Advanced BIOLOGY

Michael Kent

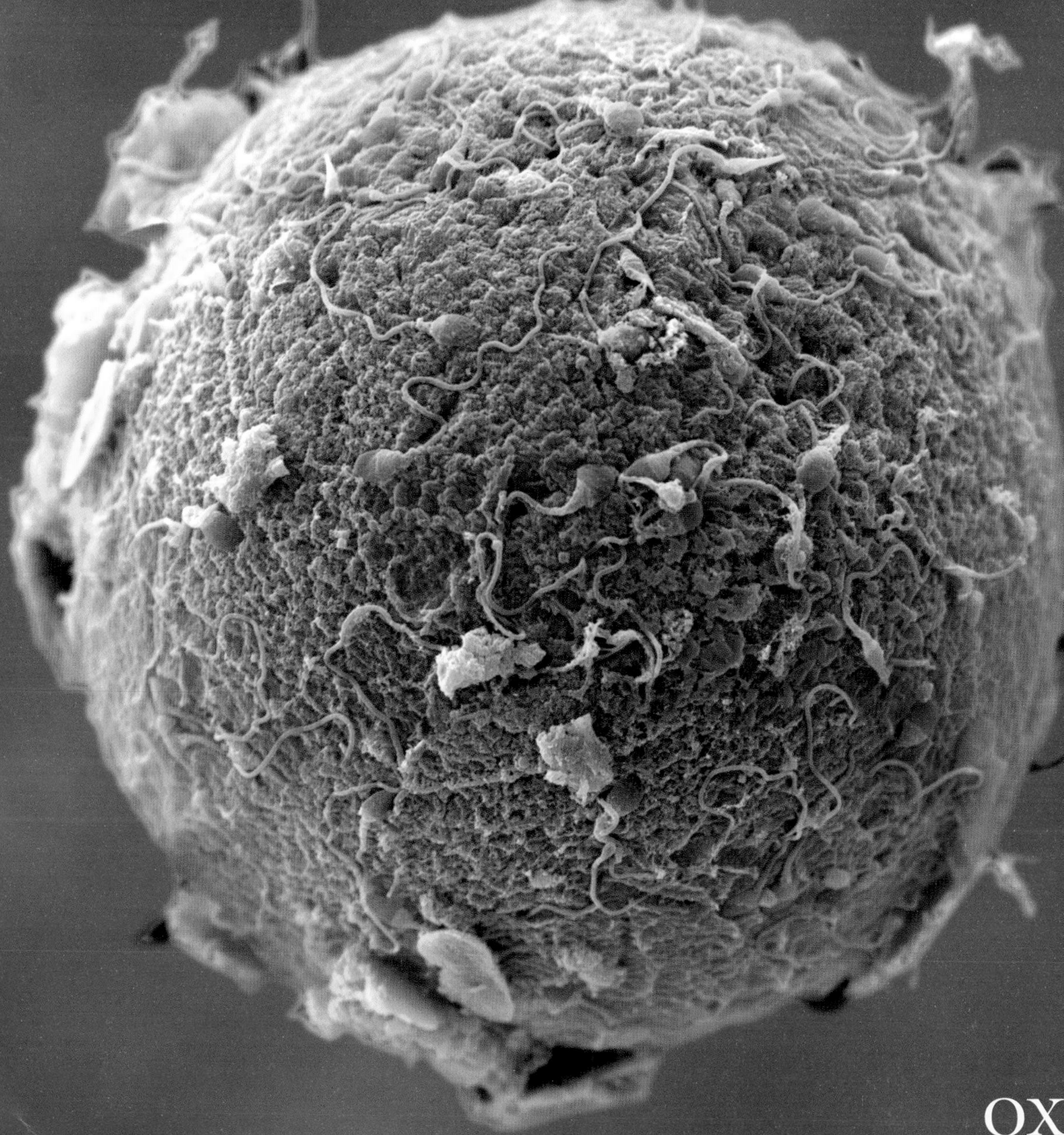

OXFORD
UNIVERSITY PRESS

OXFORD
UNIVERSITY PRESS

Great Clarendon Street, Oxford OX2 6DP

Oxford University Press is a department of the University of Oxford.
It furthers the University's objective of excellence in research, scholarship,
and education by publishing worldwide in

Oxford New York

Auckland Cape Town Dar es Salaam Hong Kong Karachi
Kuala Lumpur Madrid Melbourne Mexico City Nairobi
New Delhi Shanghai Taipei Toronto

With offices in

Argentina Austria Brazil Chile Czech Republic France Greece
Guatemala Hungary Italy Japan Poland Portugal Singapore
South Korea Switzerland Thailand Turkey Ukraine Vietnam

First published 2000
Reprinted with revisions 2013

British Library Cataloguing in Publication Data

Data available

ISBN- 978-0-19-839290-3

10 9 8

Printed in China by Sheck Wah Tong Printing Press Ltd

Paper used in the production of this book is a natural, recyclable product
made from wood grown in sustainable forests. The manufacturing process
conforms to the environmental regulations of the country of origin.

We have tried to trace and contact all copyright holders before publication. If
notified the publishers will be pleased to rectify any errors or omissions at the
earliest opportunity.

Introduction

Today's biologists are making fantastic discoveries which will affect all of our lives. These discoveries have given us the power to shape our own evolution and to determine the type of world we will live in. Recent advances, especially in genetic engineering, have dramatically affected agriculture, medicine, veterinary science, and industry, and our world view has been revolutionised by modern developments in ecology. There has never been a more exciting nor a more important time to study biology.

Advanced Biology has been written for students taking the new specifications in Advanced level and Advanced Subsidiary level biology, but it will also be useful to those of you taking the Scottish Higher Grade examination, Advanced GNVQ Science and the International Baccalaureate. It contains all the core material you will need to know, whichever board you are following. In addition, *Advanced Biology* covers all the key terms and concepts in the options offered by the different examination boards. Note, however, that within the specifications of the different boards, the units, both compulsory and optional, differ in their content and titles. Therefore, the title of a chapter or spread may not have the same heading as that in your specification. Consequently, it might be necessary to make use of the Index or the Specification Guide on the website at www.oup.com/uk/sciencegrids to look for topics; the website details the route you will need to take through *Advanced Biology* for the particular specification you are following. If you don't have convenient access to a computer, ask your teacher or lecturer for guidance.

Advanced Biology is divided into five sections. Each begins with a section-opener which includes a quotation from a famous biologist that sets the scene for the chapters which follow. Each chapter contains a series of double-page spreads, a summary, and examination questions. The double-page spread format is designed to make it easy to find topics so that they can be used at any point during the course. Each spread includes Objectives, a Fact of life box, Quick check questions, and the main text. The Objectives state clearly what you are expected to be able to do after completing the spread. The Fact of life box encourages discovery by providing interesting and thought-provoking aspects of biology. The main text covers topics included in the specifications. Some of the text and illustrations have been boxed to distinguish them from the main text. The boxes indicate that the enclosed material is either a little more difficult than the rest of the text, or that you are unlikely to need it for your examination. There is a wealth of questions, both at the end of each spread and at the end of each chapter. Used wisely, these will give you the practice you need for exam success. The Quick check questions test your recall of some of the main points in the spread, whereas the Food for thought questions will give you the opportunity to practise your problem-solving skills. Doing the Examination questions at the end of the chapter and checking your answers against those given at the end of the book will help you gain confidence for your examinations.

Although this book is primarily aimed at helping you to be successful in your examinations, I hope I have also been able to share my sense of awe at the wonder and beauty of our living world.

Author's acknowledgements

The author would like to express his sincere thanks to Michael Allaby for his advice and encouragement throughout this project; to my colleague Bill Youell for his helpful comments on the genetics section; to Jacqui Rivers for her resourcefulness in obtaining photographs of the more obscure subjects; and to the unknown readers who made such a thorough and constructive job of scrutinising the text. In addition, my heartfelt thanks go to Ruth Holmes for her stirling work as Editor during the initial stages of the book, encouraging me during some difficult moments; this book would be much poorer if it were not for her creative contribution.

Finally, I would like to thank my wife Merryn and daughter Kerris for their loving support and helpful suggestions.

Although many people contributed to this book, any mistakes or omissions remain those of the author.

Michael Kent
Wadebridge, Cornwall
2000

Introduction to update

Biological knowledge, like that of all the sciences, is characterised by change. As new discoveries are made, old ideas have to be modified or even discarded. Since 2000, biology has progressed at an amazing pace. There have been many exciting developments, particularly in DNA technology and cell biology, which have been reflected in significant changes in the examination specifications. This update of *Advanced Biology* has been made to cover those changes.

Michael Kent
Wadebridge, Cornwall
2013

Contents

Chapter 14 Reproduction and coordination in flowering plants

HEALTH AND DISEASE

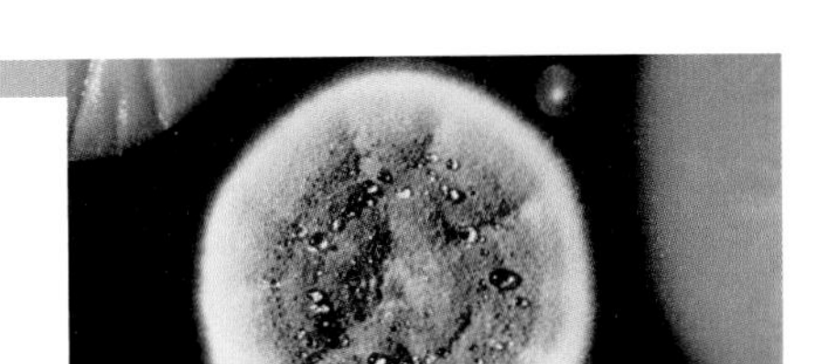

Chapter 15 Infectious diseases

Chapter 16 Non-infectious diseases

Chapter 17 Microorganisms and their applications

GENETICS AND EVOLUTION

ECOLOGY AND CONSERVATION

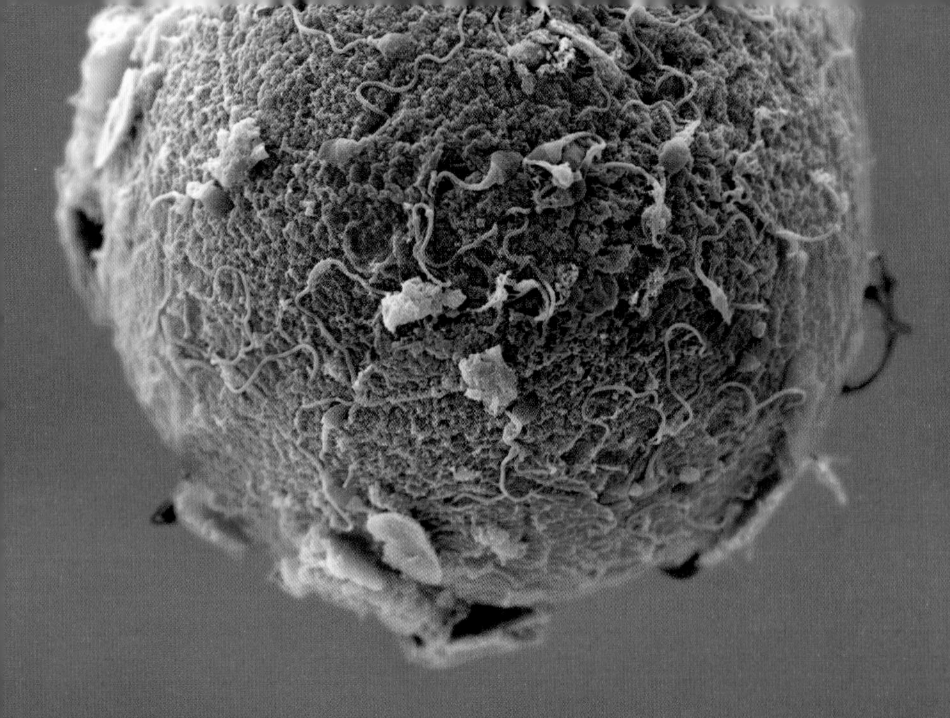

The basis of **LIFE**

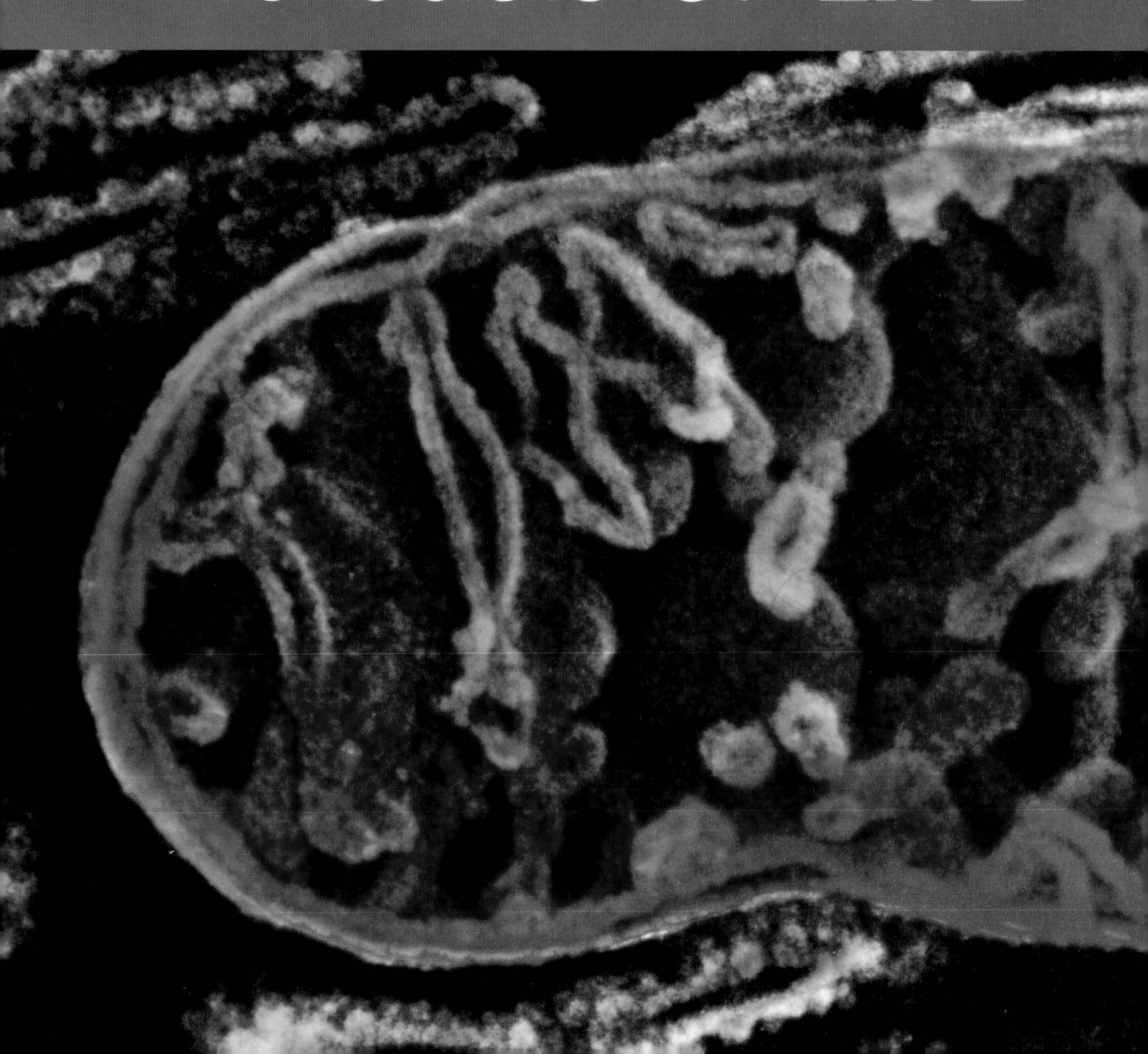

We are content to find that living things are made of elements identical with those outside the body. Indeed, we should be astounded and dismayed if anyone claimed that it were not so. Nevertheless, we shall not be satisfied that living things are therefore 'just the same' as non-living ones. We shall not be surprised to find that there is something different and even 'special' about them and particularly about their 'behaviour'.

J. Z. Young

In this part of the book, we examine the structure of life and some of the fundamental processes that characterise living things. We see that life is organised on many structural levels. Atoms combine to form complex biological molecules. These are arranged into minute structures called organelles which in turn are the components of cells. In multicellular organisms, cells are grouped into tissues and tissues into organs. Each level builds on those below it in such a way that when small components join together to form larger structures, new properties emerge that were not present in the simpler levels of organisation.

A photomicrograph of a mitochondrion, one of the organelles found in plant and animal cells.

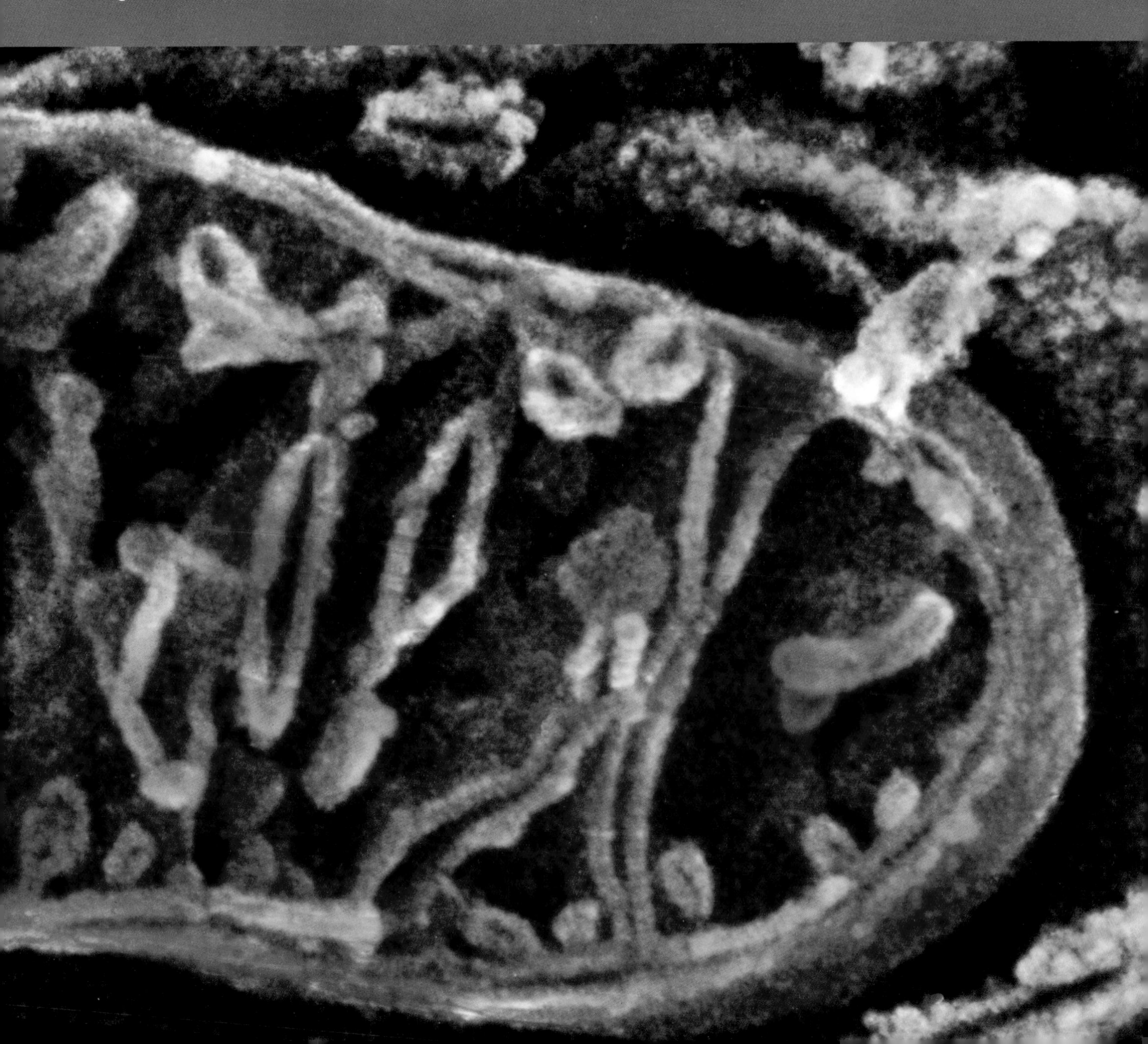

1 What is biology?

The characteristics of life

1.1

OBJECTIVES

By the end of this spread you should be able to:

- discuss the main features of living things.

Fact of life

The continued existence of life depends on reproduction, and this is perhaps the most characteristic feature of living things. Reproduction allows both continuity and change. Over countless generations this has allowed species to become well suited to their environment, and life to evolve gradually to more complex forms.

Biology is the scientific study of life. But what is life? When we see a bird on a rock it may seem obvious that the bird is alive and the rock is not, but what precisely makes the bird alive and the rock not? Throughout history, thinkers in many fields have tried to define life. Although they have failed to provide a universally accepted definition, most scientists agree that all living things share certain basic characteristics:

- Living things are made of organised structures.
- Living things reproduce.
- Living things grow and develop.
- Living things feed.
- Living things respire.
- Living things excrete their waste.
- Living things respond to their surroundings.
- Living things move.
- Living things control their internal conditions.
- Living things are able to evolve.

Non-living systems may show some of the characteristics of living things, but life is the combination of all these characteristics.

Organisation

All things are made of chemicals, but in living things the chemicals are packaged into highly organised structures. The basic structure of life is the cell. Cells themselves contain small organelles that carry out specific functions. A cell may exist on its own or in association with other cells to form tissues and organs. Because of their highly organised structure, living things are known as **organisms**.

Reproduction

Reproduction (figure 1) is the ability to produce other individuals of the same species. It may be sexual or asexual. Reproduction involves the replication of DNA. This chemical contains genetic information which determines the characteristics of an organism, including how it will grow and develop.

***Figure 1** Reproduction: a diamondback terrapin (*Malaclemys terrapin*) hatching from its egg.*

Growth and development

All organisms must grow and develop to reach the size and level of complexity required to complete their life cycle. **Growth** is a relatively permanent increase in size of an organism. It is brought about by taking in substances from the environment and incorporating them into the internal structure of the organism. Growth may be measured by increases in linear dimensions (length, height, etc.), but is best measured in terms of dry weight as this eliminates temporary changes due to intake of water which are not regarded as growth. **Development** involves a change in the shape and form of an organism as it matures. It is usually accompanied by an increase in complexity.

Feeding

Living things are continually transforming one form of energy into another to stay alive. Although energy is not destroyed during these transformations, heat is always formed. Heat is a form of energy which cannot be used to drive biological processes, so it is sometimes regarded as 'wasted energy'. Living things have to renew their energy stores periodically from their environment, to continue transforming energy and to replace the 'wasted energy'. They also have to obtain nutrients – chemicals that make up their bodies or help them carry out their biological processes. Living things acquire energy and nutrients by **feeding**, either by eating other organisms, or by making their own food out of simple inorganic chemicals using energy from sunlight or from chemical reactions.

Respiration

Living things need energy to stay alive and to do work. Although food contains energy, this is not in a directly usable form. It has to be broken down. The energy released during the breakdown is used to make ATP (adenosine triphosphate) in a process called **respiration**. ATP is an energy-rich molecule and is the only fuel that can be used directly to drive metabolic reactions in living organisms.

Figure 2 *Scientists at the Honda Motor Company have been developing a human-shaped robot equipped with artificial intelligence since the early 1990s. In 2008, one model helped conduct the Detroit Symphony Orchestra. In 2011, a new improved model was able to take a drinks order and serve it from a thermos flask.*

Excretion

The energy transformations that take place in an organism involve chemical reactions. Chemical reactions that occur in organisms are called **metabolic reactions**. Waste products are formed in these reactions, some of which are poisonous, so they must be disposed of in some way. The disposal of metabolic waste products is called **excretion**.

Responsiveness

All living things are sensitive to certain changes in their environments (stimuli) and **respond** in ways that tend to improve their chances of survival. The degree of responsiveness depends on an organism's complexity: a bacterium may be limited to simple responses, such as moving towards favourable stimuli or away from harmful ones; people can make highly sophisticated responses to a wide variety of stimuli which they may perceive either directly or with the aid of technological devices.

Movement

Responses usually involve some form of **movement**. Movement of whole organisms from one place to another is called **locomotion**. Plants and other organisms that are fixed in one place do not display locomotion, but they can move parts of their bodies. Movements of living things differ from those of non-living things by being active, energy-requiring processes arising from within cells.

Homeostasis

All living things are, to some extent, able to control their internal conditions so that their cells have a constant chemical and physical environment in which they can function effectively. The regulation and maintenance of a relatively constant set of conditions within an organism is called **homeostasis**. Homeostasis is a feature of all living systems, from a single cell to the whole biosphere (the part of Earth containing life).

Evolution

Living things are able to change into new forms of life. This **evolution** usually takes place gradually over successive generations in response to changes in the environment.

Quick check

1 What is the difference between:

a the growth of a crystal and the growth of a plant

b the movement of a cloud and the movement of an animal?

Food for thought

a You might be familiar with the mnemonic (memory aid) 'Richard Of York Gave Battle In Vain' for remembering the colours of the spectrum (red, orange, yellow, green, blue, indigo, violet). Suggest a mnemonic for the ten characteristic features of living things described in this spread. You can change the order of the features.

b The robot shown in figure 2 can move and respond, and requires energy to maintain its organisation and a constant internal environment. How would you argue that the robot is a non-living object? A robot could be made that has all the characteristic features of living things. Would it still be non-living?

1.2

OBJECTIVES

By the end of this spread you should be able to:

- describe what biologists do
- define the different levels of biological organisation
- list the main elements of a scientific method.

Fact of life

No matter how dramatic it is, any discovery must be shared before it can make a contribution to our scientific knowledge. Biologists communicate with each other mainly by means of concise reports called papers. Typically, a paper contains the aims of an investigation, a description of the method used, the results obtained, and a discussion of the significance of the results. The method is described in enough detail to allow someone else to repeat the investigation. Well over one million original papers are published in the biological sciences each year, in subjects ranging from the behavioural interactions of different animal populations to the analysis of chemical reactions taking place in cells.

What do biologists do?

Biologists study every aspect of life at every level of its organisation, from the atoms that make up biological molecules to the ecosystems that form the biosphere (figure 1).

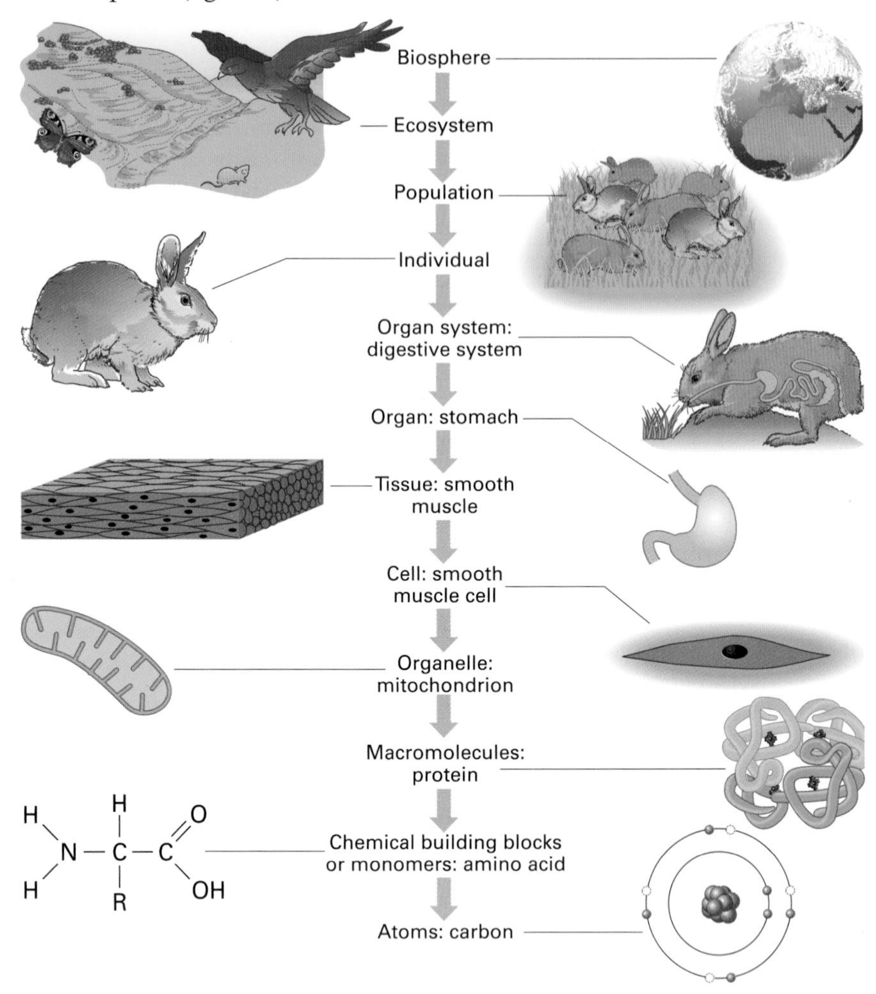

Figure 1 *Levels of biological organisation from atoms, the smallest components of living things, to the biosphere, the entire living planet.*

Aspects of biology

Modern biology is an enormous subject that has many branches. Specialists in some branches include:

- molecular biologists and biochemists who work at the chemical level, with the aim of revealing how DNA, proteins, and other molecules are involved in biological processes
- geneticists who study genes and their involvement in inheritance and development
- cell biologists who study individual cells or groups of cells, often by culturing them outside organisms; they investigate how cells interact with each other and their environment
- physiologists who find out how organ systems work in a healthy body
- pathologists who study diseased and dysfunctional organs
- ecologists who study interactions between organisms and their environment. Some focus their attention on whole organisms; others study populations, individuals of the same species living together at one location.

There are also biologists who specialise in particular groups of organisms; for example, bacteriologists study bacteria, botanists study plants, and zoologists study animals.

Biologists are employed in many fields including conservation and wildlife management, industry, medicine, health care, horticulture, agriculture, zoos, museums, information science, and marine and freshwater biology. In addition, many biologists are employed as teachers, lecturers, or research workers.

The scientific method

The definition of biology given in spread 1.1 states that it is a 'scientific study'. This distinguishes biology from other ways of studying life. However, there is no single rigid scientific method that biologists use: there are numerous ways of studying life scientifically. Nevertheless, biological investigations usually include one or more of the following key elements:

- observing: making observations and taking measurements
- questioning: asking questions about observations and posing a problem
- hypothesising: formulating a hypothesis, a statement that explains a problem and can be tested
- predicting: stating what would happen if the hypothesis were true
- testing: testing the hypothesis, usually by carrying out a controlled experiment aimed at producing data that will either support or contradict the hypothesis
- interpreting: interpreting the test results objectively and drawing conclusions that accept, modify, or reject the hypothesis.

A biologist may start an investigation by making observations or by using observations described by other biologists. Such observations may be obtained directly by the senses, such as listening to a bird song, or indirectly through instruments such as recording the song on a computer system. On the other hand, an investigation may start simply by a biologist having an idea (perhaps no more than a hunch) that something happens in a particular way, and then the idea will be tested by making observations or carrying out experiments to see if it is valid. A hypothesis is suggested and then tested in all investigations. One essential aspect of a scientific experiment is that it can be repeated by other scientists working independently.

A typical hypothesis makes a clear link between an **independent** or **manipulated** variable and a **dependent variable**. Variables are conditions or factors (such as light, temperature, or time) that can vary or may be varied. In an experiment, the independent or manipulated variable is the one that is systematically changed; the dependent variable is the effect or outcome that is measured. For example, when investigating the activity of an enzyme at different temperatures, temperature is the independent variable that is manipulated by the scientist; rate of reaction is the dependent variable that is measured at each temperature. Other variables called **controlled variables** are kept constant or controlled at set levels.

At the end of an experiment, the results must be interpreted as objectively as possible. Sometimes they are so clear that it is obvious whether they support or contradict a hypothesis. Often, however, results are variable and need statistical analysis before conclusions can be made. The conclusions may lead to the hypothesis being accepted, modified, or rejected. Even if results support a hypothesis, it is accepted only tentatively because it can never be proved completely. However, it only needs a single contrary observation to **refute** a hypothesis (prove it wrong or incomplete). A hypothesis is therefore only the best available explanation at any time. This makes biology a highly dynamic subject and not merely a collection of facts.

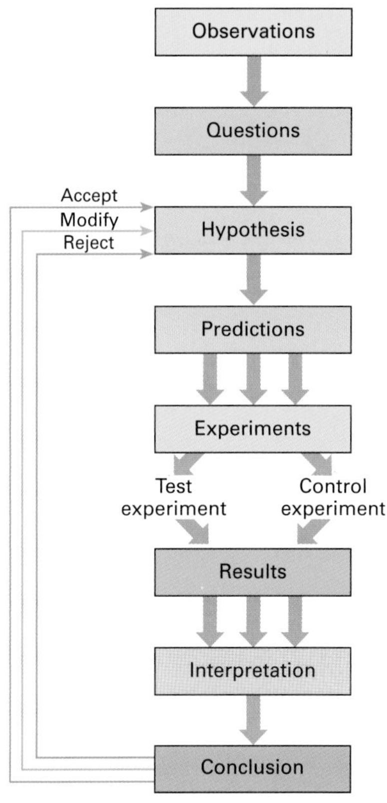

***Figure 2** A typical sequence of events in a scientific investigation.*

QUICK CHECK

1 What is the difference between a physiologist and a pathologist?

2 Which is the highest level of biological organisation on Earth?

3 In an experiment in which the rate of photosynthesis of a plant is measured at different light intensities, which is the independent (manipulated) variable and which is the dependent variable?

Food for thought

The life sciences have made an enormous contribution to human welfare, especially through their applied branches of medicine, agriculture, and biotechnology. However, an important part of understanding biology and the other sciences is realising their limitations. Science does not, for example, deal with hypotheses that are not testable. Suggest questions that might not be possible to answer using a scientific method.

2 The chemicals of life

Atoms and elements

2.1

OBJECTIVES

By the end of this spread you should be able to:

- understand that all matter is made of elements
- list the main elements found in living things
- describe the main components of an atom
- describe how isotopes can be used in biological investigations.

Fact of life

Ninety-nine per cent of all living things is made up of only four elements: hydrogen, oxygen, nitrogen, and carbon.

Even organisms as different as a giraffe and the acacia bush that it eats (figure 1) are made up of these elements.

Figure 1 A giraffe and the acacia bush that it eats are made of the same chemical elements.

The elements in living things

All parts of the Earth are made of 92 naturally occurring substances called **elements**. An element is a substance that cannot be split into simpler substances by chemical means. Living, dead, and non-living things are all made up of elements.

Only about 21 of the the 92 naturally occurring elements are important to living organisms. The most common elements in organisms are:

- carbon (C)
- oxygen (O)
- hydrogen (H)
- nitrogen (N).

Other elements occur in living organisms in smaller quantities. These include phosphorus (P), potassium (K), sulphur (S), calcium (Ca), iron (Fe), magnesium (Mg), sodium (Na), and chlorine (Cl).

Atoms

Each element is made up of **atoms**. A single atom is the smallest amount of any element that can exist. Atoms are far too small to be seen with an ordinary microscope. Many millions of atoms make this comma, and countless billions make each of us. The structures and functions of whole organisms depend on the structures of the atoms they contain.

Atoms can join together with other atoms of the same element, or with atoms of different elements, to form **molecules** (see spread 2.2). Atoms can combine in many different ways to form thousands of different materials. Combinations of different elements are called **compounds**.

Structure of the atom

The main mass of each atom is in a central **nucleus**. The nucleus is made of even smaller particles called **protons** and **neutrons**. (Hydrogen atoms are the exception. They have one proton in the nucleus but no neutrons.)

Electrons are minute particles. They move around the nucleus at very high speed within **electron shells** or **energy levels** (figure 2). Their mass is negligible compared with that of the neutron or proton.

The components of an atom have different electrical properties.

- Neutrons have no electrical charge.
- Electrons have a negative charge (–).
- Protons have a positive charge (+).

An atom with an equal number of electrons and protons has no overall electrical charge.

Ions

Ions are atoms that have gained an electrical charge.

- Some atoms can gain one or more electrons and become negatively charged.
- Others can lose one or more electrons and become positively charged.

Grouping the elements – the periodic table

The number of protons in the nucleus is different for every element. This number is called the **atomic number**. In a neutral atom the number of protons equals the number of electrons. The periodic table (figure 3) shows the elements arranged in order of atomic number. Elements with similar properties occur in the same column or **group**.

Isotopes

The atoms of any particular element always have the same number of protons but the number of neutrons can vary. The atoms may therefore have different masses. These are different versions of the same element, and are called **isotopes**.

Isotopes are identified by the **mass number** of the atom, which is the number of protons plus the number of neutrons. Different isotopes of the same element have different mass numbers, but they all have the same atomic number. For example, three isotopes of the carbon atom have mass numbers of 12, 13, and 14, but they all have an atomic number of 6. These isotopes are represented as ^{12}C, ^{13}C, and ^{14}C.

Radioactive isotopes

A few elements have isotopes that are radioactive, which means that they are unstable. They change spontaneously to form a different isotope of the same element, or a completely different element. Radiation is given out during this change.

Tracers

Isotopes of an element have the same chemical properties and so will behave in the same way inside an organism. This is useful for investigating the reactions that go on in an organism. A compound that contains a 'labelled' isotope (e.g. one that is radioactive) is used. It takes part in reactions in the organism in the same way as the unlabelled compound. The isotope and products made from it in the organism can be detected by physical measurements. For example, the radioactive isotope ^{14}C can be used to trace the path by which carbon in carbon dioxide ($^{14}CO_2$) is converted into glucose ($^{14}C_6H_{12}O_6$) during photosynthesis.

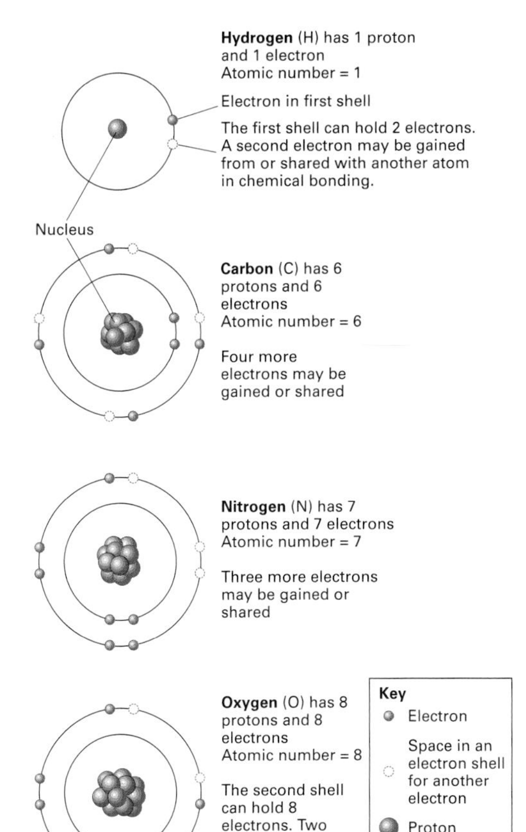

Figure 2 *Highly simplified representations of hydrogen, carbon, nitrogen, and oxygen atoms. In reality, the electron shells are not spherical. Electrons occupy charged clouds of various shapes, and there is no way of telling exactly where the electrons are at any one instant.*

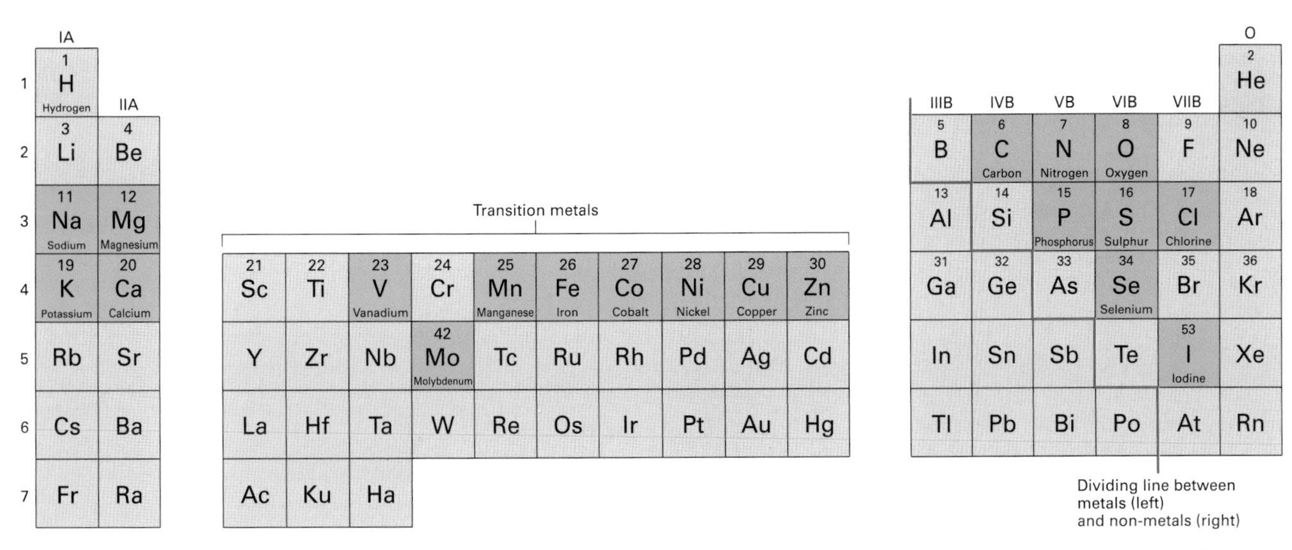

Figure 3 *In the periodic table, elements are arranged in order of atomic number. Elements in a column or groups have similar chemical properties. The main elements occurring in living organisms are shown in green.*

Quick check

1. What is the smallest possible amount of any element called?
2. Name the four most common elements found in living organisms.
3. Give the name of the relatively dense, central part of an atom.
4. Which component of an atom varies in number to form different isotopes of the same element?

Food for thought

Suggest how a biologist could test whether or not the radioactive isotope ^{14}C behaves in the same way as ^{12}C when inside a plant.

2.2

Molecules and compounds

OBJECTIVES

By the end of this spread you should be able to:

- describe how molecules and compounds are formed
- explain why carbon is uniquely suited to its role as the main element in living organisms.

Fact of life

In 1828, Friedrich Wöhler was the first person to synthesise a biological material. He synthesised urea from non-biological substances – cyanic acid and ammonia.

This showed that the substances in living organisms are made of the same chemicals as those in non-living things, and that they differ only in the way they are organised.

Atoms contain electrons, which orbit the nucleus in shells or energy levels. Isolated atoms are usually unstable because their outer electron shell is not full. They therefore form chemical bonds with other atoms so that their outer shells are filled and they reach a stable state. This can be achieved by two main types of bonding.

- In **covalent bonding** electrons are *shared* between atoms.
- In **ionic bonding** electrons are *transferred* between atoms.

Compounds are formed during chemical bonding.

The only elements in which atoms exist separately, without being bonded to other atoms, are the noble gases such as neon. The atoms of these gases have a full outer electron shell.

Chemical formula

Scientists use a chemical formula to show the composition of a compound. For example, H_2O indicates that a water molecule consists of two hydrogen atoms and one oxygen atom.

Covalent bonding: making molecules

In a covalent bond, a pair of electrons is *shared* between atoms to form a **molecule**. The atoms may be of the same element, as in the hydrogen molecule (H_2), or of different elements, forming a covalent compound. A molecule is the smallest particle into which a covalent compound can be divided without breaking up the compound.

Valency

The **valency** of an element tells you how many bonds the atoms need to form to gain a full outer shell of electrons:

- A carbon atom (C) forms 4 covalent bonds.
- An oxygen atom (O) forms 2 covalent bonds.
- A hydrogen atom (H) forms 1 covalent bond.
- A nitrogen atom (N) forms 3 covalent bonds.

For example, in a molecule of methane (CH_4) the carbon atom is linked to four hydrogen atoms by single covalent bonds (figures 1 and 2).

Bonds also act *between* molecules, pulling them together within a solid or a liquid. These bonds are much weaker than the covalent bonds holding the atoms in the molecule together. (See hydrogen bonding, spread 2.3.)

Ionic bonding: making ions

In an ionic bond, electrons are *transferred* between atoms to form **ions**. Ions have full outer electron shells and are therefore very stable. The valency of an element is the number of electrons that are donated or accepted during ionic bonding.

For example, in sodium chloride (NaCl; table salt):

- the outer electron of a sodium atom is transferred to the outer electron shell of a chlorine atom (figure 3).
- The sodium ion has a positive electric charge (+) because it has lost an electron.
- The chloride ion has a negative electrical charge (–) because it has gained an electron.
- Sodium and chlorine both have a valency of 1.

The opposite charges produce an electrostatic attraction – the ionic bond – which holds the sodium and chloride ions together in solid sodium chloride. However, when dissolved in water, the ions can move about freely, quite independently of each other. In an aqueous solution of sodium chloride, the ions are represented as $Na^+(aq)$ and $Cl^-(aq)$.

In methane, the carbon atom shares each of its 4 outer electrons with a hydrogen atom

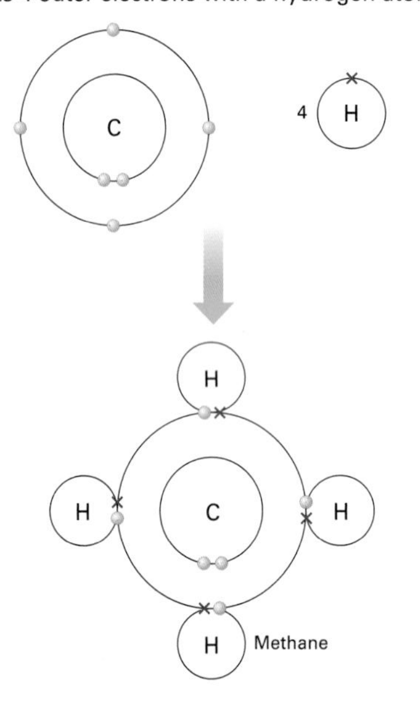

The carbon atom has now achieved stability: it has a share in 8 electrons in its outer shell. Each hydrogen atom has a share in 2 electrons: it also has a full outer shell.

Figure 1 *Covalent bonding in methane.*

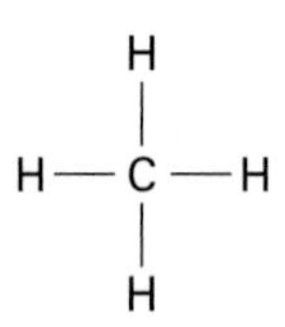

Figure 2 *Methane: each single bond between a carbon atom and a hydrogen atom is represented by a single line.*

Biological molecules

The simplest compounds may be made up of molecules that are quite small and contain just two or three atoms. However, many **biological molecules** are very large: DNA, for example, contains millions of atoms.

Biological molecules contain many carbon atoms combined with atoms of other elements. The study of these carbon-containing compounds is called **organic chemistry**. Carbon-containing compounds are not only obtained from living (organic) sources. Scientists have made many thousands of organic compounds, including pesticides and medicines.

The special property of carbon

A carbon atom is **tetravalent**, which means that it can form four covalent bonds with other atoms. These can be atoms of other elements, or other carbon atoms. The bonds around a carbon atom point towards the corners of a tetrahedron, which is a very stable arrangement.

Carbon–carbon bonds are very strong. Long chains of carbon atoms can exist in a single molecule, combined with other atoms and branched in many different ways. Hence there is an enormous variety of carbon-containing molecules, some of great complexity (figure 4).

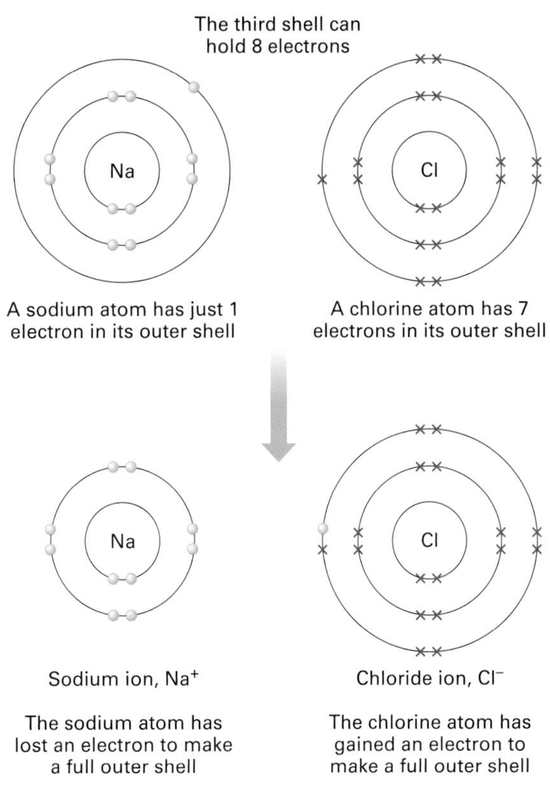

Figure 3 *Formation of sodium and chloride ions.*

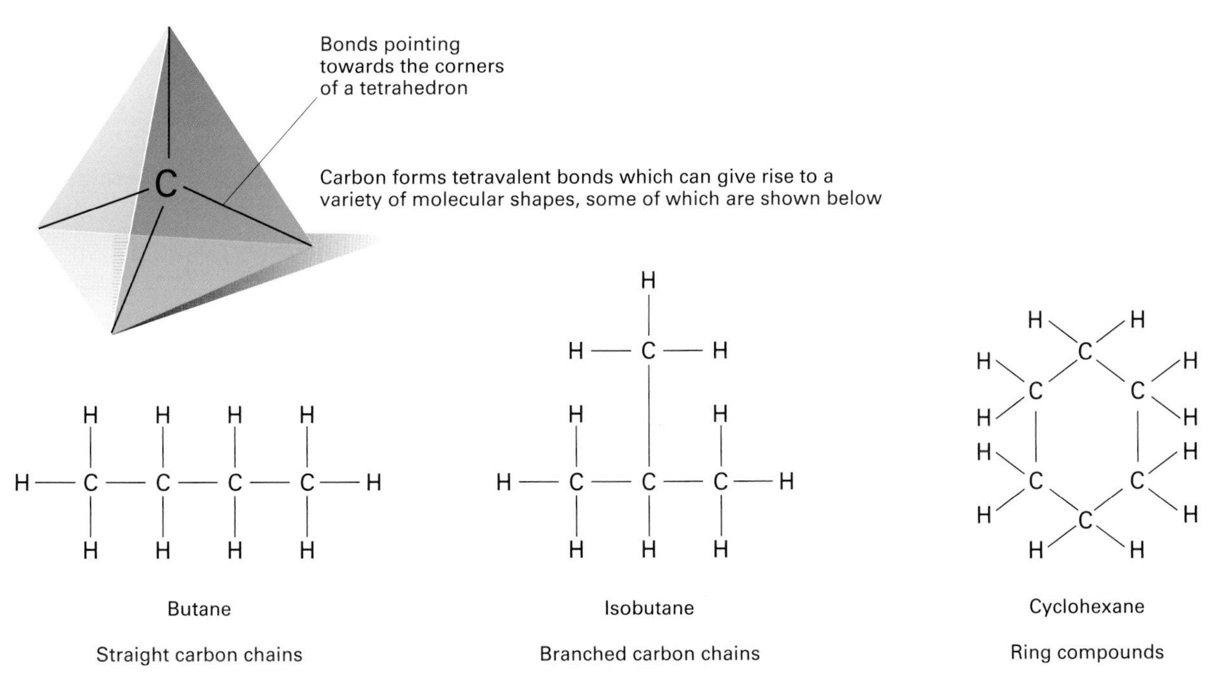

Figure 4 *Just a few examples of the variety of molecular shapes based on tetravalent bonding in carbon compounds.*

QUICK CHECK

1 Name the type of bond formed:
 a when one atom donates an electron to another atom
 b when two atoms share electrons.
2 Why is carbon referred to as tetravalent?
3 Give three reasons why carbon is uniquely suited to its role as the main element in living organisms.

Food for thought

Some science fiction writers have suggested that life on other planets could be based on an element other than carbon. Suggest why life on Earth is carbon based.

2.3

The unique properties of water

OBJECTIVES

By the end of this spread you should be able to:

- describe the physical and chemical properties of water.

Fact of life

In October 2009, NASA scientists smashed a rocket on to the dark side of the Moon creating a debris cloud 2 km high. Spectrometers on a probe that followed the rocket detected significant amounts of water vapour in the cloud. In May 2011, analysis of lunar soil brought back by the *Apollo 17* mission revealed 100 times more water than expected, supporting the idea that the Moon once had a sea the size of the Caribbean. Before the 2011 discovery it was thought that all lunar water was brought in by impacts of icy comets or watery meteorites into the Moon. If future explorations find sufficiently large quantities of water, permanent settlements on the Moon might be feasible using the water not only for drinking but also for fuel, using solar energy to convert the water into hydrogen and oxygen.

***Figure 1** An artist's impression of a manned lunar base. Such a base could support a manned mission to Mars.*

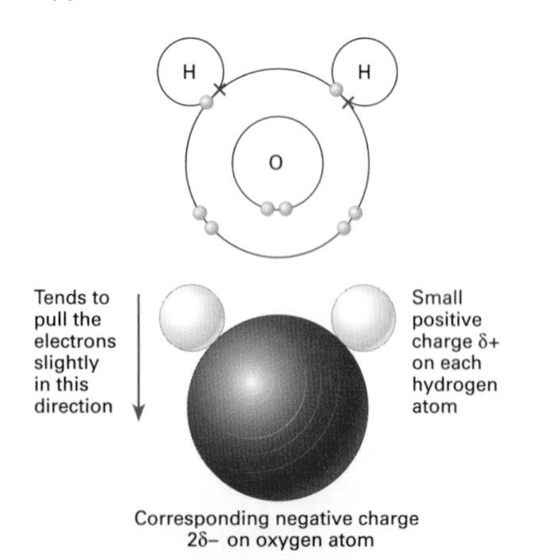

***Figure 2** The arrangement of atoms in a water molecule. The oxygen atom has a tendency to draw electrons closer to it. This gives the oxygen end of a water molecule a slightly negative charge and the hydrogen end a slightly positive charge. (δ+ and δ– are the conventional ways of representing very small charges.)*

Water is essential for life as we know it. Its unique chemical and physical properties allow it to sustain life. Organisms themselves are mainly water; the human body is about 60 per cent water.

Water molecules...

A water molecule consists of two hydrogen atoms and one oxygen atom. Each hydrogen atom shares a pair of electrons with the oxygen atom to form two covalent bonds (H—O—H). However, the electrons are not shared equally. Oxygen has a greater pull on the electrons than hydrogen: it has a greater **electronegativity**. This draws electrons away from the hydrogens slightly so that each water molecule has slightly negative and slightly positive regions (figure 2).

Molecules with charged regions are called **polar molecules**; those with two separate charged areas are called **dipolar molecules**. Water is a dipolar molecule.

...are 'sticky'

The negative and positive ends of water molecules attract each other so that the molecules tend to stick together (figure 3). These forces of attraction between water molecules are called **hydrogen bonds**. Hydrogen bonding gives water many of its unique properties.

Water is a liquid at room temperature

Compounds with molecules as small as those of water, for example hydrogen sulphide, are usually gases at room temperature. Water is a liquid at room temperature because of hydrogen bonding. One water molecule can form hydrogen bonds with up to four other water molecules.

Water is the universal solvent

The dipolar molecules make water an excellent solvent. In fact water is often called the **universal solvent** because so many things dissolve in it.

- Polar and ionic substances have an electrostatic charge, so they are attracted to the charges on water molecules. They dissolve readily in water and are called **hydrophilic** ('water-loving').
- Non-polar substances, such as oil, have no charge on their molecules. They do not dissolve readily in water and are called **hydrophobic** ('water-hating').

When a salt (an ionic compound) dissolves in water, the ions separate and layers of water molecules form around the ions (figure 4). These layers of water molecules prevent ions or polar molecules from clumping together, and keep the particles in solution.

Water forms a skin at its surface

Water molecules have a much stronger attraction to other water molecules than to molecules in the air. At an air–water interface (for example, on the surface of a pond), a water molecule on the surface forms hydrogen bonds with other water molecules around and below it, but not with air molecules above it. The unequal distribution of hydrogen bonds produces a force called **surface tension**. This causes the water surface to contract and form a surprisingly tough film or 'skin'. Surface tension is a measure of how difficult it is to stretch or break a liquid surface: water has a greater surface tension than most other liquids.

Ice floats on water

Water behaves peculiarly when cooled and is most dense at 4 °C. Consequently, the temperature at the bottom of a large body of water such as an ocean remains relatively constant at 4 °C, despite variations in the temperature of the water at the surface.

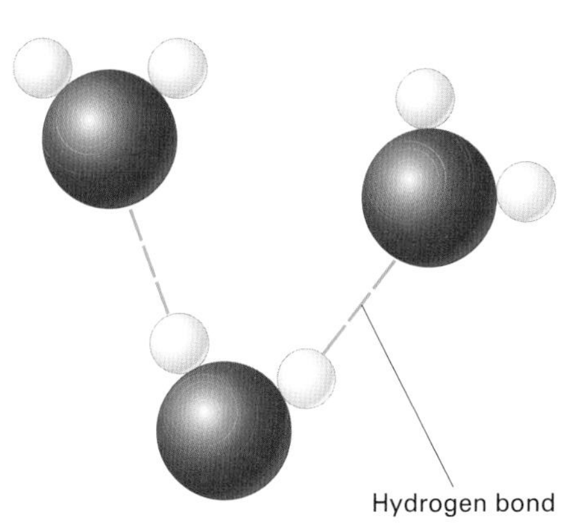

Figure 3 Water molecules have negative and positive ends so they tend to be pulled together by hydrogen bonding.

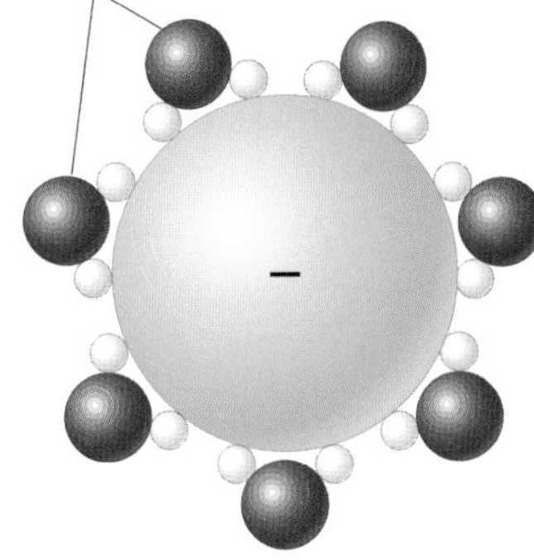

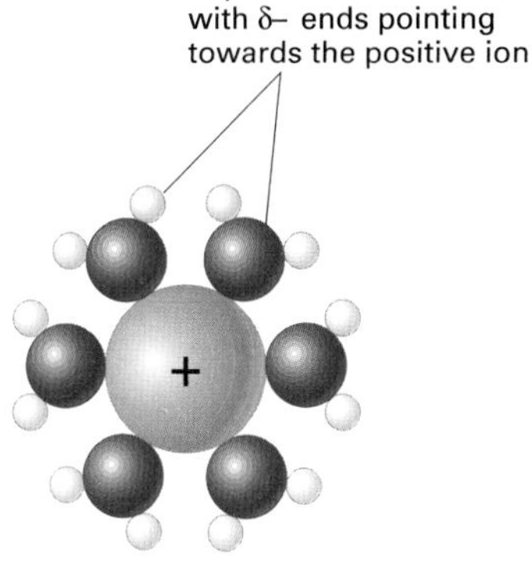

Figure 4 Water is a good solvent because its molecules form a layer around ions or molecules. In a solution of sodium chloride, the positive (hydrogen) ends of water molecules point towards the negative chloride ion, and the negative (oxygen) ends point towards the positive sodium ion.

Layers of water molecules form around many non-ionic organic substances, such as sugars, because they contain polar side-groups.

In liquid water hydrogen bonds are constantly forming and reforming. When water freezes, each molecule forms hydrogen bonds with four other molecules. This makes a rigid lattice, which holds the water molecules further apart than in liquid water. This is why water expands when it freezes. Ice is less dense than liquid water; ice floats on water. Although individual hydrogen bonds are weak, combined in their thousands they make ice as solid as rock.

Water is wet

Water tends to wet most things (that is, it sticks to them) because its molecules form hydrogen bonds with other polar substances. This attraction between molecules of different substances is called **adhesion**. The attraction between molecules of similar substances is called **cohesion**. Hydrogen bonding gives water considerable cohesive properties. Thus water molecules stick together and also stick to other surfaces, such as the sides of its container. This attraction between molecules enables water to enter and move along very narrow spaces, a process called **capillarity**.

Water and temperature

An important thermal property of water is its ability to resist temperature changes.

- Water has a high **specific heat capacity**: it needs to gain a lot of heat energy to raise its temperature. Conversely, it needs to lose a lot of heat energy to lower its temperature.
- Water has a high **latent heat of vaporisation**: a lot of heat energy is needed to evaporate it. When water evaporates from a surface, it draws heat energy out of the material underneath, creating a cooling effect.
- Water has a high **latent heat of fusion**: water at 0 °C must lose a lot of heat energy before it forms ice crystals.

Other properties of water

Other important physical properties of water include:

- It is transparent to sunlight (figure 5).
- It has a relatively high density compared with air.
- It is difficult to compress.
- It conducts electricity. (Pure water has a low conductivity, but dissolved ions make it a good conductor.)

Figure 5 The transparency of water allows plants to photosynthesise deep in the oceans.

Quick check

1 a What are hydrogen bonds?

b Explain why sugar and salt dissolve readily in water, but oil does not.

c Why is a plant leaf damaged by frost?

Food for thought

When water freezes, ice forms on the surface. Suggest what would happen if ice were denser than liquid water. What implications would this have for life on Earth?

2.4

The biological importance of water

OBJECTIVES

By the end of this spread you should be able to:

- explain the biological significance of the chemical and physical properties of water
- give examples of organisms or biological processes to illustrate the significance of these properties.

Fact of life

Dehydration (loss of water from the body) is a major hazard to most terrestrial organisms. In humans, dehydration can result in overheating, a potentially dangerous lowering of blood pressure, and a slower circulation of blood around the body. These effects can lead to heat stroke and death.

Mild dehydration commonly occurs during exercise if the amount of water lost through sweating is greater than the water taken in by drinking. This mild dehydration can cause poor performances in sport even in temperate climates. A 2 per cent loss of body weight due to water loss can lead to a 20 per cent drop in the working capacity of muscles. Dehydration reduces both the speed and strength of muscle contractions. To minimise the risk, water should be drunk at intervals during exercise and competitive sports.

Water covers more than three-quarters of the Earth's surface, and most of the planet's organisms live in water. Living organisms originated in a watery environment, and in the course of evolution they have remained dependent on water for most of their activities. The content of each living cell is mostly water. Without it, cells die or go into a state of deep inactivity.

Dehydration for survival

Water is essential to life for many reasons (see table 1). Sometimes, however, it is an advantage for living tissue to become dehydrated. Dehydration results in a state of suspended animation, so that small organisms and reproductive structures such as seeds, pollen, and eggs can survive periods of water scarcity. A very low water content also enables these structures to be small and more easily dispersed.

Some organisms dehydrate to avoid being killed during cold conditions. Examples are the tardigrades, short plump animals better known as 'water bears' which inhabit the films of water that surround terrestrial lichens and mosses (figure 1). When the temperature falls below a certain level, their body composition changes from 85 per cent water to about 3 per cent. Waterless tissues cannot freeze, so there is no danger of dehydrated tardigrades being ripped apart by ice.

Life on the surface

The surface film of a pond, resulting from water's high surface tension, is a firm elastic platform on which a wide range of surface-dwelling organisms can live. These include whirligig beetles, pond skaters, and raft spiders (figure 2) which use the surface of the water to detect vibrations caused by passing prey.

In addition to those functions mentioned in table 1, water has many other roles. For example, it provides the medium through which sperm swim during fertilisation, and it is necessary for the germination of seeds. Water also plays a major role in the regulation of pH (see spread 2.5).

Figure 1 *An electron micrograph of a tardigrade (×100).*

Figure 2 *Surface dwellers.*

Table 1 *The biological significance of the properties of water.*

Property of water	Significance for living organisms
Water is a liquid at room temperature	Provides a liquid environment inside cells and aquatic environments for organisms to live in
Water is the universal solvent: it dissolves more substances than any other common solvent	The chemical reactions inside cells happen in aqueous solution; water is also the main transport medium in organisms
Water forms a skin at its surface: it has a high surface tension	Water forms a surface film at an air–water interface; this allows some aquatic organisms such as pond skaters to land on the surface of a pond and move over it
Ice floats on water: ice is less dense than liquid water	Ice forms on the surface of a body of water and insulates the water below, allowing aquatic life to survive
Water is wet: adhesion makes it stick to polar surfaces	Along with low viscosity, this property allows capillarity so that, for example, water can move upwards through narrow channels in soil, against gravity
Water has a very high specific heat capacity	The environment inside organisms resists temperature changes; aquatic environments have relatively stable temperatures
Water has a high latent heat of vaporisation	Heat is lost from a surface when water evaporates from it; this is used as a cooling mechanism, for example sweating in mammals and transpiration in plants
Water has a high latent heat of fusion	Cell contents and aquatic habitats are slow to freeze in cold weather
Water is colourless and transparent	Transmission of sunlight enables aquatic plants to photosynthesise
Water is denser than air	Water supports organisms as large as whales; it also supports and disperses reproductive structures such as larvae and large fruits such as coconuts
Water is difficult to compress	Water is an important structural agent, acting as a skeleton (called a hydrostatic skeleton) in worms and turgid plants
Water takes part in many chemical reactions	Water is a major raw material for photosynthesis; it also takes part in many digestive reactions, breaking down food molecules by hydrolysis
Water combines with many organic molecules to form hydrated molecules	Most organic molecules occur in a hydrated form in cells; if the water is removed, their physical and chemical properties are affected
Water has a low viscosity (it flows freely: water molecules can slide easily over each other)	Water can flow freely through narrow vessels; watery solutions can act as a lubricant, for example mucus allows food to move easily down the oesophagus
Water has a high tensile strength: water columns do not break or pull apart easily	Continuous columns of water can be pulled all the way up to the top of a tree in xylem vessels during transpiration

Quick check

1 a Which property of water makes it suitable for use as a hydrostatic skeleton?

b Why is water's transparency essential for most of life on Earth?

2 Which biological process uses the latent heat of vaporisation of water?

Food for thought

Suggest what would happen to pond skaters if their pond became polluted with detergent.

2.5

OBJECTIVES

By the end of this spread you should be able to:

- describe the different components of a solution
- explain what the pH scale is
- describe the causes and effects of acid precipitation.

Fact of life

On 10 April 1974, a rainstorm at Pitlochry in Scotland had a pH of 2.4; it was more acidic than vinegar.

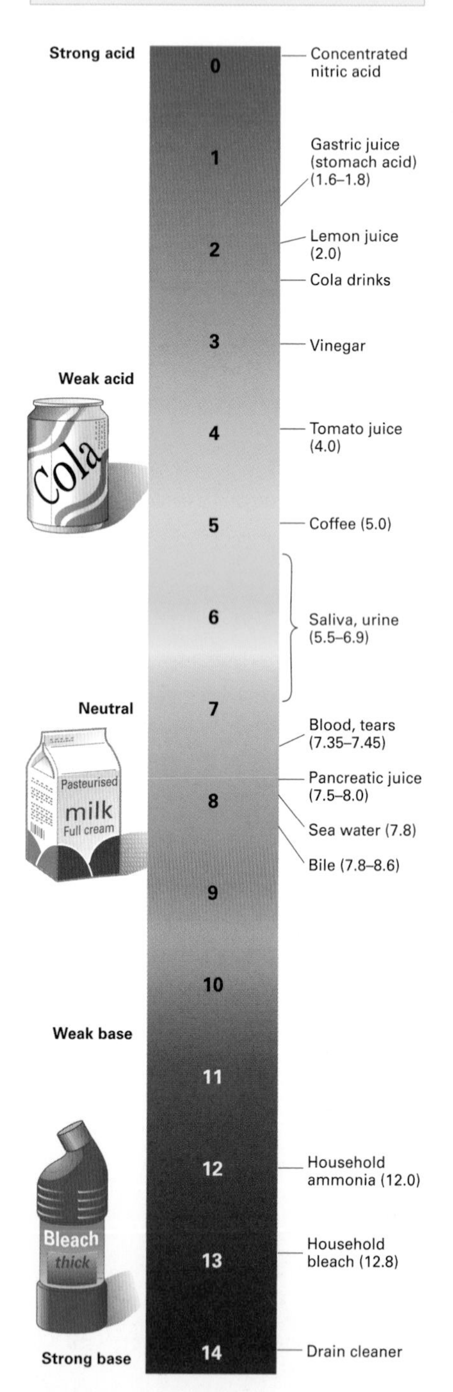

Figure 1 *The pH scale.*

THE pH OF SOLUTIONS

In solution

Almost all the processes that are essential to life take place in solution. A **solution** is a homogeneous mixture (that is, a mixture that is uniform throughout) of two or more substances in which the particles are completely dispersed. The substance of which there is more, forming the greater part of the solution, is called the **solvent**. The substance dissolved in the solvent is called the **solute**.

Acids, bases, and alkalis

In living organisms, water is the most important solvent. In spread 2.3 we discussed how water molecules are made of hydrogen and oxygen atoms. However, we did not mention that these molecules can **dissociate** or split to form protons and hydroxide ions:

$$H_2O \rightleftharpoons H^+ + OH^-$$

$$\text{Water} \rightleftharpoons \text{proton} + \text{hydroxide ion}$$

Pure water dissociates only a little, producing an equal number of protons (H^+) and hydroxide ions (OH^-), both at a concentration of 107 mol dm^{-3}. The concentration of protons in a solution is more conveniently expressed in terms of **pH**, the negative logarithm (to the base 10) of the H^+ concentration in mol dm^{-3}. Water has a pH of 7; it is neutral, neither acidic or basic. **Acids** are substances that release free protons in solution. Therefore, acidic solutions have a higher concentration of free protons than pure water and a pH less than 7. The stronger the acidity, the more protons are released and the lower the pH. **Bases** are substances that react with acids to form salts. Bases that dissolve in water are called **alkalis**. These molecules are proton acceptors, removing protons from aqueous solutions and so increasing the pH. Alkaline (or basic) solutions therefore have a higher concentration of hydroxide ions than pure water, and a pH greater than 7.

Because pH is measured on a logarithmic scale, each pH number below 7 is ten times more acidic than the next number (for example, a solution at pH 5 has a concentration of protons ten times higher than a solution at pH 6). The same is true for pH numbers above 7 (for example, a solution at pH 9 has a concentration of hydroxide ions ten times higher than a solution at pH 8). The pH scale (figure 1) extends from 0 (most acidic) to 14 (most alkaline).

Acid precipitation

The bodies of living organisms can tolerate only a limited pH range. Even a slight change in pH can be harmful because molecules in cells are very sensitive to concentrations of protons and hydroxide ions.

Rain and other forms of precipitation contain dissolved gases from the atmosphere, principally carbon dioxide and sulphur dioxide, and so naturally have a pH less than 7. **Acid precipitation** has a pH less than 5.0. The increased acidity is caused mainly by fumes from burning fossil fuels reacting with water in the atmosphere to form acids of sulphur and nitrogen (figure 2).

Effects on fish, trees, and soil

In the 1970s, rain with a pH as low as 3 was first noticed in Scandinavia, then in the north-eastern United States and south-eastern Canada, and later in northern Europe and Japan. This mixture of acids was responsible for the depletion of fish stocks in lakes and the destruction of large areas of forest. The acids react with soil minerals, releasing aluminium ions into lakes which interfere with oxygen uptake in the gills and kill the fish. The effect is so dramatic that about 700 of Norway's lakes are now completely devoid of fish. Acid precipitation has its greatest

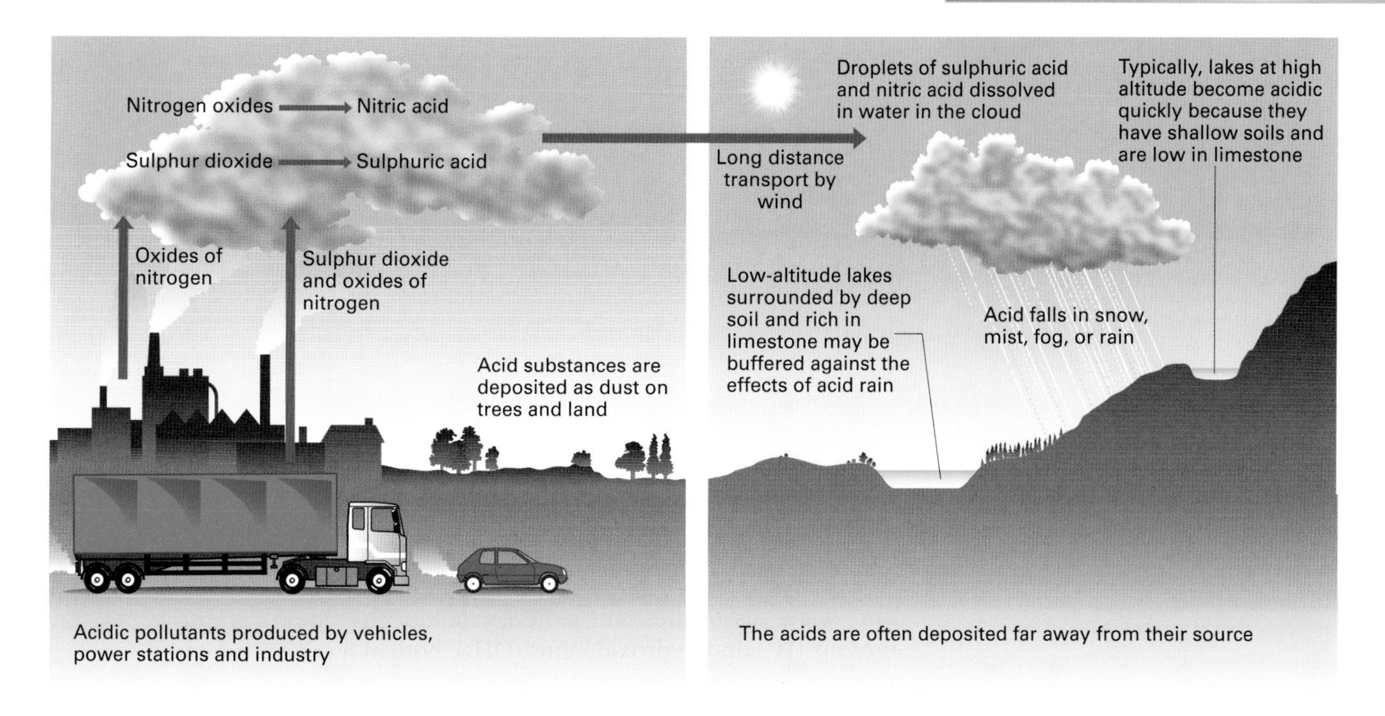

Figure 2 *Acidic gas emissions cause acid precipitation many miles from the source of the pollution.*

effect in spring when ice melts, releasing into the lakes all the pollutants accumulated during the winter. This inflow of acidic water coincides with the breeding period of many fish.

The effect of acid precipitation on trees is complex. Acid rain may affect plants directly, though only if the pH is exceptionally low. Acid mist, which clings to leaf surfaces, and airborne acid particles deposited directly on leaves probably cause more direct harm to trees than acid rain. Heavy deposits of acid cause yew and birch trees to lose their young branches, and conifers to lose their needles and become discoloured.

Acid precipitation has a dramatic effect on soils, which might explain some of the devastation in forests. As the acid precipitation trickles down through the soil, the protons interact with soil minerals. For example, they displace aluminium ions and magnesium ions, which are then dissolved in the soil water. This liberation of toxic aluminium ions can kill plants directly or make them susceptible to pests and diseases, while the dissolved magnesium ions, essential for plant growth, are leached away and lost from the soil.

International efforts are being made to reduce emissions of sulphur dioxide and other gases that contribute to acid precipitation. Unfortunately, the response of many governments has been slow because the financial cost of reducing emissions is high.

Buffers

Common aquatic plants and animals generally flourish in a fairly narrow range of pH. However, most of these organisms possess buffers which allow them to tolerate, at least for a short time, water at a pH of 6–9. A **buffer** is a substance that keeps the pH constant when small amounts of acid or alkali are added. They work to maintain a constant pH within an organism (see spread 2.9 for a fuller account of how buffers work).

Quick check

1 Copy and complete the following sentence, using the terms **solvent**, **solute**, and **solution**:

A ______ dissolves in a ______ to form a ______ .

2 How many more protons are there in a solution at pH 6 compared with an equal volume of a solution at pH 8?

3 Suggest why the term 'acid precipitation' is used in the text rather than 'acid rain'.

Food for thought

Norway is not a densely populated country, nor is it heavily industrialised. However, the acidification of its rivers and lakes has been rapid. In 1940, rivers and lakes had a pH range of 5.0–6.5; by 1976, they had a pH of 4.6–5.0, with a pH as low as 3.0 after snow melt or during dry spells. Suggest why Norway's lakes became so quickly acidified even though they are not in heavily industrialised regions.

Snow melt occurs in spring. Why does this have such a devastating effect on fish such as trout and salmon?

2.6

Carbohydrates: simple sugars

OBJECTIVES

By the end of this spread you should be able to:

- describe the main physical and chemical properties of sugars
- describe the polymerisation of monosaccharides to form polysaccharides
- discuss the functions of sugars.

Fact of life

Untreated cow's milk is not the perfect food for cats – it can cause diarrhoea and flatulence (wind)! As adults, many cats stop making lactase, the enzyme that breaks down lactose (milk sugar) into glucose and galactose. Undigested lactose therefore builds up in the gut. This encourages the growth of bacteria, which produce the unpleasant gut problems.

Carbohydrates are a group of substances that are important in many biological processes. They provide energy-rich nutrients to organisms and are used to build their body structures.

All **carbohydrates** contain carbon, hydrogen, and oxygen. They have the common formula $C_x(H_2O)_y$, where x and y are variable numbers that may be the same or different. Carbohydrates include sugars, starch, glycogen, and cellulose.

Monosaccharides

Monosaccharides are simple sugars. (The word means 'single sugar'.) They all have the general formula $(CH_2O)_n$, showing that the elements carbon, hydrogen, and oxygen are always present in the same ratio. The letter n can be any number from 3 to 7. Monosaccharides are grouped according to the value of n:

- in trioses $n = 3$
- in tetroses $n = 4$
- in pentoses $n = 5$
- in hexoses $n = 6$
- in heptoses $n = 7$.

Glucose is the best known and most abundant hexose. Its chemical formula is $C_6H_{12}O_6$, showing that each molecule contains 6 atoms of carbon, 12 atoms of hydrogen, and 6 atoms of oxygen.

The formula does not show how these atoms are structurally arranged. In fact, glucose can take up a number of different shapes. This phenomenon is called **isomerism**. Each **isomer** has the same chemical formula but a different structural formula.

Glucose has different isomers

Figure 1 shows two common isomers of glucose: alpha glucose and beta glucose. Notice that the ring structures are very similar – only the positions of the hydrogen and hydroxyl groups on carbon atom 1 are different. Imagine the glucose molecule as a three-dimensional structure. The CH_2OH group is above the ring in both alpha and beta glucose. The OH group is *below* carbon atom 1 in alpha glucose but *above* carbon atom 1 in beta glucose.

The different arrangements of glucose have important biological consequences: for example, alpha glucose molecules combine to form starch whereas beta glucose molecules combine to form cellulose (see spread 2.7). This illustrates a key concept in biology: the detailed positions of atoms in a biological molecule determine both the shape of the molecule and its function.

Other important hexoses include fructose (fruit sugar) and galactose. Important pentoses are ribose and deoxyribose (constituents of the genetic material, RNA and DNA).

Disaccharides

Figure 2 shows how two monosaccharides can combine to form a 'double sugar' or **disaccharide**. This is called a **condensation** reaction because it involves the removal of a water molecule.

- **Maltose** (malt sugar) is a disaccharide made up of two glucose molecules.
- **Sucrose** (table sugar) is made up of a glucose and a fructose.
- **Lactose** (milk sugar) is made up of a glucose and a galactose.

Disaccharides can be broken down to their constituent monosaccharides by **hydrolysis**. This is a chemical reaction involving the addition of water.

Structural formula of alpha glucose

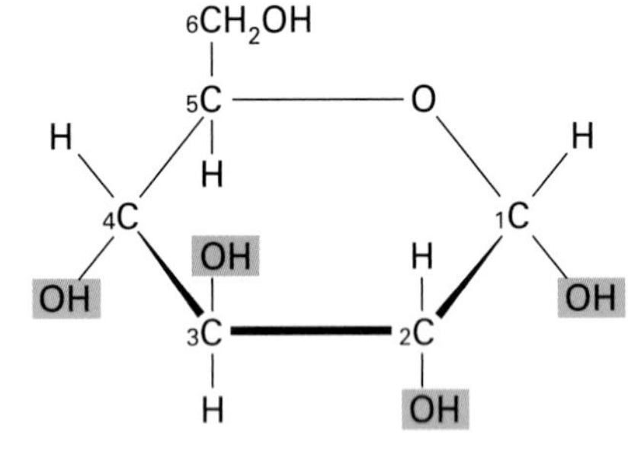

Structural formula of beta glucose

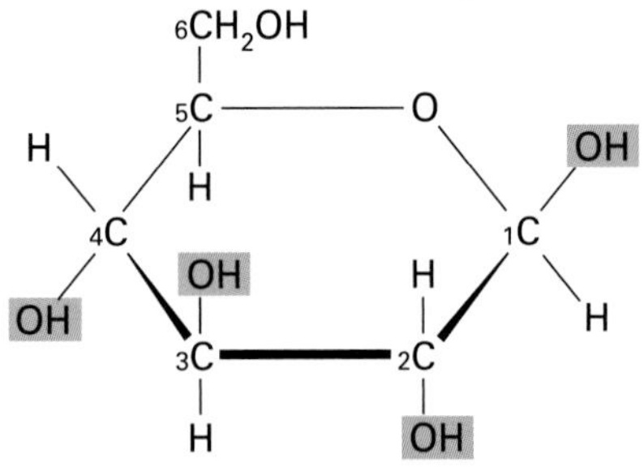

Simplified structural formula of alpha glucose

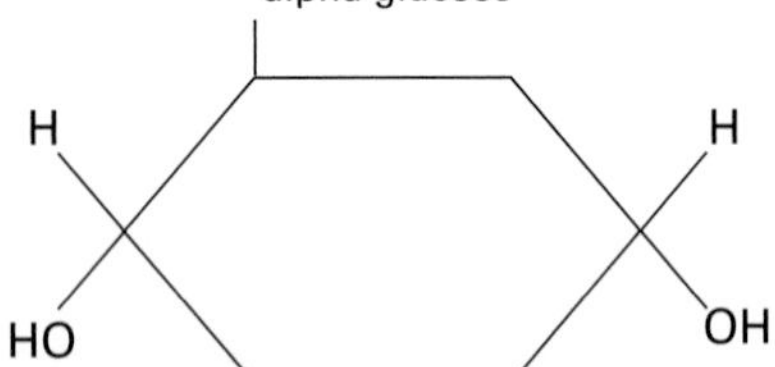

Figure 1 *The structure of glucose, $C_6H_{12}O_6$. The ring lies at right angles to the plane of the paper. The groups connected by thick lines lie in front of the groups connected with thin lines. The various side groups stick up and down at angles.*

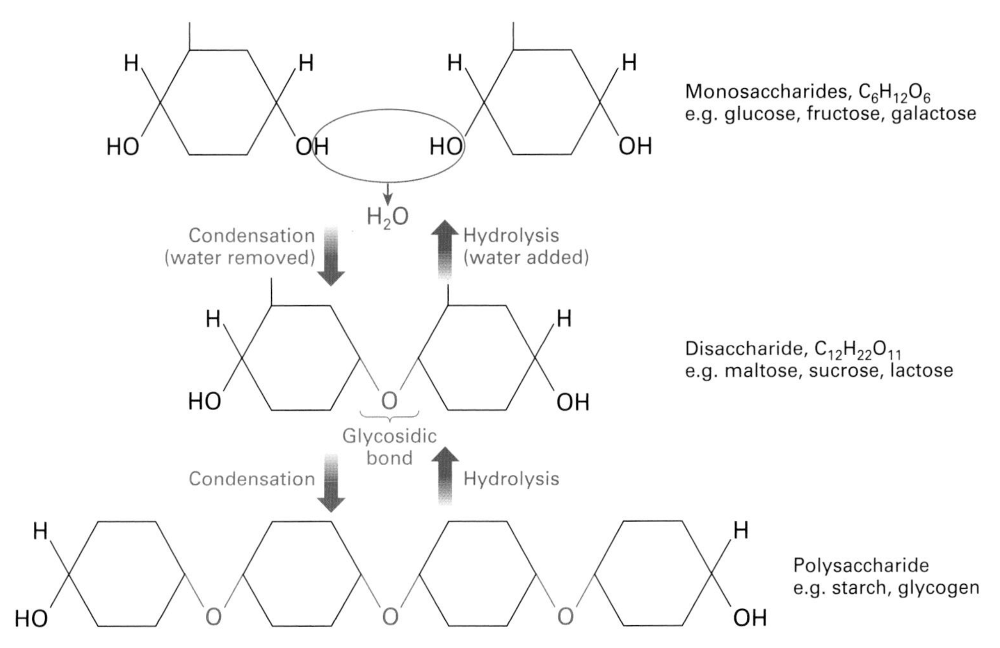

***Figure 2** The formation of disaccharides and polysaccharides by condensation (removal of water), and their breakdown to monosaccharides by hydrolysis (addition of water).*

Polysaccharides

Many monosaccharide units can be added to a disaccharide by a series of condensation reactions. A molecule with between three and ten monosaccharide units is called an **oligosaccharide**. More and more units can be added to form a very large molecule called a **polysaccharide**. This process is called **polymerisation**. Large molecules that are made up of repeating units are called **polymers**; each unit is called a **monomer**. Polysaccharides are polymers made up of monosaccharide monomers.

Monosaccharide monomers may also combine with other types of molecule to form **conjugated molecules**. Chains of monosaccharide units can combine with lipids to form glycolipids, or with proteins to form glycoproteins. These molecules are important in cell membranes.

Functions of simple sugars in living organisms

- Glucose is the major source of energy for most animals. Each gram of glucose yields approximately 16 kJ of energy when fully broken down during respiration. Glucose is also the main form in which carbohydrate is transported around the body in animals.
- Lactose is the main sugar in milk.
- Maltose is produced by the breakdown of amylose in many germinating seeds.
- In plants, carbohydrate is moved from one region to another as sucrose.

Quick check

1 What is the general formula for monosaccharides?

2 Name the bond that links the glucose monomers in a disaccharide.

3 In what form is carbohydrate transported in:

a mammals

b flowering plants?

Food for thought

Fructose is the sweetest of natural sugars, one and a half times sweeter than sucrose. Fructose is used as a sugar substitute for diabetics. Suppose a cake recipe specifies 100 g of sugar; how much fructose would you need?

Sugar in the diet

Monosaccharides, disaccharides, and oligosaccharides all have fewer than ten monosaccharide units (monomers). They are therefore classified as **sugars** – sweet-tasting crystalline substances that are soluble in water.

In the UK, the average person eats more than 1 kg of sugar each week, mostly sucrose. Sugar is a source of energy but has no other nutritional value. (It is sometimes described as 'empty calories'.) *Moderate* consumption of sugar is not harmful. However table sugar has been referred to as 'pure, white, and deadly' because eating very large amounts of it may be associated with obesity, diabetes, and heart disease. These diseases are likely to be linked to general overeating and lack of exercise, as well as eating too much sugar. Nevertheless, people who are obese or who respond adversely to high levels of sugar (including diabetics) should not eat too much sugar.

High sugar intake can also lead to tooth decay and to certain fungal infections. For example, candidosis (commonly called thrush) is a disease caused by yeast-like fungi that thrive in sugary environments.

2.7

O B J E C T I V E S

By the end of this spread you should be able to:

- describe the differences between sugars and polysaccharides
- explain how the structures of starch, glycogen, and cellulose are related to their functions.

Fact of life

Cellulose is the principal structural material of plants. As such, it is the most abundant organic compound in the world.

Carbohydrates: polysaccharides

Polysaccharides are polymers made up of many monosaccharide units (monomers), which are linked by condensation reactions (see spread 2.6). They have the general formula $(C_6H_{10}O_5)_n$. Polysaccharides are relatively insoluble in water, are not sweet, and cannot be crystallised.

The monosaccharides can join together in different ways to form chains or ring structures. Chains may be straight, helical, coils (spirals), or branched. The properties of a polysaccharide depend on the number and type of monomer it contains, and how these are joined together.

By convention, each carbon atom in a monosaccharide unit is identified by a number (see spread 2.6, figure 1).

- A condensation reaction between the hydroxyl groups at carbon 1 of one monosaccharide and carbon 4 of the other results in a bond called a **1–4 glycosidic bond**.
- If the reaction is between the hydroxyl groups on carbon 1 and carbon 6, a **1–6 glycosidic bond** is formed.
- Straight chains are formed by monomers linked by 1–4 glycosidic bonds; branched chains have one or more 1–6 glycosidic bonds (figure 1).

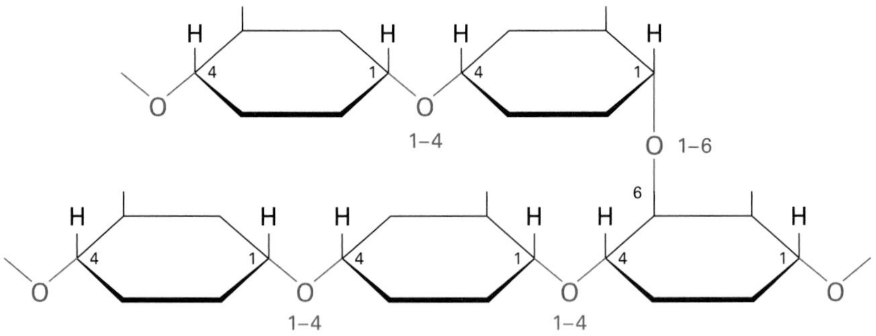

Figure 1 *Part of a glycogen molecule showing a straight-chained section of alpha glucose units interconnected by 1–4 glycosidic bonds, and a branch formed by a 1–6 glycosidic bond. In glycogen, the 1–6 branches occur at intervals of 8–10 glucose units. In amylopectin (a constituent of starch which has a similar structure to glycogen), the branches occur at intervals of 12–20 glucose units.*

Starch: energy storage in plants

Starch is a polysaccharide formed from alpha glucose units. It consists of two components: **amylose** and **amylopectin**. Both are formed by the condensation of alpha glucose units, but the units are arranged differently. In amylose, the units are linked mainly by 1–4 glycosidic bonds which form unbranched chains. Amylopectin has many more 1–6 glycosidic bonds, producing highly branched chains.

Because of its structure (figure 2) starch is compact which is ideal for a storage product. In flowering plants, starch granules are confined to double-membraned organelles called **plastids**. These may be colourless **amyloplasts** in storage structures such as potato tubers (figure 3), or green **chloroplasts** in leaves.

Starch is broken down to glucose for respiration. It is also a source of organic carbon for making other substances.

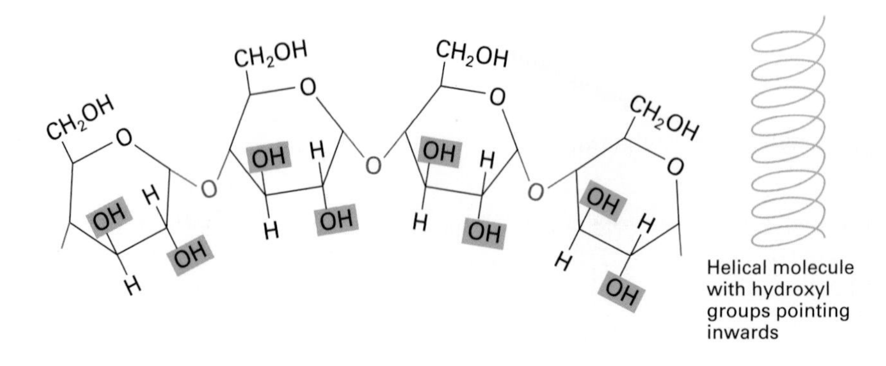

Figure 2 *The structure of a starch molecule. Amylose and amylopectin are laid down in successive rings to form starch granules. The long chains of alpha glucose units are coiled into a helix, forming a cylinder-like structure with most of the hydroxyl groups pointing inwards.*

Glycogen: energy storage in animal cells

Glycogen is also a storage polysaccharide. It is found as small granules (figure 4) and is particularly abundant in liver and muscle cells. Glycogen has a similar role and structure to starch. (It is sometimes called 'animal starch'.) However, it is much more branched because it has many more 1–6 glycosidic bonds. This makes glycogen less dense and more soluble than starch. Glycogen can be also hydrolysed readily by enzymes and is broken down more rapidly than starch. This is reflected in the higher metabolic rate in animals compared with plants.

Figure 3 *A photomicrograph of potato cells showing starch granules (×300).*

Structural materials: cellulose

Cellulose is a tough structural polysaccharide. It is the major constituent of plant cell walls. The structure of cellulose is shown in figures 5 and 6. Cellulose is completely permeable: it allows water and dissolved substances to enter and leave plant cells freely. The cells swell when they take in water by osmosis (spread 4.9) and the cell wall prevents the cells from bursting when this happens. Cellulose becomes impermeable when the gaps between fibres are filled with impermeable substances.

Unlike starch and glycogen, cellulose cannot be hydrolysed easily. Herbivores such as cows and elephants are able to digest grass because microorganisms in their guts (spread 9.10) produce **cellulase**, the enzyme that digests cellulose. Humans and most other animals do not produce cellulase so they cannot obtain the nutrient content from plant cells.

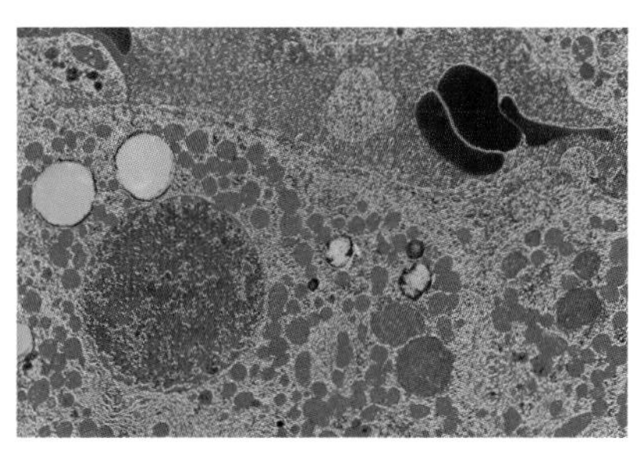

Figure 4 *A photomicrograph of a liver cell showing glycogen granules which appear as pink–red packages (×3400).*

Lignin

In wood, cellulose is further strengthened by **lignin**. Lignin is a highly complex non-carbohydrate polymer. It impregnates the cell walls of the water-transporting tubes (xylem) to form an impermeable lining, a process called lignification. Xylem cells die when they become completely impregnated or lignified. In addition to strengthening cells, lignin also helps prevent rot, infection, and decay.

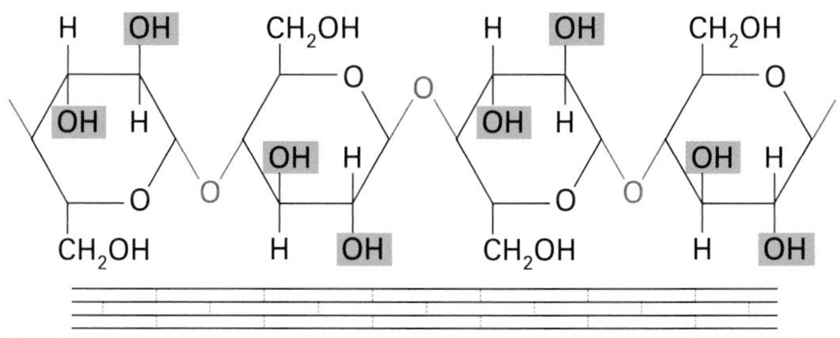

Figure 5 *The structure of a cellulose molecule. Cellulose is formed from beta glucose units linked by 1–4 glycosidic bonds. The hydroxyl groups alternate on either side of the molecule. This forms straight chains, giving cellulose a fibrous structure. Cellulose fibres are strengthened further by hydrogen bonds that link adjacent chains.*

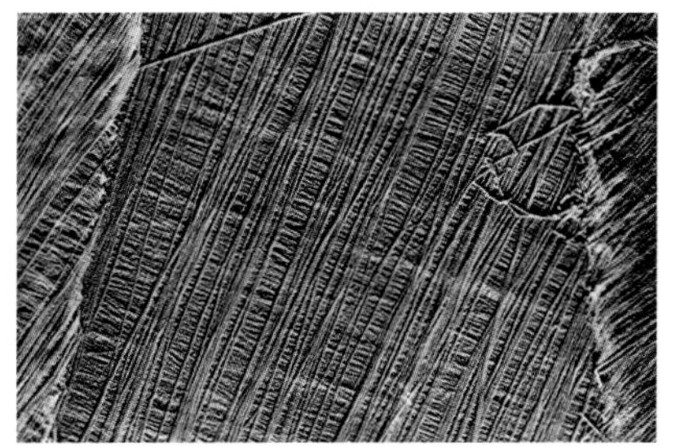

Figure 6 *An electron micrograph of a cell wall of the seaweed* Chaetomorpha. *The cell wall is made up of microfibrils which are laid down in layers. Each microfibril is a rope-like structure, about 30–40 nm in diameter, consisting of a mass of cellulose chains. Microfibrils of one layer lie roughly parallel, but at an angle to those in other layers. Some of the microfibrils are interwoven, adding strength to the cell wall.*

Quick check

1. List three differences between a sugar and a polysaccharide such as starch.
2. **a** What is the main function of starch?
   b What is the main function of cellulose?
   c Outline the essential structural differences between starch and cellulose.

Food for thought

Sea squirts are marine animals which belong to the chordates, the same phylum as ourselves. They have a larval stage which resembles a tadpole, but the adults are barrel-shaped creatures that no longer move around. Typically, sea squirts remain attached by one end to a rock and filter feed. The body of a sea squirt is covered by a special mantle called a tunic. Curiously, the principal component of most tunics is a type of cellulose called tunicine. Suggest the advantages to sea squirts of having cellulose in their tunics.

2.8

LIPIDS

OBJECTIVES

By the end of this spread you should be able to:

- describe the structure of fats and oils
- list the functions of lipids
- describe the structure of a phospholipid.

Fact of life

Worldwide, humans use 80 million tonnes of fats and oils every year, most of which are derived from plants. In addition to being used in margarines, soaps, candles, putties, printing inks, and varnishes, plant oils are also used as a fuel for vehicles. Biodiesel, oil derived from plants such as oil palms and sunflowers, is commonly used in tropical areas that do not have their own 'fossil' oils.

Lipids are a diverse group of compounds which are insoluble in water but dissolve readily in other lipids and in organic solvents such as ethanol, chloromethane, and diethyl ether. Lipids all contain carbon, hydrogen, and oxygen, though the proportion of oxygen is lower than in carbohydrates.

Fats and oils

Triglycerides (called **triacylglycerols** by chemists) are lipids made from glycerol and fatty acids. Triglycerides are also called true fats or neutral fats.

Glycerol is an alcohol that contains three carbon atoms each linked to a hydroxyl group. A **fatty acid** has a long chain of hydrogen and carbon atoms (a **hydrocarbon chain**) ending in an acidic carboxyl (—COOH) group. The carboxyl group ionises in water to release a hydrogen ion (proton), which gives fatty acids their acidic properties:

$$—COOH \rightleftharpoons —COO^- + H^+$$

A triglyceride is formed when each hydroxyl group on the glycerol molecule combines with a carboxyl group on a fatty acid molecule, in a reaction called a **condensation** (figure 1). Water is removed and an **ester bond** is formed.

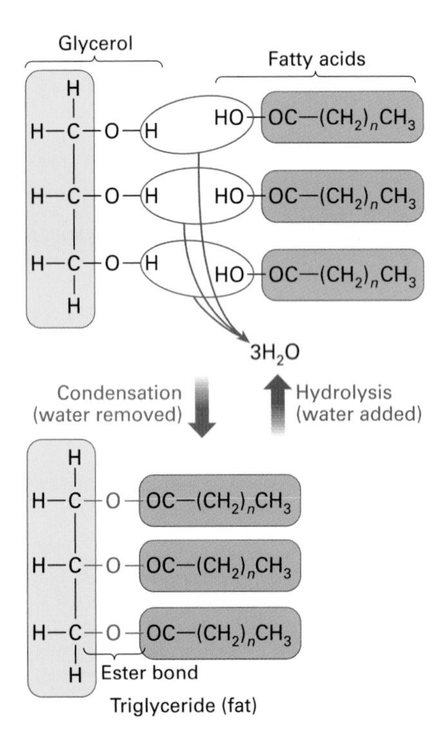

Figure 1 Formation of a triglyceride molecule from glycerol and three fatty acids. Three molecules of water are removed in the condensation reaction.

Triglycerides are sometimes referred to under the umbrella term 'fats'. Sometimes they are grouped according to their physical state: those that are liquid at room temperature (20 °C) are called oils; those that are solid are called fats. Each molecule of a fat or oil is made up of glycerol combined with three fatty acids which may all be the same, or may be different.

Fatty acids may be **saturated** or **unsaturated**. In a **saturated fatty acid**, each carbon atom in the hydrocarbon chain is linked to a carbon atom on either side and also to two hydrogen atoms. The carbon atoms are bonded to the maximum number of other atoms. Saturated fatty acids have only single bonds in the hydrocarbon chain (table 1). This makes the chain relatively straight so that triglycerides that contain only saturated fatty acids are able to pack closely together. Thus, triglycerides of saturated fatty acids tend to be solid ('fats') at room temperature.

In an **unsaturated fatty acid**, such as oleic acid (a major constituent of olive oil), the carbon atoms in the hydrocarbon chain are not bonded to the maximum number of other atoms. Two or more carbon atoms have double bonds between them (table 1). Fatty acids with one double bond are called **monounsaturated fatty acids**; those with two or more double bonds are called **polyunsaturated fatty acids**.

The atoms around a double bond may be arranged in either the *cis*-form or the *trans*-form (figure 2). Triglycerides with a high proportion of unsaturated *cis*-fatty acids tend to be oils because the *cis*-double bonds kink the hydrocarbon chain, which prevents the molecules from packing closely together. This means the attractions between the molecules are very weak, so the substance is a liquid rather than a solid. Triglycerides

Table 1 Structures of some common fatty acids. The skeletal formula shows only the carbon-to-carbon bonds in the hydrocarbon chain and the terminal carboxyl (COOH) group.

Trivial name	Systematic name	Structural formula	Skeletal formula
Stearic acid	Octadecanoic acid	$CH_3(CH_2)_{16}COOH$	COOH
Oleic acid	Octadec-*cis*-9-enoic acid	$CH_3(CH_2)_7CH{=}CH(CH_2)_7COOH$	COOH

with more unsaturated *trans*-fatty acids tend to be solids because the hydrocarbon chains are straighter, leading to properties similar to saturated fatty acids.

Functions of fats and oils in living organisms

Triglycerides have a higher proportion of hydrogen than either carbohydrates or protein, making them a more concentrated source of energy: each gram of fat or oil yields about 38 kJ, more than twice the energy yield of a gram of carbohydrate. In mammals, excess fat is laid down for storage in special connective tissue (called **adipose tissue**) under the skin.

As well as being an energy store, fats and oils have other functions, including:

- heat insulation – fat is a bad conductor of heat; mammals tend to increase their adipose tissue in winter to reduce heat loss
- shock absorption – delicate mammalian organs such as the kidneys which are vulnerable to knocks and bumps have relatively thick layers of fat around them
- buoyancy – many single-celled aquatic organisms produce an oil droplet to aid buoyancy.

Dietary fat is also a source of fatty acids (see spread 9.7) and of phospholipids, which have a number of essential functions.

Phospholipids

Phospholipids form a major part of cell membranes (see spread 4.6), including the myelin sheath around nerve fibres that allows the rapid conduction of nerve impulses. Phospholipids consist of glycerol attached to two (not three) fatty acid chains. The third hydroxyl group of glycerol combines with phosphoric acid to form a polar phosphate group.

Schematic diagrams such as figure 3 show a phospholipid as a tadpole-like structure with the phosphate group forming the head and the fatty acid chains a double tail. The polar head is strongly attracted to water: it is hydrophilic ('water-loving'). The non-polar tail, being made up of hydrocarbon chains, is oily and therefore repelled by water: it is hydrophobic ('water-hating'). The phospholipid molecule is said to be **amphipathic**: that is, one end is hydrophilic and the other end is hydrophobic.

In water, phospholipid molecules collect together in a single (monomolecular) layer with the hydrophilic heads poking into the water. In cells, both the intracellular environment and immediate external environment are watery. This causes phospholipids to form a double layer, with the hydrophobic tails pointing inwards, away from the watery environment. The phospholid bilayer gives cell membranes their fluid properties and allows lipid-soluble substances to pass easily through them (see spread 4.6).

Waxes, cholesterol, and steroids

Apart from triglycerides and phospholipids, other lipids include waxes, cholesterol, and steroids. **Waxes** are similar to triglycerides, but contain fatty acids bonded to long-chain alcohols rather than to glycerol. Waxes are usually relatively hard solids at 20 °C, providing protection and waterproofing on the surfaces of insects and leaves.

Steroids contain four rings of carbon and hydrogen atoms with various side chains. Many animal hormones are steroids, including oestrogen and testosterone (see spread 12.4), which are made from cholesterol, a lipid which shares the same four-ring structure.

Cholesterol is also a raw material for the manufacture of vitamin D. It is a vital component of mammalian cell membranes, strengthening the membranes at high body temperatures.

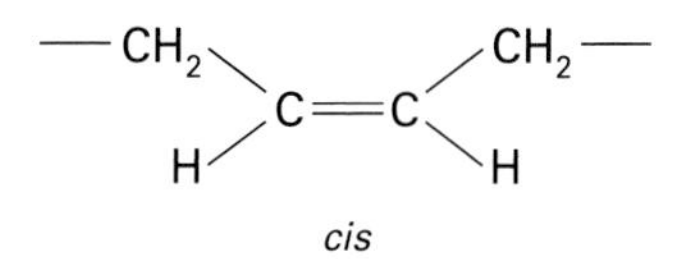

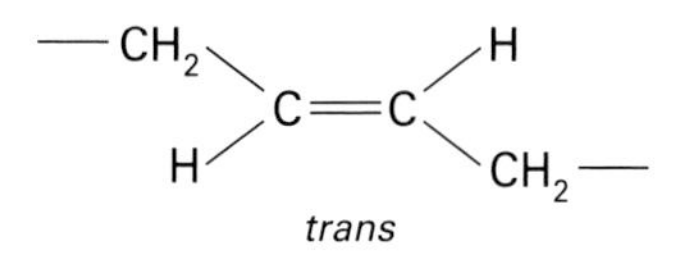

***Figure 2** The arrangement around the double bond in a* cis-*fatty acid and a* trans-*fatty acid.*

***Figure 3** Structure of a phospholipid.*

QUICK CHECK

1. Name the products of digestion of true fats.
2. Why is fat such a rich store of energy?
3. State two ways in which a phospholipid differs from a triglyceride.

Food for thought

Solid margarines are produced by treating plant oils with a stream of hydrogen in the presence of a metal catalyst such as platinum. Suggest how this process changes the plant oils.

What are the health implications of eating solid margarines rather than plant oils?

2.9 Introduction to proteins

OBJECTIVES

By the end of this spread you should be able to:

- list the main functions of proteins
- describe the structure of an amino acid
- show how amino acids polymerise
- explain how amino acids can act as buffers.

Fact of life

Simple organisms such as bacteria have about 1000 different types of protein molecule in their cells; humans probably have between 50 000 and 100 000. The largest known human protein is dystrophin. It has a relative molecular mass of 350 000 and contains about 3500 amino acids. Some people cannot make dystrophin in their cells, and this leads to muscular dystrophy, an inherited disorder characterised by muscle weakness and wasting.

Proteins are large complex biological molecules which play many diverse roles in all organisms. Proteins make up a high percentage of the structure of living things (for example, about 18 per cent of the human body is protein) and also take part in the fundamental processes that make up life. Every organism contains thousands of different kinds of proteins, each with its own unique three-dimensional structure which enables it to carry out a specific function.

Types of protein

Proteins can be grouped into seven major classes based on their functions:

- **enzymes**, biological catalysts that control biochemical reactions, for example amylase, which catalyses the digestion of starch, and ATPase, an enzyme found in all cells which catalyses the breakdown and formation of ATP
- **structural proteins** that form part of the body of an organism, for example the silk of spiders, collagen that makes up tendons and ligaments, and keratin, the major component of hair
- **signal proteins** that carry messages around the body, for example insulin, a hormone involved in controlling glucose levels in the blood
- **contractile proteins** involved in movement, for example actin and myosin, the proteins that enable muscles to contract
- **storage proteins**, for example albumen, the protein store that forms the white of eggs
- **defensive proteins**, for example blood antibodies that fight infections
- **transport proteins**, for example haemoglobin, the carrier of oxygen in the blood.

Although there are many millions of proteins, all are made from the same basic building blocks, namely approximately 20 kinds of **amino acid**.

Amino acids

All amino acids have an **amino group** (—NH_2) and a **carboxyl group** (—COOH), as shown in their general structure (figure 1).

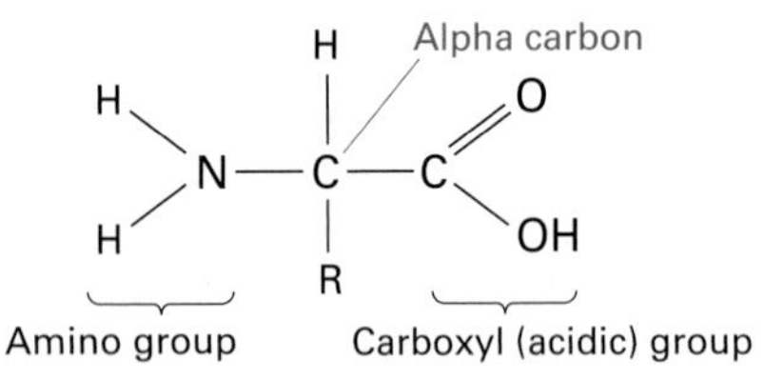

Figure 1 *General structure of an amino acid.*

Table 1 *The amino acids that make up proteins.*

Amino acid	Abbreviations		R group (side chain)	Amino acid	Abbreviations		R group (side chain)
Glycine	Gly	G	H	Aspartic acid	Asp	D	CH_2COOH
Alanine	Ala	A	CH_3	Asparagine	Asn	N	CH_2CONH_2
Valine	Val	V	$CH(CH_3)_2$	Glutamic acid	Glu	E	$(CH_2)_2COOH$
Leucine	Leu	L	$CH_2CH(CH_3)_2$	Glutamine	Gln	Q	$(CH_2)_2CONH_2$
Isoleucine	Ile	I	$CH(CH_3)CH_2CH_3$	Proline	Pro	P	H_2N O OH
Serine	Ser	S	CH_2OH	Tryptophan	Trp	W	N
Threonine	Thr	T	$CH(OH)CH_3$	Phenylalanine	Phe	F	
Lysine	Lys	K	$(CH_2)_4NH_2$	Tyrosine	Tyr	Y	OH
Arginine	Arg	R	$(CH_2)_3NHCNHNH_2$	Methionine	Met	M	$(CH_2)_2SCH_3$
Histidine	His	H	H N N	Cysteine	Cys	C	CH_2SH

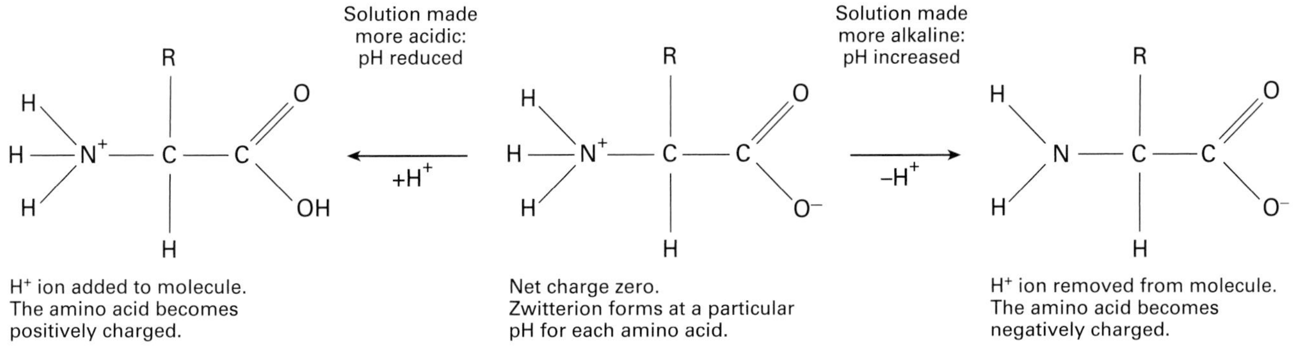

Solution made more acidic: pH reduced

Solution made more alkaline: pH increased

H^+ ion added to molecule. The amino acid becomes positively charged.

Net charge zero. Zwitterion forms at a particular pH for each amino acid.

H^+ ion removed from molecule. The amino acid becomes negatively charged.

Figure 2 *The amphoteric nature of an amino acid.*

The amino group is attached by a covalent bond to a central carbon atom called the alpha carbon. A hydrogen atom, another carbon atom, and a side chain represented by the letter R are also linked to the alpha carbon. The R group is different for each of the 20 amino acids (table 1) and it determines the specific properties of a given amino acid.

In the simplest amino acid, **glycine**, the R group is a single hydrogen atom. Several amino acids contain R groups with relatively complex ring structures. Some R groups are non-polar and hydrophobic (such as that of leucine), while others are polar and hydrophilic (such as those of cysteine and serine). Proteins containing amino acids with hydrophilic R groups can dissolve more readily in solutions within cells.

Amino acids as buffers

Amino acids are **amphoteric**; that is, they have both acidic and basic properties when they dissociate in water (figure 2). The acidic properties are derived from the carboxyl group which can donate a proton so the molecule becomes negatively charged when in an alkaline solution. The basic properties are derived from the amino group which can take up a proton so the molecule becomes positively charged in an acidic solution. At a pH that is quite specific for each amino acid, the molecule ionises so that different parts of the molecule have positively and negatively charged groups simultaneously. Ions like this with both positive and negative charges are called **zwitterions**.

The ability to donate or receive protons causes amino acid solutions to behave as **buffers**. A buffer solution tends to resist changes in pH when an acid or base is added to it. Buffer systems play an essential role in the human body, keeping the pH of the blood and of other fluids within tolerable levels.

Peptide bonds: joining amino acids together

Two amino acids can combine to form a **dipeptide** by a condensation reaction between the carboxyl group of one and the amino group of the other (figure 3). The resulting bond linking the two amino acids is called a **peptide bond**. Further amino acids can be added to either end of the dipeptide to form a **polypeptide chain**. Proteins consist of one or more polypeptide chains: they are polymers made up of amino acid monomers.

Quick check

1 What is the function of:

- **a** haemoglobin
- **b** amylase
- **c** insulin?

2 Name the four elements found in all amino acids.

3 Draw a peptide bond.

4 What is a zwitterion?

Food for thought

Over 170 amino acids are known to occur in cells and tissues. Of these, only 20 are commonly found in proteins. Suggest one function that all amino acids and proteins share.

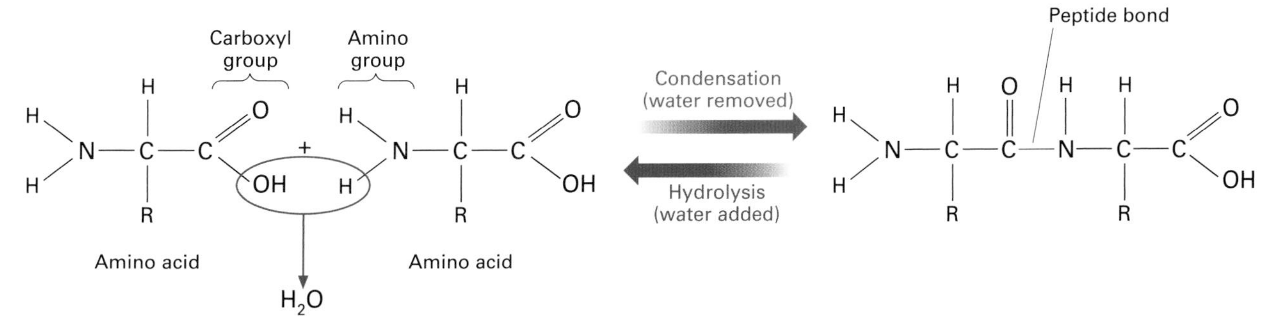

Figure 3 *The formation of a dipeptide by a condensation reaction, resulting in a peptide bond.*

2.10

OBJECTIVES

By the end of this spread you should be able to:

- describe the four levels of protein structure
- explain the significance of a protein's shape
- classify proteins according to their structure.

Fact of life

Every organism has its own particular complement of proteins. This is called its **proteome**. Different proteomes have evolved to function in different environments. The proteomes of thermophilic bacteria living in hot volcanic springs, for example, have evolved to function at temperatures in excess of 80 °C, while the proteome of ice fish living in Antarctic waters have evolved to function at sub-zero temperatures. The ability of a protein to carry out its specific function in a particular environment depends on its structure. The study of the structure and function of the complete complement of proteins of an organism is called **proteomics**.

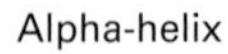

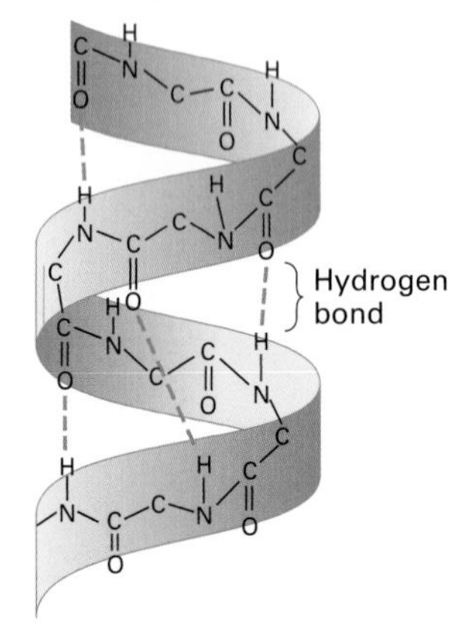

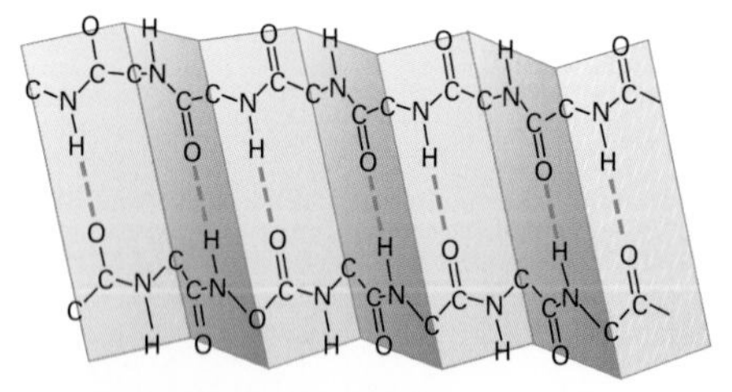

Figure 2 An alpha-helix and a beta-pleated sheet. Both secondary structures depend on hydrogen bonding of the peptide bond parts of the molecule at different areas along the polypeptide chain.

PROTEIN STRUCTURE AND FUNCTION

Proteins consist of one or more chains of amino acids (polypeptide chains) folded into a unique three-dimensional shape. The shape is determined by up to four levels of structure: the primary structure, the secondary structure, the tertiary structure, and the quaternary structure. Each level of structure determines the next. **Simple proteins** consist of amino acids only; **conjugated proteins** also contain a non-amino acid part called a **prosthetic group**.

Primary structure: the sequence of amino acids

A protein's **primary structure** is the sequence of amino acids that make up its polypeptide chain or chains. Figure 1 shows the primary structure of myoglobin, an oxygen-carrying conjugated protein consisting of one polypeptide chain of 153 amino acids, and an iron-containing prosthetic group attached to the polypeptide chain.

```
  1 GLSDGEWQLVLNVWGKVEADIPGHGQEVLIRLFKGHPETLEKFDRFKHLK
 51 SEDEMKASEDLKKHGATVLTALGGILKKKGHHEAEIKPLAQSHATKHKIP
101 VKYLEFISEAIIQVLQSKHPGDFGADAQGAMNKALELFRKDMASNYKELG
151 FQG
```

Figure 1 The primary structure of myoglobin: the sequence of 153 amino acids in the myoglobin polypeptide chain. The letters refer to the single-letter abbreviations in table 1 on spread 2.9.

In order for myoglobin or any other protein to carry out its specific function, it must contain the correct amino acids arranged in a precise order. Its ability to function may be severely disrupted if only one amino acid is out of place. For example, a particular change in a single amino acid in haemoglobin, the main oxygen-carrying blood protein, causes sickle-cell anaemia, a serious blood disorder.

Secondary structure: folding and coiling

When amino acids join up in the polypeptide chain, a variety of forces between different parts of the molecule and hydrogen bonding (see spread 2.3) cause the chain or regions of the chain either to coil into an **alpha-helix** or to fold into a **beta-pleated sheet** (figure 2). This coiling or folding is the protein's **secondary structure**. The shape of the helix or sheet is maintained by regularly spaced hydrogen bonds each formed between the —N—H group of one amino acid and the — C = O group of another amino acid in a different part of the polypeptide chain. A single polypeptide chain may have some regions coiled into an alpha-helix and others folded into beta-pleated sheets.

Tertiary structure: the 3D shape

The **tertiary structure** refers to the overall three-dimensional shape of a polypeptide chain (figure 3). Proteins are classified into two main groups on the basis of their tertiary structure:

- **Fibrous proteins** consist of parallel polypeptide chains cross-linked at intervals to form long fibres or sheets. Fibrous proteins are usually insoluble in water and physically tough, which suits them for their mainly structural functions. Fibrous proteins include collagen (a major constituent of tendons and bone), silk (which forms the threads of a spider's web), and keratin (the main component of hair).
- In **globular proteins**, the polypeptide chains are tightly folded to form a spherical shape. Many globular proteins are folded so that their hydrophobic groups are on the inside of the molecule and the hydrophilic groups face outwards, making these proteins soluble in water. Globular proteins include enzymes, antibodies, and many hormones.

The precise three-dimensional shape of a globular protein molecule determines its function: every coil and twist, bump and indentation is important. The shape is maintained by various bonds including ionic bonds (spread 2.2), hydrogen bonds (spread 2.3), disulphide bonds (figure 4), and **hydrophobic interactions** (interactions between non-polar, water-repellent groups in the protein). The tertiary structure of a protein depends on its primary structure, as the bonds holding the tertiary structure together can only form if the correct amino acids are at specific points along a polypeptide chain.

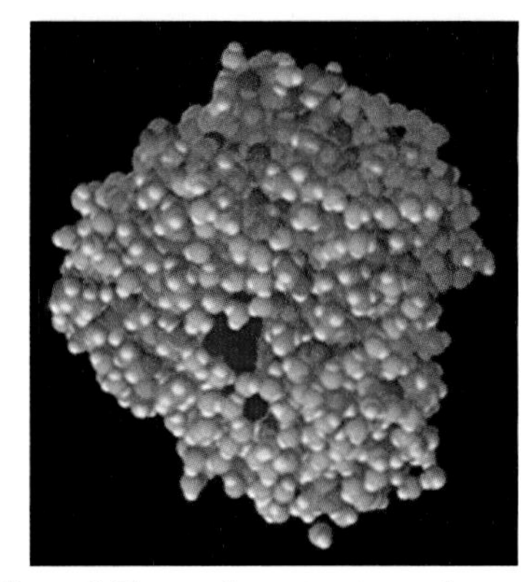

Figure 3 *The tertiary structure of a myoglobin molecule, represented as a space-filling model which shows the molecule's surface shape.*

Breaking down the tertiary structure: denaturation

If the bonds holding the protein in shape are broken, a process called **denaturation** occurs. The primary structure is retained but the polypeptide chains unravel and lose their specific shape (figure 5). As a result, denatured globular proteins lose their specific function.

Denaturation is nearly always irreversible. It can be caused by changes in pH, salt concentration, or temperature. The effects of denaturation are dramatically demonstrated every time an egg is boiled or fried: the heat quickly causes the transparent protein-containing area around the yolk to irreversibly solidify and become white and opaque.

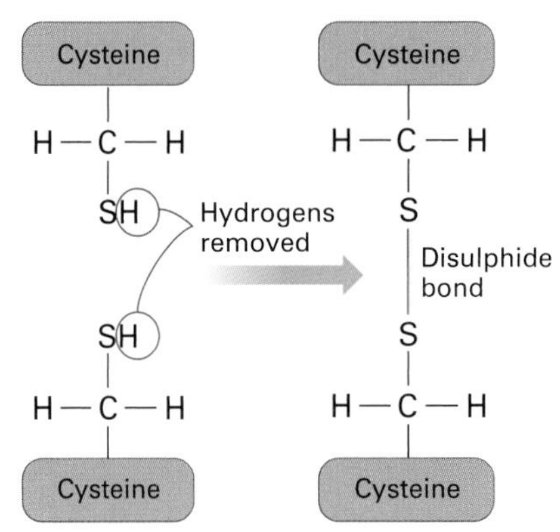

Figure 4 *A disulphide bond forms between two cysteine amino acids. It is one of the strongest and most important bonds that maintain the shape of protein molecules.*

If the protein is soluble, denaturation makes it insoluble and the protein is inactivated

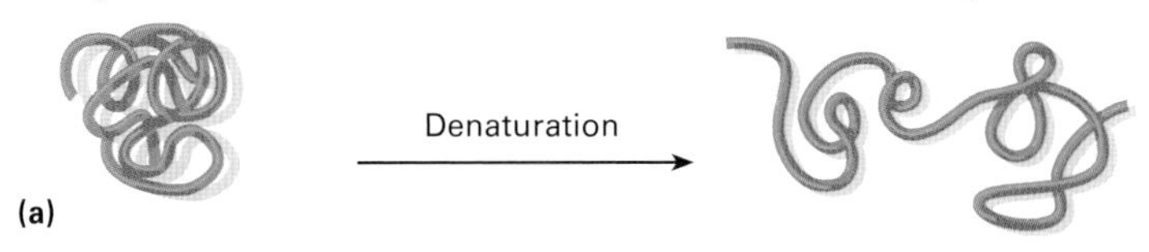

If the protein is an insoluble fibrous protein, it loses its structural strength

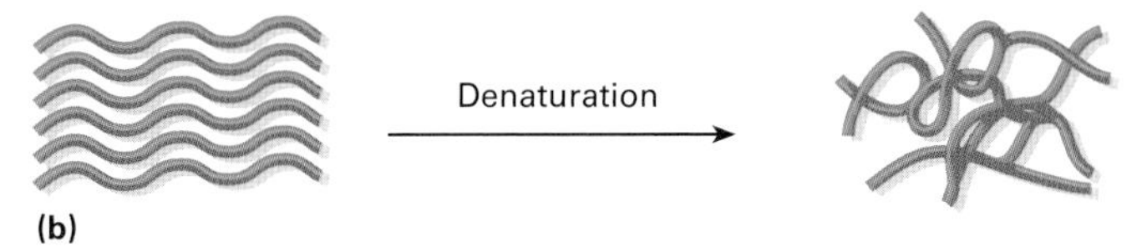

Figure 5 *Denaturation of:* ***(a)*** *a globular protein and* ***(b)*** *a fibrous protein.*

Quaternary structure: the association of polypeptide chains

Many proteins consist of more than one polypeptide chain chemically bonded to each other. The **quaternary structure** refers to the way these polypeptide chains are arranged in the protein. Haemoglobin, for example, has four polypeptide chains of two distinct types and four non-protein haem groups (figure 6), whereas **collagen**, a fibrous protein, consists of three helical polypeptides that are supercoiled to form a ropelike structure of great strength.

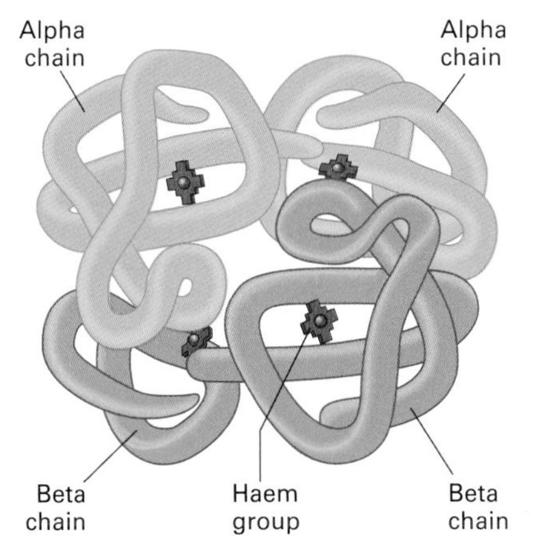

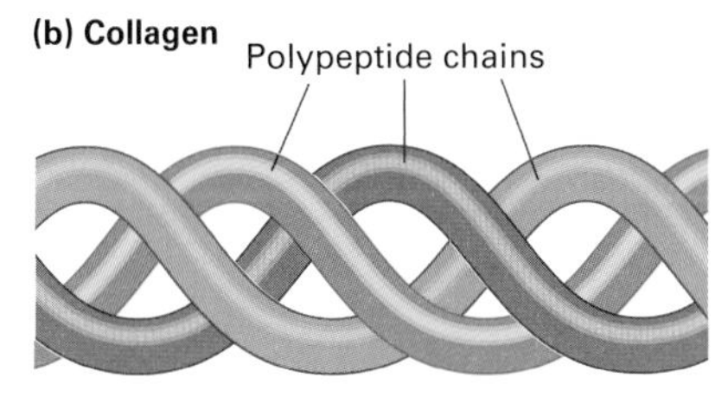

Figure 6 *The quaternary structure of* ***(a)*** *haemoglobin and* ***(b)*** *collagen. Haemoglobin has two alpha and two beta polypeptide chains, each about 140 amino acids long. The quaternary structure of collagen consists of a triple helix of three polypeptide chains wrapped around each other in a rope-like manner.*

QUICK CHECK

1. With reference to haemoglobin, distinguish between tertiary and quaternary structure.
2. What happens when a protein is exposed to excessive heat or extremes of pH?
3. Which class of protein, fibrous or globular, has mainly structural functions?

Food for thought

In humans, fats and carbohydrates are the main sources of energy. One gram of protein yields about the same amount of energy as one gram of carbohydrate. Suggest why protein normally contributes no more than 10 per cent of energy requirements, and why it has been called the 'fuel of last resort'.

2.11

NUCLEOTIDES AND NUCLEIC ACIDS

OBJECTIVES

By the end of this spread you should be able to:

- describe the main components of a nucleotide
- explain how nucleotides can combine to form a polynucleotide chain
- discuss the structure and function of ATP
- outline the structure of DNA.

Fact of life

The breakdown of ATP releases energy used to do work. A contracting muscle cell requires about two million ATP molecules per second to drive its biochemical machinery. A runner uses the equivalent of about 75 kg of ATP during a marathon race. However, the human body has only a small store of ATP (approximately 100 g in an average person). This store would be exhausted after only one second of the race if the runner could not continuously regenerate ATP from ADP and inorganic phosphate by respiration of metabolic fuels, mainly carbohydrate and fat.

Nucleotides are nitrogen-containing organic substances which play a vital role in every aspect of an organism's life. Nucleotide molecules occur singly (**mononucleotides**) or combined in numbers from two to many thousands (**polynucleotides**). Each nucleotide is made of three parts (figure 1):

- a nitrogen-containing organic **base**
- a five-carbon sugar (a **pentose**)
- one or more **phosphate** groups.

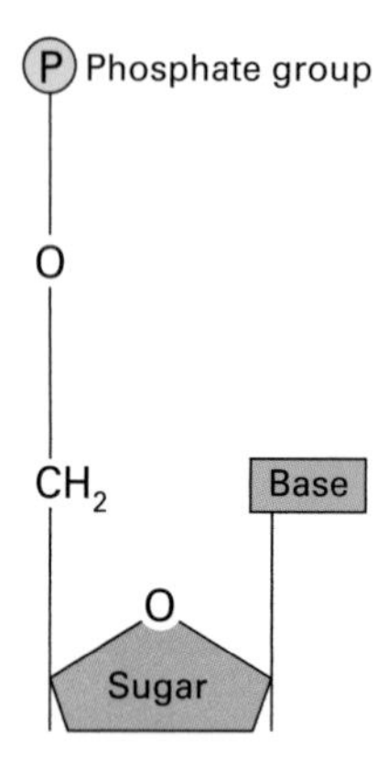

Figure 1 *The basic structure of a nucleotide.*

ATP: energy to drive reactions

Adenosine triphosphate or **ATP** is a mononucleotide. The base is **adenine** and the sugar is **ribose**. Attached to the ribose are three phosphate groups (figure 2). The covalent bond linking the second and third phosphate groups is unstable, and is easily broken by hydrolysis.

When this bond is broken a phosphate group (P_i) is removed, and ATP becomes **ADP** (**adenosine diphosphate**). At least 30 kJ of energy is released during this reaction. It is called an **exergonic** reaction because energy is released.

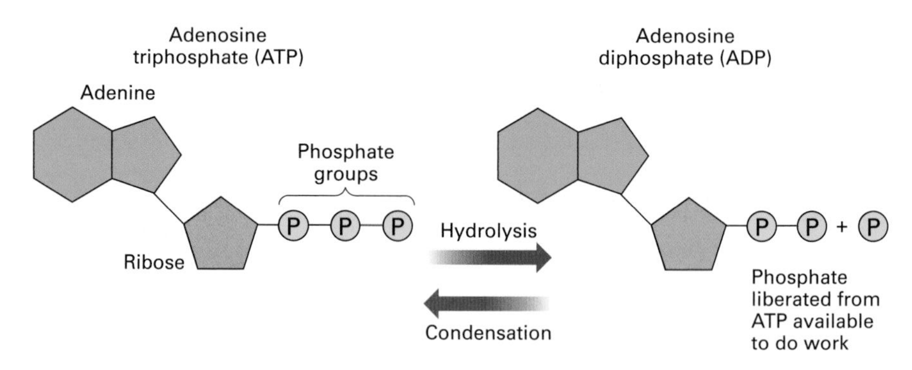

Figure 2 *The breakdown and formation of adenosine triphosphate.*

ATP can be resynthesised from ADP and inorganic phosphates by a condensation reaction. The energy needed to drive this reaction comes from respiration. The **ATP cycle**, which shows the relationship between ATP, ADP, and respiration, is outlined in figure 3.

ATP is vital because it is the main source of chemical energy for energy-consuming reactions. In cells, the exergonic breakdown of ATP is coupled to energy-consuming (**endergonic**) reactions. Phosphate groups liberated from ATP attach to other molecules (a process called **phosphorylation**). This energises the molecules, enabling them to do work.

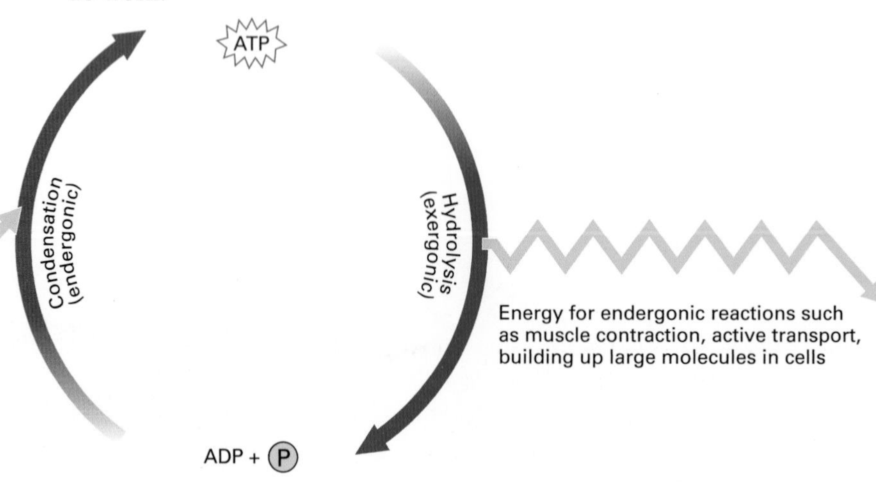

Figure 3 *The ATP cycle.*

Nucleic acids: DNA

Nucleotides can combine by condensation reactions to form long chains (figure 4). These chains are called **nucleic acids** or polynucleotides. The best known nucleic acid is **deoxyribonucleic acid** or **DNA**. DNA has been called 'the molecule of life' because it plays a key role in inheritance and protein synthesis. It influences every physical and behavioural characteristic of an organism.

The basic structure of DNA seems simple because it is made of only four types of nucleotide. However, these nucleotides can link together in many ways to form chains of infinite variety. This infinite variability enables DNA to store the genetic information for all the millions of organisms on Earth, as we shall see later.

Each DNA nucleotide has a phosphate group, a pentose sugar (deoxyribose), and one of four types of base (all of which contain nitrogen): **adenine** (A), **cytosine** (C), **guanine** (G), or **thymine** (T).

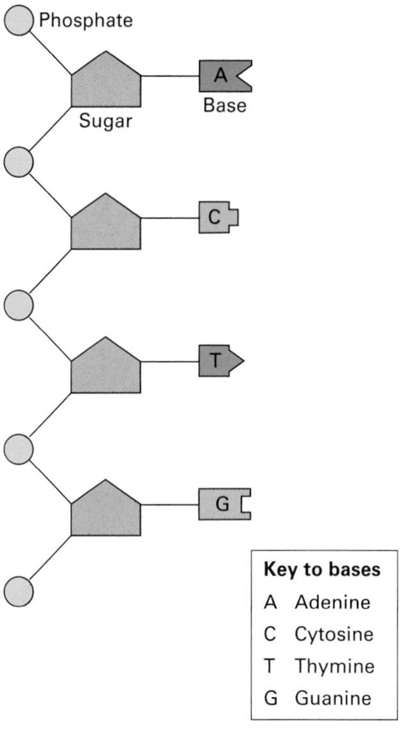

Figure 4 *Part of a polynucleotide chain.*

DNA forms a double helix

- The DNA molecule consists of two strands, each of which is a polynucleotide chain.
- Each strand has a helical (spiral) shape, so that DNA has become known as the '**double helix**'.
- The polynucleotide chains run in opposite directions and are joined by pairs of bases (figure 5).
- The bases are held together by hydrogen bonds between the hydrogen atoms of a base in one chain and the oxygen and nitrogen atoms of a base in the other chain.
- The shapes and sizes of the bases mean that nucleotides can pair up only in certain ways. Guanine always pairs with cytosine, and adenine always pairs with thymine.

The sequence of bases along the polynucleotide chain forms the genetic code, which determines the characteristics of an organism that are inherited from its parents (see spread 18.6). The sugar–phosphate backbone of DNA is the same for all organisms, from bacteria to whales, but the base sequence is extremely variable. See spread 18.1 for a more detailed account of DNA structure.

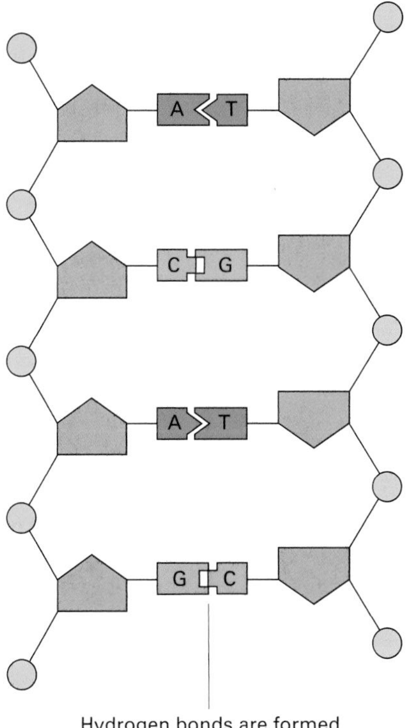

Figure 5 *Part of a DNA molecule.*

Nucleic acids: RNA

The other main nucleic acid is **ribonucleic acid** or **RNA**.

- RNA usually consists of a single polynucleotide chain that forms an alpha-helix.
- The pentose sugar in the RNA nucleotides is ribose rather than the deoxyribose that occurs in DNA nucleotides.
- RNA nucleotides contain one of four types of base: adenine, guanine, cytosine, or **uracil** (which replaces the thymine of DNA).

There are several types including messenger RNA (mRNA), transfer RNA (tRNA), and ribosomal RNA (rRNA). Each differs in length and shape, but all share the same basic structure. The different types of RNA are involved in different aspects of protein synthesis. The exact function is related to the precise shape, length, and nucleotide composition of the RNA (see spread 18.7).

Quick check

1. List the three components found in all nucleotide molecules.
2. How are the bases in a double strand of DNA held together?
3. Name three cellular activities that use ATP as an energy source.

Food for thought

Biochemists often refer to ATP as the 'energy currency of cells' because all cells use ATP as the source of energy to drive their metabolic machinery. Cells may have many different fuel molecules, but these cannot supply metabolic energy until they are converted to ATP. Suggest why cells do not use fuel molecules, such as carbohydrates and fats, directly.

Summary

Atoms of different **elements** make up the basic structure of all matter, including living things. Different elements can combine to form **molecules** and **compounds**.
Water is one of the most important molecules in biology because it has special physical and chemical properties which make it uniquely suited as an internal and external environment for life. Many substances dissolve in water to produce solutions. The **pH** of solutions varies from 0 to 14, with pure water having a pH of 7.
Acidic solutions have a pH of less than 7 whereas **basic solutions** have a pH of more than 7. **Acid rain**, which may be harmful to forests and lakes, has a pH of less than 5.6.
Carbohydrates, **lipids**, **proteins**, **nucleotides**, and **nucleic acids** are biological molecules that form most of the structure of living organisms and carry out their functions. Each biological molecule has physical and chemical properties suited to its particular function.

PRACTICE EXAM QUESTIONS

1 The diagrams below represent organic molecules. Using only the letters adjacent to the diagrams, indicate

- **a** which structure contains a peptide bond [1]
- **b** which structure contains a glycosidic bond [2]
- **c** which structure is an amino acid [1]
- **d** which structure is a nucleotide [1]
- **e** which structure is characterized by its solubility in organic solvents [1]
- **f** which structure is a monosaccharide [1]
- **g** which structures have been made as a result of a condensation reaction [2]
- **h** which structure would require 3 water molecules for complete hydrolysis. [1]

[Total 10 marks]

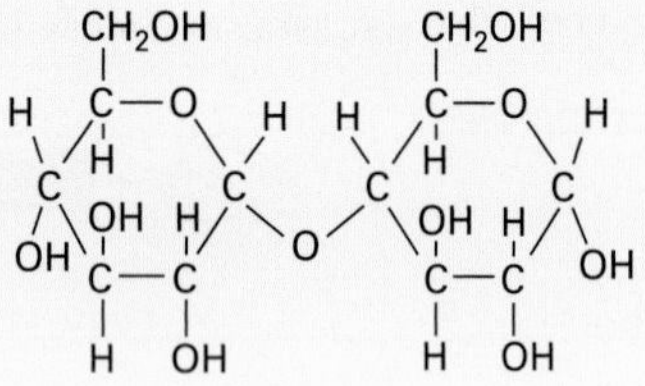

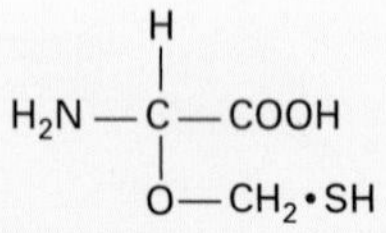

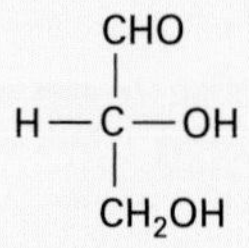

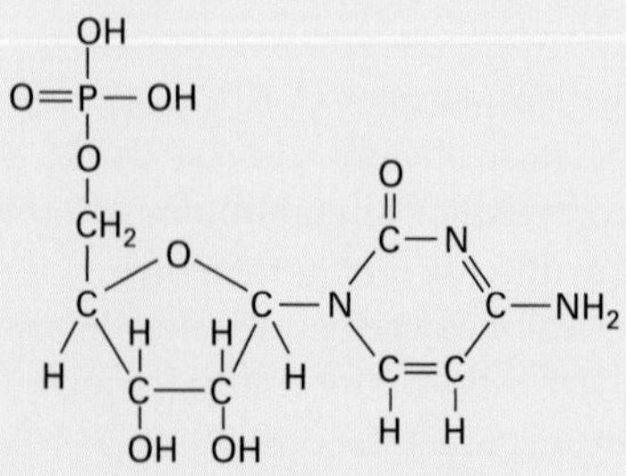

2 The diagram below shows the structure of a lipid molecule.

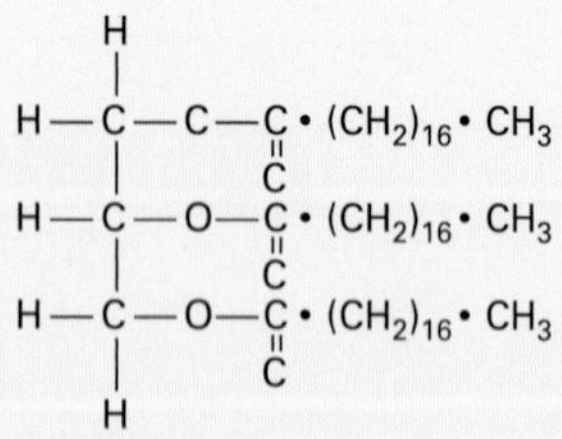

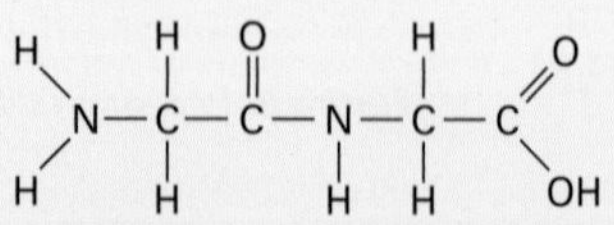

- **a i** Name the parts labelled A and B. [2]
  - **ii** Name this type of lipid. [1]
  - **iii** Name the chemical reaction used to form the bonds between A and B. [1]
- **b i** State one function of this type of lipid in living organisms. [1]
  - **ii** State one feature of the molecules of this type of lipid which makes them suitable for the function you have given. [1]

[Total 6 marks]

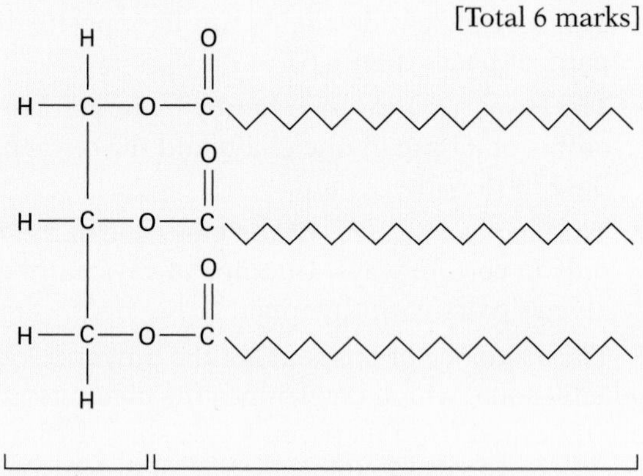

3 Starch is a polymer of glucose. It has two components, amylose and amylopectin. Amylose is a straight chain molecule of several thousand alpha glucose molecules linked by 1–4 glycosidic bonds. Amylopectin is a branched molecule formed from thousands of alpha glucose molecules linked by 1–4 glycosidic bonds with branches formed from 1–6 glycosidic bonds every 25 units (the numbers associated with the bonds refer to which carbon atoms are involved in the bond).

- **a** The diagram below represents a molecule of alpha glucose. Using a simplified structure construct a diagram to show a short length of amylopectin consisting of three glucose units in one chain linked to a branch of two units. [2]

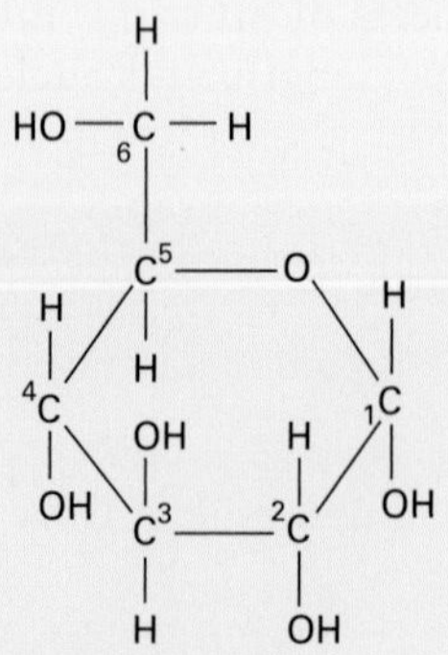

b The hydrolysis of the molecule which you have drawn would result in the production of monosaccharides.

i What is meant by the term *hydrolysis*? [2]

ii Give two different ways by which this hydrolysis could be brought about. [2]

iii Suggest why the hydrolysis of a branched molecule will be faster than a straight chain molecule. [1]

[Total 7 marks]

4 Copy and complete the table by giving four differences between cellulose and glycogen. [4]

	Cellulose	Glycogen
1		
2		
3		
4		

5 The diagrams show part of the molecular structures of two polysaccharides. The hexagonal shapes represent hexose sugars.

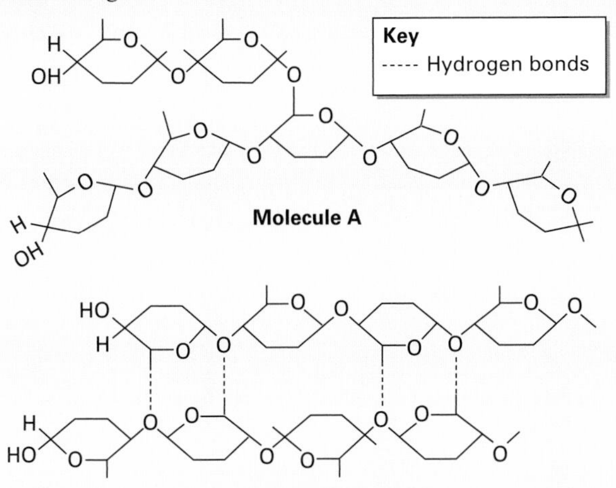

a Give the name of molecule **A**. [1]

b Give one difference between hexose sugars in molecules **A** and **B**. [1]

c Both polysaccharides contain hexose sugars joined by 1–4 glycosidic bonds.

i Explain, using an annotated diagram, how these bonds in molecule **A** are hydrolysed in the process of human digestion. [2]

ii Using information in the diagram of molecule **B**, suggest **one** reason why it cannot be digested by humans. [1]

[Total 5 marks]

6 The figure above right shows a diagram of part of a polypeptide chain. This type of twisted structure is commonly found in proteins of many different types.

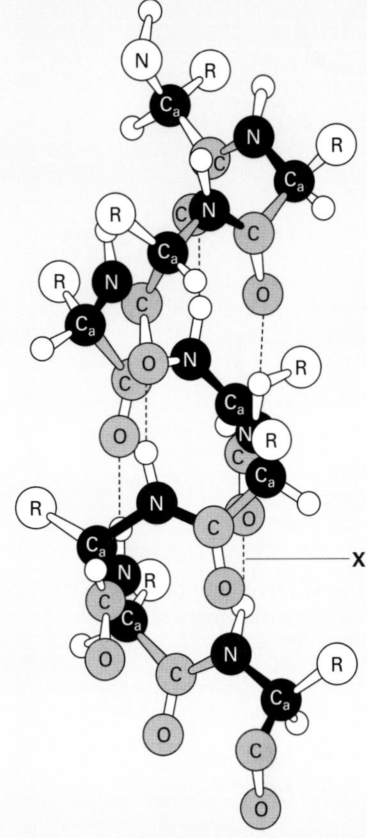

a i Name the repeating unit of a polypeptide chain. [1]

ii State the name given to the twisted structure shown in the figure. [1]

iii Identify the type of bond in the structure labelled by the letter **X**. [1]

iv Explain briefly what would happen to the polypeptide chain if it were heated to about 70 °C. [2]

v The twisted arrangement seen in the figure above is referred to as a secondary structure. Explain what is meant by the term *secondary structure*. [2]

b Another common secondary structure is known as the beta sheet. State *one* difference between the beta sheet and the structure shown in the figure above. [1]

c i Proteins can be classified as fibrous or globular. Name **one** example of each type of protein. [2]

Globular proteins such as that shown in the figure below are often described as tertiary structures. However, as indicated in the diagram, many globular proteins may also have sections of secondary structure.

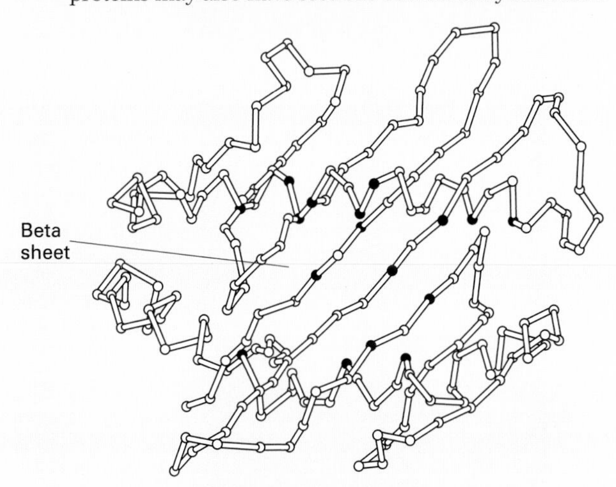

ii Explain what is meant by the term *tertiary structure*. [2]

d Monosaccharides can also be linked together to form long chain molecules called polysaccharides. Give **three** differences (other than the presence of monosaccharides or amino acids) between a polypeptide and a polysaccharide chain. [3]

[Total 15 marks]

Metabolic reactions

Energy and metabolism

3.1

OBJECTIVES

By the end of this spread you should be able to:

- understand that living organisms obey the laws of thermodynamics
- distinguish between catabolic and anabolic reactions
- explain what is meant by the activation energy of a metabolic reaction.

Fact of life

It used to be thought that all organisms on Earth obtained their energy from the Sun. However, in 1977 scientists made a stunning discovery that changed forever our understanding of life on Earth. They found a vast assemblage of weird and wonderful organisms living on the bottom of the Pacific Ocean close to hydrothermal vents (areas of underwater volcanic activity) and so deep that no light can reach them. The organisms ranged in size from microscopic bacteria to giant clams and tube worms. The bacteria, which acquire their energy from sulphur compounds released from the hot vents, are the primary food suppliers.

Types of energy

Every living organism on Earth requires energy to stay alive. **Energy** is the capacity to do work: the ability to move matter in a direction in which it would not move without an input of energy. The amount of matter may be small or large: an ion passing through a cell membrane, or a whale swimming in an ocean.

Energy exists in many forms, including chemical, electrical, nuclear, heat, light, and mechanical energy. There are two types of mechanical energy: kinetic energy and potential energy. Matter that is moving and performing work is said to have **kinetic energy** (for example, a whale rising out of the sea; figure 1). Matter that is not actually performing work, but that has the ability to do so because of its position, is said to have **potential energy** (for example, the whale at the top of its jump, when it is momentarily static). The atoms within a sugar molecule also have potential energy, because of the arrangement of their atoms. Sugar possesses **chemical energy**, the most important type of potential energy for life.

Laws of thermodynamics

Organisms survive by transforming one form of energy into another. Plants use energy from light to make glucose and starch, and animals use energy from other organisms so that they can move, grow, and carry out all their many activities. Energy that can be used to do useful work is called **free energy**.

Under most circumstances, two laws govern the energy transformations that occur in all matter, be it living, dead, or non-living. The laws are called the first and second laws of thermodynamics. (**Thermodynamics** is the study of energy transformations.)

The **first law of thermodynamics** (the **law of conservation of energy**) states that energy can be neither created nor destroyed; it can only be changed from one form into another. Therefore, when energy changes take place within an organism, the energy input always equals the energy output.

The **second law of thermodynamics** is based on the assumption that the universe is becoming increasingly disordered. This is related to the fact that when energy is transformed from one type to another, some of the energy is converted into heat. Because we are warm blooded, we tend to think that heat is useful. However, whenever heat is generated, less energy is available to do useful work. Therefore, heat energy is regarded as a low-order form of energy. The second law of thermodynamics is sometimes called the **law of entropy** because entropy is another word for the disorder or randomness in a system. (Entropy is also a measure of the unavailability of energy for useful work, a consequence of disorder.)

Organisms appear to disobey the second law because they become more ordered and complex during their lives. However, they can do this only at the expense of the environment. Organisms are 'energy parasites'; they become increasingly complex by using energy taken from their surroundings. The sum effect of life is to increase disorder in the universe.

***Figure 1** Energy is required by all organisms to carry out their functions. This humpback whale expends a great deal of energy in projecting itself out of the sea. It appears to breach the water to stun or panic shoals of fish, and to communicate to other members of its group.*

Chemical reactions and metabolism

A chemical reaction leads to a chemical change in matter. A reaction can be represented by a chemical equation. For example, the following chemical reaction takes place when oxygen and hydrogen combine to form water:

$$2H_2 + O_2 \longrightarrow 2H_2O$$

Two molecules of hydrogen ($2H_2$) and one molecule of oxygen (O_2) are the starting materials, called the **reactants** or **substrates**. Two molecules of water ($2H_2O$) are the resultant substances, called the **products**. The arrow indicates the direction of the reaction, that is from hydrogen and oxygen to water. Notice that the numbers of hydrogen atoms and of oxygen atoms on the left-hand side of the equation equal the numbers on the right-hand side. All equations are **balanced** in this way, reflecting the fact that chemical reactions do not destroy or create matter.

Cells carry out thousands of chemical reactions. The sum of all these reactions is called **cellular metabolism**. There are two main types of metabolic reaction: catabolism and anabolism. During **catabolism**, substances break down and release energy. During **anabolism**, chemical reactions take in energy to synthesise complex molecules from simple molecules. Reactions that liberate more energy than they take in are called **exergonic**; those that take in more energy than they liberate are called **endergonic**.

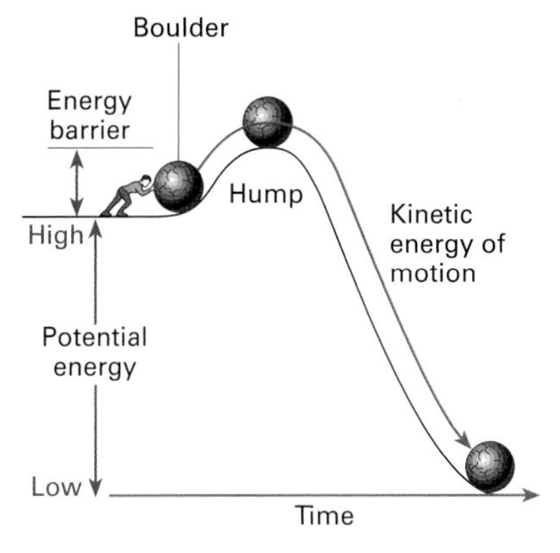

Figure 2 The boulder analogy: the potential energy of the boulder must be increased before it can roll down the hill.

Activation energy

Most molecules are in a relatively stable state and require an input of energy in order to react with each other. Chemicals in a match head, for example, do not react until the match is struck against a suitable surface. There is an **energy barrier** to the reaction. The amount of energy required to overcome this barrier and start a reaction is called the **activation energy**. Friction supplies the activation energy required to ignite the match. Once lit, the match generates much more energy, mainly as light and heat.

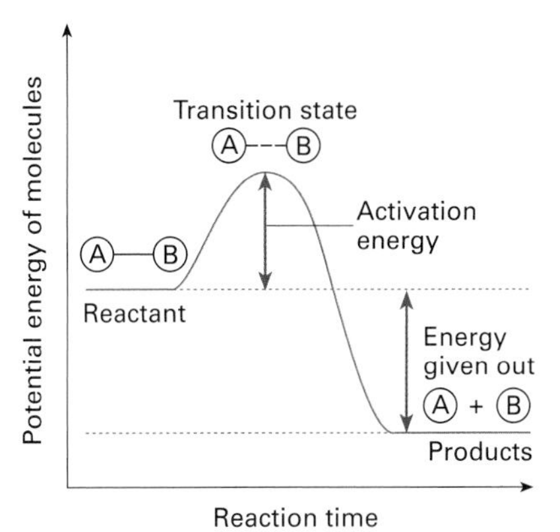

Figure 3 Potential energy of molecules during a catabolic reaction.

The energy changes of a chemical reaction can be likened to those of a boulder rolling down a hill (figure 2). The hump over which the boulder must be pushed before it can roll freely represents the energy barrier, and the energy needed to push the boulder up over this barrier is equivalent to the activation energy.

Catabolic reactions are exergonic (give out energy) because they have a small activation energy compared with the energy released during the reaction; the reaction products contain less energy than the substrates (figure 3). Anabolic reactions are endergonic because their activation energy is greater than the energy released during the reaction. The products of the reaction contain more energy than the reactants, therefore extra energy must be supplied for the reaction to proceed (figure 4).

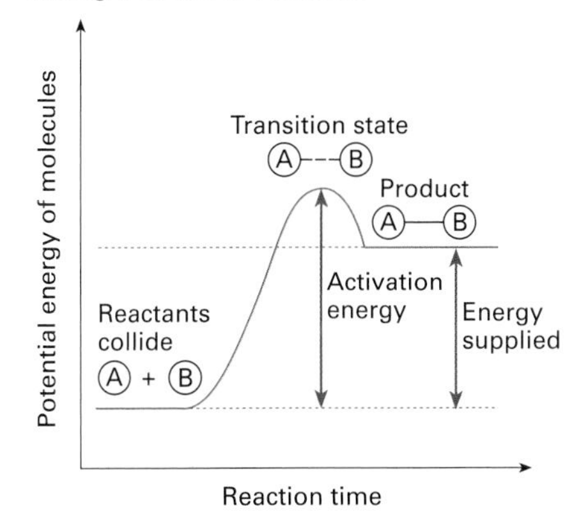

Figure 4 Potential energy of molecules during an anabolic reaction.

Quick check

1 Organisms obey the law of conservation of energy. Why then do they appear to lose energy when they transform one form of energy into another (for example, when an animal converts chemical energy into kinetic energy during locomotion)?

2 Is photosynthesis an anabolic or catabolic process?

3 What is activation energy?

Food for thought

The energy change that occurs during a chemical reaction is referred to by the measure $G^{\ominus}$, the standard free energy change. In any reaction, if $G^{\ominus}$ is negative, the reaction is exergonic, that is, it releases energy which can do work. If $G^{\ominus}$ is positive, the reaction is endergonic, that is, it requires an energy input for the reaction to take place. Suggest whether $G^{\ominus}$ is positive or negative in the following reactions:

a glucose + oxygen ⟶ carbon dioxide + water

b ADP + P_i ⟶ ATP

c water + carbon dioxide ⟶ glucose + oxygen

d $C_6H_{12}O_6$ (glucose) ⟶ $2CH_3CH(OH)COOH$ (lactate).

Most of the chemical reactions in cells are endergonic. What is their source of energy?

3.2

How enzymes work

OBJECTIVES

By the end of this spread you should be able to:

- list the main differences between enzymes and inorganic catalysts
- describe the lock-and-key theory and induced-fit theory of enzyme action.

Fact of life

A very small amount of enzyme can bring about a change in a large amount of substrate. For example, at 0 °C, the enzyme catalase has a **turnover number** of 50 000. This means that at 0 °C, one molecule of catalase can catalyse the breakdown of 50 000 molecules per second of hydrogen peroxide to water and oxygen. At human body temperature, one molecule of the enzyme catalyses the breakdown of about 600 000 molecules per second of hydrogen peroxide.

An average enzyme undergoes about 1000 reactions per second. Catalysts can speed up reaction rates by as much as a trillion (10^{12}) times. According to Dianne Gull and Bernard Brown, 'This is roughly equivalent to speeding up your life from cradle to grave so that it is completed in a single heartbeat.'

The transition state: the hump in the energy hill

Every cell carries out many thousands of metabolic reactions. Each reaction starts with the **reactants** (the molecules undergoing the reaction) and ends with a **product** or products. Metabolic reactions usually take place in a number of steps, with each intermediate step forming an intermediate molecule on the way to the final product. The intermediate molecules are less stable (hence contain more energy) than either the reactants or the products. The intermediate molecule containing the most energy is called the **transition state**. Before a reaction can get underway, there must be enough energy available to convert the reactants to the transition state. This energy is the activation energy described in spread 3.1.

Enzymes: biological catalysts

Almost every cellular reaction, whether it involves breaking things down (a catabolic reaction) or building things up (an anabolic reaction), is controlled by a biological catalyst called an **enzyme**. Enzymes can increase the rate of biochemical reactions to more than a billion times their normal rate, without the enzymes being changed themselves by the reaction. Because they can be reused, even small concentrations of enzymes can be very effective. However, like other catalysts, enzymes cannot make reactions occur which otherwise would not happen, and they do not alter the final amount of product formed, just the speed at which it is formed.

Unlike an inorganic catalyst, which might catalyse many different types of reaction, an enzyme will usually catalyse only one type of reversible reaction. Enzymes are affected by temperature and pH much more than inorganic catalysts are (spread 3.3). Each enzyme works effectively only in a limited range of temperature, pressure, and pH; inorganic catalysts often function very effectively in widely varying conditions of temperature and pressure, and extremes of pH.

Reducing the activation energy

Their ability to accelerate reactions and their high selectivity make enzymes among the most important compounds found in biology. Most enzymes catalyse reactions by combining with the substrate of a reaction to form a series of **enzyme–substrate complexes**. The complex in which the substrate binds most tightly to the enzyme corresponds to the transition state. This transition state is more stable (of lower energy) than the transition state in the uncatalysed reaction. Consequently, an enzyme reduces the activation energy of the reaction it catalyses (figure 1).

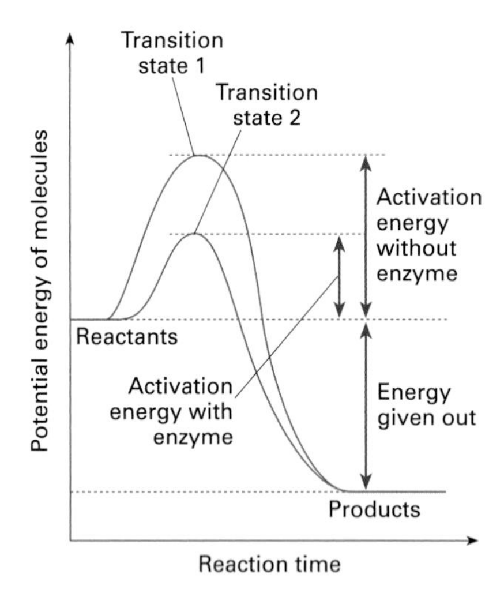

Figure 1 The lowering of the activation energy by an enzyme.

There are two main theories explaining how enzyme–substrate complexes form: the lock-and-key theory and the induced-fit theory.

The lock-and-key theory

Most enzymes are huge globular protein molecules made up of many thousands of atoms along with some metal ions. Their molecules have a very precise shape (tertiary structure) which includes a cleft or pocket called the **active site**. In the **lock-and-key theory** of enzyme action, the substrate fits into a rigid active site like a key into a lock (figure 2). This is quite a crude model of the chemical mechanism of enzyme action. Various types of bond including hydrogen bonds and ionic bonds hold the substrate in the active site to form an enzyme–substrate complex. Once the enzyme–substrate complex is formed, the enzyme can help change the substrate, either splitting it apart or linking pieces together.

In the lock-and-key theory, the shape of the substrate must fit the enzyme exactly if a reaction is to be catalysed. This explains why enzymes are specific, and why any change in enzyme shape, no matter how small, alters its effectiveness. However, it is not a totally satisfactory explanation of enzyme action. If the theory is correct, enzyme action depends on the unlikely event of randomly moving substrate molecules entering the active site in the right orientation. This would be analogous to trying to get a key in a lock by throwing it ... with your eyes shut!

The induced-fit theory

The **induced-fit theory** of enzyme action is a modified version of the lock-and-key theory. It does not depend on such precise contact being made between the substrate and the active site. In this model, the active site is able to change its shape to enfold a substrate molecule. The enzyme takes up its most effective catalytic shape after binding with the substrate. The shape of the enzyme is affected by the substrate, just as the shape of a glove is affected by the hand wearing it.

The distorted enzyme molecule in turn distorts the substrate molecule, straining or twisting the bonds. This makes the substrate less stable, reduces its potential energy, and thus lowers the activation energy of the reaction. The reaction occurs and products are formed which no longer bind to the active site, and so move away. The flexible enzyme then returns to its original shape, ready to bind the next substrate molecule (figure 3).

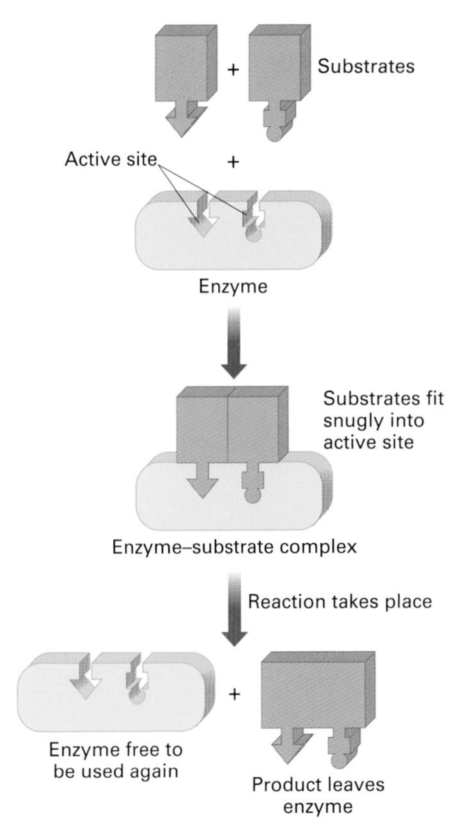

Figure 2 *Much simplified view of the lock-and-key theory of enzyme action. The reaction can go either way: the enzyme also catalyses the conversion of product to substrate.*

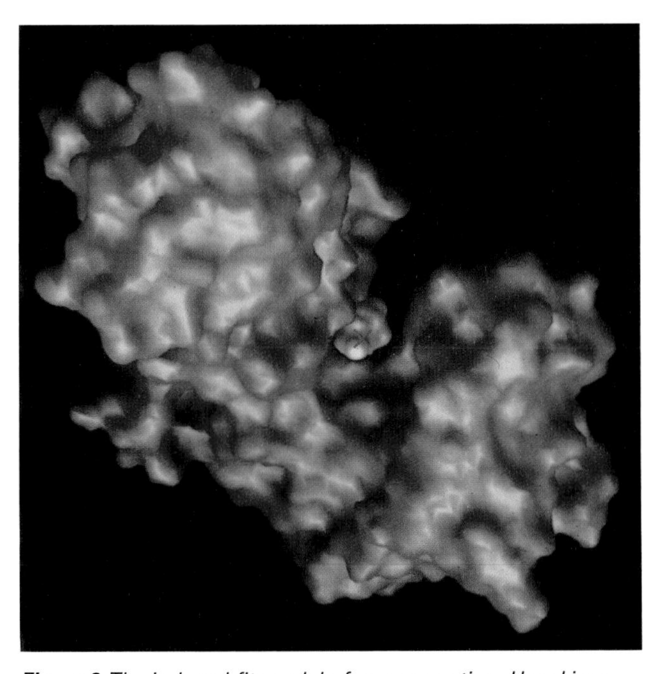

Figure 3 *The induced-fit model of enzyme action. Hexokinase is an enzyme which catalyses the phosphorylation of glucose. When the substrate, glucose (x), enters the active site it induces a slight change in the shape of the protein. This enables an enzyme–substrate complex to form.*

Quick check

1. List three differences between inorganic catalysts and enzymes.
2. Give the main difference between the lock-and-key and induced-fit theories of enzyme action.

Food for thought

The role of enzymes has been compared with matchmakers. Indeed, the Chinese word for catalyst, *tsoo mein*, means 'marriage broker'. Is this comparison appropriate? Suggest another comparison.

3.3

Factors affecting enzymes (1)

OBJECTIVES

By the end of this spread you should be able to:

- explain why enzymes affect specific types of reactions
- describe and explain the effects of enzyme concentration, substrate concentration, incubation time, temperature, and pH on enzyme-catalysed reactions.

Fact of life

Water spewing out of deep-sea volcanic vents reaches temperatures in excess of 350 °C. Certain bacteria can live in the heated surrounding water because they have evolved special thermostable enzymes that are not denatured by temperatures as high as 115 °C.

Enzymes are globular proteins. The three-dimensional shape, particularly that of the active site, determines the specificity of an enzyme and also how it is affected by other factors, such as temperature and pH.

The box at the bottom of the page explains how you can measure the initial rate of a reaction. This initial rate is used to compare the effects of different factors on an enzyme-catalysed reaction.

Enzyme specificity: which reaction does it catalyse?

- Enzymes catalyse reactions by combining with a substrate to form an enzyme–substrate complex (see spread 3.2).
- The formation of an enzyme–substrate complex can only take place if the substrate fits snugly into the active site of the enzyme.
- An enzyme will only catalyse a reaction in which the substrate and active site have complementary shapes.
- Consequently, an enzyme usually catalyses only one reaction or a closely related group of reactions.
- These reactions are usually reversible; if an enzyme can catalyse a reaction in one direction, then it will also be able to catalyse the reverse reaction.
- The degree of enzyme specificity varies. Some enzymes are so specific that they catalyse only one reaction. For example catalase catalyses only the breakdown of hydrogen peroxide (a by-product of cellular respiration) to water and oxygen. Other enzymes, particularly digestive enzymes, are a little less specific. Some lipases will break down a range of fats into fatty acids and glycerol.

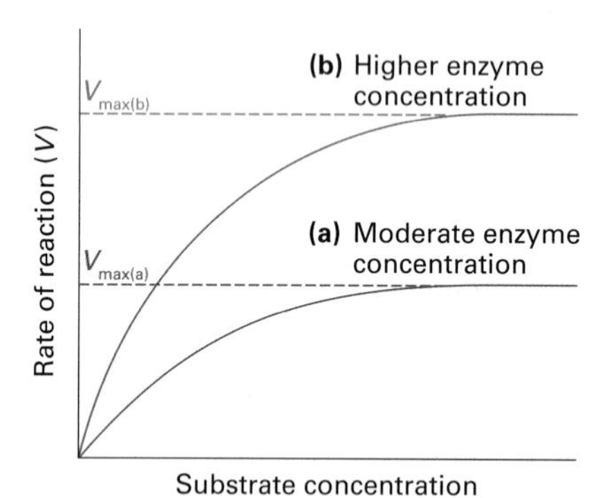

Figure 1 *Two saturation curves for an enzyme-catalysed reaction. For any given concentration of enzyme, the rate of reaction increases with increased substrate concentration until a constant rate of product formation (saturation) is reached. At saturation, all enzyme active sites are bound to the substrate and the rate of the reaction is at its maximum (V_{max}). At a higher enzyme concentration, the saturation point is raised because more active sites become available, and V_{max} is increased. The substrate concentration (usually expressed in molar units, M) that gives $\frac{1}{2}V_{max}$ is referred to as K_m.*

Substrate concentration

The rate of an enzyme-catalysed reaction increases in direct proportion to the substrate concentration until the reaction reaches a maximum rate. After the maximum all the active sites of the enzyme molecules are filled, so increasing the substrate concentration further has no effect on the rate of reaction (figure 1).

Enzyme concentration

The rate of an enzyme-catalysed reaction is directly proportional to the concentration of the enzyme present, as long as no other factors are limiting the rate (figure 1). Therefore, this relationship holds only when pH, pressure, and temperature are constant, and the substrates are present in excess concentrations. Under these conditions, the more active sites there are available, the more substrate can be converted to product.

Measuring the initial rate of reaction

Biologists often need to be able to measure the initial rate of an enzyme-catalysed reaction so that they can compare the rate under different conditions. The rate of reaction is usually expressed as the amount of substrate converted to product per unit time.

Figure 2 shows you how to calculate the initial rate of reaction from a graph of product concentration against time. First draw a line at a tangent to the beginning of the curve, as close as possible to time 0. Continue this line until you reach a convenient point along the x-axis; in this case, 30 seconds. Then read off how much product would be made if the initial reaction rate continued for 30 seconds. In this case, 20 mmol (millimoles, a measure of the number of molecules) of product would have been made in 30 seconds. This gives a rate of 0.66 mmol s^{-1} or 40 mmol min^{-1}.

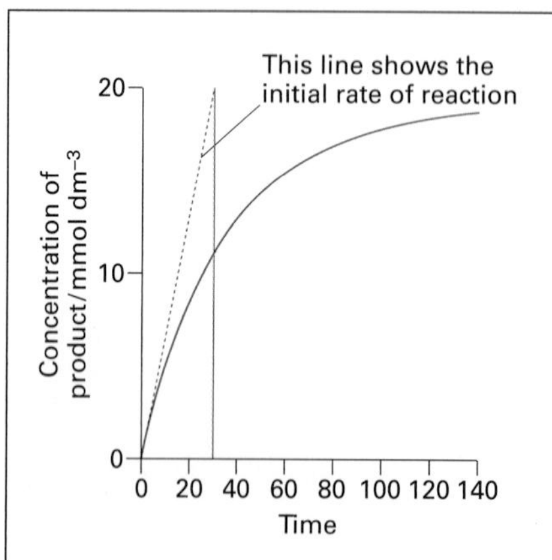

Figure 2 *Measuring the initial rate of reaction.*

Incubation time

The **incubation time** is the length of time over which a reaction has taken place. Generally, the average (not initial) rate of an enzyme-catalysed reaction decreases as the incubation time increases, even when there is excess substrate present. This decrease is probably because the enzyme gradually becomes denatured with time (spread 2.10). As the protein molecule becomes increasingly deformed, it loses its effectiveness as an enzyme.

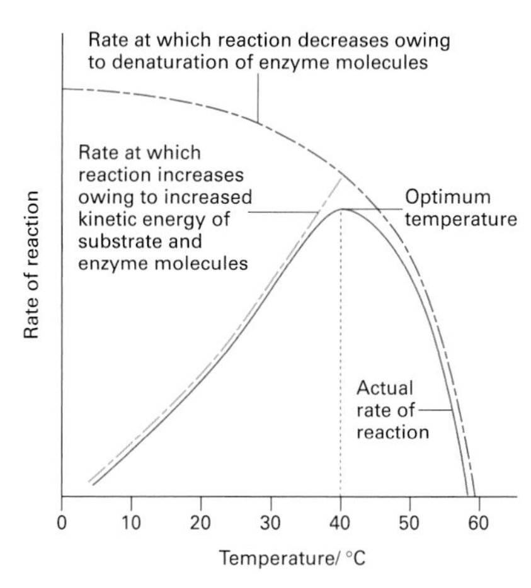

Figure 3 The effect of temperature on an enzyme-catalysed reaction. Increasing the temperature tends to speed up the reaction because the kinetic energy of the reactants increases. However the reaction also tends to slow down as the enzyme is denatured. The balance of these opposing activities is different at different temperatures:

- *Below the optimum temperature, the rate rises because the effect of the increased kinetic energy of reactants exceeds the effect of denaturation.*
- *At the optimum temperature, the rate levels off because the increase in kinetic energy of reactants is cancelled out by the denaturation of the enzyme.*
- *Above the optimum temperature, the rate falls because the effects of denaturation exceed the effects of increased kinetic energy of the reactants.*

Temperature

The rate of an enzyme-catalysed reaction increases with temperature up to a maximum, called the **optimum temperature**.

- At suboptimal temperatures, increasing temperature increases the kinetic energy of the reactants. As they move faster, they are more likely to collide and interact with each other and with the enzyme.
- The change in rate of a reaction for each 10 °C rise in temperature is called the **temperature coefficient**, Q_{10}:

$$Q_{10} = \frac{\text{rate of reaction at } x + 10\,°\text{C}}{\text{rate of reaction at } x\,°\text{C}}$$

  At suboptimal temperatures the Q_{10} for enzyme-catalysed reactions is approximately 2 (the rate doubles for each 10 °C rise in temperature).
- The rate continues to rise until it reaches a peak at the optimum temperature.
- Above this temperature, the rate usually falls dramatically. This is because the increased energy causes bonds that maintain the enzyme's shape to break, and the enzyme becomes denatured (see spread 2.10). The changed shape means that the substrate can no longer fit into the active site, and enzyme activity is lost.
- The optimum temperature for an enzyme-catalysed reaction is related to the enzyme's usual thermal environment. In humans many enzymes work best at core body temperature – about 37 °C.

Hydrogen ion concentration (pH)

Most enzymes are effective in only a narrow pH range. Within this range there will be one particular pH (the **optimum pH**) at which activity is greatest. The optimum pH for an enzyme usually matches its usual pH environment. Digestive enzymes in the stomach, for example, work best under acidic conditions, whereas those in the small intestine work best in alkaline conditions (figure 4). Deviations from the optimum pH can cause bonds to be broken (especially hydrogen bonds and ionic bonds) so that the enzyme becomes denatured. The substrate no longer fits readily into the active site to form an enzyme–substrate complex, so the enzyme becomes less effective.

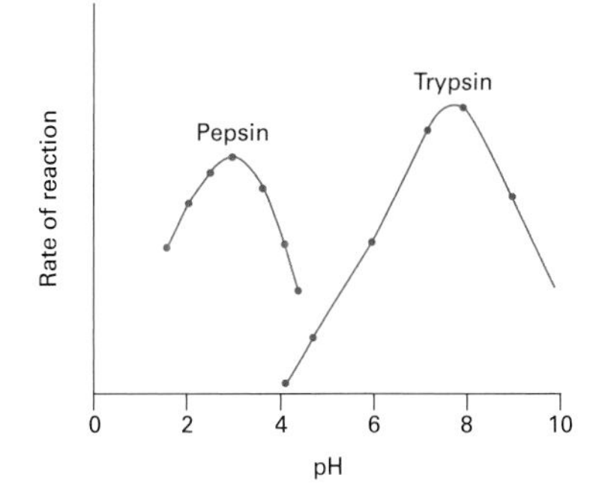

Figure 4 The effect of pH on the rate of reaction of pepsin and trypsin.

Quick check

1 Why does an enzyme catalyse only a specific reaction or a closely related group of reactions?

2 a State **two** effects of temperature on enzyme-catalysed reactions.

b What is the optimum pH of trypsin and of pepsin shown in figure 4?

c In which parts of the alimentary canal would you expect to find each of these enzymes?

Food for thought

The factors that affect enzymes can be grouped according to whether they alter the tertiary structure of an enzyme, or whether they alter the chance of an enzyme and substrate colliding (the so-called collision theory of enzyme action). Suggest to which group each of the factors in this spread belong. Are there any factors that belong to both groups?

3.4

Factors affecting enzymes (2)

OBJECTIVES

By the end of this spread you should be able to:

- explain how cofactors enable enzymes to work
- distinguish between competitive and non-competitive inhibitors
- describe an example of end-product inhibition.

Fact of life

Enzyme inhibitors can act as poisons, pesticides, antibiotics, and painkillers. Cyanide blocks the active site of cytochrome oxidase. This enzyme catalyses the oxidation of hydrogen to form water, the final product in aerobic respiration. Cyanide is therefore a respiratory poison, preventing aerobic respiration.

The pesticide malathion inhibits the action of acetylcholinesterase. This is how melathion kills insects and other pests, by disrupting the transmission of nerve impulses. Unfortunately, malathion also affects humans and other animals, but at higher doses than those required to kill insects.

The antibiotic penicillin kills bacteria by inhibiting an enzyme that is essential for the synthesis of bacterial cell walls.

Possibly the most widely used painkiller in the world is 2-ethanoyloxy-benzenecarboxylic acid, known by pharmacists for many years as acetylsalicylic acid, better known by those in pain as aspirin. Aspirin relieves pain by inhibiting an enzyme called PGHS which is required for the formation of prostaglandins. Prostaglandins are a group of compounds derived from fatty acids. They increase the sensitivity of nerve endings, making a person more aware of pain. Prostaglandins are produced after injury as part of the inflammation response.

Cofactors: activators and coenzymes

Some enzymes work only in the presence of a non-protein substance called a **cofactor**. A cofactor that is tightly bound to its enzyme is called a **prosthetic group**. Cofactors may be inorganic or organic. Inorganic cofactors are called **activators**, and these include metal atoms such as copper, iron, or zinc. Activators may attach onto the active site of the enzyme to make its shape more efficient. Organic cofactors are called **coenzymes**, and many coenzymes are vitamins or compounds made from vitamins. Some coenzymes transfer chemical groups, atoms, or electrons from one enzyme to another. For example, nicotinamide adenine dinucleotide or NAD, derived from vitamin B_3, transfers hydrogen during cellular respiration.

Inhibitors

Many substances can interfere with enzymes, reducing or even completely destroying their action. These substances are called **inhibitors**. There are two main groups of inhibitors: competitive and non-competitive. Inhibitors of either type can be either reversible or irreversible. **Reversible inhibitors** generally bind to an enzyme with weak bonds such as hydrogen bonds which are easily broken. Reversible inhibitors affect the enzyme only so long as they are attached to it. As soon as they are detached, the enzyme can function normally again. **Irreversible inhibitors** attach to an enzyme with strong covalent bonds which are difficult to break without damaging the enzyme. Consequently, the effect of an irreversible inhibitor is permanent.

A **competitive inhibitor** of a particular enzyme has a shape resembling the enzyme's normal substrate. Consequently, the inhibitor can compete with the normal substrate to occupy the active site. If the inhibitor occupies the site, it prevents the enzyme from combining with its normal substrate (figure 1). Antibiotic drugs called sulphonamides act as competitive inhibitors. Their shape resembles that of a substance called para-aminobenzoate (PAB) which is used by some harmful bacteria in the synthesis of folic acid. An enzyme catalyses the conversion of PAB into folic acid. Sulphonamides compete with PAB for the active site on this enzyme. So if sufficient sulphonamide is present, the enzyme will be inhibited and the bacteria will become deprived of folic acid and die.

The effect of the competitive inhibitor depends on its concentration compared with that of the substrate, and on how tightly the enzyme binds to the inhibitor and to the substrate. An enzyme-catalysed reaction is likely to proceed if there is more substrate than inhibitor, but will tend to become slower as the proportion of inhibitor increases.

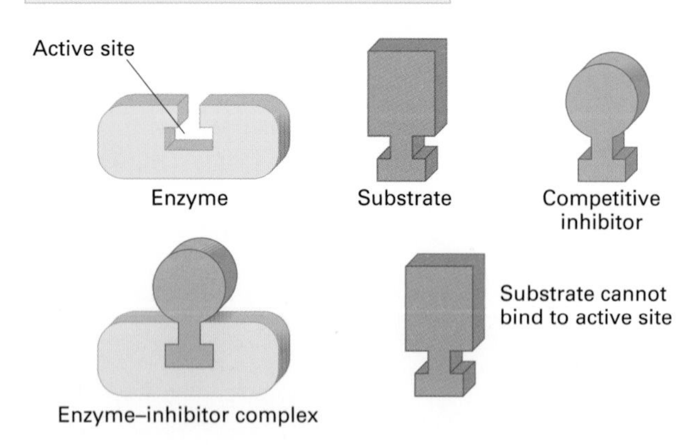

Figure 1 *A competitive inhibitor has a shape similar to that of the normal substrate. The substrate and inhibitor compete for the active site. If the inhibitor binds to the active site, it prevents the binding of the substrate.*

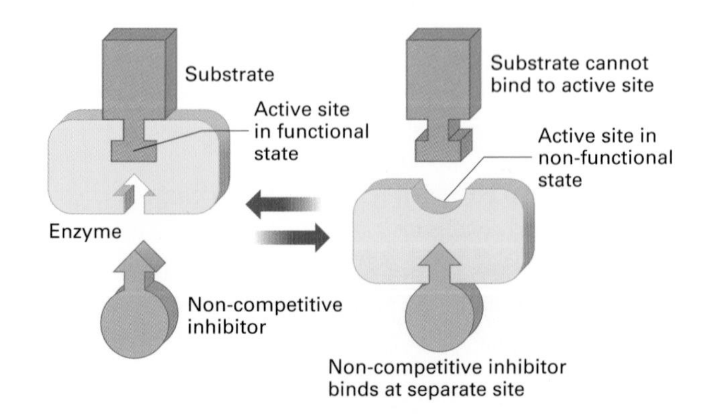

Figure 2 *A non-competitive inhibitor does not bind to the active site but attaches to an enzyme at some other place. The attachment causes the shape of the active site to change, preventing the formation of an enzyme–substrate complex.*

A **non-competitive inhibitor** does not attach to the active site, but binds with the enzyme at another site. Once attached, the non-competitive inhibitor causes the active site to change shape, preventing the normal substrate from binding there (figure 2). In most organisms, heavy metals such as mercury and cadmium act as irreversible non-competitive inhibitors, and are therefore poisonous. However, not all non-competitive inhibitors are harmful. Figure 3 shows how the end-product of a reaction can act as a non-competitive inhibitor, controlling a series of enzyme-catalysed reactions. This phenomenon is called **end-product inhibition**.

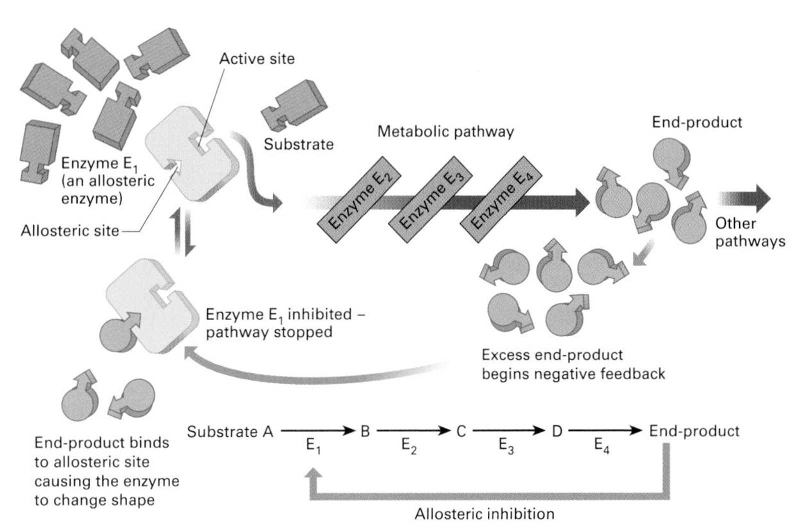

***Figure 3** End-product inhibition in a metabolic pathway. Enzyme E1 acts as a molecular switch that is turned on or off depending on the concentration of the end-product: on when the concentration is low; off when it is higher. Inhibition results from the attachment of the end-product on to the allosteric site of the enzyme, causing the enzyme to change shape so that the substrate can no longer fit onto its active site.*

End-product inhibition and metabolic control

You might be under the impression that all inhibitors are harmful substances, and indeed many enzyme inhibitors do act as pesticides and poisons. However, many other inhibitors are beneficial. Some, especially reversible inhibitors, are important regulators of cellular metabolism.

Most metabolic reactions take place in a series of steps called a **metabolic pathway**. In many cases, a metabolic pathway is regulated by the final substance produced by it. When this substance reaches a certain concentration, it acts as a non-competitive inhibitor. It binds onto one of the enzymes in the metabolic pathway (often the first one), causing the enzyme to change shape. This prevents the reaction series from progressing (figure 3). The reactions start up again when the concentration of the end-product falls to a sufficiently low level. This type of inhibition is called **end-product inhibition**. Control of a metabolic pathway by an end-product is an example of **negative feedback**.

The enzyme that is inhibited, which functions as a regulatory enzyme in the metabolic pathway, is often an **allosteric enzyme**. Allosteric enzymes have a site separate from the active site to which another substance can bind. The shape of the binding substance has to fit that of the allosteric site in much the same way as the normal substrate has to fit the active site. The binding substance may be either an inhibitor or an activator, so either slowing down or speeding up the reaction. (The word 'allosteric' means 'at another place'.)

ATP production by glycolysis (the first stage of aerobic and anaerobic respiration) is regulated by end-product inhibition of the enzyme **phosphofructinase** which functions near the start of the metabolic pathway. If there is a lot of ATP (the end-product) in the cell, this enzyme is inhibited. Respiration slows down and less ATP is produced. As ATP is used up, for example by contracting muscle cells, the inhibition stops and the reaction speeds up again.

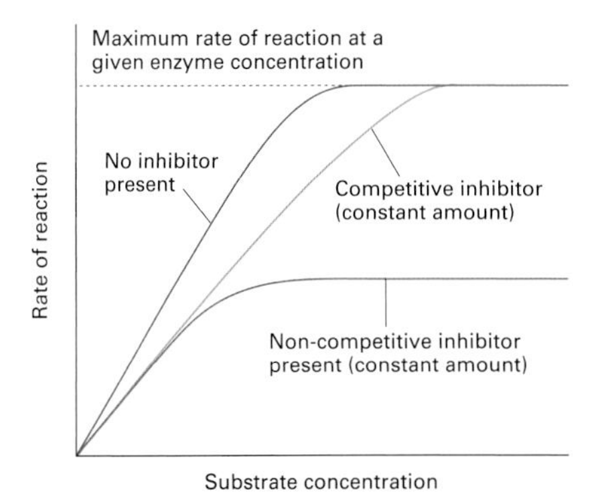

***Figure 4** The effect of a competitive inhibitor and a non-competitive inhibitor on the rate of an enzyme-catalysed reaction.*

Quick check

1 Distinguish between a cofactor and a coenzyme.

2 List the different types of enzyme inhibitor.

3 Describe how end-product inhibition works.

Food for thought

A **Michaelis curve** is a graph that shows the effect of increasing substrate concentration on the rate of reaction in the presence of a fixed amount of enzyme (spread 3.3, figure 1). Figure 4 shows that in the presence of a competitive inhibitor, the maximum possible rate can still be achieved, but only when more substrate is available to overcome the effect of the competitive inhibitor. The non-competitive inhibitor has the same effect as lowering the enzyme concentration. Suggest what the graph would look like if:

a a smaller amount of competitive inhibitor was used

b a smaller amount of non-competitive inhibitor was used.

3.5

Classification of enzymes

OBJECTIVES

By the end of this spread you should be able to:

- classify enzymes using the six main classes specified by the International Union of Biochemistry.

Fact of life

In 1897 the Buchner brothers ground yeast with sand in a mortar and extracted a lifeless juice from the mixture which was able to ferment alcohol. They called the active ingredient of this juice 'enzyme' (which means 'in yeast'). Since 1897, thousands of new enzymes have been discovered. In 1992, The International Union of Biochemistry and Molecular Biology published the sixth and current edition of *Enzyme Nomenclature* in which they list 3196 enzymes. By 2012 approximately 5000 enzymes had been named.

The main enzyme classes

1 Oxidoreductases
2 Transferases
3 Hydrolases
4 Lysases
5 Isomerases
6 Ligases

Figure 1 *Yeast produces enzymes that ferment sugar to produce alcohol. Carbon dioxide is given off in the reaction, and causes the mixture to bubble and froth.*

The six major groups of enzyme

It is difficult to estimate the total number of enzymes that exist. This is not surprising when you realise that over a thousand different reactions can take place in an individual cell, and each reaction has its own specific enzyme. Classification of these enzymes is therefore a mammoth task. The task is made even more difficult by new enzymes being discovered or synthesised every day.

Until 1964 there were several different classification systems of enzymes which led to considerable confusion. In 1964, the International Union of Biochemistry introduced a system aimed at dispelling the confusion. The system is based on the type of reaction the enzyme catalyses. There are six major classes:

1 **Oxidoreductases** catalyse redox reactions (biological oxidation and reduction reactions, see box on opposite page) by the transfer of hydrogen, oxygen, or electrons from one molecule to another, for example:

$$\text{Ethanal} + \text{NADH} + \text{H}^+ \xrightarrow{\text{alcohol dehydrogenase}} \text{ethanol} + \text{NAD}^+$$

Hydrogen is simultaneously lost from NADH and gained by ethanal. NADH is oxidised to NAD^+, and ethanal is reduced to ethanol. This particular process takes place in anaerobic respiration in yeasts and plants.

2 **Transferases** catalyse the transfer of a group from one compound to another, for example:

$$\text{Glutamic acid} + \text{pyruvic acid} \xrightarrow{\text{aminotransferase}} \text{alpha-ketoglutaric acid} + \text{alanine}$$

The R group on the amino acid, glutamic acid, is exchanged with the R group on a keto acid, pyruvic acid. A new amino acid, alanine, is formed along with a new keto acid, alpha-ketoglutaric acid. This specific type of process is called transamination and it enable us to make non-essential amino acids from the essential amino acids.

3 **Hydrolases** catalyse the splitting of a large substrate molecule into two smaller products. Water is involved in the reaction: it is a hydrolysis, for example:

$$\text{Lactose} + \text{water} \xrightarrow{\text{lactase}} \text{glucose} + \text{galactose}$$

The disaccharide lactose is broken down into two monosaccharides by the addition of water. Hydrolases enable many condensation reactions (such as the polymerisation of glucose to glycogen) and hydrolytic reactions (such as the digestion of proteins to amino acids) to take place.

4 **Lysases** catalyse the addition of a group across a double bond, for example:

$$\text{Pyruvic acid} \xrightarrow{\text{pyruvic decarboxylase}} \text{ethanal} + \text{carbon dioxide}$$

Pyruvic acid is converted into ethanal and carbon dioxide by breakage of its double bond and the addition of a new group to the 'freed' bonds. This particular reaction takes place during the fermentation of sugar by yeast. The ethanal is then converted to ethanol (alcohol).

5 **Isomerases** catalyse rearrangements within a molecule, converting one isomer to another, for example:

$$\text{Glucose 1-phosphate} \xrightarrow{\text{phosphoglucomutase}} \text{glucose 6-phosphate}$$

The position of a phosphate group in the glucose 1-phosphate molecule is changed to form the isomer glucose 6-phosphate. This reaction takes place during respiration.

6 **Ligases** catalyse bond formation between two compounds. The reaction uses energy that comes from the hydrolysis of ATP to ADP and phosphate, for example:

$$\text{Amino acid + specific tRNA} \xrightarrow{\text{aminoacyl-tRNA synthetase + ATP}} \text{amino acid tRNA complex + ADP + } P_i$$

An amino acid is joined to a tRNA molecule. This particular process is called tRNA activation and it is an essential step in protein synthesis.

Enzyme nomenclature

The naming of enzymes has suffered from a confusion similar to that of enzyme classification. Some enzymes are known by three different names: for example, an enzyme found in human saliva is known as ptyalin, salivary amylase, and alpha-1,4-glucan-4-glucanohydrolase. The first two names are referred to as **trivial names**. The first name, ptyalin, dates from before there were any agreed conventions on enzyme nomenclature. The second, salivary amylase, is based on the secretion that contains the enzyme (saliva) and the substrate on which the enzyme acts (amylose). This name also illustrates the convention of using the suffix '-ase' to indicate that it is the name of an enzyme.

The third name is the **systematic name** agreed by the International Union of Biochemistry. The name reflects the nature of the substrate and the type of reaction catalysed, with all the information based on the formal chemical equation describing the reaction.

The systematic name is used for scientific purposes when it is important to be exact in order to avoid confusion. However, it is a rather long name, difficult to remember, and clumsy. Therefore, where confusion is unlikely or exactness is not important, trivial names are still used.

Redox reactions

The term **redox** is an abbreviation of reduction–oxidation. Reduction and oxidation are two chemical processes that usually occur simultaneously in the same reaction.

Oxidation may involve the addition of oxygen or the removal of hydrogen, but it always involves a loss of electrons. Conversely, **reduction** may involve the addition of hydrogen or the removal of oxygen, but it always involves the gain of electrons.

Redox reactions play a fundamental role in many biological processes, including respiration and photosynthesis, so this is an essential concept for biology students to learn. OIL RIG is a useful mnemonic to help you remember that, in relation to electrons,

Oxidation

Is

Loss

Reduction

Is

Gain

Quick check

1 Using the classification of the International Union of Biochemistry (see box on opposite page), identify the type of enzyme that would catalyse the reactions shown below. You may use each type of enzyme once, more than once, or not at all.

a ATP + glucose ⟶ ADP + glucose 6-phosphate

b Maltose + water ⟶ glucose + glucose

c Hydrogen peroxide ⟶ water + oxygen

$$2H_2O_2 \longrightarrow 2H_2O + O_2$$

Food for thought

Abzymes and ribozymes are two groups of biological catalyst which do not fit neatly into the traditional classification of enzymes. **Abzymes** are large antibodies manufactured by biotechnologists using a technique called monoclonal antibody cloning (see spread 15.6). These antibodies are large protein molecules which fit specific substrates very precisely. They are used in many industrial and medical processes, for example, to break down blood clots and remove scar tissue. **Ribozymes** were discovered in 1983. They are sections of nucleotides which cut themselves out of nuclear RNA during the formation of messenger RNA (see spread 18.7); they have been described as autosplicing catalysts. Suggest points for and against classifying abzymes and ribozymes as enzymes.

3.6

Enzyme technology

OBJECTIVES

By the end of this spread you should be able to:

- explain why enzymes are used in industrial processes
- describe how enzymes are extracted for industrial use
- discuss the use of immobilised enzymes in industrial processes.

Fact of life

The world market for enzymes is worth many millions of pounds and is rising daily as new uses for enzymes are discovered. However, the use of enzymes in commericial and industrial processes is nothing new. They have been used for centuries in cheese-making and the leather industry, and protein-digesting enzymes were patented for use in washing powders in 1913.

Enzymes from microorganisms

For thousands of years, natural enzymes made by microorganisms (especially bacterial and fungal cells) have been used to make products such as cheeses, bread, wine, and beer. Today enzymes are used for a wide range of industrial processes (see spread 3.7). The study of industrial enzymes and their uses is called **enzyme technology**.

The use of enzymes in industrial processes has a number of advantages over using inorganic catalysts for the same processes. Unlike inorganic catalysts, most enzymes work at room temperature, atmospheric pressure, and within moderate pH ranges. Enzymes are much more specific than inorganic catalysts. Some enzymes, for example, catalyse reactions involving only one type of isomer. Consequently enzymes can be used in the food, drinks, and drugs industries to produce chemically pure products. Non-enzymatic reactions often yield a number of unwanted products with harmful side-effects.

Producing the enzyme: culturing the microbes

Microbial cells are still the most common source of industrial enzymes. Microorganisms produce enzymes inside their cells (**intracellular enzymes**) and may also secrete enzymes for action outside the cell (**extracellular enzymes**). The chosen microorganisms are usually cultured in large fermentation chambers under controlled conditions designed to maximise enzyme production. The microorganisms may have specific genes introduced into their DNA by genetic engineering so that they produce enzymes naturally made by other organisms (see spreads 18.9 and 18.10).

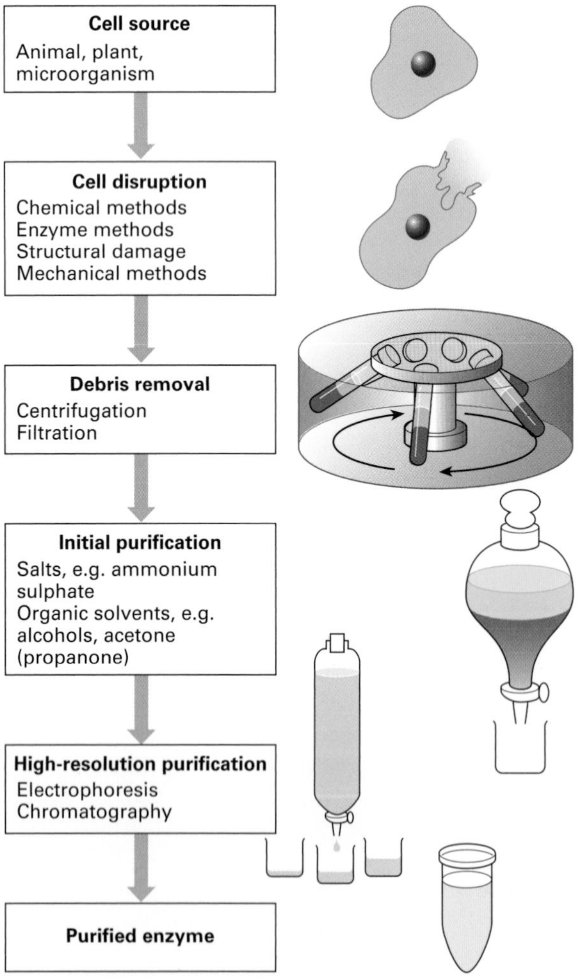

Cell source
Animal, plant, microorganism

Cell disruption
Chemical methods
Enzyme methods
Structural damage
Mechanical methods

Debris removal
Centrifugation
Filtration

Initial purification
Salts, e.g. ammonium sulphate
Organic solvents, e.g. alcohols, acetone (propanone)

High-resolution purification
Electrophoresis
Chromatography

Purified enzyme

Figure 1 *Steps in the purification of an intracellular enzyme produced by cultured microbes.*

Isolating the enzyme

Enzymes used in industry are usually needed in a pure form, so must be isolated from the microbial cells. To obtain an intracellular enzyme, the microbe cells are harvested from the culture and broken open. The resultant mixture is centrifuged to remove large cell fragments, and the enzyme is precipitated from solution by a salt (such as ammonium sulphate) or an alcohol (such as propan-2-ol). The enzyme can be purified (figure 1) by techniques such as electrophoresis or column chromatography (see appendix). Extracellular enzymes are present in the culture outside the microbial cells. They are often soluble in water, so they can be extracted from the culture medium and purified.

Using whole cells

In some circumstances it is not possible to use isolated enzymes to carry out an industrial process, and whole cells are used instead. This may happen, for example, when it is too difficult or expensive to extract and purify an intracellular enzyme, or when two or more enzymes are required to work together. When using whole cells, the substrate has to diffuse into the cells before the reaction takes place. The product may diffuse out of the cells, or the cells may be disrupted to release it. The product can then be extracted and purified.

Improving the enzyme: enzyme stability

The **stability** of an enzyme refers to its ability to retain its tertiary structure (its three-dimensional shape) so that it continues to be effective under a wide range of conditions. Enzyme stability is a key factor in the industrial use of enzymes.

Most enzymes are relatively unstable and work only within narrow ranges of temperature and pH. They quickly become denatured when subjected to unnatural environments. Many industrial processes require enzymes to work in the presence of chemicals such as organic solvents, at high temperatures, and extremes of pH, conditions which cause most enzymes to lose their shape and become inactive. It is possible to overcome this problem by taking advantage of microbes that live naturally in harsh environments.

Organisms evolve to produce enzymes that are adapted to their environmental conditions. Thermophilic bacteria living in hot volcanic springs, for example, produce **thermostable** enzymes that do not denature at high temperatures: they work effectively in the temperature range 65–75 °C. These enzymes are also resistant to organic solvents and tolerate a wide range of pH. The gene for the thermostable enzyme has been isolated from thermophilic bacteria and transferred to *Bacillus subtilis*, a microbe that can be used in industrial processes. This enables a thermostable version of the desired enzyme to be produced.

Improving the enzyme: immobilisation

Unstable enzymes may be **immobilised** by being attached to or located within an insoluble support (figure 2). Once attached, an enzyme's stability is increased, possibly because its ability to change shape is reduced.

Compared with free enzymes, immobilised enzymes have several other advantages:

- Immobilised enzymes can be recovered easily from a reaction mixture and reused over and over again.
- Because immobilised enzymes are held in an inert (unreactive) matrix, the reaction products are not contaminated by the enzyme.
- Enzymes that have been immobilised by incorporating them into a solid material such as a nylon membrane can be manipulated easily; they can, for example, be placed at a precise position in a reaction system, and they can be added or removed from a reaction system easily, giving greater control over the reaction.
- Immobilising enzymes makes continuous production of a substance easier. For example, reactants could be added to the top of a column containing immobilised enzymes. The reactants trickle down through the enzymes, and the product is collected continuously as it leaves the bottom of the column.

Table 1 shows the four main types of enzyme immobilisation. Similar methods can be used to immobilise whole cells.

Table 1 *Advantages and disadvantages of different types of immobilisation.*

Type of immobilisation	Advantages	Disadvantages
Adsorption onto an insoluble matrix (e.g. glass)	Easy to immobilise; relatively safe	Becomes detached easily; susceptible to pH and temperature changes
Covalent binding to a solid support (e.g. cellulose or nylon)	Fairly resistant to pH and temperature changes	May require the use of chemicals that can damage the enzyme
Trapping within a gel (e.g. collagen, alginate beads)	Gentle conditions help preserve the enzyme; no chemical change in the system	Enzyme may leak away; enzyme substrate has to diffuse into the gel
Encapsulation behind a selectively permeable membrane (e.g. nylon)	Little chance of the enzyme leaking away	May be difficult to bring the enzyme substrate in close contact with the enzyme; encapsulation may damage the enzyme

Adsorption onto an insoluble matrix e.g. porous glass

Enzyme

Porous glass

Covalent binding onto a solid support e.g. cellulose or nylon

Enzyme

Solid support

Entrapment within a gel e.g. collagen

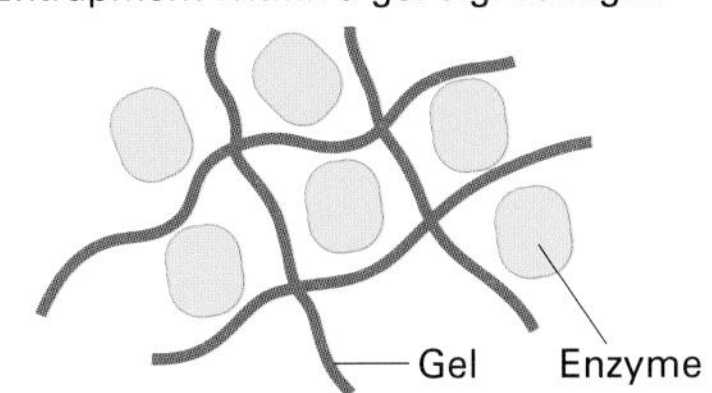

Encapsulation behind a selectively permeable membrane e.g. nylon

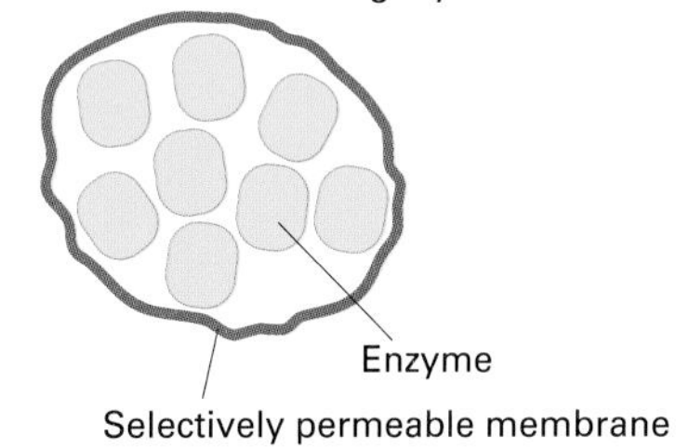

Figure 2 *Methods of enzyme immobilisation.*

QUICK CHECK

1. List three advantages and two disadvantages of using enzymes rather than inorganic catalysts in industrial processes.
2. What is the role of alcohol or ammonium sulphate during the extraction of enzymes?
3. Why is the thermostability of enzymes so important for many industrial processes?

Food for thought

Surprisingly, cats have difficulty digesting cow's milk. They lack the enzyme lactase which catalyses the breakdown of lactose into galactose and glucose. Suggest how immobilised enzymes can help.

3.7

Applications of enzyme technology

OBJECTIVES

By the end of this spread you should be able to:

- detergents
- the food industry
- biosensors.

Fact of life

The fact that the active sites of enzymes conform to the shape of specific molecules is used to detect illicit drugs. In the 1990s, a gel was developed that changes colour in the presence of heroin. The gel contains two enzymes. The first comes from the bacterium *Rhodococcus* and it breaks heroin down to morphine. The second enzyme comes from another bacterium, *Pseudomonas*, and breaks the morphine down in such a way that it causes a colourless dye to turn red.

Detergents

Because of their ability to catalyse specific reactions at normal temperatures and pressures, enzymes are of great commercial and industrial importance (table 1). Enzymes have been used for centuries in cheese-making, brewing, baking, and the leather industry. Although they were used in washing powders at the beginning of the twentieth century, it was not until the 1950s and 1960s that Novo industries of Denmark developed an enzyme stable enough for use in a general washing powder. The enzyme, **subtilisin**, was obtained from the bacterium *Bacillus subtilis*. It is a protein-digesting enzyme (protease) which is active up to 60 °C.

Table 1 *Examples of industrial applications of enzymes.*

Application	Enzyme	Use
Baking	Amylase	Catalyses breakdown of flour starch to sugar
Biofuel	Cellulases	Break down cellulose to sugars for fermentation
Biological detergents	Lipases	Remove fatty stains
Brewing	Proteases	Remove cloudiness from stored beers
Confectionary	Invertase	Helps make soft centres in chocolates
Dairy industry	Rennin	Coagulates milk in cheese manufacture
Fruit juices	Pectinases	Clarify juices
Meat industry	Trypsin	Tenderises meat and pre-digests baby foods
Paper industry	Ligninases	Break down lignin to soften paper
Rubber industry	Catalase	Helps convert latex into foam rubber

Food and drink

The food and drinks industry depends heavily on enzymes. Enzymes produced by yeast have been used for thousands of years in brewing and baking (see spread 17.10). The word ‘enzyme’ actually means ‘in yeast’. It was first used to describe the chemicals that enable yeast to convert sugar into alcohol during fermentation.

High fructose syrups contain fructose and glucose in roughly equal proportions. The high fructose syrups are in greater demand than pure glucose as food and drink sweeteners, because fructose is sweeter than glucose. Therefore, if glucose can be converted into fructose, its commercial value is increased greatly. The production of high fructose syrups by enzyme technology not only helps satisfy the sweet tooths of millions of consumers, but also helps the food industry solve a waste-disposal problem. The raw material for high fructose syrups is starch (especially corn starch in the USA), a major waste product of the food industry. The conversion of starch to high fructose syrups uses a combination of three main enzymes (figure 1): alpha amylase and glucoamylase (to convert starch to glucose), and glucose isomerase (to change glucose to its isomer, fructose). Alpha amylase and glucose isomerase are obtained from bacteria belonging to the genus *Bacillus*, and glucoamylase is obtained from the fungus *Aspergillus niger*.

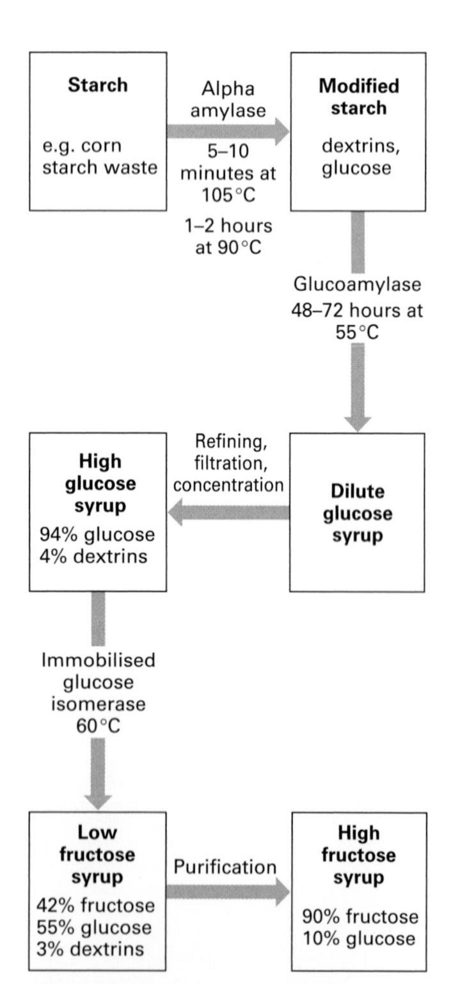

Figure 1 *The enzymes used in the production of high fructose syrup.*

Pectinases are enzymes used to clarify fruit juices. Plant cell walls contain polysaccharides called pectins, and these make fruit juice viscous and difficult to extract. The juice is also cloudy and the flavour may deteriorate. To overcome this, the pectins are hydrolysed using pectinases.

Cellulases are enzymes produced by microbes that live in the guts of ruminants such as cattle. These enzymes are used to break down lignin (see spread 2.7) in pulp so that the lignin can be removed during the manufacture of paper. Lignin discolours paper and makes it rougher. Removing the lignin is a cheaper and less environmentally damaging alternative to bleaching the pulp with chlorine.

Naturally occurring enzymes can catalyse only a limited number of reactions. Most of these enzymes are denatured by the organic solvents and high temperatures that are an integral part of many industrial processes. Existing enzymes are therefore being genetically modified to work under harsh industrial conditions and, using computer-aided design, completely new enzymes are being created to catalyse reactions that do not occur in nature.

Biosensors

The fact that enzymes bind with a specific substance is used in diagnostic tools called biosensors. A **biosensor** usually consists of a **receptor** which uses a biochemical reaction to detect a specific substance or condition, a **transducer** which converts a biochemical signal into an electrical signal, and a system for converting the electrical signal into a reading or measurement (figure 2). The receptor is usually an immobilised living cell, tissue, or enzyme (see spread 3.6).

Biosensors are used extensively in industry, forensic science, agriculture, environmental science, and medicine to monitor specific chemicals accurately and rapidly.

Some biosensors use a colour change to detect a specific substance. One common biochemical device enables diabetics to monitor their own blood glucose levels. It consists of a plastic strip with a piece of special filter paper at one end. The filter paper contains two immobilised enzymes, glucose oxidase and a peroxidase, and a colourless hydrogen donor. When the paper is dipped in blood containing glucose, the glucose oxidase catalyses the conversion of the glucose to gluconic acid and hydrogen peroxide:

$$\text{Glucose} + \text{oxygen} \xrightarrow{\text{glucose oxidase}} \text{gluconic acid} + \text{hydrogen peroxide}$$

The peroxidase enzyme then catalyses a reaction between the hydrogen peroxide produced and the colourless hydrogen donor. The product of this second reaction is a coloured compound.

$$\text{Colourless hydrogen donor} + \text{hydrogen peroxide} \xrightarrow{\text{peroxidase}} \text{water} + \text{coloured compound}$$

The type and intensity of the colour indicates the concentration of glucose in the blood.

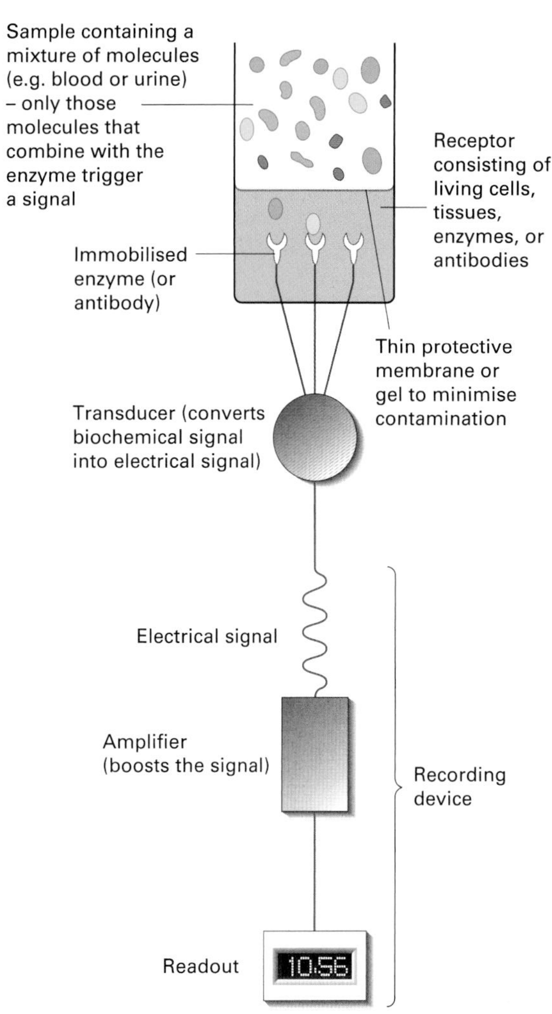

Figure 2 *Plan of a typical biosensor.*

QUICK CHECK

1 What is subtilisin?

2 Why is the commercial value of glucose increased by converting it to fructose?

3 What is the function of a transducer in a biosensor?

Food for thought

One of the major uses of enzymes is in detergents. Suggest:

a why the first biological detergents were withdrawn from the market

b which fabrics may be damaged by prolonged soaking in biological detergents

c how biological detergents affect domestic energy consumption.

Summary

Metabolism refers to the chemical reactions that take place in living organisms, There are two main types of reaction: **anabolic reactions** result in synthesis whereas **catabolic reactions** result in breakdown. Metabolic reactions obey the **laws of thermodynamics**: they neither create nor destroy energy, but when they transform energy from one form to another they always produce heat.

Before a metabolic reaction can get underway, **activation energy** is required to overcome an **energy barrier**. The amount of activation energy required depends on the type of reaction and the conditions in which it takes place. **Enzymes** are proteins which act as **biological catalysts**. They speed up specific chemical reactions by lowering the activation energy. The activity of an enzyme depends on the shape of its active site. According to the **lock-and-key theory**, the shape of the active site is rigid and fits that of its substrate precisely. In contrast, the **induced-fit theory** suggests that the active site is not rigid and that it takes up its most effective shape after binding with its substrate. In either case, the substrate fits neatly onto the active site to form an **enzyme–substrate complex**.

Factors affecting the rate of enzyme-catalysed reactions include **substrate concentration**, **enzyme concentration**, **incubation time**, **temperature**, **hydrogen ion concentration**, **cofactors**, and **inhibitors**. Extreme temperatures and pH can inactivate an enzyme by changing its **tertiary structure** so that the shape of its active site does not fit that of its substrate. This process is called **denaturation**. Cofactors are non-protein substances, inorganic **activators** or organic coenzymes that enable enzymes to work better. Inhibitors slow down enzyme-catalysed reactions. **Competitive inhibitors** compete with the normal substrate for the active site of the enzyme whereas **non-competitive inhibitors** alter the shape of the active site after binding onto another site on the enzyme. Inhibitors may be reversible or irreversible. **Irreversible inhibitors** include metabolic poisons such as cyanide. **Reversible inhibitors** include the final products of reaction pathways which regulate their own production by **end-product inhibition**.

Enzymes are classified into six major groups: **oxidoreductases**, **transferases**, **hydrolases**, **lysases**, **isomerases**, and **ligases**.

Enzymes have many industrial applications. They are used in the food and drinks industry, in detergents, and in **biosensors**. In many industrial processes, **immobilisation** of enzymes onto inert, insoluble surfaces has improved their usefulness. The study of industrial enzymes and their uses is called **enzyme technology**.

PRACTICE EXAM QUESTIONS

1 The diagram shows the tertiary structure of a molecule of the enzyme RNAase.

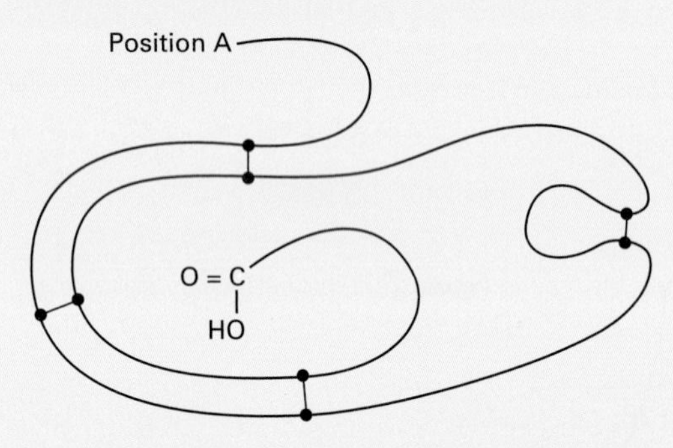

a What is the name of the chemical group found in position A? [1]

b **i** Explain what is meant by the *tertiary structure* of a protein. [1]

ii The chemical mercaptoethanol breaks disulphide bonds (bridges). Describe and explain what would happen to the enzyme activity of RNAase if it were treated with mercaptoethanol. [3]

[Total 5 marks]

2 The diagrams below illustrate one model of enzyme action.

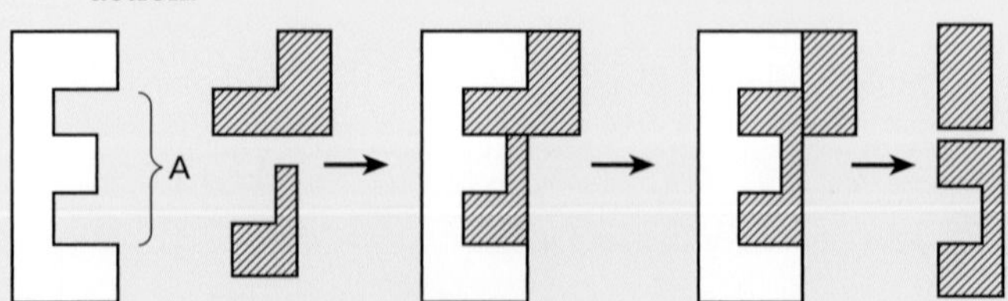

a Name the part of the enzyme labelled A. [1]

b Explain how this model can account for enzyme specificity. [2]

c With reference to this model, explain the effect of a competitive inhibitor on an enzyme-catalysed reaction. [2]

[Total 5 marks]

3 The *turnover number* of an enzyme is defined as the number of substrate molecules converted to product by one molecule of enzyme in one minute. In an experiment carried out at 20 °C the turnover number for an enzyme was found to be 2500 at the start of the experiment but dropped to 1000 after 5 minutes.

a **i** Suggest why the turnover number decreased after 5 minutes. [2]

ii How would you expect the turnover number to differ from 2500 at the start of an identical experiment but carried out at 30 °C? Explain your answer. [2]

b Explain why it would be important to have a control in the experiment at 20 °C and at 30 °C. [1]

[Total 5 marks]

4 Succinate dehydrogenase is an enzyme which catalyses the conversion of succinate to fumarate.

a Use your knowledge of enzyme structure to explain why succinate dehydrogenase catalyses this reaction only. [2]

b Malonate is an inhibitor of succinate dehydrogenase. The structural formulae of malonate and succinate are shown in the diagram.

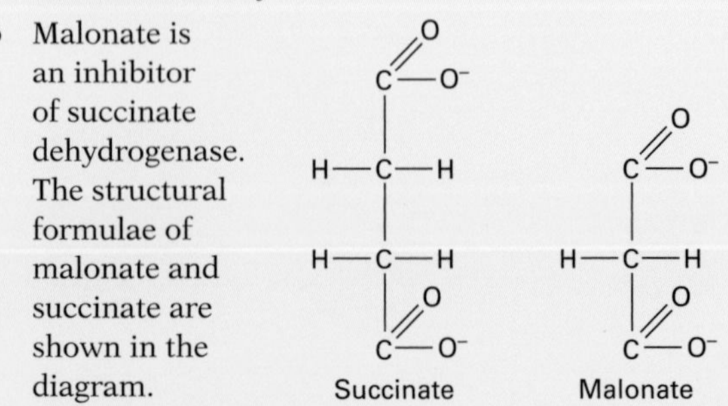

Use the information in the diagram to explain how malonate inhibits the enzyme. [2]

[Total 4 marks]

5 The graph shows the results of an investigation into the effect of a competitive inhibitor on an enzyme-controlled reaction over a range of substrate concentrations.

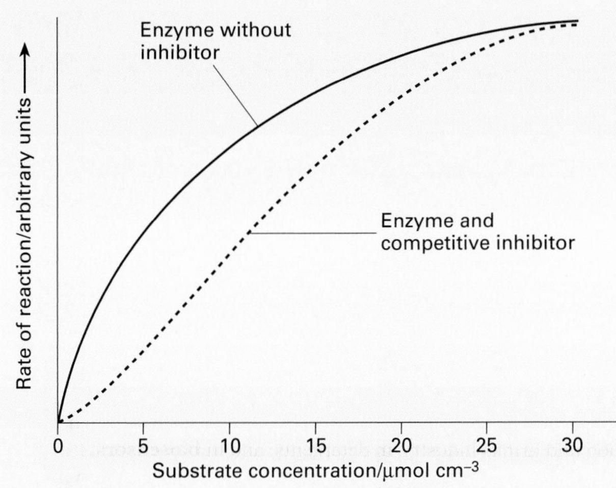

a Give **one** factor which would need to be kept constant in this investigation. [1]

b **i** Explain the difference in the rates of reaction at the substrate concentration of 10 μmol cm^{-3}. [2]

ii Explain why the rates of reaction are similar at the substrate concentration of 30 μmol cm^{-3}. [1]

c The diagram represents a metabolic pathway controlled by enzymes.

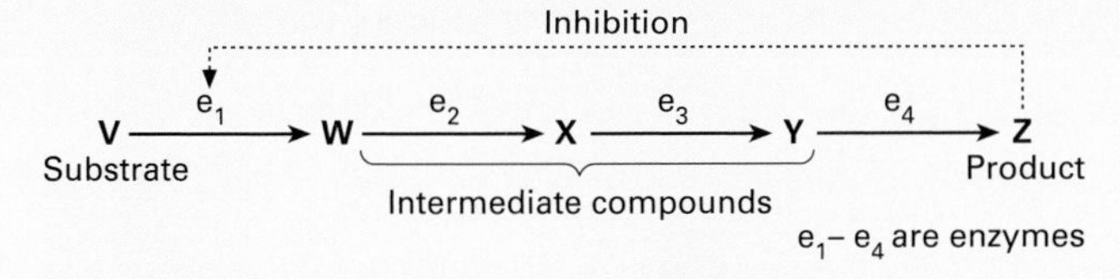

i Name the type of control mechanism which regulates production of compound **Z**. [1]

ii Explain precisely how an excess of compound **Z** will inhibit its further production. [2]

[Total 7 marks]

6 The flow diagram shows two ways in which enzymes may be produced for industrial processes.

a Explain the advantages of using immobilised enzymes. [4]

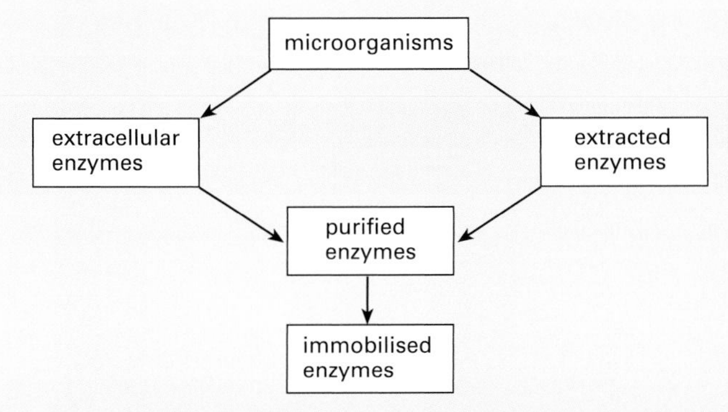

b **i** Describe **three** ways by which enzymes or cells may be immobilised. [3]

ii A bacterial enzyme used widely in the production of sweeteners for soft drinks is glucose isomerase which converts glucose to the much sweeter fructose. Outline how you would test whether different ways of immobilising this enzyme affect the rate at which it converts glucose to fructose. [3]

c Suggest why enzymes used for industrial processes may be obtained from microorganisms that are thermophilic (able to grow at high temperatures). [2]

[Total 12 marks]

7 The diagram shows the main components of a glucose biosensor.

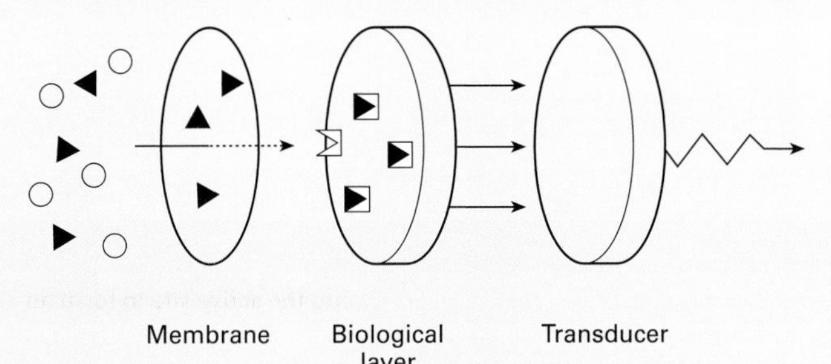

a **i** Name the receptor molecules in the biological layer. [1]

ii What is the function of these receptor molecules? [1]

b What is the function of the transducer? [1]

c A glucose biosensor can be used to measure blood glucose levels in diabetics. Suggest two advantages of using a biosensor rather than the Benedict's test in measuring blood glucose levels. [2]

[Total 5 marks]

8 *Answers should be written in continuous prose. Credit will be given for biological accuracy, the organisation and presentation of the information and the way in which the answer is expressed.*

Read the following passage.

Because they are proteins, enzymes are sensitive to changes in pH and temperature. The effects of temperature are complex but involve a balance between the movement of molecules and the structural stability of proteins. This results in an optimum rate of reaction. Inhibitors work by influencing the active site of the enzyme molecule, either directly or indirectly.

a Describe how polypeptide chains may be folded to form protein molecules. [5]

b Explain the effects of temperature on enzyme activity. [6]

c Explain how inhibitors can alter the rate of enzyme-controlled reactions by acting either directly or indirectly. [6]

Quality of language. [3]

[Total 20 marks]

9 Explain why

i enzymes which break down starch and cellulose are called hydrolases. [1]

ii amylase will break down starch but it will not break down cellulose. [3]

[Total 4 marks]

Cells

Cell theory

OBJECTIVES

By the end of this spread you should be able to:

- describe the main ideas of the cell theory
- compare the structures of animal and plant cells as seen with a light microscope.

Fact of life

Cells do not function in isolation. They have to be able to communicate with each other. Communication often involves **cell signalling**: the release by one cell of one or more substances that transmit information to other cells. Cell signalling occurs between unicellular organisms, for example prior to mating. It is also an essential feature in multicellular organisms, for example, enabling the different cells in a tissue to function in a coordinated manner and to respond appropriately to their microenvironment.

Calcium is one of many chemical signals that play an important role in regulating cellular activity and orchestrating the complexities of cellular communication. Disruption of signalling pathways in which calcium plays a key part can lead to diseases such as hypertension, cardiac arrhythmia and heart failure, cancer, and depressive illness. Cell signalling often involves complex multi-component signalling pathways that provide opportunities for feedback, signal amplification, and interactions between different signalling substances and different signalling pathways.

The discovery of cells

Cells were discovered in 1665 by the English scientist and inventor Robert Hooke. Hooke designed his own compound light microscope (see spread 4.2) to observe structures too small to be seen with the naked eye. Among the first structures he examined was a thin piece of cork (the outer surface of bark from a tree). Hooke described the cork as being made of hundreds of little boxes, giving it the appearance of a honeycomb (figure 1). He called these little boxes **cells**.

It soon became clear that virtually all living things are made of cells, and that these cells have certain features in common.

The cell theory

The concept that cells are the basic units of life became embodied in a theory called the **cell theory**,which embraces the following main ideas:

- cells form the building blocks of living organisms
- cells arise only by the division of existing cells
- cells contain inherited information which controls their activities
- the cell is the functioning unit of life; metabolism (the chemical reactions of life) takes place in cells
- given suitable conditions, cells are capable of independent existence.

A typical animal cell

Figure 2 shows the structure of a typical animal cell as seen with a light microscope.

- The cell has a **cell surface membrane** which encloses the cell contents.
- The contents consist of a central ball-shaped **nucleus** surrounded by material called **cytoplasm**.
- The nucleus contains a fibrous material called **chromatin**.
- This condenses to form **chromosomes** during cell division.
- Chromatin contains DNA, the inherited material which controls the various activities inside the cell.
- Scattered within the cytoplasm are **mitochondria**, small rod-like structures. They have been described as the 'power-houses' of the cell because they supply energy.
- Smaller dots within the cytoplasm are particles of stored food. Many consist of glycogen, which is a food storage polysaccharide.

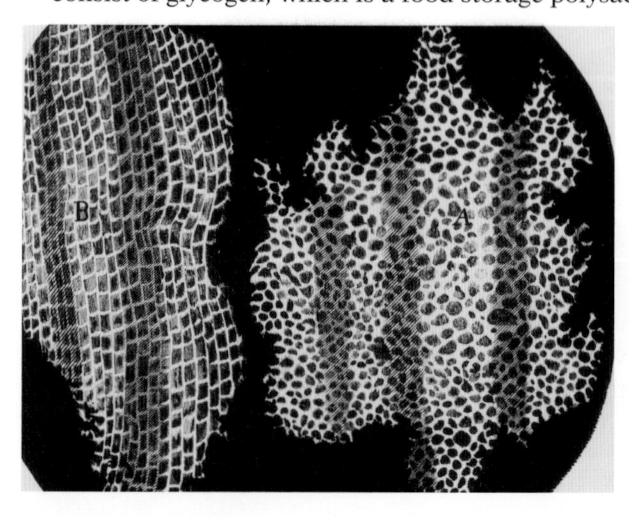

***Figure 1** Cork cells drawn by Robert Hooke for his book* Micrographia, *first published in September 1665 by the Royal Society. In the book, Hooke was the first to use the word 'cell' as a biological term.*

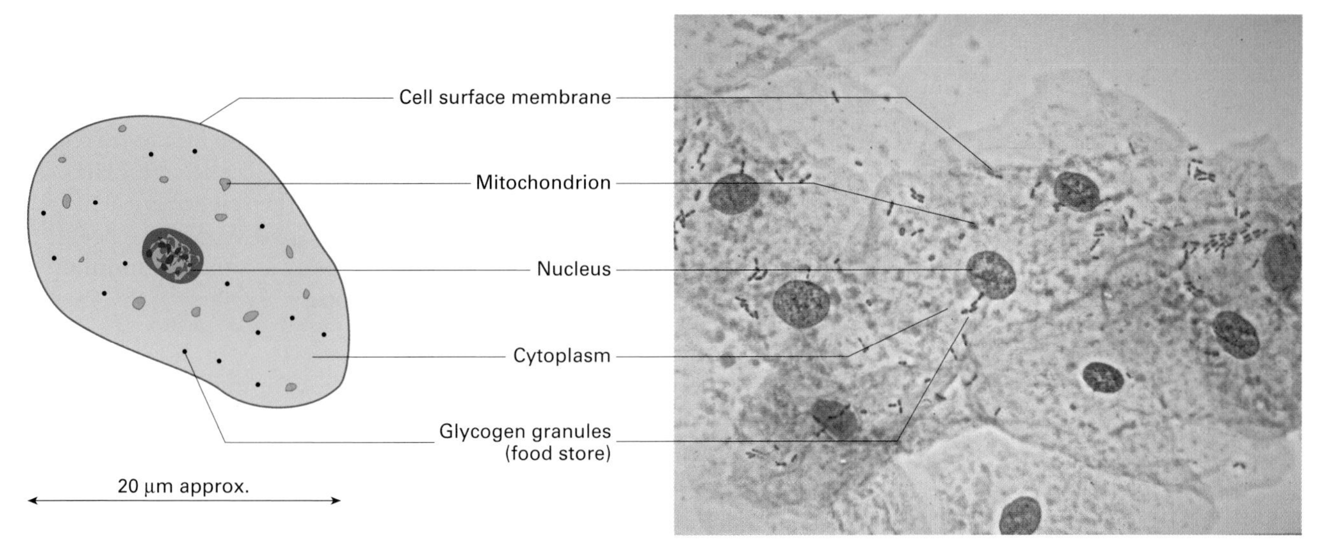

Figure 2 (a) *Drawing of a typical animal cell as seen with a light microscope. Animal cells exist in many shapes and sizes. Most share the features shown in this diagram. They may also have vacuoles, but these are not permanent features as they are in plant cells.*

Figure 2 (b) *Some cheek cells as they appear under the light microscope (×2000).*

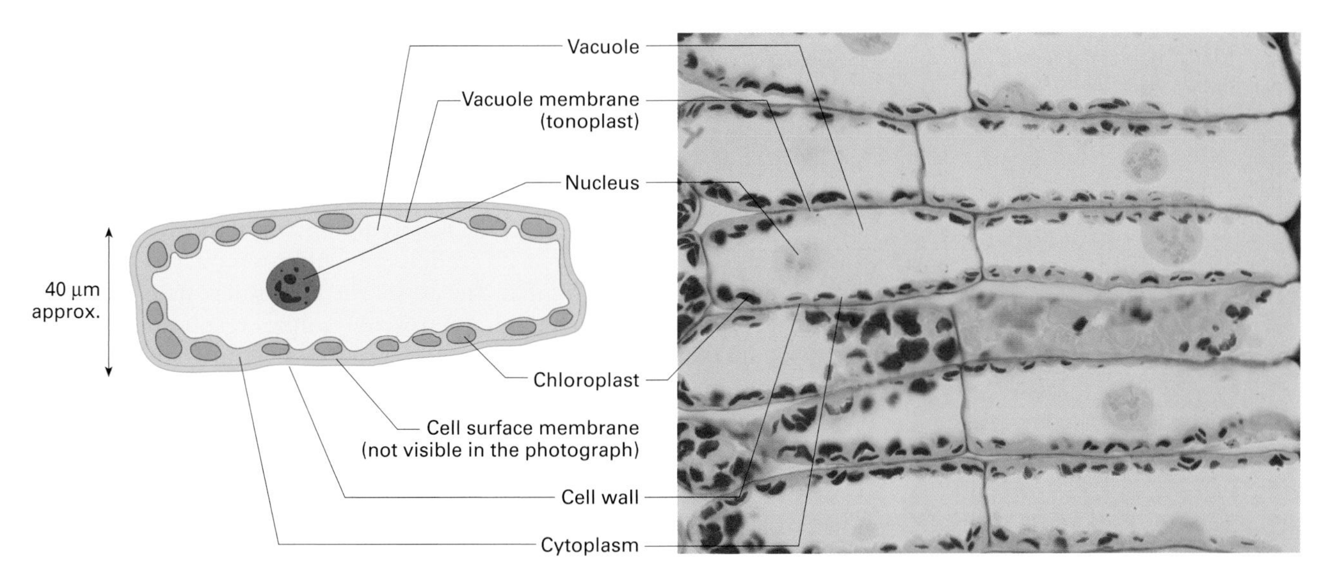

Figure 3 (a) *Drawing of a typical plant cell as seen with a light microscope. A cell wall surrounds the cell surface membrane. The cell wall is made of cellulose, a tough rubbery material which is completely permeable to water. In some plant cells, the cellulose cell wall becomes impregnated with lignin, which makes it impermeable.*

Figure 3 (b) *Some palisade cells from a leaf as they appear under the light microscope (×500). Palisade cells form a photosynthetic layer just beneath the surface of many leaves.*

A typical plant cell

Figure 3 shows the structure of a typical plant cell as seen with a light microscope. Like an animal cell, it has a cell surface membrane, cytoplasm, and a nucleus. However, plant cells differ from animal cells in several ways:

- Most plant cells have a large sap-filled cavity called the **vacuole**. Sap is a watery fluid containing salts and sugars. The vacuole is surrounded by a membrane called the **tonoplast**.
- The cytoplasm contains **starch grains**, the food storage products of plants.
- Many plant cells have **chloroplasts** in the cytoplasm. These contain the pigments used in photosynthesis. **Chlorophyll**, which is green, is the main pigment. Chloroplasts occur only in the parts of plants exposed to light – the green parts. They are absent from underground structures such as roots.

Quick check

1 Briefly state the main concept of the cell theory.

2 Using figures 2 and 3, list the features:

a that only animal cells have

b that only plant cells have

c that both animal and plant cells have.

Food for thought

Suggest why red blood cells (spread 7.8) appear to contradict the cell theory.

4.2

Microscopes

OBJECTIVES

By the end of this spread you should be able to:

- describe the main features of a light microscope and an electron microscope
- distinguish between the terms magnification and resolving power
- give the approximate size of different biological structures using an appropriate unit of measurement.

Fact of life

Since the invention of the first transmission electron microscopes (TEMs), several different types of electron microscope have been developed with special capabilities. For example, the scanning tunnelling microscope (STM) was developed in the 1980s to create images of surfaces at the atomic level. It earned its inventors Gerd Binnig and Heinrich Rohrer the Nobel Prize in Physics in 1986. The STM can achieve magnifications of more than 100 million. It has been used to view biological structures, such as DNA (figure 1), that are too small to be resolved by TEMs.

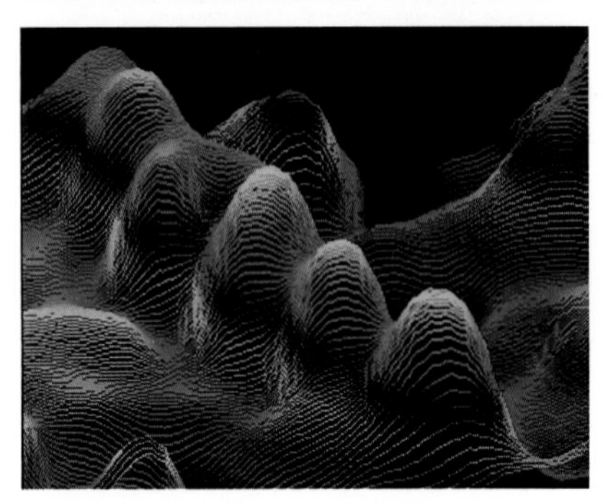

Figure 1 *A scanning tunnelling micrograph. The orange and yellow peaks represent ridges of the DNA double helix (false colour; ×2 500 000).*

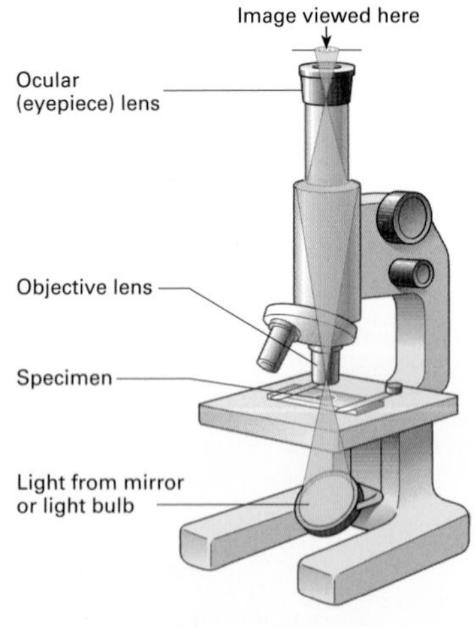

Figure 2 *A compound light microscope has two lens systems: the objective lens and the ocular or eyepiece lens.*

A microscope is used to produce a magnified image of an object or specimen. Anton van Leeuwenhoek (1632–1723) was the first to invent a microscope powerful enough to explore the world of microbes. His discoveries stimulated an explosion of interest in the scientific use of microscopes. Since the eighteenth century many new types have been invented, of which the most commonly used today are the **compound light microscope** and the **electron microscope**.

The compound light microscope

The compound light microscope is also called a light microscope or optical microscope. Light passes through a specimen (the object) and then through two sets of glass lenses, the objective lens and the ocular or eyepiece lens. The lenses refract (bend) the light to give a magnified image of the object (figure 2). The image may be projected directly into the viewer's eye or onto photographic film. A photograph taken through a light microscope is called a **photomicrograph** or light micrograph.

Magnification and resolution

The **magnification** of an instrument is the increase in the apparent size of the object. The total magnification of a compound microscope is worked out by multiplying the magnification of the objective lens by that of the ocular lens.

There is virtually no limit to the magnification produced by a light microscope; it depends on the power of the lenses used. However, above a certain magnification the image becomes blurred and it is impossible to distinguish structures lying close together. This limit of effective magnification is called the **resolving power** or **resolution** of the microscope. It is defined as the ability of a microscope to show two objects as separate. The resolving power of the light microscope is limited by the wavelength of light. A light microscope cannot resolve detail finer than 0.2 μm (see figure 3). Light microscopes can magnify objects up to about 1500 times without losing clarity.

Table 1 *Measurements and equivalents*

1 millimetre (mm)	= 10^{-3} metre (m)	= 1/1000 m
1 micrometre (μm)	= 10^{-6} metre (m)	= 1/1 000 000 m
1 nanometre (nm)	= 10^{-9} metre (m)	= 1/1 000 000 000 m
1 metre (m) = 10^3 mm = 10^6 μm = 10^9 nm; 1 kilometre (km) = 10^3 m		

Calculating magnifications

The **magnification** is the number of times larger an image is than the object (specimen). Therefore,

$$\text{Magnification} = \frac{\text{size of image}}{\text{size of object}}$$

It is essential that the same unit is used for the size of the image and the size of the object. For example, if an image measures 50 mm and the object measures 5 μm, the size of the image should be converted to μm:

$$\text{Size of image} = 50\,\text{mm} = 50\,000\,\mu\text{m}$$

$$\text{Therefore, magnification} = \frac{50\,000}{5} = 10\,000$$

Conversely, if the magnification is 50 000 times, and the size of the image is 5 mm (5000 μm), the actual size of the object is equal to:

$$\frac{\text{Size of image}}{\text{magnification}} = \frac{5000}{50\,000} = 0.1\ \mu\text{m}$$

The electron microscope

Electron microscopes use a beam of electrons instead of a beam of light. Electron beams have a much smaller wavelength than light rays, so electron microscopes have greater resolving powers and can produce much higher effective magnifications than light microscopes. Figure 3 shows the sizes of objects that can be viewed with the naked eye, the light microscope, and the electron microscope.

There are two main types of electron microscope: the **transmission electron microscope** (TEM), and the **scanning electron microscope** (SEM).

The transmission electron microscope

The TEM (Figure 4) is used to study the details of the internal structure (the ultrastructure) of cells. Extremely thin samples of the specimen are needed. To make these the specimen is supported in a resin block to prevent it collapsing during cutting, and is sliced with a diamond or glass knife. The section is then impregnated with a heavy-metal stain, such as osmium tetroxide.

As the beam passes through the specimen, electrons are absorbed by heavily stained parts but pass readily through the lightly stained parts. **Electromagnets** bend the electron beam to focus an image onto a fluorescent screen or photographic film. A photograph taken through an electron microscope is called an **electron micrograph**.

The most modern TEMs can distinguish objects as small as 0.2 nm. This means that they can produce clear images magnified up to 250 000 times. The magnification is varied by changing the strength of the electromagnets.

The scanning electron microscope

The SEM is used to produce three-dimensional images of the surface of specimens. Electrons are reflected from the surface of a specimen stained with a heavy metal. This enables the SEM to produce images of whole specimens: cells, tissues, or even organisms.

Although electron microscopes have revolutionised cell biology, they have not completely replaced light microscopes. Light microscopes are used to examine living and unstained specimens. Preparation of specimens for electron microscopy is complicated and time-consuming. Electron microscopes are very expensive and can be used only to study dead specimens stained with heavy metal, which might well produce artefacts (see spread 4.5).

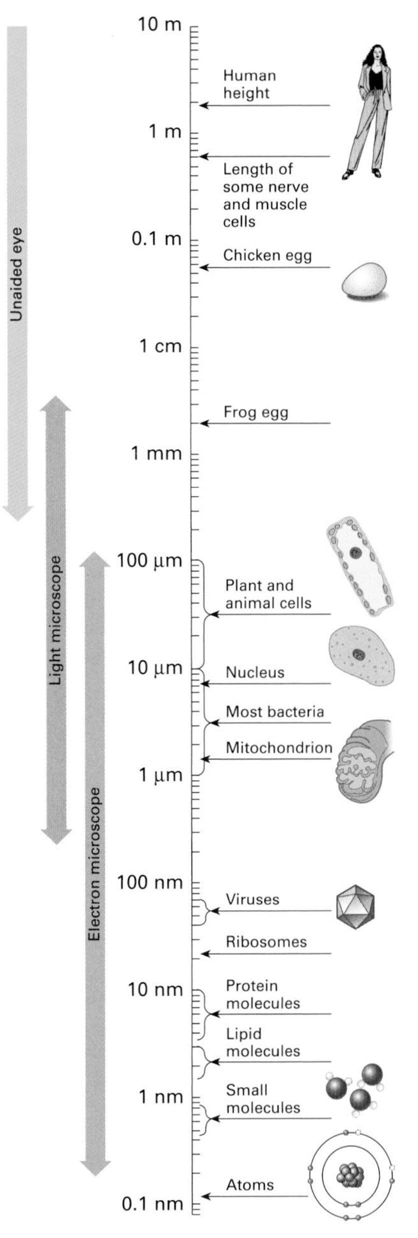

Figure 3 *The sizes of various biological structures.*

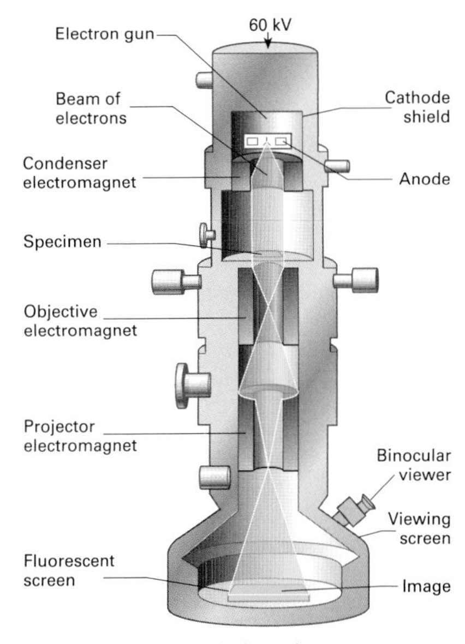

Figure 4 *A transmission electron microscope. The beam of electrons is emitted from the electron gun, a heated filament. Electromagnets bend the beam.*

Quick check

1. How is the magnification varied in:
   - **a** a light microscope
   - **b** an electron microscope?
2. Why is the resolving power of an electron microscope so much better than that of a light microscope?
3. What is the approximate size of the smallest structure that can be observed with a light microscope?

Food for thought

Suggest which unit should be used when calculating the diameter of the DNA molecule in figure 1. From the information given in the figure, what is its actual diameter? Why might there be a discrepancy between the actual diameter and that estimated from the scanning tunnelling micrograph?

4.3

THE ULTRASTRUCTURE OF ANIMAL CELLS

OBJECTIVES

By the end of this spread you should be able to:

- describe the ultrastucture of animal cells and the functions of their organelles.

Fact of life

Protein degradation is a vital function of cells. It eliminates damaged and unwanted proteins, removes excess enzymes, and helps to supply amino acids for fresh protein synthesis. **Proteasomes** are organelles that break down unwanted, abnormal, or misfolded endogenous proteins (that is, proteins synthesised inside the cell). A typical human cell contains about 30 000 proteasomes. Each proteasome is a protein complex folded to form a barrel-shaped structure open at both ends. Degradation takes place on the internal surface of the proteasome, shielded from the rest of the cell. Proteins to be degraded can gain entry into the proteasome only if they are tagged with **ubiquitin**, a small regulatory protein that has been found in nearly all eukaryotic cells. A molecular lock at one end of the 'barrel' recognises the tagged proteins and admits them for disassembly once the ubiquitin label has been removed. The peptides and amino acids formed by degradation are released from the other end of the proteasome. Proteasomes require energy from ATP to work.

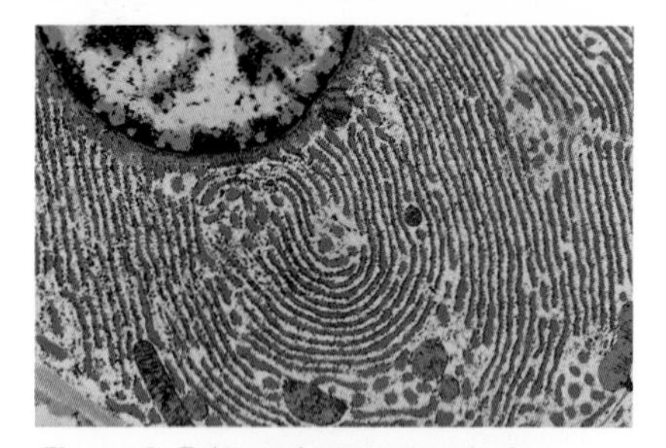

Figure 2 *False-colour transmission electron micrograph of part of the nucleus of an animal cell (×4000).*

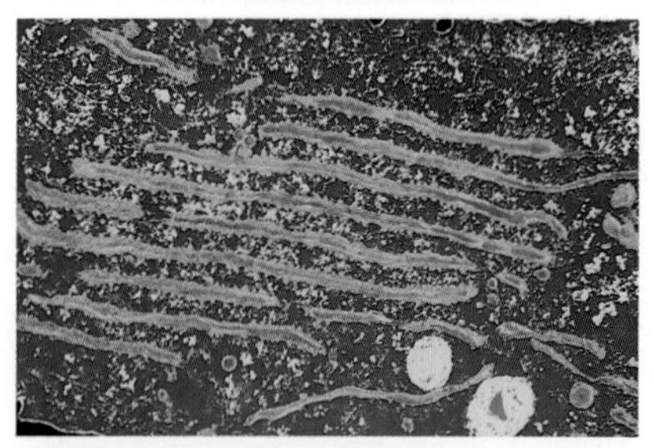

Figure 3 *False-colour transmission electron micrograph of mammalian smooth endoplasmic reticulum (×4500). (In figure 2 above, the red areas covered with black dots are rough endoplasmic reticulum.)*

A typical animal cell viewed under a light microscope appears to have a simple structure: a ball-shaped nucleus is surrounded by a relatively uniform clear cytoplasm, which is bounded by a cell surface membrane. However, the electron microscope reveals that the cell is highly complex and contains structures called **organelles** which carry out specific functions within the cell. The detailed internal structure of a cell is known as the **ultrastructure** (figure 1).

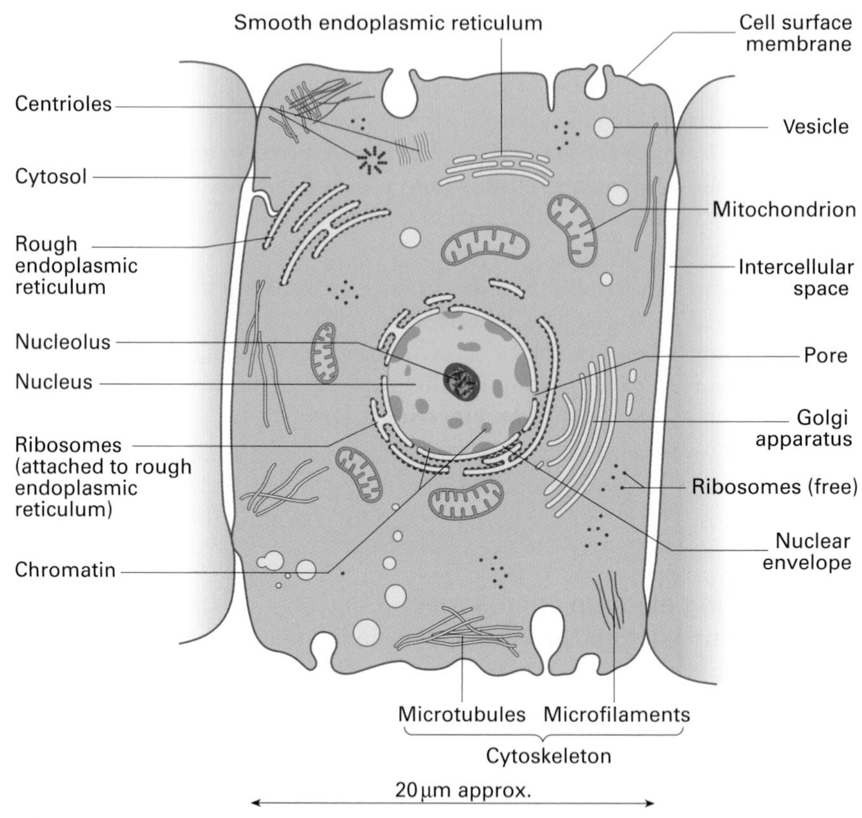

Figure 1 *A generalised animal cell showing structures visible under an electron microscope. There is an enormous variety of animal cells. The size and form of each cell is adapted to carry out specific functions. However, despite differences in detail, most animal cells have the same basic ultrastructure.*

The **cell surface membrane** is selectively permeable and controls the exchange of substances between the cell and its environment. The membrane consists mainly of lipid and protein (see spread 4.6).

The **cytoplasm** refers to all the living parts of the cell, excluding the nucleus. It consists of membrane-bound organelles and the **cytosol**, the fluid part of the cytoplasm. The cytosol contains numerous small molecules in solution and large molecules in suspension.

The **cytoskeleton** consists of microtubules, filaments, and fibres, which give the cytoplasm physical support. The cytoskeleton is also involved in cell movement.

The nucleus

The **nucleus** (figure 2) is the largest cell organelle. It is enclosed by a **nuclear envelope**, which consists of two membranes perforated by **nuclear pores**. These pores control the exchange of materials between the nucleus and cytoplasm. The nucleus contains **chromatin**, the form that chromosomes take when the cell is not dividing. Chromatin contains DNA, the molecule of inheritance which controls the activity of the cell. Two forms of chromatin are visible: **euchromatin** stains lightly and is thought to contain active DNA; **heterochromatin** stains deeply and is thought to contain inactive DNA. The nucleus also contains one or more **nucleoli**. These manufacture ribosomes (see below).

The endoplasmic reticulum and ribosomes

The **endoplasmic reticulum** (**ER**) (figure 3) is a system of flattened membrane-bound sacs forming tubes and sheets. The sacs, called **cisternae**, are full of fluid. Some regions of the ER are covered with bead-like structures called ribosomes. This type of ER is called **rough ER** (**RER**). Smooth ER (**SER**) has no ribosomes.

- The cisternae of the RER transport proteins made by the ribosomes.
- The SER synthesises, secretes, and stores carbohydrates, lipids, and other non-protein products.
- The SER of liver cells also contains enzymes which break down many chemicals.

Ribosomes are tiny organelles consisting of two subunits, one smaller than the other. Ribosomes are made of roughly equal parts of protein and RNA, and they are the sites of protein synthesis. They may be either attached to the RER or may lie free in the cytoplasm.

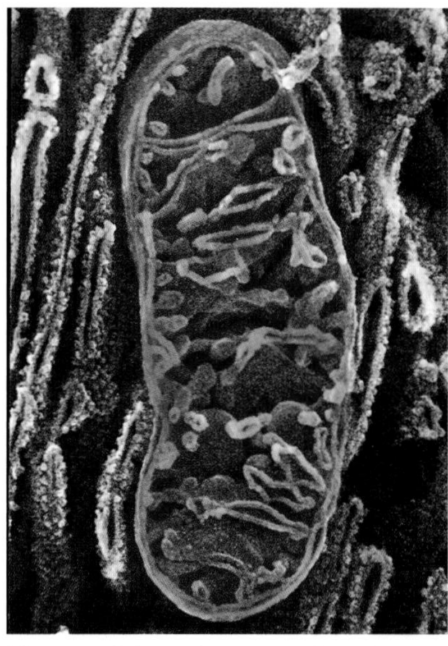

Figure 4 False-colour scanning electron micrograph of a mitochondrion (×2000).

Mitochondria

Mitochondria (figure 4) are cigar-shaped organelles, usually 1–2 μm long. They are involved in generating ATP by aerobic metabolism. Each mitochondrion is bounded by an envelope consisting of two membranes. To maximise its surface area, the inner membrane is deeply folded into shelves called **cristae**. The watery matrix inside contains DNA, ribosomes, calcium phosphate granules, and enzymes.

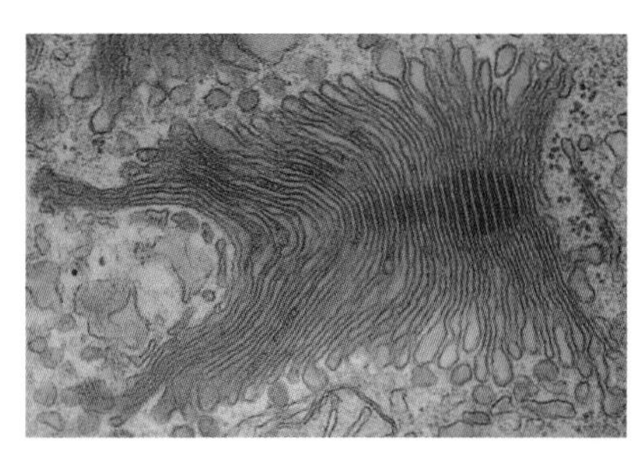

Figure 5 False-colour transmission electron micrograph of Golgi apparatus (×4000).

The Golgi apparatus and lysosomes

The **Golgi apparatus** or Golgi body (figure 5) consists of a stack of flattened membrane-bound sacs, called **cisternae**. New membrane is continuously added to one end of the Golgi apparatus and buds off as vesicles (small sacs) at the other end. The stack often forms an extensive network in animal cells. (In plant cells it is better defined and is called a dictyosome.)

The Golgi apparatus functions as a processing and packaging structure. It enables cell materials such as enzymes to be secreted from the cell in vesicles, and is involved in the formation of **lysosomes**.

Lysosomes (figure 6) contain enzymes that break down proteins derived from outside the cell, eliminate worn-out mitochondria and other redundant organelles, digest the contents of vacuoles ingested by white blood cells, and are involved in autolysis and apoptosis. **Autolysis** is the process by which a cell self-destructs when it dies or is injured. **Apoptosis** involves a genetically programmed series of biochemical events that leads to cell death (see spread 4.10). It occurs during the normal development of multicellular organisms, for example, when a tadpole loses its tail.

Other structures commonly found in animal cells include **centrioles**, **flagella**, and **cilia**. Like the cytoskeleton, these all consist of microtubules. Centrioles are involved in cell division (see spread 4.11). Cilia and flagella probably originated as locomotory structures in single-celled organisms (see spread 21.4).

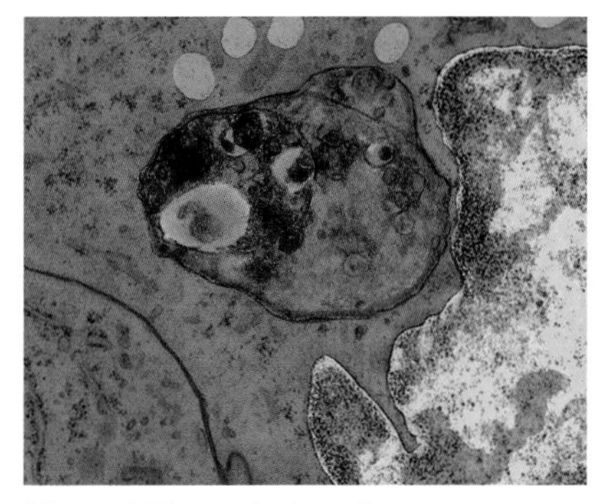

Figure 6 Transmission electron micrograph showing a lysosome (×20 000). The lysosome (red) is a simple spherical sac bound by a single membrane. It contains hydrolytic enzymes which break down intracellular materials.

Quick check

1 Which organelle:
  a carries out aerobic respiration
  b is the site of protein synthesis
  c contains hydrolytic enzymes
  d contains chromatin?

Food for thought

Organelles are defined as distinct structures in a cell in which or on which certain functions and processes are localised. They acquired their name because they were regarded as analogous to organs in a multicellular organism (organelle means 'little organ'). For example, the cell surface membrane is regarded by some biologists as an organelle because it is analogous to the skin. Suggest which parts of the human body carry out equivalent functions to each of the other organelles.

4.4

OBJECTIVES

By the end of this spread you should be able to:

- describe the ultrastructure of a typical plant cell
- describe the ultrastructure of a typical bacterial cell
- distinguish between prokaryotes and eukaryotes.

Fact of life

Mitochondria, chloroplasts, and bacteria share an uncanny resemblance to each other: they all self-replicate, and they all contain circular DNA and similar ribosomes. These striking similarities led to the theory of **endosymbiosis**. This theory suggests that mitochondria and chloroplasts evolved from bacteria. These bacteria were endosymbionts – they were whole organisms that lived inside other cells in a mutually beneficial relationship. Eventually, the relationship became permanent. The large host cell became a eukaryotic cell containing chloroplasts derived from the photosynthesising bacteria and/or mitochondria derived from the bacteria that could use oxygen for respiration. Cells having both types of bacteria became plant cells; those with only the respiring bacteria became animal cells.

It is easy to imagine how the original relationship could be mutually beneficial. The host cell could provide shelter, some protection against predators, and raw materials for its smaller partners. The ancestral chloroplasts could use sunlight and simple chemicals to provide the host with organic molecules, and the ancestral mitochondria could provide the cell with a bountiful supply of ATP.

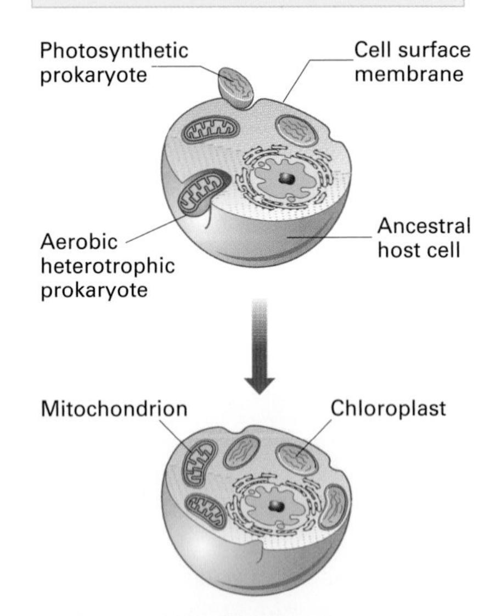

***Figure 1** Mitochondria and chloroplasts may have evolved from endosymbiotic bacteria.*

The ultrastructure of plant and bacterial cells

A typical plant cell has many features in common with a typical animal cell (figure 2). It has mitochondria, ribosomes, an endoplasmic reticulum, Golgi apparatus, and a well defined nucleus bounded by a nuclear membrane. However, there are differences between plant and animal cells:

- Each plant cell is surrounded by a rigid cellulose cell wall. This gives plant cells a more uniform and regular shape than that of animal cells.
- Animal cells have to obtain their food from an outside source. Many cells in the green parts of plants contain chloroplasts, which enable plants to make their own food by photosynthesis.
- Plant cells may contain a large fluid-filled sac called a vacuole. Vacuoles occur in some animal cells, but they are not usually permanent features.

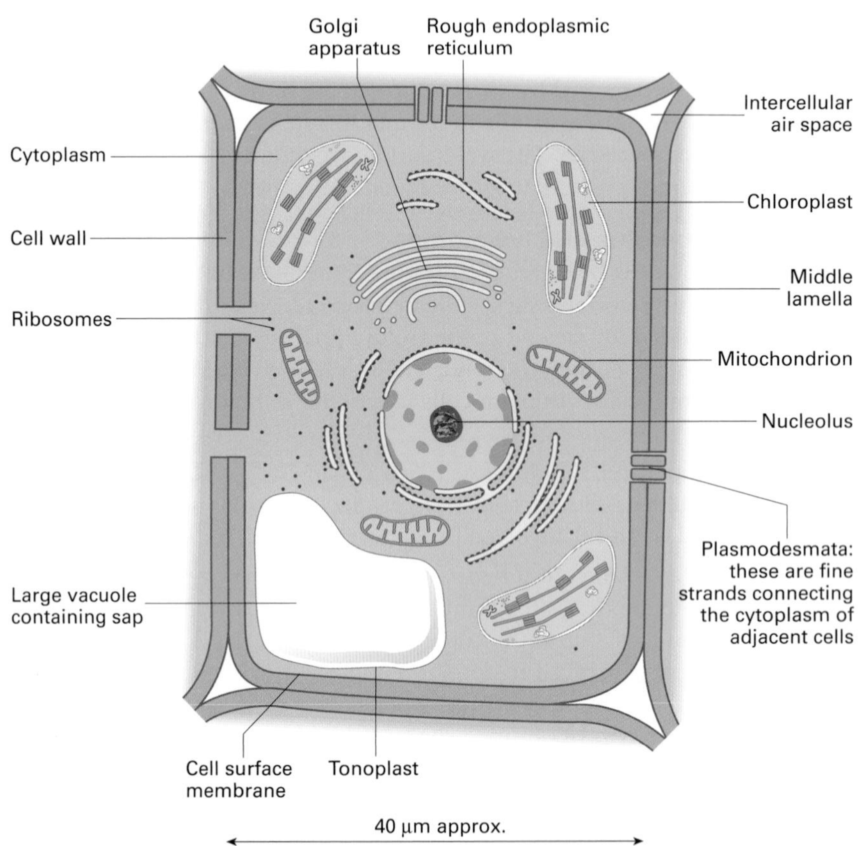

***Figure 2** Diagram of a palisade plant cell from a leaf showing structures visible under an electron microscope. Although there is no such thing as a 'typical' plant cell, palisade cells have features commonly seen in most plant cells.*

Chloroplasts, a cellulose cell wall, and a vacuole are visible under the light microscope, but the ultrastructure of a plant cell becomes clear only with the aid of an electron microscope.

A rigid **cell wall** surrounds each plant cell. This consists of cellulose microfibrils running through a matrix of other complex polysaccharides. Cell walls of neighbouring cells are cemented together by a thin layer of pectic substances (calcium and magnesium pectates) which form the **middle lamella**.

Most cell walls are not continuous but are perforated by fine membrane-lined pores through which pass cytoplasmic threads called **plasmodesmata**. These threads link the cytoplasm of two adjoining cells. Each plasmodesma has a central tubular core, usually connected at each end to the endoplasmic reticulum. Plasmodesmata enable substances to be transported easily between neighbouring cells.

Cell walls allow plant cells to become full of water without bursting, and to develop a turgor pressure which helps support the plant. Most cell walls are completely permeable to water and provide a pathway for the free flow of water and mineral salts. Some cell walls become modified by substances such as lignin. Lignin strengthens cell walls but may make them impermeable.

A **chloroplast** (spread 5.1, figure 2) is a large organelle (usually 5–10 μm long) in which photosynthesis takes place. It has an envelope of two membranes and contains a gel-like matrix called the **stroma**, and a system of membranes. The stroma contains ribosomes, DNA, and photosynthetic enzymes. It also stores lipid droplets and starch. The internal chloroplast membranes are piled up in places to form stacks called **grana**. The membranes contain photosynthetic pigments, including the green pigment **chlorophyll**, responsible for harvesting light energy for photosynthesis.

A typical plant cell has a large central **vacuole**. This is a sac bounded by a membrane called the **tonoplast**. The vacuole contains **sap**, a watery solution of various substances including sugars, mineral salts, pigments, and enzymes. When full of sap, the vacuole causes the cell surface membrane to press against the cell wall. This exerts a turgor pressure which helps to support the plant. Sometimes the vacuole functions as a lysosome.

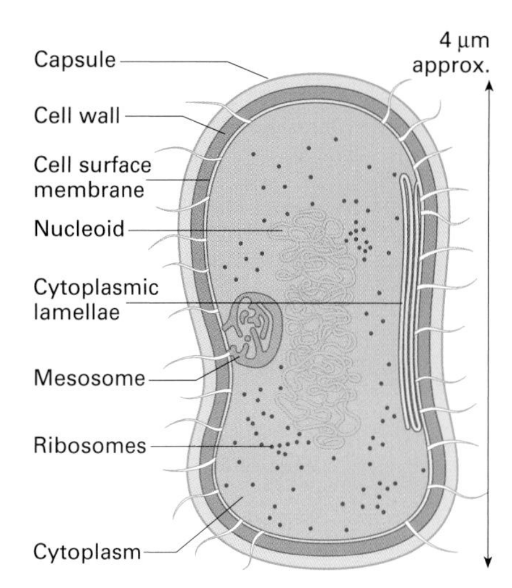

Figure 3 *A generalised bacterial cell (prokaryotic cell). In common with animal and plant cells, a typical bacterial cell has a cell surface membrane enclosing cytoplasm that contains enzymes, ribosomes, and food granules. The membrane is surrounded by a cell wall, and this may in turn be enclosed in a capsule. However, a bacterial cell lacks the high level of organisation of an animal or plant cell. It has no Golgi apparatus or endoplasmic reticulum. The flagella (if they occur) are simple; they have no complex assembly of microtubules. The genetic material of a bacterium is a single strand of DNA, usually coiled up into the centre of the cell to form a nucleoid. This nucleoid has no double-membraned nuclear envelope, so is often described as an 'ill-defined nucleus'. Some bacterial cells possess additional circular pieces of genetic material called plasmids within the cell. Respiration generally takes place on a mesosome, an infolding of the cell surface membrane, but there are no mitochondria. Photosynthesising bacterial cells such as cyanobacteria (commonly known as blue-green algae) have a special form of chlorophyll, but it is not enclosed in a double-membraned chloroplast.*

Eukaryotes and prokaryotes

Plants, animals, fungi, and protoctists (a group that includes single-celled organisms such as *Amoeba*) all have cells containing double-membraned structures and with the genetic material enclosed within a well defined nucleus. These organisms are known as **eukaryotes**, and are made up of **eukaryotic cells**. Organisms such as bacteria that lack double-membraned organelles and do not have a membrane-bound nucleus are known as **prokaryotes** (figure 3).

The biology of bacteria is covered in more detail in spreads 17.2 to 17.4, and a fuller comparison of prokaryotes and eukaryotes is given in spread 21.2. Viruses, which are neither prokaryotes nor eukaryotes, are described in spread 17.1.

Quick check

1 What are plasmodesmata?

2 a Which structure in a bacterial cell resembles a nucleus?
  b How does it differ from the nucleus of a eukaryotic cell?

3 Do double-membraned organelles occur in prokaryotes or eukaryotes?

Food for thought

A plant cell has been described as 'an animal cell crawling inside a cardboard box'. Is this an apt description?

4.5

Revealing the ultrastructure of cells

OBJECTIVES

By the end of this spread you should be able to:

- describe the nature of artefacts
- discuss the importance of freeze-fracturing
- explain how cell organelles can be isolated by cell fractionation.

Fact of life

Structures observed in electron micrographs are not always what they seem to be. When tiny vesicles surrounded by ribosomes were first noticed on electron micrographs, cell biologists thought that they had discovered a new type of organelle. They classified these structures, along with other small organelles, as **microsomes** (small bodies). However, these newly discovered microsomes turned out to be artefacts: they do not really exist. They were fragments of endoplasmic reticulum produced during the preparation of material.

The invention of the electron microscope allowed cell biologists to look at structures that were invisible under the light microscope. However, some cell biologists were sceptical: they believed that many features observed in electron micrographs were **artefacts**, that is, features not present in life.

Artefacts

An artefact may result during the preparation of a specimen. Air bubbles trapped under a coverslip are obvious artefacts produced during light microscopy.

Artefacts produced in electron microscopy are less obvious. For example, cells may be distorted during dehydration, so that they appear very different from the original living material. During the staining process, chemicals from outside the cell may appear as solid deposits which can look like intracellular structures.

Freeze-fracturing

The existence of structures in an electron micrograph can be checked by examining cells using a different technique. **Freeze-fracturing** (sometimes called freeze-etching) allows biologists to check structures observed by transmission electron microscopy (TEM).

- Living material is plunged into liquid nitrogen at –196 °C and pushed against a sharp blade in a precise way.
- The frozen tissue splits along lines of weakness, often in the middle of a membrane.
- The fractured surfaces are 'etched' with heavy metal so that the specimen can be examined by TEM.

Freeze-fracturing has confirmed the existence of structures in cells, and has also revealed new features. For example, when cell membranes fracture between the two lipid layers, tiny dots are revealed, which are thought to be protein molecules (figure 1). This provides compelling evidence that proteins form a mosaic in cell membranes (see spread 4.6).

Cell fractionation

Cell fractionation is a technique used to prepare samples of the various cell organelles so that their functions can be studied. The organelles

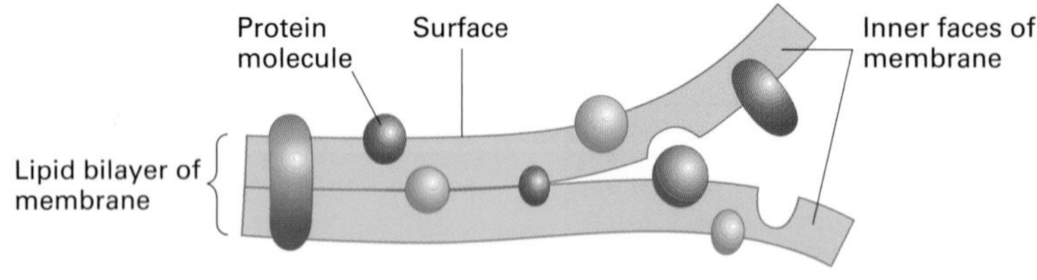

Figure 1 (a) *Freeze-fracturing. An artist's impression of how freeze-fracturing splits the bilipid layer of a cell surface membrane to reveal proteins embedded in the membrane.*

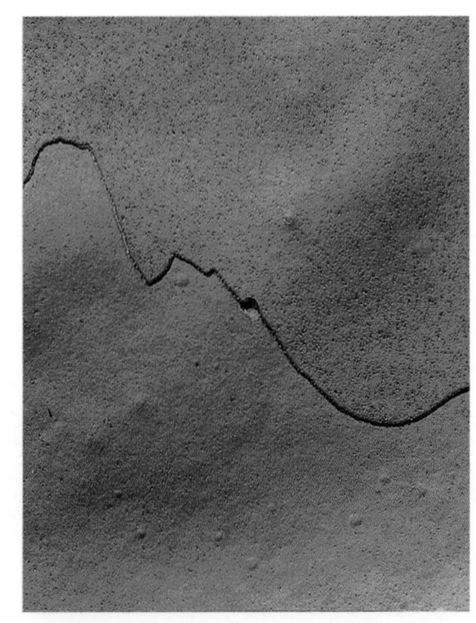

Figure 1 (b) *A scanning electron micrograph of a freeze-fractured cell surface membrane. The spherical structures are thought to be proteins.*

are separated into fractions according to their size (using **differential centrifugation**) or density (using **density gradient centrifugation**). The main steps of cell fractionation are shown in figure 2. For either method, the tissue is first cut into small pieces and then placed in a chilled, isotonic, and buffered solution.

- The temperature of the solution is kept low to slow down metabolism and minimise self-digestion of the organelles.
- The salt concentration of the solution is made isotonic so that organelles do not change volume. (A very salty solution causes organelles to shrink, whereas very watery solutions cause them to swell.)
- The solution is buffered to minimise changes in pH during the process. This prevents enzymes in the organelles from becoming denatured.

These conditions minimise the risk of damage to isolated organelles and improve the chance of organelles functioning normally during investigations.

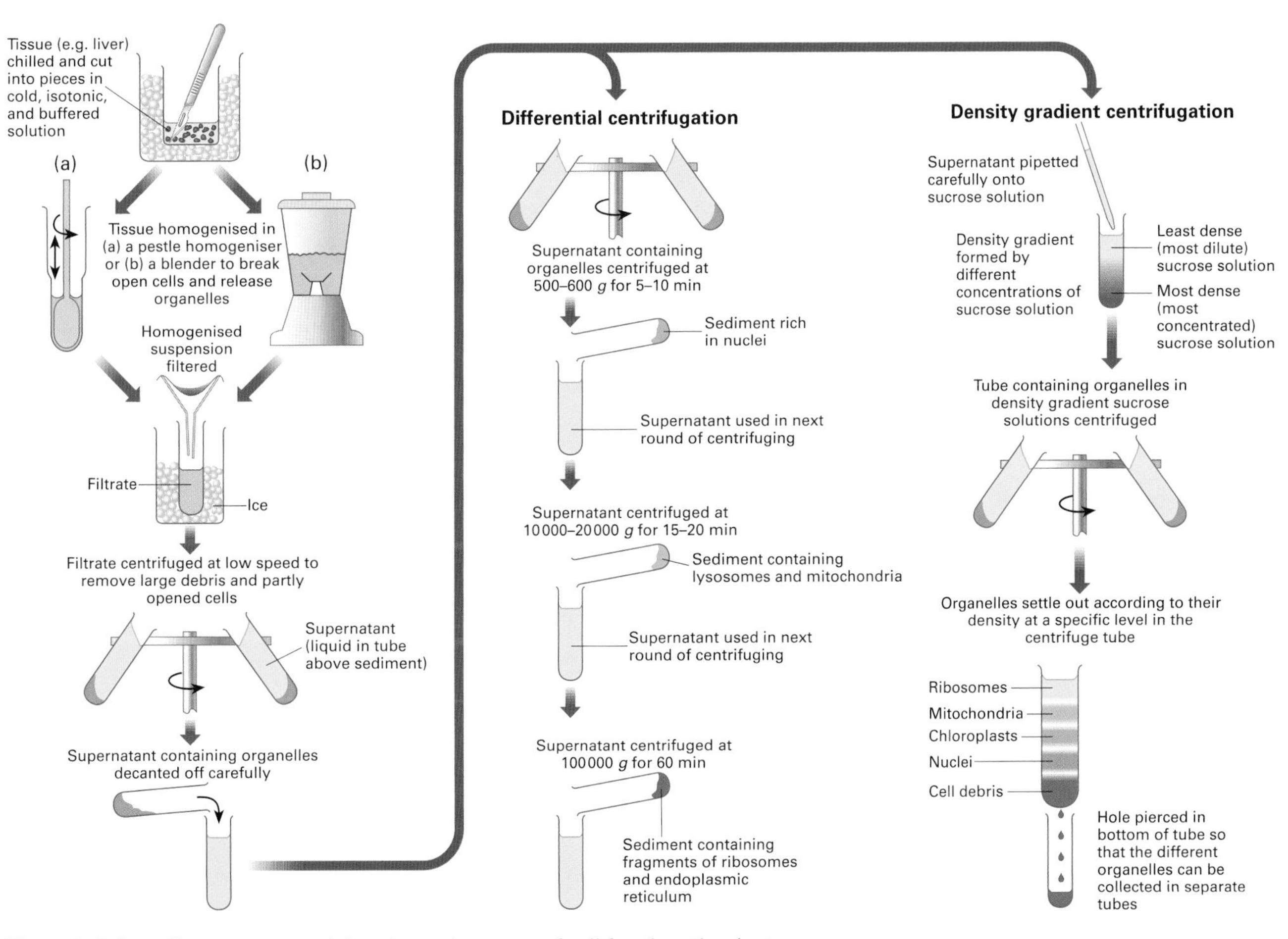

Figure 2 *A flow diagram summarising the main stages of cell fractionation by two methods: differential centrifugation and density gradient centrifugation.*

Quick check

1 Name two possible causes of artefacts in the preparation of specimens for the transmission electron microscope.

2 Which component of cell membranes, not visible in electron micrographs taken with a transmission electron microscope, is revealed by freeze-fracturing?

3 Why is tissue chilled during cell fractionation?

Food for thought

Suggest different methods you could use to ensure that a structure seen in an electron micrograph is a real structure and not an artefact.

4.6

CELL MEMBRANES

OBJECTIVES

By the end of this spread you should be able to:

- list the functions of cell membranes
- describe the fluid-mosaic structure of cell membranes
- explain the roles of the different components of cell membranes.

Fact of life

Cholesterol is infamous for contributing to heart disease. This lipid, however, is a vital component of human cell membranes, which contain almost as much cholesterol as phospholipid. Without cholesterol, our membranes would be too fluid at our relatively high body temperature, and would be liable to burst like soap bubbles.

Membranes cover the surface of every living cell, form an intricate network within the cytoplasm, and surround most cell organelles. These membranes are barely visible under a light microscope, but studies using electron microscopes, freeze-fracturing, and other modern techniques reveal that cell membranes have a complex structure and carry out a host of vital functions.

The structure of the cell membrane

Electron micrograph studies indicate that membranes are 7–8 nm wide and have three layers (figure 1). This was originally (and wrongly) interpreted as indicating a lipid layer sandwiched between two layers of protein. Biochemical analyses show that **cell surface membranes** (membranes around the outsides of cells) are approximately 45 per cent lipid, 45 per cent protein, and 10 per cent carbohydrate. Most of the lipid is **phospholipid** (see spread 2.8). Each phospholipid molecule consists of a hydrophobic tail of two fatty acids, and a hydrophilic phosphate head. In cell membranes, phospholipids arrange themselves in a layer two molecules thick (a **bilayer**), with their hydrophobic tails pointing inwards, away from the water both inside and outside the cell.

In 1972, Jonathan Singer and Garth Nicolson proposed the **fluid-mosaic model** of cell membrane structure. They proposed this model after realising that membranes must have a complex structure to carry out their many and varied activities (see below). In their model (figure 1), individual protein molecules shift and move on a fluid bilayer of phospholipid. The protein molecules are variable in structure and function, but they all contribute to the mechanical strength of membranes.

Some, called **intrinsic proteins**, span the width of the membrane; others, **extrinsic proteins**, are confined to the outer or inner surface. Some intrinsic proteins act as carrier molecules, transporting substances across the membrane, or as more passive routes for the movement of material in and out of the cell. Other intrinsic proteins are enzymes

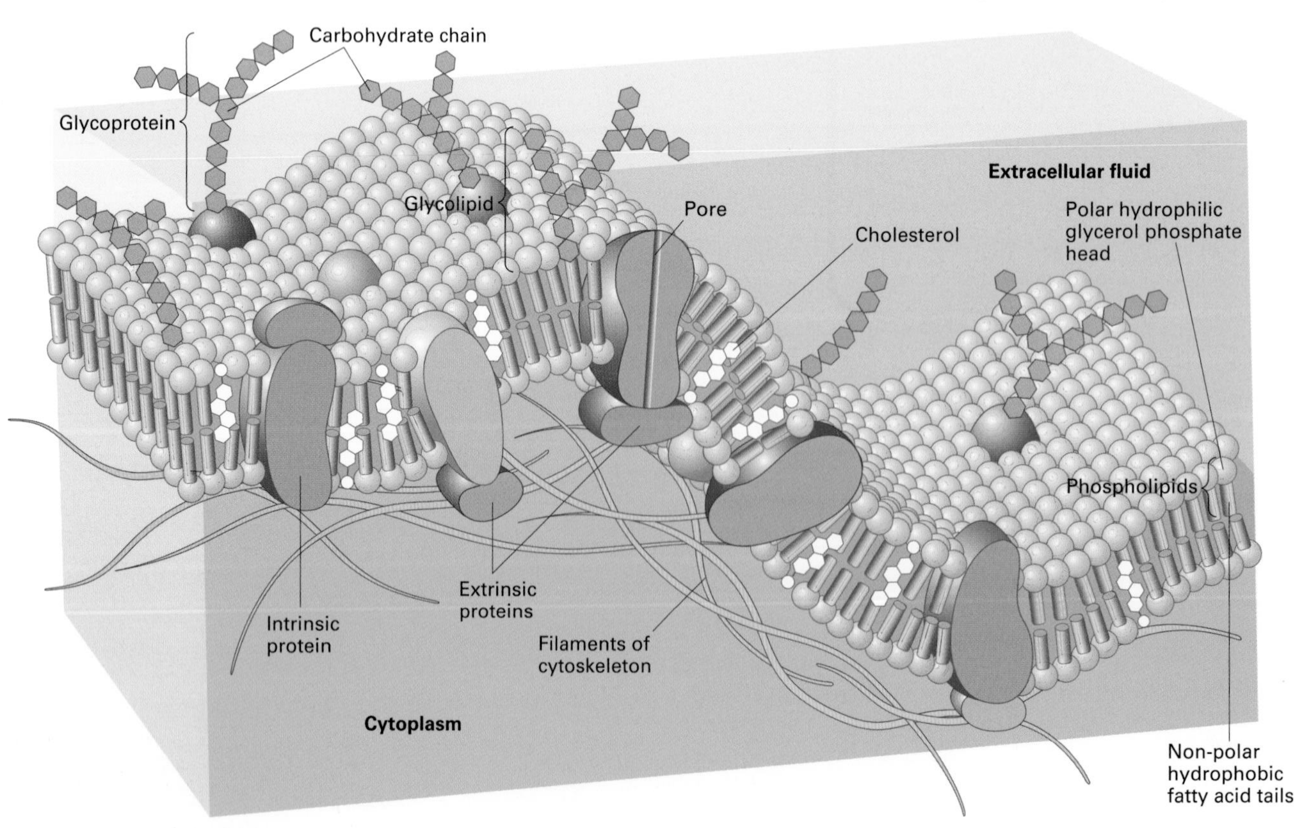

Figure 1 *Illustration of the fluid-mosaic model of the cell surface membrane.*

which catalyse specific metabolic reactions at a particular location within the cell.

Many extrinsic proteins combine with carbohydrate groups to form **glycoproteins**. The carbohydrate groups usually extend from the cell surface like antennae or feelers and act along with **glycolipids** (lipid molecules joined to a carbohydrate group) as the chemical receptors of the cell. Some proteins on the inner surface of the cell surface membrane attach onto the cytoskeleton to anchor the membrane in place.

The phospholipid bilayer is oily, giving membranes flexibility and fluidity. The phospholipids also allow the passage of certain lipid-soluble substances through the membrane.

Functions of cell surface membranes

A cell surface membrane forms the barrier between a cell and its surroundings. It has to be strong to offer mechanical support, and flexible to allow cells to move, grow, and divide. It must also be self-sealing so that the cell can divide without bursting. Some other functions of cell surface membranes are listed below along with a reference to the spread in which you can find out more about them.

- controlling the passage of materials in and out of cells (4.7 to 4.9)
- recognition of other cells (15.5)
- receptor sites for hormones and neurotransmitters (10.3 and 10.6)
- transmission of nerve impulses (10.4 and 10.5)
- insulation of nerves (10.5)

Functions of membranes inside cells

Intracellular membranes (the membranes inside cells) have a structure very similar to that of cell surface membranes. However, the proportions of the molecular components vary considerably. The membranes enveloping chloroplasts, for example, contain very little carbohydrate.

Nearly every cellular process involves intracellular membranes. A few of the functions of these membranes are given below, together with the spread containing more information about them.

- acting as a reaction surface (6.4)
- acting as an intracellular transport system (4.3)
- providing separate intracellular compartments, isolating different chemical reactions (4.3)

Quick check

1 Why must cell membranes be partially permeable?

2 Explain why the fluid-mosaic model is so called.

3 The human immunodeficiency virus which causes AIDS attacks certain white blood cells (T cells), interfering with a person's immune response. The virus attaches to specific receptor sites on the cell surface membranes of T cells. Suggest which components of the cell surface membrane act as receptor sites, and how the receptor site is specific to T cells.

Food for thought

Cholesterol is an essential component of cell surface membranes. Suggest why having cell surface membranes with too much cholesterol is harmful.

Cell membrane models

Like all scientific knowledge, our understanding of the structure of the cell membrane is tentative. Models have changed with the discovery and publication in scientific journals of new evidence that provides a better explanation of scientific observations. Early models, based on the observation that lipids and lipid-soluble substances can pass through membranes more readily than can substances insoluble in lipid, suggested that membranes are made of lipids. The discovery of protein in membranes isolated from red blood cells led Hugh Davson and James Danielli to propose a new model in 1935. In this model, cell membranes consist of a phospholipid core sandwiched between two layers of globular protein. The **Davson–Danielli** model was supported by early electron micrographs in which cell membranes appeared to consist of three layers (figure 2). Similar electron micrographs of cell membranes from a wide variety of organisms gave rise to the **unit membrane theory** which proposed that all cell membranes have this same basic structure. Detailed analysis of the electron micrographs revealed that not all membranes are identical or symmetrical. The unit membrane theory was rejected and the fluid-mosaic model proposed.

However, this is not the end of the story. The fluid-mosaic model has been refined since it was proposed in 1972 by Singer and Nicolson, and will continue to evolve as new scientific evidence comes to light.

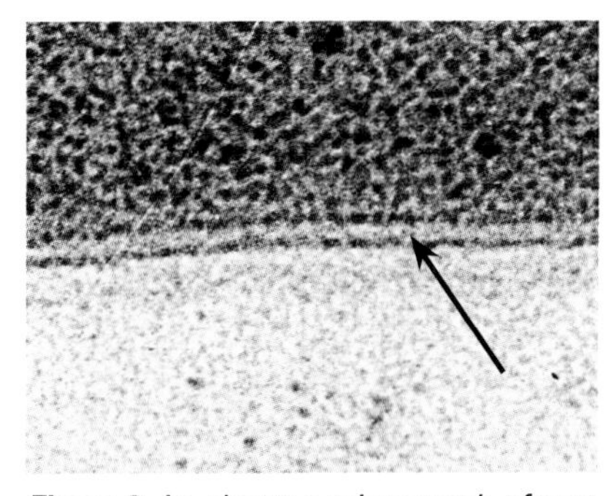

Figure 2 An electron micrograph of part of a cell. The arrow is pointing to the cell surface membrane which appears to have three layers: two dark layers separated by a lighter layer.

4.7

Diffusion

OBJECTIVES

By the end of this spread you should be able to:

- define diffusion
- list the factors that affect the rate of diffusion
- compare the surface area to volume ratio for objects of different sizes.

Fact of life

The time a substance takes to diffuse increases with the square of the average distance it has to cover. This means that although diffusion may be fast over short distances, it becomes impossibly slow over long distances. For example, it would take about three years for oxygen to diffuse the one metre or so from the lungs to the extremities of the human body. Clearly diffusion alone is not suitable for transport of gases within the human body.

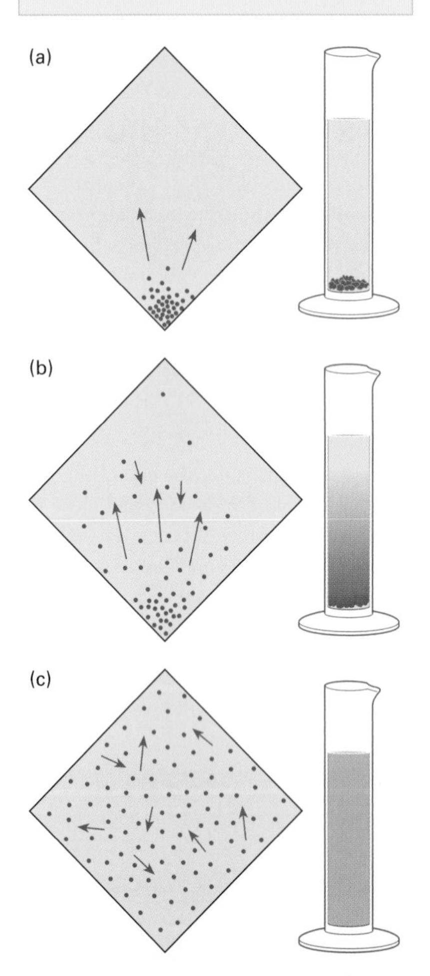

***Figure 1** A model of diffusion. Each dot represents a dye molecule moving randomly. In **(a)** and **(b)** there is a net movement from the bottom of the measuring cylinder to the top. In **(c)** an equilibrium has been established and there is no net movement; the molecules remain evenly distributed and there is an even concentration of dye throughout the cylinder.*

Moving down a concentration gradient

Particles are never still: they keep moving and bumping into each other all the time. When a drop of dye is added to water (figure 1), the molecules mix so that the dye spreads through the water even if it is not stirred. Dye molecules move in every direction, but the net effect is that they spread from where they are plentiful to where they are less plentiful.

This type of movement is called **diffusion**. It is defined as: *the net movement of molecules or ions from a region of their high concentration to a region of their lower concentration.*

Diffusion occurs in living and non-living systems wherever there is a concentration gradient. It continues until the particles are distributed evenly throughout the system. When a uniform concentration is reached, equilibrium is established. The molecules still move, but there is no net movement in any particular direction.

Diffusion is the mechanism by which gases are exchanged across respiratory surfaces, such as gills in fish and alveoli in mammals.

The rate of diffusion

The rate of diffusion in a given direction across the exchange surface:

- is directly proportional to the area of the surface
- is directly proportional to the concentration gradient
- is inversely proportional to the distance (the length of the diffusion pathway).

This is known as **Fick's law** and can be summarised simply as:

$$\text{Rate of diffusion is proportional to } \frac{\text{surface area} \times \text{difference in concentration}}{\text{length of diffusion path}}$$

Increasing the factors on the top line of the equation will make diffusion faster, while increasing that on the bottom will slow it down.

Fick's law applies to situations where there is no barrier to the movement of substances. Like the diffusion of a dye in a container of water, the diffusion of a substance into or out of a cell is a form of **passive transport**: it does not require the expenditure of energy. Diffusion through a cell membrane is affected by the nature of the membrane (for example, its permeability) and the size and type of molecule or ion diffusing through it.

Adaptation for diffusion

As would be expected from Fick's law, cellular diffusion is a very slow process unless there is a large concentration gradient over a short distance. Tissues such as those in the lungs and small intestine are especially adapted to maximise the rate of diffusion by:

- maintaining a steep concentration gradient
- having a high surface area to volume ratio
- being thin, minimising the distance over which the diffusion takes place.

Membrane permeability

Cell membranes are **partially permeable**: many substances can pass through them, but some substances cannot. Some substances, such as steroid hormones, oxygen, and carbon dioxide, dissolve readily in the inner oily phospholipid bilayer of cell membranes (see spread 4.6, figure 1). Consequently these substances diffuse readily through a cell membrane down a concentration gradient.

Facilitated diffusion

Glucose, nucleic acids, amino acids, and proteins are not soluble in lipids and do not pass readily through the phospholipid bilayer of cell membranes. Instead, these substances pass through the membrane by a process called **facilitated diffusion**.

This is the passive movement of molecules down a concentration gradient but it involves special carrier proteins in the cell membrane. These proteins may have hydrophilic channels that function as pores along which solutes can pass (figure 2). Alternatively, they may be small globular proteins that move in the membrane, shuttling their load back and forth, like a ferry carrying cars.

Facilitated diffusion of relatively large molecules such as glucose appears to be very specific and may involve a protein called a **permease**. This protein changes its shape when glucose binds to it. The solute is moved across the membrane as the protein alternates between its two shapes (figure 3). Although net movement is always down a concentration gradient, the protein can transport the solute in either direction. Like enzymes, permeases can be inhibited by substances that resemble the normal substrate.

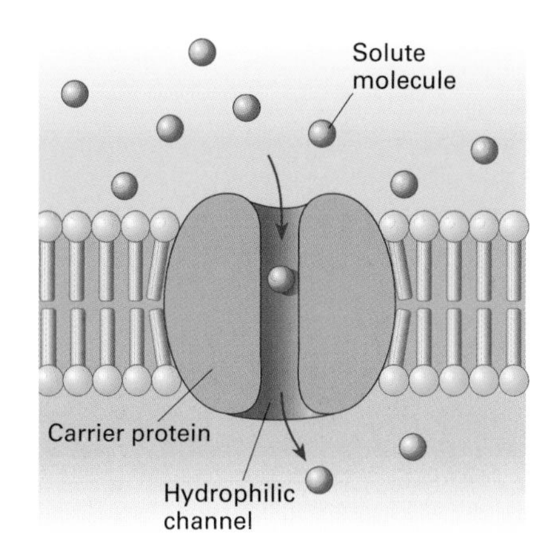

Figure 2 *Facilitated diffusion: a protein provides a functional pore in the membrane for the passage of a solute.*

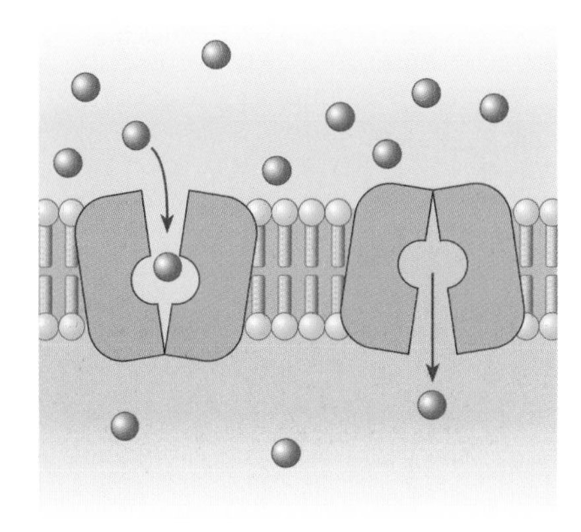

Figure 3 *Facilitated diffusion: a protein acts as a specific carrier molecule in the membrane.*

Surface area to volume relationship

Table 1 shows that as the length of the sides of a cube is increased, the ratio of its surface area to its volume decreases. A similar relationship applies to organisms: as the size of an organism increases, its surface area to volume ratio decreases.

Table 1 *Surface area to volume ratios of a cube.*

Length of side (mm)	Area of side (mm^2)	Surface area of cube (mm^2)	Volume of cube (mm^3)	Surface area to volume ratio
1	1	6	1	6:1
2	4	24	8	3:1
5	25	150	125	1.2:1
10	100	600	1000	0.6:1
100	10000	60000	1000000	0.06:1

This relationship has important consequences for organisms. Some processes such as the rate of diffusion depend on surface area; other processes such as metabolic rate depend on volume. Consequently a small organism such as the single-celled *Amoeba* can satisfy its need for oxygen by simple diffusion from its watery environment. However, simple diffusion cannot supply the oxygen needed by large organisms. It would take several years for oxygen to diffuse into the centre of a whale because its surface area to volume ratio is more than one million times smaller than that of the single-celled organism. To compensate for their relatively small body surface areas, large organisms have evolved special exchange surfaces and transport systems, such as lungs and gills to exchange respiratory gases, and blood circulatory systems to transport the gases.

Quick check

1. Define diffusion.
2. List the factors that affect the rate of diffusion of a molecule into a cell.
3. Calculate the surface area to volume ratio of a cube with sides 4 mm long.

Food for thought

Tissues involved in gaseous exchange are moist. It is a common misconception that they have to be moist in order to carry out diffusion. This is not true. The tissues are moist because they are very thin and are permeable to water and other substances. Their moistness is a result of their ability to allow diffusion, not a cause of it. Nevertheless, gills, for example, are no longer able to carry out gaseous exchange if they become dry. Suggest reasons for this loss of ability.

4.8

Active transport

OBJECTIVES

By the end of this spread you should be able to:

- describe the process of active transport
- distinguish between endocytosis and exocytosis.

Fact of life

Seaweeds can concentrate inside themselves trace metals that are present in only minute quantities in the surrounding sea. For example, *Pelvetia canaliculata, Ascophyllum nodosum*, and *Fucus spiralis* (brown seaweeds common on British rocky shores) concentrate titanium 1000–10 000 times, zinc 1000–1400 times, nickel 600–1000 times, and strontium 8–20 times.

Maintaining an internal environment

Before sending space probes to Mars, NASA asked James Lovelock to suggest features that they could use in their search for life. Lovelock was the originator of the Gaia hypothesis concerning the environmental conditions on Earth and the organisms that live there (see spread 22.1). He concluded that extraterrestrial life may not share the characteristic features of life on Earth. He proposed, however, that to stay alive, any organism would have to keep its internal environment relatively constant and different from its external environment (a process called homeostasis).

The concentrations of ions inside a cell of any Earth organism are kept at a different level from the concentrations outside (figure 1). To maintain these concentrations, a cell must import some substances and export others.

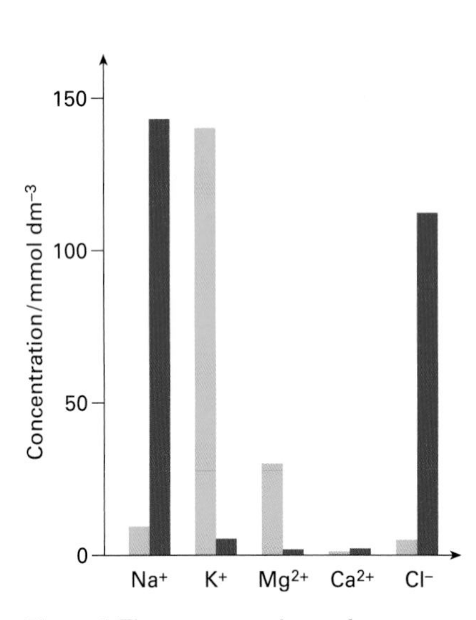

***Figure 1** The concentrations of some ions inside cells (yellow bars) and outside cells (blue bars).*

Active transport: using energy to import and export

Cells import and export their substances by **active transport**. Active transport is:

- the movement of substances,
- usually against a concentration gradient (from a region of lower to a region of higher concentration),
- across a cell membrane.
- It involves the expenditure of energy.

If a cell cannot carry out active transport, it dies as concentrations inside and outside the cell become equal.

The energy for active transport usually comes from ATP, provided by respiration in the mitochondria (see spread 2.11). Cells that carry out active transport on a large scale have an unusually high number of mitochondria and high concentrations of ATP. Active transport ceases if ATP production is prevented by metabolic poisons such as cyanide, or lack of oxygen.

How active transport happens

Several types of active transport involve membrane proteins which carry specific molecules or ions. These carriers may move:

- a single substance in a single direction (**uniport carriers**)
- two substances in the same direction (**symport carriers**)
- two substances in opposite directions (**antiport carriers**).

The precise mechanism of active transport is unclear. One hypothesis suggests that some actively transported substances can hitch a ride on a proton (hydrogen ion) in a process called **cotransport** (see box). Another

Cotransport

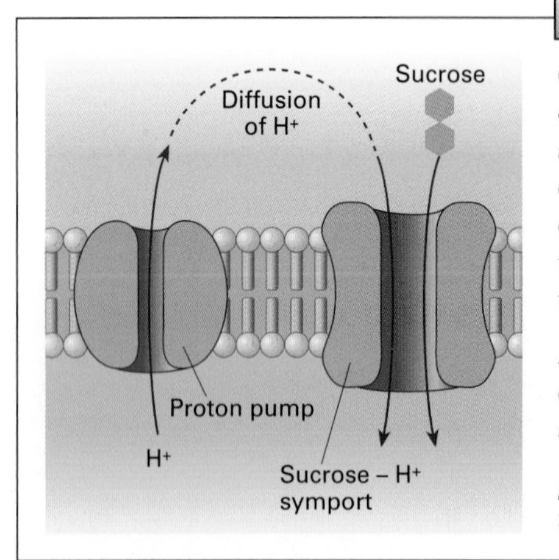

Cotransport is a relatively complex form of active transport in which the pumping of one substance indirectly drives the transport of one or more other substances against a concentration gradient. In plant cells which store sucrose, for example, an ATP-driven proton pump actively transports hydrogen ions (protons) out of the cell. This creates a concentration gradient down which the protons can diffuse. Diffusion of the protons back into the cell via a symport carrier enables sucrose to be carried into the cell at the same time (figure 2).

A similar process is used to transport glucose and amino acids into cells lining the digestive tract in mammals. In this case, absorption of the nutrients is dependent on a sodium pump.

***Figure 2** Cotransport: sucrose moves into plant cells along with protons. The protons travel down a concentration gradient set up by a proton pump.*

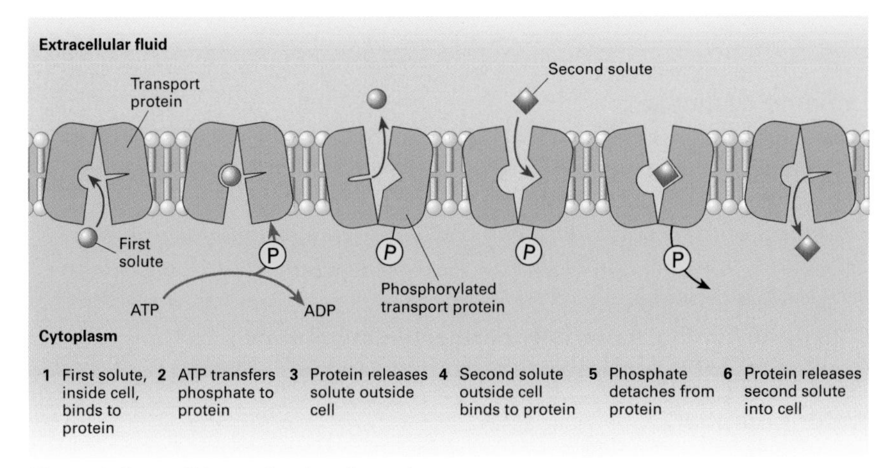

Figure 3 *A possible mechanism for active transport.*

hypothesis suggests that protein molecules change shape to transport solutes from one side of the membrane to the other (figure 3). To do this ATP is hydrolysed to ADP. The terminal phosphate group attaches directly onto the protein, causing it to change shape. Active transport systems that move two solutes together are quite common. One of the most important is the sodium–potassium pump which pumps sodium ions out of the cell and potassium ions in. This is important in generating impulses in nerve cells.

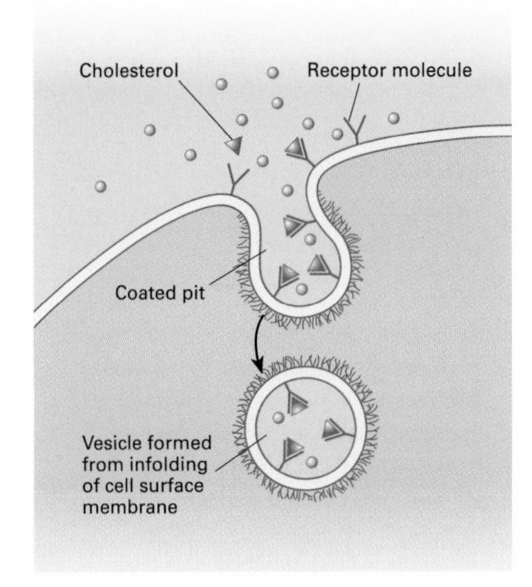

Figure 4 *Receptor-mediated endocytosis. Receptor molecules on the cell surface membrane bind with a specific substance such as cholesterol from the extracellular environment. As receptor sites are filled, the surface folds inwards until a coated vesicle finally separates from the cell surface membrane.*

Cytosis: movement with vesicles and vacuoles

Most cells can carry out **cytosis**: active transport which involves infolding or outfolding of sections of the cell surface membrane. This appears to involve the contractile proteins in cellular microfilaments and microtubules.

Cytosis may result in the bulk transport of material into a cell (**endocytosis**) or out of a cell (**exocytosis**). There are two main forms of endocytosis: **phagocytosis** and **pinocytosis**. A third form, **receptor-mediated** endocytosis, is shown in figure 4.

- In phagocytosis (cellular eating) solid substances, sometimes whole organisms, are brought inside the cell by invagination (an infolding of the cell surface membrane). A vacuole is formed, the inner surface of which is derived from the outer surface of the cell surface membrane. Many unicellular organisms such as *Amoeba* (figure 5) and certain white blood cells called phagocytes perform phagocytosis.
- Pinocytosis (cellular drinking) is a similar process, but the infoldings are much smaller. Liquids or large macromolecules such as proteins are taken in via a small vesicle.
- Exocytosis appears to be endocytosis in reverse. Vesicles and vacuoles move to the cell surface membrane, fuse with it and release their cargo to the outside world.

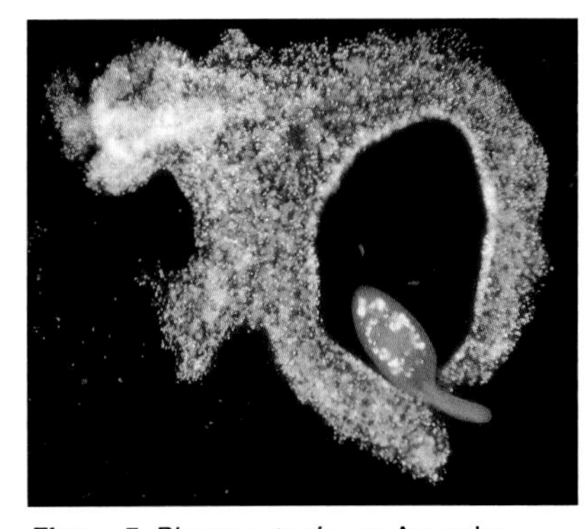

Figure 5 *Phagocytosis: an* Amoeba *engulfing its prey.*

Quick check

1 Describe active transport, mentioning ATP, concentration gradient, and oxygen consumption.

2 Suggest why active transport is affected by changes in oxygen concentration but diffusion is not.

3 Distinguish between pinocytosis and phagocytosis.

Food for thought

To study mineral uptake, a botanist used isolated roots bathed in solutions containing potassium ions and different oxygen concentrations. Suggest why the botanist found that:

a potassium ions were present in the roots even when the percentage of oxygen was zero

b the concentration of potassium ions in the roots increased rapidly with increasing oxygen concentrations up to 20 per cent

c the rate of sugar consumption by the roots rose with increases in oxygen concentration.

4.9

Osmosis

OBJECTIVES

By the end of this spread you should be able to:

- define water potential
- explain how osmosis takes place
- distinguish between isotonic, hypertonic, and hypotonic solutions and their effects on animal and plant cells.

Fact of life

Brine shrimps (*Artemia* spp) are found in tremendous numbers in salt lakes and coastal evaporation ponds (ponds used to obtain salt by evaporation of sea water). Brine shrimps can live in water with solute concentrations ranging from about 0.3 to about 300 g dm^{-3} because they can pump salts actively out of their bodies. At 300 g dm^{-3}, salt water is more than eight times saltier than normal sea water; it is so salty that the salts crystallise out of solution.

Defining osmosis

Osmosis is the process by which cells exchange water with their environment. It is a passive process similar to diffusion, but involving only water molecules. In biology, osmosis is often defined as: *the net movement of freely moving water molecules from a region of their higher concentration to a region of their lower concentration through a partially permeable membrane.*

In the definition, a **partially permeable membrane** refers to a membrane that is permeable to water and to certain solutes.

Water potential

The pressure exerted by freely moving water molecules in a system is called the **water potential**, measured in **kilopascals** (kPa). By convention, the water potential of pure water is given the value of 0 kPa. A solution with a high water potential has a high number of freely moving water molecules.

Consider a solution with a higher water potential separated from a solution with a lower water potential by a partially permeable membrane. More water molecules move from the region of higher water potential to the region of lower water potential than in the other direction (figure 1a). The movement continues until the water potentials on both sides of the membrane are equal. Water molecules continue to move in both directions after equalisation, but there is no *net* movement (figure 1c).

The effect of solutes

The water potential of a solution falls when solutes are added because water molecules cluster around the solute molecules (figure 1b). As the concentration of solutes increases, clustering also increases, further lowering the water potential.

The contribution of solutes to the water potential of a system is called the **solute potential** of that system. Because it always lowers water potential, solute potential is always negative. It becomes more negative as more solutes are added to the system.

Osmosis in animal cells

A single-celled organism living in the sea is bathed in fluid which has the same water potential as its **cytosol** (the liquid part of the cytoplasm). The sea water is **isotonic** with the cytosol because it causes no change in the volume of the cell. This is because the water potential inside and outside the cell is the same and there is no net flow of water in or out (figure 2a). A similar situation applies to organisms, such as the malarial parasite *Plasmodium*, which live in blood.

An animal cell placed in fresh water has to deal with a very dilute medium which has a higher water potential than the cytosol. The fresh water is **hypotonic** to the cytosol. It tends to enter the cell, causing the cell volume to increase. If the cell cannot eliminate the excess water, as a red blood cell cannot, it will burst (a process known as lysis) (figure 2b). However, if fresh water is the cell's natural environment, it will have a mechanism for eliminating excess water which has entered by osmosis. Amoebae that live in ponds, for example, have a **contractile vacuole** which collects and gets rid of excess water, keeping the volume of the cell constant.

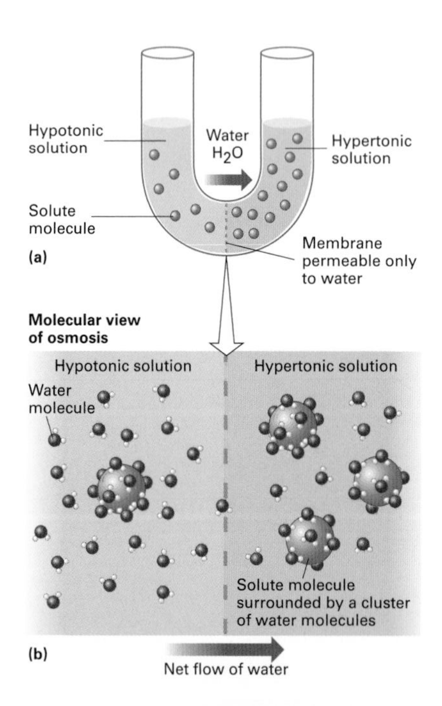

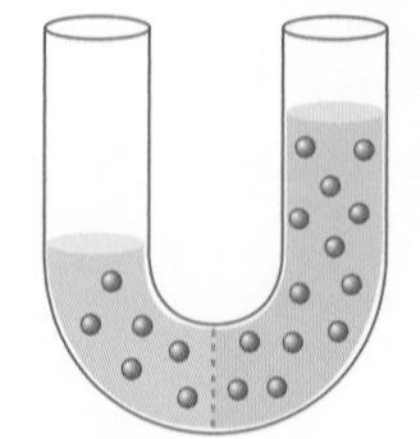

*Figure 1 Osmosis. **(a)** A hypotonic (weak) solution is separated from a hypertonic (strong) solution by a membrane that is permeable only to water. **(b)** Water molecules cluster around solute molecules (see spread 2.3) so fewer water molecules are free to move in the hypertonic solution. **(c)** Consequently there is a net movement of water molecules from the hypotonic solution to the hypertonic solution, causing the fluid level to rise on the right-hand side.*

An animal cell placed in a very salty solution has to cope with the risk of losing water. If the external solution is of a sufficiently high solute concentration, its water potential will be lower than that of the cell – it is a **hypertonic** solution. Water will tend to be drawn out of the cell, causing it to shrink and shrivel (figure 2c).

Osmosis in plant cells

Plant cells behave in the same way as animal cells in an isotonic solution: they neither gain nor lose water (figure 2a). However, because a plant cell has a rigid cellulose cell wall, it behaves differently from an animal cell in hypotonic and hypertonic solutions.

In a hypotonic solution, water enters the plant cell, filling the vacuole to capacity. The cell surface membrane pushes against the cell wall, making the cell very rigid. A cell in this condition is said to be **turgid** (figure 2b). There is no danger of the cell bursting because the pressure of the cell wall against the cell surface membrane restricts the inflow of water.

In a hypertonic solution, water moves out of a plant cell, the cell vacuole shrinks, and the cell surface membrane pulls away from the cell wall (figure 2c). The cell becomes **flaccid** because the contents are no longer pushing against the cell wall. A cell in this condition is said to be **plasmolysed**.

Pressure potential

The water potential of a plant cell is affected by both the pressure of the cell wall against the cellular contents and the solute potential of the cell. The contribution made by the cell wall is called the **pressure potential**. This always has a positive value when the cell wall is turgid, but equals 0 kPa when the cell is flaccid. Thus the water potential of a plant cell is the sum of its solute potential and pressure potential:

$$\Psi_{cell} = \Psi_s + \Psi_p$$

water potential of plant cell = solute potential + pressure potential

The water potential of an animal cell is determined primarily by its solute potential. The effect of the cell membrane is usually so small that it is ignored.

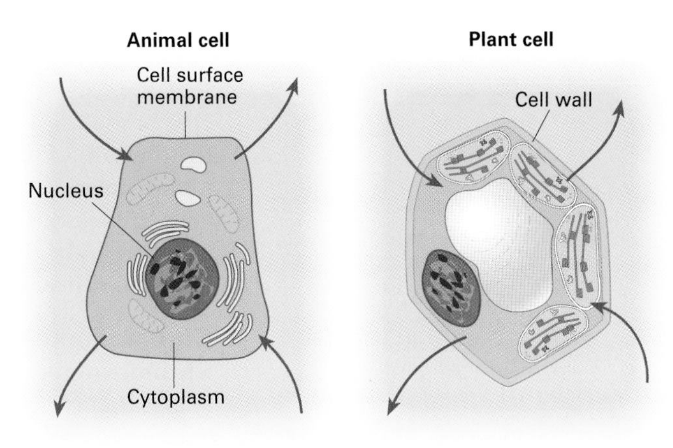

(a) Isotonic solution: no net flow of water

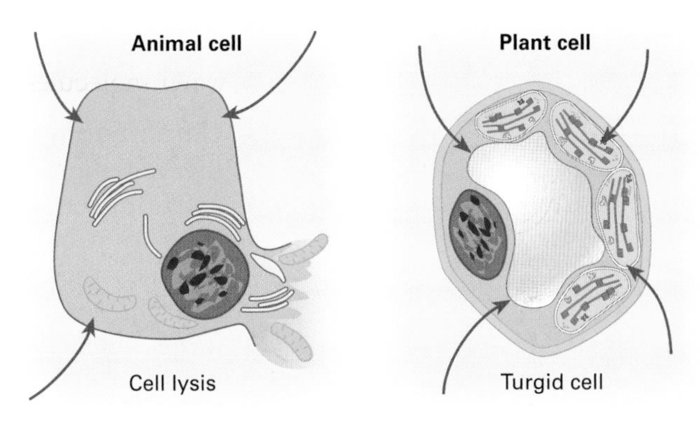

(b) Hypotonic solution: net inflow of water

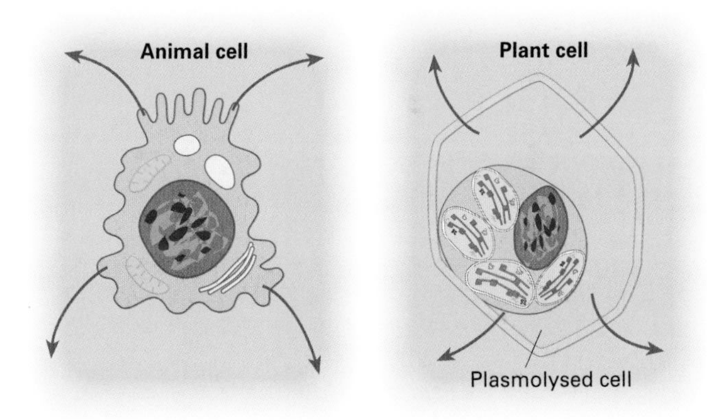

(c) Hypertonic solution: net outflow of water

Figure 2 *The effects of different solutions on animal and plant cells.*

Quick check

1 What is the water potential of pure water at standard pressure and temperature?

2 Plant cell A has a solute potential of –300 kPa and a pressure potential of 200 kPa. Lying next to it is plant cell B which has a solute potential of –400 kPa and a pressure potential of 100 kPa. Use the water potential equation to predict the direction of net movement of water in each cell.

3 What type of replacement drink should a heavily perspiring athlete take: hypertonic, isotonic, or hypotonic? Give reasons for your answer.

Food for thought

There are many different groups of amoeba. Some inhabit freshwater and marine environments; others are internal parasites which infect human intestines and cause dysentry. Suggest why contractile vacuoles appear to be absent from some marine and parasitic groups of amoeba.

4.10

Introduction to cell division

OBJECTIVES

By the end of this spread you should be able to:

- describe the main stages of the cell cycle
- distinguish between mitosis and meiosis.

Fact of life

The normal growth and development of a multicellular organism is determined by a fine balance between cell addition by mitosis and cell deletion by **apoptosis** (programmed cell death). Apoptosis controls cell quality and quantity in all multicellular organisms. It is the 'default setting' of each cell produced by mitosis. A new cell will automatically self-destruct unless signals from neighbouring cells prevent it from doing so. Apoptosis removes cells that are no longer required and infected, worn, or damaged cells that need to be replaced. Unlike uncontrolled cell death (**necrosis**), apoptosis takes place in an orderly way. Cells to be destroyed are selected by specific cell signals. Then cell metabolism is closed down and the residual debris is dismantled and recycled within the organism. In most cases, one of a number of triggers causes mitochondria to release a signalling molecule (e.g. cytochrome c) which sets in train the self-destruction programme. The mechanisms that control the cell cycle are probably linked to those that control apoptosis.

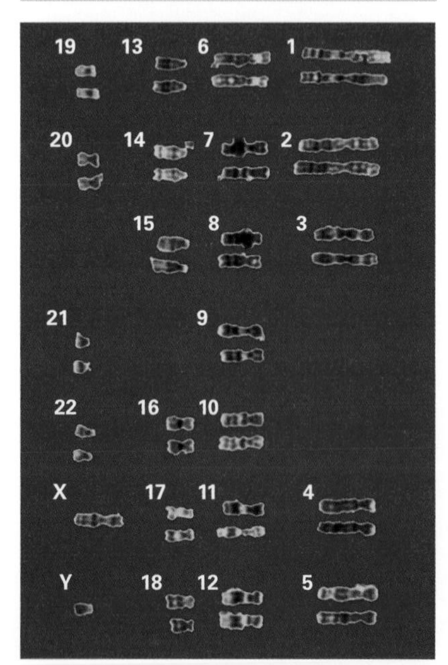

Figure 1 A karyotype: human chromosomes arranged in matching pairs. Each chromosome consists of two sister chromatids joined by a centromere. Twenty-two pairs of human chromosomes are numbered. The other pair are the sex chromosomes denoted by the letters X and Y. Usually, males have one X and one Y chromosome; females have two X chromosomes.

One of the most important concepts in biology is that cells arise only by the division of existing cells. Cell division is essential to all life. It enables a multicellular organism to grow and to replace worn out or damaged cells. It is also the basis of reproduction in every organism. Cell division starts with the division of the nucleus. There are two forms of nuclear division: **mitosis** and **meiosis**.

Chromosomes: carrying information

Chromosomes are the structures that provide continuity between one generation of cells and the next. Their name comes from the Greek: *chroma* = coloured, *soma* = body, because of their affinity for certain stains used in microscopy. Chromosomes consist of DNA, the genetic material of the cell, wrapped in protein. They become visible in the nucleus where the more dispersed chromatin existed before (see spread 4.3). Whole chromosomes can be examined microscopically after breaking a dividing cell open and staining it with a suitable dye.

Chromosomes form homologous pairs

Figure 1 shows a photomicrograph of human chromosomes cut out and arranged into matching pairs according to their size and certain other features. These are called **homologous pairs**. Apart from the sex chromosomes, both chromosomes in a pair normally contain the same genes (for example, for eye colour). However, these may be different forms, called **alleles**, of the gene (for example, one chromosome may carry the allele for green eyes and the other may carry the allele for brown eyes). Alleles are described more fully in spread 19.3.

Human cells each have 46 chromosomes (23 pairs). Other species have different numbers, for example, chimpanzee cells each have 48 (24 pairs) and cabbage plant cells each have 18 (9 pairs). One chromosome in each pair comes from the individual's mother and the other from the father.

- Cells that have the normal two sets of chromosomes are called **diploid** (usually abbreviated $2n$).
- Cells that give rise to gametes (eggs and sperm) have only one chromosome of each pair, so they have half the normal number of chromosomes. Such cells are called **haploid** (n).
- In humans, $n = 23$, so normal diploid cells have 46 chromosomes and the haploid gametes have 23 chromosomes.

Mitosis: two identical daughter cells

In **mitosis**, the nucleus divides once and produces two identical nuclei. The new daughter cells are genetically identical to the parental cell (unless their DNA has been changed in some way, for example by a mutation) (figure 2). So mitosis doubles the number of cells without changing the genetic information. New cells for growth of a multicellular organism, asexual reproduction, and wound healing, for example, are produced by mitosis.

The cell cycle

The **cell cycle** is the sequence of events that occurs between one cell division and the next. It consists of three main stages (figure 3):

1 During interphase, the cell grows, carries out its functions, and replicates its DNA (see spread 18.3). After the DNA is replicated, new protein becomes attached to it. The chromosome now consists of two strands called **sister chromatids** which contain identical genetic information. Sister chromatids are joined at some point along their length by a **centromere**. These become visible under the light microscope only during mitosis. Typically, interphase lasts for about 90 per cent of the cell cycle.

2 Nuclear division takes place during mitosis. The chromatids containing replicated DNA are separated from each other and are redistributed as chromosomes in the nuclei of the two new daughter cells.

3 In **cell division** (also called **cytokinesis**) the cytoplasm divides to form two daughter cells.

The duration of the cell cycle varies with conditions such as temperature and cell type. Mitosis of eukaryotic cells is controlled by the interaction of extracellular growth factors, such as growth hormone and insulin in humans, with intracellular cyclin-dependent kinases.

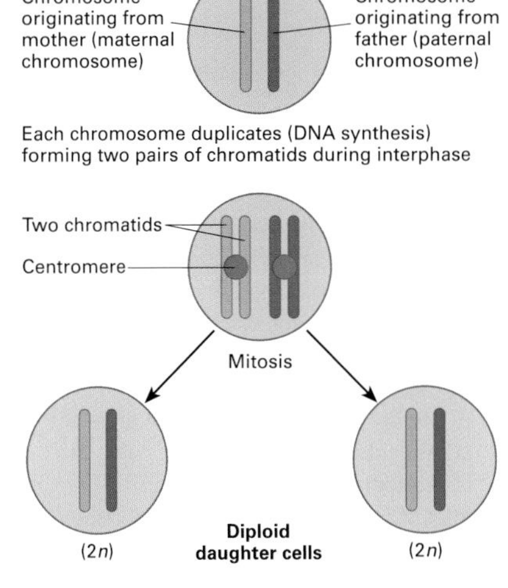

Figure 2 Outline of mitosis. Duplication of a chromosome is followed by nuclear division and cell division. Mitosis is explained in more detail in the next spread.

Meiosis: four different daughter cells

In meiosis, the nucleus divides twice. This produces four haploid nuclei (figure 4). The number of chromosomes is therefore halved during meiosis. Moreover, homologous chromosomes within a pair can exchange genetic material before being separated. The daughter cells are therefore genetically different from the parent cell (and from each other).

Meiosis is the basis of sexual reproduction, occurring at some point in the life cycle of organisms that reproduce sexually. The haploid gametes produced by meiosis fuse during fertilisation. This means that the new fertilised cell has the diploid number of chromosomes. Without meiosis in the life cycle, the number of chromosomes of a sexually reproducing species would be doubled in each generation (see chapter 12).

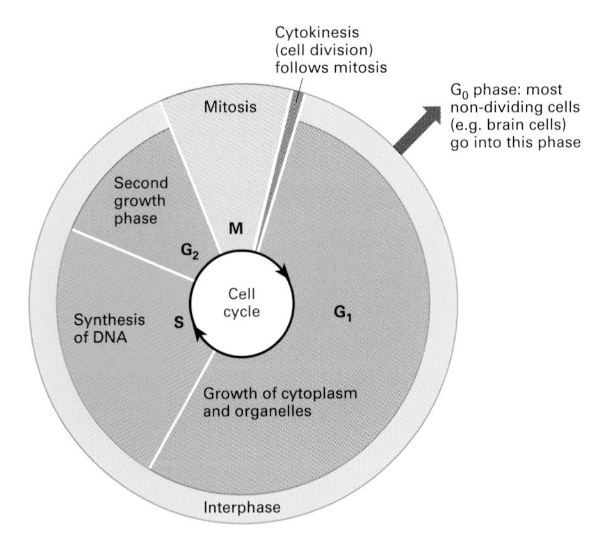

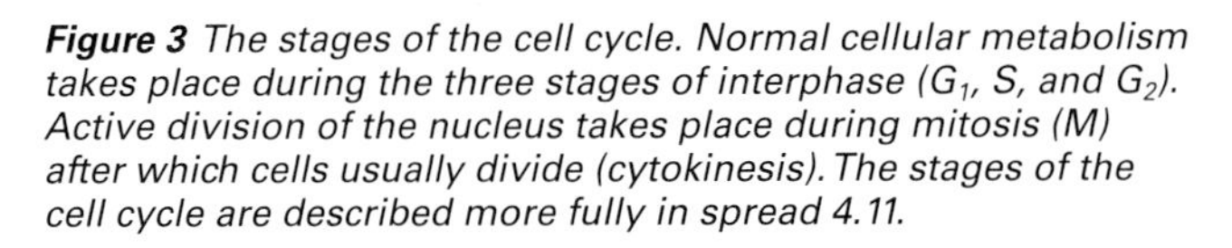

Figure 3 The stages of the cell cycle. Normal cellular metabolism takes place during the three stages of interphase (G_1, S, and G_2). Active division of the nucleus takes place during mitosis (M) after which cells usually divide (cytokinesis). The stages of the cell cycle are described more fully in spread 4.11.

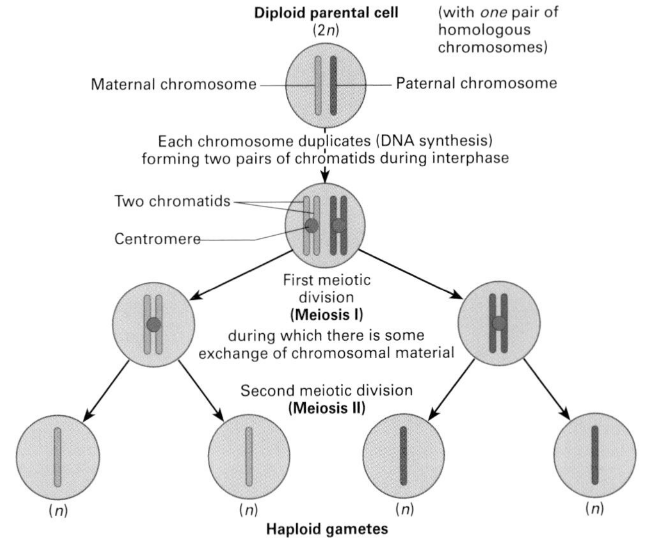

Figure 4 Outline of meiosis. Duplication of a chromosome is followed by two nuclear divisions and cell divisions. Meiosis is explained in more detail in spread 4.12.

Quick check

1 a List the main stages of mitosis, starting with interphase.

b At which stage is DNA replicated?

2 Compare mitosis and meiosis in terms of number of cell divisions and number of daughter cells.

Food for thought

Apoptosis is responsible for removing tail cells when tadpoles metamorphose into frogs, and during human embryo development for the removal of cells that form the webbing between digits. It is also responsible for destroying brain cells that have not interconnected during child development. Suggest the possible implications of this for learning a new skill such as music or a foreign language.

4.11

Mitosis

OBJECTIVES

By the end of this spread you should be able to:

• describe the main stages of mitosis.

Fact of life

When studying sea urchin cell cycles, Sir Tim Hunt discovered a new family of proteins involved in cell signalling. He named these proteins **cyclins** because their concentrations varied in a cyclical fashion during the cycle. Different cyclins control the progress of the cycle by being active at different stages. Mitosis of eukaryotic cells is regulated by the interaction of intracellular cyclin-dependent **kinases** and extracellular growth factors, such as those associated with human growth hormone and insulin. Cyclin-dependent kinases are enzymes that, when cyclin is bound to the kinase, transfer phosphate groups from high-energy donor molecules, such as ATP, to specific substrates.

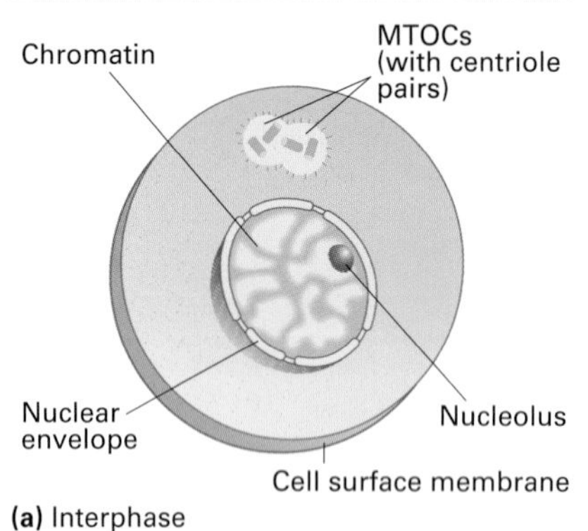

(a) Interphase

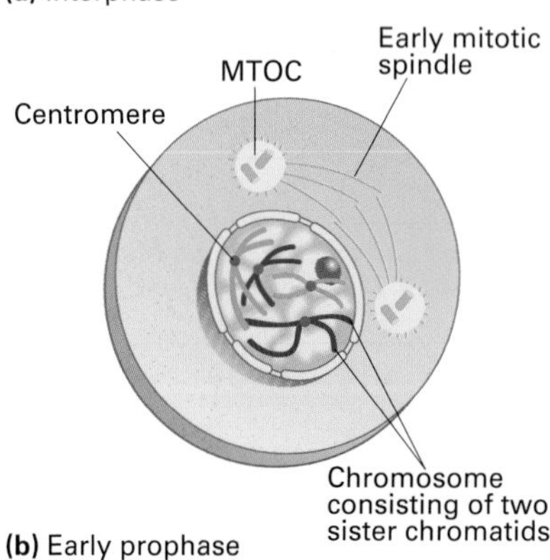

(b) Early prophase

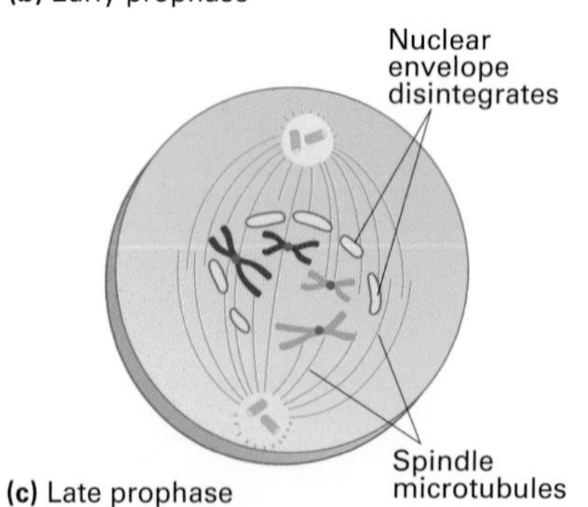

(c) Late prophase

Figure 1 *Interphase; and prophase, the first stage of mitosis.*

The stages of interphase

Mitosis is preceded by **interphase** (figure 1a). This was once called the resting stage because, under the light microscope, little seems to be happening apart from growth. In fact, interphase is a period of intense activity. DNA replicates, and energy in the form of ATP is built up ready for use during mitosis.

Interphase has three stages: G_1, S, and G_2 (see spread 4.10, figure 3).

- During stage G_1 the cell is very active, growing and carrying out its metabolic functions.
- During stage S, DNA replicates and two sister chromatids form from each chromosome.
- During G_2, the mitochondria divide and the cell continues to grow until mitosis begins. In plants, chloroplasts also divide during G_2.

Mitosis is a continuous process, but is described in four main stages: prophase, metaphase, anaphase, and telophase.

Prophase

Mitosis starts with **prophase** (figure 1b). The chromosomes condense, becoming more tightly coiled and folded and so appear shorter and fatter as prophase progresses. As soon as the chromosomes start to condense, the DNA becomes inactive. The condensation of the chromosomes into separate structures enables them to be moved easily; during interphase they are diffuse and would become entangled if they were moved about the nucleus. In the later stages of prophase pairs of sister chromatids can be seen, attached at a point called the **centromere**. The nucleoli disappear, the nuclear membrane breaks down, and a spindle apparatus appears.

The spindle apparatus

The **spindle apparatus** is made of microtubules which control the movements of the chromosomes. The microtubules originate from two **microtubule-organising centres (MTOCs)**. In animal cells each MTOC contains two barrel-shaped structures called **centrioles**. These centrioles have an internal structure similar to that of cilia and flagella. Little is known about the function of centrioles. They are absent from most plant cells and experimental destruction of animal centrioles appears to have little effect on mitosis.

By late prophase (figure 1c), the MTOCs (with or without their centrioles) have moved to opposite ends (poles) of the cell. Some microtubules of the spindle apparatus make contact with the centromeres of the chromosomes; others make contact with microtubules from the opposite pole. The chromosomes move towards the equator (an imaginary line equidistant between the poles) of the cell.

Metaphase

In **metaphase** (figure 2a) the centromeres of all the chromosomes are lined up on the equator and divide. The spindle apparatus is fully formed. One sister chromatid from each chromosome is attached by microtubules to one pole; the other is attached to the opposite pole.

Anaphase

Anaphase (figure 2b) begins with the separation of the centromeres. The sister chromatids are drawn apart to opposite poles of the cell. Once the sister chromatids are separated they are referred to as **daughter chromosomes**. The separation and movement is brought about by the shortening of the microtubules that connect the chromosomes to the poles, and the lengthening of the pole-to-pole microtubules. The poles move further apart, lengthening the cell.

Telophase

Telophase (figure 2c) begins when the two sets of daughter chromosomes have reached the two poles of the cell. The nuclear membrane and nucleoli reform, and the chromosomes become less visible under the microscope. At the end of telophase the spindle apparatus disappears. Mitosis finishes when two identical daughter nuclei are formed.

Cytokinesis

Cytokinesis (division of the cell) may occur during or after telophase, or it might not occur at all. Some cells divide their nuclei without forming new cells. For example, some skeletal muscle cells become multinucleated. In animal cells, cytokinesis involves the formation of a **cleavage furrow** which pinches the cell into two.

Plant cells divide differently. First, membrane-enclosed vesicles containing cell wall materials collect at the midline of the parent cell. The vesicles join to form a **cell plate** (figure 3). The plate grows outwards and eventually fuses with the cell surface membrane so that two distinct cell walls can be formed, dividing the daughter cells.

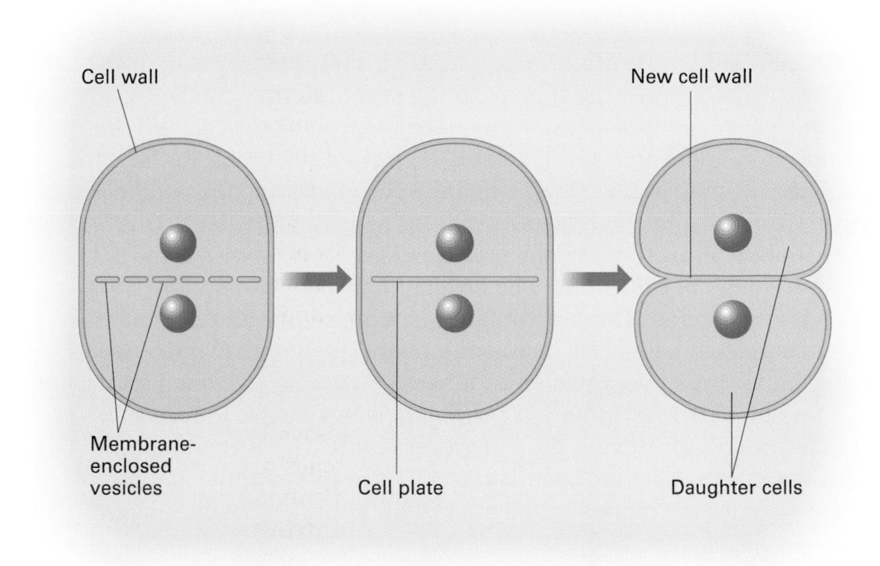

Figure 3 *Plant cell division: formation of a cell plate.*

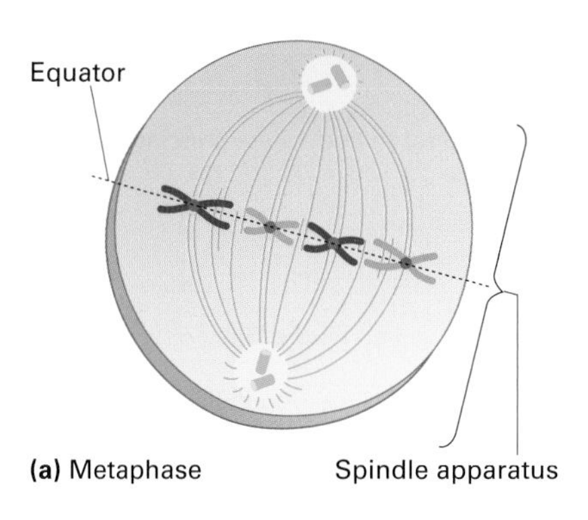

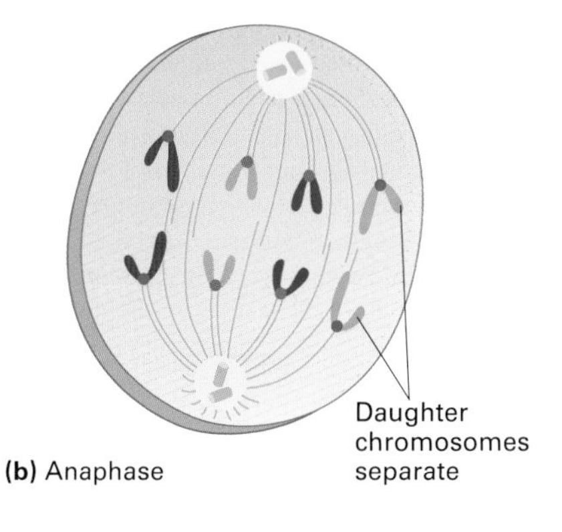

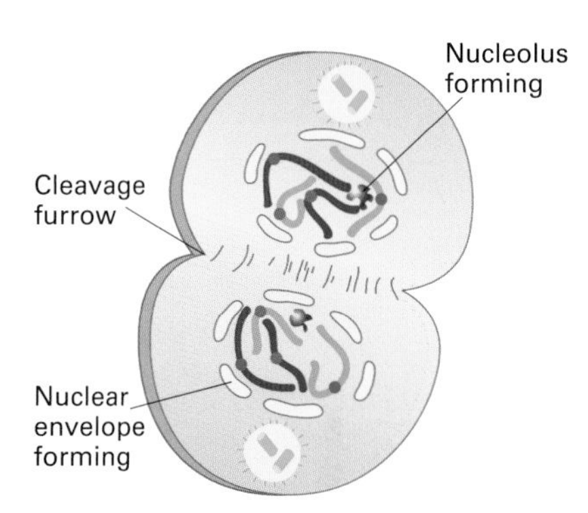

Figure 2 *Mitosis continued: metaphase, anaphase, telophase, and cytokinesis.*

QUICK CHECK

Identify which photomicrograph in figure 4 corresponds to:

a interphase
b early prophase
c late prophase
d metaphase
e anaphase
f telophase.

Food for thought

Suggest why mitosis in plants is restricted to special regions called meristems, whereas in animals there is no such restriction.

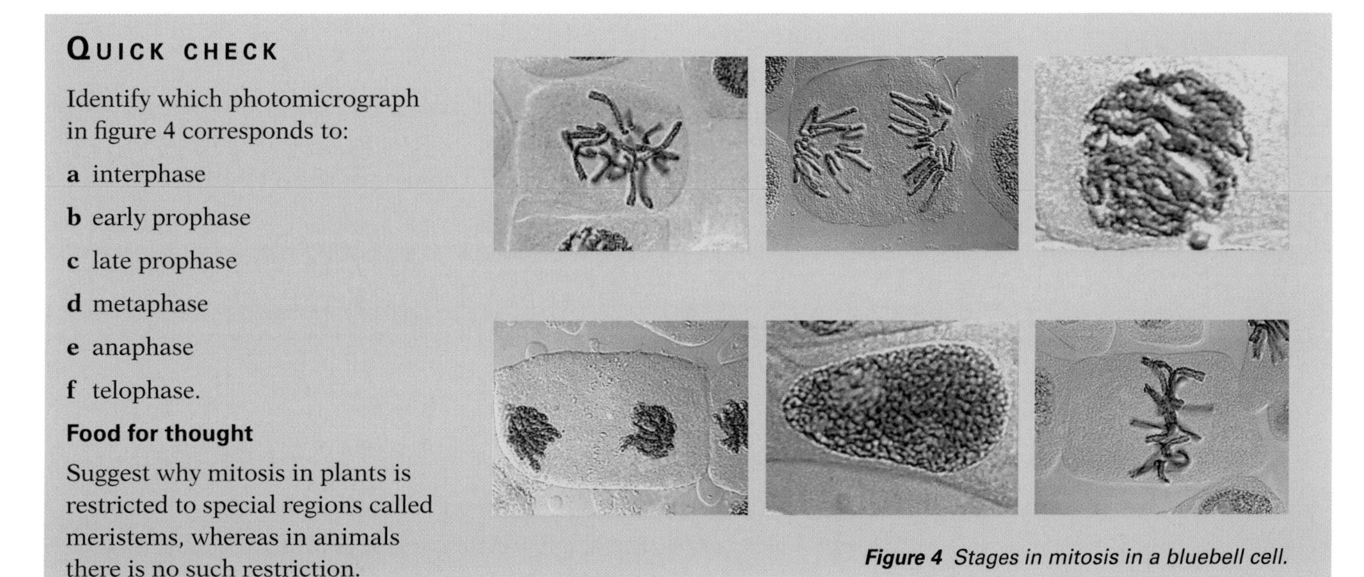

Figure 4 *Stages in mitosis in a bluebell cell.*

4.12

Meiosis

OBJECTIVES

By the end of this spread you should be able to:

- describe the main stages of meiosis.

Fact of life

During the formation of a human gamete (egg or sperm) by meiosis, only one chromosome out of each homologous pair ends up in the gamete. Which of the two chromosomes the gamete receives is a random choice. Even without chromosomes crossing over, this random segregation of the chromosomes, combined with the random choice of which gametes actually fuse, ensures a great deal of genetic diversity: 2^{23} (more than eight million) different combinations of chromosomes for any child of the same parents.

Meiosis consists of two divisions: meiosis I and meiosis II. In meiosis I (figure 1), homologous chromosomes are separated. In meiosis II (figure 2), sister chromatids are separated. During meiosis, one diploid parent cell divides to form four haploid daughter cells, genetically distinct from each other and from the parent cell. Meiosis is sometimes called reduction division because the number of chromosomes in the cell is halved.

Like mitosis, meiosis is a continuous process. However, it is described as eight main stages: prophase I, metaphase I, anaphase I, telophase I, prophase II, metaphase II, anaphase II, and telophase II. The stages are similar to those of mitosis, and the stages that share a similar name (such as prophase, prophase I, and prophase II) are superficially similar, but there are significant differences.

Meiosis I

Meiosis I is preceded by interphase, apparently identical to that preceding mitosis. DNA and organelles are duplicated, and energy is stored as ATP to be used later in meiosis.

Prophase I is a complex stage taking up almost 90 per cent of the meiotic division time. In prophase I, the chromosomes condense, the nuclear membrane breaks down, the nucleoli disappear, and the spindle apparatus appears. These features are very similar to those in prophase of mitosis. However, when the chromatids appear, homologous chromosomes come together and pair up to form a **bivalent**. This is quite unlike mitosis, where the homologous chromosomes behave independently of each other. The bivalent has four chromatids (two from each chromosome). These chromatids can intertwine and exchange segments in a process called **crossing over**. The point at which two chromatids cross over is called a **chiasma**. Crossing over changes the mixture of genetic information carried on a chromatid. It makes an important contribution to the genetic variability that sexual reproduction introduces to a species. Towards the end of prophase I, the paired chromosomes are moved by the microtubules of the spindle apparatus towards the equator of the cell.

At **metaphase I** homologous chromosomes line up side by side on the equator. The centromere of one chromosome is attached to microtubules from one pole, and the centromere of its homologue is attached to microtubules from the opposite pole. The centromere does not divide, so sister chromatids remain attached to each other.

During **anaphase I** homologous chromosomes separate and move towards opposite poles. The sister chromatids of each chromosome move as a single unit.

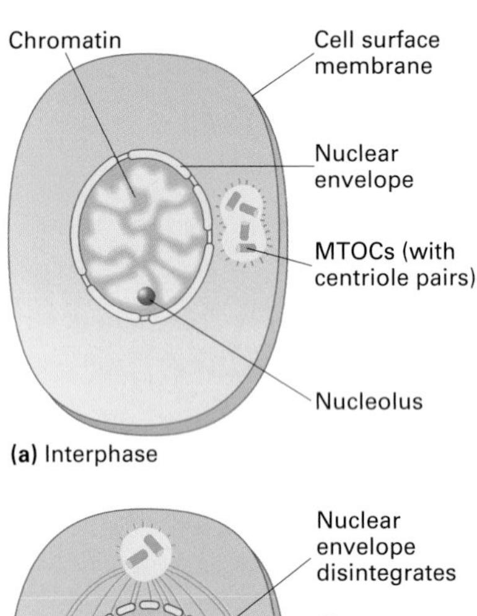

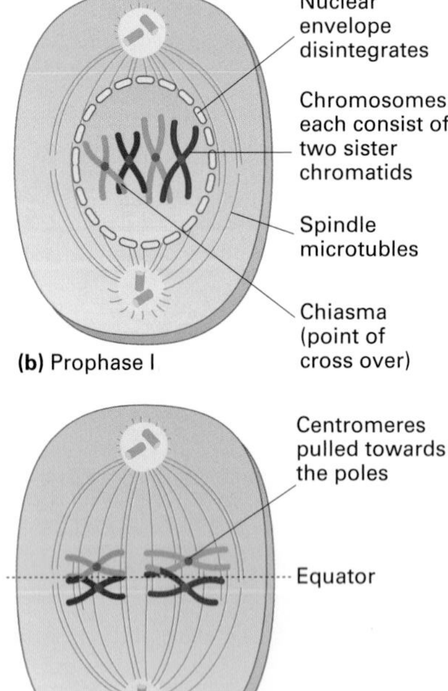

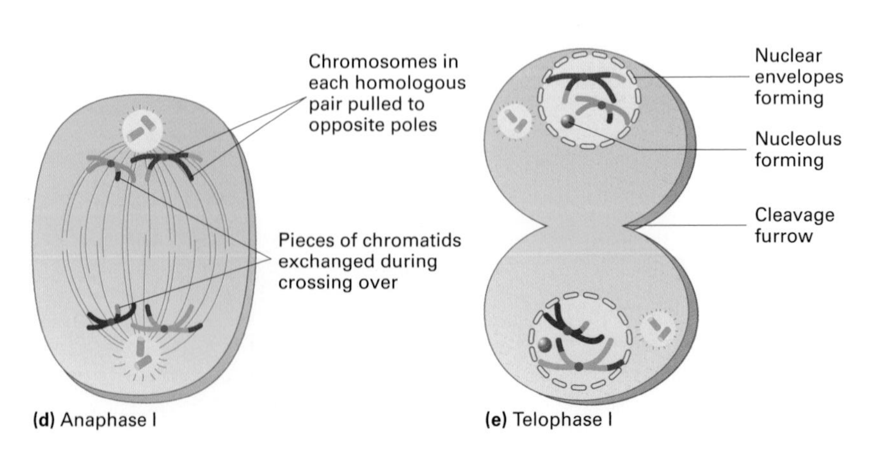

Figure 1 The stages of meiosis I.

In **telophase I**, a haploid set of chromosomes arrives at each pole, resulting in the segregation of alleles so that only one of each pair of alleles will occur in a single gamete. Each chromosome still consists of two chromatids. Cytokinesis usually coincides with telophase, and two haploid cells are formed. In some species the nuclear membrane and nucleoli reform and there is a slight delay before meiosis II. In other species prophase II follows immediately after telophase I.

Meiosis II

Meiosis II is similar to mitosis: it results in the separation of sister chromatids but involves only a haploid set of chromosomes.

Prophase II is not preceded by interphase: there is no replication of DNA. The chromosomes condense, a spindle apparatus forms and the chromosomes move to the equator of the cell.

At **metaphase II** the chromosomes line up along the equator and the centromere divides into two.

In **anaphase II** sister chromatids are separated and move to opposite poles as individual daughter chromosomes.

In **telophase II** a complete haploid set of chromosomes gathers at each pole and is encapsulated in a nuclear membrane. Nucleoli reform and cytokinesis occurs.

The implications of meiosis

The end result of the two meiotic divisions is the production of four haploid cells from a single diploid cell. During gamete formation, each of a pair of homologous chromosomes can combine with either of another pair of homologous chromosomes, resulting in independent assortment (see spread 19.2). This, along with crossing over, ensures that the four daughter cells are genetically different from the parent cell and from each other. Independent assortment, crossing over, and random fusion of the gametes all contribute to the variation of offspring produced by sexual reproduction (see spread 19.1 for an account of variation).

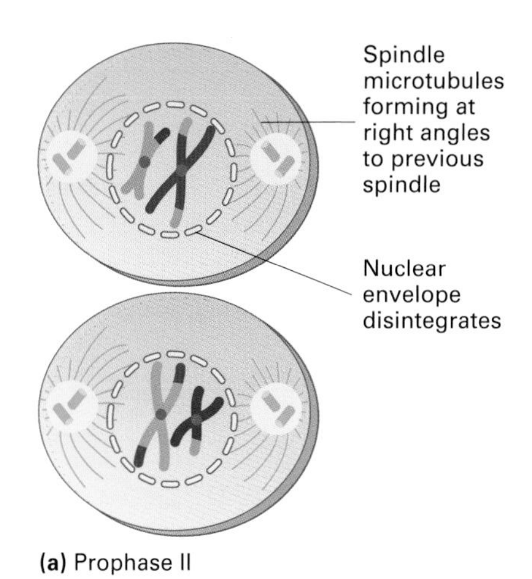

(a) Prophase II

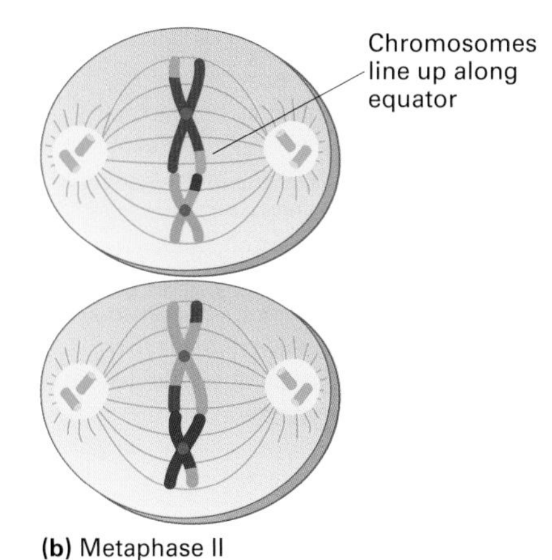

(b) Metaphase II

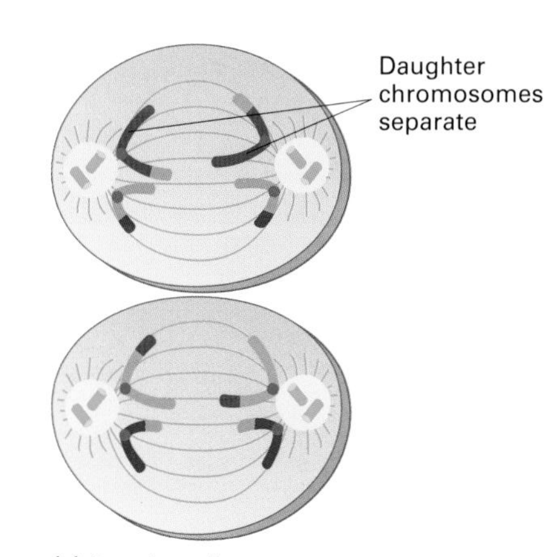

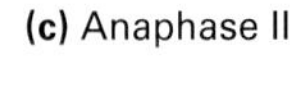

(c) Anaphase II

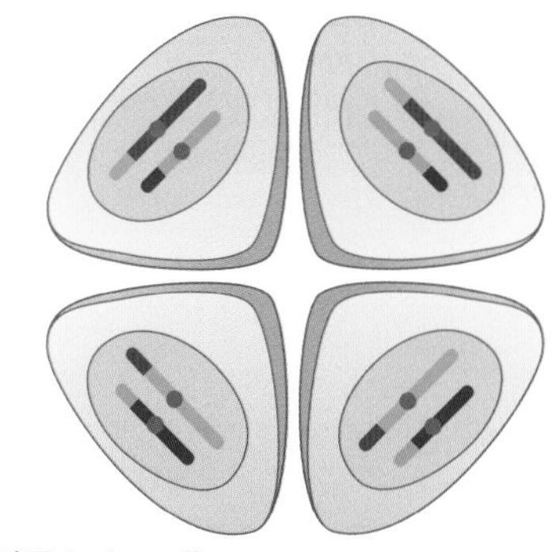

(d) Telophase II

Figure 2 *The stages of meiosis II.*

Quick check

1 Identify the stages of meiosis shown in figures 3a and b.

Figure 3 *Stages in meiosis.*

Food for thought

Cyanide is a metabolic poison that prevents ATP production in cells. If cyanide is applied to cells before cell division starts, the nucleus does not divide. However, if cyanide is applied during or after prophase, the cell continues to divide. Suggest an explanation for this.

4.13

Cells, tissues, and organs

OBJECTIVES

By the end of this spread you should be able to:

- discuss the advantages and disadvantages of being multicellular
- explain the difference between cells, tissues, and organs
- name the main types of animal and plant tissues.

Fact of life

Survival in multicellular organisms depends on cells being able to communicate with each other so that their different activities can be coordinated. Cell-to-cell communication is accomplished by cell signalling. This may be electrical or chemical. For cells in direct contact, an electrical signal may be transmitted from one cell to another. Chemical signals enable communication between cells that are some distance apart. Typically, one cell releases the chemical signal which diffuses to a target cell that has receptors to detect the stimulus. Information is then relayed along various cell-signalling pathways to activate effectors in either the nucleus or the cytoplasm. Different pathways control different stages in the life history of cells, from their formation by mitosis, through their differentiation, and finally to their death, by processes such as apoptosis. Cell-signalling pathways also regulate a wide range of adult processes such as muscle contraction, secretion, digestion, tissue repair, and information processing in neurones and sense organs.

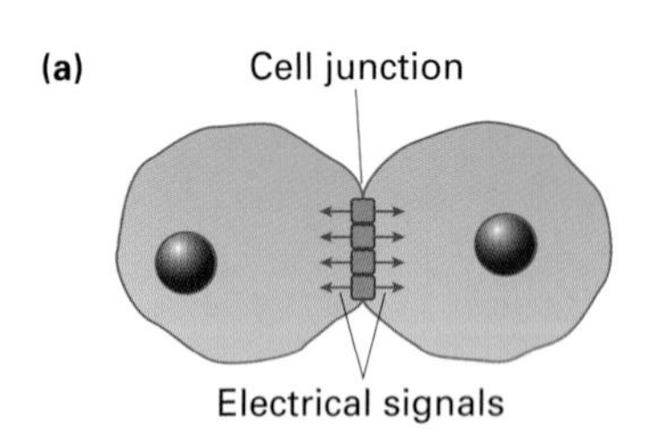

***Figure 1** Cell signalling between cells **(a)** in direct contact and **(b)** some distance apart. Chemical signals include neurotransmitters and hormones.*

There is a limit to the size of an individual cell with a single nucleus. The largest cells are just visible with the naked eye, but the majority can be seen only with the aid of a microscope. The size of the cell is limited by the ratio of the surface area of the cell surface membrane and the nuclear membrane to the cell's volume.

Surface area and volume: exchanging materials

A living cell is constantly exchanging substances with its environment. The speed at which substances can diffuse in and out depends on the cell surface area, but the amounts of those substances that are needed depend on cell volume. A cell that grew indefinitely would reach a point at which it used substances more quickly than it could obtain them. Cellular activities would then slow down or stop. (The effects of increasing size on diffusion of materials is explained more fully in spread 4.7.)

The size of the nucleus

The same limitation applies to the size of the nucleus, which exchanges substances with the cytoplasm across the nuclear membrane. If a nucleus stayed the same size while the cytoplasm grew larger, there would be problems controlling the outer regions of the cytoplasm as they became distant from the nucleus. So there appears to be an upper limit to the amount of cytoplasm that can be controlled effectively by a single nucleus. This problem can be overcome by having several nuclei in one cell – by becoming **multinucleated**. However, most large organisms have adopted another solution to the transport problems of being large. Their bodies consist of more than one cell: they have become **multicellular**.

Surface area to volume ratio

Figure 2 shows how the surface area to volume ratio changes as the number of cells is increased. A cubic cell of side 2 cm has a surface area of 24 cm^2 and a volume of 8 cm^3. If this single cell is replaced with eight cells of side 1 cm the volume remains the same but the surface area doubles.

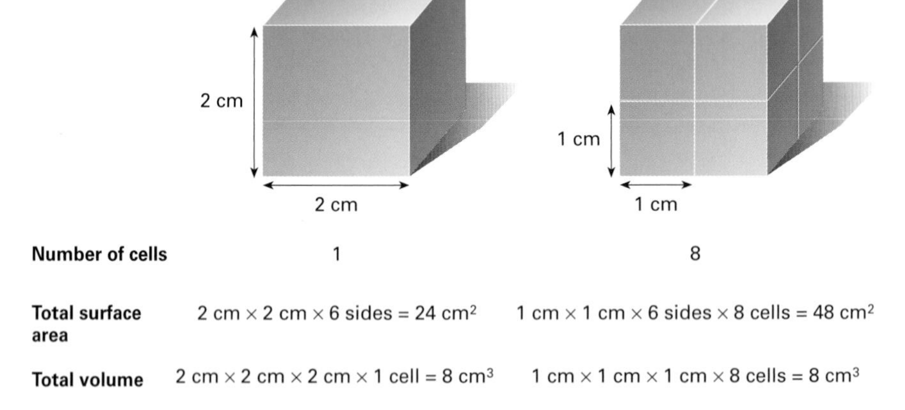

Number of cells	1	8
Total surface area	2 cm × 2 cm × 6 sides = 24 cm^2	1 cm × 1 cm × 6 sides × 8 cells = 48 cm^2
Total volume	2 cm × 2 cm × 2 cm × 1 cell = 8 cm^3	1 cm × 1 cm × 1 cm × 8 cells = 8 cm^3
Surface area: volume ratio	24:8 = 3:1	48:8 = 6:1

***Figure 2** The effect of increasing the number of cells on the surface area to volume ratio.*

Specialisation

Being multicellular not only enables organisms to be larger, it also enables their cells to become specialised. Each cell no longer has to carry out all the activities necessary to sustain life. Cells may become specialised in feeding, support, or defence. This **division of labour** among cells makes multicellular organisms more efficient and enables them to exploit modes of life not available to single-celled organisms. However, specialisation often means that individual cells lose the ability to carry out some functions. For example, a mature human red blood cell has become specialised for carrying oxygen, but its lack of a nucleus means it cannot reproduce. Cells of a multicellular organism therefore lose some, if not all, of their independence.

Tissues and organs

In most multicellular organisms, groups of cells of the same type associate together to form **tissues**. The cells in a tissue act together to perform a common function. Working as a group of cells is often more efficient than working individually.

A tissue may combine with other tissues to form an organ: for example, muscle tissue, nerve tissue, and connective tissue work together in the heart. An **organ** is a group of tissues that carries out a particular function. Figure 3 shows how cells combine to form tissues, and how tissues combine to form organs.

Human tissues

There are more than two hundred different types of cell in the human body. They combine to form a variety of tissues which:

- hold body structures together (**connective tissue**)
- line the body surface and the surfaces of organs, cavities, and tubes (**epithelial tissue**, figure 4)
- move the body or body parts (**muscle tissue**)
- communicate between different parts of the body (**nerve tissue**)
- produce eggs and sperm (**reproductive tissue**).

Plant tissues

In flowering plants, all cells are derived from unspecialised tissue (**meristematic tissue**) which has actively dividing cells. The main plant meristems are at the apex of the shoot and the tip of the root. **Non-meristematic tissues** contain specialised cells which have lost the ability to divide. **Ground tissue** includes:

- **parenchyma**, the packing tissue that fills the spaces between other tissues, which may be adapted for roles such as storage, support, or photosynthesis
- **collenchyma**, made up of cells with walls thickened with extra cellulose at the corners to provide strength and flexibility to stems and leaves
- **sclerenchyma**, tightly packed bundles of lignified fibres.

Supporting and **vascular tissue** include:

- **xylem** which transports water
- **phloem** which transports dissolved substances such as sucrose.

Protective tissue includes:

- the **epidermis** lining the surfaces of leaves and stems
- **cork** on the surfaces of stems of trees and shrubs.

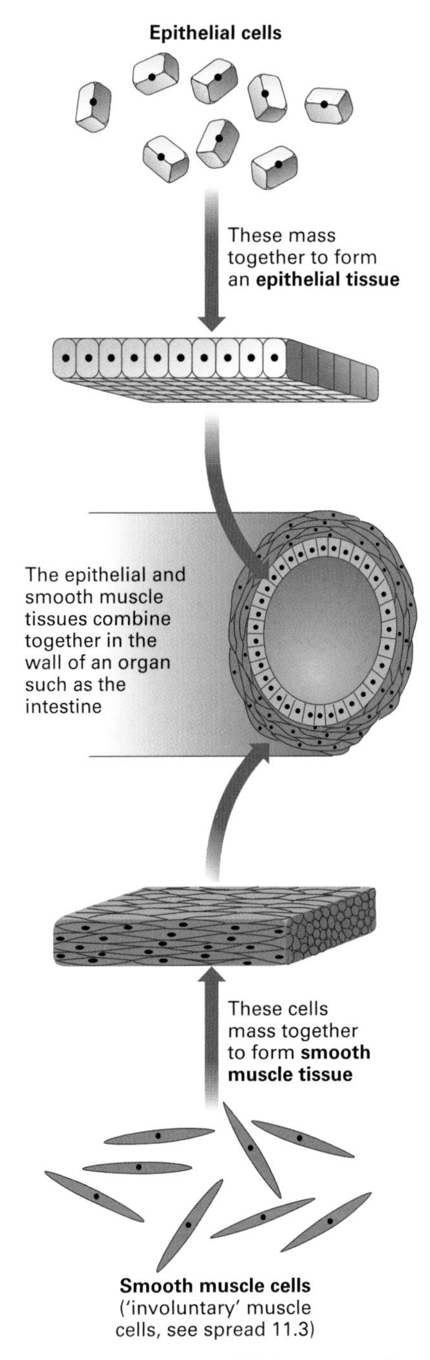

Figure 3 The relationship between cells, tissues, and organs.

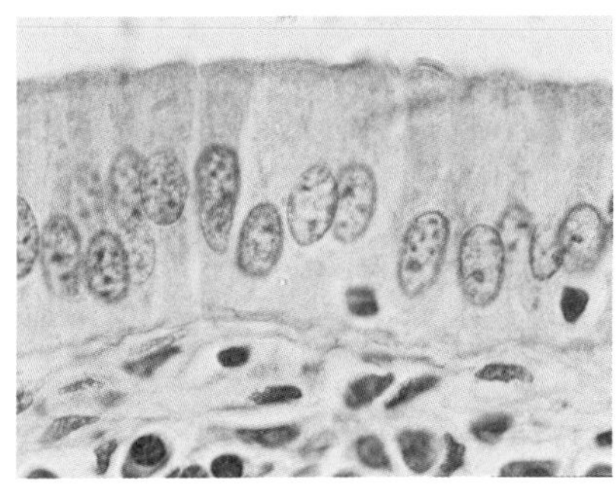

Figure 4 Light micrograph of a simple type of epithelium. This type of tissue is found on the surface of the skin and the linings of cavities and tubes within the body.

Quick check

1 Give two advantages of being multicellular.

2 What is the difference between a tissue and an organ?

3 Give one example of: **i** an undifferentiated (unspecialised) cell **ii** a tissue **iii** an organ to be found in:
a a mammal
b a plant.

Food for thought

Bearing in mind that single-celled organisms have been around longer than any other organisms on Earth, and that they make up about half of the total biomass of living things, suggest the specific advantages of being single celled.

Summary

According to the **cell theory**, cells are the basic units of life. **Prokaryotic cells** of bacteria differ from **eukaryotic cells** of animals and plants, but all cells have certain features in common.
Observations with a **light microscope** reveal that a typical animal cell consists of a **cell surface membrane** enclosing **cytoplasm** and a **nucleus**, and that typical plant cells have **cellulose cell walls**, **sap-filled vacuoles**, and **chloroplasts**. The higher **resolving power** of **electron microscopes** allows **cell ultrastructure** to be studied. **Scanning electron microscopes** are used to examine surface features of fractured or whole cells, whereas a **transmission electron microscope** reveals internal structures of thinly sectioned cells. These structures include **microtubules**, **filaments**, **fibres**, and **organelles**, such as **ribosomes**, **mitochondria**, **the Golgi apparatus**, **lysosomes**, and the **endoplasmic reticulum**. Organelles can be isolated by **cell fractionation** and their functions studied.
Cell membranes cover the surface of every cell and form a network inside many cells. In eukaryotes, a **double membrane** envelops mitochondria, chloroplasts, and the nucleus. According to the **fluid-mosaic model**, each cell membrane consists of protein molecules shifting and moving on a **bilayer of phospholipids**. The precise composition of membranes varies as do their functions.
Cell surface membranes control the passage of materials in and out of cells. Substances may move through a membrane by **diffusion**, **facilitated diffusion**, **active transport**, and **osmosis**. According to **Fick's law**, the rate of diffusion is proportional to the **surface area** and **concentration gradient**, and inversely proportional to **the length of the diffusion path**. Both active transport and facilitated diffusion involve **carrier** molecules, but only active transport requires ATP as a source of energy. Facilitated diffusion, like simple diffusion and osmosis, is a passive process.
Osmosis involves the movement of water only. Water molecules have a net movement from a high **water potential** to a low water potential through a **selectively permeable membrane**. When plant and animal cells are placed in a **hypertonic solution** the volume of their cytoplasm shrinks; plant cells **plasmolyse**, the cell membrane withdraws from the cell wall. In **hypotonic solutions**, plant cells become **turgid** whereas animal cells burst unless they have a mechanism, such as a **contractile vacuole**, to prevent the influx of water.
Cell division is essential for growth, repair, and reproduction. It starts with the division of the nucleus of which there are two main types: **mitosis**, a single division which (barring mutations) produces nuclei identical with each other and with the parent cell; and **meiosis** which involves two divisions and produces daughter cells different from each other and from their parent cells.
Some cells exist as independent organisms; others occur together to form tissues and organs in multicellular organisms. **Multicellularity** allows **specialisation** and organisms to become large. As organisms become larger, their **surface area-to-volume ratio** decreases. This has important consequences for a number of processes vital to life, including the exchange of materials between an organism and its environment.

PRACTICE EXAM QUESTIONS

1 Copy and complete the table below which compares the structures of a typical plant, animal and prokaryotic cell. Use a tick if the feature is present and a cross if it is absent. [4]

	Plant cell	Animal cell	Prokaryotic cell
nucleolus			
plasmid			
mitrochondrion			
cellulose wall			

2 The diagram below shows the structure of a liver cell as seen using an electron microscope.

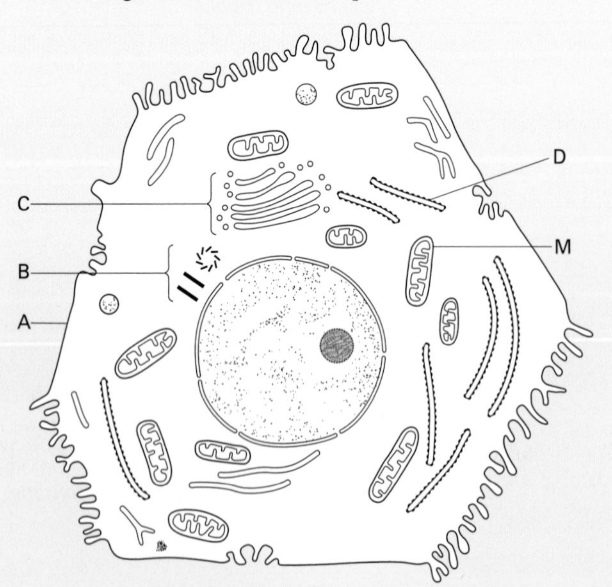

a Name the parts labelled A, B, C and D. [4]

b The magnification of this diagram is × 12 000. Calculate the actual length of the mitochondrion labelled M, giving your answer in mm. Show your working. [2]

[Total 6 marks]

3 The flow chart shows how a preparation of mitochondria may be made from liver cells.

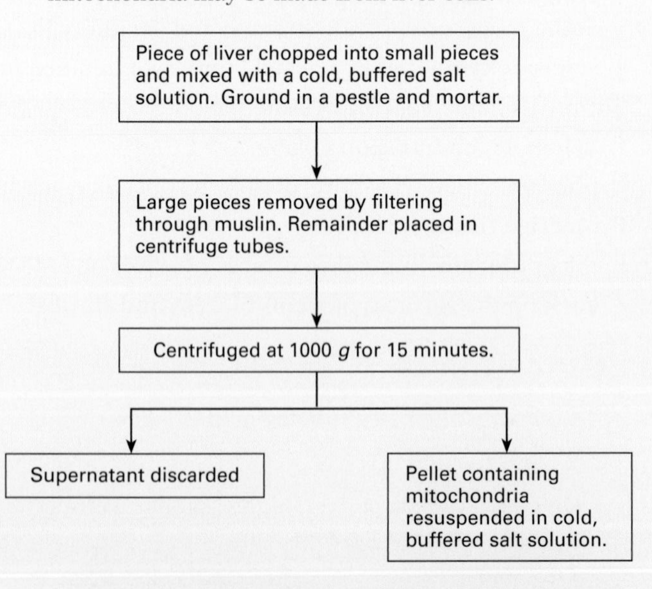

a Name one organelle that would be found in the supernatant discarded at the end of this process. [1]

b Explain why the pellet was resuspended in salt solution rather than in distilled water. [2]

c The resuspended pellet of mitochondria was mixed with pyruvate and incubated at 35 °C for 15 minutes.

Explain why, during the period of incubation, there was a fall in the amount of oxygen present in the preparation. [2]

[Total 5 marks]

4 The diagram represents a phospholipid molecule.

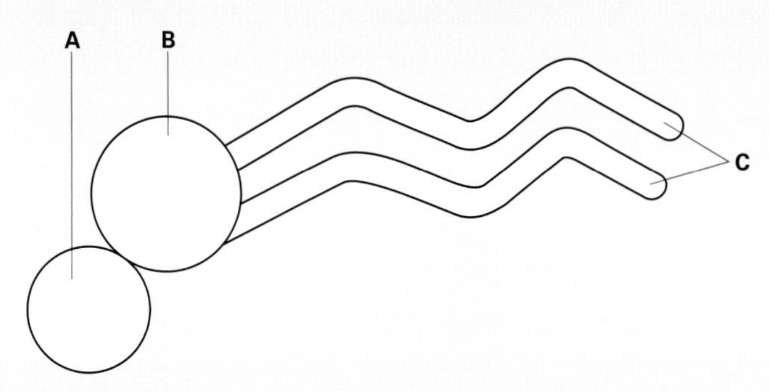

a **i** Name the parts of the molecule **A**, **B** and **C**. [1]

ii Explain how the phospholipid molecules form a double layer in a cell membrane. [2]

b Cell membranes also contain protein molecules. Give two functions of these protein molecules. [2]

[Total 5 marks]

5 The diagram below shows a plant cell, immersed in a sucrose solution. The pressure potential (ψ_p) of the cell and the solute potential (ψ_s) of the cell and of the sucrose solution are shown in the diagram.

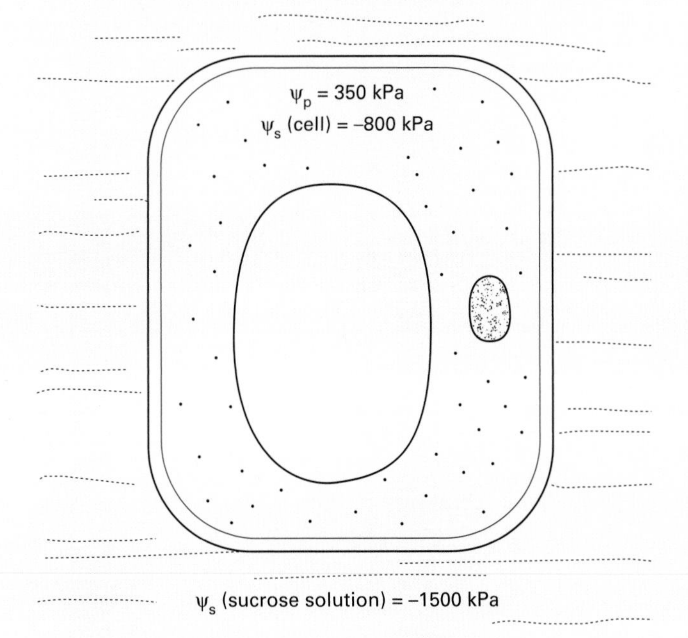

a Calculate the water potential of this cell (ψ cell). Show your working. [2]

b State whether water will move into or out of the cell. Explain your answer. [2]

c State the water potential of this cell at the point of incipient plasmolysis. Assume that changes in ψ_s (cell) are negligible. [1]

[Total 5 marks]

6 The diagram shows three adjacent plant cells.

a **i** Calculate the water potential of cell A. [1]

A
ψ_s = −10MPa
ψ_p = 3MPa
ψ = ...

B
ψ = −12MPa

C
ψ = −4MPa

ii Copy the diagram and show, by means of arrows, the direction of water movement between these cells. [1]

iii Explain why the water potential of a sucrose solution has a negative value. [2]

b An artificial membrane can be made which consists only of a lipid bilayer. The diagram below compares the permeability of such an artificial membrane with a biological cell membrane.

High permeability
Artificial membrane
Biological membrane
H_2O → ← H_2O
Glycerol → ← Glycerol
← Na^+
Na^+ →
Low permeability

i Explain why the permeability of both membranes to glycerol is the same. [1]

ii Explain why the permeability of the two membranes to sodium ions differs. [2]

[Total 7 marks]

7 In this section you are expected to answer in continuous prose, supported, where appropriate, by diagrams. You are reminded that up to two marks in this question are awarded for Quality of Language. [2]

a Write an account of the cell cycle, involving a mitotic nuclear division, highlighting the events occurring in each phase. [9]

b Describe four major differences between mitosis and meiosis. [4]

[Total 15 marks]

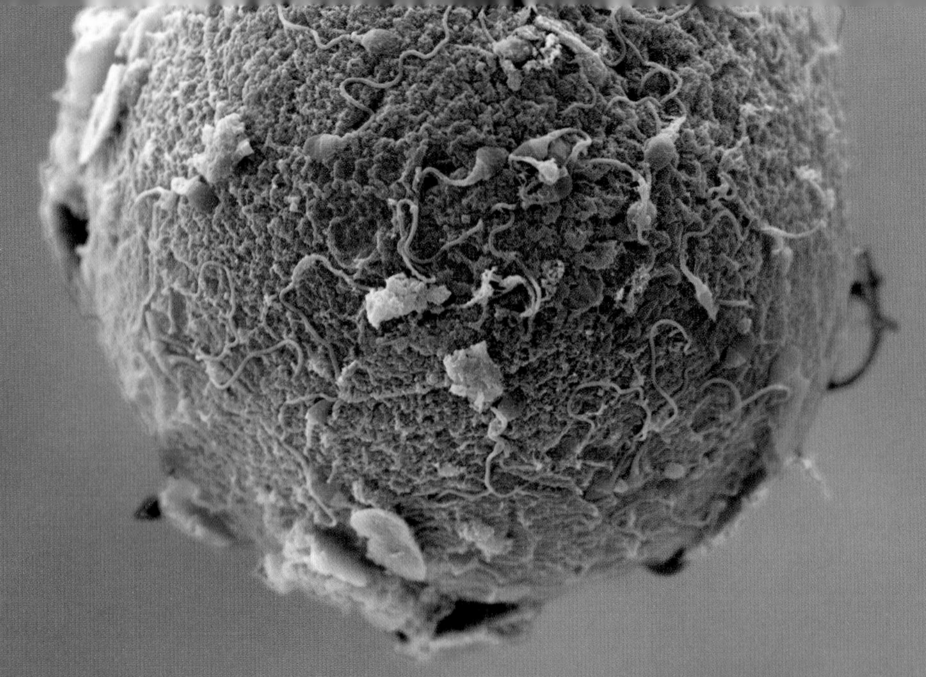

ANIMALS *and* PLANTS

When we contemplate life as the summing up of many contrivances, each useful to its possessor, in the same ways as a great mechanical invention is the summing up of the labour, the experience, the reason, and even the blunders of numerous workmen, how far more interesting – I speak from experience – does the study of natural history become.

Charles Darwin, 1859

In this section, we consider how animals and plants keep themselves alive and reproduce. Our guiding theme will be that the form of a particular structure fits its function. We will find that this 'form fits function' theme applies equally well to all levels of biological organisation, from molecules to entire organisms.

A ghost bat: the aerodynamically efficient shape of the bat's wing makes flight possible; an example of form fitting function.

Photosynthesis

Photosynthesis: an overview

5.1

OBJECTIVES

By the end of this spread you should be able to:

- describe the overall process of photosynthesis and its importance to life on Earth
- describe the structure and function of a chloroplast.

Fact of life

It has been estimated that if all the land surface of the Earth could support plants, enough food could be produced to feed 1000 billion people. Of course, this is unrealistic because not all land is suitable for growing plants, and some land is needed for urban and recreational uses. However, even if only 7 per cent of the land surface were made agriculturally productive, plants could produce enough food to support 79 billion people.

Most plants have no structures for ingesting and digesting food. They have no mouth and no alimentary canal, yet plant material is rich in carbohydrates, proteins, and fats. Instead of obtaining their food from other organisms, plants make it for themselves using simple ingredients. They are **autotrophs** ('self-feeders').

What is photosynthesis?

A typical plant takes in carbon dioxide (from the air) and water (from the soil) and builds these up into sugars and other complex substances. Oxygen is released as a waste product. The energy in the chemical bonds of the raw materials carbon dioxide and water is less than the energy in the chemical bonds of the products. Therefore the reaction is endergonic and requires an external source of free energy. This energy is supplied by sunlight that falls on the plant. A green substance, **chlorophyll**, enables the plant to trap light energy and use it to make sugars. The process of using sunlight to build up complex substances from simpler ones is called **photosynthesis**.

Photosynthesis is a complex process which takes place in a series of small steps. However, it can be summarised by the simple equation:

$$\underset{\text{raw materials}}{\underset{\text{Carbon dioxide + water}}{6CO_2 + 12H_2O}} \xrightarrow{\text{light and chlorophyll}} \underset{\text{products}}{\underset{\text{glucose + oxygen + water}}{C_6H_{12}O_6 + 6O_2 + 6H_2O}}$$

The water molecules on both sides of this equation reflect the fact that the oxygen produced by photosynthesis is derived from water. Two water molecules are required to form one oxygen molecule. The hydrogen in glucose is also derived from water. Experiments using radioactive isotopes and isolated chloroplasts, the sites of photosynthesis, suggest that there are two main stages in photosynthesis: a **light-dependent stage** in which water is broken down into hydrogen and oxygen using light energy; and a **light-independent stage** in which the hydrogen reacts with carbon dioxide to form a carbohydrate (figure 1). Water is re-formed in this reaction. The light-dependent stage happens only in the light; the light-independent stage happens both when it is light and when it is dark.

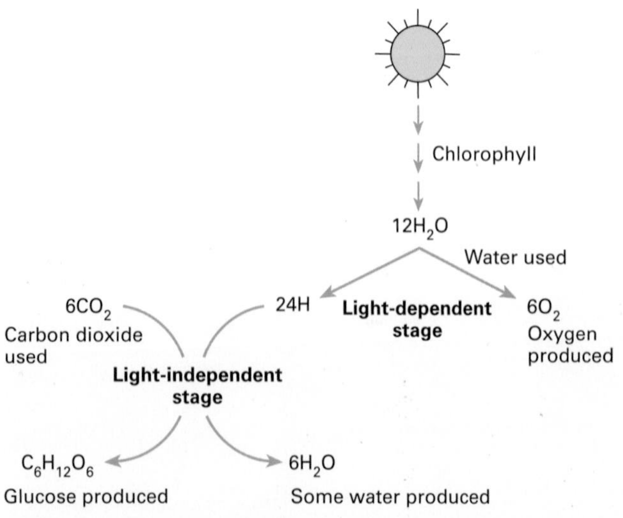

Figure 1 An outline of the two main stages of photosynthesis.

Converting glucose to other substances

The glucose formed by photosynthesis is used as the raw material for other chemical reactions. It is the main substrate used in respiration. Some of the glucose is converted to other carbohydrates: cellulose to form cell walls; sucrose to be transported to other parts of the plant; and starch for storage. Some of the glucose is combined with minerals from the soil to make proteins and other complex organic substances. Although light is needed for making glucose, it is not needed for turning the glucose into these other substances.

Photosynthesis: the basis of life

Green life has been steadily pumping out oxygen as a waste product of photosynthesis for millions of years. Some of the oxygen is used as a raw material for respiration, but most of it has accumulated in the atmosphere. So the very existence of our oxygen-rich atmosphere depends on the photosynthesising activities of green life.

Animals cannot make their own food. The only way they can obtain complex organic substances is by eating other organisms. These organisms ultimately depend on the ability of plants to harvest energy from sunlight to make food from carbon dioxide and water. Life on Earth is almost entirely solar powered.

The site of photosynthesis

Although leaves are the main sites of photosynthesis in most plants, it can take place in any part that is green. These green parts have **chloroplasts**, which contain all the biochemical machinery necessary for the light-dependent and light-independent stages of photosynthesis (figure 2).

Chloroplasts act as compartments, isolating the photosynthetic reactions from other cellular activities. Each chloroplast consists of two membranes enclosing a gelatinous matrix called the **stroma**. The stroma contains ribosomes, circular DNA, and enzymes used in photosynthesis. Suspended in the stroma are **thylakoids**. These are disc-like membrane sacs, several of which are stacked in a group to form a **granum** (plural **grana**). The space inside each thylakoid in a stack is connected with the other thylakoids in the stack, forming a continuous fluid-filled compartment called the **thylakoid space**. The thylakoid membranes contain photosynthetic pigments, including chlorophyll.

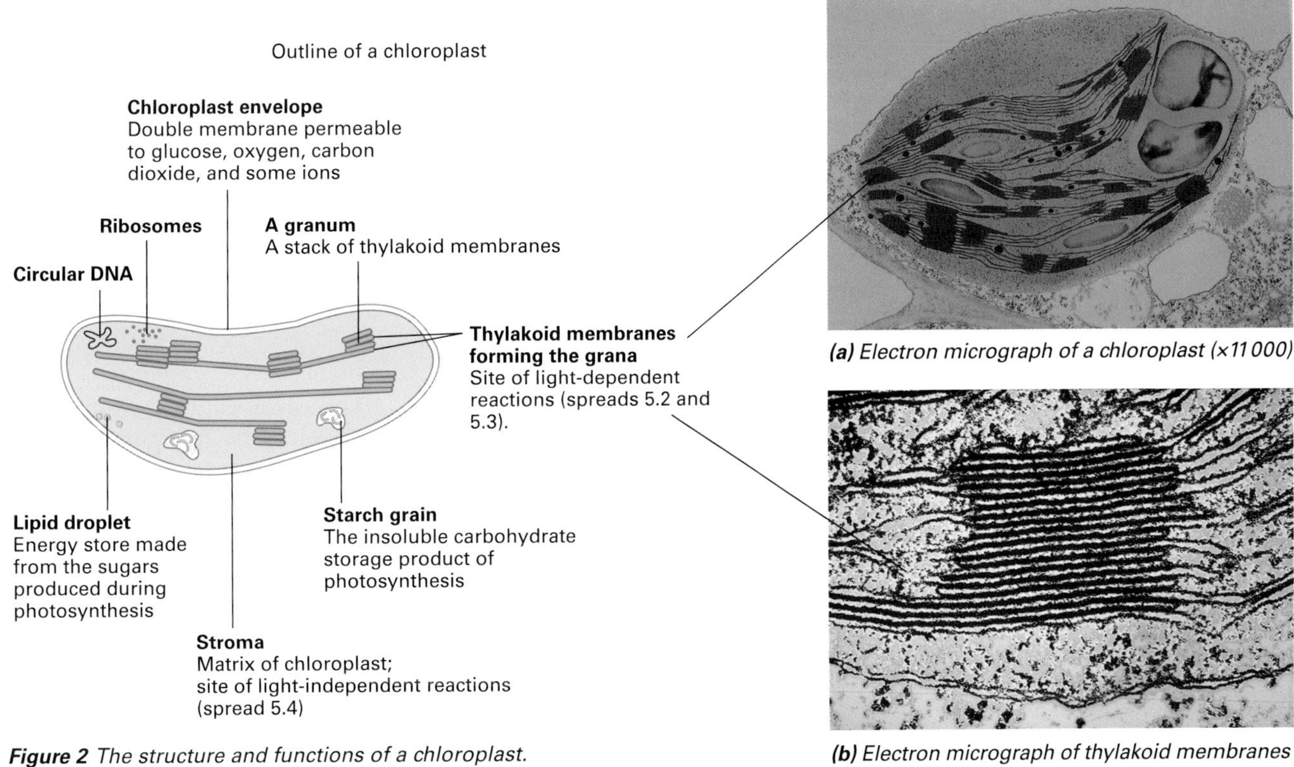

***(a)** Electron micrograph of a chloroplast (×11 000)*

***(b)** Electron micrograph of thylakoid membranes*

***Figure 2** The structure and functions of a chloroplast.*

Quick check

1 During photosynthesis, what gas is:

a a raw material

b a product?

2 Give the precise location in a typical terrestrial plant of:

a the light-dependent stage

b the light-independent stage of photosynthesis.

Food for thought

Less than one per cent of the solar energy that falls on the Earth is used by plants for photosynthesis. Suggest what happens to the other 99 per cent of solar energy.

5.2 Photosynthetic pigments

OBJECTIVES

By the end of this spread you should be able to:

- describe the function of chlorophyll
- distinguish between an action spectrum and an absorption spectrum
- discuss the function of accessory pigments.

Fact of life

Of the wide range of wavelengths in the electromagnetic spectrum (figure 1), plants can use only visible light for photosynthesis. About 48 per cent of the energy in solar radiation reaching plants is in the form of ultraviolet and infrared light, but these are not used in photosynthesis.

How light arrives at the plant

The energy that drives the light-dependent reactions of photosynthesis comes from sunlight, a form of **electromagnetic radiation**. The behaviour of light and other forms of electromagnetic radiation is described by physicists using two different models: a wave model (light is wave) and a particulate model (light is particles). The wave model is used to describe how light passes through space and how it is absorbed by photosynthetic pigments. The particulate model is used when explaining the events in photosynthesis.

Sunlight travels in space in the form of rhythmic waves similar to those made on a pond after a stone is dropped in. The distance between the crests of two adjacent waves is called the **wavelength** (measured in nanometres). Figure 1 shows the full spectrum of electromagnetic radiation produced by the Sun. The **visible light spectrum** occupies only a small part of this. Visible light consists of a range of different wavelengths which combine to give white light, but which when seen separately appear as different colours.

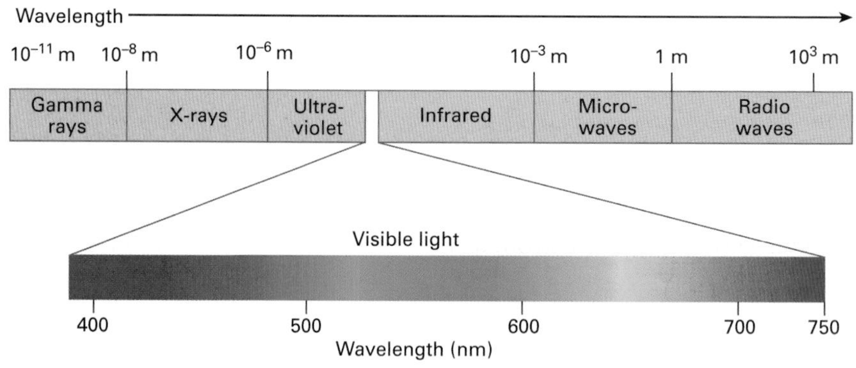

Figure 1 The electromagnetic spectrum.

As sunlight falls on a plant, some wavelengths are absorbed, while others are reflected or transmitted. The pattern of absorption and reflection depends on coloured substances called pigments, so the colour of a plant depends on the pigments it contains in its chloroplasts. Grass appears green in sunlight because grass contains chlorophyll pigments which reflect green light and absorb other wavelengths. A purple leaf has pigments that absorb lighter colours and mask the colour of chlorophyll.

Chlorophyll and other photosynthetic pigments

There are several types of chlorophyll, all containing a ring structure called a **porphyrin ring** with magnesium at the centre linked to a long hydrocarbon chain (figure 2). **Chlorophyll *a***, a type of chlorophyll common to all plants and the only one found in cyanobacteria, absorbs light mainly in the red–orange and blue–violet parts of the visible light spectrum. Chlorophyll *a* is the most important pigment in plants because it is the only one that takes a direct part in photosynthesis. Instead of chlorophyll *a*, some photosynthetic bacteria have another type of chlorophyll called **bacteriochlorophyll**, which contains manganese instead of magnesium.

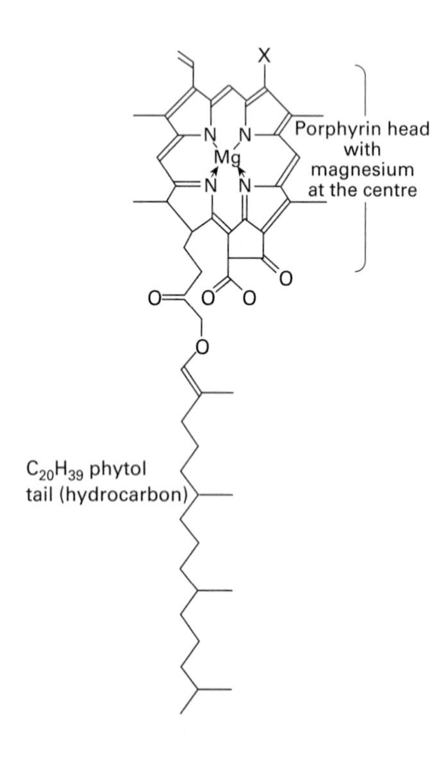

Figure 2 A chlorophyll molecule.

Apart from chlorophyll *a*, there are other photosynthetic pigments in chloroplasts called **accessory pigments**. Like chlorophyll, these act as **light-harvesting** pigments, but they do not participate directly in the light-dependent stage. Instead, they broaden the range of light wavelengths that a plant can use. The pigments absorb light of other wavelengths outside the range of chlorophyll *a* and convey the energy to chlorophyll *a* which then uses it in photosynthesis. Interestingly, none of the accessory pigments absorbs well in the green–yellow area of the spectrum, around 500–550 nm. Accessory pigments include chlorophylls other than chlorophyll *a*, carotenoids (orange, yellow, or brown pigments, which include xanthophylls and carotenes), and phaeophytin (a grey pigment).

The wavelengths of light that a particular photosynthetic pigment absorbs can be displayed in a graph called an **absorption spectrum**. The wavelengths of light that bring about photosynthesis in a particular plant can be displayed in an **action spectrum**. The absorption spectrum of the combined photosynthetic pigments in a plant coincides very closely with the action spectrum of photosynthesis for that plant (figure 3), supporting the notion that these pigments harvest light for photosynthesis.

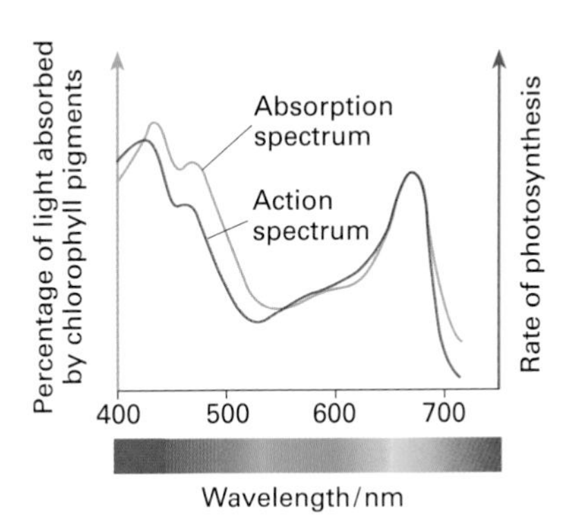

Figure 3 *Comparison of the action spectrum for photosynthesis and the absorption spectrum for plant pigments.*

How chlorophyll harnesses energy from sunlight

When sunlight is absorbed by a photosynthetic pigment, the light behaves as a series of particles called **photons**. Each photon of light contains a fixed amount of energy called a **quantum**. The size of the quantum varies with the wavelength of light: the shorter the wavelength, the larger the quantum. For example, a photon of violet light contains almost twice as much energy as a photon of red light.

When a chlorophyll molecule absorbs a photon of light, one of the pigment's electrons gains energy: its energy level is raised from its lowest most stable level (the **ground state**) to a higher level called an **excited state**. A molecule with an electron in the excited state is very unstable. In a pure solution of chlorophyll, the excess energy is usually released quickly as heat and light, and the electron reverts to its ground state almost immediately. The chlorophyll produces a reddish afterglow called a fluorescence. However, in a chloroplast the excited electron is passed on to another molecule called a **primary acceptor molecule**. This molecule is reduced (it gains an electron) as the chlorophyll is oxidised (it loses an electron: see spread 3.5).

Reduction of the primary acceptor molecule is the first stage of the light-dependent reactions. These are explained in spread 5.3.

The arrangement of pigments in the chloroplast

The photosynthetic pigment molecules are clustered in the thylakoid membranes. Each cluster is called an **antenna complex** (figure 4). The molecules in the complex are arranged so as to channel light energy to just one molecule of chlorophyll *a*. This is the only molecule that actually donates an electron to the primary acceptor molecule. Chlorophyll *a* and the primary acceptor molecule make up what is called the **reaction centre**. The reaction centre and all the other light-gathering molecules combine to form a **photosystem**, the light-harvesting unit in chloroplasts.

There are two types of photosystems: photosystem I and photosystem II. In **photosystem I**, the reaction centre is called **P700** because its chlorophyll *a* has a maximum absorption at a wavelength of 700 nm (red light). **Photosystem II** has a reaction centre called **P680** because its chlorophyll *a* has a maximum absorption at 680 nm (a more orangey red). These two photosystems usually act together in the light-dependent stage of photosynthesis.

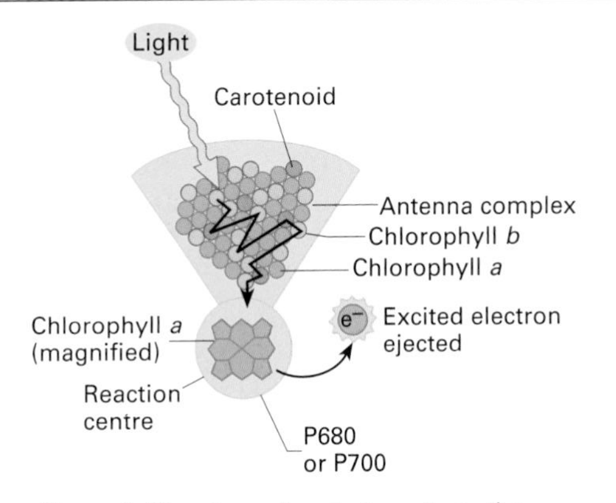

Figure 4 *Structure of a photosystem: the light-harvesting unit of the chloroplast. Light energy is trapped by pigments in the antenna complex and channelled to the chlorophyll* a *at the reaction centre.*

Quick check

1 Which colours of the spectrum does chlorophyll *a* absorb the most?

2 In the context of photosynthesis, explain the difference between an action spectrum and an absorption spectrum.

3 Which area of the spectrum is *not* absorbed well by any accessory pigments?

Food for thought

Accessory pigments absorb solar energy and pass the electrons they emit to chlorophyll *a* in the reaction centre of a photosystem (figure 4). Suggest other functions these accessory pigments might have.

5.3

The light-dependent stage

OBJECTIVES

By the end of this spread you should be able to:

- discuss what happens to chlorophyll *a* when it absorbs light energy
- explain how ATP is generated during photophosphorylation
- explain how water is broken down by photolysis.

Fact of life

ATP is the universal energy carrier in living cells. Its production during photosynthesis and respiration is a vital feature of life on Earth. In 1961 Peter Mitchell, a British biochemist, presented what is termed the **chemiosmotic hypothesis** of ATP production, for which he received a Nobel prize in 1978.

The way in which ATP is produced during photosynthesis is quite complicated (figure 1). It involves the active transport of protons from the stroma across a thylakoid membrane into the thylakoid compartment. This so-called proton pump is driven by the energy released from excited electrons as they pass down an electron transport system. Protons accumulate in the thylakoid compartment, forming a concentration gradient. They can diffuse back across the membrane into the stroma, but only through channels called **chemiosmotic channels.** These channels are formed by an enzyme complex called ATP synthase (an ATPase). As protons pass through the channels, the enzyme complex converts ADP and inorganic phosphate into ATP.

Outline of the light-dependent stage

The reactions of the light-dependent stage of photosynthesis provide energy and hydrogen. (Spread 5.4 will describe how the light-independent stage uses these to make sugars from carbon dioxide.) The light-dependent stage also produces oxygen as a waste product.

The light-dependent reactions occur in the thylakoid membranes of a chloroplast's grana (spread 5.1, figure 2). Light absorbed by chlorophyll *a* provides energy to convert ADP and inorganic phosphate (P_i) to ATP, a process called **photophosphorylation**. Light energy trapped by chlorophyll *a* is also used to split water molecules and release electrons; this process is known as **photolysis**.

Photophosphorylation and the electron transport system

Photosynthesis is initiated by the **photoactivation** of chlorophyll *a*. Light energy absorbed by the chlorophyll causes the energy level of one of the electrons in chlorophyll to be raised: the electron goes from its ground state to an excited state. In its excited or activated state, the electron leaves the chlorophyll molecule and passes to another molecule.

During the light-dependent stage, electrons in their excited state are passed along a chain of electron carrier molecules called an **electron transport system**. Most of the electron carriers in the series are proteins. Each molecule along the chain has a stronger affinity for electrons than the previous molecule; this keeps electrons moving down the chain. A redox reaction takes place during the transfer of an electron from one carrier molecule to the next. One carrier molecule donates an electron (is oxidised), while the other molecule accepts it (is reduced). $NADP^+$ is the final electron acceptor in the electron transport system. ($NADP^+$ is a dinucleotide molecule that accepts electrons and protons to form NADPH, and it is similar to NAD^+ which is described in spread 6.2.)

As electrons pass down the system in a series of redox reactions, they release energy which is used to make ATP (see Fact of life). This method of ATP production is called **photophosphorylation** because light energy is used to add a phosphate group to ADP.

Photolysis

Photolysis is the splitting of water molecules into oxygen and hydrogen by light. Like photophosphorylation, it depends on the photoactivation of chlorophyll *a*. The absorption of light causes each chlorophyll molecule to lose an electron and become positively charged. In this oxidised state, the chlorophyll becomes 'electron hungry': it needs to replace the lost electron before it can react again. By an incompletely understood process

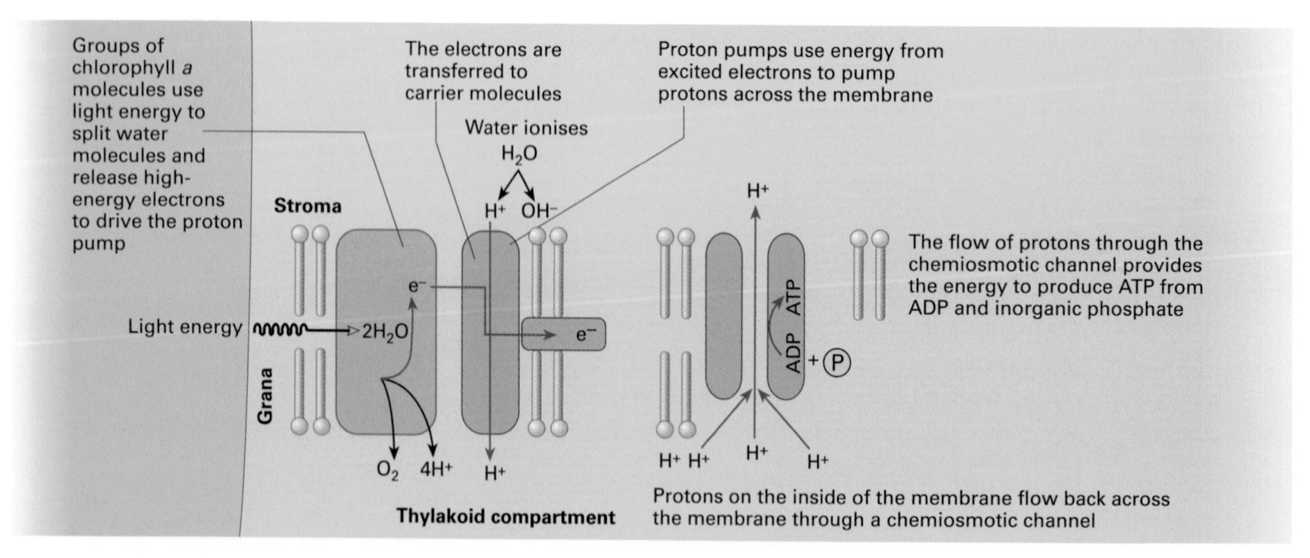

Figure 1 *The production of ATP during photosynthesis is explained by the theory of chemiosmosis.*

involving a complex of molecules, the electrons lost from chlorophyll are replaced by electrons extracted from water. It is this extraction that causes water molecules to split and produce protons and oxygen:

$$2H_2O \longrightarrow 4H^+ + 4e^- + O_2$$

The protons (H^+) are used to reduce $NADP^+$ and the oxygen is given off as a waste gas or used in respiration.

The Z scheme

Photophosphorylation and photolysis do not take place as isolated events in chloroplasts. They are intimately linked in a series of reactions which involve both photosystems (P700 and P680). These events are often summarised in a diagram called the **Z scheme** (figure 2). The *y*-axis of the diagram represents the energy level of the electrons. The Z-shaped pathway of the electrons reflects the fact that the energy level of electrons is twice boosted by the light absorbed by the photosystems and then lowered as the electrons pass down the electron transport system. $NADP^+$ is the final acceptor of the electrons. The electrons and protons reduce $NADP^+$ to NADPH. NADPH stores the electrons and protons (which together make hydrogen) until they can be transferred to carbon dioxide in the light-independent stage (spread 5.4).

__Figure 2__ The Z scheme: a diagrammatic representation of electron flow during the light-dependent stage of photosynthesis. It shows the change in energy level of the electrons in cyclic photophosphorylation and in non-cyclic photophosphorylation. Cyclic photophosphorylation is the synthesis of ATP coupled to the cyclic passage of electrons along a series of electron transport molecules to and from photosystem I (P700). During the process, water is broken down and $NADP^+$ is reduced to NADPH. Non-cyclic photophosphorylation involves the production of ATP by the non-cyclic flow of electrons from photosystem II to photosystem I.

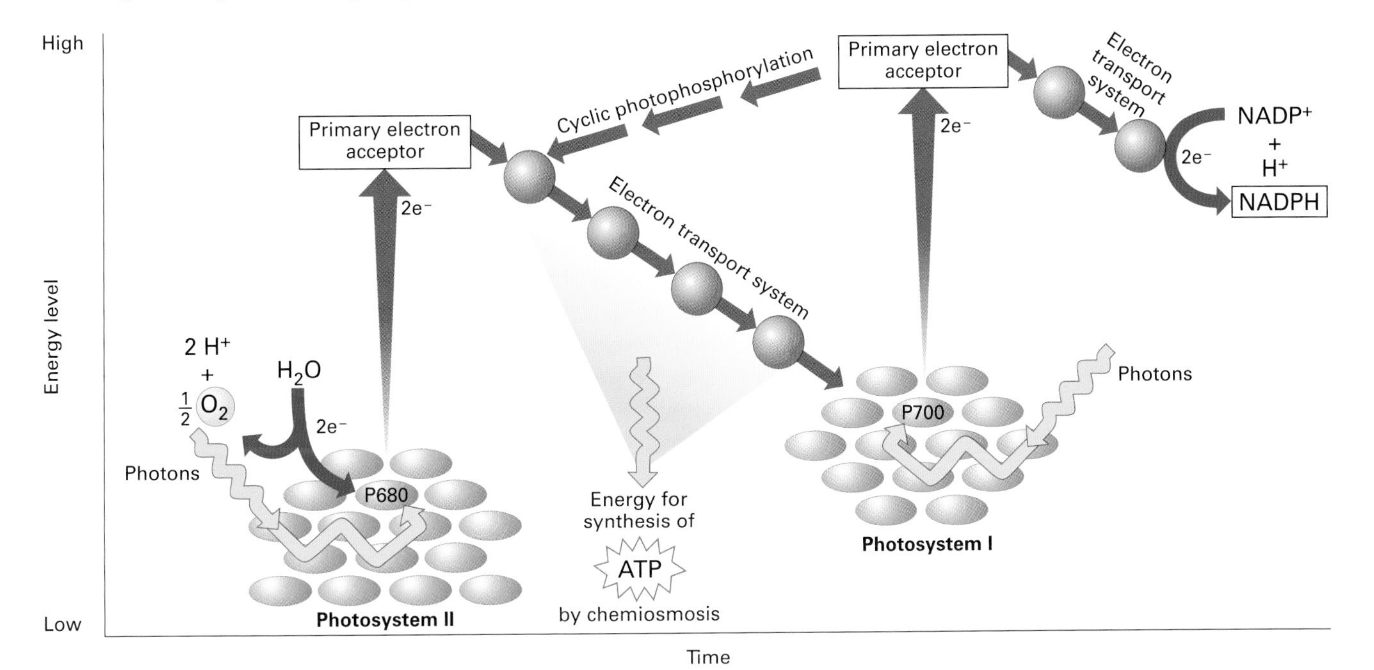

Every molecule of NADPH formed in the light-dependent stage requires two electrons from photosystem I. These electrons are replaced by those from photosystem II. This in turn regains its electrons from water.

Photophosphorylation involving the two photosystems is called **non-cyclic photophosphorylation**. An additional type, called **cyclic photophosphorylation**, occurs in green plants when there is sufficient ADP and inorganic phosphate available. Cyclic photophosphorylation involves only photosystem I. The energy from the excited electron is used to make ATP; it is not used to reduce $NADP^+$.

In summary, the light-dependent stage of photosynthesis uses energy from sunlight to make ATP, and splits water to release protons and electrons which reduce $NADP^+$ to NADPH. The ATP and NADPH are used in the next stage to synthesise carbohydrates from carbon dioxide (spread 5.4).

Quick check

1. Name the three main products of the light-dependent stage of photosynthesis.
2. What is the name of the final acceptor in the electron transport chain?
3. Which ions are produced when water molecules dissociate?

Food for thought

Suggest why pure chlorophyll fluoresces when exposed to light.

5.4 The light-independent stage

OBJECTIVES

By the end of this spread you should be able to:

- describe the Calvin cycle
- explain how carbon dioxide is fixed in a plant
- explain the role of ATP and NADPH in the light-independent stage.

Fact of life

Ribulose bisphosphate carboxylase (rubisco), an enzyme essential for photosynthesis, is very large. It is made of 16 polypeptide chains, eight of which have active sites that combine with carbon dioxide. Large amounts of rubisco occur in photosynthetic bacteria and in the stroma of chloroplasts of all green plants. Consequently, it is thought to be the most abundant enzyme in the world.

Calvin's 'lollipop apparatus'

During the light-independent stage of photosynthesis, carbon dioxide is **fixed**: it is assimilated into a plant by being converted into biochemical products in the plant. The overall reaction is the reduction of carbon dioxide to glucose. The light-independent stage takes place in the stroma of the chloroplast. Enzymes convert carbon dioxide to glucose in a series of reactions discovered by Melvin Calvin using his 'lollipop apparatus' (figure 1). The reactions use ATP and NADPH, the products of the light-dependent stage.

In Calvin's experiments, the 'lollipop' was a thin transparent vessel containing a suspension of *Chlorella*, a single-celled alga. Carbon dioxide labelled with radioactive carbon (^{14}C) was added to the suspension and light was shone on it. A sample from the suspension was collected in a tube containing boiling methanol at timed intervals from a few seconds to several minutes. The boiling methanol killed the algae quickly, inactivated their enzymes, and stopped the reactions instantaneously. The radioactive compounds that had been formed were extracted from the samples using paper chromatography and analysed by autoradiography (see appendix for a description of chromatography and autoradiography).

The Calvin cycle

In this way, Calvin was able to trace the biochemical path taken by carbon in photosynthesis. His investigations revealed that carbon dioxide is converted to glucose in a cyclical chain of reactions which is now known as the **Calvin cycle** (figure 2).

Carbon dioxide fixation involves a five-carbon compound called ribulose bisphosphate (RuBP) and an enzyme called ribulose bisphosphate carboxylase (rubisco). With the catalytic help of rubisco, RuBP combines with carbon dioxide to form an unstable six-carbon compound which splits immediately into two molecules of a three-carbon compound, glycerate 3-phosphate (GP):

$$\underset{5C}{\text{RuBP}} + \underset{1C}{\text{carbon dioxide}} \xrightarrow{\text{rubisco}} \underset{2 \times 3C}{2\ \text{GP}}$$

(where C represents a carbon atom)

GP is then reduced to triose phosphate (TP, phosphorylated three-carbon sugar) called glyceraldehyde 3-phosphate (GALP). This reduction requires hydrogen from NADPH and energy from ATP:

$$\text{ATP} \rightarrow \text{ADP} + P_i$$
$$\text{GP} \longrightarrow \text{TP (a triose phosphate, GALP)}$$
$$\text{NADPH} \rightarrow \text{NADP}^+$$

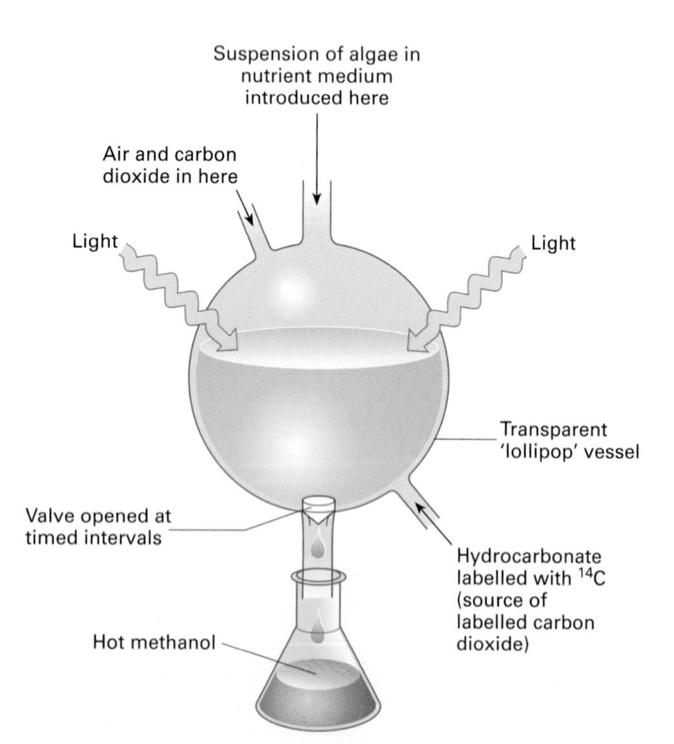

Figure 1 *The 'lollipop' apparatus used by Melvin Calvin to study the light-independent reactions of photosynthesis. Algae were cultured in the thin transparent vessel (the 'lollipop') into which hydrogencarbonate labelled with radioactive* ^{14}C *was injected. The metabolic pathway of the light-independent stage was traced by analysing the radioactive compounds formed at different times after injection of the hydrogencarbonate.*

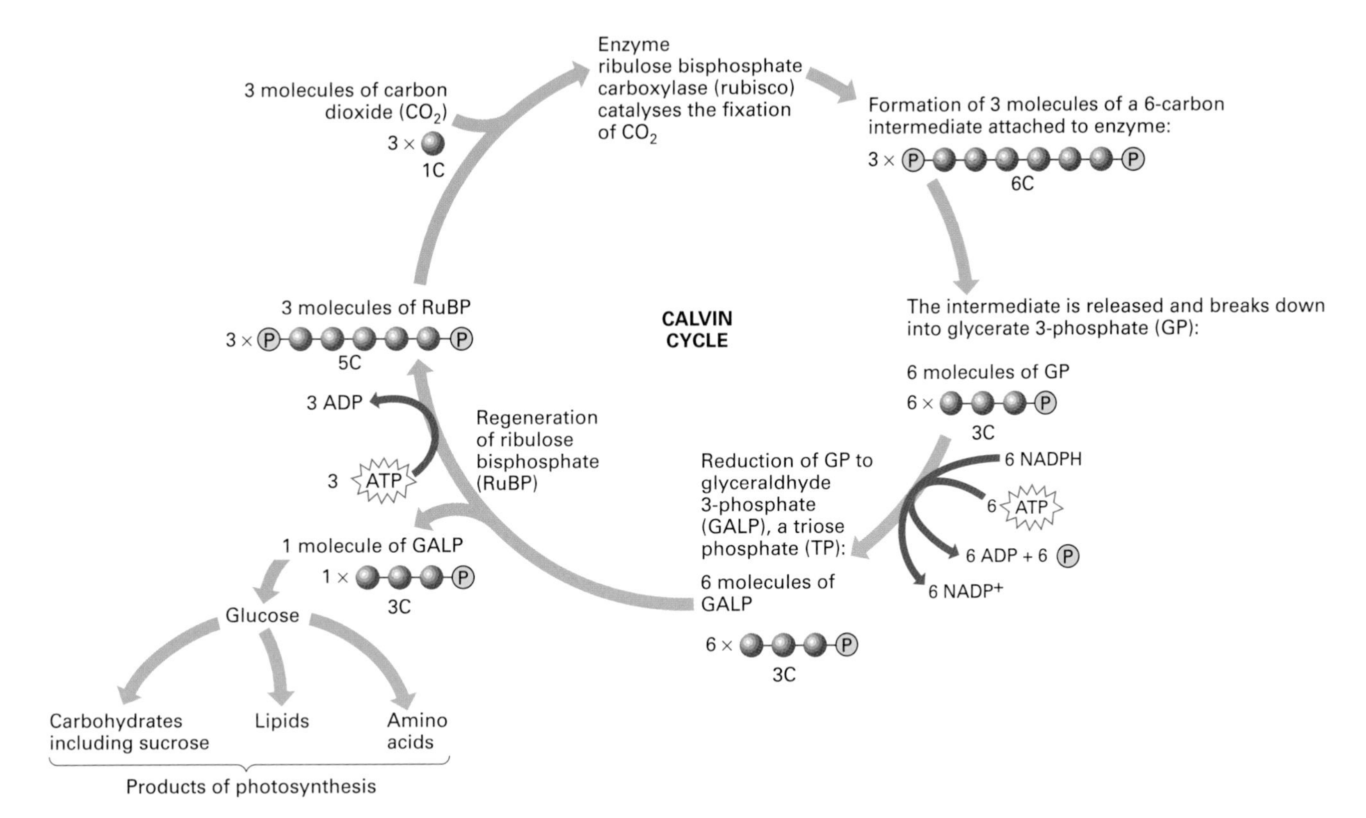

__Figure 2__ An outline of the Calvin cycle. The NADPH and ATP used in these reactions come from the light-dependent stage of photosynthesis. During the Calvin cycle, carbon dioxide is fixed and reduced to sugars from which other products can be synthesised; RuBP used for fixation is regenerated.

About one-sixth of the total GALP is then used to make glucose (which can be converted into other carboyhydrates, for example, sucrose, starch, and cellulose), amino acids, fatty acids, or glycerol. The remaining five-sixths is converted back to RuBP. This regeneration of RuBP requires phosphate from ATP.

Why is it called the light-independent stage?

The light-independent stage used to be known as the 'dark stage' because the reactions can take place in the dark if sufficient ATP and NADPH are available (for example, under experimental conditions using isolated chloroplasts). However, this name is now not thought appropriate because, under most circumstances, sufficient ATP and NADPH are available only when ATP and NADPH are being generated by the light-dependent reactions in sunlight. So although the reactions of the light-independent stage do not depend directly on light, they usually take place simultaneously with the reactions of the light-dependent stage.

Photorespiration

In bright sunlight, where there are also low carbon dioxide levels and high oxygen levels, oxygen and carbon dioxide compete for the same active site on rubisco. Consequently, in addition to catalysing the reaction between RuBP and carbon dioxide, rubisco also catalyses the oxidation of RuBP in a process called **photorespiration**. When oxidised, RuBP forms only one molecule of GP and some by-products, including glycolate (a two-carbon compound). GP enters the Calvin cycle, but the by-products are oxidised in a complex series of reactions which salvages some of the carbon dioxide. However, photorespiration is regarded as wasteful because, unlike normal respiration or photosynthesis, the by-products become oxidised without producing any ATP or NADPH. It has been shown that under some conditions, photorespiration may reduce the photosynthesising efficiency of plants by as much as 50 per cent. Some plants that live in conditions which favour photorespiration, such as sugar cane, have evolved a biochemical mechanism which overcomes this problem (see spread 5.6).

Quick check

1 Name three molecules involved in carbon dioxide fixation.

2 What is the first relatively stable product of carbon dioxide fixation?

3 What is the role of NADPH in the Calvin cycle?

4 Describe two functions of ATP in the Calvin cycle.

Food for thought

Suggest why photorespiration is not usually a problem for temperate plants.

5.5

OBJECTIVES

By the end of this spread you should be able to:

- describe the main factors affecting the rate of photosynthesis
- explain the meaning of the compensation point
- define the law of limiting factors.

Fact of life

The atmosphere contains less than 0.04 per cent carbon dioxide, yet each year plants make more than 200 billion tonnes of carbon compounds from this meagre supply of carbon dioxide.

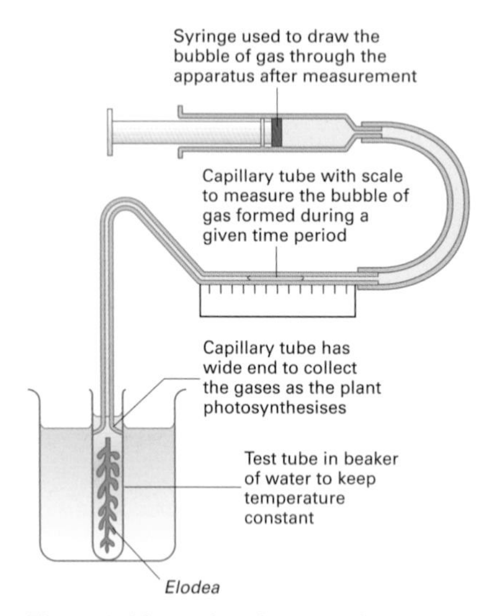

Figure 1 *Measuring the rate of photosynthesis in Canadian pondweed.*

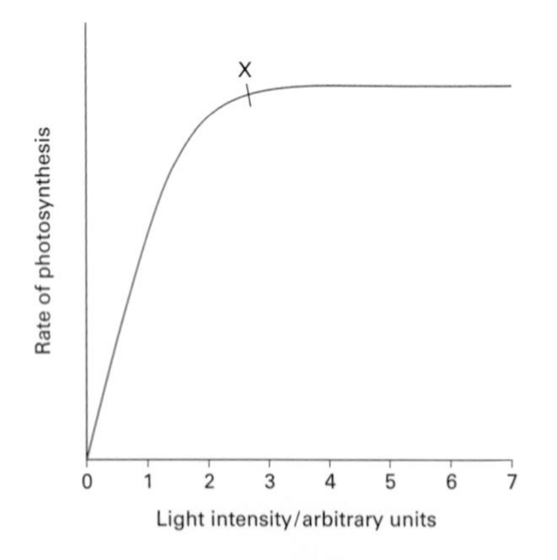

Figure 2 *The effect of light intensity on the rate of photosynthesis.*

Factors affecting the rate of photosynthesis

Measuring how fast a plant is photosynthesising

The rate of photosynthesis can be measured as the volume of carbon dioxide taken in by a plant per unit time, or as the amount of carbohydrate produced per unit time. In laboratory investigations, the rate is commonly estimated as the volume of oxygen released per unit time, which is more easily measured. A plant such as *Elodea*, Canadian pondweed (figure 1), is used in experiments such as this. However, this method does not give an accurate measure of photosynthesis. Some of the oxygen generated by photosynthesis is used by the plant for respiration. Respiration goes on all the time, even when photosynthesis is at its height. So using oxygen liberation as a measure of photosynthesis gives an underestimate of the true rate. We are actually measuring the rate of photosynthesis above a point called the **compensation point**, defined as: *the point at which the rate of photosynthesis in a plant is in exact balance with the rate of respiration, so there is no net exchange of carbon dioxide or oxygen*. The compensation point is usually related to a particular light intensity or carbon dioxide level.

Factors that affect the rate of photosynthesis

Photosynthesis is affected by many factors, both external (in the environment) and internal (inside the plant). External factors include light intensity, the wavelength of light, carbon dioxide levels, temperature, wind velocity, and water and mineral supplies. Internal factors include type and concentration of photosynthetic pigments, enzyme and water content, and leaf structure and position.

The effect of many of these factors is difficult to determine quantitatively because they interact, and they also affect other processes in the plant. For example, the importance of water to photosynthesis cannot be demonstrated easily. Simply depriving a plant of water kills it, but the cause of death may not be connected with photosynthesis. The importance of water can be demonstrated using water labelled with a heavy isotope of oxygen, ^{18}O, and tracing the isotope using an instrument called a mass spectrometer which can measure the masses of atoms. One batch of *Chlorella* (green algae) is placed in water in which the oxygen atoms have been replaced by the heavy isotope. Then a second batch of *Chlorella* in unlabelled water is given a supply of carbon dioxide labelled with ^{18}O. Only the first batch of *Chlorella* gives off oxygen labelled with ^{18}O, confirming that the oxygen formed in photosynthesis comes only from water, not from carbon dioxide.

Light intensity, carbon dioxide concentration, and temperature are three external factors that are relatively easy to manipulate. Consequently they have been the focus of many investigations on photosynthesis.

Light intensity

The rate of photosynthesis is directly proportional to light intensity. Figure 2 shows the response of a typical whole plant to changes in light intensity. The graph levels off at point X because the photosynthetic pigments have become saturated with light, and some other factor (for example, availability of carbon dioxide, or amount of chlorophyll) stops the reaction from going any faster. Very high light intensities may actually damage some plants, reducing their ability to photosynthesise.

The **light compensation point** (the light intensity at which the rate of photosynthesis is exactly balanced by the rate of respiration) varies for different plants. Two major groups have been identified: **sun plants** and **shade plants** (figure 3). Sun plants include most temperate trees, such as oak. They photosynthesise best at high light intensities. Shade plants include those of the shrub layer, such as ferns. Their light compensation point is relatively low, but they cannot photosynthesise very efficiently at high light intensities. Consequently sun plants outcompete shade plants at high light intensities.

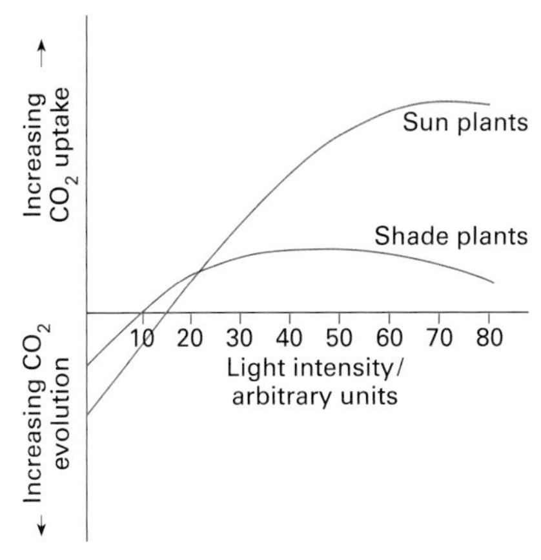

Figure 3 *The response of sun plants and shade plants to changes in light intensity.*

Carbon dioxide levels

The average carbon dioxide content of the atmosphere is about 0.04 per cent. As long as there is no other factor limiting photosynthesis, an increase in carbon dioxide concentration up to 0.5 per cent usually results in an increase in the rate of photosynthesis. However, concentrations above 0.1 per cent can damage leaves. Therefore the optimum concentration of carbon dioxide is probably just under 0.1 per cent (figure 4). In dense, warm, and well-lit vegetation, low levels of carbon dioxide often limit the rate of photosynthesis. Growers of greenhouse tomatoes recognise this and provide a carbon dioxide enriched atmosphere for their plants.

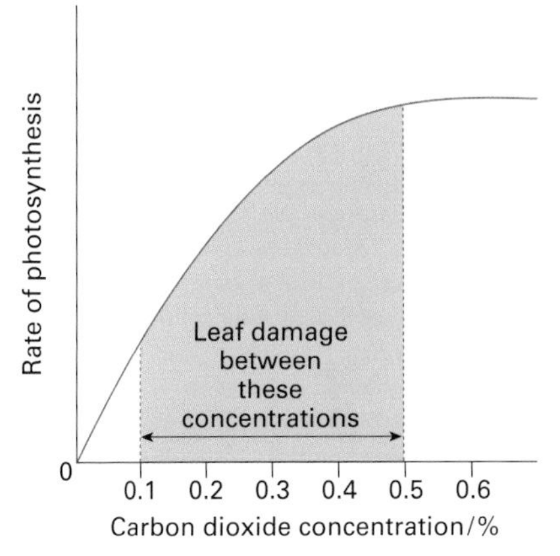

Figure 4 *The effect of carbon dioxide level on the rate of photosynthesis.*

Temperature

Changes in temperature have little effect on the reactions of the light-dependent stage because these are driven by light, not heat. However, the reactions of the Calvin cycle are catalysed by enzymes which, like all enzymes, are sensitive to temperature. The effect of temperature on these reactions is similar to its effects on other enzymes. The optimum temperature varies for each species, but many temperate plants have an optimum temperature between 25 °C and 30 °C.

Law of limiting factors

So far we have looked at the effects of isolated factors. However, under natural conditions plants are subjected to many factors simultaneously. Figure 5 shows the simultaneous effect of carbon dioxide level, temperature, and light intensity. Above a light intensity of about 3 units, the factor limiting the rate of photosynthesis in **A** and **B** is carbon dioxide concentration (indicated by the significantly higher rate of photosynthesis in **C**); temperature is limiting the rate of photosynthesis in **C**; and a factor other than light intensity is beginning to limit the rate of photosynthesis in **D**. As you can see, the rate is determined by the least optimal factor: the factor that is furthest from its optimum value. This factor is called the **limiting factor** and the figure illustrates **the law of limiting factors**. This law states that: *when a physiological process depends on more than one essential factor being favourable, its rate at any given moment is limited by the factor at its least favourable value and by that factor alone.* When other factors are kept constant, an improvement in the value of the limiting factor leads to an increase in the rate of the process. Conversely, when the rate of the process does not increase in response to an improvement in an important factor, some other factor is limiting the process. For a process to go at its maximum rate, *all* factors must be at their optimum level.

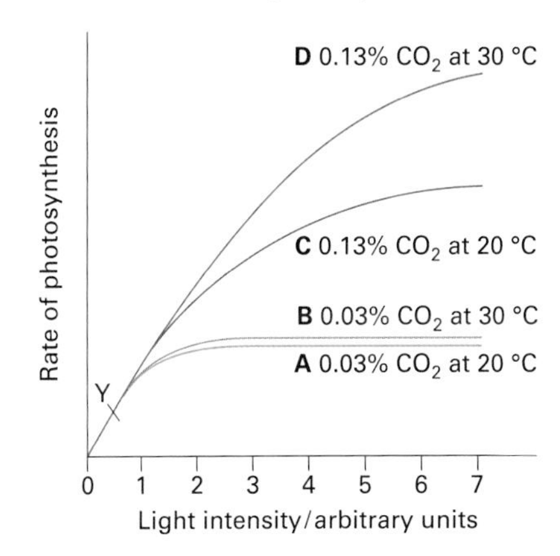

Figure 5 *The interaction of different factors and their effect on the rate of photosynthesis.*

QUICK CHECK

1 Why is it difficult to demonstrate the importance of water to photosynthesis?

2 How does the light compensation point of a shade plant differ from that of a sun plant?

3 In figure 5, which factor is limiting the rate of photosynthesis at point Y?

Food for thought

Greenhouses (glasshouses) can be used to grow plants in controlled environmental conditions. Carbon dioxide can be enhanced by burning fossil fuels, and temperature and light can be controlled by automatic heating, ventilation, and lighting. Tomatoes and wheat grown in glasshouses have far greater yields than those grown outdoors. Suggest why tomatoes are grown commercially in greenhouses whereas wheat is not.

5.6

OBJECTIVES

By the end of this spread you should be able to:

- distinguish between C_3 and C_4 plants
- explain the advantages and disadvantages of crassulacean acid metabolism (CAM)
- give examples of C_3, C_4, and CAM plants.

Fact of life

Plants are not very efficient at harnessing energy from the sunlight they receive. Under the most carefully controlled laboratory conditions plants can reach 25% efficiency but on cloudy days the natural photosynthetic efficiency of most individual plants is about 0.1%. The annual winter evening primrose, *Oenothera claviformis*, has the highest natural photosynthetic efficiency at 8%, closely followed by sugar cane at 7%.

Photosynthesis in different climates

Green plants thrive in environments ranging from hot and dry equatorial regions to freezing-cold polar regions. Their success depends on their adaptability. To survive and breed, each plant has had to evolve specific adaptations to cope with the demands of its particular environment. These adaptations include ways of fixing carbon dioxide.

C_3 plants: fixing directly into the Calvin cycle

C_3 plants fix carbon dioxide directly into the Calvin cycle as the three-carbon compound glycerate 3-phosphate (GP). Common and widely distributed, they include some of our most important crop plants such as wheat, soya beans, and rice. C_3 plants function efficiently in temperate conditions. However, they suffer two major disadvantages in hot, dry environments.

First, to obtain sufficient carbon dioxide, C_3 plants must open their stomata (small pores in their leaves). Unfortunately, when stomata are open, they not only allow carbon dioxide to enter the plant, but also allow water to escape. So in hot dry conditions C_3 plants have to either cease photosynthesising or run the risk of wilting and dying.

The second disadvantage relates to the ability of ribulose bisphosphate carboxylase (rubisco) to combine with oxygen. Rubisco is the enzyme that catalyses carbon dioxide fixation. On a hot, sunny day, carbon dioxide concentrations around photosynthesising cells decrease, because a large proportion of the carbon dioxide is being used up in photosynthesis. In these conditions, rubisco combines with oxygen rather than carbon dioxide in a process called **photorespiration**. The process results in the loss of fixed carbon dioxide from the plant, reducing photosynthetic efficiency and plant growth. Unlike photosynthesis, photorespiration does not produce sugar molecules; and unlike respiration, it yields no ATP. As much as half of the carbon dioxide fixed in the Calvin cycle may be released by photorespiration. Therefore, in hot, arid conditions, or in conditions where carbon dioxide levels are low, C_3 plants do not grow well.

C_4 plants: the Hatch–Slack pathway

C_4 plants, such as sugar cane and maize, have evolved a special metabolic adaptation which reduces photorespiration. They do not use ribulose bisphosphate (RuBP) to fix carbon dioxide directly into the Calvin cycle. Instead, they use phosphoenolpyruvate (PEP) to fix carbon dioxide as a four-carbon compound, **oxaloacetate**. The reaction is catalysed by phosphoenolpyruvate carboxylase (PEP carboxylase). This enzyme cannot combine with oxygen. Consequently C_4 plants can continue to fix carbon dioxide even when its concentration is very low.

The leaves of C_4 plants are specially adapted to carry out this initial fixation. A ring of large closely packed cells called the **bundle sheath** surrounds the leaf veins. Surrounding the bundle sheath is a smaller ring of mesophyll cells (figure 1). This distinctive arrangement is called **Kranz anatomy** and can be used to identify C_4 plants ('Kranz' means crown or halo and refers to the two distinctive rings). The initial fixation of carbon dioxide into oxaloacetate takes place in the small ring of mesophyll cells. Then the oxaloacetate is converted to malate, another four-carbon compound. Malate is transported into the bundle sheath cells where it releases carbon dioxide. Once released, the carbon dioxide is reassimilated by RuBP and enters the Calvin cycle in the same way as described for C_3 plants. The metabolic pathway that transports carbon dioxide into the bundle sheath cells is called the **Hatch–Slack pathway** (figure 1b). As a result of this pathway, the concentration of carbon dioxide in the bundle sheath cells is 20 to 120 times higher than normal.

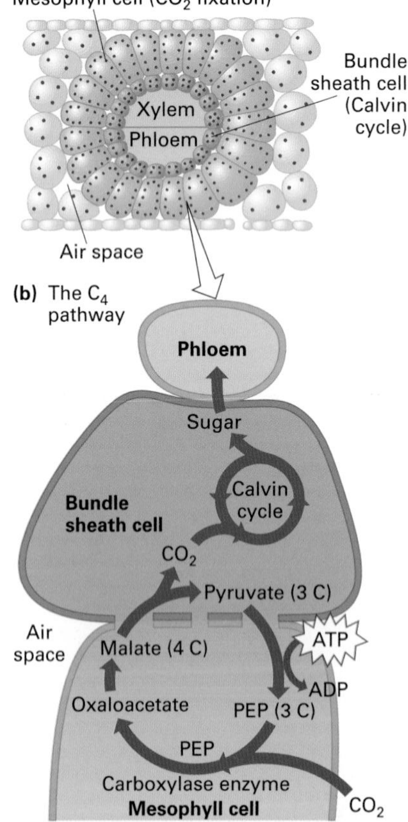

Figure 1 The metabolism of C_4 plants. (a) Kranz anatomy of C_4 plants showing the two rings of cells formed by the mesophyll cells and bundle sheath cells. (b) The metabolic pathway by which carbon dioxide is initially fixed in a mesophyll cell and then passed on to the bundle sheath cell in which the carbon dioxide is converted to sugar via the Calvin cycle.

C_4 plants have two main advantages in hot, dry environments. First, because PEP carboxylase has a high affinity for carbon dioxide and does not combine with oxygen, C_4 plants can continue to photosynthesise even when their stomata are closed for long periods. This reduces water loss and photorespiration. C_4 plants need only about half as much water as C_3 plants for photosynthesis. Secondly, because high carbon dioxide concentrations can be maintained in the bundle sheath cells, C_4 plants can increase their photosynthetic efficiency (figure 2a).

These adaptations enable C_4 plants to outcompete C_3 plants in hot and very sunny conditions, but not in temperate conditions. Fewer than 0.5 per cent of plant species are C_4 plants, yet they include economically important crops such as maize, sugar cane, and millet.

CAM plants: photosynthesising in the desert

Although C_4 plants are well adapted to occasional periods of drought, they cannot cope well with desert conditions. A group of plants including cacti and pineapples have evolved a third type of carbon dioxide fixation which enables them to survive in very dry climates. These plants are called **CAM plants**. CAM is an abbreviation for crassulacean acid metabolism, a type of metabolism first observed in the family of plants called Crassulaceae (which includes the stonecrops, fleshy-leaved plants that will grow on rocks and walls).

CAM plants conserve water by only opening their stomata at night. During the night, they fix carbon dioxide into oxaloacetate which is converted into malate. This acts as a carbon dioxide storage compound. During the day, malate releases carbon dioxide into the Calvin cycle. This allows photosynthesis to take place on hot, dry, sunny days, even though the stomata are closed (figure 2b).

CAM plants conserve water very well and are able to survive in extremely dry conditions, but CAM plants do not photosynthesise very efficiently. Most are very slow growing. Where there is plenty of water, CAM plants cannot compete well with C_3 or C_4 plants.

CAM plants and C_4 plants have a similar metabolism: carbon dioxide is first fixed into a four-carbon intermediate before it enters the Calvin cycle. However, in CAM plants the initial fixation and the Calvin cycle occur at separate times, whereas in C_4 plants the initial fixation and the Calvin cycle are separated structurally but both occur during the day. C_4 plants live in hot, very sunny, and periodically dry environments but where lack of water is rarely a limiting factor (partly because the plants can reduce water losses due to their C_4 metabolism) and annual rainfall is high (typically, tropical rainforest-type climates); CAM plants are desert plants that live in areas of very low annual rainfall. Note that C_3, C_4, and CAM plants all eventually use the Calvin cycle to make glucose from carbon dioxide.

Figure 2 *C_4 and CAM plants.* ***(a)*** *Sugar cane is a C_4 plant: photorespiration is avoided by a mechanism for transporting carbon dioxide into the bundle sheath cells.* ***(b)*** *Pineapple is a CAM plant: it fixes carbon dioxide at night so stomata can stay closed during the day.*

Quick check

1 a Name two C_3 plants.
  b Why is sugar cane called a C_4 plant?

2 When do CAM plants fix carbon dioxide?

3 Suggest which type of carbon dioxide fixation (C_3, C_4, or CAM) is most efficient:
  a if it is hot and sunny and the carbon dioxide level is low, but water is freely available
  b in hot, dry and sunny climates where stomata are closed
  c in bright light and temperate regions where there is an ample water supply.

Food for thought

If C_4 plants have a greater photosynthetic efficiency than C_3 plants, suggest why all plants do not have C_4 metabolism.

Summary

Photosynthesis is the process by which light energy is used to build up complex organic substances from carbon dioxide and water. It has two main stages: the **light-dependent stage** and the **light-independent stage**. In plants, the light-dependent stage takes place on the **grana** of **chloroplasts**. Light is absorbed by a number of **photosynthetic pigments**, but only **chlorophyll *a*** participates directly in the reactions. These result in the breakdown of water (**photolysis**), and the formation of ATP, NADPH, and oxygen. The light-independent stage takes place in the **stroma**. Carbon dioxide is fixed by **ribulose bisphosphate** (RuBP) to **glycerate phosphate** (GP) which is then reduced, using hydrogen from NADPH and energy from ATP formed during the light-dependent stage, to **glyceraldehyde phosphate** (GALP). GALP is a **triose phosphate**. It is used to make glucose and to regenerate RuBP. The series of reactions by which carbon dioxide is fixed and RuBP regenerated is called the **Calvin cycle**.

The rate of photosynthesis is affected by a number of factors including **light intensity**, **carbon dioxide levels**, and **temperature**. The rate is determined by the factor furthest from its optimum level (the **limiting factor**). The light intensity at which the rate of photosynthesis equals the rate of respiration is called the light **compensation point**.
Different species of plants are adapted to photosynthesise in different climates. Temperate plants usually fix carbon dioxide directly into GP, a 3-carbon compound, and are called **C_3 plants**. Many tropical plants living in wet conditions fix carbon dioxide as a 4-carbon compound and are called **C_4 plants**. Their special metabolism allows them to avoid photorespiration which reduces photosynthetic efficiency. Some desert plants are called CAM plants because they have a special metabolism, called **crassulacean acid metabolism**, which allows them to fix carbon dioxide at night and keep their stomata closed during the day to conserve water.

Practice exam questions

1 The figure below shows bubbles of gas coming from the cut stem of an illuminated inverted shoot of the aquatic plant *Elodea*.

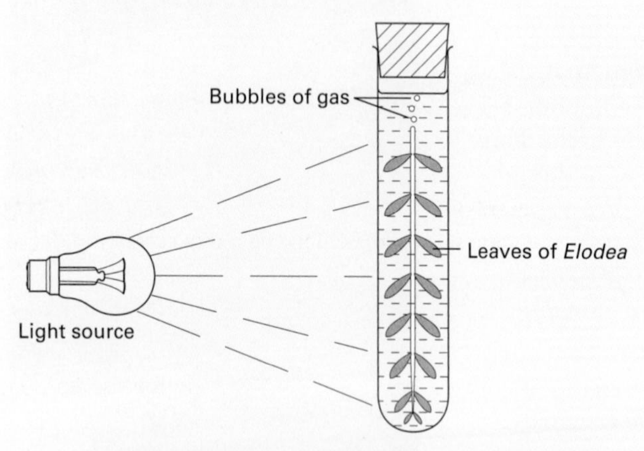

a State **three** environmental factors, in addition to light intensity, that could affect the rate of bubble production in *Elodea*. [3]

b Explain briefly how the oxygen in the gas bubbles is produced. [3]

c The intensity of the light falling on *Elodea* is proportional to $\frac{1}{d^2}$ (where d is the distance between the lamp and the inverted shoot).

If a lamp, at 10 cm from the shoot, produces a reading of 20 bubbles per minute, how many bubbles per minute would you expect to count if the lamp is placed at 5 cm from the shoot? Show your working. [2]

d The figure above right represents several different components of a thylakoid.

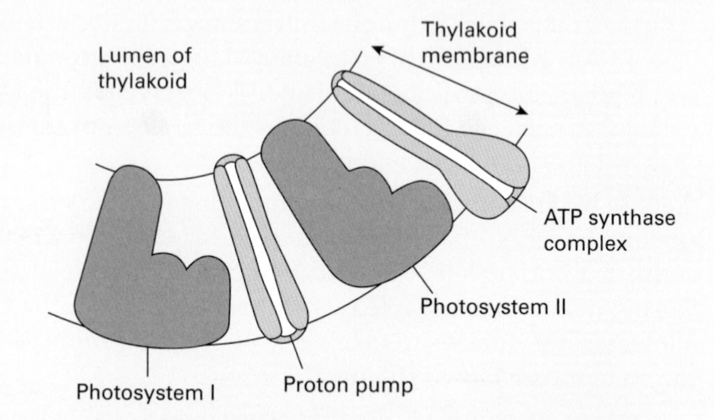

State precisely in which component(s) of a thylakoid each of the following processes takes place:

i photolysis

ii absorption of light of wavelength 680 nm

iii cyclic photophosphorylation

iv non-cyclic photophosphorylation

v the reduction of NADP+ to NADPH. [5]

e What is a proton pump and what is its role in photosynthesis? [3]

[Total 16 marks]

2 Calvin investigated the pathway by which carbon dioxide is converted to organic compounds during photosynthesis. He used the apparatus shown in the diagram. The apparatus contained cells of a unicellular alga. While the apparatus was in a dark room, Calvin supplied the algal cells with carbon dioxide containing radioactive carbon. The contents of the apparatus were thoroughly mixed, then a light was switched on. At five-second intervals he released a few of the cells into hot alcohol, which killed the cells very quickly.

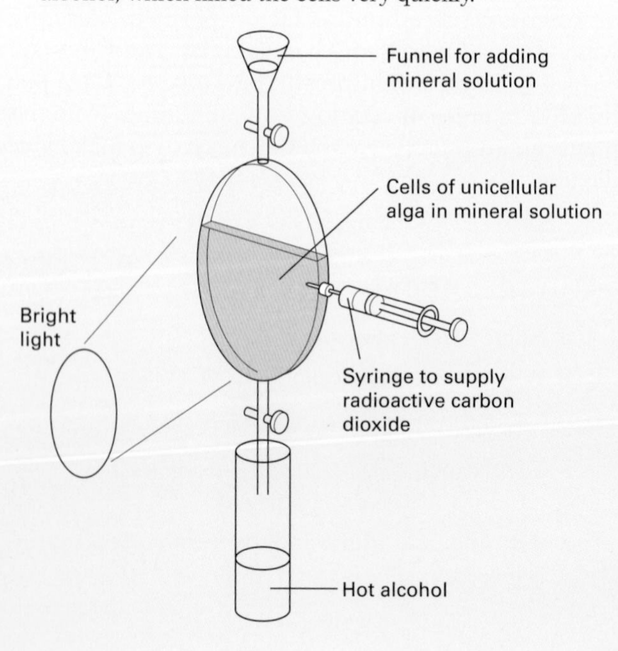

a Suggest

i why the part of the apparatus containing the algal cells is thin;

ii how it would be possible to prevent the light source from heating the algal cells;

iii why it was necessary for the algal cells to be killed very quickly. [3]

Calvin homogenised the killed algal cells and carried out two-way paper chromatography. This technique involves running the chromatogram with one solvent, then turning the paper through 90° and running it with a different solvent.
By using this technique, Calvin was able to investigate the organic compounds produced during photosynthesis. The diagram shows the chromatograms he obtained. The spots are those containing radioactive compounds.

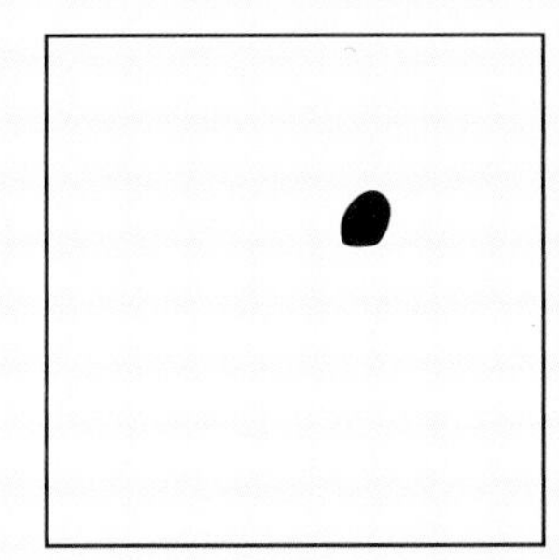
After 5 seconds

After 15 seconds

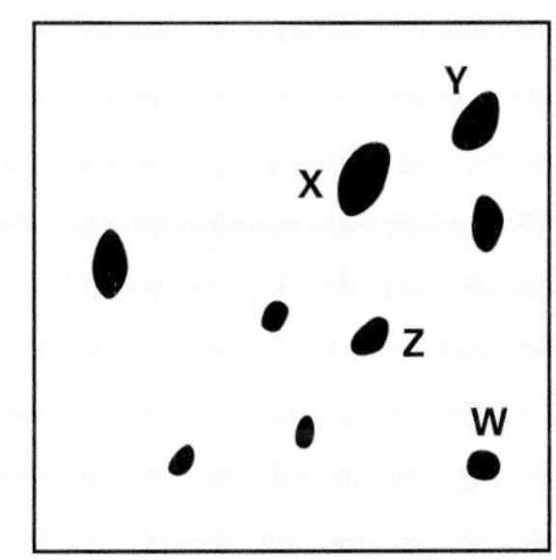

After 60 seconds

b Suggest how the compound present in each spot on the chromatogram might be identified. [1]

c **i** Name the compound first formed in the pathway in which carbon dioxide is converted to an organic compound. [1]

ii Which of the spots, **W**, **X**, **Y** or **Z** contained this compound? [1]

iii Explain your answer to **c ii**. [1]

[Total 7 marks]

3 **a** Describe how the structure and distribution of chloroplasts ensures the efficient trapping of light by leaves. [10]

b The synthesis of ATP is driven by a flow of protons. Describe where and how the required proton gradient is produced in the chloroplast. [10]

[Total 20 marks]

4 The graph below shows how the wavelength of light affects the rate of photosynthesis.

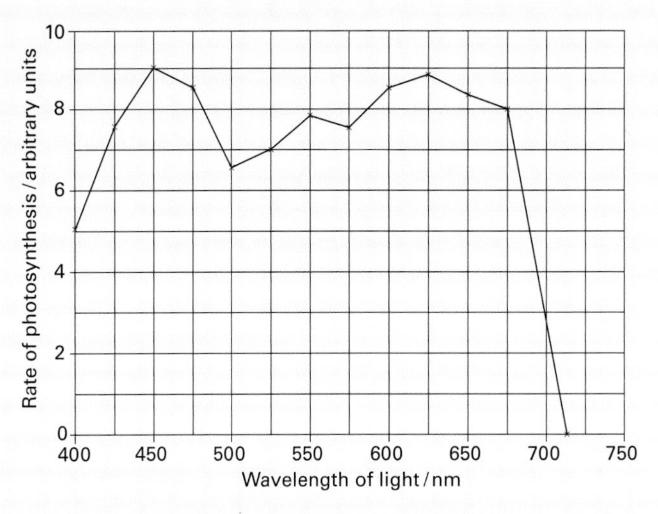

a **i** What name is given to the relationship between wavelength of light and the rate of photosynthesis, as shown by this graph? [1]

ii From the graph, state the optimum wavelength of light for photosynthesis. [1]

iii Explain the effect on the rate of photosynthesis of varying the wavelength of light from 550 to 700 nm. [2]

b **i** Name the stage of photosynthesis that produces oxygen. [1]

ii State where in a chloroplast this stage would occur. [1]

[Total 6 marks]

5 **a** The diagram summarises the light-independent reaction of photosynthesis.

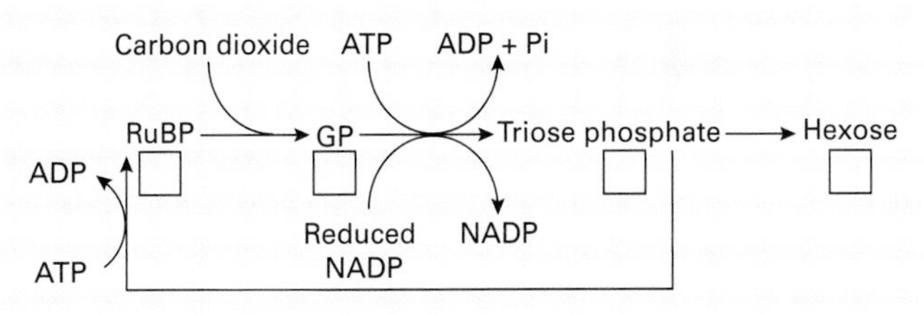

i Copy and complete the **four** boxes to show the number of carbon atoms in a molecule of each substance. [1]

ii Where in the chloroplast does the light-independent reaction take place? [1]

iii Explain why the amount of GP increases after a photosynthesising plant has been in darkness for a short time. [2]

b Describe the role of water in the light-dependent reaction of photosynthesis. [2]

[Total 6 marks]

Respiration

Cellular respiration: an overview

6.1

OBJECTIVES

By the end of this spread you should be able to:

- distinguish between the biological meaning of the word respiration and its everyday meaning
- explain the significance of cellular respiration
- distinguish between anaerobic and aerobic respiration.

Fact of life

Cyanobacteria (blue-green bacteria) related to those shown in figure 1 had a profound effect on the evolution of life on Earth. They were the first organisms to be able to photosynthesise. The oxygen they liberated into the atmosphere was deadly to anaerobes. Aerobic respiration is thought to have evolved as a means of coping with the oxygen-rich atmosphere created by primitive cyanobacteria.

What is respiration?

Most people use the term 'respiration' to mean breathing. Biologists use the term in quite a different way. To biologists, **respiration** is defined as: *the process by which the energy in food molecules is made available for an organism to do biological work.* To avoid confusion, this process is sometimes called **cellular respiration**. Cellular respiration occurs in every living cell: it is the only way a cell can obtain energy in a usable form.

All living things need usable energy to move, grow, reproduce, and repair damaged structures. They obtain this energy from food. The food is either made by the organism (for example, by photosynthesis) or it is obtained from other organisms (for example, by predation).

Producing ATP

As we saw in spread 2.11, **adenosine triphosphate** (ATP) is the form of chemical energy that is used to fuel energy-consuming (endergonic) biological activities. The fact that all living organisms use ATP is one of the unifying features of life. Without ATP an organism will quickly die, yet all organisms store little ATP. At any one moment you, for example, have only enough ATP to keep yourself alive for one second. Therefore you have to produce ATP continuously to survive. You do this by cellular respiration.

During cellular respiration the energy in food molecules is transferred to molecules of ATP. There are two main types of cellular respiration: anaerobic respiration and aerobic respiration.

Anaerobic respiration

Hundreds of millions of years ago, at the very dawn of life, conditions on Earth were quite different from now. The atmosphere was very thin and consisted of swirling clouds of water vapour and methane, ammonia, and carbon dioxide gases. It contained very little, if any, oxygen. We would not survive in such an atmosphere without special breathing apparatus. So the first organisms transferred energy from food to ATP without using oxygen. This process is called **anaerobic respiration** (respiration without oxygen).

Many organisms respire anaerobically today. Habitats such as stagnant ponds and deep sediments have no oxygen. Bacteria living in these habitats are **obligate anaerobes**; that is, they can only respire anaerobically.

Some organisms such as yeast and *Escherichia coli* (a bacterium that lives in human intestines) are **facultative anaerobes**. They use anaerobic respiration only when there is a lack of oxygen.

Anaerobes generate ATP by only partly breaking down food molecules, rather than breaking them down fully to carbon dioxide. Glucose is broken down into pyruvate in a process called **glycolysis**. The pyruvate may then follow one of two main anaerobic metabolic pathways. Plants and fungi convert pyruvate to ethanol and carbon dioxide in a process called **alcoholic fermentation**. Animals convert pyruvate to lactate in a process called **lactate fermentation**. These two types of anaerobic respiration may be summarised by the following equations:

In yeast: $C_6H_{12}O_6$ (glucose) $\longrightarrow$ $2C_2H_5OH$ (ethanol) + $2CO_2$ (carbon dioxide)

In animals: $C_6H_{12}O_6$ (glucose) $\longrightarrow$ $2C_3H_6O_3$ (lactate)

Glycolysis and fermentation are discussed more fully in spread 6.2.

The evolution of aerobic organisms

The first organisms on Earth are thought to have obtained their food by engulfing carbon compounds that were present in the primordial seas. Other forms of life gradually evolved which could use the energy from sunlight to make their own food. These early photosynthetic organisms had a profound effect on all life to come, for by photosynthesising they pumped masses of oxygen into the atmosphere. It is difficult to comprehend that the oxygen in today's atmosphere is the accumulated waste of millions of years of photosynthesis.

To anaerobic organisms, oxygen is as toxic as cyanide. To survive in an oxygen-rich environment, anaerobes had to evolve a means of disposing of this (to them) deadly gas. **Aerobic respiration** (respiration with oxygen) is thought to have evolved as a means of getting rid of oxygen.

Aerobic respiration

Aerobic respiration is often summarised by the equation:

$$\underset{\text{glucose}}{C_6H_{12}O_6} + \underset{\text{oxygen}}{6O_2} \longrightarrow \underset{\text{carbon dioxide}}{6CO_2} + \underset{\text{water}}{6H_2O}$$

The equation implies that glucose is broken down in a single step, but this is a gross simplification. If glucose is oxidised in this way outside the cell, it burns and creates a flame. Cells have to oxidise glucose in a much more controlled manner so that the heat generated does not destroy them.

Aerobic respiration is a complex process in which food molecules are broken down in a series of steps. During the breakdown, energy is released which is used to synthesise ATP from ADP and inorganic phosphate. The heat produced during respiration does not usually cause organisms to burn, because its release is spread over many biochemical reactions. However, fires associated with bacteria in decomposing haystacks remind us of the dramatic consequences of not allowing the heat from respiration to escape.

There are three major stages in aerobic respiration: glycolysis, the Krebs cycle, and the electron transport system (a chain of electron acceptors similar to that in the light-dependent stages of photosynthesis). These are the topics of the next three spreads.

Figure 1 *Cyanobacteria blanketing a pond. The silvery beads are bubbles of gas, mainly oxygen, the waste product of their photosynthesis.*

Figure 2 *Polar bears scavenging on refuse at a dump in Arctic Canada.*

Quick check

1 Define cellular respiration in terms of the type of energy it generates.

2 Why does ATP have to be regenerated continuously in the body?

3 **a** In what type of environment do anaerobes thrive?

b Distinguish between an obligate anaerobe and a facultative anaerobe.

Food for thought

Humans have influenced the environmental conditions even in the remote reaches of the Arctic (figure 2). Suggest why you would expect to find anaerobic organisms living in the mud under the refuse.

6.2

Glycolysis and fermentation

OBJECTIVES

By the end of this spread you should be able to:

- describe the main stages of glycolysis
- explain how ATP is generated during glycolysis
- distinguish between alcoholic fermentation and lactate fermentation.

Fact of life

Anaerobic respiration occurs in aerobes, such as ourselves, when there is insufficient oxygen to meet the energy demands of cells. An adult human usually produces about 150 g of lactate per day. But during heavy exercise, over 30 g of lactate can be produced in minutes. The production of large amounts of lactate is associated with muscle fatigue, often experienced as cramps.

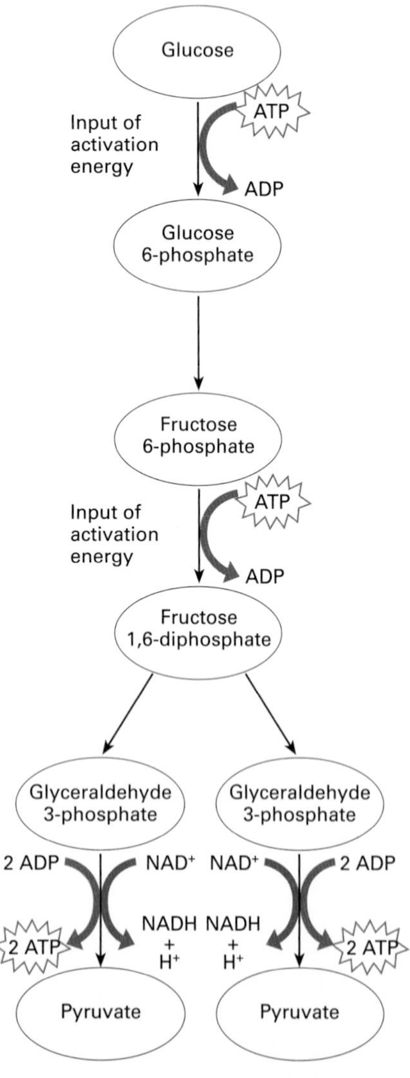

Figure 1 An outline of the main stages of glycolysis, in which one molecule of glucose is broken down to two molecules of pyruvate.

Glycolysis: sugar splitting

Glycolysis ('sugar splitting') is the first stage of both aerobic and anaerobic respiration. It takes place with or without oxygen, in the cytosol (the fluid part of the cytoplasm).

As figure 1 shows, glycolysis starts with one molecule of glucose and ends with two molecules of pyruvate (pyruvic acid), a three-carbon compound. During the process, a nucleotide called nicotinamide adenine dinucleotide (NAD, see box) is reduced (it has hydrogen added to it) and there is a net production of two molecules of ATP. Although this is a relatively small energy yield, glycolysis takes little time to complete and does not require oxygen. It can therefore provide energy immediately.

Activation: adding phosphate

Glycolysis takes place as a series of reactions. Before glycolysis can even start, glucose has to be given some **activation energy** (spread 3.1). This is provided by two molecules of ATP which break down to ADP and inorganic phosphate. The glucose is energised by the addition of phosphate to it, first to form glucose 6-phosphate, then to form fructose 1, 6-diphosphate. So far the cell has invested two molecules of ATP, with no return.

Fructose 1, 6-diphosphate then splits to form two molecules of glyceraldehyde 3-phosphate (GALP).

The reduction of NAD^+ and the synthesis of ATP

Next, each GALP molecule is converted by a series of reactions to pyruvate. The conversion involves the removal of hydrogen (oxidation). This reaction is coupled with the production of ATP and the reduction of NAD^+. Each redox reaction generates enough energy to synthesise two molecules of ATP. Therefore, for each molecule of glucose broken down during glycolysis, four molecules of ATP are produced. However, as two molecules of ATP are used as activation energy, glycolysis produces a net gain of two ATP molecules for each glucose molecule. Overall, glycolysis produces

- 4 molecules of ATP (but 2 are used as activation energy)
- 2 molecules of pyruvate
- 2 molecules of NADH.

If oxygen is available, the **link reaction** converts pyruvate to acetylcoenzyme A which enters the Krebs cycle (spread 6.3), and NADH can feed electrons into the electron transport system (spread 6.4).

Regenerating NAD^+ with or without oxygen

The NADH produced during glycolysis must be oxidised back to NAD^+ to be used again in glycolysis. If this did not happen, NAD^+ would soon run out and the production of ATP would halt.

If oxygen is available, NAD^+ is regenerated when NADH releases hydrogen into the mitochondria. The hydrogen enters the electron transport system and generates about six more molecules of ATP (see spread 6.4). However, if oxygen is unavailable, NAD^+ is regenerated by fermentation, a process in which no more ATP molecules are generated.

Fermentation

In the absence of oxygen, pyruvate can follow one of two metabolic pathways: alcoholic fermentation or lactate (lactic acid) fermentation.

Alcoholic fermentation occurs in plants and yeasts which are respiring anaerobically. Pyruvate is first converted to ethanal by **decarboxylation** (removal of carbon dioxide). The carbon dioxide is released as a gas. This feature of fermentation is taken advantage of in bread-making. Ethanal is then reduced to ethanol using hydrogen from

NADH (figure 2a). Ethanol is another product of yeast fermentation that has been exploited for thousands of years (see spread 17.10).

Lactate fermentation occurs in animals that are respiring anaerobically. The pyruvate from glycolysis is converted in a single step to lactate (lactic acid). The reaction is catalysed by lactate dehydrogenase and the process requires hydrogen from NADH (figure 2b).

Glycolysis transfers only a small proportion of the energy in glucose to ATP. Anaerobic organisms have to satisfy their energy needs by glycolysis, despite its low level of ATP production. However, aerobic organisms can use oxygen to release a much greater proportion of the energy in glucose to make many more ATP molecules via the Krebs cycle, the next stage of aerobic respiration, and ultimately via the electron transport system.

Figure 2 Fermentation: the fate of pyruvate and NADH in anaerobic conditions: (a) in plants and yeast; (b) in animals.

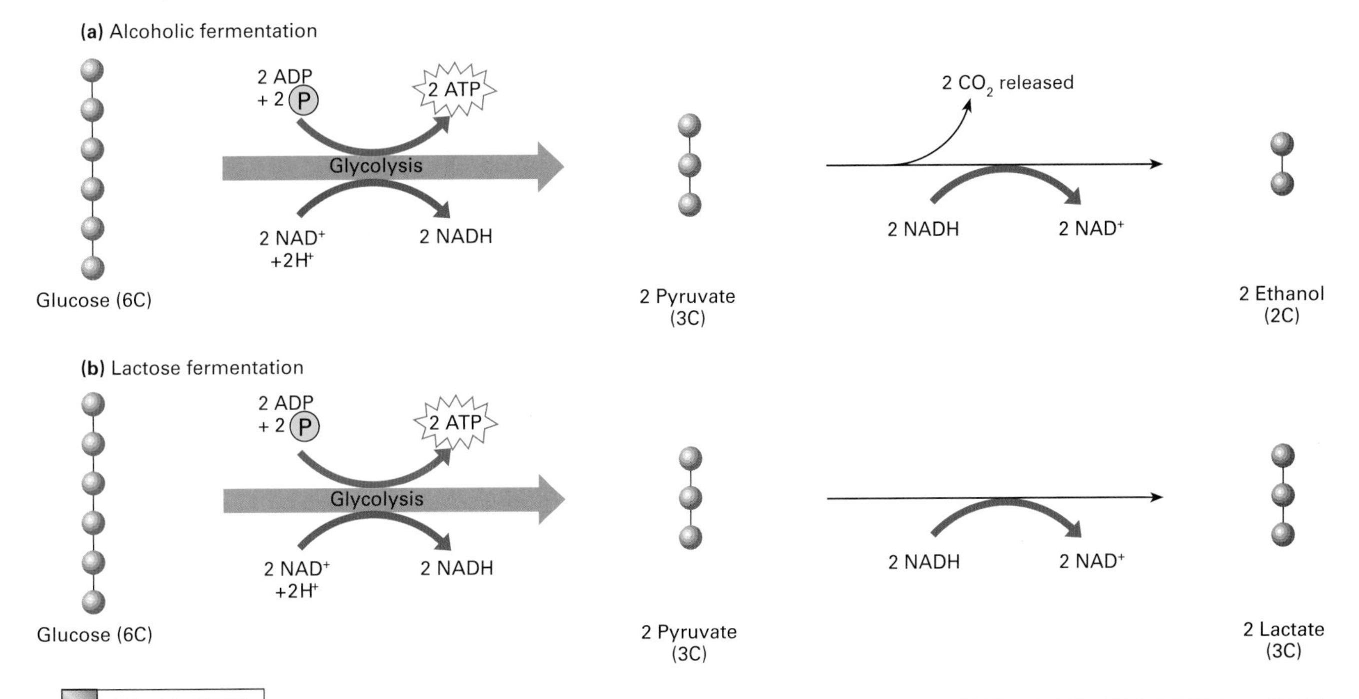

What is NAD?

During glycolysis, NAD (nicotinamide adenine dinucleotide) plays a pivotal role. When its oxidised form (NAD^+) is converted into its reduced form (NADH), two molecules of ATP are produced (figure 3).

The addition of an inorganic phosphate group (P_i) to ADP (or any other molecule) is called **phosphorylation**. In glycolysis, the phosphate group is obtained from a substrate (a molecule other than ATP, ADP, or AMP) and the process is called **substrate-level phosphorylation**.

NAD is an organic molecule formed from two nucleotides. It is synthesised in our bodies from vitamin B_3 (also called niacin, nicotinic acid, or nicotinamide). A dietary deficiency of this vitamin results in a disease called **pellagra**. The provision of energy in the brain is impaired in sufferers of pellagra, resulting in nervous disorders. NAD is regarded as a coenzyme because it assists dehydrogenases, enzymes that catalyse the removal of hydrogen from a substrate.

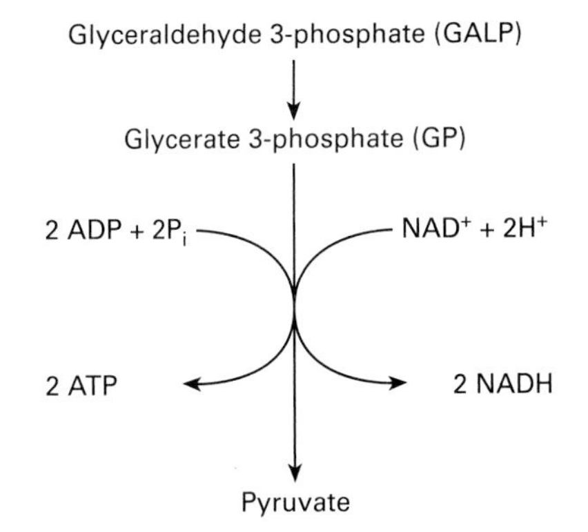

Figure 3 The reduction of NAD^+ is coupled with phosphorylation during glycolysis.

Quick check

1 In which part of the cell does glycolysis take place?

2 How many ATP molecules are produced (gross) during glycolysis of one glucose molecule?

3 In which type of fermentation is carbon dioxide released, alcoholic fermentation or lactate fermentation?

Food for thought

In animals, pyruvate can be regenerated from lactate when oxygen is available. Suggest why plants cannot regenerate pyruvate from ethanol.

6.3

THE KREBS CYCLE

OBJECTIVES

By the end of this spread you should be able to:

- describe the main processes of the Krebs cycle
- define the term oxidative decarboxylation.

Fact of life

Glucose is not the only substrate of aerobic respiration. Substances produced by the metabolism of other carbohydrates, and of proteins and lipids, can also be respired. The different products are fed into the aerobic respiration pathway at different points. Fatty acids, for example, bypass glycolysis and are fed directly into the Krebs cycle. They are converted to acetyl-CoA by a process called **beta oxidation**. Essentially, this involves two-carbon fragments of acetyl-CoA being split off from a fatty acid so that the long hydrocarbon chain is shortened by two carbon atoms at a time. Each acetyl-CoA fragment enters the Krebs cycle in the usual manner.

A fatty acid can generate a great deal of energy. Oxidation of one molecule of stearic acid, for example, can produce 147 ATP molecules. This makes fats a very important metabolic fuel, contributing at least half the normal energy requirements of resting skeletal muscle, heart muscle, liver, and kidneys.

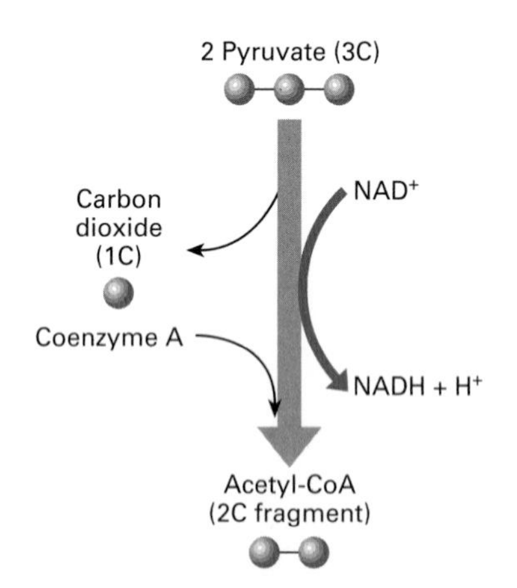

Figure 1 ***The link reaction**: the conversion by oxidative decarboxylation of pyruvate to acetylcoenzyme A.*

The **Krebs cycle** is the central phase of the aerobic respiration of glucose. The cycle was named after Sir Hans Krebs who discovered its main components.

The link reaction

If oxygen is available, a reaction takes place that links glycolysis to the Krebs cycle. The three-carbon pyruvate formed at the end of glycolysis enters the mitochondria by active transport and is converted into **acetylcoenzyme A** (acetyl-CoA, a two-carbon fragment). The conversion involves the removal of carbon dioxide (decarboxylation) and hydrogen (oxidation) from the pyruvate. The process by which this link reaction takes place is called oxidative decarboxylation (figure 1).

In the mitochondria, the two-carbon acetyl component of acetyl-CoA diffuses to the site of the Krebs cycle. The coenzyme A helps the acetyl fragment to enter the cycle, and then is detached and can be recycled (figure 2). Two acetyl fragments enter the Krebs cycle for each molecule of glucose that entered glycolysis.

Oxidative decarboxylation and the reduction of NAD and FAD

The two-carbon acetyl fragment combines with a four-carbon compound, oxaloacetate, to form a six-carbon compound, citrate. Then, by a series of reactions (the main ones of which are shown in figure 2), citrate is progressively broken down to reform oxaloacetate.

The breakdown again involves oxidative decarboxylation. Two molecules of carbon dioxide are produced in the cycle for every acetyl fragment that enters. The carbon dioxide is excreted as a waste gas. Hydrogen is stripped off the organic molecules as they progress around the cycle. The hydrogen is taken up by either NAD or **flavine adenine dinucleotide** (**FAD**), another molecule formed by the combination of two nucleotides.

Synthesising ATP

These redox reactions are exergonic. They release enough energy to synthesise one molecule of ATP for each turn of the cycle. This synthesis of ATP is called **substrate-level phosphorylation** because the addition of a phosphate group to ADP is coupled with the exergonic breakdown of a high-energy substrate molecule.

Each molecule of glucose that enters glycolysis generates two molecules of ATP in the Krebs cycle, because one molecule of glucose gives two molecules of acetyl-CoA. This immediately doubles the ATP production of glycolysis. However, more importantly, the release of hydrogen during the Krebs cycle provides the reducing power to generate many more ATP molecules in the next stage of aerobic respiration, the electron transport system.

The net effects of the Krebs cycle

To summarise: during each revolution of the Krebs cycle, two molecules of carbon dioxide are released, NAD and FAD are reduced, and one molecule of ATP is generated. In addition, oxaloacetate is regenerated so that, if more acetyl-CoA is available, the cycle can start again.

In addition to its respiratory role, the Krebs cycle is also the centre for metabolic interconversions. For example, it is a valuable source of compounds used to manufacture chlorophyll, fatty acids, and amino acids.

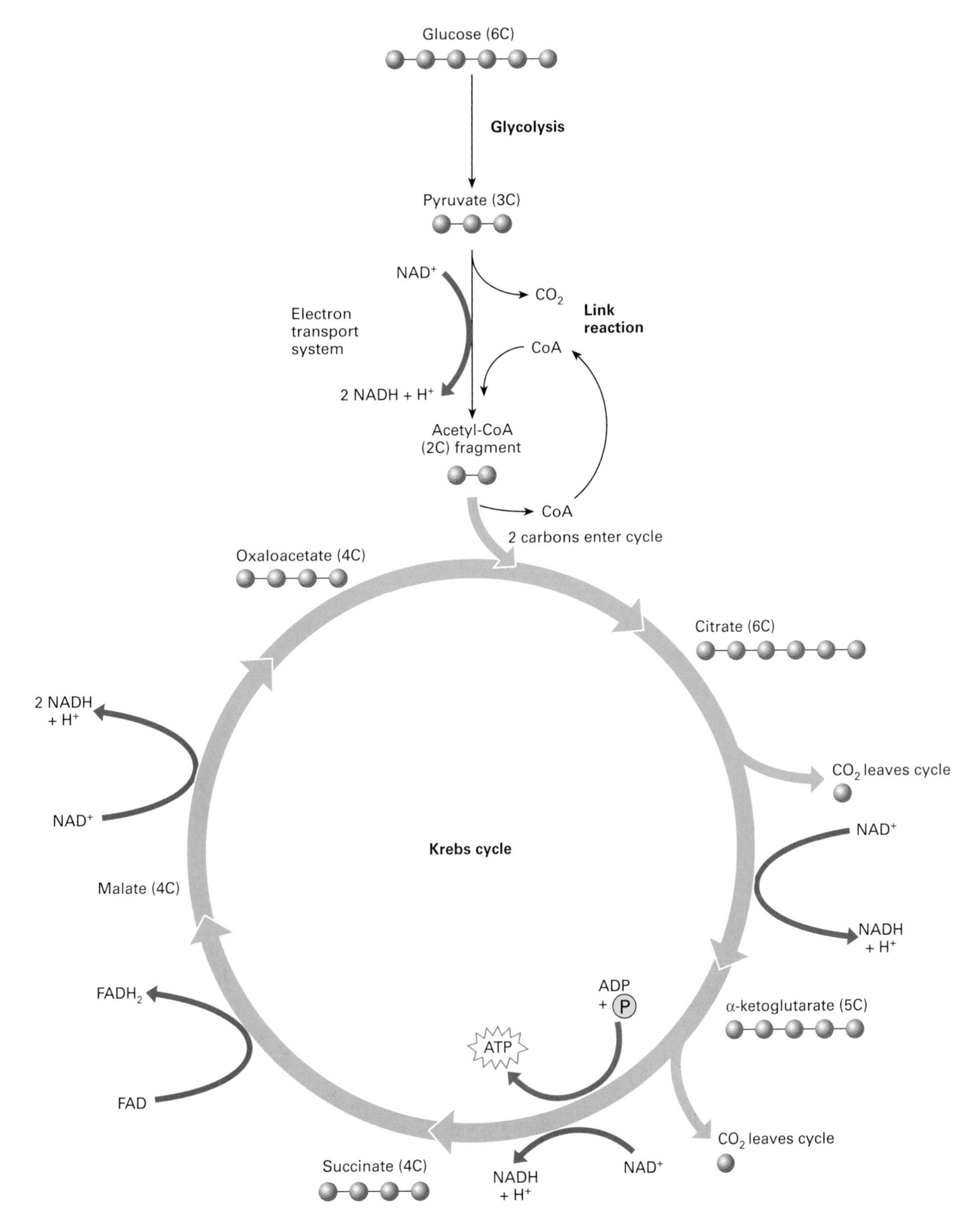

***Figure 2** An outline of the Krebs cycle, in which two molecules of ATP are generated for each glucose molecule that entered glycolysis.*

Quick check

1 a In which part of the cell does the Krebs cycle take place?

b How many ATP molecules are generated by each revolution of the Krebs cycle?

2 The Krebs cycle involves oxidative decarboxylation. What does this mean?

Food for thought

What do you think is the significance of the Krebs cycle being a cyclical process?

6.4

THE ELECTRON TRANSPORT SYSTEM

OBJECTIVES

By the end of this spread you should be able to:

- describe the electron transport system used in aerobic respiration
- explain the meaning of mitochondrial oxidative phosphorylation
- discuss how the structure of a mitochondrion is related to its function.

Fact of life

Flavine adenine dinucleotide (FAD) has a similar function to NAD. Both are involved as coenzymes in the electron transport system, helping the enzyme dehydrogenase to transfer hydrogen from one molecule to another. Whereas NAD is made from niacin (also known as nicotinic acid or vitamin B_3), FAD is made from riboflavin (vitamin B_2). Riboflavin is obtained from a wide variety of foods including yeast, liver, eggs, and milk. It is quickly decomposed by heat, and when exposed to light, it is broken down into a substance called lumiflavin. Milk in bottles left in sunlight therefore loses much of its riboflavin. Riboflavin deficiency causes ariboflavinosis, characterised by cracked skin and eye problems.

The main purpose of the Krebs cycle (spread 6.3) is to feed electrons into the next stage of aerobic respiration. The hydrogen atoms carried by reduced NAD and FAD molecules are the source of these electrons, which are passed along a chain of carrier molecules called the **electron transport system** (**ETS**). This is similar to the electron transport system in the light-dependent stage of photosynthesis (spread 5.3). For each turn of the Krebs cycle, 3 molecules of NADH and 1 molecule of $FADH_2$ are generated. As the Krebs cycle turns twice for each molecule of glucose respired, 6 molecules of NADH and 2 molecules of $FADH_2$ are produced for each molecule of glucose.

A chain of redox reactions

NADH and $FADH_2$ pass on electrons when they donate hydrogen to the next carrier in the system, so that a redox reaction takes place (for example, NADH becomes oxidised by losing hydrogen and its electron, whereas the next carrier becomes reduced by gaining hydrogen and the electron). At one point in the ETS, hydrogen atoms split to form protons and electrons. The ETS concludes with the reduction of oxygen (the final acceptor of electrons and protons) which forms water (figure 1).

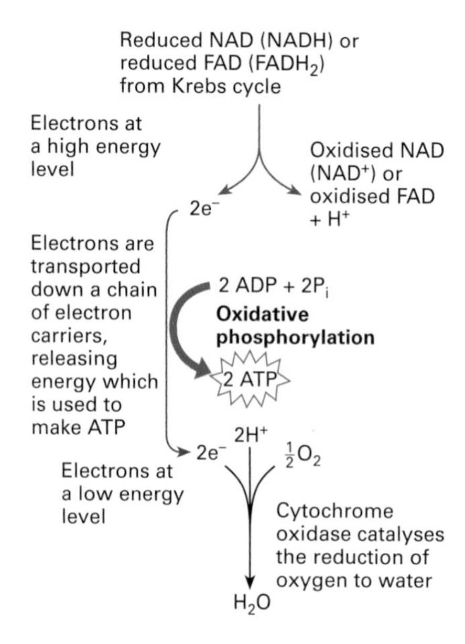

Figure 1 *An outline of the electron transport system (ETS). This system liberates energy which is used to synthesise ATP by oxidative phosphorylation. Note that oxygen is the final electron and proton acceptor. Without oxygen, this type of respiration cannot take place.*

The respiratory ETS includes **cytochromes**, iron-containing pigmented molecules which give cells rich in mitochondria a brownish colour. The cytochromes are embedded in the inner membranes of mitochondria, where they form an electron transport chain within the ETS. Each redox reaction in the ETS releases energy which can be used to synthesise ATP. The oxygen-dependent synthesis of ATP within mitchondria using energy released from redox reactions is called **oxidative phosphorylation** (compare with photophosphorylation in photosynthesis, spread 5.3). ADP is phosphorylated (has a phosphate group added) and the energy comes from the oxidation reactions of the ETS. The box explains how, according to the chemiosmotic theory, ATP is synthesised.

The final redox reaction is catalysed by **cytochrome oxidase**. Some metabolic poisons such as cyanide inhibit the action of this enzyme, with potentially fatal results.

The site of the ETS

The redox reactions happen on the inner membranes of mitochondria (figure 2) in which the electron carriers are embedded. The inner membrane is highly folded into structures called cristae which increase the surface area on which the reactions can take place.

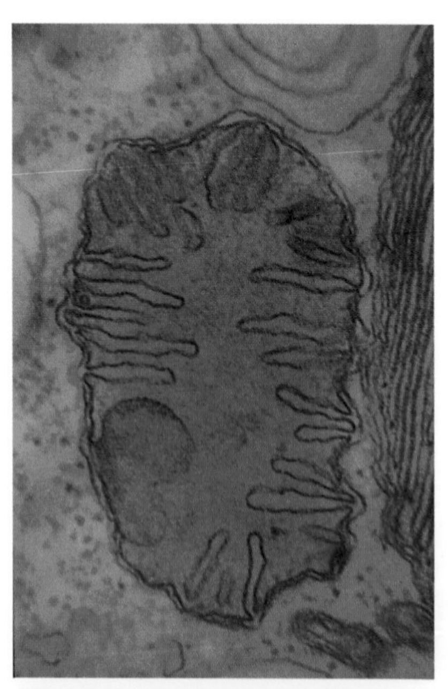

Figure 2 *Transmission electron micrograph of a mitchondrion (x2000). The inner membrane is the site of most ATP production in cells. The cristae increase the surface area of the inner membrane on which are sited the stalked particles, the location of the ATP synthase (an ATPase enzyme).*

Chemiosmotic theory: how ATP is synthesised

On spread 5.3 we saw how the chemiosmotic theory explains how free energy released in the ETS in the chloroplast is converted to chemical energy in ATP. The same theory explains the production of ATP in the mitochondrion. Energy released from the electron transport system is again used to pump protons (H^+), this time from the mitochondrial matrix into a compartment between the inner and outer mitochondrial membranes. The protons accumulate so that a steep concentration gradient develops between the compartment and the matrix. The inner membrane is generally impermeable to protons, which can only diffuse back into the matrix through chemiosmotic channels. These are located in special structures in the inner membrane called **stalked particles**. The energy associated with the movement of protons through the chemiosmotic channels drives the synthesis of ATP. The phosphorylation of ATP is catalysed by the enzyme ATP synthase (an ATPase) in the bulbous ends of stalked particles:

$$ADP + P_i \xrightarrow{\text{ATP synthase}} ATP$$

The ATP synthase in mitochondria has a structure and function almost identical to that in chloroplasts.

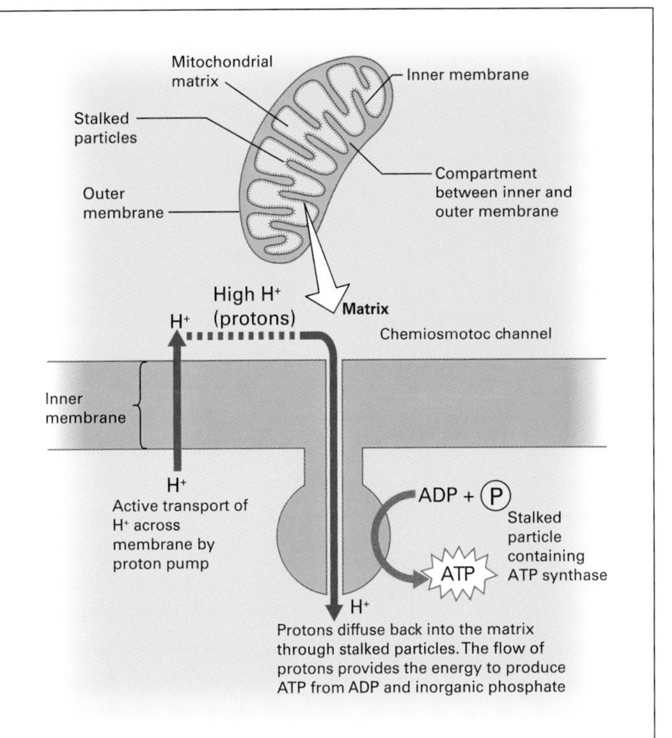

Figure 3 *The chemiosmotic theory revisited: how ATP is synthesised using the energy released in the ETS.*

Efficiency of ATP production

Theoretically, the complete aerobic respiration of each molecule of glucose results in a net production of about 38 ATP molecules. Estimates vary according to the tissue being studied, and there is also some uncertainty about the number of ATP molecules produced from each NADH. Nevertheless, most experts agree that when oxygen is available, a cell can harvest as much as 40 per cent of the potential energy within one glucose molecule; most of this production depends on the activities of the ETS. A much smaller percentage yield is provided by anaerobic respiration.

If oxygen is unavailable, the ETS cannot work. This in turn stops the Krebs cycle, because the reduced NAD and FAD cannot be re-oxidised. Consequently, anaerobic respiration results in a gross production of only 4 molecules of ATP for each molecule of glucose, with a net production of only 2 molecules of ATP.

Quick check

1 Precisely where in the cell is the electron transport system located?

2 Why is the process that results from the passage of electrons along the electron transport system called oxidative phosphorylation?

3 What is the significance of the inner mitchondrial membrane being highly folded?

Food for thought

ATP synthase, the enzyme that plays an essential role in chemiosmosis, consists of two subunits. One subunit occurs above a membrane; the other within it. It is thought that the subunits were originally independent and had their own functions, but then became coupled together and gained new functionality. Suggest why this coupling must have occurred early in the evolution of life.

6.5

MEASURING RESPIRATION

OBJECTIVES

By the end of this spread you should be able to:

- describe how to measure the oxygen consumed and carbon dioxide produced by an organism
- calculate respiratory quotients
- explain how respiratory quotients can be used to determine the respiratory substrate.

Fact of life

A person's maximal oxygen consumption is a good indicator of his or her level of aerobic fitness (stamina). Maximal oxygen consumption is often estimated as VO_2 max, the maximum volume of oxygen consumed per kilogram body mass per minute ($cm^3\ kg^{-1}\ min^{-1}$). The average VO_2 max for a typical 20-year-old female is 32–38 $cm^3\ kg^{-1}\ min^{-1}$, and that of a typical 20-year-old man is 36–44 $cm^3\ kg^{-1}\ min^{-1}$. The highest recorded VO_2 max is 94 $cm^3\ kg^{-1}\ min^{-1}$, for a Norwegian champion cross-country skier. Aerobic (endurance) training can improve a person's VO_2 max by 15–20 per cent or more.

Figure 1 Cross-country skiing is one of the most demanding sports. Successful competitors usually have exceptionally high maximal oxygen consumptions.

The rate of respiration

Respirometers are devices that measure the rate of respiration, typically by measuring oxygen consumption and carbon dioxide output. Respirometers vary in complexity. Some incorporate sophisticated computer technology, automatically measuring the volumes of gases exchanged and drawing off small samples to analyse the proportions of oxygen and carbon dioxide in the gases. **Spirometers** such as that shown in figure 2 on spread 7.2 are commonly used to measure the oxygen consumption of a human subject under different conditions (for example, different states of activity). Spirometers can also be used to measure the depth and frequency of breathing. The respiration rates of small organisms can be measured using a simple respirometer such as the one shown in figure 2 below.

In this particular respirometer, woodlice have been placed in a boiling tube which is connected to a U-tube. The U-tube acts as a manometer (a device for measuring pressure changes). The other end of the U-tube is connected to a control tube which is treated in exactly the same way as the first tube, except that it has no woodlice but instead glass beads which take up the same volume as the woodlice. The two boiling tubes (but not the manometer) are kept in a water bath at a constant temperature. The U-tube contains a coloured liquid which moves according to the pressure exerted on it by the gases in the two boiling tubes. Both tubes contain potassium hydroxide solution which absorbs any carbon dioxide produced.

When the woodlice respire aerobically, they consume oxygen, which causes the liquid to move in the U-tube in the direction of the arrows. The rate of oxygen consumption can be estimated by timing how long it takes for the liquid to rise through a certain height.

The experiment can be repeated by replacing the potassium hydroxide solution with water. Comparing the changes in manometer liquid level with and without potassium hydroxide solution gives an estimate of carbon dioxide output. Such respirometer measurements of oxygen consumption and carbon dioxide production can be used to measure the respiratory quotient (see below).

If the internal radius of the manometer tube is known, the volumes of gases can be calculated using the equation:

$$\text{Volume} = \pi r^2 h$$

where r is the internal radius of the tube and h is the distance moved by the liquid.

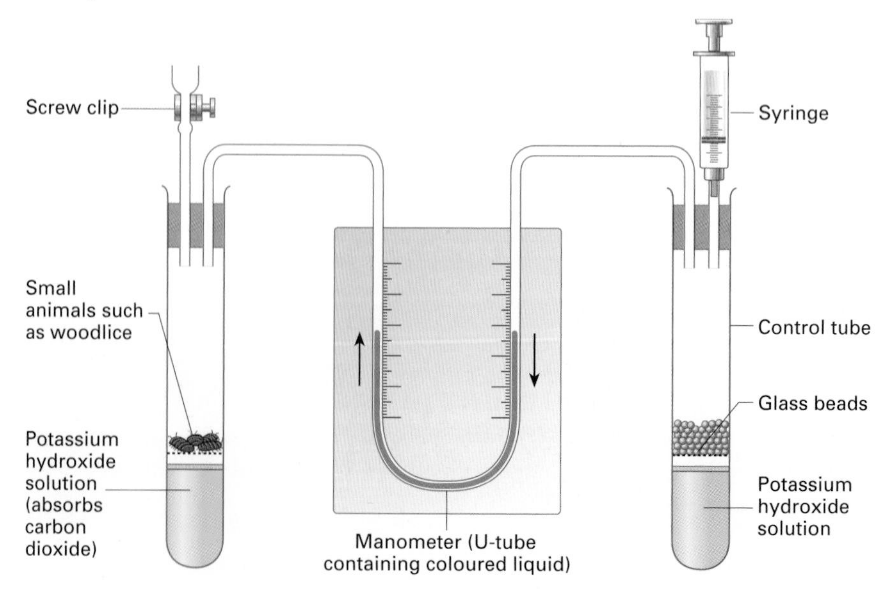

Figure 2 Simple respirometer used to measure respiration in small animals.

Respiratory quotient

The **respiratory quotient** (**RQ**) is the ratio of the volume of carbon dioxide produced to the volume of oxygen used in respiration during the same period of time. The RQ is often assumed to equal the ratio carbon dioxide expired: oxygen inspired during a given time.

$$RQ = \frac{\text{volume of carbon dioxide given out}}{\text{volume of oxygen taken in}}$$

The RQ is important because it can indicate what kind of substrate is being respired, and whether the respiration is aerobic or anaerobic. A look at the equation for the aerobic respiration of glucose (spread 6.1) shows that for each molecule of glucose respired, six molecules of carbon dioxide are produced and six molecules of oxygen are consumed. As each molecule of gas occupies the same volume, this would give an RQ of 1.0. It is possible to work out the RQs for fat and protein by looking at their respiratory equations. The figures are shown in table 1. An RQ greater than 1.0 indicates anaerobic respiration; respiration that is completely anaerobic has an RQ of infinity.

Table 1 *The respiratory quotients for different respiratory substrates.*

Respiratory substrate	RQ
Carbohydrate	1.0
Protein	about 0.9
Fat	about 0.7

Figure 3 *Respiration during germination. During the early stages of seed germination (spread 14.6), it is difficult for oxygen to penetrate the seed coat. At this stage, the RQ is about 3 to 4. Later, when the seed coat is shed, it becomes easier for oxygen to reach respiring tissues so the RQ falls. Eventually, seeds with large carbohydrate stores have an RQ around 1.0 and those with large lipid stores have RQs of 0.7 to 0.8.*

Quick check

1 The results of an experiment using a simple respirometer are given in table 2. In both cases, the manometer fluid moved in the direction shown by the arrows in figure 2.
Use these data to calculate the RQ.

Table 2 *Results of respirometer experiments.*

Solution in experimental tube	Volume change ($mm^3\ h^{-1}$)
Potassium hydroxide	40
Water	4

2 Using the following equation, calculate the RQ for the complete aerobic respiration of oleic acid (a fatty acid found in olive oil):

$$2C_{18}H_{34}O_2 + 51O_2 \longrightarrow 36CO_2 + 34H_2O$$

3 From your calculation of the RQ in question 1, state which substrate is being respired.

Food for thought

Strictly speaking, the ratio of the volume of carbon dioxide expired by an organism to the volume of oxygen inspired should be referred to as the **respiratory exchange ratio** (gas exchange ratio or RER), not the respiratory quotient. This is because the RQ refers to gaseous exchange resulting only from cellular respiration. Under certain circumstances, this may not equal the RER because expired air may contain gases from non-respiratory sources. In mammals during rest, the RER usually equals the RQ, but during exercise, expired air contains carbon dioxide and oxygen derived from stores within the body, and the RER is not the same as the RQ. A similar situation arises in plants because of the effects of photosynthesis.

Suggest reasons for the following observations. A plant that uses glucose as a substrate for aerobic metabolism has an RQ of 1.0, but under certain circumstances it may have:

a no respiratory exchange ratio (RER)

b an RER equal to 1.0

c an RER of more than 1.0.

Summary

Cellular respiration is the process by which organic fuels (mainly carbohydrates and fats) are broken down to make **ATP**, the only form of energy that organisms can use directly. Respiration may be **anaerobic** (without oxygen) or **aerobic** (with oxygen). Both forms start with glycolysis in the cytoplasm, outside mitochondria. Glycolysis involves the partial breakdown of glucose to **pyruvat**e. It requires two molecules of ATP as **activation energy**, and results in the reduction of **NAD⁺** to **NADH** and the production of 4 molecules of ATP (giving a net production of 2 molecules). If oxygen is unavailable, the pyruvate is converted to **lactic acid** in animals, and **alcohol and carbon dioxide** in plants and fungi. This **lactic acid fermentation** and **alcoholic fermentation** allows NAD⁺ to be regenerated. If oxygen is available, pyruvate is converted to **acetylcoenzyme A** which is fed into the Krebs cycle in mitochondria.

During the **Krebs cycle**, the two-carbon acetylcoenzyme A combines with a four-carbon compound to form a six-carbon organic acid. This six-carbon compound is the starting point for a cycle of reactions which results in the generation of ATP, hydrogen, and carbon dioxide. The hydrogen is used by the **electron transport system** to generate ATP by **oxidative phosphorylation**. This involves a series of **redox reactions** that finishes with the addition of oxygen to hydrogen to form water. Aerobic respiration results in the complete breakdown of glucose to carbon dioxide and water with the generation of about 38 molecules of ATP. The ratio of the volume of oxygen consumed to the volume of carbon dioxide released during respiration is called the **respiratory quotient** (**RQ**). It can be measured using a **respirometer** or a **spirometer**. The RQ can be used to determine the type of respiration and respiratory substrate.

PRACTICE EXAM QUESTIONS

1 The diagram below shows some of the stages in anaerobic respiration in a muscle.

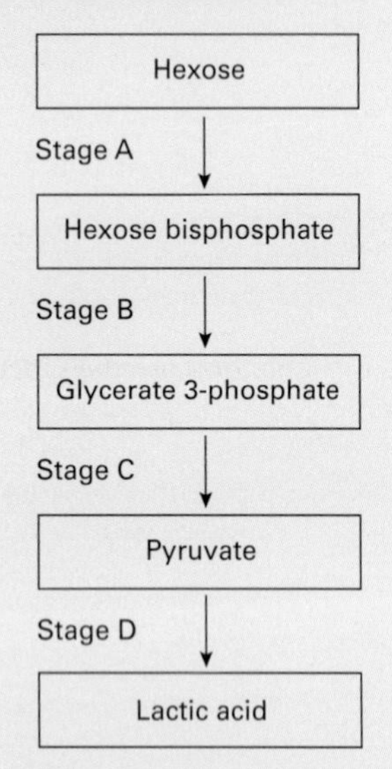

a **i** Name the process shown by stages A to C. [1]

ii State where in a cell this process occurs. [1]

b **i** Give *two* uses of ATP in cells. [2]

ii At which of the stages shown in the diagram is ATP used? [1]

c NADH + H⁺ is a reduced coenzyme which is involved in anaerobic respiration.
At which of the stages shown is NADH + H⁺ oxidised? [1]

[Total 6 marks]

2 **a** The inner membrane of a mitochondrion is folded to form cristae. Suggest how this folding is an adaptation to the function of a mitochondrion. [2]

b The equation represents oxidation of a lipid.

$$C_{57}H_{104}O_6 + 80O_2 \longrightarrow 57CO_2 + 52H_2O + \text{Energy}$$

Use the equation to calculate the Respiratory Quotient (R.Q.) of this lipid. Show your working. [1]

c Suggest an explanation for each of the following:

i the R.Q. of germinating maize grains is 1; [1]

ii the R.Q. of a normal, healthy person varies over a 24-hour period. [1]

[Total 5 marks]

3 **a** Copy and complete the table with a tick if the statement is true about the particular stage of respiration.

Stage of respiration	Statement		
	ATP produced	**Carbon dioxide produced**	**Reduced NAD converted to NAD**
Glucose → pyruvate			
Pyruvate → acetylcoenzyme A			
Krebs cycle			
Electron carrier chain			

[4]

b The graph shows the volume of oxygen consumed and the volume of carbon dioxide released by a small animal over a 15-minute period.

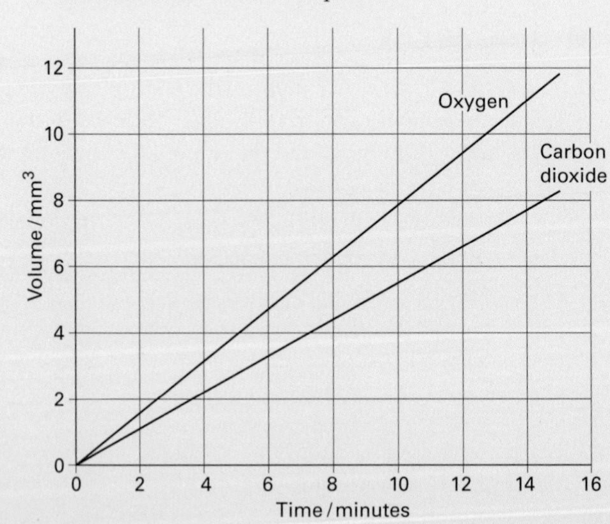

Showing your working in each case, use the graph to calculate:

i the rate of oxygen consumption; [1]

ii the respiratory quotient (R.Q.) of this animal over the 15-minute period. [2]

4 Mitochondria are organelles that are thought to have entered into a mutualistic existence with cells many millions of years ago. They are surrounded by a double membrane system and are usually 0.5 µm – 1.0 µm in length. The inner membrane is folded into structures called cristae and the space inside is called the matrix. Mitochondria possess proteins called cytochromes. They also contain small circles of DNA which code for a few of the mitochondrial proteins. Mitochondria have the biochemical machinery needed for carrying out some protein synthesis.

Liver cells have between 1000 and 2000 mitochondria which can occupy up to 20% of the cell volume.

a Use information in the passage to answer the following:

i Outline **two** pieces of evidence suggesting that mitochondria originally existed outside eukaryotic cells as free living organisms. [2]

ii Suggest why mitochondria can no longer exist as free living organisms. [1]

iii Suggest why liver cells have so many mitochondria. [1]

iv Explain the significance of the folding of the inner mitochondrial membrane. [2]

b The figure below represents the main stages of aerobic respiration.

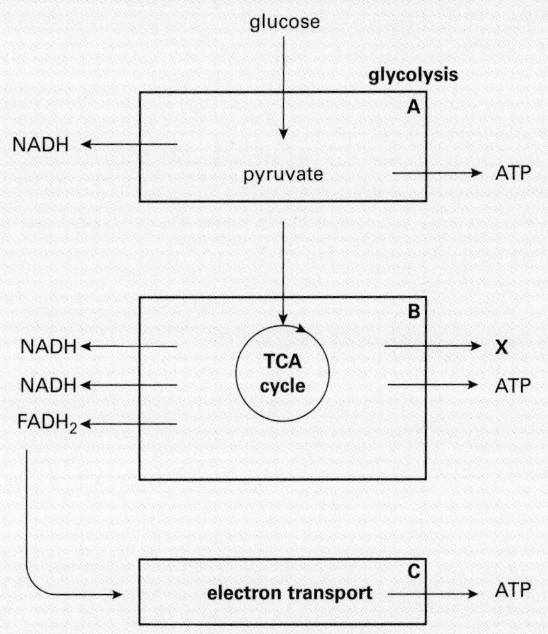

i State precisely where the reactions in boxes **A**, **B** and **C** occur in the cell. [3]

ii What substance is **X**? [1]

iii A total of 38 molecules of ATP is formed during the complete breakdown of one molecule of glucose. State how many molecules are formed at each of the stages **A**, **B** and **C**. [2]

iv If glucose is burned, the energy transferred as heat and light is 2881 kJ mol^{-1}. In the reactions described above, some of the energy is retained in the form of ATP. The energy 'trapped' in ATP is 50 kJ mol^{-1}. Calculate the percentage of the energy made available from the breakdown of one molecule of glucose which is retained for biological reactions in the cell. Show your working. [3]

[Total 15 marks]

5 a The diagram below shows a section through a mitochondrion as seen using an electron microscope.

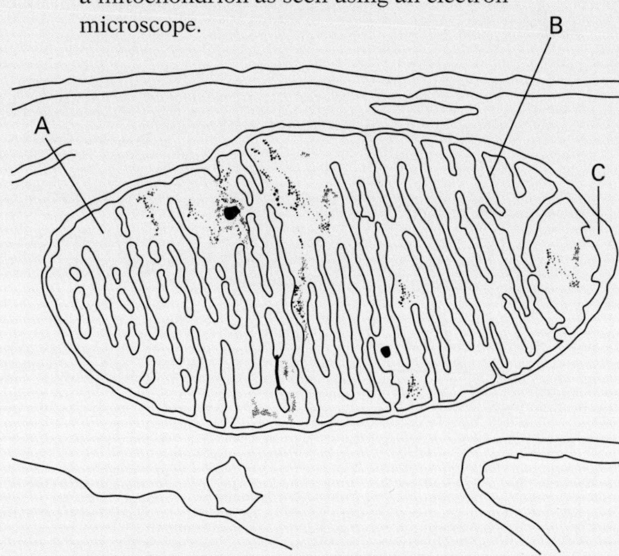

i Name the parts labelled A, B and C. [3]

ii Sketch the diagram. By means of an arrow, show the location of the electron transport system. [1]

iii The magnification of this diagram is ×70 000. Calculate the actual length of the mitochondrion, giving your answer in suitable units. Show your working. [2]

b Active mitochondria can be isolated from liver cells. If these mitochondria are then incubated in a buffer solution containing a substrate, such as succinate, dissolved oxygen will be used by the mitochondria. The concentration of dissolved oxygen in the buffer solution can be measured using an electrode.

An experiment was carried out in which a suspension of active mitochondria was incubated in a buffer solution containing succinate, an intermediate of the Krebs cycle. The concentration of dissolved oxygen was measured every minute for five minutes. A solution containing sodium azide was then added to this preparation and the concentration of dissolved oxygen was measured for a further five minutes. Sodium azide combines with cytochromes and prevents electron transport.

The results are shown in the graph below.

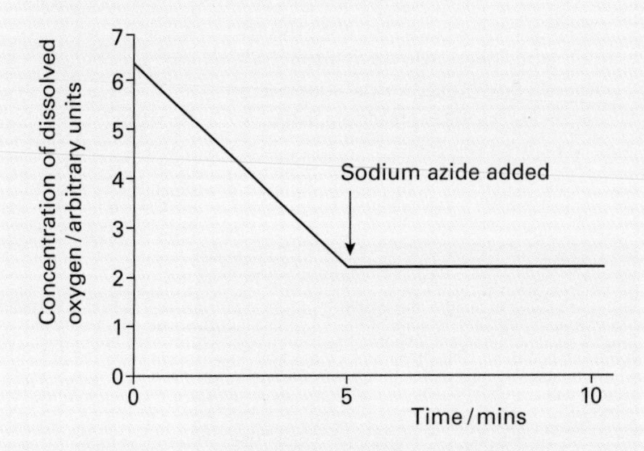

i Explain why the concentration of oxygen decreased during the first five minutes. [2]

ii Suggest what effect the addition of sodium azide will have on the production of ATP and give an explanation for your answer. [3]

[Total 11 marks]

7 Gaseous exchange and transport in mammals

Gaseous exchange systems

7.1

OBJECTIVES

By the end of this spread you should be able to:

- describe the main structures of the mammalian respiratory system
- explain the importance of lungs to mammals.

Fact of life

Although lungs are the main organs of gaseous exchange among terrestrial vertebrates, and gills among aquatic vertebrates, some vertebrates breathe through their skin. These include amphibians, sea snakes, and amphibious fish. They all live in moist habitats and have skin that is well supplied with blood vessels (highly vascularised). The common rocky shore fish *Blennius pholis*, for example, can survive several hours out of water as long as its body surface is kept moist. Adult frogs usually breathe through their lungs and skin, but the South American frog *Telmatobius culeus*, which lives permanently submerged in the waters of Lake Titacaca, breathes exclusively through the skin. The surface area of its skin is enlarged by loose folds. In mammals, little if any oxygen is taken up across the skin. However, some bats can lose carbon dioxide through their skin. These bats have large, thin, hairless, and highly vascularised wing membranes which provide a large surface area (figure 1).

***Figure 1** The Australian ghost bat excretes carbon dioxide through its skin as well as through its lungs. At 27 °C, as much as 11.5 per cent of the carbon dioxide excreted by a bat can be lost through the skin, but this figure drops to 0.4 per cent at 18 °C.*

Cellular (internal) respiration is the process within cells that generates usable energy in the form of ATP (spread 6.1). Mammals obtain most of their energy from the aerobic respiration of food molecules (chapter 6), although they can respire anaerobically for short periods.

- Oxygen is essential for aerobic respiration. It is obtained from the air and is transported in the blood.
- Carbon dioxide is produced as a waste product of respiration. It is transported in the blood, and is breathed out into the air.
- A **gaseous exchange system** (sometimes called external respiration) is therefore needed to exchange oxygen and carbon dioxide between the air and the blood.
- In mammals this gaseous exchange takes place in the **lungs**.
- For gaseous exchange to happen there must be a continous flow of air into and out of the lungs. This is called **ventilation**. It is achieved by the breathing movements of the chest.

Why do mammals need a gaseous exchange system?

All the cells of a mammal, even those deep in the centre of body, need a continous supply of oxygen and must be able to eliminate carbon dioxide.

Most animals have evolved as land animals. Their skin is adapted to conserve water and is unsuitable as a surface for gaseous exchange. In addition, mammals are too large to rely on simple diffusion to exchange gases. It would take too long for oxygen to reach the central cells from the skin (see spreads 4.7 and 4.13). Mammals have evolved specialised internal organs – the lungs – to enable them to exchange gases but without losing too much water.

The alveoli are the site of gaseous exchange

The lungs consist of a system of tubes of ever-decreasing size (figure 2) which end in microscopic bulbous sacs called **alveoli**. This is where gaseous exchange takes place. The alveoli have the following adaptations for efficient gas exchange (spread 7.3):

- very thin walls
- a moist inner surface
- a huge combined surface area
- a rich blood supply – each alveolus is surrounded by capillaries.

The lungs are deep within the chest. This minimises water loss from the alveolar surface, but means that air must be moved in and out of the lungs so that the alveoli are in contact with a constantly changing supply of air (spread 7.2).

The human respiratory system

The human resiratory system is shown in figure 2.

- Air enters the human airway through either the nose or the mouth. Air entering via the nose cavity is filtered by hairs in the nasal passages, warmed by contact with the tissues in the nasal cavity, and moistened by cells in the **mucous membrane** of the nasal cavity. Mucous membranes line much of the airway. These membranes contain globlet cells (figure 2) which secrete **mucus**, a slimy material rich in glycoproteins. If an irritating substance is breathed in, this can stimulate a sneeze. The irritant is ejected at speeds exceeding 160 km h^{-1}.

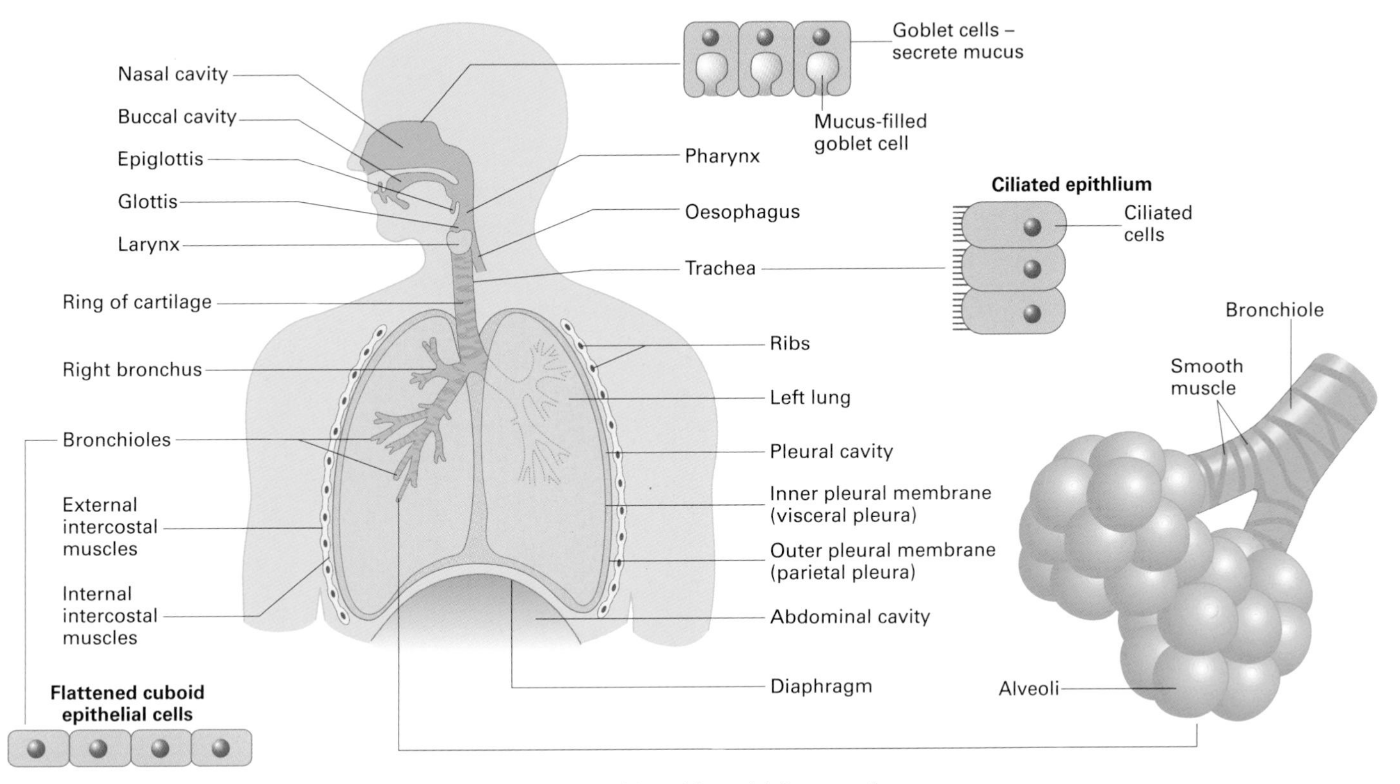

***Figure 2** The human respiratory system. The trachea, bronchi, and bronchioles contain **smooth muscle** which can contract to constrict the tubes. They also contain **elastic fibres** which allow the tubes to expand during inhalation and contribute to the lungs' elastic recoil during exhalation.*

- Air entering the respiratory system through the mouth into the **buccal cavity** is not warmed or moistened as much as air passing through the nose, and it is not filtered at all.
- The nasal and buccal cavities lead into the **pharynx**, a tube that conducts both food and air. When food is swallowed a flap of tissue called the **epiglottis** closes over the **glottis**. This is a reflex action which prevents food from going down the trachea (windpipe).
- The **larynx** is a box-shaped structure just above the trachea. The flow of air in and out of the respiratory system makes the **vocal cords** (folds of mucous membrane) vibrate, producing sounds. The sound is varied by changing the position and tension of the cords.
- Air passes through the larynx into the **trachea**. This single tube forms the major airway. The trachea is held open by horseshoe-shaped rings of cartilage. Without them the trachea would collapse during breathing out, when the external atmospheric pressure is higher than the pressure inside the trachea. The gaps in the cartilage rings allow the trachea to be flexible so that food can pass easily down the oesophagus which runs behind the trachea.
- The trachea is lined with mucous membrane containing **ciliated epithelium**. The epithelial cells have microscopic hair-like extensions called **cilia**. These beat in a wave-like manner, moving mucus, dust, and microorganisms upwards and out of the lungs.
- The trachea subdivides into two main branches, the right and left **bronchus**. The bronchi are narrower than the trachea but have a similar structure. Each bronchus divides repeatedly into smaller tubes called **bronchioles**. Larger bronchioles are lined by complete rings of cartilage which collapse quite easily. The smallest bronchioles have no cartilage (this enables them to constrict completely) and are lined with **flattened cuboidal epithelium**. Some gaseous exchange takes place through these bronchioles.
- The bronchioles lead ultimately to numerous alveoli. This is where most of the gaseous exchange takes place.

Quick check

1 List the structures through which air passes on its way from the nose to the alveoli.

2 Give two reasons why mammals need lungs, rather than exchanging gases through the skin.

Food for thought

Ventilation is a form of **mass transport** (also known as mass flow) in which particles of matter are transferred in bulk from one place to another by being carried in a fluid. Nicotine within cigarettes can stop cilia from beating. Suggest how this can affect the ventilation of the lungs.

7.2

OBJECTIVES

By the end of this spread you should be able to:

- explain how air gets in and out of the lungs
- define the lung volumes.

Fact of life

The lung volume of a mammal makes up about 5 per cent of the body volume, irrespective of body size. Approximately the same proportion of the body is taken up by the lungs in a whale and a shrew.

VENTILATION

In mammals, **ventilation** (breathing) is the flow of air into and out of the lungs. It results in the mass transport of gases, and liquid or solid particles suspended in air, to and from the respiratory exchange surfaces in the alveoli. An efficient ventilation mechanism is essential for survival.

The mechanism of ventilation

Gases flow from regions of high pressure to regions of low pressure. When the total gas pressure inside the alveoli is equal to the pressure of the surrounding atmosphere, no movement of gas is possible. For **inhalation** (breathing in or inspiration) to occur, the gas pressure in the alveoli must be less than that in the atmosphere. For **exhalation** (breathing out or expiration), the gas pressure in the alveoli must be greater than that in the atmosphere.

Inhalation

Humans inhale by enlarging the thoracic (chest) cavity, which enlarges the lungs as well. This reduces the gas pressure in the alveoli, creating a pressure gradient which draws air into the lungs.

The expansion of the thoracic cavity is brought about by the combined movements of the rib cage (ribs and sternum) and diaphragm. The rib cage is moved by the two sets of intercostal muscles between the ribs (figure 1a).

- During inhalation the **external intercostal muscles** contract. This draws the rib cage upwards and outwards and causes the sternum to move outwards and forwards.
- The diaphragm also contracts: the central portion of this sheet of muscle moves downwards.

The lungs cannot expand on their own. The **inner pleural membrane**, which covers the surface of the lungs, is closely linked to the **outer pleural membrane**, which lines the inside of the thorax. Only a thin layer of fluid separates the two. Expansion of the thorax therefore causes the lungs to expand.

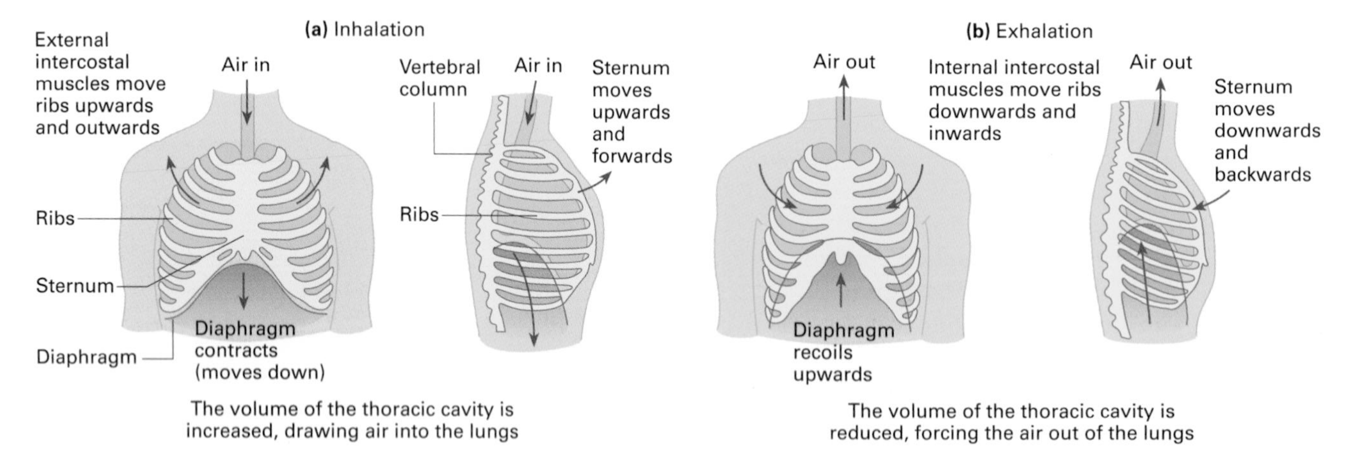

***Figure 1** Movements that ventilate the lungs.*

Inhalation requires considerable work. The active contraction of the intercostal muscles and the diaphragm has to provide enough force to overcome a series of resistances. These include:

- the recoil of elastic tissue of the lungs and thorax (their resistance to being stretched)
- the frictional resistance of air as it passes through the hundreds of thousands of small bronchioles leading into the alveoli
- the resistance created by surface tension at the fluid–gas interfaces in the alveoli.

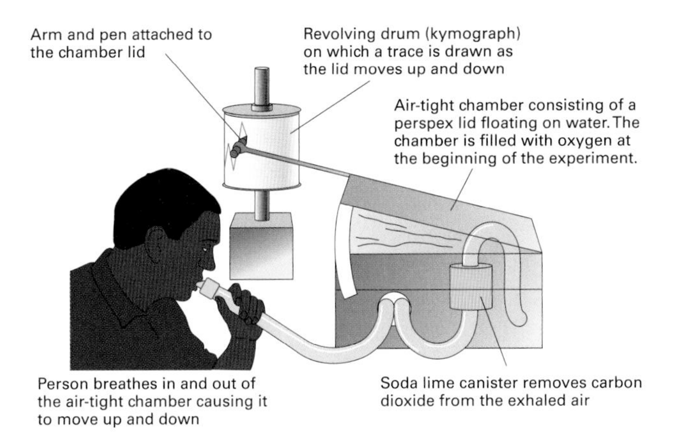

Figure 2 *A spirometer records the volume of air breathed in and out.*

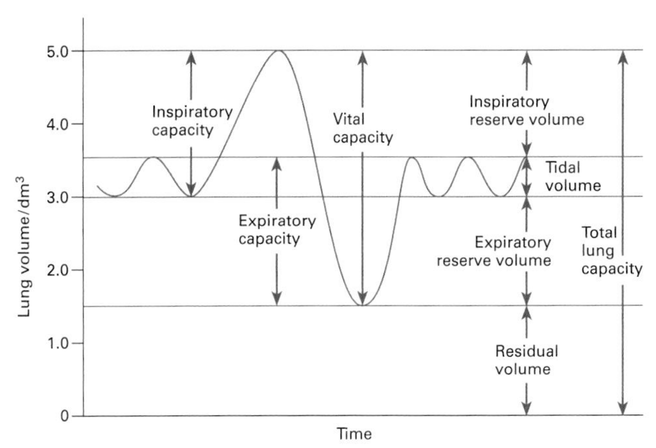

Figure 3 *Typical human lung volumes measured with a spirometer.*

Exhalation

Exhalation is usually a relatively passive process (figure 1b). Enlargement of the thorax during inhalation stretches the tissues of the thorax and lungs, which recoil naturally during exhalation. The diaphragm also returns to its resting position by elastic recoil. The muscles are contracted actively during exhalation only during high rates of ventilation or when there is an obstruction in the airway. The **internal intercostal muscles**, which act in opposition to the external intercostal muscles, draw the rib cage downwards and inwards.

Lung volumes

The volume of air inhaled and exhaled can be measured using a **spirometer** (figures 2 and 3).

- The volume of air breathed in or out of the lungs per breath is called the **tidal volume**. It is about 0.5 dm^3 at rest but varies with each individual; it increases during exercise.
- The **vital capacity** is the maximum volume of air that can be forcibly expired after a maximal intake of air. It varies between 3.5 and 6.0 dm^3 depending on the size and fitness of the person.
- The volume of air remaining in the lungs at the end of a maximal expiration is called the **residual volume** (typically about 1.5 dm^3). Inhaled air mixes with residual air, keeping the levels of gases in the alveoli relatively constant.
- In addition to these lung volumes, physiologists are also interested in the **dead space**. This is the volume of air taken into the lungs which does not take part in gaseous exchange.

Regulation of the ventilation rate

The rate at which a person breathes is called the **ventilation rate**, often expressed as the volume of air breathed per minute (i.e. the **minute ventilation**):

Ventilation rate = tidal volume × number of breaths per minute

Changes in the circulation of blood and the ventilation rate ensure that the blood always supplies sufficient oxygen to meet the needs of the tissues. During exercise, for example, breathing becomes both deeper and faster, so that the ventilation rate increases – to as much as 200 dm^3 per minute in a top-class athlete.

Respiratory centres in the hindbrain (the medulla oblongata, see spread 10.13) control the rate and depth of breathing. They have an intrinsic rhythmic activity which keeps the movements of ventilation going automatically. The basic rhythm of breathing is modified by inputs from stretch receptors in the bronchi, other receptors sensitive to carbon dioxide levels in the blood, and higher centres of the brain. This ensures that the ventilation rate meets the demands of specific situations (figure 4).

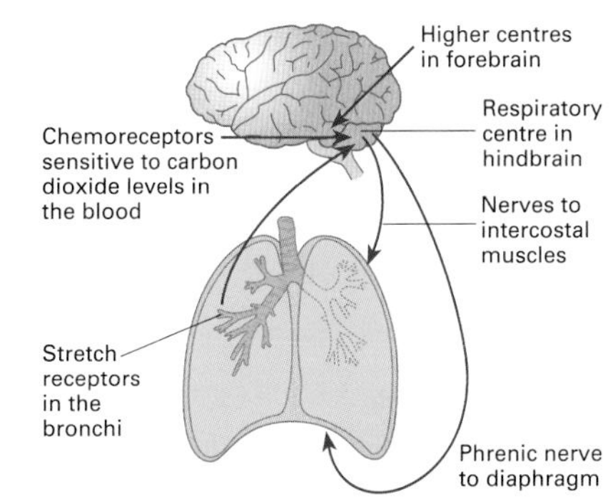

Figure 4 *The regulation of ventilation in mammals. The respiratory centre receives information from various sources and sends impulses which cause the inhalation muscles to contract rhythmically.*

Quick check

1 Which muscles cause the ribs to move during:
  a inhalation
  b exhalation?

2 What name is given to the volume of air inhaled or exhaled per breath?

Food for thought

The trachea, bronchi, and their branches are not especially adapted for gaseous exchange. At the end of exhalation these tubes contain 'used' air. The volume of this 'used' air, called the **dead space**, is about 0.15 dm^3. When a person inhales, this 'used' air is drawn back into the lungs. Therefore, out of 0.5 dm^3 of air inhaled, only about 0.35 dm^3 of fresh air enters the lungs. The dead space thus makes up about one-third of the tidal volume at rest. Suggest how the proportion of dead space to tidal volume changes during exercise.

7.3

Gaseous exchange in the alveoli

OBJECTIVES

By the end of this spread you should be able to:

- describe the structure and function of alveoli
- explain how they are adapted to maximise the rate of diffusion.

Fact of life

The composition of gas in human alveoli remains remarkably constant at about 15 per cent oxygen and 5 per cent carbon dioxide during all states of breathing. This is because during inhalation, fresh air mixes with air already in the lungs, resulting in only about one-fifth of the air being renewed at each breath.

Minimising evaporation

Gaseous exchange (the diffusion of oxygen into the body and carbon dioxide out of it) takes place mainly in the **alveoli**. Each alveolus is a tiny sac with a wall just one cell thick. Alveoli vary in diameter from 75 μm to 300 μm. Fluid from the cytoplasm passes through the cell surface membrane onto the surface of alveolar cells, keeping the alveoli moist. However, it is essential that these membranes do not dry out, otherwise gases would not be able to diffuse across them. Mammals minimise water loss by having their lungs inside the body. Here the alveoli can be kept moist without losing too much water during breathing (figure 1a).

The details of diffusion

- The blood is the transport medium that carries oxygen and carbon dioxide between the lungs and the body cells. These gases are exchanged between the air and the blood in the alveoli.
- Alveoli are in close contact with a vast network of blood capillaries (figure 1b); each alveolus has its own blood supply.
- Gases dissolve in the fluid on the cell surface membrane and diffuse through the thin walls of the alveolus and its neighbouring capillaries into the blood. Oxygen enters the blood in this way. Carbon dioxide leaves the blood and diffuses into the air in the alveolus.

Composition of alveolar air

The gas in alveoli does not have the same composition as atmospheric air. Each breath brings a fresh supply of air into the lungs, which mixes with the gases already in the respiratory tract (the residual volume). In normal quiet breathing, about 0.5 dm^3 of air is drawn into the lungs with each breath. Of this, about 0.35 dm^3 reaches the alveoli and mixes with the residual volume of gas (about 1.5 dm^3). It is this mixture of gases that forms the **alveolar air** involved in gaseous exchange. Table 1 shows the

Table 1 *Composition of inspired (atmospheric), alveolar, and expired air. Note that expired air still contains quite a high proportion of oxygen, enough to save someone's life by mouth-to-mouth resuscitation.*

	Percentage composition by volume		
Gas	**Inspired air**	**Alveolar air**	**Expired air**
Oxygen	20.95	13.8	16.4
Carbon dioxide	0.04	5.5	4.0
Nitrogen	79.01	80.7	79.6

Figure 1 *The structure and position of alveoli, and gaseous exchange by diffusion across the wall of an alveolus. By the time the blood leaves the alveolus, it is almost saturated with oxygen and has lost much of its carbon dioxide.*

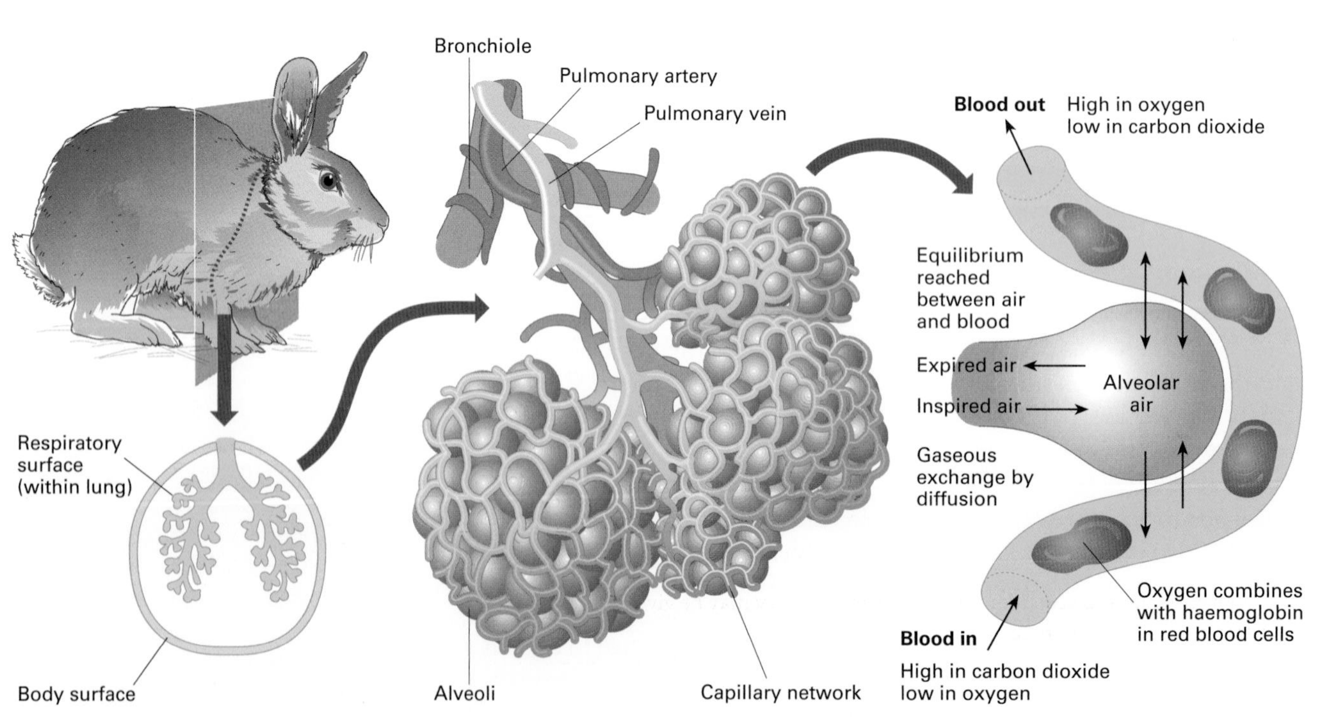

relative proportions of the main respiratory gases at various points of the respiratory tract.

In comparison with alveolar air, the blood arriving in the lungs from the rest of the body has a relatively high concentration of carbon dioxide and a relatively low concentration of oxygen. Both gases diffuse down their concentration gradients and equalise between the blood and the air. The blood leaving the lungs therefore has a similar composition to that of expired air (figure 1c).

Partial gas pressures

Table 1 gives the percentage composition of gases. However, **partial pressures** are usually used to compare the proportions of gases in a mixture. The partial pressure of a gas in a mixture of gases is the pressure exerted by that gas. It is usually measured in kilopascals (kPa). For example, at sea level the total atmospheric pressure is 101.3 kPa. The atmosphere contains 21 per cent oxygen, which therefore has a partial pressure of $\frac{21}{100} \times 101.3$ kPa, that is 21.3 kPa. The partial pressure of oxygen is abbreviated as $p(O_2)$.

Maximising the rate of diffusion

Diffusion between the blood and the alveoli obeys Fick's law (see spread 4.7). The rate of diffusion is maximised by a number of adaptations.

- Good ventilation and an efficient circulatory system maintain steep concentration gradients. With each breath, air carrying carbon dioxide is expelled from the lungs and fresh supplies of oxygen are brought in. With each heart beat, blood high in oxygen is transported to all the body's tissues and blood high in carbon dioxide is returned to the lungs.
- The large number of alveoli provide a very large surface area. There are about 300 million alveoli in a pair of human lungs. Spread out they would fill half a tennis court, about 50 times the total surface area of the skin.
- The alveolar and capillary walls are very thin and close together, minimising the diffusion path. The alveoli are lined by **squamous epithelium**. The cells of this epithelium are flattened so the alveolar membrane is only about 0.2 μm thick. Capillary walls are also thin and the lumen is so narrow that red blood cells have to squeeze their way through. This brings haemoglobin (the protein that transports most of the oxygen around the body) very close to alveolar air. The combined effect of the large surface area of the lungs and the very thin alveolar membranes allows very high rates of gaseous exchange by diffusion.

Lung surfactant

Alveoli must be kept open for their extensive surface area to be used for gaseous exchange. However, the inner surface of the alveoli is lined with a watery film which exerts a surface tension. If there was nothing between this film and the alveolar air, the alveoli would collapse as the water molecules stick together, especially when air is exhaled. To prevent this, a mixture of phospholipid molecules called **lung surfactant** occupies the space between the watery film on the inner surface of the alveoli and the air within them. Lung surfactant reduces the surface tension so that the alveoli remain open and available for gaseous exchange (figure 2).

Lung surfactant is especially important at birth when a baby's lungs must inflate to take the first breath of its life. This first breath is 15 to 20 times more difficult to take than subsequent breaths. However, once expanded the alveoli become lined with surfactant, making breathing much easier. Premature babies may not have enough surfactant to keep their alveoli expanded. They may be given synthetic surfactant to help them breathe more easily.

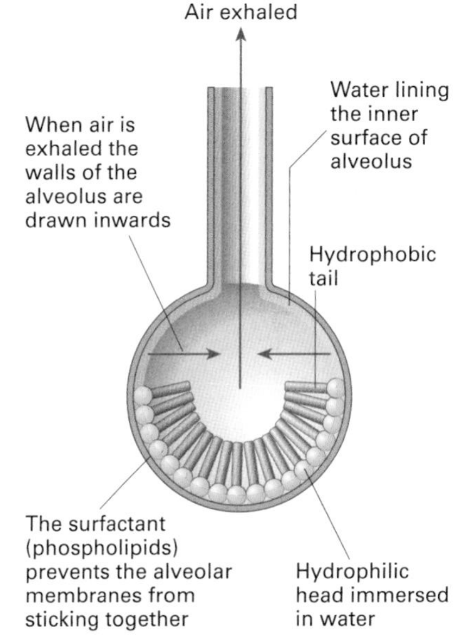

Figure 2 *Surfactant prevents the alveoli collapsing during exhalation.*

Quick check

1. List the adaptations of alveoli that make them suitable for gaseous exchange.
2. If the air contains 0.04 per cent carbon dioxide, calculate $p(CO_2)$. Assume that the total pressure of the air is 101.3 kPa.

Food for thought

Suggest why each cubic centimetre volume of a frog lung has a total gaseous exchange surface of about 20 cm^2, whereas a cubic centimetre volume of a mouse lung has a gaseous exchange surface of about 800 cm^2.

7.4

The cardiovascular system

OBJECTIVES

By the end of this spread you should be able to:

- explain what is meant by mass flow
- describe an open circulatory system and a closed circulatory system
- discuss the structure and functions of arteries, capillaries, and veins.

Fact of life

William Harvey was physician to King Charles I of England. Harvey calculated that the heart beats more than 1000 times in 30 minutes, and that each beat of the heart pumps about 60 g of blood. He concluded that the body does not take in sufficient fluid to replace this blood, so it cannot be used up or lost from the body. Therefore the blood that leaves the heart must come back to it. He suggested that blood circulates the body in closed vessels. He stated: *'I frequently and seriously bethought me, and long revolved in my mind, what might be the quantity of blood ... I began to think whether there might not be a motion, as it were, in a circle.'*

Figure 1 *William Harvey (1578–1657). His skilful dissections combined with accurate observations and careful reasoning led him to the idea that blood is enclosed in vessels and circulates the body – a closed circulatory system.*

Why do mammals need a cardiovascular system?

Mammals are relatively large and highly organised, with different body regions adapted to carry out different functions. For example, cells in the small intestine are adapted to absorb food; those in the alveolus are adapted for gaseous exchange. These specialised regions depend on each other: intestinal cells need oxygen acquired by the alveoli; the alveoli need nutrients acquired by the small intestine. This interdependence demands an efficient transport system.

Substances can move by diffusion across exchange surfaces (such as the gut and the alveolar walls), and in and out of individual cells. However, mammals are too large to rely solely on diffusion to move materials throughout their bodies. Over long distances substances are transported more efficiently by **mass flow**: the bulk movement of substances from one area to another due to differences in pressure. The **blood vascular system** provides the most important means of mass flow in mammals. Many substances are carried in the blood, which is pumped in special vessels continuously and fairly rapidly to all parts of the body.

Open circulatory systems

Some small animals have an **open circulatory system** in which blood is not confined to vessels. For example, in arthropods blood flows freely over tissues, through spaces known collectively as the **haemocoel**. Blood flow is slow and at low pressure with little control over its distribution.

Closed circulatory systems

Large mammals have a **closed circulatory system**, which consists of the heart, arteries, arterioles (narrow thin-walled arteries), capillaries, venules (small veins), and veins. Figure 2 shows the relationship between these vessels and compares their structures.

- **Arteries** carry blood away from the heart.
- **Capillaries** are microscopic thin-walled structures made only of **squamous endothelium** (a layer of flattened epithelial cells); they form networks in various parts of the body.
- Veins carry blood towards the heart.

In a closed circulatory system, the blood can be at a much higher pressure than in an open circulatory system (important, for example, for ultrafiltration in the kidney) and there is much more control over its distribution. Vessels can be widened (**vasodilation**) or narrowed (**vasoconstriction**) by contraction of smooth muscle. This allows blood to be shunted to areas where it is needed. For example, during exercise blood is shunted from the intestine to respiring skeletal muscles.

The double circulatory system of mammals

Mammals have a **double circulatory system** (figure 3). This means that in each complete circuit of the body blood flows through the heart twice.

- The **pulmonary circulation** transports blood between the heart and the lungs; the **systemic circulation** carries blood between the heart and all other parts of the body.
- Each organ has a major artery which supplies it with blood and most have a major vein taking blood back to the heart. Exceptions are the stomach and small intestines. The blood from these organs is carried by the **hepatic portal vein** to the liver where the digested food molecules are processed. The blood is transported back to the heart by the hepatic vein.

The single circulatory system of fish

Fish have a **single circulatory system** – blood flows through the heart only once for each complete circuit of the body. Blood is pumped from

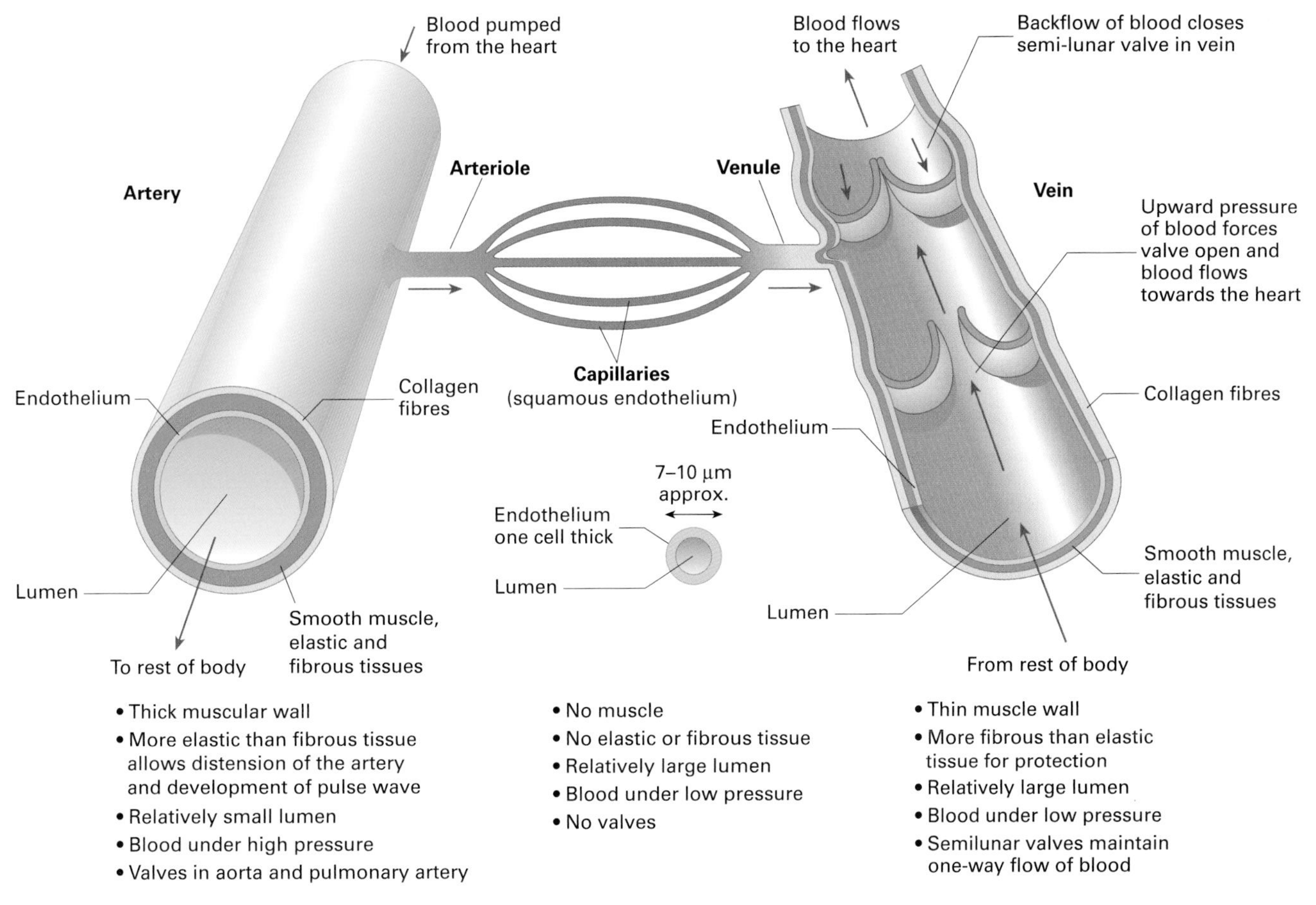

Figure 2 *Schematic diagram comparing an artery, a capillary, and a vein.*

the heart to the gills, and then flows directly to the rest of the body. The pressure drops dramatically as blood leaves the gills, so the blood flow to vital organs is both slower and at a lower pressure than in a double circulatory system. A single circulatory system would be unsuitable for mammals because their kidneys cannot function efficiently at low pressures.

Moving the blood through the system

As in any mass flow system, the blood vascular system transports fluid from high-pressure areas to low-pressure areas – down pressure gradients. The pressure gradients are produced in three main ways:

- by the pumping action of the heart
- contractions of skeletal muscles squeeze blood along veins
- inspiratory movements of the thorax reduce the pressure inside the thoracic cavity, which helps to draw blood back to the heart.

Valves prevent backflow and ensure that the blood flows in one direction only.

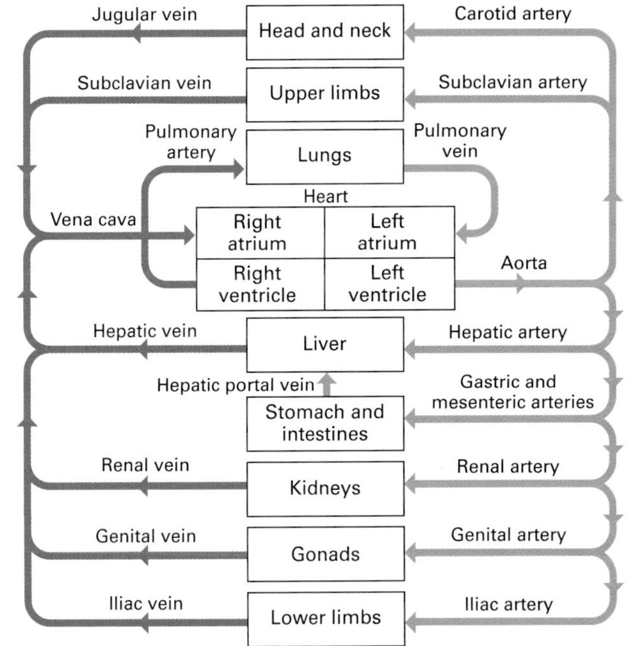

Figure 3 *Schematic diagram of the mammalian double circulatory system.*

Quick check

1 Define mass flow.

2 Give two advantages of a closed double circulatory system.

3 Why is the thickness of a blood vessel not a good indicator of whether it is an artery or a vein?

Food for thought

Suggest what would happen if a valve in a vein in the leg became weakened. What might cause the weakening?

7.5

STRUCTURE OF THE HEART

OBJECTIVES

By the end of this spread you should be able to:

- describe the structure and function of a mammalian heart.

Fact of life

The septum (the inner wall dividing the right and left sides of the heart) is not complete until after birth. During embryonic development, blood is oxygenated by the placenta, rather than the lungs. Most blood therefore bypasses the lungs and goes directly from the pulmonary artery to the aorta through a small vessel called the **ductus arteriosus.** Usually the vessel closes completely within the first few weeks after birth. However, sometimes the closure is incomplete and a person is said to have a 'hole in the heart' or, medically, an atrial septal defect. If large, the hole can cause severe health problems because the blood does not become fully oxygenated. Surgical closure of the hole (the so-called 'hole in the heart' operation) usually resolves such problems.

The mammalian heart is a remarkable organ. From birth to death it pumps blood continuously around the body. In a healthy person, the heart pumps continuously 24 hours a day without tiring. The heart of a resting person pumps more than 10 000 dm^3 of blood per day, enough to fill a small swimming pool.

The pumping action of the heart is of vital importance as the circulation of blood supports all the living cells of the body. If the human brain is deprived of blood, an individual loses consciousness within three to five seconds; after 15 to 20 seconds the body begins to go into convulsions; after only a few minutes, parts of the brain may become permanently damaged.

The human heart

The human heart (figure 1) is about the size of a clenched fist and lies in the chest cavity between the two lungs. It is encapsulated by a double layer of tough inelastic membranes which form the **pericardium**. A fluid (the pericardial fluid) is secreted between the membranes allowing them to move easily over each other. The pericardium protects the heart from overexpansion caused by elastic recoil when it is beating very fast.

The walls of the heart consist mainly of a special type of muscle called **cardiac muscle** (spread 11.3, figure 1). Cardiac muscle is found only in the heart. Unlike other muscle, it never fatigues. However, it does not tolerate lack of oxygen or lack of nutrients and soon dies if its supply of blood is cut off. (See spread 11.3 for a comparison of cardiac muscle with other types of muscle.)

The heart is divided into a left side and a right side separated by the **septum**. The septum becomes rigid just before the heart contracts, so that it serves as a fulcrum for the action of heart muscle. Each side of the heart has two chambers: an **atrium** which receives blood from the veins, and a **ventricle** which pumps blood into the arteries.

Figure 1 *The structure of the human heart.*

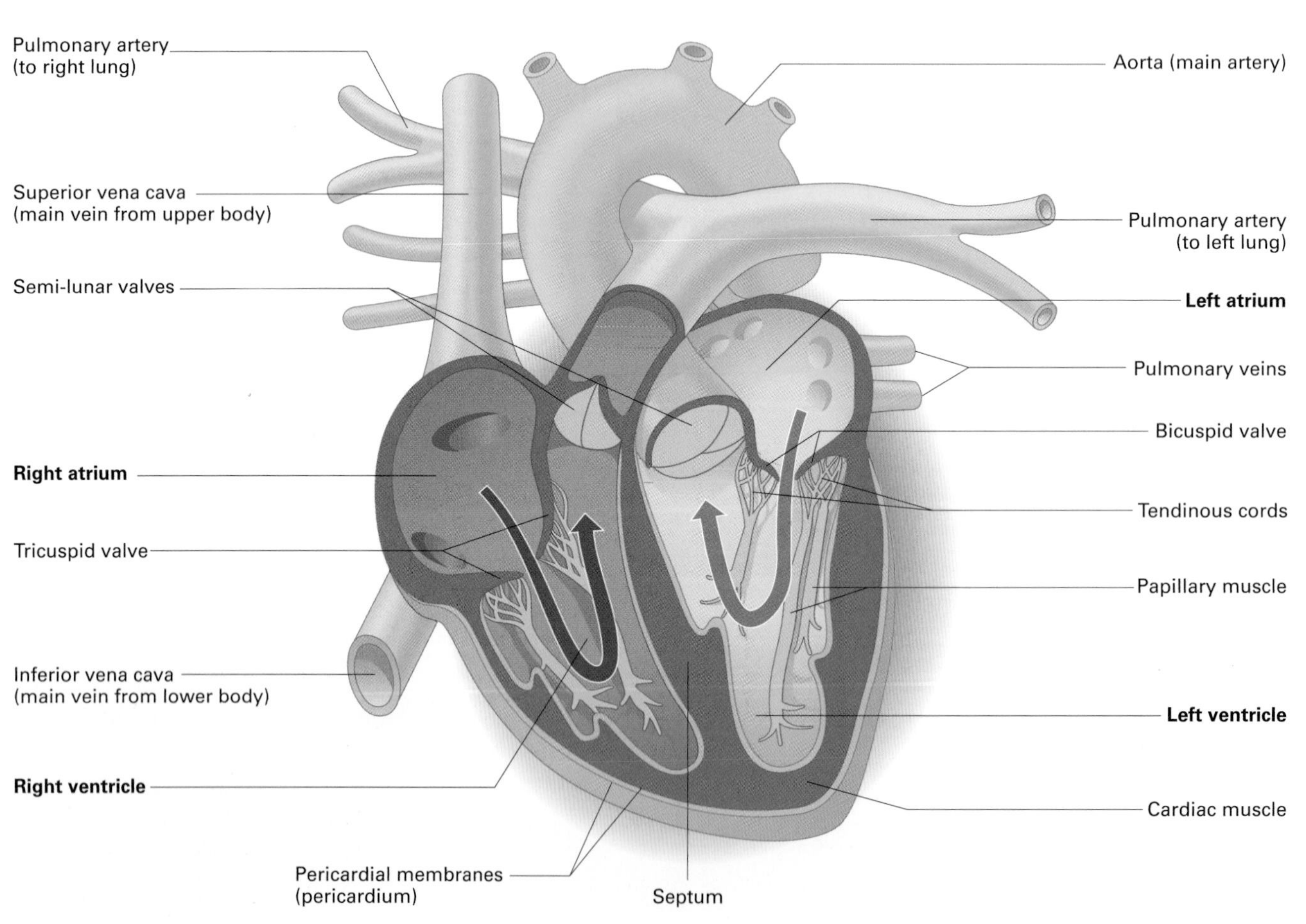

Deoxygenated blood from the systemic veins enters the **right atrium** and is passed through the **tricuspid valve** into the **right ventricle**. This contracts and pumps blood through the **pulmonary artery** into the lungs. Oxygenated blood returns through the **pulmonary vein** into the **left atrium** and then, through the **bicuspid valve**, into the **left ventricle**. This has a very thick muscular wall which enables it to contract strongly and exert sufficient pressure to pump blood into the aorta and all the way around the body.

The role of the valves

The one-way flow of blood is maintained by **valves**. The tricuspid valve between the right atrium and right ventricle has three flaps; the bicuspid valve has two flaps. These valves allow blood to flow freely from atria to ventricles, but when the blood pressure in the ventricles exceeds that in the atria, the valves prevent backflow from ventricles to atria. A sound is produced at the precise moment the valves shut; this is the 'lub' of the 'lub-dub' sounds heard when a stethoscope is placed on the chest. **Tendinous cords** attached to special muscles (**papillary muscles**) prevent the valves from turning inside out. The papillary muscles do not move the valves, they just increase the tension of the tendinous cords so that they can resist the powerful back pressure of blood. **Semi-lunar valves** (half-moon shaped valves) prevent backflow in the pulmonary artery and dorsal aorta. Closure of these valves produces the 'dub' sound heard through a stethoscope. The venae cavae constrict during each heartbeat so that blood does not flow back into the veins.

The heart's blood supply

Because the heart is a very active organ, it has a high demand for oxygen and nutrients. Some of the oxygenated blood leaving the left ventricle goes directly to the heart through the **coronary arteries**, of which there are usually three. These arteries branch out to supply the thick heart muscle with nutrients and oxygen (figure 2). Disease of these arteries can lead to them becoming blocked, resulting in a heart attack (see spread 16.7).

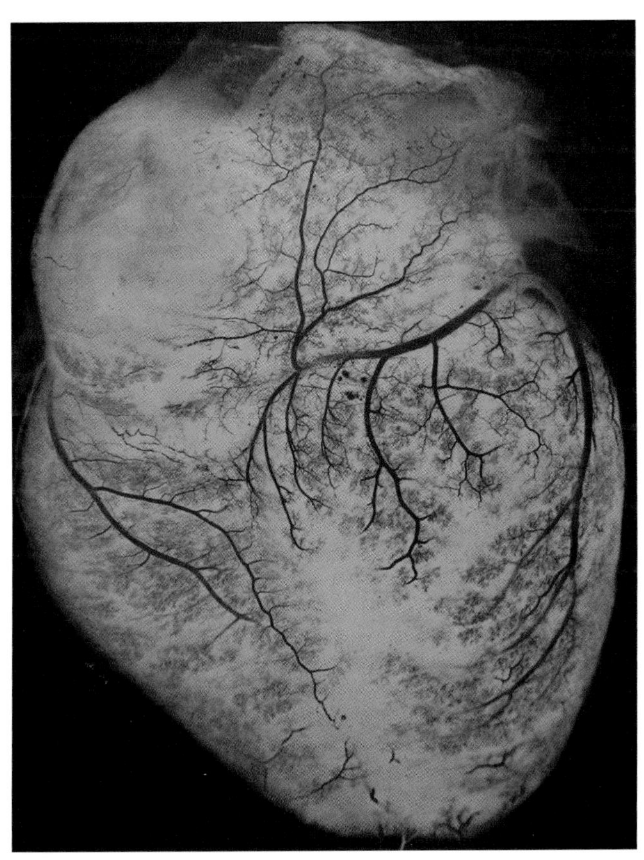

Figure 2 *The coronary arteries supply nutrients and oxygen to the working heart muscle.*

Quick check

1 Draw a flow diagram showing all the main structures through which a red blood cell moves on its passage from the superior vena cava to the dorsal aorta.

2 Which chamber of the heart has the most muscular wall? Give reasons for your answer.

Food for thought

Endurance athletes commonly have a condition known as 'athlete's heart': the heart muscle becomes enlarged as a result of regular training; the volume of the chamber of the left ventricle increases; and the muscular wall of the left ventricle becomes thicker. The condition is not detrimental to health because the blood supply to the heart is also improved. Suggest why an enlarged heart in an inactive person may indicate heart disease, such as a faulty valve.

7.6

The cardiac cycle

OBJECTIVES

By the end of this spread you should be able to:

- describe the main events of the cardiac cycle
- explain how a heartbeat is initiated.

Fact of life

The heart rate of a mammalian species is inversely proportional to its size. An elephant has a resting heart rate of about 25 beats per minute, whereas a shrew has a resting heart rate of over 600 beats per minute.

(a) Atrial systole

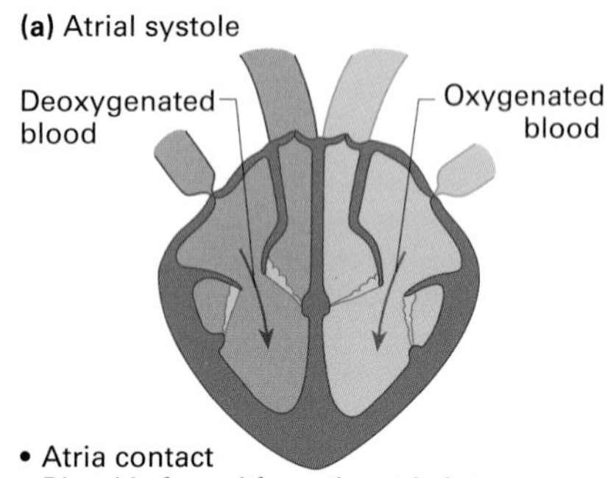

- Atria contact
- Blood is forced from the atria into the ventricles

(b) Ventricular systole

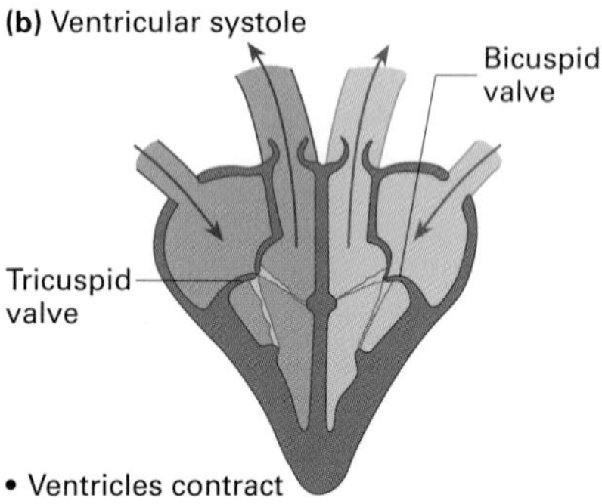

- Ventricles contract
- Blood is forced into the arteries
- Bicuspid and tricuspid valves prevent blood flowing back into the atria
- Atria start to fill with blood again

(c) Diastole

- Heart relaxes and fills with blood from the veins
- Semi-lunar valves prevent blood entering through the arteries

***Figure 1** Main events in a cardiac cycle: **(a)** atrial systole (followed by a **pause**) **(b)** ventricular systole **(c)** diastole.*

Cardiac muscle

The heart consists mainly of **cardiac muscle**, which has a very peculiar property: if an isolated cardiac muscle cell, a group of cells, or even the whole heart, is provided with oxygen, nutrients, and the right mixture of mineral salts, it will contract spontaneously without nervous or hormonal stimulation. These contractions originate within the muscle, and are called **myogenic** contractions.

The cardiac muscle cells work together to produce the heartbeat. This is a sequence of events called the **cardiac cycle** (figure 1).

The pacemaker

The cycle starts at the **sinoatrial node** (**SA node**). This is a small patch of tissue that has its own inherent (natural) rhythm of contraction. The rhythm is made faster or slower by nervous impulses and hormones. The SA node is sometimes called the **pacemaker** because it determines the rate of contraction of the rest of the cardiac muscle.

The SA node generates waves of electrical impulses called **cardiac impulses** over the atria. When the impulses reach the junction between the atria and ventricles, they are delayed by the **atrioventricular septum**. The delay gives time for the wave of contraction to pass over the whole of both atria. The atrioventricular septum is a layer of non-conducting connective tissue. It completely separates the atria from the ventricles except for a region in the right atrium called the **atrioventricular node** (**AV node**).

The AV node is the only route of transmission for the cardiac impulse. It acts as a second pacemaker region, relaying the cardiac impulse to the ventricles. If the SA node malfunctions, the AV node takes over as pacemaker of the whole heart, but it has a slower inherent rhythm.

Systole (contraction)

The cardiac impulse is relayed from the AV node over the ventricles through a bundle of specialised muscle fibres called the **bundle of His** or **atrioventricular bundle**. This bundle branches into other fibres called **Purkyne fibres** (formerly called Purkinje fibres). The cardiac impulse passing through the Purkyne fibres causes a wave of contraction. It starts at the apex of the heart, and passes rapidly over the ventricles (**ventricular systole**). Regions close to the AV node have thin Purkyne fibres which carry impulses more slowly than the thick fibres that supply more distant parts of the ventricles. This ensures that all parts of the ventricles contract more or less simultaneously.

Diastole (relaxation)

The whole heart relaxes after ventricular systole, allowing blood from the veins to fill up the heart. This stage of the cardiac cycle is called **diastole**.

Pressure changes in the heart

Figure 2 shows the pressure changes that take place on the left side of the heart during a cardiac cycle. Note that the maximal pressure in the left ventricle is far higher than that in the atrium. This is because the ventricle has to work hard to pump blood to all parts of the body (except the lungs). The atrium only pumps blood into the ventricle.

Electrical changes in the heart: the ECG

Cardiac muscle contracts as a result of electrical stimulation. A wave of electrical charge is initiated in the pacemaker region and spreads across the heart. This generates electrical currents in the body fluids around the heart. The currents can be detected on the body surface using recording electrodes. Electrical signals can then be shown on a cathode ray oscilloscope or a chart recorder. The record produced by this procedure is called an **electrocardiogram** (**ECG**).

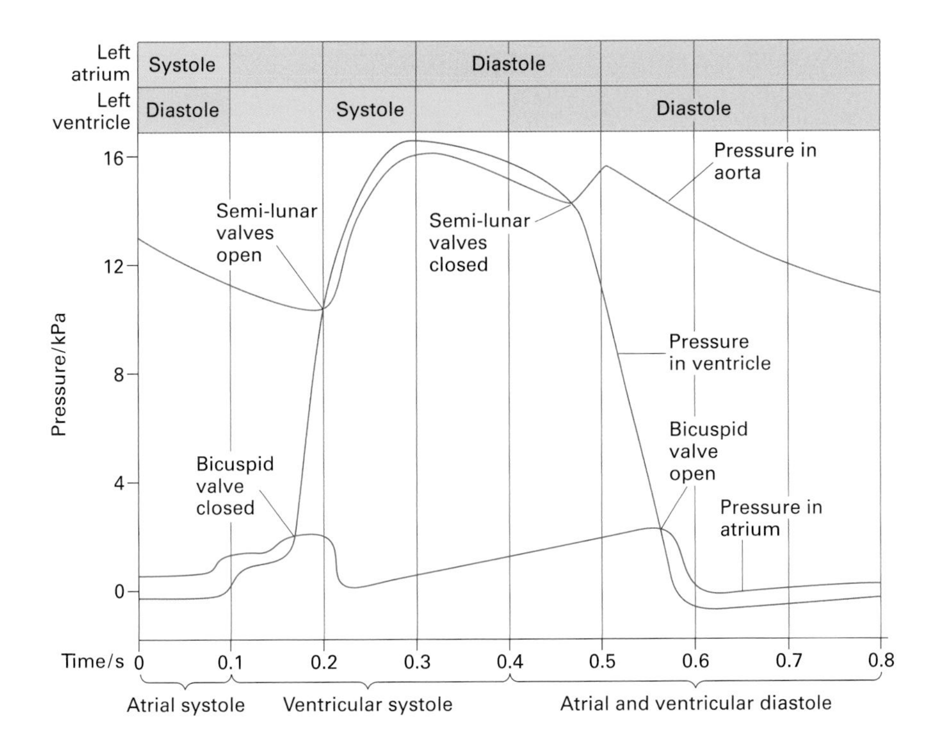

Figure 2 *Pressure changes in the left side of the heart during the cardiac cycle.*

Note that closure of the valves is a passive process. It depends on the relative pressures either side of the valve. For example, the atrioventricular valves (bicuspid and tricuspid valves) close when the pressure in the ventricles is higher than the pressure in the atrium. They open when the pressure is higher in the atrium than in the ventricle.

A typical ECG consists of characteristic waves which correspond to particular events in the cardiac cycle (figure 3).

- The P wave is caused by atrial systole.
- The QRS wave is caused by ventricular systole.
- The T wave coincides with ventricular diastole.
- The heart rate can be calculated from the interval between one P wave and the next.

If disease disrupts the heart's conduction system the ECG is changed. ECGs are therefore used to diagnose cardiovascular disease. The doctor can tell what has happened to the heart from the pattern of the ECG.

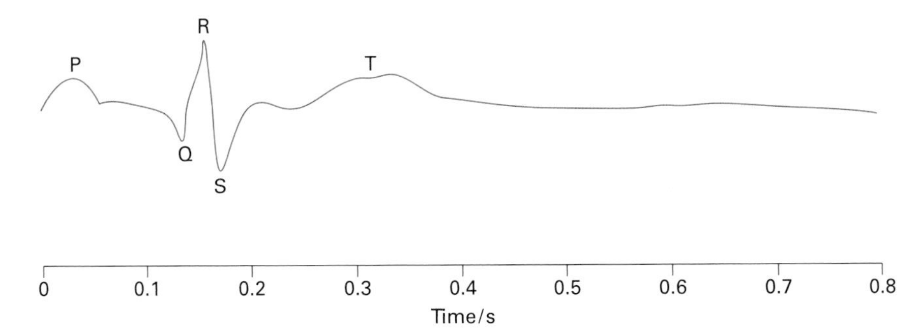

Figure 3 *An electrocardiogram (ECG) shows the electrical activity in the heart. This trace shows a single cardiac cycle in a healthy heart.*

Quick check

1 a List the four stages of the cardiac cycle in the order in which they occur.

b Using figure 2, estimate the duration of ventricular systole.

2 State the precise location of the main pacemaker of the heart.

Food for thought

Figure 2 shows the pressure changes on only the left side of the heart. Suggest how the pressures will change on the right side of the heart during the cardiac cycle. In what ways will the changes be similar to those on the left side; in what ways will they be different?

7.7

The control of the heart

OBJECTIVES

By the end of this spread you should be able to:

- define cardiac output
- describe how to measure the pulse rate
- explain how the action of the heart is controlled.

Fact of life

In some people, the conduction of electrical impulses from the natural pacemaker of the heart (the SA node) is impaired. This condition, known as heart block, can be remedied by inserting an artificial pacemaker under the muscle in the upper thorax (figure 1). The battery-powered pacemaker stimulates the heart through a wire connected to the ventricle or the heart lining. New pacemakers fitted with complex electronic circuits can sense changes in breathing, body temperature, and movement so that the heart rate is adjusted accordingly.

Cardiac output

The amount of blood pumped around the body by the heart depends on two factors: how quickly the heart is beating and the amount of blood it pumps out per beat. These two factors give the **cardiac output**, which is defined as the product of **stroke volume** (the volume of blood leaving the left ventricle with each beat) and **heart rate** (number of heart beats per minute).

$$\text{Cardiac output} = \text{stroke volume} \times \text{heart rate}$$

In humans, typical values for a sedentary male at rest are a stroke volume of 75 cm^3 and a heart rate of 70 beats per minute. When exercising, both the stroke volume and the heart rate increase significantly. The heart rate also varies considerably from one individual to another. Males tend to have lower heart rates than females, and well trained endurance athletes tend to have very low resting heart rates, often below 40 beats per minute. An endurance athlete develops strong heart muscles and has a high stroke volume, so that the cardiac output at rest is about the same as that of a sedentary person with a higher heart rate.

The heart rate is often measured by taking the pulse. In most circumstances pulse rate and heart rate are identical, but this is not so if the heartbeat becomes irregular. The pulse is actually a wave of pressure that passes along the arteries, causing them to expand and recoil rhythmically. The pulse can be felt in any artery that passes over a bone close to the body surface (figure 2).

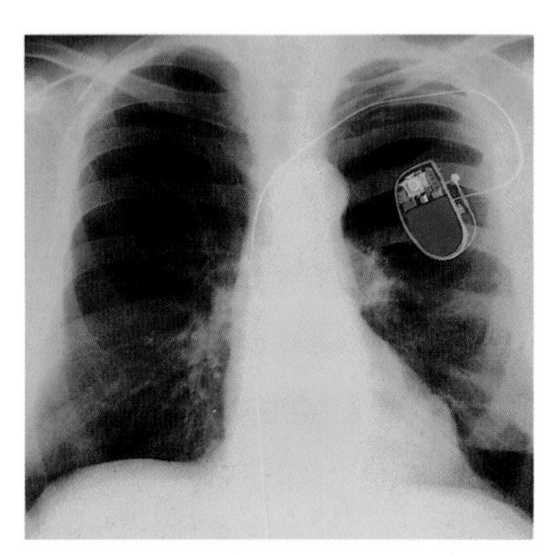

Figure 1 An artificial pacemaker fitted into the chest cavity. The pacemaker sends regular electrical pulses which help keep the heart beating regularly. Having a pacemaker fitted is one of the most common types of heart surgery. In 2010 in England, more than 40 000 people had a pacemaker fitted.

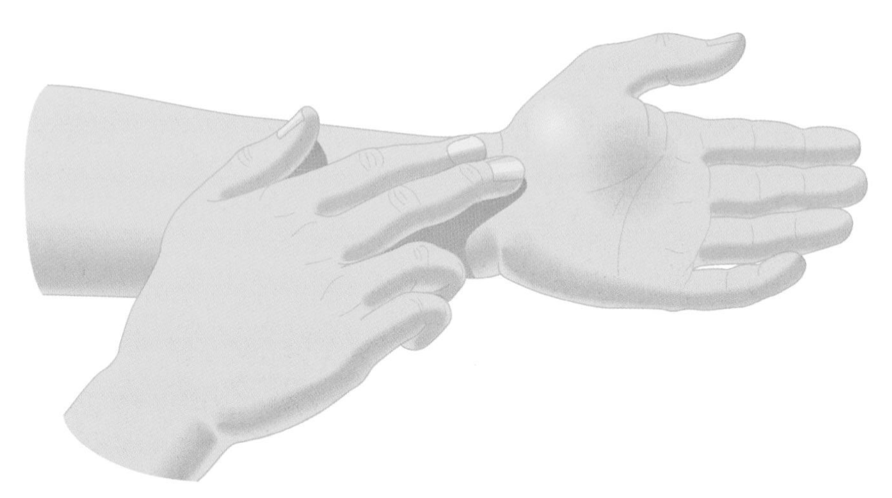

Figure 2 One of the best sites for checking the pulse is directly above the base of the thumb on the underside of the wrist. This is called the ***radial pulse*** *because it is where the radial artery passes alongside the radius. To count the pulse, press the fingertips of your middle and index fingers gently but firmly against the radial artery. It usually takes a few seconds to become aware of the pulse, and a few seconds more to become sensitive to the rhythm. Once this is established, an accurate count can be made over a 10- or 15-second period, and the pulse rate per minute calculated.*

Control of cardiac output

As cardiac output is the product of stroke volume and heart rate, a change in either of these will affect the amount of blood pumped out of the heart each minute.

Although heart muscle has its own inherent rhythm, the heart rate is carefully regulated by the nervous and hormonal systems so that the cardiac output can adapt to the demands of a particular situation. Sensory receptors in the walls of the heart chambers and some blood vessels (e.g. the **carotid** and **aortic sinuses**) are sensitive to changes in blood pressure. These receptors convey information to the cardiorespiratory centre in the medulla oblongata at the base of the brain (see chapter 10). A branch of the vagus nerve, part of the parasympathetic nervous system, leads directly to the sinoatrial node. The vagus is an inhibitory nerve, and impulses from it slow the heart rate (figure 3).

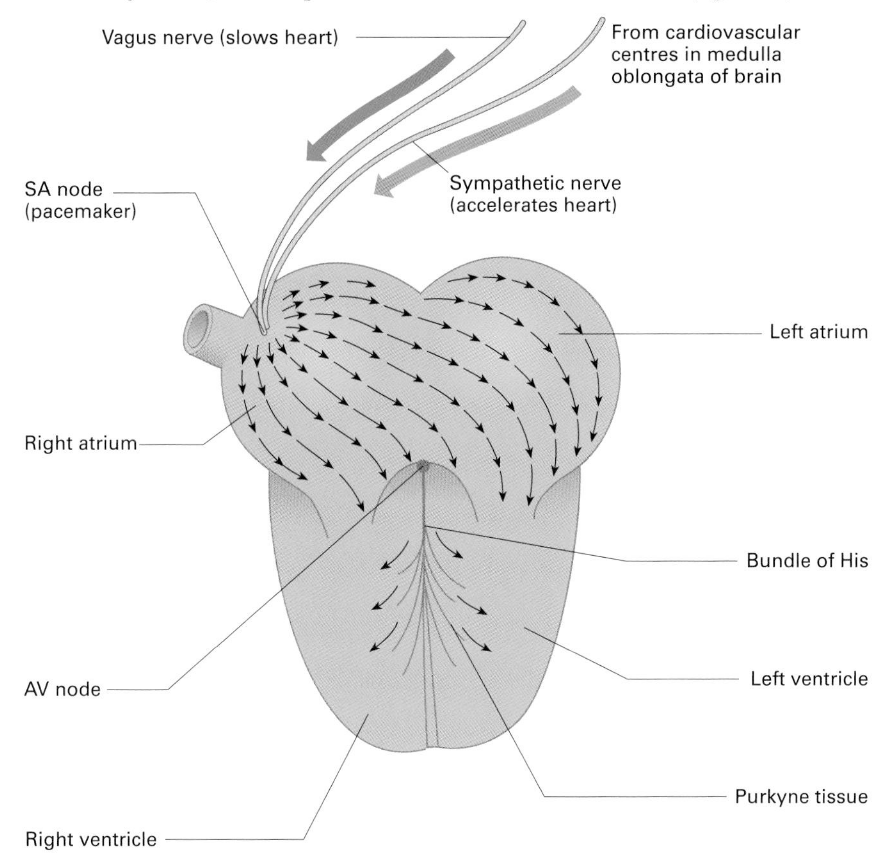

Figure 3 *The cardiac output is modified by the parasympathetic and sympathetic nervous systems.*

Branches of the sympathetic nerve have the opposite effect on the heart: impulses from these nerves increase the heart rate. At times of excitement or danger, the sympathetic nervous system also stimulates the release of adrenaline from the adrenal glands. Adrenaline increases both the strength and speed of cardiac contractions (that is, it increases both the stroke volume and the heart rate).

Cardiac output varies with the volume of blood returning to the heart (the **venous return**). When the venous return is high, the walls of the right atrium are stretched and the heart beats faster (the **Bainbridge reflex**). A high venous return also stretches the walls of the left ventricle, causing the ventricles to contract more strongly, giving a greater stroke volume (the **Frank–Starling effect**). These responses enable the heart to adjust the strength and rate of its contractions according to the volume of blood passing through it at any given time.

Quick check

1 a Calculate the cardiac output of a heart when the stroke volume is 60 cm^3 and the heart rate is 75 beats per minute.

b Calculate the volume of blood ejected from this heart in one hour.

2 Where is the radial pulse taken?

3 Suggest what will happen to the heart rate if the vagus nerve is cut.

Food for thought

Suggest what would happen if the nerve supply to the heart were removed totally.

7.8

OBJECTIVES

By the end of this spread you should be able to:

- list the main components of blood
- discuss the functions of blood
- describe the process of blood clotting.

Fact of life

In adult mammals, most blood cells derive from special **haematopoietic stem cells** in the bone marrow. (The word 'haematopoietic' comes from two Greek words that mean 'blood' and 'to make'.) These stem cells are undifferentiated and have the ability to divide and give rise to any of the different types of blood cell. After a blood cell matures and differentiates to become specialised, it loses its **multipotency** (its ability to form blood cells of any other type). A healthy adult human produces millions of new blood cells each day just to maintain the number in the circulatory system. The biology of haematopoietic stem cells is now so well understood that transplants are used routinely to treat patients with cancers and other disorders of the blood and immune systems.

Blood

Blood makes up about one-twelfth of the body mass of a mammal. The average human has 4–6 dm^3 of blood circulating around the body. Blood consists of cellular components suspended in a fluid called **plasma**. The cellular components are red blood cells, white blood cells, and platelets. They can be separated from the plasma by centrifugation (figure 1).

Red blood cells, white blood cells, and platelets

The red blood cells or **erythrocytes** of most mammals are round discs, concave on each side (they are biconcave) and have no nucleus. The size varies between species: human red blood cells, for example, are about 8 µm in diameter. They are filled with a red protein called **haemoglobin**. The main function of haemoglobin is to carry oxygen (spread 7.9).

Human red blood cells live for only about 120 days. During that time each cell will have made many thousands of journeys around the body. Because of their short life span, red blood cells have to be replaced continuously, mainly by cells in the bone marrow.

Several different types of white blood cell (**leukocytes**) are involved in defence. Two types, **neutrophils** and **monocytes**, can move freely around the body, even penetrating deep into bone. These cells are also known as **phagocytes** because they engulf foreign substances and other cells such as bacteria. (For more detail on the structure and function of white blood cells, see spread 15.5.)

Platelets are fragments of cells broken off from large cells in the bone marrow. They play an important role in blood clotting.

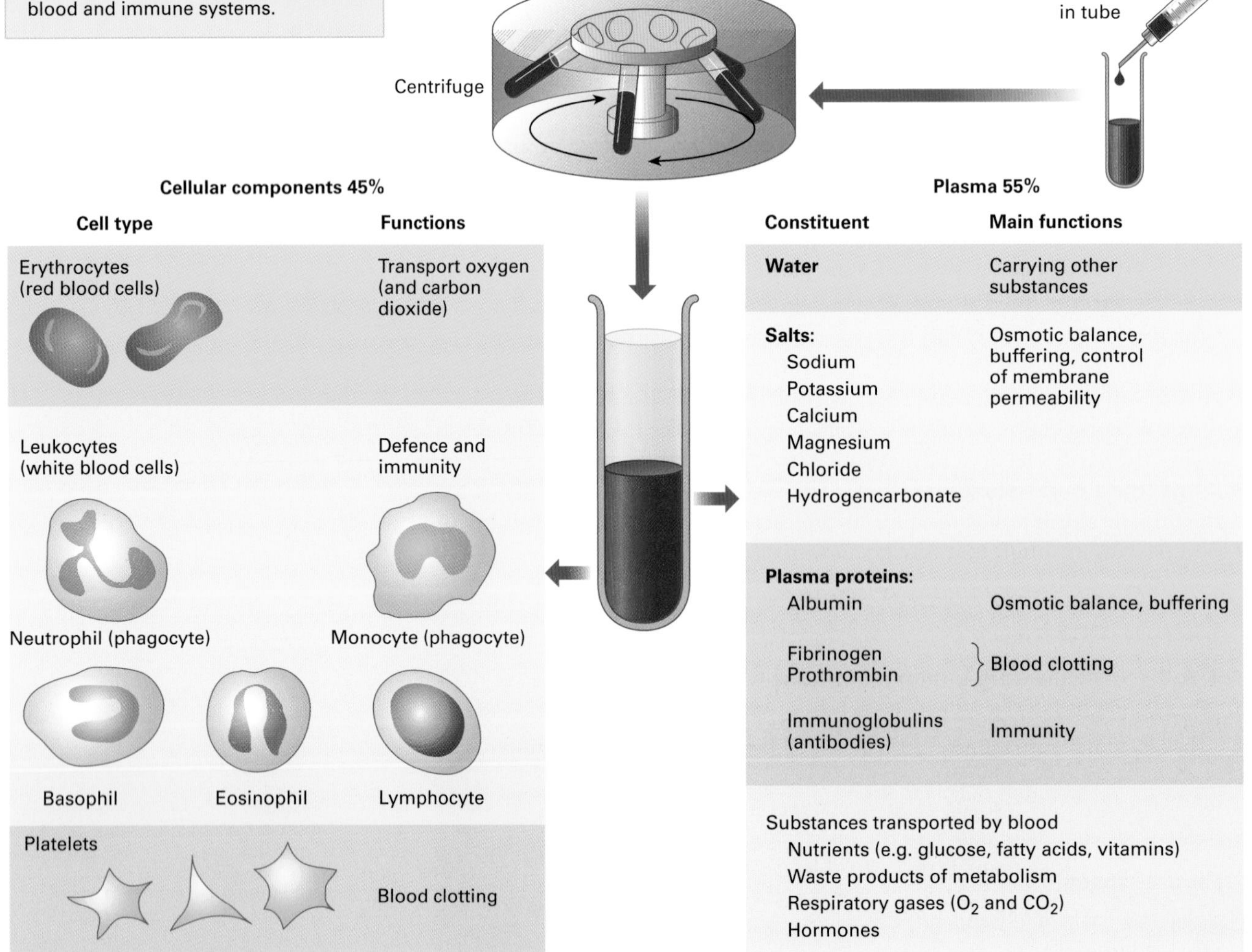

***Figure 1** The major components of blood can be separated by centrifugation.*

Counting blood cells

The number of cellular components in each cubic millimetre of blood can be estimated using a special slide called a **haemocytometer** (figure 2). The slide holds a known volume of a liquid within an area divided up into squares. The cellular components within these squares are counted, and the total number of each component per cubic millimetre estimated. In each cubic millimetre of blood there are between 5 and 6 million red blood cells, 5000 to 10 000 white blood cells, and 250 000 to 400 000 platelets.

Plasma and blood clotting

The straw-coloured plasma is mainly water, but is slightly denser than pure water because it contains many dissolved substances. The plasma is the main transport medium in the body. In addition to many chemicals, it also transports heat from hot regions of the body to cooler regions.

Plasma proteins are involved in buffering (keeping body fluids at a constant pH) and defence against injury and disease. Fibrinogen plays a key role in blood clotting. Removal of this protein from plasma produces **serum**, which does not clot.

If a blood vessel is ruptured, it is vital to stem the flow of blood. Most mammals can lose up to one-third of their blood without any long-lasting damage, but if a mammal loses more than half of its blood there is little chance of survival.

Blood clotting minimises blood loss following injury. The blood coagulates to form a solid plug (clot) made of cells trapped in a fibrous network (figure 3). The clot prevents further blood loss, reduces the risk of pathogens (harmful microorganisms) entering the body, and provides a framework for the repair of damaged tissue.

Blood clotting involves a complex series of biochemical reactions. If a blood vessel is damaged, collagen fibres in the vessel wall become exposed to blood. Platelets stick rapidly to the exposed collagen fibres. The platelets release **thromboplastins** (clotting factors). These clotting factors, with the help of calcium and vitamin K, convert **prothrombin** (an inactive plasma protein) to **thrombin** (an active plasma protein). Thrombin acts as an enzyme, catalysing the conversion of soluble **fibrinogen** into insoluble **fibrin**. Fibrin forms the network of fibres that traps blood cells and debris to form the clot.

Blood does not normally clot in intact blood vessels because of the action of a number of anticoagulants such as heparin circulating in the bloodstream. Intact endothelium (the inner lining of blood vessels) also produces molecules which inhibit clotting. Blood clots quickly when exposed to air because of the absence of an endothelium and a lack of anticoagulants.

Quick check

1 Distinguish between plasma and serum.

2 Name two proteins mentioned in this spread which help to buffer the blood.

3 What is the function of fibrinogen?

Food for thought

Mammalian red blood cells lack nuclei, but the red cells of all other vertebrates (fish, amphibians, reptiles, and birds) have nuclei, and almost all are oval. The significance of whether red blood cells have a nucleus is not understood. However, non-mammalian red blood cells are sometimes over 100 times larger than those of mammals. Suggest *possible* advantages and disadvantages of red blood cells having a nucleus.

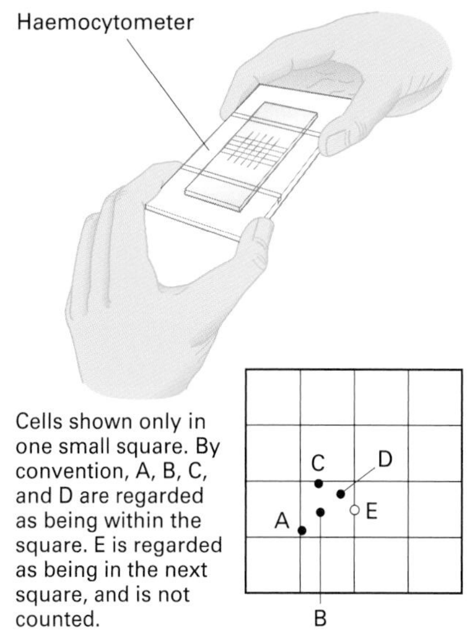

Figure 2 A haemocytometer used to estimate the number of blood cells and other cells. By knowing the depth of the liquid on the slide (0.1 mm) and the area of each small square (0.0025 mm²) it is possible to estimate the density of the cells on a haemocytometer. By convention, only cells in a square and touching its top or left-hand side are counted (that is, cells ABCD). Therefore the density of the cells in the square is 4 per 0.00025 mm³ or 16 000 per mm³.

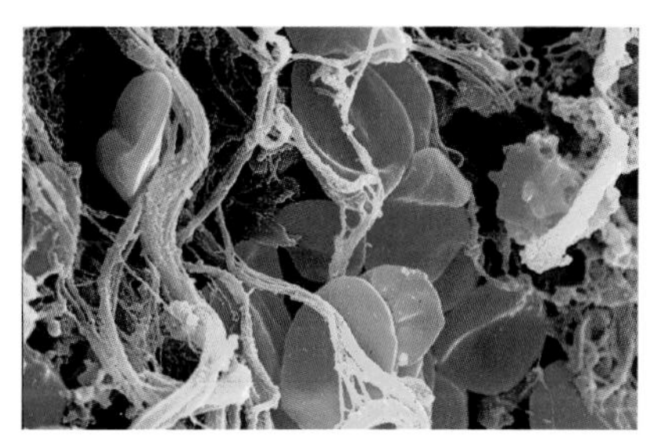

Electron micrograph showing red blood cells trapped in a fibrous mesh. (x2000)

Platelets in contact with collagen fibres exposed in damaged tissue
↓
Thromboplastins (clotting factors)
Calcium ions → ← Vitamin K
↓
Prothrombin (inactive) → Thrombin (active)
↓
Fibrinogen (soluble) → Fibrin (insoluble)
↓

Figure 3 Blood clotting: the clotting factors act in series, each factor being converted into its activated form which then activates the next factor in the pathway. This produces a cascade effect which has been called a 'biochemical amplification' because an initial small reaction leads to a large final reaction. Lack of clotting factors can cause excessive bleeding, as in haemophilia A (caused by lack of clotting Factor VIII) and haemophilia B or Christmas disease (caused by a deficiency of clotting Factor IX).

7.9 Haemoglobin

OBJECTIVES

By the end of this spread you should be able to:

- describe the structure of haemoglobin
- explain how haemoglobin transports oxygen
- interpret oxygen dissociation curves for haemoglobin.

Fact of life

Respiratory pigments are not confined to the animal kingdom: leguminous plants such as beans have a red oxygen-binding protein pigment resembling haemoglobin. It is called leghaemoglobin and it occurs in root nodules where it takes up oxygen and maintains an anaerobic environment for nitrogen-fixing bacteria, which are anaerobic.

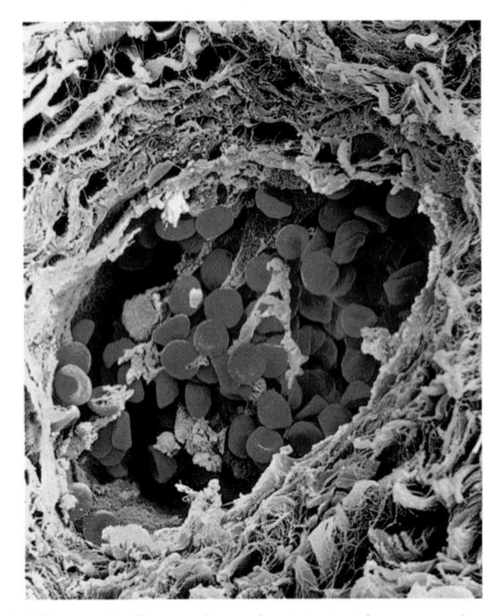

***Figure 1** Scanning electron micrograph of a small blood vessel containing red blood cells.*

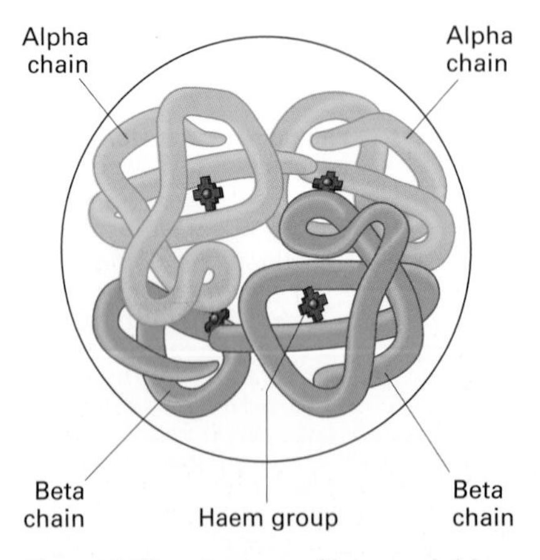

***Figure 2** The structure of haemoglobin.*

Carrying oxygen

Blood is the main transport medium for respiratory gases. Most of the carbon dioxide excreted by cells is transported in solution as hydrogencarbonate ions (see below). Oxygen does not dissolve well in water and only a very small amount (no more than 20 cm^3 in humans) is carried in solution. Most of the oxygen supplying mammalian cells is carried around the body by **haemoglobin**. The total haemoglobin content of blood is about 750 g, which is normally confined within red blood cells (figure 1).

The structure of haemoglobin

- Haemoglobin is a conjugated protein (figure 2).
- The protein part (called **globin**) consists of four polypeptide chains. These chains are of two types called alpha and beta. They are about the same length (about 140 amino acids) but have slightly different compositions.
- Each chain is combined with a non-protein prosthetic group called **haem**. Haem consists of an atom of iron enclosed in a ring structure.
- Each haem group can combine with one molecule of oxygen. This process is called oxygenation. (It is not the same as oxidation, because the iron does not lose any electrons and is not chemically oxidised.) Each molecule of haemoglobin can therefore combine with a maximum of four molecules of oxygen.

Oxygen dissociation curves

The degree of oxygenation of haemoglobin is determined by the partial pressure of oxygen $p(O_2)$ in the immediate surroundings (see spread 7.3). If $p(O_2)$ is low (as in the capillaries of respiring tissues), haemoglobin carries a relatively small amount of oxygen; if $p(O_2)$ is high (as in the alveolar capillaries), haemoglobin becomes almost saturated with oxygen.

An oxygen dissociation curve shows the degree of haemoglobin saturation with oxygen plotted against different values of $p(O_2)$. The curve (figure 3) is an S- or sigmoid shape.

- At $p(O_2)$ close to zero there is no oxygen bound to the haemoglobin.
- At low $p(O_2)$, the polypeptide chains are tightly bound together, making it difficult for an oxygen molecule to gain access to the iron atoms, and the curve rises only gently.
- As one molecule of oxygen becomes bound to one haem group, the polypeptide chains open up, exposing the other three haem groups to oxygen. This makes it much easier for them to become oxygenated, and the curve rises steeply.
- At very high $p(O_2)$, the haemoglobin becomes saturated and the curve levels off.
- The oxygen dissociation curve is steep within the narrow range of $p(O_2)$s in the body. The percentage saturation of haemoglobin in this range varies greatly for small changes in $p(O_2)$. This means that oxygen readily combines with haemoglobin in the lungs, where $p(O_2)$ is higher, and is readily released from haemoglobin in the tissues, where $p(O_2)$ is lower.

Carbon dioxide and the Bohr shift

Unloading of oxygen in the capillaries of tissues is helped by the relatively high concentrations of carbon dioxide produced by cellular respiration. Carbon dioxide reduces the affinity of haemoglobin for oxygen at $p(O_2)$s in the body. This therefore shifts the oxygen dissociation curve to the right. This effect is known as the **Bohr shift** or Bohr effect (figure 4).

The Bohr shift results from the way in which carbon dioxide is transported in the blood. About 5 per cent of carbon dioxide is carried in solution as molecular carbon dioxide, and a small percentage combines with amino groups in haemoglobin to form **carbaminohaemoglobin**. However, carbon dioxide is carried mainly in solution as hydrogencarbonate ions. Figure 5 shows that most hydrogencarbonate ions are formed by a series of reactions in red blood cells.

- First carbon dioxide (CO_2) diffuses into red blood cells where it is converted into carbonic acid (H_2CO_3). This reaction is catalysed by the enzyme **carbonic anhydrase**.
- Carbonic acid dissociates, forming protons (H^+) and hydrogencarbonate ions (HCO_3^-).
- The hydrogencarbonate ions diffuse out of the cell. They are transported in solution in the plasma.
- Chloride ions (Cl^-) diffuse inwards from the plasma to maintain electrical neutrality. This process is called the **chloride shift**.
- The protons left inside the cell are mopped up by haemoglobin to form **haemoglobinic acid** (HHb). This forces the haemoglobin to release its oxygen load, hence the Bohr shift.
- By taking up excess protons haemoglobin is acting as a buffer. This is important in preventing the blood from becoming too acidic.

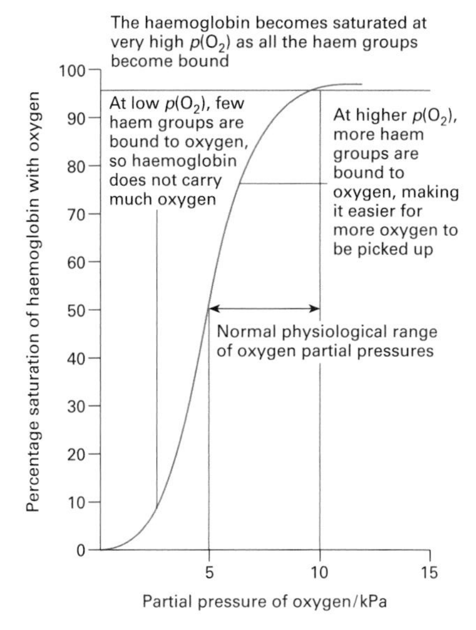

***Figure 3** Oxygen dissociation curve for adult haemoglobin.*

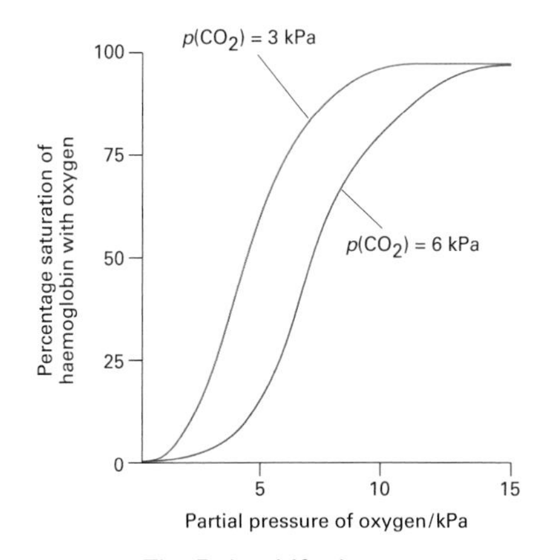

***Figure 4** The Bohr shift: the movement of the oxygen dissociation curve to the right at higher p(CO_2) shows that carbon dioxide reduces the affinity of haemoglobin for oxygen (within the normal physiological range of oxygen partial pressures.)*

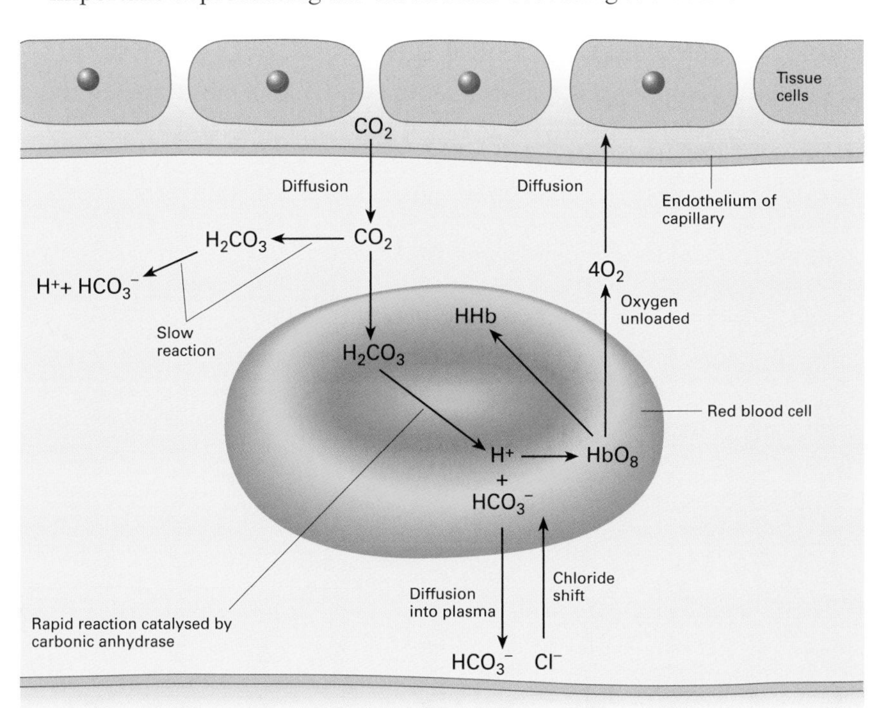

***Figure 5** Formation of hydrogencarbonate ions and the transport of carbon dioxide in the blood.*

Quick check

1 Why is haemoglobin called a conjugated protein?

2 How many oxygen molecules can each haemoglobin molecule transport?

3 What is the effect of high carbon dioxide concentrations on the oxygen dissociation curve of haemoglobin?

Food for thought

Suggest explanations for the following observations:

a Lugworms and midge larvae called bloodworms that live in mud have haemoglobin with an oxygen dissociation curve to the left of human haemoglobin.

b The haemoglobin of a pigeon has an oxygen dissociation curve well to the right of human haemoglobin.

c Human haemoglobin is confined within cells and has a relative molecular mass of about 68 000, whereas the haemoglobin of the earthworm *Lumbricus* is free in the plasma and has a relative molecular mass of about 2 946 000.

7.10

THE EFFECTS OF EXERCISE

OBJECTIVES

By the end of this spread you should be able to:

- describe the effects of exercise on respiration and on the circulation.

Fact of life

Roger Bannister was the first to break the four-minute mile barrier at Iffley Road, Oxford, on 6 May 1954. The photograph (figure 1) shows that he was totally exhausted at the end of the run, both mentally and physically, having used all his reserves of energy to gain the world record.

Figure 1 *Roger Bannister completing the first four-minute mile at Iffley Road, Oxford.*

In an all-out effort, gasping for breath and with heart pounding, an athlete generates energy at more than 12 times the resting rate (figure 1). To obtain this amount of energy, the athlete has to have a very efficient means of acquiring oxygen from the air, transporting it to active muscles, and using it in cellular respiration. The oxygen is required to respire food molecules so that ATP can be synthesised.

Exercise and breathing

During exhausting exercise lasting a few minutes, the rate of breathing may rise to more than 30 breaths per minute and ventilation of the lungs may exceed 120 dm^3 per minute. All the respiratory muscles, the internal and external intercostal muscles and the diaphragm, work as hard as they can to draw air in and out of the lungs. A substantial proportion of the energy produced by cellular respiration is used in these ventilatory movements.

During a sprint, an athlete cannot obtain sufficient oxygen to meet his or her energy requirements by aerobic respiration alone. Therefore the athlete has to generate ATP anaerobically. The difference between the theoretical oxygen requirement and the volume of oxygen actually obtained is called the **oxygen deficit**. Lactate and hydrogen ions form as waste products of anaerobic respiration (spread 6.2). Their accumulation in the blood leads to muscle fatigue and limits the contribution that anaerobic respiration can make to an activity.

At the end of the sprint, the sprinter breathes much more quickly and deeply, and consumes more oxygen than normal (figure 2). This process is commonly referred to as paying off the **oxygen debt**. However, exercise physiologists point out that this term gives a false impression: it implies that oxygen has somehow been 'borrowed' from a store during exercise, and must be replaced. Although a small amount of oxygen *is* 'borrowed' from oxymyoglobin during exercise (see opposite), reoxygenating myoglobin uses only a small proportion of the extra oxygen consumed after exercise. A better description of the extra oxygen consumed is **recovery oxygen**, since most of the oxygen is used to restore the body to its pre-exercise condition. In addition to replacing oxymyoglobin stores, the extra oxygen is used to repair tissue and to convert lactate to glucose which is then stored as glycogen.

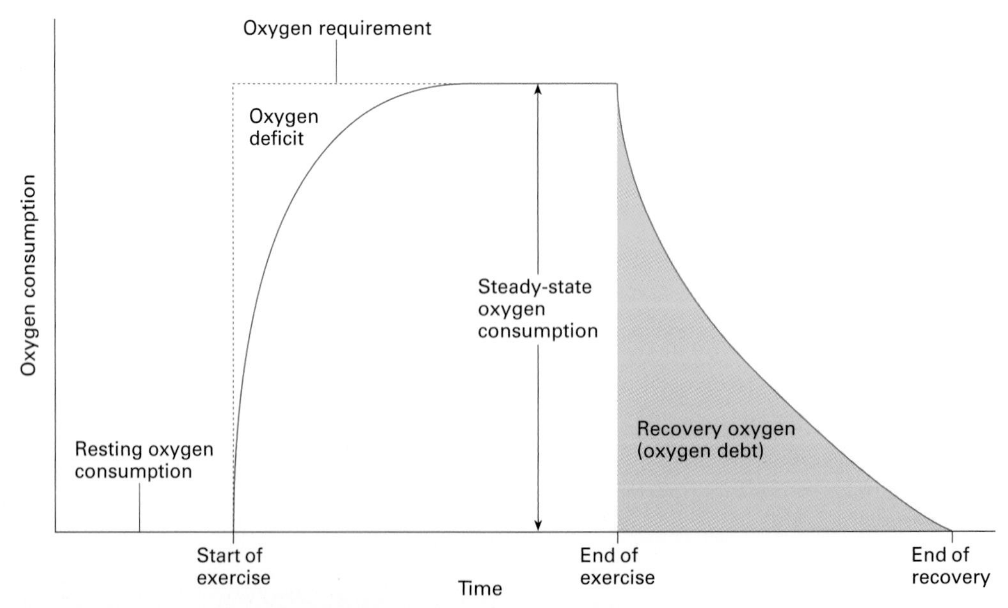

Figure 2 *Oxygen consumption and exercise. The oxygen deficit arises because the body's oxygen requirements cannot be met by breathing. Steady-state oxygen consumption occurs when the oxygen requirements and oxygen supply are in equilibrium: aerobic metabolism can meet the energy demands of the body. The recovery oxygen restores the body to its pre-exercise condition.*

Exercise and circulation

Exercise requires a good supply of oxygenated blood to sustain active muscles. This is achieved by increasing the cardiac output and by pumping proportionately more of the body's blood to the muscles. Even the anticipation of exercise increases the stroke volume and heart rate. At full speed, a sprinter's heart may beat more than 200 times per minute. The heart rate remains high for a while after exercise.

During exercise, the skeletal muscles receive proportionately more of the body's blood by a process called **shunting**. By dilating blood vessels in one part of the body and constricting those in other parts, blood is diverted to active muscles from, for example, the intestines. This is one reason why it is unwise to exercise after a big meal. Whereas the blood supply to the intestines and skeletal muscles varies with the level of exercise, the blood suppy to the vital organs, especially the brain, heart, and kidneys, remains relatively constant.

When the blood reaches muscle tissue, the maximum amount of oxygen is given up by the blood. Carbon dioxide excreted by the respiring cells causes haemoglobin to release more oxygen because of the Bohr effect. The high temperatures in and around very active muscles cause even more oxygen to be unloaded (oxygen dissociation increases with temperature). At high workloads, muscle is able to extract a much higher percentage of oxygen from blood than at low workloads.

Aerobic exercise

Aerobic exercise is any repetitive, rhythmical, relatively low-intensity exercise using large muscle groups. Aerobic exercise increases the body's demands for oxygen and adds to the workload of the heart and lungs, strengthening the cardiovascular system. To obtain lasting benefit, exercise should have a frequency of at least three days per week, an intensity of between 60 and 80% maximum heart rate (HR_{max} = 220 – age in years) and a duration of 20–60 minutes of continuous activity.

The role of myoglobin

When muscles are very active, $p(O_2)$ becomes very low and the muscle takes advantage of an additional store of oxygen. This store is in the form of **oxymyoglobin**, the oxygenated form of myoglobin, a respiratory protein found in skeletal muscles. Myoglobin is similar to haemoglobin, but has only one polypeptide chain with a single haem group. Each myoglobin molecule can therefore carry only one oxygen molecule. Myoglobin has a hyperbola-shaped oxygen dissociation curve (figure 3) showing that it combines very readily with oxygen but releases it only when $p(O_2)$ is very low.

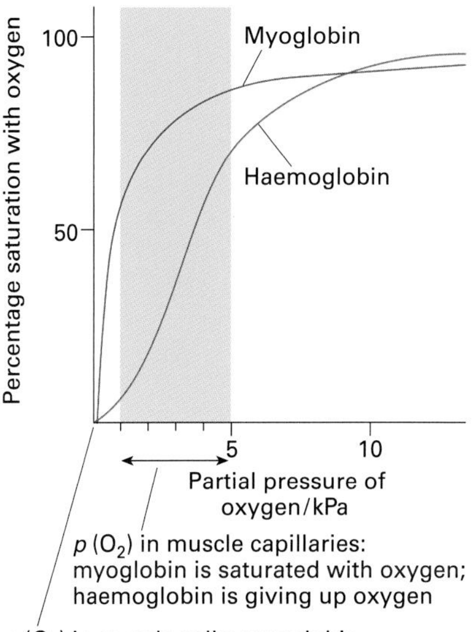

***Figure 3** Oxygen dissociation curves for myoglobin and haemoglobin.*

The advantages of regular exercise

Regular aerobic exercise (typified by brisk walking, cross-country skiing, long-distance running, swimming, and cycling) has profound long-term effects on respiration and circulation. The ventilation mechanism becomes more efficient. Respiratory muscles become stronger, the blood supply to the lungs is increased, and the ability of blood to take in oxygen from the alveoli is improved. The blood volume and total number of red blood cells are increased. The heart becomes enlarged (a harmless condition called **athlete's heart**) and the resting pulse rate is lowered, indicating a more efficient oxygen transport system.

Regular exercise also improves the ability of cells to generate and use energy because the concentration of respiratory enzymes and number of mitochondria are increased.

Quick check

1 a List three effects of exercise on:

i the respiratory system

ii the cardiovascular system.

b Give three factors that cause oxygen to be unloaded from blood supplying active muscle tissue.

Food for thought

During exercise, blood is shunted (redistributed) from one part of the body to another. For example, the blood supply to the muscles increases many times, whereas that to the liver and intestines decreases. Suggest (with reasons) what might happen to the blood supply during exercise to:

a the brain

b the heart

c the kidneys.

7.11

Tissue fluid and lymph

OBJECTIVES

By the end of this spread you should be able to:

- explain how tissue fluid and lymph are formed
- discuss the relationship between blood, tissue fluid, and lymph.

Fact of life

Lymph vessels can be blocked by a number of factors. This can result in a build-up of lymph and tissue fluid which can cause massive swelling of tissues. The effects of blocked lymph vessels can be seen in people suffering from elephantiasis. This condition is caused by infection with a parasitic nematode worm, *Wucheria bancrofti*, which most commonly affects the legs (figure 1).

W. bancrofti played an important part in the history of parasitology: Patrick Manson, a British doctor, discovered in 1878 that a mosquito was required in the worm's life cycle. This was the first proof that an internal parasite could be transmitted by an insect.

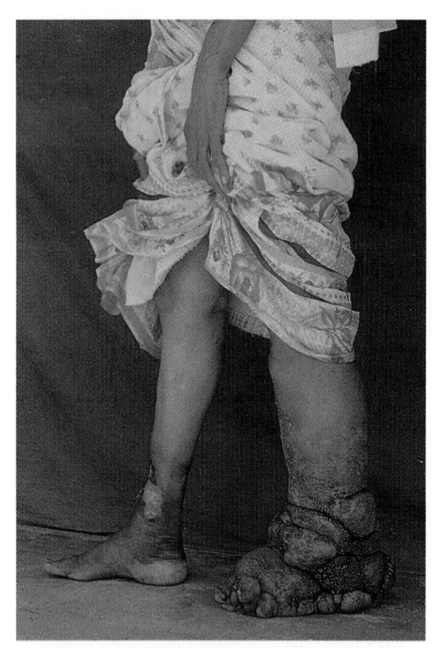

Figure 1 *A severe case of elephantiasis.*

Tissue fluid: the medium surrounding body cells

Mammalian cells are bathed in a fluid called **tissue fluid**. This fluid forms the immediate environment of the cells. It is the source of water, oxygen, and nutrients. It also receives excretory substances from the cells. The chemical composition, temperature, and pH of the tissue fluid are very important to the cells. If the conditions are unsuitable, a cell may not function efficiently, or it may even die.

Tissue fluid is derived from blood

Blood at the arterial end of a capillary is under high pressure. This pressure is sufficient to force water and small molecules out through the endothelium (the capillary lining) in a process called **ultrafiltration**. The fluid seeps into the interstitial spaces (the spaces between cells).

Tissue fluid normally contains solutes in concentrations similar to those in blood plasma. However, it has little if any proteins and no red blood cells. Plasma proteins such as albumin, globulin, and fibrinogen are too large to filter through the endothelium (although small amounts may leak through very thin capillary walls). The proteins remaining within the capillary exert an osmotic effect. This tends to keep some fluid in the capillary, countering the ultrafiltration pressure. Similar effects operate in the interstitial spaces, so that the movement of fluid out of or into the capillaries is determined by the relative pressures in the tissue fluid and the blood. The total pressure at the arterial end of the capillaries is usually greater than that in the tissue fluid: water and salts therefore enter the tissue fluid. Blood pressure is much reduced at the venous end of capillaries, and the reverse process takes place: water, salts, and waste products flow back into the capillary (figure 2).

The lymphatic system

Not all the tissue fluid circulates back into the venous ends of capillaries. Some of it drains into a second circulation system: the **lymphatic system**. In humans, the lymphatic system is an extensive network of vessels that resemble veins (figure 3). The colourless or pale yellow fluid within the lymphatic system is called **lymph**. Lymph has a composition similar to tissue fluid, but it contains more fatty substances. It also contains more protein and white blood cells because of the activity of cells within the lymphatic system.

The largest organ of the lymphatic system is the **spleen**, located to the left of and just behind the stomach. The spleen stores an emergency supply of blood and also contains white blood cells. Other lymphatic organs include the adenoids, tonsils, appendix, and vessels within the bone marrow. The **thymus** is prominent in infants but usually degenerates in early childhood. It is important in the maturation of certain **lymphocytes** called T cells, which play a major role in the immune system. Along the larger lymph vessels are numerous sac-like organs called **lymph nodes**.

Defence against disease

The lymph nodes help to protect the body against disease in several ways.

- They produce and store white blood cells called lymphocytes.
- When foreign substances invade the body, these white blood cells multiply rapidly and become very active, engulfing bacteria and secreting antibodies to kill them. The accumulation of white blood cells and dead bacteria can cause the lymph nodes to become swollen and tender.
- The lymph nodes also filter out foreign particles, bacteria, and dead tissue before these enter the bloodstream.

The role of the lymphatic system in the immune response is described in more detail in chapter 15.

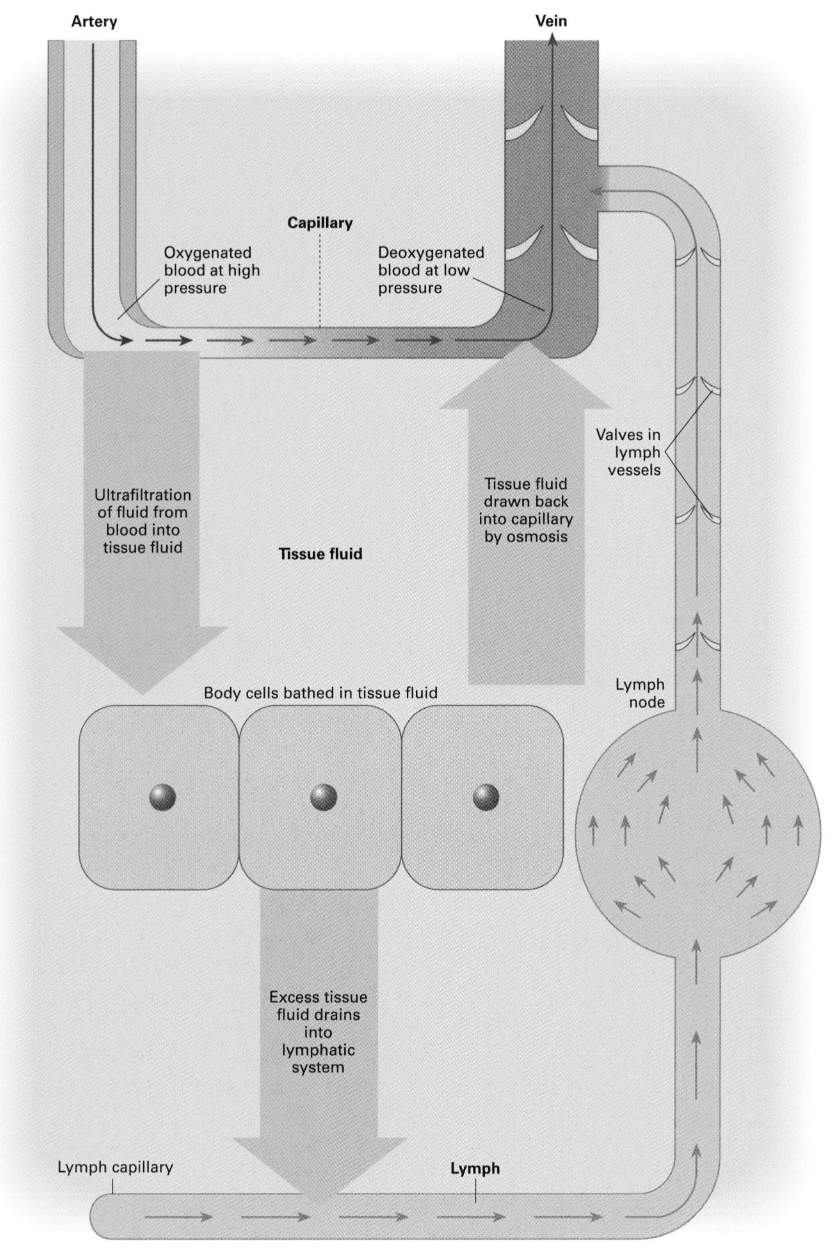

Figure 2 *The formation of tissue fluid and lymph from blood plasma.*

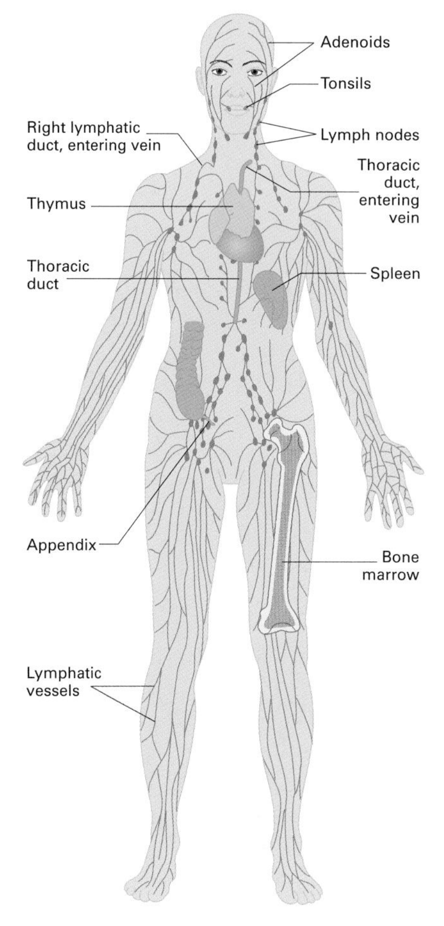

Figure 3 *The human lymphatic system.*

Transport of tissue fluid

The primary function of the lymphatic system is to carry excess tissue fluid back to the blood. This is essential for creating a suitable environment for cells, and for keeping conditions in the blood constant. The lymphatic system also transports substances such as fatty acids and some enzymes and hormones. The **lacteals** into which fatty acids are absorbed in the small intestine are small branches of the lymphatic system. The lymphatic system is also the main route by which cholesterol makes its way into the blood.

The lymphatic system is a closed system

Like the blood circulatory system, the lymphatic system is a closed system. Fluid enters through the walls of lymphatic vessels, which can stretch to hold more fluid as necessary. The lymphatic system has no heart to pump lymph around the body. Lymphatic vessels can contract, but the circulation of lymph depends mainly on external pressures: lymph is squeezed along vessels by the pressure associated with breathing, intestinal movements, and movements of skeletal muscles. These pressures enable the lymphatic system to return fluid sluggishly back to the veins at a rate of about 30 cm^3 per minute. As in veins, backflow is prevented by one-way valves. Blockage of the lymphatic system causes the body to swell with excess fluid, a condition called **oedema**.

Quick check

1 By which process does fluid leave the blood and enter the tissue fluid?

2 Which components of the blood do not enter the tissue fluid?

Food for thought

Suggest ways in which tissue fluid and lymph are:

a similar

b different.

Summary

The **gaseous exchange system** of mammals consists of the mouth and nose which leads via the pharynx through a series of tubes to the **lungs**. **Gaseous exchange** takes place in the lungs by **diffusion** through the surface of millions of small, very thin, highly vascularised air-filled sacs called **alveoli**. **Ventilation** of the lungs is brought about the action of **intercostal muscles** and the **diaphragm** which change the volume of the chest cavity. The rate and depth of breathing is controlled by **respiratory centres** in the hind brain. The volume of air inhaled and exhaled can be measured using a **spirometer**.
Mammals have a **cardiovascular system** consisting of the **heart** and a **closed circulatory system** of vessels through which blood passes. The heart is a **double pump** with four chambers. Valves ensure that blood passes in one direction through the heart. The heart is **myogenic** and has its own inherent rhythm of beating, but this can be modified by the action of nerve impulses and hormones. The series of events of each heartbeat is called the **cardiac cycle**. It consists of **atrial systole**, a **pause**, **ventricular systole**, and **diastole**. Each heartbeat starts at the **sinoatrial node** which acts as a **pacemaker**. **Blood** is a complex tissue with many functions. It consists of **red blood cells**, **white blood cells**, **platelets**, and **plasma**.
Red blood cells contain **haemoglobin**, involved in the **transport** of oxygen and carbon dioxide. The oxygen-carrying properties of haemoglobin are reflected by its **oxygen dissociation curves**. White blood cells are concerned mainly with **defence**. Platelets are involved in **blood clotting**. Plasma is a watery fluid which acts as the **transport** medium for many substances and heat. Plasma forced through capillary walls forms **tissue fluid**. Excess tissue fluid drains into the **lymph**. Both the respiratory and cardiovascular systems are affected by **exercise**.

Practice exam questions

1 a The figure below shows the lungs in their respective pleural cavities.

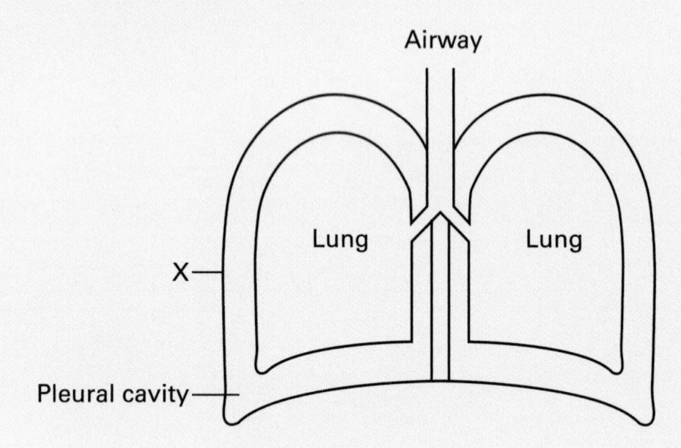

i Sketch on a copy of the figure below the appearance of the lungs if the wall of the thorax were punctured at X. [1]

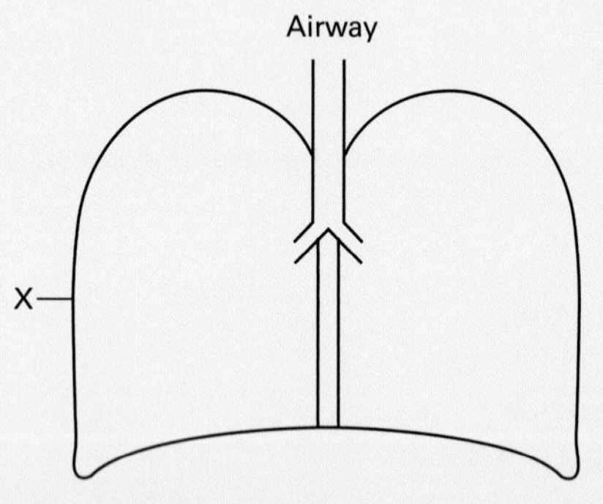

ii Explain the reasons for the altered appearance you sketched in the figure above. [2]

b The figure above right shows the oxygen dissociation curve for haemoglobin under different conditions. Curve A is the normal curve; curves B and C represent the dissociation curves in conditions of a high level of carbon dioxide or a high level of toxic carbon monoxide.

i Explain briefly why the dissociation curves have a sigmoid or 'S' shape. [2]

ii Identify which curve is due to the presence of a high level of carbon dioxide. [1]

iii Explain any physiological advantages resulting from the altered dissociation curve in the presence of a high level of carbon dioxide. [3]

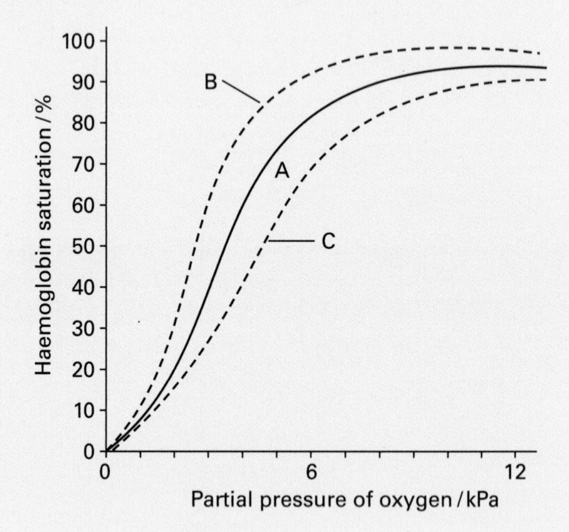

iv Suggest how the effect of carbon monoxide on the oxygen dissociation curve shown contributes to the toxic effect of the gas. [1]

[Total 10 marks]

2 *Answers should be written in continuous prose. Credit will be given for biological accuracy, the organisation and presentation of the information and the way in which the answer is expressed.*

Read the following passage.

The mammalian heart is a double pump adapted to forcing blood, at the same rate but at different pressures, along the two systems of a double circulation. High pressure in the systemic (body) circulation has evolved with lower pressure in the pulmonary (lung) circulation and a low pressure lymphatic circulation. Each heart beat is controlled by a wave of electrical excitation. In turn, the cardiac output of the heart adapts to meet the body's needs and is influenced by nervous and hormonal control.

a The heart forces blood at the same rate but at different pressures along the two systems of a double circulation (lines 1 and 2). Explain how the mechanism that controls each heart beat, and the structure of the heart, enable it to do this. [6]

b Describe the part played by hormones and the nervous system in controlling heart rate. [7]

c Describe how lymph is formed. [4]

Quality of language [3]

[Total 20 marks]

3 The diagram shows a vertical section through a human heart. The arrows represent the direction of movement of the electrical activity which starts muscle contraction.

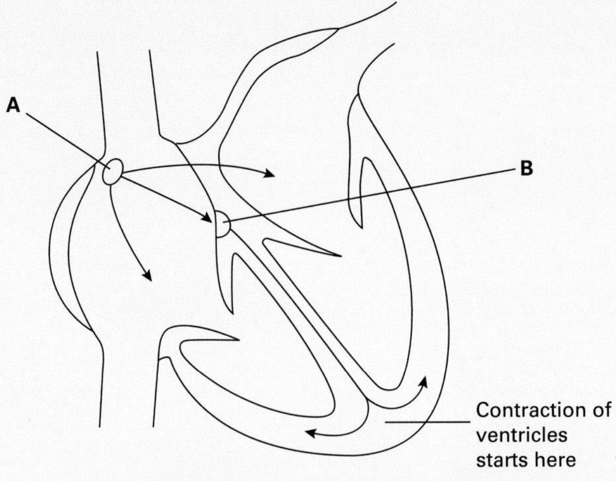

a Name structure **A**. [1]

b Explain why each of the following is important in the pumping of blood through the heart.

i There is a slight delay in the passage of electrical activity that takes place at point **B**. [1]

ii The contraction of the ventricles starts at the base. [1]

c Describe how stimulation of the cardiovascular centre in the medulla may result in an increase in heart rate. [2]

[Total 5 marks]

4 **a** Outline the main *chemical* components of mammalian blood plasma, briefly indicating **one** function of **each**. [6]

b Describe fully how mammalian blood picks up and transports respiratory gases. [10]

c Explain briefly how the blood capillaries exchange **other** materials with the cells of the body's tissues. [4]

[Total 20 marks]

5 The diagram shows two oxyhaemoglobin dissociation curves.

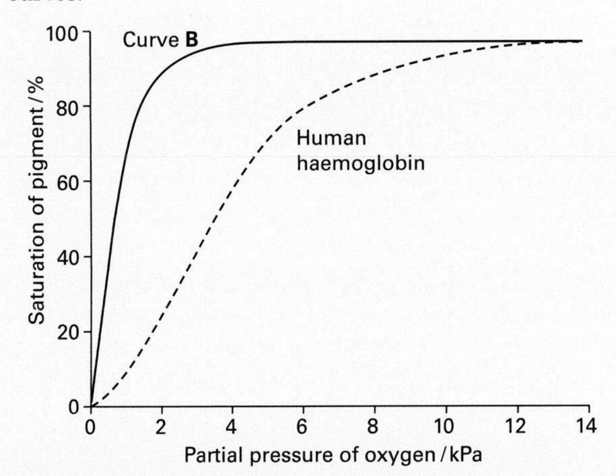

a Use the graph to explain why human haemoglobin:

i is saturated with oxygen in the lungs; [1]

ii releases oxygen when it reaches the tissues. [1]

b Explain what causes human oxyhaemoglobin to give up a greater proportion of the oxygen it carries during vigorous exercise. [2]

c Lugworms live in deep sand on sea shores. When the tide is in, they obtain oxygen by pumping water over their gills. At low tide there is no water to pump over the gills. Curve **B** on the graph shows the dissociation curve of lugworm oxyhaemoglobin. Explain the advantage of haemoglobin of this type to the lugworm. [2]

[Total 6 marks]

6 The graph below shows the oxygen dissociation curve of haemoglobin from a mammal at two different temperatures (38 °C and 43 °C).

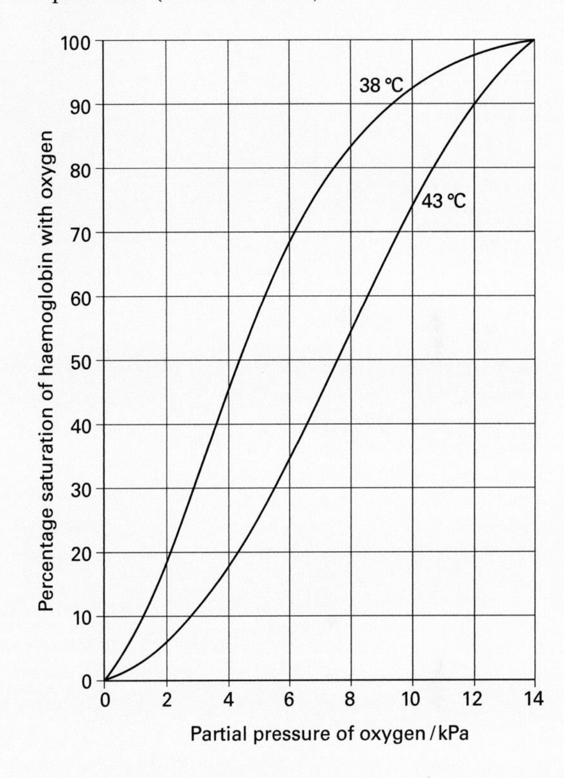

a **i** From the graph find the percentage saturation of haemoglobin in blood from an area of the body where the temperature is 43 °C and the partial pressure of oxygen is 4 kPa. [1]

ii Blood that is fully (100%) saturated with oxygen carries 105 cm^3 of oxygen in 1 dm^3 (litre) of blood. [1]

Calculate the volume of oxygen released from 1 dm^3 of blood when blood that has become 90% saturated at 38 °C reaches a part of the body where the temperature is 43 °C and the partial pressure of oxygen is 4 kPa. Show your working. [3]

b Suggest how this effect of temperature on the oxygen dissociation curve of haemoglobin might be advantageous to the mammal. [2]

[Total 6 marks]

Homeostasis

Homeostatic control systems

OBJECTIVES

By the end of this spread you should be able to:

- define homeostasis
- describe the main components of a homeostatic control system
- distinguish between positive and negative feedback.

Fact of life

Our ability to maintain a constant internal environment enables humans to survive in a greater range of habitats than any other animals on Earth. By using artificial homeostatic devices to support our physiological processes, we are able to go from the depths of the oceans into outer space (figure 1).

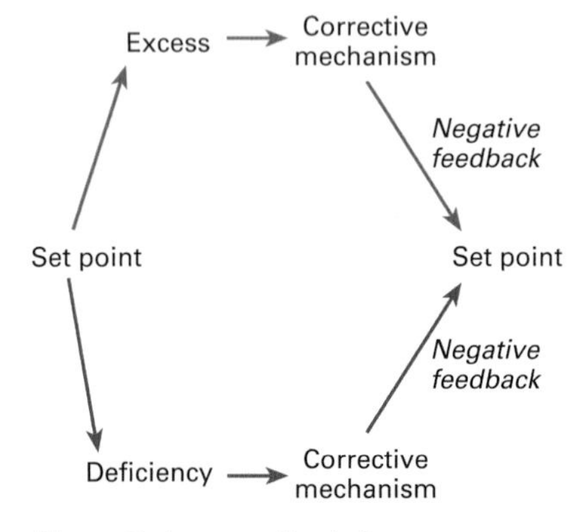

Figure 2 *A generalised diagram showing how negative feedback keeps a system in a stable condition. A change from the usual level of a factor (the set point for that factor) triggers a corrective mechanism which restores the factor to its usual level.*

Figure 1 *Artificial homeostatic devices enable humans to go underwater and into space.*

A steady state

In order to survive, an organism has to be able to keep its internal environment within tolerable limits. The internal environment of a multicellular organism is the tissue fluid bathing the cells. Keeping the conditions of the tissue fluid such as pH, temperature, and salt content at a relatively constant level is called **homeostasis** (*homoios* means the same; *stasis* means standing).

Negative feedback

Homeostasis is a characteristic of living things, but the term is used for any system, biological or non-biological, which is in a steady state.
In biological systems, homeostasis is usually achieved by a process called **negative feedback** (figures 2 and 3). A change in the level of an internal factor causes effectors to restore the internal environment to its original level. For example, an increase in the internal body temperature causes the body to lose more heat; a decrease in body temperature causes the body to generate more heat (see spread 8.9). This type of system, in which a change in the level of a factor triggers a corrective mechanism, is called a self-adjusting system.

Thermostatic control: an example of a homeostatic mechanism

A thermostat is a simple example of a non-biological control system. The output being controlled is room temperature. The desired temperature is set on a control panel. A thermosensor detects the actual room temperature and sends information to a control box. The control box acts as a comparator, comparing the actual room temperature with the set temperature. If the actual temperature deviates from the set point, the control box starts up the appropriate corrective mechanism, switching a heater off if the temperature is above the set point, or switching it on if it is below the set point.

Positive feedback

Sometimes a homeostatic mechanism breaks down and negative feedback does not occur. Deviations from the set point are not corrected. Even worse, the deviations may be made larger. This results in a process called **positive feedback** in which a small change in output causes further change in the same direction (figure 4).

Control mechanisms

All homeostatic control mechanisms that use negative feedback, whether they be physical or biological, share the same components (figure 3).

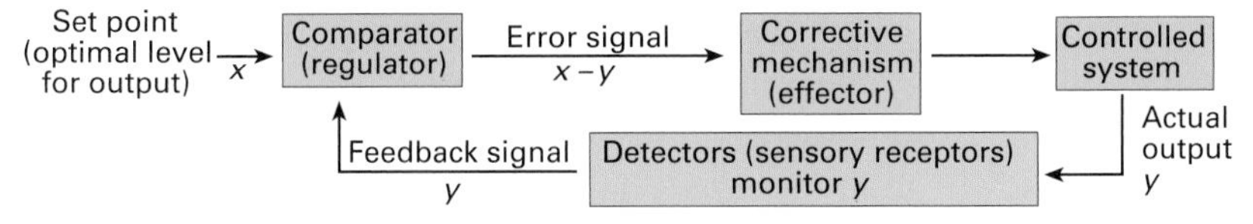

Figure 3 *A simple negative feedback system.*

They all have an **output** (for example, an internal factor such as blood temperature) that is controlled, and a **set point** (also called the norm or reference point, given the value x in the diagram). In a physiological process, the set point is usually determined genetically and is the desired or optimal physiological state for the output. The control mechanism also has **detectors** (sensory receptors in physiological systems) to monitor the actual output (given the value y in the diagram). A **comparator** (sometimes called a regulator in physiological systems) compares the actual output with the set point. The comparator produces some sort of **error signal** which conveys information to the corrective mechanism about the difference between the set point and the actual output ($x - y$). In physiological systems, the error signal is usually in the form of nerve impulses or hormones, and the corrective mechanism may include one or more effectors which restore the output to its set point. In some physiological processes (such as thermoregulation; see spread 8.9) separate but coordinated mechanisms control deviations in different directions from the set point (for example, rises or falls in body temperature), giving a greater degree of control.

The corrective mechanism described above is the key component of homeostatic control: it varies the output y so that this output can be brought back to the set point. Homeostasis is a dynamic process: it works by making continual adjustments to compensate for fluctuations of output. It is therefore more accurate to describe such systems as being in a **steady state** or in **dynamic equilibrium** rather than being constant.

Positive feedback is usually harmful because it tends to produce unstable conditions. For example, when the negative feedback mechanisms in mammalian temperature regulation break down, a rise in body temperature can spiral upwards and threaten death (spread 8.10). However, in certain circumstances positive feedback can be useful. It is the basis of oxytocin release during childbirth (spread 12.9), and occurs in nerves in which a small stimulus can bring about a large response.

Homeostasis in mammals

In 1859, Claude Bernard, a French physiologist, was the first to recognise the importance of homeostasis in mammals. He studied variations in glucose concentrations in the blood of dogs. He found that the concentrations remained remarkably stable despite dramatic variations in diet. For example, dogs recently fed meat or sugar-rich food had similar glucose concentrations to a starving dog. He concluded that dogs and other mammals must have a control mechanism that keeps their internal environment constant. (The following two spreads describe the homeostatic control of blood glucose concentration in more detail.)

The ability of mammals to maintain a stable internal environment makes them independent of changing external conditions, enabling them to exploit a wide range of habitats. For example, our highly developed internal mechanisms keep our body temperature close to 37 °C in spite of wide variations in environmental temperature. This allows us to remain active in these different environments, whereas other animals such as lizards have limited powers of thermoregulation and cannot function over such a wide range of environmental temperature.

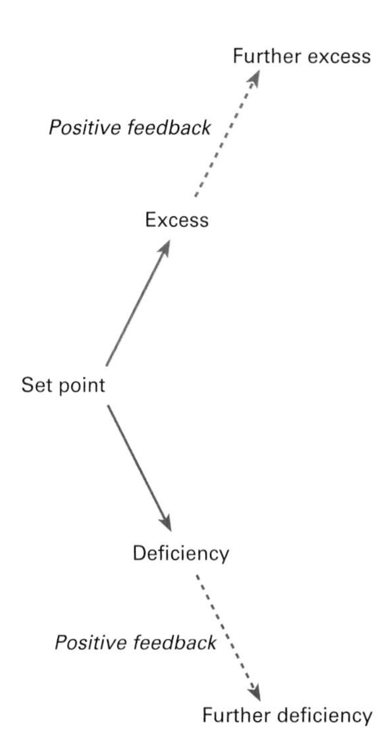

Figure 4 *Positive feedback.*

Quick check

1 Define homeostasis.

2 What is the function of:
  a detectors
  b comparators
  c effectors
  within homeostatic control systems?

3 What type of feedback system is involved in bringing the temperature of an overheated person back to normal?

Food for thought

Positive feedback is generally harmful because it tends to amplify changes in a factor that needs to be kept constant. Suggest biological examples (physiological, behavioural, or ecological) in which positive feedback might be beneficial.

8.2

OBJECTIVES

By the end of this spread you should be able to:

- describe how mammalian blood glucose concentrations are maintained at a relatively constant level
- explain the roles of insulin, glucagon, and other hormones in the control of blood glucose concentrations.

Fact of life

Most cells can use several high-energy compounds as respiratory substrates, including a variety of carbohydrates, fats, and proteins. But the brain cells of mammals can respire only glucose.

Insulin secretion

The process by which an increase in blood glucose concentration causes beta cells to release insulin is complex (figure 1). Glucose transporter molecules facilitate the diffusion of glucose into beta cells. The glucose is phosphorylated, effectively trapping it within the cell, and then broken down by glycolysis to create ATP. An increase in the ATP:ADP ratio causes ATP-sensitive potassium (K^+) channels to close. This prevents K^+ from being shunted out of the cell and results in depolarisation due to a rise in the positive charge inside the cell. The change in electrical charge activates voltage-gated calcium (Ca^{2+}) channels which transport Ca^{2+} into the cell. The abrupt increase in intracellular calcium triggers the release of insulin from storage granules into the blood.

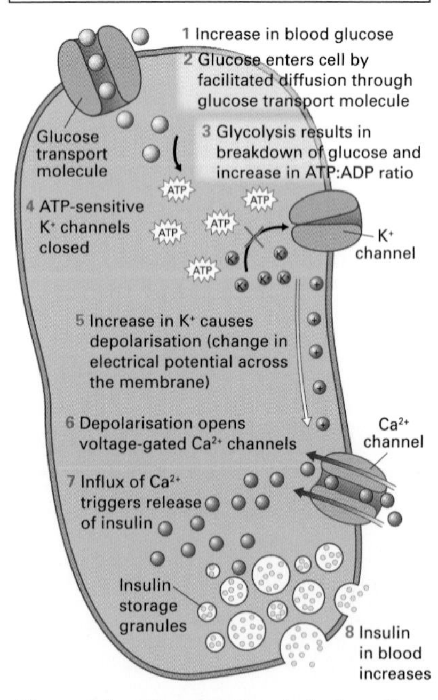

***Figure 1** Insulin secretion in beta cells.*

CONTROL OF BLOOD GLUCOSE CONCENTRATION

The importance of glucose

Glucose is the starting point for the manufacture of many organic compounds within the body, and it is a major fuel for cellular respiration. This simple, highly soluble sugar is transported in solution in blood plasma.

Glucose is so vital that mammals have evolved a complicated system for maintaining blood glucose concentrations at a steady state. The normal human blood glucose concentration is about 90 mg of glucose per 100 cm^3 of blood. This set point is genetically determined.

Two interacting mechanisms control blood glucose concentration: one compensates for levels that are too high; the other compensates for levels that are too low. **Insulin** acts as the error signal or regulating chemical for the first mechanism. **Glucagon** acts as the main (but not only) error signal or regulating chemical for the second mechanism (figure 2).

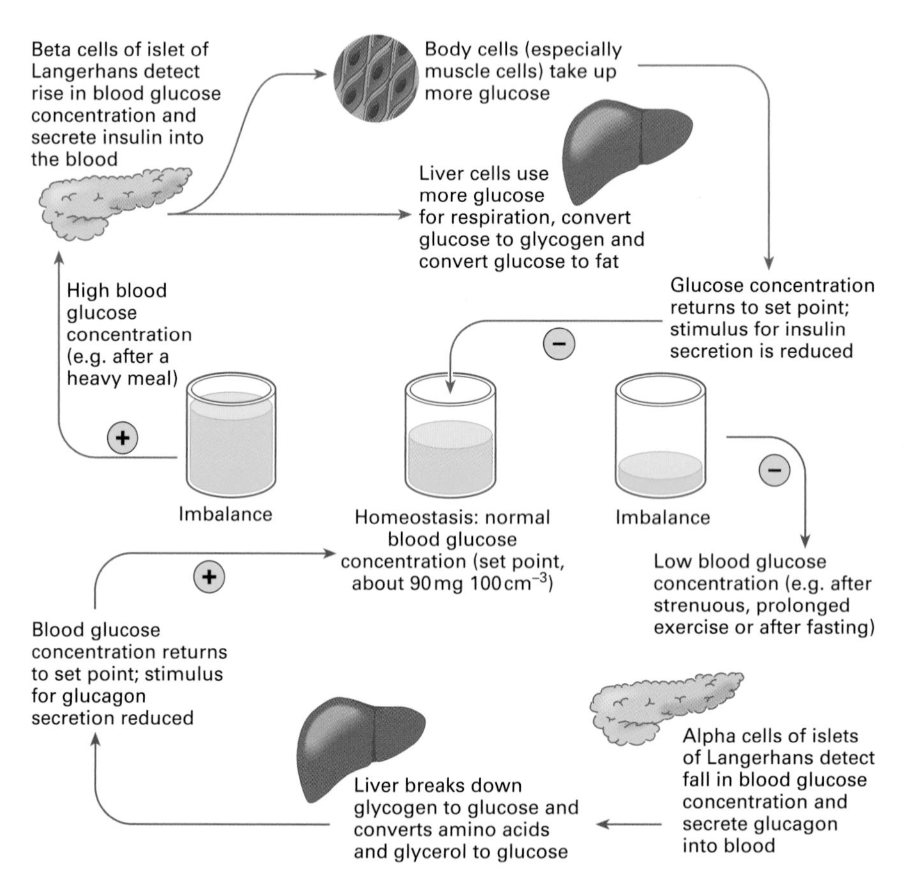

***Figure 2** Homeostatic control of blood glucose concentration.*

Insulin

Insulin is a small protein consisting of only 51 amino acids. The amino acids form two polypeptide chains linked together by covalent (disulphide) bonds. Insulin is secreted by special cells called beta cells in the **islets of Langerhans**, endocrine tissue within the pancreas. When the blood glucose concentration rises above the set point, more insulin is secreted from the pancreas (figure 1).

In order to act on cells, insulin molecules bind to an exposed glycoprotein receptor on the cell surface membranes, in much the same way as a substrate binds to an enzyme (figure 2). In some as yet unknown way, the membrane-bound insulin brings about a number of cellular responses which reduce the blood glucose concentration. The responses include changes in both cell surface membrane permeability and enzyme activity which lead to four major effects:

1 an increase in the uptake of glucose and amino acids into cells
2 an increase in the rate of cellular respiration and the use of glucose as a respiratory substrate
3 an increase in the rate of conversion of glucose to fat in adipose (fat-storing) cells
4 an increase in the rate of conversion of glucose to glycogen in liver and muscle cells (**glycogenesis**).

In addition to its role as an error signal in the control of blood glucose, insulin is also an anabolic (body-building) hormone: it promotes the synthesis of large molecules and plays a key role in cell growth. Insulin appears to affect all cells, but muscle, fat, and liver cells are its main targets.

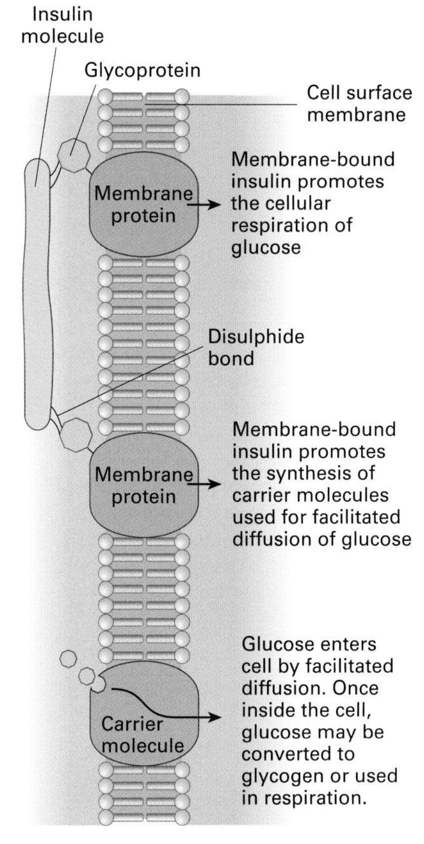

***Figure 3** Schematic diagram of suggested mechanism of action of insulin.*

Glucagon

Glucagon is a protein made of a single polypeptide chain with 29 amino acids. It is secreted by **alpha cells** in the islets of Langerhans. When the blood glucose concentration falls below the set point, the alpha cells secrete glucagon. Glucagon activates phosphorylase (an enzyme in the liver) which catalyses the breakdown of glycogen to glucose (**glycogenolysis**). Glucagon also increases the conversion of amino acids and glycerol into glucose 6-phosphate. This synthesis of glucose from non-carbohydrate sources is called **gluconeogenesis**.

The interaction of glucose control mechanisms

In addition to glucagon, at least four other hormones can also increase blood glucose concentrations. For example, in times of acute stress or excitement, **adrenaline** is secreted. This causes the breakdown of glycogen in the liver, boosting blood glucose concentrations. When glycogen stores in the liver become exhausted, **cortisol** is secreted by the adrenal glands. Cortisol promotes liver cells to convert amino acids and glycerol into glucose.

The system for controlling blood glucose concentrations is self regulating: the blood glucose concentration itself determines the relative amounts of insulin and glucagon secreted, and these hormones alter the blood glucose concentration so that it remains relatively stable. The two corrective mechanisms regulated by insulin and glucagon act in opposition to each other or **antagonistically** (figure 4). They provide a much more sensitive control system than one that relies on only one set of corrective mechanisms. Cells in the islets of Langerhans are thought to act as detectors and comparators for both sets of corrective mechanisms.

The control systems usually keep blood glucose concentrations close to the norm. However this system, like any other system, can break down. If, for example, the pancreas cannot secrete insulin, or if target cells lose their responsiveness to insulin, the blood glucose concentration can reach dangerously high levels. The resulting condition is called **diabetes mellitus**, and is the subject of the next spread.

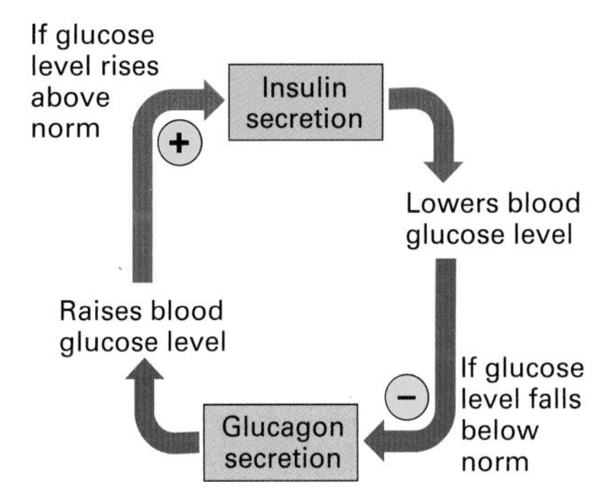

***Figure 4** The antagonistic actions of insulin and glucagon.*

Quick check

1 Explain briefly the advantage of having two corrective mechanisms that regulate blood glucose concentrations, rather than one.

2 Under what conditions is glucagon secretion increased?

Food for thought

The homeostatic control mechanism of blood glucose concentration involves several components:

- sensory receptors to monitor the actual blood glucose concentration
- a comparator to compare the actual concentration with the set point
- several error signals to regulate the corrective mechanisms
- two corrective mechanisms which use a number of effectors to keep blood glucose concentration constant.

Suggest how shortcomings in each of these different components might influence the efficiency of the overall process.

8.3

Diabetes mellitus

OBJECTIVES

By the end of this spread you should be able to:

- describe the causes, symptoms, and treatments of the two main types of diabetes mellitus.

Fact of life

The discovery that insulin can be used to treat diabetes has enabled millions of people with diabetes worldwide to enjoy a full and active life. Many top class sportspeople, including the former Olympic rower Sir Steve Redgrave, are diabetics. Sir Steve was diagnosed with diabetes at the age of 35, just prior to the Sydney Olympic Games in which he won his fifth gold medal.

Figure 1 *Sir Steve Redgrave in action in the Sydney Olympic Games.*

Diabetes mellitus is a metabolic disorder caused by a lack of insulin or a loss of responsiveness to insulin. It can result from incomplete development of, damage to, or disease of the islets of Langerhans, the endocrine portion of the pancreas which secretes insulin (figure 2).

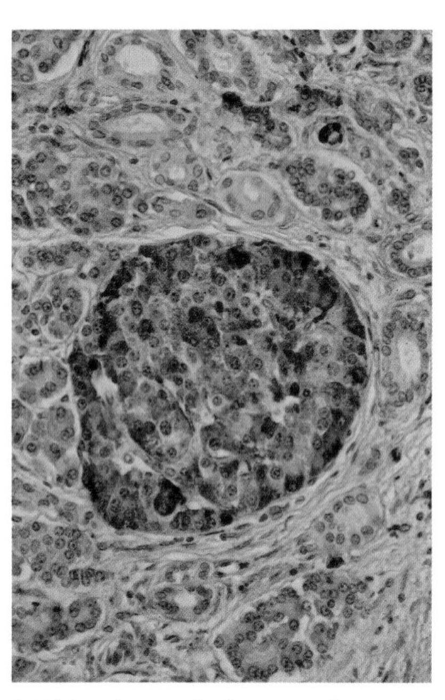

(a) Light micrograph of pancreas tissues

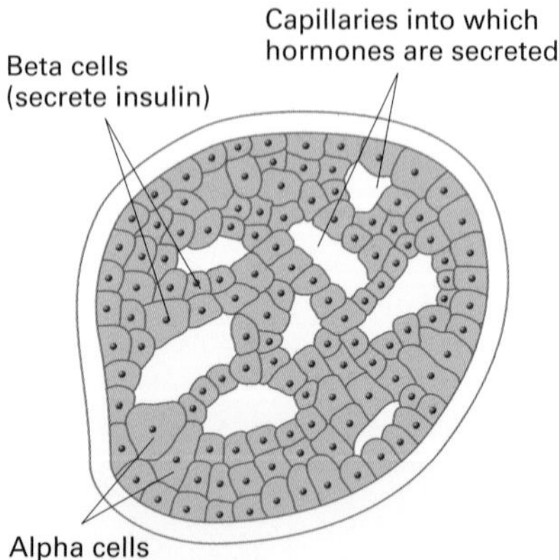

(b) Detail of an islet of Langerhans

Figure 2 *Structure of an islet of Langerhans. The pancreas secretes digestive enzymes as well as endocrine hormones.*

Diagnosing diabetes

Diabetes mellitus is characterised by the excretion of large amounts of sugary urine (diabetes mellitus actually means 'sweet fountain'). The blood glucose concentration becomes so high that the kidney is unable to reabsorb all the glucose filtered into its tubules back into the blood. Consequently glucose is excreted in the urine (normally, urine contains no glucose). The high blood glucose is called **hyperglycaemia**, and the presence of glucose in the urine is called **glycosuria**; both are signs of diabetes mellitus. Other diagnostic features include the patient complaining of a lack of energy, a craving for sweet foods, and persistent thirst. The main diagnostic test is a **glucose tolerance test** in which the patient swallows a sugar solution and a doctor measures the blood glucose concentration at intervals. Graphs of the results from a diabetic and a person with normal glucose metabolism are distinctly different (figure 3).

In addition to disrupting the homeostatic control of blood glucose, diabetes mellitus has other harmful effects. Insulin acts as an anabolic (body-building) hormone, therefore lack of it causes weight loss and muscle wasting. Although the blood of diabetics is rich in glucose, their carbohydrate metabolism is disrupted and their cells are unable to use glucose as a major source of fuel. Consequently, they have to resort to metabolising fat as a source of energy. In severe cases, the products of this fat metabolism can have harmful effects.

Types I and II

There are two forms of diabetes mellitus with different causes. **Type I diabetes** (also known as insulin-dependent diabetes or juvenile-onset diabetes) usually occurs suddenly in childhood. It appears to be an autoimmune disease: cells from the immune system attack beta cells in the islets of Langerhans, destroying a person's ability to secrete insulin. **Type II diabetes** (also known as insulin-independent diabetes or late-onset diabetes) usually occurs later in life. It is often caused by a gradual loss in the responsiveness of cells to insulin, but it can also be due to an insulin deficiency.

Treating type I diabetes

Most diabetics with type I diabetes are treated with insulin injections (insulin cannot be taken orally because, being a protein, it would be digested in the alimentary canal). Before insulin became available about 75 years ago, type I diabetes would have meant death by a slow and progressive wasting. Nowadays, with proper use of insulin and careful management of diet and exercise, diabetics can lead a perfectly normal life.

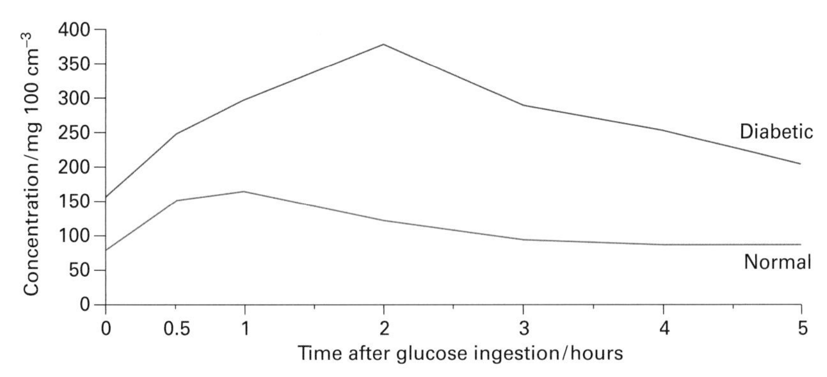

***Figure 3** The results of a glucose tolerance test in a person with normal glucose metabolism and a diabetic.*

Insulin used to be obtainable only from non-human sources (mainly pigs and cows), but recombinant DNA technology has made human insulin available to diabetics (see spread 18.10). In the short term, insulin from non-human sources cannot usually be distinguished from human insulin, but non-human insulin is not absolutely pure and prolonged use may lead to immunological reactions which can damage health. The immune system identifies the non-human insulin as foreign and attacks it with antibodies. The insulin is either destroyed or coated so that it cannot bind to receptor sites on target cells.

Care has to be taken with the doses of insulin, because an overdose causes too much glucose to be withdrawn from the blood, reducing the blood glucose concentration below the set point (a condition called **hypoglycaemia**). Glucose is the main fuel for brain cells, and a lack of glucose can lead rapidly to unconsciousness and, in extreme circumstances, to coma and even death. This is why a diabetic should be given sugar if found unconscious.

All diabetics need to monitor their blood glucose levels. Easy-to-use biosensors and dipsticks are available for this. Diabetics also need to manage their diet and levels of exercise very carefully. They should avoid eating too much sugar or going for long periods without food.

Treating type II diabetes

Most type II diabetics can control their blood glucose concentration by carefully regulating their diet and exercise. These diabetics tend to become less responsive to insulin as they age. This can be offset if necessary by injecting insulin.

Type II diabetes is increasing worldwide, and primarily in well fed communities. It is less common among communities that have low-fat and high-fibre diets, and eat complex carbohydrates that are not quickly broken down to glucose in the body. Type II diabetes is linked with high-fat diets and obesity. One of the the highest incidences is among a group of people who live in a district of Port Moresby, the capital of New Guinea. In the past, fat was usually in short supply, so it is believed that these people evolved an ability to store fat in times of plenty. This may have led to them develop diabetes when they adopted a western, high-fat diet. As people live longer, type II diabetes is becoming more common and the demand for insulin is increasing annually. The ability to produce human insulin by genetic engineering allows us to treat the millions of people worldwide who have diabetes.

Stem cell therapy and diabetes

Theoretically, stem cells can be induced to make any tissue type and the newly made tissue used to treat a human disorder, including type I diabetes. In 2009, the *Daily Telegraph* reported that stem cell transplants 'have freed patients with type I diabetes of daily insulin injections'. The report was based on a clinical trial involving 23 patients with newly diagnosed type I diabetes, before the pancreas was irreversibly damaged and before the onset of any complications due to raised blood glucose levels. Of the 23 patients, 20 remained free from insulin injections for at least 7–31 months. Side effects included pneumonia in 2 patients, endocrine problems in 3 patients, and low sperm counts in 9 patients. There were no deaths. These stem cell transplants apparently worked by 'resetting' the immune system so that the body stopped attacking the pancreas. The researchers used **haematopoietic stem cell transplantation** (HSCT) in which stem cells derived from the patient's own bone marrow were collected from the blood. Chemotherapy was used to partly destroy the patient's own bone marrow cells and immunosuppressant drugs administered to stop the immune system attacking the pancreas. *NHS Choices*, a website designed to help NHS patients make informed health choices, concluded that the study illustrated a promising approach to treating type I diabetes if the disease was detected early enough and if patients were prepared to accept the risk of side effects. However, the study involved only a small number of selected patients and so was not randomised. Also, there was no control group for comparison. The final conclusion was that 'Randomised trials to test the new treatment against current care in a larger group of patients will help establish whether this is truly a 'cure for diabetes' or simply a way of prolonging insulin production by a few years.'

Quick check

1 What is the main cause of type I diabetes ?

2 What is the main treatment for type I diabetes?

3 What type of diet is most closely associated with the development of type II diabetes?

Food for thought

The **blood glucose pool** is the total amount of glucose in the blood at any one time. Suggest why some people may become **hypoglycaemic** (have a reduced blood glucose concentration) after eating a large sugar-rich meal.

8.4

The liver

O B J E C T I V E S

By the end of this spread you should be able to:

- discuss how the structure of the liver is related to its functions
- describe in outline the ornithine cycle.

Fact of life

In the USA, it is estimated that 10–15 per cent of adults have gallstones. Gallstones can exist for many years without any symptoms, but when they block the bile duct they can cause considerable pain. The high incidence in the USA is linked to a fatty diet. Gallstones are usually formed in the gall bladder from cholesterol, sometimes with bile pigments and calcium salts.

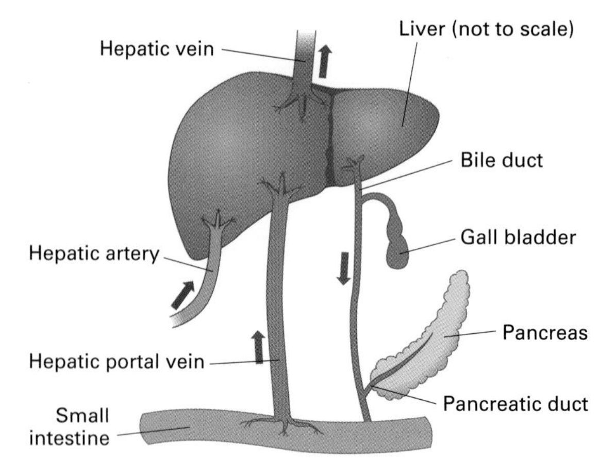

***Figure 1** Schematic diagram of the liver showing the blood vessels and bile duct.*

The role of the liver

The liver is the largest organ in the human abdomen (figure 1). It lies on the right-hand side of the body, close to the diaphragm. It plays a central role in metabolism, regulating the levels of a wide range of chemicals in the blood and preventing harmful substances from reaching chemically sensitive organs such as the brain.

To carry out its functions, the liver has a double blood supply. The **hepatic artery** delivers oxygenated blood so that liver cells can generate energy by aerobic respiration to carry out all their energy-demanding functions. The **hepatic portal vein** takes all the blood from the intestines to the liver. This enables the liver to process substances absorbed from the gut before they enter the general circulation. Poisons, for example, can be made harmless by detoxification before they damage the brain. Similarly, the lymphatic vessels from the gut carry fatty substances to the liver for processing before these substances go to the rest of the body.

Lobules and hepatocytes

The liver is made up of **lobules** (figure 2). In each lobule, blood from the hepatic portal vein and hepatic artery spreads out and flows slowly through **sinusoids** (blood-filled spaces) past each **hepatocyte** (liver cell). These are the sites of the many chemical reactions carried out by the liver. Under the light microscope hepatocytes appear relatively uniform, undistinguished, and unspecialised cells. The electron microscope, however, reveals their specialised ultrastructure. They have an exceptionally high density of mitochondria to provide sufficient ATP for all their energy-consuming functions, and microvilli (on the cell surface membrane bordering blood sinusoids) to increase the surface area for uptake of substances from the blood.

The liver acts as the body's chemical-processing factory, carrying out many functions including:

- storage of minerals (including iron) and the vitamins A, D, and B_{12}
- destruction of red blood cells (the lobules are patrolled by star-shaped white blood cells called Kupffer cells which ingest worn out red blood cells)
- manufacture of plasma proteins (such as albumins) and blood clotting agents (including fibrinogen)
- detoxification of poisons: all mammals are likely to be exposed to poisonous chemicals at some time, and the liver has evolved a group of enzymes which break these chemicals down into less harmful products. An example is the enzyme catalase, which catalyses the breakdown of hydrogen peroxide to water and oxygen.

The major roles of the liver in carbohydrate, fat, and protein metabolism are outlined below.

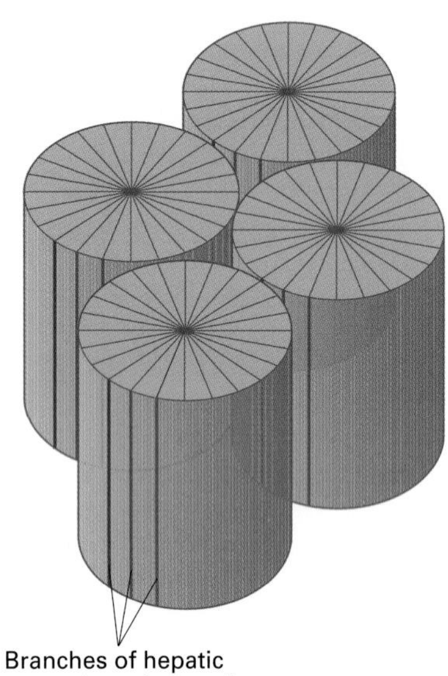

(a) Liver lobules

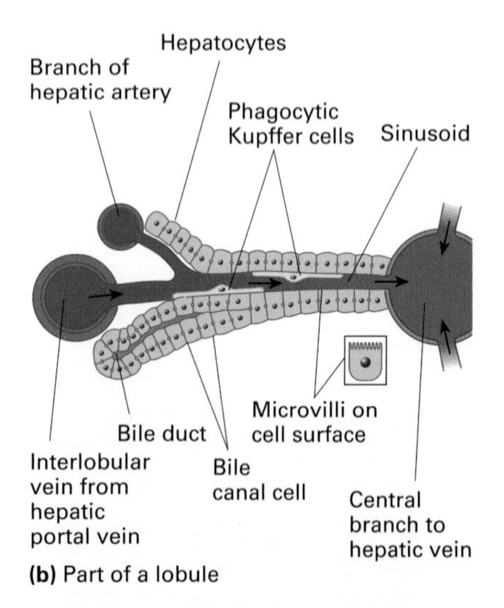

(b) Part of a lobule

***Figure 2 (a)** Liver lobules. **(b)** The arrangement of cells in part of a lobule.*

Carbohydrate metabolism

The liver helps to regulate blood glucose levels (see also spread 8.2). The glucose concentration of blood leaving the liver may be kept the same as that entering the liver, it may be reduced, or it may be increased. Glucose can be added to the blood by converting glycogen to glucose (**glycogenolysis**), or by converting non-carbohydrate substances such as amino acids and glycerol into glucose (**gluconeogenesis**). Glucose can be removed from the blood by storing the glucose as glycogen (**glycogenesis**), converting the glucose to fat, or using glucose as a fuel for cellular respiration.

Fat metabolism

Liver cells process fatty acids so they can be transported and deposited in the body. The liver also regulates the amounts of phospholipids and cholesterol, synthesising them or eliminating them as required. Excess cholesterol and phospholipids are eliminated in the bile.

The liver produces copious quantities of bile, which aids the digestion and absorption of fats. **Bile** is a thick yellow liquid containing water, bile salts, bile pigments, bile acids, inorganic ions, and cholesterol. Bile salts and bile acids help to break down fat into small droplets. This process, called **emulsification**, provides a larger surface area for digestion by enzymes, and aids absorption. Bile pigments give faeces their characteristic colour. The pigments are derived mainly from breakdown products of haemoglobin, released from defunct red blood cells.

Bile is transported via bile canaliculi (small channels in the liver lobules) to the bile duct. Humans produce about 0.5 dm^3 of bile per day which is passed to a storage organ, the **gall bladder**, before it is secreted into the small intestine. Although no one can live without a liver, it is quite possible to survive without a gall bladder. Many people suffering from gallstones (hard masses of cholesterol and minerals) have their gall bladders removed. Some mammals, such as rats and horses, lack a gall bladder and always secrete the bile directly into the small intestine.

Protein metabolism

In addition to manufacturing valuable proteins, the liver is also involved in **nitrogenous excretion**. The body cannot store excess proteins or amino acids. If more protein is eaten than is used to build up muscle or as a fuel for cellular respiration, the excess has to be eliminated. The non-nitrogenous parts of amino acids can be converted to fats and carbohydrates that can be used immediately or stored. However, the nitrogen in the amino group has to be eliminated from the body because it forms toxic products. The first step is **deamination**, removal of the amino group from the amino acid to form ammonia. Ammonia is very toxic and very soluble; only aquatic animals can use it as an excretory product. Terrestrial animals convert it to a less toxic and less soluble substance both to avoid its harmful effects and to conserve water. Mammals convert ammonia to **urea** by combining ammonia with carbon dioxide in a series of enzyme-catalysed reactions called the **ornithine cycle** or urea cycle (figure 3). Animals that excrete urea are called **ureotelic**. Urea is transported in the blood to the kidneys where the blood is filtered and the urea passes out into the urine (spread 8.5).

Heat production

The liver has a very high metabolic rate. This has led people to think that it is always a heat exporter, generating heat to keep the rest of the body warm. However, experiments with rats indicate that, except in cold environments, the temperature of blood entering the liver is about the same as that of blood leaving it. This indicates that exothermic (heat-releasing) and endothermic (heat-consuming) reactions in the liver usually cancel each other out. In cold conditions, it seems that mammals generate extra heat by increasing the rate of exothermic reactions in many parts of the body, not only the liver.

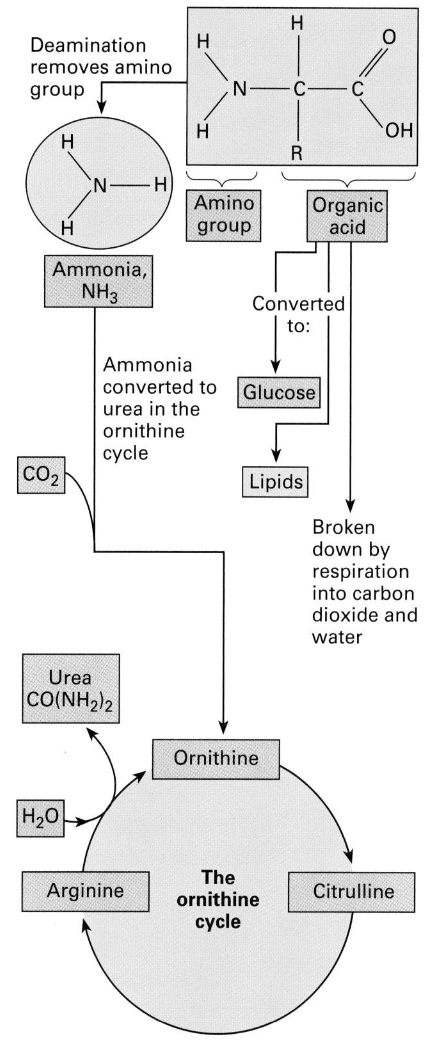

Two molecules of ammonia enter the ornithine cycle for each molecule of urea produced. Arginine is hydrolysed to urea in a reaction catalysed by the enzyme arginase.

Figure 3 *Nitrogen excretion: outline of deamination. The ornithine cycle converts ammonia, NH_3, into urea.*

Quick check

1 Explain the importance of the liver's dual blood supply.

2 Study figure 3. What are the three substances that enter the ornithine cycle to synthesise urea?

Food for thought

Suggest *two* different reasons why a person with gallstones might be put on a low fat diet. How might gallstones be removed without using surgery?

8.5

OBJECTIVES

By the end of this spread you should be able to:

- discuss the main functions of the kidney
- describe the structure of the kidney
- describe how kidney transplants are performed.

Fact of life

There are over a million nephrons (tubules) in a human kidney. As each nephron is about 5 cm long, this adds up to about 5 km of kidney tubules.

Structure and functions of the kidneys

Mammalian kidneys play a central role in homeostasis, helping to eliminate waste from the blood and to regulate the salt and water content of the body. In this spread we are going to focus on the structure and functions of human kidneys. The kidneys of other mammals have the same basic structure but, as we shall see (spread 22.14), they differ in detail.

Excretion and osmoregulation

The kidneys function as vital filtration and purification organs. Each day, up to 2000 dm^3 of blood pass through them. During its passage, unwanted substances are eliminated from the blood and essential ones are retained. The end result is the production of urine consisting of variable amounts of water, urea (the nitrogenous waste synthesised in the liver; spread 8.4), and other waste substances. The elimination of metabolic waste is called **excretion**. The production of urine containing variable amounts of water allows the kidneys to control the water content of the body, a process called **osmoregulation**.

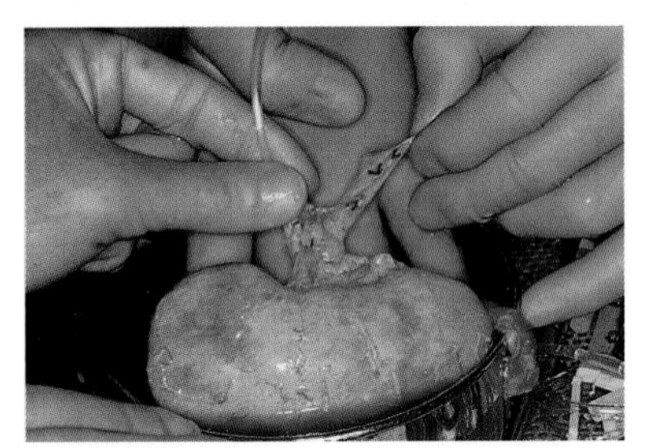

Figure 1 Preparation of a kidney for a transplant operation.

Kidney failure

Kidney failure is a common disease affecting tens of thousands of people each year. It is possible to live if one kidney fails, but failure of both is fatal if untreated. Treatment of kidney failure may involve kidney dialysis or transplantation.

Dialysis is a method of separating particles of different size in a liquid mixture by passing the mixture through a partially permeable membrane. In kidney dialysis, blood is filtered either through the peritoneum (a natural membrane in the abdomen of the patient) or through an artificial membrane in a 'kidney machine'. In either case, waste products of small particle size (e.g. urea and mineral salts) are removed from blood, which retains large particles (such as blood cells and proteins). In **transplantation**, a kidney from a donor is implanted in the lower abdomen near the groin and connected to the recipient's blood supply and bladder. The failed kidneys are usually left in place. As with other forms of transplantation, tissue rejection was a major problem when kidney transplants were first performed in the 1960s (see spread 15.8 for a general discussion of the problems of transplantation). However, nearly 2000 transplants are now carried out each year in the UK. There is a high survival rate and most of the problems of tissue rejection have been overcome.

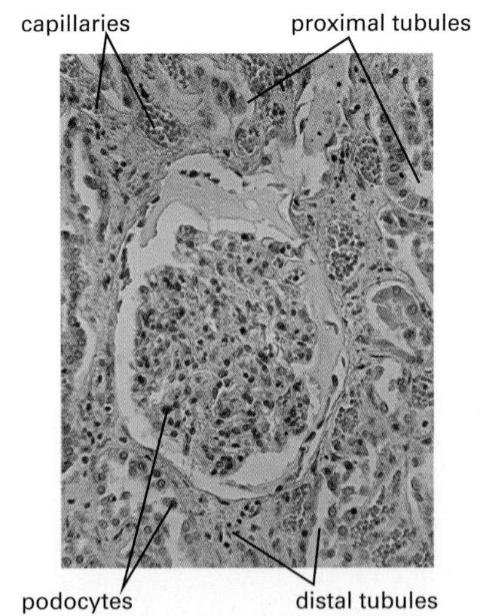

Figure 2 Histology of part of the kidney revealed in a photomicrograph of a transverse section.

Structure of the kidney

Humans have two kidneys which lie just in front of the twelfth ribs at the back of the abdominal cavity (figure 3a). The kidneys are held in place by layers of fat; loss of these fat layers can cause a kidney to slip and buckle the ureter (the large tube from which urine drains into the bladder), restricting the flow of urine to the bladder. Kidneys have a good blood supply: a **renal artery** supplying oxygenated blood and nutrients to the kidney, and a **renal vein** carrying away filtered blood to the heart. Cut in half, a kidney can be seen to consist of an outer **cortex** and an inner **medulla** which leads to the **pelvis** and then the **ureter** (figure 3b). Examination of a section of the kidney with a microscope reveals that it consists of minute tubules, each of which is called a **nephron** (figure 3d). All the nephrons have a similar structure and carry out the same functions.

- The **Bowman's capsule** acts as an ultrafiltration unit, filtering the blood and separating the large particles (which stay in the blood vessels) from the small ones (which pass into the nephron) (spread 8.6).
- The **proximal convoluted tubule** is mainly concerned with selective reabsorption so that valuable substances such as glucose are taken back into the blood and are not lost in the urine (spread 8.6).

- The **loop of Henlé** acts as a countercurrent exchange mechanism, creating a low water potential (high solute content) in the medulla of the kidney, so that water can be reabsorbed by osmosis (spread 8.7).
- The **distal convoluted tubule** and collecting ducts are concerned with osmoregulation, varying the amount of water reabsorbed into the blood (spread 8.8).

The nephron is also involved in regulating the salt content of the blood, blood pressure, and pH of the urine.

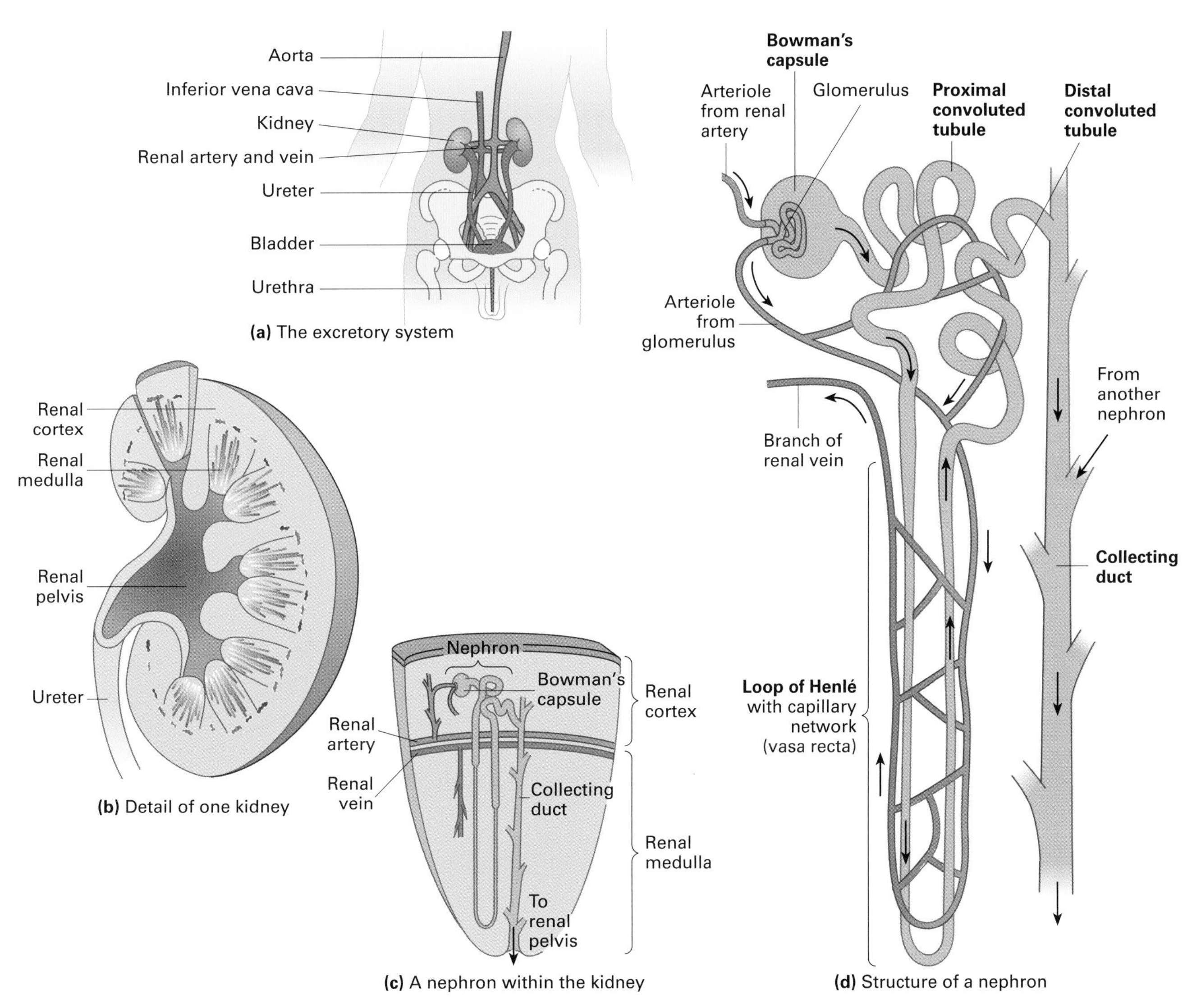

***Figure 3** Structure of the human kidney.*

Quick check

1 Define:

a excretion

b osmoregulation.

2 List the three main regions of the nephron, starting where fluid enters the nephron and ending where it flows into the collecting duct.

3 In kidney transplants, in which part of the body is a kidney usually implanted?

Food for thought

Kidney transplants are far cheaper than kidney dialysis, and they give the patient much greater independence. However, the demand for transplants is about double the supply of donor kidneys. Kidney transplants depend on being able to obtain fresh, healthy kidneys in which the tissue is alive. Sometimes close living relatives act as donors (a person can live normally with just one kidney), but in about 35 per cent of cases the kidney comes from a person who has been recently certified 'brain dead'. Suggest ethical and moral issues raised by kidney dialysis and kidney transplantation.

8.6

OBJECTIVES

By the end of this spread you should be able to:

- describe the process of ultrafiltration in the glomerulus
- explain how glucose and water are reabsorbed from the proximal convoluted tubule.

Fact of life

Normally, proteins are too large to pass through the kidney and into the urine. However, during intense lengthy exercise the blood pressure can become very high, forcing a small amount of protein into the urine. This condition, called **athletic pseudonephritis**, mimics the symptoms of a very serious kidney disease called glomerulonephritis. However, athletic pseudonephritis is a temporary condition that clears completely within three days of rest.

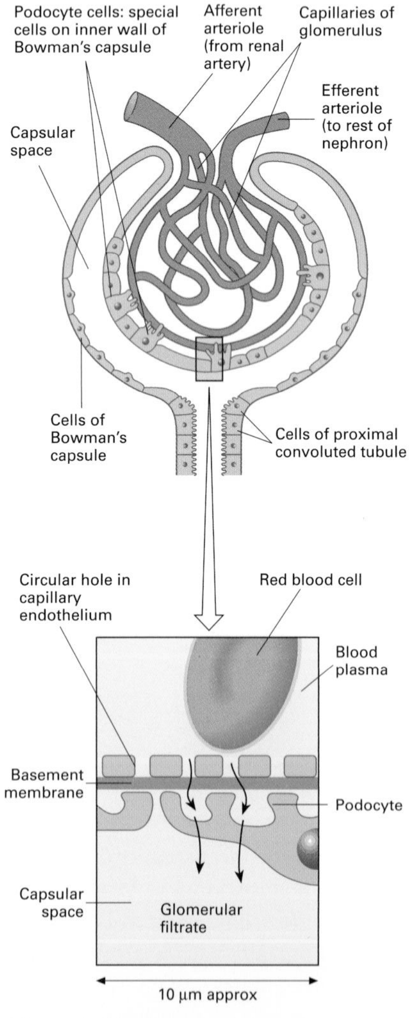

***Figure 1** Ultrafiltration takes place through the glomerulus. Only small molecules can pass through the basement membrane of the capillaries forming the glomerulus. These molecules make up the glomerular filtrate.*

THE PROXIMAL CONVOLUTED TUBULE

Ultrafiltration: the formation of the glomerular filtrate

The first task of the nephron is **ultrafiltration**: the separation of large molecules from small ones by a very fine filter. This takes place in the **glomerulus**, a knot of capillaries surrounded by the Bowman's capsule at the beginning of a nephron. A **basement membrane**, a very fine and delicate membrane made of mucopolysaccharide and protein fibres, separates the glomerulus from the capsule. The Bowman's capsule (or **renal capsule**) has special cells called **podocytes** with numerous foot-like processes that grip on to the basement membrane.

Blood enters the glomerulus at high pressure (this is necessary for filtration) through the afferent arteriole from the renal artery. The pressure is kept high because the arteriole leaving the glomerulus is narrower than the one entering it, thus creating a damming effect. The pressure forces some water and solutes through the barriers between the capillary and the nephron: the filtrate passes through the endothelium (the layer of thin flattened cells lining a capillary) and the basement membrane, and down through the spaces between the podocytes into the capsular space (figure 1). The endothelium and podocytes act as coarse filters, but it is the continuous basement membrane that acts as the fine filter, preventing the passage of large molecules into the nephron.

At the point of entering the nephron, the filtrate is called **glomerular filtrate**. The glomerular filtrate has a solute composition similar to that of blood, but it normally lacks red blood cells and proteins, which are too large to pass through the basement membrane. In principle, the formation of glomerular filtrate is similar to the formation of tissue fluid (spread 7.11), but the large surface area of the glomerulus and the high pressure of its blood supply mean that a greater proportion of the blood passes into the filtrate.

The rate of ultrafiltration is high: about 125 cm^3 of filtrate enters the nephron each minute. This is much more fluid than the body can afford to lose. Because substances pass through the glomerulus according to their particle size and not according to their usefulness, the glomerular filtrate contains valuable substances such as glucose and other solutes too precious to lose in the urine. The rest of the nephron is concerned with reabsorbing these useful substances and eliminating those which are either in excess of requirements (such as water and mineral salts) or toxic (such as urea).

Selective reabsorption of glucose

Glucose entering the glomerular filtrate is reabsorbed into the blood by active transport in the proximal convoluted tubule, the first and longest region of the nephron. The proximal convoluted tubule is lined with a single layer of epithelial cells, well adapted for reabsorption (figure 2). The cell surface membranes facing the lumen of the tubule are highly folded into microvilli. These microvilli provide a large surface area for the movement of solute molecules into the cell. The proximal convoluted tubule is bathed in tissue fluid and is never far from a capillary. The cell surface membranes nearest the tissue fluid are highly indented to form numerous intercellular spaces. A basement membrane (a thin, delicate membrane formed outside the cell) covers the outside of the tubule. The membrane is thought to contain protein molecules which carry specific substances into the cell from the tissue fluid, and others out of the cell into the tissue fluid.

Glucose reabsorption takes place by active transport, which needs energy provided by ATP. Proximal convoluted tubule cells have many mitochondria to generate ATP by aerobic respiration. Normally, all the glucose is reabsorbed from the proximal convoluted tubule. However, despite the high capacity of tubule cells to generate ATP, there is a limit to the amount of glucose that can be reabsorbed. If the blood glucose concentration (and, therefore, the glucose concentration in the glomerular filtrate) is higher than a critical level called the **renal threshold**, glucose appears in the urine. This is why untreated diabetics often produce a sugary urine.

Reabsorption of water and other solutes

In addition to glucose, the proximal convoluted tubule reabsorbs amino acids, vitamins, hormones, sodium ions and other soluble minerals, and water. Small protein molecules that have managed to squeeze their way into the tubule are reabsorbed in **pinocytotic vesicles** (see spread 4.8). Sodium is actively transported into the tissue fluid outside the tubule and water follows passively by osmosis. In humans, about 85 per cent of the water in the proximal convoluted tubule is always reabsorbed. The loop of Henlé, distal convoluted tubule, and collecting ducts are also involved in water reabsorption, but the amount reabsorbed from these structures varies according to the state of the person (see spread 8.8).

Surprisingly, about 55 per cent of the urea (the major excretory product that the kidney has to eliminate) also moves out of the nephron into the tissue fluid and blood. This happens because of the small size of the urea molecule and its ability to pass freely through cell surface membranes. It diffuses down a concentration gradient from the nephron into the tissue fluid and blood, where it remains in circulation. The 45 per cent left in the tubule is excreted in the urine.

By the time the fluid leaves the proximal convoluted tubule and the loop of Henlé, it has lost about 85 per cent of its water, all of its glucose, and some of its salt. This is not, however, the end of selective reabsorption, since a variable amount of water and salts are reabsorbed along the rest of the nephron.

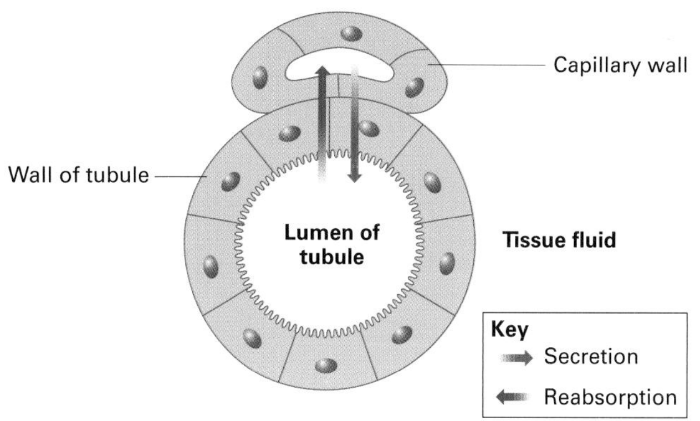

(a) Cross-section of proximal convoluted tubule

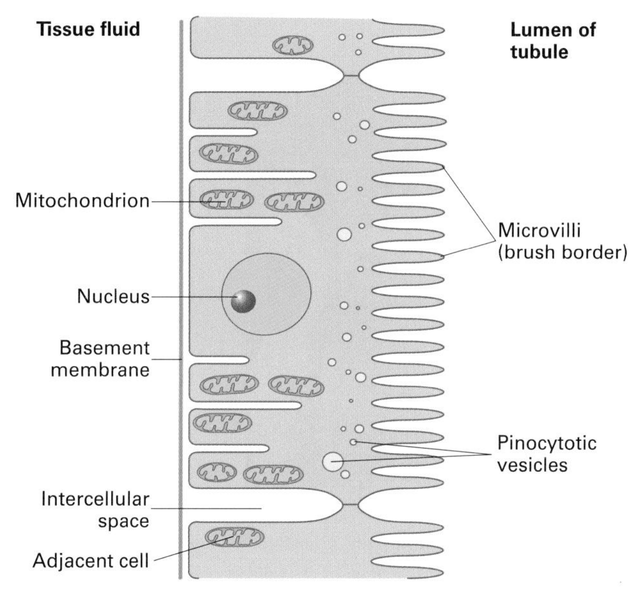

(b) Tubule epithelial cell as seen under the electron microscope

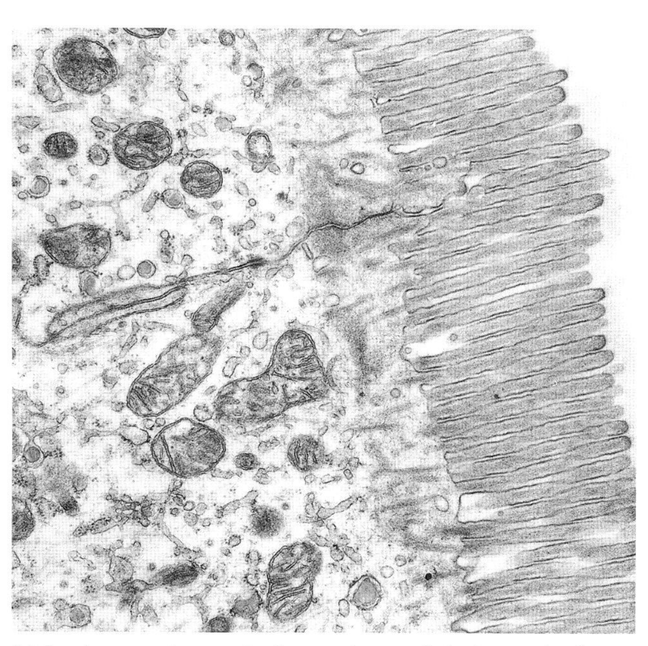

(c) An electron micrograph of parts of two cells in the proximal convoluted tubule (magnification ×5000). Microvilli and a large number of mitochondria show that it is well adapted to its function of selective reabsorption.

Figure 2 *Structure of the proximal (first) convoluted tubule.*

Quick check

1 What is the concentration of glucose in the glomerular filtrate relative to the concentration in the blood?

2 a Explain how the cells lining the proximal convoluted tubules are adapted to give a large surface area.

b Suggest why the proximal tubule epithelial cells have a high density of mitochrondria.

Food for thought

At first sight, mammalian excretion appears to be very inefficient: ultrafiltration selects substances only according to their size. A lot of energy has to be expended to reabsorb useful substances from the glomerular filtrate back into the blood. Suggest why mammalian kidneys do not selectively excrete useless and harmful substances into the nephron, leaving the useful ones in the blood.

8.7

The loop of Henlé

OBJECTIVES

By the end of this spread you should be able to:

- explain the principle of countercurrent exchange
- describe the role of the loop of Henlé in producing salty tissue fluid in the kidney medulla.

Fact of life

Countercurrent exchange of heat in the blood vessels going to and from the feet of an Emperor penguin minimises heat losses when it is standing on Antarctic ice (figure 1).

Figure 1 *Emperor penguins are adapted to minimise heat losses through their feet.*

The loop of Henlé appears to function as a countercurrent exchange mechanism that increases the solute concentration in the medulla of the kidney. By doing so, it creates an osmotic gradient along which water can be withdrawn from the collecting duct if circumstances demand it.

The principle of countercurrent exchange

A **countercurrent exchange mechanism** involves the exchange of heat or materials (such as gases, water, and solutes) between fluids flowing in opposite directions ('countercurrent') in two systems. This type of exchange mechanism is used in power stations to maximise the exchange of heat between cold water entering a generator and steam leaving the generator. Inside the generator, the cold water is heated to turn it to steam which turns the turbines. The steam then leaves the generator. The outgoing steam needs to be cooled as much as possible before releasing it into the environment, and the incoming water needs to be heated by the outgoing steam so that it can be more easily converted to steam to drive the turbines. Figure 2 shows that if the incoming water and outgoing steam flow in the same direction, the temperatures of the two systems become virtually the same, but neither is particularly low or particularly high. On the other hand, when the water and steam flow in opposite directions, the final temperatures are very different: the incoming water becomes almost as hot as the steam, and the outgoing steam is almost as cool as the water before it enters the generator. Countercurrent exchange therefore enables the incoming water to extract much more heat from the outgoing steam than does a system using parallel flow.

Countercurrent multiplier

In the loop of Henlé, a countercurrent exchange mechanism is combined with the active secretion of solutes. A system that uses this combined type of exchange is called a **countercurrent multiplier**. It enables the loop of Henlé to create a very high concentration gradient between the tissue fluid and blood in the medulla of the kidney and the urine in the collecting ducts.

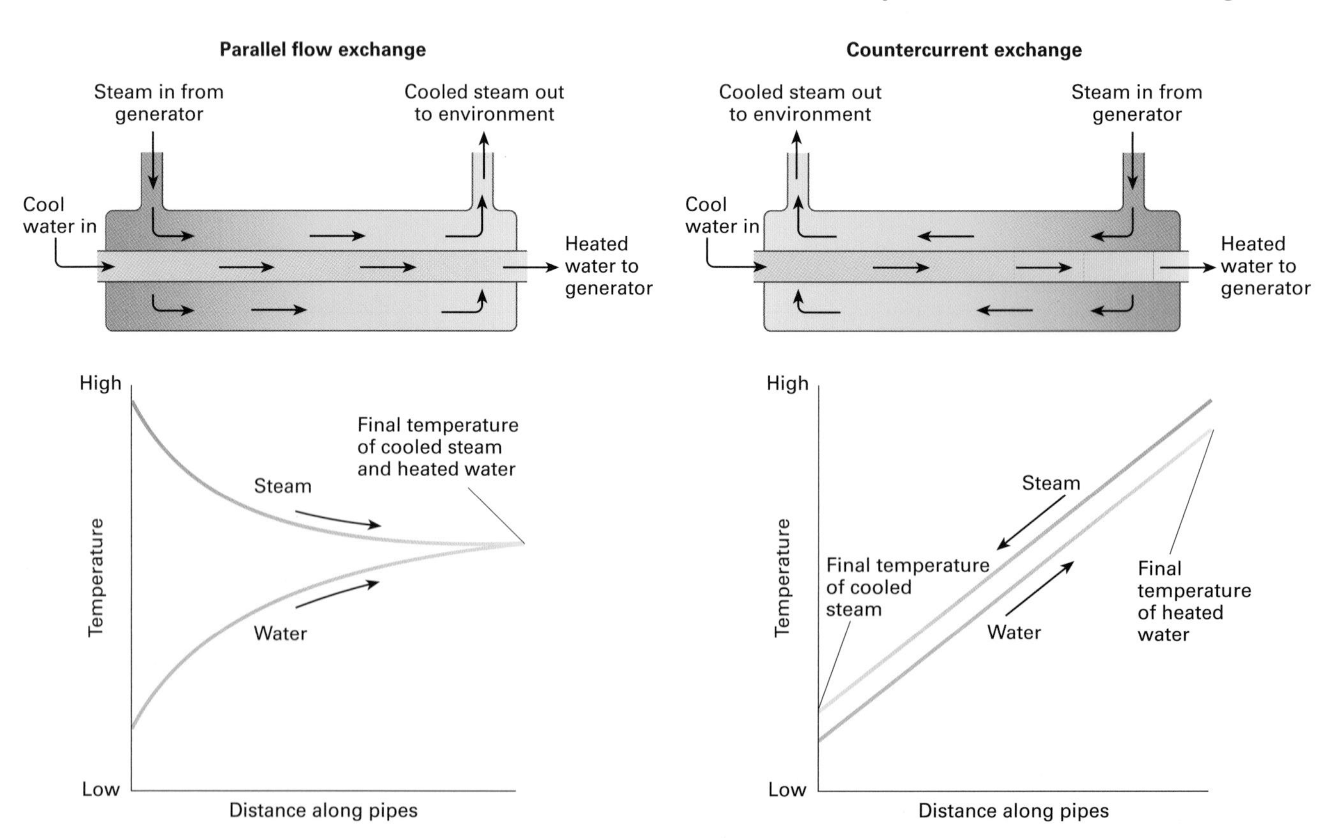

Figure 2 *Comparison of parallel flow and countercurrent mechanism of exchange.*

Before discussing the countercurrent multiplier, we need to look again at the arrangement of the loop of Henlé and its neighbouring structures. As you can see in spread 8.5, figures 2c and d, the loop of Henlé is connected at one end to the proximal (first) convoluted tubule and at the other end to the distal (second) convoluted tubule. Between these two tubules, the loop of Henlé first descends deep into the medulla and then after a hairpin bend ascends into the cortex again. Throughout its course, it is surrounded by fine looped blood vessels called the **vasa recta**. These vessels carry blood from the glomerulus to the renal vein.

Exchange using a countercurrent multiplier is a dynamic process, taking place along the whole length of the loop simultaneously. Salt (sodium and chloride ions) is pumped actively out of the ascending limb into the surrounding tissue fluids. This creates an osmotic gradient which draws water out of the descending limb into the medulla (the ascending limb is impermeable to water). At any one level of the loop of Henlé, this results in the fluid in the descending limb having a slightly higher salt concentration than that in the adjacent ascending limb. The process continues down the length of the loop so that this concentrating effect is multiplied. The countercurrent multiplier means that the fluid in and around the loop of Henlé becomes saltier as it goes down the loop, is saltiest at the hairpin bend and becomes less salty as it goes up the ascending limb (figure 3). For animals of the same size, the final salt concentration that can be produced in the tissue fluid depends on the length of the loop: the longer the loop, the higher the final salt concentration.

The vasa recta supplies oxygen and nutrients to the loop of Henlé. In addition, its looped arrangement is important because it ensures that, by simple countercurrent exchange, the incoming blood gets first saltier as it passes down into the medulla and then less salty as it leaves for the cortex. This allows the tissues of the medulla to retain their high salt concentration, rather than the salt being carried away in the blood to other parts of the body.

The high salt concentration of tissues in the medulla of the kidney creates a concentration gradient between the tissue fluid there and the urine within collecting ducts. As we shall see in the next spread, under certain circumstances, this concentration gradient allows water to be reabsorbed from the urine so that water losses can be minimised.

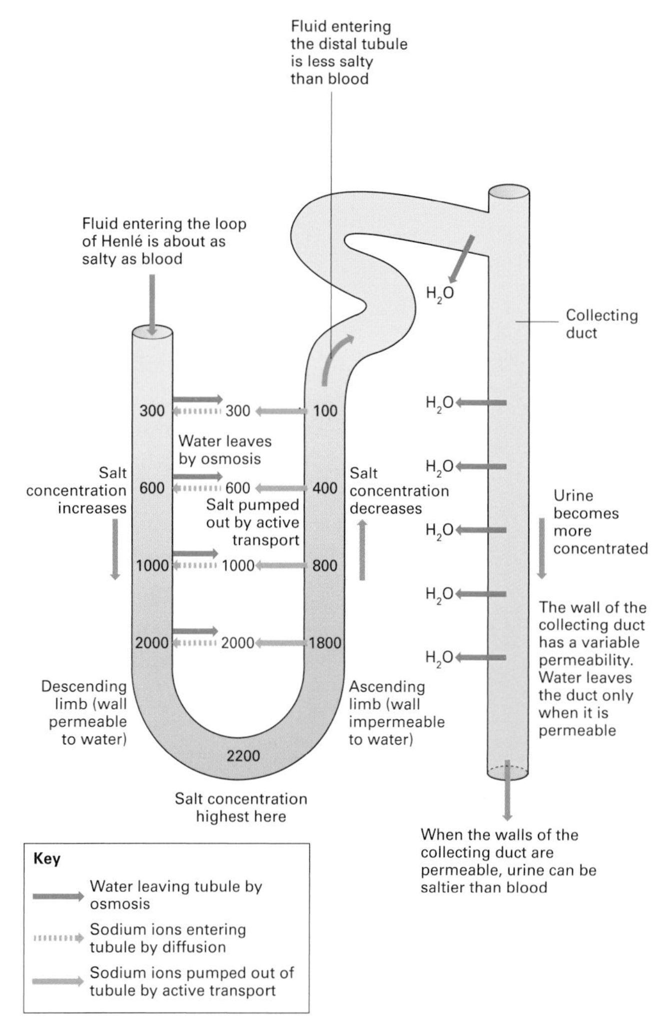

Figure 3 *Countercurrent multiplier mechanism in the loop of Henlé.*

Quick check

1 What is the difference between a simple countercurrent exchange system and a countercurrent multiplier?

2 Why is it important that the ascending limb of the loop of Henlé is impermeable to water?

Food for thought

Suggest why rats that live in deserts have a longer loop of Henlé than rats that live in habitats with a plentiful water supply.

8.8

OBJECTIVES

By the end of this spread you should be able to:

- define osmoregulation
- explain how the volume of urine is controlled by ADH
- list the other homeostatic functions of the kidney.

Fact of life

Diabetes is generally taken to mean a disorder of carbohydrate metabolism, often related to lack of insulin. This is **diabetes mellitus**; there is also another form of diabetes called diabetes insipidus. **Diabetes insipidus** is a rare metabolic disorder caused by a deficiency of antidiuretic hormone (ADH). The patient produces large quantities of dilute urine and is constantly thirsty. The disorder is treated with ADH.

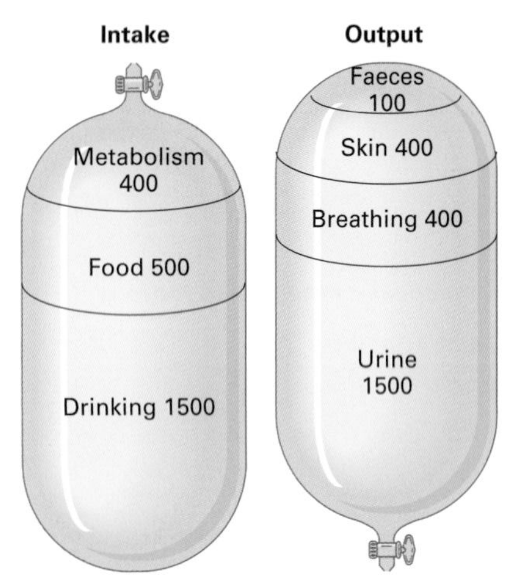

***Figure 1** Water balance in a typical human adult with a mass of 70 kg. The values are the volumes in cm^3 of water intake and output in 24 hours.*

The distal convoluted tubule and collecting ducts

Gaining and losing water

The body of mammals is about 65 per cent water. About a third of this water is extracellular (outside the cells in tissue fluid and blood); the rest is intracellular. There is a continuous exchange of water among the fluid compartments of the body, and between the mammal and the outside world. Water is gained from drinking, eating food, and breaking down food during respiration (for example, the complete oxidation of one molecule of glucose yields six molecules of water). Water is lost from the body by sweating, breathing, in the urine, and in the faeces. In humans, as much as 5 dm^3 of sweat per hour can be lost in extremely hot and dry conditions. Every time you exhale you lose water because expired air is fully saturated with water while inspired air is usually not. Only a small amount of water is normally lost in the faeces but during gastrointestinal disorders (for example, diarrhoea) this loss can be very high.

Despite all this movement of water, the water content of a mammal is maintained at a remarkably constant level from day to day (figure 1). Keeping the volume of fluid in the body constant depends on maintaining the water potential of the blood relatively constant. This process is called **osmoregulation**, and is achieved mainly by regulating the volume of urine production from the kidneys.

Reabsorption of water

In humans, about 180 dm^3 of water passes from the blood into the Bowman's capsules each day. Approximately 85 per cent of this water is automatically reabsorbed from the proximal convoluted tubules, irrespective of the state of the person. This reabsorption is by the passive process of osmosis which happens because a concentration gradient is created by the active transport of sodium ions out of the tubule into the surrounding tissue fluid and blood (spread 8.6).

Fluid leaving the proximal convoluted tubule still contains a substantial amount of water. It is largely the job of the distal convoluted tubules and collecting ducts to determine how much of this water is excreted in the urine. The loop of Henlé provides the concentration gradient along which water can be reabsorbed from these structures if necessary.

In order to conserve water in times of water shortage or heavy water losses, mammals can produce a **hypertonic urine** (a urine with a higher concentration of solutes than the blood). The production of a hypertonic urine depends on two main conditions:

1. a concentration gradient so that water can move by osmosis out of the distal convoluted tubules and collecting ducts into the blood
2. tubules and collecting ducts that are permeable to water.

We have already seen how the countercurrent multiplier mechanism creates an osmotic gradient between the tissue fluid in the kidney medulla and the urine in the collecting ducts (spread 8.7). Fluid leaves the loop of Henlé with a water potential greater than that of blood plasma so there is a concentration gradient between fluid in the distal convoluted tubules and the surrounding tissue fluid. This concentration gradient is enhanced when salts are actively pumped out of the distal convoluted tubule (for example, as a part of salt regulation). The amount of water reabsorbed from the distal convoluted tubules and collecting ducts depends on their permeability. This is controlled by a negative feedback mechanism involving a hormone called **antidiuretic hormone** (**ADH**). ADH increases the water permeability of the distal convoluted tubules and collecting ducts, allowing more water to pass out of the filtrate and be conserved in the body.

Osmoreceptors in the hypothalamus of the brain (spread 10.14) monitor the water potential of the blood. If the water potential is higher than normal (the blood is too dilute), less ADH is released from the posterior part of the pituitary gland, the collecting ducts remain impermeable to water, and water is excreted in a **hypotonic urine** (a urine with a lower concentration of solutes than the blood). If the water potential is lower than normal (the blood is too concentrated), more ADH is released and the collecting ducts become permeable to water, the urine is hypertonic, and water is conserved (figure 2).

Release of urine (micturition)

After leaving the collecting ducts, the urine flows continuously down the ureters and into the bladder. The bladder is connected to the outside environment via the urethra. Urine release to the environment (**micturition**) depends on how full the bladder is, and also on a number of muscles. When the bladder is empty, nerve impulses from the sympathetic nervous system (part of the autonomic nervous system: see spread 10.7) cause its muscular wall to relax and close its opening into the urethra by contracting an internal sphincter muscle. When the bladder is full, nerve impulses from the parasympathetic nervous system cause the internal sphincter to relax and the bladder wall to contract, and urine enters the urethra. However, the urine can be prevented from being released by contraction of an external sphincter muscle which is under voluntary control.

Urine contains any chemical particles that are small enough to pass through the filter of the glomerulus and that are not completely reabsorbed in the tubules. Therefore, midstream urine samples are used to detect illicit drugs and chemicals associated with pregnancy.

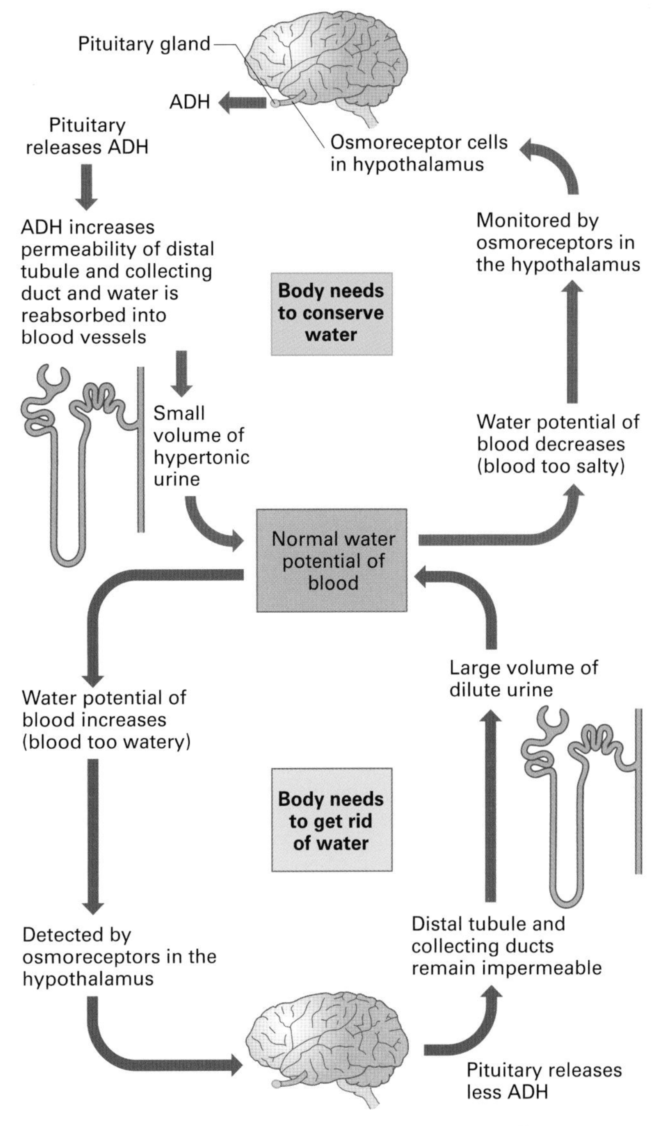

Figure 2 *The role of the kidney in osmoregulation. The volume of urine is controlled by a negative feedback mechanism involving antidiuretic hormone (ADH).*

Regulation of salts, blood pressure, and pH

Variable amounts of salts, water, protons, and hydroxide ions are reabsorbed from the distal convoluted tubule to control the salt concentration, blood pressure, and pH of the blood. If the salt concentration in the body is too low, sodium ions may be secreted actively out of the tubule with chloride ions following. If the tubule walls are permeable, water will also leave by osmosis.

The regulation of salt and blood pressure are closely connected. If the sodium ion concentration in the blood is low, then the blood water potential increases (becomes less negative) and water moves by osmosis into the tissues. This results in a slight lowering of blood pressure, monitored by cells which are in contact with the afferent vessels going into the glomerulus. If blood pressure is below the set point, these cells stimulate the secretion of **renin**, an enzyme that reacts with a blood protein to produce **angiotensin**. When angiotensin reaches the adrenal cortex, it stimulates the secretion of **aldosterone**. Aldosterone accelerates the active transport of sodium ions from the distal convoluted tubule and collecting ducts to the tissue fluid and blood. Water follows by osmosis, and the blood pressure rises until it returns to the set point. If the blood pressure is higher than the set point, negative feedback reduces the secretion of aldosterone.

The distal convoluted tubule also helps regulate blood pH by varying the pH of urine. If the blood pH is too low, more protons are excreted. If the blood pH is too high, more hydroxide ions are excreted. By these processes, the pH of urine may be as low as 4 or as high as 9.

Quick check

1 **a** Define osmoregulation.
  b Why is it important for terrestrial mammals to be able to produce a hypertonic urine?

2 Is a dehydrated person likely to produce more or less ADH?

3 List the homeostatic functions of the kidneys mentioned in this spread.

Food for thought

Suggest why strenuous physical activity is often associated with an increase in the volume of urine production, even when the activity causes water loss due to sweating.

8.9

Temperature regulation in animals

OBJECTIVES

By the end of this spread you should be able to:

- distinguish between ectotherms and endotherms
- discuss the advantages and disadvantages of maintaining a constant body temperature
- describe the main ways by which heat can be exchanged between an organism and its environment.

Fact of life

According to **Allen's rule**, as the climate becomes warmer mammals of the same species tend to become more rounded and have shorter limbs and appendages (such as ears). According to **Bergmann's rule**, as the climate becomes colder they also tend to become larger. However, these 'rules' are often broken: many mammals have special adaptations for heat conservation that are more important than size and shape.

(a) Morning

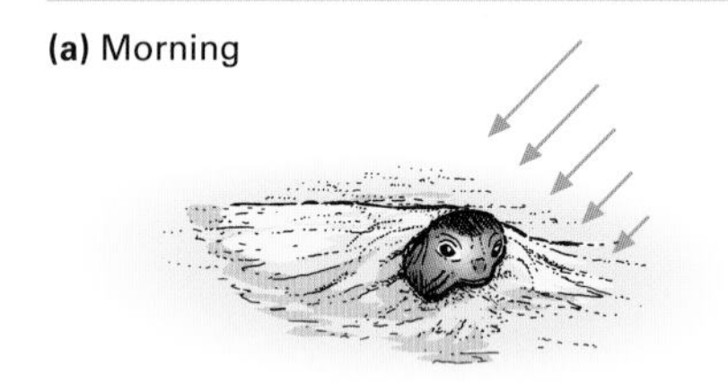

(b) Noon

(c) Afternoon

Figure 1 The earless lizard of south-western USA regulates its temperature by its behaviour: (a) The morning sun warms blood in the head protruding from the sand. The rest of the body remains hidden until the lizard is warm enough to be active. (b) At noon the lizard seeks shelter from the hot sun. (c) It emerges in the afternoon when it is cooler, and lies parallel to the sun's rays.

Living in extremes of temperature

Life exists in an amazingly wide range of thermal environments, from sub-zero polar seas to the bubbling waters of hot volcanic springs. Individual species have evolved a diverse array of mechanisms to cope with extreme conditions. However, a particular organism can only withstand a wide range of thermal environments if it can maintain its internal body temperature at a relatively constant level. This is mainly because enzyme-catalysed metabolic reactions work efficiently only within a limited temperature range (spread 3.3). If the temperature is too high, enzymes become denatured, disrupting metabolic reactions. If the temperature is too low, metabolic reactions take place too slowly to maintain an active life. At extreme temperatures macromolecules may even change their state: for example, phospholipids become very fluid at very high temperatures and solid at very low temperatures.

Thermoregulation

The process by which an animal regulates its temperature is called **thermoregulation**. Animals that can maintain a stable body temperature are sometimes called **homoiotherms** or warm blooded. Animals with a body temperature that is more or less the same as that of the environment are sometimes called **poikilotherms** or cold blooded. However, these terms can be misleading; for example, a hibernating European hedgehog (regarded as warm blooded) maintains its body at a low temperature (about 6 °C), whereas some fish and reptiles (commonly regarded as cold blooded) can maintain their bodies at a relatively high temperature (about 30°C) for long periods. To avoid these problems, many biologists use the terms **ectotherms** and **endotherms** when describing thermoregulation of animals.

An ectotherm has a body temperature that changes with the environmental temperature (for example, an earthworm), or it uses mainly behavioural control mechanisms to regulate its internal body temperature. Reptiles can keep their body temperature relatively constant by using external sources of heat gain and loss, for example, by moving in and out of the sun (figure 1). Most animals except birds and mammals are ectotherms.

All mammals and birds are endotherms. They control their body temperature independently of the environment using internal physiological control mechanisms as well as behavioural ones. However, not all the body of a mammal is kept at a constant temperature, only the **body core**. This consists of the vital organs of the chest and abdomen, and the brain. The skin and tissue close to the body surface are always cooler than the core because it is through these structures that heat is exchanged with the environment (figure 2).

Generally, endotherms can remain active over a far wider range of environmental temperatures than can ectotherms. Within certain limits, endotherms are free to migrate long distances and maintain high rates of activity in all sorts of weather. This allows them to capture and kill ectothermic prey or escape from ectothermic predators. However, maintaining a body temperature different from that of the external environment requires a great deal of energy for metabolism. Consequently, although being endothermic has freed mammals and birds from fluctuations in environmental temperatures, it has made them slaves to their stomachs. They require much more food than ectotherms of equivalent size. A pygmy shrew, for example, consumes about its own mass of food per day whereas a cockroach of the same size can go for days without a meal.

Exchanging heat with the environment

To maintain a constant body temperature, an animal's heat gains must equal its heat losses. In warm-blooded mammals, heat is gained mainly by metabolism and this is a function of body volume. Heat is lost, however, through the body surface and this is a function of surface area. Physiological responses to actively either heat or cool the body are energy demanding. As mammals increase in size their surface area to volume ratio decreases (spread 4.7). Consequently, for mammals with the same body shape and with no special adaptations to conserve or lose heat, a larger body size increases the danger of overheating in a hot environment whereas a smaller body size increases the danger of losing too much body heat in a cold environment.

Figure 3 shows that heat exchange involves radiation, convection, conduction, and evaporative cooling.

- **Radiation** is the transfer of heat as infrared waves. The amount of heat radiated by a body is proportional to the temperature difference between the body and its surroundings.
- In **convection**, heat is transferred by fluid molecules (air or water) moving in a current. The faster the rate of fluid movement, the faster the rate of heat transfer.
- **Conduction** is the transfer of heat by physical contact between two bodies.
- **Evaporative cooling** occurs when water changes to water vapour. The evaporation of 1 gram of water requires the loss of 2.45 kJ from a body.

Spread 8.10 discusses how mammalian thermoregulation combines behavioural and physiological control mechanisms.

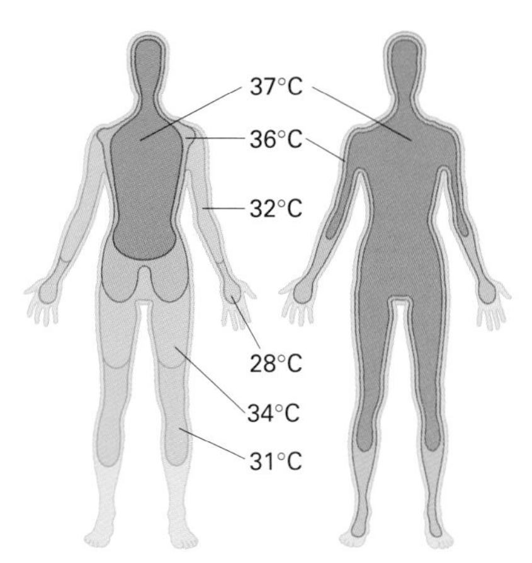

Figure 2 *Temperature distribution in the body of a human in environments at 20 °C (left) and 35 °C (right). The lines are isotherms indicating regions of equal temperature. The core temperature is the dark red area.*

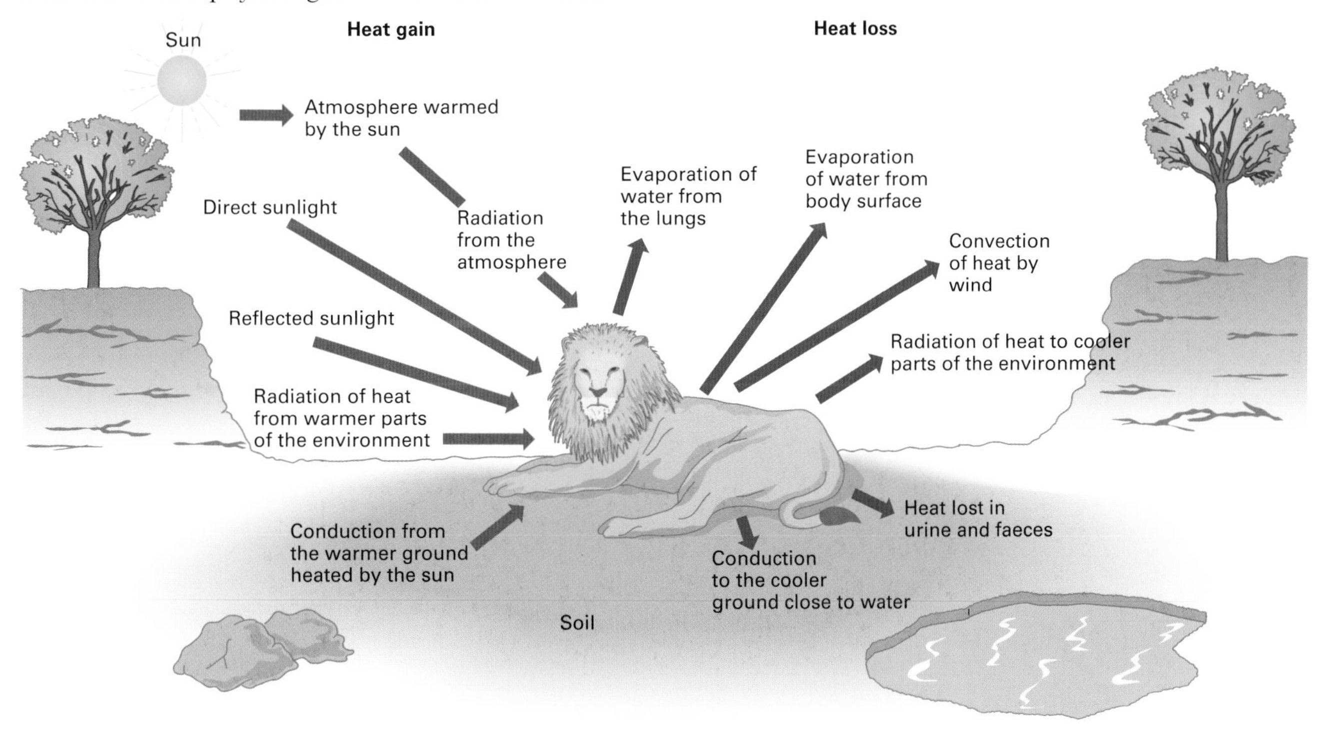

Figure 3 *Energy exchanges between a lion and its environment.*

Quick check

1. Why is the lizard shown in figure 1 called an ectotherm?
2. Why does a mammal have to eat much more food than a reptile of equivalent size?
3. List the three main ways in which heat can be transferred from the environment to an animal.

Food for thought

Some animals have been called **heterotherms** because they do not fit neatly into the two main categories of thermoregulation (endotherms and ectotherms). For example, the powerful, streamlined skipjack tuna (*Katsuwonus pelamis*) can maintain its swimming muscle at a temperature as much as 14 °C higher than the water in which it swims. Suggest how the fish might achieve this.

8.10

Controlling body temperature in humans

OBJECTIVES

By the end of this spread you should be able to:

- describe the mechanisms of human thermoregulation.

Fact of life

In humans, the brain and the organs of the chest and abdomen produce 72 per cent of body heat, even though they make up only 8 per cent of the body mass.

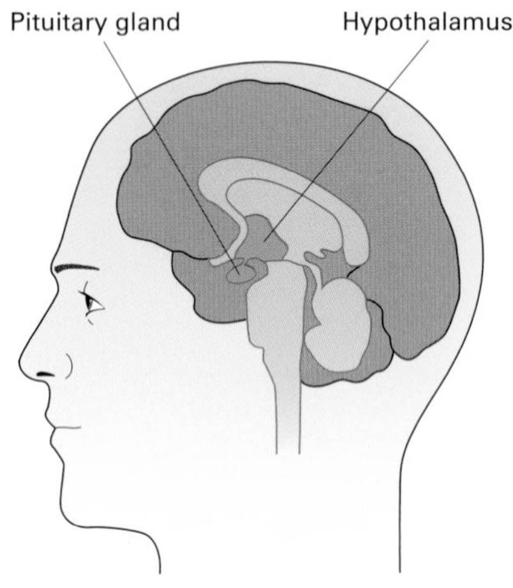

Figure 1 *The position of the hypothalamus in the brain.*

There are two interacting homeostatic systems that control human body temperature: one regulates the temperature of the skin surface; the other regulates the temperature of the body core (the vital organs in the abdominal cavity and the chest cavity, and the brain).

Regulation of skin temperature

The system that regulates the skin surface temperature is more obvious to us because we are conscious of most of its main components: the set point, detectors, comparator, and corrective mechanism.

- The set point is the preferred skin temperature, the temperature at which the person feels comfortable.
- The detectors are thermoreceptors in the skin. Although there is disagreement about the nature of these thermoreceptors, most physiologists believe that **heat receptors** detect increases in skin temperature while **cold receptors** detect decreases in skin temperature. These thermoreceptors can only detect changes in skin temperature; they do not give any information about the actual temperature.
- The cortex of the brain (the outer area, responsible for forming our conscious thoughts and feelings) acts as the comparator. If we feel too hot or too cold, we may decide to move to a cooler or warmer area, remove or add clothing, or take some other voluntary action which brings our skin temperature back to its norm.
- The error signals consist of nerve impulses supplying voluntary skeletal muscle.
- Behavioural responses act as the corrective mechanism.

Regulation of body core temperature

We are not conscious of our second thermoregulatory system. It works mainly by autonomic (involuntary) physiological responses. The set point of this system is the optimal body core temperature, about 37 °C. This is genetically determined but can be temporarily shifted. For example, during certain bacterial infections, substances known as pyrogens raise the set point, causing a fever. Pyrogens may be toxins produced by bacteria, or they may be secreted from white blood cells to raise the body temperature and stimulate the body's defence responses.

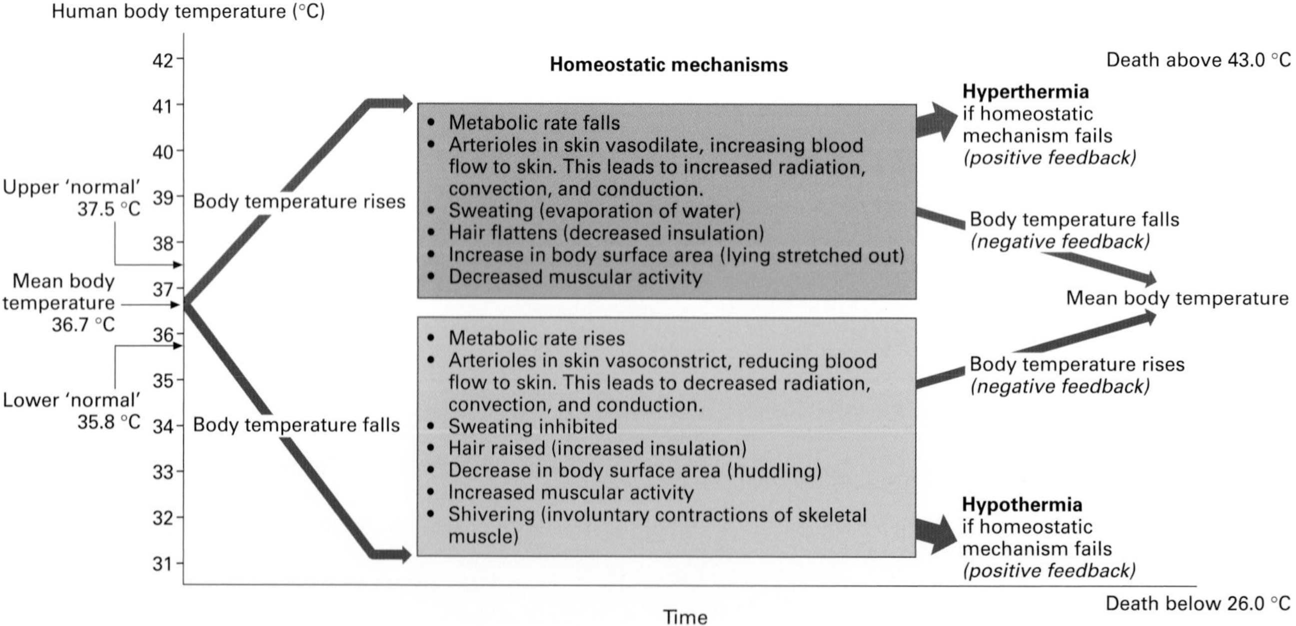

Figure 2 *Homeostatic control of body temperatures.*

The detectors are thermoreceptors in the **hypothalamus**, a small part of the brain just above the pituitary gland (figure 1). The hypothalamus is strategically placed for monitoring the temperature of the blood supplying the nervous tissue of the brain. This tissue is especially sensitive to temperature fluctuations; blood temperatures that are either too high or too low can cause mental derangement.

The hypothalamus contains two thermoreceptor centres. A **heat loss centre** in the anterior hypothalamus is activated by increases in blood temperature. It uses nerve impulses and hormones as the error signals to activate responses that increase heat loss from the body so that the core temperature can be brought back down to its set point. A **heat gain centre** in the posterior hypothalamus is activated by decreases in blood temperature. This uses error signals to initiate a variety of corrective mechanisms which conserve body heat and raise the blood temperature (figure 2).

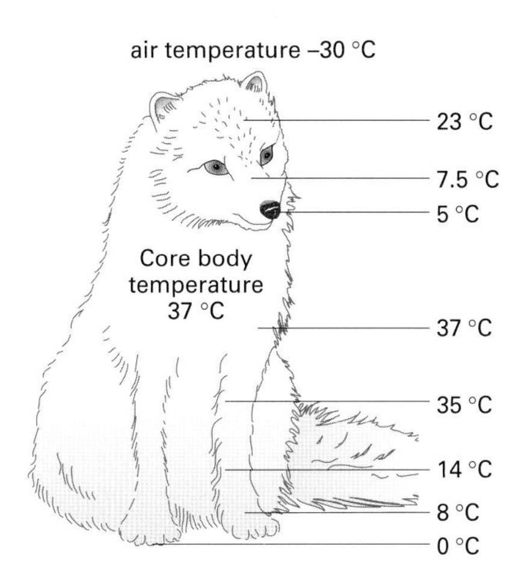

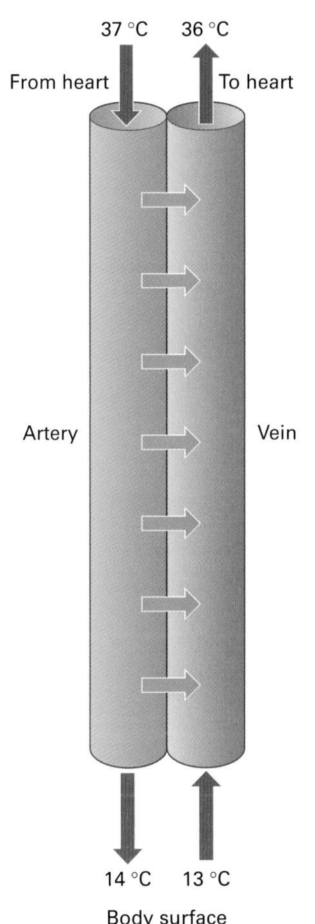

Figure 3 *Some arctic mammals conserve body heat using a countercurrent heat exchange mechanism. The body surface chills more easily than the core. The countercurrent exchange between warm blood coming from the interior and cool blood at the surface ensures that blood returning to the heart is warmed, so heat is retained in the core.*

Hyperthermia and hypothermia

If the core temperature rises above about 41 °C, the thermoregulatory mechanisms usually break down, positive feedback occurs, and a person goes into a state of **hyperthermia** and suffers **heat stroke**. The skin becomes hot, dry, and flushed, and the patient shows signs of mental confusion and loss of muscular coordination. If the body temperature is not reduced quickly (for example, by loosening clothing, fanning, and tepid sponging), the condition can be fatal. Death usually occurs when the core temperature reaches 43 °C. It is especially important that pregnant women avoid hyperthermia because of the adverse effects of high temperatures on fetal devlopment; maternal temperatures should not exceed 38 °C. Pregnant women, especially those who exercise, are advised to take plenty of fluids to reduce the risk of dehydration and assist cooling.

Thermoregulation also usually fails if the core temperature falls below 32 °C, when a person may go into a state of **hypothermia**. The pulse becomes progressively weaker and the patient becomes increasingly irrational and sluggish. If the patient is not warmed in a controlled manner, the core temperature will continue to fall to such a low level that he or she goes into a coma and may die.

Severe cases of both hypothermia and hyperthermia require expert medical attention.

Interaction of the two regulatory systems

The two thermoregulatory systems do not work in isolation; they interact. It is thought that the skin receptors pass information to the hypothalamus about changes in the environmental temperature and set up responses to compensate before the core temperature starts to change. In most situations, both systems work together to control body temperature.

The skin plays a key role in thermoregulation. This and other functions of the skin are discussed in spread 8.11.

Quick check

1 a Distinguish between hyperthermia and hypothermia.

b Suggest why a person who has dry skin in a hot environment might be in danger.

Food for thought

Compare the arctic fox in figure 3 with the desert fox in spread 10.9, figure 1. Suggest how the ears may play a significant part in the thermoregulation of both of these animals. In what other ways do these two foxes differ?

8.11

Temperature regulation and the skin

OBJECTIVES

By the end of this spread you should be able to:

- describe the structure of mammalian skin
- discuss the functions of mammalian skin, especially its role in thermoregulation.

Fact of life

One of the most striking features of humans is our nakedness. The only other mammals of comparable nudity are elephants, pigs, a few rodents that live underground, and aquatic mammals (figure 1). Our nakedness is due to the tiny size of our hairs rather than their small number. Nakedness probably evolved together with our high number of sweat glands. This kept our ancestors cool when hunting large animals in a hot climate (long hairs would interfere with water evaporation).

Figure 1 The nakedness of the dolphin's epidermis provides it with a smooth surface for swimming underwater. External hair or fur would slow its progress through the water, reducing the advantages gained by its streamlined body shape.

The skin is the first line of defence and the first point of contact between the body and the outside world. Taken as a whole, it is the largest organ in the body. Mammalian skin is a complex mixture of tissue and cell types which carry out a wide variety of functions that include thermoregulation; protection against foreign bodies, mechanical damage, and solar radiation; energy storage; and production of vitamin D.

Structure of mammalian skin

The skin of mammals shows considerable differences between species. It may be tough and leathery like an elephant's, or as soft as a human baby's, or hardened into flat protective plates as in the armadillo. It may have thick hair like an arctic fox's, or thin hair like that of the domestic pig, or hair adapted as spikes like that of a porcupine. Despite variations in outward appearance, the skin of all mammals shares the same general structure (figure 2).

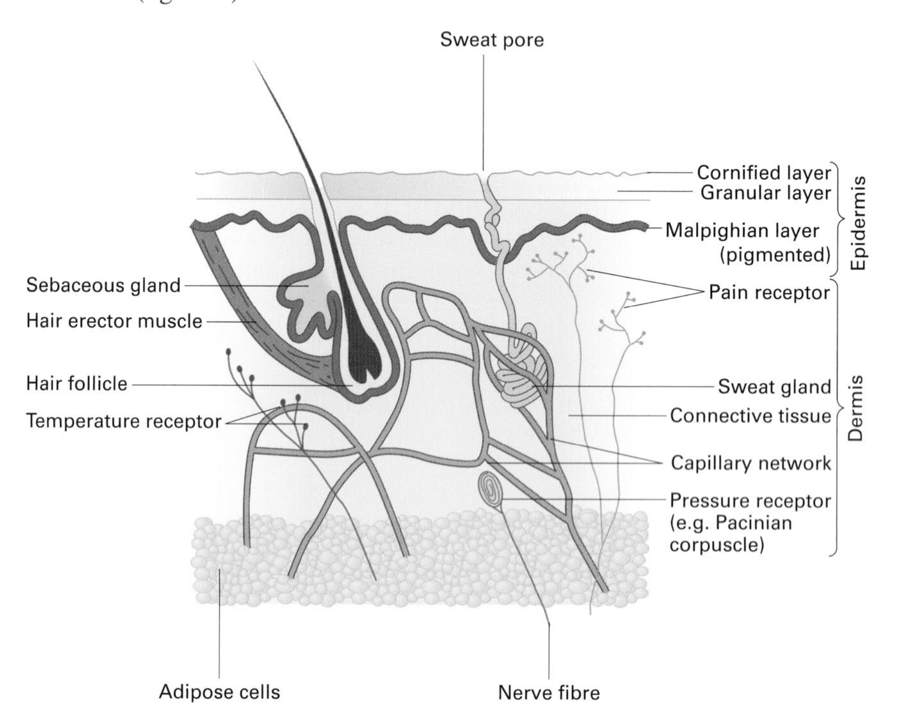

Figure 2 Vertical section through mammalian skin.

Essentially, the skin is a two-layered organ made up of an outer surface layer, the **epidermis**, and an inner layer, the **dermis**.

The surface epidermis layer consists of dead cells. These cells swell when wet and many of them are shed daily. The outer layer of the epidermis becomes cornified or keratinised (the cytoplasm of the outermost cells is replaced by a fibrous protein, keratin). The extent of cornification depends on the amount of friction on the skin (manual labourers, for example, generally have harder and thicker skin than office workers). Living epidermal cells beneath the cornified layer are continually dividing to replace dead cells lost from the surface. **Sweat glands** and **sebaceous glands** release their secretions through pores in the epidermis onto the skin surface. The innermost layer of the epidermis contains the **Malpighian layer**, separated from the underlying dermis by a basement membrane. The Malpighian layer forms special cells called melanocytes which produce melanin, the pigment that gives human skin its dark colour and protects underlying cells from ultraviolet light.

The dermis is well supplied with receptor cells sensitive to temperature, pressure, touch, hair movements, and pain. These receptors are connected via nerves to the brain. The dermis contains collagen fibres and elastic fibres which give the skin toughness and flexibility, and fat cells which store energy and provide mechanical as well as thermal insulation.

The dermis contains sweat glands and blood vessels, both of which play an important part in thermoregulation. The watery sweat secreted by sweat glands evaporates from the skin surface, cooling the body in times of heat stress. The capillaries in the dermis are structured so that the blood can be shunted to different parts of the skin to help control temperature:

- blood is shunted to the surface so that heat can be radiated away from the body when the body becomes too hot
- blood remains confined to deep layers, insulated by fat, when the body needs to retain heat (figure 3).

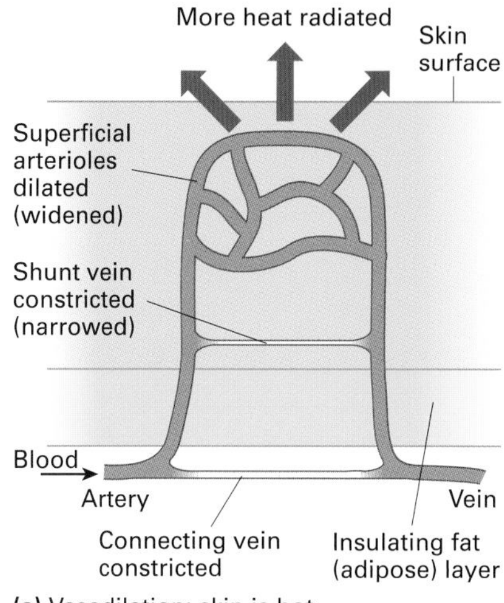

(a) Vasodilation: skin is hot

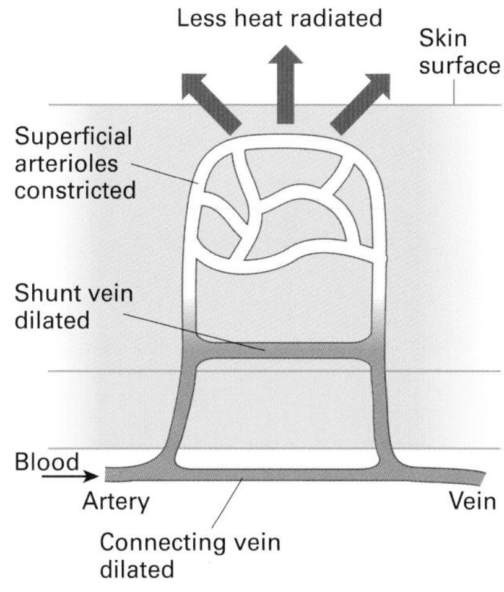

(b) Vasoconstriction: skin is cold

Figure 3 *Thermoregulation by the blood vessels of the skin.*

Summary of the main functions of skin

The skin carries out a wide range of functions. They include:

- protection against:
  - mechanical damage (epidermis becomes hardened by friction)
  - ultraviolet light (melanin acts as a barrier to ultraviolet light)
  - microorganisms (dead cells act as a barrier; sebum secreted by sebaceous glands has a bactericidal action)
  - predators (defensive structures such as claws and horns can be used for attack as well as defence)
  - high water loss or water gain (for example, oily sebum secreted onto the skin surface helps waterproof the skin)
- thermoregulation (by varying amount of sweating, thickness and position of hair, thickness of adipose layer, and blood supply; see spread 8.10)
- detection of stimuli (receptors sensitive to pain, touch, heat, cold, light, pressure, and heavy pressure)
- communication (chemicals called pheromones are released by specialised sweat glands; see spread 12.5)
- camouflage (by colour of hair)
- production of vitamin D (the dermis contains lipids called sterols which are converted by ultraviolet light into vitamin D).

QUICK CHECK

1 List three structures in the skin that play a part in thermoregulation.

2 Explain what is meant by 'shunting'.

Food for thought

Brown fat is a special type of heat-producing fatty tissue which owes its colour to a high density of mitochondria. The mitochondria generate heat rather than ATP, making a significant contribution to non-shivering thermogenesis (metabolic heat production). Brown fat has its own nerve supply and a good blood supply, enabling it to produce heat more quickly than ordinary fat when a mammal becomes cold. Brown fat is very important to some mammals which break it down to generate heat when coming out of hibernation. In humans, brown fat is more abundant in babies than adults. Suggest why brown fat may make a significant contribution to metabolic heat production in babies, but is probably not so important in adults.

Summary

Homeostasis is the process by which conditions within a system (e.g. tissue fluid) are kept at a steady state. In biological systems, homeostasis is usually achieved by **negative feedback**; breakdown of homeostasis often results in **positive feedback**. The homeostatic mechanism maintaining blood glucose concentrations involves **insulin** and **glucagon** produced in the **Islets of Langerhans**. These hormones affect, among other things, the interconversion of glucose and glycogen in the liver. Lack of insulin (or insulin insensitivity) can lead to **diabetes mellitus**.

The **liver** has many functions in addition to its involvement in the control of blood glucose levels. It is the site of amino acid breakdown by **deamination** and the formation of **urea** in the **ornithine cycle**. **Excretion** of urea takes place in the **kidneys** which are also involved in **osmoregulation**. The kidney consists of millions of **nephrons**. **Ultrafiltration** through the **glomerulus** in the **Bowman's capsule** produces a **glomerular filtrate**. This passes into the proximal convoluted tubule where useful substances, including glucose and water, are reabsorbed. The fluid then passes into the **loop of Henlé** where a **countercurrent multiplier** creates a high concentration of salts in kidney medulla. The fluid then enters the **distal convoluted tubule** and **collecting duct** where, under the control of **antidiuretic hormone** (ADH), a variable amount of water is reabsorbed.

Temperature regulation in animals varies from those that conform to the environmental temperature, to those that maintain a constant body temperature in a wide range of environmental temperatures. Humans are **endothermic homoiotherms**, controlling their **body core** temperature by physiological mechanisms. Several structures in **mammalian skin**, such as **hair**, **sweat glands**, **blood capillaries**, and **adipose tissue** play an important part in temperature regulation.

PRACTICE EXAM QUESTIONS

1 A person fasted overnight and then swallowed 75 g of glucose. The graph shows the resulting changes in the concentrations of insulin and glucose in the blood.

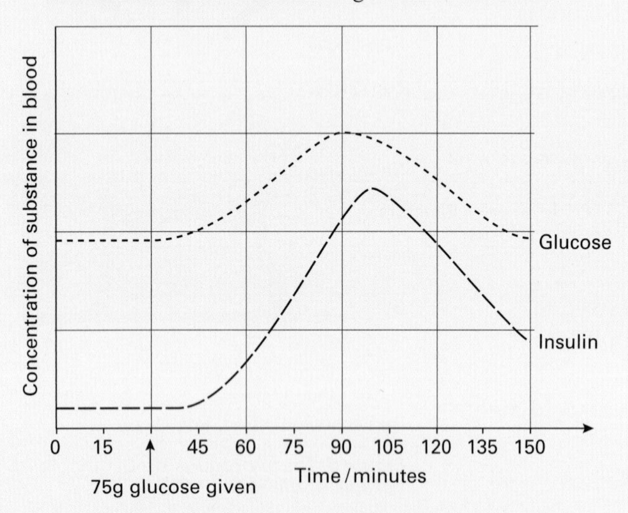

a **i** Explain the relationship between the concentrations of glucose and insulin in the blood in the first 30 minutes after the glucose was swallowed. [2]

ii Use information from the graph to explain what is meant by the term negative feedback. [1]

b Explain why the concentration of glucagon in the blood rises during exercise while that of insulin falls. [2]

[Total 5 marks]

2 Diabetes mellitus is a complex disease which occurs in two main forms, insulin dependent (type 1) and non-insulin dependent (type II). Type I diabetes is sometimes called juvenile onset diabetes and people affected require frequent injections of insulin. The cells which produce the insulin appear to have been destroyed by the body's immune system. Type II diabetes can occur at any age but characteristically in adulthood. It can often be found in people who are obese from over-eating. Type II diabetes appears to be the result of a failure to respond to insulin which may be the result of reduced numbers of insulin receptors. In many cases this form of the disease may be controlled by dietary means. For both types of diabetes, exercise is also often beneficial.

a Which cells are destroyed in *type I diabetes*? [1]

b Outline the role of insulin in the control of blood glucose levels. [3]

c Why does insulin have to be taken by injection rather than by swallowing a tablet? [1]

d Suggest what sort of molecules insulin receptors are and state where they would be found. [2]

e **i** Suggest why dietary management might be used for patients with *type II diabetes*. [2]

ii Suggest why exercise may be beneficial for both types of diabetes. [1]

[Total 10 marks]

3 The graph shows the concentration of the solutes in the fluid in the different regions of a nephron from a human kidney. Curve **A** shows the concentration in the presence of the hormone, ADH. Curve **B** shows the concentration when no ADH is present.

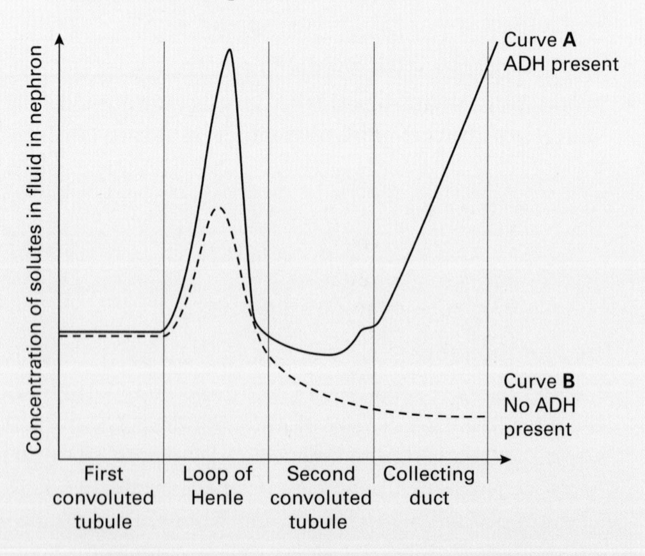

a Although solutes such as glucose are reabsorbed in the first convoluted tubule, the concentration of these solutes does not change as the fluid flows along this part. Explain why. [1]

b What causes the concentration of the solutes to increase then decrease as the fluid flows through the loop of Henlé? [4]

c Explain the difference in concentration of the fluid in the collecting duct in curves **A** and **B**. [2]

[Total 7 marks]

4 Copy the following account of kidney function, then write on the dotted lines the most appropriate word or words to complete the account.

In the kidney, the renal artery branches to form many smaller arterioles, each of which divides further to form a knot of capillaries called a ....

Here, small molecules such as ....
and .... are forced into the cavity of the Bowman's (renal) capsule by the process of .... Selective reabsorption takes place in the nephron. In the proximal convoluted tubule all the .... is reabsorbed.

In the ascending limb of the loop of Henlé, ions are actively pumped out of the nephron. This causes .... to be drawn out of the collecting duct.

[Total 6 marks]

5 The diagram shows the way in which temperature is regulated in the body of a mammal.

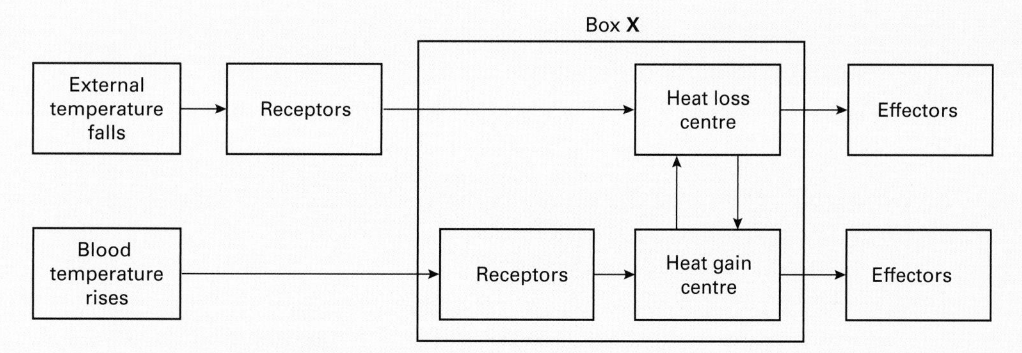

a Which part of the brain is represented by box **X**? [1]

b **i** How does the heat loss centre control the effectors which lower the body temperature? [1]

ii Explain how blood vessels can act as effectors and lower the body temperature. [3]

[Total 5 marks]

6 **a** What is meant by the term *homeostasis*? [1]

b With reference to suitable examples, distinguish between *endothermy* and *ectothermy*. [4]

[Total 5 marks]

9 Heterotrophic nutrition

Heterotrophs

9.1

OBJECTIVES

By the end of this spread you should be able to:

- describe the main forms of heterotrophic nutrition.

Fact of life

In 1953, Stanley Miller and Harold Urey showed how organic molecules might have evolved from simple molecules in the primordial seas, leading to the evolution of life on Earth. They simulated the atmospheric conditions of the early lifeless Earth using electrical discharges, warm water, and the gases methane, ammonia, and hydrogen. A condenser 'rained' water back into a flask filled with liquid which represented the ancient seas. After only one week organic molecules (including amino acids) formed, changing the water from clear to murky brown (figure 1).

The origin of life

More than three thousand million years ago, the oceans of the Earth are believed to have been cauldrons of hot bubbling water energised by fierce lightning storms, irradiated with ultraviolet light, and bombarded by spewing volcanic ash and molten lava. The water in these oceanic basins was so rich in dissolved minerals that it has been called 'primordial soup'. Biochemists believe that these ancient waters became the sites of intense chemical activity, resulting in the formation of complex carbon molecules including sugars and amino acids. Laboratory simulations of these early conditions seem to support this notion of the origin of organic molecules. The interaction of these chemicals over millions of years gradually resulted in more complex and larger organic molecules. It is possible that some chemicals brought by meteorites from outer space contributed to this process. Eventually a new chemical appeared, a nucleic acid which was capable of carrying instructions for the manufacture of proteins, and also capable of replicating itself. When this occurred, groups of accumulated chemicals could reproduce themselves and life began on Earth.

Nutrition in early organisms

It is unlikely that the first living organisms made their own food. The biochemical apparatus to synthesise food is much more complicated than that needed to feed on ready-made organic matter. The first organisms probably consumed the various carbon compounds that were present in the primordial seas. Organisms such as these, that cannot synthesise organic compounds from inorganic substances, are called **heterotrophs**.

As the populations of heterotrophs grew, their source of food would have been gradually exhausted and competition for nutrients would have become increasingly fierce. Any organism that could tap a new source of food would have a competitive advantage. This situation set the scene for the evolution of **autotrophs**: organisms able to manufacture their own food, either by chemosynthesis or by photosynthesis.

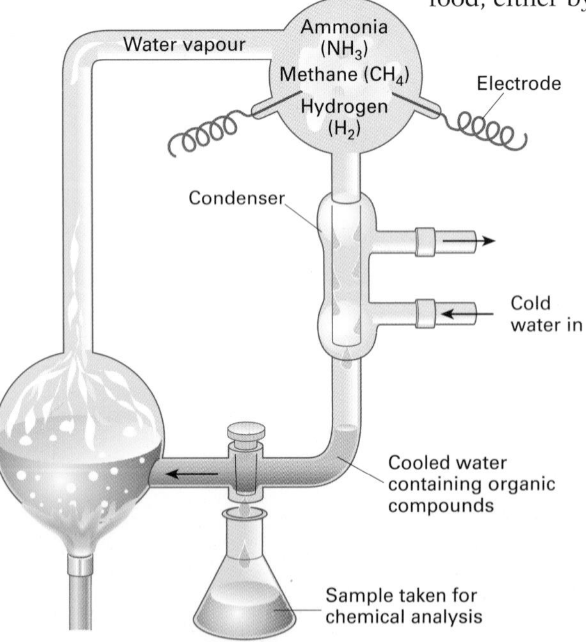

(a) Miller and Urey's apparatus

(b) A scientist recreating the Miller–Urey experiment that showed how organic molecules might have been first formed on Earth. The flask containing a mixture of water, hydrogen, ammonia, and methane has an electric field applied across it and is exposed to ultraviolet light to simulate the conditions on Earth before life evolved.

Figure 1 *Miller and Urey's experiment, which showed how organic molecules might have evolved in primordial seas.*

Autotrophs and heterotrophs now existed together, giving new possibilities for nutritional strategies. The first herbivores to evolve would have been predatory heterotrophs that feed on autotrophs. Carnivores would also have evolved, organisms that feed on other heterotrophs rather than relying on the presence of free organic molecules in their environment. So the first balanced ecosystem came into being, with all the basic nutritional types represented. Today heterotrophs include all animals and fungi, and many bacteria and single-celled organisms.

Types of heterotrophs

There are three main forms of heterotrophic nutrition: **saprobiontic nutrition** (sometimes called saprotrophic nutrition or saprophytic nutrition), **parasitic nutrition**, and **holozoic nutrition**.

Saprobionts feed on dead and decaying organic matter. They include mushrooms and toadstools which secrete enzymes onto their food and digest it before absorbing the products into their bodies.

Parasitic organisms feed on living tissue. Unlike a predator, a parasite does not kill its host in order to obtain nourishment from it.

Holozoic heterotrophs feed on solid organic matter. They can be classified according to whether their diet consists of plants, animals, or both. **Herbivores** feed almost exclusively on plants, although they may consume some animals along with the plant material; examples include sheep, cows, and rabbits. **Carnivores** feed mainly on other animals; examples include lions, dogs, and cats. **Omnivores** feed on both plants and animals; examples include humans, bears, badgers, and pigs.

Heterotrophs can also be classified according to the relative size of the particles of food they ingest. **Macrophagus feeders** eat relatively large particles of food, while **microphagus feeders** eat small particles. Many microphagus feeders, such as cockles and mussels, filter plankton from water. **Liquid feeders** consume a variety of plant and animal juice or soft tissues. Some liquid feeders are free living; bees and humming-birds, for example, consume the nectar of plants; the garden spider (*Aranea diadema*) injects its struggling prey with enzymes, and sucks up the liquefied digested products. Other liquid feeders include **external parasites**, attacking the surface of their living host and sucking up juices from their bodies. For example, mosquitoes suck the blood of mammals, and aphids suck up the sap of plants. **Internal parasites**, such as liver flukes and tapeworms (spread 21.11), are usually bathed in nutrients and absorb their food already digested by the host.

Although we tend to classify heterotrophs into groups for convenience, heterotrophic nutrition is very varied. Each species has its own precise nutritional requirements and its own way of obtaining food (figure 2).

Bracket fungus.

Zebra.

Frog.

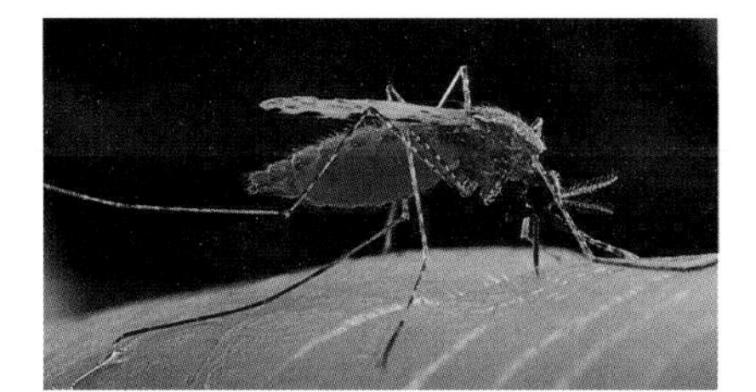

Mosquito.

Housefly.

Fanworm.

__Figure 2__ Different types of heterotrophic nutrition.

Quick check

1 Study figure 2 and identify which type of heterotrophic nutrition each of these organisms is showing.

Food for thought

What types of nutrition are illustrated by a spider which traps a small fly in a web, injects the fly with a poison and a digestive fluid, and then sucks up the liquefied juices of the fly?

9.2 HOLOZOIC NUTRITION IN MAMMALS

OBJECTIVES

By the end of this spread you should be able to:

- describe the main stages of holozoic nutrition in mammals.

Fact of life

Weighing up to 146 tonnes (1 tonne = 1000 kg), the blue whale (*Balaenoptera musculus*) is the largest mammal that has ever existed. These gigantic creatures feed selectively on small shrimp-like planktonic crustacea called krill. During the summer, each whale can take in over 4 tonnes of krill per day.

In mammals, holozoic nutrition involves five main processes:

- ingestion
- digestion
- absorption
- assimilation
- egestion.

These stages can be seen in figure 1.

Ingestion

Ingestion is the act of eating: taking food into the gut, where it is processed. The humpback whale ingests enormous amounts of food (usually small fish, and crustacea called krill). It is a filter feeder, straining its food out from the water using a baleen, a very large comb-like array of plates on either side of the upper jaw. After the food has been sorted mechanically in the baleen, it is swallowed whole and passes into the stomach to be first mechanically and then chemically broken down, a process called **digestion**.

Digestion

Chemical digestion takes place as a series of hydrolytic reactions in different regions of the gut, each with its own specific types of enzymes. Digestion breaks down the large biological molecules in food into small molecules: polysaccharides are broken down to monosaccharides, fats into glycerol and fatty acids, proteins into amino acids, and nucleic acids into nucleotides. The whale can use these small molecules as building blocks for the synthesis of its own tissue.

Absorption and egestion

After the fish and krill have been broken down sufficiently, **absorption** takes place. Useful digested products and other soluble substances such as vitamins and minerals are transported across the cells lining the gut wall into the bloodstream and lymph. Undigested food is eliminated from the gut through the anus. This process is called **egestion**.

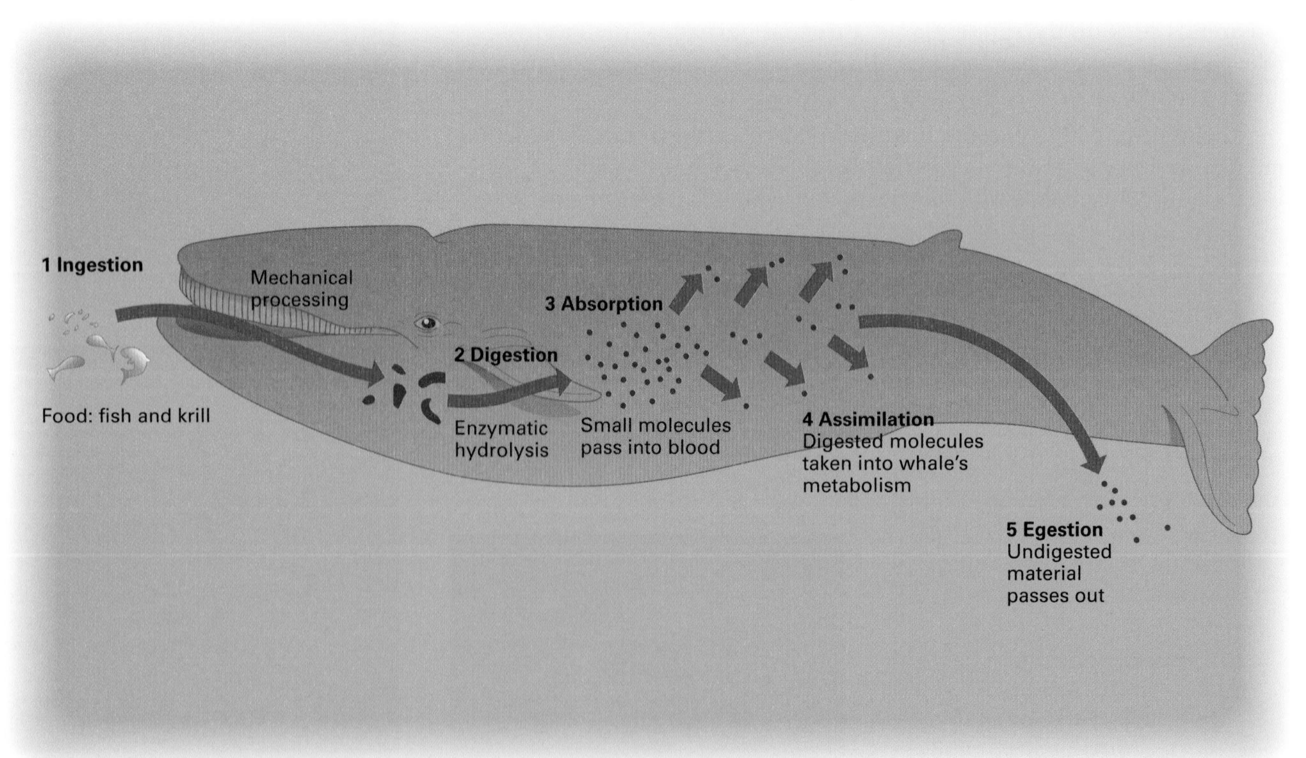

Figure 1 *The five main processes of holozoic nutrition in a mammal.*

Assimilation

After absorption, digested food molecules are carried by the bloodstream to the whale's body cells. The molecules may be stored for future use, broken down further by respiration to generate energy, or used by cells to maintain good health, for growth, or to repair body tissues. (A humpback whale can be more than 16 m long and can weigh up to 65 000 kg.) The process of incorporating and using digested food molecules in the body is called **assimilation**.

The human alimentary canal

Like whales and all holozoic mammals, humans have an **alimentary canal**, a tubular passage or gut divided into specific regions where different digestive processes take place (figure 2). The main parts of the alimentary canal are the mouth (oral cavity or buccal cavity), tongue, pharynx, oesophagus, stomach, small intestine, large intestine, rectum, and anus. Other organs outside the alimentary canal that contribute to digestion include the salivary glands, pancreas, and liver. These organs secrete digestive juices which enter the alimentary canal through ducts. Bile, the digestive secretion of the liver, is stored in the gall bladder before it passes to the small intestine.

The wall of the alimentary canal has the same basic structure throughout its length (figure 3). The **serosa** (serous membrane), a thin protective and supportive layer of epithelial and connective tissue, lines the outer wall of the gut. The **mucosa** containing epithelial tissue lines the inner wall. The tissue is glandular, secreting chemicals into the gut lumen to ease the passage of food or to digest it. The type of secretion varies according to the gut region. In some regions mucus and specific enzymes are secreted; in others only mucus.

The shape of the mucosa also varies. In many regions it is highly folded into structures called **villi** which increase the surface area for digestion and absorption. All parts of the gut contain the **muscularis externa**: two layers of smooth muscle, one layer of circular muscle and one layer of longitudinal muscle. These contract to propel food along the canal by **peristalsis** (waves of rhythmic contraction). Another layer of smooth muscle, the **muscularis mucosa**, forms the boundary between the mucosa and submucosa. Its muscle fibres are oriented in different directions so that their contraction can promote continuous agitation of the glands and mucosa lining the lumen of the gut. The agitation probably improves secretion, digestion, and absorption.

It usually takes about 12 hours for food to travel the full length of the gut, although this transit time varies enormously. A fatty diet, for example, tends to extend transit times.

QUICK CHECK

1 Define the five main processes of holozoic nutrition in mammals in the order in which they occur.

Food for thought

Suggest how egestion is different from excretion.

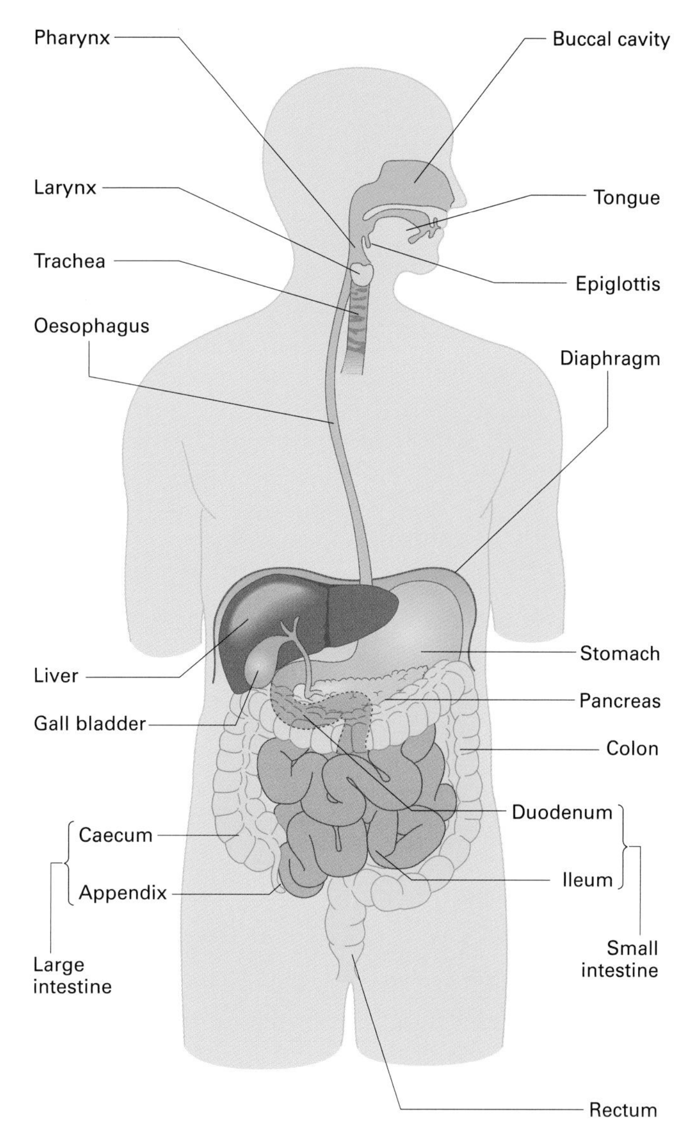

Figure 2 *The human alimentary canal, and other organs involved in digestion.*

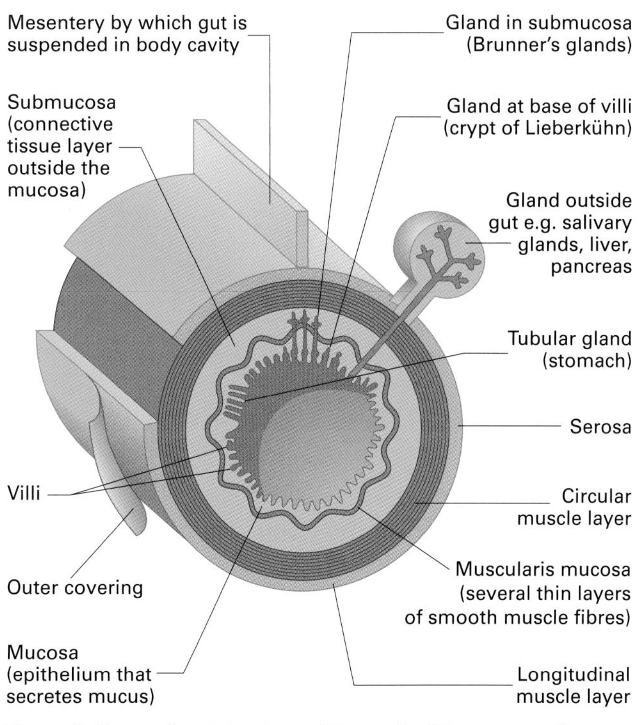

Figure 3 *Generalised structure of the wall of the mammalian gut, showing the main layers.*

9.3

Biting, chewing, and swallowing

OBJECTIVES

By the end of this spread you should be able to:

- describe the functions of the tongue and teeth in human nutrition
- explain how mechanical and chemical digestion starts in the buccal cavity
- describe how swallowing takes place.

Fact of life

Giraffes often feed on the thorny leaves and shoot tips of acacia trees. The long neck is an adaptation that enables giraffes to feed on parts of the tree that shorter creatures cannot reach. The length of its black tongue is almost as impressive as that of its neck: the tongue can be extended up to 46 cm out of its mouth to gather food. The heavily grooved mouth of a giraffe, combined with its ability to secrete large quantities of sticky saliva, allows it to compress and swallow its thorny food. A well developed swallowing reflex and very strong oesophageal muscles ensure that food has a smooth journey from pharynx to stomach.

Jo is hungry. The stomach feels empty and even the merest mention of food sets the salivary glands flowing. Jo likes cheese, so the sight and smell of a freshly prepared cheese and salad sandwich increases the salivation, making the mouth water in anticipation of food. Gurgling noises from the abdomen indicate that other parts of the gut are preparing to digest a meal by secreting digestive juices. These initial responses are controlled by nerve reflexes, either simple reflexes in which the stimulus of food in the mouth causes the salivation response, or conditioned reflexes caused by stimuli which Jo has learned to associate with food. The nature of simple reflexes and conditioned reflexes are discussed in more detail in spreads 11.8 and 11.10.

Taking a bite

Jo uses blade-like **incisor teeth** to take a bite of the sandwich. The **canines** tear the bite free from the sandwich and the tongue pushes it into the **buccal cavity** (figure 1). In the buccal cavity the food is tasted, chewed, lubricated, and moulded into a **bolus** (a rounded ball) before being swallowed.

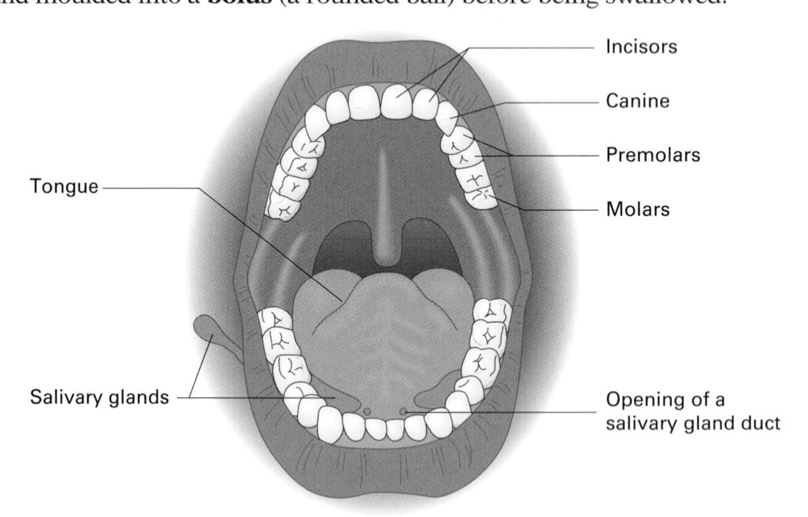

Figure 1 *The mouth and buccal cavity.*

The tongue is the organ of **taste**. It has receptor cells called **taste buds** (figure 2). Four basic tastes, sweet, salt, sour, and bitter have been widely recognised for hundreds of years. In the 1980s, **umami** was recognised as the fifth basic taste. Umami receptors respond to L-glutamate and are responsible for meaty tastes.

The pleasant taste of the sandwich, reinforced by the appetising smells released from the food as it is chewed, causes the **salivary glands** to secrete even more saliva. The **saliva** consists mainly of water and mucoproteins which lubricate the food. It also contains an enzyme, buffers, and antibacterial agents. The enzyme is **salivary amylase** which hydrolyses starch to maltose. This makes the bread taste sweeter as it is chewed. The buffers moderate the effects of acids in foods, which might damage the teeth. The antibacterial agents kill some bacteria before they enter the stomach.

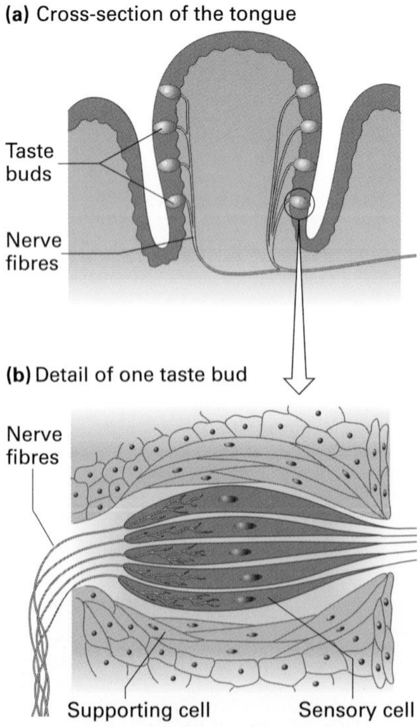

(a) Cross-section of the tongue

Taste buds

Nerve fibres

(b) Detail of one taste bud

Nerve fibres

Supporting cell

Sensory cell

Figure 2 *The tongue and taste buds.*

Chewing the food

The **premolars** and **molars** mechanically digest the food, grinding and crushing it into smaller particles so that it can be chemically digested and swallowed more easily. The teeth are not inert objects but are living structures that can manipulate and feel the food as it is being chewed. Each tooth has its own supply of nerves and blood vessels to give it sensitivity and keep it alive (figure 3).

In addition to tasting the food, the highly muscular tongue manipulates the food as it is being chewed. It moulds the food into a bolus which is passed backwards to the pharynx where it is swallowed.

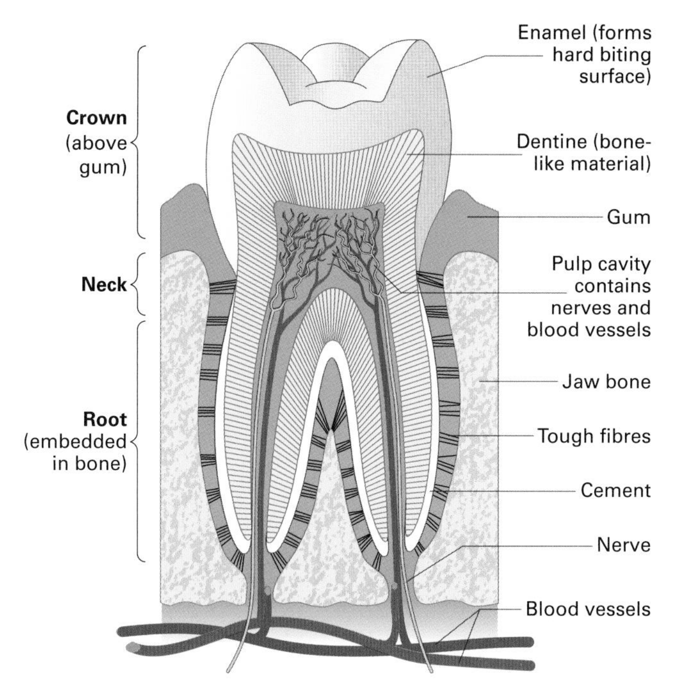

Figure 3 Structure of a human molar tooth.

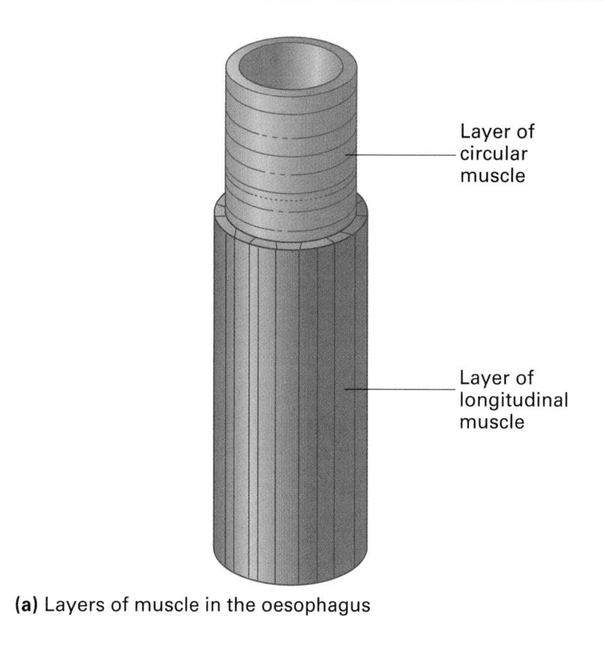

(a) Layers of muscle in the oesophagus

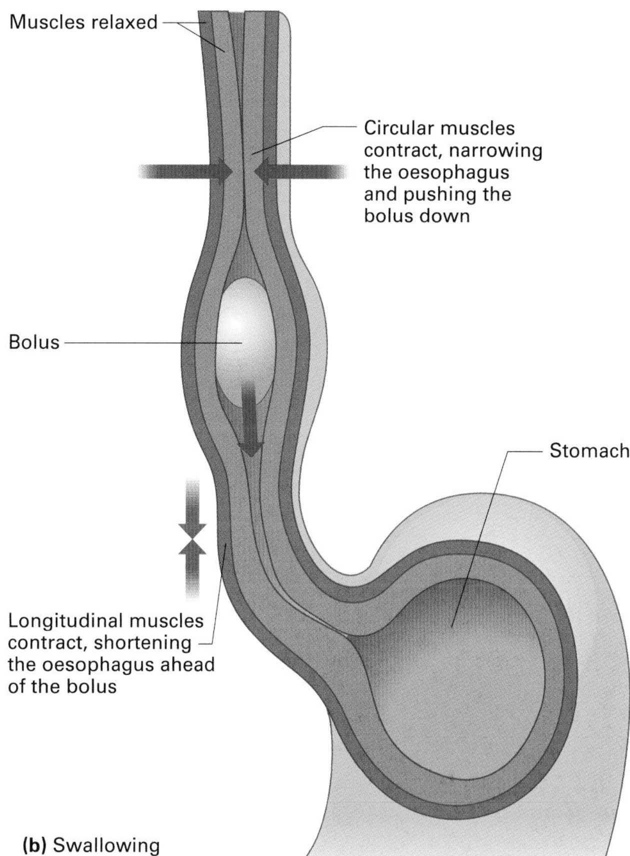

(b) Swallowing

Figure 4 *The mechanism of swallowing.*

Swallowing

When Jo swallows the bolus, a series of reflex actions takes place. The larynx moves upwards and pushes the **epiglottis**, a small flap of cartilage, over the opening of the trachea, preventing the bolus entering the trachea which would cause Jo to choke. After the bolus has entered the oesophagus, the larynx moves downwards and the trachea opens. The movements of the larynx are visible on Jo's throat as the Adam's apple (another name for the larynx) bobs up and down.

The **oesophagus** contains two layers of smooth muscle which contract involuntarily during the swallowing reflex to move the food quickly to the stomach (figure 4). The inner layer consists of **circular muscle**. When this contracts it makes the oesophagus narrower (constricted) and longer. The outer layer consists of **longitudinal muscle**. When this contracts it makes the oesophagus wider (dilated) and shorter. These muscles work antagonistically: when the circular muscles contract the longitudinal muscles relax, and vice versa. Their actions combine to produce the wave-like contractions of peristalsis that travel down the oesophagus.

Peristalsis squeezes the bolus of food down the oesophagus until it enters the stomach, which is discussed in the next spread.

Quick check

1 Give two functions of the tongue during feeding.

2 Name the enzyme in human saliva and describe its action.

3 Describe what happens to the oesophagus when its circular muscle contracts.

Food for thought

Suggest reasons for the following observations:

a It is difficult to swallow when there is not much saliva in the buccal cavity.

b When food touches the back of the pharynx, it is difficult not to swallow.

c The stomach may start to secrete gastric juices before any food reaches it.

9.4

OBJECTIVES

By the end of this spread you should be able to:

- describe the process of digestion in the stomach
- explain how gastric secretions are controlled
- explain how the stomach empties its contents into the duodenum.

Fact of life

When fully stretched, a human stomach can contain about 5 dm^3 of food or drink. The stomach is not usually thought of as an organ of absorption. However, very small lipid-soluble molecules such as alcohol can pass through the phospholipid bilayer of cell membranes and enter the bloodstream. Aspirin is also lipid soluble in acidic conditions and can therefore be absorbed directly from the stomach into the bloodstream.

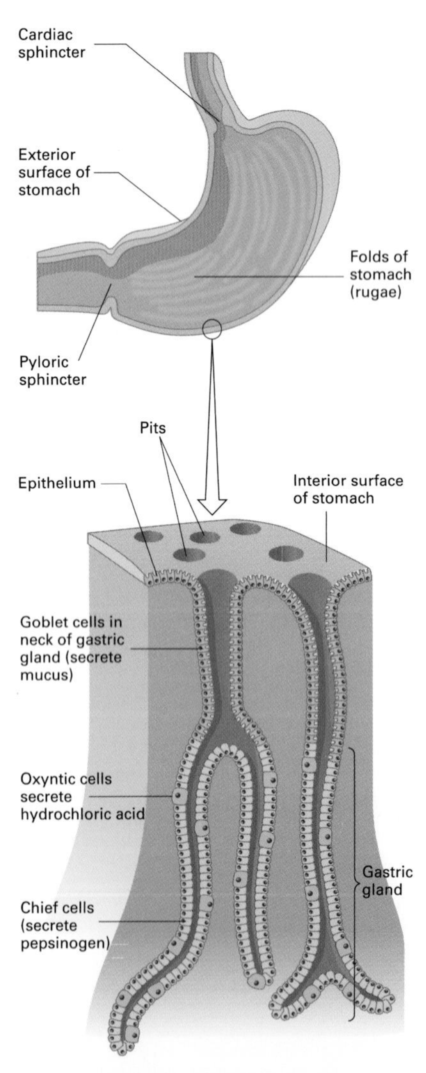

Figure 1 *The mammalian stomach and its tubular gastric glands.*

THE STOMACH

In spread 9.3 we saw how Jo ate a cheese and salad sandwich. Chemical digestion of the sandwich started in the buccal cavity, where starch was partly broken down by salivary amylase to maltose. In this spread, we consider why chemical digestion is necessary, and the role of the stomach in digesting protein and preparing the food for its final breakdown in the small intestine.

The need for digestion

The chemicals within the cheese sandwich are going to be incorporated into the fabric of Jo's body. After Jo has eaten the sandwich, the food molecules could become part of Jo's legs, liver, or even brain. They could be stored for future use, or they could be used as a fuel for cellular respiration.

The sandwich contains all three of the main groups of **macronutrients**: carbohydrates, lipids, and proteins. Macronutrients are food groups that are required in large amounts. However, the chemicals in the sandwich are not in a form that Jo's body can use. First, the proteins and carbohydrates are polymers, and the fats are also molecules that are too large to pass easily into cells. Secondly, most of the large molecules in the cheese and salad sandwich are different from those in Jo's body. The large food molecules have to be broken down into monomers and other small molecules which can then be used as building blocks for Jo's own particular types of large molecules.

The process of breaking food down into small molecules that can be absorbed is called **digestion**. Chemical digestion involves the hydrolysis of polysaccharides, lipids, proteins, and nucleic acids. Enzymes speed up the hydrolysis. Minerals and vitamins do not have to be digested. Their molecules are small enough to be absorbed directly across the gut wall.

Digestion in the stomach

The **stomach** is a highly elastic and muscular organ which can expand easily to hold a large meal. The stomach wall is highly folded and is dotted with **pits** leading to tubular **gastric glands** which secrete gastric juices (figure 1). **Gastric juices** consist mainly of water, along with three other main types of secretions:

- **mucus** from **goblet cells** (so called because they are shaped like a wine glass)
- **hydrochloric acid** from **oxyntic cells** (also called parietal cells)
- **pepsinogen** from **chief cells** (also called peptic or zymogen cells).

Pepsinogen is a precursor, an inactive form of an enzyme which is converted by hydrochloric acid to **pepsin**, the active enzyme. Pepsin hydrolyses large polypeptide chains into smaller polypeptides. Pepsin is an **endopeptidase**: it breaks the peptide links within a polypeptide chain, but does not break the terminal peptide links and therefore produces almost no individual amino acids. Pepsin works most efficiently in acidic conditions.

A fourth secretion of the gastric glands, **prorennin**, is important in young mammals (see Food for thought).

Continual movements of the muscular stomach wall churn the food, mixing it with gastric juices and helping to break it down mechanically. The combination of chemical and mechanical digestion reduces the food into a uniform creamy paste called **acid chyme**.

Control of gastric secretion

Nerves and hormones control the secretion of gastric juices so that they are produced only when food is either about to be eaten or is in the gut. The presence of active enzymes in an empty stomach can lead to autolysis (self-digestion) of the stomach. Control occurs in three distinct phases:

Peptic ulcers

A peptic ulcer is a breach in the lining of the stomach or the first part of the duodenum. The stomach lining is normally covered with a layer of mucin which protects it from the hydrochloric acid and pepsin inside the stomach. It is thought that an ulcer is caused by excessive secretions of hydrochloric acid and pepsin, or by mechanical damage to the stomach wall. The breach in the lining may develop into a hole. An ulcer can be very painful and is easily infected. For a long time the disease was attributed to stress and poor dietary habits which result in oversecretion of acids. Traditional treatments include the use of antacids to control the gastric acids and encourage the ulcer to heal. In the 1980s the disease was linked to the presence of *Helicobacter pylori* (a species of bacterium) in the gut. The bacteria seem to cause the stomach lining to lose its protective coat of mucin. Some patients have been successfully treated with antibiotics which eliminate the bacteria: their ulcers have disappeared and there has been no recurrence. Before the discovery of *H. pylori* in stomachs, the stomach was thought to be too hostile an environment for any bacteria to survive. It is estimated that about 50% of the world's population is infected with *H. pylori*; in about 10% of these, peptic ulcers develop. Evidence also links *H. pylori* to certain types of stomach cancer.

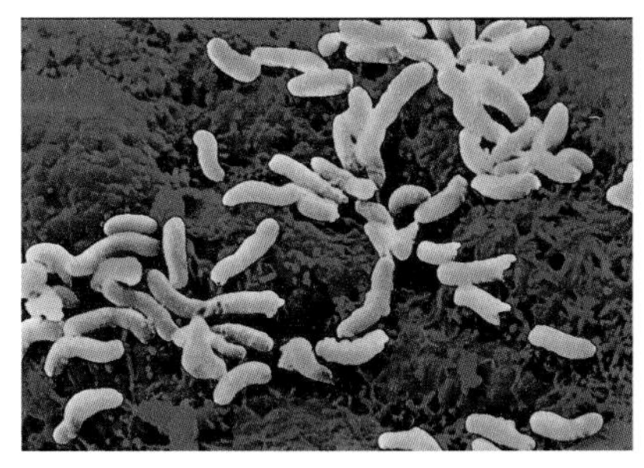

***Figure 2** Scanning electron micrograph of* Helicobacter pylori *(x3500) in the stomach.*

1 **Nervous phase**: the sight, smell, or taste of food triggers a reflex in which nerve impulses relayed from the brain cause the gastric glands to release their secretions.
2 **Gastric phase**: substances from food in the stomach stimulate endocrine cells in the stomach wall to secrete a hormone called **gastrin**. The gastrin has a direct effect on adjacent gastric glands, increasing gastric sections, and it is also carried to all other gastric glands in the bloodstream.
3 **Intestinal phase**: partially digested food in the duodenum stimulates the release of another hormone, gastric inhibitory peptide (enterogastrone), from the duodenal lining. This hormone inhibits gastric gland secretion and reduces movements of the stomach muscles.

Sphincters: opening and closing

The stomach has two sets of sphincter muscles (rings of smooth muscle that can open and close) at its openings. The **cardiac sphincter** guards the opening from the oesophagus, and the **pyloric sphincter** guards the opening to the small intestine. The sphincters are closed most of the time. The cardiac sphincter opens when food is swallowed, and the pyloric sphincter opens briefly to let acid chyme into the small intestine. Occasionally the cardiac sphincter remains open for a moment after swallowing and allows acid chyme to flow up from the stomach into the oesophagus, causing a burning sensation called 'heartburn'. This sphincter also opens during vomiting when the direction of peristalsis reverses, bringing the food back up to the buccal cavity and, usually, out through the mouth.

Food that is not too fatty stays in the stomach for about four hours. The pyloric sphincter regulates the outflow of the food into the duodenum. The sphincter muscle relaxes briefly in response to a number of factors in the stomach. These factors include the presence of acid chyme (detected by chemical receptors in the stomach wall), gastrin, and food (detected by pressure receptors when the food stretches the stomach wall). When the pyloric sphincter is relaxed, chyme enters the duodenum a little at a time.

The duodenum is the first region of the small intestine, the subject of our next spread.

Quick check

1 List the functions of hydrochloric acid in the stomach.
2 Where are these secretions produced:
   a gastrin
   b gastric inhibitory peptide?
3 What prevents food that has entered the stomach from the oesophagus from immediately going into the duodenum?

Food for thought

Prorennin is the precursor of **rennin**, an enzyme that catalyses the conversion of caseinogen (a soluble milk protein) to casein (an insoluble milk protein). Like pepsinogen, prorennin is activated by hydrochloric acid. The conversion of caseinogen to casein results in milk coagulating, delaying the passage of milk proteins into the duodenum. Suggest why relatively large amounts of prorennin are present in the stomachs of young mammals, though it may be absent in adults.

9.5

The intestines

OBJECTIVES

By the end of this spread you should be able to:

- relate the structure of the duodenum to its function
- interpret a table showing the main enzymes and digestive secretions found in the human gut
- discuss the role of the large intestine.

Fact of life

The human small intestine acquired its name because of its small width, not because of its lack of length. On the contrary, it is the longest part of the gut and may be up to 6 m long. Its great length combined with the complex folding of its inner surface provides a large surface area (about 250 m^2) for its functions of digestion and absorption of food.

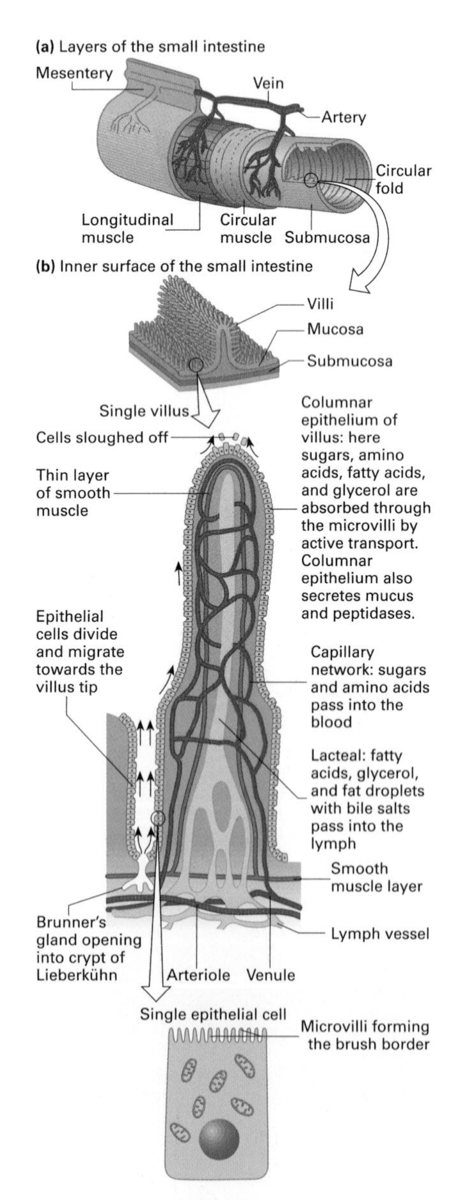

***Figure 1** The small intestine.*

Food that has been pulverised and semi-digested in the buccal cavity and stomach enters the small intestine via the pyloric sphincter as an acidic, creamy paste called **chyme**. Digestion is completed in the small intestine, which is also the major site of absorption. Indigestible food components pass into the large intestine to be expelled through the anus.

Digestion in the duodenum

The first part of the small intestine, the **duodenum**, receives chyme from the stomach, and secretions from the gall bladder and pancreas (see spread 8.4, figure 1). The duodenum is the site of intense chemical activity. All four types of macromolecules (carbohydrates, fats, proteins, and nucleic acids) are digested by specific hydrolytic enzymes with the help of other substances (table 1). Some of the secretions are produced in the pancreas, and others are produced by intestinal cells.

The digestive juices present in the duodenum contain **proteases**, protein-digesting enzymes which include both **exopeptidases** and **endopeptidases**. **Aminopeptidase** is an exopeptidase that acts on terminal peptide bonds at the amino end of a polypeptide chain, whereas **carboxypeptidases** hydrolyse the peptide bond at the carboxyl end. Endopeptidases such as **trypsin** catalyse the hydrolysis of internal peptide bonds. These enzymes may be very specific, acting on bonds linking two particular amino acids.

Hydrogencarbonate ions are secreted by the pancreas and by **Brunner's glands** in the duodenum. The alkaline ions neutralise the acidic chyme and provide an optimal pH for the action of enzymes in the small intestine. Brunner's glands lead to **crypts of Lieberkuhn**, tubular invaginations of glandular epithelia at the base of villi. Actively dividing **stem cells** occupy a region close to the base of each crypt. They replace cells continually lost from the crypt and villi. **Paneth cells** in the base of each crypt secrete a variety of antimicrobial agents to protect the stem cells from infection. **Bile**, produced by liver cells, transforms large lumps of fat into tiny droplets. This process, called **emulsification**, increases the surface area of the fats, making it much easier for lipases to digest them.

The secretion of intestinal and pancreatic juices is triggered by acid chyme entering the duodenum. The chyme also triggers the secretion of three hormones from the intestinal wall: enterogastrone, cholecystokinin (formerly called pancreozymin), and secretin. These hormones circulate in the blood.

- When **enterogastrone** reaches the stomach lining, it inhibits the secretion of hydrochloric acid.
- **Cholecystokinin** stimulates pancreatic cells to secrete digestive juices rich in enzymes. It also induces the gall bladder to release bile into the duodenum.
- **Secretin** stimulates the pancreas to secrete digestive juices rich in hydrogencarbonate ions.

Absorption in the ileum

The ileum, the second part of the small intestine, is the most important site of absorption in the gut. Its structure is well suited to this function (figure 1). Convolutions, folds, and twists in the intestinal wall give it a huge surface area for the absorption of nutrients. The finger-like projections in the intestinal wall, the **villi**, consist of epithelial cells which themselves have microscopic folds called **microvilli** which project into the lumen of the small intestine. These collectively form the **brush border** and they add further to the surface area available for absorption. The epithelial cells also have large numbers of mitochondria which provide the energy for active transport of nutrients that need to be

Table 1 *The main secretions associated with the small intestine.*

Secretion	Source	Function in small intestine (for enzyme-catalysed reactions: substrate ⟶ product)
Bile salts	Liver	Emulsify fats
Sodium hydrogencarbonate	Liver (bile): pancreas	Neutralises stomach acid
Trypsinogen	Pancreas	Converted to trypsin in duodenum
Nuclease	Pancreas	Nucleic acids ⟶ nucleotides
Peptidases	Pancreas	Peptides ⟶ amino acids
Pancreatic amylase	Pancreas	Starch ⟶ maltose
Chymotrypsin	Pancreas	Proteins ⟶ small polypeptides
Lipase	Pancreas	Triglycerides ⟶ fatty acids + glycerol
Enterokinase	Small intestine	Trypsinogen ⟶ trypsin; trypsin converts proteins ⟶ peptides at pH7–8
Maltase	Small intestine	Maltose ⟶ glucose
Sucrase	Small intestine	Sucrose ⟶ fructose + glucose
Lactase	Small intestine	Lactose ⟶ glucose + galactose
Peptidases	Small intestine	Peptides ⟶ amino acids
Nucleotidases	Small intestine	Nucleotides ⟶ bases, sugars + phosphates
Mucus	Small intestine	Protection and lubrication
Water	Liver, pancreas, and small intestine	Essential for hydrolytic breakdown of food; transport medium for secretions

absorbed against a concentration gradient. Each villus has a good blood supply to carry sugars, amino acids, minerals, and other water-soluble substances (including vitamins A, C, and E) to the liver for processing. **Lacteal**s (lymph capillaries) transport fat-soluble substances.

The wall of the small intestine is very muscular. The coordinated actions of its circular and longitudinal muscles create peristaltic waves which move the chyme along the intestines. After about five hours in the small intestine, the liquid chyme enters the **large intestine** (colon) through a T-shaped junction guarded by a sphincter muscle (figure 2).

Reabsorption in the large intestine

One arm of the junction between the small and large intestines leads into a blind pouch called the **caecum**. The caecum terminates in a worm-like extension called the **appendix**. The appendix does not seem to have any important function: its removal (for example, after appendicitis) has no long-term harmful consequences. However, the appendix harbours a mass of white blood cells, so it might play a part in immunity.

After most of the nutrients have been absorbed in the small intestine, the intestinal contents, now called **faeces**, pass slowly along the large intestine. The main function of the large intestine is to reabsorb water into the bloodstream, making the faeces less bulky. This is a vital function; about 7 dm^3 of water from drinks and from internal watery secretions enter the gut every day. Without water reabsorption we would soon become severely dehydrated with potentially fatal consequences.

After passing through the large intestine, the faeces are stored in the **rectum**. As the faeces accumulate, pressure increases in the rectum, causing a desire to defaecate. Adults expel the faeces by voluntarily opening the anal sphincter; infants have to learn to do this. Although a lot of water has been extracted from faeces, water still makes up about 65–80 per cent of their bulk. The rest of the faecal dry matter consists of intestinal secretions (including bile, which is responsible for much of the colour of faeces), cells sloughed off the gut wall, bacteria, and undigested food. Most of the colonic bacteria (bacteria in the large intestine) are dead, but some (for example, *Escherichia coli*) produce important vitamins such as vitamin K, biotin, folic acid, and some B vitamins. Only the indigestible parts of food, such as cellulose, contribute to the faeces.

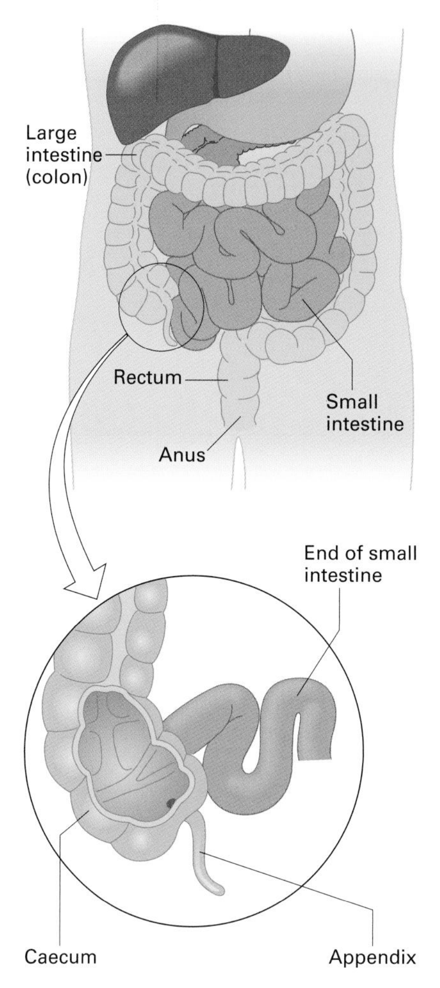

Figure 2 *The large intestine.*

Quick check

1 Distinguish between microvilli and villi.

2 What are the final digestion products of:

a starch

b lipids

c proteins?

3 What is the main function of the large intestine (colon)?

Food for thought

Salivary secretion is mainly controlled by nerve reflexes (spread 9.3); gastric secretion is controlled by a combination of nerve reflexes and hormonal mechanisms (spread 9.4); pancreatic secretions are controlled mainly by hormonal mechanisms. Suggest the significance of this progressive switch from neural to endocrine control from saliva to pancreatic juice.

9.6

Energy from food

OBJECTIVES

By the end of this spread you should be able to:

- explain why we need food
- list the components of a balanced diet
- describe how a person's energy expenditure can be estimated.

Fact of life

The daily energy consumption of an elephant, the largest terrestrial mammal, is about a million times greater than that of a shrew, the smallest mammal. However, gram for gram, the energy consumption of an elephant is only about one per cent of that of the shrew. This high energy consumption per gram body mass is associated with the higher surface area to volume ratio of the shrew.

Energy budgets

An energy budget shows the amount of energy in food consumed against the amount of energy expended over a given period of time. In a balanced diet, the total amount of energy consumed equals the total energy expended. The following table shows some examples of the energy expended by a typical 25-year-old woman office worker with a mass of 62 kg.

Activity	Average energy expenditure (kJ min^{-1})
Everyday activities	
Sitting, eating	5
Standing, cooking	9
Washing and dressing	9
Walking moderately quickly	15
Walking up and down stairs	28
Work and recreation	
Office, sitting	6
Office, walking slowly	11
Dancing	19
Jogging	27

Males expend about 10 per cent more energy than females of the same body mass and age when performing a comparable activity. Adolescents demand extra energy for growth and repair. After maturity, energy requirements generally decline with age.

A balanced diet

Like all other mammals, we need food to provide energy for us to do work and keep ourselves warm, for growth and repair of tissues, and to maintain health.

To satisfy these needs we must have a **balanced diet**. The food we eat each day should contain the correct proportions of carbohydrates, fats, proteins, fibre, minerals, vitamins, and water. Our **energy budget** should also be balanced: we should consume food with the same amount of energy as we expend. An unbalanced diet in which we consume too few or too many calories, or in which nutrients are deficient, in excess, or in the wrong proportion, leads to **malnutrition**. Obesity, excessive weight loss, and vitamin deficiency diseases are different manifestations of **malnutrition**.

Food as fuel

Energy in the form of ATP is generated by the respiration of molecules obtained from our food. Carbohydrates and fats are the main energy nutrients, but dietary protein and alcohol can also be used as metabolic fuels. During prolonged exercise, for example, protein can supply 5–10 per cent of energy requirements.

The energy content of a particular food can be estimated from burning a known mass of the food completely in oxygen in an apparatus called a **bomb calorimeter**. The heat given out can be measured, providing an idea of the energy that food liberates when respired. The total heat generated from one gram of food is its **energy-producing value**. When pure fat is respired, it generates more than twice as much energy as proteins or carbohydrates. Each gram of fat provides about 37.6 kJ, protein and carbohydrate about 16.7 kJ per gram, and alcohol about 29.3 kJ per gram.

Joules and calories

Even though the **joule** (**J**) should be used in all scientific accounts as the standard unit of energy, work, and heat, the **calorie** is still often used to describe energy values of food in general literature. One calorie is the heat required to raise the temperature of 1 g of water through 1°C (more precisely, from 14.5 °C to 15.5 °C). One calorie equals 4.18 joules. These units are very small in relation to the energy used by a person, so it is usual to use kilocalories (kcal) or kilojoules (kJ), units 1000 times larger. Kilocalories are often referred to as Calories (note the uppercase C).

Growth, pregnancy, and lactation

Although an increase in physical activity is the most obvious factor in increasing energy expenditure (see table opposite), additional energy is also needed during periods of growth, pregnancy, and lactation (breast-feeding). A newborn infant uses about one-third of its energy consumption in forming new tissues. In the final months of pregnancy, a mother requires about 800 kJ extra each day to meet the demands of the developing baby and to lay down fat for lactation. During lactation, a nursing mother requires an additional daily intake of 1900–2500 kJ.

Energy expenditure

A person's daily energy requirements depend on sex, age, size, level of physical activity, and metabolic rate. The **basal metabolic rate** (**BMR**) refers to the minimum amount of energy needed to keep the body alive at absolute rest. Because it is difficult to ensure a subject is at absolute rest, the BMR is usually estimated as the amount of energy used by a person resting quietly after at least 8 hours' sleep and 12 hours after the last meal. The BMR for the average healthy adult is usually between about

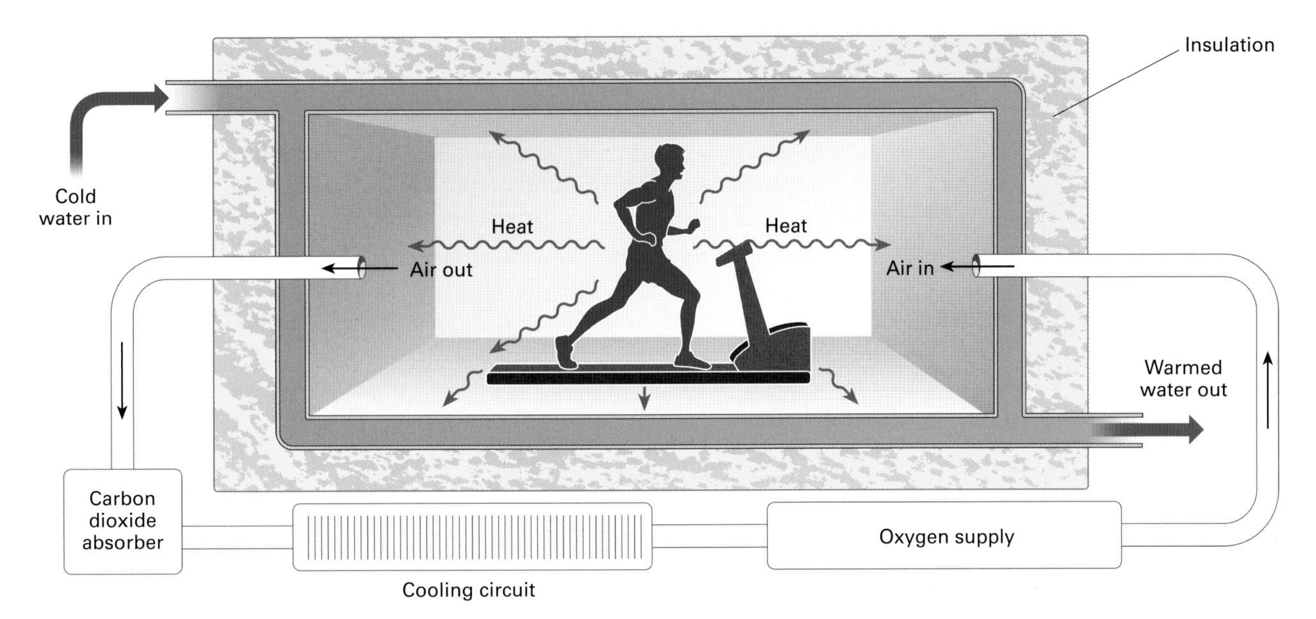

***Figure 1** A calorimeter used to measure energy consumption by direct calorimetry.*

5000 and 7500 kJ per day. It may, however, be as low as 2900 kJ. The BMR usually remains constant for each individual but varies widely from person to person.

Short-term energy expenditures are greatly affected by changes in activity levels (table 1). However, people rarely keep up intense activity for long periods of the day and the average daily activity level is quite low for most people. A person's daily activity level is indicated by his or her **physical activity level** (**PAL**). This is calculated as:

$$\text{PAL} = \frac{\text{total energy expended over 24 hours}}{\text{BMR over 24 hours}}$$

The average PAL for the adult UK population is 1.4, indicating that BMR is the larger component of an average person's daily energy expenditure. According to the World Health Organization, the desirable PAL for good health is 1.7; in economically developed countries, only 13 per cent of women and 22 per cent of men achieve this level of activity.

Measuring energy expenditure

A person's energy expenditure can be measured using direct calorimetry or indirect calorimetry. **Direct calorimetry** is simple theoretically but difficult and expensive in practice. The subject is placed in a specially designed chamber and the amount of heat he or she generates is estimated by measuring the increase in temperature of water circulating the chamber (figure 1). This method is based on the assumption that all the energy expended by a person is ultimately converted to heat.

Indirect calorimetry involves measuring the oxygen consumption of the subject (figure 2). Each cubic decimetre of oxygen used by the person releases a certain amount of energy when each food type is completely broken down by aerobic respiration. This is known as the **heat equivalence** of the food type. One cubic decimetre of oxygen releases 20.9 kJ of energy from carbohydrates; 19.65 kJ from fats; and 18.8 kJ from proteins. The type of food being respired by a person at any given time is reflected by the respiratory quotient (RQ):

$$\text{RQ} = \frac{\text{volume of carbon dioxide liberated}}{\text{volume of oxygen consumed}}$$

Respiration of glucose gives an RQ of 1.0; proteins give an RQ of about 0.9; and fats give an RQ of 7.0–8.0. By combining information about the heat equivalence and RQ, it is possible to estimate a person's energy expenditure from his or her oxygen consumption.

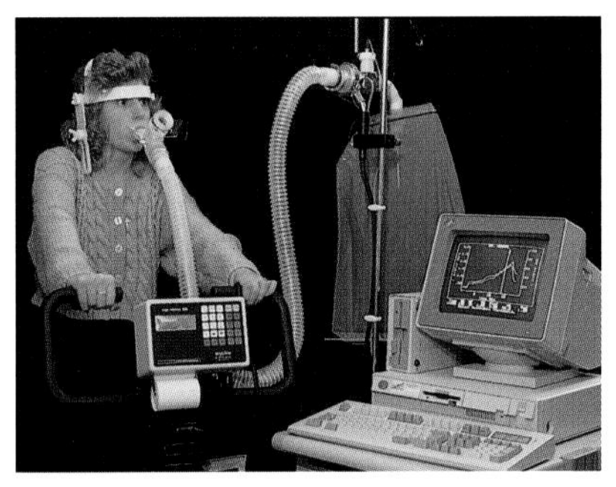

***Figure 2** Equipment used in indirect calorimetry to measure oxygen consumption and RQ.*

Quick check

1. State which type of nutrient has the highest energy-producing value per gram.
2. List the components of a balanced diet.
3. Calculate the BMR of a person who has a daily energy expenditure of 12 500 kJ and a PAL of 1.6.

Food for thought

Some women eat a lot more energy-producing foods during pregnancy, with the explanation that they are 'eating for two'. Suggest why government guidelines recommend that only pregnant mothers in the final trimester (the final third) of pregnancy need to increase their intake of energy-producing foods.

9.7

OBJECTIVES

By the end of this spread you should be able to:

- discuss the role of carbohydrates, fats, proteins, fibre, and water in the diet.

Fact of life

Obesity is associated with a number of illnesses such as heart disease. However, the Inuit people living in the Arctic are typically obese but enjoy robust good health and strong hearts. This apparent paradox is thought to be due to eating large amounts of fish oils which contain a group of fatty acids called omega-3 fatty acids (for example, arachidonic acid and eicosapentaenoic acid).

These unsaturated fatty acids increase the clotting time of blood, ensuring that blood flows smoothly through the vessels and reducing the risk of heart disease. Omega-3 fatty acids also seem to lower blood cholesterol levels and prevent the arteries from becoming clogged with cholesterol plaques.

THE COMPONENTS OF A BALANCED DIET

Carbohydrates, fats (lipids), and proteins are called **macronutrients** because they are required in relatively large amounts in the diet. The chemistry of each of these nutrients is given in chapter 2.

A **balanced diet** combines the different macronutrients, along with fibre, minerals, and vitamins, in the proportions recommended for good health (figure 1). Minerals and vitamins are explored in the next spread.

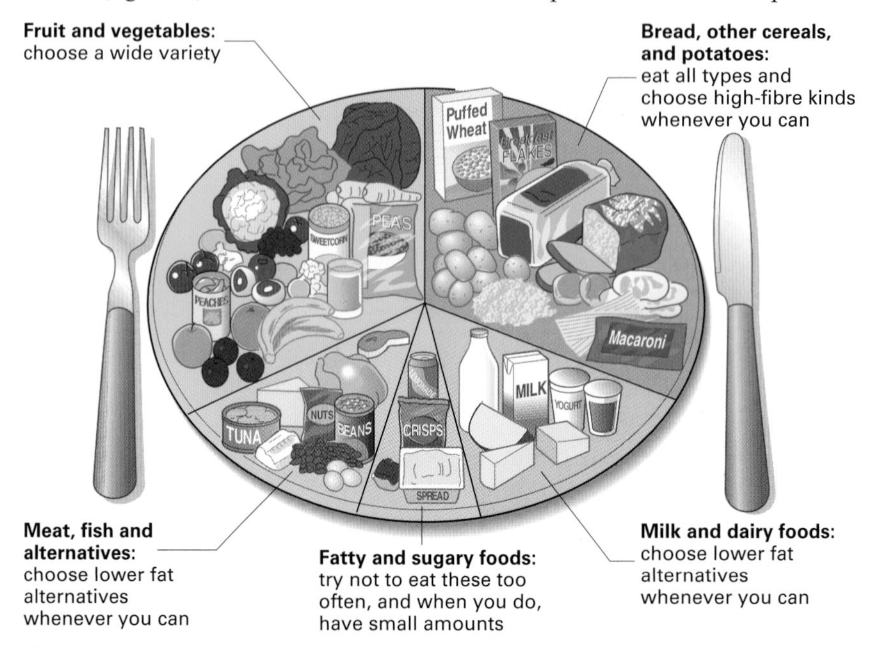

Figure 1 *The 'food plate' was devised by UK government nutritionists as a means of promoting a balanced diet. Fruit and vegetables provide mainly roughage, carbohydrates, minerals, and vitamins; bread, cereals, and potatoes are rich in energy-giving carbohydrates; meat and fish supply protein (as does cheese); and milk and dairy products provide calcium. People are recommended to limit their intake of fatty foods and sweets.*

Carbohydrates

Carbohydrates are the most used form of fuel in the body. Ideally, about two-thirds of our energy should be derived from carbohydrates and the rest from fats. Although carbohydrates are stored in the body as glycogen, glucose is the main respiratory substrate. The brain requires a constant supply of glucose and is very sensitive to deficiencies. This is one reason why blood glucose levels are kept relatively constant (spread 8.2).

Muscles use glycogen as their metabolic fuel. A limited amount of glycogen can be stored in the body (about 8400 kJ of energy; this is equivalent to the energy needed to run about 32 km). Athletes preparing for an endurance activity sometimes boost their muscle glycogen stores by a procedure known by sportspeople as '**carboloading**' or **glycogen loading**. This involves reducing the level of training in the week before the activity, and eating a high-carbohydrate diet for the three days before the start of the activity. Pasta parties have become traditional events the night before a marathon.

In addition to being important as fuels, carbohydrates also have other functions in the body. For example, chains of monosaccharide units combine with proteins and lipids to form glycoproteins and glycolipids in cell membranes (spread 4.6).

Lipids

Lipids (fats and oils) have generally had a bad press. They are often assumed to have no value except as a fuel and as mechanical and thermal insulation. However, lipids are a major component of cell membranes, and we would soon become ill if we ate no lipids. Certain **essential fatty**

acids such as linoleic acid have vital functions and cannot be made in the body. They are used, for example, in the synthesis of prostaglandins and leukotrienes, hormone-like substances that help to control blood pressure and play a role in the body's immune response. Essential fatty acids are obtained from foods such as oil-rich fish, nuts, oil seeds and their products (for example, sunflower oil). Lack of essential fatty acids may lead to hyperactivity, reduced growth, and even death.

Proteins

Proteins are used as fuel, especially during starvation and fasting, but their main function is as a source of amino acids. These are used to synthesise new proteins for growth and repair. Twenty amino acids are needed by humans, 11 of which can be synthesised by the body and are termed **non-essential amino acids**. The remaining nine amino acids must be obtained from the diet and are called **essential amino acids**. One of these, histidine, was once thought to be essential only for infants and children. However, since 1975 it has been recognised as an essential nutrient for adults as well.

The body cannot store excess amino acids, so we need a regular supply of protein in our diet. The amount varies with age (table 1). Animal products such as meat, eggs, milk, and cheese are good sources of proteins because they contain all the essential amino acids. They are called **complete proteins**. Most plant proteins are incomplete because they lack one or more essential amino acid. Nevertheless, all the amino acids can be obtained if a variety of plant foods is eaten. Figure 2 shows that by eating a mixture of beans and corn, for example, a vegetarian can obtain all nine essential amino acids.

Malnutrition due solely to protein deficiency is very rare, but protein deficiency combined with a lack of energy-giving nutrients is relatively common in some parts of the world. Children suffering from such a condition are likely to be retarded mentally and physically; many do not survive into adulthood.

Fibre

Dietary fibre, the indigestible part of plant food (mainly cellulose), is not a nutrient because it is not absorbed and assimilated into the body; it passes out of the gut in the faeces. However, dieticians recommend that about 25–50g of fibre should be eaten each day to ensure good health. Fibre aids peristalsis, assisting the passage of food through the gut, and reducing the risk of intestinal disorders such as constipation. Western diets are notoriously deficient in fibre. This deficiency has been linked to a number of diseases, including certain cancers. Good sources of fibre include wholegrain cereals, fruits, vegetables (especially beans), nuts, and whole baked potatoes.

Water

Although water provides no energy, it is our most important nutrient (see spread 2.4). For our bodies to work efficiently we have to maintain a water balance: too little or too much water can be harmful.

Failure to replace water results in dehydration. Even slight losses of body water can adversely affect physical performance: a loss of 1 per cent of body water, for example, results in a 2 per cent reduction in muscle efficiency. Water loss causes the heart rate to spiral upwards. A loss of 6 per cent of total body water is serious; a loss of more than 10 per cent can be fatal.

Drinking too much water in a short time can lead to water intoxication. Water dilutes the tissue fluid and inflates cells. This can cause metabolic disorders, nausea, vomiting, muscular cramps, and, in extreme cases, death. In February 1995, a 20-year-old man from Derby drank himself to death after taking a single tablet of the drug ecstasy. It was estimated that he drank 26 pints of water which, according to a pathologist, caused brain damage.

The amount of water a person needs depends on water losses, but the World Health Organization recommends at least eight cups each day.

***Table 1** Recommended daily intake of protein according to age and sex.*

Age (years)	Protein (g) Males	Protein (g) Females
1	30	27
5–6	43	42
9–11	56	51
15–17	72	53
18–64	72	54
65–74	60	47
75+	54	42
Pregnancy	–	60
Lactation	–	69

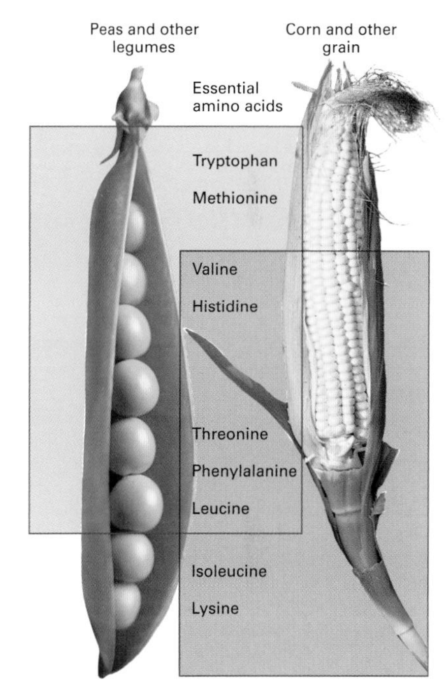

***Figure 2** Essential amino acids obtained from grains and legumes.*

Quick check

1 a What are macronutrients?
  b Name one essential fatty acid.
  c Explain the difference between a complete protein and an incomplete protein.

Food for thought

Milk is often described as the perfect food. Although this is true for newborn babies feeding on human milk, suggest why cow's milk is not the perfect food for people of any age.

9.8

OBJECTIVES

By the end of this spread you should be able to:

- discuss the role of minerals and vitamins in the diet.

Fact of life

Although most people are aware that vitamin deficiencies are harmful, they do not realise that very high doses of niacin, folic acid, and vitamins A, B_6, D, and K are known to be toxic. Excessive consumption of vitamin A, for example, can cause the legs to swell and bones to change structure. Excessive consumption of niacin (vitamin B_3) can cause liver damage, while vitamin D in excess can cause brain and kidney damage. In extreme cases hypervitaminosis (a disorder resulting from consuming excessive vitamins) can be fatal, but this is exceedingly rare.

Minerals and vitamins

Minerals

Using modern techniques, all minerals can be detected in the human body, but most are just 'passing through' and are not incorporated into body tissues or used in metabolic reactions. However, some of these minerals are essential components of a balanced diet. Table 1 shows 16 minerals that are known to be essential because lack of them can cause a deficiency disease: for example, lack of iron can cause anaemia, and lack of calcium can lead to osteomalacia or osteoporosis (spreads 16.4 and 16.10). Other minerals such as vanadium and selenium are believed to play an important role in the body, but not all nutritionists are convinced that they are essential.

The major minerals or **macrominerals** are required in relatively large amounts. The recommended daily adult intake of calcium, for example, is 700 mg. This contrasts with **microminerals** or **trace elements** which are required in very small amounts. A daily intake of only 0.06 mg of selenium, for example, is sufficient to maintain health.

Vitamins

Vitamins are a group of non-protein organic compounds required in minute amounts for the maintenance of good health and efficient metabolism. So far, 13 compounds have been classed as true vitamins; that is, a deficiency of the compound results in a deficiency disease with characteristic symptoms (see spread 16.10 for a discussion of diseases associated with a deficiency of vitamins A and D).

Table 1 *Minerals known to be essential for good health.*

Mineral	Major food sources	Functions
Macrominerals		
Calcium	Dairy foods, eggs, green vegetables	Formation of bones and teeth; blood clotting; enzyme activator
Chloride	Table salt	Formation of hydrochloric acid in stomach; maintenance of water balance; maintenance of anion/cation balance
Iron	Meat, liver, legumes, watercress	Component of haemoglobin and myoglobin needed for oxygen transport in blood; component of cytochromes used in respiration
Magnesium	Meat and leafy green vegetables	Formation of bones and teeth; constituent of coenzymes
Phosphorus	Meat, fish, dairy foods, eggs, vegetables	Involved in transfer of energy in ATP; formation of bones and teeth; component of DNA and RNA; important role in muscle and nerve activity
Potassium	Meat, fruit (especially bananas), vegetables	Needed for correct functioning of heart; important role in muscle and nerve activity; maintenance of anion/cation balance
Sodium	Table salt	Needed for muscle and nerve activity; maintenance of anion/cation balance
Sulphur	Proteins from many sources	Components of certain amino acids; needed for muscle growth
Zinc	Seafoods, yeast products	Activator of many enzymes; helps heal wounds
Microminerals		
Chromium	Liver, meat, cheese, yeast	Improves ability of insulin to convert glucose to glycogen
Cobalt	Liver, meat, milk	Component of vitamin B_{12}; activator of some enzymes
Copper	Liver, meat, fish	Required for synthesis of haemoglobin and bone; enzyme activator
Fluoride	Many water supplies	Component of bones and teeth; helps resist tooth decay
Iodide	Fish, shellfish, seaweeds, iodised table salt	Component of the hormone thyroxine
Manganese	Liver, kidneys, fruit, nuts, legumes, tea, coffee	Bone formation; enzyme activator
Molybdenum	Liver and cereal crops	Essential for functioning of enzymes involved in DNA metabolism

***Table 2** Vitamins: their sources and functions*

Mineral	Major food sources	Functions	Symptoms of deficiency
Vitamin A (retinol)	Oranges, green vegetables, carrots	Maintenance of mucous membranes of eye and respiratory tract; component of visual pigments	Night blindness; increased risk of infections, especially of the mucous membranes
Vitamin B_1 (thiamin)	Liver, lean meat, eggs, unpolished rice, yeast extract	Helps release energy from carbohydrates	Disease beri beri (decreased appetite, gastrointestinal disturbance; nerve and muscle disorders)
Vitamin B_2 (riboflavin)	Wholegrain cereals, meat, liver, yeast extract	Component of FAD, a coenzyme in respiratory metabolism	Disease ariboflavinosis (skin lesions, such as cracks in the cornersof the mouth; blurred vision)
Vitamin B_3 (niacin or nicotinic acid)	Meat, fish, unpolished rice, nuts, yeast extract	Component of coenzymes NAD and NADP; important in respiration	Disease pellagra (skin and gut lesions, muscle weakness, loss of appetite)
Vitamin B_6 (pyridoxine)	Meats, vegetables, wholegrain cereals, yeast extracts	Coenzyme to over 60 enzymes in protein metabolism	Nervous irritability, anaemia, sores in eyes and mouth
Vitamin B_{12} (cobalamin)	Liver, fish, dairy products	Coenzyme in DNA metabolism; needed for maturation of red blood cells	Anaemia and weight loss, muscle twitching, irritability
Folate (folic acid)	Green vegetables, nuts, yeast extract, liver, pulses	Coenzyme in DNA synthesis and amino acid metabolism	Anaemia and gastrointestinal problems
Vitamin B_5 (pantothenic acid)	Most foods, but especially liver, legumes	Component of vitamin A; carbohydrate and fat metabolism	Fatigue, numbness, tingling in the hands and feet ('burning feet syndrome')
Biotin	Legumes, meats	Coenzyme in fat, protein, and carbohydrate metabolism	Scaly skin, pallor, loss of appetite, nerve and muscle disorders
Vitamin C (ascorbic acid)	Citrus fruits, dark green vegetables, potatoes	Used in collagen synthesis (e.g. bones, cartilage, and gums); antioxidant; improves iron absorption	Joint pains, poor wound healing, increased susceptibility to infection; scurvy (bleeding gums, anaemia, red spots in the skin, degeneration of muscle and cartilage)
Vitamin D (antirachitic factor)	Oily fish, dairy products, egg yolk (also made by action of sunlight on lipids in the skin)	Calcium metabolism and formation of bone	Rickets in children (bones do not harden and become deformed); osteomalacia in adults (bones soften)
Vitamin E (tocopherols)	Wheatgerm oil, sunflower oil, peanuts	Antioxidant, prevents damage to phospholipids in cell membranes	Possibly anaemia; sterility in rats, but no effect on human fertility
Vitamin K	Green vegetables, tea, liver	Important in blood clotting (needed to manufacture prothrombin)	Defective blood clotting resulting in easy bruising, excessive bleeding

Vitamins A, D, E, and K are fat soluble; the others are water soluble. Beta-carotene (a nutrient converted in the body to vitamin A) and vitamins C and E are **antioxidants**: chemicals that mop up unstable products of metabolism called free radicals which can damage cells. There is an increasing body of evidence showing that antioxidants provide some protection against certain types of cancer (for example, cancer of the bowel and bladder) and against cardiovascular disease. Vitamin E is added as an antioxidant to some margarines (for example, margarine made from sunflower oils) to prevent the margarine from going rancid.

QUICK CHECK

1 Using table 1, list the minerals required for the following body functions:

a as a component of haemoglobin

b bone formation (there is more than one mineral required)

c as a component of thyroxine.

Food for thought

Many governments have recognised the importance of diet in relation to their nation's health and economic prosperity. As well as providing guidelines about the amounts of nutrients people should eat, some have made more general recommendations aimed at reducing diet-related diseases (for example, to reduce fat consumption to less than 35 per cent of energy intake). Analyse your own eating habits and suggest improvements you could make to reduce your own risk of diet-related diseases.

9.9

OBJECTIVES

By the end of this spread you should be able to:

- discuss the adaptations of a carnivore to its diet.

Fact of life

Lions share their feeding ranges with leopards, cheetahs, wild dogs, and spotted hyenas. But only the lion regularly kills prey larger than about 250 kg. Most kills are made by female lions which weigh less than 200 kg. An adult female lion requires about 5 kg of meat per day and an adult male about 7 kg.

The skull, jaws, teeth, and muscles of a lion's head

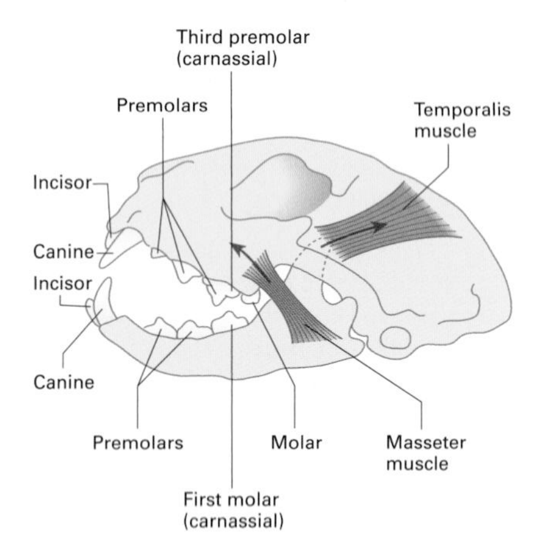

Dental formula of lion

i $\frac{3}{3}$	c $\frac{1}{1}$	pm $\frac{3}{2}$	m $\frac{1}{1}$

Top row shows number of each kind of tooth on each side of upper jaw
Bottom row shows number of each kind of tooth on each side of the lower jaw

Key

i = incisor pm = premolar
c = canine m = molar

Action of the carnassial teeth

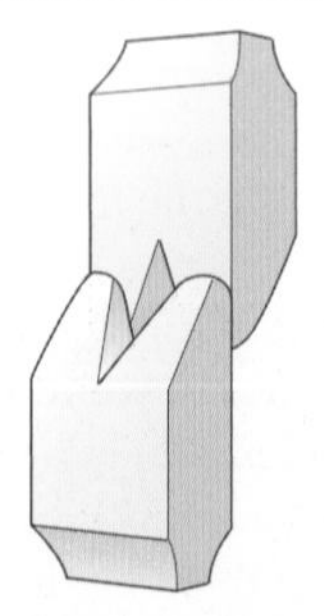

The carnassial teeth in the upper and lower jaws slide past each other like blades of a knife, easily slicing through meat.

Figure 1 *Adaptations of a lion's head to a carnivorous diet.*

CARNIVORES

What is a carnivore?

The word **carnivore** is applied generally to any flesh-eating animal. More specifically, it refers to a group of mammals that have one characteristic in common: they have all evolved from an ancestor whose last upper premolar and first lower molar teeth are adapted to shear flesh. These four **carnassial teeth** have sharp tips, high cusps, and jagged edges that fit together as perfectly as the cutting surfaces of a pair of scissors. They are admirably suited to slicing up pieces of meat (figure 1).

Today's representatives of this mammalian group are classified as the order Carnivora. This is a diverse group that includes cats, dogs, bears, weasels, racoons, hyenas, and the cat-like civets (a family that, in some classifications, includes the mongooses). Representatives of the Carnivora vary greatly in size (the largest, the polar bear, is 25 000 times the size of the smallest, the weasel), shape, agility, social behaviour, and even diet. You would be forgiven for thinking that all carnivores are meat eaters, but some species have a predominantly vegetarian diet. Many living Carnivora are adapted to an omnivorous diet or even largely a vegetarian diet. The giant panda, for example, lives in mountainous forests of China searching placidly for its immobile prey, the bamboo. Its carnassial teeth have lost their pointed tips and jagged edges and have reverted to grinding surfaces.

How a lion is adapted to its diet

We are going to examine in more detail the adaptations of one carnivore, the lion.

Because of its strength and majestic appearance, the lion has been called the 'king of beasts'. It is a true carnivore. Although lions are opportunistic feeders that eat small rodents, hares, and even reptiles, their diet comprises mainly of animals weighing 50–500 kg. These animals are much easier to digest than plants, but they are much harder to catch. The predatory behaviour of a lion hunting a zebra is much more complex than that of a herbivore such as a cow grazing a field.

Hunting

Lions are social animals living in groups called **prides**. The females in a pride do much of the hunting. Their acutely evolved senses of sight and hearing (smell is of little importance to a lion) enable them to find and stalk their prey. The lion's eyes are large and much closer together than those of herbivores, sacrificing field of vision for clear sight (acuity) and stereoscopic vision (see spread 10.12). Its ears are also large and funnel sound efficiently to the inner ear, so that the lion is aware of even the slightest sound. A lion's skeleton and musculature are adapted for a quick attack. Lions are not endurance athletes; they are sprinters with a sprinter's muscles which become fatigued quickly. A lion outpaces a zebra over a short distance, especially if it takes the zebra by surprise, but the zebra easily outruns the lion in an extended chase.

Once the target is in range, the lion must kill its prey. This is a dangerous time for the lion; a well placed kick from a healthy adult zebra could do serious damage. Although lions are more likely to kill healthy prey than are the other carnivores, they usually select a weak straggler to attack. Humans are easy prey, being neither swift nor strong; attacks on humans are well documented.

Usually several lions hunt together, fanning out and partially encircling the target, cutting off possible escape routes. When within about 10 m of its prey, a lion charges and grabs or slaps its prey on the flank before the prey can run away. Typically only one in four attacks is successful. Once knocked down, the prey has little chance of escape. Large prey, such as a zebra, are usually suffocated by a bite to the throat or by clamping the muzzle shut.

Figure 2 *A pride of lions sharing a meal caught by the lionesses in the pride. The lionesses will eat only after the male has finished.*

Killing and eating

The lion's skull is a good killing and dismembering machine: its jaws are short and its muscles powerful (figure 1). The massive **temporalis muscle** enables the large pointed canine teeth to deliver sufficient pressure to suffocate its prey. The **masseter muscle** provides the force needed to cut flesh, and grind and cut meat and gristle when the jaws are almost shut. All members of the cat family have another set of weapons: lethal retractile claws which can grip or slash through the thickest skin.

Backward-curved horny papillae (clusters of taste buds projecting from the surface of the tongue) cover the upper surface of a lion's tongue. These hold on to the meat as it is being cut by the teeth, and make it easier to manipulate and swallow the meat.

Apart from adaptations of the teeth and tongue in the buccal cavity, the alimentary canal of lions is also adapted for digesting meat. The stomach is muscular, and the intestines are short with only a very small caecum. Meat is easier to digest than plant material, so the intestines do not have to be so long in a carnivore as in a herbivore.

Unlike herbivores, which spend most of their time grazing and chewing their food, only a small proportion of a lion's time is taken up with hunting and eating. A diet of easily digestible highly nutritious meat leaves lions plenty of time to rest and socialise.

Quick check

1 What is the function in a lion of:

a carnassial teeth

b the large temporalis muscle

c backward-curving horny papillae on the tongue?

Food for thought

Pride males may survive almost exclusively on kills made by females. Suggest why adult males rarely take part in hunts, yet they readily defend their territory and females from other males.

9.10

HERBIVOROUS MAMMALS

OBJECTIVES

By the end of this spread you should be able to:

- discuss the adaptations of a herbivorous mammal to its diet.

Fact of life

Elephants are non-ruminant herbivores with a hind-gut digestive system similar to that of horses. Being large (a bull elephant may weigh up to 6000 kg), elephants have to eat huge amounts of food in order to sustain themselves. They cannot rely on energy-rich fruits and nuts because these are often in short supply. Elephants therefore eat anything that is available, including very poor-quality vegetation. Consequently, an adult elephant consumes about 150 kg of food each day. However, because of its poor quality, half the food leaves the body undigested.

All herbivorous mammals, from a tiny rodent weighing a few grams to a massive elephant weighing 6000 kg, share the same problem: plant nutrients are contained within cells guarded by thick cellulose cell walls. In order to obtain the maximum nutritional value from plants, the cell walls have to be ripped apart mechanically and the cellulose has to be digested chemically. Chewing with specially adapted teeth performs the first function; microbes that produce the enzyme cellulase in the gut perform the second.

Feeding

Large herbivores such as deer, sheep, and cows spend a long time grazing to obtain sufficient nutrients to survive. While grazing, herbivores have to be acutely sensitive to signs of danger. Grazing animals have to look up periodically to scan the environment for possible threats of attack. Typically, their eyes are positioned laterally to give them a wide field of view but limited stereoscopic vision (see spread 10.12; rabbits have a total binocular visual field of 360°, but a frontal stereoscopic vision of only 20°). An advantage of grazing in herds is that guard duties are shared and individuals can spend more time feeding. Many herbivore species have evolved visual alarm responses which they use at the first hint of danger. Some rabbits and deer have white tails which are especially conspicuous when the animal is running, and serve as a danger signal to other members of the group. If given sufficient warning, most large herbivores can outrun their predators.

Mechanical digestion

Herbivores have jaws and dentition adapted to acquiring and mechanically digesting (grinding) plant food (figure 1). Some grass eaters such as rabbits have sharp incisors with chisel-like edges so they can crop grass. Their incisors, unlike those of carnivores, continue to grow throughout their lives. They are kept short by being worn away continually by the grass. If, however, the upper and lower incisors are out of alignment, their unrestricted growth can have fatal consequences. In sheep and cattle, the upper incisors have been replaced by a **horny pad**. This pad is used in conjunction with the highly mobile tongue to grab grass and, with a twist of the head, wrench a mouthful out of the ground.

The back teeth of herbivores are adapted as grinding mills so that the plant cell walls are mechanically disrupted and the digestible contents released (figure 1). The **premolars** and **molars** have broad ridged surfaces which can break plant material down to fine particles. The **temporalis** and **masseter muscles** move the lower jaw across the upper jaw, either backwards and forwards (as in an elephant), sideways (as in cattle), or in a rotary movement (as in horses). The direction of the ridges on the surfaces of the teeth reflect the type of movement and maximise their grinding action (for example, in elephants, the ridges go across the width of the molars).

Some grass eaters retain the **canine teeth** as defensive weapons; others have lost them. The gap created by their absence is called the **diastema**. It allows the tongue to manipulate food and to keep freshly cropped grass separate from cud (partly digested grass).

Chemical digestion: gut microbes

Herbivores require the enzyme **cellulase** to break down cellulose cell walls. However, they cannot produce cellulase themselves, so they harbour in their guts large communities of microbes (bacteria and other single-celled organisms) that can. The microbes hydrolyse cellulose and use the breakdown products (sometimes combined with substances obtained from the host's mucus and saliva) to make organic molecules such as fatty acids and amino acids. Herbivores may obtain their nutrients directly from these breakdown products and organic molecules, or they may obtain their nutrients by digesting the microbes.

The skull, jaws, teeth, and muscles of a sheep's head

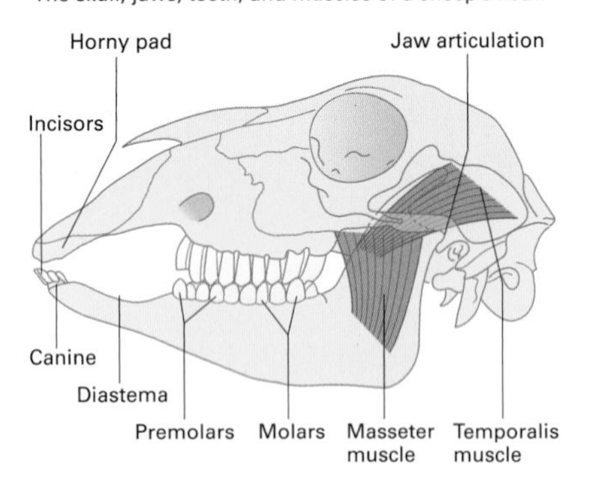

Dental formula of a sheep

i $\frac{0}{3}$ c $\frac{0}{1}$ pm $\frac{3}{3}$ m $\frac{3}{3}$

Top row shows number of each kind of tooth on each side of upper jaw
Bottom row shows number of each kind of tooth on each side of the lower jaw

Key

i = incisor pm = premolar
c = canine m = molar

Action of the molar teeth

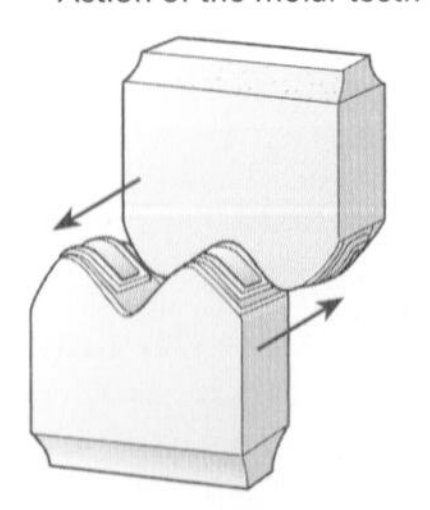

The molars slide over each other and grind

***Figure 1** Adaptations of a sheep's head to a herbivorous diet.*

The microbes usually live in chambers specially adapted as fermenters. The conditions within the fermenters, such as pH and temperature, are at an optimal level for microbial growth. Horses, elephants, and some other herbivores house their cellulose-digesting microbes in the colon and large caecum. These **hind-gut fermenters** can absorb only a little of the fermentation products in the caecum and colon. The fermentation products do not pass through the small intestine which is much better adapted for absorption so a high proportion of the products are lost in the faeces.

Rabbits have cellulose-digesting bacteria in their caecum and appendix (figure 2). The products of digestion cannot be absorbed in the large intestine. To overcome this disadvantage, rabbits produce two batches of faeces. The faeces in the first batch are soft and watery, and are eaten again; these faeces are usually produced at night. Reingesting these faeces gives the rabbit the chance to absorb the products of bacterial digestion as they pass through the gut. The second batch of faeces is drier and more compact, and is not reingested.

Ruminants, such as deer, cattle, sheep, and goats, have a more efficient digestive system that includes a complex four-chambered stomach (figure 3). When grass is swallowed for the first time, microbes in the first two chambers, the **rumen** and **reticulum**, convert the cellulose in the plant cell walls to **cellobiose** (a disaccharide made up of two beta glucose monomers) and beta glucose. Microbes use these digestion products and substances obtained from the host's mucus or saliva to make fatty acids and amino acids, some of which are absorbed through the stomach wall. The partly digested grass (now called cud) is regurgitated to the buccal cavity to be chewed again. This act, called **rumination** or 'chewing the cud', softens and helps break down plant fibres. When the cud is swallowed again, it re-enters the rumen and reticulum for further microbial digestion. Then it goes into the omasum which acts as a strainer, retaining large particles requiring further breakdown, and passing a trickle of small particles into the **abomasum** (sometimes called the 'true stomach', because the first three chambers are really enlargements of the oesophagus). The food then passes through the gut to be digested in the normal mammalian way.

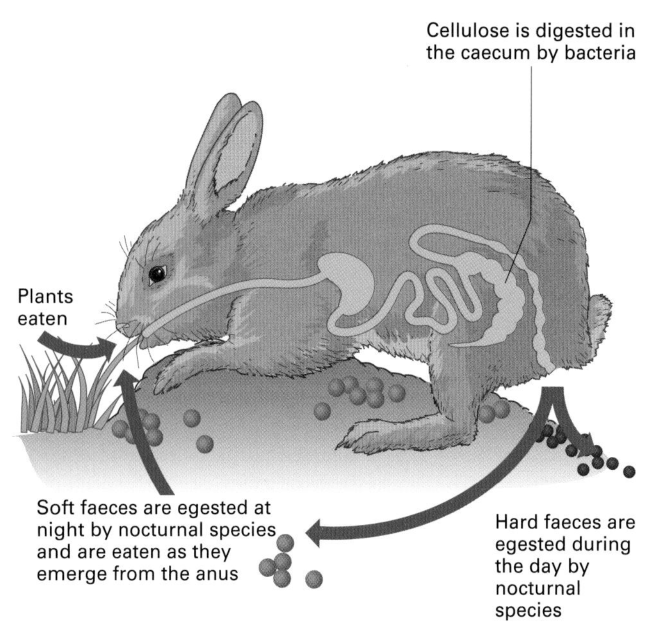

Figure 2 *Digestion of cellulose by a rabbit.*

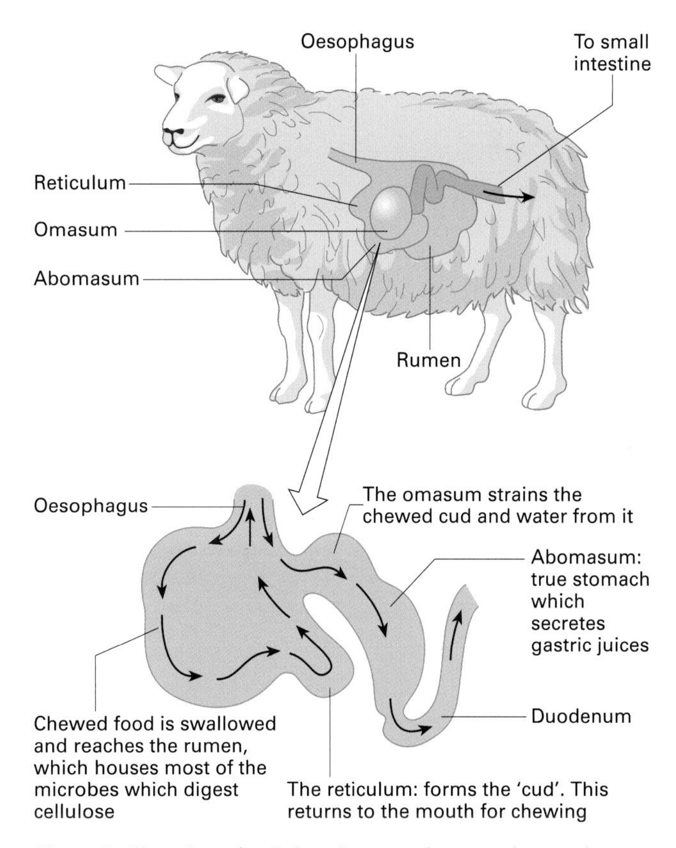

Figure 3 *Digestion of cellulose by a ruminant such as a sheep.*

Quick check

1 a What is the function of the diastema present in some herbivores?

b In which chambers of ruminants do microbes convert cellulose to cellobiose?

Food for thought

Ruminants excrete significant amounts of urea in their saliva. Suggest how this non-protein source of nitrogen helps ruminants survive on diets poor in protein.

Summary

Heterotrophs cannot make their own food; they have to obtain organic molecules by eating other organisms or their products. Heterotrophic nutrition may be **saprobiontic**, **parasitic**, or **holozoic**. In mammals, holozoic nutrition involves **ingestion**, **digestion**, **absorption**, **assimilation**, and **egestion**. Ingestion involves **biting**, **chewing**, and **swallowing**. Once in the **stomach**, food is immersed in gastric secretions and broken down chemically. The **gastric secretions** are under nervous and hormonal control to ensure that they are produced only when there is food in the stomach or about to enter it. The **pyloric sphincter** regulates the outflow of food into the **duodenum** and **ileum** of the **small intestine** where digestion, accelerated by a number of **hydrolytic enzymes**, and absorption take place. **Villi** and **microvilli** provide a very large surface area for absorption. After passing through the small intestine, the intestinal contents, now called **faeces**, enter the **large intestine** where water is reabsorbed and the faeces made less bulky.

A **balanced diet** provides all the nutrients required for growth and repair; and to maintain health. Heterotrophs also obtain **energy** from food. The energy content of each food item can be measured using a **bomb calorimeter**. Energy expenditure of a person can be measured using **direct calorimetry** or **indirect calorimetry**. It varies with sex, age, size, and level of activity. The **basal metabolic rate** (**BMR**) is the minimum energy expenditure needed to stay alive. Components of a **balanced diet** include **carbohydrates**, **lipids**, **proteins**, **fibre**, **water**, **minerals**, and **vitamins**.
The teeth, jaw muscles, gut, and feeding behaviour of **carnivores** are adapted to eating flesh whereas those of **herbivorous mammals** are adapted to eating vegetation. In particular, the guts of herbivores harbour bacteria which enable the animals to deal with cellulose cell walls.

PRACTICE EXAM QUESTIONS

1 The diagrams show different types of teeth from three different mammals.

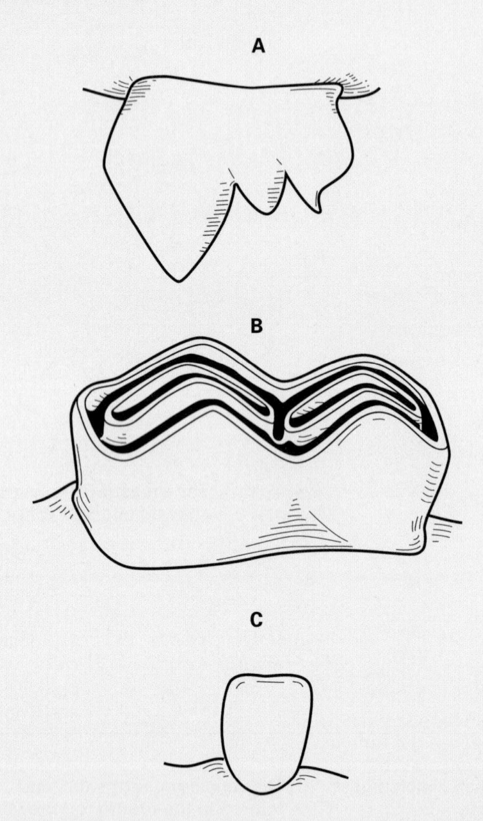

Copy the table below and complete the boxes to show:

i the type of tooth. [3]

ii one example of a mammal in which this tooth is found, [3]

iii the function of each type of tooth. [3]

	Tooth		
	A	B	C
Type of tooth			
Named mammal			
Tooth function			

[Total 9 marks]

2 The table below refers to some enzymes involved in the digestion of carbohydrates in the human digestive system. Copy and complete the table by writing the correct word or words in the empty boxes.

Name of enzyme	Site of production	Products of reaction
	Wall of intestine	Glucose + galactose
Sucrase		
	Pancreas	Maltose

[Total 4 marks]

3 The flow chart represents the breakdown of starch in the human gut.

$$\text{starch} \xrightarrow{\text{amylase}} \text{maltose} \xrightarrow{\text{maltase}} \text{glucose}$$

a Name two organs which produce amylase in humans. [1]

b Describe how the release of amylase from each of these organs is controlled. [3]

c Describe the precise location of maltase in the human gut. [2]

[Total 6 marks]

4 The small intestine in humans is 5 to 6 metres long. Along its length it is thrown into circular folds each approximately 10 mm high. The mucosa is modified to create finger-like processes known as villi, each of which is between 0.5 and 1.5 mm tall and contains capillaries and a lacteal. The columnar epithelial cells making up the surface of the villi are covered in microvilli. These projections, 1 μm long, incorporate several enzymes including exopeptidases and carbohydrases such as maltase. The epithelial cells are joined together by 'tight junctions', ridges which span the space between the adjacent membranes and match complementary grooves in the other membrane. These not only link the two cells involved, but also obliterate the intercellular space.

a Describe, in detail, the function of the capillaries and lacteal in each villus. [4]

b Explain how the structure of the wall of the small intestine is adapted for its functions. [6]

c Amylase is secreted into the lumen of the gut, but maltase is attached to the surface of the epithelial cells. Suggest the importance of this difference. [2]

[Total 12 marks]

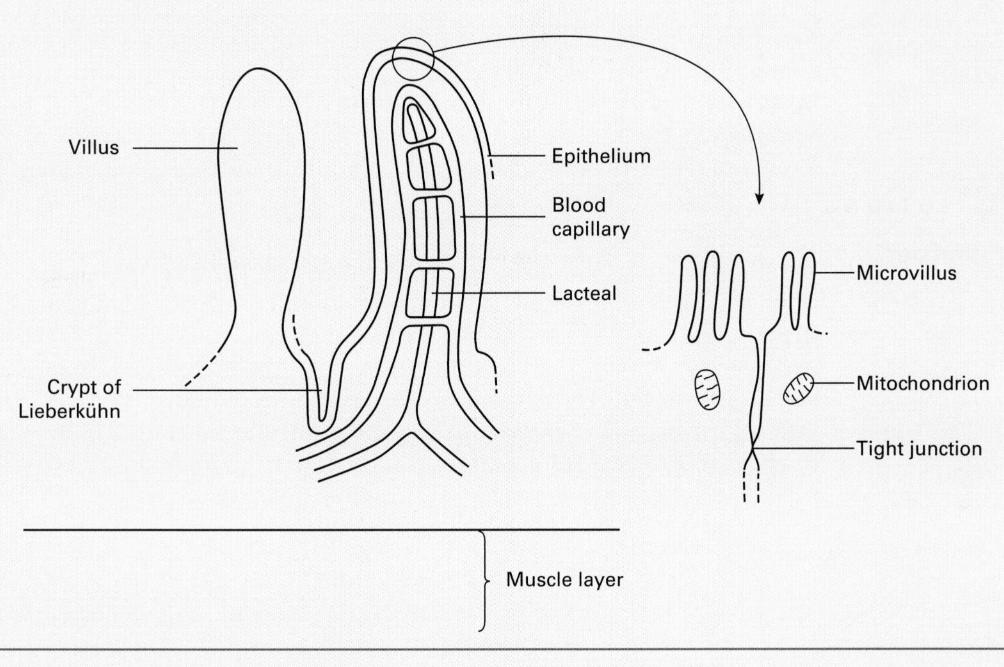

5 The diagram below shows a part of a transverse section through the ileum as seen using a low magnification with a light microscope.

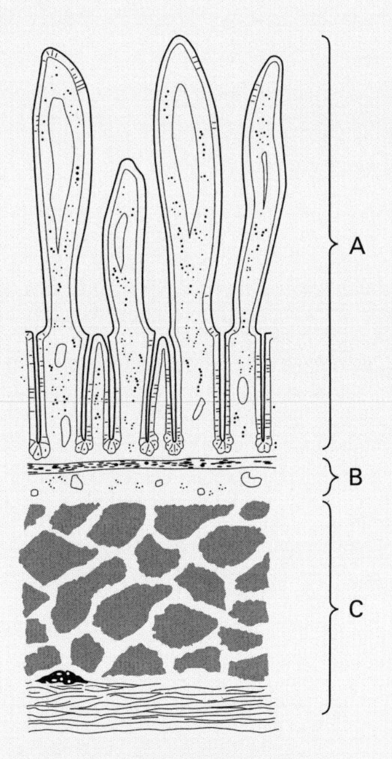

a Name the parts labelled A, B and C. [3]

b Describe the function of part C. [2]

[Total 5 marks]

10 Nervous and hormonal coordination

Nerves and hormones

10.1

OBJECTIVES

By the end of this spread you should be able to:

- explain how information is transferred in a multicellular animal
- compare nervous systems with endocrine systems.

Fact of life

Most nerve fibres are very thin (less than 10 μm in diameter), but the giant nerve fibre of a squid may be more than 1 mm across.

Sensitivity: responding to stimuli

All living organisms must be able to detect changes in their environment and respond appropriately. Changes in the environment are called **stimuli** (singular: stimulus). A stimulus may be in either the external environment (outside the organism) or the internal environment (inside the organism). **Sensitivity**, the ability to respond appropriately to stimuli, is one of the characteristic features of life. Each organism has its own specific type of sensitivity that improves its chances of survival. A single-celled amoeba, for example, can move away from a harmful stimulus such as very bright light, and move towards a favourable stimulus such as food molecules, but it can only distinguish between a limited number of different stimuli.

In an amoeba, the detection of the stimulus and the response to the stimulus must both take place in a single cell. However, in large multicellular animals such as mammals, stimuli are detected in **sense organs**, and organs that respond are called **effectors**. The sense organs and effectors may be in quite different parts of the body. In addition, responses usually involve the coordinated actions of many different parts of the body. To achieve this coordination, one part of the body must be able to pass information to another part. In mammals there are two major systems that convey information: the nervous system and the endocrine system. Both involve **cell signalling**.

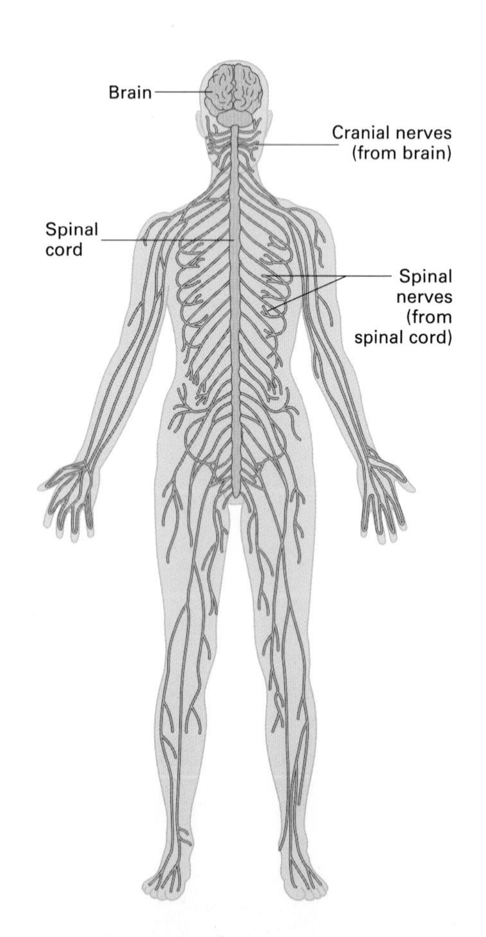

***Figure 1** General plan of the human nervous system. The brain and spinal cord form the central nervous system; the cranial nerves and spinal nerves form the peripheral nervous system.*

The nervous system

Nervous systems range from the simple nerve nets of jellyfish and sea anemones, which have no brain and relatively few interconnections, to the nervous system of humans (figure 1), with brains of staggering complexity. The human brain contains many millions of cells, each of which may communicate with thousands of other nerve cells. Their interconnections form circuits which enable us to control our muscles, think, remember, and even study our own brains.

All animal nervous systems are fast-acting communication systems containing nerve cells, **neurones**, which convey information as nerve impulses (electrochemical cell signals, see spread 10.4). Neurones take various forms but each has a **cell body**, containing a nucleus, and **nerve fibres**, long extensions that transmit nerve impulses rapidly from one part of the body to another. Fibres carrying impulses away from the cell body are called **axons**; those carrying impulses towards the cell body are called **dendrons**. Apart from the main nerve fibre, there may be small dendrons (**dendrites**) extending from the cell body.

In mammals, **sensory neurones** carry messages from peripheral sense organs to a **central nervous system** (CNS) consisting of the brain and spinal cord. The CNS acts as an integration centre and processes information from many sources. **Motor neurones** convey instructions from the CNS to effector organs (mainly muscles and glands).

A mammalian motor neurone (figure 2) can convey information rapidly over considerable distances; for example, a single nerve impulse may be transmitted from the spinal cord to the feet in a few milliseconds. These fast-conducting neurones are enclosed along most of their length by a thick insulating material called the **myelin sheath** (figure 3). The myelin sheath is produced by special supporting cells called **Schwann cells**. The sheath is essentially a series of cell membranes, each produced by a Schwann cell and wrapped many times around the axon. Gaps between the membranes of each Schwann cell, called the **nodes of Ranvier**, are the key to the fast transmission of nerve impulses (see spread 10.5).

Fast transmission enables mammals to respond almost instantaneously to stimuli. Nerve impulses can be directed along the nerve fibres to specific points in the body so that responses can be very localised.

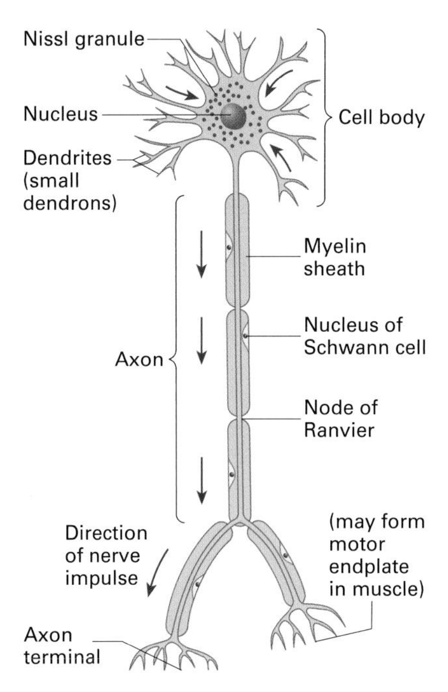

Figure 2 *Structure of a motor neurone.*

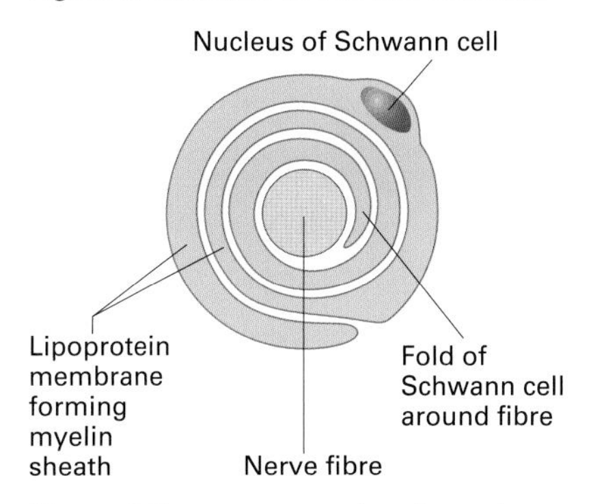

Figure 3 *Transverse section through an axon showing the myelin sheath. Only vertebrate nerve fibres have a myelin sheath.*

The endocrine system

Typically, the nervous system conveys information rapidly between specific locations so that quick responses can be made. In contrast, the **endocrine system** uses chemical signals to convey information from one source to many destinations to bring about long-lasting responses.

The endocrine system consists of a number of glands that secrete **hormones** (the chemical signals, usually proteins or steroids). The glands of the endocrine system are called endocrine glands or ductless glands because they secrete their hormones directly into the bloodstream (in contrast to **exocrine glands**, such as those of the pancreas, which pass their secretions into ducts). Once inside a blood vessel, a hormone is carried in the bloodstream so that it can reach almost any cell in the body. However, each hormone has its own **target cells** on which it acts. Therefore, although all the hormones are transported together in the bloodstream, each has its own specific effect on the body. In some cases, a target cell has specific receptor molecules on its cell surface membrane which bind the hormone molecule. Once bound onto the membrane, the hormone brings about its response.

Endocrine glands occur at strategic points around the body. Their hormones regulate a wide range of activities, including blood glucose concentration, gastric secretion, heart rate, metabolism, growth rate, reproduction, and water balance (spread 10.2).

Quick check

1 What is an axon?

2 In what form is information conveyed in:

a the nervous system

b the endocrine system?

Food for thought

Squids can escape from danger because they have giant nerve fibres. These fibres can conduct nerve impulses very rapidly, since speed of conduction is directly related to the diameter of the fibre. Squids have nerve fibres of normal diameter to control their slow cruising movements, but giant nerve fibres control their rapid escape response. When danger threatens, giant nerve fibres carry information from the brain down the body, causing circular muscles to contract and force a jet of water out of the body, enabling the squid to make a quick backward escape (figure 4).
Suggest why squids have giant nerve fibres only for rapid escape responses. Why do mammals not require giant nerve fibres?

Figure 4 *The squid* Loligo vulgaris *inhabits coastal waters and shallow seas in large numbers. Its giant nerve fibres enable it to respond quickly to danger.*

10.2

The mammalian endocrine system

OBJECTIVES

By the end of this spread you should be able to:

- locate the major endocrine organs in a mammal's body
- describe the main effects of hormones.

Fact of life

Endocrinology (the study of hormones) has been revolutionised in recent years by modern techniques which enable doctors to detect the type and amount of hormones (and other substances) present in the blood in extremely low concentrations. For example, radioimmunoassays can detect the hormone gastrin in a concentration as low as 0.1 picogram per cm^3 of blood plasma (1 picogram = 10^{-12} grams).

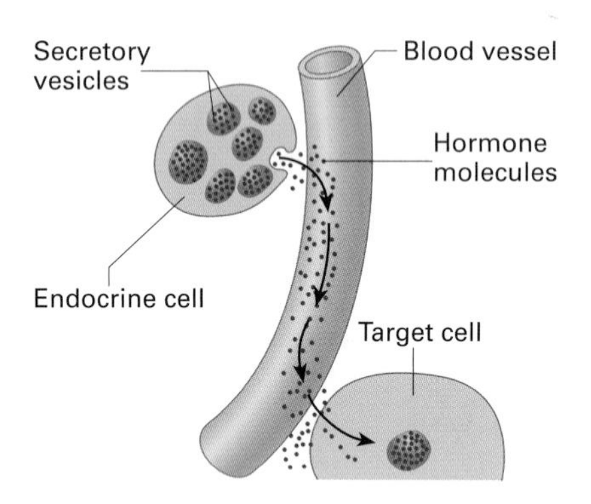

Figure 1 A hormone from an endocrine cell is secreted into the blood, which carries it to its target cell.

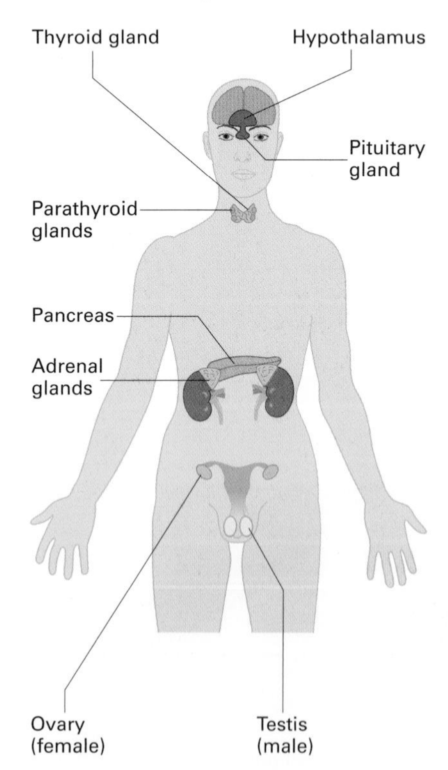

Figure 2 The major human endocrine glands.

The mammalian **endocrine system** includes all the cells that secrete hormones into the bloodstream (figure 1). A **hormone** is a regulatory chemical that is transported by the blood to its target cells in some other part of the body. It affects its target cells by attaching onto specific receptor molecules either on the surface of or within the target cells.

Glands

The endocrine system includes specialised individual nerve cells called **neurosecretory cells** (spread 10.14), as well as glands. There are more than a dozen endocrine glands in the mammalian body (figure 2). Collectively, they secrete more than 50 hormones; the main ones are shown in table 1. Several glands have more than one function: for example, the pancreas produces digestive enzymes as well as hormones, and the ovaries produce ova as well as oestrogen and progesterone.

Hormones

Hormones are often effective in very small concentrations. The average concentration of insulin in the blood, for example, is 10^{-5} g cm^{-3}. Most hormones are either excreted in the urine or broken down by enzymes quite quickly (half the insulin released by the pancreas is broken down within 9 minutes). Many hormones, such as those regulating growth and metabolism, exert their effects over a very long time because they are secreted as a steady trickle into the bloodstream.

The target cells of some hormones are very specific. For example, thyroid stimulating hormone secreted by the pituitary gland only affects the activity of the thyroid gland. However, because hormones are transported in the blood, they can reach all parts of the body and their effects can be widespread. They are therefore especially effective in controlling whole-body activities such as growth and metabolism.

Control of endocrine glands

Specific stimuli regulate the release of hormones from endocrine glands into the bloodstream. Secretion from some glands (for example, the adrenal medulla) is controlled by direct nervous stimulation. Secretion from many other glands is regulated by negative feedback mechanisms involving hormones or some other chemicals. For example, insulin secretion from the islets of Langerhans is regulated by blood glucose concentrations (see spread 8.2).

The chemical structure of hormones is varied, but most are proteins or steroids. Despite the diversity of hormones, they appear to have two main methods of action. These are described in spread 10.3.

Quick check

1 Precisely where in the body is antidiuretic hormone made?

2 According to table 1, name the hormone that:

a stimulates secretion of milk

b raises basal metabolic rate

c promotes development of male secondary sexual characteristics.

Food for thought

With reference to negative feedback mechanisms, suggest the significance of hormones such as insulin being broken down quickly in the body.

Table 1 *The main human hormones and their sites of secretion.*

Site of secretion	Hormone	Type of chemical	Main functions
Pituitary gland: posterior lobe (releases hormones made by hypothalamus)	Oxytocin	Polypeptide	Stimulates milk flow from the mammary glands and contraction of uterus at birth
	Antidiuretic hormone (ADH)	Polypeptide	Promotes reabsorption of water in kidneys
Pituitary gland: anterior lobe	Growth hormone (GH)	Protein	Stimulates growth (especially of bones)
	Prolactin	Protein	Stimulates mammary glands to secrete milk
	Follicle stimulating hormone (FSH)	Glycoprotein	Stimulates growth of Graafian follicle and oestrogen secretion in ovaries; stimulates testosterone secretion in testes
	Luteinising hormone (LH)	Glycoprotein	Stimulates oocyte maturation, ovulation, and formation of corpus luteum in ovaries; stimulates testosterone production in testes
	Thyroid stimulating hormone (TSH)	Polypeptide	Regulates growth of thyroid gland and secretion of its hormones
	Adrenocorticotrophic hormone (ACTH)	Polypeptide	Stimulates growth of adrenal cortex and release of adrenocortical steroids; stimulates lipid breakdown and release of fatty acids from fat cells
Thyroid gland	Thyroid hormone (thyroxine and triiodothyronine)	Derived from tyrosine (amino acid) and contains iodine group	Raises basal metabolic rate; regulates tissue growth and development (especially of nervous and skeletal tissue)
	Calcitonin	Polypeptide	Lowers blood calcium levels
Parathyroid glands	Parathyroid hormone	Polypeptide	Raises blood calcium levels; regulates bone formation; essential for normal skeletal growth
Adrenal cortex	Aldosterone	Steroid	Regulates concentration of sodium and potassium ions in the blood
	Cortisol	Steroid	Suppresses immune response to combat stress
	Androgens	Steroid	Promote development of testes and male secondary sexual characteristics
Adrenal medulla	Adrenaline	Amine	Prepares body for action, e.g. increases heart rate and metabolic rate
Pancreas (beta cells of islets of Langerhans)	Insulin	Protein	Lowers blood glucose concentration
Pancreas (alpha cells of islets of Langerhans)	Glucagon	Polypeptide	Raises blood glucose concentration
Ovaries	Oestrogens	Steroid	Promote development and maintenance of female secondary sexual characteristics and growth and function of the ovaries; control menstrual cycle and pregnancy
	Progesterone	Steroid	Controls menstrual cycle and pregnancy
Testes	Androgens (e.g. testosterone)	Steroid	Promote development of testes and male secondary sexual characteristics
Stomach	Gastrin	Polypeptide	Stimulates secretion of gastric juices
Duodenum	Enterogastrone	Polypeptide	Inhibits secretion of gastric juices; reduces stomach mobility
	Cholecystokinin	Polypeptide	Stimulates secretion of pancreatic juices; inhibits gastric secretions

10.3

Hormonal mechanisms

OBJECTIVES

By the end of this spread you should be able to:

- describe the main mechanisms by which hormones produce their effects on target cells.

Fact of life

In 1994, Alfred Gilman and Martin Rodbell won the Nobel Prize in Medicine for their pioneering work on **cell signalling**. They discovered **G-proteins**. These transmembrane proteins function as **transducers** by translating a signal on the surface of a cell into a response inside the cell. They are molecular switches, turning a response 'on' when the cell signal is present and 'off' when it is absent. In addition to being involved in cellular responses to adrenaline (figure 2) and other hormones, they also play an important part in a host of other signalling pathways including those for vision, smell, and cell division. Malfunctions of G-proteins have been linked to diseases as diverse as cholera and cancer.

Hormones are transported in the bloodstream and can therefore reach almost any cell in the body. However, each hormone only affects cells that have specific receptor molecules onto which the hormone can bind. The binding of a hormone to its receptor molecule is similar to the binding of a substrate to its enzyme; both depend on complementary shapes. The mechanism by which a hormone acts depends mainly on whether its receptor molecule is within the target cell or on its surface.

Binding to receptors inside the cell: steroids

Oestrogen and other steroids exhibit a type of **ligand–receptor interaction** by binding to protein receptor molecules inside their target cell. Because it is lipid soluble, a steroid can pass easily through the phospholipid of cell membranes and into the cytoplasm. Here it binds to a receptor protein which carries the steroid into the nucleus. In the nucleus, the hormone–receptor complex attaches to a section of DNA and regulates the expression of a gene which carries the instructions for the manufacture of a specific polypeptide (see chapter 18). The hormone can therefore bring about a specific response by changing the amount of polypeptide synthesised by the cell (figure 1). Oestrogen regulates different genes in different target cells, bringing about specific responses in different parts of the body.

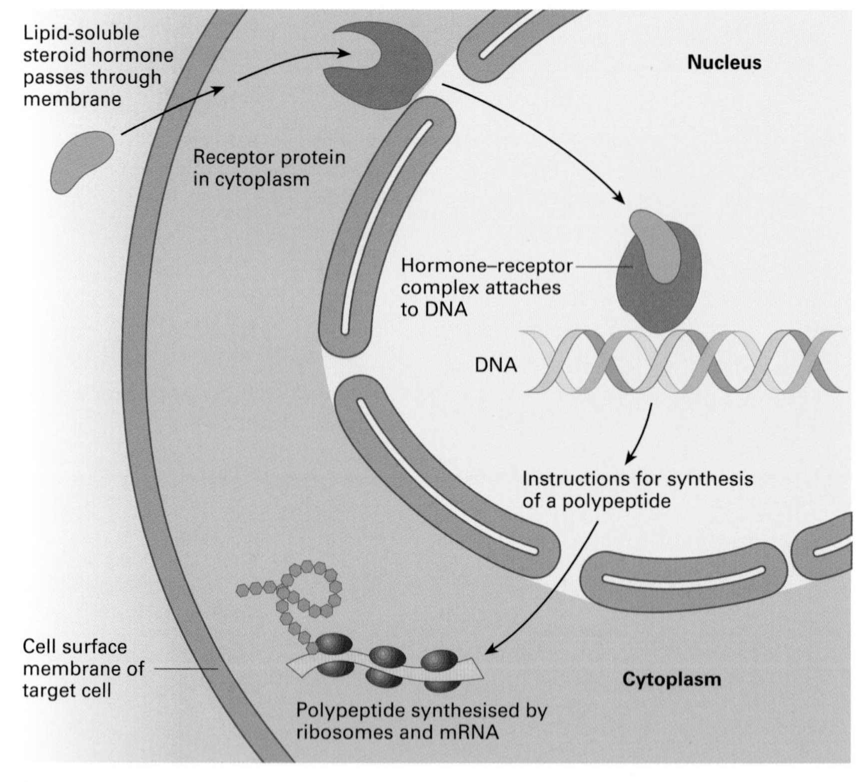

Figure 1 *The action of a steroid hormone such as oestrogen on its target cell. Because it is transported in the blood, oestrogen can reach any cell in the body. In cells that have oestrogen receptors onto which it can bind, oestrogen acts as a cell signalling molecule, triggering transcription and the synthesis of specific proteins (see spread 18.7).*

Binding to receptors on the cell surface membrane: adrenaline

Hormones that attach onto cell surface receptor molecules tend to be small hydrophilic molecules. They include adrenaline, glucagon, parathyroid hormone, and insulin. These hormones affect their target cells by a **signal-transduction mechanism**. The attachment of the hormone onto its receptor molecule in the cell surface membrane acts as a signal, initiating a series of molecular changes that bring about a specific response inside the cell. The signal-transduction mechanism was first described for the action of adrenaline on muscle and liver cells.

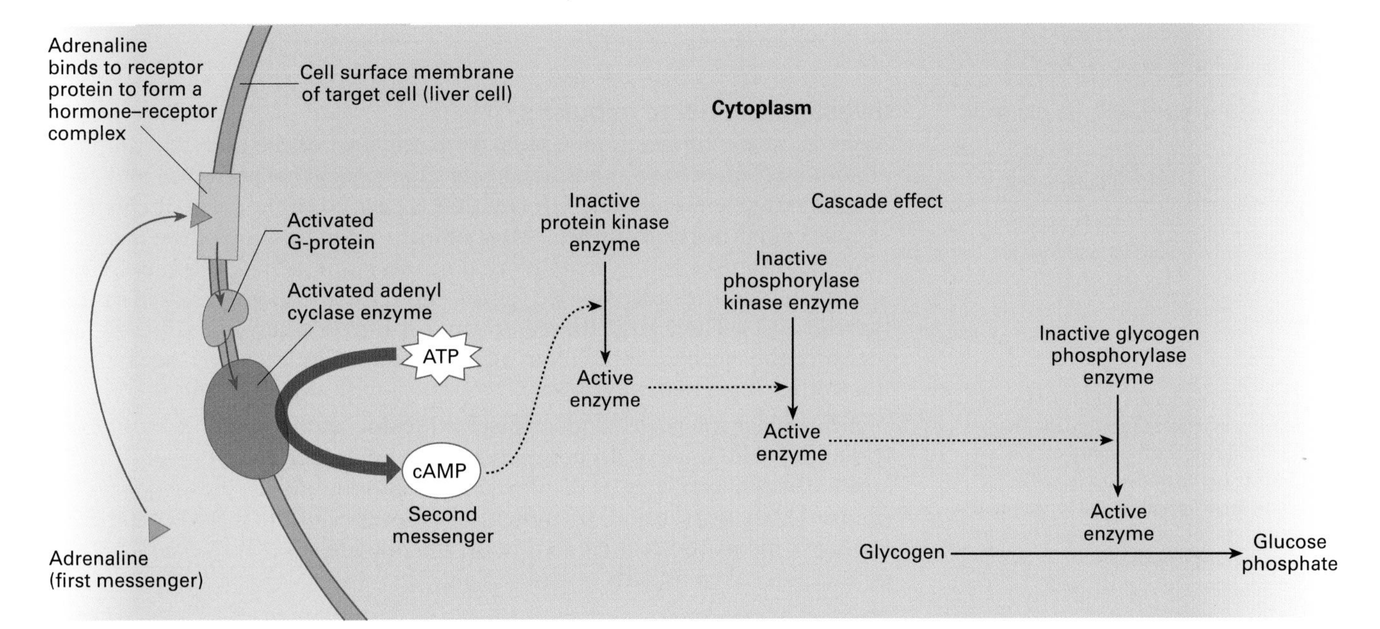

***Figure 2** The mode of action of adrenaline on a liver cell.*

Figure 2 shows how adrenaline catalyses the breakdown of glycogen in liver cells. Adrenaline molecules bind onto receptor molecules in the cell surface membrane. The formation of a hormone–receptor complex activates a protein in the membrane. The protein, called a **G-protein** (short for **guanine nucleotide-binding protein**), acts as a relay between the hormone–receptor complex and an enzyme. This enzyme, **adenyl cyclase**, catalyses the conversion of ATP to **cAMP** (**cyclic adenine monophosphate**) within the liver cell. cAMP is a small molecule which acts inside the cell as a **second messenger** (the first messenger is the hormone). It diffuses into liver cells where it initiates a complex chain reaction. First, it activates a **kinase** (a phosphate-transferring enzyme) by changing the enzyme's shape. The activated enzyme then catalyses a reaction which activates another enzyme, and so on, until the activation of glycogen phosphorylase catalyses the conversion of glycogen to glucose phosphate. This complex chain reaction produces a **cascade effect** which amplifies the response to the hormone. Throughout the chain, each enzyme molecule activates many molecules of its substrate, which become the next enzyme in the chain. Thus a small signal (a few molecules of adrenaline on the cell surface) results in the production of a very large response (a very large number of molecules of glucose phosphate inside the cell).

Adrenaline affects many types of cells in the body and can bring about a variety of responses. The precise response depends on which specific enzyme is activated by cAMP.

Several surface-acting hormones have a mechanism similar to adrenaline and work through cAMP. Others, such as insulin, also use a signal-transduction mechanism, but involving another second messenger.

The 'fight or flight' response

When confronted by a real (or even perceived) danger, we commonly experience a feeling of exhilaration and power combined with fear. This feeling is brought about by a surge of adrenaline secreted from the adrenal medulla into the bloodstream. Adrenaline is a stimulant that has wide-ranging effects on muscles, the circulatory system, and carbohydrate metabolism. An adrenaline surge prepares the body for action by what is commonly called the **'fight or flight response'**. It increases heart rate, depth and rate of breathing, metabolic rate, and the conversion of glycogen to glucose. It also improves the force of muscle actions and delays the onset of fatigue.

Quick check

1 a Name two hormones that enter their target cells.

b Name two hormones that attach onto surface receptor molecules on their target cells.

c Why is cyclic AMP referred to as a 'second messenger'?

Food for thought

Hormones and enzymes are similar in that they are both effective in very small amounts and are not consumed in the many metabolic processes they affect. Suggest how hormones and enzymes differ.

10.4

Setting up a nerve impulse

OBJECTIVES

By the end of this spread you should be able to:

- explain how a resting potential is maintained
- explain how an action potential is generated.

Fact of life

Puffer fish produce a highly potent neurotoxin called tetrodotoxin. This selectively blocks the entry of sodium ions into nerve and muscle cells during an action potential, preventing the generation of nerve impulses and muscle contractions.

Investigating nerve impulses

Nerves convey information rapidly from one part of the body to another, enabling animals to respond quickly to changes in their external and internal environments. The information is carried in the form of electrical signals called **nerve impulses**. Most of our understanding of the nature of nerve impulses comes from work done on giant axons of squids. These are the nerve fibres responsible for the rapid escape movements of squids (spread 10.1). Their large diameter (up to 1 mm) makes it possible to measure the electrical activity in a giant axon when it is at rest and when it is conveying a nerve impulse.

A fine glass microelectrode is inserted inside an axon, and the voltage (potential difference; p.d.) between it and a reference electrode on the surface of the axon can be displayed on a cathode ray oscilloscope (figure 1). By convention, the potential difference of the inside of the cell is always measured relative to that on the outside, so that the outside potential is taken as zero.

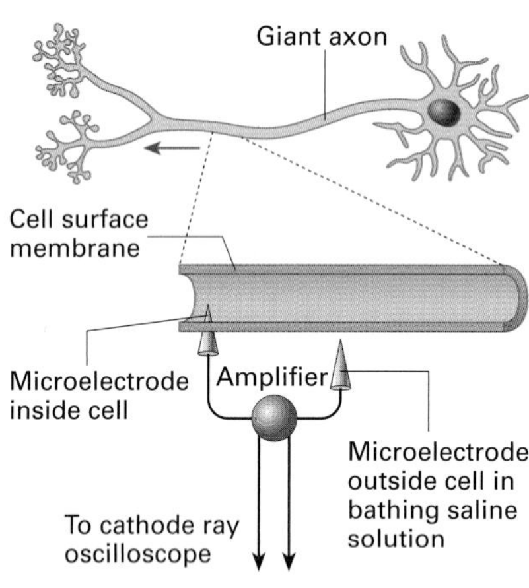

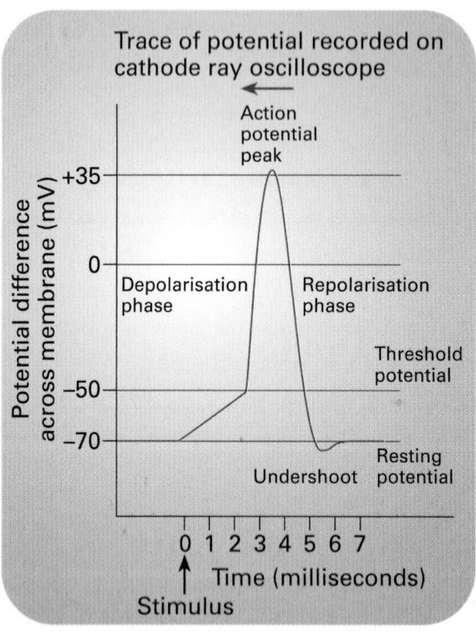

Key
Direction in which nerve impulse travels

Figure 1 Measuring the potential difference across the membrane of a giant axon.

Resting potential

A resting neurone is so called because it does not convey a nerve impulse, not because it is inactive. On the contrary, a resting neurone expends much energy in maintaining a potential difference across its membrane. This is called the **resting potential** and measures about –70 millivolts.

During the resting potential, the inside of the neurone is negative relative to the outside because of an unequal distribution of charged ions. On the outside, sodium ions (Na^+), chloride ions (Cl^-), and calcium ions (Ca^{2+}) are present in higher concentrations than inside the cell. By contrast, the inside of the cell has a higher concentration of potassium ions (K^+) and organic anions (negative ions).

This unequal distribution of ions results from a combination of active transport and diffusion of sodium and potassium ions across the cell membrane, and the inability of large organic anions to pass out of the cell (figure 2). A **sodium–potassium pump** actively transports sodium ions out of the neurone and potassium ions in. For every three sodium ions pumped out, only two potassium ions are pumped inwards. On its own, this would result in only a slight potential difference across the membrane. However, this difference is amplified by the membrane being about 50 times more permeable to potassium ions than to sodium ions. Potassium ions are able to diffuse freely back out of the cell down their concentration gradient, but the sodium ions diffuse back into the cell only very slowly. This creates a negative electrical charge inside compared with outside. Without active transport, an equilibrium would eventually be reached and there would be no potential difference across the membrane.

Action potential

A nerve impulse occurs when the resting potential across the membrane of a neurone has a sufficiently high stimulus. A **stimulus** is any disturbance in the external or internal environment which changes the potential difference across a membrane. The stimulus may be chemical, mechanical, thermal, or electrical, or it may be a change in light intensity.

The recording on the cathode ray oscilloscope in figure 1 shows the effects of a stimulus on a giant axon. When the stimulus is applied, the axon becomes **depolarised**; that is, the inside becomes temporarily less negative. If the stimulus is strong enough (if it exceeds the **threshold level**), an **action potential** occurs. There is a complete reversal of the charge across the nerve cell: the interior becomes positively charged relative to the outside. Typically, the action potential reaches a peak of about +35 millivolts. The potential difference then drops back down, undershoots the resting potential and finally returns to it. The return of the potential difference towards the resting potential is called **repolarisation**. The entire action potential takes about 7 milliseconds. Although figure 1 refers specifically to a giant axon, its general features apply to all animal neurones.

Ion channels and action potentials

The action potential results from the changes in the permeability of cell membranes to ions. At rest, the membrane permeability of a nerve fibre is thought to depend on **ion channels** through which specific ions can move. An ion channel consists of a protein molecule spanning the membrane, with a pore through the centre. Sodium ions move through one type of channel and potassium ions through another. There are many more of these ion channels for potassium than for sodium, therefore at rest the membrane permeability to potassium ions is much greater than that to sodium ions (figure 2).

During an action potential, special ion channels control ion movements across the membrane. These channels are believed to have voltage-sensitive gates that open and close in response to voltage changes, and are therefore called **voltage-gated ion channels** (figure 3).

During the resting potential, the voltage-gated sodium and potassium ion channels are closed. When a stimulus is applied, sodium ion channels open rapidly, sodium ions move in, and the inside becomes more positive. If the stimulus reaches the threshold level, an action potential occurs. When the action potential reaches its peak, the sodium ion channels close slowly and potassium ion channels open slowly. Sodium ions stop moving into the cell but potassium ions diffuse more rapidly out. These changes cause the potential difference to drop. When the membrane returns to its resting potential, potassium ion channels close, but because they do this slowly, the potential dips below the resting level. Finally, when the potassium ion channels are closed, the membrane returns to its resting condition.

So far, we have examined how an action potential is generated at the point of stimulation. However, this is only the first step in the propagation of a nerve impulse along a neurone. These localised action potentials are converted into nerve impulses which transmit information from one part of a neurone to another neurone or to an effector such as a muscle or a gland.

Action potentials obey the **all-or-none law**. This means that no matter how strong the stimulus, the size of an action potential is always the same. Therefore, information about the strength of a stimulus is carried along a nerve fibre not as variations in the size of nerve impulses, but by changes in their frequency. The next spread discusses these points more fully.

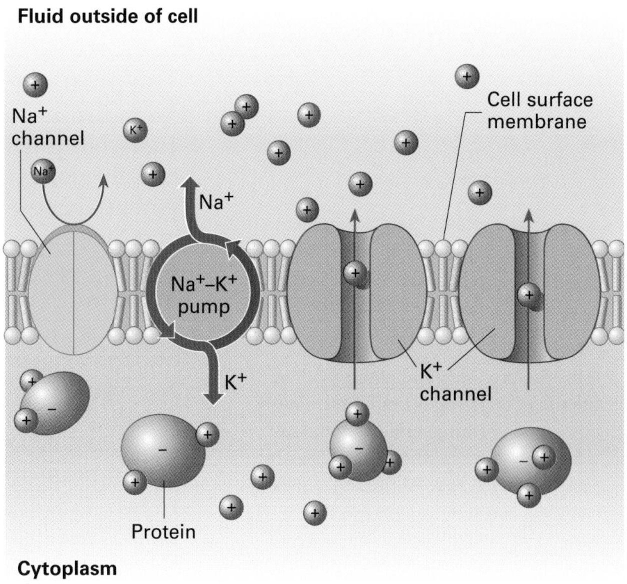

Figure 2 *The generation of a resting potential. The resting potential is brought about by an unequal distribution of ions either side of a cell surface membrane. Note that the potassium ion channel shown is always open, whereas the sodium ion channel shown opens only during an action potential.*

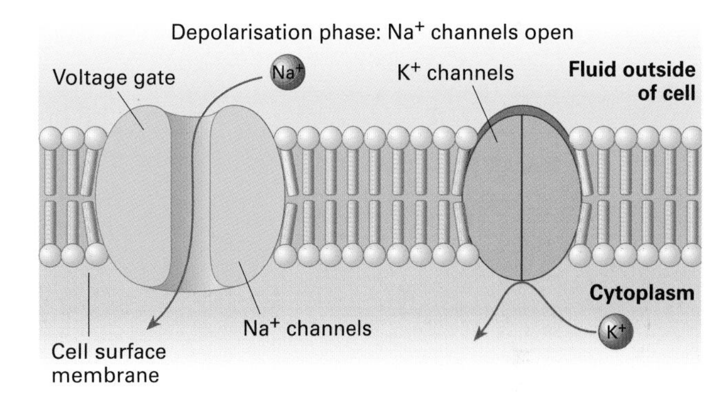

Figure 3 *Voltage-gated ion channels and the action potential.*

Quick check

1 What are the main factors that determine the resting potential of a neurone?

2 State whether
 i the voltage-gated potassium ion channels
 ii the voltage-gated sodium ion channels
 in a neurone membrane are open or closed:
 a during the resting phase
 b during the depolarisation phase of the action potential
 c in the repolarisation phase
 d during the undershoot.

Food for thought

Suggest why depolarisation is not a good word for describing what happens during the transmission of an action potential.

10.5 TRANSMISSION OF A NERVE IMPULSE

OBJECTIVES

By the end of this spread you should be able to:

- explain how a nerve impulse is propagated along a neurone
- discuss the role of the myelin sheath in the transmission of nerve impulses.

Fact of life

Information conveyed along a nerve fibre is even simpler than information conveyed in Morse code. Instead of a series of dashes and dots, there are just dots, each represented by a nerve impulse. Nerve impulses are all the same size. Therefore the only way of varying information about a stimulus is to alter the frequency, the number of impulses per second. Frequencies higher than 1000 impulses per second have been measured in some mammal fibres, but most human fibres conduct at frequencies less than 100 per second.

The propagation of a nerve impulse

An individual action potential is a short-lived, localised event (spread 10.4), but it leads to the transmission of a **nerve impulse** which travels rapidly along a nerve fibre. The impulse forms because the action potential causes a small current to flow in the cytoplasm of a nerve fibre and the extracellular fluid. In an unmyelinated neurone (one that has no myelin sheath), this local current acts as the stimulus for the next part of the nerve membrane, causing further depolarisation, and so on along the nerve fibre (figure 1). Thus, a nerve impulse is propagated as a wave of depolarisation, with one portion of the fibre repolarising (returning to the resting potential) as the next portion depolarises. The action potential remains exactly the same size as it travels along the neurone.

Transmission in a myelinated nerve fibre: saltatory conduction

In myelinated nerve fibres, the myelin sheath acts as an effective electrical insulator. Consequently the local flow of current can only be set up between adjacent nodes of Ranvier. There is no myelin sheath at these nodes and so here the nerve fibre membrane is exposed to the extracellular fluid. Also, there are large numbers of sodium ion channels at the nodes of Ranvier; in the myelinated parts of a nerve fibre there are few. The nodes are about 1 mm apart. The local current set up by the depolarisation of one node depolarises the next node, and so on, so the nerve impulse 'leaps' from node to node (figure 2). This type of nerve impulse transmission is called **saltatory conduction** (*saltus* is the Latin verb 'to leap' or 'to

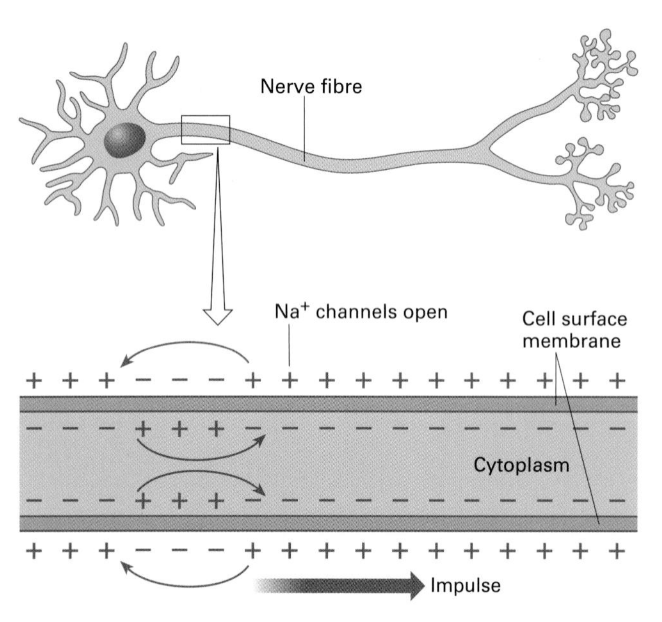

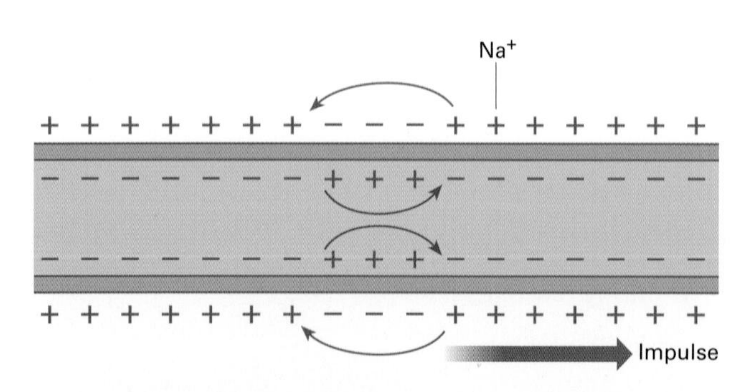

Figure 1 The transmission of a nerve impulse in an unmyelinated fibre.

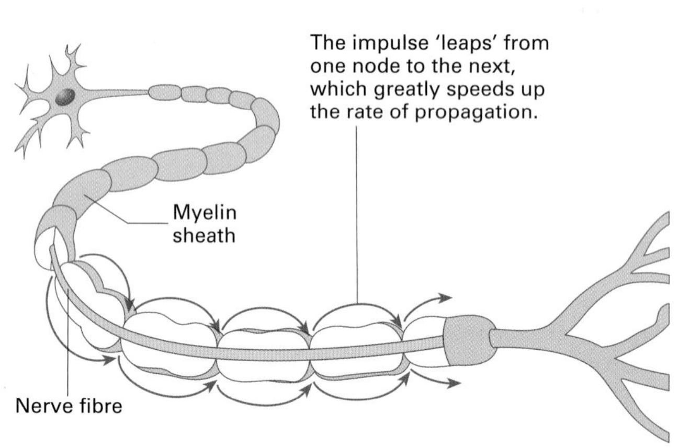

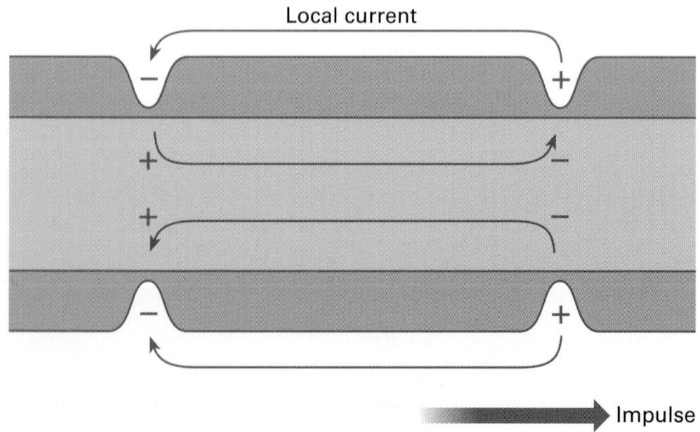

Figure 2 *Saltatory conduction along myelinated nerve fibres: action potentials occur only at the nodes of Ranvier because local currents in one node reach the next node, starting a new action potential there. Thus the impulse jumps from node to node.*

jump'). Saltatory conduction transmits nerve impulses at speeds up to 100 m s^{-1}, making myelinated fibres particularly suited for transmitting impulses over long distances. Vertebrate unmyelinated fibres conduct at up to about 1 m s^{-1}, so these fibres occur only for conduction over short distances (e.g. in the brain) or where fast conduction is not necessary. The conduction speed of all fibres, myelinated and unmyelinated, increases with temperature until the body temperature gets so high that enzymes are denatured and membranes are damaged.

All or nothing and frequency coding

In both unmyelinated and myelinated nerve fibres, a nerve impulse is an **all-or-none event**. A neurone will not transmit an impulse unless given a stimulus of a certain strength. Once this **threshold** level of stimulation is reached, an action potential will be set up. It does not make any difference how far above the threshold level the stimulus is: the action potential is either set up, if the stimulus reaches the threshold, or it is not, if it does not. Increasing the stimulus above the threshold does not cause a larger action potential because all nerve impulses in a particular neurone have the same amplitude. How do we know how strong a stimulus is? Nerve impulses do not vary in size, so this information is coded by the number of impulses generated in a given time. In this **frequency coding**, the rate of impulse generation may increase with the strength of the stimulus, for example.

Frequency coding would provide limited information about changes in the environment if sense organs were connected to only one neurone. However, sense organs usually have many neurones that transmit information to the brain or another part of the body. Different neurones have different threshold levels. Therefore, further information about the nature of a stimulus is provided by the number, type, and location of the neurones affected by the stimulus.

The refractory period

Neurones can generate nerve impulses over a wide range of frequencies, from one or a few per second to more than 100 per second. The frequency is limited because during a brief period following the action potential called the **absolute refractory period**, the nerve fibre is completely inexcitable so cannot propagate another impulse. For another short period afterwards, only a stimulus above the normal threshold level sets up another action potential; this is called the **relative refractory period**.

Usually a nerve fibre is stimulated at one of its ends and an impulse travels in one direction only. It does not, for example, turn round and go backwards half-way along the fibre. This is because of the refractory period: once a region of nerve fibre has been depolarised it cannot be excited again, but the region immediately ahead of the depolarisation can be, so the action potential travels in that direction. Synapses (the special junctions between one neurone and another cell) ensure that the impulse flows in only one direction along a series of nerve fibres. Synaptic transmission is the subject of the next spread.

Multiple sclerosis

Multiple sclerosis (MS) is a major neurological disease that results in the gradual deterioration (demyelination) of the myelin sheath around neurones of the brain and spinal cord. In demyelinated regions, hardened scars or plaques (scleroses) replace the sheath. These scars interfere with the transmission of nerve impulses, slowing them down and causing a gradual loss of motor activity. The symptoms, including loss of coordination, blurred vision, and weakness, depend on the location of the demyelination.

The precise cause of MS is unknown, but evidence strongly suggests that proteins within the myelin sheath are attacked by certain enzymes or by cells from the body's own immune system. The sheath consists of about 70 per cent phospholipids and 30 per cent proteins. One theory suggests that MS appears in people who were infected by the measles virus during childhood. According to the theory, the virus lies dormant in the myelin sheath until it emerges later in life, coated with protein from myelin. The immune system attacks the virus and the myelin proteins, destroying the myelin sheath.

MS is more common in temperate regions than tropical regions, and it affects proportionately more men than women. In some people the condition progresses unrelentingly, but in most there are intermittent periods of relapse and remission. Although the disease is progressive and disability increases, only about one in ten MS sufferers is confined to a wheelchair. As yet there is no cure for MS, but the symptoms can be relieved by appropriate therapy, and the rate of relapse may be reduced by steroids and other drugs.

Quick check

1 Typically, what happens to:
  a the frequency
  b the amplitude (size)
  of an action potential when a stimulus is increased above the threshold level?

2 What is the name of the supporting cell that produces the myelin sheath?

Food for thought

Suggest why birds and mammals can remain active within a wide range of thermal environments, while most amphibians or reptiles become sluggish in cold environments and respond quickly only when their bodies have been warmed by, for example, the Sun.

10.6

Synapses

OBJECTIVES

By the end of this spread you should be able to:

- explain how information passes across a synapse from one neurone to another, or from a neurone to its effector.

Fact of life

A typical postsynaptic cell (figure 1) receives information from hundreds or even thousands of presynaptic neurones. The numerous synaptic connections allow the cell to combine different sources of information before responding. Its response will depend on the sum of all the excitatory and inhibitory postsynaptic potentials produced by spatial and temporal summation.

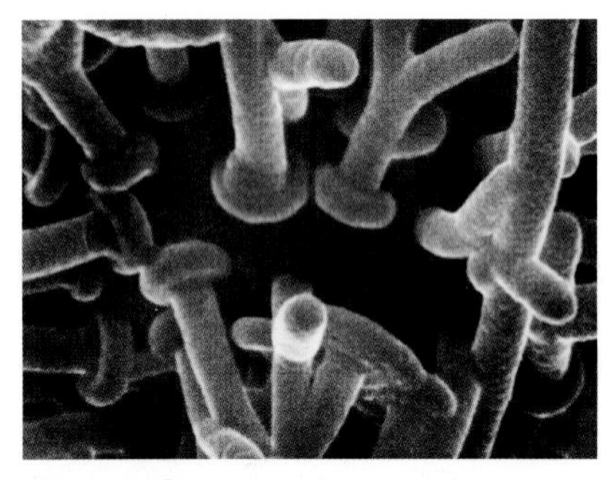

Figure 1 Scanning electron micrograph of presynaptic neurones terminating in synaptic knobs, which send information to postsynaptic cells (×5000).

We have already discussed how a stimulus generates an action potential (spread 10.4), and how this is propagated as a nerve impulse along a nerve fibre (spread 10.5). In this spread we examine how information in the nervous system passes from one cell to another; or, as one physiologist phrased it, how neurones 'talk' to each other and to their effectors. Information transfer occurs across a **synapse**, the junction between a neurone and another cell, either by the passage of electrical signals or more commonly by chemical signals called **neurotransmitters**.

Chemical synapses

Most nerve pathways have cells that communicate with each other by means of neurotransmitters. The cells are not in direct contact with each other. They join at synapses which have a **synaptic cleft**, a small gap of about 20 nm. A neurotransmitter is released from one membrane in the synapse (the **presynaptic membrane**), diffuses across the synaptic cleft, and binds onto receptors on the other membrane (the **postsynaptic membrane**). The postsynaptic cell may be either another neurone or a cell in an effector organ, such as a muscle or gland.

Different neurones release different types of neurotransmitters. **Excitatory presynaptic cells** release neurotransmitters that make the postsynaptic membrane more excitable and more likely to generate nerve impulses. **Inhibitory presynaptic cells** release neurotransmitters that make the postsynaptic membrane less excitable and less likely to transmit an impulse.

Excitatory synapses

Figure 3 shows the various events that take place in a chemical synapse which uses acetylcholine as the neurotransmitter. Acetylcholine is synthesised within the presynaptic knob and stored in special organelles called **synaptic vesicles**. When an action potential reaches the presynaptic membrane, it depolarises the membrane, that is, makes the membrane less negative than at rest. This depolarisation triggers the opening of **calcium ion channels** in the presynaptic membrane. Calcium ions diffuse into the presynaptic knob, causing the presynaptic vesicles containing acetylcholine to migrate and fuse with the presynaptic membrane. The acetylcholine is released into the synaptic cleft and diffuses across the synapse. Then it binds to specific protein receptor molecules on the postsynaptic membrane, a process known as **receptor activation**.

Acetylcholine acts as an excitatory neurotransmitter at synapses connecting motor neurones with skeletal muscle. Receptor activation causes sodium ion channels to open, making the membrane more permeable and producing a **graded potential**. If enough acetylcholine is released, the graded potential may become large enough to generate an action potential.

Electrical synapses

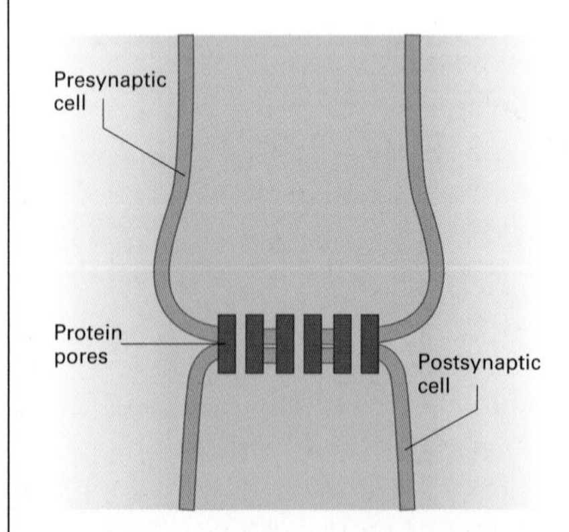

In an electrical synapse, the presynaptic and postsynaptic membranes are close together, with pores made of protein linking the cytoplasm of the two cells. This enables electrical signals to jump relatively unchanged from the presynaptic cell to the postsynaptic cell. Transmission is very fast because there is no intervening chemical, therefore electrical synapses are found where responses have to be very quick. Electrical synapses are found in pathways that control the escape responses of some animals. In mammals, they occur in nerve pathways which control rapid eye movements; they are also common in the heart.

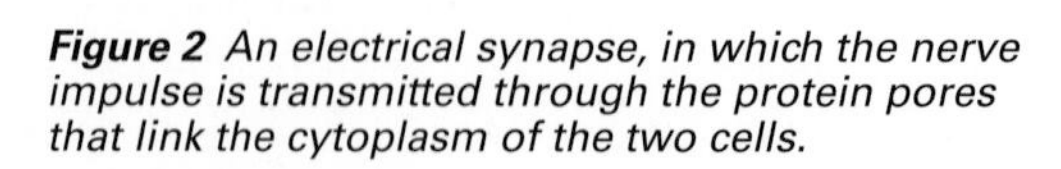

Figure 2 An electrical synapse, in which the nerve impulse is transmitted through the protein pores that link the cytoplasm of the two cells.

If an excitatory synapse receives a continuous stream of action potentials at high frequency, eventually transmission across the synapse stops. This is because the neurotransmitter cannot be resynthesised fast enough and it runs out. The synapse becomes **fatigued**.

Inhibitory synapses

Receptor activation by inhibitory neurotransmitters causes other effects on postsynaptic membranes. Usually it opens chloride ion channels which makes the postsynaptic membrane more negative than normal (it becomes **hyperpolarised**) and less likely to depolarise sufficiently to generate an action potential.

Recycling the neurotransmitter

After a neurotransmitter has affected the postsynaptic membrane, it is removed. This prevents further, unwanted effects. Acetylcholine, for example, dissociates from its receptor and is broken down by the enzyme **cholinesterase**. The breakdown products diffuse back into the presynaptic knob where they are resynthesised into acetylcholine, using energy from ATP. (The high density of mitochondria in the synaptic knobs ensures that there is plenty of ATP available for the synthesis of the neurotransmitter.)

Combining the effects of more than one impulse

Most postsynaptic membranes have many synaptic connections and only produce an action potential when excited through several synapses simultaneously. In this case, a single synapse does not release enough neurotransmitter to start an action potential, but an action potential is fired when sufficient neurotransmitter builds up from several synapses acting together. Thus the graded potentials of several synapses can combine to generate an action potential. This is called **spatial summation**.

A postsynaptic membrane may fail to generate an action potential after a single impulse reaches the presynaptic membrane, but does so when two or more impulses arrive in quick succession. In this case, the graded potentials produced by successive impulses add together, generating an action potential in the postsynaptic neurone. This is called **temporal summation**.

Transmission of nerve impulses through chemical synapses has several advantages:

- it enables information from different parts of the nervous system to be integrated
- it provides a mechanism for filtering out trivial or non-essential information
- it ensures that nerve impulses are one-way, passing only from presynaptic membranes to the postsynaptic membranes
- it allows the synapses to act as switches, so that nerve impulses can pass along one of several discrete pathways in the nervous system.

Neurotransmitters that act like adrenaline

Noradrenaline and dopamine are neurotransmitters called **neurohormones** because they act like hormones. **Noradrenaline**, released by axons in the sympathetic nervous system, has effects similar to those of adrenaline (spread 10.3). **Dopamine** is secreted in the brain where it helps regulate many types of behaviour, motivation, and voluntary movements. Adrenaline, noradrenaline, and dopamine are chemically related and affect their target cells via G-protein-coupled receptor molecules (see spread 10.3). One receptor, **DRD4** (dopamine receptor D_4), has been implicated in schizophrenia, bipolar disorders, and **Parkinson's disease**, a condition characterised by a loss of ability to execute smooth, controlled movements. DRD4 is the target for drugs such as **L-DOPA** (levodopamine). L-DOPA is a precursor of dopamine administered to treat Parkinson's disease because, unlike dopamine, it can cross the **blood–brain barrier** (a network of capillaries around the brain which are specially modified to prevent substances in the blood from diffusing freely into the brain).

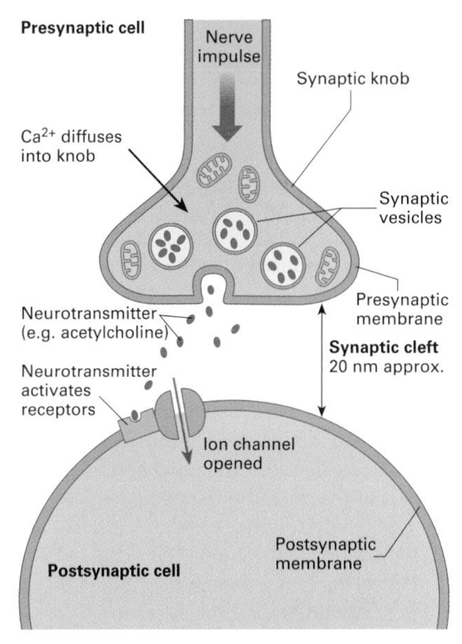

Figure 3 *A chemical synapse, in which the neurotransmitter changes the excitability of the postsynaptic membrane.*

QUICK CHECK

1 a Explain why synaptic knobs have a high density of mitochondria.

b Which mineral ion causes synaptic vesicles to fuse with the presynaptic membrane?

c Which neurotransmitter is used in synapses between neurones and skeletal muscle?

Food for thought

Some drugs have an **antagonistic** effect on neurotransmitters, reducing their action, whereas other drugs are **agonists**, amplifying the effect of a particular neurotransmitter. Some nerve gases act by inhibiting the action of cholinesterase at nerve–muscle junctions. Suggest what effect they have on muscle contractions. Are they agonists or antagonists?

10.7 THE MAMMALIAN NERVOUS SYSTEM

OBJECTIVES

By the end of this spread you should be able to:

- describe the main features of a mammal's nervous system
- discuss the functions of the peripheral nervous system.

Fact of life

The spinal cord is an extension of the brain, acting as an information highway between the head and the rest of the body. Spinal nerves link the spinal cord to other parts of the body. If the spinal cord is severed in an injury, the closer the point of damage to the brain, the more severe the disability. For example, a break near the neck may result in paralysis from the neck down.

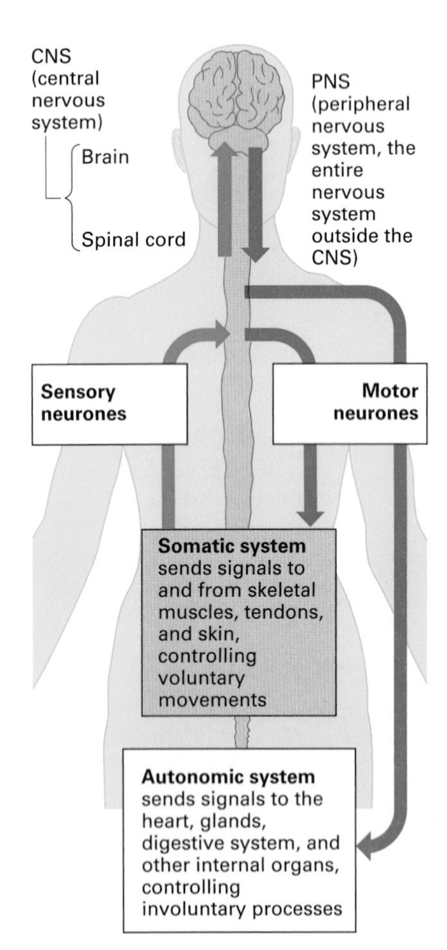

Figure 1 Organisation of the mammalian nervous system.

The mammalian nervous system is very complicated. It not only enables mammals to perform highly complex mental and physical activities at high speed, but it also acts as the main control system for internal physiological processes. It can function well as a control system because it is highly cephalised and centralised. **Cephalisation** is the concentration of nervous tissue at the head end; **centralisation** is the presence of a central nervous system as distinct from a peripheral nervous system (figure 1).

Organisation: central and peripheral

The **central nervous system** (**CNS**) consists of a prominent **brain** enclosed in a skull, and a **spinal cord** (a cylinder of tissue which runs from the brain down the back). The brain integrates information from many internal and external sense organs, stores information in memory centres, controls our emotions and intellect, and acts as the main command centre, sending instructions through the spinal cord to muscles and organs throughout the body. The structure and function of the brain is described in more detail in spread 10.13. The spinal cord acts as a relay centre between the brain, peripheral sense organs, and peripheral effectors (figure 2).

The **peripheral nervous system** consists of all the nerves and associated nervous tissue outside the brain and spinal cord. A **nerve** is a thread-like structure containing a bundle of neurone fibres (axons and dendrons). A single nerve may contain both sensory and motor neurones. The peripheral nervous system also contains **ganglia**. Each ganglion is a mass of nervous tissue containing many cell bodies and synapses enclosed in a connective tissue sheath.

The peripheral nervous system forms a vast communication network of sensory pathways and motor pathways. Sensory pathways transmit impulses from external and internal sensory receptors to the CNS. Motor pathways carry impulses from the CNS to muscles and glands, the effector organs.

The somatic (voluntary) nervous system

The peripheral nervous system is divided into the somatic nervous system and the autonomic nervous system. The **somatic nervous system** includes sensory neurones transmitting information from peripheral sensory receptors to the CNS, and motor neurones which send nerve impulses to skeletal muscles. It is sometimes called the **voluntary nervous system** because many of its actions are under conscious control. The relationship between nerves and skeletal muscles is dealt with in spread 11.4.

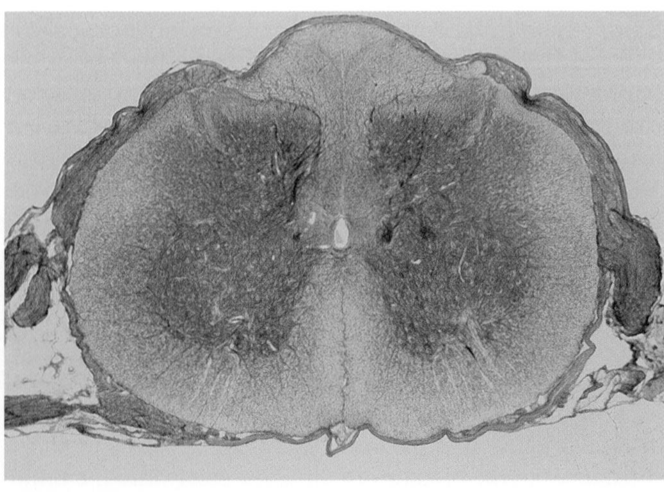

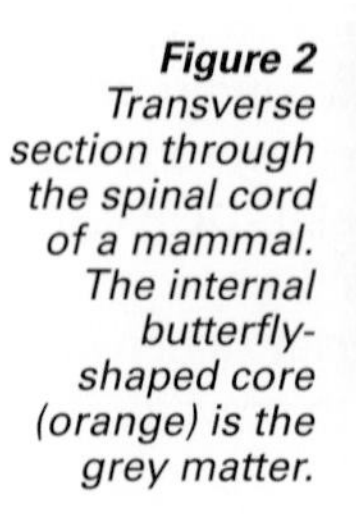

Figure 2 *Transverse section through the spinal cord of a mammal. The internal butterfly-shaped core (orange) is the grey matter.*

The autonomic (involuntary) nervous system

The **autonomic nervous system** is said to be **involuntary** because it enables internal organs to function properly, usually without our conscious control. The autonomic system has two divisions, the **parasympathetic nervous system** and the **sympathetic nervous system** (figure 3), which generally have opposing actions (that is, many of their actions are antagonistic).

The parasympathetic nervous system maintains the normal functioning of the body in non-threatening situations, and helps keep the body in a relaxed, unstressed condition. For example, stimulation of parasympathetic nerves reduces heart rate, constricts the pupils, and constricts bronchioles, but dilates blood vessels leading to the gut. Parasympathetic nerves usually have cholinergic nerve endings (that is, they use acetylcholine as a neurotransmitter).

The sympathetic nervous system brings about physiological responses to threatening situations (so-called 'fight-or-flight' responses). For example, stimulation of sympathetic nerves increases heart rate, dilates the pupils, and constricts blood vessels going to the gut. Most sympathetic nerves have adrenergic nerve endings (that is, they use noradrenaline as a neurotransmitter). The effects of noradrenaline on its target cells are similar to those of adrenaline in preparing the body for intense, energy-consuming activities such as fighting, or fleeing from danger.

Interaction of the parasympathetic and sympathetic systems

A mammal is rarely fully relaxed or fully active. The parasympathetic and sympathetic divisions work antagonistically so that the mammal's activity level is appropriate to a particular situation. Most mammalian organs (the liver and adrenal medulla are exceptions) are supplied with both parasympathetic and sympathetic neurones. Opposing signals from the two types of neurones adjust each organ's activity to a suitable level.

Quick check

1 a Which structures make up the central nervous system?

b Distinguish between the somatic nervous system and the peripheral nervous system.

2 Which part of the autonomic nervous system prepares the body for 'fight or flight'?

Food for thought

Suggest why the following actions, brought about by stimulation of the sympathetic nervous system, adrenaline, or drugs that mimic adrenaline, prepare a person for activity:

a dilation of the pupils

b inhibition of gut movements

c dilation of bronchioles.

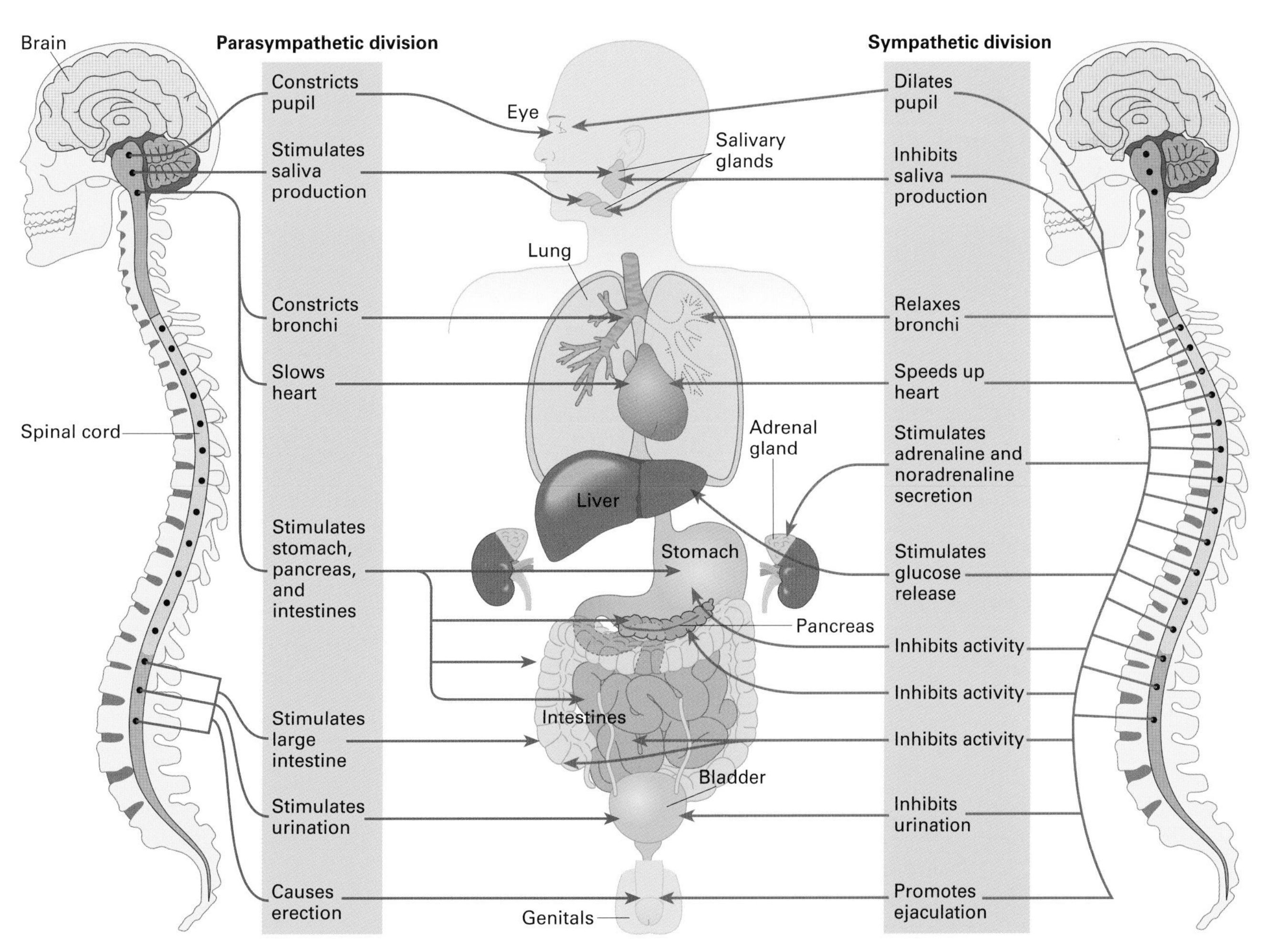

Figure 3 *The functions of the parasympathetic and sympathetic nervous systems.*

10.8

OBJECTIVES

By the end of this spread you should be able to:

- describe the main types of sensory receptors
- discuss the main functions of the sensory system
- distinguish between an action potential and a generator potential
- explain the significance of sensory adaptation.

Fact of life

A male silk moth (*Bombyx*) can pick up the scent of a female several kilometres away. The male responds to as little as one molecule of sex attractant per 10^{15} air molecules.

ANIMAL SENSES

A typical animal must be able to find its food, select a mate, and avoid predators. It must also be able to keep its internal environment within tolerable limits. These and other activities depend on the animal's ability to gather information about what is happening inside and outside its body.

Sensing environmental changes

The physical and chemical conditions in an animal's internal and external environments are continually changing. A change that can be detected is called a **stimulus**. To some extent, all animal cells are sensitive to stimuli, but some cells called **receptors** have become especially sensitive to particular stimuli. There is a huge number of environmental variables that an animal could sense. However, each species has evolved receptors only to environmental variables that have an appreciable effect on its chances of survival. Thus, humans can sense all the colours of the rainbow but can sense neither infrared nor ultraviolet light.

Classifying receptors

Receptors are commonly classified according to the type of stimulus energy they detect. The main types are:

- **mechanoreceptors** which detect changes in mechanical energy, such as movements, pressures, tensions, gravity, and sound waves
- **chemoreceptors** which detect chemical stimuli, for example, through taste and smell
- **thermoreceptors** which detect temperature changes
- **electroreceptors** which detect electrical fields
- **photoreceptors** which detect light and other forms of electromagnetic radiation.

Receptors can also be classified according to their structure. Simple receptors, known as **primary receptors**, consist of a single neurone, one end of which is sensitive to a particular type of stimulus (figure 1). A primary receptor gathers sensory information and transmits it to another neurone or an effector. For example, **Pacinian corpuscles** (figure 4) are mechanoreceptors located in the skin (spread 8.11), tendons, joints, and muscles. Their ends consist of concentric rings of connective tissue. Application of pressure against the connective tissue deforms stretch-mediated sodium ion channels in the cell surface membrane, causing an influx of sodium ions which leads to a generator potential. A **secondary receptor** is more complex. It consists of a modified epithelial cell which is sensitive to a particular type of stimulus. The cell senses changes and passes this information on to a neurone which transmits it as nerve impulses. Sense organs are complex stimulus-gathering structures consisting of grouped sensory receptors. In many sense organs, several receptors make synaptic connections with a single receptor neurone. This is called **convergence**, and it increases the sensitivity of the sense organ. Thus, an environmental change may be too weak to generate a nerve impulse when it stimulates a single receptor, but it will generate a nerve impulse if it stimulates several receptors simultaneously. This effect is similar to summation in synapses (spread 10.6).

A third classification of receptors is based on the source of stimulation: **exteroceptors** respond to stimuli outside the body; **interoceptors** respond to stimuli inside the body; and **proprioceptors** respond specifically to changes of joint angle and the amount of tension in muscles.

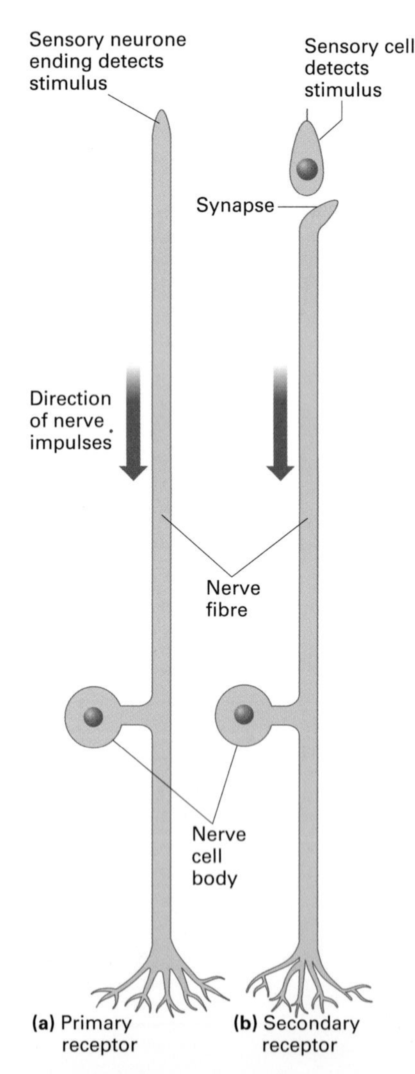

Figure 1 *Two types of receptor cells:* ***(a)*** *A primary receptor consisting of a single neurone, the end of which is sensitive to a particular stimulus.* ***(b)*** *A secondary receptor consisting of a modified epithelial cell that detects a particular stimulus, and a sensory neurone that transmits nerve impulses.*

Sensory systems

Receptors are the first component of a sensory system, which has three major functions:

1 **transduction:** receptors gather sensory information and then convert it into a form of information that can be used by the animal (nerve impulses)

2 **transmission:** sensory neurones transmit nerve impulses from the receptors to the central nervous system (CNS)
3 **processing:** the CNS processes the information so that appropriate responses can be made to environmental changes.

A receptor converts the energy from the stimulus into an electrical potential that is proportional to the stimulus intensity. This graded electrical potential is known as the **receptor potential** or **generator potential**. If the stimulus is sufficiently high (that is, above a critical threshold level) the graded potential is high enough to fire an action potential. If the stimulus is beneath the threshold, no action potential results (figure 2).

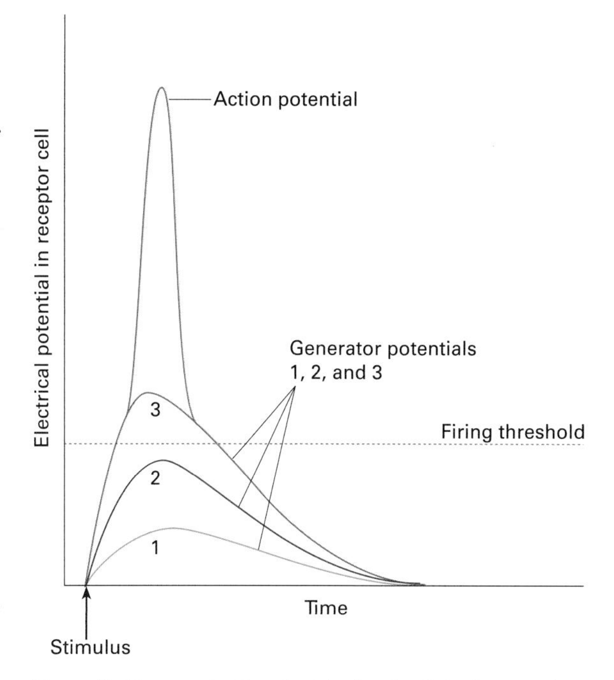

Figure 2 *Changes in the electrical potential of a receptor when stimulated by three separate stimuli. Only the third stimulus produces a generator potential high enough to start an action potential.*

Sensory adaptation

Receptors are adapted to detect potentially harmful or beneficial changes in the environment. When given an unchanging stimulus, most receptors stop responding so that the sensory system does not become overloaded with unnecessary or irrelevant information. Loss of responsiveness is brought about by a process called **sensory adaptation** (figure 3). An unchanging stimulus results in a decline in the generator potentials produced by sensory receptors. Consequently, the nerve impulses transmitted in sensory neurones become less frequent and may eventually stop. The mechanism of sensory adaptation is not fully understood, but it involves changes in the membranes of receptor cells. Sensory adaptation explains why, for example, a person becomes insensitive to the touch of clothing on skin. Even a hair shirt becomes tolerable after wearing it for a long time!

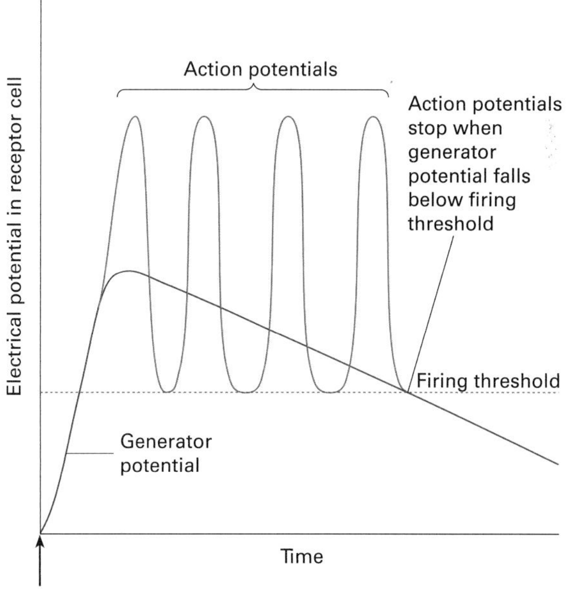

Figure 3 *Sensory adaptation: the generator potential gradually declines in response to a constant stimulus. When it falls below the threshold, the action potentials stop.*

Transferring information

After gathering and transducing the stimuli, the sensory system transmits information about the stimulus to the CNS and effectors. The frequency of nerve impulses propagated along a sensory neurone usually gives information about stimulus strength. The transfer of information is rarely direct. In mammals, much of the sensory information goes to sensory projection areas in the brain where information processing takes place (spread 10.13).

Quick check

1 Distinguish between a primary receptor and a secondary receptor.

2 Which type of receptor detects changes in the internal environment of the body?

3 What is the main difference between a generator potential and an action potential?

Food for thought

Pacinian corpuscles in the skin are mechanoreceptors. Each consists of a single sensory neurone, the ending of which is surrounded by concentric rings of connective tissue (figure 4). Deformation of the rings bends the nerve ending, which sends nerve impulses to the brain, creating the sense of being touched. Rods and cones are photoreceptors in the retina of the eye (spread 10.12) which, when stimulated by light, send nerve impulses to the brain that create the sense of seeing something. Suggest how these special sensory receptors can create different sensations.

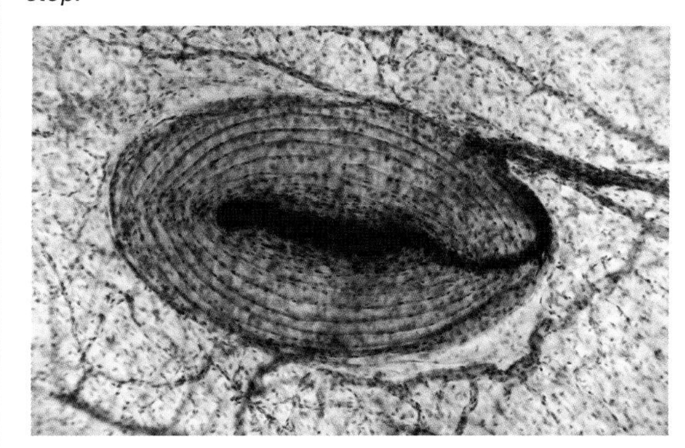

Figure 4 *A Pacinian corpuscle.*

10.9

Structure of the human ear

OBJECTIVES

By the end of this spread you should be able to:

- describe the structure of the human ear and the functions of its main parts.

Fact of life

The fennec or desert fox (figure 1) hunts insects, lizards, and birds at night. Its large cone-shaped ears help it locate its prey by directing and concentrating sound waves into the ear canal. Like many other mammals (including some people) this fox can also move its pinnae, enabling it to locate the source of sounds without moving its head. Sound location depends on the difference in time between sound reaching one ear and then the other.

Figure 1 *The fennec (*Fennecus zerda*) lives in the deserts of North Africa and Arabia.*

The human ear is a complex sensory organ that enables us to hear sounds, detect body movements, and maintain balance. Sound is the sensation we feel when structures in the ear are acted upon by pressure waves (vibrations) in the air within a limited frequency range, about 20 hertz to 20 000 hertz. (The hertz, Hz, is the SI unit of frequency, indicating the number of cycles per second.) For the purposes of this and the next spread, we shall call the air vibrations within this frequency range **sound waves** and the sensation we feel **sound**.

The ear has three main parts: an air-filled **outer ear**, an air-filled **middle ear**, and a fluid-filled **inner ear** (figure 2). The structure of each part is closely related to its functions.

The outer ear: funnelling sound waves to the ear drum

The outer ear consists of a flap (called the **pinna** or **auricle**) which funnels sound waves from the air into a canal called the **auditory canal** or ear canal. Some people share the ability of other mammals such as dogs and cats to move the pinna towards sources of sound, but in most people the pinna is fixed. The sound waves travel down the auditory canal to its inner end where they hit a tough membrane called the **ear drum** or **tympanum**. The walls of the auditory canal produce wax which keeps the ear drum soft and supple.

The middle ear: transmission of vibrations by the ossicles

On being hit by the sound waves, the ear drum vibrates and transmits its vibrations to three ossicles (small bones) in the middle ear: the **hammer** (**malleus**), the **anvil** (**incus**), and the **stirrup** (**stapes**). The ossicles form a lever system transferring vibrations from the ear drum to the inner ear. The ossicles reduce the amplitude of the vibrations but amplify the pressure they produce more than 20 times. Two small muscles attach the ossicles to the skull. During high intensity sounds, these make the ossicles stiffer and less sensitive to sound waves, protecting receptor cells (hair cells) in the inner ear from over-stimulation.

If the ear is subjected frequently to loud sounds, the ossicles may become worn down by vibrating too much. They then become less effective at transmitting sound waves, and the hearing becomes less sensitive. Operators of noisy machinery may suffer permanent loss of hearing unless they wear protective ear pads.

A tube called the **Eustachian tube** connects the middle ear to the back of the throat. This allows air to pass in and out of the middle air to equalise pressure on either side of the tympanum during swallowing. The Eustachian tube enables your ears to 'pop' when changing altitude during an aeroplane flight or deep-sea diving, for example.

The inner ear: transduction of vibrations to nerve impulses

The inner ear is made up of two main parts: the **cochlea** and the **vestibular apparatus**. The entire inner ear is filled with fluid called **endolymph** and is separated from the skull wall by another fluid called **perilymph**.

The cochlea is responsible for our sense of hearing. Vibrations of the stapes are transmitted to the cochlea through a flexible membrane called the **oval window** (**fenestra ovalis**). When the oval window vibrates it sets up pressure waves in the fluid in the cochlea, a long three-chambered coiled tube.

The cochlea is a complicated structure containing special mechanoreceptors called **hair cells**. These vibrate in response to pressure waves in the cochlear fluid, and generate nerve impulses which are passed along the auditory nerve to the brain. (Spread 10.10 describes the transduction of sound waves more fully.)

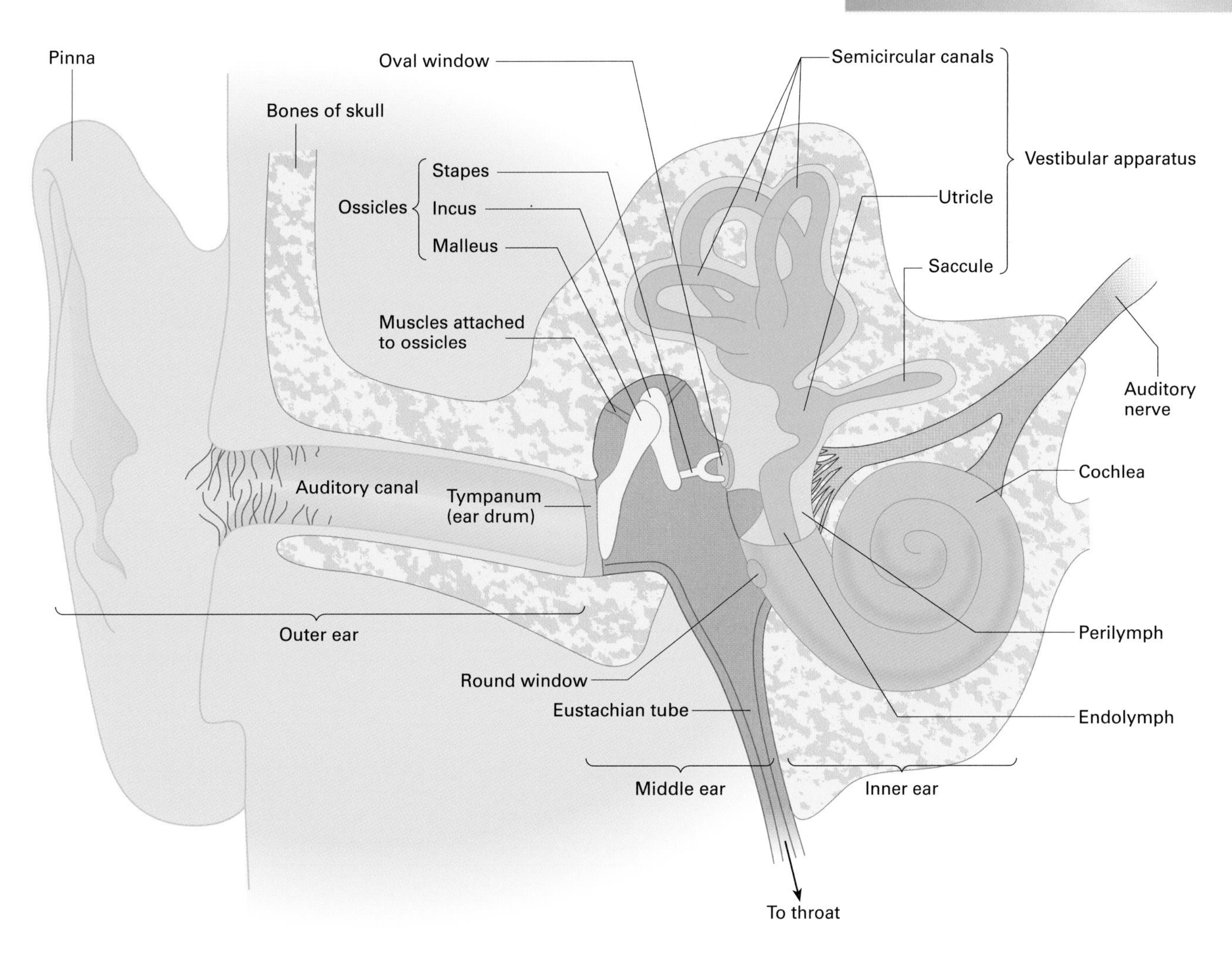

Figure 2 *Structure of the human ear.*

Pressure waves in the cochlear fluid finally reach the **round window** (**fenestra rotundis**). The energy from the waves dissipates as they push against the round window and its flexible membrane bulges into the middle ear.

In addition to the cochlea, the inner ear also contains the vestibular apparatus composed of the **semicircular canals** and two membranous, fluid-filled pouches, the **utricle** and the **saccule**. Spread 10.10 explains how mechanoreceptors in these structures detect head movements and position, helping us to maintain balance and produce smooth, coordinated movements.

Quick check

1 a In which part of the ear are the organs of balance?

b Which bone is pressed against the tympanum (ear drum)?

c Which structure equalises the pressure on either side of the ear drum?

Food for thought

Suggest why whales and other marine mammals do not have external ears (pinnae).

10.10

OBJECTIVES

By the end of this spread you should be able to:

- explain how mechanoreceptors enable sounds to be heard and balance to be maintained.

Fact of life

Humpback whales communicate with each other by complex 'songs' made up of repeated patterns of sounds which have been described as snores, chirps, groans, yups, ees, and oos. Some of their sounds can be detected by hydrophones at a range of more than 185 km. The prime function of the song appears to be sexual. The changing frequencies of the song may be used to broadcast information about range and bearing so that whales can find each other.

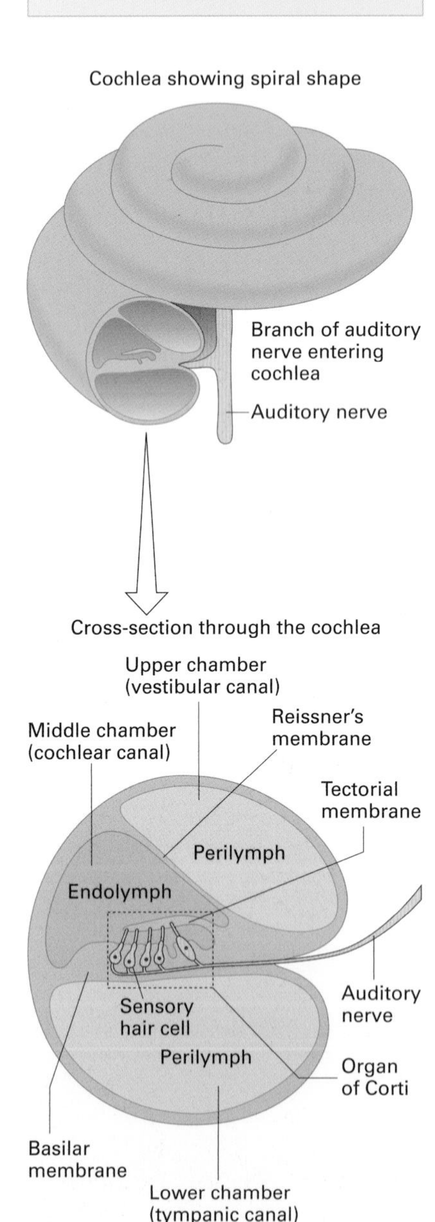

***Figure 1** The cochlea and the organ of Corti.*

MECHANORECEPTORS IN THE HUMAN EAR

Hearing, the maintenance of balance, and the detection of body movements all depend on receptors in the inner ear that detect mechanical stimuli. In the sense of hearing, these stimuli are pressure waves (vibrations) in the air, which are commonly called sound waves because they produce a sensation that we call sound. The ear also uses special mechanoreceptors to maintain balance. These receptors are sensitive to movement and gravity.

The cochlea and the organ of Corti

The cochlea is a three-chambered, spirally wound, fluid-filled tube resembling a snail's shell. The upper chamber (the **vestibular canal**) and the lower chamber (the **tympanic canal**) communicate with each other via a small hole called the **helicotrema** at the tip of the cochlea. The central chamber (the **cochlear canal**) contains the **organ of Corti** (figure 1). This has **hair cells** which are the mechanoreceptors that transduce sound waves into nerve impulses. The fluid in the upper and lower chambers is **perilymph**, rich in sodium ions; the fluid surrounding the organ of Corti in the middle chamber is **endolymph**, rich in potassium ions.

The hair cells are modified ciliated cells. Their bases are embedded in the **basilar membrane** which forms the floor of the middle chamber, and their hair-like heads are embedded in a rigid membranous flap, the **tectorial membrane**. **Reissner's membrane** (also called the **vestibular membrane**) separates the cochlear canal from the vestibular canal.

When vibrations are transmitted into the cochlea from the ossicles, pressure waves pass along the upper chamber (the vestibular canal) and cause the Reissner's membrane to vibrate. These vibrations are transmitted to the endolymph, causing the basilar membrane in the organ of Corti to vibrate. As the basilar membrane moves, the hair cells are bent against the rigid tectorial membrane. This produces a generator potential which, when it reaches a threshold level, is transmitted as a nerve impulse along the auditory nerve to the midbrain (figure 2).

Hearing different sounds

Association neurones (see spread 11.8) carry information about the frequency, duration, intensity, and direction of sound waves to the hearing centres in the cortex of the brain. The cortex processes this information so that we can not only distinguish between very similar notes at many different volumes, but we can also distinguish between the same note played on different instruments.

The human ear can detect a wide range of sound waves which vary in pitch (frequency) and loudness or intensity (amplitude). Frequency is measured in hertz (Hz), cycles per second. Loudness is measured in decibels (dB). The decibel scale is a logarithmic scale in which the faintest sound that can be heard is given a value of zero decibels (0 dB); the loudest tolerable sound is about 120 dB, one million times as loud as faintest detectable sound.

It is thought that the ear can discriminate between different audible sound waves because of the different responsiveness of hair cells along the length of the cochlea. At the base of the cochlea, the basilar membrane is narrow, thin, and rigid. Its hair cells respond more readily to high frequencies which are interpreted by the brain as high-pitched sounds. In the apex of the cochlea, the basilar membrane is wider and less rigid. It contains hair cells which are more sensitive to low-frequency pressure waves; movements of these are interpreted as low-pitched sounds. Responsiveness to high-pitched sounds decreases with age.

Perception of intensity or loudness is thought to depend on the number of neurones that are activated and the frequency of their impulses. The ear

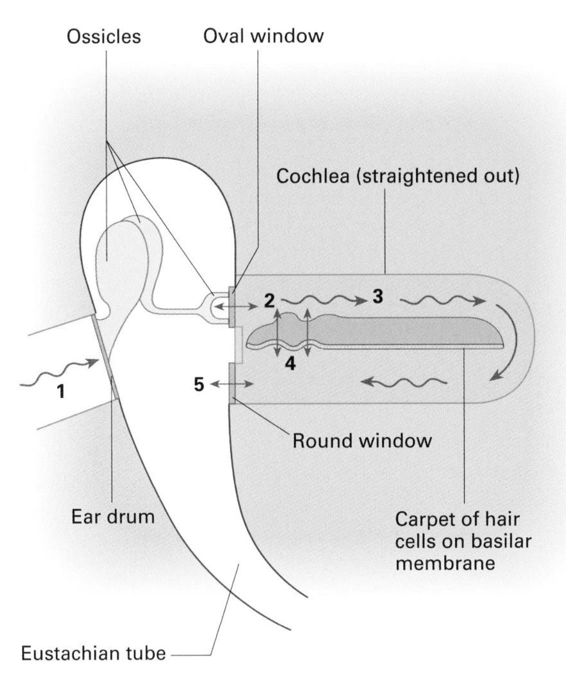

Figure 2 *Diagram showing how the ear transduces sound waves into nerve impulses. The cochlea is shown as a straight tube for simplicity.*

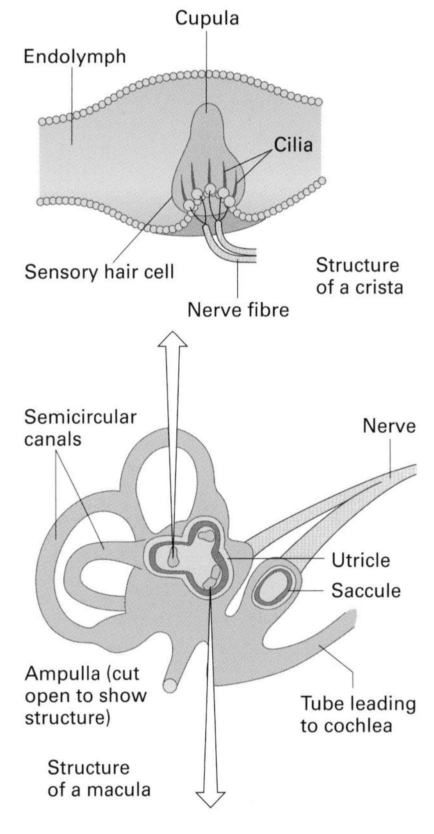

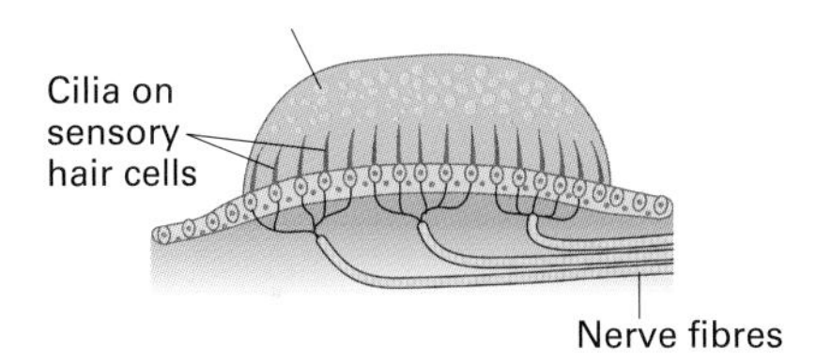

Figure 3 *The vestibular apparatus showing the semicircular canals, a crista, and a macula.*

is more sensitive to different levels of loudness at some frequencies than at others. Its maximum loudness sensitivity is at about 1000–3000 Hz. It is probably no coincidence that this corresponds to the frequency range of normal human speech. The ear is completely insensitive to sound waves above 20 000 Hz.

The vestibular apparatus and the sense of balance

Our sense of balance and information about position and movement come from the **vestibular apparatus** in the inner ear. The vestibular apparatus consists of the semicircular canals, containing organs called cristae, and sacs called the saccule and utricle, containing organs called maculae.

The inner ear has three **semicircular canals**, each lying in a different plane at right angles to the other two (figure 3). This arrangement means that changes of movement (acceleration or deceleration) in any direction can be detected. Each canal is filled with fluid (endolymph) and has a small swelling (an **ampulla**) containing a sense organ. The sense organ, called a **crista**, has hair cells embedded in a gelatinous structure called a **cupula**. The cupula moves in response to movements of the endolymph within the canal, bending the cilia that project from the hair cells. When the cilia bend, they produce generator potentials which, if they reach the threshold level, elicit nerve impulses that are transmitted to the brain.

The **saccule** and **utricle**, the two sacs in the vestibular apparatus, contain sense organs called **maculae** which are lined with hair cells surrounded by a jelly-like fluid (figure 3). Fine granules of calcium carbonate, called **otoliths**, float in the fluid. The granules move in response to changes in body position, and bend the cilia on the hair cells. If the cilia bend in one direction, the frequency of nerve impulses increases; bending the cilia in the other direction decreases the frequency.

The information from the vestibular apparatus is transmitted to several parts of the central nervous system: to the spinal cord where body position can be adjusted by reflex actions; to the cerebellum, where it is processed with other sources of information to produce smooth, coordinated movements; and to higher centres of the brain which are involved in the control of eye movements (the vestibular apparatus works with the eyes to maintain balance).

Quick check

1 What is:

a the function of the oval window

b the difference between pitch and intensity of sound

c the significance of the three semicircular canals being in different planes?

Food for thought

Loud bangs can damage the sensitive hairs in the inner ear, despite the stiffening of the ear ossicles by contraction of the two muscles that attach the ossicles to the skull. Suggest why gunners learn to open their mouths as well as cover their ears when firing cannons.

10.11

STRUCTURE OF THE HUMAN EYE

OBJECTIVES

By the end of this spread you should be able to:

- describe the structure of the human eye.

Fact of life

As a person ages, the lens gradually hardens and becomes less able to change shape. This may result in **presbyopia**, an inability to focus on near objects. Whereas a young child can usually focus clearly on objects as close as 6.3 cm, many people over 50 years cannot focus clearly on objects closer than 40 cm. Presbyopia is treated by the use of converging lenses. Elderly people are also more likely to suffer from **cataracts** which cause blurred vision, due to the lens becoming more opaque with age. Cataracts are treated by surgical removal of the lens. A contact lens, appropriate spectacles, or an **intraocular lens** (a lens implanted in the eye) may be used to compensate for the missing lens. Other eye defects include **myopia** (near-sightedness), corrected by diverging lenses, **hypermetropia** (far-sightedness) corrected by converging lenses (figure 1), and **astigmatism** (in which the eye cannot focus on vertical and horizontal lines simultaneously), corrected by wearing cylindrical lenses.

The **eye** is a complex light-sensitive organ that enables us to distinguish minute variations of shape, colour, brightness, and distance. The function of the eye is to transduce light (visible frequencies of electromagnetic radiation) into patterns of nerve impulses. These are transmitted to the brain, where the actual process of seeing is performed.

Outside the eye

The eyes are situated in cavities called **orbits** which enclose and protect all but the front parts. The exposed front surface of each eye is protected by a thin membrane called the **conjunctiva** and by eyelids, eyelashes, and lachrymal (tear) glands. Hairy **eyebrows** and **eyelashes** help protect the eyes from strong sunlight and prevent sweat from the forehead irritating the eyes. They also contribute greatly to facial expressions.

When a mammal blinks, the **eyelids** act like windscreen wipers on a car and keep the front surface of the eye free from dust and dirt. Tears secreted from the **lachrymal glands** in the upper and outer part of each eye contribute to this cleaning process. They also prevent the eyes from drying out and contain a bactericide (called **lysozyme**) which reduces the risk of infection.

The eyes are never stationary. Even when the head is completely still, small movements of six **extrinsic muscles** produce a constantly changing pattern of light. The muscles also rotate the eye to follow moving objects and to direct attention to a chosen object. Both eyes work together and are always directed at the same spot.

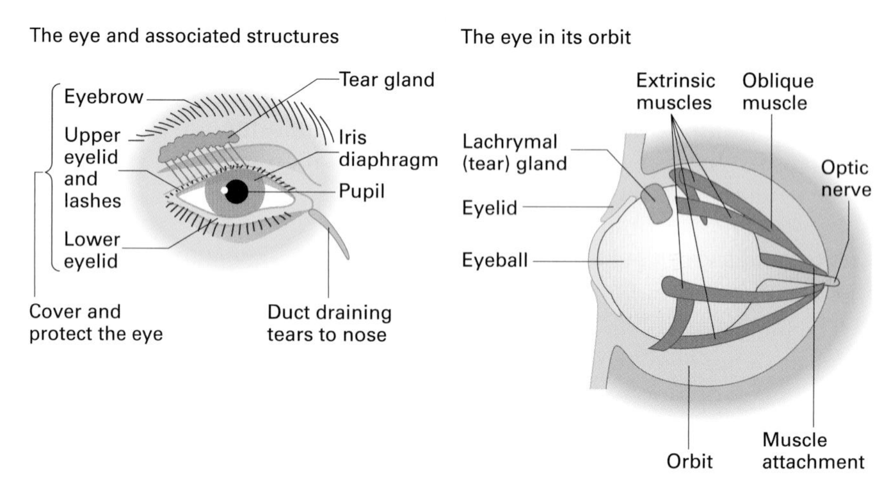

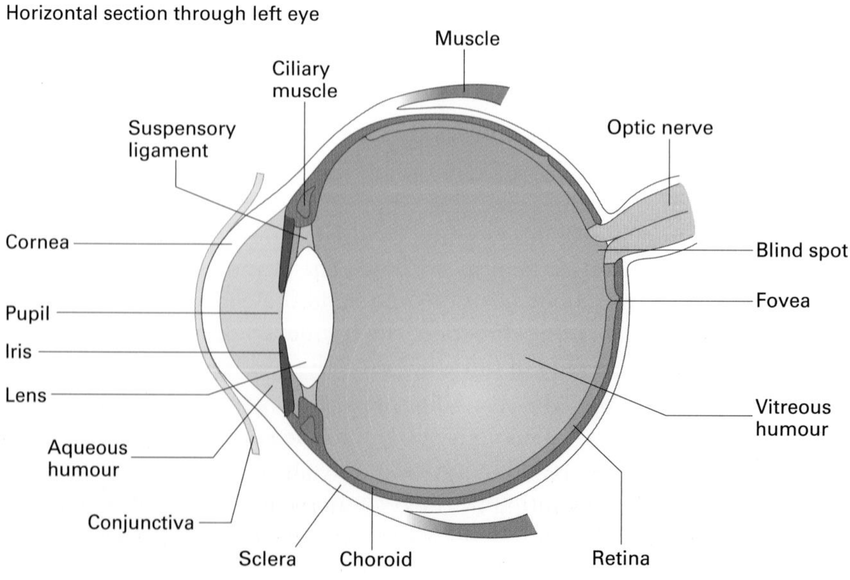

Figure 2 Different views of the human eye.

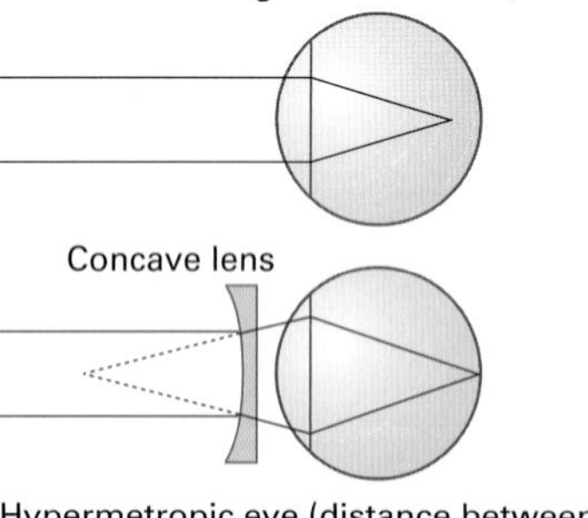

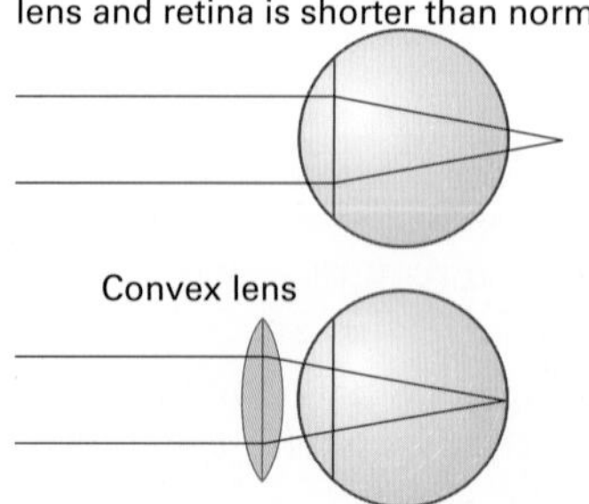

Figure 1 The correction of myopia and hypermetropia.

Inside the eye

The eyeball is a roughly spherical structure approximately 2.5 cm in diameter (figure 2). The outer part is composed of three layers of tissue: an outer protective coating called the **sclera**; a middle vascular layer called the **choroid**; and an innermost layer of photoreceptors (light-sensitive cells) called the **retina**.

Light enters the eye through the **cornea**, a transparent curved structure continuous with the sclera. The cornea is a tough, five-layered membrane that bends (refracts) light waves as they pass through it, so that the light is focused onto the photoreceptor cells at the back of the eyeball.

Behind the cornea is a chamber filled with **aqueous humour**, a clear watery fluid similar in composition to blood plasma. It separates the cornea from the lens.

The role of the lens: accommodation

Although in terrestrial mammals light waves are refracted mostly by the cornea, the **lens** can vary the amount of refraction and thus fine tune the focusing of light on the back of the eye. The lens is a flattened crystalline sphere made of a large number of transparent fibrous cells arranged in layers. It is connected by ligaments to radial and circular **ciliary muscles** which, with their surrounding tissue, form the **ciliary body**. Figure 3 shows how the ciliary muscles work antagonistically to change the shape and thus the focal length of the lens for focusing on near and distant objects. This adjustment process is called **accommodation**.

The iris: adjusting the size of the pupil

The ciliary body is actually the front edge of the choroid layer, which has many blood vessels that supply the eye with nutrients and oxygen. The choroid is also full of a dark brown pigment which prevents internal reflections of light inside the eyeball.

The **iris** is another extension of the choroid layer and is continuous with the ciliary body. Named after the Greek for rainbow, the iris gives colour to the eye. It consists of pigmented radial and circular muscles around a central opening called the **pupil**. The muscles control the size of the pupil, as shown in figure 4. In humans, the pupil is circular when both dilated and constricted. However, other mammals show much variation of pupil shape when constricted: for example, in cats the dilated circular pupils become a slit when constricted.

Behind the lens

The main body of the eye behind the lens is filled with a transparent, jelly-like substance called the **vitreous humour**. The pressure of the vitreous humour gives shape and firmness to the back of the eye.

The cornea, aqueous humour, lens, and vitreous humour all play a part in focusing an image onto the retina which covers the back of the eye. Information about the image is transmitted by the optic nerve to the brain. Spread 10.12 explains how photoreceptors in the retina transduce the energy from light into nerve impulses to produce the sensation of sight.

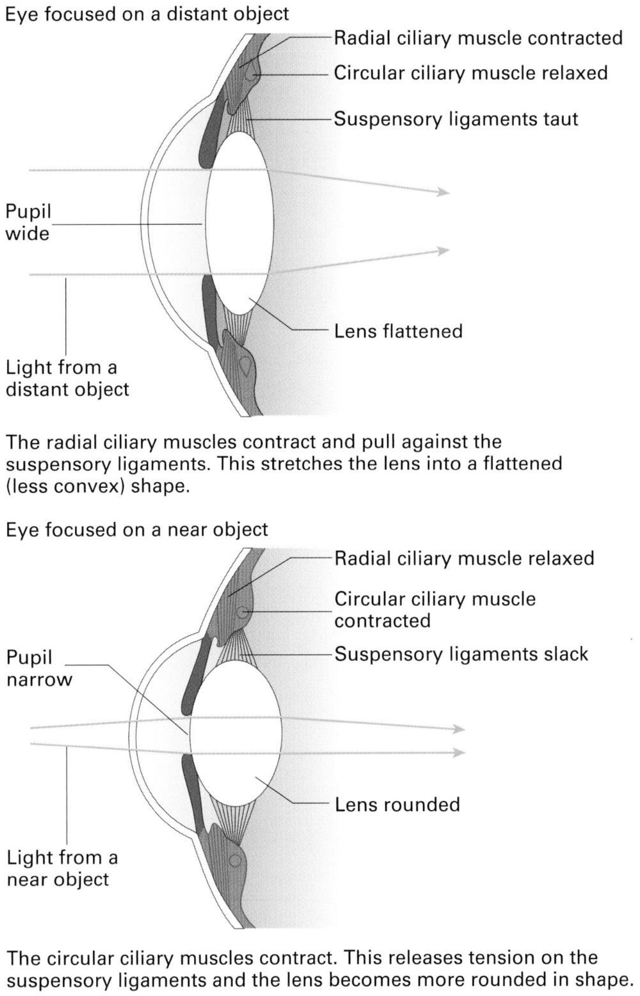

Eye focused on a distant object

Radial ciliary muscle contracted

Circular ciliary muscle relaxed

Suspensory ligaments taut

Pupil wide

Lens flattened

Light from a distant object

The radial ciliary muscles contract and pull against the suspensory ligaments. This stretches the lens into a flattened (less convex) shape.

Eye focused on a near object

Radial ciliary muscle relaxed

Circular ciliary muscle contracted

Suspensory ligaments slack

Pupil narrow

Lens rounded

Light from a near object

The circular ciliary muscles contract. This releases tension on the suspensory ligaments and the lens becomes more rounded in shape.

Figure 3 Accommodation: adjusting the lens to focus light onto the retina from both distant and near objects. In this diagram, the light is shown bending only at the lens for simplicity: in reality the light is also bent by the cornea.

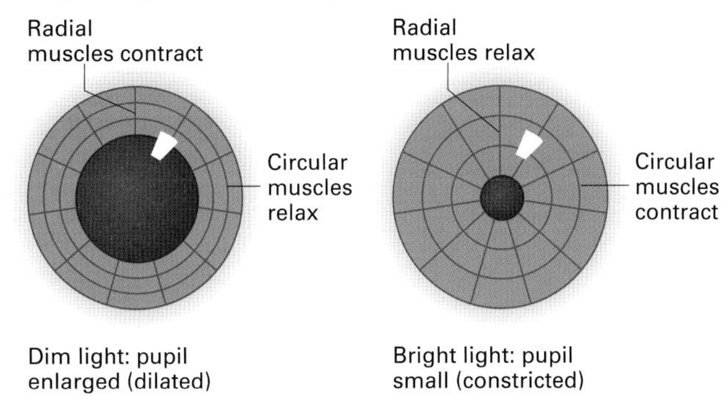

Figure 4 The response of the iris to dim and bright light conditions.

QUICK CHECK

1 a What is the function of the cornea?

b What is the shape of the lens when the eye is focused on a near object?

c Which structure controls the size of the pupil?

Food for thought

In terrestrial vertebrates, tear glands secrete liquid to lubricate the surface of the eye and keep it clean. Suggest how tear glands in reptiles such as sea snakes and brackish water terrapins help adapt them to their habitat.

10.12

The retina and vision

OBJECTIVES

By the end of this spread you should be able to:

- describe the structure of the retina
- explain how rod cells transduce light energy into nerve impulses
- explain how cone cells produce colour vision
- discuss the significance of binocular vision.

Fact of life

In very low light intensities, the dark pigment between the retina and choroid coat is completely withdrawn. In cats and some other mammals, this exposes a reflective layer of protein and crystals of guanine in the choroid coat called the **tapetum lucidum**. Any light that passes through the retina without being absorbed by the photoreceptors is reflected back again by this layer, and may stimulate the photoreceptors second time around. The tapetum lucidum increases the sensitivity of the eyes to dim light and explains why cats' eyes give bright reflections when the light from a car shines into them at night.

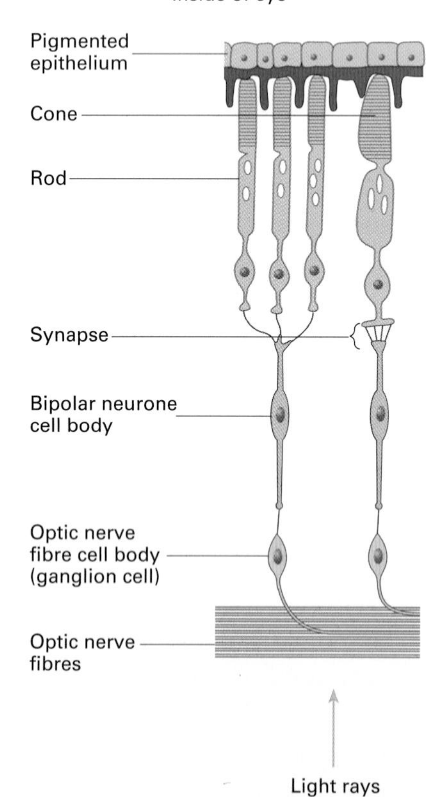

***Figure 1** Simplified diagram of part of the retina. Notice that light has to pass through the optic nerve fibres before it reaches the photoreceptor cells which are at the back of the retina.*

The **retina** is a photosensitive layer at the back of the eye. It contains two types of photoreceptors called **rods** and **cones**, and their nerve fibres. In the human eye there are about 12 million rods and 7 million cones.

The structure of the retina: rods and cones

A single-celled layer containing the dark pigment melanin separates the retina from the choroid coat. The pigment can migrate up and down in the gaps between the rod and cone cells, helping the eye to adapt to different light intensities (see Fact of life).

Figure 1 shows the arrangement of rods and cones in the retina. Rods and cones are secondary receptors: some cones have their own bipolar neurone (a neurone with a process leaving the cell body at either end), but up to 150 rods may have to share one bipolar neurone. This enables cones to provide much more sensory information, but they require about 1000 times more light than rods before they stimulate their bipolar neurone to fire an action potential.

Rods are very sensitive to low light intensities because of **convergence** (see spread 10.8); the many rods sharing the same nerve fibre are able to detect light from a relatively wide area and combine its stimulatory effects. This increased sensitivity to low intensities is at the expense of **visual acuity** (the ability to discriminate fine details). Consequently, in dim light when the cones are not functioning, the rods provide only enough information for an ill-defined image that lacks colour. Also, because rods are the only type of photoreceptor at the outermost edge of the retina, something seen 'out of the corner of the eye' lacks detail. However, the rods are very sensitive to changes in light intensity caused, for example, by something moving; this ability is potentially vital because that 'something' could be an enemy or predator.

The fovea and blind spot

Over most of the retina, the rods and cones are buried under a layer of blood vessels and nerve fibres which lead into the optic nerve. However, the **fovea**, a small depression in the retina opposite the lens, consists only of cones. There is only a thin layer of nerve fibres here, and no capillaries. This means that light falling on the fovea produces a clear well-defined visual image in colour. When a person wants to examine the details of an object, the eyes move automatically so that light from the object falls on the fovea.

Sensory information from rods and cones passes to the brain via the optic nerve, a bundle of hundreds of thousands of nerve fibres. The point of exit of the optic nerve and blood vessels from the eye is called the **blind spot** because it is devoid of photoreceptors and insensitive to light.

Transduction of light energy by rods

Vesicles in rod cells (figure 2) contain **rhodopsin**, a photosensitive pigment made of the protein **opsin** combined with **retinal** derived from vitamin A. In the dark, rods are depolarised: cation channels (mainly sodium channels) are opened. Depolarisation results in the release of a synaptic transmitter which inhibits the bipolar neurones, preventing them from transmitting action potentials along the optic nerve fibres. Light causes retinal to change shape and break apart from opsin, a process called **bleaching**. This leads to cation channels becoming closed and the rod cell becomes hyperpolarised (the cell membrane potential becomes more negative than normal). No inhibitory synaptic transmitter is released and the bipolar cell is able to generate an action potential which is transmitted in the optic nerve. Activation of a rod cell by light is therefore a **hyperpolarisation** (inhibition) of the cell.

Our eyes adapt to different levels of brightness by varying their sensitivity.

A long period (more than about 30 minutes) in the dark results in **dark adaptation**: photosensitive pigments are formed faster than they are broken down, increasing sensitivity to light. A prolonged period in bright light results in **light adaptation**: the photosensitive pigments are broken down faster than they are formed, reducing sensitivity to light.

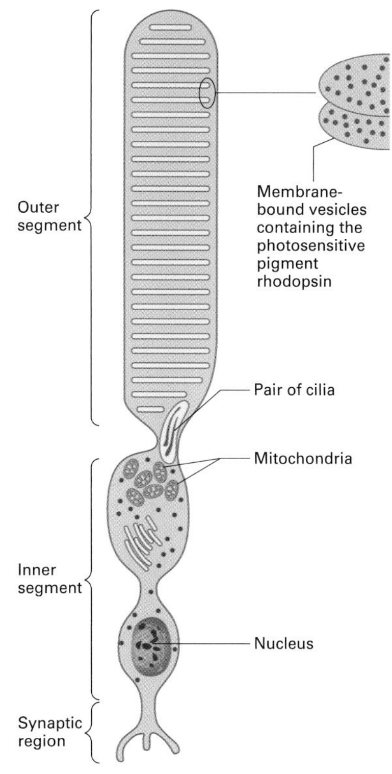

***Figure 2** A rod cell.*

Colour vision and cones

Cones are responsible for colour vision. They are thought to work on the same principle as rods. However, there are three types of cone, each containing a different form of the pigment **iodopsin** which breaks down only in bright light. The three pigments together are sensitive over the visible spectrum, but one is most sensitive to blue light, one to green light, and the other to red light. According to the **trichromatic theory**, different colours are perceived by mixing the information from the different types of cone. An alternative theory of colour vision, the **retinex** theory, suggests that the brain cortex as well as the retina is involved in colour perception. This would explain why we usually perceive a particular object as being the same colour under different types of illumination.

Central processing: how we see

The visual cortex of the brain is responsible for sorting out all the sensory information from the retina and integrating it with other information so that we can 'see' an object. One illustration of this processing is the fact that the lens produces an inverted image on the retina; the brain turns it the 'right way up'. Optical illusions are also evidence of central processing; in these cases the brain is fooled into perceiving something that is not present in the environment.

Stereoscopic vision: combining two images

Having two eyes (**binocular vision**) is better than having one for the following reasons:

- it gives a larger field of vision
- a defect in one eye does not result in blindness
- in animals with two forward-facing eyes, it provides the potential for stereoscopic vision.

Stereoscopic vision depends on each eye being able to look at the same object from a slightly different perspective. The visual centre in the brain combines the two views to make a three-dimensional image. Stereoscopic vision provides information about the sizes and shapes of objects and enables distances to be judged accurately. However, because the eyes have to be relatively close together for stereoscopic vision, the field of vision is relatively small. Mammalian predators tend to have well developed stereoscopic vision, while herbivores tend to have eyes wide apart, sacrificing stereoscopic vision for a wide field of view (figure 3).

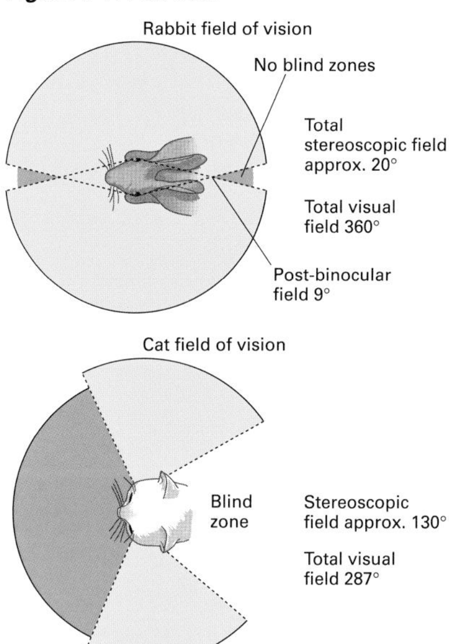

***Figure 3** Stereoscopic vision in a rabbit and a cat. The rabbit is a herbivore: it has a wide field of vision but a small stereoscopic field. The cat is a predator: it has sacrificed a wide field of vision for good stereoscopic vision.*

QUICK CHECK

1. Which type of photoreceptors occur in the fovea?
2. Briefly explain what happens to rhodopsin when it absorbs light.
3. According to the trichromatic theory of colour vision, to which colours of light are the three different types of cone sensitive?
4. Distinguish between binocular vision and stereoscopic vision.

Food for thought

Suggest why:

a the flicker on a cinema screen is visible out of the corner of your eye but not when you look directly at the screen

b objects that appear brightly coloured in a well-lit room appear in dim light to be either white or black, or various shades of grey

c on first entering a dimly lit room it is difficult to see objects clearly, but they gradually become more clearly visible.

10.13

OBJECTIVES

By the end of this spread you should be able to:

- identify the main regions of the brain and discuss their main functions.

Fact of life

Until recently, the most common way of studying brains was by dissecting tissue on bodies undergoing autopsy or on patients undergoing surgery. It is now possible to study brains without wielding a scalpel. **Magnetic resonance imaging (MRI)** is a non-invasive technique used to generate images of the brain of a living person. High frequency radio waves are absorbed and transmitted by the water molecules in tissues placed in a strong magnetic field. A computer 'maps' out variations in the signals to produce a three-dimensional image of the brain (figure 1). Oxygenated blood causes the image to become brighter. These changes in image brightness are the basis of a special MRI procedure called **functional magnetic resonance imaging** (fMRI). This allows a record to be made of the rapid changes in blood flow that follow the activation of a particular area of brain tissue. fMRI can therefore be used to study brain function as well as structure.

The human brain

The human brain is a truly amazing organ. Weighing less than 1.5 kg, this spongy mass of nerve tissue is the master control centre for the whole body. It is the seat of our intelligence, emotions, and memory. It has direct control over moving body parts connected to it by nerves. Through the hypothalamus, it has indirect control over other body parts, such as the endocrine glands, connected to it by the circulatory system (spread 10.14).

The structure of the brain

The brain is surrounded by three protective membranes called **meninges**. Inflammation of these membranes due to a bacterial or viral infection is called **meningitis**. The bulk of the brain consists of spaces called **ventricles** which are continuous with the central canal of the spinal cord. These spaces are filled with **cerebrospinal fluid** which supplies the many millions of neurones in the central nervous system (CNS) with respiratory gases, nutrients, white blood cells, and hormones. The cerebrospinal fluid also fills the spaces between the meninges, cushioning the CNS against mechanical disturbances such as knocks and bumps.

Underneath the meninges, the CNS contains non-excitable **neuroglial cells** and two distinct areas of excitable nervous tissue called **white matter** and **grey matter**. White matter consists mainly of nerve fibres; grey matter is mainly nerve cell bodies. In the mammalian brain most of the grey matter is in the cerebral cortex, the superficial layer of the cerebrum. Neuroglial cells do not conduct nerve impulses; they have a supportive and protective role in the CNS.

The structure of the brain is shown in figure 1.

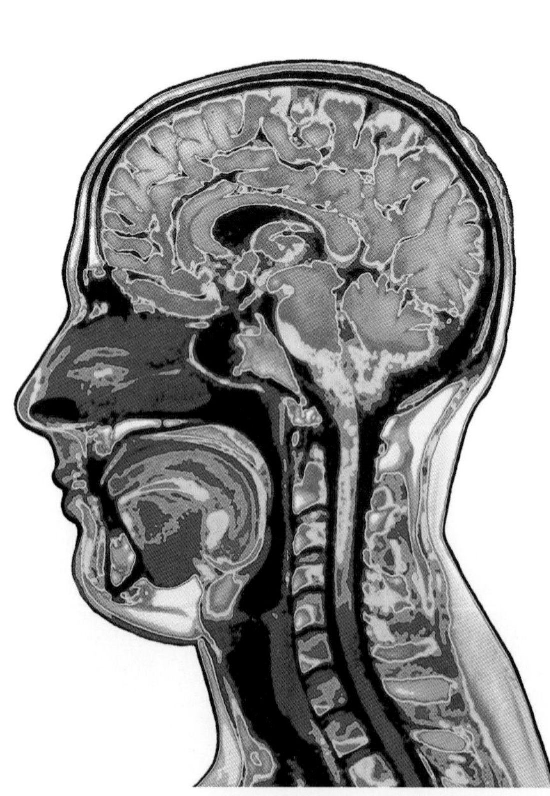

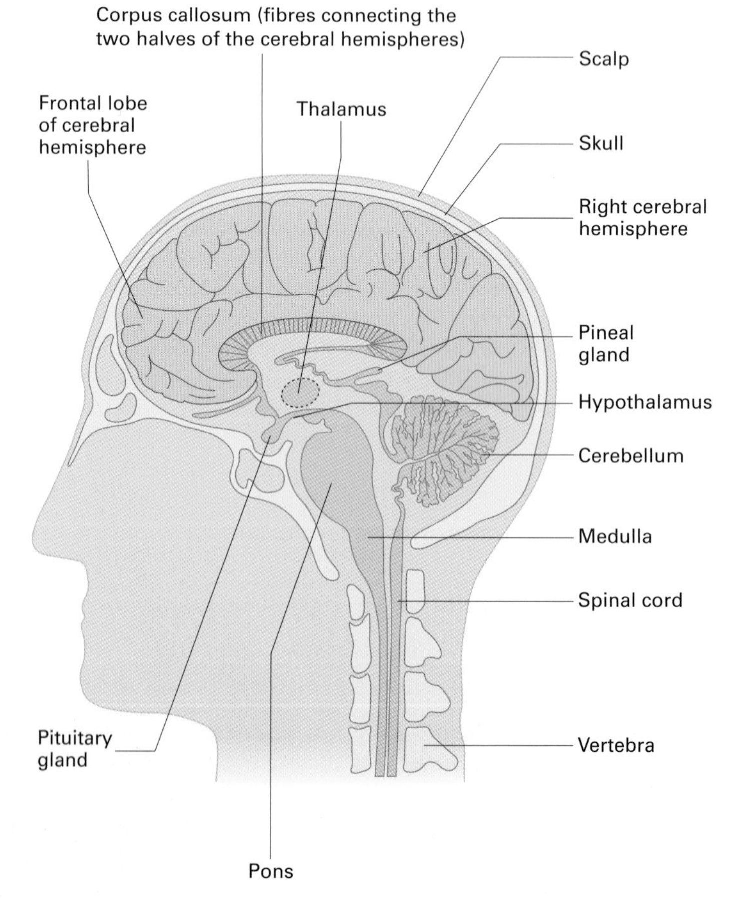

Figure 1 MRI scan of the brain, and the structure it reveals.

Functions of the brain

The **hindbrain** consists of three parts: the medulla oblongata, the pons (or pons Varolii) and the cerebellum. The **medulla oblongata** contains the neural centres of the autonomic nervous system which control vital physiological processes such as breathing rate and blood pressure. The **pons** (which means 'bridge' in Latin):

- relays messages between the forebrain and cerebellum
- contains the **pneumotaxic centre** which regulates the change from inspiration to expiration when breathing
- is involved in sensory analysis, and controlling movement and posture, swallowing, sleep, and the level of consciousness.

The **cerebellum** is a large and complex lobe at the back of the brain. It receives sensory information from the organs of balance in the ears, and from proprioceptors in muscles and tendons. It works in close association with the cerebral cortex to coordinate movement, posture, and balance.

The **midbrain** links the hindbrain with the forebrain. It has centres involved in voluntary movements and visual and auditory reflexes which, for example, enable a person to follow a moving object and locate a sound.

The **forebrain** includes the cerebrum, the thalamus, and the hypothalamus. The **cerebrum** forms the bulk of the brain and consists of two **cerebral hemispheres**, each a symmetrical rounded mass. The cerebral hemispheres have a deep fissure between them, but are connected by a bridge of fibres known as the **corpus callosum** which enables information and learning to be transferred from one hemisphere to the other. Small clusters of nerve cells called basal ganglia occur under the corpus callosum. They have dopamine-sensitive receptors, including DRD4, which play an important role in motor coordination. When damaged (for example, by **Parkinson's disease**) motor coordination is disturbed (see also spread 10.6).

The **cortex**, a superficial layer of the cerebrum only 2–3 mm deep, is the largest and most complex part of the brain. It contains thousands of millions of neurones, each with many synaptic connections. Consequently, there is an immeasurable number of potential nervous pathways in the brain. The pathways that are actually used are determined genetically and by experience (learning). These pathways enable us to have conscious thoughts and actions. They also produce our most distinctive human characteristics: the ability to solve problems, imagine abstract concepts, and display artistic talent and appreciation. The cortex has four discrete lobes which have a number of motor and sensory functional areas as well as association areas (figure 2). The association areas make up most of the cerebral cortex. One large association area in the frontal lobe is a site of higher mental activities: it integrates information from other areas so that we can think, make decisions, and use language. The visual association area interprets, processes, and stores visual information. The auditory association area performs the same function for sound waves.

Much of human personality, learning, and memory seems to be associated with the **limbic system**. This consists of several interconnecting areas, including parts of the **hippocampus** (so named because it is shaped like a sea-horse), the thalamus, and the hypothalamus in the base of the brain. The hippocampus interacts with another area of the cortex, the prefrontal cortex, and is involved in complex learning, reasoning, and personality. The **thalamus** relays signals from various sensory systems to the brain. It is also involved in the perception of pain and pleasure. The **hypothalamus** provides a link between the brain and the endocrine system (spread 10.14).

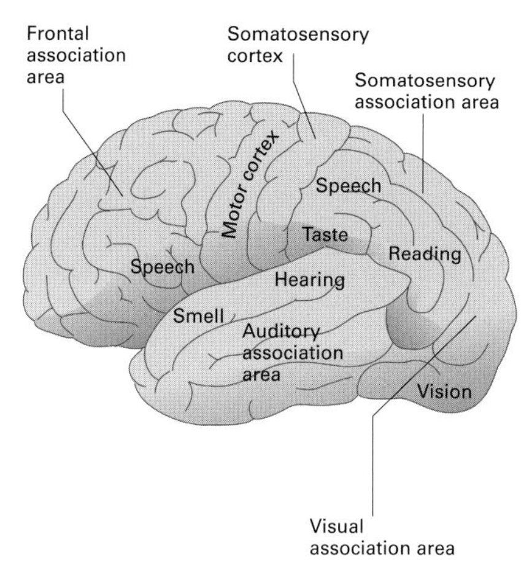

***Figure 2** The main functional areas of the cortex of the left cerebral hemispheres. The motor and sensory areas have been identified through information gained from accidents, surgery, and experimental stimulation. The **sensory areas** receive sensory impulses from the sense organs in different parts of the body. **Motor areas** send instructions along motor neurones from the brain to voluntary muscles. The **association areas** integrate sensory information.*

Quick check

1 a What are the functions of the meninges?

b Why does a person die immediately if the medulla oblongata is destroyed?

c Which part of the brain is concerned with conscious activities?

Food for thought

A stroke is an interruption of the blood flow to the brain. It can permanently damage brain tissue, and often leads to paralysis. The degree of paralysis depends on the region of the brain affected and the severity of the interruption to the blood supply. Suggest why stroke patients often regain some control over paralysed arms and legs after regular and frequent physiotherapy.

10.14

OBJECTIVES

By the end of this spread you should be able to:

- describe how the hypothalamus provides a link between the nervous system and the endocrine system
- explain how the hypothalamus affects the metabolic rate of mammals
- describe the structure and function of the thyroid gland.

Fact of life

The remedy for an underactive thyroid gland used to be to take one fried sheep's thyroid weekly with blackcurrant jelly. Today, patients are given carefully regulated quantities of thyroid hormone.

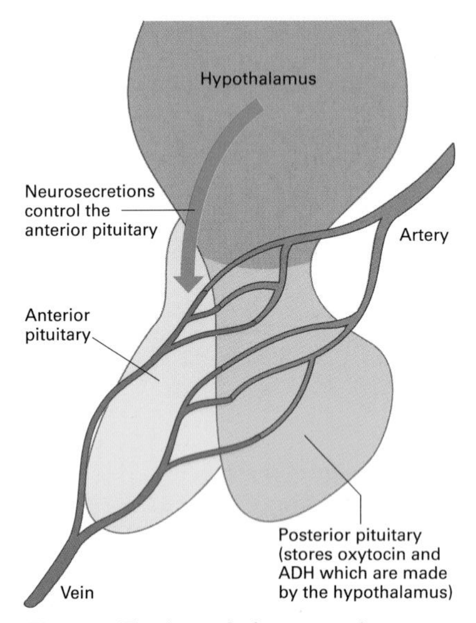

***Figure 1** The hypothalamus and pituitary gland.*

THE HYPOTHALAMUS, PITUITARY, AND THYROID GLAND

The hypothalamus: integrated homeostatic control

The nervous and endocrine systems do not work in isolation. The **hypothalamus**, a small organ in the lower brain (spread 10.13, figure 1), integrates the activities of the two systems to control a number of vital homeostatic processes. The hypothalamus is attached by numerous nerves to the brain and is exceptionally well supplied with blood from the rest of the body. It therefore receives a lot of information about conditions inside the body and in the external environment. It uses this information to regulate physiological processes such as thirst, appetite, and sleep. It also uses the information to regulate the activity of many endocrine glands. It does this through its effects on the pituitary gland.

The posterior pituitary: oxytocin and ADH

The **pituitary gland** (figure 1) consists of two distinct parts: a posterior lobe and an anterior lobe. The posterior lobe is really an extension of the hypothalamus. It stores two hormones made in the hypothalamus: **oxytocin**, which induces contraction of the uterine wall; and **antidiuretic hormone** (**ADH**), which regulates water reabsorption in the kidneys (see spread 8.8). These hormones are released into the bloodstream when the posterior lobe is stimulated by nerve impulses from the hypothalamus.

The anterior pituitary: the 'conductor of the endocrine orchestra'

The anterior lobe of the pituitary consists of endocrine tissue that secretes hormones which regulate the activity of many other glands. Because of this regulatory ability, the pituitary has been called the 'conductor of the endocrine orchestra'. However, the anterior pituitary works under the directions of the hypothalamus. The hypothalamus produces **releasing factors** that stimulate the release of hormones from the anterior pituitary, and **inhibiting factors** that slow down or stop the release of hormones from the anterior pituitary.

The chemical messengers produced by the hypothalamus are often referred to as **neurosecretions** because they are secreted from special neurones. Unlike neurotransmitters, they go into a blood vessel, not across a synapse (figure 2).

The relationship between the nervous and endocrine systems can be illustrated by the effects of the hypothalamus on the thyroid gland.

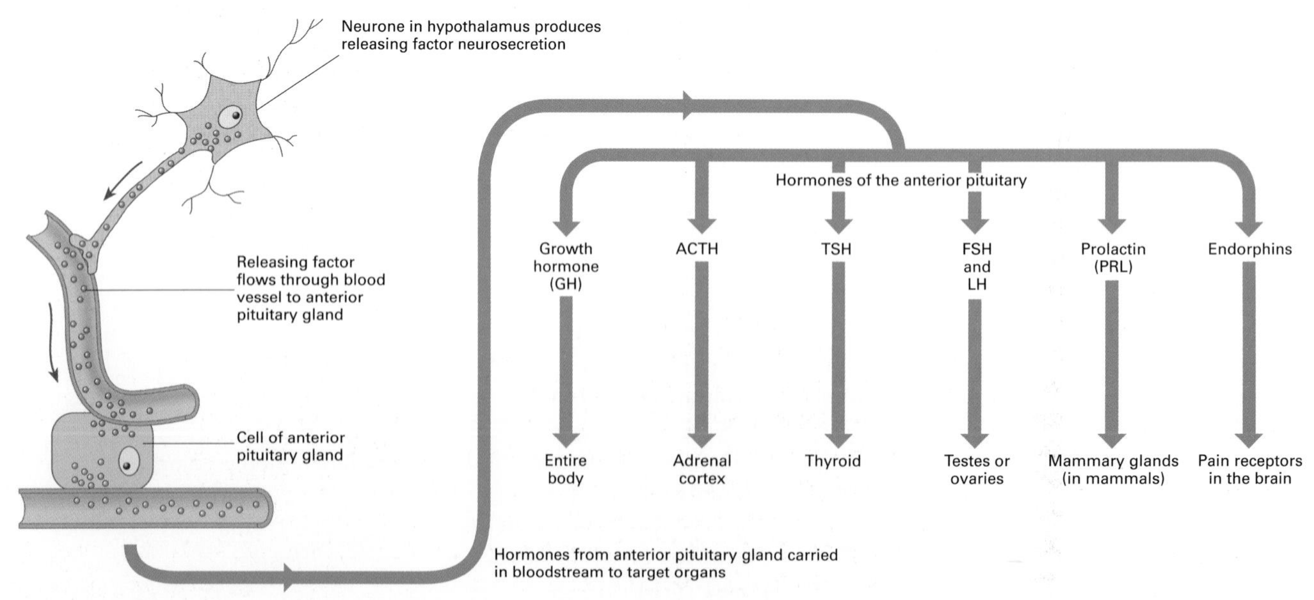

***Figure 2** Neurosecretions from the hypothalamus control the secretory activity of the anterior pituitary gland and therefore affect many other endocrine glands indirectly.*

Control of the thyroid gland

The **thyroid gland** lies in the neck, overlying the trachea close to the larynx. The gland consists of numerous follicles in close association with blood capillaries (figure 3). It secretes **calcitonin** and **thyroid hormone**.

Calcitonin helps to regulate the calcium ion concentration in the blood. Thyroid hormone is composed of at least two iodine-containing chemicals, the most abundant of which is **thyroxine** or tetraiodothyronine (the other is triiodothyronine). Thyroxine controls the metabolic rate and stimulates respiration of glucose and fats.

Thyroxine is stored in the lumen of the follicles as an inactive precursor, **thyroglobulin**, a glycoprotein containing the amino acid tyrosine. When required, a protease breaks down the thyroglobulin making its tyrosine available to bind with iodine and form thyroxine. Thyroxine, unlike thyroglobulin, is able to pass into the bloodstream.

The regulation of thyroxine secretion involves negative feedback (figure 4). The hypothalamus produces **thyrotrophin releasing factor (TRF)** which passes to the anterior pituitary along blood vessels.

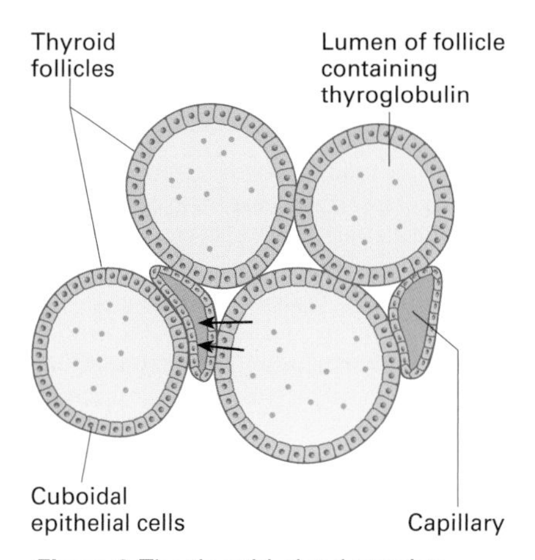

***Figure 3** The thyroid gland consists of numerous follicles which store thyroglobulin in their cavities. When required, thyroglobulin is converted to thyroxine which enters one of the many capillaries in the thyroid gland.*

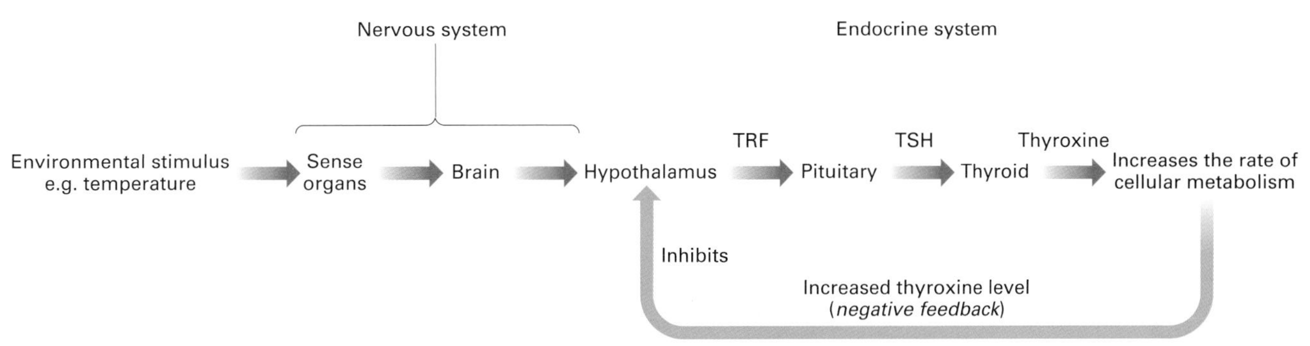

***Figure 4** Control of thyroxine secretion by negative feedback mechanisms. The hypothalamus coordinates the nervous and endocrine systems, in this case, gathering information from internal and external sense organs to set the level of cellular metabolism.*

TRF stimulates the release of **thyroid stimulating hormone (TSH)** from the anterior pituitary. In turn, TSH stimulates the secretion of thyroxine. When the thyroxine concentration exceeds a certain level (the norm) in the blood, it inhibits TRF production from the hypothalamus and TSH production from the pituitary. Thus the production of thyroxine is reduced until its concentration in the blood drops below the norm. However, information from the brain and sense organs can set a new norm. For example, in many mammals the norm is increased during a severe bout of cold weather: the hypothalamus releases more TRF, the anterior pituitary secretes more TSH, and the thyroid secretes more thyroxine until it reaches its new norm. The metabolic rate is increased, generating more body heat to compensate for the cold weather.

By this complex interaction between different parts of the nervous system and endocrine system, the rate of cellular metabolism can be adjusted to the environmental conditions, but be kept steady from moment to moment.

Disorders of the thyroid

The cellular metabolic rate affects many vital processes including growth and development. Disorders of the thyroid gland can therefore have significant effects. Overactivity (**hyperthyroidism**) is characterised by a raised body temperature, hyperactivity, increased irritability, and restlessness. Underactivity (**hypothyroidism**) in adults is characterised by a reduced alertness and increased body mass (myxoedema). In children, hypothyroidism results in mental, sexual, and physical retardation. Hypothyroidism can result from a thyroid gland disorder or, more rarely, from lack of dietary iodine. It is commonly treated by taking hormone-replacement tablets, usually laevothyroxine to replace thyroxine.

Quick check

1. What is the name given to chemical messengers produced by neurones but carried in the blood?
2. Name the hormone from the hypothalamus that stimulates the release of thyroid stimulating hormone from the pituitary gland.
3. Name two processes affected by thyroxine.

Food for thought

The pituitary gland has been called *the* 'master gland' in the body because its secretions coordinate the activity of many other endocrine glands. Suggest why the hypothalamus might be more deserving of the title.

Summary

The **nervous system** and **endocrine system** convey information which coordinates the activities of animals. The nervous system contains **neurones** which transmit **nerve impulses** rapidly from one specific location to another so that responses can be localised. Key features of neurones are **cell bodies**, **axons**, **dendrons**, and, in myelinated fibres, **Schwann cells**, **myelin sheaths**, and **nodes of Ranvier**.

The **endocrine system** consists of **endocrine glands** which produce **hormones**. The hormones are transported in the blood stream, usually in very low concentrations, and can affect target cells in different parts of the body; the responses are usually slow. Hormonal secretion is usually regulated by **negative feedback mechanisms** involving other hormones, but it may also be regulated by direct nerve stimulation. Hormones such as **steroids** affect a target cell by binding to a **receptor** inside the cell, whereas others such as **adrenaline** affect a target cell by binding onto a receptor in the cell surface membrane and use a **second messenger**, such as **cAMP**, to affect processes inside the cell.

A neurone not transmitting a nerve impulse has a **resting potential**. An above-threshold stimulus causes **depolarisation** and an **action potential** which obeys the **all-or-none-law**. Nerve impulses are propagated as a wave of depolarisation along a nerve fibre. Conduction speed is increased in myelinated fibres by **saltatory conduction**. Information passes from one nerve to another, or between a nerve and an effector (e.g. a muscle) via **synapses** which may be excitatory or inhibitory, depending on the type of **neurotransmitter** used. The nervous system of mammals is divided into a **central nervous system** and **peripheral nervous system** which can be further divided into the **somatic nervous system** and **autonomic nervous system**. The autonomic nervous system has two main divisions: the **parasympathetic nervous system** which helps keep the body in a relaxed state, and the **sympathetic nervous system** which prepares the body for action.

Animal **sense organs** contain **receptors** which detect **stimuli**. **Mechanoreceptors** in the **human ear** enable us to hear different sounds, detect body movements, and maintain balance. Photoreceptors in the retina of the **human eye** transduce light energy into nerve impulses, enabling us to see. Information from the eyes and ears goes to the **brain** where it is processed. The brain is the major control centre for the whole body. It contains the **hypothalamus** which provides a link between the nervous and endocrine systems.

Practice exam questions

1 The diagram summarises the way in which adrenaline can control a chemical reaction in a liver cell.

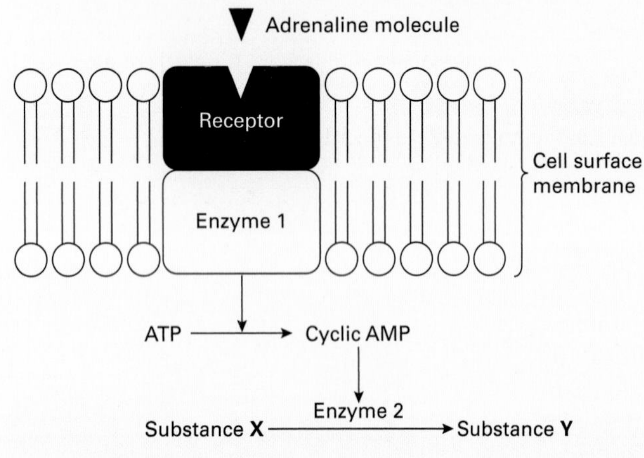

a Describe the function of cyclic AMP in this process. [1]

b Give **one** example of a chemical reaction in a liver cell which is controlled by adrenaline by naming:

i substance **X**; [1]

ii substance **Y**. [1]

c Use the diagram to explain:

i why adrenaline may affect some cells and not others; [1]

ii how a single molecule of adrenaline may cause this cell to produce a large amount of substance **Y**. [2]

[Total 6 marks]

2 The diagram below shows a nerve cell or neurone.

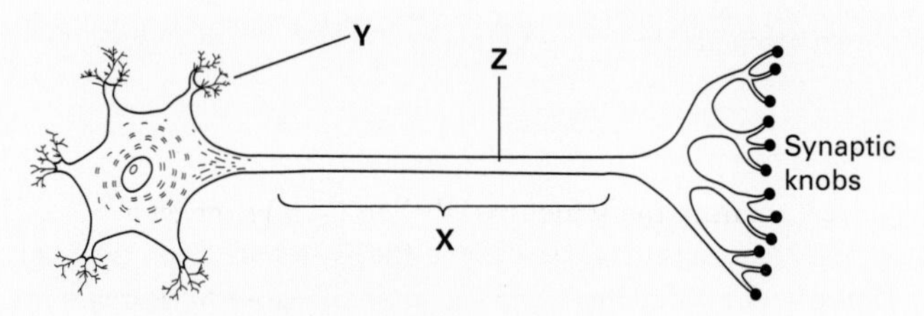

a Name the type of neurone shown. [1]

b Name the structures labelled **X** and **Y**. [2]

c A nerve impulse can be initiated by stimulation with a microelectrode. What would be the effect of stimulation at point **Z**? [1]

d The synaptic knobs release a chemical transmitter, acetylcholine. Nerve gases prevent the breakdown of this chemical. From this information suggest

i one early symptom of nerve gas poisoning, [1]

ii one reason for this observed symptom. [1]

[Total 6 marks]

3 The graph shows the changes in permeability of the cell surface membrane of an axon to sodium and potassium ions during an action potential.

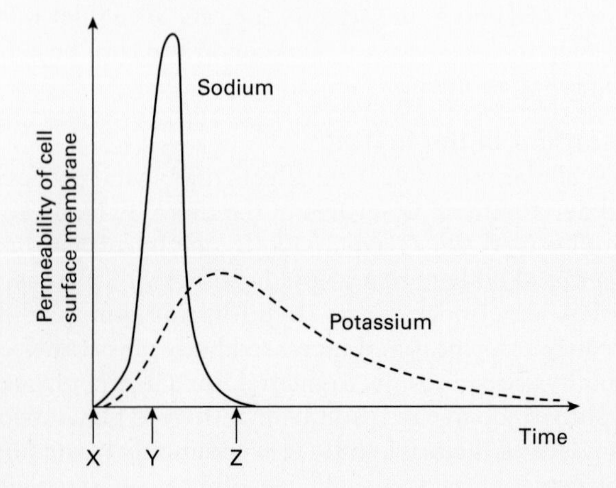

a Explain how the events which take place between times X and Y on the graph can lead to a change in the potential difference across the membrane. [2]

b What happens to the potential difference across the membrane between times Y and Z? [1]

c Explain why a nerve impulse travels faster in a myelinated neurone than in a non-myelinated one. [2]

[Total 5 marks]

4 The diagram below represents the structures visible at a synapse with the aid of electron microscopy.

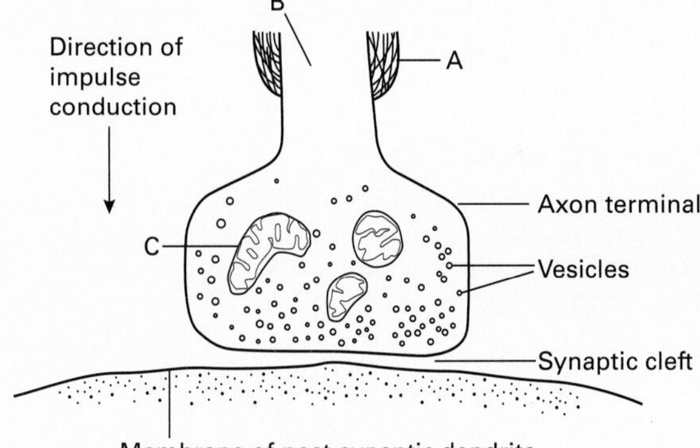

a Identify the structures labelled A and B. [1]

b Name the chemical found in the numerous vesicles that occur in the synaptic knob. [1]

c Identify the structure labelled C and suggest a reason for its presence in the synaptic knob. [2]

d A powerful hydrolytic enzyme is found in the synaptic cleft. What is its function in normal synaptic transmission? [2]

[Total 6 marks]

5 a The list describes the main stages in the process by which information is transmitted across a cholinergic synapse.

Cholinergic synapse

- *An action potential arrives at a synaptic knob of a presynaptic neurone.*
- *This causes .... ions to enter the synaptic knob.*
- *Vesicles move to the .... membrane.*
- *A neurotransmitter called .... is released into the synaptic cleft.*
- *This moves across the cleft by a process known as ....*
- *The neurotransmitter combines with a .... on the postsynaptic membrane.*
- *Influx of .... ions causes local depolarisation and an action potential is set up in the postsynaptic neurone.* [3]

i Copy the list. Using the correct scientific terms, add the words that have been omitted.

ii Explain what happens to the neurotransmitter after it has passed information across a cholinergic synapse. [3]

b Some nerves, especially those of the sympathetic nervous system, produce noradrenaline in their synaptic vesicles. Name this type of synapse. [1]

[Total 7 marks]

6 The diagram represents an enlarged section of part of the retina and choroid of a human eye.

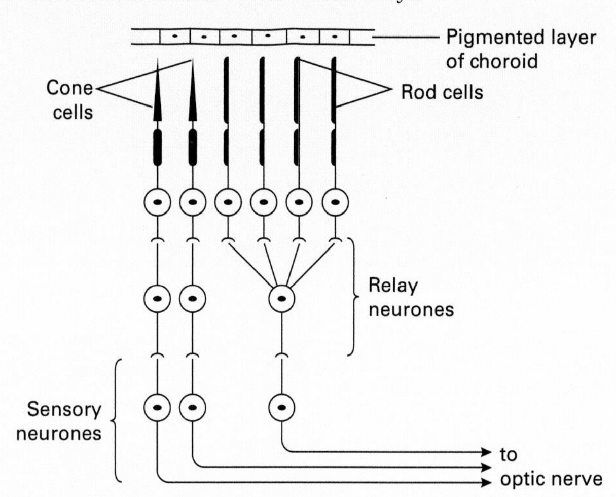

a i Draw an arrow on a sketch of the diagram to show the direction in which light passes through the retina.

ii Suggest a function of the black pigment which occurs in the choroid layer of the eye. [2]

b Use information in the diagram to explain how a person is able to:

i see light of low intensity;

ii see in great detail in bright light. [4]

[Total 6 marks]

7 Copy and complete the following table by stating which region of the brain controls each of the functions listed.

FUNCTION	REGION OF BRAIN
Osmoregulation	
Control of posture	
Modification of heart rate	

[3]

8 The diagram below shows a human brain seen from the right side.

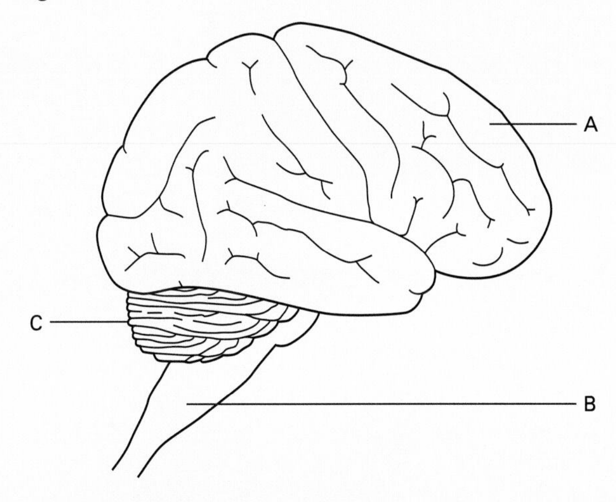

a Name the parts labelled A, B and C. [3]

b Give two functions of the part labelled B. [2]

[Total 5 marks]

11 Locomotion and behaviour

Types of animal skeleton

OBJECTIVES

By the end of this spread you should be able to:

- describe the three main types of animal skeleton
- discuss the functions of skeletons
- state and discuss the advantages and disadvantages of exoskeletons.

Fact of life

All cells are supported internally by a skeletal system of microtubules. Figure 1 is a photograph of cells taken using a technique called immunofluorescence, in which the cytoskeletal proteins in the cells are coated with antibodies carrying a fluorescent dye. When the cells are illuminated with ultraviolet light, the whole of the cytoskeletal frameworks show up brightly against the dark background.

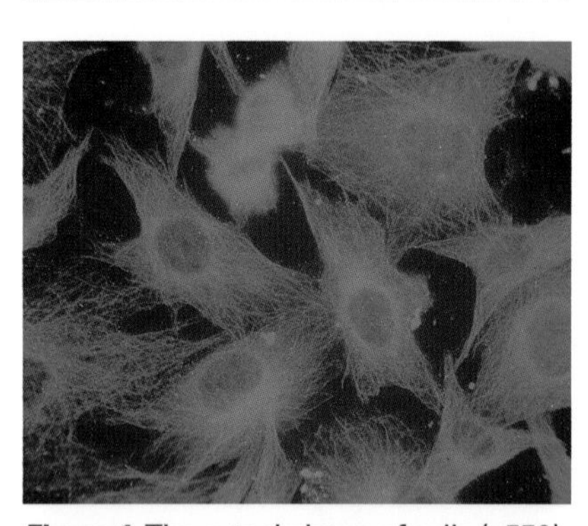

***Figure 1** The cytoskeleton of cells (×550).*

Without a skeleton to support it, an animal would be a jelly-like mass of unsupported cells. Terrestrial animals (animals that live on land) rely on a skeleton even more than aquatic animals, which are given some support by the water surrounding them.

A **skeleton** is any firm structure that gives mechanical support to the body and protection to the softer parts of the organism. Single-celled organisms (and also the cells of multicellular organisms) have a **cytoskeleton**. Multicellular animals also have either a **hydrostatic skeleton**, an **exoskeleton**, or an **endoskeleton**.

Cytoskeleton: support in the cell

Cells have an internal framework, the cytoskeleton, composed of protein microfilaments and microtubules (figure 1). This gives the cell its shape and provides support for cellular extensions such as microvilli. In some cells, the cytoskeleton is involved in intracellular transport. Microtubules going down a nerve fibre, for example, carry material from the cell body to the terminals at the end of the fibre.

The cytoskeleton is a dynamic structure. Its microtubules are continually being disassembled, reassembled, and rearranged to support the ever-changing shape of a cell. Some microtubules contain **actin**, a contractile protein also found in muscle cells. These microtubules appear to be involved in many processes including the movement of chromosomes during mitosis and meiosis, phagocytosis, and the movement of secretory vesicles to the cell surface membrane.

Hydrostatic skeleton: supporting with water

A hydrostatic skeleton is formed by internal watery fluids contained within confined spaces in the body.

Water has three important properties that make it suitable to act as a skeleton:

- it is relatively incompressible and can make a soft-walled structure rigid, so that muscles can act against it
- it can transmit pressure changes equally in all directions
- it has a low viscosity which enables its shape to change with the shape of its container as the water moves from one structure part to another.

Hydrostatic skeletons are used in many soft-bodied organisms for locomotion. The organisms include *Hydra*, sea anemones and other members of the Cnidaria (spread 21.9), and earthworms.

The earthworm's hydrostatic skeleton is formed in a fluid-filled space called the **coelom** (figure 2). The coelom is surrounded by two antagonistic layers of muscle, circular muscle and longitudinal muscle, which contract to cause pressure changes in the coelom. The earthworm's body is divided into segments, so individual parts of the body can change in shape, resulting in different pressures in different parts of the coelom. When the circular muscles contract, segments become thinner and longer. When the longitudinal muscles contract, segments become fatter and shorter.

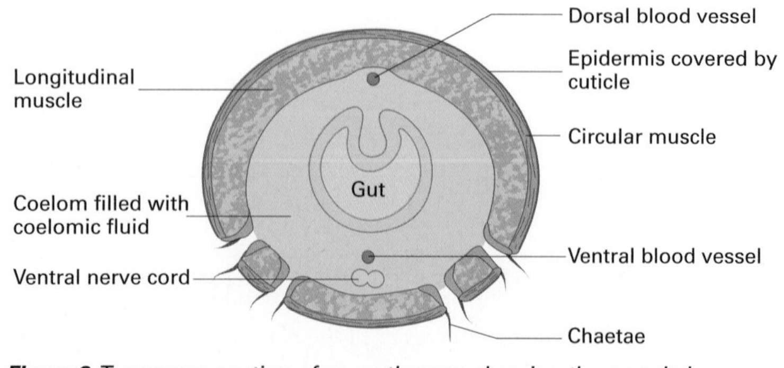

***Figure 2** Transverse section of an earthworm showing the muscle layers and the coelom.*

During locomotion, the circular and longitudinal muscles contract in turn to produce peristaltic waves along the body, beginning at the front end and working backwards. **Chaetae** (small, stiff bristles sticking out of the side of the body) anchor parts of the body to the ground so that other parts can be pushed away from or pulled towards these points (figure 3).

Hydrostatic skeletons are also used to extend soft structures, such as the siphon of a mollusc and the penis of a mammal.

Exoskeleton: support outside the body

An exoskeleton lies outside the muscles that are attached to it. Exoskeletons are found in all arthropods, including insects and crustacea.

The exoskeleton of an insect covers the body surface and is called a **cuticle**. It is a non-living structure secreted by cells in the epidermis which lie beneath it. The cuticle has two main layers: an outer **epicuticle**, and an inner **procuticle**. The epicuticle is very thin and is covered with wax. Its main function is to prevent water loss. The procuticle can be further divided into two layers: an outer rigid **exocuticle** and an inner more flexible **endocuticle**. Both contain **chitin**, a very tough structural polysaccharide.

Figure 4 shows how an antagonistic pair of muscles moves the jointed parts of an insect. The muscles are attached to **apodemes**, special projections on the inner surface of the cuticle. The hard epicuticle and rigid exocuticle are absent from the joints, leaving only the flexible endocuticle. This allows joints to bend easily. Although each joint can move in one plane only, successive joints in an insect's limb bend in different planes, giving a wide range of movements. This joint mobility is particularly important for insects that fly (spread 21.14).

The jointed exoskeletons of arthropods provide an excellent means of locomotion. They also offer good support and protection for small animals. However, they are unsuitable for large animals because of the surface area to volume relationship. As an animal increases in size, its body volume and therefore body mass increase by the power of 3, but its surface area only goes up by the power of 2 (see spread 4.7). The strength of an exoskeleton depends on its surface area, so a large animal would need a disproportionately large and unwieldy exoskeleton to support its body mass.

Arthropod exoskeletons also suffer the serious disadvantage of not being alive, so they are incapable of growth. Consequently, arthropods have to shed their exoskeleton periodically (a process called **moulting** or **ecdysis**) in order to increase in size. During moulting, arthropods are particularly vulnerable to predation.

Endoskeleton: support from the inside

Endoskeletons are typified by the bony skeletons of mammals. These are described in the next spread.

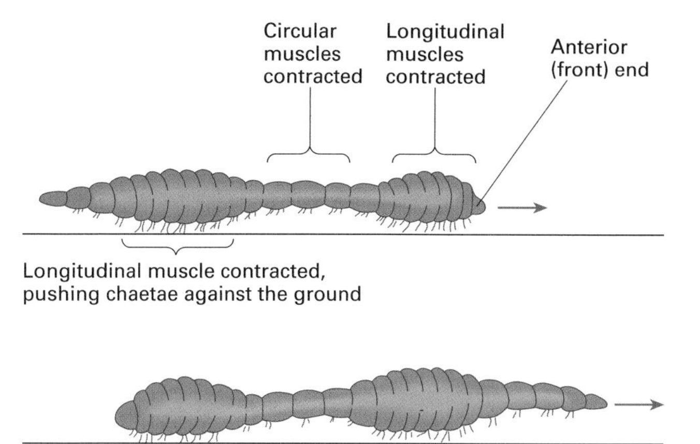

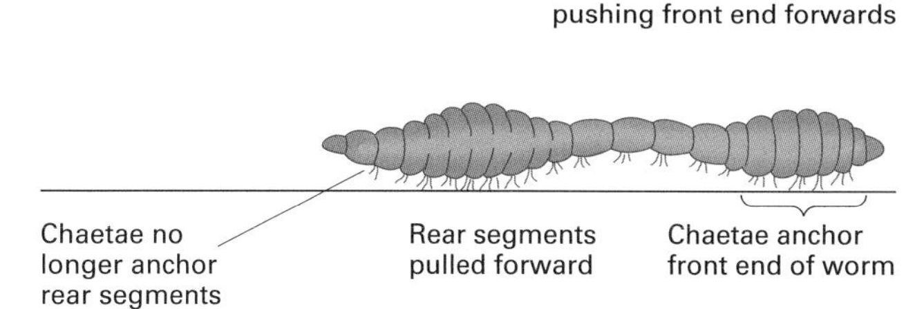

Figure 3 *Locomotion in an earthworm.*

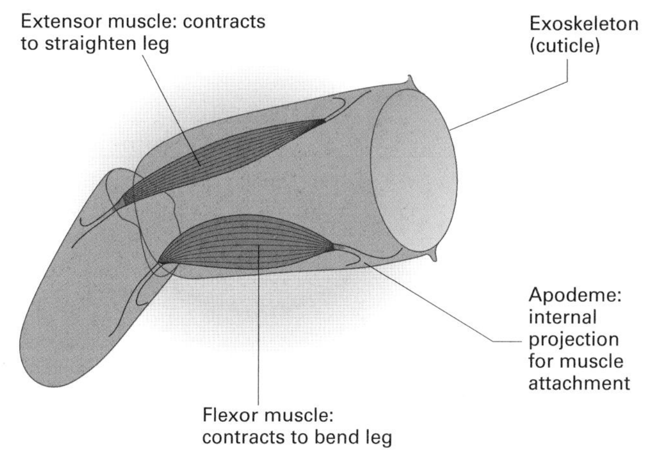

Figure 4 *One joint of an insect's limb showing the attachment of its antagonistic muscles.*

Quick check

1 The cytoskeleton is a dynamic structure, continually changing its shape. Which one of the three main types of animal skeleton shares this property?

2 List three functions of the exoskeleton of an arthropod.

3 Exoskeletons are an efficient means of supporting small animals. Explain briefly why an animal the size of a dog would require a disproportionately large and unwieldy exoskeleton to support its body mass.

Food for thought

One of the major disadvantages of having an exoskeleton is the need to moult. Suggest the problems an arthropod such as a crab or insect faces during the process of moulting, and how the animal might overcome or minimise these problems.

11.2

VERTEBRATE ENDOSKELETONS

OBJECTIVES

By the end of this spread you should be able to:

- describe the main functions of a vertebrate endoskeleton
- distinguish between cartilage and bone
- describe the histology of compact bone
- list the main parts of the vertebral column.

Fact of life

The human skeleton consists of about 200 bones and makes up approximately 20 per cent of body mass.

All vertebrates have an **endoskeleton**: a rigid framework of bone and cartilage within the body tissue to which muscles are attached. The functions of the vertebrate endoskeleton are that it maintains body shape, supports soft tissues, and acts as a system of levers for locomotion. Some parts manufacture blood cells and store minerals, such as calcium salts.

Cartilage and bone

Vertebrate skeletons are made of living material that can grow with the animal. In all vertebrate embryos, the main skeletal material is **cartilage**. This is a firm, flexible material which can support great weight. In some vertebrates (for example, sharks) cartilage is the main skeletal material for adults too. In mammals and most other vertebrates it is largely replaced by **bone**, a firmer and denser material. In adult mammals cartilage occurs where firmness combined with flexibility is required, for example, at the ends of bones, in the external ear, at the end of the nose, and in the stiffening rings of the trachea and larynx.

Cartilage and bone are forms of **connective tissue**, which is tissue largely consisting of the material between cells (the **intercellular matrix**). Cartilage has relatively few cells which lie in cavities called **lacunae** embedded in an intercellular matrix of fibres and non-living material (figure 1).

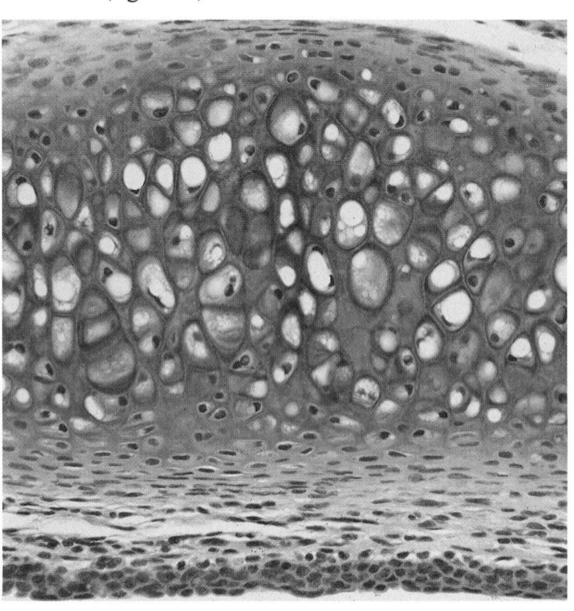

Figure 1 Photomicrograph of hyaline cartilage, the most common type of cartilage. The cartilage cells (chondrocytes) lie inside spaces called lacunae. The hyaline matrix contains fibres of collagen.

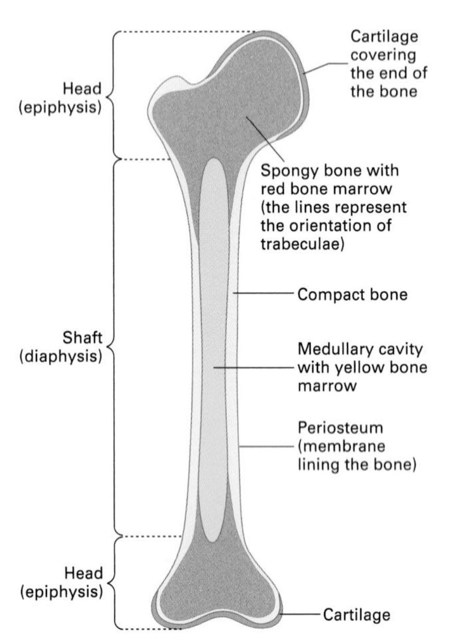

***Figure 2** Structure of a long bone: vertical section through the femur.*

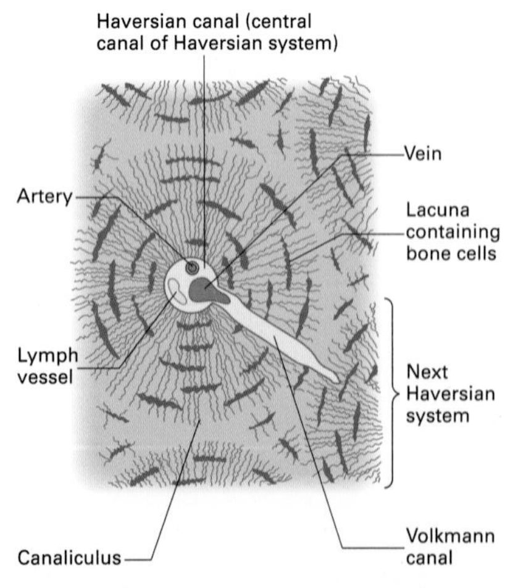

***Figure 3** The microscopic structure of compact bone showing the Haversian system.*

Bone has a hard, relatively rigid matrix, with many collagen fibres. The matrix contains a relatively large amount of water and is impregnated with inorganic salts such as calcium carbonate and calcium phosphate. The inorganic material may make up up to 65 per cent of the dry mass of bone. Like cartilage, bone has few cells which are located in lacunae in the matrix.

Types of bone

Some bony material is called **spongy bone** because it consists of a network of hardened bars called **trabeculae** between which are large spaces filled with **red bone marrow**, a soft tissue made by cells that also form blood cells and lymph cells. **Compact bone** is another type of bone; this is an almost continuous mass of cells and matrix, and has only microscopic spaces. Compact bone is harder and much denser than spongy bone. It forms the main bulk of the shafts of typical long bones, such as those of the upper arm and thigh (figure 2). In these bones, the compact bone surrounds a large central cavity called the **medullary cavity**. This cavity is filled with **yellow bone marrow**, which produces white blood cells and is more fatty than the red bone marrow filling the spaces in spongy bone.

Combinations of spongy bone, compact bone, and cavities allow bones to be both strong and hard, but also light. Bone is particularly resistant to compressive (squashing) forces so it is well suited to supporting the bodies of terrestrial vertebrates. It is, however, relatively weak under tensile (pulling) forces.

The structure of compact bone

Compact bone is made of structural units called **Haversian systems** (figure 3). Each system forms an irregular cylinder with layers of matrix called **lamellae**. Dotted throughout the matrix are bone cells in their lacunae. Non-dividing and inactive bone cells are called **osteocytes**; dividing bone cells that can lay down a new matrix are called **osteoblasts**; and bone cells that can reduce the size of bones by reabsorbing the matrix are called **osteoclasts**.

Osteocytes are connected to each other and to a central canal (the **Haversian canal**) by radiating **canaliculi** (microscopic channels) filled with tissue fluid. The canaliculi transport material to and from blood vessels in the Haversian canal, which also contains nerves, blood vessels, and lymph vessels. The nerves and vessels are carried down from the bone surface to the Haversian systems by Volkmann canals.

The human skeleton

The human skeleton is typical of mammalian skeletons (figure 4). It can be divided into two components: an **axial skeleton**, which is the skull and vertebral column with the rib cage; and the **appendicular skeleton**, which includes the bones of the paired limbs and the pectoral and pelvic girdles.

The vertebral column is a series of small bones forming a flexible and supportive structure down the back. The bones, called **vertebrae**, occur in five regions, as shown in figure 4. Between the vertebrae are intervertebral discs of cartilage. The discs have a hard outer surface but a softer, gelatinous centre which cushions the vertebrae during locomotion.

Quick check

1 List the main functions of a vertebrate skeleton.

2 Name two features which distinguish cartilage from bone.

3 List three structures contained within a Haversian canal.

4 What is the total number of bones in a typical human vertebral column?

Food for thought

Bone is living tissue. Human adult long bones, such as the femur, do not lengthen. However, according to Wolff's law, changes in the function of bones are followed by changes in the internal structure and external shape of the bone. In a mature bone, the bone elements are built up or broken down and its mass increases, or decreases, according to the mechanical demands imposed on it. Suggest how this might affect the bones of people:

a travelling in space

b having an active lifestyle

c having a sedentary lifestyle.

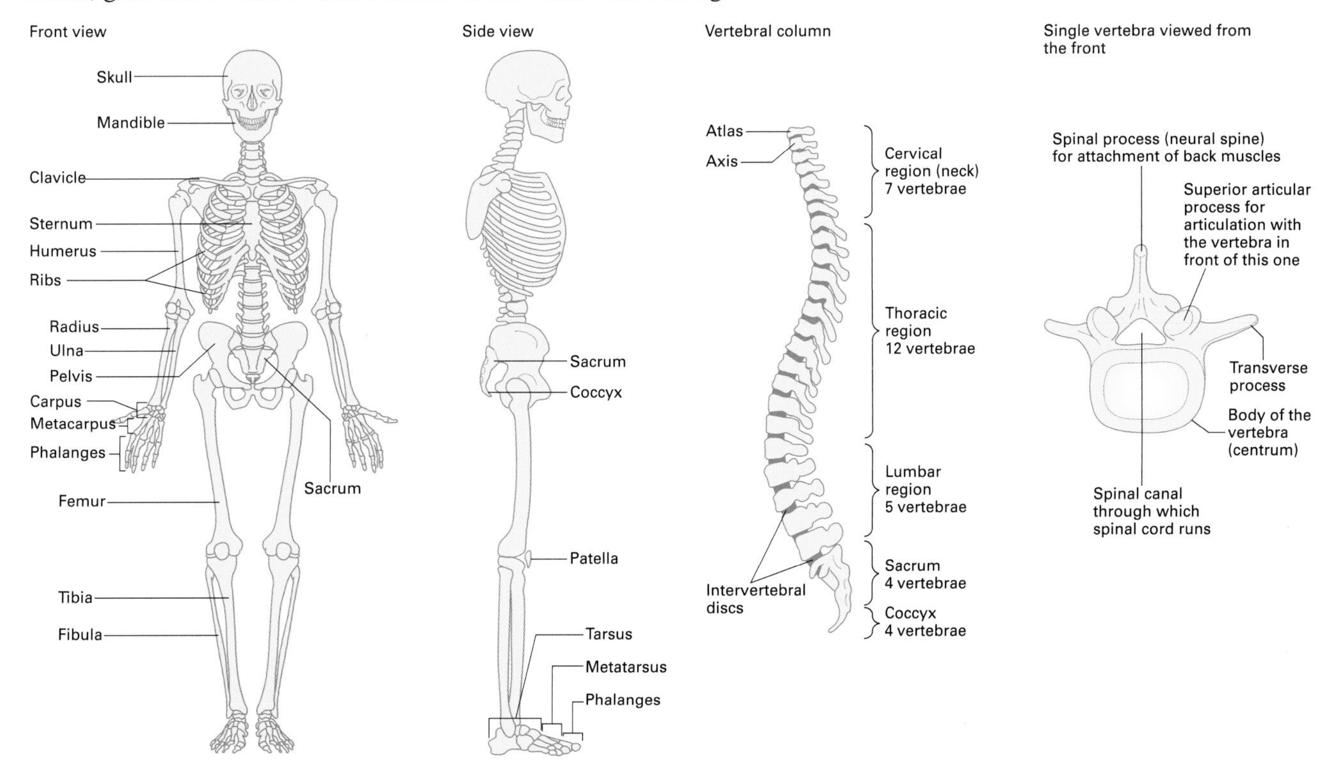

Figure 4 *The human skeleton.*

11.3

Muscles and joints

OBJECTIVES

By the end of this spread you should be able to:

- describe the main types of mammalian muscle
- describe the features of a synovial joint
- explain the role of antagonistic muscles.

Fact of life

The elastic components in the feet and ankles of running animals store energy after impact with the ground, increasing running efficiency. In humans, the Achilles tendon acts as a spring, stretching by about 6% of its length and returning over 90% of its stored energy. It has been estimated that the efficiency of conversion of chemical energy to mechanical energy is increased from 25% to 40% or more by the elasticity of tendons.

Muscle tissue consists of long cells called **muscle fibres** which have the remarkable ability to change their length and produce tension (pull). This is the basis of the main function of muscle: to move the whole body or parts of it.

There are three main types of muscle: cardiac muscle, smooth muscle, and skeletal muscle (figure 1).

Cardiac muscle

Cardiac muscle forms the bulk of the heart (spread 7.5). Cardiac muscle fibres appear lightly striped or striated under the light microscope and are branched, forming a net-like arrangement. Each cell has one nucleus and the cells are joined to each other by tight, folded junctions called **intercalated discs**. Cardiac muscle is very odd in that it is **myogenic**: it contracts rhythmically even when removed from the body. This inherent rhythm is the basis of the heartbeat (spread 7.6). Cardiac muscle is unique in never suffering fatigue. However, its muscle fibres will stop contracting if they are starved of oxygen or bathed in fluid of the wrong chemical composition.

Smooth muscle

Smooth muscle consists of numerous slender, spindle-shaped muscle fibres, each with a single nucleus. Smooth muscle fibres occur in many parts of the body, sometimes singly or in small groups (for example, in the skin), but often massed together in sheets, bundles, or rings (for example, in the alimentary canal and in the lining of the uterus). Rings of smooth muscle fibres are called **sphincter muscles**. Sphincter muscles occur in the gut: the pyloric sphincter controls the passage of food from the stomach to the duodenum, and the anal sphincter controls the elimination of faeces from the body. Small sphincter muscles surround some blood vessels, helping to regulate blood pressure and control the distribution of blood.

When massed together, smooth muscle fibres usually contract slowly and rhythmically, and they are slow to fatigue.

Smooth muscle is sometimes called **involuntary muscle** because its contractions are usually controlled unconsciously. However, humans can learn to control many smooth muscles, such as those of the anal sphincter. After intense training, a few people can even control the sphincter muscles around blood vessels so that they do not bleed if a pin is stuck into their hand. Some dentists use hypnosis on haemophiliac patients who bleed a lot during surgery, and here an appropriate hypnotic suggestion controls the sphincter muscles and reduces bleeding.

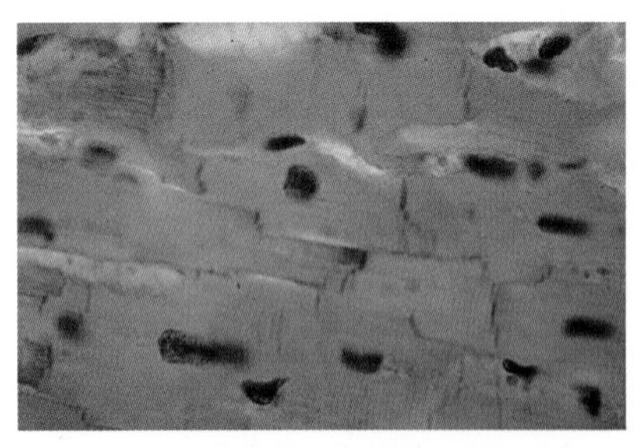

(a)

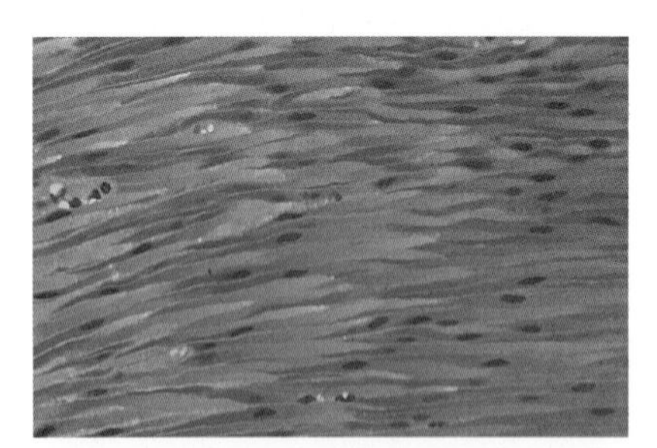

(b)

(c)

***Figure 1** Photomicrographs of: **(a)** cardiac muscle; **(b)** smooth muscle; **(c)** skeletal muscle.*

Skeletal muscle

Skeletal muscle is so named because it is attached to bones. It is responsible for the voluntary movements of the body. When skeletal muscles contract, they pull bones closer together or allow body parts to resist external forces. Skeletal muscle is also called striated muscle as it appears striped or striated under a microscope.

Each muscle consists of thousands of contractile muscle fibres wrapped in connective tissue, supplied with blood vessels and nerve fibres. Individual skeletal muscle fibres have many nuclei.

Vertebrate joints and locomotion

Skeletal muscles are attached to at least two bones by **tendons**. These are tough, strong strands of connective tissue: mainly collagen fibres, providing strength, along with elastin fibres, providing elasticity. The two bones are connected at a **joint** and held in place by **ligaments**. The contraction of a muscle moves one bone relative to the other. The **insertion**, the tendon attachment point furthest from the body mid-line, usually moves while the **origin**, the tendon attachment point closest to the body mid-line, is fixed.

A muscle can only shorten and pull on a bone. Once shortened, it can be extended back to its original length only by the action of another muscle. This is illustrated by movements of the arm (figure 2). The biceps at the front of the arm and the triceps at the back act as an **antagonistic pair** of muscles, that is, the muscles work against each other, moving the arm in opposite directions.

Joints can be classified into three types according to their structure and the amount of movement they permit.

- **Fibrous joints** (for example, sutures between the bones of the skull) hold bones tightly together by connective tissue consisting of short, tough fibres. They permit no movement.
- **Cartilaginous joints** are held together by cartilage. They include intervertebral discs which hold vertebrae together by fibrous cartilage.
- **Synovial joints** contain a cavity filled with fluid and they move freely in one or more planes (figure 3 and table 1).

***Table 1** The main parts of a synovial joint and their functions.*

Structure	Main functions
Ligament	Binds bone to bone and keeps joint stable
Tendon	Binds muscle to bone and keeps joint stable
Joint capsule	Encloses the joint cavity
Synovial fluid	Acts as a lubricant, reducing friction; supplies nutrients; is load bearing
Synovial membrane	Secretes synovial fluid
Articular cartilage	Provides a smooth surface for joint movement; absorbs mechanical shocks; spreads forces

Types of synovial joint

There are five main types of synovial joint:

1 **ball-and-socket joints** (for example, between the femur and pelvic girdle in the hip) allow movement in all three planes

2 **hinge joints** (for example, between the ulna and humerus in the elbow) allow movement mainly in one plane (bending)

3 **pivot joints** (for example, that between the radius and ulna) allow rotation around one axis only

4 **gliding joints** (for example, between the shoulder blade and collar bone) allow only sliding movement

5 **saddle joints** (for example, between the carpal and metacarpal bones in the wrist) allow rotation about two separate axes.

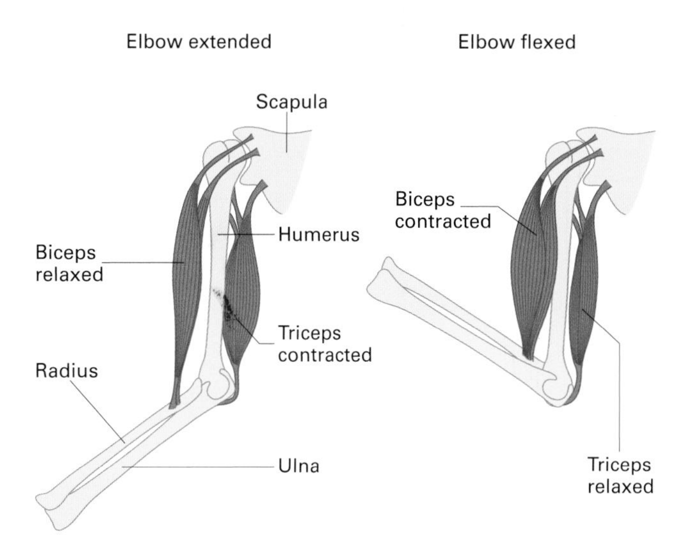

***Figure 2** The antagonistic muscles of the upper arm.*

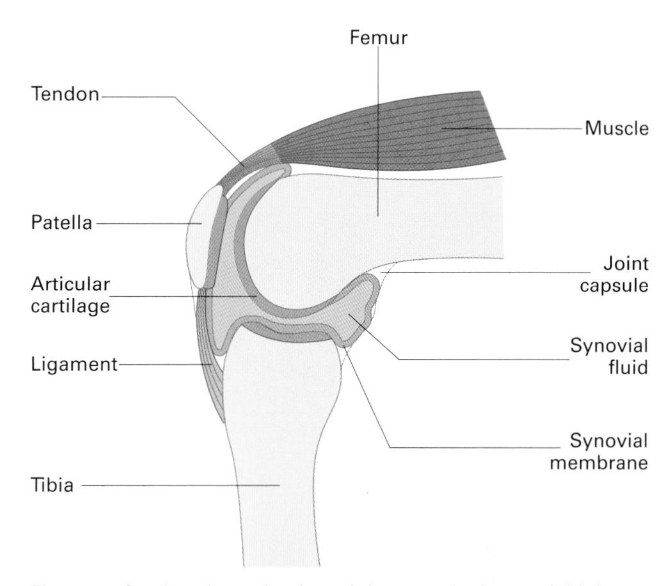

***Figure 3** Section through a knee joint: a typical synovial joint. The functions of its component parts are shown in table 1.*

Quick check

1 List the three main types of muscle in order according to their tendency to fatigue, starting with the muscle type that fatigues least.

2 What is the main function of articular cartilage in a synovial joint?

3 Name:

a the flexor

b the extensor

that form the antagonistic pair of muscles in the upper arm.

Food for thought

According to McCutchen's Weeping Lubrication Theory, synovial fluid is squeezed in and out of the articular cartilage at points of contact during load-bearing movements. Suggest the significance of this and the implications it has for people who have a sedentary lifestyle.

11.4

Skeletal muscle

OBJECTIVES

By the end of this spread you should be able to:

- describe the structure of skeletal muscle
- explain the function of a motor unit
- distinguish between slow-twitch and fast-twitch muscle fibres.

Fact of life

Body-builders train regularly with heavy weights and have a special diet to develop the size and appearance of muscles. Good training routines can reduce fat levels, and increase flexibility and strength. However, body-builders sometimes sacrifice fitness and health for rapid improvements in appearance by using poor diet and training routines, supplemented with anabolic steroids (body-building chemicals).

Even though our muscles may not be as large and well formed as those of Arnold Schwarzenegger (figure 1), we all have about 600 muscles that share a similar internal structure. In this spread we are going to look at a typical muscle, the biceps or upper arm muscle which bulges when the forearm is flexed (figure 2).

Looking at the biceps

The biceps (figure 2) is a large fleshy organ covered by a sheath of connective tissue. It is a spindle-shaped muscle attached to the scapula by two tendons and to the radius by one tendon. The biceps is richly supplied with blood vessels: arteries and veins branch repeatedly into numerous capillaries, forming vast networks in and around the muscle, so that all parts of the muscle have a good supply of oxygen and nutrients. To make muscles look bigger, body-builders repeatedly contract and relax their muscles to increase the volume of blood flowing through them, a procedure called muscle pumping.

Nerves containing both sensory and motor neurones enter the muscle along with the blood vessels. The nerves branch many times to reach all parts of the muscle. Motor fibres coming from the central nervous system (brain and spinal cord) control the amount of tension in the muscle. Sensory neurones carry information from pain receptors and **proprioceptors** in the muscle to the central nervous system. Proprioceptors monitor the level of tension in the muscle and provide information about the orientation and movement of body parts.

Figure 1 Arnold Schwarzenegger (right). He later became a film actor and then Governor for California.

Muscle fibres in the biceps

Under a light microscope we can see that the biceps is made up of many thousands of small cells called **muscle fibres** (figure 3). Like other cells, a muscle fibre has cytoplasm (called **sarcoplasm**), an internal membrane system (the **sarcoplasmic reticulum**), and a cell surface membrane (the **sarcolemma**). However, it is peculiar in having many nuclei (being multinucleated) and being very long. Three or four blood capillaries surround each muscle fibre in a sedentary person, while up to seven capillaries supply each fibre in a trained athlete.

At this level of magnification, we can see that the axon of each motor neurone branches to supply up to 150 muscle fibres. All the muscle fibres served by the same motor neurone are called a **motor unit** because they work as a unit, contracting or relaxing at the same time. The motor unit is the basic functional unit of skeletal muscle. Muscles that perform delicate movements may have as few as one muscle fibre per motor unit. Muscles that perform heavy work such as the quadriceps in the thigh may have more than 150 muscle fibres in a motor unit.

Each branch of an axon terminates at a plate-like structure called a **neuromuscular junction** (also called a myoneural junction or motor endplate). A neuromuscular junction forms the neurone-to-muscle synapse, the connection between a neurone and a muscle fibre. As with neurone-to-neurone synapses (spread 10.6), the neuromuscular junction has a small gap between the neurone and muscle fibre called a synaptic cleft. The nerve impulse is transmitted across the cleft by the neurotransmitter **acetylcholine**.

Slow-twitch and fast-twitch fibres

Muscle sections can be stained with special dyes that show two main types of muscle fibre. They have different colours and contract at different speeds. The two types of muscle fibres are **slow-twitch** or **red fibres** and **fast-twitch** or **white fibres**.

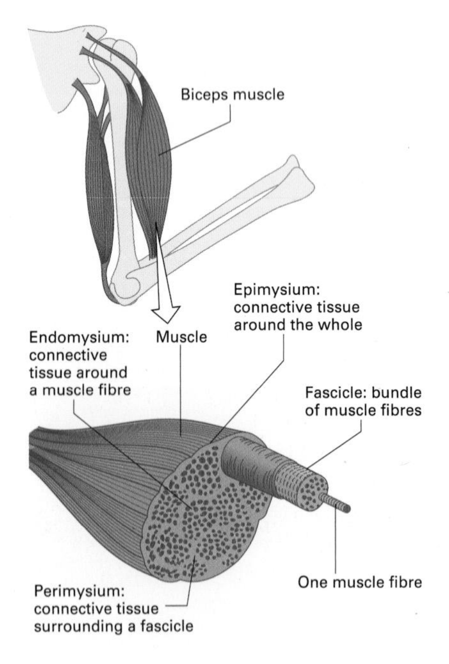

Figure 2 The biceps muscle: a spindle-shaped muscle made up of contractile muscle fibres covered with connective tissue.

Slow-twitch fibres are adapted to function over long periods. They respire aerobically to avoid the build-up of lactic acid which would quickly fatigue them. They have their own metabolic fuel (muscle glycogen) but, because they are aerobic, they can also use the almost limitless supply of fat stores in the body. Their high content of myoglobin and good blood supply means that they can obtain sufficient oxygen, and their high density of mitochondria ensures that they can use oxygen efficiently to generate large amounts of ATP. However, a disadvantage of aerobic metabolism is that they cannot generate ATP at a very fast rate, therefore slow-twitch fibres are not very powerful.

Fast-twitch fibres are adapted for short bursts of explosive action. They generate ATP quickly and anaerobically from stores of a high-energy compound, **creatine phosphate** (CP), and by lactate fermentation (spread 6.2). When CP breaks down it releases energy and phosphate ions which can be used to make ATP for up to 10 seconds of activity; CP is regenerated during aerobic respiration. Although fast-twitch fibres have a relatively low myoglobin content and a small number of mitochondria, they are rich in the enzymes required for anaerobic respiration. However, during lactate fermentation, they can only use glycogen as a fuel, and the lactate and hydrogen ions they produce makes them fatigue quickly.

Most people have roughly equal numbers of slow- and fast-twitch fibres, but the proportion varies in trained athletes: endurance athletes tend to have more slow-twitch fibres while power athletes tend to have more fast-twitch fibres.

Muscle fibres under the electron microscope

Looking at a single muscle fibre under an electron microscope reveals that it is made up of a bundle of smaller fibres called **myofibrils**. Skeletal muscle appears striated (striped) under a microscope because of the combination of these myofibrils causing alternate light and dark bands (figure 4). A myofibril consists of repeating units called **sarcomeres**.

A sarcomere is the region between two dark lines called **Z lines**. The sarcomere is the fundamental unit of action of a muscle fibre. It contains two kinds of filament called thin filaments and thick filaments. A **thin filament** is made up of a double strand of the protein **actin**, along with one strand of a regulatory protein, coiled around each other. Each **thick filament** is made up of parallel strands of the protein **myosin**. The thin and thick filaments are arranged within the sarcomere in a way that produces bands of light and dark in electron micrographs. The broad dark band (also known as the **A band**) consists of thick filaments interspersed with thin filaments, except in a central region (the **H zone**) where only thick filaments occur. The light band (the **I band**) contains only thin filaments along with the proteins in the Z line that connect adjacent thin filaments.

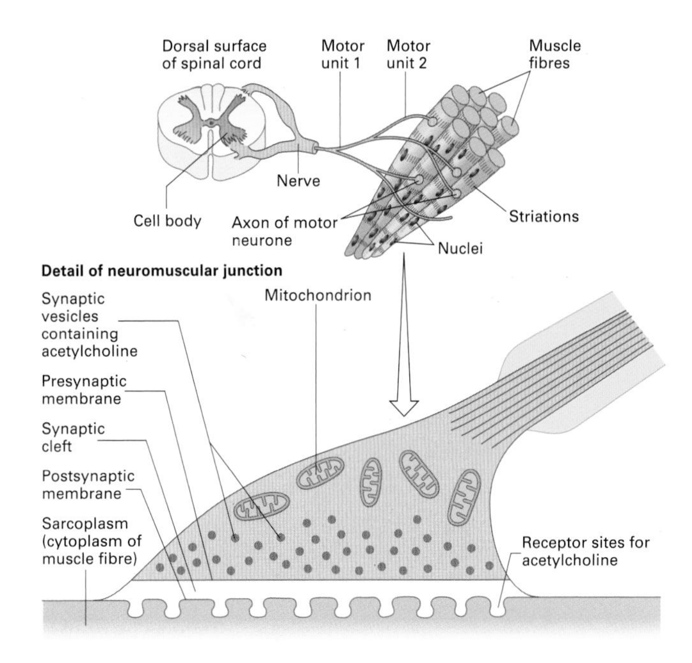

Figure 3 *Muscle fibres and the motor nerves that supply them.*

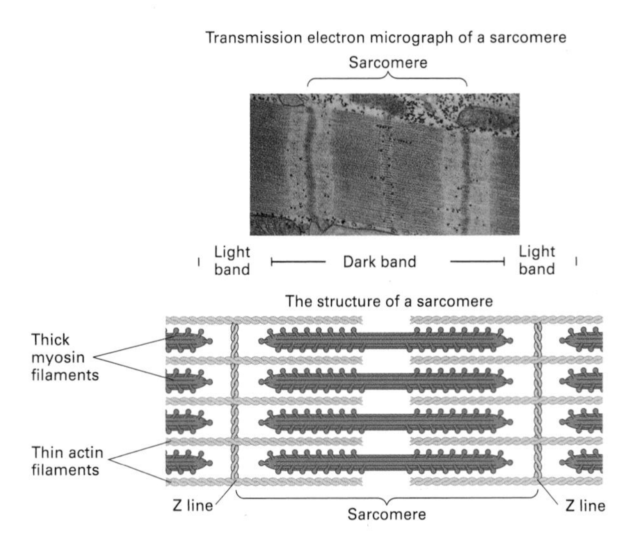

Figure 4 *A sarcomere within a muscle fibre as seen under the electron microscope.*

QUICK CHECK

1 What is a sarcomere?

2 What major difference would you expect between motor units that perform delicate movements and those that perform powerful movements?

3 Which will have more fast-twitch muscle fibres and which will have more slow-twitch muscle fibres:

a the arm muscles of a shot-putter

b the leg muscles of a cross-country skier?

Food for thought

You can improve your strength with regular load-bearing exercises (for example, push-ups, sit-ups, and weight training). The training brings about changes in muscles, nerves, and blood vessels. Suggest what these changes might include.

11.5

How muscle fibres contract

OBJECTIVES

By the end of this spread you should be able to:

- describe the sliding filament theory of muscle contraction
- distinguish between temporal summation and muscle fibre recruitment.

Fact of life

Muscle makes up to 40 per cent of human body mass, and a single muscle may have as many as 400 000 muscle fibres.

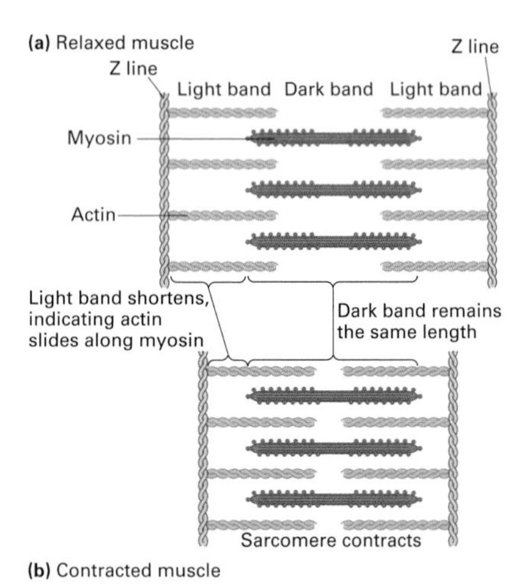

***Figure 1** A sarcomere in **(a)** relaxed **(b)** contracted muscle.*

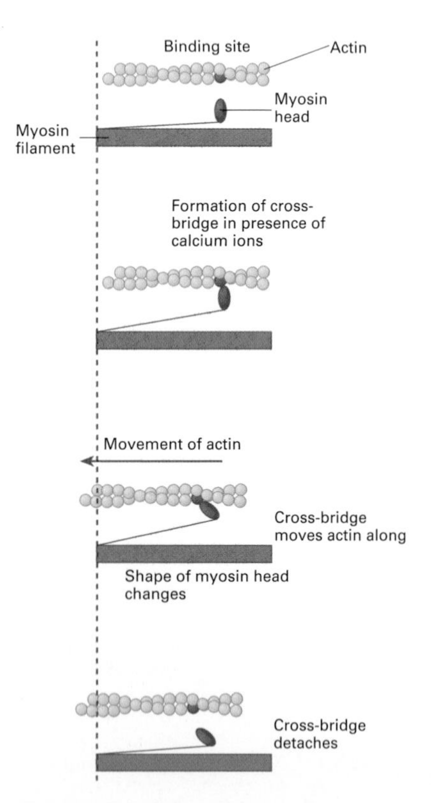

***Figure 2** The sliding filament theory. Cross-bridges attach and reattach 50 to 100 times per second, but only if ATP is available.*

In the previous spread we saw that a single skeletal muscle is made up of many muscle fibres (muscle cells), each of which is divided into many myofibrils. Myofibrils contain sarcomeres, repeating units of thick and thin filaments. These filaments are the contractile apparatus of muscles.

The sliding filament theory

In the 1950s, Andrew Fielding Huxley and other researchers used the electron microscope to study sarcomeres of relaxed and contracted muscles (figure 1). It appears that during contraction, the Z lines and the thin filaments slide towards the middle of the sarcomere. The sarcomere shortens, but the lengths of the thick and thin filaments do not change.

From these studies, Huxley proposed the **sliding filament theory** of contraction. According to the theory, a sarcomere shortens when its thin filaments slide along its thick filaments (figure 2). High-magnification electron micrographs show that the myosin filaments are rod shaped with a globular end (called the **myosin head**). The head can form a cross-bridge with actin, the protein in the thin filaments. When attached to actin, a myosin head can change shape and slide the actin further along the myosin. During a muscle contraction, these cross-bridges are formed and broken down repeatedly up to 100 times per second, causing the sarcomere to shorten by a ratchet-like mechanism. The mechanism can only shorten a sarcomere. It cannot actively return the sarcomere to its original length; muscle elongation is usually brought about by the action of antagonistic muscles (spread 11.3, figure 2). Calcium ions are required for cross-bridges to form and the breakdown of ATP provides the energy needed by the ratchet mechanism. The combined actions of millions of sarcomeres can contract a whole muscle to about half its resting length. Contraction is started by a nerve impulse which triggers the availability of calcium ions and ATP.

The role of calcium ions

When a muscle is at rest, calcium ions are not present in the sarcoplasm because they are stored in the **sarcoplasmic reticulum**, fine membrane-bound channels in the muscle fibres. In the absence of calcium ions in the sarcoplasm, **tropomyosin** (a protein in thin filaments) prevents the myosin heads from attaching onto actin by blocking the binding sites. When a muscle is stimulated sufficiently by nerve impulses, calcium ions are released from the sarcoplasmic reticulum and combine with **troponin** (another protein in thin filaments), causing the tropomyosin to change shape and unblock the binding sites.

Calcium ions are released from the sarcoplasmic reticulum at the end of a sequence of events which begins when an action potential reaches a neuromuscular junction. When the action potential arrives, calcium ion channels are opened in the membrane of the nerve fibre and calcium ions diffuse into the synaptic cleft. This causes synaptic vesicles to move to the junction membrane and fuse with it, releasing acetylcholine into the synaptic cleft. The acetylcholine diffuses across the cleft and attaches onto receptor molecules on the muscle fibre membrane. This attachment of acetylcholine changes the membrane permeability, and leads to a localised depolarisation of the postsynaptic membrane called a **graded potential**. Unlike an action potential (spread 10.4), a graded potential does not obey the all-or-none law: its amplitude increases with the intensity of a stimulus. When the graded potential exceeds the threshold level, an action potential sweeps across the muscle fibre and passes into membranous tubules called **T tubules** or transverse tubules that fold inwards from the sarcolemma. Where the T tubules make contact with the sarcoplasmic reticulum, the action potential causes the sarcoplasmic reticulum to release calcium ions into the sarcoplasm. Calcium ions spread through the sarcoplasm, enabling myosin heads to bind onto actin. Energy from the breakdown of ATP enables the heads to take up a new position.

When action potentials stop arriving, calcium is actively pumped back into the sarcoplasmic reticulum, tropomyosin blocks the myosin head binding sites on the actin, and the muscle relaxes.

Contraction of a whole muscle

A single motor unit (one or more muscle fibres served by the same motor neurone) obeys the all-or-none law: it either contracts completely or not at all. However, a whole muscle can produce graded responses by two mechanisms called temporal summation and muscle fibre recruitment.

Temporal summation can be demonstrated by stimulating an isolated muscle electrically (figure 3). If the isolated muscle is given a single electrical stimulus, it will produce a **simple twitch**. If a second stimulus is given before the first twitch is over, the muscle tensions will add together to produce a greater response. If the rate of stimulation is fast enough, the twitches will fuse to produce a smooth sustained contraction called **tetanus**. In normal situations, muscles produce smooth tetanic contractions rather than the jerky movements of muscle twitches.

In **muscle fibre recruitment**, the amount of tension produced in a muscle is altered by changing the number of motor units activated. As more force is needed, more muscle fibres are stimulated. Slow-twitch fibres are recruited for small muscular forces and more and more fast-twitch fibres are recruited as the force reaches maximum levels (figure 4).

Under normal conditions, the nervous system does not recruit all of the available fibres even during maximal voluntary efforts. This prevents damage to muscles and tendons. The potential power of human muscles has been demonstrated many times by people exerting apparently superhuman efforts when confronted with a crisis. For example, apparently weak grandmothers have been known to lift a car to release a trapped grandchild.

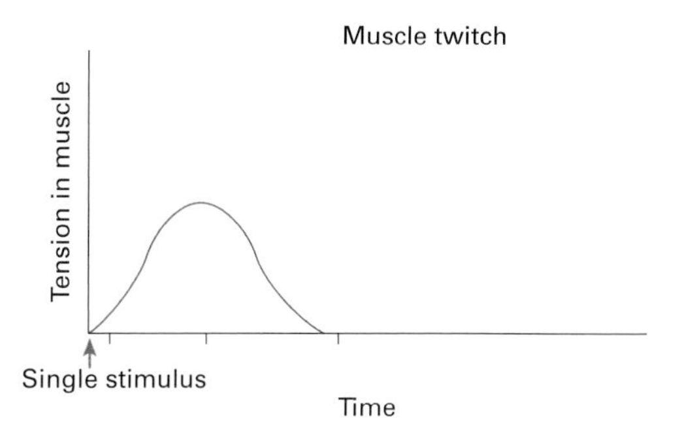

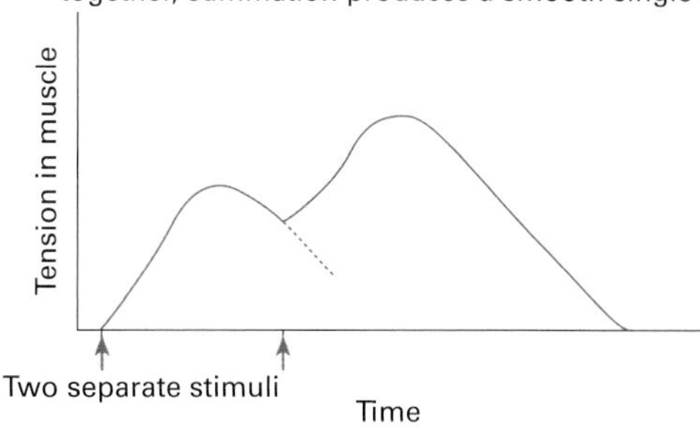

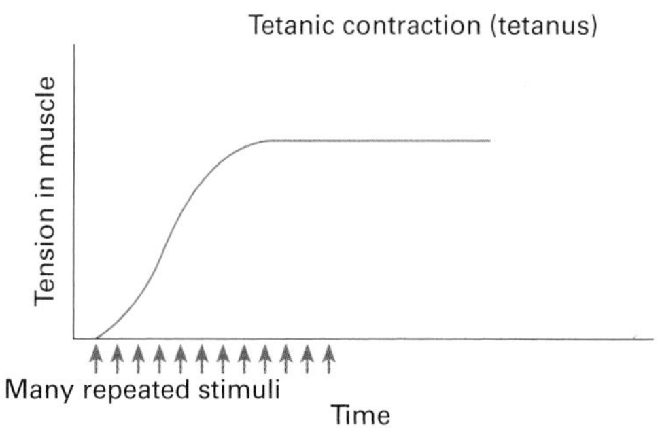

Figure 3 *Contraction of an isolated muscle induced by electrical stimulation, showing temporal summation.*

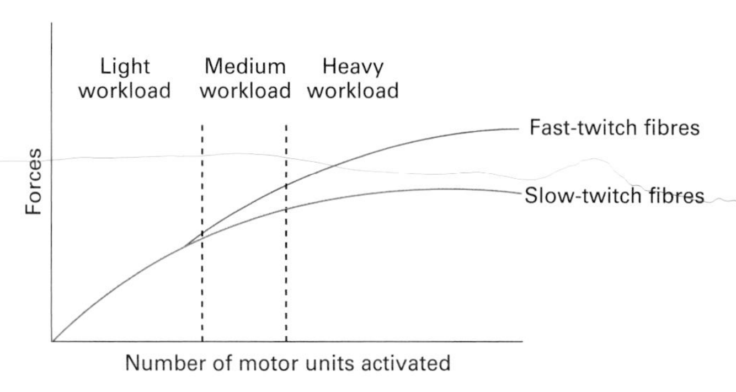

Figure 4 *Muscle fibre recruitment.*

QUICK CHECK

1 What is the role of:
 a calcium
 b tropomyosin
 c acetylcholine
 d ATP

 in the sliding filament theory of muscle action?

2 Distinguish between temporal summation and muscle fibre recruitment.

Food for thought

The ratchet mechanism hypothesis of muscle contraction involves attachment and detachment of the myosin cross-bridge onto actin. This mechanism requires energy supplied by ATP. The muscles of a dead person are often very stiff: a condition known as **rigor mortis**. What does this suggest about the position of the cross-bridges in the muscles of a dead person, and about the precise role of ATP?

11.6

Swimming and flying

OBJECTIVES

By the end of this spread you should be able to:

- describe the forces that need to be overcome before locomotion can take place
- explain how fish swim
- explain how birds fly.

Fact of life

The speed at which a fish swims is generally related to body size. The spectacular swordfish can grow up to 4.9 m long and is probably the fastest fish, reaching speeds up to 100 km h^{-1} (over 60 m.p.h.). The function of its sword is unclear; it might aid streamlining or it might be used to strike at schools of prey.

A few animals, such as filter-feeding barnacles, marine worms, and suckling babies, can stay in one place and wait for food to come to them, but most of us have to actively find our own food. To do this we must be able to move from one place to another. This is called **locomotion**. It is an active process requiring the expenditure of energy. To move around an animal needs to overcome two major forces: gravity and friction. The relative importance of these two forces varies according to the environment in which the animal lives and moves. Vertebrates move in water, on land, and in the air. They have evolved a diverse variety of skeletons, muscle systems, and nervous control mechanisms to provide support and allow locomotion in these different environments. In this spread, we shall look briefly at adaptations that enable fish to swim in water, and birds to fly.

Swimming

Fish are among the best swimmers in the animal world. A fish has little problem overcoming gravity because most of its body weight is supported by water. Overcoming friction is a greater problem because water is a relatively viscous and dense medium (800 times denser than air) and offers considerable resistance to a body moving through it. To minimise this resistance, many fish have sleek, streamlined body shapes.

A fish generates forces that propel it forwards by passing S-shaped waves down its body. These waves can be clearly seen in eels (figure 1a). The waves result from contractions of **myotomes**, W-shaped muscle blocks that make up each side of the fish's body. Alternating waves of contraction and relaxation pass down the myotomes on either side of the body from head to tail. They cause different parts of the body including the tail fin to be swept from side to side, pushing water backwards and sideways. This causes reaction forces to push the fish's body forwards. In most bony fish, the waves of contraction passing down the body are much less obvious than in eels because most of the forces they generate are transmitted to the sideways movements of the tail (figure 1b).

In addition to moving the fish forwards, contractions of muscles in the body wall also produce forces that cause unwanted movements (figure 2). These movements lead to lower swimming efficiency, and fish have evolved structures to reduce such movements. Unwanted sideways movements (called **yawing**) are resisted by the vertical dorsal and ventral fins and, in some fish, by a large head. **Rolling** (movement around the transverse plane) is also resisted by the dorsal and ventral fins, and **pitching** (movement in the vertical plane) is resisted mainly by the paired pectoral fins which act as hydrofoils.

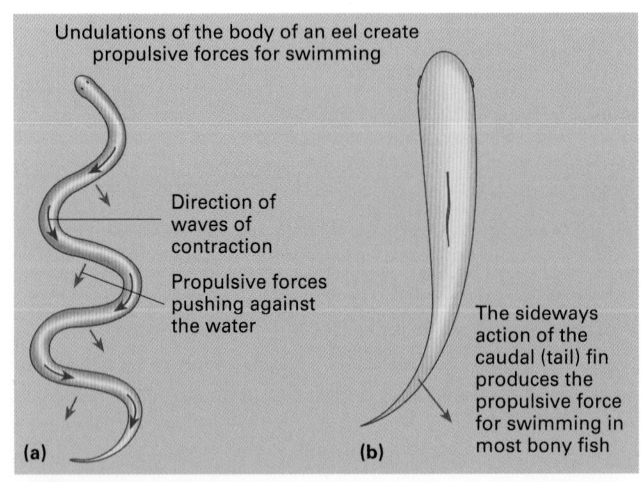

Figure 1 (a) During swimming, the long body of an eel is thrown into S-shaped waves which produce locomotory forces. (b) In a typical bony fish, contractions of the body muscles are transmitted mainly to the tail, where propulsive forces are generated.

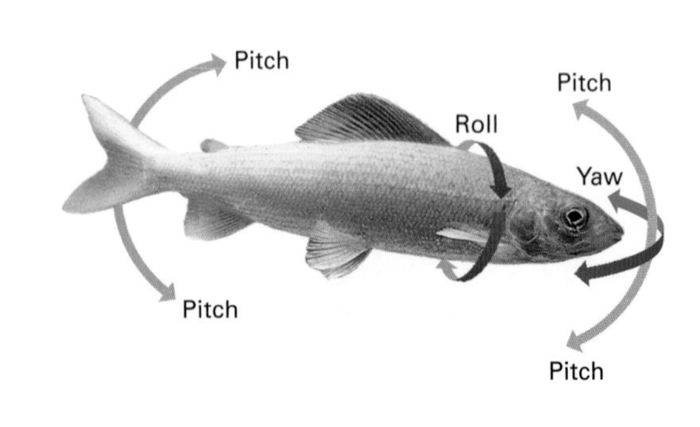

Figure 2 Some of the undesirable movements in a fish.

Many bony fish have acquired a **swim bladder** that acts as a source of buoyancy. This allows them to hover in mid water. Fish without a swim bladder sink to the bottom when they stop swimming. The swim bladder is a gas-filled sac which has developed as an outgrowth of the alimentary canal. A fish can control the amount of gas in its swim bladder to change its buoyancy. When neutrally buoyant, the fish has the same relative density as the surrounding water so it neither sinks nor rises. The swim bladder frees the pectoral fins from their function as hydrofoils, enabling them to be used as brakes, rudders, or (in flying fish) even as wings.

Flying

For a bird to fly, it must be able to generate enough lift to overcome the downward pull of gravity, and enough thrust to overcome the drag forces that resist forward motion. It also has to contend with the low density of air which means the air provides little support and offers little resistance against which limbs can act.

The wings of a bird are highly modified forelimbs that act as **aerofoils** (figure 3a). The leading edge of a wing is thicker than its trailing edge. It also has a slightly convex (domed) upper surface, and a concave (hollow) or flat lower surface. This shape makes the air passing over a wing travel further than the air passing under it. This means that the air pressure under a wing is greater than the air pressure above it. In this way, the wing generates lift when it moves through air, or when air moves over it.

Air can be made to pass over wings by gliding or soaring, and by flapping. **Gliding** and **soaring** are forms of **unpowered flight** which use natural air currents. Gliding birds usually have long wings with a large surface area to take maximum advantage of air currents. Some gliding birds such as eagles live on mountainsides, others such as gulls live on cliffs, so that they can make use of upcurrents and thermals in the air. **Flapping** is used in **powered flight**; the birds flap their wings to create movement of air over themselves. Flapping is not a simple up and down motion, but involves moving the wings in a figure-of-eight pattern. The wing angle is changed throughout the movement to provide the required amount of thrust and lift, and to minimise drag. Flapping requires the action of two sets of powerful flight muscles which work antagonistically, one set producing the downstroke of the wing, the other set producing the upstroke. The sternum of a bird is very large and has a deep keel to provide a surface for the attachment of these muscles (figure 3c).

Birds have a low body weight, which reduces the energy they have to expend in flight. The skull is thin and lacks teeth, the long bones are hollow and filled with air, and the body surface is covered in several types of very light **feathers** (horny outgrowths of the skin epidermis, made mainly of keratin). **Flight feathers** in the wings are strengthened to resist strong air currents; **contour feathers** covering the body are smooth to improve streamlining; and **down feathers** next to the skin are fluffy to provide high levels of insulation.

(a) Aerofoil shape

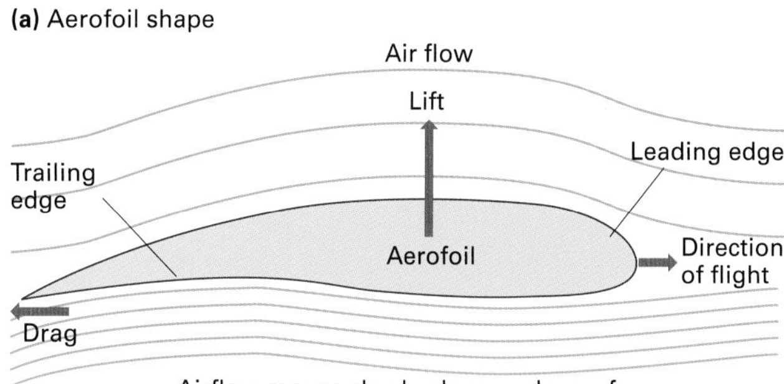

(b) A bald eagle in flight

(c) Sternum and flight muscles

Pectoralis tendon wraps around a groove to inset onto the upper surface of the humerus
humerus
Wing aerofoil giving lift
Sternum
Contractions of pectoralis major pull wing down
Contractions of pectoralis minor pull wing up
Deep keel of sternum

Figure 3 *Some aspects of bird flight.* ***(a)*** *The wing of a bird acts as an aerofoil.* ***(b)*** *A bald eagle in flight.* ***(c)*** *The smaller group of breast muscles (pectoralis minor) produces the upstroke of the wings; the larger muscles (pectoralis major) produce the downstroke, generating lift and forward movement. During the upward recovery stroke, the wing is flexed at the elbow and wrist to reduce air resistance.*

Quick check

1 What are the two major forces that an organism has to overcome in order to move from one place to another?

2 Name the muscle blocks responsible for locomotion in a fish.

3 When a bird is in flight, is air pressure greater on the upper or lower surface of the wing? Explain the significance of this.

Food for thought

Suggest why:

a in most marine fish the swim bladder makes up about 5% of the body volume, whereas it makes up 7% of the body volume of freshwater fish

b some fast-swimming fish, such as tuna and mackerel, have no swim bladder.

c What other types of bony fish are likely to lack a swim bladder?

11.7 WALKING AND RUNNING

OBJECTIVES

By the end of this spread you should be able to:

- distinguish between walking and running in humans
- describe the locomotion of a quadruped
- explain how body posture is maintained.

Fact of life

The cheetah has a rate of acceleration comparable to that of a high-performance sports car. It can reach speeds of up to 110 km h^{-1} (about 70 m.p.h.), enabling it to outrun all other animals over short distances.

Locomotion in humans

Humans have been ingenious in inventing devices that transport them effortlessly from one place to another. Nevertheless, we still use walking as our main form of active locomotion. Walking involves a cyclical sequence of leg movements (figure 1). The body is propelled forwards when the ball of the foot is pressed against the ground and when the hamstrings at the back of the leg retract the thigh. The main muscles responsible for walking are shown in figure 2. Arm movements usually accompany the leg movements: the left arm swings forwards with the right leg, and the right arm swings forwards with the left leg.

When the foot of a walker makes initial contact with the ground, the ankle is flexed (bent), and when the foot leaves the ground, the ankle is extended. Each foot is in contact with the ground for more than half the stride and sometimes both feet are on the ground together. During walking, at least one foot is in contact with the ground at all times. During running, each foot is on the ground for less than half the stride and both feet may be off the ground at the same time. Because of the shorter contact time, the foot has to exert considerably more force during the retraction phase while running than while walking. Also, the limbs go through a greater range of movement during running. While walking, the legs are kept relatively straight so that leg length is fairly constant, whereas in running the leg is flexed (bent) much more so that longer strides can be taken.

The feet of a runner hit the ground much harder than those of a walker. On impact, the tendons act as springs, storing energy which is released as elastic recoil when the leg is moved backwards. Above a critical speed, this reduces the energy cost of running and makes it more efficient than walking. Below that critical speed, walking is more efficient.

Locomotion in quadrupeds

A **quadruped** is a mammal that walks on four legs. During slow movements, a quadruped would fall over as soon as one leg was lifted off the ground unless the other three legs provided a sufficiently stable triangle of support. This is achieved by shifting the centre of gravity. A balanced triangle is maintained by lifting each leg off the ground in a definite order (right forefoot, left hindfoot, left forefoot, right hindfoot, and so on) and by lifting only one leg at a time. When off the ground, each leg goes through a sequence of movements similar to those of a human walker, with flexor and extensor muscles generating propulsive forces during retraction.

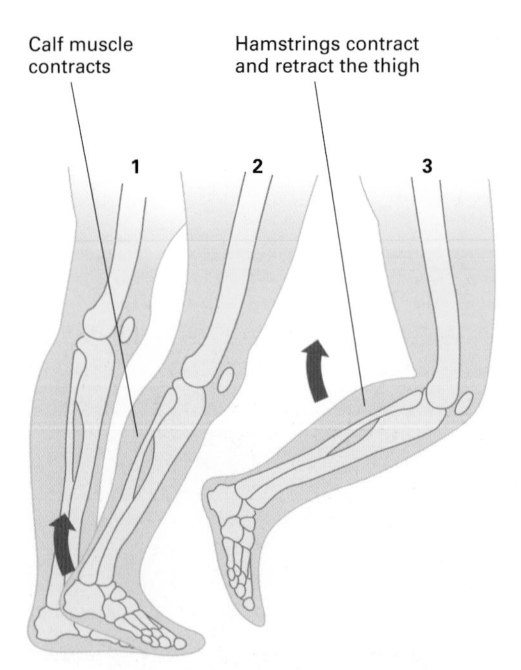

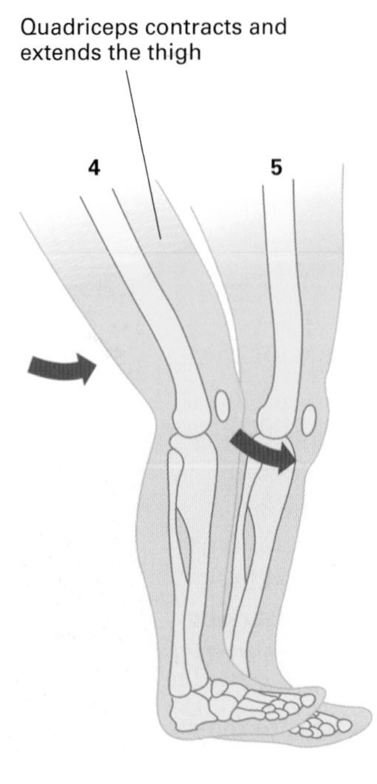

Figure 1 *Walking: the sequence of movements of the right leg during a single stride. From a standing position* ***(1)****, the calf muscle contracts and raises the heel* ***(2)****. In doing so, it exerts a forward thrust by pushing the ball of the foot against the ground. The hamstrings exert more forward thrust as they contract to bend the knee slightly and raise the leg* ***(3)****. As the foot loses contact with the ground, the weight of the body is supported by the left leg which is still in contact with the ground. Next, the quadriceps contracts to extend the leg and draw it forwards* ***(4)****. When extension is completed, the foot regains contact with the ground, heel first* ***(5)****. The weight can now be supported on the right leg, and the whole sequence can be repeated with the left leg.*

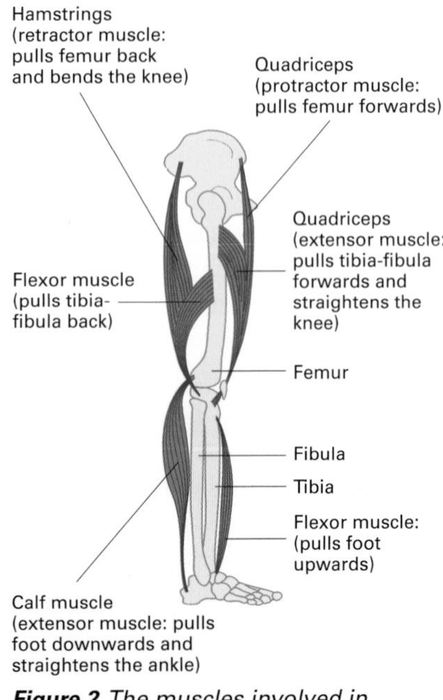

Figure 2 *The muscles involved in walking.*

Figure 3 A cheetah running flat out swivels its shoulder blades and flexes its spine to increase the length of its stride and increase its speed. The long tail acts as a balancing organ to help maintain stability. Even when running at top speed towards its prey, its head maintains the same relative position with eyes remaining focused on a single point.

In rapid locomotion, all legs may be off the ground simultaneously, but only for a short period because limb movements are so rapid. A cheetah, for example, running at speeds exceeding 110 km h^{-1} (about 70 m.p.h.) appears almost to be in flight.

Running speed is a product of the stride length and the cadence (the rate of strides). When running fast, a cheetah's highly flexible back maximises its stride length (figure 3). Running on its toes also contributes to the cheetah's high speed.

Mammals that move on their toes are called **digitigrades**. Those such as ourselves and bears, which move on the soles of their feet, are called **plantigrades**. Hooved mammals that touch the ground only with the tips of their digits are called **unguligrades**.

An animal's stride rate is limited by how quickly its muscles can contract and how soon it fatigues. However, the speed of movement in a digitigrade or unguligrade can be increased as these animals can use their toes and hoofs as extra limb joints when running. This means they can use different muscles to move different joints of the leg in the same direction at the same time, increasing their rate of movement. The evolutionary climax of the unguligrade limb is seen in the long slender legs of a horse. Each hoof is derived from a single toe (the third) enabling a horse to run with both speed and endurance.

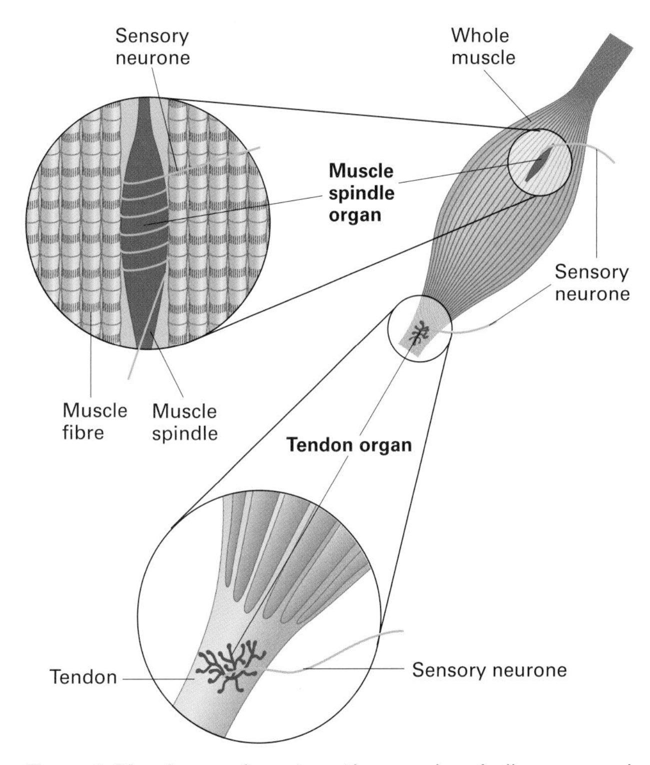

Figure 4 Muscle proprioceptors: the muscle spindle organ and the tendon organ.

Posture

Muscles do not usually completely relax: at least a few fibres are always contracted in order to create a slight muscle tension called **muscle tone**. Muscle tone keeps muscles primed for action. It also helps an animal maintain a particular **body posture** or position of the body when not completely supported by some other object. To maintain an upright body posture, an animal has to hold certain joints steady by partially contracting antagonistic pairs of muscles around those joints. Posture is said to be good when it can be maintained with very little effort. For example, when standing, a person's head, backbone, hips, and legs should be aligned so that the person is balanced with the feet planted firmly on the ground. In this position, muscles carry little weight and simply keep the body balanced.

In order to maintain body posture and bring about smooth coordinated movements, the central nervous system requires information about the tension in muscles in different parts of the body and their positions. Two types of **stretch receptor** called **muscle spindle organs** and **tendon organs** provide this information (figure 4). These internal sense organs or **proprioceptors** are present in all skeletal muscles. They monitor muscle tension continually so that adjustments can be made to maintain a particular body posture or to carry out a particular body movement. Muscle spindle organs are also involved in the stretch reflex (see spread 11.8).

Quick check

1. What is the main difference between walking and running?
2. Give two ways in which a cheetah maximises its stride length when running flat out.
3. Give two functions of muscle tone.

Food for thought

Backache is second only to the common cold as a reason for absence from work in highly industrialised countries. Suggest how poor posture could increase the risk of developing backache. Explain how abdominal exercises such as sit-ups can reduce the risk of backache when performed properly (that is, slowly from a legs-bent position with the soles of the feet planted on the ground and the small of the back in contact with the floor throughout the exercise) but can harm the back if performed incorrectly (for example, forcibly from a lying position with legs straight out and hands on the back of the head).

11.8

Behaviour: simple responses

OBJECTIVES

By the end of this spread you should be able to:

- describe the components of a simple reflex action
- distinguish between a reflex action and a fixed action pattern
- discuss how kineses and taxes can orientate animals to favourable environments.

Fact of life

Doctors test the knee-jerk reflex (also called the patellar reflex) by tapping the tendon just below the kneecap (figure 1). In a healthy patient, the tap should cause a reflex contraction of the quadriceps muscle in the upper thigh so that the leg kicks. In this reflex pathway, a stretch receptor in the tendon is stimulated by the tap and a sensory neurone carries a nerve impulse to the spinal cord. Here the impulse passes across a synapse to a motor neurone, which runs from the spinal cord to the quadriceps, the leg extensor. Contraction of this muscle brings about the knee-jerk response. Failure of the leg to respond to the tap may indicate disease or damage to one or more components of the reflex arc.

A reflex contraction of a muscle in response to its being stretched is called a **stretch reflex**. Stretch reflexes play an important role in maintaining body posture.

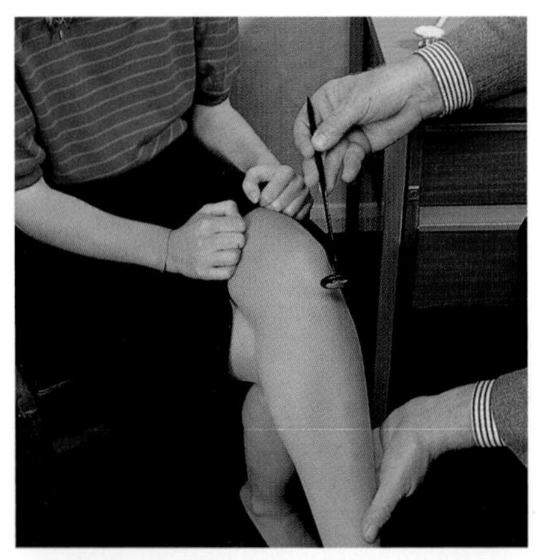

Figure 1 *The knee-jerk reflex.*

Animal behaviour is the observable responses of an animal to the environment around it. These responses usually involve movements of the whole body or part of it, but may include other changes. The reproductive behaviour of a male three-spined stickleback (spread 11.13), for example, includes its colour changes as well as its courtship movements. Unlike tropic movements of plants (spread 14.11), most behavioural responses are reversible.

Reflex actions

A **reflex action** is the simplest type of animal behaviour. It is an automatic and involuntary response characterised by being stereotyped (a particular stimulus always evokes the same response). Many reflex actions, such as escape responses (spread 10.1) and the withdrawal of a hand from a hot object, are protective. All reflex actions are **innate**: they are unlearned, genetically determined responses that use inherited nervous pathways.

Figure 1 shows a doctor testing the **knee-jerk reflex** of a patient. The simple nervous pathway that transmits information rapidly from receptor to effector is called a **reflex arc**. This type of pathway is adapted for responses that need to happen as quickly as possible.

Reflex arcs vary in complexity. The knee-jerk reflex arc is **monosynaptic** because the sensory neurone has a single synaptic connection with the motor neurone. In some other reflex arcs, such as that shown in figure 2, there is another neurone in the spinal cord between the sensory neurone and motor neurone, called an **interneurone**. This is also called a **relay neurone** or an **association neurone**, because it makes synaptic connections with nerves that pass upwards into the association areas of the brain. These connections enable the brain to influence reflex actions, though the brain can only modify a reflex action; it cannot stop it completely, because information is transmitted along the reflex arc very quickly

Even the knee-jerk reflex is not as simple as it might seem. When the stretch receptors within the quadriceps are stimulated, in addition to bringing about the stretch reflex they also send information to antagonistic muscles ensuring that they are relaxed. This process is called **reciprocal inhibition**.

Stereotyped behaviour

Stereotyped behaviour occurs when the same response is given to the same stimulus on different occasions. Reflex actions are examples involving simple nerve pathways. More complex examples, such as the courtship dance of sticklebacks (spread 11.13), the suckling response of new born babies, foraging in *Drosophila*, and the food-begging response of gull chicks (figure 3) have more complex nervous pathways that can bring about a specific sequence of coordinated movements called **fixed action patterns** (FAPs).

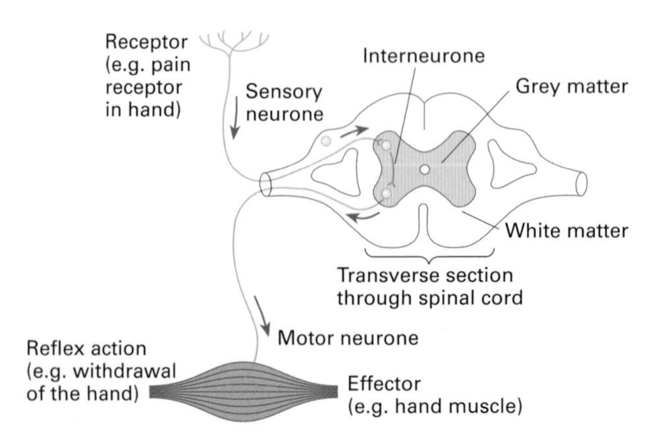

Figure 2 *A simple reflex arc consisting of three neurones: a sensory neurone, an interneurone, and a motor neurone.*

The simplest FAPs are largely innate and genetically determined responses to specific stimuli called **sign stimuli**. The presence of food, for example, acts as a sign stimulus for foraging in the fruit fly *Drosophila*. Food can trigger one of two foraging strategies called 'rover' and 'sitter'. The FAPs of 'rovers' cause the flies to traverse a large area, while those of 'sitters' causes them to cover a small area. Whether a fly is a 'rover' or 'sitter' is determined by the alleles of a single gene. In this and many other types of insect behaviour, a sign stimulus activates nervous pathways which make muscles move without any decision-making in the brain. However, stereotyped behaviour incorporating FAPs is generally more complex than simple reflex actions and can be modified by experience. For example, the begging response of a gull chick is triggered by the red spot on its parent's beak, but experiments show that the response varies with the size of the spot and other features of the beak (figure 3). Similarly, the calls of many birds are greatly influenced during their early development by the songs of nearby members of the same species.

Kineses and taxes

Kineses and taxes are forms of behaviour involving the locomotion of organisms or cells in response to a specific external stimulus. They are both forms of behaviour which help to keep an individual animal, such as a woodlouse, in a favourable environment (figure 4).

A **kinesis** is a random movement in which the rate of movement is related to the intensity of the stimulus, but not to its direction. There are two main types: **orthokinesis**, which involves changes in speed of movement; and **klinokinesis**, which involves changes in the rate of turning (the lower the rate of turning, the more likely it is that the animal will leave an unfavourable area). In a choice chamber, a container divided into dry and damp areas, woodlice tend to move faster and turn less in the dry areas. This tends to orientate them towards the damp areas.

A **taxis** is a movement in response to the direction of a stimulus. Movements towards a stimulus are positive; those away from a stimulus are negative. Woodlice generally show negative phototaxis: they move away from light and tend to be found in shaded or dark places, such as under stones and leaves.

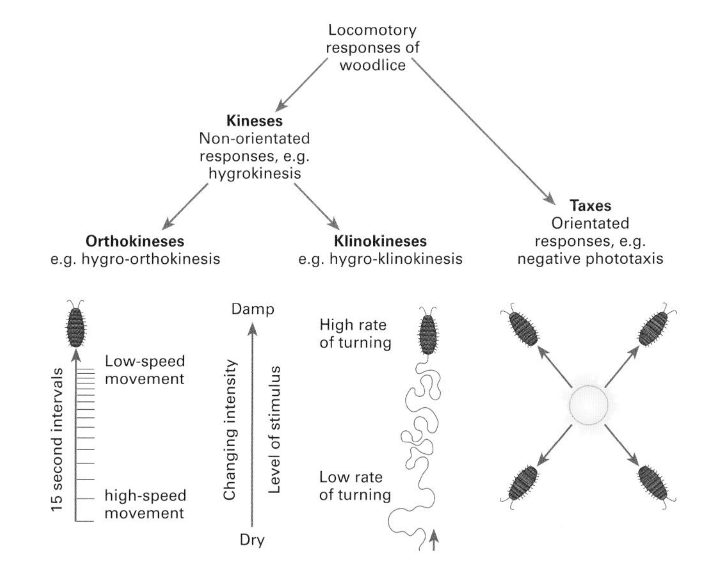

(a) Normal feeding of gull chick, pecking at its mother's beak.

Spot colour	Percentage effectiveness
Red	100%
Black	86%
Blue	85%
White	71%
No spot	30%

(b) Experiments with hand-held cardboard models

Figure 3 *Feeding behaviour of gull chicks.* ***(a)*** *Normal feeding behaviour of a chick on the nest. It pecks at the red spot on the parent's beak to stimulate the parent to regurgitate food.* ***(b)*** *Laboratory experiments using hand-held cardboard models of a gull's head compared the number of pecks at each model with the number of pecks at the red spot model. The feeding response involves fixed action patterns.*

Quick check

1. List the main components of a reflex arc.
2. Distinguish between a reflex action and a fixed action pattern.
3. Name the type of behaviour that produces a movement towards a directional light source.

Food for thought

Ground-nesting birds such as gulls and geese retrieve eggs when they roll out of the nest. Egg retrieval is a fixed action pattern (FAP) triggered by the sight of the egg rolling away. Experiments show that the larger the egg, the stronger the response. An abnormally large model egg may provide a **supernormal stimulus** and trigger a greater than normal response. Using this information, suggest why a willow warbler may feed a cuckoo chick in preference to its own offspring.

Figure 4 *Orientation behaviour of woodlice in choice chambers. Woodlice respond to humidity levels by changing their rate of turning (hygro-klinokinesis) and their speed of movement (hygro-orthokinesis). They also move away from light (negative phototaxis).*

11.9

OBJECTIVES

By the end of this spread you should be able to:

- discuss the contribution of innate behaviour and learned behaviour to an animal's behaviour
- explain the value of habituation
- give examples of imprinting and explain its significance.

Fact of life

A police horse is able to ignore the loud noises of traffic and a bustling crowd because it has been habituated to noises as part of its training.

The 'nature versus nurture' debate

In psychology, the 'nature versus nurture' debate concerns the relative importance of inheritance ('nature') and environment ('nurture') in determining a behaviour. Most modern psychologists agree that a behaviour is not usually determined by nature *or* nurture, but by an interaction between the two. **Twin studies** can help to distinguish between genetic and environmental influences on the development of behaviour. Typically, such studies involve comparing the behaviour of genetically identical twins that have been brought up in distinctly different environments. Behavioural differences between the identical twins are assumed to be determined mainly by environmental factors. The effects of genetic factors on behaviour can be studied by comparing the variations in behaviour in a large sample of identical and non-identical twins brought up in comparable environments. If the similarity between the behaviour of identical twins is more than that for non-identical twins, the behaviour is assumed to be determined mainly by genetic factors. **Cross-cultural studies**, in which the behaviour of people in different cultural environments is compared, can also help identify the effects of environment on the development of behaviour. These studies are characterised by the sample being sufficiently large for statistical analysis to show relationships or lack of relationships between the behavioural traits and type of culture being examined.

Learning: habituation and imprinting

Figure 1 *Mounted policeman clearing the crowds invading the pitch at the first FA Cup Final played at Wembley in 1923.*

Learning is the capacity to record specific experiences and to modify behaviour in the light of those experiences. Animals that learn can adjust their behaviour to suit specific environments and can change their behaviour as the environment changes. Animals that rely entirely on stereotyped, genetically determined behaviour patterns cannot adapt quickly to environmental changes; their behaviour changes only as a result of slowly acting genetic mutations over many generations.

Most forms of animal behaviour observed in nature are a variable mixture of innate behaviour and learned behaviour. Even the relatively stereotyped food-begging behaviour of a gull chick is modified by experience. Inexperienced gull chicks instinctively peck at conspicuous objects, but quickly learn to associate pecking accurately at their parents' beaks with obtaining food. Animals such as insects, with a short life span spent in a relatively unvarying environment, rely heavily on genetically determined stereotyped forms of behaviour. Animals with long life spans in variable environments would not survive long enough to reproduce unless they could adapt their behaviour to the changing environment.

Learning occurs when an animal's behaviour becomes consistently modified as a result of experience (not as a result of growth or fatigue). The behavioural modifications and types of experience vary with the type of learning. The simplest type of learning is called habituation.

Habituation

Habituation is learning in which repeated exposure to a stimulus results in decreased responsiveness. A sudden loud noise, for example, causes a horse to bolt on initial exposure, but if the horse is subjected to repeated noises, the responsiveness decreases (figure 1).

A common feature of habituation is that the response reappears if the stimulus is not given for a long period of time. Thus, a previously habituated horse will bolt again if it hears a noise after being kept in a very quiet environment for a few weeks. If the horse is exposed regularly to noises again, it will become habituated to them more rapidly than previously.

When an animal is habituated to one stimulus, it will usually treat another similar stimulus in the same way. Thus, if a horse has been habituated to one particular noise, it will tend to habituate to a new noise quickly. This phenomenon is called **generalisation**, in which a learned response is brought about by stimuli other than the ones the animal is used to.

In some circumstances, loss of responsiveness to a stimulus may result from **fatigue** or **sensory adaptation** rather than habituation. However, fatigue is not the cause if after the loss of one response, the same muscles can be used in another activity. In a similar way, sensory adaptation is not the cause of the loss of responsiveness to a particular stimulus if the same stimulus can bring about a different response. Once habituated to a stimulus, an animal still senses it but has learned to ignore it.

Habituation is a very widespread and important form of learning. It enables an animal to avoid wasting time and energy responding to harmless stimuli that do not threaten its survival and reproduction.

Imprinting

Imprinting is a form of learning that occurs during a brief, genetically determined **critical period** in the lives of animals, usually shortly after birth. A particular stimulus becomes permanently associated with a particular response, as when young birds follow any large moving object. The behaviour is irreversible and influences future patterns of social behaviour.

Konrad Lorenz (an Austrian zoologist who gained a Nobel Prize for his studies of animal behaviour) demonstrated imprinting in greylag geese. He divided a batch of eggs into two groups. One group was reared normally by the mother, and the goslings showed normal behaviour, following the mother around when young (gaining food, shelter, and protection from her) and growing up to mate and interact with other geese. The other group was isolated from the mother and hatched in an incubator. For the first few hours after hatching, the goslings spent their time with Lorenz. Imprinting occurred and the goslings treated Lorenz as their mother, following him in preference to their natural mother or to other geese (figure 2). Such goslings became socially dysfunctional in adult life; they continued to prefer Lorenz to other geese and even attempted to mate with humans.

Other examples of imprinting include newly hatched salmon, which use this process to learn the complex mixture of odours associated with the stream in which they hatch. This early imprinting helps adult salmon to find their home stream after their migratory swim to the sea and back.

Imprinting usually involves a young animal learning the characteristics of its parents so that it can recognise its parents and other members of its species. Imprinting also occurs in adult animals, enabling them to recognise their own offspring. For example, shortly after giving birth, a mother goat is sensitive to the smell of her kid for about one hour. During this critical period, a few minutes' contact with any kid is sufficient for it to be accepted as her own. If she is separated from all kids during that first hour, even her own kid will be rejected and will not be allowed to suckle.

Imprinting tends to occur in animals in which social bonds (with parents, a family group, or members of the opposite sex) are important.

Quick check

1 How does innate behaviour differ from learning?

2 Name two causes of loss of responsiveness apart from habituation.

3 In what ways might the normal imprinting of goslings be important for their survival?

Food for thought

Habituation is a common form of learning in all animals, including ourselves. Describe some circumstances in which you might become habituated. Suggest the value of such habituation. Describe circumstances in which habituation could prove to be dangerous.

Figure 2 *A group of goslings that had imprinted on Konrad Lorenz follow him as if he is their mother.*

Critical periods in human development

The role of critical periods and imprinting in human development is controversial. This is especially so for **sexual imprinting** in which experiences during early development can determine future sexual behaviour. For example, it is well documented that non-human adults can bias young animals to choose future mates that resemble their parents, but such findings are only beginning to emerge from studies of humans. Unlike our understanding of imprinting in non-human animals, we do not know the precise mechanism of imprinting-like phenomena in humans, nor do we know to what extent imprinting during early development contributes to adult human behaviour. Much better understood are the critical periods in which normal human sensory development depends on having particular sensory inputs. For example, **visual acuity,** the acuteness or clearness of vision, depends not only on the sharpness of the retinal focus within the eye in which defects can be corrected by wearing glasses, but also on the ability of the brain to interpret the incoming sensory information. There are sensitive periods, called **critical windows**, during which infants are more vulnerable to vision defects and require normal visual input for different visual abilities. These critical windows vary. For **grating acuity** (the ability to distinguish the elements of a fine grating composed of alternating dark and light stripes or squares) the window is from birth to at least 5 years old, whereas it is up to approximately 10 years for **Snellen acuity** (the ability to distinguish letters of decreasing size arranged in lines on a chart commonly used in eye tests), and at least to early teenage years for peripheral vision.

11.10

Conditioning and latent learning

OBJECTIVES

By the end of this spread you should be able to:

- distinguish between classical conditioning and operant conditioning
- discuss the significance of latent learning.

Fact of life

The fire salamander shown in figure 1 might appear to be endangering itself with its conspicuous coloration, but the colours act as a warning. Predators soon learn to associate the bright colours with the salamander's ability to squirt a nerve poison from glands on its back. After one taste of the toxic and unpalatable secretions or one squirt in the eye, a predator is unlikely to regard any similarly coloured animal as a potential meal again.

Figure 1 *A fire salamander, which, when attacked by a predator, can squirt streams of nerve poison from glands on its back.*

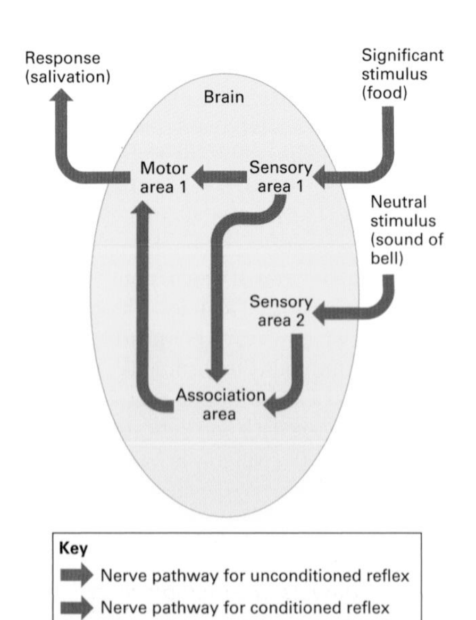

Figure 2 *A new nerve pathway developed by classical conditioning.*

Many biologists believe **conditioning** is the most important learning process for animals in the wild. Conditioning (also called **association** or **associative learning**) happens when an animal learns to associate a particular behaviour pattern with a reward or punishment. Most predators learn quickly to associate certain kinds of prey with unpleasant experiences (figure 1). Animal behaviourists recognise two main experimental circumstances in which conditioning occurs: classical conditioning and operant conditioning.

Classical conditioning (Pavlovian conditioning)

Classical conditioning occurs when a behaviour that is normally triggered by a certain stimulus comes to be triggered by a substitute stimulus which previously had no effect on the behaviour. It was first demonstrated in dogs by Ivan Pavlov in 1902. A hungry dog salivates when given food. Taking great care to exclude any other stimuli, Pavlov presented a dog with a neutral stimulus such as the sound of a bell five seconds before giving it food. After presenting the combined stimuli of bell and food five or six times, the dog associated the bell with the food and salivated when the bell was rung without food being given. The dog had learned to respond to a new stimulus which was previously neutral. In this type of experiment, the behaviour that associates a neutral stimulus (the bell) with a significant stimulus (the food) is called a **conditioned reflex** and the animal is said to have become **conditioned** to the bell. This form of learning requires the development of a new nervous pathway in which the association area in the cortex of the brain makes a link between two different types of stimuli and gives the same response to them (figure 2).

In Pavlov's experiment, the unconditioned stimulus (food) was pleasant and acted as a reward or **positive reinforcement** when the association between it and the bell was being established. Classical conditioning can also be established using unpleasant stimuli which act as a punishment or **negative reinforcement**. For example, a cow that touches an electric fence and receives a mild shock soon learns to associate the fence with the shock and is less likely to approach the fence in future.

Operant conditioning (instrumental conditioning)

Operant conditioning differs from classical conditioning in a number of ways. Whereas in classical conditioning the animal learns to associate a particular stimulus with a reinforcer (the unconditioned stimulus), in **operant conditioning** an animal learns to associate a particular behavioural act with a reinforcer. Also, in classical conditioning the delivery of reward or punishment is controlled by a person (for example, an experimenter or animal trainer); in operant conditioning, the animal's own behaviour determines whether or not a reward appears.

Burrhus Frederic Skinner studied operant conditioning extensively in an apparatus of his own design called a Skinner box. In one study, a hungry rat learned to associate pressing a lever with obtaining food (figure 3). The behavioural act which causes the lever to be pressed appears to be generated spontaneously; the rat may press the lever accidentally, or it may press it because it is hungry and actively sniffing for food. This type of learning is also called **instrumental conditioning**, because the animal's own spontaneously generated behaviour has been instrumental in obtaining food. Operant conditioning is thought to be the predominant learning process found among animals, including humans. For example, a bird may associate turning over a leaf with the discovery of an insect, a dog can quickly associate retrieving a stick with obtaining a food reward, and a young child can be conditioned to display good manners by the use of appropriate rewards and punishments.

Operant conditioning is also demonstrated in **trial-and-error** learning in a maze. Mazes vary from a simple T- or Y-shaped tube to a complicated series of pathways. The animal is confronted with one or more choices of direction to make. If it chooses a particular direction or series of directions, it is given a reward or avoids a punishment. A single attempt at completing the maze is called a trial. Each choice of direction which leads to a dead end, away from a reward, or towards a punishment is recorded as an error. An animal is regarded as having 'learned' the maze when it can repeatedly pass through the maze without making a wrong turn. Rats are famous for learning complicated mazes very quickly when given a reward, improving both speed and accuracy with each successive trial.

Latent learning

Rats are very inquisitive creatures. Even when given no reward, a rat will explore a new maze for its own sake, excitedly entering those parts of the maze that give the maximum chance of exploration. This exploratory behaviour is thought to result in a form of learning called **latent learning** which has no apparent reward except encountering novel situations. Hungry rats allowed to explore a maze with no reward seem to learn the maze so that when food is provided as a reward they quickly find the shortest route to it. They require significantly fewer trials to learn the maze than do hungry rats who are unfamiliar with the maze (figure 4). Exploratory behaviour and latent learning are probably vitally important to small mammals, enabling them to become familiar with their home territory so that if danger threatens, for example from a predator, they can make a quick escape and find a refuge.

In most natural situations, learning is much more complex than in the controlled circumstances of a laboratory. A bee visiting one particular flower repeatedly, for example, learns to associate the colour and scent of the flower with its nectar. It will also learn the relative position of the flower to its hive, and at what time of day the flower is producing most nectar. Thus its foraging behaviour is a result of a complex interaction between innate behaviour and different types of learning.

Quick check

1 What are the main distinctions between classical conditioning and operant conditioning?

2 How does latent learning help a rat in its natural habitat?

Food for thought

Some animal behaviourists believe that children are best educated, like animals in a Skinner box, by instrumental conditioning that involves doing things to gain rewards or not doing things to avoid punishment. Suggest arguments for and against this type of teaching.

Figure 3 *A Skinner box. B. F. Skinner designed this type of box to study operant conditioning in small animals. In this experiment, when a rat presses a lever a small pellet of food (a positive reinforcement) is given to it in a cup. Skinner boxes have also been used for experiments to study the effects of negative reinforcements and punishments on learning. In the UK, before any such experiments are conducted, they must conform to an ethical framework enshrined legally in the Animals (Scientific Procedures) Act 1986. The Act requires that proposals for research involving the use of animals must be fully assessed in terms of any harm to the animals. A detailed examination of the particular procedures must be made, including the types and numbers of animal used. The potential stress and psychological harm to the animals must be identified. These 'costs' are then weighed against the potential benefits of the investigation.*

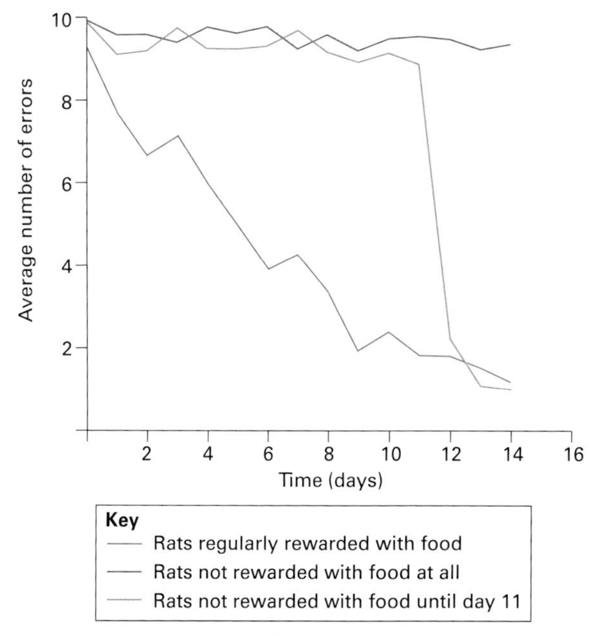

Figure 4 *Experimental results from rats in a multiple maze showing latent learning.*

11.11

OBJECTIVES

By the end of this spread you should be able to:

- describe an example of imitation
- explain how insight learning differs from trial-and-error learning
- discuss the basis of memory.

Fact of life

African hunting dogs (*Lycaon pictus*, also called African wild dogs) are specialised cooperative predators. They hunt, rest, travel, and reproduce in packs of 2–20 dogs. When not staying in one place to look after pups, they cover 2–50 km each day hunting for food. Their home ranges average 1500 km^2. The adult males and adult females have separate dominance hierarchies (see spread 11.13). Usually, only the dominant male and female mate. Unlike most other animals, however, the females compete to breed. The African hunting dog is listed as a species threatened with extinction: the world population is probably less than 10000.

Figure 1 *A pack of African wild dogs (*Lycaon pictus*) mounting a coordinated attack on a wildebeest in Serengeti, Tanzania.*

Imitation, insight, and memory

Imitation

Imitation is a form of **social learning** which may explain the rapid spread of certain types of animal behaviour from one area to another. It involves copying the behaviour of another animal, usually a member of the same species. For example, cream-stealing behaviour by birds such as bluetits is known to have started in one area of Britain when metal- or cardboard-capped milk bottles were first delivered to customers' doorsteps. It is thought that one or two enterprising tits learned to steal the cream at the tops of the bottles by pecking at bottle caps, and this behaviour spread quickly to other areas because birds copied one another. Unlike imprinting which may incorporate copying (spread 11.9), imitation is not confined to a critical period in the animal's life.

Imitation is an important component of many complex behaviour patterns in mammals. African hunting dogs seem to learn their skills through a combination of genetic preprogramming, trial-and-error learning, and imitation. They hunt in packs with each pack appearing to prefer a particular prey and a particular hunting method. Most packs avoid healthy large animals such as zebra and wildebeest because they can be dangerous. However, a few packs hunt large animals by mounting closely coordinated attacks, one dog grabbing the victim's lip and another its tail, while the rest disembowel it (figure 1). For these packs, zebra-hunting is a tradition learned by each generation imitating the behaviour of its predecessor.

Insight

In figure 2, a dog is presented with a situation called a **detour problem**. The lead is long enough to reach the food, but only if the dog first walks away from the food and around the pole immediately behind the dog. Dogs have great difficulty solving this type of problem. Although they might eventually reach the food by trial and error, they tend rather to strain on the lead, unable to reach the food. Chimpanzees, on the other hand, usually solve detour problems immediately without resorting to trial and error. They appear to assess the situation and use their reasoning ability to solve the problem. This type of learning is called **insight**.

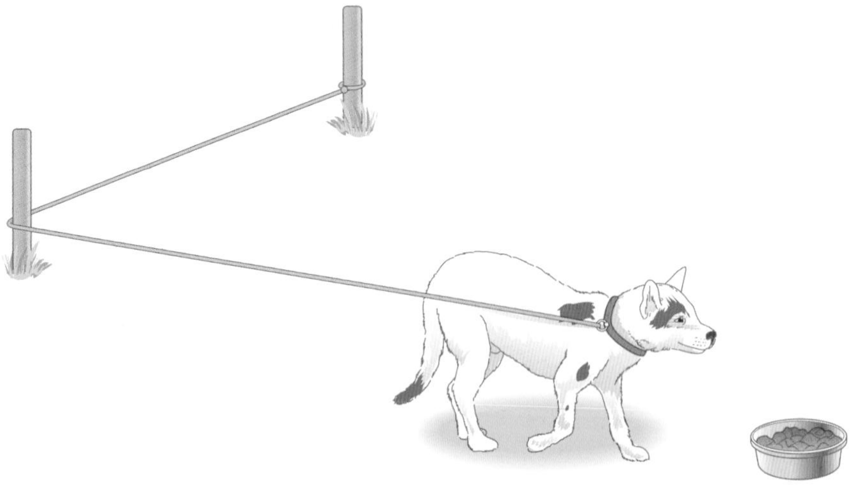

Figure 2 *Detour problem: the dog will continue to pull on the leash until, by trial and error, it eventually finds itself the correct side of the poles. It cannot solve this problem by insight.*

Wolfgang Kohler performed a series of classic experiments demonstrating that chimpanzees are good at solving many different problems by insight. For example, they readily learn to use a stick to gather in a piece of fruit that is out of reach behind bars (figure 3). To do this requires the chimpanzee to appreciate the relationship between the size of the stick and the distance between it and the banana. In another problem, a chimpanzee was placed in a room with several boxes on the floor. A bunch of bananas was hung out of reach, high above its head. The chimp apparently sized up the situation and then stacked the boxes on top of each other to reach the bananas (figure 4).

Figure 3 Insight learning in chimps: a chimp using a stick to rake in a banana that is out of reach.

Figure 4 A chimp obtaining a banana after stacking up boxes.

In insight learning, animals seem to solve problems by internal mental processes which cannot be directly observed. Among humans, problem-solving usually involves **cognition**, a mental process which includes conscious thinking, self-awareness, judgement, and language. Some scientists believe that the term cognition should be confined to humans because we cannot infer the mental state of an animal from its actions. If we make such inferences, there is a danger of **anthropomorphism** (falsely attributing human motivations, feelings, and thoughts to other animals). Most biologists take great care to avoid anthropomorphism, not because they believe that the mental processes of animals are completely different from those of humans, but because they cannot be certain whether or to what extent a non-human animal can think and feel as a human does.

Memory

In order to learn, an animal has to be able to record specific experiences and to modify its behaviour in the light of those experiences (spread 11.10). The ability to store and recall experiences is called **memory**. Memory is shown by a learned response continuing over a period of time.

Researchers have suggested at least two separate mechanisms for storing information: a **short-term memory** which lasts only a few minutes, and a **long-term memory** which can last for many years. Short-term memory is easily disrupted whereas long-term memory is relatively permanent and resistant to change. There is also evidence that short-term memory can be overloaded but that long-term memory has an infinite capacity. Information appears to pass into the short-term memory before it can be stored in the long-term memory.

In humans, the transfer of memory from the short-term memory to the long-term memory is made easier by:

- rehearsal (for example, practising techniques or revising study notes)
- positive emotional state (you learn better if you are feeling alert and well motivated)
- linking the new information to information already in the long-term store (it is easier to add to your understanding than to learn something completely new).

Understanding how nerve networks in the brain store, retrieve, and use memories is one of the most challenging aspects of modern biology. There are a number of competing hypotheses. The process may involve chemical changes (for example, changes in RNA composition in the neurones), or changes in the synaptic interconnections between neurones: some synapses may be weakened, others may be strengthened, or entirely new ones may be formed. However, no theory has gained general acceptance yet. It is likely that learning and memory involve more than one set of mechanisms for storing different types of information.

Quick check

1 Name the kind of learning that involves copying the behaviour of another animal.

2 Explain briefly the difference between the way a dog may solve a detour problem and the way a chimpanzee usually solves the same type of problem.

3 How can the transfer of short-term memory to long-term memory be enhanced?

Food for thought

Dr Jane Goodall has become famous for her studies of chimpanzees in their natural habitats in East Africa. She has described a wide range of complex behaviour which has convinced her that chimpanzees are truly cognitive beings. Some scientists object to the use of the term cognition to describe the thought processes of non-human animals. How do you think the contrasting viewpoints about chimpanzee cognition might affect the use of such animals in experiments?

11.12

Social behaviour in insects

OBJECTIVES

By the end of this spread you should be able to:

- discuss the advantages and disadvantages to insects of living in societies
- describe how ants use pheromones as a means of communication
- describe how foraging honeybees communicate using a dance.

Fact of life

Termite mounds may be over 7.6 m high (figure 1). The nest is underground and contains separate chambers for larvae, fungi (which termites eat), and the royal cell where the queen lives. The mound itself consists of a central chimney and side chimneys through which air circulates to help maintain the temperature inside the nest.

Figure 1 *A termite mound in the Northern Territory, Australia.*

Social behaviour involves interactions between two or more animals of the same species. It is most clearly seen in animals which organise themselves into highly structured social groups called **societies**. Members of an animal society interact strongly, influencing the behaviour of one another. They are not just a group or aggregation of individuals living in the same place.

Social life brings certain advantages and disadvantages. The **advantages** include:

- better protection against predators, such as improved detection or escape systems
- better use of and defence of limited resources
- increased feeding efficiency
- increased reproductive efficiency
- increased survival of offspring through communal feeding and protection
- saving of energy by endothermic animals as a result of being close together
- saving of energy by moving fish and birds which can take advantage of vortices (whirling movements in the water or air) created by others in the group.

Disadvantages of living in a society include:

- increased competition for water, space, food, mates, and other resources
- increased susceptibility to disease and parasites
- higher risk of being harvested by humans
- higher risk of predation on young by cannibalistic neighbours.

Social behaviour has evolved where the advantages of a social life exceed the disadvantages. It hinges on the ability of animals to communicate with one another in order to influence each other's behaviour.

Communication

Communication is the transfer of information from one animal to another. Communication takes many forms, but always involves one animal producing a signal or **sign stimulus** that can be detected by another. A sign stimulus may be seen, felt, heard, a chemical smelled or tasted, or a combination of these. The simplest type of communication uses a sign stimulus which releases a fixed action pattern (spread 11.8). Communication in a society is often complex, combining different sign stimuli with intricate learned or preprogrammed behaviours.

Insect societies

True societies are found in two orders of insects, the Isoptera or termites, and the Hymenoptera which includes ants, bees, and wasps. These societies show four main characteristics: cooperative care of their young, overlapping generations, division of labour, and communication among their members.

A typical insect colony contains many members inhabiting a nest or hive. A beehive may have more than 75 000 individuals, a wood ant nest may contain more than 300 000 members, and over a million termites may live in a large mound (figure 1). However, all members of a colony are closely related as they are all the offspring of one fertile female, the queen.

The members are divided into different groups called **castes**, each with a specific task. One caste might care for the young, while others find food, defend the colony, or remove dead members. An individual's caste may be determined by a number of factors, including its sex, nutrition, and the temperature of the nest or hive. An individual may change its caste as it ages or if the needs of the colony change. The evolutionary success of insect societies is largely due to the division of labour, made possible by the different castes. The actions of all individuals within a colony are so well integrated that such a colony is often called a 'superorganism'.

Communication: pheromones and waggle dances

Workers are the largest caste within an insect colony. They are sterile (non-reproductive) individuals working on behalf of the whole colony. Their activities are controlled by communication between each other. Ants and termites live mostly in the dark, so the workers communicate tactically (by touch) and chemically by signals called pheromones. A **pheromone** is a substance (a scent) secreted by one organism that stimulates a physiological or behavioural response in another individual of the same species. On food-seeking expeditions, worker ants lay down scent trails which can be followed by other workers. Pheromones act as sexual attractants in many species of insects, and have also been identified in birds, fish, and mammals.

Bees also communicate using pheromones. If, for example, a queen leaves a hive, workers detect her absence chemically and hastily build a chamber in which a new queen can develop. Foraging European honeybees communicate to other workers by means of a dance performed on a vertical surface inside the hive, or on the floor at the hive entrance. If a new food source is less than about 70 m from the hive, a foraging bee performs a **round dance** which gives no directional or distance information about the food source. If the food is more than about 70 m from the hive, the returning bee does a figure-of-eight **waggle dance**. The dance gives information about the new food source, including its distance and direction from the hive (figure 2). A dancing bee usually has pollen and nectar on its body. It regurgitates nectar, emits a buzzing noise, vibrates its wings, and waggles its abdomen from side to side. Fellow workers learn the dance by touching and following the dancer. They probably use all the tactile, sound, and chemical signals associated with the dance to discover the precise location and type of food.

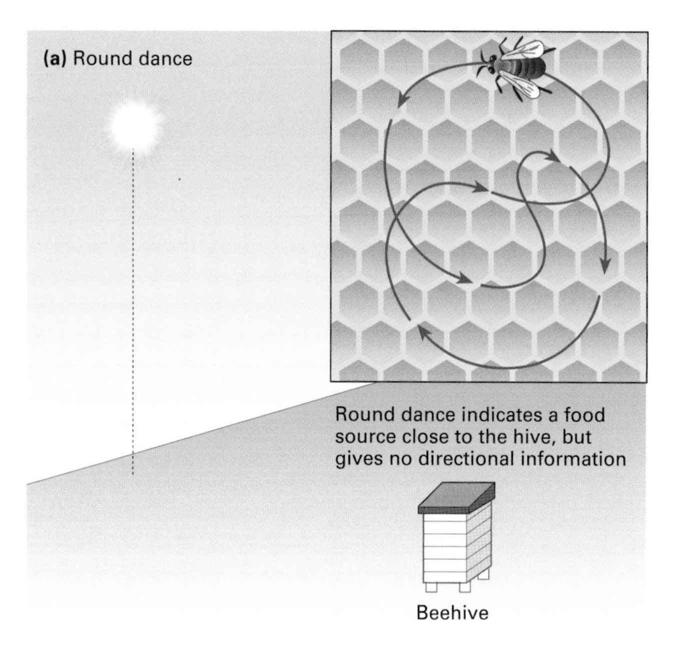

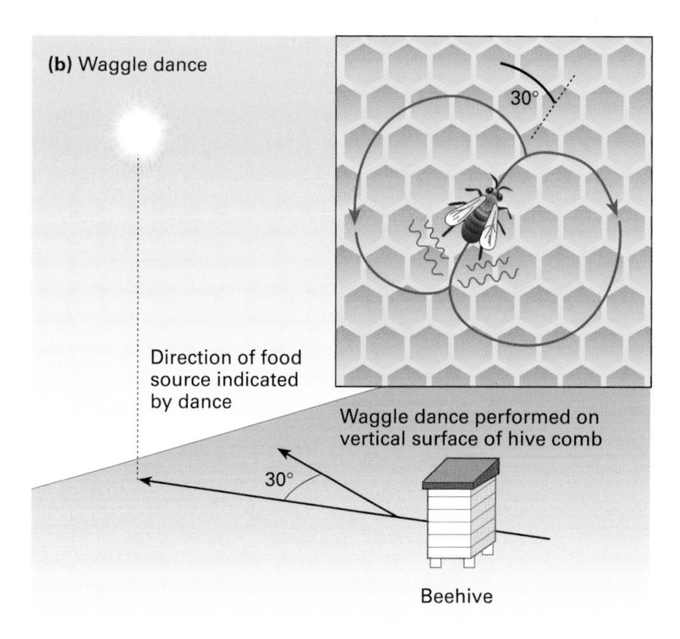

*Figure 2 Bee dances. **(a)** A round dance indicates that the food source is near the hive. **(b)** A waggle dance gives information about the distance and direction of the food. According to Karl von Frisch, the first scientist to describe these dances, the distance is indicated by the duration of each dance and the number of abdominal waggles performed per dance. The direction of the food is indicated by the angle of the straight run in the dance in relation to the vertical surface of the hive. If the run is up the hive, the food is towards the Sun; if the run is down the hive, the food is located away from the Sun. In the example illustrated, the food is 30° to the right of the Sun.*

Quick check

1 What is the difference between an insect society and an aggregation of insects?

2 What are pheromones?

3 What information is given by:

a a round dance

b a waggle dance?

Food for thought

When a foraging bee returns to its hive it uses not only the waggle dance to pass on information about a new food source, but also other forms of communication. Suggest what information would be useful to the other bees in the hive, and the different ways in which this information might be communicated.

11.13

OBJECTIVES

By the end of this spread you should be able to:

- explain why courtship behaviour is often highly ritualised
- describe how birds and mammals maintain their territory
- discuss the significance of dominance hierarchies.

Fact of life

In red deer (*Cervus elephus*), size really does matter. Generally, the bigger the stag, the higher it is in the dominance hierarchy. Usually, rutting contests become bloody only between stags of similar size. A hind's position in the dominance hierarchy is also related to her size; this affects the sex ratio of her offspring. Larger hinds have a higher rank and produce significantly more male calves than female calves.

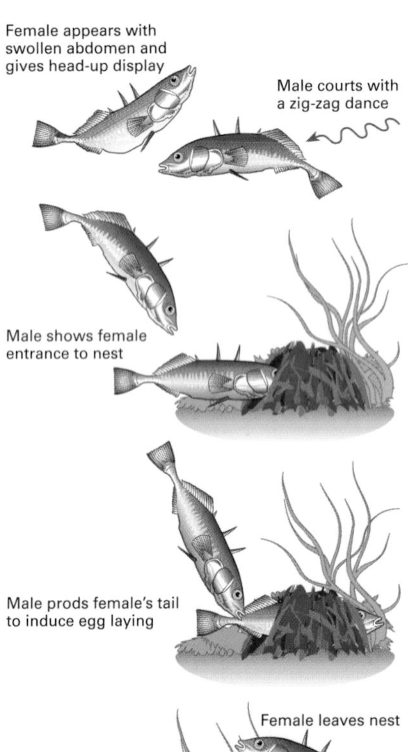

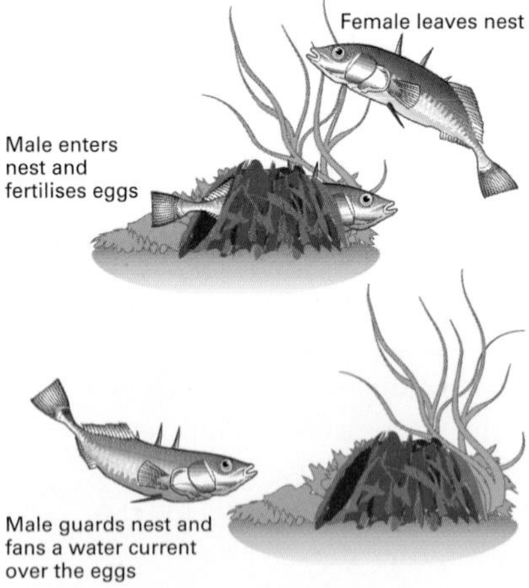

***Figure 1** The courtship dance of sticklebacks.*

Courtship, territoriality, and dominance hierarchies

In this spread we shall look at three aspects of social behaviour in animals.

Courtship in sticklebacks

Courtship, the specialised behaviour that precedes the fertilisation of eggs by a male, is highly ritualised in most animals: it follows a set pattern in different individuals of the species. Courtship is well illustrated in the three-spined stickleback (*Gasterosteus aculeatus*), a small fish that inhabits freshwater ditches, streams, and ponds.

During the breeding season, male sticklebacks leave their school of fish to stake out a territory, aggressively fending off intruders. In his territory, the male builds a nest of weeds glued together by a sticky substance from his kidneys. After completing the nest, the male's underside suddenly changes colour from a dull grey to a bright red. Then he begins to court a gravid female (one that has grown shiny and bulky, showing she is carrying eggs). Whenever a gravid female enters his territory, the male and female perform a highly ritualised courtship dance (figure 1). A successful courtship results in the female laying her eggs in the nest about one minute after the start of the zig-zag courtship dance. The male enters the nest, fertilises the eggs, and chases the female away. After this, he seeks another partner. He may court as many as five females successfully. Then his colour darkens and his energies are devoted to looking after the eggs, fanning water over them (to improve the supply of oxygen), and protecting them from predators. Even after they have hatched, the father looks after his brood until, a few days later, they become independent and join the young of other broods.

The sex life of sticklebacks is mainly controlled by fixed action patterns (spread 11.8) and is therefore instinctive and automatic. The male's zig-zag dance, for example, is triggered by the sight of a female's swollen abdomen which acts as a sign stimulus (spread 11.12). The dance proceeds like a chain reaction, each stage providing the stimulus for the next one.

As a female and male approach each other during the reproductive period, there is a conflict between sexual attraction and attack or escape. The courtship ritual seems to be the means by which aggression is reduced. The zig-zag dance may have evolved from the male alternately attacking the female and leading her to the nest.

Courtship is also important because it brings the sexes together when they are fertile; it ensures that male and female reproductive behaviours are synchronised so that egg-laying and fertilisation by the sperm happens at approximately the same time; and it is highly species specific, ensuring that breeding only takes place between members of the same species.

Territoriality

Although commonly associated with birds and mammals, territoriality is widespread in the animal kingdom. Even sedentary animals, such as some species of sea anemone and limpets, are territorial. **Territoriality** is the defence of an area occupied relatively exclusively by an animal or group of animals. The area is defended by clear acts of aggression or signals. The aggression is usually highly ritualised and rarely results in injury. Signals may be visual, vocal, or chemical, for example:

- fiddler crabs wave their claws
- male robins take up a threat posture
- many species of bird proclaim their territory in song
- toads croak to keep others away
- bull alligators roar
- the cat family use urine as a chemical signal
- hippopotamuses use their solid wastes as signals.

A territory (the defended area) may be used for shelter, gathering food, courting, mating, nesting, or building a home. Some animals maintain a large territory for all these functions; others may defend a relatively small area for only one or a few of these functions. A flightless cormorant on the Galapagos Islands, for example, defends only the area within its reach when sitting on its nest. Some animals hold a territory only in the breeding season, while others keep one all year round (a pair of tawny owls, *Strix aluco*, defends a single territory for the whole of their adult lives).

Territoriality serves several functions. It spreads the population out over the available resources such as food (the amount of food often determines the size of a territory); it leaves animals free from disturbance during pair formation; and, because it increases the distance between individuals, it reduces the risk of infections spreading.

Figure 2 *Territoriality: the regular spacing between gannet nests is a good indicator that this cliff area is divided into territories.*

Dominance hierarchy

A **dominance hierarchy** is the social ranking of each member in a social group. The group is organised so that higher-ranking individuals are dominant over lower-ranking ones. Most dominance hierarchies are linear; they have no members of equal rank. For example, among farmyard chickens, the alpha (top-ranked) chicken is dominant over all others; another chicken dominates all but the top chicken; and so on. Dominance hierarchies are maintained by aggressive behaviour. If a new batch of chickens is brought together, the chickens chase and peck each other until a dominance hierarchy or pecking order is established (figure 3). In an established hierarchy, each chicken can peck only those below it.

Figure 3 *The pecking order of members of a new flock of hens is established by a series of individual combats.*

Most animals use aggression as a means of establishing their status. However, this aggression is expressed in highly ritualised fights, and rarely results in bloodshed. Red deer (*Cervus elaphus*) and several other mammals occasionally establish their status by impressive charging displays (figure 4). Once established, the dominance hierarchy is comparatively stable and is maintained mainly by threat postures. This makes life within the group generally peaceful.

An animal's status determines its right to valuable resources such as food and a mate. The higher the status, the freer the access to a resource. Dominance hierarchies ensure that resources are distributed with the minimum of fuss and that the strongest individuals have the best chance to pass on their genes.

Figure 4 *Rutting in red deer: during the rut (the mating period that coincides with oestrus in most hinds) a roaring contest leads to stags fighting for possession of hinds. Once a dominance hierarchy is established most contests end after the roaring stage and do not reach the fight stage.*

Quick check

1. Which stimulus acts as a sign stimulus initiating the zig-zag dance of stickleback males?
2. How is the territory of, for example, a bird usually established and maintained?
3. What is the name given to the animal at the top of a dominance hierarchy?

Food for thought

African hunting dogs (*Lycaon pictus*) are highly aggressive predators that feed on a wide range of prey including wildebeest (spread 11.11, figure 1). Suggest the advantages and disadvantages to a dog of:

a living in a pack
b having a low position within the dominance hierarchy of the pack
c being part of a pack that holds a territory.

11.14

Behavioural rhythms and biological clocks

OBJECTIVES

By the end of this spread you should be able to:

- describe the main types of behavioural rhythms in animals
- discuss how behavioural rhythms are regulated
- explain the significance of behavioural rhythms.

Fact of life

The pineal gland is a peculiar organ that may have evolved from an ancient third eye. Fossil ostracoderms, extinct fish that are thought to have been our remote ancestors, have holes for three eyes. Some present-day vertebrates have retained this third eye. In the New Zealand reptile *Sphenodon punctatus* (the tuatara, figure 1), this eye even has a lens and its own retina.

Figure 1 *The tuatara, a New Zealand reptile with a third eye.*

In humans, the pineal gland is a pea-sized organ consisting of a small mass of nerve tissue located deep in the brain (spread 10.13, figure 1). It is not exposed to light, but has nervous connections with the retina so it can receive information about changing light and dark cycles indirectly. It secretes two hormones, melatonin and serotonin, which may play a role in the behavioural rhythms of mammals. Serotonin secretion is stimulated by light. Melatonin secretion is inhibited by exposure to light so that melatonin levels rise during the night and fall during the day. Melatonin is thought to affect the development of the gonads and may be involved in the onset of puberty. The pineal gland is connected to the suprachiasmatic nuclei (SCN) in the hypothalamus. The SCN and pineal gland probably interact to form the biological clock.

Regular patterns of behaviour

Many patterns of behaviour, such as sleep, feeding, and drinking, recur at regular intervals and are known as **behavioural rhythms**. The time between each cycle in a rhythm can vary from minutes to years. For example, *Arenicola marina* (polychaete worms that live in U-shaped burrows on muddy shores) feed every 6 or 7 minutes. Many mammalian predators have a **circadian rhythm** (an approximately 24-hour rhythm) of eating during the day and resting at night. In contrast, most small desert mammals are nocturnal, hiding in their burrows during the day to avoid the blistering heat of the desert sun and gathering their food at night when it is cooler.

Fiddler crabs have a twice-daily **tidal rhythm** of activity: they forage when the tide is out and seek refuge when the tide is in. Some intertidal animals (animals that live between the high-water and low-water marks) have a cycle of activity corresponding to the 28.5-day lunar cycle of the Moon's movement around the Earth. The grunion (*Leuresthes tenuis*) is a small fish that spawns on Californian beaches in May and October, when the tides are at their highest.

Many birds and mammals have **circannual rhythms** (rhythms of approximately one year) of breeding, migration, and hibernation. Some migration cycles are longer: the wandering albatross (*Diomedea exulans*) continually circles the southern seas in an easterly direction, returning to its nesting island once every two years.

Regulation of behavioural rhythms

Behavioural rhythms enable animals to use environmental resources at the safest and most profitable times. The rhythms correspond to regular variations in environmental factors such as water availability, temperature, and light intensity. It might appear that the **exogenous** (external) environmental factors are driving the behavioural cycles. However, many rhythms are maintained even when an animal is placed in an unchanging environment that gives no indication about time. These rhythms must be regulated by some sort of **endogenous** (internal) mechanism.

Circadian (daily) rhythms may continue in the absence of an external time cue, though the start of activity may vary by a small amount each day (rhythms with an exact 24-hour cycle are called **diurnal rhythms**). For example, in its natural environment, the mosquito *Anopheles gambiae* (one of the most important carriers of malaria) is active mainly at night, with a peak of flight activity at dusk. In the uniform darkness of an experimental chamber, the mosquito's 'dusk' peak is repeated for several days, but it has a rhythm of close to 23 hours (figure 2). Therefore, after a few days, the peak activity no longer coincides with dusk in the environment. In its natural environment, light (an exogenous factor) seems to adjust the mosquito's rhythm to a more precise 24-hour cycle.

Studies on numerous other animals confirm that circadian rhythms have an endogenous timing mechanism or **biological clock** which is synchronised with external cycles by an exogenous cue. The exogenous cue is sometimes called the *zeitgeber* (German for 'time giver'). The most common zeitgeber is light.

The timing of rhythmical behaviour with a longer cycle, such as a circannual (yearly) migration, is determined partly by hormonal changes linked directly to exogenous factors such as day length. The timing may also involve a biological clock, but it is difficult to test this experimentally as the animals would have to be kept in constant conditions for months or even years.

Shorter rhythms are probably regulated mainly by an endogenous mechanism and may be little affected by exogenous factors. The feeding behaviour of the lugworm, for example, appears to be controlled exclusively by an endogenous mechanism.

The biological clock

The exact nature of the biological clock is the subject of much active research. It probably involves a structure that can function as a pacemaker. This structure appears to be different in different animal groups. In lugworms, for example, the pharynx acts as a pacemaker for feeding rhythms, and transmits its effects through the ventral nerve cord. In mammals, two clusters of neurones in the hypothalamus called the suprachiasmatic nuclei (SCN) seem to act as the major clock. The SCN contain many important behavioural centres, receive nerve signals directly from the retina, and have connections with the pineal gland (see Fact of life) and pituitary gland (spread 10.14). They are therefore in an ideal position to regulate endocrine and behavioural activity in response to changing periods of light and dark.

Human behavioural rhythms

Behavioural rhythms affect humans as well as other animals. For example, the 28-day female menstrual cycle involves changes in the ovaries, uterus, and breasts (spread 12.4), and usually also results in considerable variations in body mass, total body water, and metabolic rate. These variations may have a dramatic effect on a woman's sense of well-being. Circadian rhythms also include changes in body temperature: this tends to be at its lowest in the early hours of the morning (incidentally, this is when most people die); it rises soon after getting up; and it reaches a peak in the late afternoon or early evening.

Sports scientists can use a knowledge of circadian rhythms to maximise the physical performance of athletes: several studies indicate, for example, that runners, swimmers, and cyclists perform best in the afternoon and early evening. Business executives can minimise the effects of 'jet lag' when moving from one time zone to another by resting after their long journey to give their biological clocks a chance to be reset. Sufferers of **seasonal affective disorder** (SAD, a feeling of depression and lethargy associated with a reduction in the number of daylight hours in autumn and winter) may be able to speed up their recovery by exposing themselves to a bright artificial light for one or two hours each day. This is thought to reduce the secretion of melatonin, a hormone linked to the disorder.

Quick check

1 Define and give an example of a zeitgeber.

2 Which part of the human brain probably contains the major biological clock?

Food for thought

In addition to business executives and sports coaches, human circadian rhythms are of interest to astronauts and military commanders. Why might these last two groups be interested in circadian rhythms? How do you think researchers study circardian rhythms in humans?

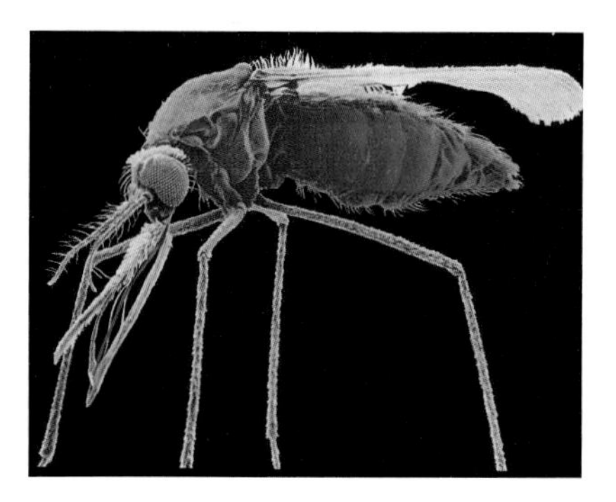

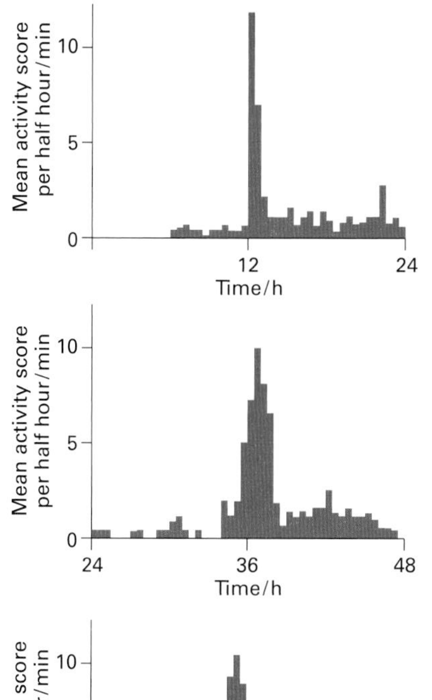

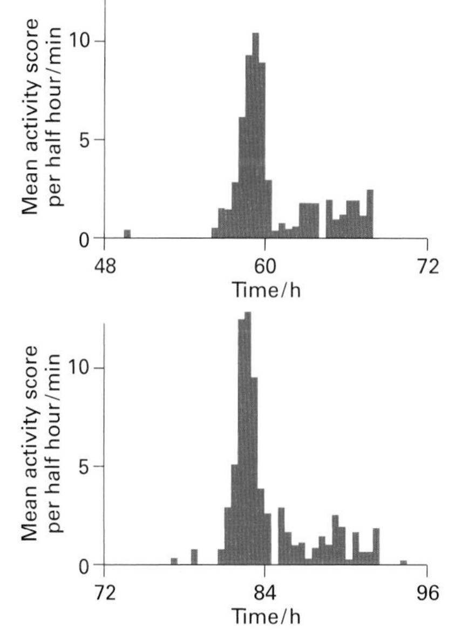

***Figure 2** The effect of uniform conditions on rhythms of flight activity in the mosquito* Anopheles gambiae.

11.15

Bird migration

OBJECTIVES

By the end of this spread you should be able to:

- distinguish between migration and dispersion
- discuss the advantages of bird migration
- discuss how birds navigate.

Fact of life

The longest migration of any known animal is that of the Arctic tern. Each year, it travels up to 5000 miles (40 000 km) on its round trip from the Arctic to the Antarctic, and back again.

What are bird migrations?

Bird migrations are among the most spectacular, the most incredible, and the most researched types of animal behaviour. They have always fascinated scientists and non-scientists. Some birds travel thousands of miles to return to the same breeding ground or even the same nesting box. Unlike dispersion, or weather movements undertaken by species such as lapwings in hard weather, true bird migration involves regular movements between fairly well defined breeding and wintering areas and it is confined to definite seasons; also, migrations are often performed by entire populations.

Half of all bird species migrate, yet the journey of each is unique. Large birds generally migrate by day whereas most small songbirds migrate at night to avoid predators. Swallows and housemartins are exceptions; although small, these birds can migrate relatively safely during the day because of their superior flying skills. Birds such as the sedge warbler (figure 2), which make exceptionally long journeys, may migrate by day and night. Whooper swans fly both by day and night when there is moonlight or clear sky but land on the sea in conditions of low cloud and poor visibility, and also in darkness when there is heavy cloud cover and no moon.

Why do birds migrate?

Most British migrants (for example sedge warblers, cuckoos, swallows, swifts, nightingales, and housemartins) breed in Britain in the summer when days are long, food is plentiful and the temperature high; they migrate to Africa and the Middle East in the winter, avoiding the worst of the British weather. Barnacle geese and Brent geese, on the other hand, breed in the Arctic during the summer and overwinter in northern Britain and Europe.

According to one theory, the timing of a migration is controlled by changes in the photoperiod (spread 14.12) through its effects on the production of gonadal hormones. An increasing amount of light and lengthening of the day stimulates the development of the gonads, while a decrease has the reverse effect.

No matter when they occur or in which direction they take place, migrations move a species between two different types of habitat, each of which is more favourable at different times. The productivity of each habitat is seasonal. The low-productivity period is insufficient to support the population, so it has to migrate. The main purpose of migration,

Bird ringing

Each year the legs of hundreds of thousands of birds are ringed or banded by scientists. The technique consists simply of placing around the bird's leg a ring carrying a unique serial number (figure 1) and an address to which its discovery should be reported. In addition, colour codes may be used so that birds can be identified through binoculars without having to recapture them. The rings are made of a light alloy which does not harm or hinder the birds in any way. To ensure the birds come to no harm, only highly skilled ornithologists who have undergone thorough training are permitted to carry out ringing on their own. Ringing has enabled ornithologists to greatly increase their understanding of migration routes, the timing of migration, and wintering areas used by birds. The data that is gathered is very important in telling us about what a bird species needs in order to survive. This information is essential for formulating a sound conservation policy.

Figure 1 *A bird being ringed.*

therefore, is to ensure an adequate food supply at all times, but it also enables migrants to avoid seasonal bad weather. By making their arduous journeys, migrating birds are able to utilise food in parts of the world where they might not be able to live permanently. For example, in winter, Brent geese can make the most of the milder weather and abundant food found in British wetlands and rain-soaked farm fields. In the summer, the long days of northern latitudes enable them to feed and raise their young around the clock in the presence of relatively few predators.

The remarkable sedge warbler, *Acrocephalas schoenobaenus*

Sedge warblers (figure 2) are the commonest and most widely distributed of the European marshland warblers. From about March to September, they inhabit dense vegetation along waterways and lakesides, as well as dense hedgerows and thickets. They perform one of the most extraordinary migrations, leaving Britain in autumn to spend winter in the African savannah grasslands, south of the Sahara desert, and returning in spring. Their long journey requires a phenomenal amount of strength and stamina. Like most migrants, sedge warblers store body fat before their journey. A warbler normally weighs about 10.5 g, but after fattening on insects for two to three weeks before its migration it may weigh more than 22 g. If heavy enough, it will fly at about 40 kilometres per hour (25 m.p.h.) for up to 70 hours to make its journey in a single marathon flight. By the time it reaches its destination it could weigh as little as 10.5 g, less than half its starting weight. There is little margin for error: below 9.5 g, death is inevitable. If not fat enough, or if conditions are unfavourable during a migration, sedge warblers usually stop to refuel before they make the dangerous Sahara crossing.

Figure 2 *A sedge warbler.*

How do birds navigate?

There appears to be no simple answer to the question of how birds find their way on immense journeys. Although in some species, such as geese, old birds may show young birds the way, the ability of many species to navigate is instinctive and is not dependent on learning or copying. This is shown by a comparison between the timing of migrations of old and young birds. For example, cuckoos migrate alone, well after the departure of their true parents whom they never meet; in many passerines (perching birds and songbirds), young, new migrants often set off before their elders.

Distinctive smells, wind direction, and topography may all play a part in navigation, but to migrate long distances over trackless stretches of sea, birds probably use a more precise navigational aid.

Daytime-migrants may use the position of the sun while night-migrants may use the position of the moon and stars. Many, if not all, birds are able to navigate by monitoring the Earth's magnetic field. This has been shown by fitting experimental groups with slender rods of iron. Those with rods magnetised in such a way as to obscure the Earth's magnetic field become disorientated, while the migrations of those with unmagnetised bars are unaffected. Microscopic grains of an iron oxide, magnetite, found in the brains of some birds may be part of the sensing system.

In order to compensate for changes in the position of the sun at different times of the day or the stars at different times of the night, birds must be able to tell the time. A biological clock (spread 14.12) would enable them to do this. The effect on navigation of resetting the biological clock has been studied using caged birds. When they are in a migratory state, they flutter in the direction they want to go. If the clock is reset by artificially altering the day length, the direction in which they flutter changes in the expected manner. Similarly, caged night-migrants vary their flutter direction according to the artificial night sky in a planetarium.

QUICK CHECK

1 How does migration differ from dispersion?

2 State the main advantage of seasonal migrations.

3 Explain why cuckoos cannot learn to navigate from their parents.

Food for thought

Bird migrations are not simple responses to lack of food or inclement conditions. Many birds leave Britain to fly south as early as the end of summer, when the weather is fine and plenty of food is still available. Others migrate even though they would be able to find enough food and are hardy enough to survive the winter in Britain. Bearing these observations in mind, suggest how migrations originated.

Summary

There are several types of **animal skeleton**, including tubular **cytoskeletons** within cells, watery **hydrostatic skeletons**, **exoskeletons** containing chitin, and **endoskeletons**.
Vertebrate endoskeletons are made of bone and/or cartilage.
Bone varies in size and shape, and it may be compact or spongy.
The human skeleton has a combination of bony and cartilaginous structures divided into two main components: the **axial skeleton** and **appendicular skeleton**.
Movement of the body or body parts requires **muscles and joints**. There are three main types of muscle in mammals: cardiac, smooth, and skeletal. A joint is the junction between two bones. Freely moving joints are called **synovial joints**.
Movement of a joint in opposite directions is brought about by antagonistic pairs of muscles. **Skeletal muscle** is attached to **bones** by **tendons**. Each skeletal muscle is made up of thousands of **muscle fibres**. They may be **slow-twitch** or **fast-twitch** fibres.
Under the electron microscope, muscle fibres can be seen to be made of repeating units called **sarcomeres**. Huxley proposed the **sliding filament theory** to explain how muscle fibres contract.
Locomotion resulting from contractions of skeletal muscle includes **swimming**, **flying**, **walking**, and **running**.

Behaviour refers to observable responses of an organism to changes in the environment. Simple responses include **reflex actions**, **stereotyped behaviour** composed of **fixed action patterns** triggered by **sign stimuli**, and **kineses** and **taxes**.
Learning allows behaviour to be modified by experience. Types of learning include **habituation**, **imprinting**, **conditioning**, **latent learning**, and **social learning** by **imitation**, for example.
In order to learn, an animal has to have a **memory** to store and recall information. **Social behaviour** relies on **communication** between individuals. This may involve **pheromones** or specific movements, as in the **waggle dance** of bees. **Courtship displays**, **territoriality**, and the formation of **dominance hierarchies** are common forms of social behaviour. Patterns of behaviour, such as courtship and feeding, often occur rhythmically.
Behavioural rhythms may be **circadian**, **diurnal**, **tidal**, or **circannual**. They may be regulated by environmental factors that affect an internal timing mechanism called a **biological clock**.

PRACTICE EXAM QUESTIONS

1 The diagram represents a longitudinal section through part of a striated muscle.

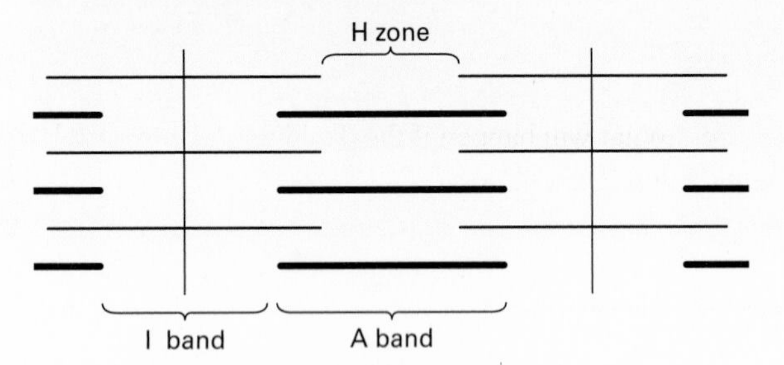

a The diagram shows the A band, the I band and the H zone. Which one or more of these:

i contains actin but not myosin; [1]

ii shortens when the muscle contracts? [1]

b Describe the part played by each of the following in muscle contraction:

i ATP; [2]

ii calcium ions. [1]

[Total 5 marks]

2 The diagram shows the apparatus used to investigate muscle contraction following stimulation of a nerve. As the muscle contracts it causes a lever to move. The lever produces a trace on a sheet of paper moving at a constant speed. The trace shown was obtained after a single stimulus.

a i Calculate the length of the delay between the stimulus being applied (**X**) and the muscle starting to contract (**Y**). [1]

ii Explain what causes this delay. [2]

b The nerve is kept moist with a solution containing sodium ions. Describe the part played by sodium ions in a nerve impulse. [2]

c In this investigation, the nerve impulse spreads in both directions along the nerve from the point of stimulation. Explain why a nerve impulse normally goes in only one direction in a living animal. [1]

[Total 6 marks]

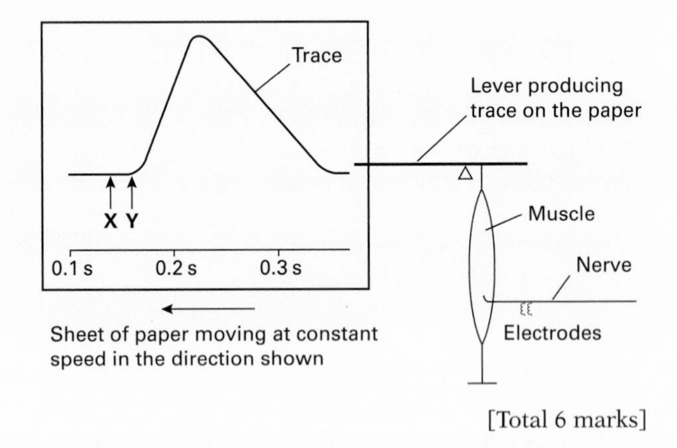

[Total 6 marks]

3 The diagram shows the structure of a nerve-muscle junction together with part of the associated muscle.

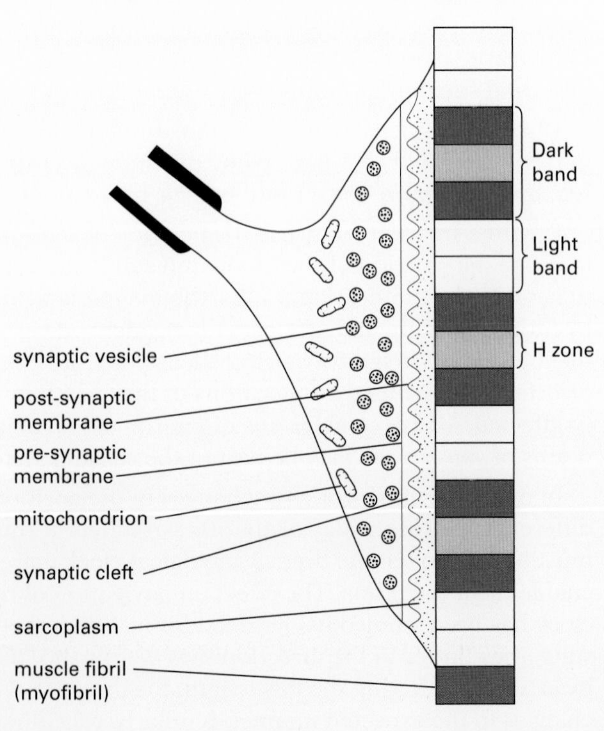

a Describe how transmission of information occurs across the nerve-muscle junction when an impulse arrives at the pre-synaptic membrane. [5]

b **i** What causes the banding pattern seen in the muscle fibril? [3]

ii How and why will the banding pattern change when the muscle fibril contracts? [4]

[Total 12 marks]

4 The ability of rats to learn the route out of a maze was investigated. Three groups of rats were used. Each rat in group **A** was put in the maze and when it found its way out it was given one pellet of food. The time taken for each rat to get through the maze was measured and the mean speed for each group of rats was calculated. Each rat in group **A** was put through the maze 20 times. The rats in the other two groups were treated in exactly the same way, except that each rat in group **B** was rewarded with 16 pellets of food and each rat in group **C** with 50 pellets.

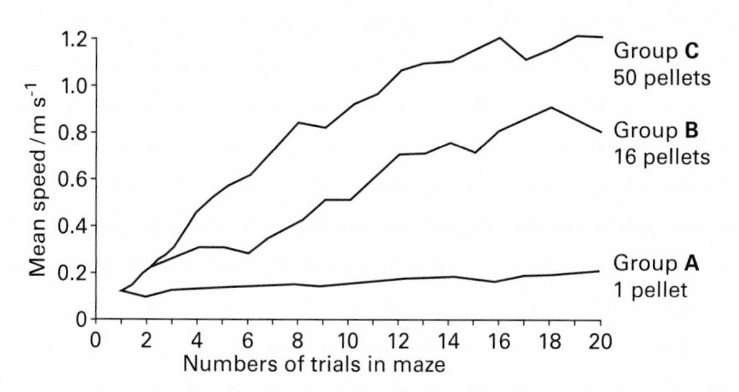

a Describe *one* piece of evidence that the rats were able to learn their way through the maze. [1]

b Suggest and explain what type of learning behaviour was shown by the rats in these experiments. [2]

c Describe and explain the effect that the amount of foot had on the behaviour of the rats. [2]

[Total 5 marks]

5 Robins have individual territories during the autumn and early winter, but from early in the new year pairs of birds begin to share the same territory, which they maintain throughout the spring and summer.

a Describe and explain the advantages of territorial behaviour, with reference to the behaviour of the robin throughout the year. [7]

b The territories are usually defended by song and displays, which often involve exhibiting the red breast as much as possible to any intruding robin. Fighting is sometimes involved, especially when the territory is being established. Suggest why robins usually defend their territories by song and display rather than fighting. [3]

c A robin shows aggressive behaviour towards models of adult robins and bunches of red feathers placed within its territory. However, models of young robins which do not have a red breast, are not threatened. What can be deduced from these observations about the factors controlling the aggressive behaviour of robins? [2]

[Total 12 marks]

6 In a series of experiments on ducklings, the effectiveness of imprinting was measured at various times after hatching. In each of these experiments, the ducklings had the opportunity to follow a model duck. The figure below represents the percentage of ducklings at a particular age which followed the model.

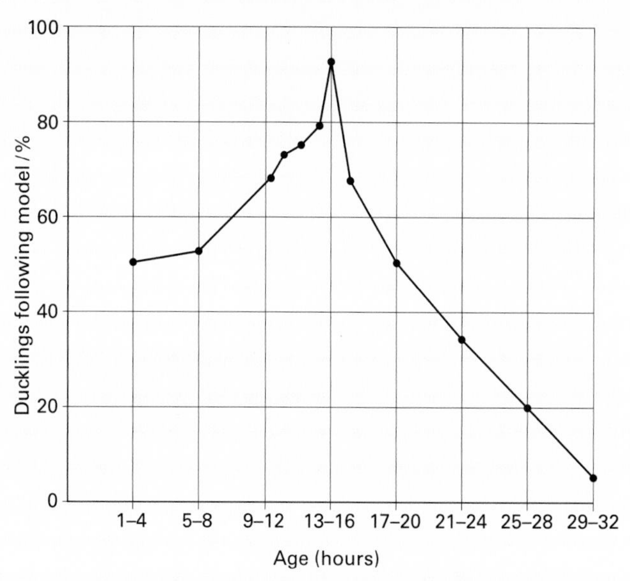

a Explain what is meant by *imprinting*. [2]

b What conclusions can be drawn from these results? [2]

c What will happen if the ducklings have no adult to imprint on? [1]

[Total 5 marks]

7 Young chaffinches which have been hand-reared from eggs and isolated from other chaffinches produce songs of the same length and containing the same notes as wild chaffinches, but the phrasing is abnormal. These birds never develop a normal song pattern, even if they are later exposed to songs of wild chaffinches. Young chaffinches caught just after leaving the nest and isolated from other chaffinches develop an almost normal song pattern. If these chaffinches are exposed to the songs of adult chaffinches when they are eleven months old they develop a completely normal song pattern, similar to the adult birds around them.

a Use the information in this passage to describe and illustrate the main features of

i innate behaviour;

ii imprinting. [8]

b Suggest why it is important for chaffinches to develop a normal song pattern. [4]

[Total 12 marks]

8 **a** Define the term 'behaviour' and describe the difference between 'instinctive' and 'leamed' behaviour in animals. [8]

b With example(s) of instinctive behaviour explain fully the way in which the response is brought about. [6]

c Discuss, with examples, the advantages and disadvantages of this type of behaviour to individuals and to the animal species. [6]

[Total 20 marks]

12 Animal reproduction

Reproduction

12.1

OBJECTIVES

By the end of this spread you should be able to:

- describe the main types of asexual reproduction in animals
- compare the advantages and disadvantages of asexual reproduction and sexual reproduction.

Fact of life

The rate of reproduction achieved by asexual reproduction can be astronomical. For example, in favourable conditions some bacterial cells can divide into two every 20 minutes. This means that in 24 hours a single cell could give rise to a population exceeding 4000 million million (4×10^{15}), assuming all the offspring survived.

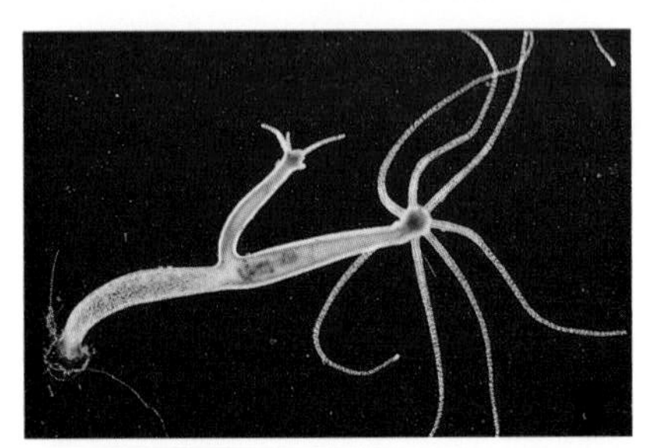

Figure 1 *Budding in* Hydra.

Figure 2 *Parthenogenesis in aphids.*

Reproduction is an essential feature of life on Earth. Every individual dies: even if it survives the dangers of predation, disease, and natural disaster, its homeostatic mechanisms will eventually wear out. However, though individuals die, a species continues as long its members produce offspring that live long enough to have offspring of their own. The ability of life to colonise new habitats and to survive in changing conditions depends on offspring resembling their parents, but not being *exact* copies. Reproduction therefore has to be able to provide both continuity and change.

There are two main types of reproduction: **asexual reproduction** and **sexual reproduction**.

Asexual reproduction: single parents

Asexual reproduction involves a single parent producing one or more cells by mitosis (spread 4.11). These cells develop into offspring which (barring mutations) are genetically identical with each other and with the parent. An advantage of asexual reproduction is that time and energy are not spent seeking a receptive mate. It can also be an efficient method of producing large numbers of offspring quickly, enabling organisms to take advantage of favourable environmental conditions. The passing on of unchanged genetic information in asexual reproduction may also be an advantage, but only if the environment is stable. Lack of variation among the offspring makes it difficult for a species to adjust to a changing environment.

There are several types of asexual reproduction including vegetative propagation (spread 14.7), cloning (spread 18.10), fission, budding, and fragmentation.

Fission is the splitting of an individual into two individuals (binary fission) or more than two individuals (**multiple fission**) of approximately equal size. It is a common form of reproduction in bacteria and single-celled organisms, and also occurs in some sea anemones.

Budding (figure 1) is common among Cnidaria, the phylum that includes *Hydra*, sea firs, sea anemones, and jellyfish. The buds (outgrowths of the body wall) of *Hydra* eventually separate from the parent to lead an independent life. Sea firs, such as *Obelia*, have a stage in their life cycle when they live in colonies attached to other objects such as seaweeds. The colonies form from buds that have not detached and the individual offspring remain connected by a common gut cavity.

Fragmentation is an extreme form of regeneration. Many invertebrates can regenerate lost body parts: an earthworm can regrow its head, a starfish its arms, and a crab its legs. Some invertebrates have developed this ability to such a high level that they use it to reproduce asexually. A planarian (a free-living flatworm), for example, constricts in the middle to split into two pieces. The missing ends are regenerated to form two individuals.

Sexual reproduction: two parents

Sexual reproduction generally involves two parents. Typically, the offspring result from the fusion of two haploid cells (cells with nuclei containing only one set of chromosomes, spread 4.10) called **gametes**. In some organisms there is no apparent difference between gametes, but in most animals there are separate male and female organisms that produce gametes which are quite different: the female gamete is usually a relatively large and immobile cell containing food reserves; the male gamete is usually much smaller and swims to the female gamete. Meiosis (reduction division, spread 4.12) takes place at some stage of the life cycle of sexually reproducing organisms to produce the gametes. Meiosis and fertilisation produce new combinations of genes, giving rise to offspring that are genetically different from their parents and from each other.

By increasing the genetic variability of a species, sexual reproduction enables the species to adapt to new environmental conditions. Offsetting this major benefit is the cost in terms of the energy and time required to find a mate. The reproductive strategy that a species adopts (asexual or sexual) depends on the relative merits of each method for the species' particular lifestyle and environment. Some animals reproduce only asexually or sexually; others can use either method at different times or under different conditions. Some sexually reproducing animals, such as the dunnock (figure 4), have a highly variable mating system.

Parthenogenesis: unfertilised gametes

An alternation of asexual and sexual phases occurs in the life cycle of most aphids (small insects which feed on plant sap; greenfly are one type of aphid). Aphids produce two types of gamete depending on the environmental conditions and the time of year. In temperate regions, all aphids are asexually reproducing females during the spring and summer (figure 2). Their gametes develop into small aphids without being fertilised. This process, called **parthenogenesis**, enables the whole population to produce young. It allows the population to grow very rapidly so that the species can take maximum advantage of plentiful supplies of food and favourable environmental conditions. Towards the end of autumn, food becomes scarcer and the weather less favourable. The parthenogenetic females produce young of both sexes. Mating between sexual males and females results in fertilised eggs that can survive the winter. The eggs lie dormant in the soil or on the host plant until the following spring when the cycle starts again.

Some fish, amphibians, and reptiles reproduce only by parthenogenesis, and have no sexual reproduction. For example, whiptail lizards (*Cnemidophorus uniparens*) are sexless; all individuals have the potential to produce eggs which can develop without fertilisation. However, more eggs are produced if the lizards imitate the courtship behaviour of sexually reproducing species. During the breeding season, whiptail lizards form 'mating' pairs. One lizard assumes the male role and the other the female role as they go through the motions of sexual reproduction (figure 3).

Figure 3 *These two whiptail lizards reproduce asexually, but they go through the motions of mating.*

Monogamy, polyandry, and polygyny

Figure 4 *Cloacal pecking in the dunnock (*Prunella modularis*).*

The **dunnock** or hedge sparrow, a small, rather drab brown-grey bird, has a variable mating system. Although essentially monogamous (one female mates with one male) or polyandrous (one female mates with two males), it may also be polygynous (one male mates with two females) or even polygynandrous (two to three males mate with two to four females). Male reproductive success (defined as the passing of a male's genes onto the next generation) is least in polyandry, intermediate in monogamy, and greatest in polygyny. Females have the greatest reproductive success in polyandry. In polyandrous dunnocks, sperm from the two or more males compete to fertilise the eggs of a single female. In an attempt to reduce the number of competing sperm and increase his chances of paternity, each male pecks at the female's cloaca during courtship to stimulate her to eject the sperm of other males that have recently mated with her. This pecking is rarely 100 per cent successful: DNA fingerprinting reveals that chicks within the same brood often have different fathers. Males provide paternal care in proportion to their mating success. Therefore, one result of polyandry is that it is common to see chicks in the same brood being cared for by more than one male. Polyandry is most common in areas where food is less readily available. The extra paternal care probably gives the chicks a better chance of survival especially in food-poor environments.

Quick check

1. List three types of asexual reproduction.
2. Give two advantages of asexual reproduction.

Food for thought

One of the major disadvantages of asexual reproduction is low variability of offspring. Many species use only asexual reproduction, but their offspring are not all clones. Suggest how variability comes about in these asexually reproducing species.

12.2

The mammalian reproductive system

OBJECTIVES

By the end of this spread you should be able to:

- describe the reproductive system of a mammalian male
- describe the reproductive system of a mammalian female
- state where the male gametes are produced and how they leave the body
- state what happens to an egg cell when it leaves an ovary.

Fact of life

Whereas female primates (including women) have a single central uterus which usually supports one embryo at a time, some placental mammals (including mice, rats, and rabbits) have two long, separate uterine arms. Each elongated uterus can support many embryos at once, each with its own placenta.

With the exception of artificial cloning (spread 18.10), mammals reproduce only sexually. Sexual reproduction has been a potent force in their evolution and continued existence. Apart from offspring from the same fertilised egg (such as identical twins), every individual born by sexual reproduction is unique. This results in a population of varied individuals, enabling the species to adapt to changes in the environment. However, this capacity for change does not come cheaply. Much energy and time is expended in mating, and only one partner actually produces offspring. Also, parents perfectly adapted to their environment are unlikely to leave offspring that are equally well adapted to the same environment. Nevertheless, the advantages of sex in mammals outweigh the disadvantages and they have developed highly effective reproductive systems which ensure its success.

Reproduction in humans

The human reproductive system is typical of many mammals. Therefore, by examining it in this and subsequent spreads (to spread 12.10) we can learn much about the reproductive systems of other mammals. Most human adults can easily be distinguished as either male or female. The male's reproductive function is to produce large quantities of small motile gametes (**spermatozoa**, commonly called sperm), and discharge them from his body into the female reproductive tract. Each discharge of sperm is called an **ejaculation**. A female's reproductive function is to produce female gametes (egg cells), receive the sperm, provide protection and nourishment for the development of the fertilised egg cell into a baby, and then expel it from her body when it is ready to be born.

Although structurally and functionally very different, the male reproductive system (figure 1) and the female reproductive system (figure 2) have some features in common. Both sexes have:

- a pair of gamete-producing organs, the **gonads** (ovaries or testes)
- a system of ducts which connects the gonads to other parts of the body and the outside world
- structures of **copulation** (the act of mating, when sperm from the male are transferred to the female).

In common with other terrestrial animals, the copulating structures and ducts have evolved to bring the gametes together without them having to pass through the air or over the land.

The male reproductive system

In the male, the **testes** (two spherical structures) produce sperm by spermatogenesis (spread 12.3). Sperm form in **seminiferous tubules** of which there are about 1000 in each testis. The total length of the tubules is over 500 m, enabling the tubules to produce 200–500 million sperm for each ejaculation. As men are able to produce sperm for the whole of their adult lives, the total production of sperm can be astronomical. The testes hang outside the abdominal cavity in a sac called the **scrotum** so that the sperm do not overheat. The temperature in the abdominal cavity is often too high for sperm to develop properly: sperm develop most efficiently at 34 °C.

A system of tubes carries sperm from the testes to the penis. The sperm pass from the seminiferous tubules through the **epididymis** and **vas deferens** (which also act as sperm stores), into the **urethra**. The urethra acts as the exit route for both sperm and urine. During the journey of the sperm from testis to urethra, accessory organs add secretions to the sperm. The **prostate gland** secretes a milky white fluid that neutralises the acidity of any traces of urine and helps protect sperm from acidic secretions in the female body. The **seminal vesicles** secrete a thick, clear

Side view

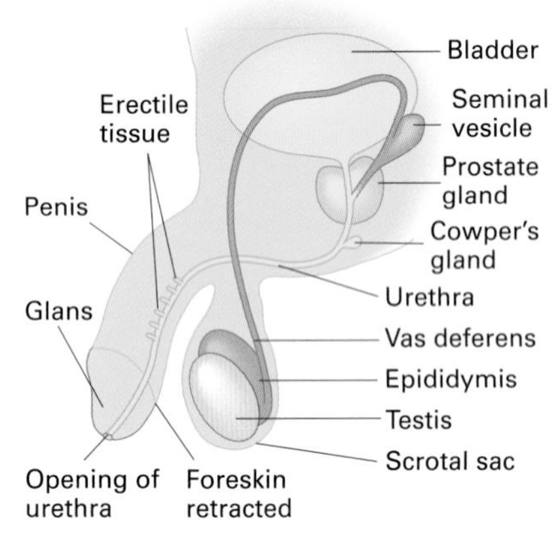

Front view

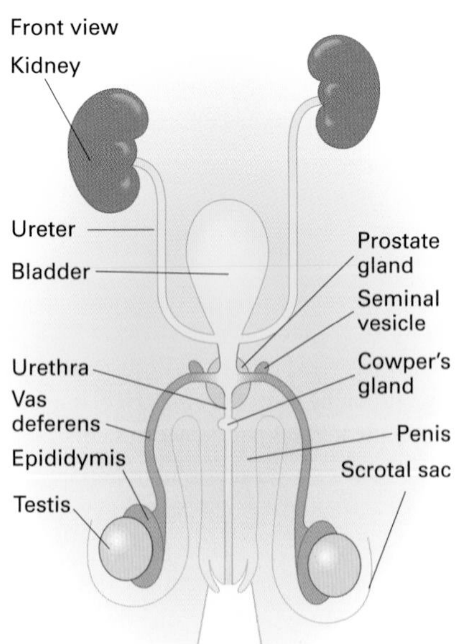

Figure 1 *The human male reproductive system.*

fluid that nourishes the sperm, and the Cowper's glands (bulbourethral glands) secrete a lubricant. The sperm, along with fluid from the prostate gland, seminal vesicles, and Cowper's glands, make up the **semen** (**seminal fluid**) that is discharged during an orgasm. For semen to be ejaculated into the female, the penis has to become stiff (a process called **erection**) so that it can be inserted into the **vagina** (the canal leading from the cervix to the external opening of the female's reproductive system). This act of **copulation** or **coitus** is discussed more fully in spread 12.5.

The female reproductive system

A woman's **ovaries** are structures about 2.5 cm long containing small sac-like structures called **follicles**. The wall of each follicle surrounds and nourishes an **oocyte** which undergoes meiotic division to form an egg cell. The process by which egg cells are formed is called oogenesis, and is discussed in the next spread. A woman is born with about 200 000 to 400 000 follicles, but only 200 to 400 complete their development: the rest degenerate.

Starting at puberty, a woman usually sheds one (but sometimes more) egg cells from her ovaries every 28 days or so. The shedding of an egg cell is called **ovulation**, and is controlled by the interaction of a number of hormones.

When an egg cell leaves the ovary, it passes across a small space and is caught up in the **infundibulum**, the funnel-shaped end of the **oviduct** (Fallopian tube) which leads to the **uterus**. Cilia sweep the egg cell towards the uterus. If fertilisation occurs, it usually takes place high up in the oviduct. A fertilised egg cell, the **zygote**, develops into a multicellular **embryo** which passes to the uterus.

The embryo embeds itself in the **endometrium**, a glandular layer inside the muscular uterine wall, which is richly supplied with blood vessels. A placenta develops to supply the embryo with food and oxygen, and the embryo continues to develop in the uterus until birth. An unfertilised egg cell degenerates and leaves the body during **menstruation**, the shedding of the endometrium that happens approximately each 28 days in a fertile adult female who is not pregnant.

The uterus opens into the muscular tube of the **vagina** through a small ring of muscle, the **cervix**. The vagina wall is lined with smooth muscle and an epithelium that secretes mucus. The vagina is the receiver of semen and the exit passage for a baby. It opens to the outside world through the **vulva**, the collective name for the external genital organs which play an important role in copulation. The urethra has a separate opening just anterior to the vulva.

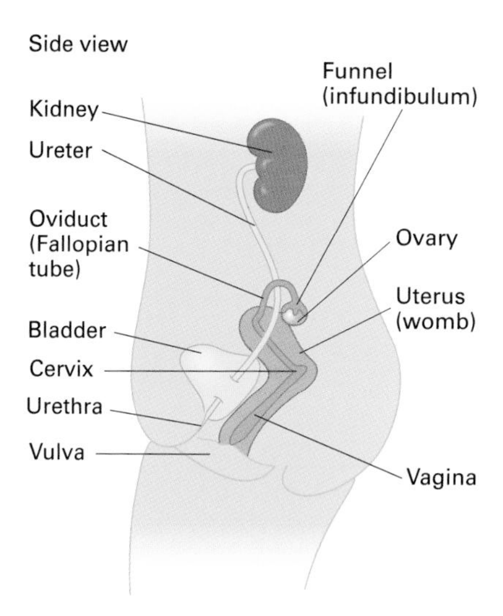

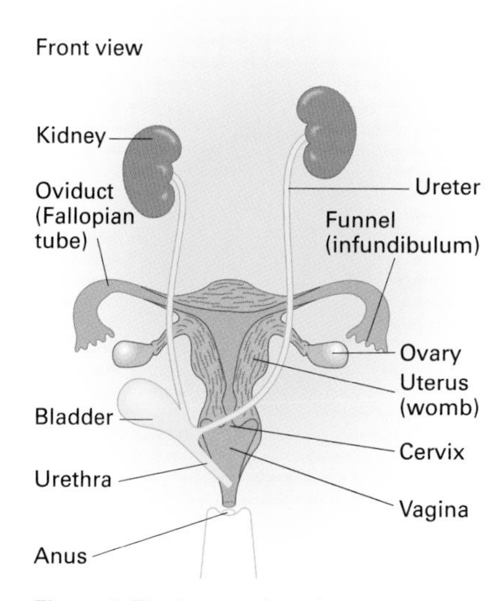

Figure 2 *The human female reproductive system.*

Quick check

1 What is the function of the prostate gland?

2 In which ovarian structures do oocytes develop?

3 List the sequence of structures through which a sperm passes from its site of production to the site of fertilisation of an ovum.

4 How do egg cells move from the ovary to the uterus?

Food for thought

The female urinogenital system is usually arranged as in figure 2. However, very occasionally, the two ovaries lead into two separate uteri and two vaginae. Women with this rare type of reproductive system can have two simultaneous pregnancies. Suggest how the type of pregnancies that occur with this rare type of reproductive system differ from those in typical reproductive systems.

12.3

OBJECTIVES

By the end of this spread you should be able to:

- describe the structure of the mammalian testis
- describe the structure of the mammalian ovary
- explain how spermatozoa are produced
- explain how oocytes are produced.

Fact of life

Drugs may disrupt the normal process of sperm formation. Marijuana, tobacco, and excessive alcohol have all been linked to the increased production of abnormal sperm (for example, two-headed or multiple-tailed sperm).

GAMETOGENESIS: THE PRODUCTION OF GAMETES

Gametes are haploid cells formed by sexually reproducing organisms. In mammals, there are two distinct types of gamete: **spermatozoa** (sperm), formed in the testes of the male; and **oocytes** (egg cells), formed in the ovaries of the female. During fertilisation (spread 12.5), a spermatozoon fuses with an oocyte to form a diploid **zygote**, the first cell of a new individual.

Spermatogenesis

Figure 1 shows the structure of a mammalian testis. It is made up of a system of interconnecting tubes which lead into the vas deferens. The tubes are enclosed within a fibrous capsule covered with a sac of skin, the scrotum. **Spermatogenesis** (spermatozoa formation) takes place in the densely coiled **seminiferous tubules**. The seminiferous tubules contain two types of cell: **germ cells** undergo the two divisions of meiosis to form the spermatozoa, while **Sertoli cells** act as nurse cells ensuring, for example, that the germ cells have adequate nourishment.

Seminiferous tubules provide a vast area for spermatogenesis so that they can produce huge numbers of spermatozoa. The youngest cells in the tubule (**primordial germ cells**) are diploid (have two sets of chromosomes) and are situated around the outside of the tube. As the cells multiply, grow, and divide meiotically, they are carried inwards. By the end of meiosis their chromosome number has been halved and they have become haploid **spermatids**. Spermatids attach to Sertoli cells and mature into spermatozoa.

Maturation from spermatid to spermatozoan involves **differentiation**, the development of specialised structures within the cell. A centriole develops into a tail or flagellum. The Golgi body is converted into the **acrosome**, an organelle at the head of the cell which contains hydrolytic enzymes to penetrate the egg cell. The nucleus becomes condensed and elongated. Mitochondria replicate and group around the flagellum so that they can provide it with energy for swimming.

The production of spermatozoa is regulated by sex hormones made in **Leydig cells**. These cells lie in spaces between the seminiferous tubules. Under the influence of luteinising hormone (a gonadotrophic hormone released from the anterior pituitary gland) they secrete steroids including **testosterone**, the main male sex hormone. Testosterone stimulates the development of **secondary sexual characteristics** at puberty (for example, beard growth), gonad development, and the production of spermatozoa.

Figure 1 *Spermatogenesis in a mammalian testis.*

Section through a testis

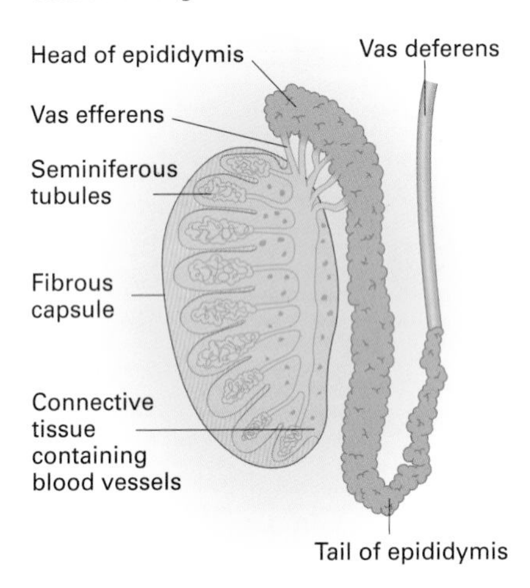

The stages of spermatozoa formation

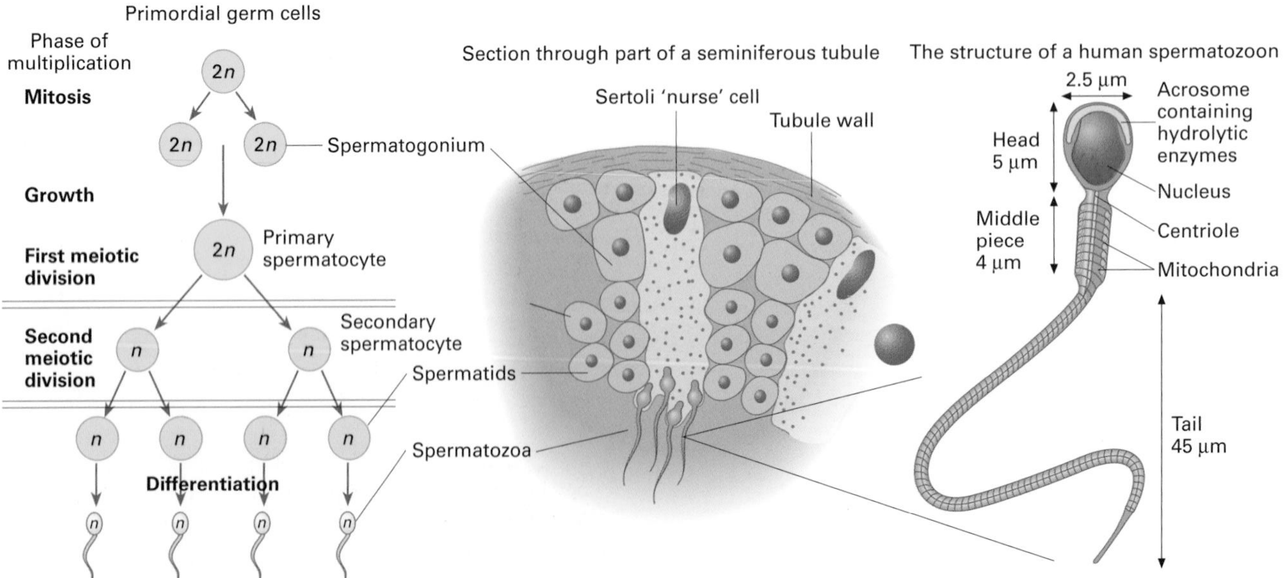

After their formation, spermatozoa pass into the epididymis, a long convoluted tube. It is so long (about 7 m) that it takes a sperm up to 14 days to pass along it. The end of the epididymis stores sperm until they are ejaculated.

Oogenesis

Oogenesis (the formation of oocytes, figure 2) takes place in the ovaries. In the human female, each ovary is roughly the same size and shape as an almond. The ovaries produce only a few hundred oocytes during the lifetime of a female mammal. The development of oocytes takes place in **follicles** situated in the outer region of the ovary. Each follicle is a small sac-like structure consisting of a ball of cells surrounding a single germ cell, as shown in figure 3.

A female is born with a finite number of follicles. Each follicle contains a **primary oocyte** which developed from primordial germ cells and oogonia before birth. Nothing much happens to the primary oocytes between birth and puberty. At puberty, a regular cycle of events starts to take place, regulated by hormones. At the start of one cycle, **follicle stimulating hormone** (**FSH**) stimulates the cells around a follicle to multiply. A maturing follicle secretes **oestrogen** which inhibits the secretion of FSH, usually ensuring that only one follicle develops at a time. The diploid primary oocyte in the follicle undergoes the first meiotic division to form haploid cells. The division is not equal; it results in a smaller cell and a larger cell. The smaller cell is called a **polar body** which degenerates and leaves the female's body. The larger cell becomes a **secondary oocyte** in which the second meiotic division is not yet complete.

The secondary oocyte starts to grow in the follicle. A thick covering, the **zona pellucida**, is secreted round it, and fluid is secreted into small pools between follicular cells. This relatively large, mature follicle is now called an **ovarian follicle** (Graafian follicle).

At intervals of approximately 28 days in women, **ovulation** occurs: the Graafian follicle merges with the wall of the ovary, the ovary wall ruptures, and the secondary oocyte is released into the oviduct. Ovulation is triggered by **luteinising hormone** (**LH**), secreted by the pituitary gland. The second division of meiosis is completed only if a sperm penetrates the secondary oocyte, fertilising it. At this point the oocyte is called an **ovum**. Unfertilised oocytes are released from the body during menstruation.

After ovulation, the remains of the Graafian follicle are converted into a glandular mass of yellow tissue called a **corpus luteum**. The corpus luteum secretes **progesterone**, a hormone which stimulates the growth of the uterus and the development of a placenta. In the absence of fertilisation, the corpus luteum degenerates before another follicle starts to develop.

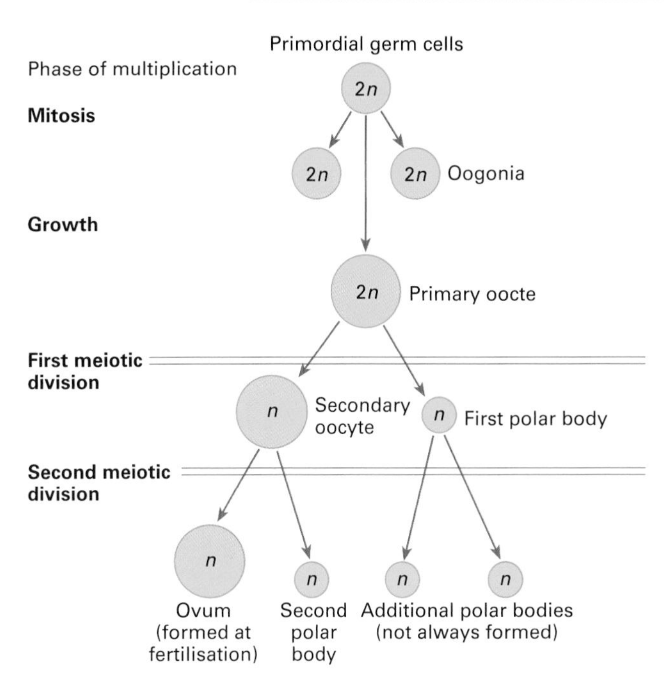

Figure 2 The stages of oogenesis.

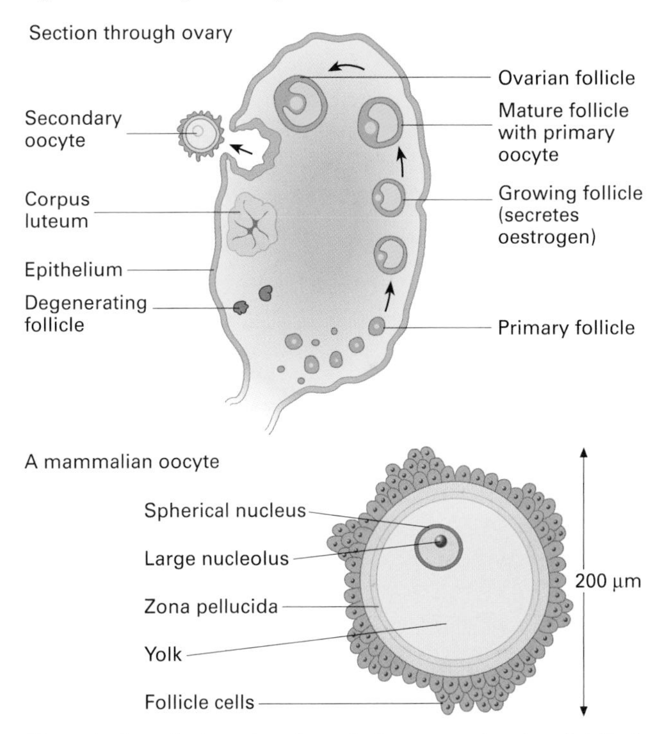

Figure 3 A vertical section through the ovary, showing detail of a secondary oocyte.

Quick check

1 Where within the testis does meiosis take place?

2 From what structures do primary oocytes develop?

3 Which cells nourish spermatids as they develop into spermatozoa?

4 How many secondary oocytes are usually formed from one primordial germ cell?

Food for thought

The testis has a rich blood supply. Suggest why this is necessary. In which part of the testis would you expect to find blood vessels?

12.4

Mammalian reproductive cycles

OBJECTIVES

By the end of this spread you should be able to:

- distinguish between an oestrus cycle and a menstrual cycle
- describe the main events in the menstrual cycle
- explain how hormones interact to regulate the menstrual cycle.

Fact of life

Menopause (when ovulation and menstruation cease) happens in most women between the ages of 45 and 54 years. At menopause, the lowering of oestrogen and progesterone levels can cause unpleasant symptoms such as hot flushes, depression and anxiety, insomnia, and bone loss. The bone loss, which may be as much as 7 per cent each year, can lead to osteoporosis (see spread 16.4). To minimise or delay the effects of menopause, many women are prescribed hormone replacement therapy (HRT). They take low doses of oestrogen and progesterone. Although HRT has proven beneficial, it is not suitable for all women.

Premenstrual tensions

As many as 90 per cent of women suffer from premenstrual tension (PMT). The disruptive emotional and physical symptoms (such as appetite changes, irritability, and headaches) that precede menstruation may last up to two weeks or longer. The symptoms vary in intensity but can be so severe that work and domestic harmony are threatened. Physical changes associated with PMT are an increase in breast size, increased fluid and salt retention with associated weight gain, abdominal distension and an increase in skin pigmentation. There is probably no single cause of PMT. It appears, however, to be linked to a hormonal imbalance, particularly of progesterone. Sufferers are generally advised to reduce their intake of saturated fats and salty foods, increase their consumption of fruit and green vegetables, and drink plenty of watery fluids.

Mammals are unique in feeding milk to their young. Many mammals are placental mammals, which have long pregnancies during which a developing embryo is completely dependent on its mother for food and protection. The costs of a long pregnancy and lactation (secretion of milk) are high. Consequently, many mammals living in seasonal environments have reproductive cycles which ensure that the young are born when food is abundant. In red deer, for example, the sexual cycle occurs only during one restricted season of the year, called the **breeding season**. Dogs and cats have two or three breeding seasons per year. The timing of the breeding season is regulated by a biological clock which is probably adjusted by seasonal changes in day length (spread 11.14).

Within the breeding season, a female mammal goes through a cycle which includes a period of heightened sexual activity (called **oestrus** or 'being on heat') when she becomes sexually receptive and attractive to males. Oestrus occurs just before ovulation, when the female is most fertile.

Humans and their primate relatives are exceptional among mammals: they are sexually receptive throughout the year; they have the potential to mate at any time; and they menstruate. The uterus lining in all mammals undergoes a similar pattern of thickening during a reproductive cycle. However, if fertilisation does not occur, the uterine lining of primates breaks down and is discharged with blood through the vagina, whereas the uterine lining of non-menstruating mammals is reabsorbed and there is no extensive bleeding. The discharge of blood and uterine lining is called **menstruation**.

The reproductive cycle of non-primate mammals is known as the **oestrus cycle** because oestrus is the most prominent event in it. The primate reproductive cycle is called the **menstrual cycle** because of the prominence of menstruation.

The human menstrual cycle

In human females, the menstrual cycle lasts approximately 28 days. A regular sequence of changes is controlled by the interaction of several hormones (figure 1 and table 1). The events are divided into three phases: the follicular phase, the ovulatory phase, and the luteal phase.

Follicular phase

The menstrual cycle begins when blood is first discharged from the uterus. From day 1 to about day 5, the uterine lining continues to break down and is discharged along with varying amounts of blood. A day before menstruation occurs, the hypothalamus produces **gonadotrophin-releasing hormone (GnRH)**. GnRH triggers the secretion of **follicle stimulating hormone (FSH)** from the anterior pituitary gland. FSH triggers the development of one or more follicles in the ovary. As a follicle grows in size, it secretes increasing amounts of **oestrogens** (a group of steroid hormones). Oestrogens stimulate the repair and growth of the uterine lining, and the growth of milk-producing tissue in the mammary glands. They inhibit further production of FSH, so that usually only one follicle matures at a time (negative feedback: FSH stimulates the production of oestrogens, and oestrogens inhibit the production of FSH). Oestrogens also stimulate the anterior pituitary gland to secrete **luteinising hormone (LH)**. Although most of the LH is stored in the anterior pituitary gland during this phase, some LH is released into the blood and stimulates the mature ovarian follicle to produce another hormone, **progesterone**.

Ovulatory phase

At about day 13 to day 15, the level of oestrogens increases rapidly. Whereas a slow rise in oestrogen levels during the follicular stage inhibits GnRH production, at a critical high level oestrogen stimulates the hypothalamus to secrete more GnRH. GnRH secretion is also enhanced by the small

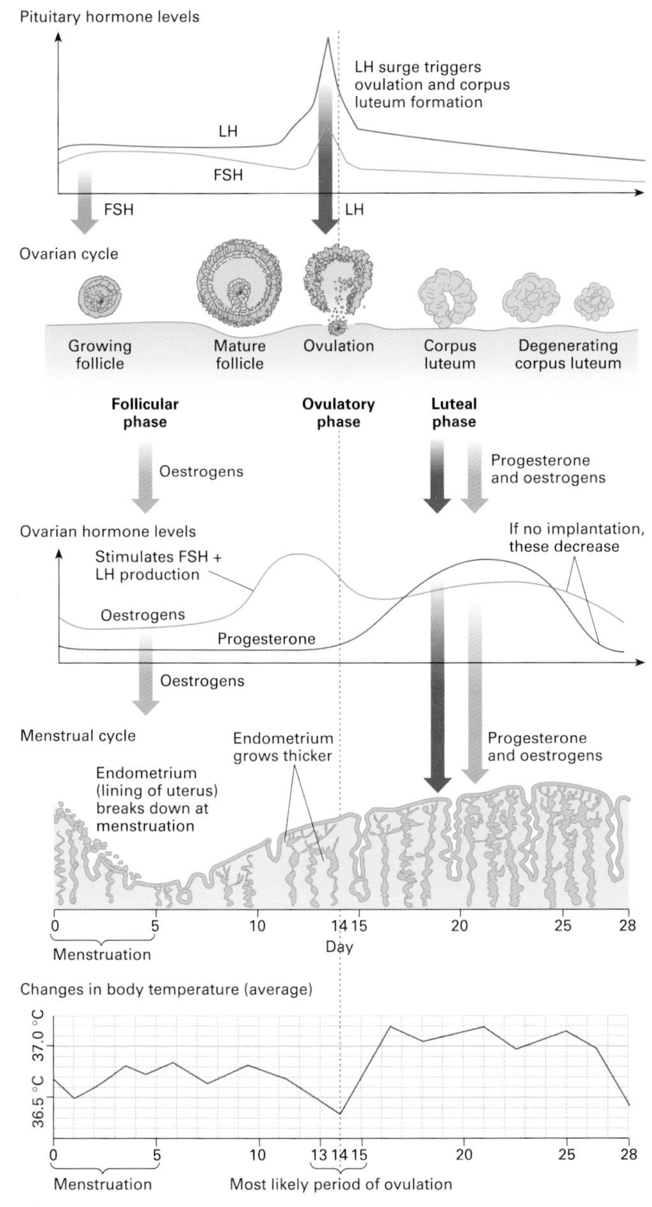

Figure 1 *The human menstrual cycle.*

Table 1 *Hormones regulating the menstrual cycle.*

Hormone	Site of secretion	Major effects and sites of action
GnRH	Hypothalamus	Stimulates the anterior pituitary to secrete LH and FSH
FSH	Anterior pituitary	Stimulates growth of follicle in the ovary
LH	Anterior pituitary	Acts on ovaries to stimulate growth of follicle and promotes ovulation; promotes development of corpus luteum
Oestrogens	Ovarian follicle	Act on the anterior pituitary at low levels to inhibit FSH, and at a critical high level to stimulate release of LH and FSH; promotes repair and growth of uterine lining
Progesterone	Corpus luteum	High levels inhibit FSH secretion; a sharp decrease, along with a decrease of oestrogen, triggers menstruation

amounts of progesterone in the blood. GnRH causes a temporary increase in FSH and a sudden release of LH from the pituitary gland (called the **ovulatory surge**). The surge of LH and FSH brings about **ovulation**, the release of a secondary oocyte from an ovarian follicle. Immediately after ovulation, a woman is fertile and can conceive a baby if she has sexual intercourse, or if sperm are already present in her oviducts.

Luteal phase

After ovulation, the remains of the ovarian follicle form the **corpus luteum**. This yellow glandular tissue secretes large amounts of progesterone and smaller amounts of oestrogens. These hormones stimulate further development of the mammary glands and uterus in anticipation of pregnancy.

If the oocyte is not fertilised within about 36 hours of being shed into the oviduct, it dies. Following this, the corpus luteum gets smaller so progesterone and oestrogen secretion is reduced. At day 28, about 14 days after ovulation, lack of progesterone brings about another menstruation and the cycle starts again. Menstruation occurs in females from puberty to **menopause**, after which there are no more fertile follicles so follicular development and ovulation cease.

If the oocyte is fertilised, the zygote is propelled towards the uterus. During its journey it forms an embryo which becomes implanted in the uterine lining (spread 12.8). A placenta forms between the embryo and mother; this produces human chorionic gonadotrophin (HCG) which stops the degeneration of the corpus luteum, and menstruation is suspended.

Hormonal control of sexual reproduction in men

In men, the pituitary gland supplies a regular secretion of the hormones LH and FSH to maintain a continuous production of sperm in the testes. These hormones are the same ones that regulate follicular development in females. Males can mate at any time and, assuming they are fertile, their sperm could fertilise an egg cell at any time.

Quick check

1 What is the main difference between an oestrus cycle and a menstrual cycle?

2 What significant event happens:

  a at day 0

  b between day 13 and day 15 of the menstrual cycle?

3 The decrease of which hormones trigger menstruation?

Food for thought

There are alarming reports of high levels of oestrogens (or chemicals with very similar effects) in lakes and rivers used as sources of drinking water. These chemicals are thought to cause male alligators to develop tiny, useless penises, and fish to become sterile. Why do you think levels of oestrogens have increased in drinking water? Suggest possible effects of these high levels of oestrogens on humans.

12.5

OBJECTIVES

By the end of this spread you should be able to:

- describe how mammals mate
- explain how a sperm enters and fertilises an ovum
- explain how only one sperm fertilises an ovum.

Fact of life

Humans and several other primates are exceptional among mammals in having no distinct mating seasons; females are potentially receptive to males throughout the year. The ability to have sexual intercourse at any time of year strengthens the bond between partners. In humans, the pleasure experienced by both partners during orgasm (the climax of sexual intercourse, characterised by rhythmic contractions of the reproductive structures of both partners and ejaculation by the male) is also important in strengthening the bond.

Figure 1 Like most mammals, elephants mate from the rear. The male has a very large penis (about 1.5 m long, and weighing about 27 kg). Nevertheless, it is only just long enough to enter the female's vagina. For most mammals, mating is a dangerous time because they are vulnerable to attack. However, because of their large size, elephants have few enemies to fear.

Mating and fertilisation in mammals

Mammals reproduce sexually by **internal fertilisation**. The male sexual organ is pushed into the vagina and ejaculation takes place inside the female. This act of mating ensures that sperm are introduced into the reproductive tract without contact with air. Mating in mammals is usually preceded and accompanied by highly ritualised courtship behaviour. Mammalian courtship is similar to that of other animals (spread 11.13) in that it brings the partners together for a period of time before mating occurs.

Courtship

Although its primary function is to attract a mate, courtship has several other functions:

- it ensures the mate is of the right species and sex
- it usually brings partners together when they are sexually most responsive and fertile (that is, during oestrus of the female; see spread 12.4)
- it heightens sexual responsiveness and suppresses other responses, such as flight and aggression, so that the partners can come into close contact and copulate
- in many mammals, courtship also establishes a close relationship called the **pair bond** between the partners which may keep the partners together while the young are brought up, or even for life.

The first stage of courtship is attracting a mate and signalling of sexual status and readiness. Many female mammals use chemicals called **pheromones** not only to attract mates but also to signal the fact that they are in oestrus and can conceive. A bitch on heat, for example, secretes pheromones in her urine and produces a strong scent from all over her body. This causes dogs to exhibit intense courtship behaviour. In species that live in herds or flocks, pheromones may also be used by a male to induce sexual readiness in a mate with whom he lives all the time. Odorous substances in the urine of male rats and sheep, for example, initiate oestrus in the female. In pigs, the boar's urine, sweat, and breath contain pheromones that cause the sow to adopt a mating posture (remaining still, ears erect, back straight) if her oestrous cycle is at its peak. If not, she is unaffected and walks away. Pig breeders can buy artificial pig pheromone to test a sow's sexual responsiveness. Sexual pheromones may also occur in humans, but their precise role and even their existence is hotly debated by scientists. Nevertheless, the pharmaceutical industry spends millions of pounds each year trying to convince us that perfumes, at least, are potent sexual attractants.

Copulation

Successful courtship leads to copulation and fertilisation. For all mammals, fertilisation of the female's oocytes by male sperm must take place deep in the female's body (internal fertilisation). To achieve this, male mammals have a penis which becomes erect at the moment of mating for insertion in the female's vagina. Women have a small structure called the **clitoris** which protrudes just in front of the urethra. Like the male penis, the clitoris contains erectile tissue but, unlike the penis, the clitoris lacks a reproductive opening. Erection of the penis is brought about by hydraulic action (the penis becomes gorged with blood which acts as a hydrostatic skeleton; spread 11.1) or by combining hydraulic and muscular actions. In most male mammals (but excluding men), a bone called the **os penis** gives the penis extra support. During mating, most male mammals stand and enter the female from behind (figure 1). Only whales, humans, and, on rare occasions, orang-utans, gorillas, and chimpanzees mate face to face. This is thought to play an important role in establishing long-term pair-bond relationships.

In men, an erection (essential for the insertion of the penis into the vagina) occurs as a result of sexual arousal. In women, sexual arousal causes the clitoris to become swollen and stimulates the secretion of mucus which lubricates the vagina during intercourse. At the peak of arousal, a complex combination of muscular actions brings about an **ejaculation** in the male. Secretions are forced from the accessory glands into the vas deferens; the bladder sphincter closes, preventing urine from entering the urethra; sperm are expelled from the epididymis into the vas deferens; a sphincter muscle at the base of the penis relaxes, admitting semen into the penis; and then a series of muscle contractions around the base of the penis and along the urethra propel semen out of the body.

Fertilisation

Human sperm have to spend at least seven hours in the female before they can fertilise an oocyte. During this time they are activated by **capacitation** (a poorly understood process by which sperm are made capable of fertilisation by secretions from the female genital tract). After capacitation, they are able to swim more vigorously and release chemicals to penetrate the oocyte. They swim towards the oocyte, nourished first by fructose within the semen and then by nutrients in vaginal secretions. Uterine contractions and the movement of cilia in the oviduct draw the sperm to their destination. Millions of sperm may start the journey to an oocyte in the oviduct, and several hundred may crowd around the cell, but only one will eventually fuse with it.

With the help of the enzyme **hyaluronidase**, sperm burrow through the follicle cells surrounding the oocyte, and into the cloudy outer zone of the oocyte. Contact of sperm with the **zona pellucida** (a thick transparent jelly-like coat on the cell surface membrane of an oocyte) triggers the **acrosome reaction** (figure 2). Acrosomal enzymes, probably released from many sperm, digest the zona pellucida. The first sperm to wriggle its way through the weakened membrane and touch the cell surface membrane of the oocyte initiates a number of instantaneous events. The oocyte completes its second meiotic division to form a haploid gamete, the **ovum**, and the **second polar body** (spread 12.3). The second polar body degenerates and the ovum becomes fertilised. The electrical charge across the ovum membrane changes, keeping other sperm out temporarily until a tough **fertilisation membrane** forms to prevent their entry permanently: this is called the **cortical reaction**. Meanwhile, after successfully penetrating the oocyte membrane, the sperm loses its tail and its head moves towards the nucleus of the ovum. The head swells and releases its chromosomes, which join with those of the ovum to form a diploid cell called the **zygote**. The fusion of sperm chromosomes with ovum chromosomes is called **fertilisation** or **conception**.

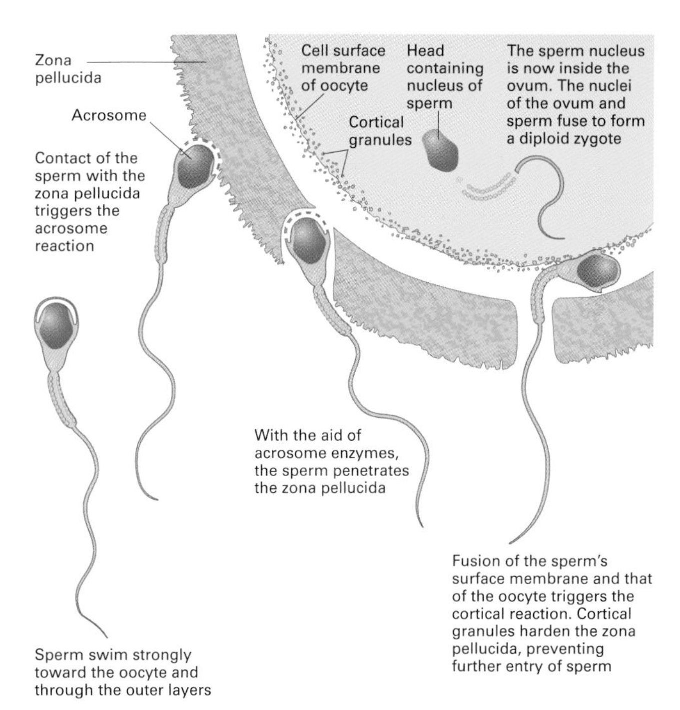

Figure 2 *Fertilisation: the acrosome reaction.*

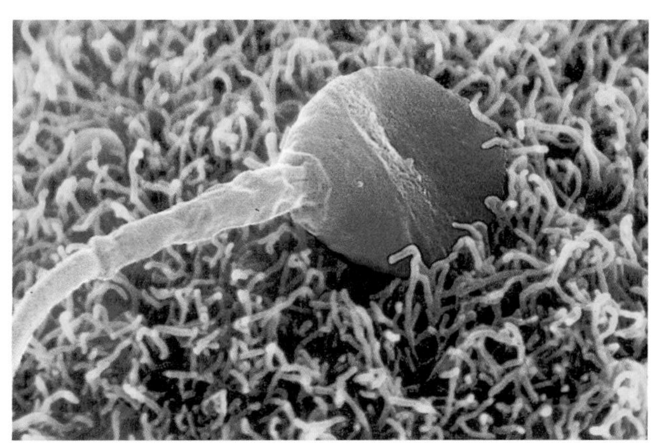

Figure 3 *False-coloured scanning electron micrograph of a human sperm entering an ovum prior to fertilisation (×6000).*

Quick check

1 Why is internal fertilisation carried out by mammals?

2 Why are sperm unable to fertilise an oocyte until they have spent at least seven hours in the female?

3 What is the function of:

a the acrosome reaction

b the cortical reaction?

Food for thought

According to reports made in November 1997 by the National Academy of Sciences and the National Institutes of Health, human sperm counts are falling at an alarming rate. Throughout the world, average sperm counts have almost halved in the last 50 years, with the reduction being greatest in the United States and western Europe. Why do you think sperm counts have fallen? Suggest what biological and social effects this might have.

12.6

Birth control

OBJECTIVES

By the end of this spread you should be able to:

- describe the main types of birth control techniques
- discuss the advantages and disadvantages of different birth control methods.

Fact of life

In 1978, the first so-called 'test-tube baby' was born. Test-tube babies grow in a mother's uterus, but are conceived outside the body by a process called IVF (*in vitro* fertilisation: literally 'in glass fertilisation'). IVF is commonly used to treat infertility caused by blockage of a woman's oviducts, or some other factor preventing her conceiving. The mother-to-be is given a synthetic hormone such as clomiphene, which opposes the action of oestrogen. At the right dosage, clomiphene stimulates ovulation and causes a number of oocytes to mature at the same time. Several oocytes are removed and placed in a laboratory dish (not a test tube), mixed with sperm from the partner, and incubated until the zygote develops to the blastocyst stage (see spread 12.7). The blastocyst is implanted in the mother's uterus, and the pregnancy is allowed to continue as normal.

Birth control is a part of the reproductive strategy of many animals. Examples of abstinence from intercourse, abortion, and infanticide can be found among both human and non-human societies. If a female bobcat, for example, cannot find food easily, she produces only one or two kittens instead of the usual three. However, only humans can manipulate their own reproductive cycles to prevent the fertilisation of an egg cell and its development into a baby.

Contraception and birth control

Contraception is the use of devices or methods which act against conception (the fertilisation of an egg cell by a sperm; spread 12.5). Other birth control methods prevent a fertilised egg cell from developing into a baby. There are many methods of birth control. Some prevent sperm from reaching the egg cell, and others interfere with the reproductive cycle, preventing ovulation or implantation. Such methods may involve the use of physical devices, chemical manipulation, or behavioural techniques.

Chemical contraception takes the form of birth control pills. The most commonly used types contain combinations of synthetic oestrogen-like and progesterone-like substances. Taking these hormones stops the secretion of GnRH, FSH, and LH so that neither follicle development (an oestrogen effect) nor ovulation (a progesterone effect) occurs.

Each birth control method has its advantages and disadvantages (table 1) and each poses its own ethical dilemmas, especially for members of religious faiths that prohibit certain types of birth control. However, our global population is growing exponentially and birth control has the potential to limit this growth. Some people argue that birth control programmes are essential to prevent a population explosion that could trigger worldwide starvation.

Increasing fertility

In addition to birth control, we also have the ability to increase fertility by various techniques. These include artificial insemination, fertility drugs, and *in vitro* fertilisation.

- In **artificial insemination**, sperm from an anonymous donor is inserted artificially through the cervix of the mother-to-be.
- **Fertility drugs** are synthetic chemicals that stimulate ovulation: they either provide gonadotrophins such as FSH which stimulate the development of follicles, or they provide a chemical which inhibits the natural production of oestrogens. Lack of oestrogen (the normal FSH inhibitor) results in more FSH being produced, which stimulates follicular development.
- ***In vitro* fertilisation** involves the fertilisation of egg cells outside the body which are then artificially implanted in the uterus to produce 'test-tube babies'.

Paradoxically, some scientists argue that fertility improvement techniques are as vital to our future survival as birth control. This is because they fear that pollutants accumulating in the environment are decreasing male fertility. If we did not have the means to increase our fertility, our population could plummet to such low levels that our social structures and very existence would be threatened.

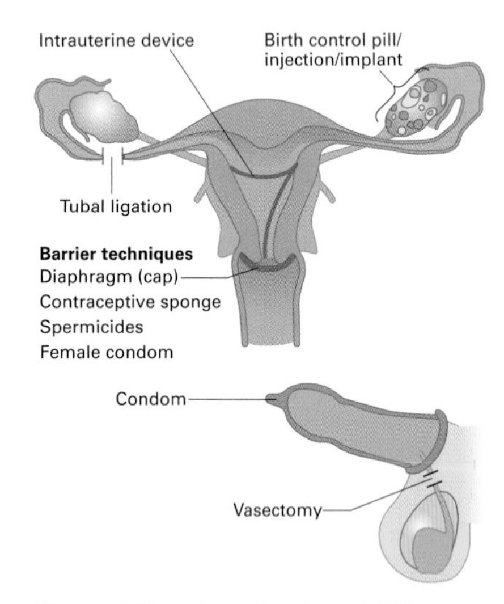

Figure 1 The sites of action of different birth control methods.

Quick check

1 Name the birth control method that is used after sexual intercourse.

2 Which type of birth control method offers some protection against sexually transmitted diseases?

Food for thought

Contraception is a highly controversial subject, especially when proposed as a means of limiting large population increases. Suggest ethical arguments for and against mass contraception programmes to control population size.

Table 1 Summary of birth control methods.

Method	How it works	Advantage	Disadvantages	Comments
Barriers preventing sperm from reaching egg cell	**Male condom** A thin latex sheath worn over an erect penis; at ejaculation, semen collects in the tip **Female condom** A thin latex sheath with two elasticated rings; one is inserted into the vagina, the other remains outside	Offer some protection against sexually transmitted diseases; no serious side-effects; reliable and easy to obtain	Can break or leak; cannot be used spontaneously; may lessen the enjoyment of intercourse	Protection improved if combined with spermicide; give some protection against sexually transmitted diseases
	Diaphragm (cap) A dome-shaped sheet of thin rubber with an elasticated rim; inserted into the vagina, it lies over the cervix and blocks entry of sperm to the uterus.	Reliable and inexpensive	Must be fitted by a doctor initially; can be messy	Usually used with a spermicide
	Spermicide Cream, foam, or gel inserted into the vagina. Kills sperm and blocks entry of sperm to uterus	Inexpensive; easy to obtain and use	Not reliable on its own	Usually used with another barrier method
	Sponge Polyurethane sponge impregnated with spermicide fitted over cervix	Easy to use	Not reliable	Needs to be fitted 24 h before intercourse and left in place at least 6 h after intercourse
Hormone manipulation interfering with ovulation and/or implantation	**Combination pill** Contains synthetic oestrogens and progesterone; prevents ovulation and implantation	Does not interrupt spontaneity; reliable if taken regularly; lowers the risk of some cancers	May increase risk of thrombosis; occasionally causes side-effects such as nausea and weight gain	Hormonal manipulation provides no protection against sexually transmitted diseases
	Injection contraceptive Prevents ovulation; thickens cervical mucus	Easy to use; one injection lasts 3 months	May cause menstrual changes and weight gain	
	Contraceptive implant Implant placed under the skin releases hormones which prevent ovulation	Long lasting (each implant may last up to 5 years); very reliable	Can cause menstrual irregularities	
	'Morning-after pill' Contains high levels of oestrogen to prevent implantation	Can be used after rather than before intercourse	High dose of oestrogen can produce unpleasant side-effects	Not suitable for regular use
Behavioural	**Rhythm method** Couple refrain from intercourse during fertile periods of menstrual cycle	Natural; no cost	Difficult to do; unreliable	Special kits are available for monitoring fertile period, increasing reliability
	Withdrawal Removal of penis from vagina before ejaculation	No cost	Difficult to do; unreliable	
Surgical	**Vasectomy** Vasa deferentia of male cut and tied off to prevent passage of sperm to the penis	A single operation gives a permanent effect; high reliability	May be irreversible	
	Ligation of oviducts Oviducts are cut and tied off	A single operation gives a permanent effect; high reliability	Usually irreversible	
Other	**Intrauterine device** (loop; coil) Plastic or copper device that prevents implantation	Allows sexual intercourse to be spontaneous	Can cause discomfort; can be displaced	Requires expert insertion

12.7

OBJECTIVES

By the end of this spread you should be able to:

- define implantation
- distinguish between a human embryo and fetus
- describe how a human embryo develops.

Fact of life

The innermost cells of a blastocyst are called **embryonic stem cells** (ESCs). They can replicate indefinitely and are **pluripotent** (able to generate any human cell type except those needed to support and develop a fetus in the womb). Therefore, ESCs could replace any compatible cell type destroyed or damaged by disease or injury. However, isolation of ESCs destroys the embryo; this raises major ethical issues. Cells similar to ESCs called 'cord-blood-derived embryonic-like stem cells' or CBEs are found in the umbilical cord. CBEs can be isolated, stored, and replicated without any of the ethical dilemmas associated with ESCs. Cord-blood banks have been established around the world in which CBEs from a newborn baby are stored. These CBEs can be used by the donor if a disease strikes in later life or by a patient with matching tissue. CBEs have already been used to treat leukaemia and immune system diseases.

Implantation and embryonic development

One reason for the biological success of humans is that they spend a long period of early development in the mother's uterus. Here, developing offspring are nurtured and protected during their most vulnerable stage. The period between conception and birth (the **gestation period** or pregnancy) is about nine months (40 weeks), much longer than for any other animal of comparable size. Elephants have gestation periods of 22 months and sperm whales of 14–15 months, but they are much bigger.

Cell division and implantation

After fertilisation in the upper part of the oviduct (spread 12.5), a human **zygote** (fertilised ovum) starts to move towards the uterus. It is swept along by peristaltic contractions of the oviduct and movements of cilia lining the oviduct. At the start of its journey, the zygote goes through the first stage of embryonic development called **cleavage**. Cleavage involves a special kind of cell cycle in which cells divide repeatedly without increasing the amount of cytoplasm; the interphase is not long enough for growth to occur. By the time it reaches the uterus (about 4 days after conception), the zygote has developed into a solid mass of cells called a **morula**. This continues to divide to form a hollow fluid-filled ball of about 100 cells called a **blastula**.

About 7 days after conception, when it makes contact with the wall of the uterus, the blastula becomes a **blastocyst**. The blastocyst appears to burrow its way into the uterine lining by the action of **trophoblast cells** in the outer layer of the blastocyst. These cells have finger-like projections (**trophoblastic villi**) which penetrate the lining of the uterus. The trophoblast cells digest uterine cells and obtain nourishment from them. The blastocyst soon becomes firmly embedded among the glands and blood vessels lining the uterus. This process, called **implantation**, is usually completed about 11 or 12 days after conception (figure 1).

After implantation the trophoblast, now called the **chorion**, develops a blood circulation and continues growing into the uterus. The chorion comes into contact with uterine blood vessels from which it gains nourishment. The blood vessels of the chorion and mother grow and expand together to form the **placenta** (spread 12.8). About 4 weeks after conception, the blood circulation of the chorion becomes linked to that of the developing embryo. From then until the end of pregnancy, the embryo derives its life support from its mother via blood vessels connected to the placenta. Despite its total dependence on its mother, a developing baby is an individual, genetically distinct from its mother.

By about 9 days after conception, a complex embryo called a **gastrula** develops from the blastocyst by a process called **gastrulation**. Cells migrate to different areas, forming the gut cavity and the three main layers of the body. These layers are the **endoderm** (inner layer), **mesoderm** (middle layer), and **ectoderm** (outer layer). After gastrulation, the fate of the cells within each layer is **determined**: even before there are any visible signs of differentiation, the cells are irreversibly committed to a particular development pathway (table 1).

In some way not yet fully understood, information about a determined cell's development pathway is passed on to its daughter cells. This information specifies the type of cell that the daughter cells will become, even if they are removed to a different environment.

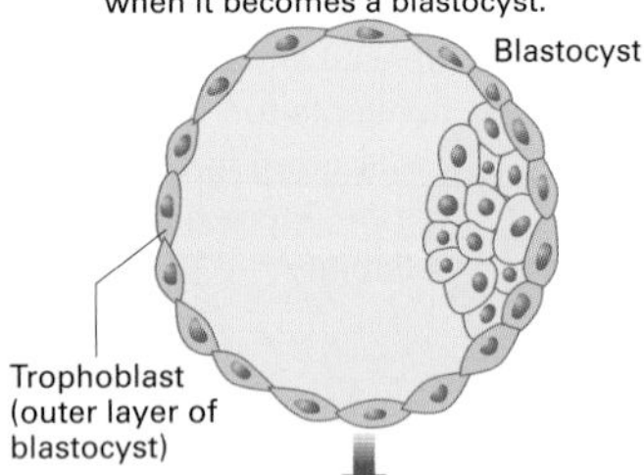

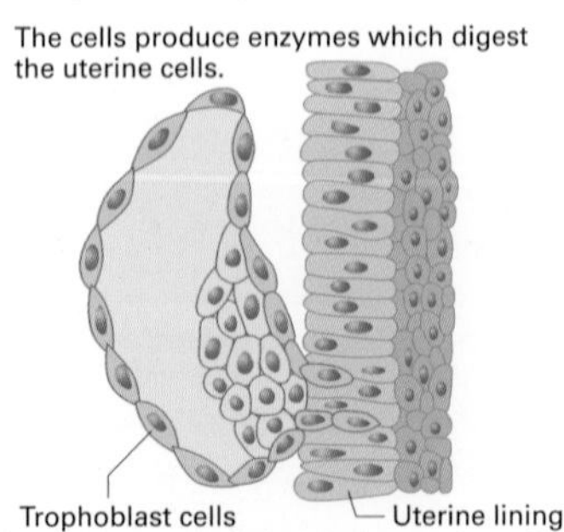

Figure 1 Implantation.

From embryo to fetus

After determination, embryonic cells produce tissues and organs (figure 2). Some cells grow inwards to form a line of tissue known as the **primitive streak** from which the **notochord** (the rod of cells from which the spinal column forms) develops. The gut and other internal organs, including the nerve cord, begin to appear. The front end of the nerve cord is thickened, indicating where the brain will develop. The heart also forms early in the embryo. Soon, muscles and buds (which will become limbs) appear. By 8 weeks after conception, the developing baby becomes clearly human and is called a **fetus**. It has a well developed head and limbs, and bones are beginning to form. Twenty or so weeks after conception, sex organs develop and the mother can feel the fetus in her womb. Within eight more weeks, the fetus grows downy hair and eyelashes, and can open its eyes. The baby has developed to the point at which it can survive if born prematurely, as long as it is given special medical care. During the last third of pregnancy (24–40 weeks after conception), the fetus's circulatory and respiratory systems undergo changes so that the newborn baby will be able to breathe air. Its muscles thicken and its bones harden, so that it will cope with the full force of gravity. It also develops the ability to regulate its own temperature. As the time of birth approaches, the baby becomes less active. Approximately 40 weeks after conception the baby, about 50 cm long and weighing 2.7–4.5 kg, is born (spread 12.9).

Table 1 *The fate of cells within each of the three layers of the gastrula.*

Endoderm becomes	Mesoderm becomes	Ectoderm becomes
Gut lining	Notochord	Brain
Oesophagus	Vertebrae	Spinal cord
Stomach	Muscle	Sense organs
Intestines	Dermis	Epidermis
Lungs	Kidneys	Adrenal medulla
Thyroid	Sex organs	
Pancreas	Tendons	
Liver	Ligaments	
	Limb bones	
	Circulatory system	

Quick check

1 How does a blastocyst burrow its way into the uterine lining during implantation?

2 What is the name given to an embryo when it reaches 8 weeks old?

3 From which body layer in the embryo does the brain develop?

Food for thought

It is now technologically possible to save babies born prematurely at 16 weeks. Suggest why it costs a great deal of money (hundreds of thousands of dollars per infant in the USA) to save these babies, and why many of the surviving infants have mental and physical disabilities.

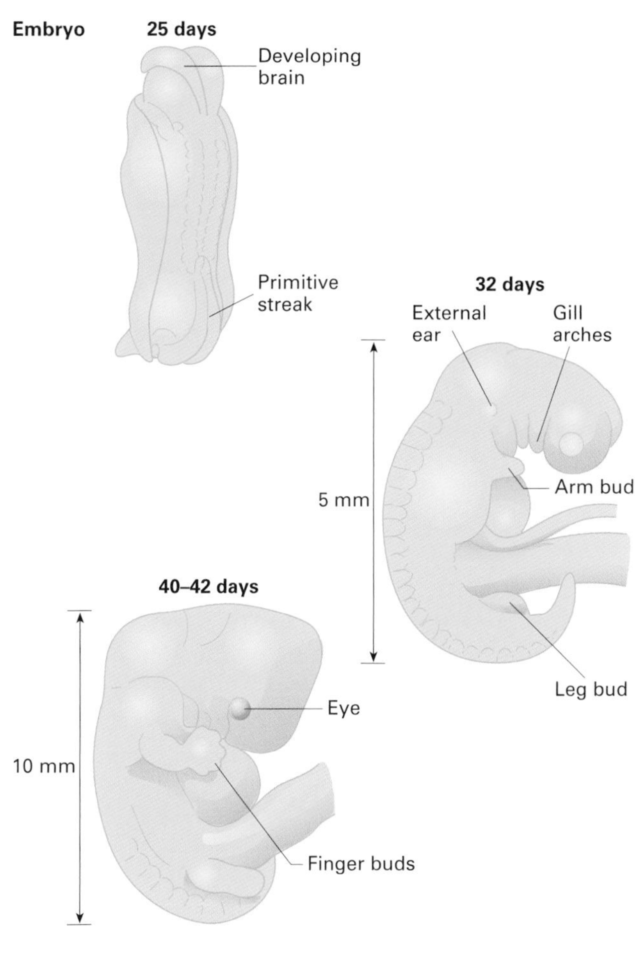

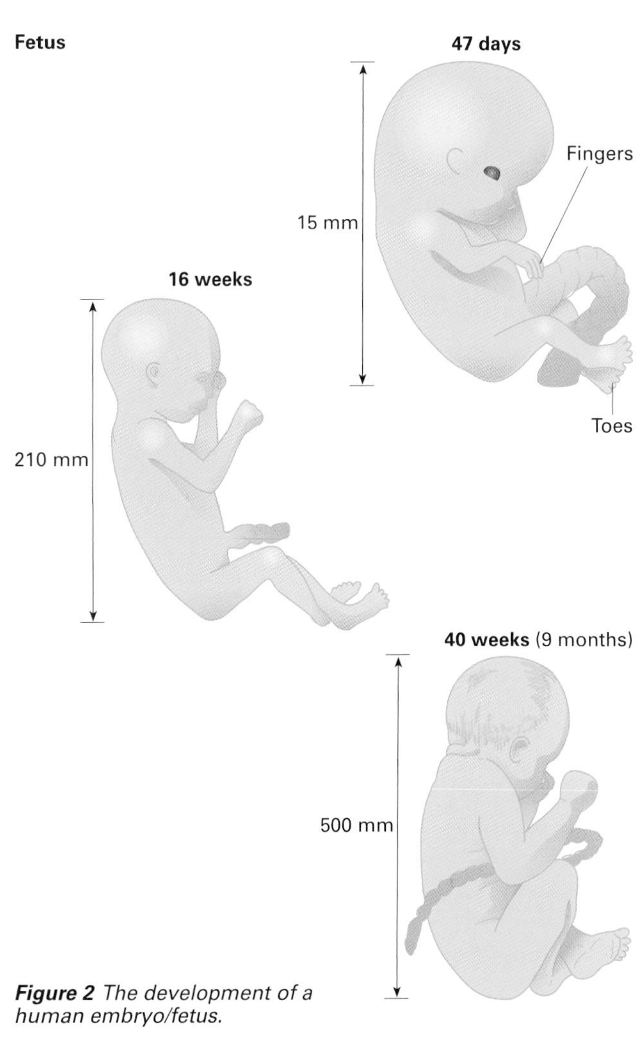

Figure 2 *The development of a human embryo/fetus.*

12.8

Pregnancy and the placenta

OBJECTIVES

By the end of this spread you should be able to:

- describe the role of the extraembryonic membranes
- explain how the placenta forms
- discuss the functions of the placenta.

Fact of life

During pregnancy, a hormone called human chorionic gonadotrophin (HCG) is released from the placenta, accumulates in the bloodstream, and is released in the urine. Detection of this hormone in the urine is the basis of some home pregnancy test kits.

HCG is a glycoprotein. This means that specific monoclonal antibodies can be produced which bind with it (see spread 15.6). In the test kit, these specific antibodies are immobilised on a urine dipstick and tagged with a blue latex. When the dipstick is dipped in urine, the hormone attaches onto a latex-tagged antibody. The HCG bound to the antibody moves up the strip and combines with anti-HCG antibodies (figure 1), producing a clearly visible result in the test window.

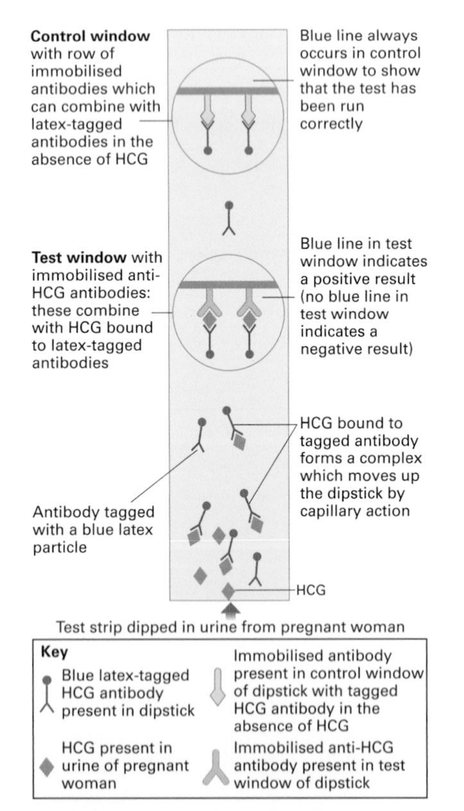

***Figure 1** If HCG is in the urine, it forms a complex with the blue tagged antibodies present in the stick. The HCG–antibody complex rises up the stick and binds to anti-HCG antibodies in the test window, producing a visible blue line indicating pregnancy.*

Pregnant female mammals have a **placenta** which provides remarkably close contact between a mother and her unborn infant. The placenta consists of a double layer of spongy vascular tissue formed in the lining of the uterus after the implantation of the embryo.

The extraembryonic membranes

After implantation, the embryo produces four **extraembryonic membranes** (membranes outside the embryo): the amnion, the yolk sac, the chorion, and the allantois (figure 2). The **amnion** is the innermost membrane. It lines a cavity which surrounds the embryo. The cavity is filled with fluid (**amniotic fluid**) in which the embryo is suspended and cushioned against mechanical damage. The **yolk sac** has no obvious function in humans and becomes buried in the placenta. The **chorion** is the outermost membrane derived from trophoblast cells (spread 12.7). It forms villi which develop into the fetal side of the placenta; the maternal side is derived from the highly vascularised uterine lining. Within three weeks of implantation, the **allantois**, a sac-like outgrowth from the embryonic gut, fuses with chorionic villi (finger-like outgrowths of the chorion) in the uterine wall. The fused structure forms the embryo's part of the placenta. As an embryo grows, the allantois forms the **umbilical cord**, a tube which carries embryonic blood vessels to and from the chorionic villi.

Circulation in the fetus

Blood flow through the umbilical cord and embryo is maintained by the pumping action of the embryo's heart. However, fetal circulation is distinctly different from that in a newborn baby because gaseous exchange is carried out by the placenta, not the fetal lungs. Oxygenated fetal blood from the placenta is carried via the umbilical vein to the right atrium. Most of this blood is forced into the left side of the heart through a hole called the **foramen ovale** and through a special fetal vessel, the **ductus arteriosus**. At birth, the foramen ovale and ductus arteriosus close so that blood can flow through the lungs (spread 12.9).

The placenta

The blood vessels of the allantochorion lie close to the maternal blood vessels in the placenta. The endometrium (uterus lining) breaks down in the region of the chorionic villi so the villi are bathed in maternal blood (figure 3). This allows ready exchange of substances but does not allow blood from the two individuals to mix. The separation of maternal and fetal blood is one of the most significant features of the placenta. Without it, the mother's immune system would reject the baby like any other foreign material. Although the allantochorion is also genetically distinct from the mother and should be rejected, it appears to be coated with molecules which prevent this from happening.

The chorionic villi provide a vast surface area for the exchange of nutritional, respiratory, and excretory substances between the circulatory systems of mother and developing baby. A human placenta, for example, has more than 48 km of villi. Substances are exchanged by diffusion, facilitated diffusion, and active transport. Substances absorbed into the fetus from the mother's blood include oxygen, nutrients (including water), some hormones, and certain antibodies which give the fetus some resistance to disease. Fetal haemoglobin has a greater affinity for oxygen than maternal haemoglobin (figure 4), ensuring maximum efficiency of oxygen uptake. Carbon dioxide, urea, and other waste products are excreted into the maternal blood.

The placenta presents a partial barrier to the passage of harmful substances from mother to infant. Some microorganisms cannot pass into the fetus. However, viruses (such as HIV), nicotine from cigarettes, addictive drugs (such as heroin), and alcohol can cross the placenta and harm the fetus.

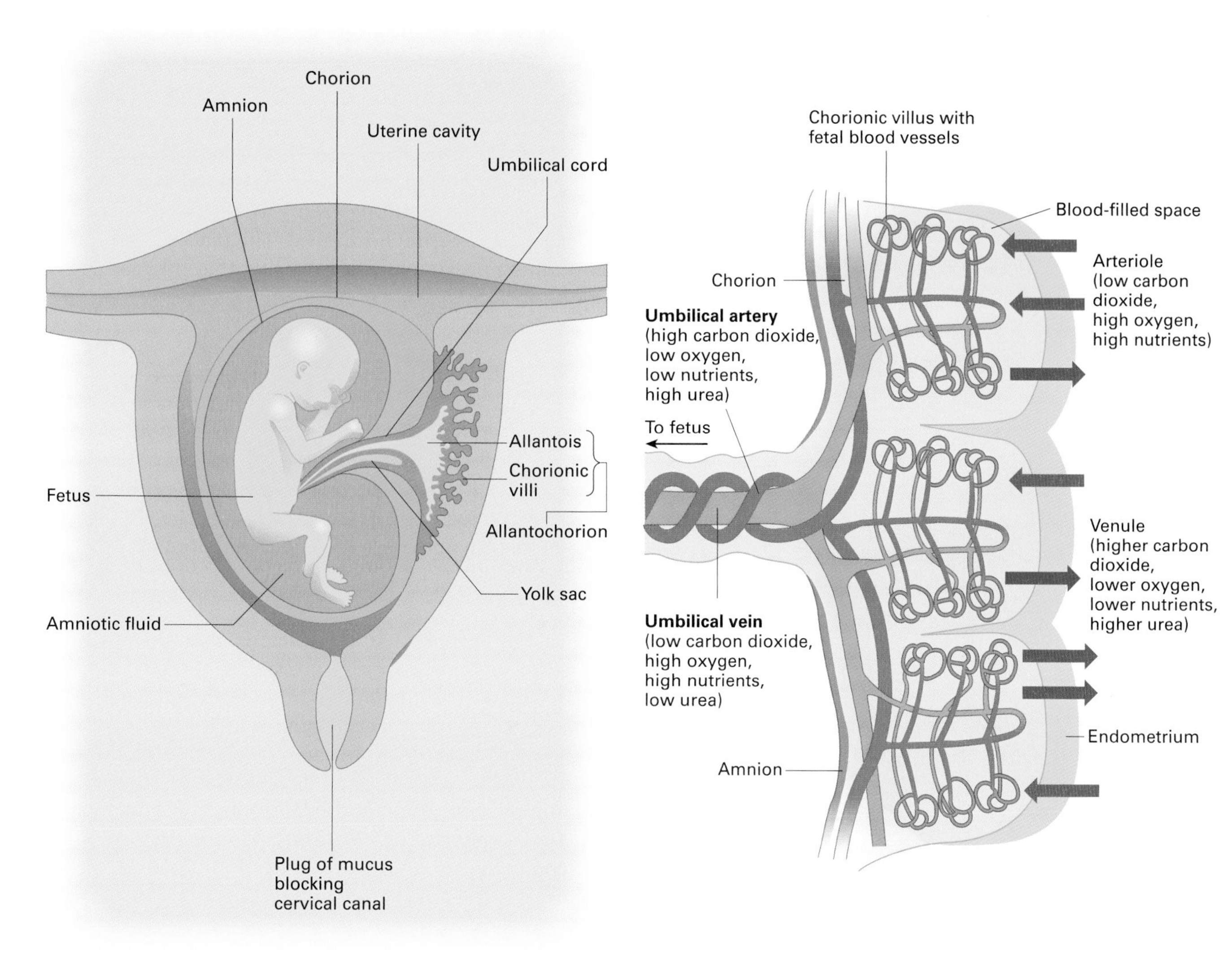

***Figure 2** A fetus at about ten weeks within the amniotic cavity.*

***Figure 3** The structure of part of the placenta.*

In addition to being an exchange surface, the placenta also acts as a temporary endocrine organ, producing various hormones involved in the maintenance of pregnancy. These hormones include HCG, progesterone, and oestrogen.

After birth

The protection and nourishment given to the developing fetus during the long gestation period of mammals means that offspring can be born well developed. Some mammals are able to move around and find food relatively soon after birth: for example, 20 minutes after birth a giraffe can stand, and within hours begins to suckle. Giraffes start to feed on vegetation when two to three weeks old. Humans are exceptional in having a very long period of development between birth and independence (spread 12.10).

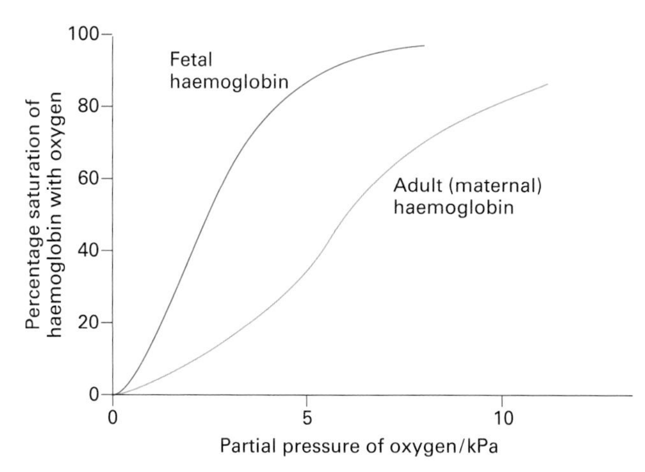

***Figure 4** The oxygen dissociation curve of fetal haemoglobin lies to the left of that of maternal haemoglobin. This ensures that oxygen passes from the mother into the fetus (see spread 7.9 for a discussion of oxygen dissociation curves).*

Quick check

1 List the extraembryonic membranes.

2 Which membranes form the fetal side of the placenta?

3 List three functions of the placenta.

Food for thought

Some people are born with a congenital defect called a 'hole in the heart'. Suggest the most likely cause of this, its location, its effects, and how it can be treated.

12.9

Birth

OBJECTIVES

By the end of this spread you should be able to:

- describe the role of the placenta in maintaining pregnancy
- discuss how birth is initiated
- describe the main stages of birth.

Fact of life

The birth of a primate can be a lengthy process. A newborn primate usually emerges head first so that it can breathe before the process is complete.

Parturition, the act of giving birth, is one of the most dramatic and dangerous events in the life of an organism. Its timing is crucial: it must not be too early or the baby will not be developed sufficiently to begin life in the outside world; it must not be too late, or the baby will be too big to make a safe exit from its mother. The timing seems to be determined by a complex interplay of fetal and maternal hormones. The placenta plays an important role in maintaining pregnancy and ensuring that pregnancy is not terminated too soon.

The role of the placenta

During the first few months of pregnancy, the placenta plays an increasingly important role in producing reproductive hormones. It secretes **human chorionic gonadotrophin** (HCG), a peptide hormone which prolongs the activity of the corpus luteum until four months into the pregnancy. During this time, the corpus luteum is responsible for maintaining pregnancy by producing oestrogens and progesterone. Without HCG, the corpus luteum atrophies (degenerates), inducing menstruation. The presence of HCG in the urine of a woman is the basis of most pregnancy tests.

After the corpus luteum has degenerated, the placenta takes over the role of producing oestrogens and progesterone, preventing menstruation and ovulation. Oestrogens also stimulate the growth of the mammary glands, increase the size of uterine muscle cells, and trigger the formation of receptors to the hormone oxytocin in the uterine muscles. Progesterone also stimulates the growth of the mammary glands, inhibits the contraction of uterine muscles, and inhibits the release of **prolactin** (a hormone that stimulates milk production; see spread 12.10). **Relaxin**, another hormone released by the placenta, relaxes the connective tissue in the bones of the pelvic girdle and enlarges the cervix in preparation for birth.

As pregnancy nears its end, the amount of oestrogen relative to progesterone rises (figure 1). These changes help to initiate birth. A few days before birth, the baby usually shifts around in the womb so that its head is close to the cervix.

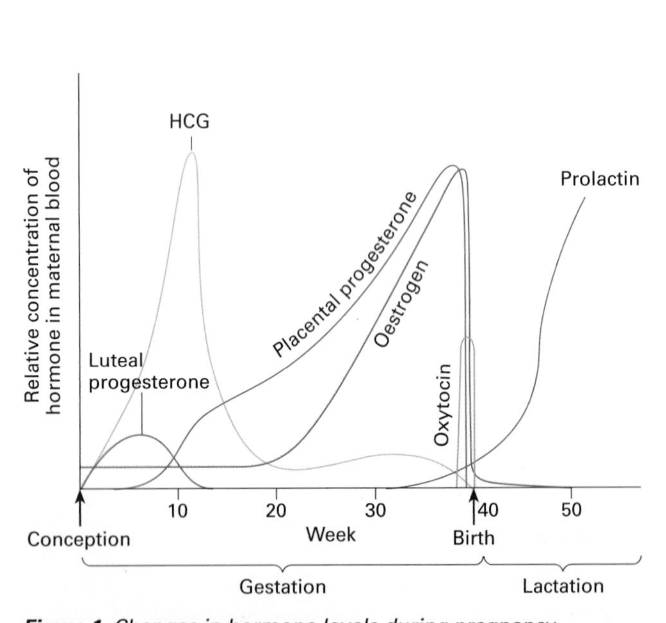

Figure 1 *Changes in hormone levels during pregnancy.*

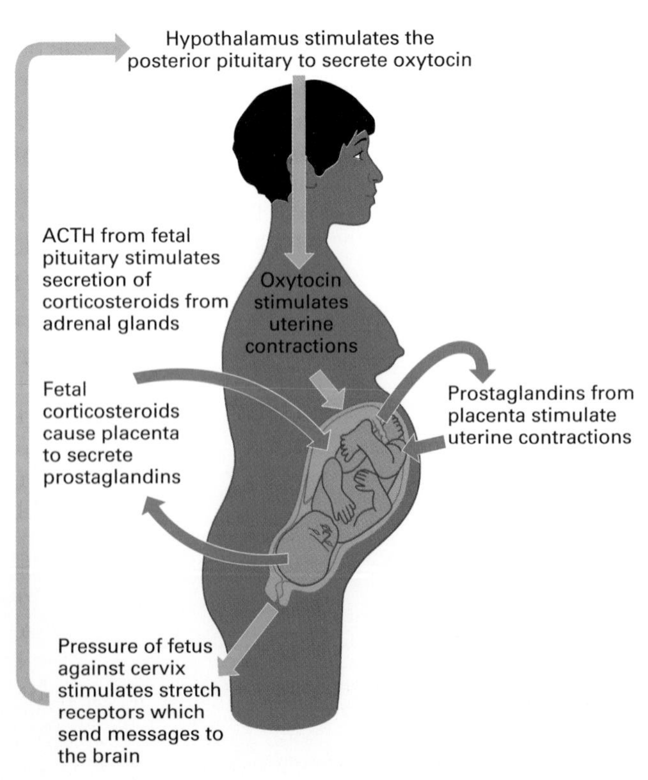

Figure 2 *Hormonal control of uterine contractions.*

Initiation of birth

There are a number of different hypotheses to explain the initiation of birth. All involve the hormone oxytocin causing uterine muscles to contract. Figure 2 summarises one hypothesis. It suggests that the fetal pituitary gland secretes **adrenocorticotrophic hormone (ACTH)** which stimulates fetal adrenal glands to secrete **corticosteroids**. The steroids pass into blood sinuses in the placenta and cause maternal cells to produce substances called **prostaglandins**. These neurosecretory substances act as local hormones and are thought to cause the muscles of the uterus to contract. At the same time, pressure of the fetal head against the cervix stimulates stretch receptors. The receptors send information to the mother's brain, causing the release of **oxytocin**. This hormone is known to cause uterine muscles to contract, but only if the muscles have specific oxytocin receptors. Oxytocin receptors form in the muscles during the later stages of pregnancy in response to high levels of oestrogen.

The levels of prostaglandins and oxytocin build up during labour. The uterine contractions are weak at first but gradually increase in frequency and intensity, causing labour pains. The physical and emotional stresses associated with labour stimulate the release of more oxytocin and prostaglandins, creating a positive feedback mechanism. Eventually, the contractions are so strong that they push the baby out of the uterus.

The process of birth

In humans, birth occurs in three main stages (figure 3). First, the cervix expands so that the baby's head can pass into the vagina. The expansion pushes out the plug of mucus in the cervix, and amniotic fluid is usually released from the uterus. During the second stage, called **delivery**, the uterine contractions are strong and very close together. The baby usually adopts a face-down position and is pushed through the vagina and out of the mother's body. In the third stage, further uterine contractions push the umbilical cord and placenta (afterbirth) out of the mother's body.

When a baby is born, it leaves the watery environment of the uterus where all its needs were met by its mother and enters a hostile world where it must start to fend for itself. One of the first things it must do is convert to breathing air.

In the uterus, a fetus obtains oxygen from its placenta. Its lungs do not work and they receive only a little oxygen (spread 12.8). At birth, the placenta stops functioning and its blood supply is cut off. This triggers a rapid alteration in the circulation, ensuring that blood is diverted to the lungs as they inflate with air for the first time. An increase in blood pressure in the left atrium causes a flap of tissue to be pressed against the foramen ovale so that blood no longer flows directly from the right to the left side of the heart. A few hours later the ductus arteriosus closes. From then on, all the blood from the right ventricle flows to the lungs.

The first breath of the baby is usually accompanied by crying. The crying does not indicate distress but is a reflex action which ensures that the lungs can be fully inflated and that the baby establishes regular breathing.

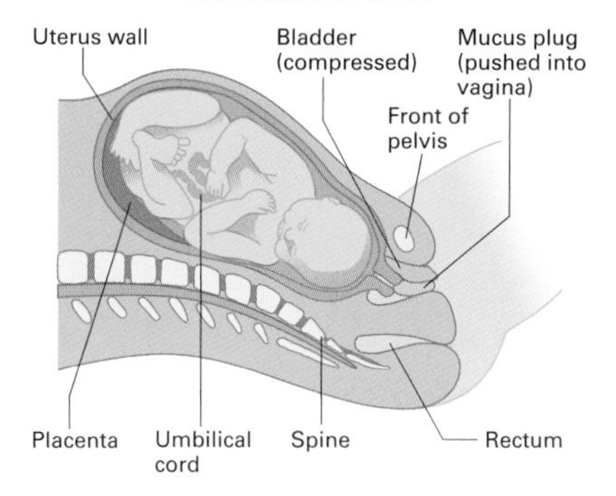

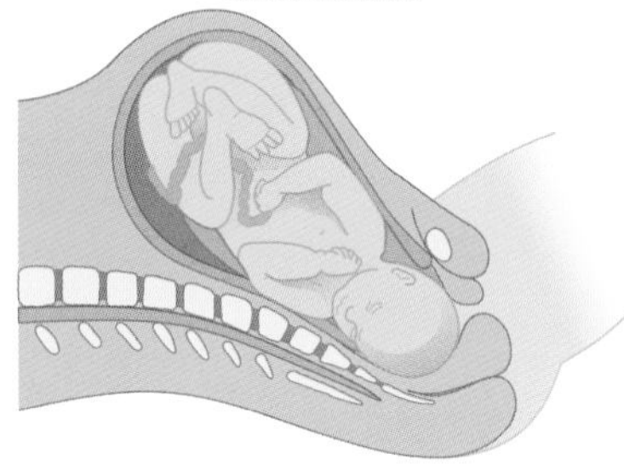

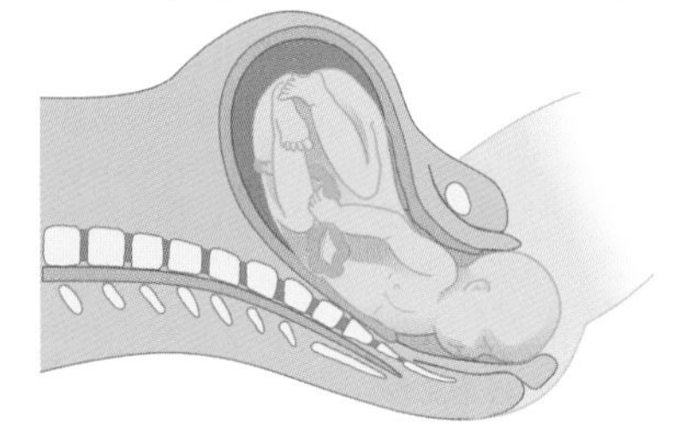

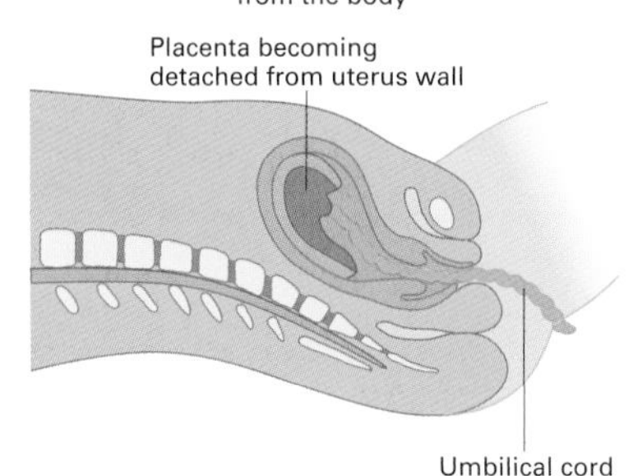

Figure 3 *Birth of a human baby.*

Quick check

1 List three hormones secreted by the placenta during pregnancy.

2 What is the role during pregnancy of:

a relaxin

b oxytocin?

3 A phrase commonly used to describe the first stage of birth is 'breaking of the waters'. To what do you think this refers?

Food for thought

Most mammalian mothers, herbivores as well as carnivores, eat the afterbirth. Explain what the afterbirth consists of and suggest reasons for eating it.

12.10

OBJECTIVES

By the end of this spread you should be able to:

- discuss the significance of parental care in mammals
- explain how lactation is controlled.

Fact of life

During lactation (breast-feeding), all the nutritional needs of an infant are supplied by the mother. Dieticians recommend that a mother eats more during the later stages of pregnancy so that she can build up an extra 2 kg or so of fat which may be drawn upon during lactation. In addition, they recommend that she consumes an extra 1900–2400 kJ per day from the start of lactation for three months. If her diet does not contain sufficient energy and nutrients, the mother's own stores of nutrients may be reduced to satisfy the requirements of the infant.

Parental care

Parental care includes all the activities of animals that maintain and support their own offspring or the offspring of near relatives.

Animals show the complete spectrum of parental care, from none at all to providing extensive support for years. A cod, for example, provides no parental care after producing fertilised eggs. It produces millions of eggs, but few will survive to reproduce. In contrast to the cod, mammals invest much time and energy caring for their young, and primates are among the most caring of all parents. Many primates have single births and nurture their offspring for several years. After the early development of the embryos/fetus in the uterus, they prepare a home for the young, provide food (milk) for the newly born infant by lactation, and teach their offspring the skills needed to fend for itself.

Lactation

Whereas the placenta supports the developing embryo/fetus in the uterus, lactation nourishes the baby after birth.

Lactation, the secretion of milk from mammary glands, is a form of postnatal care unique to mammals. All baby mammals rely on milk for their early survival after birth. **Mammary glands** are composed mainly of fat tissue and a complex system of ducts (figure 1). Each duct ends in milk-secreting cells surrounded by muscle cells which help to pump the milk to the nipples. Under the nipples, the ducts widen into pouches for milk storage.

Lactation depends on the action of two hormones: **prolactin** which promotes milk production in the glandular cells; and **oxytocin** which stimulates the ejection of milk from the nipples.

Prolactin is secreted by the anterior pituitary gland of the mother during pregnancy, but its effects are inhibited by progesterone and oestrogens. After birth, decreasing levels of progesterone and oestrogen

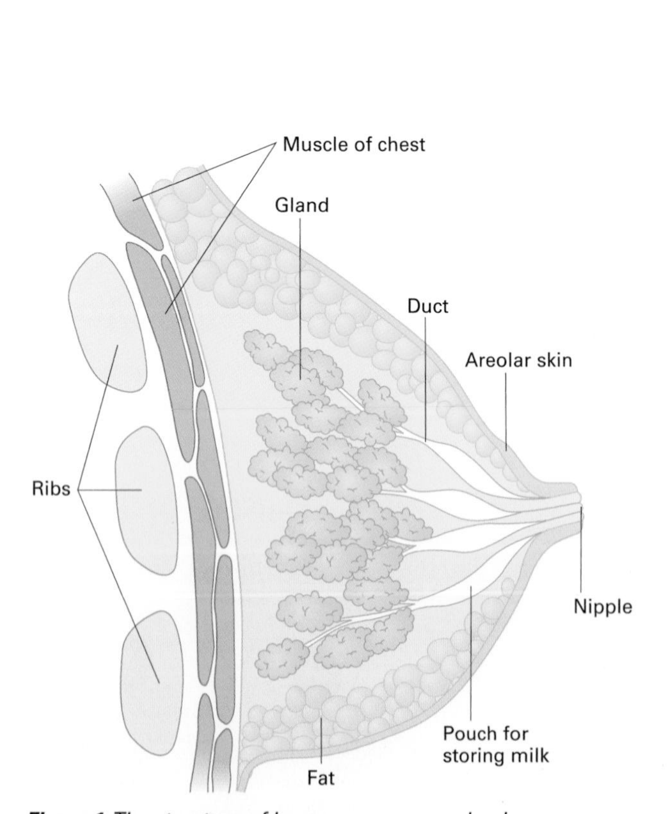

Figure 1 *The structure of human mammary glands.*

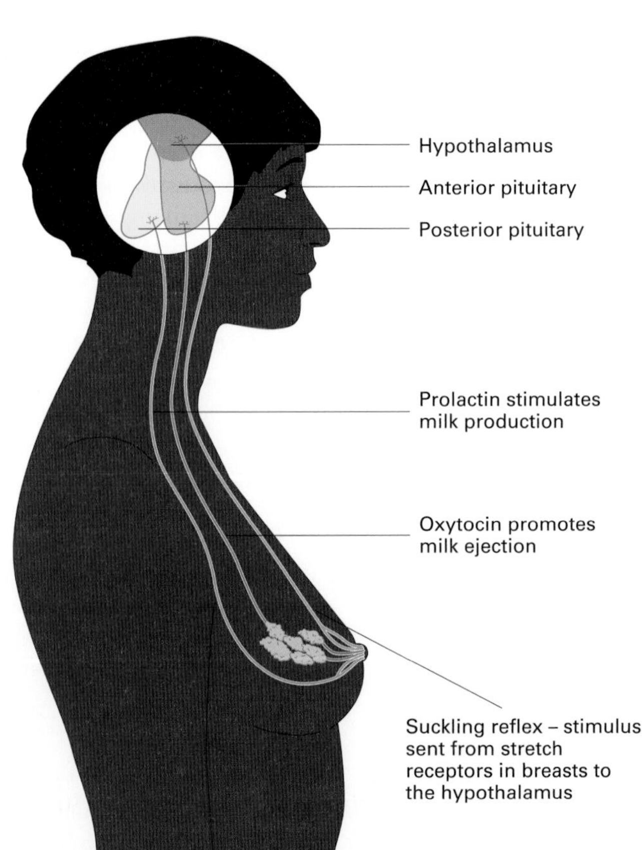

Figure 2 *The production and ejection of milk.*

allow prolactin to promote milk production. Prolactin levels increase temporarily every time a baby suckles from its mother due to a reflex action involving mechanoreceptors in the breasts (figure 2). In this way, the supply of milk is regulated to match the demands of the baby. The more a baby feeds, the greater the prolactin secretion and milk production. This is another example of positive feedback playing a beneficial part in a physiological process. Suckling also triggers the release of oxytocin, which ensures that milk flows freely to the nipple. Milk production stops soon after a mother ceases breast-feeding, but can go on for years if nursing is continued.

Suckling also inhibits the secretion of luteinising hormone (LH) and follicle stimulating hormone (FSH) from the anterior pituitary gland. This phenomenon is called **lactational anovulation** because it suppresses follicular development and ovulation. This is not a reliable method of contraception, but frequent suckling does appear to delay pregnancies in some nomadic tribes. For example, women in the !Kung, a Kalahari tribe that practises no other form of contraception, have on average only four or five children each.

By suckling their young, mammalian mothers provide their offspring with a highly nutritious fluid. Milk is rich in fats (including those required for the proper development of brain tissue), proteins, minerals, vitamins, and milk sugar (lactose). The first fluid that the baby suckles from its mother's mammary glands is a special milk called **colostrum**. This is a yellowish fluid containing antibodies which provide some immunity against infection.

A suckling infant's growth and welfare depend not on its own ability to gain resources, but on its mother's. Parental care means that the infant does not have to find its own food until it reaches an age and size such that it can compete on more or less equal terms with adults. Lactation also postpones the need for teeth until much of the jaw is complete. This allows the development of the complex fitting together (occlusion) of teeth, typical of many mammals, and essential for efficient gnawing, biting, and chewing.

Other aspects of parental care

In addition to providing nutrition, most mammalian parents also protect their young. Rodent parents build nests to provide warmth and shelter; dogs retrieve wandering pups, picking them up by the scruff of the neck, and carry them to safety (figure 3); and even apparently docile mammals such as the antelope and elk become surprisingly aggressive when protecting their young against attacks from predators.

In social mammals, parental care involves much more than feeding and protecting the infant, because the young must learn to respond to others and fit into the social structures around them. This is especially true of primates, whose infants need early experiences with their parents in order to learn to fend for themselves, and to learn to live effectively in social units. Parental (especially maternal) care and, in particular, the quality of the mother–infant bond can have a profound effect on the subsequent development of the infant's behaviour and its ability to learn to communicate effectively.

Human infants are dependent on their parents for longer than any other mammals. During its early days, the infant is totally reliant on its parents for food, warmth, and protection. Parental care also involves the teaching and learning of language. Within one year, an infant can usually speak a few words; by the age of two, most children have a vocabulary of more than 200 words. Infants learn language mainly by listening to parents and mimicking them. Parents also teach their children many other social skills. In this way, they provide an additional form of inheritance called **cultural** inheritance. Unlike genetic inheritance, our cultural inheritance is passed on deliberately and is absorbed consciously.

Figure 3 *A female African hunting dog (*Lycaon pictus*) carrying her pup away from danger.*

Quick check

1 Why does prolactin not stimulate milk production during pregnancy?

2 Some lactating mothers have plenty of milk, but it does not flow freely to the nipple. They therefore have difficulty in breast-feeding their babies. Suggest which hormone they might lack.

Food for thought

Although breast-feeding and bottle-feeding both provide babies with the same proportions of major nutrients, they differ in a number of ways. Breast milk, for example, has far more long-chain polyunsaturated fat than formula milk. A study of 926 premature babies found that those fed on breast milk had significantly higher mental development (as determined by standardised intelligence tests) by the time they reached seven years old than those given formula milk. How might the availability of long-chain polyunsaturates help the mental development of premature babies? Suggest other ways in which breast-feeding and bottle-feeding might differ, and how they might affect the social, mental, and physical development of a baby.

12.11

OBJECTIVES

By the end of this spread you should be able to:

- distinguish between growth and development
- describe a sigmoid growth curve
- distinguish between continuous growth and discontinuous growth in animals.

Fact of life

One of the most astounding marvels of biology is how a single fertilised cell gives rise to a whole complex organism such as ourselves. We now understand that **cell signals** that switch genes on and off play a crucial role. Cell signals from a group of genes called **homeotic genes** are especially important. They were first discovered in fruit flies and subsequently found in all fungi, plants, and animals. They are called homeotic genes because they all have a similar structure (homeo = the same). They function as molecular architects for specific adult body plans. One particular sequence of DNA in homeotic genes acts as a master switch, determining the fate of cells during growth and development. The sequence is 180 bases long and is called **homeobox** (**Hox**). It translates into a protein of 60 amino acids which binds to DNA, switching transcription on or off and thereby controlling gene expression.

Growth and development

Defining growth and development

Growth is a fundamental characteristic of life. In biology it means more than just 'becoming bigger'. It is a quantitative concept that generally refers to a relatively permanent and irreversible change in size usually accompanied by a change in dry mass, solid matter, and the amount of cytoplasm in an organism. Growth can be negative or positive. An organism that changes size by filling itself or emptying itself of water is not, in the biological sense, growing. Also, daily fluctuations in mass that accompany feeding and activity are not regarded as growth.

Development refers to the progressive changes that take place in an animal from conception to adulthood. Growth and development commonly occur together and determine the overall change in shape of an organism, a process called **morphogenesis**.

Growth and development in animals

The growth and development of animals begin with the embryo. Typically, there are three major stages:

- **cleavage** – the division of the zygote without an increase in mass into a **morula**, a ball consisting of many daughter cells
- **gastrulation** – the rearrangement of cells into distinct layers (the ectoderm, mesoderm, and endoderm of vertebrates)
- **organogenesis** – the development of tissues and organs.

The zygote and the morula are **totipotent**: they have the potential to generate any of the different types of cell found in the animal. In humans, after cleavage the cells form a hollow fluid-filled ball called the blastocyst (figure 1). Its innermost cells are called **embryonic stem cells** because they are **pluripotent**: they have the potential to generate any type of cell found in an adult except those needed to support and feed the fetus.

Organogenesis involves a process called **differentiation**, the development of specialised types of cell to carry out specific functions. Differentiation involves biochemical as well as structural changes. It enables particular types of cell to carry out specific functions more efficiently. However, most specialised cells lose their pluripotency and become **unipotent**: they are able to generate only cells of their own type. Because they lack a nucleus, human red blood cells lose even this ability.

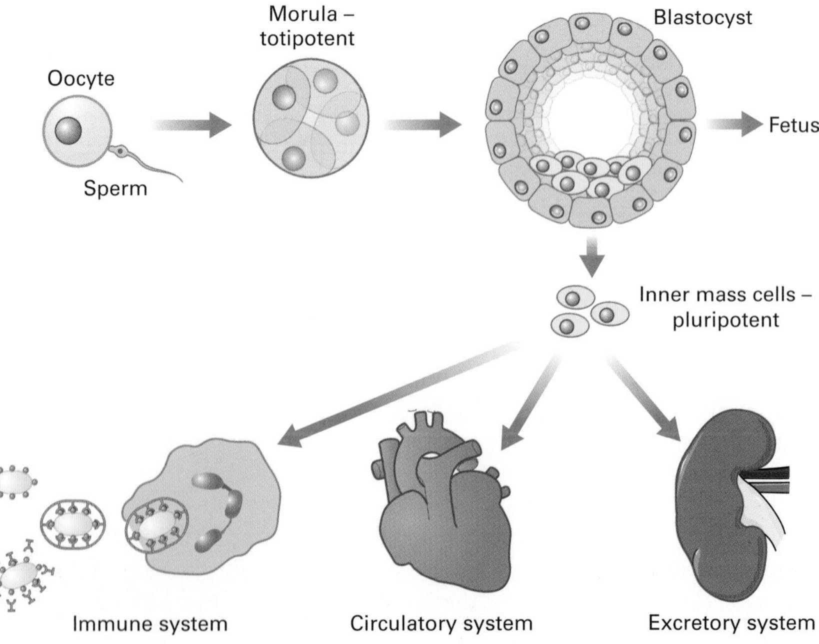

Figure 1 Sources of pluripotent cells in a developing embryo.

Not all cells become differentiated. Some remain unspecialised and retain their potential to generate other cells, even in adult animals. These cells are called somatic stem cells or **adult stem cells** and are found throughout the body in juveniles and adults. They are essential for repairing damaged or diseased tissue, and to maintain regenerative organs such as blood, skin, or intestinal tissues, which have a continuous turnover of cells. A stem cell is able to go through many mitotic cell divisions while remaining undifferentiated. Some adult stem cells are **multipotent**: they can generate progeny of several distinct cell types. Others are **oligopotent**: they can generate cells of only a few types. In humans, the blood and bone marrow are particularly rich sources of adult stem cells and are used for stem cell therapies.

Cell signalling, the ability of cells to transmit and receive information, is the basis of normal development. Signalling pathways between neighbouring cells determine their fate, what type of cell they are to become, and even whether or not they are to survive. Programmed cell death (**apoptosis**), which is regulated by cell signals, is an essential part of growth and development. Without apoptosis, tissues would continue to grow uncontrollably, as in a tumour.

Growth patterns

Although the growth of each individual results from a unique interaction between its environment and its genetic make-up, many organisms share similar basic patterns of growth. These can be expressed in graphs called growth curves in which increases in growth are plotted against time. The sigmoid growth curve (spread 22.6) describes the pattern of growth of many populations and individual organs, as well as whole organisms.

Humans have two periods of rapid growth (called growth spurts) between which there is a period of steady growth. One growth spurt occurs in early infancy and the other in adolescence (figure 2). Human organs grow at different rates from the entire body (figure 3). This is called **allometric growth**. It contrasts with **isometric growth** in fish, for example, in which organs appear to grow at the same rate as the rest of the body.

Mammals continue growing from birth to maturity, although their growth rate changes. Arthropods, on the other hand, usually have a number of periods of extremely rapid growth followed by periods in which there appears to be little or no growth. This striking growth pattern is called **discontinuous growth** or **intermittent growth** (figure 4). It happens because arthropods have to shed their inelastic exoskeletons or moult before they can grow (spread 21.13).

Mammalian growth ceases at maturity. The growth curve flattens out or even declines prior to the animal's death. Many other animals continue to grow throughout their lives. This pattern of unlimited growth is found among invertebrates, fish, and reptiles.

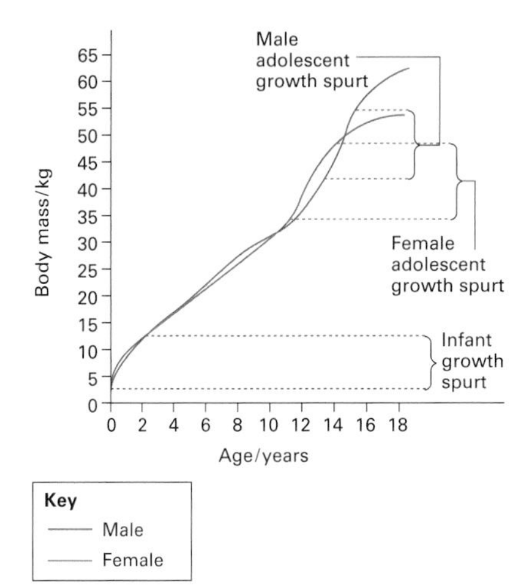

Figure 2 *Human growth curve.*

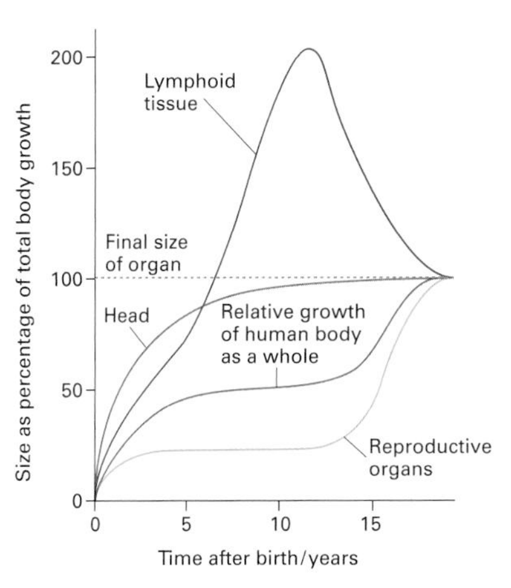

Figure 3 *Relative growth rates of lymphoid tissue, head, and reproductive organs in the human body.*

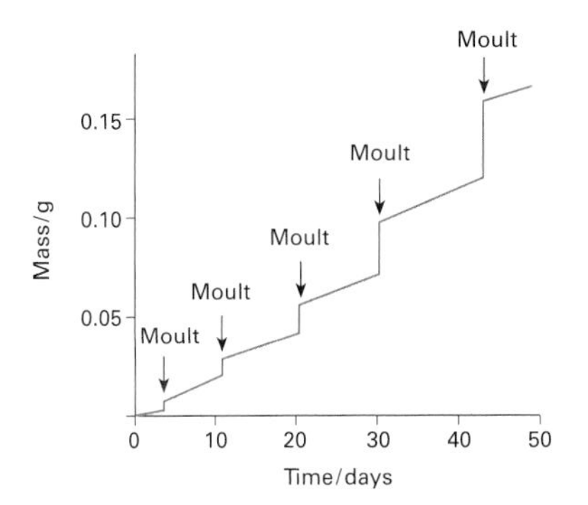

Figure 4 *Growth curve of an insect showing discontinuous growth.*

Quick check

1 Distinguish between growth and development.

2 Are embryonic stem cells totipotent, pluripotent, or unipotent?

3 Which one of each of the following pairs of terms applies to human growth:

a continuous/discontinuous

b allometric/isometric

c limited/unlimited?

Food for thought

Figure 3 shows that different human tissues have different growth curves. Suggest possible advantages of these differences.

Summary

Reproduction may be sexual or asexual. **Asexual reproduction** results in offspring being identical with each other and with the parent animal (barring mutations). There are several types including **fission**, **budding**, and **fragmentation. Sexual reproduction** results in offspring being different to each other and their parents. Animals which reproduce sexually have cells which divide by **meiosis** at some stage of their life cycle. Asexual reproduction might involve **parthenogenesis**.
The **mammalian reproductive system** includes gamete-producing gonads: the **testes** in males and **ovaries** in females. **Gametogenesis** (gamete production) involves mitotic and meiotic divisions. During **spermatogenesis**, **spermatozoa** develop in the **seminiferous tubules**; during **oogenesis**, **oocytes** develop in the **ovarian follicles. Oestrus cycles** and **menstrual cycles** are two types of **mammalian reproductive cycles**.
The human menstrual cycle is regulated by several hormones, including **gonadotrophin-releasing hormone**, **follicle stimulating hormone**, **luteinising hormone**, **oestrogens**, and **progesterones**.
In mammals, **fertilisation** is **internal** and is usually preceded by **courtship** and **copulation.** When in the female reproductive tract, sperm are activated by **capacitation**. They burrow their way into the oocyte using hydrolytic enzymes released from the **acrosome**. The **cortical reaction** results in the formation of a **fertilisation membrane** which prevents entry of more than one sperm. A variety of **contraception** techniques may be used to prevent fertilisation. The chance of fertilisation can be increased by **artificial insemination**, ***in vitro* fertilisation**, or the use of **fertility drugs**.
After fertilisation, the **zygote** develops into an **embryo**. **Implantation** of the embryo occurs in the wall of the uterus where a **placenta** develops. During pregnancy, the placenta provides an exchange surface between **fetus** and mother. **Parturition**, the act of giving **birth**, is triggered by the interaction of a number of hormonal and nervous stimuli. **Parental care** is very well developed in mammals. In addition to protecting the baby and providing it with food by **lactation**, mammalian parents generally have a profound effect on the social development of their offspring.
Growth and development of individual animals begins with the embryo. Whole organisms and organs often have a **sigmoid growth curve**. Different organs may grow at different rates (**allometric growth**) or at the same rate (**isometric growth**), and growth may be **continuous** or **discontinuous**, **limited** or **unlimited**.

PRACTICE EXAM QUESTIONS

1 The diagram shows a section through part of a human testis.

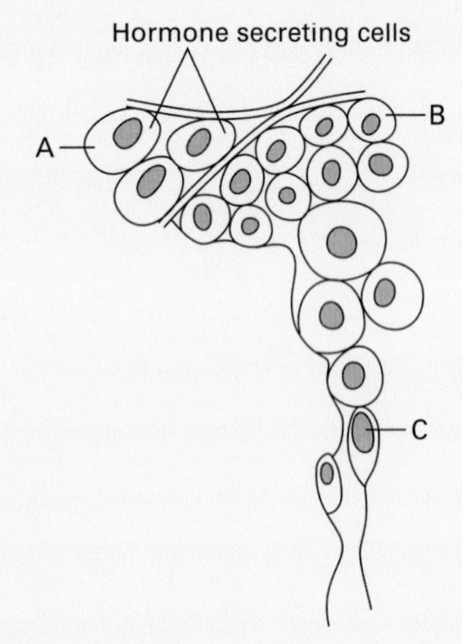

a Giving a reason for your answer in each case, what percentage of each of the following types of cell would you expect to contain a Y chromosome:

i type A cells; [1]

ii type C cells? [1]

b Many of the type B cells will undergo mitosis. Explain the importance of mitosis in a mature testis. [1]

c Give two ways in which cell division results in the type C cells being genetically different from one another. [2]

[Total 5 marks]

2 The diagram shows some stages in the formation of a mammalian egg cell.

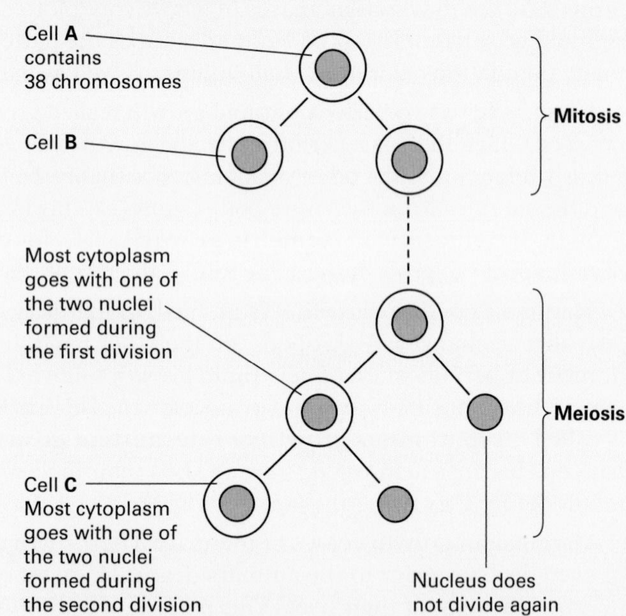

a How many chromosomes will there be in

i cell **B**; [1]

ii cell **C**? [1]

b Suggest **one** advantage in the way in which the cytoplasm divides during meiosis. [1]

c Describe and explain **two** ways in which the events of meiosis cause the egg cells to be genetically different from one another. [4]

[Total 7 marks]

3 a Describe the similarities and differences between male and female gametes. [3]

b Explain two ways in which the placenta is adapted to provide a developing fetus with its nutrients. [4]

[Total 7 marks]

4 **a** The mammalian oestrous cycle is controlled by hormones secreted from the pituitary gland and from the ovaries.
Describe the roles of the following hormones in the control of this cycle.

i The pituitary hormones FSH and LH. [5]

ii The ovarian hormones oestrogen and progesterone. [5]

b Rats can reproduce at any time of the year. Some other mammals, for example deer, have a specific breeding season. Suggest **one** advantage to each animal of its particular pattern of breeding. [2]

[Total 12 marks]

5 The graph shows the effect of light on the period of oestrus in two groups of Suffolk ewes (female sheep). Ewes in the control group were subjected to the normal seasonal variation in day length. Part way through the first year of the experiment, ewes in the experimental group were subjected to a reversed day-length cycle. The period of oestrus of each group of ewes is shown as solid bands above the graph.

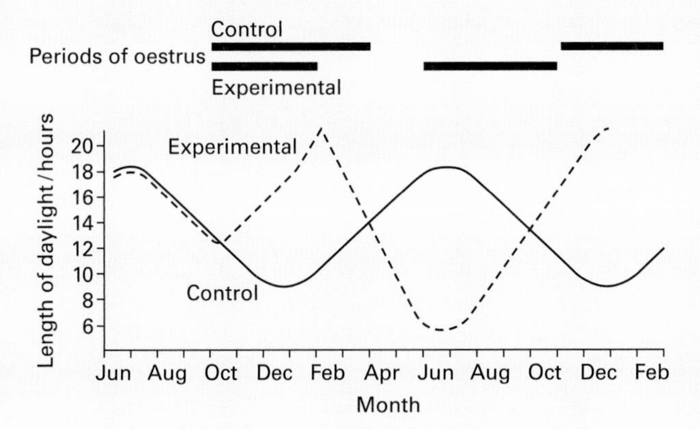

a **i** During what time of year did oestrus normally occur? [1]

ii Suggest **one** advantage of this pattern of oestrus to the ewes in the control group. [2]

b What does the experiment suggest about the regulation of the onset of oestrus in ewes? [2]

[Total 5 marks]

6 The table below refers to four hormones associated with the human menstrual cycle.
Copy the table. If the statement is correct, place a tick (✓) in the appropriate box and if the statement is incorrect place a cross (✗) in the appropriate box.

Hormone	Secreted by ovaries	Reaches highest level in blood before ovulation
Follicle stimulated hormone (FSH)		
Luteinising hormone (LH)		
Oestrogen		
Progesterone		

[Total 4 marks]

7 In this section, answers should be illustrated by large, clearly labelled diagrams wherever suitable.
Your answer must be in continuous prose, where appropriate.
Up to 4 additional marks in this section are awarded for quality of English.
Your answer must be set out in sections **a**, **b** etc., as indicated in the question.

a **i** Describe the structure of the human placenta. [6]

ii Outline the transport mechanisms involved in the exchange of substances between mother and fetus. [6]

b Explain how smoking, by a pregnant woman, may damage the unborn baby. [4]

[Total 16 marks]

13 Structure and transport in plants

The leaf

13.1

OBJECTIVES

By the end of this spread you should be able to:

- describe the structure of a dicotyledonous leaf
- describe the functions of different tissues in a leaf
- distinguish between parenchyma, collenchyma, and sclerenchyma.

Fact of life

The longest leaves belong to palm trees. Those of the palm *Raphia ruffia* (from which raffia fibres are obtained) may reach over 22 m long.

The leaf is the main site of photosynthesis, the process by which green plants manufacture their own food (chapter 5). Figure 1 shows the structure of a typical dicotyledonous leaf. (The **dicotyledons** form a major group of flowering plants in which the seeds have two embryonic leaves; see spread 21.7.) The **lamina** or blade of a leaf is flat and thin. Its shape provides a large surface area for absorption of light and carbon dioxide. The leaf is attached to a stem or branch by a **leaf stalk** or petiole. The stalk holds the leaf in a position such that its surface is exposed to the maximum amount of light. From the stalk, a main **vein** leads down the leaf with side veins branching out on either side. These veins connect the leaf to the rest of the plant, bringing the leaf some of the raw materials required for photosynthesis, and carrying products of photosynthesis away from it. The veins also provide mechanical support, maintaining the flat shape of the leaf. The stem and branches raise the leaves above the ground so they are exposed to the light. On many plants the leaves are arranged on branches in such a way that they do not shade one another.

The tissues of a leaf

In common with stems (spread 13.2) and roots (spread 13.3), leaves are made up of three main types of tissue: **epidermal tissue**, **vascular tissue**, and **ground tissue**. Each tissue forms a continuous system throughout the plant.

The epidermis covers and protects the leaves. It is the first line of defence against physical damage, infection, and being eaten. The upper epidermis consists of one or more layers of rectangular cells. In terrestrial plants, these epidermal cells secrete a waxy coating called the **cuticle**. The waxy cuticle is waterproof, minimising water loss from the surface of the leaf. It is often thicker on the upper surface, making this surface appear more shiny than the lower surface.

***Figure 1** The structure of a dicotyledonous leaf.*

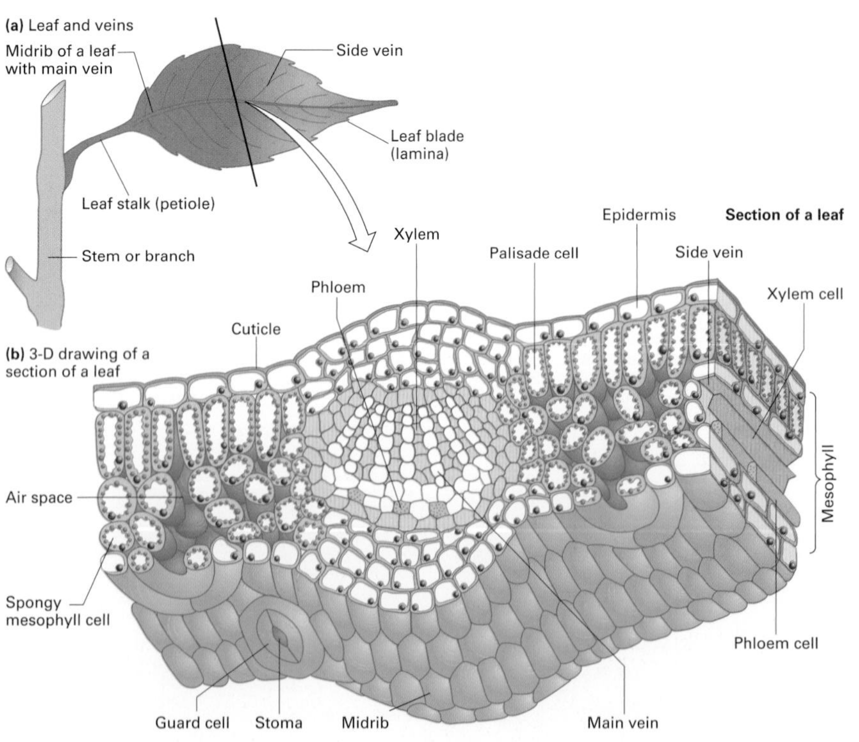

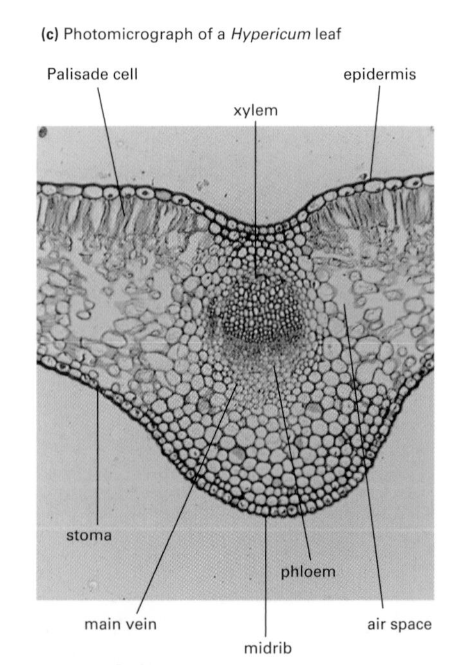

(Adapted from original artwork by Don Mackean: *GCSE Biology, second edition*, John Murray, 1995)

The epidermis is perforated by microscopic pores called **stomata**. Stomata allow carbon dioxide and oxygen to gain easy access into the plant, but also allow water to escape. Each stoma is flanked by a pair of **guard cells** that regulate the size of the pore, closing it in times of water stress (spread 13.7). Water is more likely to be lost from the upper surface of a leaf because it is more exposed to sunlight. The upper surface usually has fewer stomata than the lower surface; this minimises water loss.

The **vascular tissue** consists of veins adapted to transport liquid substances around the plant, and it is made up of vascular bundles, groups of vessels running from the root up the stem and to the leaves. **Xylem** (figure 2) forms the upper part of a vascular bundle in the leaf, bringing water and mineral salts to the leaf. **Phloem** forms the lower part of a bundle, transporting sucrose and other products of photosynthesis away from the leaf.

Ground tissue is all the tissue in a plant other than the epidermis, reproductive tissue, and vascular tissue. It makes up the bulk of a leaf and consists mainly of parenchyma cells reinforced by collenchyma and sclerenchyma.

The cells of the ground tissue

Parenchyma cells are the least specialised of plant cells (figure 2a); they are characterised by having intercellular air spaces which vary in size. Parenchyma cells are regarded as the basic cells from which other cells have evolved. Parenchyma cells form the packing tissue of plants, and include the palisade cells and spongy mesophyll cells shown in figure 1 which make up the main photosynthesising tissue in the leaf.

Palisade cells are a dense green colour due to the numerous chloroplasts they contain. These cells are packed tightly together in a regular arrangement near the upper surface of the leaf so they obtain the maximum exposure to light. The chloroplasts can move around inside the cells according to the amount of light available. If it is a dull day, they are often clustered at the tops of the cells, in the best position to trap light; in very sunny conditions, they may be grouped towards the bottoms of the cells to avoid being overexposed to light.

The **spongy mesophyll** is the chief site of gaseous exchange in the leaf. It consists of rounded or sausage-shaped cells with fewer chloroplasts than palisade cells. The cells are loosely arranged and between each of them are air spaces connecting the mesophyll with the stomata.

Collenchyma and sclerenchyma (figure 2b, c) make up tissues that have a supportive, structural role in plants. In leaves, these cells are common around the vascular bundles (especially the midrib) and at the leaf tips. **Collenchyma** cells are elongated and have unevenly thickened cell walls with extra cellulose in the corners of the cells. There are two main types of **sclerenchyma: fibres** are very elongated and have very thick cell walls impregnated with lignin; **sclereids** (or stone cells) are more spherical in shape. Both types of sclerenchyma cells are specialised for support. Fibres in particular have great tensile strength and do not break easily when stretched. Mature sclerenchyma cells are dead because they are enclosed in a complete layer of lignin which is impermeable to water.

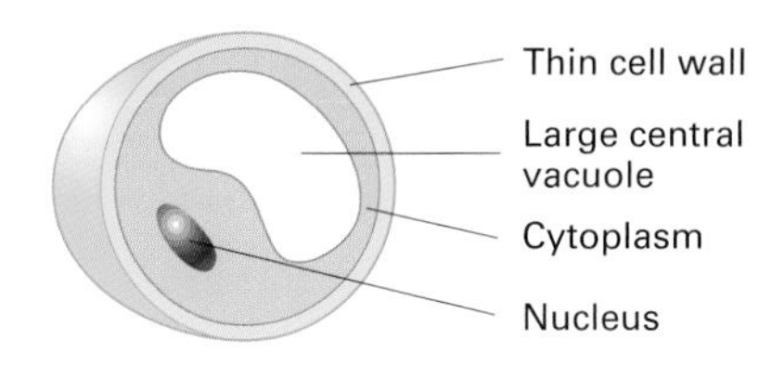

(a) Parenchyma

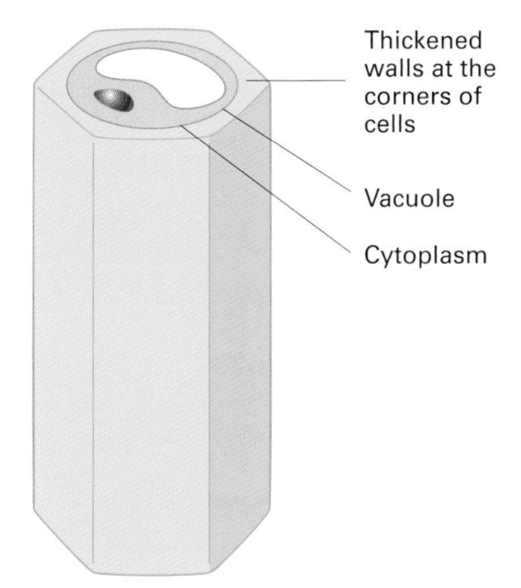

(b) Collenchyma

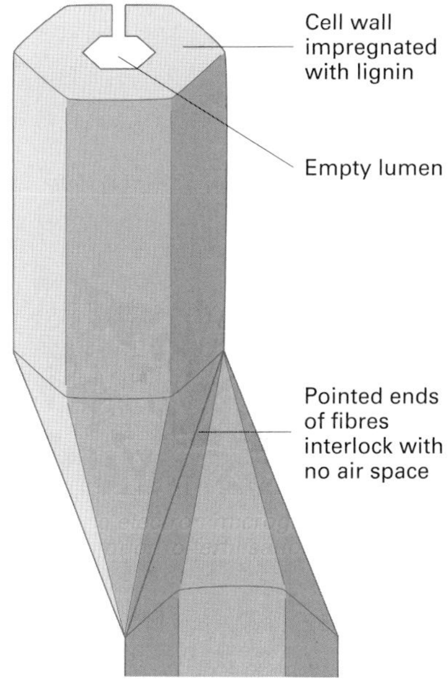

(c) Sclerenchyma: fibre

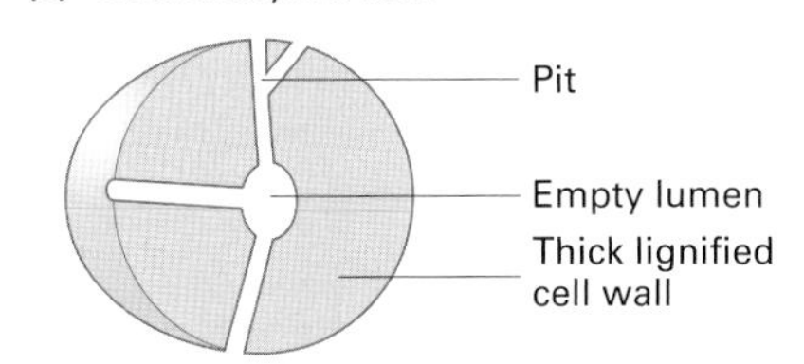

(d) Sclerenchyma: sclereid

Figure 2 *The main types of cells in the ground tissue of leaves.*

QUICK CHECK

1 a Which structure forms a waterproof layer on the surface of a leaf?

b Why is this structure thicker on the upper surface of the leaf than on the lower surface?

2 What is the main function of palisade cells?

3 How does collenchyma differ from sclerenchyma?

Food for thought

Leaves from different species have an enormous variety of size, shape, and structure. In addition to being adapted to absorbing light for photosynthesis, to what other factors might leaves be adapted?

13.2

The stem

OBJECTIVES

By the end of this spread you should be able to:

- describe the structure of a dicotyledonous stem
- state the major functions of stems
- explain how different tissues contribute to the mechanical support of stems.

Fact of life

Stems of wood plants have a layer of protective tissue called cork just below the epidermis. Cork is made of dead cells coated with a waxy, waterproof substance (suberin). The exceptionally thick cork layer of the cork oak (*Quercus suberi*) is removed for commercial use. If cork formed a complete layer, the cells in the stem would die because they would not be able to exchange respiratory gases with the environment. However, slit-like openings called **lenticels** (figure 1) develop in cork to allow gaseous exchange.

Figure 1 *Photomicrograph of an elder stem. A lenticel perforates the cork layer of the stem, allowing gaseous exchange to take place between the stem cells and the environment.*

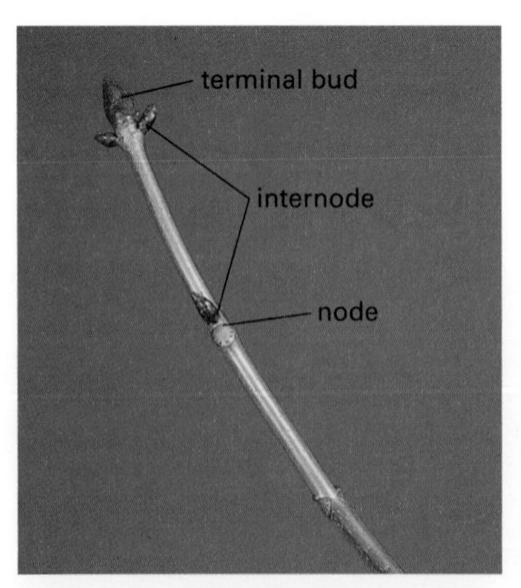

Figure 2 *A woody twig: an example of a stem showing nodes and internodes of a horse chestnut (*Aesculus hippocastanum*).*

Functions of a stem

The stems of most plants are first and foremost organs of support. They lift terrestrial plants above the ground, raising their leaves towards the Sun and holding them in the best position to gain optimum exposure to light and carbon dioxide. They also hold flowers and fruit in positions that allow efficient pollination and seed dispersal.

As well as support, stems have three other major functions:

- they transport materials from one part of the plant to another
- they produce new living tissue to replace cells that die and to make new growth
- they store food and water.

The stem as a plant organ

The attachment site of a leaf or bud on a stem is called a **node**, and the portion between nodes is called an **internode**. Most stems point upwards from the ground and are easily distinguished from other plant organs. Some stems, however, have an unusual shape or location which makes them more difficult to identify. Potato tubers, for example, appear root-like, but they are actually swollen underground stems specialised for food storage. All stems, of whatever size, shape, or location, are distinguishable as such by the presence of nodes and internodes (figure 2).

The tissues and cells of a stem

Figure 3 shows a dicotyledonous, non-woody (herbaceous) stem. The epidermis is like that of a leaf: a single layer of cells perforated by stomata. The epidermis helps maintain the shape of the stem. It is covered with a waxy cuticle to reduce water loss. In woody stems of trees and bushes, the epidermis is replaced by **bark** consisting of many layers of dead cells. Bark is penetrated by small pores called **lenticels**, through which gaseous exchange takes place. The lenticels usually appear as raised spots surrounded by a powdery and impermeable material.

Just inside the epidermis, a layer of collenchyma gives both support and flexibility to the stem. Some collenchyma cells contain chloroplasts which make the stem appear green.

The inner parts of the stems of most non-woody plants consist of vascular bundles embedded in undifferentiated parenchyma cells. When fully inflated with water (turgid), the parenchyma cells press against the epidermis and collenchyma, strengthening the stem. The stems of trees and bushes are supported not by parenchyma but by rigid woody tissue which makes up the bulk of these stems. The woody tissue consists of xylem and associated cells such as fibres formed by a process called secondary growth. Spread 14.9 explains how new wood is added outside the old wood each growing season to form annual growth rings, visible in transverse sections of the stems of trees and shrubs.

Vascular tissue in the stem consists of bundles containing phloem and xylem surrounded by protective and strengthening **sclerenchyma fibres** (figure 4). The xylem is located towards the inside of the stem and the phloem towards the outside. The tough rigid vascular bundles embedded in softer turgid parenchyma tissues have been likened to reinforced concrete, in which rigid steel girders are embedded in softer concrete. This arrangement gives the stem strength and flexibility, making it well suited to resisting sideways bending in strong winds. The vascular bundles of dicotyledonous plants are arranged in a ring pattern around the outside of the stem, while in monocotyledons such as cacti the vascular bundles are scattered throughout the stem.

The stem centre is called the **pith**. It may consist of parenchyma cells for storage, or it may be devoid of cells, in which case it is called a **pith cavity**.

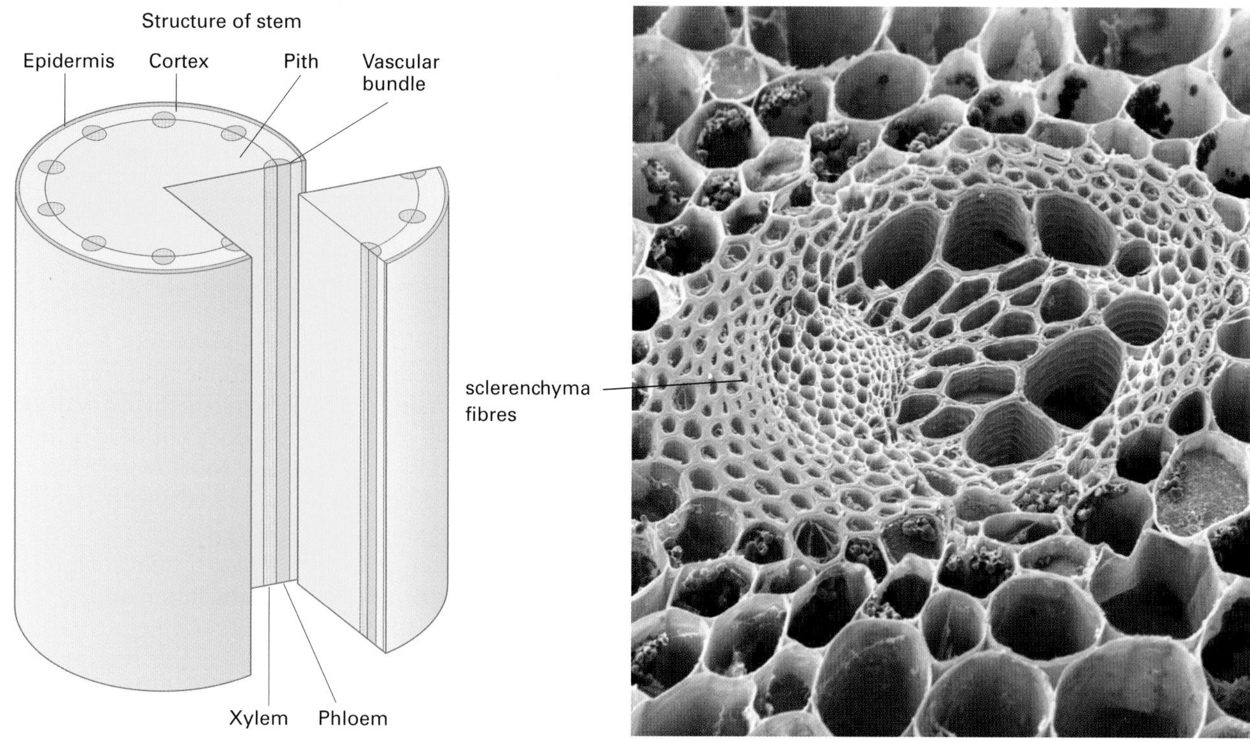

Figure 3 *The structure of a non-woody dicotyledonous stem.*

Figure 4 *Coloured scanning electron micrograph (SEM) of a transverse section through part of a buttercup stem showing sclerenchyma fibres in a vascular bundle.*

Plant fibres

Fibres are a vital part of a plant's anatomy. They hold plants together, providing strength, support, and resilience against wind, wear, and tear. Typically, plant fibres are long, spindle-shaped structures, tapered or blunt at both ends. They are usually arranged in groups. Those in the xylem (xylary fibres) are xylem vessels. Those outside the xylem (extraxylary fibres) are mainly sclerenchyma fibres. Like all plant cells, fibres have a cellulose cell wall made of microfibrils which contribute to a plant's fibrous nature (spread 2.7). Cotton is composed mainly of cellulose that forms twisted ribbon-shaped fibres which burst out the seeds of the cotton plant. Most fibres are strengthened by lignin. Trees and shrubs are strengthened further by secondary thickening. This results in secondary xylem vessels becoming thickened with extra layers of lignin and cellulose to form dense woody tissue (spread 14.9). Traditionally, plant fibres have been used in many products including fabrics, ropes, and cordage. Recently, there has been great interest in using plant-based products (for example, fibres from bamboo shoots and rice husks mixed with an amino acid resin) to replace oil-based plastics. The tensile strength of plant fibres can be tested by fixing a fibre at both ends and applying a force (e.g. suspended masses) onto its centre until it breaks. Fibres from celery can be obtained easily by peeling. Those from nettles can be extracted by soaking the stem in water for about a week to rot away the soft parts (this process is called **retting**). After soaking, xylem vessels and sclerenchyma fibres can be peeled away from the central pith and dried before testing.

Figure 5 *An exotic garden display of several different species of cacti.*

Quick check

1 What distinguishes stems from other plant structures?

2 List the four main functions of stems.

3 How do parenchyma cells help support herbaceous stems?

Food for thought

Cacti (figure 5) live in hot dry American deserts. To conserve water and deter herbivores, their leaves lose their photosynthesising function and are modified into spines. Nevertheless, cacti may lose as much as 20 per cent of their tissue fluids in a severe drought. Suggest how the stem is adapted to:

a carry out photosynthesis

b minimise water losses and minimise the effect on the plant of water losses.

13.3

THE ROOT

OBJECTIVES

By the end of this spread you should be able to:

- describe the structure of a typical dicotyledonous root
- describe the functions of root tissues.

Fact of life

The roots of some desert plants grow very deeply to tap underground water supplies. The taproot of a mesquite, a small tree found in the tropical deserts of central America, was discovered 53 m underground by engineers. In contrast, a particular yam called hottentot bread (*Dioscorea elephantipes*) overcomes water scarcity by storing water in an immense underground water-filled root tuber that may weigh up to 320 kg.

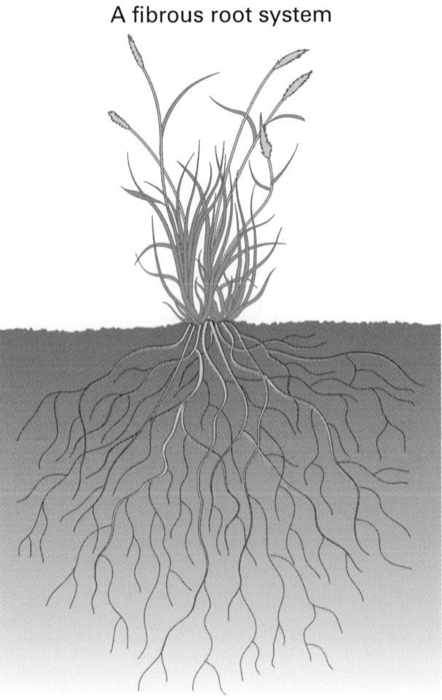

Figure 1 *A taproot system and a fibrous root system.*

Usually hidden from sight, roots are vital to the life of terrestrial plants. For many plants that live in seasonal environments, the roots are the only structures to survive the winter. Roots usually grow underground to obtain water and mineral salts. Other functions of roots include support, storage, transport, and production of new cells. Roots also absorb oxygen from the soil.

Fibrous roots and taproots

The roots of a plant usually form either a **taproot system** or a **fibrous root system** (figure 1). In a taproot system, the radicle (the first root to emerge from a seed) forms the major root that persists throughout the life of the plant. In a fibrous root system, the radicle does not persist for long after germination but is replaced by **adventitious roots** (roots that grow from stems and leaves). Dicotyledons such as trees and shrubs usually have taproot systems; monocotyledons such as grasses usually have fibrous root systems. A mature root system can be extremely complex and the combined length of the roots can measure many kilometres.

The root as a plant organ

Although roots usually grow underground, they cannot always be distinguished from stems on the basis of their being under the soil; some plants have roots wholly above the ground, and others have underground stems. However, no matter what their location, shape, or size, roots differ externally from stems in not bearing leaves or buds and by having tips covered by a layer called the **root cap** or calyptra. The root cap is a thin thimble-shaped layer of cells which protects the growth region (**root meristem**) just behind the root cap from abrasion as the root grows and pushes its way between soil particles. Cells of the root cap are constantly worn away as they push through the soil: rapidly growing roots lose as many as 10 000 cells per day. As the cells of the root cap are sloughed off, they release a polysaccharide slime called **mucigel** which protects the root tips from drying out. Mucigel also lubricates the root tip as it forces its way through the soil.

The tissue and cells of a root

Figure 2 shows the internal structure of a root. Starting from the outside and working inwards, it has first an **epidermis**: an outer layer of small thin-walled cells which have no waxy cuticle. In one region of the root, specialised epidermal cells called root hair cells grow outwards to form **root hairs**, which are thin walled and tube shaped. The root hairs are in intimate contact with the soil and provide a very large surface area for the uptake of not only water and mineral salts but also oxygen. Root growth and root metabolism are very active processes, requiring a good supply of oxygen. Roots in heavily compacted soil (such as under a path) or waterlogged soil quickly die. Soil oxygen is as vital for root health as soil water.

The zone in which root hairs occur is called the **piliferous layer**, lying just behind the root tip. Further up the root the piliferous layer is replaced by other cells which may become corky due to the presence of suberin (a waxy substance).

The outer region of the root between the epidermis and vascular bundle is called the **cortex**. It is made up of parenchyma cells which may store starch grains. There are large air spaces between the parenchyma cells, making it easier for oxygen to diffuse into the root.

The inner boundary of the cortex is marked by the **endodermis**. This is a single layer of cells which forms a ring around the centrally placed vascular bundles. The radial (side) and end walls of endodermal cells are lined with suberin to form the **Casparian strip**. This is thought to influence the route by which water enters the vascular bundles from the root cortex (spread 13.6).

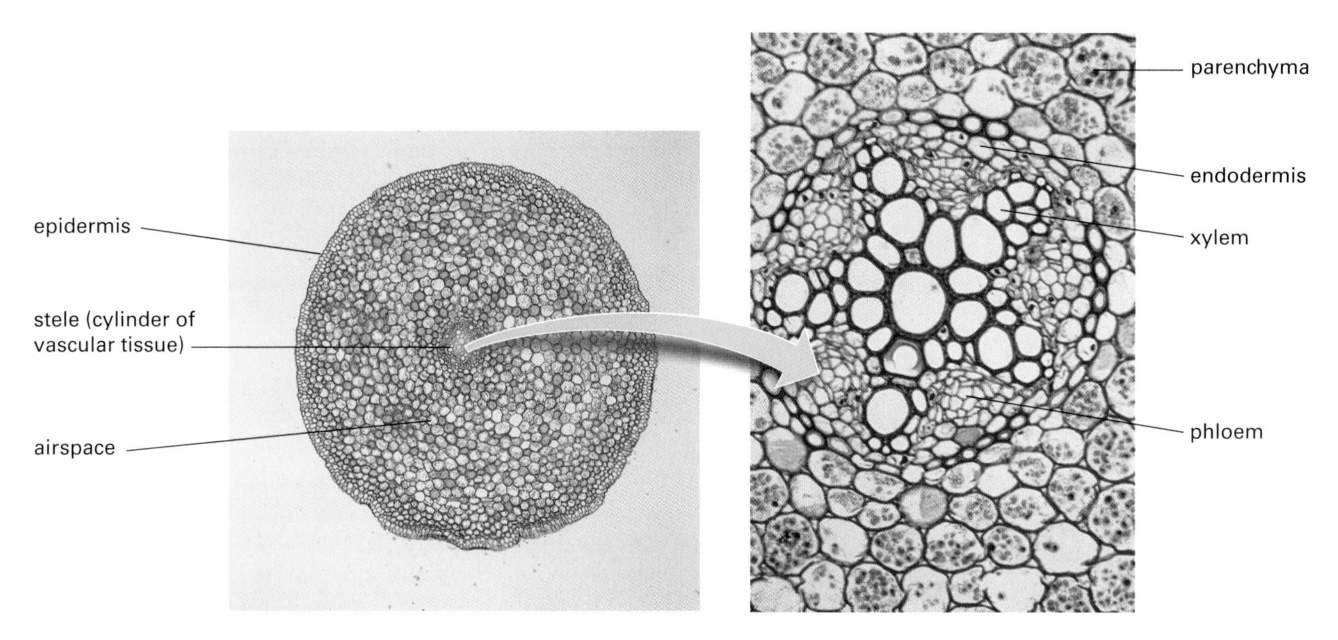

***Figure 2** The structure of a dicotyledonous root.*

Immediately inside the endodermis, the **pericycle**, an indistinct layer of lignified (woody) sclerenchyma, provides some extra mechanical support. The pericycle contains unspecialised cells from which lateral roots develop.

Internally, roots differ from stems in having their vascular tissue located in a central region called a **vascular cylinder** or **stele**. The vascular bundles are enclosed within the endodermis and pericycle, and their central location enables them to withstand pulling forces on the roots when the shoot is blown around by the wind. In many plants, the water-conducting xylem radiates out from the centre of the vascular cylinder like the spokes of a wheel. The food-conducting phloem fills the areas between the spokes.

Longitudinally, roots are divided into regions of cell division, cell growth, and cell differentiation (figure 3). Mitosis and cell division (spread 4.11) occur in the root meristem. Behind the meristem, in the region of elongation, cells expand and push the root tip through the soil. Further back, the cells differentiate into specialised types, and it is in this region of differentiation that root hairs first develop. Above this region, fully developed root tissues occur.

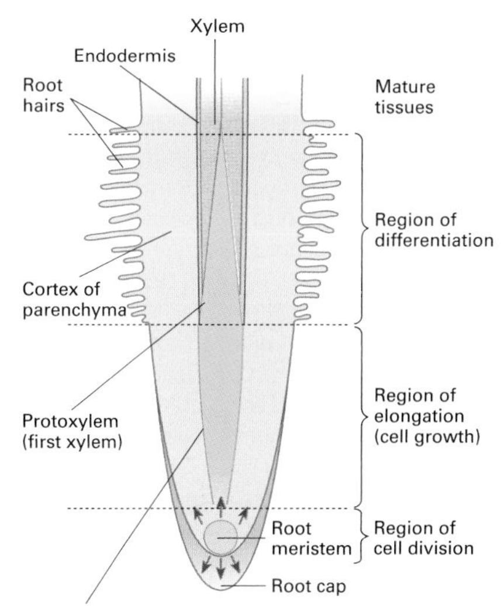

***Figure 3** Longitudinal section through a typical root.*

Quick check

1 Describe the structure of root hair cells.

2 State the function of:

a root hairs

b the root cap

c root xylem.

Food for thought

Mangrove swamps are humid and hot places pervaded by the smell of decay. Each tidal inflow of water deposits more glutinous mud. The white mangrove tree inhabits this inhospitable environment. Why do you think its roots stick out of the mud (figure 4)? Why are these structures so important to mangrove trees?

***Figure 4** Roots of white mangrove trees protruding from the mud.*

13.4

Transport: xylem and phloem

OBJECTIVES

By the end of this spread you should be able to:

- discuss the need for transport systems in plants
- distinguish between tracheids and xylem vessels in xylem tissue
- describe the functional unit of phloem.

Fact of life

Infections of xylem vessels by bacteria, fungi, or nematodes can cause potentially devastating **vascular wilt diseases.** Typical symptoms include **wilting** and death of leaves, which harms the whole plant and can lead to death. **Panama disease** is a wilt disease of bananas caused by spores of the fungus *Fusarium oxysporum* becoming lodged in xylem vessels. When the spores germinate, hyphae grow and produce a further batch of spores which spread rapidly through the whole xylem system. Water transport is blocked. Leaves yellow, wilt, and die. Eventually the whole shoot dies, leaving only the rhizome. Although the rhizome produces suckers in the next growing season which can form new shoots, these shoots are also diseased. In the 1940s and 1950s, Panama disease threatened to wipe out commercial banana plantations in the whole of the Caribbean and Central America. The plantations were saved only by replacing susceptible bananas with a resistant type that had not previously been in commercial production. This cultivar has remained resistant to the present day.

Figure 1 *The giant redwood (*Sequioa sempervirens*) is the tallest tree in the world. It may live more than 200 years and grow nearly 120 m high. Redwood forests were widespread in the age of the dinosaurs but today their natural range is restricted to a narrow coastal belt from California to southern Oregon. Redwood cultivars have been widely planted elsewhere, including the UK.*

Growing tall

All green plants depend on light energy from the Sun to power photosynthesis. Because of this dependency, there is intense competition among plants for exposure to light, especially in forests. An ability to grow tall is therefore a great advantage: the tallest plant in a forest can expose its leaves to light unobstructed while smaller plants risk being overshadowed by their neighbours and deprived of light. Tall plants can also disperse pollen, seeds, and fruit further than their smaller competitors.

These advantages of height have resulted in the evolution of trees of truly magnificent proportions. The giant redwood tree, for example, can grow to a height of over 100 m (figure 1). Although its great height gives the redwood access to the best light, it also poses considerable problems. The tree's photosynthesising leaves require a constant supply of water and mineral salts, but the only source of these is the soil, 100 m below. Its roots have to probe deep into the soil to quench its insatiable demand for water, but in order to grow the roots need to use food manufactured in the distant leaves. Tall plants therefore have extensive vascular systems which transport water and nutrients. The gases carbon dioxide and oxygen are transported in and out of even the largest plant by diffusion.

Flowering plants have evolved two separate vascular systems: xylem transports water and dissolved mineral salts; and phloem transports organic substances. Both these transport systems involve mass flow (spread 13.10).

Xylem: tracheids and vessels

Xylem is a vascular tissue made up of cells working together to carry out two main functions: support and transport. Xylem contains two types of water-conducting cells: **tracheids** and **vessels**. Other cells in the vascular tissue include parenchyma, collenchyma, and sclerenchyma (spread 13.1), which are not involved in transport but contribute to support.

Tracheids and vessels are elongated cells arranged end to end (figure 2). During development their cell walls become impregnated with lignin (spread 2.7), making them woody and impermeable. Mature tracheids and vessels are hollow and dead. The walls of these water-conducting cells are perforated by pits, which allow water and salts to pass sideways between the cells.

Tracheids have tapered sloping end walls that contain cellulose-lined pits which allow water to pass from cell to cell. Xylem vessels are usually shorter and fatter than tracheids. Their end walls break down so that the cells join end to end to form a continuous hollow tube. This arrangement allows water to flow upwards relatively unimpeded from one vessel cell to another.

The first xylem vessels to develop in a plant are known collectively as **protoxylem**. They have cell walls in which lignin is laid down in distinctive rings or spirals. This allows the cells to expand and lengthen between the lignin. Later in the plant's life, **metaxylem** vessels develop which are larger and more extensively lignified than protoxylem.

Transporting water

The water-conducting role of xylem can be demonstrated in a number of ways. For example, if the cut end of a leafy shoot is placed in a dye (such as eosin) and sections of the shoot are examined after a few hours, only the xylem is found to be stained with the dye (red with eosin). Also, if a ring of tissue is removed from a stem or trunk, water continues to flow upwards as long as only the phloem has been destroyed. Water ceases to flow if the xylem is removed, destroyed, or blocked. However, metabolic poisons, such as cyanide, do not stop the upward flow of water in a shoot.

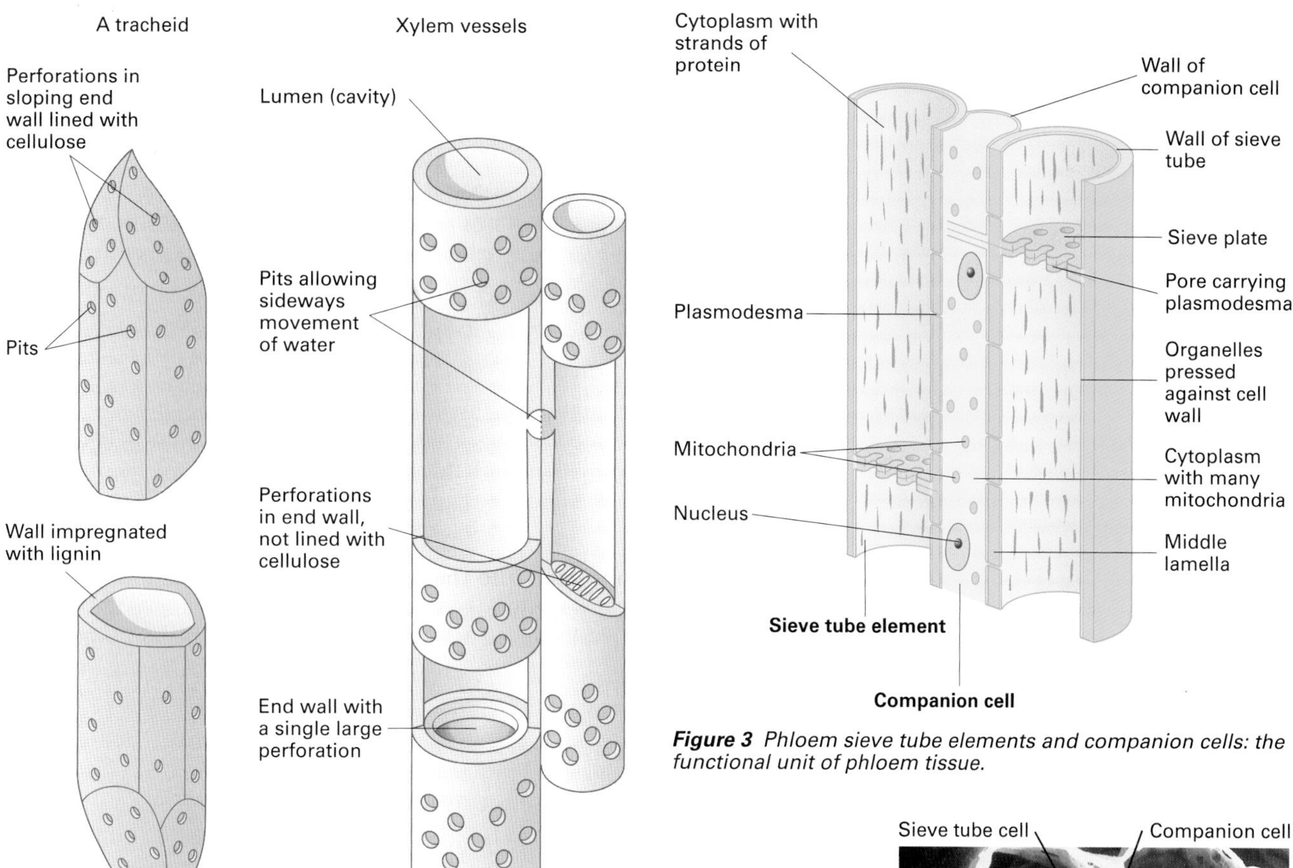

Figure 2 *The structure of tracheids and vessels, two kinds of xylem cell.*

Figure 3 *Phloem sieve tube elements and companion cells: the functional unit of phloem tissue.*

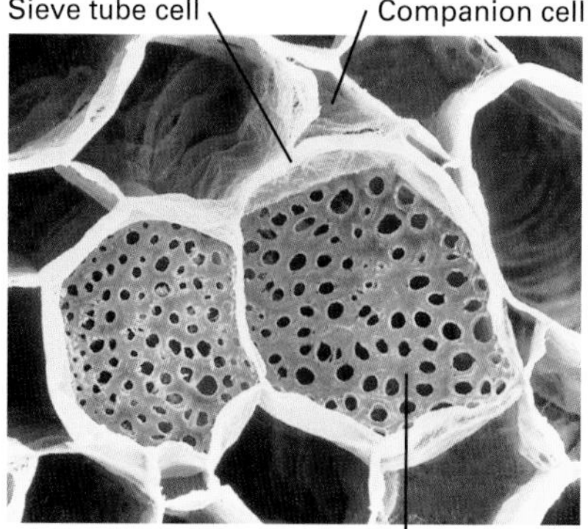

Figure 4 *Photomicrograph of a transverse section through phloem tissue showing sieve plates.*

Water movement must therefore be a passive process. Even though xylem cells are dead, they are the plant's water transport system. Spreads 13.5 and 13.6 describe the different mechanisms by which water moves in the plant, and spread 13.8 discusses mineral absorption and transport.

Phloem: sieve tube elements and companion cells

Phloem is a mixed tissue which may contain a variety of cell types, including parenchyma, sclereids, fibres, sieve tube elements, and companion cells. Organic substances (such as amino acids, sucrose, and plant hormones or auxins) are transported along the phloem **sieve tubes**. Each sieve tube is a cylindrical column of **sieve tube elements** joined end to end (figure 3). The end walls are perforated by pores carrying enlarged plasmodesmata (cytoplasmic strands passing through the cell walls) forming structures called **sieve plates**, so named because they look very much like a sieve. When mature, sieve tube elements have no nucleus and their cytoplasm is pushed to the sides of the cell. They are kept alive and supported in their function by **companion cells**.
A typical companion cell has a cytoplasm containing the usual organelles, but it has an unusually thin cell wall and its vacuole, if present, is small and difficult to detect. It also has an exceptionally high density of mitochondria, indicating that it is very active metabolically. Companion cells are intimately connected to sieve tube elements via numerous plasmodesmata. Sieve tubes and companion cells form the functional unit for **translocation**, the transport of organic solutes in the phloem. Sieve tubes cannot work without their companion cells.

Organic substances are translocated downwards from the leaves to storage organs and, later, upwards from storage organs to the growing regions. Translocation is discussed in more detail in spread 13.9.

Quick check

1 By what process does gaseous exchange take place in even the largest plants?

2 How do tracheids differ from xylem vessels?

3 What is the functional unit of translocation in the phloem?

Food for thought

The redwoods (figure 1) tower above other plants in the forests of Sierra Nevada. They are clearly the dominant tree within the forest. If being tall is such a great advantage, why don't all forests contain trees this tall?

13.5 Water transport

OBJECTIVES

By the end of this spread you should be able to:

- distinguish between transpiration and evaporation
- discuss the causes of transpiration
- describe the cohesion-tension theory of water movement.

Fact of life

The volume of water transpired from a plant can be phenomenal: in one growing season, a single corn (maize) plant can transpire more than 450 dm^3 of water, and in one day, a single oak tree can transpire more than 700 dm^3 of water. Transpiration can have profound effects on the local weather and is a major factor in maintaining the water cycle in tropical rainforests.

From the highest part of its shoot to its deepest root, a terrestrial plant contains a continuous stream of flowing water called the **transpiration stream** (figure 1). A current of water and dissolved mineral salts passes from the soil, through the roots, up the stem system via xylem vessels, and finally the water passes out of the shoot, usually through the leaves. How quickly the transpiration stream flows depends on the rate at which water is lost from the plant. This loss of water from a living plant is called **transpiration**.

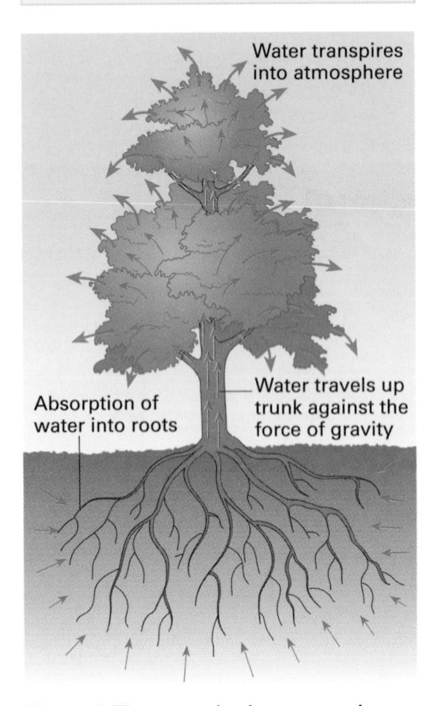

***Figure 1** The transpiration stream in a tree.*

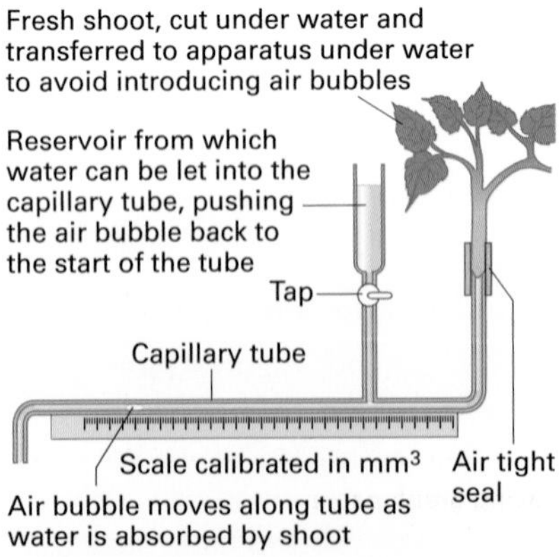

Measuring transpiration

Transpiration is easily demonstrated by placing a transparent plastic bag over a potted plant which has its soil covered by a plastic sheet. Water vapour (identified using cobalt II chloride paper, which turns from blue to pink in the presence of water) quickly builds up inside the bag.

Measuring transpiration rates directly can be difficult, but they are often estimated using a simple potometer (figure 2). The potometer actually measures water uptake. Estimates of transpiration assume that the water lost by transpiration equals the water taken in by the plant.

Fresh shoot, cut under water and transferred to apparatus under water to avoid introducing air bubbles

Reservoir from which water can be let into the capillary tube, pushing the air bubble back to the start of the tube

Tap

Capillary tube

Scale calibrated in mm^3

Air tight seal

Air bubble moves along tube as water is absorbed by shoot

***Figure 2** A potometer measures the uptake of water, which is taken to equal the water transpired by the plant.*

Experiments with a potometer show that transpiration is similar in many ways to evaporation of water from a physical system. The rates of both processes are increased by an increase in temperature, a decrease in humidity, and an increase in wind speed. All of these factors increase the rate of water loss from a wet surface. However, transpiration differs from evaporation in that it occurs through living tissue and is affected by a plant's activities, especially photosynthesis, and on internal as well as external factors. The most important internal factor affecting transpiration is how open the stomata are. Typically, stomata open in response to increases in light intensity so that carbon dioxide can diffuse easily into photosynthesising leaves (see spread 13.7). Transpiration rates increase substantially when the stomata are open. Therefore, transpiration is affected by light intensity whereas evaporation from a physical system is not.

Functions of transpiration

Water is an essential raw material for photosynthesis. It is therefore easy to fall into the trap of thinking that the main function of transpiration is to ensure that photosynthesising leaves have a continuous supply of water. However, transpiration rates are far too great to be explained simply in terms of the plant's need of water for photosynthesis and growth: less than one per cent of the water in the transpiration stream is used by the plant. Most of the water merely passes through the plant and is never incorporated into its structure or involved in its metabolism.

Most plant physiologists have concluded that transpiration is the unavoidable consequence of having aerial structures which allow gaseous exchange for photosynthesis and respiration. Water escapes from a plant when gases enter and leave it. As such, transpiration has been called a necessary evil. Nevertheless, transpiration does have some beneficial effects. Like the evaporation of sweat from the skin of a mammal, it helps to keep the plant cool on hot, sunny days. (However, the very low transpiration rates in plants living in hot and humid conditions, such as parts of some tropical rainforests, would suggest that this cooling by transpiration is not essential.) Transpiration also plays a role in the absorption and movement of minerals and ions from the roots to the different parts of the shoot. This is one of the topics in spread 13.8.

Sites of transpiration

The leaves are the main sites of water loss in a typical, fully leafed plant. About 90% of transpiration takes place through the stomata and 10% through the cuticle (the amount varies with the thickness of the waxy cuticle). Minute amounts of water are usually lost through other parts of the shoot. After leaf fall, however, lenticels (small pores in the stem which allow gaseous exchange) are the main site of water loss in deciduous plants.

Mechanism of water movement up a tall plant

The transpiration stream in tall plants is unlike a stream of water flowing over the ground in that it flows uphill. The force of gravity is overcome by stronger forces generated in the leaves during transpiration. The whole process is driven by energy from the Sun. Heat causes water on the external surface of mesophyll cells to evaporate and the water vapour diffuses from the plant (usually through stomata; spread 13.7). The loss of water from a mesophyll cell decreases the water potential of that cell. This causes water to move into the mesophyll cell by osmosis from an adjacent cell. A water potential gradient becomes established, causing water to flow from the xylem along a chain of cells to the outermost mesophyll cell (figure 3).

According to the **cohesion–tension theory**, a continuous column of water exists in the xylem from the deepest root to the tip of the shoot. Because of the unique cohesive properties of water (spread 2.4), the loss of water from the xylem creates a tension (a pulling force) in the water column. The tension is transmitted all the way down to the roots, stretching the entire water column so that, in heavily transpiring shoots, the diameter of the stem is reduced. Hydrogen bonds between water molecules hold the water molecules together giving the water great tensile strength. This prevents the water column from breaking apart as it is pulled upwards. Adhesion (the forces of attraction between unlike molecules) between the water molecules and the walls of the water-conducting vessels prevents the water from slipping back down. Therefore, by a combination of tension, cohesion, and adhesion, water moves up the stem.

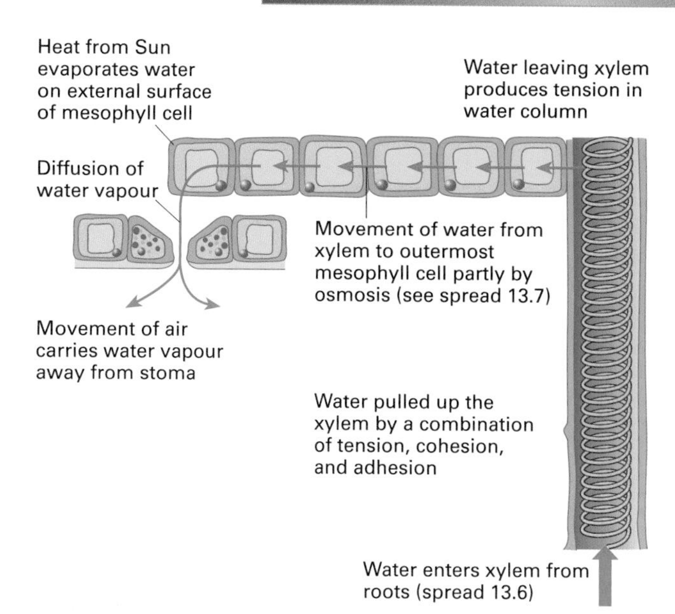

Figure 3 *Water movement up a stem and across to the mesophyll cells.*

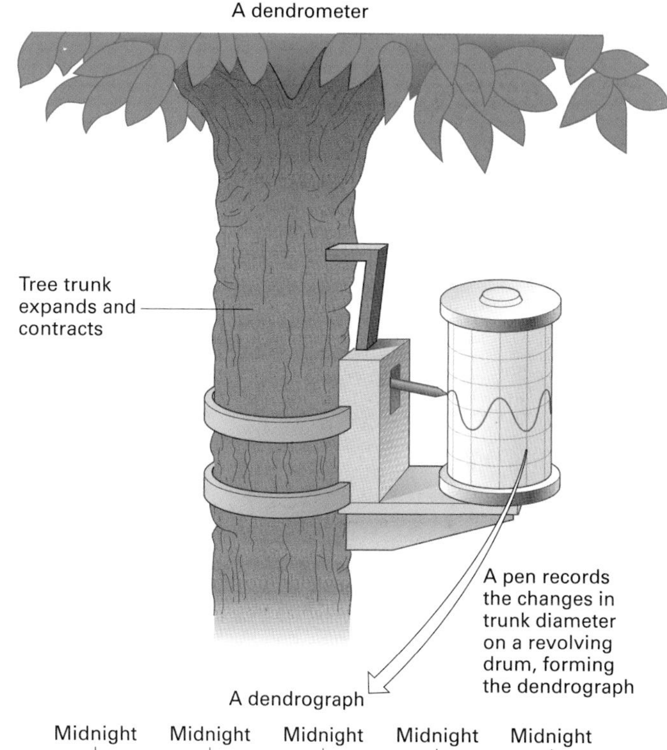

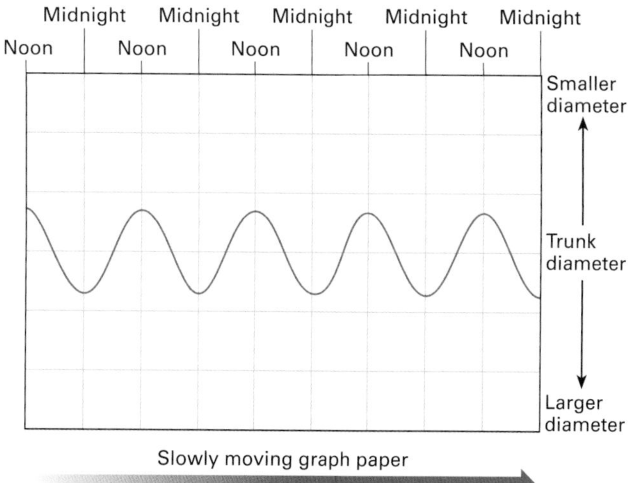

Figure 4 *A dendrometer measures the diameter of a tree trunk by recording a dendrograph: a tracing of the contraction and expansion of the trunk.*

Quick check

1. Name one environmental factor that affects transpiration but not evaporation.
2. Give the main reason why transpiration is thought to take place.
3. Name the forces that prevent the column of water in the xylem from breaking apart when it is pulled upwards in the trunk of a tall tree.

Food for thought

Typically, during the growing season, tree trunks expand at night and contract during the day (figure 4). Trunk girth often reaches a maximum at midnight and a minimum at noon. Suggest reasons for these typical changes. Suggest circumstances that might result in the diameter of trees *increasing* at noon.

13.6

Water uptake

OBJECTIVES

By the end of this spread you should be able to:

- describe how water moves into the root hair cells
- discuss the contribution of root pressure and capillarity to water uptake and movement in the xylem
- distinguish between the apoplast route and symplast route of water movement.

Fact of life

In their search for water, the roots of large trees extend sideways and downwards (figure 1), sometimes to depths far exceeding the height of the tree.

***Figure 1** The Chinese banyan tree (*Ficus retusa*) has extensive aerial roots.*

***Figure 2** Radish seedling with a dense growth of hairs.*

One vital adaptation to life on land is a plant's ability to absorb enough water to replace transpiration losses. If water losses exceed water gains, a plant will go into a water deficit, it will wilt, and eventually it will die.

Plants have evolved root systems that are as extensive and carefully arranged as their leaves. Even in well watered soils, a plant has to compete with other plants for water. A plant therefore needs a wide network of roots close to the surface so that it can absorb rainwater before its competitors take it all.

Two mechanisms of water uptake

Water is absorbed mainly, but not exclusively, by **root hairs** (figure 2). These are cellular extensions of special epidermal cells (sometimes called **root hair cells**) just behind the root tip. The root hairs come into intimate contact with soil particles and soil water. They increase the surface area of the roots enormously. Water uptake into the root hairs involves two main mechanisms which may or may not occur simultaneously.

The first mechanism happens when more water is transpired than is being absorbed. The loss of water from a shoot produces tension so that the pressure of the water in the xylem is lower than atmospheric pressure. This tension is transmitted all the way down the plant and (as long as water is freely available) pulls water into the roots passively. Under such conditions, water can even be pulled into dead roots.

The second mechanism of water absorption is thought to result from a combination of osmotic forces and active transport in the roots. Active transport occurs only in living plants. Instead of water being pulled up the stem by tension, it is pushed up by pressure generated in the roots.

This mechanism depends on an osmotic gradient being created in the roots. Dissolved substances build up in epidermal cells by a combination of diffusion and active transport (spread 13.8). This accumulation of solutes gives the root hair cells a lower water potential than that of the water in the soil. Consequently, water enters a root hair cell by osmosis, increasing its water potential above that of its neighbours. Water then moves on to the next cell by osmosis, and so on across the root to the endodermis.

It is thought that endodermal cells actively secrete solutes into the xylem. This creates a concentration gradient sufficiently high to draw water up the stem against gravity. The pressure of water in the xylem created in this way is called **root pressure**. It can be demonstrated in a plant that has been cut just above the root. Water seeps out of the cut stem, usually at a pressure of between 100 and 200 kPa, but the pressure can reach 800 kPa. The pressure is high enough to account for water movement up many relatively short plants, but not for water movement up large trees. Root pressure only occurs in living plants. It is inhibited by metabolic poisons (such as cyanide) and lack of oxygen.

Root pressure is thought to be responsible for water oozing out of leaves: a process called **guttation**. This occurs at night or in very humid conditions. It is common in many rainforest plants and is often seen at the tips of the leaves of grasses. Guttation ensures that water and dissolved minerals continue to move up a plant when transpiration rates are low.

Plant physiologists are unsure of the relative importance of the two methods of water uptake, but the first mechanism probably dominates when transpiration is fast and the second when transpiration is slow.

Capillarity (the spontaneous movement of water along fine tubes due to surface tension effects; see spread 2.3) may also contribute to water movement up the stem. It can occur along the fine xylem vessels and tracheids (which have a lumen diameter of between 0.01 and 0.2 mm), but the forces generated by capillarity are not sufficient to explain root pressure. Also, capillarity is a passive process which occurs in non-living and dead systems.

Movement of water from the root to the xylem

There is also some doubt about the route that water takes from the root hair cells to the endodermis. There are three possible routes (figure 3); each has its supporters.

In the **apoplast route**, water passes freely through cellulose cell walls from one cell to another. Up to half the volume of a cell wall may be taken up by water molecules. Movement of water through this route could be an entirely passive process resulting from the tension created by transpiration. As water is pulled up the xylem, the cohesive forces between water molecules would ensure that water is drawn across adjacent cell walls.

In the **symplast route**, water diffuses along a water potential gradient through the cytoplasm of adjacent cells. The cytoplasm of adjacent cells is interconnected by cytoplasmic strands called plasmodesmata which pass through pores in the cellulose cell walls. The plasmodesmata and cytoplasm form a continuous pathway for water movement.

In the **vacuolar route**, water moves along the same water potential gradient as in the symplast route, but through the vacuoles as well as the cytoplasm.

Whatever route it takes to get there, once water reaches the endodermis, it is forced to go through the living part of the cell. The impermeable Casparian strips prevent water and its dissolved mineral salts from entering the xylem via the cellulose cell walls. This seems to play an important role in mineral uptake, and will be dealt with in more detail in spread 13.8.

Quick check

1 In a plant that is taking up water in its roots, is the water potential in the soil higher or lower than that in the root hair cells?

2 Which of the following processes take place only in living roots:

a capillarity

b osmosis

c root pressure?

3 Explain why it is not possible for water to follow the apoplast route through cells of the endodermis.

Food for thought

Root hairs are exceedingly delicate, whether on a tree or a seedling (figure 2). When transplanting young trees from a nursery to a woodland, damage to the roots and root hairs often leads to the death of the tree. What precautions might be taken to minimise this damage and increase the success of moving young trees?

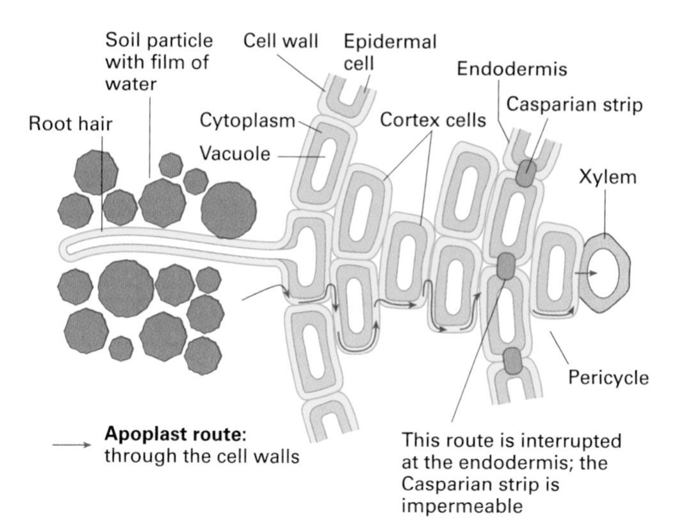

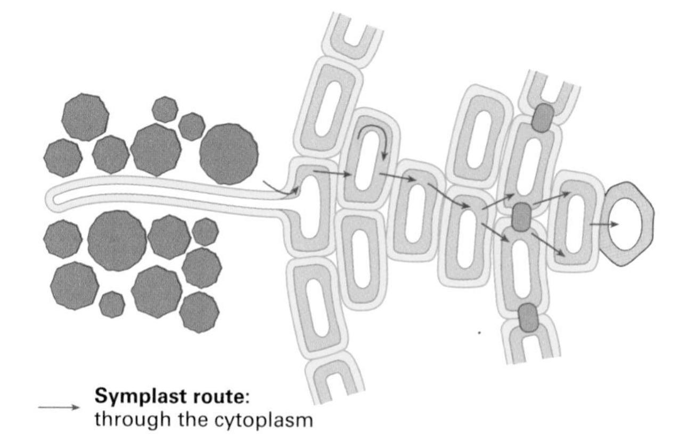

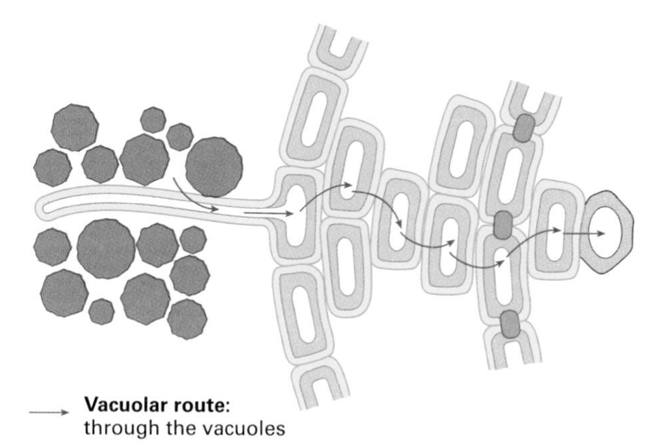

Figure 3 The possible routes that water can take from the soil to the xylem in the root.

13.7

STOMATA

OBJECTIVES

By the end of this spread you should be able to:

- describe how carbon dioxide gets from the atmosphere to photosynthesising cells
- discuss the conflict between photosynthesis and transpiration
- explain how stomata open and close.

Fact of life

In the leaves of many trees, the number of stomata on the lower epidermis far exceeds the number on the upper epidermis. In species of oak, for example, the lower epidermis has about 45 000 stomata per cm^2 whereas the upper epidermis has none.

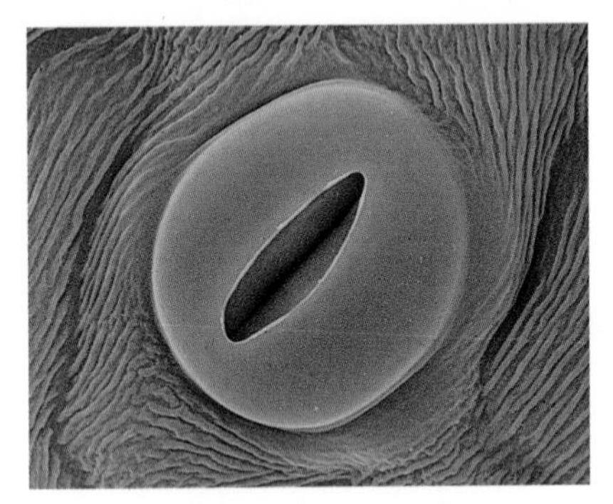

Figure 1 A false-colour scanning electron micrograph of a stoma on a sepal of the primula flower (×600).

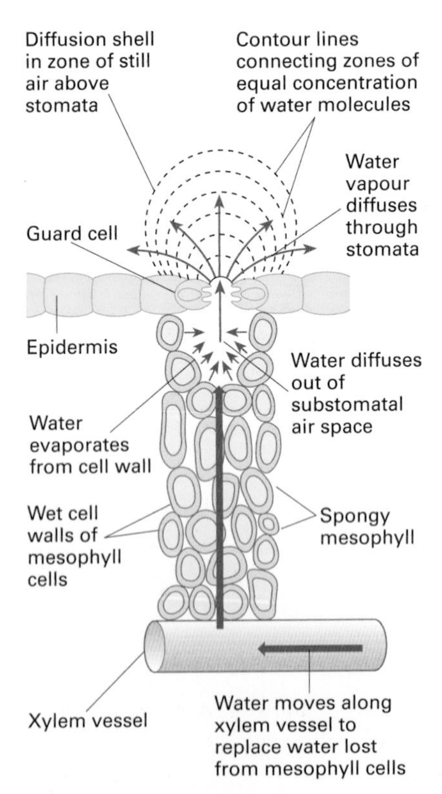

Figure 2 Diffusion of water from mesophyll cells out of the leaf.

Stomata are microscopic pores in the epidermis of the leaves and stems of terrestrial plants (figure 1). In this spread we are going to see how, by controlling the aperture of stomata, plants can compromise between maximising gaseous exchange and minimising water losses. When stomata are open, atmospheric gases (carbon dioxide for photosynthesis and oxygen for respiration) can be exchanged between the plant and the air. However, when stomata are open, water can leave the plant by transpiration (spread 13.5).

Carbon dioxide uptake through stomata

When stomata are open, they connect the air spaces inside a leaf (called **substomatal air spaces**) with the atmosphere. This means that carbon dioxide and oxygen can diffuse quickly into the plant. An unobstructed diffusion route is especially important for carbon dioxide, because it is present in a low concentration in the atmosphere (0.035 per cent), so concentration differences between the plant and the atmosphere are small (spread 4.7). In well watered plants exposed to sunlight, the stomata open to allow mesophyll cells to take up carbon dioxide. The cells use the carbon dioxide in photosynthesis, creating a concentration gradient from the atmosphere to the chloroplasts of the mesophyll cells down which carbon dioxide diffuses. Carbon dioxide diffuses much faster in air than in water, so it is an advantage for mesophyll cells to be in contact with the air spaces in leaves. However, a large surface area for carbon dioxide absorption also provides a large area for the loss of water.

Water losses through stomata

Heat from the Sun evaporates water on the surface of mesophyll cells and the water vapour formed diffuses into the substomatal air spaces. The water lost from the mesophyll cells is replaced by water from the xylem. This water moves to the mesophyll cells along the apoplast, symplast, or vacuolar route, as described for water movement in the root (spread 13.6). In leaves, the apoplast route is thought to be the most important and the vacuolar route the least important.

The water that evaporates into the substomatal air spaces diffuses out of the leaf through the stomata. Water vapour diffuses from a high water potential inside the leaf to a lower water potential outside. A layer of stationary air called the **diffusion shell** lies immediately next to the leaf. The thickness of this layer depends on a number of factors including the size, shape, and hairiness of the leaf, and also wind speed. Water vapour has to diffuse through the diffusion shell (figure 2) before being carried away in moving air by mass flow. Any factor that reduces the thickness of the diffusion shell or increases the water potential gradient tends to increase the rate of diffusion of water from a leaf. Air movements, for example, disrupt diffusion shells by carrying water vapour away from a leaf surface. An increase in temperature accelerates the evaporation of water from mesophyll cells, boosting the concentration of water vapour in the substomatal air spaces. An increase in temperature also lowers the relative humidity of the air outside the leaf. Both these events result in a steeper water potential gradient between the atmosphere and the substomatal air spaces. However, the effects of these factors on transpiration rates are complicated because the stomatal aperture can vary, changing the area through which diffusion can take place.

Mechanism of opening of stomata

Figure 3 shows a stoma flanked by a pair of specialised epidermal cells called **guard cells**. Unlike other epidermal cells, guard cells have chloroplasts and are kidney shaped when turgid. The guard cells vary the size of aperture of the stoma by changing their shape with changes in turgidity. The cellulose cell walls that surround the stoma are thicker and less elastic than those in contact with the epidermis. As the guard cells inflate with water (become more turgid), the increased pressure inside causes the cells to expand. Because of their greater flexibility, the walls furthest from the stoma aperture expand more than the thicker walls that line the stoma opening. This causes the guard cells to bend into a semicircular shape, enlarging the stoma opening. When the guard cells lose water and become less turgid, the stoma closes.

Plant physiologists are certain that stomatal aperture varies as a result of changes in the turgidity of the guard cells. They are less certain about how these changes are brought about, though the following observations have been made:

- Most stomata open during the day and close at night.
- Some stomata show a circadian (daily) rhythm of opening and closing even when kept in constant conditions.
- Stomata generally close when a plant suffers water stress (for example, when transpiration exceeds water absorption).
- The stomata of some desert plants close during the day and open at night.

Plants can therefore vary the stomatal aperture. This allows a compromise between the need to conserve water and the need to exchange gases for photosynthesis.

Molecular explanations for stomatal opening

Experiments with radioactively labelled isotopes and metabolic poisons suggest that stomatal opening involves the active transport of potassium ions into the guard cells. The build-up of potassium ions lowers the water potential of the guard cells and draws water into the cells by osmosis (spread 4.9), the cells become turgid and the stoma opens. When active transport is not taking place, potassium diffuses out of the guard cells, water leaves by osmosis, the cells become flaccid, and the stoma closes.

A number of factors may trigger this active transport of potassium ions. Light may start it by activating ATPase in the guard cells. The ATPase increases the ATP produced by photosynthesising chloroplasts in the guard cells, and the ATP provides the energy for active transport.

For a plant under water stress, its need to conserve water is greater than its need to obtain carbon dioxide for photosynthesis. Under these conditions a plant secretes **abscisic acid** (**ABA**). This is a chemical messenger which causes stomata to close. It is thought that ABA triggers a metabolic pump which actively secretes potassium ions out of guard cells, causing the cells to lose water and become flaccid.

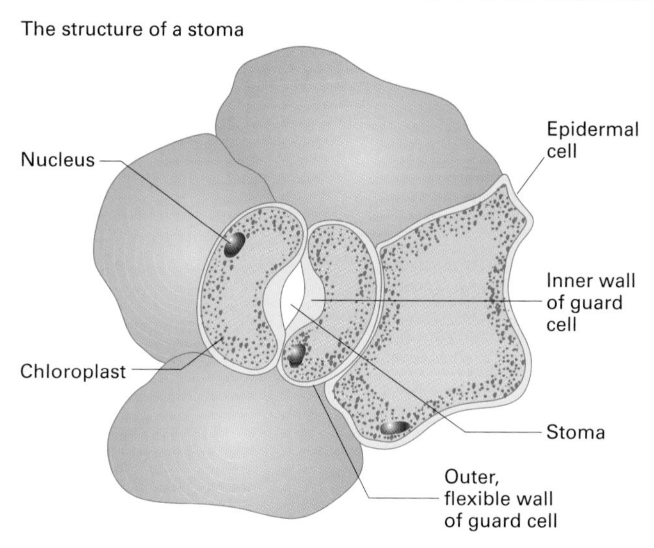

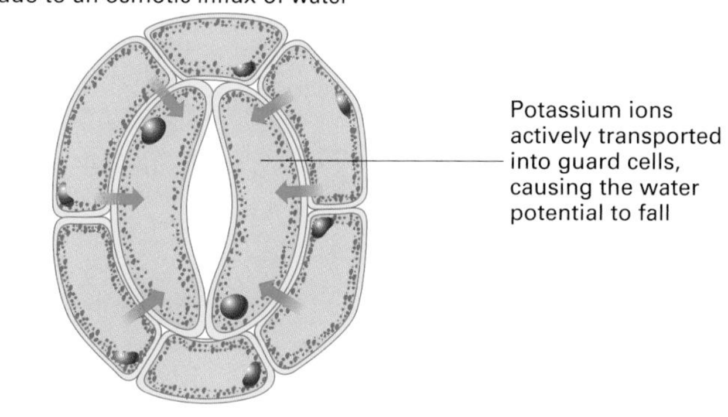

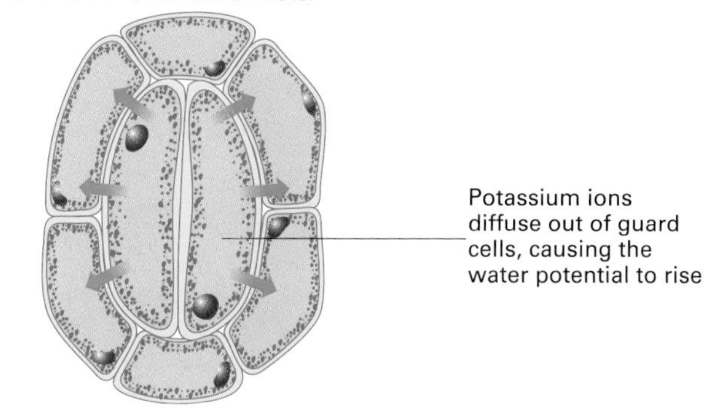

Figure 3 *The stomatal mechanism.*

Quick check

1. By what process does carbon dioxide enter the photosynthesising cells of a leaf?
2. Plants have to replace the water they lose in transpiration. How do some temperate plants cut down transpiration losses in winter?
3. Assuming that the stomatal mechanism involves active transport, describe what will happen to a stoma flanked by guard cells that have been given a metabolic poison.

Food for thought

Plants open their stomata to allow carbon dioxide to diffuse easily into photosynthesising cells. Why do they not have to keep their stomata open all the time to allow oxygen to diffuse into respiring cells?

13.8

Uptake of mineral ions by roots

OBJECTIVES

By the end of this spread you should be able to:

- describe how mineral ions are absorbed in terrestrial plants
- describe how mineral ions are transported in terrestrial plants
- list the main macronutrients and micronutrients of plants.

Fact of life

The mineral ion requirements of plants and the effects of mineral deficiencies can be determined by growing plants in different culture solutions. The culture solutions are aerated to ensure that the roots have oxygen for cellular respiration. The ion of a particular element, such as potassium, is omitted from what is thought to be a complete culture medium. The growth of the plant in an incomplete medium (lacking only the potassium, for example) is compared with the growth of a plant in a complete medium. Common symptoms of mineral deficiency are chlorosis and stunted growth. In chlorosis, synthesis of chlorophyll is inhibited, resulting in pale yellow leaves.

Mineral elements required by plants

In addition to the carbon, hydrogen, and oxygen obtained from carbohydrates made by photosynthesis, plants require a variety of mineral elements. More than 50 elements have been extracted from plants, though not all of them are believed to be essential to the plant. To a certain extent, the composition of a plant reflects the soil in which it grows. Plants growing on mine wastes, for example, contain gold, tin, and silver. To determine which mineral elements are actually essential, plants are grown in solutions containing different combinations of minerals (figure 1). Such experiments show that (as well as carbon, hydrogen, and oxygen) all plants require at least 14 mineral elements to grow and maintain health (table 1). Mineral elements required in large amounts are called **macronutrients**. They are calcium, magnesium, nitrogen, phosphorus, potassium, and sulphur. Elements required in small amounts are called **micronutrients** or **trace** elements. They are boron, chlorine, copper, iron, manganese, molybdenum, nickel, and zinc.

Some plants need other mineral elements: for example, sodium is required by certain desert shrubs (such as *Atriplex*) which have C_4 metabolism (spread 5.6), but is not essential for most species.

How mineral ions enter the plant

Plants take up mineral elements in salts, which are made up of ions. When salts are dissolved in water, these ions can move freely and can enter the plant in a number of ways. In terrestrial flowering plants, most mineral ions are absorbed into root cells relatively independently of water, by a combination of diffusion and active transport. The water that is absorbed through cell walls to the endodermis (spread 13.6, figure 3) may contain mineral ions, which therefore move into the plant along with the water by mass flow.

If the concentration of a mineral ion is higher in the soil water than in the cytoplasm of an epidermal root cell, the mineral ion will diffuse passively and unselectively through the cell surface membrane into the cell. If the concentration of an ion is lower in soil water than in the cytoplasm, root cells may selectively pump the ion into the cytoplasm (see also spread 4.8). Some mineral ions are taken in by specific carrier molecules and stored in plant cells so that their concentration is much higher than in the soil water. Absorption against this strong concentration gradient requires a considerable expenditure of energy. Plants actively absorbing minerals therefore use large amounts of oxygen to generate ATP by respiration. In experimental studies on temperate plants, increasing the temperature around roots by 10–30 °C increased the rate of respiration and also the rate of mineral absorption (diffusion is relatively insensitive to temperature changes). Metabolic poisons inhibit active mineral ion uptake.

Transport of mineral ions within the plant

After being taken into epidermal cells by either diffusion or active transport, mineral ions diffuse down a concentration gradient to the endodermis. In order to enter the xylem and be transported out of the roots into stems, water and dissolved mineral ions have to pass through the cytoplasm of cells in the endodermis: the impermeable **Casparian strips** (figure 2) prevent the water and dissolved minerals ions from passing along the endodermal cell walls. From the cytoplasm in the endodermis, the mineral ions may diffuse into the xylem or they may be transported actively into it; this active transport of mineral ions from the cytoplasm of endodermal cells is thought to be the basis of root pressure (spread 13.6).

Once in the xylem, mineral ions are carried to other parts of the plant by mass flow in the transpiration stream. Experiments with radioactive tracers show that some mineral ions can be transferred to the phloem and carried there too.

When mineral ions have reached the sites where they are used, they either diffuse or are actively transported into the cells that need them. The chief sites of mineral use are the regions of plant growth (for example, meristems, young leaves, flowers, and fruits) and storage organs.

Figure 1 The upper grapevine leaf has been grown in a culture solution that lacks potassium. The lower one has been grown in a solution with potassium and is healthy.

Table 1 Essential mineral elements for plants. The table shows the 14 elements taken up as mineral ions from the soil or surrounding water of plants. The other three elements known to be essential (carbon, hydrogen, and oxygen) are obtained as carbohydrate by photosynthesis.

Mineral element	Available to plants as	Major functions
Macronutrients		
Calcium (Ca)	Calcium ions in soil or surrounding water	Important in formation of cell wall; regulates many cellular responses to environmental stimuli; activates some coenzymes
Magnesium (Mg)	Magnesium ions in soil or surrounding water	Component of chlorophyll; cofactor for many enzymes
Nitrogen (N)	Ammonium and nitrates in soil or surrounding water; nitrogen-fixing bacteria	Component of proteins, nucleic acids, some plant growth substances, and coenzymes
Phosphorus (P)	Phosphate ions in soil or surrounding water	Component of nucleic acids, phospholipids, ATP, and several coenzymes, including NADP
Potassium (K)	Potassium ions in soil or surrounding water	Cofactor in protein synthesis; involved in water and ion balance; involved in the stomatal mechanism (spread 13.7)
Sulphur (S)	Sulphate in soil or surrounding water	Component of proteins and some coenzymes (e.g. coenzyme A)
Macronutrients		
Boron (B)	Probably as borate ions in soil or surrounding water	Required for normal cell division in meristems; cofactor in chlorophyll synthesis
Chlorine (Cl)	Chloride ions in soil or surrounding water	Required for photolysis (splitting) of water molecules during photosynthesis; involved in water and ion balance
Copper (Cu)	Copper ions in soil or surrounding water	Component of electron carriers in photosynthesis and respiration; component of enzyme involved in synthesis of lignin
Iron (Fe)	Iron ions in soil or surrounding water	Component of cytochromes (electron carriers in respiration and photosynthesis); component of an intermediate molecule formed during chlorophyll synthesis
Manganese (Mn)	Manganese ions	Activator of some enzymes; involved in formation of some amino acids
Molybdenum (Mb)	Molybdenate ions	Essential for nitrogen fixation; cofactor in conversion of nitrate to nitrite
Nickel (Ni)	Nickel ions	Involved in nitrogen metabolism
Zinc (Zn)	Zinc ions in soil or surrounding water	Constituent of a variety of enzymes; deficiency prevents expansion of leaves

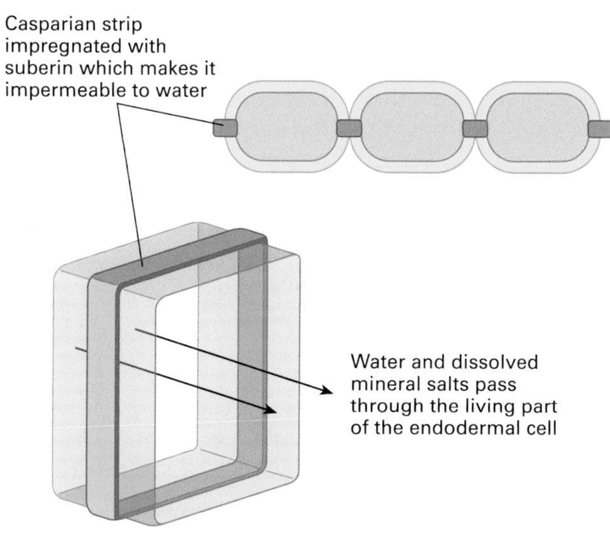

***Figure 2** The Casparian strip forces water and mineral ions to pass through the living part of endodermal cells.*

Quick check

1 How are mineral ions absorbed by plants? List three ways.

2 Which structures transport mineral ions in plant stems?

3 Which element, in addition to carbon, hydrogen, oxygen, and nitrogen, forms an essential part of chlorophyll?

Food for thought

Hydroponic farming involves growing plants on nutrient solutions, similar to the method shown in figure 1, but on a much bigger scale. Crops such as tomatoes and lettuces are planted in gravel or sand to provide mechanical support, and a nutrient solution flows over the roots. Suggest the advantages and disadvantages of hydroponics compared with greenhouse cultivation in soil.

13.9

TRANSLOCATION OF ORGANIC SUBSTANCES

OBJECTIVES

By the end of this spread you should be able to:

- discuss the evidence for organic substances being translocated in the phloem
- describe the composition of phloem sap
- explain how aphids can be used to collect samples of phloem sap.

Fact of life

An intriguing triangular relationship has evolved between certain species of plants, aphids, and ants. The aphids (greenfly and blackfly) obtain their food by sucking sap from the phloem of the plant. The sap is so rich in nutrients that the aphid has to egest excess sugars as a 'honeydew' droplet from the anus (figure 2a). The ants feed on the droplet. In return for their 'honeydew' reward, the ants, using their potent stings, protect the aphids from predators. At the same time, the ants also protect the plants against herbivores other than the aphid.

Sources and sinks

Flowering plants are complex multicellular organisms made up of many different organic substances which carry out a wide range of functions in different parts of the plant. A plant's survival depends on its ability to transport organic substances from their **sources** where they are made to **sinks** where they are used. Leaves are the main sources in plants, and growth centres are the main sinks. Storage organs such as potato tubers act as sources when substances within them are broken down and released to other parts of the plant; they act as sinks when materials are brought to them to be stored.

The transport of a substance in solution within a plant is called **translocation**. The phloem is thought to be involved in the translocation of organic substances, and evidence for this comes from ringing experiments, investigations involving sap-sucking organisms, and experiments with radioactive tracers.

Ringing experiments

In a ringing experiment, a complete ring of bark is removed from a tree trunk or stem. The phloem lies just beneath the bark, and is removed with it. The tissue just above the ring swells, whereas that below the ring tends to wither (figure 1). Analysis of the liquid within the swollen tissue shows that sugars and other organic solutes have accumulated above the ring. This suggests that removal of the phloem interrupts the downward movement of these substances, and that the tissues below the ring wither due to lack of nutrients. If the main trunk of a tree is ringed, the whole tree will eventually die. Ringing does not remove the xylem, which lies deeper within the stem. Conduction of water is therefore not affected by ringing. Ringing experiments help explain why trees can be killed by African elephants, which strip off bark with their trunks, and grey squirrels, which gnaw the bark.

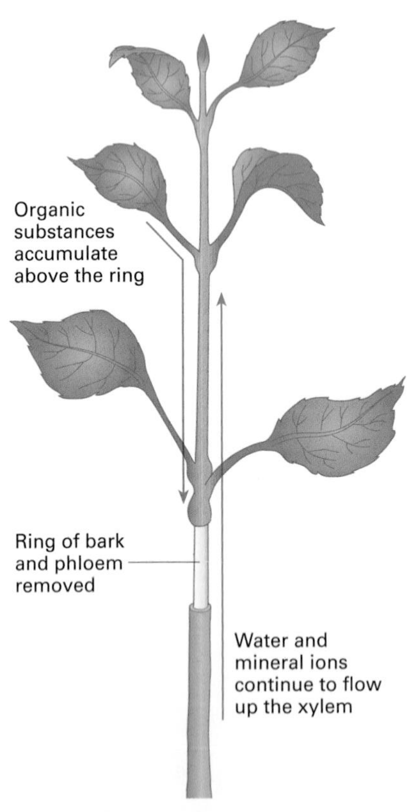

Figure 1 *Results of a ringing experiment.*

Sap-sucking organisms

Aphids and some other insects feed on the sap within phloem sieve tubes. They probe a leaf or stem with their needle-like mouthparts, called **stylets**, and pierce their way into sieve tubes to extract the sap (figure 2). If the head of a feeding aphid (anaesthetised with carbon dioxide) is detached from the mouthparts, sap continues to exude from the exposed end of the stylet. The liquid may continue to flow for days, and it can be collected and analysed. It is found to contain sugars and other organic substances. Looking at sections through the stem under the microscope shows that the stylet has penetrated a sieve tube. The closer the stylet is to a sugar source, such as photosynthesising leaves or an organ releasing its store of sugar, the faster the sap flows out of the cut end of the stylet, and the higher its sugar concentration. The flow of sap stops if the phloem is killed or poisoned with a respiratory inhibitor. These results show that the exudate from the cut end of a stylet is sap from phloem. They also support the idea that sap moves in the phloem from source to sink because of pressure generated at the source end by the active pumping of sugar into sieve tubes (see spread 13.10).

Radioactive tracers

A radioactive isotope of carbon, ^{14}C, can be used to trace the carbohydrates formed during photosynthesis. A plant is exposed to an atmosphere containing carbon dioxide labelled with ^{14}C. Carbohydrates are synthesised using the radioactively labelled carbon and their movement in the shoot can be traced using photographic film. The film blackens where it is exposed to the radioactive isotope, revealing the location of the carbohydrates. This process is called **autoradiography**.

(a) An aphid feeding on a plant stem

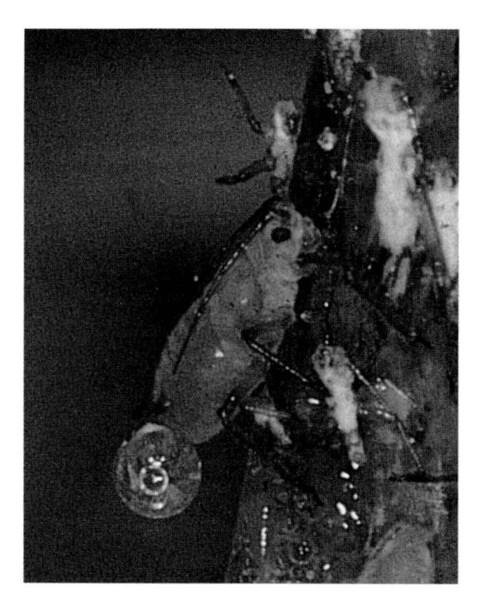

(b) Photomicrograph of the mouthparts (stylet) of an aphid in a sieve tube

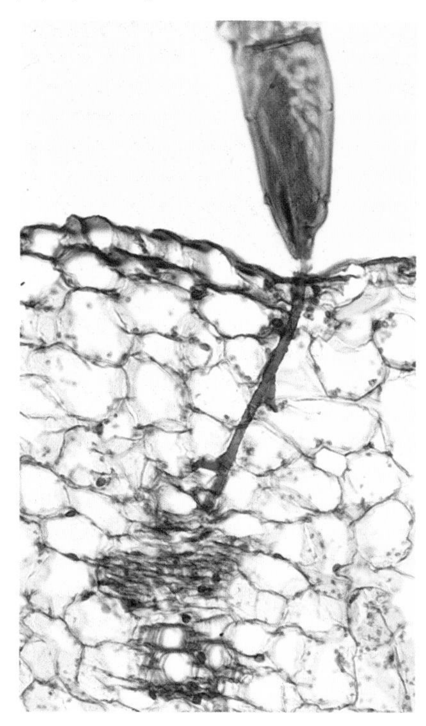

(c) Diagram showing the source of the sap exuded from a sieve tube

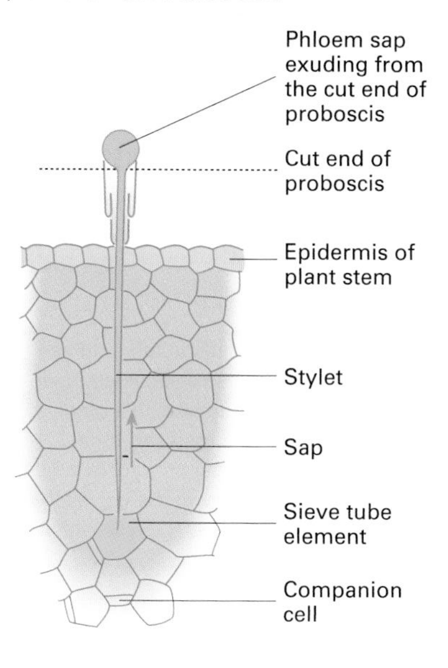

Figure 2 *Using an aphid to obtain sap from sieve tubes.*

The blackened areas correspond to the position of the phloem in the stem, indicating that the organic substances synthesised in the leaf are translocated in the phloem.

Composition of phloem sap

The main substance carried in solution in sieve tubes is the disaccharide sucrose. The concentration of sucrose is particularly high in the sieve tubes of sugar cane. Sucrose is an ideal form in which to transport carbohydrate as it is very soluble and also unreactive. Once it has reached its destination (the sink), it can be hydrolysed into glucose and fructose, stored, or used as a raw material for the manufacture of more complex organic substances. The phloem sap also contains amino acids and other nitrogenous compounds, plant growth substances, and mineral salts (transferred from xylem; see spread 13.8).

Quick check

1 When a ring of bark is removed from a stem, which vascular tissue is also removed?

2 Apart from water, what is the main substance found in phloem sap?

3 How do the experiments with aphid mouthparts show that a sieve tube is at a higher pressure than the air?

Food for thought

Suggest why stripping a thin layer of bark all the way around a tree will kill the tree, but a deep gouge along one side will not.

13.10

Mass flow and pressure flow

OBJECTIVES

By the end of this spread you should be able to:

- describe the main features of the translocation of organic substances
- describe the pressure flow hypothesis.

Fact of life

In one growing season, up to 250 kg of sucrose can be translocated down the trunk of a large tree. In a giant redwood tree, sucrose and other organic solutes may be translocated more than 100 m from the leaves to the roots.

Ringing experiments, investigations using aphid mouthparts, and radioactive tracing experiments have revealed the following about the translocation of organic substances in stems:

- Translocation takes place mainly in the sieve tubes of phloem.
- It is an active process which takes place only in living tissue.
- It is inhibited by metabolic poisons and lack of oxygen.
- Large amounts of organic substances can be translocated.
- Its rate varies for different organic substances (between 20 and 600 cm per hour).
- Different organic substances are translocated up and down the stem simultaneously (but not necessarily in the same sieve tube).
- Translocation takes place over great distances.
- In temperate plants, translocation increases with increases in temperature of between 10 °C and 30 °C.

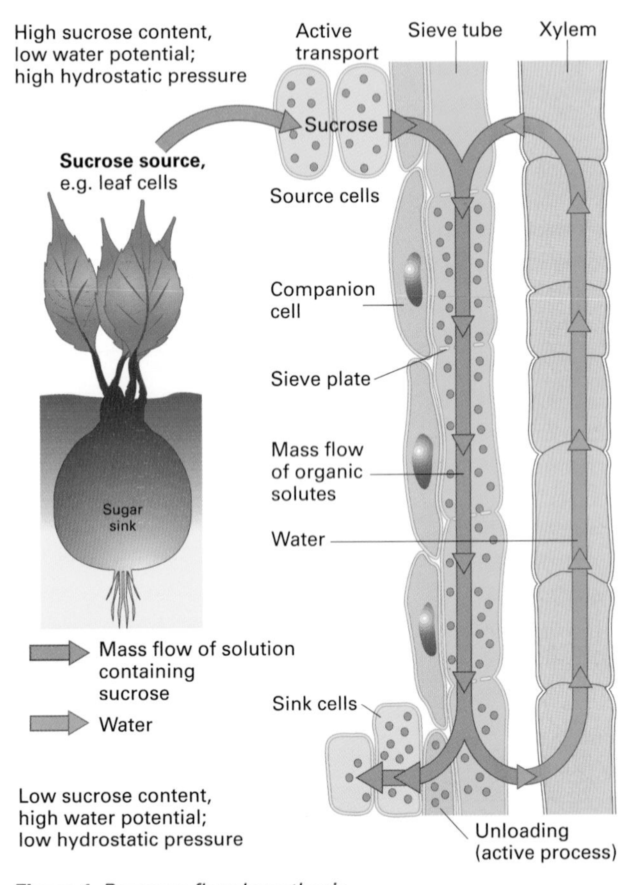

Figure 1 Pressure flow hypothesis.

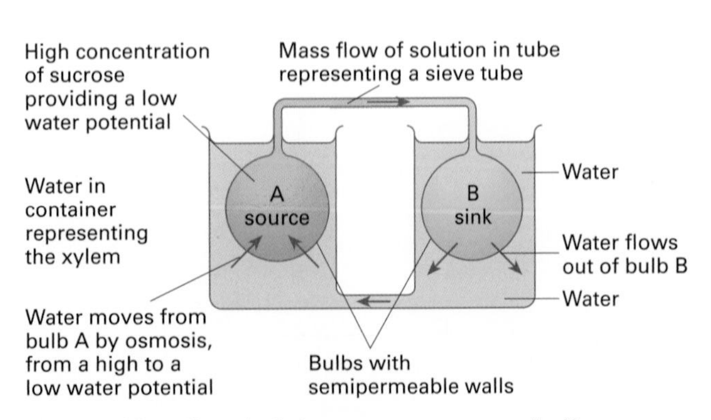

Figure 2 Mass flow: bulb A represents source cells (for example, in leaves) and bulb B sink cells (for example, in roots).

It is not clear how translocation of organic substances takes place, but it is not by diffusion. Diffusion is far too slow to account for the fast rates of translocation and the large amounts translocated. Also, substances that cannot diffuse (such as viruses) are translocated in the phloem. Several hypotheses have been made about how substances are translocated, and the most popular one is called the **pressure flow hypothesis**, summarised in figure 1. According to this hypothesis, translocation involves a combination of active transport and mass flow, and takes place in three stages.

The pressure flow hypothesis

The first stage occurs at a source (such as a leaf) in which a sieve tube is loaded with sucrose and other organic solutes. The loading is carried out by companion cells and involves an active transport mechanism. This explains how the sucrose concentration in the sieve tubes may be as high as 30% while the concentration in the leaf cells is only 0.5%. It also explains why translocation will only take place in living cells. Modified companion cells, called **transfer cells**, appear to be involved in loading sucrose into sieve tubes by cotransport (spread 4.8, figure 2), but transfer cells have not been found in all plants. Transfer cells have many internal projections of the cell wall which give the cell surface membrane a very large area for transport. These cells also have exceptionally large numbers of mitochondria which generate the energy required for active transport.

In the second stage, the sap is translocated in the stem from source to sink by mass flow. **Mass flow** is a purely physical process demonstrated by Ernst Münch in a famous experiment in 1930. He tried to model the translocation system of a plant (figure 2). In mass flow, fluid flows from a region of high hydrostatic pressure to a region of low hydrostatic pressure. As the fluid flows, it carries different substances in it, in much the same way as a stream carries suspended and dissolved materials from one place to another.

According to the pressure flow hypothesis, a high hydrostatic pressure in a sieve tube is produced by an osmotic influx of water. This happens because of the accumulation of sucrose during loading.

One criticism of the pressure flow hypothesis is that substances cannot flow in opposite directions in the same sieve tube. However, phloem contains numerous sieve tubes and different solutes could travel in opposite directions at the same time by travelling in two different sieve tubes with different sources and sinks. Translocation in different sieve tubes could also explain how different solutes can move in the same direction at different rates.

In the third stage of the pressure flow hypothesis, sucrose and other organic solutes are unloaded at the sink where they may be used or converted into a storage product. Water follows the solutes out of the sieve tube, reducing the hydrostatic pressure and maintaining the pressure gradient from source to sink. Unloading is thought to be an active process requiring living companion cells and using energy.

Problems with the pressure flow hypothesis

The pressure flow hypothesis explains why translocation takes place only in living phloem, and how the concentration of sucrose can be much higher in the sieve tubes than in photosynthesising leaf cells. However, the hypothesis is incomplete and leaves much still to be investigated. No one is sure how loading and unloading take place. There is also some uncertainty about the function of sieve plates and the cytoplasm in sieve tube elements, which appear to be unnecessary obstacles to the flow of phloem. One suggestion is that sieve plates provide mechanical support to the sieve tubes, preventing them from buckling or splitting when they carry sap at high hydrostatic pressure. Sieve plates and sieve tube cytoplasm may also be involved in repairing leaky sieve tubes, or isolating a sieve tube that has burst open. **Phloem protein (P protein)**, a fibrous protein in the sieve tube, is thought to seal off a burst sieve tube element. Nevertheless, no one knows how the cytoplasm of the sieve tube element is isolated from the sap and what determines whether nutrients are unloaded or loaded.

There is enough uncertainty about the pressure flow hypothesis for some plant physiologists to believe another mechanism is involved. Contractile proteins recently discovered in sieve tubes may help to generate the forces needed to move solutes. The proteins are related to actin and myosin, responsible for the muscle contractions of animals. It has also been suggested that **cytoplasmic streaming** may play a part in translocation (see box). Translocation remains a subject of dispute and, as one plant physiologist stated, there are almost as many explanations as there are research workers.

Cytoplasmic streaming

If you look at living plant cells with the aid of a microscope, you may see the chloroplasts moving around the cytoplasm in a process called cyclosis. A similar process, called **cytoplasmic streaming**, is thought by some researchers to be responsible for translocation of organic substances in individual sieve tube elements, with active transport conveying the substances across the sieve plate either upwards or downwards in the sieve tube (figure 3). Although the hypothesis neatly explains the two-way flow of substances, cytoplasmic streaming has been observed in immature sieve tubes only, and is not fast enough to account for phloem transport.

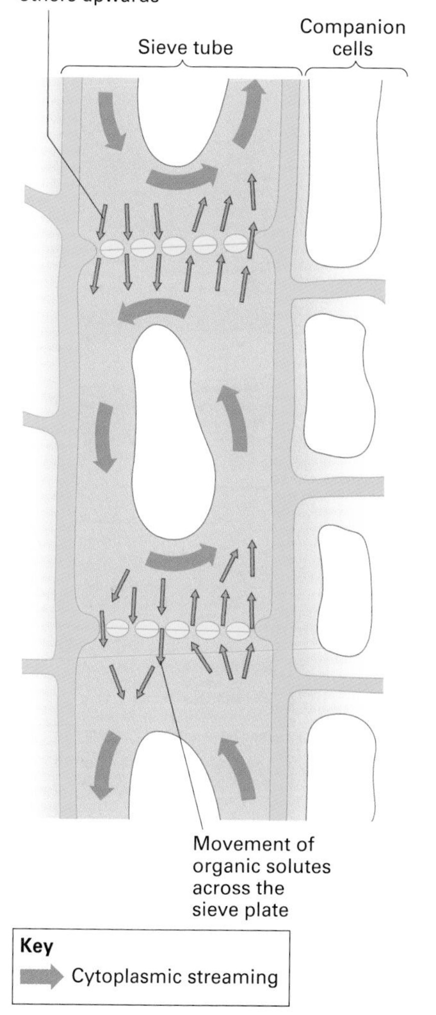

Figure 3 *Cytoplasmic streaming.*

QUICK CHECK

1 Explain why, according to the pressure flow hypothesis, translocation can only take place in living phloem.

2 According to the pressure flow hypothesis, how do organic substances move in opposite directions in the phloem?

Food for thought

Münch carried out the experiment shown in figure 2 to support his theory that translocation is a purely physical process that can be explained by mass flow. Explain why mass flow eventually stops in the model.

Münch argued that mass flow would not stop in a living plant. Suggest the basis of this argument.

Summary

The structure of a **dicotyledonous leaf** is well adapted for **photosynthesis**. The **epidermis** covers and protects the leaf, **stomata** and **spongy mesophyll** allow gas exchange, **vascular tissue** transports material to and from the leaf, and **palisade cells** are the main sites for photosynthesis. The **stem** connects leaves to the roots. It supports the plant, transports materials, produces new cells for growth, and may store food and water. It contains tissue adapted to mechanical support and transport of water and mineral salts. The roots anchor the plant in the soil and are adapted to take up water and mineral salts.
The vascular tissue contains **xylem** which transports water and mineral salts, and **phloem** which translocates organic solutes. Water is transported in the **transpiration stream**. Stomata are the main site of **transpiration**. According to the **cohesion-tension theory**, water moves up the stem by a combination of tension, cohesion, and adhesion. **Water uptake** occurs mainly through root hairs. In living roots, it usually involves a combination of **active transport** and **osmosis**. Water may take the **apoplast**, **symplast**, and **vacuolar routes** to the **endodermis**. Active transport of salts from the endodermis into the xylem may cause a **root pressure** which forces water up the stem. Water escapes from leaves through stomata flanked by **guard cells** which can alter the stomatal aperture to regulate the amount of water lost. Minerals are taken up by a combination of **diffusion** and **active transport**. Essential minerals include **macronutrients** and **micronutrients**.
Translocation of organic substances occurs from **sources** to **sinks**. **Ringing experiments**, sometimes involving radioactive tracers, and studies of **sap-sucking insects**, show that translocation takes place in the **sieve tubes** of phloem. According to the **pressure flow hypothesis**, translocation involves a combination of **active transport** and **mass flow**.

PRACTICE EXAM QUESTIONS

1 The diagram shows the main pathways by which water moves through a plant.

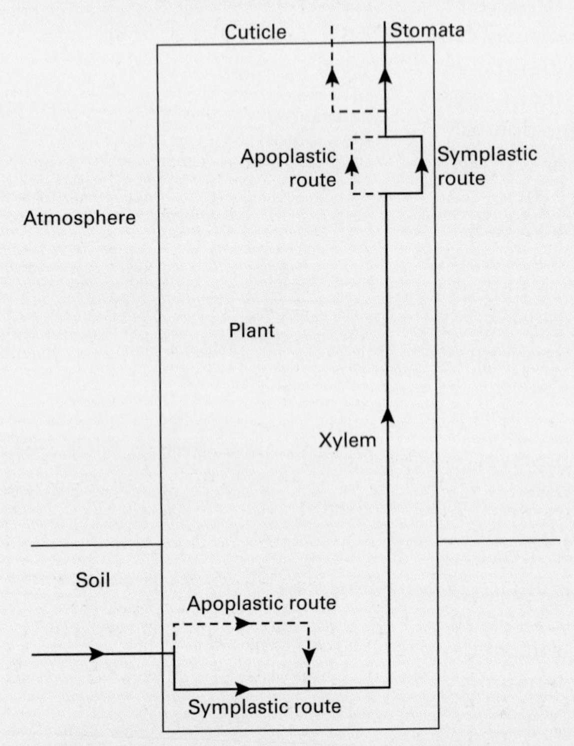

a Explain how water moves from the soil into the xylem of the root. [6]

b Explain how water moves through the xylem in the stem of the plant. [5]

c Explain how the structure of a leaf allows efficient gas exchange but also limits water loss. [6]

[Total 17 marks]

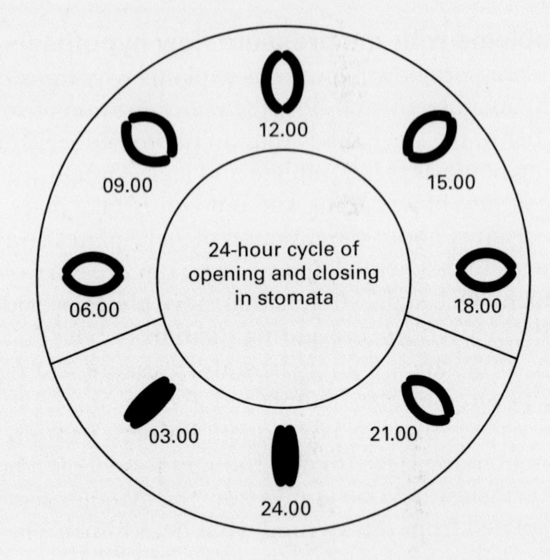

2 The drawing above shows a 24-hour cycle for the opening and closing of stomata from the same plant.

a Explain how this cycle of opening and closing of stomata is advantageous to the plant. [2]

b The diagram shows the potassium (K^+) ion concentrations in the cells around an open and closed stoma in *Commelina*. The concentrations are in arbitrary units.

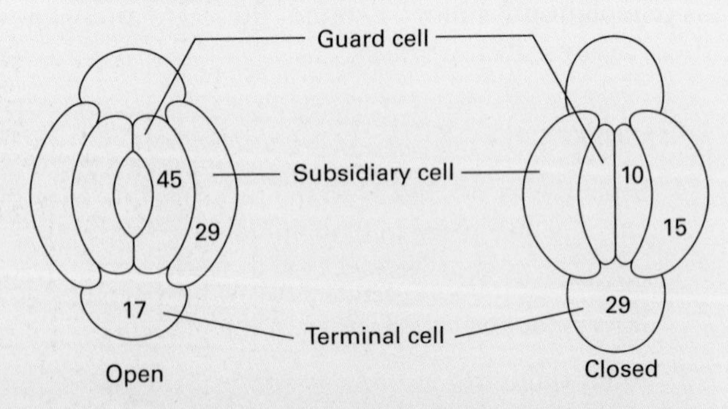

i Explain how the movement of K^+ ions accounts for the opening of the stoma. [3]

ii Explain how K^+ ions are moved against a concentration gradient. [2]

[Total 7 marks]

3 An investigation was carried out into the relationship between the rate of water absorption and the rate of transpiration in sunflower plants at various times of the day. The results are shown in the diagram below.

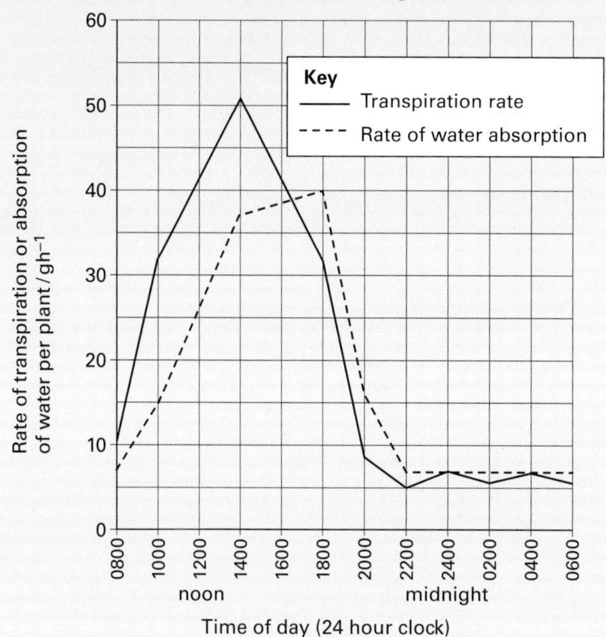

a i Describe the changes in the rate of transpiration that took place during the experiment. [3]

ii Suggest why these changes occurred. [3]

b Comment on the relationship between the rate of transpiration and the rate of water absorption during the experiment. [2]

c Describe a simple method that you could use in the laboratory to measure the rate of transpiration of a flowering plant. [4]

[Total 12 marks]

4 The diagram below shows a longitudinal section of two cells of phloem tissue in a plant stem.

a Name the cells labelled A and B in the diagram. [2]

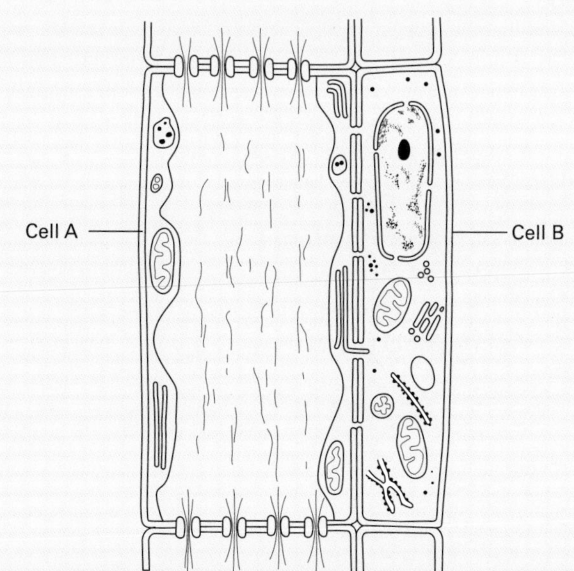

b i State the function of phloem in a plant. [1]

ii Describe how aphids can be used to investigate the function of phloem. [3]

[Total 6 marks]

5 The diagram below shows some of the cells involved in the loss of water from part of a leaf.

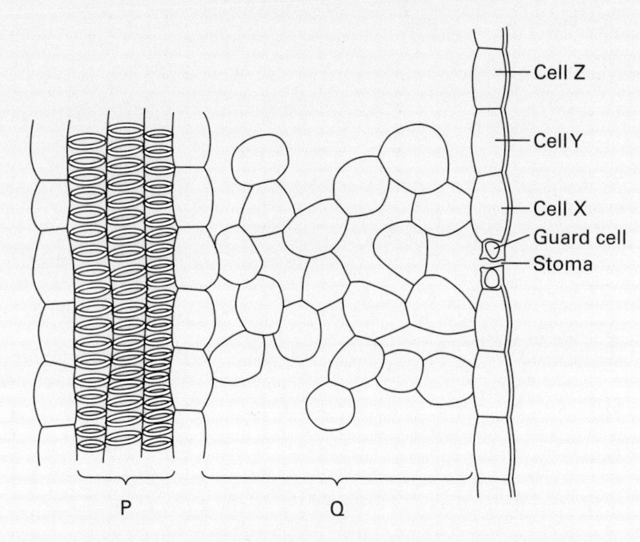

a Name the tissues labelled P and Q on the diagram. [2]

b The table below shows the concentrations of potassium ions in some of the cells shown in the diagram when the stoma is open and when the stoma is closed.

Cell	Concentration of potassium ions / arbitrary units	
	Stoma closed	Stoma open
Guard cell	95	448
Cell X	157	293
Cell Y	199	98
Cell Z	448	73

i Describe the changes that take place in the concentrations of potassium ions in cells X, Y and Z when the stoma opens. [2]

ii Explain how these changes in potassium ion concentration are related to the mechanism for the opening of the stoma. [3]

[Total 7 marks]

6 a Describe fully the passage of water from the soil to the xylem tissue of plant roots. [4]

b Explain how this route is also used to carry nitrogen into the plant. [4]

c Indicate the function of nitrogen in plants and the consequences if the nitrogen supply is inadequate.[2]

[Total 10 marks]

7 Outline the functions of the stomata of a leaf. Give an illustrated account of the structure and function of the stomata, describing the theories which have been put forward to explain their opening and closing. [10]

[Total 10 marks]

Reproduction and coordination in flowering plants

Flower structure

OBJECTIVES

By the end of this spread you should be able to:

- describe the structures of typical flowers
- explain the functions of flower parts.

Fact of life

The largest flower in the world belongs to the genus *Rafflesia*. It is a parasitic plant living on jungle floors in south-east Asia. The plant spends most of its life as a mesh of filaments hidden within liana (vine) roots. It flowers once every ten years, emerging from the underground roots like a giant cabbage. The monstrous bloom of one Sumatran species measures over a metre across. Its leafless, warty, maroon flower is not pretty, but its smell is even worse: it gives out a stench similar to rotting flesh, to attract swarms of pollinating flies.

Figure 1 Rafflesia, *the world's biggest flower.*

Flowers bring an almost infinite variety of shapes, colours, and scents to gardens, hedgerows, and forests. They are among the most beautiful of living structures, giving immense pleasure to millions of people. In addition to their aesthetic value, flowers are of major importance in food production. However, flowers evolved not for our benefit, but as the main reproductive structures of angiosperms, a division of seed-bearing plants. The beauty of angiosperms serves one purpose: the reproduction of new individuals. The flower is the centre of the sexual reproductive cycle and is the structure in which meiosis and fertilisation take place.

There are two main groups of angiosperms: dicotyledons and monocotyledons. Their classification is discussed in chapter 21. In this spread, we shall examine their flowers.

Dicotyledons

The flowers of dicotyledons may be radially symmetrical like buttercups and the typical flower in figure 2, or bilaterally symmetrical like the member of the pea family shown in figure 3. They vary greatly in size, colour, and shape, but the flower of a dicotyledon is typically radially symmetrical, insect pollinated, and produces both male and female gametes (figure 2). It has four types of organs: **sepals**, **petals**, **stamens**, and one or more **carpels**. These organs are formed from a series of modified leaves arranged in concentric rings called **whorls**. From the outside to the centre, the whorls are:

- the **calyx**, made up of the leaf-like sepals
- the **corolla**, formed from petals (the calyx and the corolla together form the **perianth**)
- the **androecium**, consisting of the stamens
- the **gynoecium**, made of one or more carpels.

Monocotyledons

The flower of a typical monocotyledon (figure 4) is wind pollinated. It has the same organs as a typical dicotyledonous flower but they are arranged differently and lack the coloured petals, scent, and nectary associated with insect pollination. In addition, the stamens hang downwards so that they can release pollen easily into the wind, and the stigma is large and feathery to gather the pollen.

The reproductive cycle of flowering plants

In both dicotyledons and monocotyledons, the flower starts the sexual reproductive cycle (figure 5) which involves the following process:

- the production of special reproductive cells
- pollination
- fertilisation
- fruit and seed development
- fruit and seed dispersal
- seed germination.

These topics are dealt with in the following spreads.

Carpel: a flower's female sexual organ consisting of an ovary, style, and stigma. The ovary forms the fruit after fertilisation. The style joins the ovary to the stigma. The stigma has a sticky surface to which pollen grains adhere. When several carpels occur fused together, the structure is called a **pistil**.

Stigma

Style

Ovary

Ovule: contains the egg cell which develops into a seed after fertilisation.

Sepal: may be green or coloured like the petals. Green sepals can photosynthesise, but their main function is to enclose the flower bud and protect it before it blossoms.

Flower stalk

Petals: are often brightly coloured and scented to attract pollinating insects. Some petals have markings called guide lines leading down to a nectary.

The stamens form the male organs of the flower. Each stamen consists of a filament (stalk) which supports an anther. An anther has four sacs containing pollen grains in which the male gametes develop.

Anther

Filament

Stamen

Nectary: a cup-shaped glandular structure at the base of the petal which secretes nectar. As an insect probes with its mouthparts to suck up the nectar, it may pick up or deposit pollen.

Receptacle: the structure at the top of the flower stalk which supports the flower.

Figure 2 *A typical flower of a dicotyledon: it is radially symmetrical and pollinated by insects.*

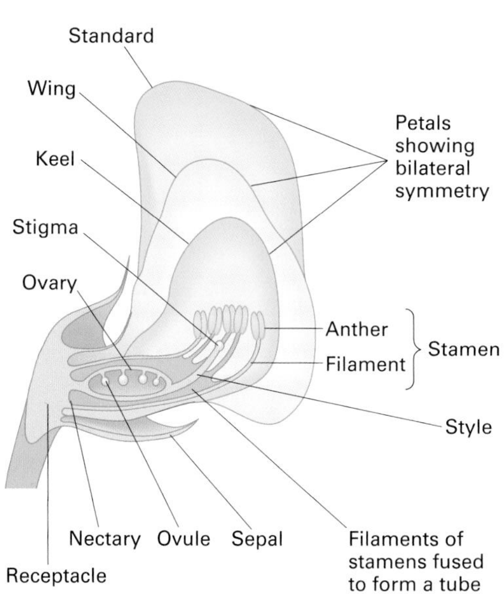

Figure 3 *A bilaterally symmetrical dicotyledonous flower: a member of the pea family.*

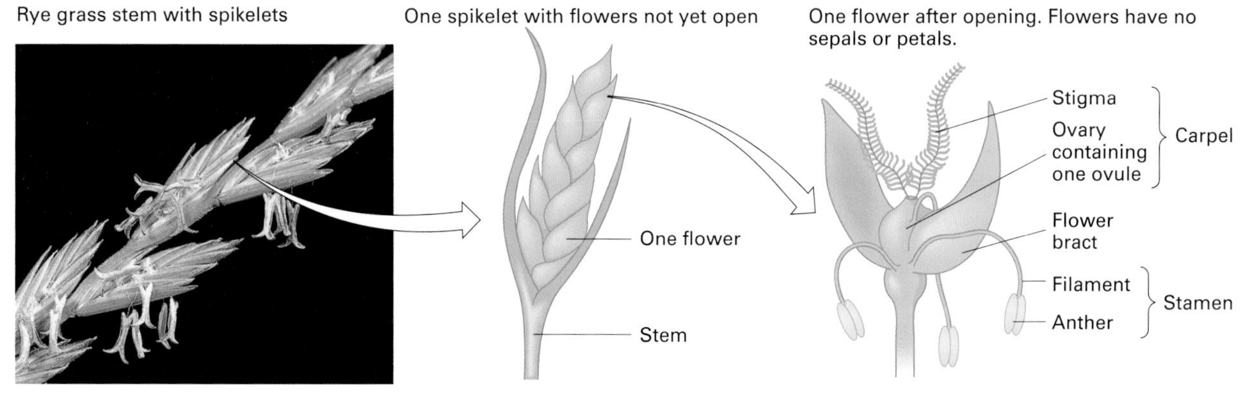

Figure 4 *The flowers of rye grass (a monocotyledonous plant).*

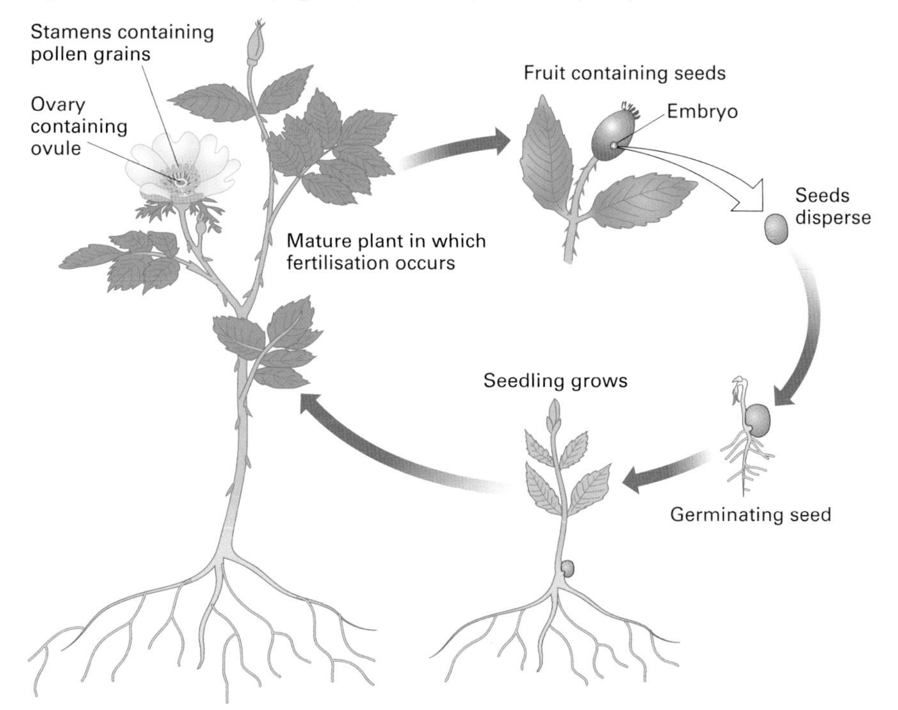

Figure 5 *The life cycle of a typical flowering plant.*

Quick check

1 a Which structures make up:

- **i** the perianth
- **ii** the corolla
- **iii** the androecium
- **iv** the gynoecium?

b What is the difference between a carpel and a pistil?

2 What is the main function of sepals?

Food for thought

The flower of *Rafflesia* (figure 1) is enormous. Suggest advantages of its large size and its use of chemicals to attract pollinating insects in a jungle habitat.

14.2

OBJECTIVES

By the end of this spread you should be able to:

- explain how pollen grains are formed in the anthers
- explain how an embryo sac and egg cell form.

Fact of life

A single birch catkin can contain five and a half million pollen grains.

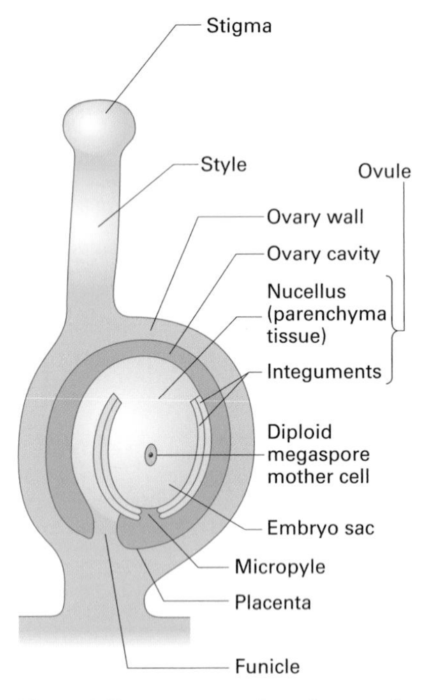

Figure 1 *Transverse section of a carpel.*

Gametogenesis in flowering plants

A central feature of sexual reproduction is **gametogenesis**, the production of gametes. Spread 12.3 deals with gametogenesis in mammals. In this spread, we are going to examine gametogenesis in a flowering plant.

Formation of an egg cell

The female part of the plant (the gynoecium) consists of a carpel or carpels which may occur separately or fused together in some way. A typical carpel has a stigma which receives pollen and a long style which leads down to a swollen structure, the ovary. The ovary contains one or more ovules. Ovules eventually develop into the seeds.

Each ovule (figure 1) is made up of an oval ball-shaped **nucellus** of parenchyma cells surrounded by one or two protective layers called **integuments**. At the apex of the ovule is a small opening through the integuments called the **micropyle**. The ovule develops from tissues within the ovary called the **placenta** and it remains attached to the placenta by a stalk called the **funicle**.

One of the cells within the nucellus is the **megaspore mother cell**, so called because it is a large cell which gives rise to other cells. Figure 2 outlines one way, generally regarded as typical, by which the megaspore mother cell develops into a megaspore, then into an embryo sac in which an **egg cell** (also called an **ovum**) forms. The megaspore mother cell first divides by meiosis to produce a row of four haploid cells called **megaspores**. Three of the megaspores usually degenerate, whereas one grows. This enlarged megaspore now develops into a mature embryo sac. Its nucleus divides by mitosis to form eight haploid cells which have no cell walls. Two of the cells, called **polar nuclei**, may fuse to form a diploid central cell in some plants. One of the nuclei close to the micropyle forms the **egg cell** and is flanked by two cells called **synergids**. These are thought to be involved in providing nutrition and support. The other three cells, the **antipodal cells**, take up a position opposite the egg cell.

Formation of male gametes

The male part of the plant (the androecium) consists of stamens. Each stamen has an anther connected to the rest of the flower by a filament. The stalk-like filament contains vascular tissue which supplies food and water to the anther. An anther contains four **pollen sacs** in which pollen grains develop from **pollen mother cells** (figure 3).

Figure 4 shows how each pollen mother cell undergoes meiosis to form four **pollen cells**, from which the male gametes develop. Immediately after meiosis, the four pollen cells are grouped together to form a **tetrad**. After the pollen cells separate, each develops into a pollen grain with a thick outer wall (the exine) and a thin inner wall (the intine). The wall has thin areas called apertures out of which a pollen tube can grow. In a

Figure 2 *Formation of an embryo sac.*

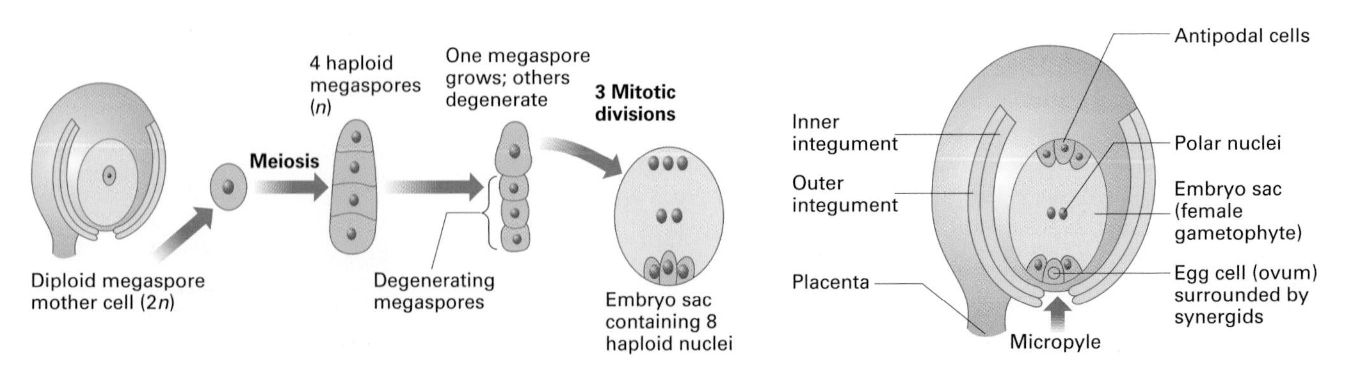

mature pollen grain, the haploid nucleus has divided mitotically to form a **generative nucleus** and a **tube nucleus**. After pollination, the generative nucleus again divides mitotically, giving rise to two male gametes.

Pollen grains range in size from 20 nm to 250 nm (about the size of dust particles). They are produced in astronomical numbers: in a large alder wood, for example, many billions of grains will be produced and each catkin alone may produce several million pollen gains (spread 14.3, figure 2).

Pollen grains come in an amazing diversity of shape. The outer wall of each grain often has a highly sculptured pattern characteristic of a species or genus. The wall has a waterproof lining of a substance so stable and resistant to decay that it may remain intact for tens of thousand of years. The fact that the walls of pollen grains last so long and that they have characteristic shapes enables scientists to chart the ecological history of an area by analysing the pollen in sediments. Pollen grains from successive layers of sediment from a lake, for example, can be extracted, identified, and counted to reveal the colonisation, development, and extinction of different species of plant. The study of pollen, called **palynology**, is also of interest to medical scientists because pollen can trigger allergic diseases of the respiratory tract, such as asthma.

Pollen grains are collected by some animals such as honeybees as food. Pollen grains are quite nutritious, containing variable proportions of protein, carbohydrate, minerals, and water.

Pollen grains as palaeothermometers

A cubic centimetre of mud can contain as many as 100 000 pollen grains. They are dead and empty, yet many retain their distinctive shapes, allowing them to be identified. Many trees appear to be sensitive to air temperatures. Analysing the pollen in sediments which have been dated in some way (for example, by radiocarbon dating) can therefore serve as a palaeothermometer, an indicator of ancient temperatures. Such an analysis is used to reconstruct the climatic history of the Earth. The results reveal that the climate of almost every region has been different from the climates of today. During the glacial period, even the Sahara desert was cooler than now and sufficiently wet to hold lakes.

Quick check

1 What is a tetrad?

2 In which structure does the embryo sac develop?

Food for thought

Palynology is a well established discipline that has generated valuable information about the relationship between past organisms and the environment in which they lived. Which features of pollen grains make them convenient objects to study?

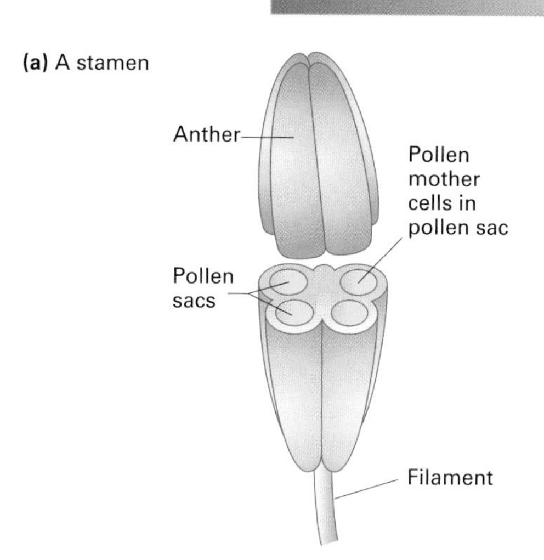

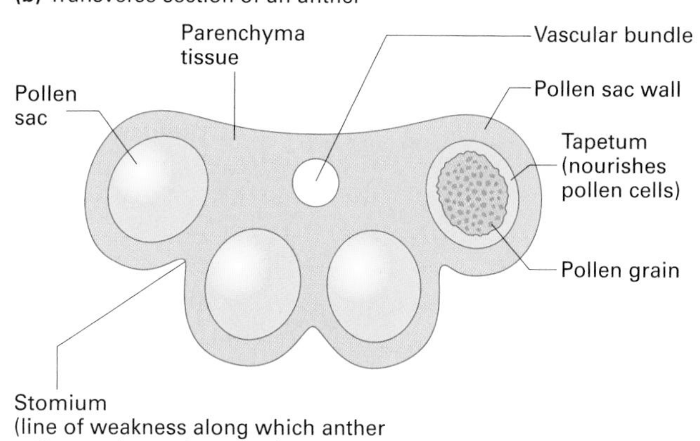

Figure 3 Transverse section of an anther.

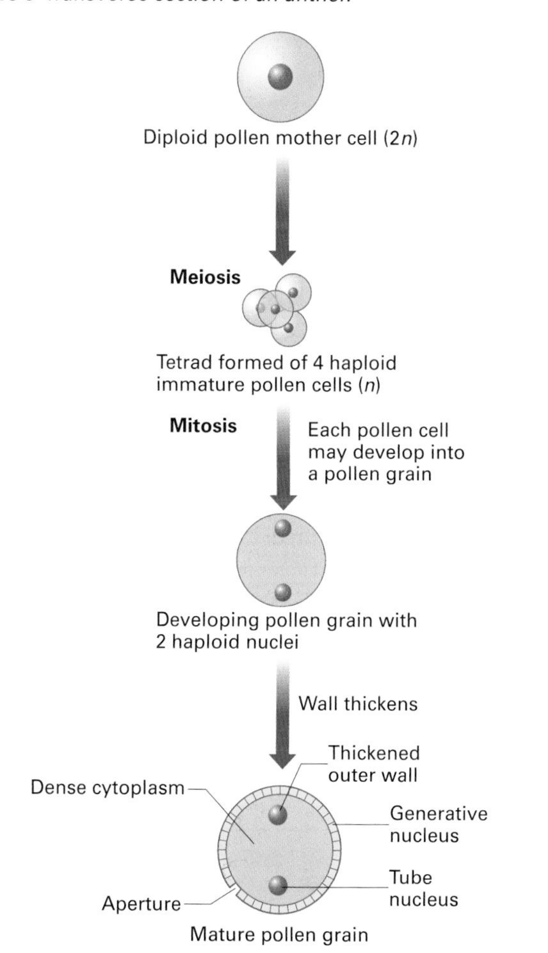

Figure 4 Formation of pollen grains.

14.3

Pollination

OBJECTIVES

By the end of this spread you should be able to:

- distinguish between cross pollination and self pollination
- discuss the process of coevolution in insects and plants
- compare the flowers of insect-pollinated plants and wind-pollinated plants.

Fact of life

Wind can transport pollen through enormous distances. The tiny grains from wind-pollinated trees have been found as high as 6000 m and can be carried by the wind as far as 3000 miles from the parent tree.

Pollen grains develop in the anthers of flowers. When mature, the anther wall splits open (a process called **dehiscence**) and the pollen is shed. Each pollen grain contains tissue which will develop into the male gametes. Pollen grains are a wonderfully efficient adaptation to life on land. The male gametes of aquatic plants can swim unprotected in a watery environment, but those of most terrestrial plants have to overcome the problems of travelling through air. The hard waterproof outer coat of a pollen grain protects the male tissue from drying out. The shape of the pollen grain and devices such as spikes or grooves in the surface of the outer wall aid the dispersal of the pollen (figure 1).

Transfer by insects or wind

Pollination is the transfer of pollen from the stamens to the stigma. Pollen grains cannot move independently, and their main means of transport is by animals or wind. Animal pollinators include humming-birds, bats, and honey possums (small marsupials that feed on nectar), but by far the most common animal pollinators are insects. Insect-pollinated flowers are called **entomophilous** flowers. Wind-pollinated flowers are called **anemophilous** flowers. Table 1 summarises the main differences between these two types of flowers, and figure 2 shows an example of each.

Adaptations to insect pollination

Most entomophilous plants provide some kind of reward to lure insects into the flower. This usually takes the form of pollen, nectar, or both. The presence of the reward must be advertised. Most insect-pollinated plants have large petals that appear brightly coloured to insects (flowers that appear white to us are brightly coloured when viewed under ultraviolet light to which insects are sensitive). The flowers are often heavily scented to attract pollinators and have marks on their petals which guide the pollinator to the nectary. In getting to the nectary, the insect has to brush past the stigma, the anthers, or both (figure 3).

Orchids have evolved a range of enticements that provide no obvious reward for their pollinators. Some have flowers that mimic the appearance of a female bee; they may also release a scent that resembles the female pheromone and acts as a sexual attractant. When a male bee, lured by the promise of sex, attempts to copulate with the flower, he picks up pollen. He will deposit this pollen on the next orchid which deceives him into trying to mate with it.

Coevolution

Many other plants have also evolved adaptations that lure certain types of insect to their flowers. These adaptations make pollen transfer more

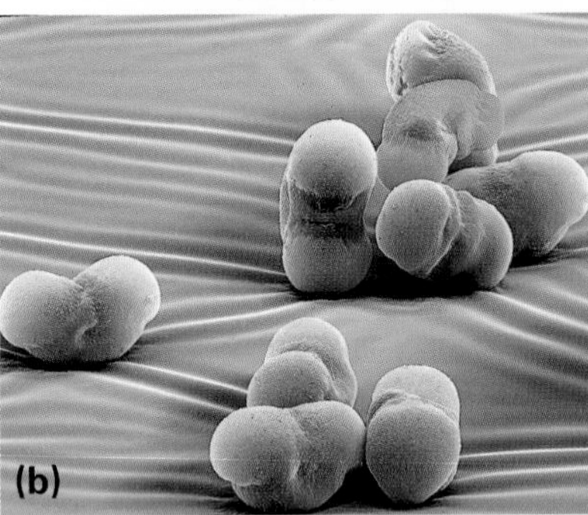

***Figure 1 (a)** Pollen from an insect-pollinated plant (daisy, ×1000). The grains are relatively large and spiky and can attach to the body of an insect.*
***(b)** Pollen from a wind-pollinated plant (silver fir, ×270). The grains are very small, smooth, and can easily be carried by the wind.*

***Figure 2 (a)** An insect-pollinated flower: wild sage. **(b)** A wind-pollinated flower: silver birch.*

efficient because it is more likely that one insect will visit different flowers of the same species, so reducing the risk of pollen being wasted on the flower of another species. Similarly, particular insects have evolved behavioural and structural adaptations which enable them to obtain the maximum reward from a visit to a flower. In many cases these adaptations have evolved together so that particular plants and insects became adapted to each other, each shaping the other's structure. This process, called **coevolution**, can lead to a total interdependence, as in the relationship between a yucca plant and the female yucca moth. The plant is totally dependent on the moth because the moth is the only insect with the correctly shaped mouthparts to extract pollen from it. The moth is totally dependent on the plant because it lays its eggs in one of the ovules, which later becomes the only source of food for its emerging caterpillars. If either partner becomes extinct, the other will be unable to reproduce.

Self pollination and cross pollination

There are two main types of pollination: **self pollination**, which is the transfer of pollen from the stamens to the stigma in the same flower; and **cross pollination**, which is the transfer of pollen from the stamens of one flower to the stigma of a different flower.

Cross pollination has the distinct advantage of providing more genetic variation. When a male gamete from one plant fertilises the egg cell of a flower on another plant, the zygote contains a new mixture of chromosomes from the two parents. Mechanisms favouring cross pollinations include:

- being dioecious, that is, having separate male and female plants (for example, willow, *Salix* sp.)
- being monoecious, that is, having separate male and female flowers on one plant (for example, hazel, *Corylus avellana*)
- anthers and stigmas maturing at different times. If anthers mature first, this is called **protandry** (for example, white dead-nettle, *Lamium aibum*); if stigmas mature first, this is protogyny (for example, bluebell, *Endymion non-scriptus*)
- structural adaptations which make self pollination unlikely; for example, in primroses the anthers and stigma are arranged so that cross pollination between pin-eyed and thrum-eyed flowers is favoured (figure 3).

Sometimes self pollination occurs but a **self-incompatibility mechanism** ensures that the pollen tube does not develop unless it has a different genetic composition from that of the stigma. This prevents self fertilisation.

Table 1 *Summary of differences between typical insect-pollinated (entomophilous) and wind-pollinated (anemophilous) flowers.*

Insect-pollinated flowers	Wind-pollinated flowers
Conspicuous flowers with large brightly coloured petals	Inconspicuous flowers with small drab petals or no petals
Scented	Not scented
Nectaries present	Nectaries absent
Relatively low pollen production	Very high pollen production
Stamens enclosed in flower so that insect brushes past them	Stamens pendulous (dangle outside the flower)
Stigmas relatively small and sticky	Stigmas large, feathery, and hang outside flower
Pollen relatively large and often spiny	Pollen small and usually smooth
Plants often single or in small groups	Plants often in dense groups covering a large area
Very complex structural modifications to attract insects	Relatively simple flower

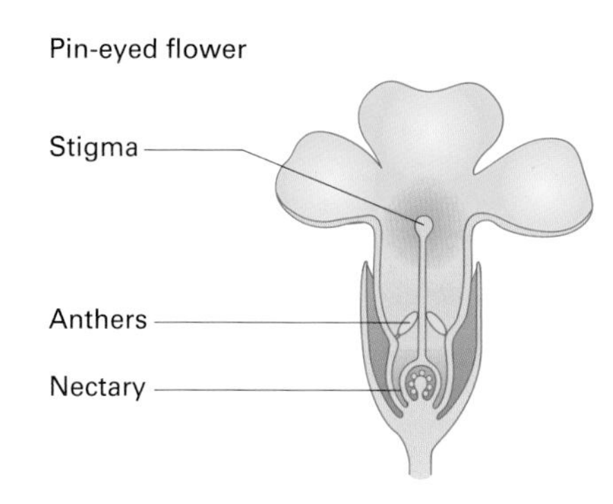

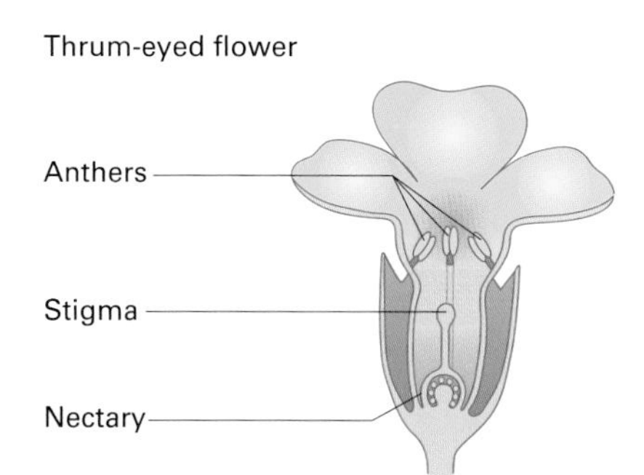

Figure 3 *Two types of primrose flower (*Primula vulgaris*). As a bee collects nectar from the nectary, it receives pollen on parts of its body which brush against the anthers. When it visits other flowers it will therefore tend to pollinate stigmas which are at the same height as the anthers from which the pollen was collected. Thus it will tend to transfer pollen from pin-eyed flowers to thrum-eyed flowers, and vice versa.*

Quick check

1 What is the main advantage of cross pollination?

2 Explain why the relationship between bees and flowers is described as mutually beneficial.

3 Why are the stamens of wind-pollinated plants and insect-pollinated plants different?

Food for thought

Many plants have devices that increase the chances of cross pollination but allow self pollination and self fertilisation to occur if cross pollination does not take place. Some uncommon species living in harsh environments with plants separated by long distances use self pollination and self fertilisation as their *only* reproductive strategy. Suggest advantages and disadvantages to these plants of this reproductive strategy.

14.4

OBJECTIVES

By the end of this spread you should be able to:

- distinguish between pollination and fertilisation
- describe the process of double fertilisation
- explain how seeds and fruits develop.

Fact of life

Between pollination and fertilisation, a pollen grain has to germinate, grow a tube the length of the style, and then the male gametes have to pass down the pollen tube to fertilise the egg cell and polar nuclei in the ovule. The time taken for this varies greatly: in geraniums the whole process can be completed in a few hours; in an orchid, it may take several months.

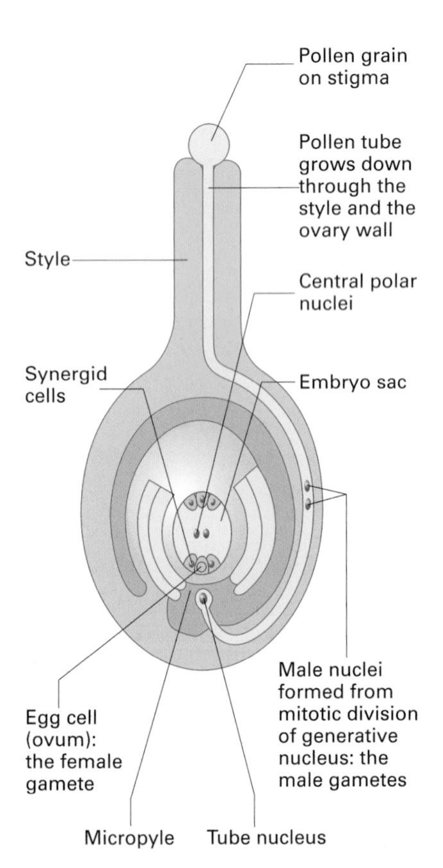

Figure 1 *Growth of the pollen tube.*

Fertilisation and seed and fruit development

Pollination and fertilisation

Pollination (the transfer of pollen from stamens to stigma, spread 14.3) does not always lead to fertilisation (the fusion of male and female gametes). The fate of a pollen grain after it has been transferred to a stigma depends on whether it is compatible with the stigma. A compatible pollen grain absorbs water from the stigma by osmosis and swells. A pollen tube develops out of the swelling and grows through an aperture in the outer coat of the pollen grain. A sugar solution secreted by the stigma may provide nourishment for the germinating pollen grain.

After the pollen tube has emerged, it grows down through the style and ovary wall, digesting its way through the tissue by secreting hydrolytic enzymes at its tip (figure 1). The growth is controlled by the pollen tube nucleus in the tip of the tube and it may be directed by chemicals secreted by the synergid cells in the embryo sac.

During the growth of the pollen tube, the haploid generative nucleus of the pollen grain divides by mitosis to form two haploid male nuclei. These are the male gametes, and they move down the pollen tube and eventually fuse with the polar nuclei and the egg cell (ovum). (In some plants, the polar nuclei fuse before fertilisation to form a diploid nucleus.) When the pollen tube reaches the micropyle, it penetrates into the embryo sac. The tube nucleus disintegrates and the tip of the pollen tube bursts, leaving a clear passage for the two male gametes to move into the embryo sac. One male gamete fuses with the polar nuclei, producing a **triploid nucleus** (it has three sets of chromosomes) which develops into the **endosperm** (food store) of the seed. The other male gamete fuses with the egg cell, forming a **zygote**. This **double fertilisation** is unique to angiosperms (flowering plants). In many cases, more than one ovule is present in a gynoecium. Each must be fertilised by a separate pollen grain, possibly even from a different plant, if it is to develop into a seed.

Development of the seed

Soon after fertilisation the ovule develops into a seed. The outer integument of the ovule becomes the **testa** (seed coat). The endosperm nucleus divides rapidly by mitosis to form a food store. (Double fertilisation therefore ensures that a food store is formed only in fertilised ovules.) In **endospermic seeds**, such as the castor oil seed, the endosperm continues to grow and is kept as the seed's food store. In **non-endospermic seeds**, such as shepherd's purse, the endosperm becomes absorbed by the developing cotyledons (embryonic leaves) which provide the food store for the germinating seed.

While the endosperm is forming, the zygote divides to form a column of cells. The innermost cell of this column develops into the embryo. Figure 2 shows the embryonic development of shepherd's purse (*Capsella bursa-pastoris*). This has two cotyledons and is non-endospermic. Nutrients for the growing embryo and food store are supplied by the disintegration of surrounding nucellus cells in the embryo sac. Food is also supplied by the parent plant through the vascular bundles of the funicle (the stalk connecting the ovule to the parent).

Formation of the fruit

While the seeds are developing, the ovary wall grows to let them expand. The mature ovary becomes the **fruit**, the outer wall of which is known as the **pericarp**. Fruit structure and function varies greatly among angiosperms, but most fruits are adapted to protect the seeds and to help their dispersal. When detached from the parent plant, a fruit has two scars, one formed by the withered style, the other showing its attachment point with the receptacle. A detached seed, on the other hand, has only one scar formed from the remains of the funicle.

The role of the seed

During the final stages of seed development, the water content of the seed is reduced. Partial dehydration helps the seed to remain dormant until it germinates: it greatly reduces metabolic activity and protects the seed from damage which might be caused by freezing and thawing.

The above account of seed and fruit development relates to angiosperms, but seed-bearing plants (spermatophytes) also include the non-flowering gymnosperms (conifers and their relatives). The evolution of seeds was a major factor in the successful colonisation of land by these plants. A seed contains a multicellular embryo, a new plant in miniature, surrounded by a protective coat to shield it from adverse environmental conditions and biological attack. In particular, the seed protects the embryo from desiccation in deserts or frozen tundra. Desert seeds can remain in dry sand for decades, awaiting a shower of rain; arctic lupin seeds remain dormant in cold conditions for even longer periods, and will germinate when the conditions become warmer.

Seeds can be very wasteful when compared with asexual (vegetative) reproduction. A seed-bearing plant has to invest a lot of energy in the formation of seeds, and few of them reach suitable places for germination. Most seeds are lost during **dispersal**, when they are spread away from the parent plant by methods which depend on variable factors such as wind, animals, and water. To ensure that the species survives, spermatophytes have to produce large amounts of seeds.

The seed is adapted to remain dormant until conditions are suitable for germination and the emergence of a new plant. When conditions are right, the food store and the ready-made tissues of the embryo enable the new plant to grow very quickly (spread 14.6). Plant growth substances (spread 14.10) control the development of the fruit and seed.

QUICK CHECK

1 Distinguish between pollination and fertilisation.

2 Explain how fertilisation in angiosperms is unique.

3 Name the parts of the shepherd's purse seed which will develop into:

a the root

b the shoot

c the leaves.

Food for thought

Only spermatophytes produce seeds, and only angiosperms produce fruits. Ferns and mosses produce **spores** rather than seeds. Spores are small unicellular structures from which a new organism arises. Suggest the advantages and disadvantages of using spores as reproductive bodies rather than seeds.

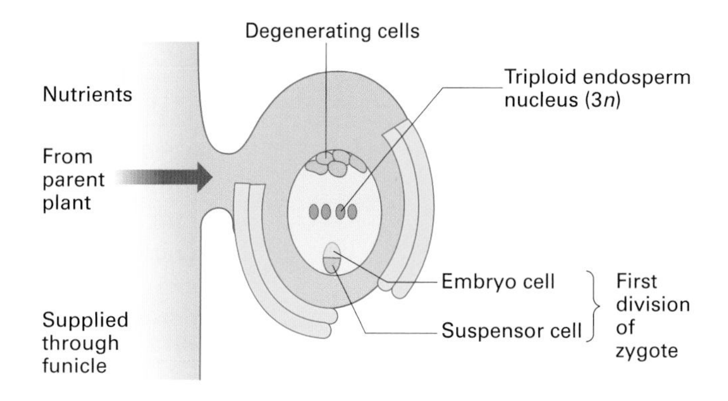

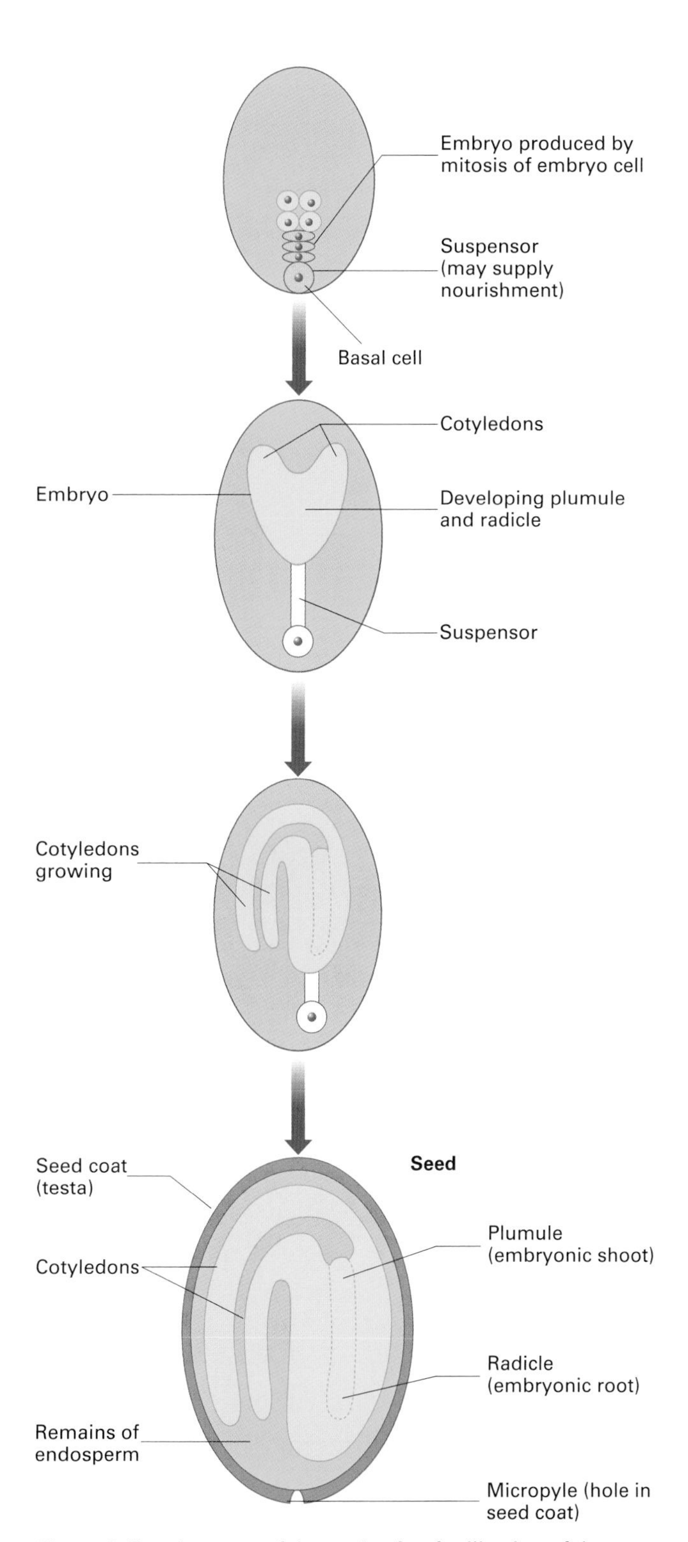

***Figure 2** Development of the ovule after fertilisation of the egg cell in shepherd's purse (*Capsella bursa-pastoris*).*

14.5

Seed dormancy and dispersal

OBJECTIVES

By the end of this spread you should be able to:

- give examples of how seeds are dispersed by explosive devices, wind, water, and animals
- discuss the biological significance of dormancy
- explain how dormancy is maintained.

Fact of life

It has been suggested that a tree native to the island of Mauritius, *Calvaris major*, is now extinct because its very thick seeds could be scarified and triggered out of dormancy only by passing through the gizzard of a dodo. According to this suggestion, the tree coevolved so closely with the dodo that extinction of the bird resulted in the demise of the tree.

Figure 1 *As its seeds ripen, the Mediterranean squirting cucumber inflates its seed pods with so much slimy fluid that they explode, burst off their stalk, and shoot through the air as far as 7 metres. During the flight of each seed pod, seeds spill out from a hole in its base like the exhaust gases of a space rocket.*

An adult plant rooted in the soil relies on the dispersal (spreading) of its seeds in order to colonise new areas. Dispersal also enables a seed to germinate and grow without having to compete with a tougher and taller parent for light, water, and nutrients. Most plants invest a great deal of energy in fruit production, not only to protect seeds but also to help disperse them over long distances.

Mechanisms of seed dispersal

Seed dispersal mechanisms range from the simple to the weird and wonderful.

- Members of the pea family and Mediterranean squirting cucumber (figure 1) use **explosive** devices.
- When ripe, the fruits of poppies form a **censer mechanism** like an incense shaker which releases seeds when it is shaken in the wind.
- Dandelion seeds have radiating threads that form a **parachute** so that they can be carried long distances by even the gentlest breeze.
- European maple, sycamore, and many other seeds of woodland trees use **wing** structures which allow them to be carried away from their parent.
- The birdcage plant does not use the wind to lift and carry it through the air, but instead it uses the wind to push its seed-carrying body along desert sands (figure 2).
- Seeds of the coconut palm are adapted for sea travel: they have a hard fibrous shell and rich supply of food and water.

Most succulent and brightly coloured fruits are designed to be eaten by birds and mammals. The taste and nutrition from the pulpy part of the fruit acts as a reward, enticing the animal to feed on more fruit and disperse the seeds. The seeds are carried in the animal's gut until they pass out with the faeces, a convenient source of nutrients for a young seedling. The seeds escape digestion in the gut because their testa is hardened with lignin and cutin.

Many plants, such as goosegrass, burdock, and the vicious-looking African grapple plant (figure 3), develop hooks from the withered style, receptacle, or ovary wall. These hooks allow the seeds to hitch a ride on the bodies of animals. Mistletoe seeds gain free air travel by sticking themselves onto a bird's body with a gluey secretion.

Dormancy

Supplied with water, oxygen, and a favourable temperature, the seeds of some plants (for example, groundsel) are ready to germinate as soon as they are shed from the parent plant. If conditions are unfavourable, these seeds remain in a state of dormancy awaiting better conditions. During dormancy, metabolism is reduced to the absolute minimum for survival.

Most plants delay germination even if the external conditions are favourable. Dormancy improves the chance of a seedling growing to maturity: for example, temperate seeds shed in the autumn may delay germination until the following spring; this reduces the risk of being frozen to death in winter. An extended period of dormancy also gives seeds a greater opportunity to be moved away from the parent plant.

The mechanisms that keep a seed dormant differ from species to species. Some seeds have an impermeable layer in the seed coat which prevents the entry of water or oxygen. An impermeable coat may be removed in the following ways:

- by the action of water
- by physical removal (scarification), for example, by the action of bacteria in the soil or the mechanical abrasion of a bird's gizzard

Figure 2 *The birdcage plant inhabits the Californian desert. The shifting sands can change a shady, moist area in which a parent plant becomes fertilised into a hot cauldron, killing the parent. Even though dead, the plant, transformed into a seed-carrying cage, continues its parental duties. The rootless cage is pushed to new areas by the wind. Seeds are shed as the cage moves, but most spill out when it comes to rest. The birdcage plant's last resting place is likely to be the side of a dune, where there is a good chance of shade and moisture for the seeds.*

Figure 3 *The seed pod of the African grapple plant can inflict serious injury to the animal that transports it. When, for example, it is trodden on by an elephant, one or more of its sharp hook-like barbs penetrates the sole of the foot. The seed pod stays embedded in the foot until the pounding action of the elephant's foot on the ground wears away the bottom hooks, arms, and seed casing and allows the seeds to break out.*

- by chemical removal during a seed's passage through the gut of an animal
- by the action of fire, which burns the seed coat away.

Horticulturists and gardeners may bring about scarification to speed up germination by gently sandpapering a seed or pricking a hole in it.

Many seeds and fruits contain **chemical inhibitors** which prevent germination. The relative concentrations of the plant growth substances (spread 14.10) abscisic acid, gibberellin, and cytokinin appear to play an important role in keeping a seed dormant. In desert plants, the chemical inhibitors are usually removed by the seed being soaked in water for a period of time.

Seeds of plants living in seasonal environments commonly require **prechilling** (exposure to a cold period) before germinating. This dormancy-breaking method is common among cereals and members of the rose family. It means that the seeds are less likely to germinate in the autumn or winter.

In some plants, such as certain varieties of lettuce, germination is stimulated after a period of exposure to light; in a few others, it is inhibited by light, so that the seeds germinate only after a period of time buried in the soil.

The physiological mechanisms by which seeds time their exposure to low temperature and particular conditions of light are not clear. They may include changes in permeability of the seed coat, changes in the relative concentrations of plant growth substances, and, for light-sensitive plants, phytochromes (see spread 14.12).

QUICK CHECK

1 Give one example of a plant that uses each of the following dispersal mechanisms:

a an explosive device which works by being inflated with water

b a winged seed lifted by air currents

c a buoyant seed carried by sea currents

d a gluey substance which sticks the seed to an animal.

2 Briefly explain why a seed may remain dormant even when the environmental conditions are favourable for germination.

3 State an advantage gained by a seed in which germination is inhibited by light.

Food for thought

Some plants and the animals that eat their fruits have coevolved a relationship in which each partner is highly adapted to the other. How might this normally mutually beneficial relationship become detrimental to either partner?

14.6

Seed germination

OBJECTIVES

By the end of this spread you should be able to:

- state the conditions required for germination
- describe hypogeal germination in a broad bean
- outline epigeal germination in a sunflower
- explain the role of the coleoptile in the germination of most monocotyledons.

Fact of life

Arctic lupin seeds estimated to be 10000 years old have germinated after being excavated from Alaskan permafrost (permanently frozen ground), taken to a laboratory, watered and warmed.

The seed occupies a unique position in the life cycle of a flowering plant. In its dehydrated and inactive state it is able to tolerate conditions so extreme that viable seeds can be found deep in permafrost or on the surface of sun-baked desert sands. The embryo wrapped inside the tough impermeable seed coat (the **testa**) is a new plant in miniature. Equipped with a food store and all the necessary genetic information, it has the potential to become an independent member of the next generation of plants. However, this potential remains unfulfilled until it breaks out of its state of dormancy.

When dormancy is broken and conditions are favourable, seed germination begins. During **germination** the almost inert, usually drab, and often tiny seed goes into a state of frenetic activity which transforms it into a young seedling, dynamic, colourful, and much larger. Although the passage of time usually has little effect on a dormant seed, time is of the essence during germination. The new plant must grow quickly to reach a new energy source, the light, before its store of food runs out.

The start of germination: taking in water

The following account of germination refers specifically to the broad bean (figure 1). Although the details differ, the seeds of all plant species require similar conditions for germination: water, oxygen, and a favourable temperature. Some species also need an appropriate level of light.

Germination of the broad bean starts with a rapid intake of water, usually through the micropyle (a tiny hole in the seed coat). The tissue beneath the hard seed coat contains hydrophilic substances which attract water from the surrounding soil. This process, called **imbibition**, rapidly saturates this outer layer of tissue. Tissue deeper inside the seed is dehydrated and so has a very low water potential, therefore water moves quickly into these cells by osmosis. The increase in water content enables metabolism to speed up inside the seed, fuelled at first by anaerobic respiration of food stores. However, energy demands are high so the germinating seed needs to switch quickly to aerobic respiration, hence the need for oxygen during germination.

The food stores can be used as an energy source only in the presence of water, which is needed for hydrolytic enzymes to convert insoluble storage polymers (polysaccharides, fats, and proteins) into soluble products. These can be translocated and used in respiration and the synthesis of new tissue. The enzymes need a favourable temperature to work efficiently. Imbibition of water also causes the embryo to release hormones that stimulate the rapid mobilisation of food stores.

Figure 1 *Hypogeal germination in a broad bean.*

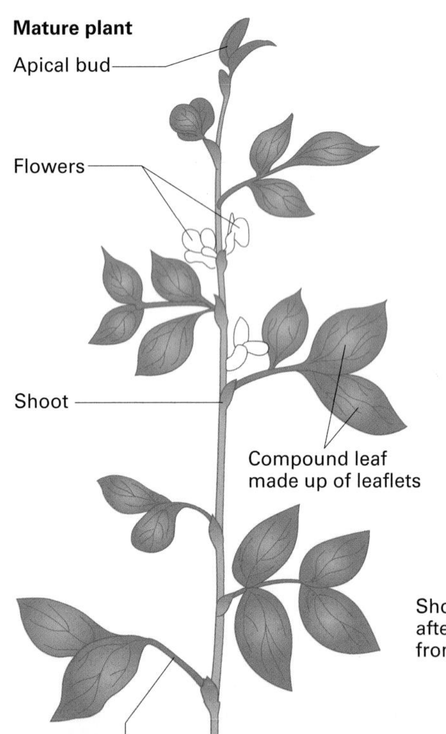

Growth of the seedling

An increase in metabolism allows the seed to start growing. First the **radicle** (embryonic root) swells, bursts through the seed coat by turgor pressure, and grows downwards into the soil. As it grows, it quickly develops a root system with lateral roots and root hairs. This ensures that the young seedling is safely anchored in the soil and is able to gain water and nutrients for its growth. Once the radicle is established, the **plumule** (embryonic shoot) emerges and grows upwards out of the soil.

The broad bean shoot shows **hypogeal germination**: the two cotyledons (embryonic leaves) remain underground. Mitotic division of cells in the **epicotyl** (the area just above the cotyledons) pushes the plumule out of the soil. The plumule adopts a hooked shape to protect its delicate tip as it grows through the soil. The tip acts as the main growth region (meristem) of the shoot after the leaves have grown and started photosynthesising.

In the sunflower (figure 2), the two cotyledons are pushed out of the soil by growth of the **hypocotyl** (an area just below the cotyledons). This is called **epigeal germination**. During the growth of the plumule through the soil, the tip of the plumule is tucked safely between the two cotyledons.

In most monocotyledons, such as grasses and cereals, a sheath called the **coleoptile** surrounds the plumule and the first leaves, protecting them as they emerge through the soil into the light (figure 3).

Once a seedling is able to photosynthesise, germination of the seed is over and the life of the young plant begins.

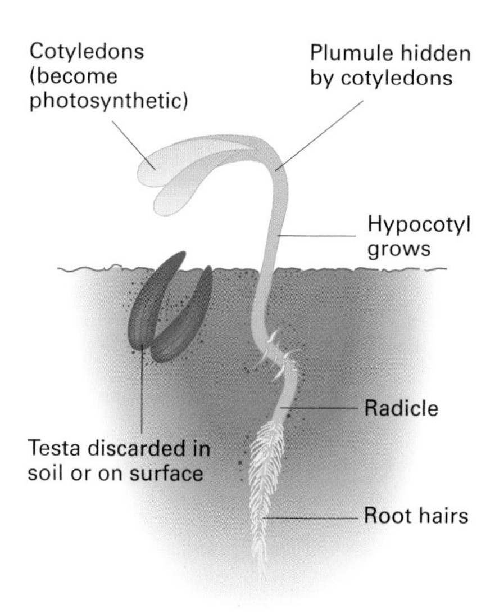

Figure 2 *Epigeal germination, for example in the sunflower and the French bean.*

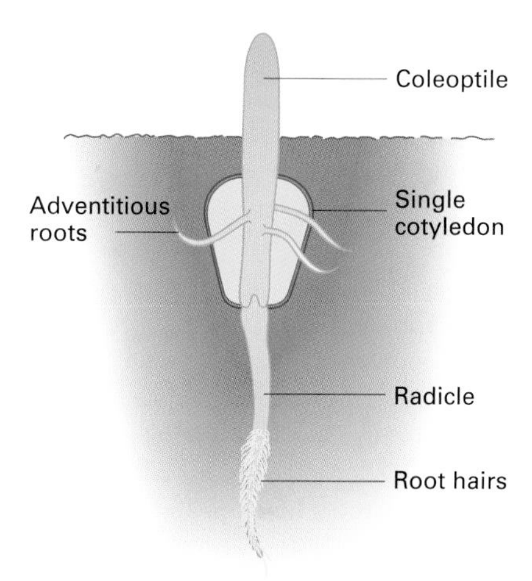

Figure 3 *Germination of a typical monocotyledon such as maize.*

The importance of seed research

Seed research is of immense importance to the human species. We rely on seeds and the seed products of cereals and legumes (peas and beans) for much of our nutritional needs. The cereals (barley, maize, millet, oats, rice, rye, sorghum, and wheat) supply about half of our protein and energy requirements. Despite our dependence on these seeds, we have much to learn about their formation, dormancy, and germination. Many questions remain unanswered or only partly answered. For example, how do some seeds (so-called **orthodox seeds**) survive dehydration whereas others (called **recalcitrant seeds**) do not? Precisely what triggers the end of dormancy? What are the specific molecular mechanisms that control cell elongation and differentiation during germination? New cellular techniques including recombinant DNA technology (spread 18.9) enable scientists to unravel the secrets of the seed's remarkable properties. This research will enable us to keep seeds successfully in seed banks, which are becoming increasingly important as natural habitats are being destroyed because of our progressive need for more space. Seeds kept dormant in a seed bank may one day prove essential for our survival.

Quick check

1 Give two reasons why water is required for germination.

2 Explain why a seed will not germinate unless the temperature is favourable.

3 Distinguish between hypogeal and epigeal germination.

4 State the role of the coleoptile in monocotyledons.

Food for thought

There has been considerable excitement in recent years about the effects of space travel on seed germination. Now that the prospect of colonising new planets is no longer complete science fantasy, seed germination in space could be of much more than academic interest. Since the late 1980s, scientists have been conducting seed germination experiments in spacecraft. How might the conditions in a spacecraft differ from those on Earth, and how do you think they might affect seed germination? Design an experiment that you could carry out if ever you travelled in space.

14.7

Vegetative propagation

OBJECTIVES

By the end of this spread you should be able to:

- discuss the advantages and disadvantages of vegetative propagation
- describe the main types of vegetative propagation
- explain the significance of perennating organs.

Fact of life

Most grasses propagate asexually by sprouting shoots and roots from underground runners. A small patch of grass can spread in this way to cover an acre or more.

Seeds without sex

Sexual reproduction in flowering plants produces seeds, but seeds can also be produced asexually. Some plants, such as many common species of dandelion (*Taraxacum* spp.), produce seeds without fertilisation. This type of reproduction is known as **apomixis**. It can occur in many ways but always results in offspring genetically identical with the parent. In dandelions, the bright yellow flowers attract insect pollinators, but the pollination is only a trigger for seed formation; it does not lead to fertilisation.

Apomixis ensures that a parent plant that is well adapted to a specific environment produces equally well adapted offspring. Only asexual reproduction can do this; sexual reproduction reshuffles genes. Dormancy and seed dispersal enable apomictic daughter plants to leave the specific habitat of the parent plant. Therefore, as long as there are similar habitats elsewhere, apomictic daughter plants have a good chance of surviving even if the parent's habitat is destroyed.

The role of asexual reproduction

Sexual reproduction provides the genetic variety that plants need in order to adapt to ever-changing environments. But this variety comes at a high price. Parent plants must invest a lot of energy in pollen and seeds but with little chance of producing large numbers of viable offspring. Asexual reproduction is much less costly and gives a much better rate of return. It usually produces many offspring that are genetically identical with the parent, but this is no disadvantage if the offspring can find a habitat which is similar to the parent's. In fact, only asexual reproduction ensures that a plant that is well adapted to a specific environment will pass on its genetic secrets of success to the next generation.

The lack of variation associated with asexual reproduction may be a disadvantage to long-term survival in a changing world. Most flowering plants which, in evolutionary terms, have been around for a long time, adopt a mixed reproductive strategy: they use sexual reproduction, which produces variety, as well as asexual reproduction, which is more efficient.

Fragmentation and regeneration

Vegetative propagation is the most common form of asexual reproduction in flowering plants. It generally involves **fragmentation**, the separation of parts of the parent plant, and **regeneration** of those parts into new individuals. It is such a commonplace event that we tend to take it for granted. However, if a similar process happened in a person it would make headline news throughout the world (imagine the sensation created by a headline such as 'Child grows from man's finger!').

Vegetative propagation is not usually a haphazard process relying on accidental detachment of a plant part. It usually involves the development of highly specialised structures which bear little resemblance to the parent organ from which they developed. As table 1 shows, almost any plant organ may be adapted to vegetative propagation.

Perennation

In many seasonal plants, the organ involved in vegetative propagation is also used as a **perennating organ**; that is, an organ that enables the plant to survive the harsh conditions of winter. In temperate regions, the aerial parts of most **herbaceous perennials** (non-woody plants which live for several years) die back in the autumn. Their perennating organs, often swollen with food stores built up in the previous growing season, lie buried under a blanket of soil, protected from the worst of the winter cold. Perennating organs often fragment to produce the new generation in the following year. The food store enables the offspring to grow quickly as soon as light and temperature conditions become favourable. This is especially important for small woodland plants, such as bluebells, which must grow and flower before trees come into leaf and plunge them into shade.

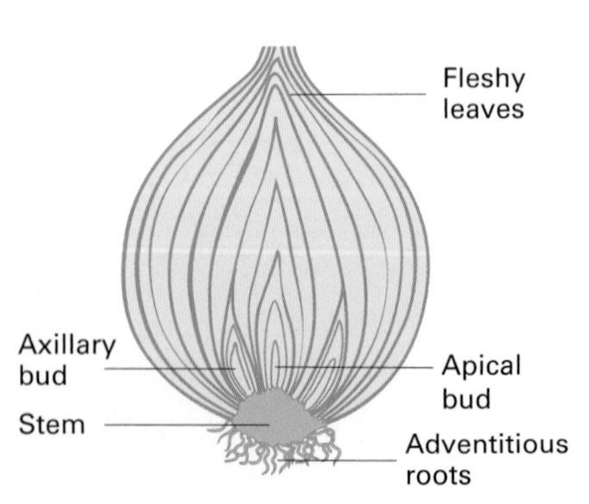

***Figure 1** Longitudinal section of an onion bulb.*

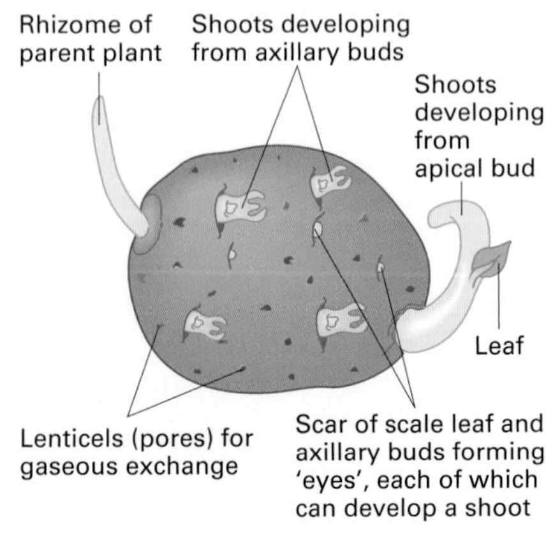

***Figure 2** Well developed potato tuber.*

***Table 1** Organs of vegetative propagation. An asterisk (*) shows organs that may also be used for perennation. The table refers to **axillary buds** and **apical buds** from which new plants may develop. An axillary bud occurs in the axil of a leaf, that is, the angle between the upper side of a leaf and the stem. An apical bud occurs at the tip of a shoot.*

Vegetative structure	Example	Description	Comments on mechanism of vegetative propagation
Bulb* (figure 1)	Daffodil (*Narcissus*) Onion (*Allium*) Tulip (*Tulipa*)	Short stem with underground fleshy leaves	Each apical bud and axillary bud can develop into a new plant.
Corm*	*Crocus* *Gladiolus*	Enlarged, swollen, and usually rounded base of stem; covered externally with thin scales or leaves	Axillary buds may develop into new plants.
Offset	Water hyacinth (*Eichhornia*)	Short thickened horizontal branch of stem with tuft of leaves at apex and small roots below	Similar to runner, but shorter and stouter. Often breaks away from parent to form new daughter plant.
Rhizome*	*Iris*, mint (*Mentha*), couch grass (*Agropyron repens*)	Thick horizontal stem usually growing underground	Sends out shoots above and roots below to give rise to new plants.
Runner	Strawberry (*Fragaria*) and creeping buttercup (*Ranunculus repens*)	Thin lateral stem on the soil surface	A number of stems usually radiate from the parent plant. New plants develop along the runner where it puts down adventitious roots. Runner decays when the daughter plant becomes established.
Stolon	Blackberry (*Rubus*)	Slender stem	Stem grows diagonally upwards then bends to the ground, putting roots out at the tip and producing a bud which grows into a daughter plant.
Sucker	*Chrysanthemum* Raspberry (*Rubus idaeus*)	Branch of a stem	Initially grows underground then emerges to form a new plant, at first nourished by the parent plant.
Tiller	Grasses	Side shoot arising at ground level	Tillers normally remain connected to form a plant that can cover a very wide area. Occasionally tillers become detached and give rise to new individuals.
Tuber* (stem) (figure 2)	Potato (*Solanum tuberosum*)	Swollen stem	A single parent plant may produce many tubers. Each tuber may have several axillary buds, each of which can develop into a new plant, enabling very rapid asexual reproduction.
Tuber* (root)	*Dahlia*	Swollen fibrous root	New plants develop from axillary buds at the base of the old stem.

Quick check

1 Give one advantage and one disadvantage of vegetative propagation .

2 Distinguish between:

a a bulb and a corm

b a stolon and a runner

c a stem tuber and a rhizome.

3 Explain why perennating organs are particularly advantageous to small flowering plants such as bluebells that live in temperate woodlands.

Food for thought

Despite the disadvantages of asexual reproduction, some plants adopt it as their only means of reproduction. Although the offspring usually inhabit the same type of environment as the parent plant, some asexually reproducing species adapt to a changing environment. Suggest how they might do this.

14.8

OBJECTIVES

By the end of this spread you should be able to:

- describe how cuttings can be made
- explain how plants can be reproduced by micropropagation.

Fact of life

English elms, trees that once graced a large proportion of the English countryside, are sterile and propagate vegetatively. This occurs naturally by root suckers which can produce a new shoot up to 50 metres away from the parent plant, or commercially by taking cuttings. DNA fingerprints have revealed that all native English elms are probably descendants of a single field elm belonging to the variable species *Ulmus minor.* Because they are clones and genetically identical, English elms are highly vulnerable to infections. In 1967, a deadly form of Dutch elm disease (a blight caused by fungal spores carried by the bark beetle) arrived in Britain. Between 1967 and the late 1990s, the disease spread through the countryside and wiped out 25 million elm trees.

Figure 1 *Taking cuttings: a traditional method of vegetative propagation. A stem of busy lizzie (*Impatiens *sp.) was cut and placed in water for a week or two, and roots formed from the cut end, resulting in a new plant.*

Figure 2 *The methods of grafting. Traditional grafting involved knife-work and binding up; a newer technique uses a stamping machine to cut a mortice and tenon in the scion and stock which lock together to form a new plant. In the 1870s the entire European stock of vines was nearly destroyed by a root-eating pest,* Phylloxera vastatrix. *Disaster was averted by pulling up the European vines and grafting the shoots onto rooted cuttings of North American vines that are resistant to attack.*

COMMERCIAL APPLICATIONS OF VEGETATIVE PROPAGATION

Selective breeding and vegetative propagation

The cabbages, cauliflower, kale, Brussels sprouts, and broccoli that grace a greengrocer's stall all originated from a single species of wild mustard. Early varieties of these vegetables were produced by farmers selecting and breeding together plants with specific characteristics: for example, large fleshy flower clusters for cauliflowers, a large terminal bud for cabbages, and leaves for kale. Traditional forms of this **selective breeding** are usually slow and costly.

Once a plant with the desired characteristics has been bred, commercial growers usually use vegetative propagation (spread 14.7) to produce more of the same. Maintaining the same characteristics through seed production often involves a complex genetic breeding programme which needs to be managed carefully for each generation. In contrast, vegetative propagation ensures that the desirable qualities of a plant will be maintained from one generation to the next with little effort on the part of the grower. Barring a few mutations, this form of reproduction produces **clones**, plants genetically identical with each other and their parent. Vegetative propagation has other advantages to the commercial plant grower. It allows plants to be produced quickly, in very large numbers, and with the minimum expense and effort. It also allows the grower to produce offspring from sterile **hybrids** formed from a cross between two different species. Many commercial flowers, trees, and shrubs are sterile: they may be very healthy, but the only way of propagating new plants from them is vegetatively.

Cuttings and grafts

One of the most commonly used methods of vegetative propagation is taking a **cutting** (figure 1). A small piece of a healthy plant is removed and kept in water or compost until roots grow. Sometimes a **plant growth substance** (an auxin) is added to speed up root development. When the roots emerge, the plant is grown in soil in the normal way. The cutting may be a leaf, a healthy young twig, or a stem tip.

Grafting (figure 2) is a similar technique but in this case the cutting (a young twig or shoot referred to as the **scion**) is inserted into a branch or stem of another plant (referred to as the **stock**). Two plants are often bound together with tape or raffia to help them join together, and the joint is covered with wax to prevent evaporation and infection. Grafting is common in gardening and commercial growing, and is the mainstay of the wine industry in many regions of the world. The only way of producing certain vines is by grafting a scion chosen for its grape production onto a stock chosen for its efficient root system and its resistance to disease. The quality of the vine is determined by the genetic make-up of the scion and is not diminished by grafting.

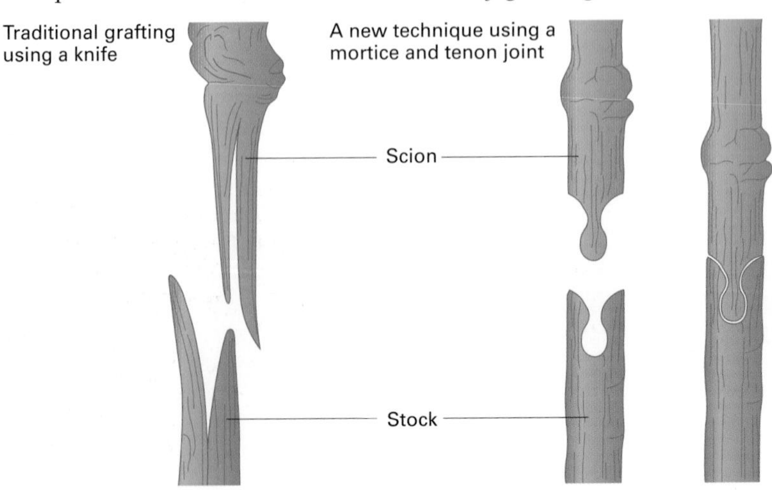

Micropropagation

Micropropagation (sometimes called tissue culture propagation or cloning) refers to test-tube methods of culturing whole plants asexually from very small pieces of tissue (called **explants**) cut from the parent (figure 3).

It was realised at the start of the twentieth century that each non-reproducing body cell (**somatic cell**) of a plant has the potential to form a whole plant (a phenomenon called **totipotency**). However, the conditions that allow the cell to fulfil this potential were not discovered until the 1950s. These conditions are:

- a source of energy
- water
- a suitable nutrient medium, including mineral salts, vitamins, and growth regulators (spread 14.10)
- a suitable temperature (usually 15–30 °C)
- suitable light levels
- sterile conditions, which must be maintained until an independent plant is formed in order to exclude microorganisms that would find the culture medium an ideal place in which to grow.

The cultured cells divide by mitosis to form a **callus** (a hard mass of undifferentiated cells). By altering the balance of the growth regulators, cells in the callus can be made to differentiate into roots and shoots. The plantlet which develops is grown in another medium, usually agar based, to support the growing shoots, and then transferred to soil. The plants are grown in greenhouses under special environmental conditions until they are sufficiently robust for the market.

Some plants develop a well defined embryo-like structure during micropropagation. This structure is called a **somatic embryo** to distinguish it from the embryo in a seed. Somatic embryos have been encapsulated with nutrients in a polysaccharide gel to make artificial seeds.

By micropropagation, a single plant can produce thousands of clones, all genetically identical with each other and the parent plant. The commercial applications of micropropagation are already great: it is a viable alternative to conventional vegetative propagation for many plants. It allows enormous numbers of plants to be produced in a small space at a time convenient to the grower. It is an especially useful technique for propagating rare plants which cannot be propagated by traditional methods.

When combined with genetic engineering (spread 18.9), the potential uses of micropropagation are almost mind-boggling. However, micropropagation tempts growers to create **monocultures** (large areas of land with a single plant variety), giving an increased susceptibility to diseases and pests which can be transmitted rapidly from one plant to another. Also, some scientists fear that long-term micropropagation may result in genetically unstable plants, or in plants that are usually fertile becoming sterile. In the 1970s, micropropagated oil palms introduced into Malaysia turned out to be sterile. Although this particular problem has now been resolved, it is a warning that, in the future, we must use micropropagation with care.

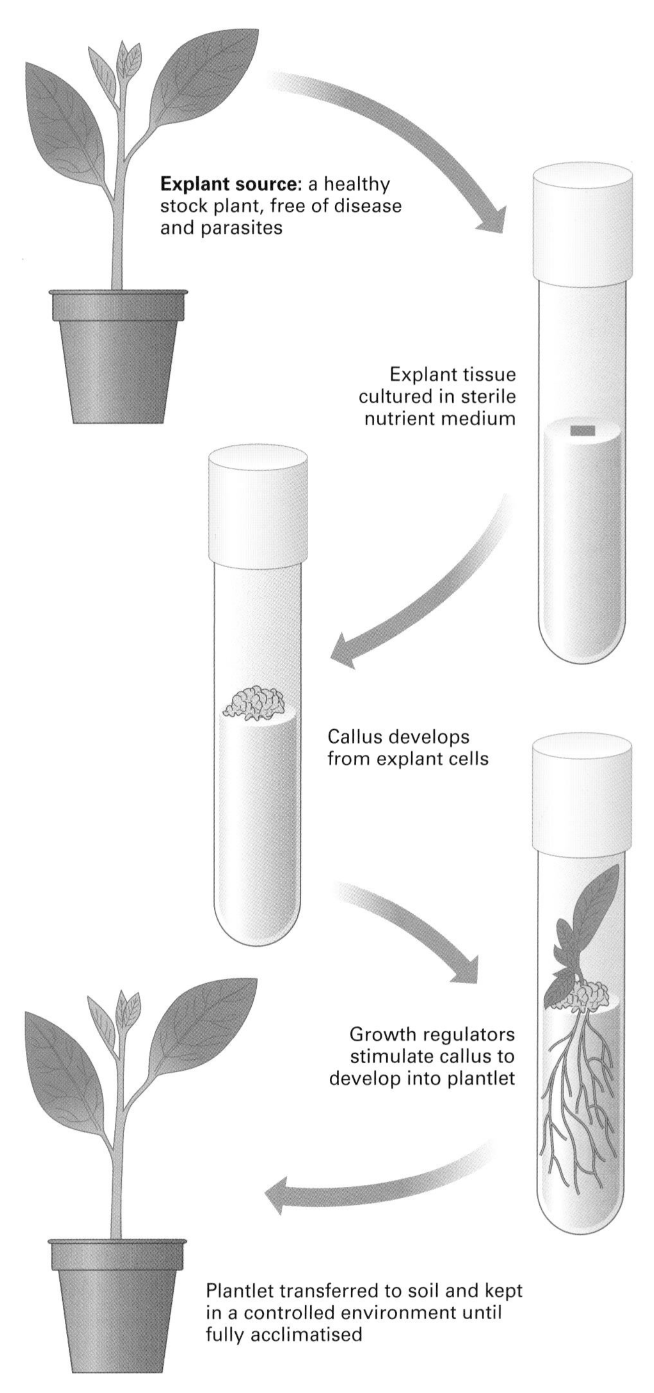

Figure 3 *An outline of micropropagation. The conditions vary for each type of explant propagated. The culture medium and environmental conditions are usually changed several times during the process. Extreme care is taken at all times to keep things sterile.*

Quick check

1 Give three advantages to agriculturists and horticulturists of propagating plants asexually.

2 What is a cutting?

3 What is totipotency, and why is it important for micropropagation?

Food for thought

Discuss the advantages and disadvantages of the agricultural practice of growing clones of a plant as monocultures. Suggest how the advantages could be maximised and the disadvantages minimised.

14.9

Plant growth

OBJECTIVES

By the end of this spread you should be able to:

- distinguish between primary growth and secondary growth
- explain how secondary growth takes place
- discuss the significance of secondary growth.

Fact of life

The oldest known plants are a group of bristlecone pines growing in the White Mountains of eastern California (figure 1), estimated to be over 4600 years old.

Figure 1 *Bristlecone pines of California, thought to be the oldest plants on Earth (see above).*

Differentiation of plant cells

Differentiation is the process by which distinct cell types are formed and become different from each other. A typical plant has about 12 basic cell types to carry out its daily functions. Differentiated cells do not usually divide unless induced to do so (for example, by micropropagation techniques, spread 14.8). However, some differentiated cells, such as parenchyma cells, can be transformed into another cell type given the right environmental or chemical signals. Others, such as xylem vessels which undergo cell death as a part of their differentiation, can never divide or be transformed.

Plants can grow to a great height. Giant redwoods (spread 13.4) tower above all other living organisms, reaching a height of over 100 m. Plants can also live to a ripe old age (figure 1). Plants grow mainly by producing new cells. After the earliest stages of development, the division of plant cells happens only in regions called **meristems**. There are two main types of meristem: apical meristems and lateral meristems. They give rise to two different kinds of growth: primary growth and secondary growth.

Primary growth: getting longer

Primary growth involves **apical meristems**, growth regions just behind the root tips and shoot tips. Primary growth is mainly concerned with making the plant longer. The roots elongate to penetrate the soil, and the shoots elongate to reach sunlight. The meristematic cells divide by mitosis (spread 4.11). Daughter cells may stay in the meristem, continuing to divide and increasing the number of cells in the shoot and root tip. However, a plant only gets longer when a daughter cell stops dividing and becomes elongated by taking in water. The cell wall is stretched and vacuoles form during the process. As the cell enlarges, the vacuoles may join to form one large central vacuole. After this growth by elongation, cells differentiate into their final specialised form. This may involve internal changes as well as the formation of a characteristic cell wall. The wall of a xylem cell, for example, may become spirally thickened with lignin.

Many herbaceous (non-woody) plants are formed only by primary growth. In general, because cell division and cell expansion happens only at the apical meristems, their size is limited because they cannot expand sideways (laterally). These plants also tend to be short lived because cells formed by primary growth usually have a life span of only 3–5 years. Some herbaceous plants called **annuals** complete their whole life cycle in a year. During unfavourable periods (winter for many plants), they survive only as seeds. As with many animals, the growth curves of annuals and many other herbaceous plants are usually sigmoid (S-shaped), and growth is limited (spread 12.11).

Secondary growth: getting wider

Trees are **perennial** plants, normally living for longer than two years. Trees can reach a great size and age because they can expand sideways by secondary growth as well as upwards by primary growth. **Secondary growth** or secondary thickening occurs in all trees and shrubs, and many herbaceous dicotyledons. The thickening of stems and roots originates in two **lateral meristems** called the **vascular cambium** and the **cork cambium**. Both these meristems contain undifferentiated cells that have retained their ability to divide by mitosis.

The **vascular cambium** is a cylindrical layer of actively dividing cells within and between the vascular bundles. After the cambium cells divide, they expand and differentiate. They form cylinders of secondary phloem towards the outside of the stem and of secondary xylem towards the inside of the stem (figure 2). As secondary growth continues, phloem cells are pushed outwards, collapse, and eventually die. But as they are continually replaced by new phloem, their death does not affect the vitality of the whole plant. Secondary xylem also changes with age: it becomes strengthened with lignin and cellulose to form dense woody tissue. **Wood** (secondary xylem), like the hard skeletons of land animals, is an evolutionary adaptation to life on land, providing the mechanical support that allows trees like the giant redwood to grow to such magnificent heights.

All the tissue external to the vascular bundle is called **bark**. It includes secondary phloem and also **cork cambium**, a layer of mitotically dividing cells which produces cork cells. The cork cambium is found just beneath the epidermis. Mature cork cells are dead and have thick walls impregnated with **suberin**. This is a waxy substance which makes the cork waterproof. Cork cells in the bark minimise water losses from the stem, and also protect the underlying living tissue from weathering and infection. As the stem thickens and the internal tissues expand, cork, cork cambium, and dead phloem tissue are pushed outwards and shed. New cork cambium develops from parenchyma cells in the living secondary phloem. Like moulting in arthropods, the shedding of cork during growth allows internal tissues to expand.

Because of their ability to thicken laterally, most large trees show **unlimited growth**; that is, they can grow continuously throughout their lives. They restrict flowering to some of their lateral branches, allowing growth to continue at the apical meristems of the rest of the branches. In contrast, annuals and **biennials** (plants with a two-year life cycle) die after flowering. Even if they did not die they would be unable to grow because all their apical meristems are converted into flowers.

Although growth in trees and shrubs is continuous, the growth rate of a plant may vary considerably. Generally, it is faster when the tree is young and when conditions are favourable, and slower as the tree reaches old age and when conditions are unfavourable. Large xylem elements are formed when conditions are favourable for fast growth (such as in the spring) and small ones are formed when growth slows (such as in the late summer and autumn). Consequently, variations of growth rate in a tree are recorded in the trunk as **growth rings** (figure 3). In non-seasonal regions, such as tropical rainforests, trees may not show annual growth rings.

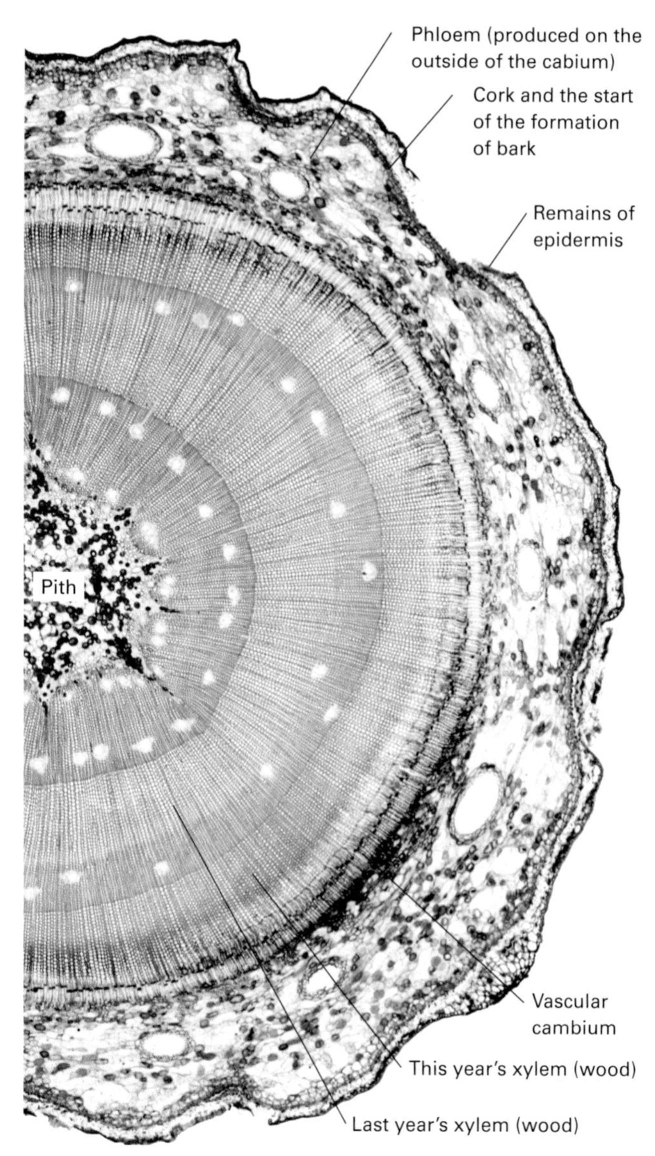

Figure 2 *Secondary growth of vascular cambium.*

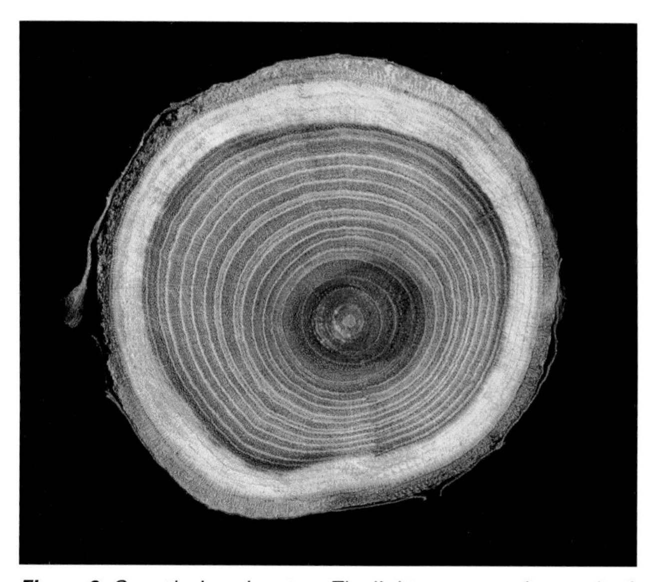

Figure 3 *Growth rings in a tree. The light areas are the result of larger xylem cells formed in the spring and summer. The dark areas result from tightly packed xylem formed in the autumn when growth is slower.*

Quick check

1. List the differences between primary growth and secondary growth.
2. What are the main meristematic regions of a tree?
3. Explain briefly what wood is and its major function.

Food for thought

The age of a living tree can be estimated by boring a hole into its trunk, penetrating to the heart of the tree, and extracting a core containing the sequence of growth rings. Why do you think this process is particularly difficult for the bristlecone pine trees (figure 1)? Suggest the inaccuracies involved in estimating their age by such a method.

14.10

Plant growth substances

OBJECTIVES

By the end of this spread you should be able to:

- discuss the similarities and differences between plant growth substances and animal hormones
- list five categories of plant growth substances and describe their main functions
- discuss the importance of interactions between plant growth substances.

Fact of life

The seedlings of rice plants infected with a fungus called *Gibberella fujikoroi* have been called 'foolish seedlings', because they appear to invest all their energy in growing tall rather than in producing seeds. They become so tall that they usually topple over under their own weight and die before reaching maturity. Infections of the fungus have threatened rice crops in Japan for centuries, so Japanese scientists have been keen to learn how to treat the disease. In the 1920s, they found that the fungus secreted chemicals into the rice plants. They named the chemicals gibberellins. They discovered that gibberellins occur naturally in plants and act as important growth promoters, but that infections of *G. fujikoroi* give rice an overdose of these chemicals.

***Figure 1** 'Foolish seedlings': the rice plants are infected with the fungus* Gibberella fujikoroi.

Plant growth substances are chemicals that occur naturally in plants at extremely low concentrations. They act as cell signals, regulating many aspects of plant growth, plant development, and plant responses to environmental stimuli (for example, in phototropism and geotropism; spread 14.11). There are five main groups of plant growth substances: **auxins**, **gibberellins**, **cytokinins**, **abscisic acid**, and **ethene** (ethylene).

Auxins

Indole-3-acetic acid (IAA) is the primary natural auxin. Others exist but their roles are unknown. IAA is synthesised mainly in the apical meristems. Its major effect is to promote primary growth by increasing the rate of cell elongation. IAA is also involved in many other plant responses, including **apical dominance**, in which relatively high concentrations of IAA in the stem apex prevent the development of lateral branches near the apex, resulting in a single long stem, and also growth responses to light and gravity (see spread 14.11). IAA interacts with ethene to bring about abscission (leaf fall): as levels of IAA decrease during the autumn, the abscission layer becomes more sensitive to ethene which triggers the production of hydrolytic enzymes such as cellulase that cause the leaf to become detached. Other processes in which IAA plays a key part are root development, and fruit growth and development. Synthetic auxins are used to accelerate root development in cuttings and as herbicides (see Food for thought).

Gibberellins

Gibberellins are another group of growth promoters. Their major role is to stimulate the growth of shoots (by promoting growth of the internodal regions) and leaves. Gibberellins are formed in young leaves around the growing tip and possibly in the roots of some plants. Like auxins, they promote cell elongation (see Fact of life). But unlike auxins, they can stimulate rapid growth in plants with certain inherited forms of dwarfism. The signalling pathway that gibberellins use to stimulate plant growth involves **DELLA proteins**, a group of gibberellin-sensitive factors that act as growth repressors. They act as **transcription blocking factors** that bind to DNA sites on genes that code for proteins essential for growth. By blocking transcription, the DNA is not transcribed to mRNA, proteins are not produced, and growth stops. In the presence of gibberellins the DELLA proteins normally break down, inhibition is removed, and growth is promoted. But a mutation of the DELLA genes can reduce the sensitivity of DELLA proteins to gibberellins and lead to dwarfism. **Dwarf wheats**, which have made an important contribution to increasing food production in the twentieth century, are the result of such a mutation. Gibberellins also play a role in the germination of some seeds. In barley seeds, for example, gibberellins stimulate the synthesis of alpha amylase. This enzyme breaks down starch to produce sugars for the germinating seed.

Cytokinins

Cytokinins are the third main group of growth promoters. They are synthesised in the roots and transported to other parts of the plant. When combined with auxins, they stimulate cell division. In some plants, cytokinins also seem to slow down **senescence** or ageing (the sequence of irreversible changes that eventually results in death).

Abscisic acid

In contrast to the three groups of plant growth promoters just described, abscisic acid (ABA) is a powerful growth inhibitor, often acting antagonistically to the growth promoters. It is synthesised in chloroplasts throughout the shoot but is particularly concentrated in the leaves, fruits, and seeds. ABA appears to promote dormancy in some seeds, and it stimulates the closing of stomata. The name abscisic acid derives from the belief that it was the direct cause of abscission (leaf fall). This is now known not to be the case.

Ethene and newly discovered substances

Ethene is a gas released from ripening fruits, nodes of stems, ageing leaves, and flowers. It is involved in seed dormancy, fruit ripening, and leaf abscission.

In recent years, new substances have been discovered that affect plant growth. They include brassinosteroids, which stimulate nutrient storage when conditions are favourable; salicylates; and jasmonates. No doubt, many more are waiting to be discovered.

The action of plant growth substances

Plant growth substances are sometimes called plant hormones because, like animal hormones, they are chemicals that carry information from one place to another, and they help regulate responses to environmental stimuli. However, unlike animal hormones, plant growth substances are not manufactured in special organs. Instead, they are made by cells in many different parts of the plant. It is often difficult to determine their precise sites of production because they occur in such small concentrations. Again, unlike animal hormones, plant growth substances are not always moved away from their site of production to their target cells. Ethene, for example, usually acts on the tissues from which it is released. Plant growth substances that are transported to their target cells from their site of production in another part of the plant move by mass flow in solution in phloem sap and in the transpiration stream of the xylem. They also move by diffusion and active transport.

Much confusion surrounds the effects of plant growth substances. An account in one textbook often appears to contradict that in another. Much of this confusion arises from the complex interactions between different plant growth substances. For example, **synergism** occurs when two or more plant growth substances interact to give a greater effect than the sum of their individual actions. In **antagonism**, on the other hand, two or more substances may interact to reduce each other's effects. ABA and gibberellins, for example, work antagonistically: ABA causes dormancy in buds and gibberellins break the dormancy. Some substances may even interact to give new effects.

The confusion is made worse by the fact that the effects of a plant growth substance depend on its concentration, or on the tissue being acted on, or on the developmental stage of the plant. For example, a high concentration of one hormone may inhibit the growth of roots but stimulate the growth of shoots. In addition, the effects often vary in different species. Ethene promotes the development of leaves, flowers, and roots in some species, but inhibits it in others.

In nature, a plant contains a cocktail of plant growth substances. It is not just one compound that produces a given response, but rather the balance of concentrations of various growth substances.

Quick check

1. How do plant growth substances differ from animal hormones?
2. List the five categories of plant growth substances and give one major effect on plants for each.
3. Distinguish between antagonistic and synergistic effects of two or more plant growth substances.

Food for thought

Synthetic auxins, such as 2,4-dichlorophenoxyacetic acid (2,4-D) and 2,4,5-trichlorophenoxyacetic acid (2,4,5-T) are commonly used as herbicides because of their differential effects on dicotyledons and monocotyledons. The herbicides contain a greater concentration of synthetic auxin than the amount of auxin naturally produced within a plant. Dicotyledons are killed by much lower concentrations than monocotyledons. Synthetic auxins are therefore used to control broad-leaved weeds (mainly dicotyledons) on lawns and in cereal crops. Although the exact mechanism of action is not known (it probably involves interference with DNA), the weedkiller causes dicotyledons to elongate so fast that they cannot sustain their own growth and they die. Study figure 2, and suggest which feature of dicotyledons might make them more sensitive to synthetic auxins.

***Figure 2** Synthetic auxins used to kill weeds in an area of grasses.*

14.11

Plant movements

OBJECTIVES

By the end of this spread you should be able to:

- distinguish between tropisms and nastic movements
- explain how phototropic responses are brought about in coleoptiles
- explain what gravitropic responses are.

Fact of life

Plant growth substances are present in such low concentrations that about 20 000 tonnes of coleoptile tips would be required to obtain one gram of indoleacetic acid (IAA).

Descriptions of large terrestrial plants moving about from one place to another are the stuff of science fiction. Nevertheless, plants do move in response to environmental stimuli, mainly by moving a body part.

Types of plant movement

There are many types of plant movement. Three types, described below, are hygroscopic movements, nastic movements, and sleep movements.

- **Hygroscopic movements** are responses to changes in moisture levels shown by certain non-living plant organs, such as fruits which explode when dry to disperse their seeds.
- **Nastic movements** are movements by plant organs in response to external stimuli such as touch or changes in temperature or light level. The movement is independent of the direction of the stimulus (if it has one). Nastic movements are usually brought about by differential growth (different parts of the plant growing at different rates). However, the quick curling response of the sensitive plant (*Mimosa*) to touch is caused by rapid changes in cell turgidity, possibly triggered by minute electrical currents similar to those in the nerves of animals (spread 10.4).
- **Sleep movements** are changes in position at night of leaves and petals for example. Sleep movements happen in circadian (daily) rhythms (spread 11.14), but they are usually triggered by changes in light level and temperature.

The most important plant movements are tropisms. A **tropism** is a movement of one part of the body in response to a stimulus. However, unlike nastic movements, the direction of movement depends on the direction of the stimulus. Most tropic movements are slow and they are usually brought about by differential growth. Growth of a body part towards a stimulus source is a **positive tropism**; growth of a body part away from a stimulus source is a **negative tropism**. Tropisms include **gravitropisms** (movements towards or away from gravity; figure 1) and phototropisms.

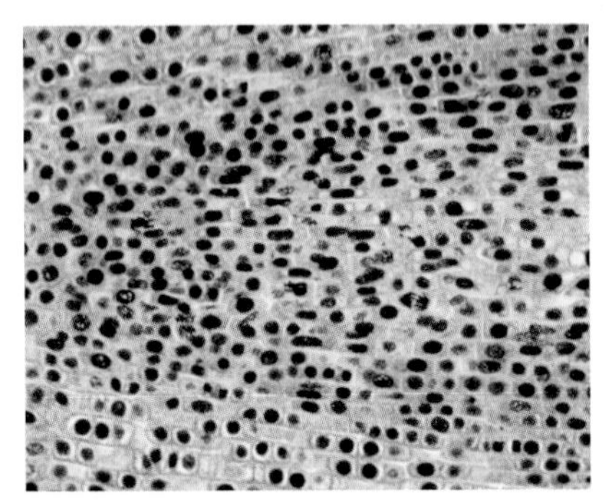

Figure 1 *Root tip cells (×300). Statocytes are root tip cells that store dense granules of starch called statoliths. These are thought to enable roots to respond positively to gravity. Gravity pulls the starch granules to the lowest point in the cell. The position of the granules may determine the distribution of plant growth substances in the root, causing the root to grow downwards. Auxins, abscisic acid, and gibberellins are probably involved in this positive geotropism.*

Phototropism

Phototropisms are plant growth movement in response to the direction of light. The stems of plants generally grow towards a directed light source. This **positive phototropism** is an adaptive response, directing shoots towards the sunlight they need for photosynthesis. Many roots show **negative phototropism**, bending away from a light source, but some roots are insensitive to light.

Figure 2 *Some early experiments carried out on coleoptiles.*

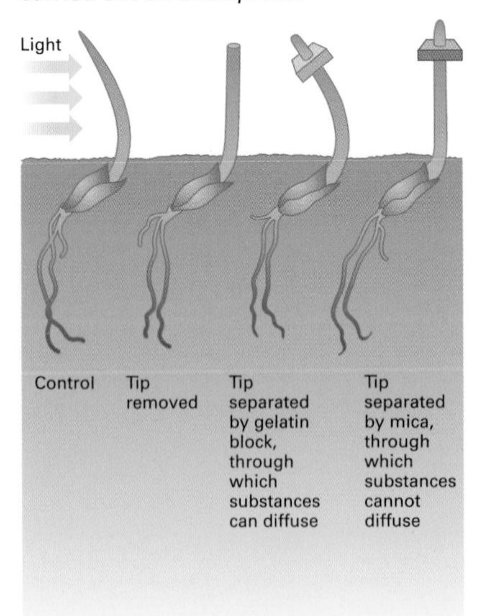

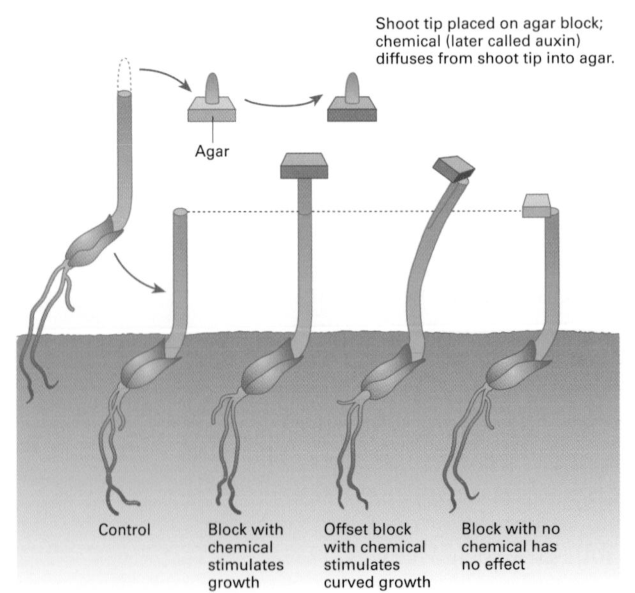

Figure 3 *Some of the experiments that show that phototropisms involve a chemical messenger. The agar blocks collect auxin which diffuses from the cut ends of shoot tips. By placing the blocks in various positions on coleoptiles with their tips cut off, it has been shown that the auxin stimulates cell growth. If more auxin is made to pass down one side of a coleoptile, that side will grow quicker than the other, causing the coleoptile to curve.*

Most of our knowledge about phototropisms has been gained from experiments on coleoptiles, the protective sheaths around the germinating shoots of grasses. Coleoptiles have been used because their response to light is easy to observe, they are small, and they are easy to grow in large numbers.

If a grass seedling is exposed to light from one side, it grows towards that light. Microscopic examination reveals that the cells on the shaded side of the coleoptile are significantly longer than those on the lit side. It seems that coleoptiles bend towards light by differential growth: cells on the shaded side elongate faster and become longer than those on the lit side. Early experiments were carried out by Charles Darwin. These involved whole coleoptiles, decapitated coleoptiles, and coleoptiles shaded in various ways. Peter Boysen-Jensen used mica to separate the tip of the coleoptile (figure 2). Later experiments carried out by Fritz Went used agar blocks to collect a chemical diffusing downwards from shoot tips (figure 3). The experiments showed that a receptor in the coleoptile tip receives information from the light source and regulates the release of a chemical messenger which carries information down the coleoptile to growing cells. Within limits, cell elongation increases with increased concentration of the chemical. In the 1930s, the chemical messenger responsible for phototropic responses was extracted and given the name **auxin** (from the Greek *auxein*, to increase). Auxin was later found to consist mainly of indole-3-acetic acid (IAA).

Coleoptile tips appear to produce equal amounts of auxin in the dark or the light, but its distribution can vary. It seems that when light strikes one side of a coleoptile, a receptor (probably a flavin, such as the coenzyme FAD derived from vitamin B_2, which absorbs light in the blue end of the spectrum) triggers the redistribution of auxin so that more travels down the shaded side of the coleoptile. The increased auxin concentration causes cells on the shaded side to grow longer than those on the lit side. Precisely how auxin stimulates cell elongation is unknown. One suggestion is that it triggers an increase in protein synthesis. Another suggestion is that auxin causes target cells to secrete protons (hydrogen ions) into their primary cell walls. The resulting increase in acidity is thought to weaken the bonds between cellulose microfibrils, allowing the cell wall to expand when the cell takes in water.

Care should be taken about generalising from these experiments because coleoptiles are formed only in the grass family; they do not exist in any dicotyledons. Some plants without coleoptiles appear to have a similar mechanism for phototropism, but others are believed to function differently.

Quick check

1 What is the main difference between nastic responses and tropic responses?

2 Tropic responses can be analysed in terms of the following:

Stimulus → receptor → transmission → effector → response

According to information in this spread, if a coleoptile is stimulated by a light source from one side and responds by bending towards that light source, what is:

a the receptor

b the effector of the response?

c How is transmission (communication) between receptor and effector achieved?

3 Which organelles in root cells are thought to form part of the receptor mechanism for gravitropisms?

Food for thought

To the relief of prospective space travellers, experiments conducted on seeds in space (mentioned in spread 14.6) yielded viable plants. But these 'extraterrestrial' organisms differed from Earth-bound relatives in one major respect: they did not grow straight; roots and shoots grew in all directions. Explain what might have caused this to happen and suggest ways of producing 'straight' plants in space.

14.12

OBJECTIVES

By the end of this spread you should be able to:

- describe current views about the photoperiodic control of flowering
- discuss the nature and function of phytochromes.

Fact of life

By controlling the length of uninterrupted darkness in greenhouses, horticulturalists can guarantee a supply of flowers such as chrysanthemums at any time of the year (figure 1).

PHOTOPERIODISM

Many of the activities of plants occur at specific times of the day or year. The sensitive plant (*Mimosa*) curls its leaves inwards every night, which minimises transpiration losses. Sleep movements such as this happen rhythmically even if the plant is kept under constant conditions of light and temperature. Flowers of insect-pollinated plants open at particular times of the day which coincide with the foraging activities of their insect pollinators. Plants in temperate regions of the world germinate and flower at certain times of the year and take advantage of particular environmental conditions.

All plants have a **biological clock** which enables them to anticipate certain events, rather than to be controlled by them. In animals, biological clocks are clusters of nerve cells located in specific parts of the body (spread 11.14). Plants have no nervous tissue, so the mechanism of their biological clock must be different. Furthermore, plants seem to have more than one mechanism to control different processes. In this spread, we are going to examine how plants use rhythmic changes in light to detect the time of year.

The relative length of day and night varies with the time of year and is called the **photoperiod**. The response of a plant to changes in day length or night length is called **photoperiodism**. One of the most studied examples is flowering.

Photoperiodic control of flowering

Plants whose flowering is triggered by the photoperiod are often placed into two main groups. **Short-day plants**, such as chrysanthemums and strawberries, generally flower in late summer, autumn, winter, or early

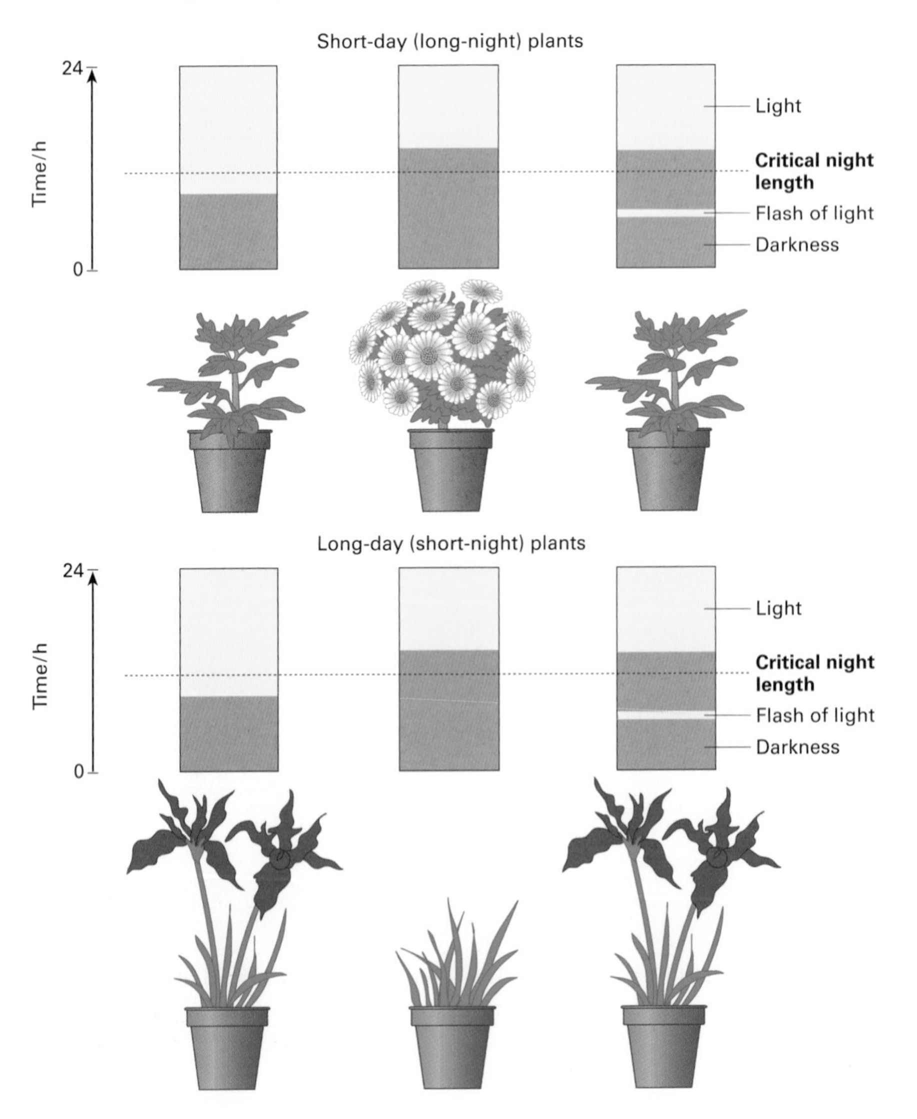

Figure 1 *These experiments show that it is the length of uninterrupted darkness at night that controls flowering, rather than the duration of exposure to light.*

spring when light periods are short. **Long-day plants**, such as lettuce, clover, and many cereals, usually flower in late spring or early summer, when the days become longer and the nights shorter. It should be borne in mind that the flowering times of different species are very variable and the two groups merge into each other. Moreover, the flowering of some plants, including tomatoes and cucumber, appear to be unaffected by the photoperiod. These plants are called **day-neutral plants**.

Experiments conducted since the 1940s (figure 1) demonstrate that the photoperiodic control of flowering depends on the length of unbroken darkness at night, not the length of exposure to the sun during the day. For example, if short-day plants have their long night interrupted by flashes of light, they do not flower. (This is one way that growers delay flowering of, for example, poinsettias and chrysanthemums, so that they are available at Christmas time.) Therefore, short-day plants should be called long-night plants, and long-day plants should be called short-night plants. However, the former names became so well established in botany textbooks before the 1940s that they have been retained.

Phytochrome and the detection of night length

Photoperiodism requires a mechanism for detecting the length of darkness. The best-known candidate for this job is a pale-blue light-sensitive protein called **phytochrome**. It exists in two interconvertible forms: P_r, which absorbs red light (light with a wavelength of 660 nm), and P_{fr}, which absorbs far-red light (light with a wavelength of 730 nm). Phytochrome is synthesised in its P_r form. Experiments with specific wavelengths of light show that P_r is converted into P_{fr} when exposed to red light, and that P_{fr} is converted back into P_r when exposed to far-red light or when in darkness. During the day, there is more P_{fr} than P_r because sunlight has a higher proportion of red light than far-red light. During the darkness of night, P_{fr} is gradually converted back to P_r (figure 2). Thus the relative amounts of the two forms of phytochrome could indicate day length. It is thought that the balance between the two forms of phytochrome controls flowering in short-day and long-day plants. A build-up of P_{fr} could stimulate flowering in long-day plants, whereas in short-day plants, P_{fr} could inhibit flowering. The above description of how phytochromes affect flowering is rather vague because no one is certain about the mechanism.

If phytochromes are involved, their detection of the photoperiod takes place in the leaves. If only one leaf of a short-day or long-day plant is exposed to normal daily routines of light and darkness and the rest of the plant including all the flower buds is kept in constant darkness, the plant will still flower normally. As the flowering response takes place in the flower buds, the plant must have some means of transmitting information from leaves to flower buds. A chemical messenger called **florigen** has been suggested. The evidence for a chemical messenger is compelling. If, for example, two plants with severed stems are grafted together at one stem, and one plant is exposed to the appropriate photoperiod to induce flowering while the other is not, both plants flower. Sometimes this works even if one plant is a short-day and the other a long-day plant.

If florigen exists, it might be secreted in response to the relative amounts of the two forms of phytochrome in the leaves, and then carried in the vascular system to the flower buds. In long-day plants, high levels of P_{fr} or low levels of P_r may stimulate florigen secretion, whereas in short-day plants, florigen may be secreted when P_{fr} levels are low or when P_r levels are high. However, until florigen has been isolated, these suggestions remain highly speculative.

Phytochrome has been found in all species of plants and algae examined to date. In addition to its probable role in flowering, it may also play an important part in other physiological responses, such as phototropisms, germination, and stomatal opening.

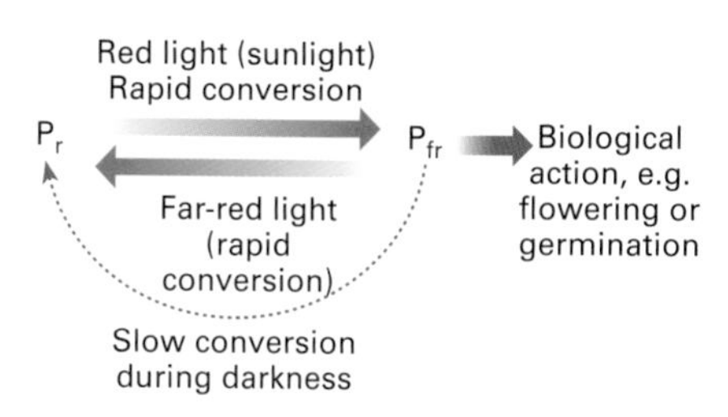

Figure 2 The interconversions between the two forms of phytochrome, P_r and P_{fr}.

Figure 3 Small woodland plants, such as this wild strawberry (Fragaria vesca), are often short-day plants.

Quick check

1 What is the photoperiod?

2 Give the possible role of phytochrome in the photoperiodic control of flowering.

3 Which form of phytochrome builds up in leaves exposed to light during the day?

Food for thought

Plants belonging to the same species often have very specific flowering times. Small flowering plants such as wild strawberries (figure 3) living in temperate woodlands are often short-day plants. Suggest advantages to the plant of evolving this flowering strategy.

Summary

A typical **dicotyledonous flower** has **sepals**, **petals**, **stamens**, and **carpels**. **Gametogenesis** occurs in **pollen grains** produced in **anthers** and **embryo sacs** produced in **ovules**. **Pollination** may be by wind, water, or animals. **Coevolution** has resulted in some insects pollinating specific plants and a complete interdependence between insect and plant. Many plants have adaptations which reduce the chance of **self pollination** and increase the chance of **cross pollination**. Flowering plants are unique in having **double fertilisation**. After **fertilisation**, a **triploid nucleus** develops into the **endosperm** and the **diploid nucleus** of the **zygote** develops into the **embryo** within the **seed**. **Seeds** develop from the **ovules**, and the **fruits** from mature **ovaries**. Seeds act as the **dispersal phase** in the life of the plant and may remain **dormant** until environmental conditions are suitable for germination. **Seed germination** starts with an intake of water followed by activation of **hydrolytic enzymes**. Broad beans have **hypogeal germination** whereas sunflowers have **epigeal germination**.

Flowering plants reproduce asexually by **vegetative propagation**. In many plants, organs involved in vegetative propagation are also **perennating organs**. Vegetative propagation is achieved commercially, for example, by taking **cuttings**, making **grafts**, or by **micropropagation** (tissue culture or cloning).
Plant growth occurs in **meristems**. **Primary growth** involves **apical meristems** whereas **secondary growth** or **secondary thickening** occurs in **lateral meristems**. Plant growth and development are regulated by **plant growth substances**, mainly **auxins**, **gibberellins**, **cytokinins**, **abscisic acid**, and **ethene**. Plant growth substances are also involved in plant movements. Phototropisms, for example, are brought about by **auxins**. **Flowering** may or may not be regulated by the **photoperiod**: **short-day plants** flower when the period of uninterrupted darkness is long whereas **long-day plants** flower when the period of uninterrupted darkness is short. Flowering in **day-neutral plants** is unaffected by the photoperiod. Photoperiodic control of flowering involves two interconvertible forms of **phytochrome** and, possibly, **florigen**.

PRACTICE EXAM QUESTIONS

1 The figure shows the life cycle of a flowering plant.

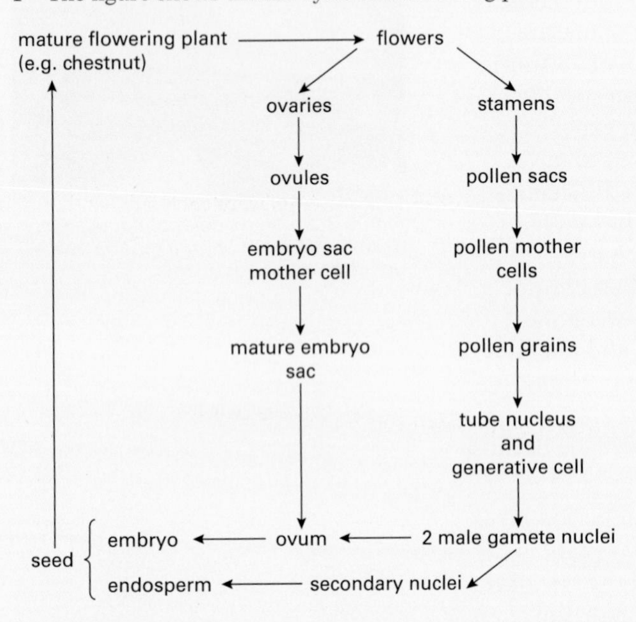

a Which stage(s) in the life cycle shown in the figure are the:

i sporophyte? [1]

ii gametophytes? [2]

iii spores? [3]

b How does this life cycle show alternation of generations? [2]

c On the figure above, at which points does meiosis occur? [2]

d In this life cycle there is said to be 'double fertilisation'. What is meant by *double fertilisation?* [2]

e Name **two** organisms which exhibit alternation of generations in which the gametophyte generation is dominant. [2]

[Total 14 marks]

2 The diagram below shows a germinating pollen grain and a mature ovule from a flower of the Papilionaceae. Some nuclei have been labelled.

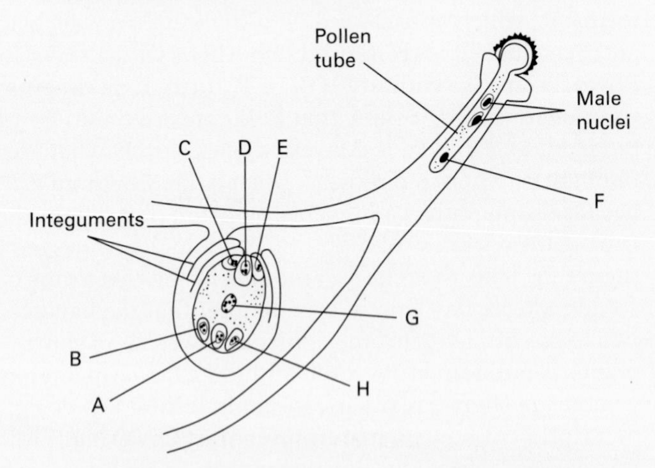

a Describe how pollination usually occurs in the Papilionaccae. [2]

b Give the letter of the nucleus which fuses with a male nucleus to form each of the following:
The zygote
The endosperm. [2]

c Describe one mechanism which prevents self-fertilisation in flowering plants. [2]

[Total 6 marks]

3 Some barley grains were soaked in water and allowed to germinate. The graph on the next page shows changes that took place in amylase activity in these germinating barley grains during the following 12 days.

a Calculate the rate of increase of amylase activity in the germinating barley grains between day 2 and day 6. Show your working. [2]

b Describe and explain the function of amylase in a germinating barley grain. [3]

c Suggest reasons for the change in amylase activity between day 0 and day 8. [4]

d The rate of amylase activity falls between day 8 and day 12. Suggest why a high rate of amylase activity is not required at this stage of growth. [2]

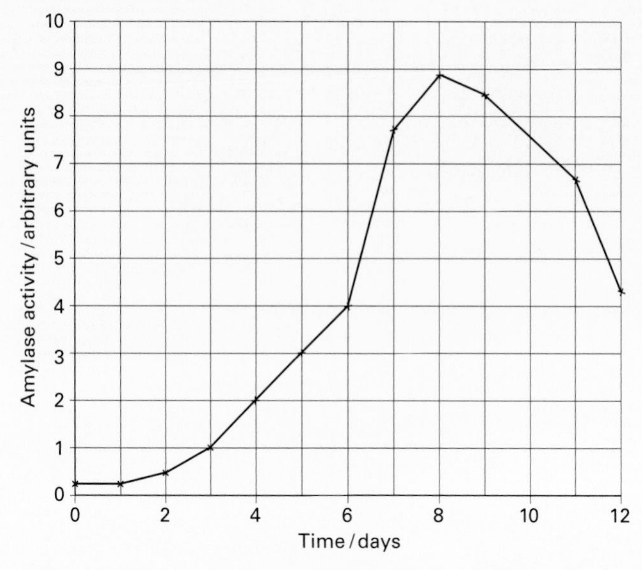

[Total 11 marks]

4 The diagram shows how the movement of auxin through plant shoots was investigated. Shoots were removed from young plants and each was placed on a block of agar. A second block of agar was placed in the side of each shoot where a portion of its tissue had been cut away. Samples of shoots treated in this way were left for several hours. Half the shoots were left in darkness and half were left with a light source from one side.
The concentration of auxin collected in the two blocks of agar from each shoot was measured. The table shows the mean percentage of auxin found in the two blocks of agar from each shoot.

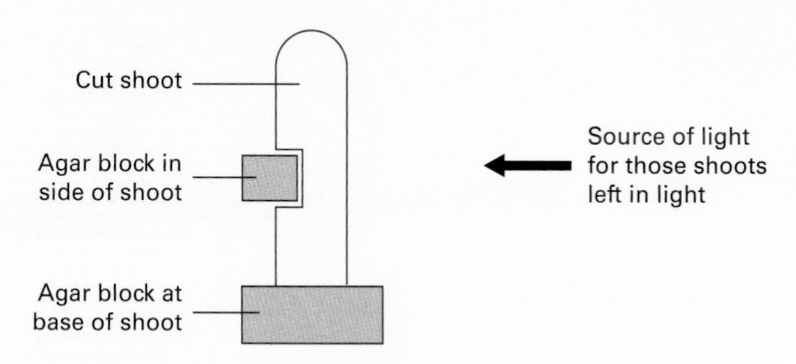

Location of block	Mean percentage of auxin collected in the two agar blocks from each shoot	
	Shoot in dark	Shoot lit from side
Side of shoot	27	35
Base of shoot	73	65
Both blocks	100	100

a What do the results of this investigation show about the movement of auxin? [2]

b Explain how the movement of auxin shown in this experiment enables intact plants to photosynthesise efficiently. [3]

[Total 5 marks]

5 The graph shows the response of oat seedlings to various concentrations of externally applied auxin.

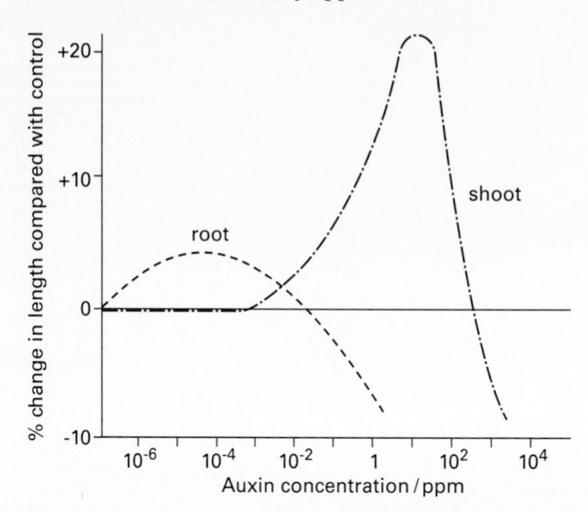

a Give two differences between the responses of shoots and roots shown by the graph. [2]

b Suggest a suitable control of this investigation. [1]

c Describe how the action of auxin on cells of the shoot promotes growth. [2]

d Explain why auxins may be used as selective weedkillers in cereal crops. [2]

[Total 7 marks]

6 Two groups of short day plants were each subjected to one of two different treatments of light and dark periods as shown in the diagram.

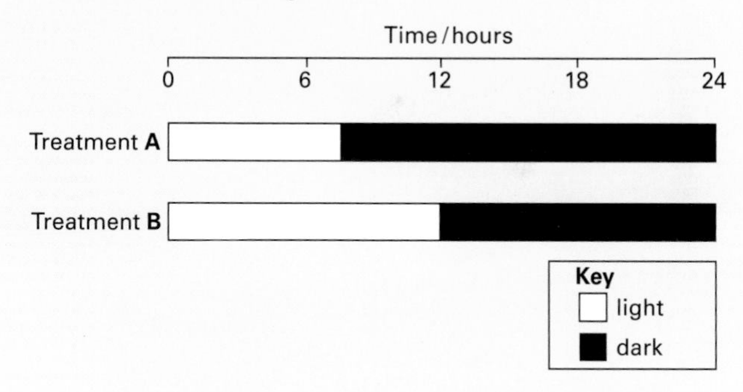

a Which treatment would you expect to trigger a flowering response? Give a reason for your answer. [2]

b In a second series of treatments, two groups of short day plants were each subjected to long dark periods, as in Treatment A, interrupted by a short period of light.

One group was exposed to a short period of red light and the second group exposed to a short period of far-red light.
What flowering response would you expect for each group? Give a reason for your answer in terms of phytochrome conversions.

i red light interruption

ii far-red interruption. [2]

[Total 4 marks]

HEALTH *and* DISEASE

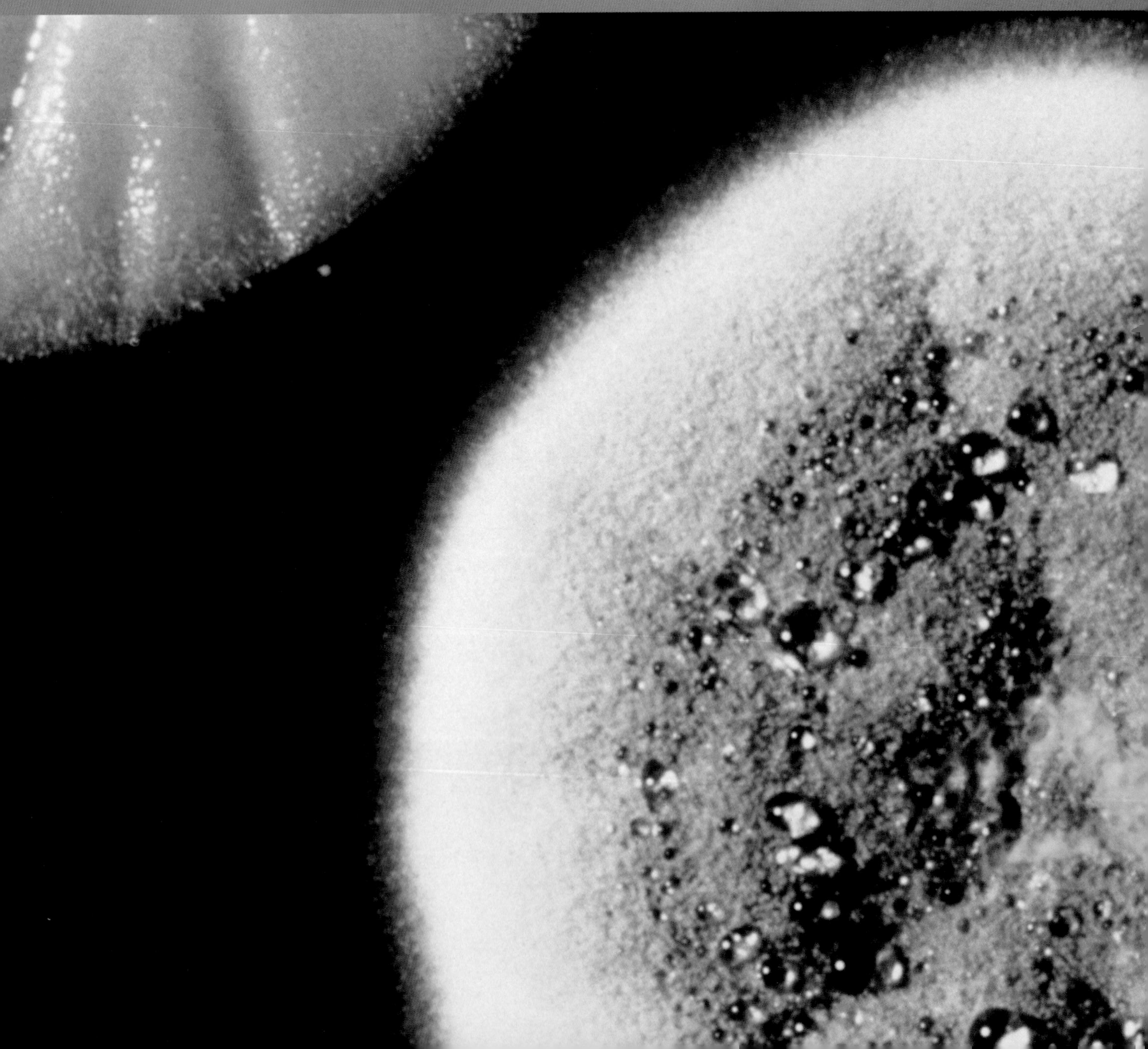

Look to your health; and if you have it, praise God, and value it next to a good conscience; for health is the second blessing that we mortals are capable of; a blessing that money cannot buy.

Izaak Walton (1593–1683)

Soldiers have rarely won wars. They more often mop up after the barrage of epidemics. And typhus, with its brothers and sisters – plague, cholera, typhoid and dysentery – has decided more campaigns than Caesar, Hannibal, Napoleon and all the generals of history.

Hans Zinsser

This section of the book deals with a wide range of diseases, both infectious and non-infectious. You will learn that although some microorganisms cause disease, others are used in its treatment or as sources of beneficial products. You will also learn how the human body responds to disease, how your lifestyle can affect your chances of contracting a disease, and how diseases can be treated.

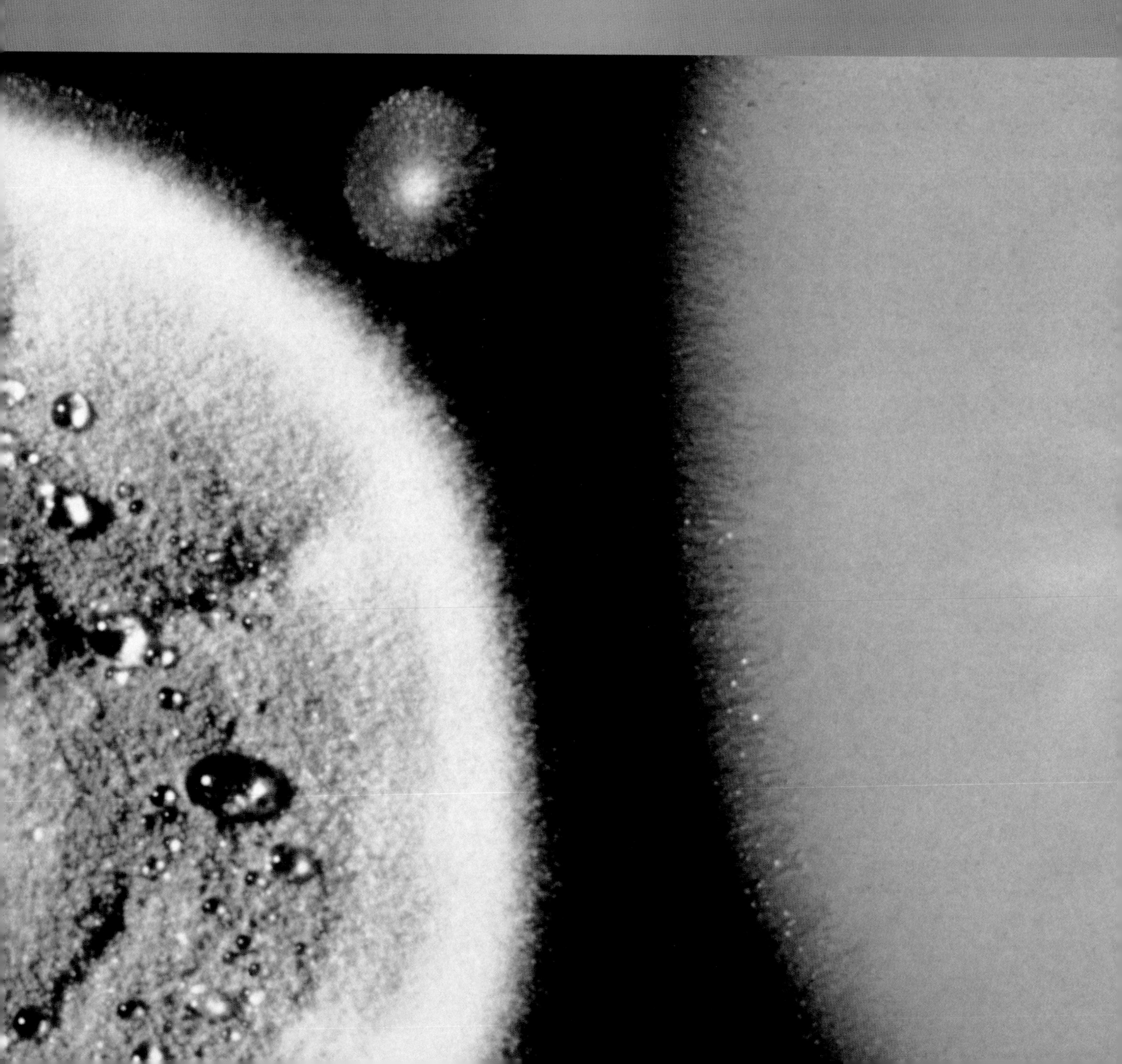

The mould growing on the petri dish is Penicillium, *from which the antibiotic penicillin is obtained. Penicillin, first used in 1943, revolutionised the treatment of disease.*

15 Infectious diseases

Human health and disease

15.1

OBJECTIVES

By the end of this spread you should be able to:

- discuss what is meant by health and disease
- identify different categories of disease and give an example of each
- discuss how global patterns of disease are studied.

Fact of life

During a sneeze, enormous numbers of moist droplets are propelled into the air (figure 1). Each droplet is about 10 μm in diameter and shoots out at about 100 m s^{-1} (more than 200 m.p.h.). Droplet nuclei (small droplets less than 4 μm in diameter, formed by evaporation of the initial droplets) can remain airborne for hours or even days and can travel long distances. Airborne droplets can carry pathogens that may be inhaled and cause infection in a susceptible person (table 1). The blast from an unstifled sneeze can be so powerful that even a surgical mask may not prevent inhalation.

What is health?

Good health is more than just freedom from disease and infirmity. The term includes mental and social as well as physical well-being. A healthy person is able to use all of his or her mental, spiritual, and physical resources to maximise the chance of survival, to live a happy and fulfilling life, and to help dependants and society. Lifestyle can have a great effect on human health, particularly with regard to sexually transmitted diseases such as AIDS and non-infectious diseases such as diabetes (spread 8.3), cancer (spreads 16.5 and 16.6), and coronary heart disease (spreads 16.7 and 16.8). Changes in lifestyle, for example by drinking less alcohol, eating a more balanced diet, stopping smoking and taking more exercise, can reduce the risk of contracting these diseases.

Types of disease

A **disease** is often defined medically as any physical or mental disorder or malfunction with a characteristic set of signs and symptoms. This excludes disorders resulting from physical trauma (for example, a leg broken during a football match). **Signs** are indications of a disease that can be observed by a doctor. **Symptoms** are indications of disease perceived by a patient.

There are many different ways of classifying diseases. The classification system in table 2 places diseases into eight categories. Some authorities also include a ninth category, **self-inflicted diseases**, for diseases such as anorexia and alcoholism that appear to be self induced. Diseases may also be classified simply into single-factor diseases or multifactorial diseases, according to the number of causes:

- **Single-factor diseases** have a single cause and usually involve a specific organ or tissue.
- **Multifactorial diseases**, such as those of the heart and blood vessels, have many causes.

Another simple classification is as acute and chronic diseases:

- **Acute diseases** develop rapidly and usually last a short time.
- **Chronic diseases** are usually slow to develop and last a long time.

Table 1 *Some diseases that can be transmitted in airborne droplets.*

Pathogen	Type of microorganism	Disease
Varicella	Viruses	Chickenpox
Influenza virus		Influenza ('flu)
Morbillivirus (or Rubeola)		Measles
Rubella		German measles
Poliomyelitis		Polio
Bordetella pertussis	Bacteria	Whooping cough
Cornyebacterium diphtheriae		Diphtheria
Mycobacterium tuberculosis		Tuberculosis (TB)
Candida spp.	Fungus	Fungal infections, including thrush

Figure 1 *'Coughs and sneezes spread diseases'.*

Studying patterns of disease: epidemiology

The study of patterns of disease and of the various factors that affect the spread of disease is called **epidemiology**.

Epidemiologists have been described as disease detectives. They try to discover the factors that cause a disease and to develop methods to prevent its spread. The main clues they use come from data about the number of people in a particular area (such as a city, a region of a country, or a whole country) affected by specific diseases, and the number who have died. The data are commonly expressed as incidence or morbidity rates, and mortality rates.

The **incidence** or **morbidity** rate is the number of new cases of a disease in a given population occurring during a specific period (a week, month, or year). It is calculated as:

$$\text{Incidence or morbidity rate} = \frac{\text{number of new cases of a given disease}}{\text{number of individuals in the population}}$$

To find how many cases of a disease are new, this calculation requires information about the **prevalence rate**. This is the total number of individuals infected in a population at any one time, no matter when the disease began. Information about prevalence rates is very useful in its own right.

The **mortality rate** of a disease may be estimated for a whole population irrespective of whether they have the disease or not:

$$\text{Mortality rate} = \frac{\text{number of deaths due to a given disease}}{\text{number of individuals in the population}}$$

Alternatively, it may be calculated using only those people who have the disease:

$$\text{Mortality rate} = \frac{\text{number of deaths due to a given disease}}{\text{number of people with the same disease}}$$

Epidemiological studies are used to identify whether a disease is endemic, epidemic, or pandemic:

- An **endemic** disease is always present in a population.
- An **epidemic** is a disease that spreads rapidly, suddenly, and unexpectedly to affect many people. The term does not include expected rises, for example, the increased incidence of influenza that happens every winter in temperate regions.
- A **pandemic** disease affects people over very large areas, such as a continent or even the whole world. AIDS and TB are pandemic at present.

Epidemiological information helps public health workers direct health care efforts effectively to control the spread of infectious diseases. For example, a sudden increase in the morbidity rate for a disease in a particular region may warn that immediate action is needed to prevent a future rise in the number of deaths from the disease. Effective and quick action may also prevent the disease spreading to other areas.

***Table 2** Classification of diseases into eight categories. Note that a disease may belong to more than one category: for example, sickle-cell anaemia is a non-infectious disease and an inherited disease.*

Category	Examples	Comments
Infectious disease	Cholera; malaria	A disease caused by an invading organism or virus transmitted from person to person
Non-infectious disease	Stroke; sickle-cell anaemia	Any disease that cannot be transmitted from one person to another
Inherited disease	Cystic fibrosis; sickle-cell anaemia; haemophilia	A disease caused by a genetic fault that may be passed from parents to children
Degenerative disease	Arthritis	A gradual decline in function, often associated with ageing
Social disease	Alcoholism	Drug dependence, often induced by social pressures and social behaviour
Mental illness	Schizophrenia; anorexia	Disorder of the mind which may or may not have a physical or chemical cause
Eating disorder	Anorexia; obesity	A disease caused by undereating or overeating
Deficiency disease	Scurvy; rickets	A disease caused by a poor diet lacking one or more essential nutrients (spread 16.10)

Quick check

1 Explain in what ways health is not just the absence of disease.

2 Which category of disease is transmitted from person to person?

3 Distinguish between morbidity and mortality.

Food for thought

Epidemiological studies have revealed great differences in the prevalence of certain diseases between economically well developed countries and countries that are less developed. Which categories of disease do you think are more prevalent in the well developed countries, and which in the less developed countries? Suggest reasons for these differences.

15.2

SMALLPOX AND AIDS

OBJECTIVES

By the end of this spread you should be able to:

- describe the nature of smallpox and explain how it has been eradicated
- describe the cause of AIDS
- explain how AIDS is transmitted
- discuss how the spread of AIDS can be controlled.

Fact of life

In 2011, 1.7 million people died of AIDS/HIV worldwide and an estimated 34.2 million people were living with the disease.

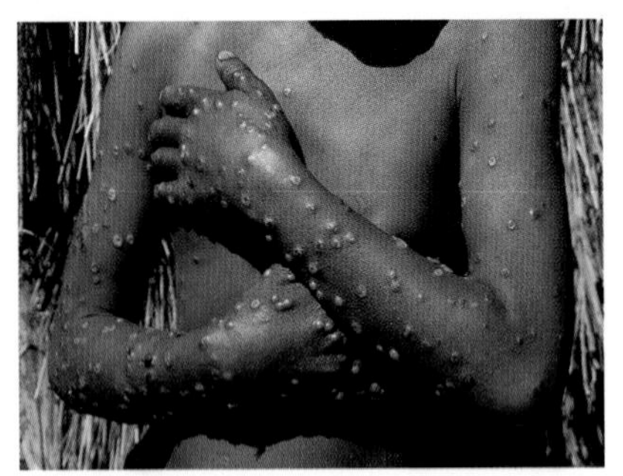

Figure 1 *The arms of a person infected with smallpox.*

The eradication of smallpox

One of the great success stories of twentieth-century medicine has been the eradication of **smallpox** as a disease. Smallpox was a horrible disfiguring viral disease which, until the 1960s, affected more than 15 million people annually in 33 different countries. In 1956 the WHO launched a campaign to wipe it out. Large populations were vaccinated and people with the disease were isolated. Everyone within a certain area of an outbreak (a spontaneous occurence of the disease) was vaccinated in a process called **ring vaccination**. In 1977, the last case of smallpox was reported in Somalia. At the time of writing the virus still exists locked up in laboratories in the USA and Russia.

Smallpox was an acute, highly infectious disease transmitted by direct contact. Obvious symptoms of the disease were red spots on the face, trunk, and extremities that changed to pea-sized blisters and became filled with pus (pustules) (figure 1). The pustules were so excruciatingly irritating that sufferers had to be prevented from tearing at their flesh. About 20–30 per cent of those with the disease died; many who recovered were blinded or permanently disfigured.

The eradication of smallpox was successful because:

- The smallpox virus is stable (see spread 17.1 for details of viruses). During the eradication programme it did not mutate, so the same vaccine could be used throughout the programme.
- The smallpox virus does not linger in the body after infection, nor does it infect other animals, so it cannot remain hidden anywhere.
- The living but non-virulent vaccine used was highly effective and easy to administer by a scratching technique (see spread 15.7 for an account of vaccination).
- It was easy to identify people with the disease, so isolation and ring vaccination were effective.

With the concerted efforts of many nations and a great deal of money, the fight against smallpox appears to have been won, but that against AIDS is still raging.

AIDS (acquired immune deficiency syndrome)

AIDS is characterised by a suppression of the immune system leading to the development of a number of rare infectious diseases. It is caused by a virus known as **HIV** (**human immunodeficiency virus**). The disease was first identified in Los Angeles in 1981, but probably originated in central Africa in the 1950s. It is believed that an HIV-related virus in African green monkeys somehow entered humans, mutated, and evolved into HIV. Since the discovery of the virus, HIV and AIDS have spread across the world to become a scourge of our times (figure 2). It is estimated that between 1990 and 2010, the number of people infected with HIV rose from 8 million to about 34 million. Since the beginning of the epidemic, about 30 million people have died from AIDS-related diseases.

HIV is transmitted mainly sexually: homosexually or heterosexually. The virus can pass from infected semen or vaginal fluids to the blood of the new host through damaged tissue in the rectum, vagina, or penis. It can also be transmitted via infected blood or blood products (for example, through transfusion of infected blood, or through contaminated needles used by drug takers), and from an infected mother to her child across the placenta or as a result of breast-feeding. In 2009, there were about 2.5 million children living with HIV/AIDS; most were infected from their mothers. Fortunately, HIV is a fragile virus and does not survive well outside the body. Therefore, ordinary social contact with HIV-positive people presents no risk of infection.

The action of HIV

HIV is potentially deadly because it attacks particular white blood cells (**helper T cells**) which are essential components of the body's immune system (spread 15.5). HIV is a retrovirus which incorporates its genetic material into that of its host cell (spread 17.1). HIV may remain dormant in a cell, with its DNA being replicated each time the host cell divides. At this stage, a person with HIV may be symptomless. This condition may last for several or even many years. However, when active, HIV uses the host's cellular machinery to manufacture new viruses. The new viruses burst out of the host cell and eventually kill it. Then they find new host cells to infect. When this happens, the body's immune system is suppressed. The HIV patient develops full-blown AIDS and may succumb to a number of diseases (such as thrush, pneumonia, and a rare type of skin cancer called Kaposi's sarcoma) that would normally be fought off by the immune system. Death may result from one of these opportunistic diseases (usually cancer or pneumonia) or from a pre-existing disease, such as tuberculosis. There is no known cure for AIDS: modern drugs slow its progress, but cannot stop it.

There are three main reasons why HIV is proving such a difficult enemy to defeat. First, it destroys helper T cells, the very cells that should be defending the body against it (figure 3). Secondly, HIV can remain hidden in cells for months or years. While inactive, it cannot be targeted and destroyed. Thirdly, HIV is extraordinarily variable. Cells in our defence system identify infective agents by the shapes of antigens on their protein coats. HIV escapes detection by frequently changing the shape of its antigens.

Although there is no known cure, antiretroviral therapy is bringing about a steady decline in the number of new infections and reducing significantly the number of AIDS-related deaths. Anteretroviral therapy is a special form of chemotherapy in which the patient takes a combination of two or more antiretroviral drugs, including **reverse transcriptase inhibitors** and **protease inhibitors**, designed to keep the amount of HIV in the body low.

Education programmes are also contributing to the decline by informing people about the risks of unprotected sex and the sharing of needles. People who have sex with different partners are told that wearing condoms reduces the risk of infection but does not eliminate it completely. Only abstinence or completely monogamous relationships between uninfected individuals are completely safe. Partners in a new relationship can check their HIV status by having a blood test. This usually involves screening the blood for particular antibodies. Those with the virus (who are HIV positive) should take great care not to spread the disease.

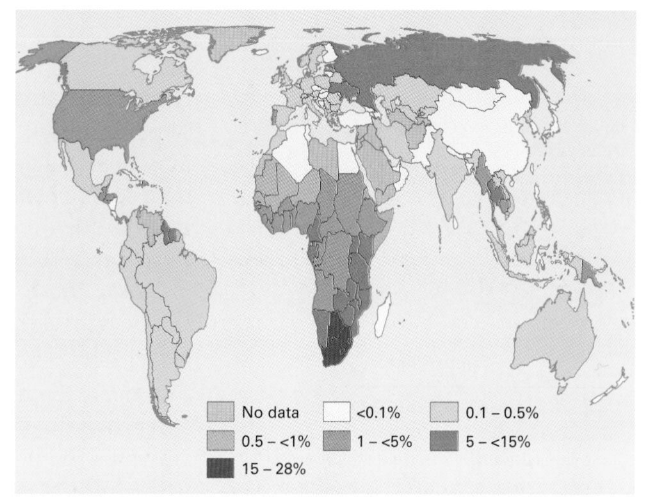

Figure 2 *Global distribution of AIDS (source: Global Report: UNAID report on the Global AIDS epidemic 2010).*

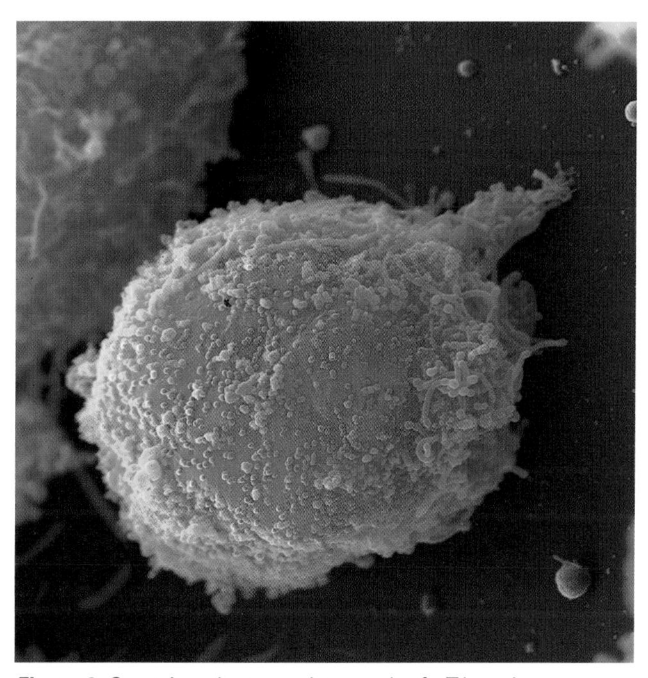

Figure 3 *Scanning electron micrograph of a T-lymphocyte white blood cell (orange) infected with HIV (×6000). In full-blown AIDS, the viruses (blue dots) replicate continuously and eventually kill the host cell.*

QUICK CHECK

1. Explain what is meant by ring vaccination.
2. What is the difference between HIV and AIDS?
3. State the main ways in which HIV is transmitted.
4. List the reasons given in this spread why AIDS is more difficult to eradicate than smallpox.

Food for thought

The smallpox virus is not extinct: it still exists in laboratories. Suggest arguments in favour of:

a retaining the virus

b eliminating it altogether.

15.3

Bacterial diseases

OBJECTIVES

By the end of this spread you should be able to describe the causes, mode of transmission, and control of:

- salmonellosis
- cholera
- tuberculosis (TB).

Fact of life

Robert Koch (1843–1910), a German doctor, was the first to demonstrate the role of bacteria in disease when he infected healthy mice with anthrax by injecting them with material that contained the disease.

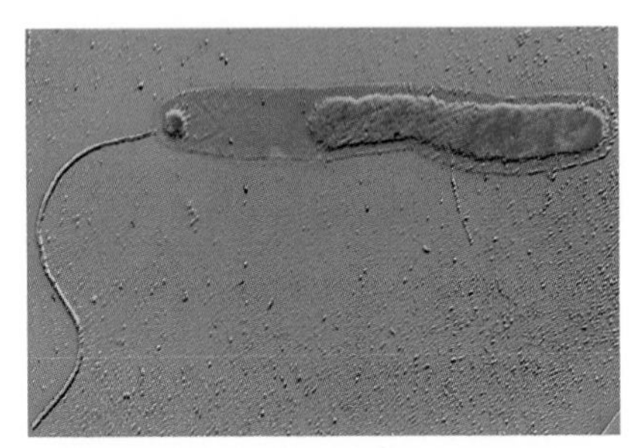

Figure 1 False-colour scanning electron micrograph of Vibrio cholerae *(×50 000), a small comma-shaped bacterium with a single flagellum, which causes cholera.*

Oral rehydration solution

Oral rehydration solution (ORS) is an aqueous solution of electrolytes (mineral salts) and glucose given as a drink to patients severely dehydrated by vomiting or diarrhoea. Oral rehydration therapy takes advantage of the sodium–glucose carrier proteins found in the columnar epithelial cells lining the small intestine. When both glucose and sodium are present, they bind to the protein and are co-transported into the cytoplasm. Absorption of water follows by osmosis. In addition to water and glucose, the ORS contains sodium, potassium, and bicarbonate salts to replace those lost by diarrhoea or vomiting. Although simple and cheap, it has been estimated that ORS has saved over 50 million lives in the past 25 years.

Harmless and pathogenic bacteria

The human body is home to many millions of bacteria. These bacteria may take nutrients from us and, when present in unusually large numbers, cause rashes and mild discomfort. However, they rarely do any significant damage; indeed, those in the large intestine benefit us by fermenting undigested foods and releasing vitamin K and biotin in the process. These harmless bacteria that normally live on and in us should not be confused with 'germs', those disease-causing (**pathogenic**) bacteria about which bleach manufacturers are forever warning us.

A bacterial disease is caused by entry of bacteria into a host which can grow and flourish there, causing harm to the host. Such an invasion activates the host's immune system (the biological defence mechanism). If the immune system is unsuccessful in fighting the infection, the host sickens and may die.

Salmonellosis

One of the most common routes of bacterial transmission is by taking in contaminated water or food. Pathogenic bacteria belonging to the genus *Salmonella* enter via this route, causing a form of food poisoning called **salmonellosis**. The initial source of infection are *Salmonella* in the intestinal tracts of animals. The bacteria enter humans in contaminated food, such as meat and poultry and their products, raw eggs, and unpasteurised milk. Salmonellosis is an acute intestinal infection with an **incubation time** (the interval between exposure to infection and the appearance of the first symptoms) of between 8 and 48 hours. The symptoms are abdominal pain, cramps, diarrhoea, nausea, vomiting, and fever. The disease usually lasts 2–5 days, but may go on for several weeks.

There are many different types of *Salmonella*. *Salmonella typhi*, the bacterium that causes typhoid or enteric fever, produces an endotoxin (a toxin released only after the death of the bacterium). Some strains of *Salmonella enteriditis*, the species that commonly causes simple food poisoning, appear to irritate the gut simply by the large numbers of bacterial cells present. They produce large amounts of gas by fermenting glucose. At the height of infection, as many as one billion *Salmonella* bacteria may be found in each gram of faeces. Most patients with food poisoning recover with rehydration treatment using a salt/sugar solution. In some cases intravenous rehydration is required.

Cholera

Cholera is another water-borne and food-borne disease. Like salmonellosis, it is an intestinal infection. It is also treatable and avoidable, yet it kills over 100 000 people each year. The bacterium that causes the disease, *Vibrio cholerae* (figure 1), is prevalent in regions of the world where sanitation is poor, clean water is unavailable, or food is contaminated. Transmission is mainly by eating food or drinking water contaminated with faeces. The incubation period is short (1–5 days). The bacteria release an enterotoxin called choleragen, a poison that affects the lining of the intestines specifically, causing vomiting and large quantities of watery diarrhoea (described graphically as 'rice water'). A great deal of water may be lost from the body (10–15 dm^3 per day in a very severe infection). This can quickly lead to severe dehydration and death if treatment is not given promptly: without treatment, over 50 per cent of cholera patients may die. However, effective treatment is cheap and simple. Those who are able to drink (80–90 per cent of patients) can be treated with a solution of salts and glucose, a process called **oral rehydration therapy**. Most of those too ill to drink recover if given an intravenous drip. In severe cases, an antibiotic such as tetracycline is often effective.

The prevention and control of both cholera and salmonellosis depend on:

- the supply of clean and safe drinking water (a number one priority)
- hygienic disposal of faeces
- hygienic handling and preparation of food (for example, correct refrigeration, cooking food, and eating food while it is still hot)
- preventing contamination from flies and other vectors.

Although these measures are usually simple to carry out in economically developed regions of the world, they are often difficult to achieve following natural disasters or in poor, overcrowded communities.

Tuberculosis

Tuberculosis (TB) is caused by an airborne bacterium, *Mycobacterium tuberculosis* (figure 2). Coughs, sneezes, and even speaking can spread the disease quickly from an infected person to another person, especially in areas of social deprivation and overcrowding. Currently, about one-third of the world's population is infected with TB. In 2010, 8.8 million people fell ill with TB and 1.4 million died from the disease. Over 95 per cent of deaths occurred in low- or middle-income countries. Since 1993, TB has been regarded by the WHO as a global emergency.

M. tuberculosis is a rod-shaped bacterium that is referred to as **invasive** because it enters and spreads through tissues, most commonly those of the lungs. An infection may result in an immediate disease, disease later in life, or no disease at all. The reason for the different outcomes is unclear, but it appears to depend on the state of a person's immunity. TB is an **opportunistic** infection, striking people with a depressed immunity (for example, it is the leading cause of death of HIV-positive people; see spread 15.2). A patient has small rounded lesions called **tubercles**, and suffers weight loss and muscle wasting (hence the common name of the disease in Victorian times, consumption).

In 1944, streptomycin was the first anti-TB drug to be used successfully to treat a TB patient. Until then, TB was called the 'White Plague' and, because it was untreateable, it was regarded by many as the most dreaded enemy of humankind. Although these first drugs were effective, they had to be taken for about 6 months. Many patients failed to complete their treatment properly, enabling drug-resistant TB strains to emerge. These strains have become more common. Multidrug-resistant TB (MDR-TB) is even resistant to isoniazid and rifampicin, two very powerful anti-TB drugs. While MDR-TB is generally treatable, it requires extensive chemotherapy (up to 2 years of treatment) which includes a cocktail of expensive anti-TB drugs that produce more side-effects. In 2006 extensively drug-resistant (XDR) TB began to appear. This is even more difficult to treat.

(a) Transmission and site of infection

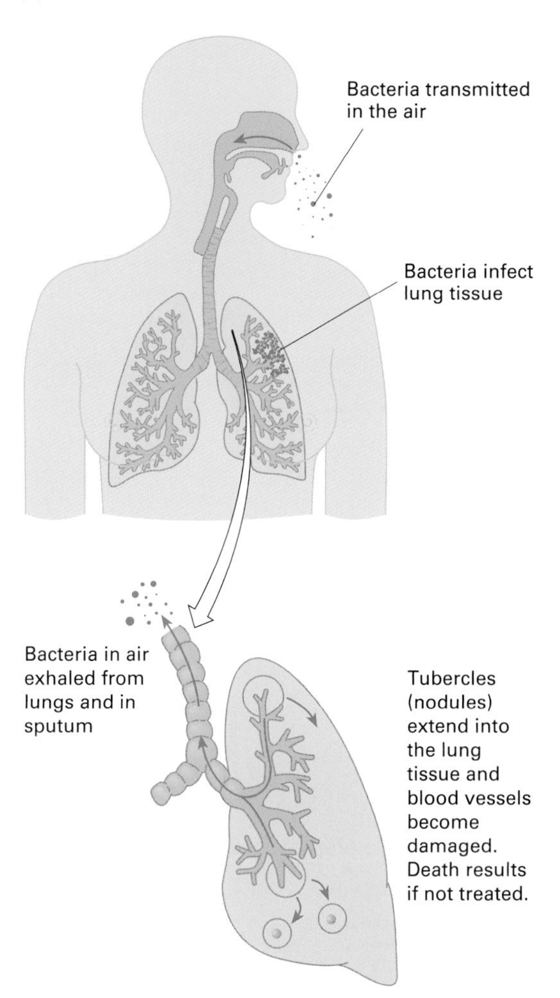

(b) *Mycobacterium tuberculosis*

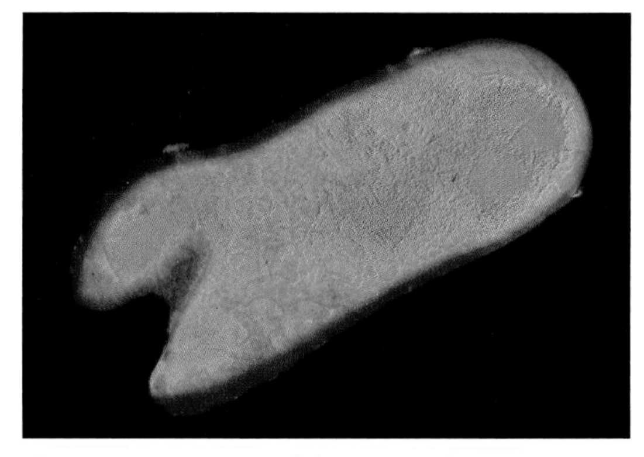

***Figure 2** Tuberculosis. **(a)** Its route of transmission and site of infection. **(b)** The causative agent,* Mycobacterium tuberculosis, *as seen with a transmission electron microscope (×60 000). The wall of this specimen is splitting, probably because it is being killed by antibiotics.*

Quick check

1 What is the main way in which salmonellosis is spread?

2 a What is an enterotoxin?
  b Which of the following organisms produces an enterotoxin: *Salmonella*, *Vibrio cholerae*, *Mycobacterium tuberculosis*?

3 What makes tuberculosis highly infectious?

Food for thought

Despite an intense vaccination programme and the availability of antibacterial drugs, TB continues to pose serious health problems. In particular, there has been a recent increase of TB in the economically developed countries in western Europe and North America. Suggest reasons for this.

15.4 Malaria and sleeping sickness

OBJECTIVES

By the end of this spread you should be able to:

- explain why malaria is so difficult to control
- describe the cause of sleeping sickness
- explain why sleeping sickness is almost impossible to eradicate in some countries.

Fact of life

In 2009, there were an estimated 225 million cases of malaria worldwide. The word 'malaria' originated in the Middle Ages when Italians referred to the bad air of fever-producing swamp areas as 'malaria'.

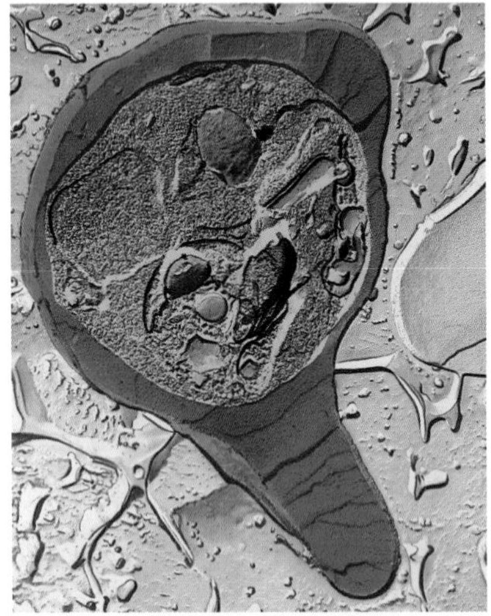

Figure 1 *Coloured freeze-fractured transmission electron micrograph of a section through a red blood cell infected with malarial parasites (*Plasmodium *sp). The blood cell appears pear-shaped. Single-celled malarial parasites (green) are in the swollen part of the blood cell.*

Malaria

References to periodic fevers and chills in ancient Hindu and Chinese writings suggest that **malaria** has plagued humans since the dawn of civilisation. Malaria probably contributed to the fall of the ancient Greek and Roman empires. It has also influenced the outcome of many military campaigns in tropical areas. According to a history of the US Army Medical Corps, the Pacific and Asiatic campaigns of the Second World War would have been impossible if malaria had not been controlled.

The life cycle of *Plasmodium*

Malaria is caused by a single-celled parasite belonging to the genus *Plasmodium* (figure 1). It enters the bloodstream through the bite of an infected female *Anopheles* mosquito (the adult male mosquito does not feed). Once in the human body, the parasites make their way to the liver, where they multiply. After an incubation period of 12 days to 10 months (depending on the species of *Plasmodium*) the parasites return to the bloodstream and invade healthy red blood cells, in which they grow and multiply rapidly. (The parasite has a very active aerobic metabolism and cannot grow in unhealthy cells, which have a low binding capacity for oxygen; see sickle-cell anaemia, spread 16.1.) The red blood cells burst, releasing the parasites which infect other cells. This causes a short bout of shivering, fever, and sweating. The life cycle of *Plasmodium* is shown in figure 2. At the height of the fever, body temperature can exceed 40 °C, a dangerously high temperature, and the spleen can become inflamed, which may cause it to rupture. The loss of healthy red blood cells leads to anaemia. People who are continually infected over a period of years may become immune to malaria. However, many infants die from the disease before they can develop an immunity.

Only female *Anopheles* mosquitoes feed on human blood; they obtain protein from the blood for their developing eggs. When a mosquito feeds on a malaria patient, it takes up blood infected with *Plasmodium* into its stomach. *Plasmodium* then invades the mosquito's salivary glands. The next time the mosquito sucks up another blood meal through its needle-like mouthparts, it can inject *Plasmodium* into a new human host.

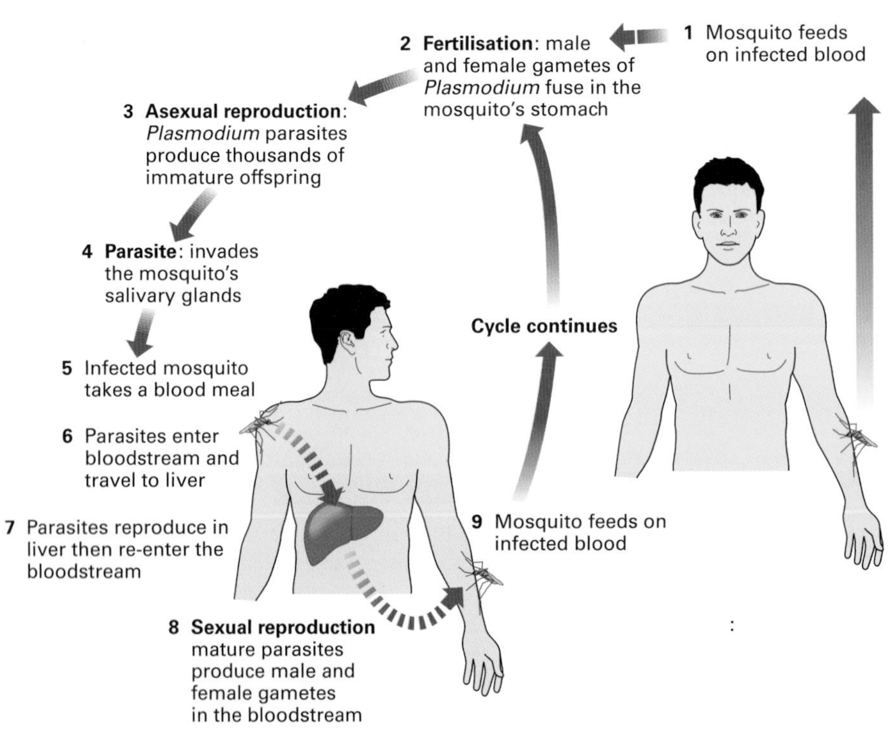

Figure 2 *The life cycle of* Plasmodium*, the causative agent of malaria.*

Controlling malaria

Malaria control measures include eradicating the **vector** (the carrier of the disease, in this case the female *Anopheles* mosquito); preventing mosquitoes biting people; and using drugs to prevent infections. The weak link in the life cycle of *Anopheles* is its aquatic air-breathing larvae. Draining marshes, covering the surfaces of ponds and lakes with oil (to prevent the larvae from breathing), and introducing larvae-eating fish or larvae-killing bacteria (*Bacillus thuringiensis*) have all been used to attack the larvae. Adult mosquitoes have been killed by insecticide sprays such as DDT, but many mosquitoes have developed a resistance to such sprays. Using mosquito nets and insect repellents reduces the risk of being bitten by a mosquito. Anti-malarial drugs such as quinine and chloroquine are used to treat infected people, but new drugs have to be developed continually because drug-resistant strains of *Plasmodium* have evolved.

Behind the bare statistics telling of the millions of people who die each year of malaria lies the untold misery of individuals whose lives have been devastated by this parasite. There could be few greater achievements for medicine than to control malaria. So far, measures to eradicate the vector from the world have been unsuccessful. Many scientists are now directing their efforts at attacking *Plasmodium* by developing vaccines and potent anti-malarial drugs using genetic engineering.

Trypanosomiasis

Trypanosomiasis (known as **sleeping sickness** in Africa, and **Chagas' disease** in South America) is another disease that spreads misery among people living in tropical regions. This debilitating and sometimes lethal disease is caused by single-celled organisms belonging to the genus *Trypanosoma*, known as trypanosomes.

Trypanosomes enter the human bloodstream either via a blood-sucking insect vector or are transmitted in the faeces of an insect vector. Vectors include kissing bugs (bugs that bite the lips and cheeks), sand flies, and tsetse flies (figure 3). Symptoms of infection appear one to two weeks after being bitten, when the victim develops intermittent fever and anaemia. The immune system tries to destroy the parasite, but first it has to detect and respond to proteins (antigens, see spread 15.5) on the surface coat of the trypanosomes. Before all the trypanosomes have been eliminated, their coat changes so that the immune system has to go through a new process of detection and response. During this time, the trypanosomes are able to multiply rapidly (figure 4). The immune system responds again, to be met with the same result, until the system becomes exhausted. In humans, death is usually caused by a secondary infection with which the immune system is unable to cope.

Treatment of trypanosomiasis

In Africa and South America, sleeping sickness often limits the economic and social development of whole communities. Trypanosomes live in many wild animals as well as in humans. These animals act as a huge reservoir of infection, making it virtually impossible to control the disease in many affected countries. Until recently, the most effective treatments have been arsenic-based compounds or DNA/protein binding drugs which are so toxic that they are themselves sometimes lethal. Several new drugs are being developed which target particular areas of the parasite's metabolism, but some of them (such as eflornithine which has been described as a 'miraculous cure') are not used because they are prohibitively expensive. One new drug prevents trypanosomes from changing the antigens on their coats. This would give the immune system a chance to recognise and destroy all the trypanosomes, and improve the effectiveness of anti-trypanosome vaccines.

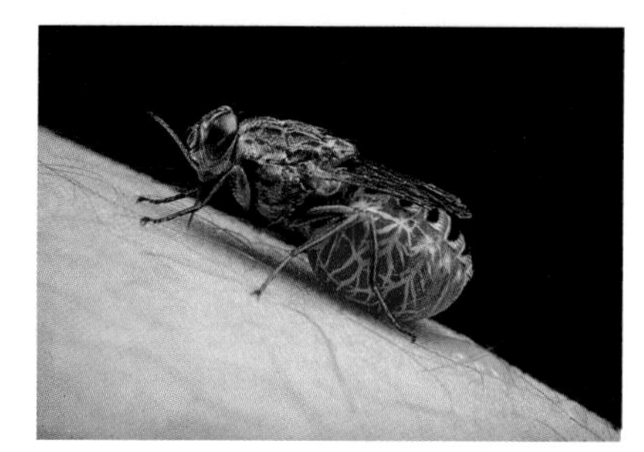

Figure 3 *A tsetse fly adding more blood to its already inflated stomach. In the process of feeding on human blood, it can transmit trypanosomes.*

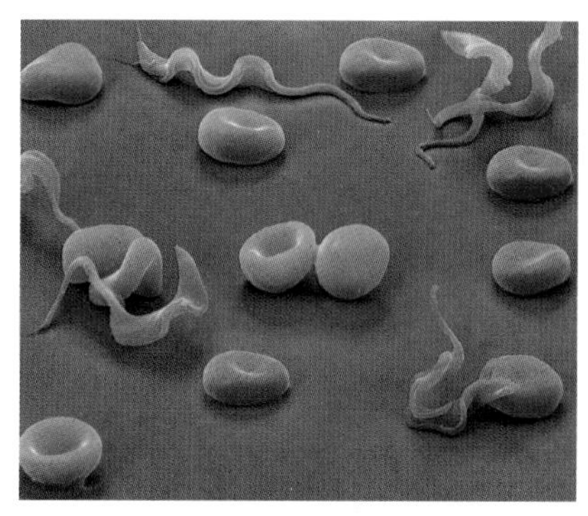

Figure 4 *Trypanosomes in the blood of a patient with sleeping sickness (×1200).*

Quick check

1. For malaria, name:
   a the causative agent
   b the vector.
2. Explain the function of mosquito nets.
3. Name the causative agent of sleeping sickness.
4. Explain how trypanosomes avoid recognition by the host's immune system.

Food for thought

In 1955 the WHO began an expensive worldwide malaria eradication programme. It failed to remove this parasite from most of the tropical regions of the world (spread 20.6). In 1976, the programme finally collapsed. Many countries are continuing eradication programmes in highly populated areas, but others have abandoned malaria control programmes completely. Today, malarial infections are rising. Suggest reasons why malaria continues to cause misery and death.

The immune system

OBJECTIVES

By the end of this spread you should be able to:

- distinguish between the cell-mediated and humoral responses
- describe antigen–antibody reactions
- explain how the primary immune response differs from the secondary response.

Fact of life

According to current estimates, each human can synthesise more than 10 million different types of antibody.

Prostaglandins

Prostaglandins are modified fatty acids derived from plasma membranes. They were first discovered in semen as secretions from the prostate gland (hence their name). Prostaglandins in semen stimulate the uterine wall to contract, helping to convey sperm to an egg. Prostaglandins secreted by the placenta make nearby muscle cells of the uterus more excitable, helping to induce labour during childbirth. More generally, prostaglandins act as local regulators of the mammalian immune system. During an infection or after physical trauma, various prostaglandins help induce a fever and inflammation, and intensify feelings of pain. These unpleasant effects act as an alarm. Ibuprofen and aspirin inhibit the secretion of prostaglandins.

The immune system

The **immune system** can deal appropriately with thousands of different infective agents, saving us from certain death from disease. It is also involved in the recognition and rejection of foreign cells and tissues, as in organ transplantation and blood transfusion (spread 15.8).

The non-specific and specific immune systems

The immune system has two main components: the primary, innate **non-specific immune system** and the adaptive or **specific immune system**. **Primary defences** are present from birth. They are quick acting and effective against a very wide range of pathogens and foreign substances. They always give the same type of response. For example, when the skin (the first natural barrier to infection) is irritated by a foreign substance such as poison from a stinging nettle, **mast cells** (large mobile cells) release **histamine** which causes an acute **inflammation response** involving pain, heat, redness, swelling, and sometimes loss of function of the affected part of the body. **Phagocytes** (large white blood cells which engulf foreign material), **complement proteins** (several blood proteins that contribute to the breakdown and removal of pathogens), and **cytokines** (small protein molecules) are also part of the non-specific immune system, as are **lysozyme**-containing tears and mucus, and acid-containing stomach juices (lysozyme is an enzyme that damages bacterial cell walls).

The **specific immune system** includes cells and proteins within the blood and lymph which attack, disarm, destroy, and remove foreign bodies. It responds slowly and is effective only against specific pathogens. However, the response is speeded up on repeated infection with the same pathogen (a phenomenon called **immunological memory**). A central feature of the specific immune system is that every cell has a complex of molecules, usually proteins and glycoproteins, called **antigens** on its surface membrane. Antigens also occur in non-cellular substances such as snake venom. Antigens and their receptors enable a cell to distinguish between friend and foe or, in the language of immunologists, between self and non-self. The immune system is normally tolerant of the body's own antigens (**self antigens**) and does not organise an attack against them. However, breakdown of the recognition system can lead to **autoimmune diseases** such as **myasthenia gravis** ('grave muscle weakness', an uncommon condition that causes certain voluntary muscles to weaken, usually those that control eye and eyelid movements), AIDS (spread 15.2), and rheumatoid arthritis (spread 16.4), which result in self-destruction of body parts.

There are two main types of specific **immune response**: the humoral response and the cell-mediated response. Both are activated by antigens from foreign organisms and substances to which the body has not become adapted (**non-self antigens**).

***Figure 1** Some types of antigen–antibody reaction.*

(a) Opsonins are antibodies that bind with antigens to act as markers so that phagocytes can recognise foreign cells and destroy them.

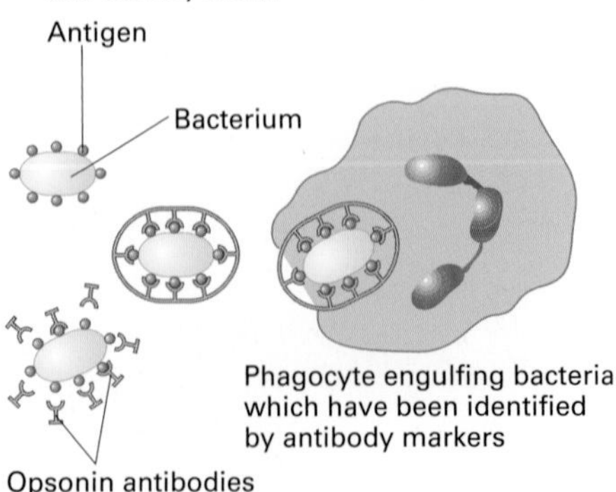

(b) Agglutinins are antibodies that bind to antigens, causing foreign substances to clump together. This prevents them from entering cells and reproducing.

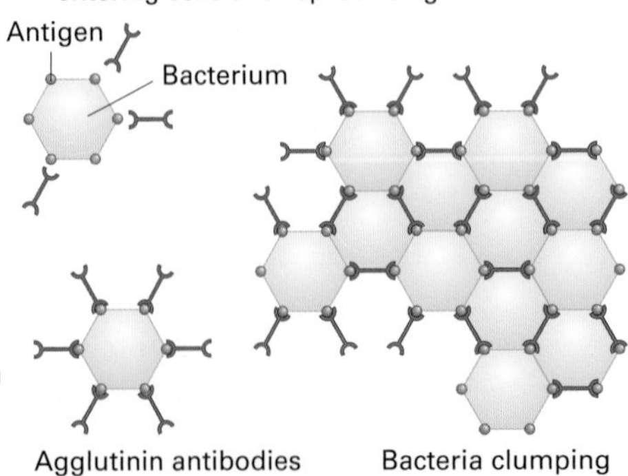

(c) Lysins are antibodies that bind to antigens and cause a foreign cell to rupture or disintegrate.

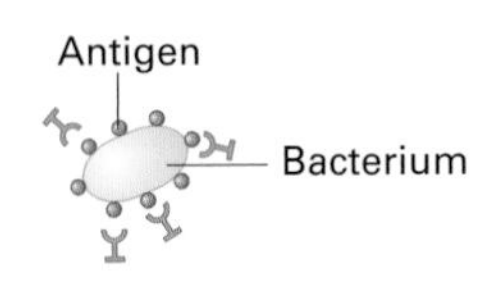

Lysin antibodies binding with antigens

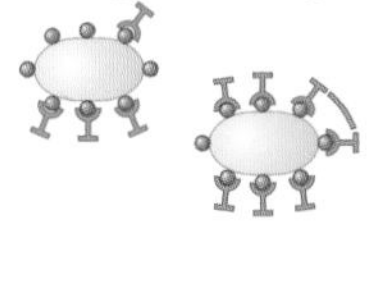

Lysis (disintegration) of bacterium

The humoral response: B cells

The **humoral response** uses soluble antibodies in the blood and lymph (the term 'humoral' means 'of or related to body fluids'). An **antibody** is a type of globular protein that reacts with a specific antigen (spread 15.6). There are several types of reaction (figure 1), all of which depend on the ability of an antibody to bind to an antigen. This binding only happens when the antigen and antibody molecules have complementary shapes.

Antibodies are produced by small lymphocytes (white blood cells, spread 7.8) called B lymphocytes (which mature in bone marrow) or simply **B cells**. There are countless types of B cell, each of which can produce a specific antibody in response to a particular antigen. When it meets this antigen, a B cell divides mitotically and after several generations differentiates into **plasma cells**. All plasma cells formed from one type of B cell secrete the same antibody.

When confronting an antigen for the first time, B cells produce **memory cells** as well as plasma cells. The response of the immune system to an antigen it meets for the first time is called the **primary response**. The primary response is usually slow, taking days or even weeks to recruit enough plasma cells to bring an infection under control. However, the body responds very rapidly to a second invasion. Memory cells are involved in this **secondary response** in an as yet imperfectly understood mechanism (figure 2, spread 15.7).

Figure 2 Killer T cells attacking a tumour cell (x300). It was identified as 'foreign' because it has non-self antigens on its cell surface.

The cell-mediated response: T cells

In the **cell-mediated response**, cells rather than antibodies are produced which are specific to the antigens on the invading pathogens or foreign substances. The cells are lymphocytes called **T cells** (which mature in the thymus). Of the several groups of T cells, **helper T cells** appear to have a key role. Each helper T cell has particular receptor proteins on its cell surface that fit a specific antigen. When it meets this antigen, a helper T cell divides mitotically. Some of the resulting cells remain in the blood and lymph as memory cells; others carry out a variety of functions. They activate other cells in the immune system, including:

- **macrophages**, large phagocytotic white blood cells which destroy pathogens; they are derived from monocytes and differ from lymphocytes in having an indented nucleus, spread 7.8
- **killer T cells** which attack body cells that have been antigenically altered (for example, cells injected by viruses or cancer cells as in figure 2); they also attack large pathogens such as unicellular parasites. Killer T cells demolish their opponents by punching holes through their cell surface membranes so that the cell contents spill out (figure 3)
- **suppressor T cells** which are involved in the general regulation of the immune system by 'switching off' or suppressing immune responses when appropriate. They probably do this by interacting in an antigen-specific manner with helper T cells and/or B lymphocytes.

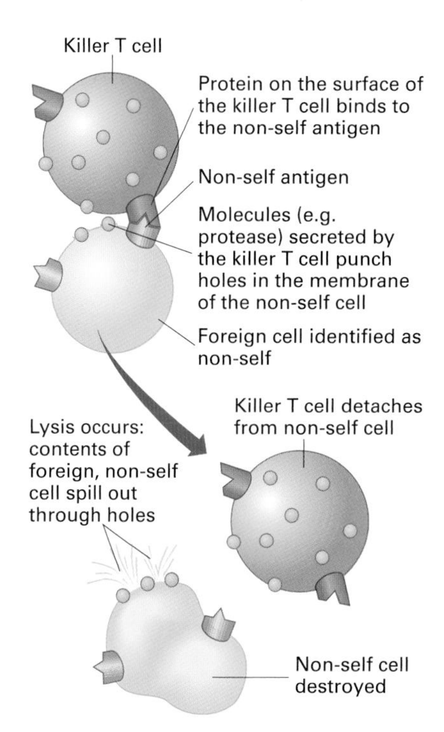

Figure 3 A killer T cell punching holes in a target cell.

Communication in the immune system

Helper T cells also activate B cells, showing that the humoral and cell-mediated responses work together to resist disease. These cells and others within the immune system communicate with each other by means of small protein molecules called **cytokines**. Cytokines are small protein molecules which function as activators and inhibitors of **cell signalling pathways** for the growth, differentiation, and behaviour of cells within the immune system. **Interferons** are a group of cytokines that modulate immune responses to viruses, bacteria, and cancers. One type, interferon alpha, regulates the action of genes that produce proteins needed for growth. It is used in medicines to boost the immune system and reduce the growth of tumour cells.

Quick check

1 Draw a flow diagram comparing the cell-mediated response and the humoral response to a foreign antigen.

2 With reference to cell-mediated responses, what is the significance of a body cell being 'antigenically altered' by, for example, a virus infection?

Food for thought

All antibodies are proteins. Suggest which property or properties of proteins have favoured the evolution of antibodies within this particular group of biological molecules.

15.6

ANTIBODIES

OBJECTIVES

By the end of this spread you should be able to:

- explain how antibodies are formed from B cells
- explain how monoclonal antibodies are made
- discuss the applications of monoclonal antibodies.

Fact of life

Monoclonal antibodies can help save the lives of cancer patients. The patient in figure 1 has been injected with radioactively labelled monoclonal antibodies that recognise an antigen known to be present on the surface of a tumour in the ovary. The antibody circulates in the blood until it meets the antigen, whereupon it binds to the antigen. The red, pink and white regions in the upper abdomen indicate that the cancer has spread from the ovary to the spleen.

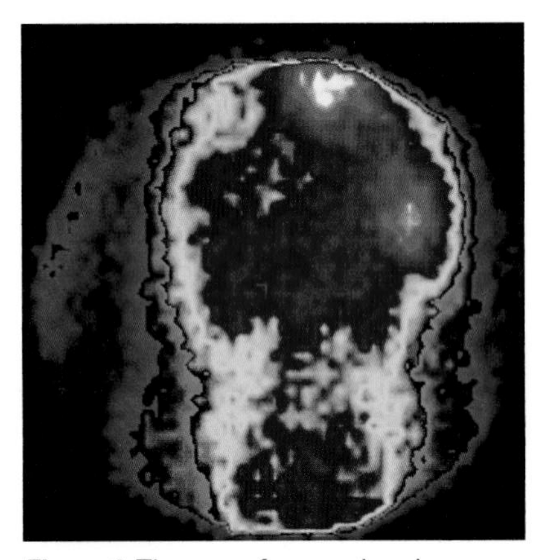

Figure 1 *The use of monoclonal antibodies to locate cancer cells in a patient (see text above).*

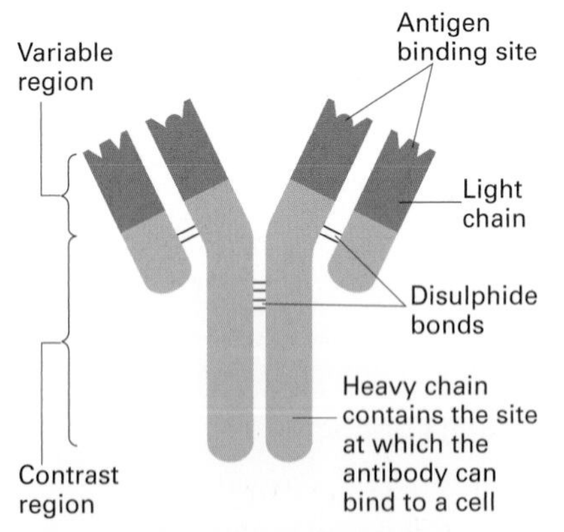

Figure 2 *The basic structure of an antibody such as the monomeric immunoglobin IgG. An antibody molecule has four polypeptide chains, two light chains and two heavy chains. Each chain has a variable region which forms the antigen binding site.*

When attacked by disease-causing organisms, the body retaliates by waging biochemical warfare. The most important weapons in the body's armoury are **antibodies**: specialised protein molecules that bind specifically to antigens, their target molecules (figure 2).

The B cells of the immune system can manufacture antibodies with an almost infinite variety of shapes and sizes to fit any antigen that enters the body.

How B cells produce antibodies

B cells have glycoprotein receptors on their cell surface membranes known as **immunoglobulins**. B cells contain multiple copies of the genes for these immunoglobulins. As each B cell matures in the bone marrow, these genes recombine randomly to produce a new and unique gene giving the particular immunoglobulins on that cell. This is how such a great variety of B cells with highly specific immunoglobulins is produced: at least one is likely to fit any particular antigen. When the B cell meets this antigen, it produces large amounts of its immunoglobulin and releases it into the blood. When released into the blood, immunoglobulins act as **antibodies.**

Every B cell has the capacity to make large amounts of its own specific antibody. However, it does this only when it meets an antigen that fits neatly onto its immunoglobulins. The signal for a B cell to go into action does not come from the antigen itself, but from **helper T cells** that are specific for the same antigen. When activated, a B cell divides rapidly by mitosis to produce **plasma cells**. These cells are genetically identical (clones) and produce the same kind of antibody. Each plasma cell is capable of producing up to 2000 identical antibody molecules per second so that the combined output of all members of a clone can be phenomenal. The process by which a particular antigen promotes the production of a specific antibody is called **clonal selection**.

By a combination of B cell variability and clonal selection, the immune system can defend the body against new attacks from previously unknown enemies. However, the body cannot produce much antibody initially, so the primary response is usually slow. If the delay between exposure to a new antigen and the body's response to deal with it is too long, a lethal antigen may be able to kill before the body can deal with it. This is why a massive dose of snake venom can kill a victim; antibodies may be present, but not in sufficient numbers to neutralise the venom. Secondary responses are usually much faster than the primary response because memory cells (formed from the same B cell clone as the cells that produce antibodies) multiply into plasma cells which can produce large amounts of antibody very quickly (spread 15.7).

Monoclonal antibodies

Until 1975, plasma cells would only manufacture antibodies inside the body. Attempts to culture plasma cells outside an organism failed because the cells are unable to divide. In 1975, Georges Kohler and Cesar Milstein discovered a new method of culturing cells which produce a pure and highly specific antibody known as a **monoclonal antibody**. They fused mouse spleen cells (the spleen is an organ rich in lymphocytes) with certain myeloma (cancer) cells from a mouse or rat. Myeloma cells can divide, unlike the lymphocytes. Some of the resulting hybrids developed into a clone of cells called a **hybridoma**. A small but significant fraction of the hybridomas obtained in this way combined the antibody-producing capacity of plasma cells with the tumour cells' ability to reproduce. A desired clone could then be isolated and used for the industrial production of antibodies (figures 3 and 4). In 1984, Kohler and Milstein were awarded a Nobel prize for their discovery, which has had an enormous impact on many aspects of medicine and other fields.

Uses of monoclonal antibodies

Since their discovery in 1975, **monoclonal antibodies** (mAbs) have transformed biological research and created a multi-billion-pound industry. They are used in pregnancy test kits (spread 12.8), in AIDS-testing kits, and in many other diagnostic and therapeutic tools. However, the initial method of manufacture generated mouse mAbs that could provoke an immune response if given to a human. In the 1980s, to reduce the risk of rejection, recombinant DNA technology was used to humanise mAbs by combining mouse DNA encoding the binding portion of a monoclonal antibody with human antibody-producing DNA. More recently, attempts have been made to reduce the risk of rejection even further by using transgenic mice that are genetically engineered to produce fully human antibodies.

mAbs are used in medicine to locate diseased cells (figure 1), to carry drugs to specific target cells, and to treat a wide variety of diseases including cancer, cardiovascular disease, inflammatory diseases, multiple sclerosis, and viral infection. Anti-cancer mAbs may work by making a diseased cell more visible to the immune system; by blocking chemical cell signals which stimulate growth in cancerous cells; by stopping new blood vessels forming around a tumour, thereby starving the tumour of oxygen and nutrients; or by delivering anti-cancer drugs to cancer cells.

Quick check

1 What is an immunoglobulin?

2 Why is it not possible to use lymphocytes or plasma cells as commercial producers of specific antibodies?

3 How does 'humanisation' of monoclonal antibodies increase their potential use?

Food for thought

Monoclonal antibodies are used as 'magic bullets' in the fight against cancer using a technique called **ADEPT** (antibody-directed enzyme pro-drug therapy). The patient is injected first with a monoclonal antibody which is loaded with an enzyme. The antibody is designed to recognise specific antigens and, as an antibody–enzyme complex, bind to tumour cells. The patient is then injected with a pro-drug (an inactive form of a drug). When it comes into contact with the antibody–enzyme complex the pro-drug is activated into a cytotoxic drug (a drug that destroys cells). In this way it is hoped that tumour cells are killed without damaging normal cells. Suggest why, although the concept is sound, progress has been slow in developing such monoclonal antibody therapies.

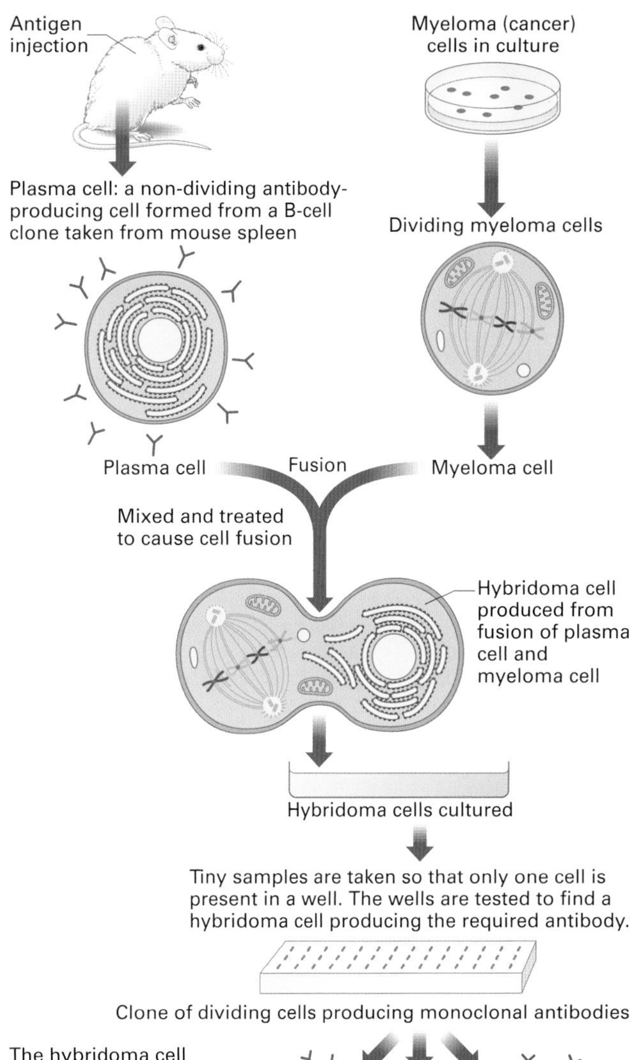

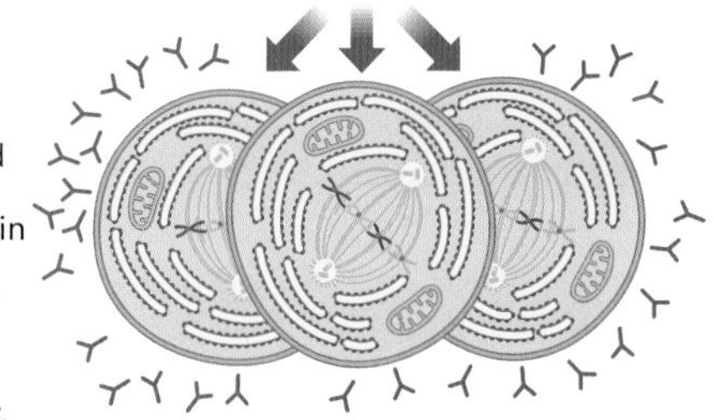

Figure 3 *Manufacture of monoclonal antibodies.*

Figure 4 *Commercial production of monoclonal antibodies in fermenters.*

15.7

Immunity and vaccinations

OBJECTIVES

By the end of this spread you should be able to:

- distinguish between natural active immunity and artificial active immunity
- give one example of artificial passive immunity and one of natural passive immunity
- explain why vaccinations are effective against some diseases but not others.

Fact of life

Edward Jenner was the first person to publish successful clinical trials of vaccinations. Stimulated by the claims of a milk maid that she could not catch smallpox because she had cowpox, Jenner began inoculating people by scratching their skins and introducing material from cowpox lesions into the scratches. The cowpox virus is closely related to the smallpox virus: it has similar antigens but much milder symptoms. Cowpox caused these symptoms in Jenner's patients, but this was a small price to pay for the immunity it gave against the dreaded and often fatal smallpox disease. By 1800, the practice was a common feature of a doctor's duties (figure 1). It became known as vaccination after *vacca*, the Latin word for cow. Subsequently, Louis Pasteur found that some other weakened pathogens could also give immunity if inoculated into a patient. In honour of Jenner, Pasteur gave the name 'vaccination' to all such preparations.

Figure 1 *A nineteenth-century painting of a doctor vaccinating a child.*

Modern advances in medicine have caused infant mortality rates to plummet. Artificial forms of immunity can be provided to reinforce natural immunity, making babies resistant to infectious diseases which once reaped a deadly harvest. Measles illustrates the effectiveness of this preventative medicine: the disease has been tamed in most economically developed regions of the world, but remains a killer in areas lacking modern medical facilities.

Measles

Measles is a highly contagious disease found throughout the world. It is caused by the morbillo virus which enters the body through the respiratory tract or conjunctiva of the eyes and replicates inside human cells. Symptoms of a distinctive rash and fever develop about 8–14 days after the initial infection. In otherwise healthy children the disease clears, usually with no complications, after about 10 days. Permanent immunity follows the disease. This type of immunity is called **natural active immunity**: the term 'active' means that the body makes its own antibodies in response to the antigens. Memory cells (spread 15.5) provide an immunological memory so that antibodies can be produced more quickly during a secondary response (figure 2). Very rarely, the measles virus attacks the central nervous system, causing it to degenerate. This complication and others are more frequent if the patient is already weakened by malnourishment or another disease. Measles is a major cause of infant mortality in many overcrowded and poor parts of the world.

Vaccinations: artificial active immunity

The potential dangers of measles make it medically worthwhile to **immunise** children: to activate their immune responses artificially before infection can occur. Morbillo virus antigens are injected into the body (a process commonly known as **vaccination**). The resulting immune response is called **artificial active immunity**, which is similar to the immunity following a natural infection. The aim of vaccination is to give active immunity without causing symptoms of the disease. Edward Jenner is believed to have been the first to use such a vaccination in the late eighteenth century (see Fact of life). One type of vaccination contains live organisms that have been weakened so they cause only very mild symptoms, but their antigens still trigger a primary response. Weakening of pathogens (referred to as **attenuation**) can be natural (as in cowpox) or it may be carried out artificially, for example, by culturing the pathogen repeatedly at high or low temperatures, or without oxygen, until suitable mutants occur. The Sabin poliomyelitis vaccine that is taken orally contains attenuated viruses.

Another type of vaccination uses viruses or bacteria that have been killed (for example, by heating or by adding chemicals such as formaldehyde) so that they can still trigger an immune response but cannot induce the disease. The Salk poliomyelitis vaccine is of this type.

A third type of vaccination uses inactivated toxins called **toxoids**. The toxoids are harmless, but trigger the same type of immune response as the normal toxins produced in particular bacterial diseases. Toxoids are used in vaccines against diphtheria and tetanus. Genetic engineering (spread 18.10) has enabled medical scientists to produce specific toxoids and vaccines free from substances that produce harmful side-effects.

Vaccination strategies and types of immunity

Ideally, vaccinations for highly contagious diseases such as measles should be administered to a large number of people at the same time. This provides a general immunity in the population called **herd immunity**. Vaccinations are not equally effective on all people. In fact, some people may not respond at all, perhaps because they have inherited a defective

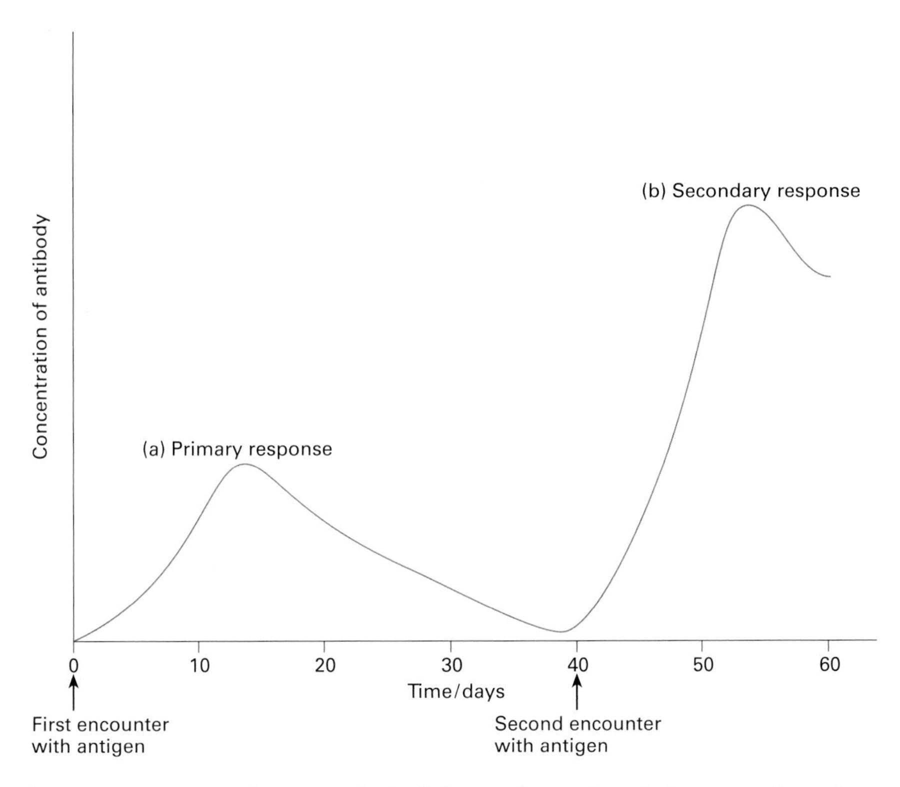

Figure 2 *Comparison of antibodies produced to the same antigen during a primary response and a secondary response.* ***(a)*** *Primary response. Typically, one B cell is activated by a specific antigen that has a shape complementary to the immunoglobulins on its cell surface. This B cell divides many times to produce a clone of cells, all of which can produce identical antibodies against the antigen. Some of the B cells become memory cells.* ***(b)*** *Secondary response. On being exposed to the same antigen a second time, the memory cells formed in the primary response divide to produce a large population of B cells. In turn, each of these can produce clones of B cells that can make antibodies to destroy the antigen or the organism to which it belongs. The secondary response is faster and greater than the primary response because it starts with a large number of memory cells rather than a single B cell.*

immune system, or because their defences have already been weakened by disease or malnutrition. Herd immunity offers them some protection because it reduces the risk of encountering the infectious agent.

Some diseases are difficult to vaccinate against. The influenza virus is particularly awkward because it mutates regularly, changing its surface antigens. Novel strains can also evolve when genes from bird and pig 'flu combine with those in the human influenza virus. Our immune systems can recognise small changes (known as **antigenic drift**) but cannot cope with large changes (**antigenic shift**). The influenza virus evades attack by **antigenic variation**; other pathogens avoid the immune system by **antigenic concealment**. *Plasmodium* (the cause of malaria) lives on the intestinal wall bathed in digestive juices concealing it from antibodies in the blood and lymph.

Active immunity, whether natural or artificial, is not immediate: there is a delay between infection and a full immune response. During this delay, toxins released by disease-causing organisms (for example, by *Clostridium tetani* which causes tetanus) or toxins introduced directly into the body (for example, by a snake bite) could kill. A person with a wound that might be infected with *C. tetani* or who has been bitten by a poisonous snake is therefore given ready-made antibodies against the toxin. This type of immunity is called **artificial passive immunity**.

Another type of immunity called **natural passive immunity** occurs when a fetus is still in the uterus. Maternal antibodies cross the placenta into fetal blood. They may protect the baby, for example against measles, during the first few months after birth. Infants also acquire some passive immunity from the antibodies in colostrum (spread 12.10). In all cases, passive immunity is only temporary because the antibodies are eventually broken down in the spleen and liver. There is no immunological memory as the baby did not make the antibodies for itself.

The battle against smallpox seems to have been won; that against measles is well underway but some diseases such as cholera are resisting our best attempts to eliminate them. However, these pathogens may succumb to other biochemical weapons we use in our fight against disease: antibiotics and chemotherapy (spread 15.9).

Quick check

1. Which type of immunity involves the body making its own antibodies in response to an antigen?
2. Explain why passive immunity is only temporary.
3. Why are influenza vaccines sometimes not effective?

Food for thought

The common cold is caused by more than 100 different strains of rhinovirus. Suggest why no effective vaccine has been developed against the cold virus, and why a person can catch a cold several times in the space of a few months.

15.8

Transplants and transfusions

OBJECTIVES

By the end of this spread you should be able to:

- describe different types of transplant
- explain why certain blood groups are incompatible and cannot be used in blood transfusions.

Fact of life

Valves from a pig heart, made immunologically inert by a tanning process, are used routinely to repair defective human heart valves. Mechanically, a pig's heart has the potential to replace a defective human heart: it is very similar in size and structure to a human heart, and has the capacity to pump blood around the human body. But the hearts are genetically very different. Therefore, whole-heart **xenotransplants** (transplants from one species to another) can be successful only if rejection is overcome. Genetic engineering has now made this possible. Human genes can be incorporated into a pig's heart as it is developing, making it acceptable to a human recipient. Such hearts from **transgenic pigs** could be the solution to the critical shortage in human donor organs. However, xenotransplantation raises many ethical and biological concerns. One of the greatest is that new infectious diseases to which humans have no defence could be transmitted from xenotransplants into their human hosts.

The HLA system

The **HLA system** (Human Leucocyte Antigen system) consists of a series of four gene families (designated A, B, C, and D) that code for proteins on the surface of nucleated cells. Individuals inherit from each parent one allele or set of alleles for each subdivision of the HLA system. Two individuals with identical HLA types are histocompatible (tissue from either individual will be tolerated by the immune system of the other because both individuals have similar antigens on their cell surface membranes). Allograft success requires a minimum of HLA differences between donor and recipient tissues.

Types of transplant

Transplants of kidneys, hearts, livers, and other organs have given many thousands of people a new lease of life. Medical technology has improved so much in recent years that the number of successful transplants has increased several fold. This has led to a demand for transplantations far exceeding the supply of donor organs and tissue. The short supply of donor material, coupled with biotechnological advances, has led increasingly to the implantation of tissues from non-human species. Transplants from one species to another (for example, from a pig to a human) are called **xenografts**. Other forms of transplant include:

- **allografts**, in which tissue is grafted from one individual to a genetically different individual of the same species
- **autografts**, in which tissue is grafted from one part of the body to another on the same individual
- **isografts**, in which tissue is grafted from one individual to a genetically identical individual (for example, a twin).

Overcoming rejection

Allografts and xenografts consist of foreign tissue with cells containing non-self antigens (see spread 15.5). Consequently, left to its own devices, the body would launch an antibody attack against most of these tissues and reject them. Some transplants, however, such as corneal and heart-valve transplants, rarely trigger an immune response no matter what the source of the tissue. Rejection is also not a problem with autografts and isografts because these tissues are genetically identical to host tissue.

The risk of allograft rejection is minimised by matching the tissues as closely as possible (see box below left) and by suppressing the immune response. Monoclonal antibodies (spread 15.6) are often used to identify the types of antigens present in donor tissue and improve the matching. Matching is more likely among close relatives than between non-relatives. A cocktail of drugs known as **immunosuppressants** is often used to inhibit the activity of all cells in the immune system. Another form of immunosuppression involves irradiating bone marrow and lymph tissues with X-rays so that the immune response is reduced. However, both these forms of suppression disable the body's natural immune system to such an extent that the patient is very vulnerable to infections. One way of reducing this problem is to destroy only the T cells required to activate the immune response against the grafted tissue. Recently, monoclonal antibodies have been developed which suppress only T cells, leaving the rest of the immune system to protect the patient against disease.

Blood transfusion and ABO blood groups

Blood transfused from one person to another can be regarded as a special kind of allograft because blood is a fluid tissue. It is a much simpler allograft than many other tissues because, in most cases, rejection or acceptance involves only two antigens. These antigens are the basis of a type of blood grouping called the **ABO blood system**. There are four blood groups within this system: A, B, AB, and O. For a successful blood transfusion, the blood of the donor and recipient must be compatible, that is, they must be able to mix without **agglutination** (the clumping together of cells). Blood compatibility depends on antigens on the surface of red blood cells (figure 1), and antibodies in the blood plasma. The two types of antigen on the red blood cells are A and B (figure 1). There are also two types of antibody in the plasma: anti-A and anti-B. These occur in the blood groups as shown in table 1.

Anti-A plasma agglutinates red blood cells with A antigen; anti-B plasma agglutinates red blood cells with B antigen (figure 2). In blood transfusions, mixing A antigens with anti-A antibodies or B antigens with anti-B antibodies should be avoided or the blood will agglutinate. However, in most blood transfusions only a relatively small quantity of donated blood is transferred to the recipient. The recipient's red blood cells are not clumped because the donor's antibodies are quickly diluted in the plasma, but the donor's red blood cells are clumped by the recipient's antibodies. Therefore, blood can usually be transfused safely as shown in table 1.

Table 1 *The ABO blood group system.*

Blood group	Antigens (red blood cells)	Antibodies (serum)	Can donate blood to	Can receive blood from
O	None	Anti-A, anti-B	All groups	O
A	A	Anti-B	A and AB	A and O
B	B	Anti-A	B and AB	B and O
AB	A and B	None	AB	All groups

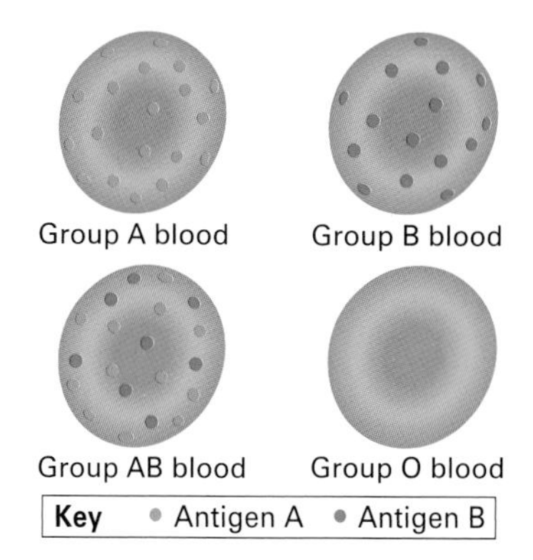

Figure 1 *Red blood cells from each of the four main blood groups showing the antigens on their surfaces.*

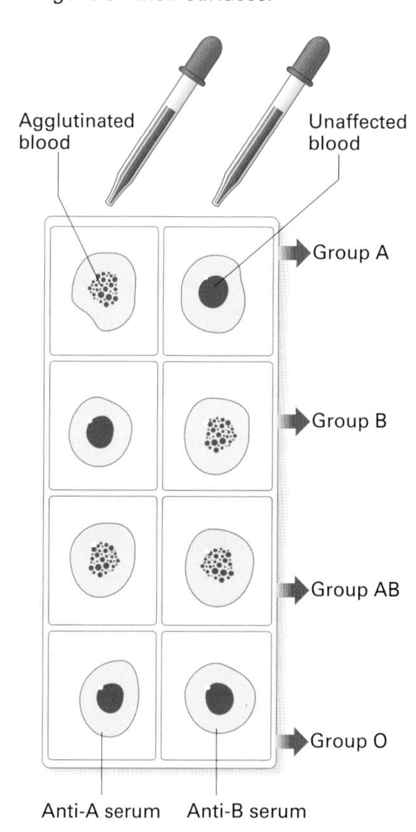

Figure 2 *Determination of blood group. A person's blood group can be determined by the response of the person's blood to specific antibodies. A sample of blood is taken and two drops put onto a white tile. To one drop of blood is added serum containing anti-A antibodies, and to the other drop of blood is added serum containing anti-B antibodies. All possible results are shown.*

The rhesus factor

There are several other blood group systems in addition to the ABO system which affect the compatibility of blood. The most important is the **rhesus blood group system**. This system involves a group of antigens known as the **rhesus factor**. Most people have the rhesus factor and are called rhesus positive (Rh-positive). People who lack the factor are rhesus negative (Rh-negative). Rh-negative people can receive one transfusion of blood from a Rh-positive donor without being harmed because they have no antibodies against the rhesus factor. However, exposure to the rhesus factor causes their B cells to produce antibodies against the rhesus factor (a process called **sensitisation**). In a subsequent transfusion, the antibodies agglutinate Rh-positive blood. However, Rh-negative blood can be transfused into Rh-positive people any number of times.

The rhesus factor can also cause problems if a Rh-negative mother has more than one Rh-positive baby. During pregnancy, Rh-positive cells may pass from fetus to mother, perhaps through a leaky placenta or during birth, and cause the mother to produce antibodies against subsequent Rh-positive fetuses. This danger can be avoided if a Rh-negative mother is injected with drugs that suppress her production of rhesus antibodies.

No problems arise between a Rh-negative fetus and a Rh-positive mother. Some red blood cells with rhesus factor may enter the fetus, but the amount of antibody produced by the fetus is too small to have any significant effect on the mother.

Quick check

1 In which types of transplant is tissue rejection:

a not a problem

b a problem?

2 a Universal donors are people who have blood with no A or B antigens. To which blood group do they belong?

b Universal recipients are people who have no anti-A or anti-B antibodies in their blood plasma. To which blood group do they belong?

Food for thought

Genetic engineering has made xenotransplants technically possible. Thousands of patients with defective hearts, livers, and kidneys die each year because donor organs are not available to them. Xenotransplants could solve the problem of lack of donors. Nevertheless, many medical scientists are extremely worried that xenotransplants could increase the risk of infectious disease not only to the patient, but also to the human population as a whole. Suggest the scientific basis of these worries.

15.9

ANTIBIOTICS

OBJECTIVES

By the end of this spread you should be able to:

- discuss the meaning of the term antibiotic
- explain how antibiotics work
- discuss the reasons for antibiotic resistance.

Fact of life

About 20% of the human population are long-term carriers of *Staphylococcus aureus* (*S. aureus*) (figure 1). The bacterium is usually found on the skin of carriers and seems to cause no major problems. But if it enters the body, under the skin or into the lungs, it can cause dangerous infections.

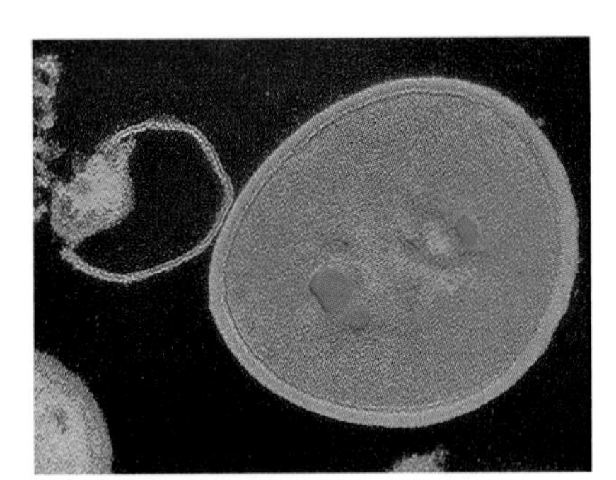

***Figure 1** An electron micrograph (×30 000) showing a complete, undamaged* Staphylococcus aureus *along with one being destroyed by antibiotics and undergoing lysis.*

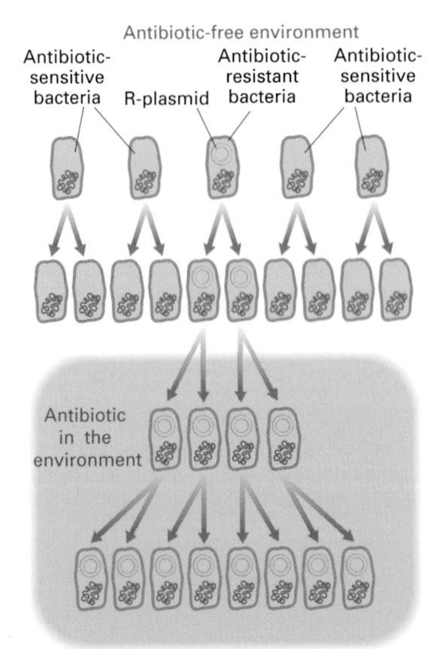

***Figure 2** Vertical gene transfer: antibiotic resistance genes form by spontaneous mutations and are transferred to offspring by DNA replication. Exposure to antibiotics kills bacteria without the resistance gene while those with the gene thrive to pass their resistance to the next generation.*

When a potentially deadly pathogen has penetrated our first line of defence and entered the body, then avoided detection and elimination by our immune system, antibiotics may be our last chance of survival.

What is an antibiotic?

An **antibiotic** is a substance produced by one microorganism that is capable of destroying or inhibiting the growth of another microorganism. Antibiotics are regarded as secondary metabolites because they are not essential for the growth or reproduction of the organism that produces them. Their formation is thought to give a selective advantage to the microorganism (antibiotics produced by *Penicillium* might, for example, act as growth inhibitors to competing microorganisms). Antibiotics that destroy (kill) microorganisms are described as **biocidal** or **bactericidal**; those that inhibit the growth and reproduction of microorganisms are **biostatic** or **bacteriostatic**. Antibiotics are not effective against viruses.

Only microorganisms (mainly bacteria and a few fungi) produce antibiotics, so substances produced by plants and animals or synthesised by humans are not strictly antibiotics. However, the term is generally used more loosely to include a wide range of chemicals which damage pathogens. Thus antibiotics generally include chemicals produced naturally by microorganisms, animals and plants; chemicals such as penicillins which are produced naturally but altered chemically to make them more effective; and chemicals such as chloramphenicol which were originally produced naturally but which are now entirely synthetic.

Antibiotics work by interfering with some essential metabolic function of microorganisms, such as cell wall synthesis, protein synthesis, nucleic acid synthesis, or cell membrane function.

Mechanisms of antibiotic action

Penicillin, vancomycin, and cephalosporins are among the most commonly used antibiotics. All weaken bacterial cell walls. These walls are made up of long chains of peptidoglycan molecules. The antibiotics inhibit the synthesis of peptide links which bind the molecules together. The cell wall becomes so weak that the bacterium bursts (lysis). Unfortunately, the antibiotics only work on bacteria that are growing; they have no effect on bacteria lying dormant in the body, waiting for a suitable time to reproduce.

Streptomycin, erythromycin, tetracyclines, and chloramphenicol either inhibit protein synthesis or promote the synthesis of abnormal proteins in bacterial cells. The antibiotics bind to bacterial ribosomes (the sites of protein synthesis), but do not affect eukaryotic ribosomes (including those of humans), which are larger.

Anthracyclines and rifampicin disrupt the synthesis of nucleic acids. Anthracyclines do this by inhibiting DNA replication, whereas rifampicin prevents transcription (the copying of DNA into mRNA; see spread 18.7).

Some antibiotics kill or destroy pathogens by disrupting cell membranes. For example, fungal cell surface membranes are normally stabilised by ergosterol, which has the same function as cholesterol in animals. The antibiotic amphotericin B binds to ergosterols, distorting the lipid bilayer. It is thought that this allows the contents of the cell to leak out, with fatal consequences for the pathogen.

Selection of antibiotics

The best antibiotic to choose to cure a human disease is one that specifically attacks the pathogen cells without harming the patient. Unselective antibiotics may act on beneficial as well as harmful microorganisms. The least desirable antibiotic is one which harms a patient. Sometimes, however, there is only a choice of the lesser of two evils: an antibiotic which cures a disease but which may harm the patient; or a disease which is eventually likely to kill the patient.

In the past, streptomycin saved many tuberculosis (TB) patients from a lingering death, but it had the unfortunate side-effect of attacking nerves responsible for hearing and balance. Consequently, many patients cured of TB became completely deaf. New anti-TB drugs have now been developed which do not have this harmful side-effect.

The demand for more effective treatments that are harmless to humans is one reason for the continual search for new antibiotics. Another reason is the very worrying ability of pathogens to evolve antibiotic-resistant strains.

Antibiotic resistance

Antibiotic resistance is a serious problem throughout the world. It has evolved in *Vibrio cholerae* and *Mycobacterium tuberculosis* (the causative agents of cholera and tuberculosis, respectively; spread 15.3) and in *S. aureus*. *S. aureus* is a very common bacterium, usually harmless when inhabiting the skin, but potentially lethal if it enters the bloodstream. It produces an **exotoxin**, a toxic protein released into a bacterium's surroundings as a result of its normal metabolism. Some strains of *S. aureus* have become resistant to several antibiotics (see box below).

S. aureus has probably become antibiotic resistant by a combination of **vertical gene transmission** (figure 2) and **horizontal gene transmission** (figure 3). In vertical gene transmission, random mutations (spread 19.1) result in many different offspring. By chance, some of them will be more resistant to a particular antibiotic than others. Exposure to doses of antibiotic that kill some but not all of the bacteria has provided a situation in which natural selection takes place. The least-resistant bacteria are killed; the most-resistant survive to pass on their resistance to the next generation. If this process is repeated enough times, a very resistant strain will evolve into the most common type of *S. aureus*.

Antibiotic resistance may also be linked to the widespread use of antibiotics in animal feedstuffs, especially in the USA where more than half the livestock is fed antibiotics to improve growth rates. Antibiotic-resistant genes are often carried in plasmids (small self-replicating circles of DNA) which can pass from one bacterium strain or species to another by horizontal gene transmission. Genes for resistance to tetracycline have been tracked from the guts of farm animals to quite different bacteria in the human gut.

With wise use of antibiotics, we may be able to slow the development of antibiotic resistance. Nevertheless, for doctors to continue to be successful in calming a raging typhoid fever in a child or halting an epidemic of cholera, scientists must continue their search for new varieties of antibiotics and discover how to maintain the effectiveness of older varieties.

The incentive to discover new antibiotics is not merely to help the human race; antibiotics are big business, with the world market worth many billions of pounds. It is to be hoped that the lure of money does not distract medical scientists from their prime aim: the eradication of suffering due to infectious diseases.

MRSA: a major problem of antibiotic resistance

In 1946, when penicillin first came into general use, almost all strains of *S. aureus* were penicillin sensitive. Today, most hospital strains of the microbe have evolved a resistance to penicillin; some, referred to as **MRSA** (methicillin-resistant *Staphylococcus aureus* or multi-resistant *Staphylococcus aureus*) are resistant to several antibiotics, including methicillin. Like other strains, MRSA can be carried on the skin and can contaminate dust particles. Therefore stringent control measures are required to prevent its spread, especially in hospitals where patients with weakened immune systems are at greater risk of infection than the general public. Infection controls include **contact isolation**, in which everyone in contact with a patient (or in contact with anything the patient touches) is careful about cleaning hands before entering and when leaving the hospital. In some circumstances, it is necessary to move patients who are carriers of MRSA to an isolation unit for their medical care.

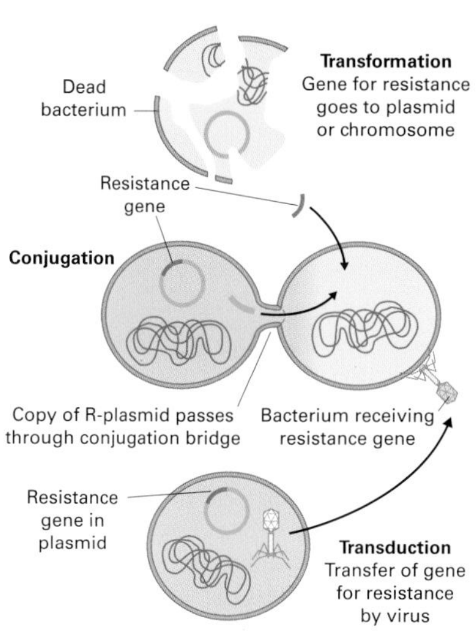

***Figure 3** Horizontal gene transfer (HGT) in which antibiotic resistance is passed from one bacterium to another. The diagram shows **conjugation**, probably the most common type. It occurs when two closely related bacteria make direct cell-to-cell contact and one bacterium transfers small circular pieces of self-replicating DNA called plasmids to the other bacterium through a cellular extension called a conjugation bridge or pilus (see spread 17.3). HGT may also occur by **transformation**, when bits of DNA present in the environment due to death and lysis of bacteria are taken up by another bacterium, and by **transduction**, when bacteriophages (bacterial viruses) transfer a section of DNA from one bacterium to another closely related bacterium.*

Quick check

1 Distinguish between biocidal and biostatic antibiotics.

2 List four ways antibiotics can harm microorganisms.

3 By what method has *Staphylococcus* probably acquired its resistance to antibiotics?

Food for thought

A Japanese strain of *Staphylococcus aureus* has evolved resistance to all forms of antibiotics. Suggest a set of guidelines that would minimise antibiotic resistance spreading within Japan and to other countries.

15.10

Emerging infectious diseases

OBJECTIVES

By the end of this spread you should be able to:

- explain what a prion is
- discuss the cause of new-variant CJD.

Fact of life

The Ebola virus (figure 1) was first identified in 1976 after significant outbreaks of haemorrhagic fever occurred in Zaire (now the northern Democratic Republic of the Congo) and Nzara in southern Sudan. It had one of the highest case fatalities (88 per cent) of any human pathogenic virus.

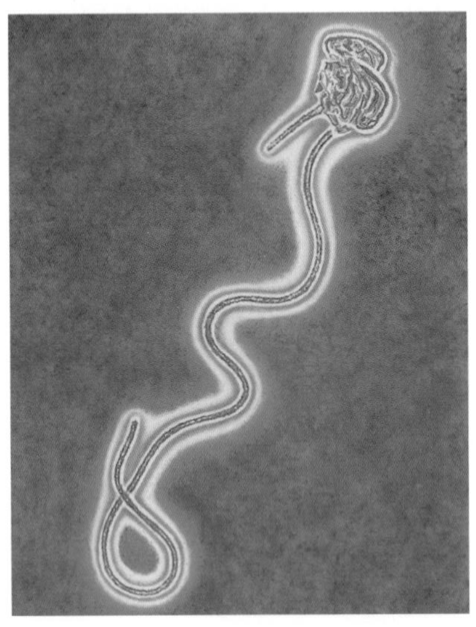

Figure 1 *Scanning electron micrograph of Ebola haemorrhagic virus (×20000).*

The war against disease is fought on many fronts and against many different enemies. Many diseases, such as smallpox, have been brought under control, but new emerging infectious diseases threaten our health and survival. An **emerging infectious disease** (**EID**) is an infectious disease that has been detected for the first time in a population or that has been known for some time but has rapidly increased in incidence or geographic range in the last three or four decades, or threatens to do so. EIDs include the following diseases.

- ***Borrelia burgdorferi*** is a spirochaete bacterium first detected in 1982 and identified as the cause of Lyme disease, a disease transmitted by ticks; symptoms typically include fever but secondary complications can lead to chest pains, neurological conditions, and arthritis.
- ***E. coli* O157:H7** is a potentially deadly strain of *Escherichia coli*, the common gut bacterium, first detected in 1982. It is often transmitted through contaminated food, and has caused outbreaks of haemolytic diseases (diseases associated with the breakdown of red blood cells).
- **Ebola virus:** discovered in 1976, the Ebola virus is thread-like in structure (figure 1) and infects primates, causing haemorrhagic fever (characterised by severe internal bleeding). There is no vaccine for Ebola.
- **Hepatitis C:** first described in the 1980s, this is the most common blood-borne hepatitis worldwide. The virus is transmitted in blood or body fluids, most commonly through shared contaminated needles.
- **HIV/AIDS:** we have already seen that HIV (spread 15.2) is proving difficult to contain and to treat.
- **Influenza A(H5N1) virus** is the most pathogenic strain of avian or bird 'flu which has spread in Asia since 2003 and reached Europe in 2005.
- ***Legionella pneumophila*** is the bacterium that causes legionnaires' disease, which can lead to collapse of respiration due to a massive pneumonia; the bacterium thrives in aqueous environments and outbreaks have been linked to poorly maintained air conditioning systems.

Another example of an emerging infectious disease is the new variant of **Creutzfeldt–Jakob disease** (nvCJD), which was first described in 1996.

Creutzfeldt–Jakob disease (CJD)

There are several forms of CJD. Sporadic CJD (sCJD) is the most common. It seems to occur spontaneously by some unknown cause, usually in people aged over 40. Iatrogenic CJD (iCJD) occurs when someone is infected with CJD through medical or surgical treatment. It is very rare: in the UK there was only one death in 2009. Inherited prion disease (IPD) is another rare disease caused by inheriting a faulty gene that produces prions, proteins in the brain (see below). There were four deaths from IPD in the UK in 2009. New variant (nvCJD) mainly affects people in their twenties. The number of cases peaked in 2000 and is now declining. There were two confirmed deaths from nvCJD in the UK in 2009.

All CJDs are forms of transmissible spongiform encephalopathies (TSEs) which include scrapie in sheep, bovine spongiform encephalopathy (BSE) in cattle, and kuru (a human TSE, formerly associated with cannibalism among the Fore people of New Guinea; the disease died out when cannibalism ceased). In all these diseases, numerous tiny holes develop in the brain so that, under the microscope, the tissue appears spongy. In humans, TSEs are characterised by a progressive loss of memory and coordination, mood changes, and loss of ability to move and speak. Individuals affected by CJD usually die within 6 months of the disease appearing, often through pneumonia. Sadly there is no known cure, although there are many drugs that can alleviate distress.

TSEs, including CJDs, are caused by an infectious protein in the brain called a **prion**. Normal prions probably play a part in long-term memory. However, prions can sometimes lose their shape. A build-up of these defective prions has been linked to Alzheimer's disease. Infectious prions are contorted prions that enter the brain and cause normal prions to contort as well (figure 2).

BSE and nvCJD

In 1986, BSE (known more dramatically as 'mad cow disease') was a new form of prion-associated disease that occurred in cattle. Between 1986 and 1996, hundreds of thousands of infected cattle were eaten by Europeans, particularly the British. During this period nvCJD appeared in humans. This variant differs from the classic type of CJD in that, among other things, it tends to attack younger people. Between 1990 and 1996, there were 23 confirmed cases of nvCJD. Its discovery provoked intense media interest in the UK and Europe because it confirmed many people's suspicions that nvCJD is the human variant of BSE.

Since the link between nvCJD and BSE was confirmed, strict controls have been in place to stop BSE entering the human food chain and the number of cattle infected with BSE has dropped dramatically. These controls include a ban on animal-derived feed supplements to cattle and the removal of all parts of an animal's carcass that could be infected with BSE.

In the UK, after four cases of nvCJD were linked to blood transfusions, there have also been strict controls to minimise the risk of blood supplies being contaminated. These controls include not accepting blood donations from people who have received a blood transfusion since 1980, and removing white blood cells, which may carry the greatest risk of transmitting CJD, from all blood used for transfusions.

The emergence of new diseases that resist antibiotics and chemotherapy reminds us that we must keep a constant watch against pathogenic microorganisms. The complacency that developed with the introduction of antibiotics bringing about miracle-like cures is misplaced. They are useful weapons but antibiotics alone will not win the war against disease. With increased international air travel and world trading of foods, lethal diseases can spread within days or even hours from one continent to another. Emergent infectious diseases are a global threat that will require all nations to work together to fight them effectively.

Quick check

1 What is a prion?

2 Some researchers believe that CJD and other transmissible spongiform encephalopathies (TSEs) are caused by a **virino**, a tiny piece of nucleic acid coated with prion protein. It is suggested that the virino nucleic acid interacts with the host cells in some way to cause the disease. What piece of evidence in the above account indicates that virinos are not the causative agent of TSEs?

Food for thought

Suggest some experiments that might be useful in determining what prions are and how they cause disease.

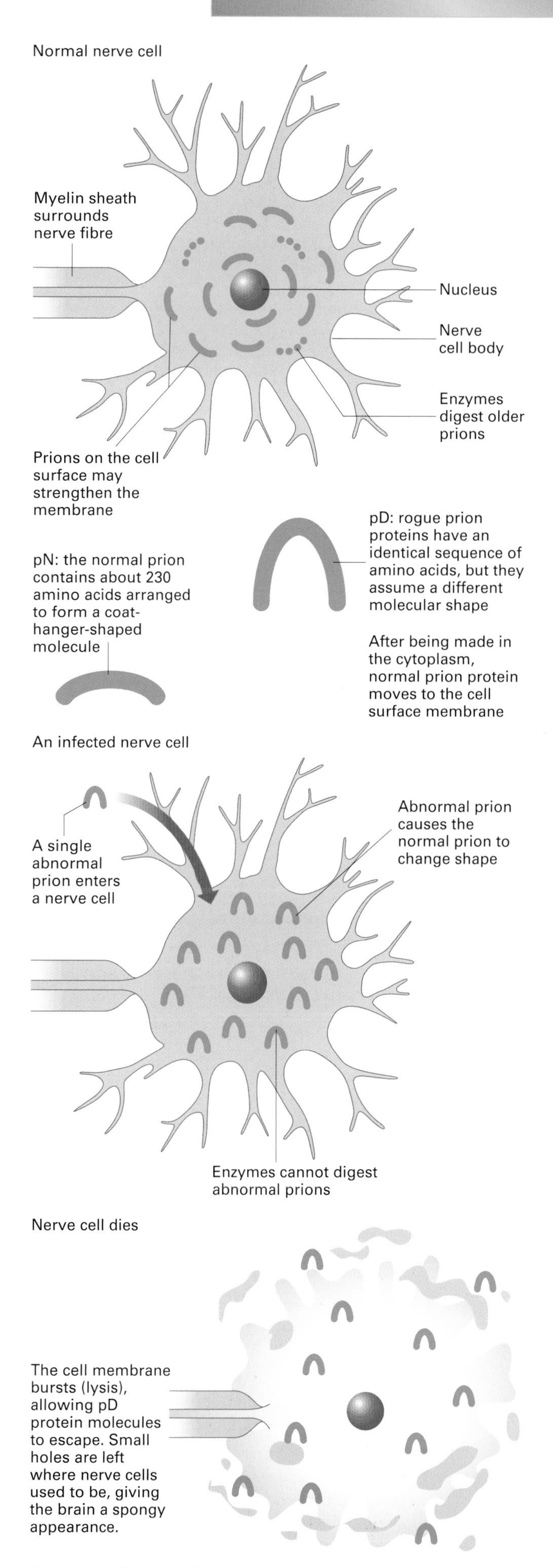

Figure 2 *CJD: how prions may infect the brain.*

Summary

Health includes mental, spiritual, and social well-being as well as physical well-being. **Diseases** have defined signs and symptoms. There are many types including **infectious diseases, non-infectious diseases, degenerative diseases, social diseases, mental illness, eating disorders**, and **deficiency diseases**. **Epidemiology** is the study of patterns of disease. **Smallpox** and **AIDS** are caused by viruses. Bacterial diseases include **salmonellosis, cholera**, and **tuberculosis**. **Malaria** and **sleeping sickness** are caused by single-celled organisms transmitted by insect **vectors**.

In mammals, the **immune system** provides the major line of defence against disease. Immune responses may be **cell-mediated** or **humoral** and involve **antigen–antibody reactions**. **Antibodies** are produced by B cells. **Monoclonal antibodies** can be manufactured using a hybrid between a B cell and a **myeloma**; the resulting cell is called a **hybridoma**. Monoclonal antibodies are used in pregnancy test kits and as diagnostic tools for a number of medical conditions, including cancer. **Immunity** against specific diseases can be provided by **vaccination**. When organ **transplants** are made, immune responses are sometimes reduced by **immunosuppressants**.

Antibody–antigen reactions are avoided in **blood transfusions** by ensuring compatibility between donor and recipient blood. **Antibiotics** destroy or inhibit the growth of microorganisms. Antibiotic resistance is becoming a problem with some microorganisms, such as *Vibrio cholerae* and *Staphylococcus aureus*. **Creutzfeldt-Jakob disease** (CJD) is unaffected by antibiotics; it is thought to be caused by **prions**.

PRACTICE EXAM QUESTIONS

1 a Copy and complete the table below to show which of the five statements about disease transmission and control apply to malaria, cholera, AIDS and tuberculosis (TB). In each box, use a tick (✓) to show that the statement applies and a cross (✗) if it does not. [5]

statement	*maleria*	*cholera*	*AIDS*	*TB*
causative agent is a bacterium				
causative agent is blood-bourne				
countrolled by pasteurisation of milk				
sexually transmitted				
transmitted by a vector				

b Explain how someone can be diagnosed as HIV positive, but not show any symptoms of AIDS. [1]

c Explain why TB has become the major cause of death for people who are HIV positive. [2]

d Explain why care must be taken in selecting antibiotics for treating cases of TB. [2]

[Total 11 marks]

2 a Outline one function of each of the following in defending the human body against bacterial infection.
Skin
Tear fluid
Gastric juice [3]

b In vaccination against tuberculosis (TB), children are injected with a weakened strain of TB bacteria.

i Explain how this procedure can result in long-term defence against TB. [4]

ii Suggest why sufferers from AIDS may contract TB, even though they have been vaccinated against it. [2]

c The body's defence system may cause problems if a blood transfusion is carried out without pre-testing for the compatibility of the blood groups of the donor and the recipient.

i Explain clearly the causes of failure of a transfusion of blood of Group A into an individual of Group O. [4]

Among North American Indians, 75% of the population belong to Group O and the remainder to group A.

ii Calculate the percentage risk of failure of random transfusion (without pre-testing of blood groups) in this population. [1]

[Total 14 marks]

3 Granulocytes are phagocytic white blood cells. The diagram shows a granulocyte magnified 10 000 times.

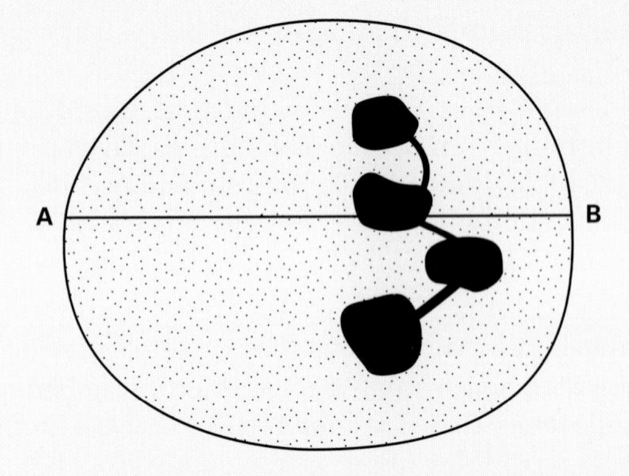

a Calculate the actual diameter **AB** of the granulocyte in micrometres. Show your working. [2]

b Give **one** major difference between the structure of this cell and a lymphocyte. [1]

c The area of a local infection contains large numbers of dead granulocytes.

i Explain the function of granulocytes at the arm of infection. [2]

ii Suggest **one** cause of the death of large number of granulocytes at the area of infection. [1]

[Total 6 marks]

4 **a** Describe the role of T cells in the immune response. [2]

b The graph below shows the development of an infection with human immunodeficiency virus (HIV) over a period of 10 years. Changes in the number of a type of T lymphocyte, T4 cells, are also shown.

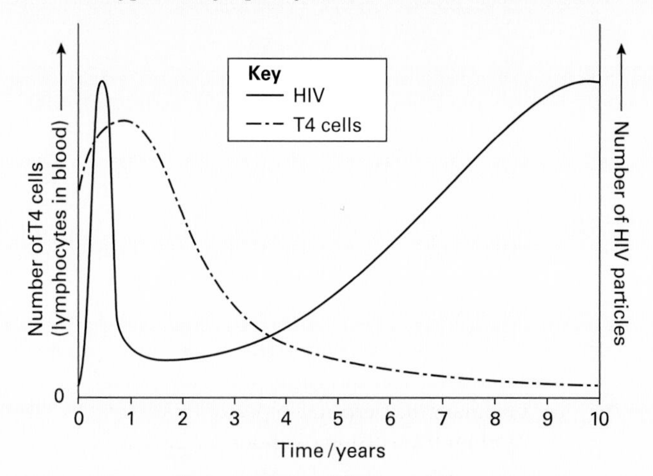

i Describe the changes in the number of HIV particles over the 10 year period. [3]

ii Describe two ways in which the curve for the T4 cells differs from that for the HIV particles. [2]

iii Suggest an explanation for the differences between the number of T4 cells in the blood and the number of HIV particles over the period. [3]

[Total 11 marks]

5 **a** Give an account of the life cycle of the parasite which causes malaria in human beings. [10]

b Indicate **five** different methods used to control malaria. [5]

c Comment on the disadvantages associated with techniques used to control malaria. [5]

[Total 20 marks]

6 Copy the following passage on monoclonal antibodies and then write on the dotted lines the most appropriate word or words to complete the passage.

Monoclonal antibodies are produced by cells called hybridomas.

First, a mouse is exposed to a particular …………………… which stimulates the production of large numbers of …………………… cells which secrete a particular antibody. These mouse cells are then fused with …………………… to produce hybridomas. The hybridomas divide to produce a culture of identical cells thus forming a …………………… The monoclonal antibodies produced by these cells are collected and concentrated before being used, for example, in diagnosing ……………………

[Total 5 marks]

7 **a** Describe how active immunity to a viral infection can be acquired naturally. [3]

b State *two* ways in which passive immunity may be acquired naturally by a young child. [2]

[Total 5 marks]

8 Give an account of the production and use of antibiotics.

[Total 10 marks]

16 Non-infectious diseases

INHERITED DISEASES

16.1

OBJECTIVES

By the end of this spread you should be able to:

- distinguish between an inherited disease and a congenital malformation
- explain the importance of genetic predisposition to common non-communicable diseases
- describe the relationship between sickle-cell anaemia and malaria.

Fact of life

One of the most common inherited diseases is primary **haemochromatosis**: an autosomal genetic disorder of a gene (known as the HFE gene) on chromosome 6 which encodes for a transmembrane protein (the HFE protein) which interacts with another small protein to regulate iron absorption. Defective HFE genes result in too much iron being absorbed from iron-rich food and laid down in body tissues, particularly in the liver. Symptoms of haemochromatosis usually manifest themselves in men aged between 30 and 40 years and women over 50. They include liver swelling and bronzing of the skin. Over time, the liver can become damaged by **iron overload**. One in about 200 people of northern European origin have this condition.

Figure 1 Woody Guthrie, a famous American folk singer who had Huntington's disease.

The best way to ensure a long life free of any major disease is to choose the right parents. If your parents live to a ripe old age, then you are more likely to. If your parents die young of, for example, a heart attack, then you have a higher risk of suffering the same fate.

The interaction between genes and the environment

The chance of a person succumbing to disease depends on an interplay between genes (the carriers of inherited information) and environment (the conditions under which the person lives, including such factors as diet and exercise). Which component, genes or environment, has the greater effect will depend on the particular disease and other circumstances.

Diseases determined almost entirely by genetic factors which can be passed from parent to offspring are called **inherited diseases**. For example, cystic fibrosis (CF) is a common genetic disorder that occurs worldwide. The genetics of this and some other inherited diseases are dealt with in chapter 19. Many inherited diseases, such as CF, sickle-cell anaemia, phenylketonuria, and haemophilia, are present from birth (and even before). Others, such as Huntington's disease (see box), do not appear until later in life, possibly after a person has had children.

All defects that are present at birth are called **congenital malformations** (**CMs**). Although some are inherited, and others are caused by environmental factors, the precise cause of many CMs is uncertain. Environmental factors which might disrupt fetal development include drugs. Thalidomide, for example, taken by pregnant women for morning sickness was found to cause fetal abnormalities such as deformed limbs (thalidomide is no longer used as an anti-morning sickness drug). Other substances that can cross the placenta and cause fetal malformations include alcohol and also viruses such as the rubella virus which causes German measles.

Many non-communicable diseases develop well after birth, and have many causes (are multifactorial). Epidemiological research has shown that they tend to occur more frequently among members of the same family, suggesting that some individuals have a genetic make-up which makes them more likely to suffer a particular disease (they are **genetically predisposed** to the disease). Common diseases with a genetic predisposition include atherosclerosis (narrowing of the arteries), coronary heart disease, hypertension (high blood pressure), diabetes mellitus, some cancers, rheumatic illnesses, and some mental illnesses. Genetic predisposition is regarded as the single most important risk factor for coronary heart disease (spread 16.7). According to the World Health Organization: *it is highly probable that genetic approaches to the prevention of common diseases will emerge as one of the dominant strategies for the improvement of health*.

Huntington's disease: a ticking time-bomb

Huntington's disease is a serious inherited disease which usually shows itself between the ages of 35 and 45. It is caused by a single mutation which causes nerves to degenerate. Brain cells die, causing behavioural changes, loss of mental powers, and uncontrollable movements. It is a dominantly inherited disease; consequently, children of a parent with Huntington's disease have a 50 per cent chance of inheriting it. About 100 000 people worldwide have Huntington's disease; 3000 of them live in the United Kingdom.

For infectious diseases, the environment (especially exposure to the pathogen) is often of prime importance in determining whether someone will suffer the disease. However, inherited characteristics can also affect the person's ability to resist and tolerate an infectious disease. The interaction between genetic and environmental factors is shown dramatically in the relationship between sickle-cell anaemia and malaria.

Sickle-cell anaemia

Sickle-cell anaemia is an inherited disease caused by a single gene mutation. (An account of how genes determine inherited characteristics is given in chapter 18.) Two genes are responsible for determining the type of haemoglobin we have: one for the pair of alpha chains and the other for the pair of beta chains (spread 7.9). Research shows that sickle-cell anaemia results from a mutation in just one nucleotide base, causing a difference in one amino acid between the abnormal beta chain and the normal beta chain.

The gene responsible for sickle-cell anaemia has two possible forms or **alleles**. The sickle-cell allele is given the symbol $\mathbf{Hb^S}$; the normal allele is $\mathbf{Hb^A}$. As each person usually has two alleles for a particular gene, three combinations are possible: $\mathbf{Hb^AHb^A}$; $\mathbf{Hb^AHb^S}$; and $\mathbf{Hb^SHb^S}$.

Only $\mathbf{Hb^SHb^S}$ individuals have the full-blown version of sickle-cell anaemia. At low oxygen partial pressures (for example, in active muscles), a large percentage of the red blood cells of people with sickle-cell anaemia become sickle shaped and unable to carry oxygen efficiently (figure 2). The distorted cells tend to form clots and block blood vessels. If the sickle-shaped cells reach the spleen, they are usually removed from circulation, causing anaemia and jaundice. There is no satisfactory treatment for the disease, which can kill, particularly during childhood, though some patients can live to 60–70 years.

$\mathbf{Hb^AHb^S}$ individuals have a much lower tendency to form sickle-shaped red blood cells. These individuals are said to have **sickle-cell trait**. The trait sometimes causes mild anaemia, particularly when oxygen partial pressures are low (for example, at high altitude or when exercising very vigorously in hot environments), but it usually causes no adverse symptoms. People with sickle-cell trait can lead normal and active lives.

Sickle-cell anaemia is endemic in parts of Africa, the Middle East, and India, where malaria is prevalent. In these areas, the $\mathbf{Hb^S}$ allele is far more common than in other areas because it gives some resistance against malaria. In people with sickle-cell anaemia or sickle-cell trait, red blood cells have a low binding capacity for oxygen. Because the malarial parasite has an active aerobic metabolism, requiring a plentiful supply of oxygen, it cannot grow and reproduce within these red blood cells.

The relationship between sickle-cell anaemia and malaria shows how the development of disease is affected by a complex interaction between inherited and environmental factors.

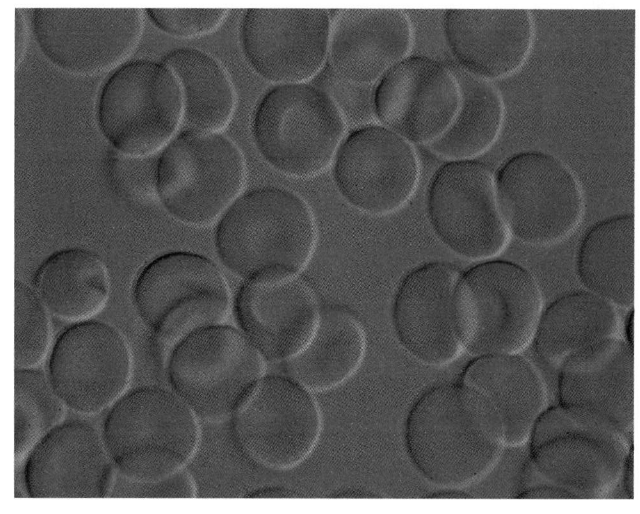

(a)

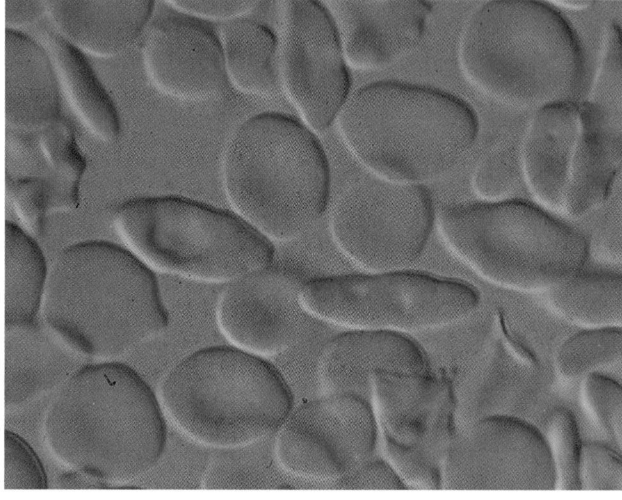

(b)

Figure 2 ***(a)*** *Normal red blood cells (×2000).* ***(b)*** *Abnormal red blood cells from a person with sickle-cell anaemia (×2000).*

Quick check

1 Give one example of an inherited disease that is *not* congenital.

2 Name one non-communicable disease for which genetic predisposition is known to play a large part in its development.

3 Under what conditions do the red blood cells of people with sickle-cell anaemia and, to a lesser extent, those with sickle-cell trait become sickle shaped?

Food for thought

The human genome project (spread 19.6) aims to identify all the human genes and map their chromosome locations. When completed, it will be possible to evaluate the genetic predisposition of an individual to particular diseases. Suggest the advantages and disadvantages of being able to carry out such evaluations. What use do you think insurance companies might make of such information?

16.2

Allergies

OBJECTIVES

By the end of this spread you should be able to:

- discuss the role of immunoglobulins in allergies
- distinguish between generalised and localised allergic reactions
- discuss the causes, symptoms, and treatment of asthma and hay fever.

Fact of life

The Ancient Greeks recognised asthma more than 2000 years ago, naming it 'the panting attack'. Asthma is one of the most common ailments in the UK; in 2012, there were 4 million people with this non-infectious disease. For most, symptoms are mild. But for some it's a serious condition: three people a day die from asthma in the UK.

Figure 1 Scanning electron micrograph of a house dust mite (×125). House dust mites can cause allergic responses such as asthma, dermatitis (inflammation of the skin), or an inflamed, runny nose. These tiny mites are virtually invisible, being about 0.3 mm long. They feed on human skin that gathers in bedclothes, carpets, and mattresses. Their faeces, which are about the size of pollen grains (10–40 μm in diameter), contain glycoproteins and small proteins that act as allergens: they may trigger an allergic reaction in some sensitive people. According to the World Health Organization, 45–85% of asthmatics are allergic to mites, compared with 5–30% of non-asthmatic people.

What is an allergy?

An **allergy** is a response to an antigen that, in most people, has no effect. The response damages tissues in the body. Allergies happen in people who are **hypersensitive** to the antigen. They develop the allergic response the second or subsequent times they come into contact with the antigen. An antigen that causes an allergic response is called an **allergen**.

The role of mast cells

In the 1960s, scientists discovered that an allergic reaction is an exaggerated immune response (spread 15.5). On the first exposure to an allergen, B cells differentiate into plasma cells that (with the help of T cells) produce a specific type of immunoglobulin (antibody) called **IgE** (figure 2). Individuals with allergies have a genetic predisposition to produce IgE. Once produced, IgE binds to **mast cells**. These cells are derived from the same cells that form blood cells. They have a large nucleus and they are often amoeboid (they look and move like the unicellular organism *Amoeba*).

On a second exposure to the allergen, the allergen attaches to the IgE on the sensitised mast cells. This causes the lysis (breakdown) of mast cells which release chemicals such as histamine, serotonin, and heparin. These chemicals cause the symptoms of the allergic reaction.

Generalised responses

An allergy may be generalised or localised. A generalised allergic reaction is potentially fatal. In hypersensitive individuals, the allergen triggers the immediate release of large amounts of chemicals from mast cells. The chemicals act on smooth muscle cells, typically causing the bronchioles to constrict, restricting the passage of air in and out of the lungs (see spread 7.1). The chemicals also cause dilation of arterioles (the small vessels leading from arteries into capillaries), greatly reducing arterial blood pressure. Within a few minutes following contact with the allergen, the combined effect of these reactions can kill an individual from asphyxiation and circulatory shock. Treatment is by immediate injection of adrenaline. Allergens that can trigger such a generalised allergic response include nuts, drugs (for example, penicillin), and insect venom from the stings of bees, hornets, and wasps.

Localised responses: hay fever and asthma

Examples of localised allergic reactions include hay fever and bronchial asthma. **Hay fever** (**allergic rhinitis**) involves the upper airway. The first exposure to the allergen (commonly pollen or the faeces of house dust mites) sensitises mast cells in the mucous membranes of the upper airway, especially the nasal cavity. Subsequent exposure to the allergen typically causes itchy and watery eyes, congested nasal passages, coughing and sneezing. Hay fever can often be relieved with antihistamine drugs.

Asthma (from a Greek word meaning 'panting') may be brought on by exercise or lung infections, but the most common form results from an allergic reaction in which mast cells in the lower part of the airway release their chemicals. These chemicals cause the bronchioles to constrict and the alveoli to become full of fluid and mucus. This leads to coughing, wheezing, and difficulty in breathing. Common allergens include pollen, fur, feathers, and house dust. In the UK, the number of asthmatics has risen dramatically in recent years so that asthma is now one of the most common childhood ailments.

Treatment for asthma includes gentle reassurance to reduce anxiety (asthmatic attacks are made worse by anxiety); bronchodilators (drugs related to adrenaline which dilate the bronchioles, figure 3); and steroids which reduce the inflammatory response.

It is vital that asthmatics who use bronchodilators keep them handy for quick use. In the UK, about 1500 people die of asthma attacks each year. Many of these deaths could probably have been avoided if bronchodilators had been readily available.

Lung function tests

Asthmatics have difficulty expelling air from the lungs. They tend to have a lower **forced expiratory volume$_{1.0}$** ($FEV_{1.0}$) than non-asthmatics of the same age ($FEV_{1.0}$ tends to decrease with age). The $FEV_{1.0}$ is the volume of air forcibly exhaled in the first second after a maximal inhalation. It is one of the lung function tests used in the diagnosis of asthma and other lung diseases. Another lung function test measures **peak expiratory flow** using a peak flow meter. The meter measures how quickly you can exhale after taking a deep breath.

Figure 3 *An asthma patient using a bronchodilator drug.*

Quick check

1. To which cells are immunoglobulins (IgE) attached during an allergic reaction?
2. Explain how a generalised allergic reaction can result in circulatory shock.
3. Name a localised allergic reaction of the upper respiratory tract.

Food for thought

Bronchodilators such as salbutamol are sympathomimetic drugs (they mimic or copy the effects of the sympathetic nervous system; see spread 10.7), with effects similar to adrenaline. Suggest why Olympic athletes suffering from asthma are permitted to use salbutamol, but they are not allowed to take *any* sympathomimetic drugs by injection.

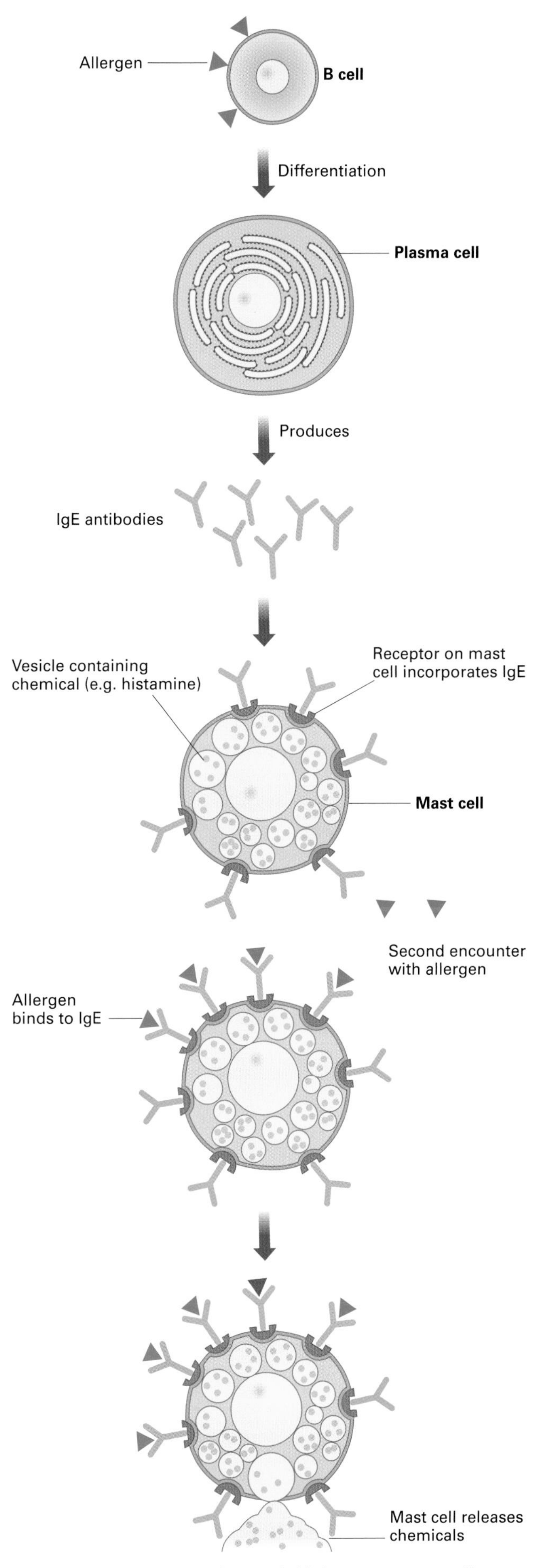

Figure 2 *An allergic reaction: on initial exposure to an allergen, IgE antibodies attach to mast cells. Upon second exposure, the allergen combines with the antibody and stimulates the mast cells to release their potent chemicals, causing the allergic reaction (e.g. hay fever or asthma).*

16.3

Ageing and Alzheimer's disease

OBJECTIVES

By the end of this spread you should be able to:

- discuss how the human body changes with age
- describe the changes that take place in the brain with age
- discuss the nature of Alzheimer's disease.

Fact of life

In 2010, there were about 10 million people over 65 in the UK. This is expected to increase to 15.5 million by 2030, and 19 million by 2050.

You may be delighted to know that the chance of reaching a 'ripe old age' is higher now than it has ever been. As our understanding of ageing increases, we will be able to extend our lives even further. Ageing and longevity (length of life) appear to have a strong genetic component:

- Children of long-lived parents tend to have long lives.
- The normal ageing process can be speeded up by gene defects such as Werner's syndrome (see box).
- Different animals have different life expectancies (figure 1).

Because of this genetic involvement in ageing, we might soon be able to manipulate our genes to improve our life expectancy. Although an understanding of the genetic basis of longevity may give us the 'secret of eternal youth', the goal of most gerontologists (biologists who study ageing) is to improve the quality of later life, not to extend it indefinitely.

The effects of ageing: senescence

As we grow older we tend to become physically weaker: muscle mass decreases, and bones become more brittle and break more easily (see spread 16.4). The physiological processes that maintain our organs and systems become less sensitive, less accurate, and slower: reaction times become slower, maximum heart rate is lowered by about one beat for each year a person ages, and the resting heart rate and basal metabolic rate (BMR) decline (after the age of 60, BMR slows down at a rate of up to 3 per cent each year). The likelihood of suffering disability and disease increases (as the immune system becomes less efficient) and, eventually, we are confronted with a situation to which we can no longer respond effectively and we die. The complex of ageing processes that eventually lead to death is called **senescence**. Senescence affects every part of our body, including the brain.

Nerve cells in the brain either cannot regenerate at all or do so only slowly. Dead and damaged cells are therefore not replaced as quickly as they are lost. The number of nerve cells declines with age and the brain becomes smaller. It is estimated that the brain of a 90-year-old has about 10 per cent fewer nerve cells than that of a 30-year-old. You might think, therefore, that brain function would inevitably decline with age also. However, this is not necessarily so. In the absence of senile dementia, although some memory loss commonly occurs, certain types of intellectual skill can actually increase with age. Most biologists believe that a decrease in brain function is not an inevitable consequence of ageing. It is now thought that memory loss, personality changes, and other malfunctions could be due to a disease process which might be treated or even prevented.

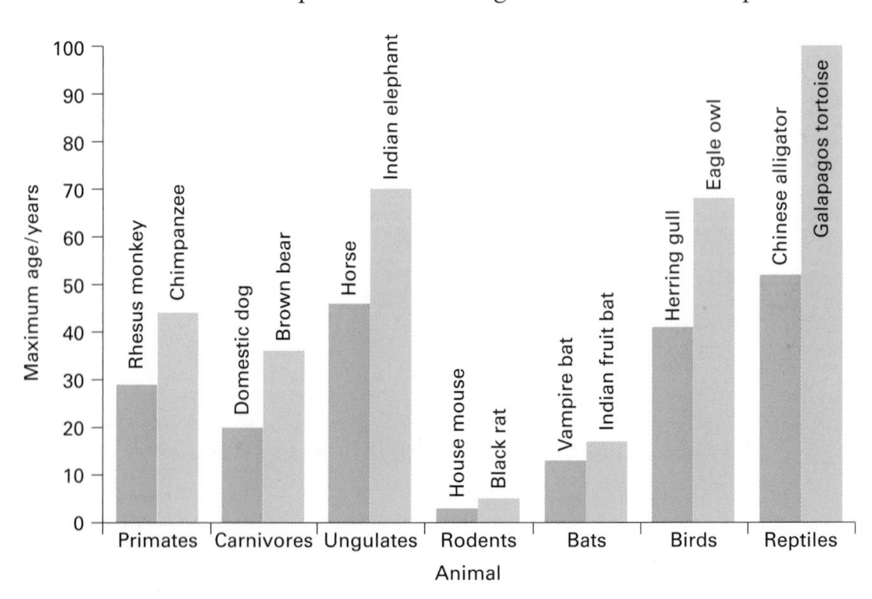

Figure 1 *The varied maximum lifespans of different animals suggest that longevity is partly determined by genes.*

Senile dementia and Alzheimer's disease

Senile dementia is a loss of intellectual faculties which usually begins after the age of about 65 years. It is characterised by memory loss, a gradual decline in reasoning, a tendency to be confused easily, and personality changes. It is not itself a disease, but rather the symptom of a variety of diseases of which Alzheimer's disease is the most common. In this disease there is a gradual build-up of protein which forms deposits called plaques outside cells, and tangles within the nerve cells which disrupt brain function (figure 2). The plaques consist of an abnormal protein called beta-amyloid which has between 39 and 42 amino acids. Normally, **prions** (spread 15.10) help regulate the production of beta-amyloid and prevent the formation of tangled plaques. In Alzheimer's patients, the areas most affected by plaques are the cerebral cortex and the hippocampus; both are involved in memory, and the hippocampus is the seat of personality. The brain of an Alzheimer's patient also shrinks significantly.

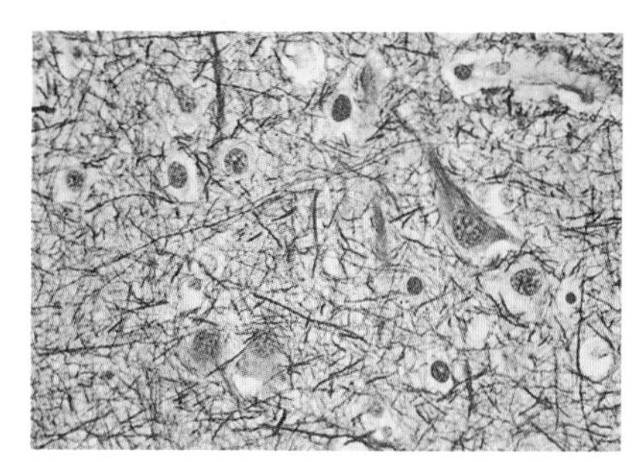

***Figure 2** Plaques (patches of beta-amyloid) and tangled nerve fibres due to accumulated tau in the brain of an Alzheimer's sufferer.*

Causes of Alzheimer's disease

A variety of genetic and environmental factors contribute to the development of Alzheimer's disease. Although it is mainly a disease of old age, it can also strike younger people. Those with a family history of the disease are at increased risk, indicating a strong genetic component. Very rarely patients with early-onset Alzheimer's disease have gene mutations in chromosome 21 (the chromosome involved in Down's syndrome). This mutation affects the production of beta-amyloid proteins. Other genetic abnormalities have been found in both young and old Alzheimer's patients.

Alzheimer's disease is the fourth biggest killer in economically developed countries after heart disease, cancer, and stroke. It not only kills, it can also strip a person of the will to live and spreads misery among loved ones. Although symptoms can be controlled to a variable extent, for example, by antidepressants such as monoamine oxidase-B inhibitors, there is no known cure. Advances in stem cell research have given hopes for an effective treatment in the future. **Stem cells** have the ability to turn into any cell type, including nerve cells in the brain. Unfortunately, Alzheimer's disease is caused by the cell death of many different groups of cells in several areas of the brain, making stem cell therapies challenging. It is very difficult to ensure that transplanted cells will have all the cell signals needed to become integrated into a brain damaged by Alzheimer's disease. Nevertheless, stem cells may be used to delay the development of Alzheimer's disease. Genetically modified stem cells have the potential to carry substances to the Alzheimer brain that stop nerve cells from dying and stimulate the function of existing cells. Stem cell therapy offers great potential for other diseases such as Parkinson's disease, in which dementia may develop due to the malfunctioning of well known cell types.

Delaying the inevitable

You cannot halt ageing. It starts at conception and ends in death. However, research indicates that you can increase your chance of living a long and healthy life by taking regular exercise and eating sensibly. Although everyone suffers a steady deterioration in fitness with each decade, the deterioration in active people is much less than in those who are inactive. Eating food such as fruits and vegetables that are rich in antioxidants (for example, vitamins A, C, and E) may also slow down the ageing process by mopping up free radicals (spread 16.6). Many scientists believe that these are one of the main environmental cause of ageing. Women approaching menopause may also be able to delay some of the adverse effects of ageing by taking hormone replacement therapy (see spread 12.4).

Accelerated ageing

About one in 10 million people will get old before their time because of a rare genetic disorder called **Werner's syndrome**. The first signs are failure to go through the usual adolescent growth spurt. Soon after this, hair goes prematurely grey, cataracts develop in the eyes, the skin becomes less flexible, and diabetes mellitus, osteoporosis, thickening of the arteries, and tumours develop. These degenerations normally occur much later in life. The life expectancy of someone with Werner's syndrome is only about 45 years, even with considerable medical assistance.

Quick check

1 What is the name given to the complex of ageing processes that eventually leads to death?

2 State the major change that takes place in the brain with age.

3 Which type of protein associated with Alzheimer's disease accumulates outside nerve cells in the brain?

Food for thought

Laboratory mice and rats kept on energy-restricted diets have been found to live longer and have fewer age-related diseases than those fed on unrestricted diets. Suggest reasons for these results. How applicable do you think these results might be to humans?

16.4

Osteoporosis and osteoarthritis

OBJECTIVES

By the end of this spread you should be able to:

- discuss how the development of osteoporosis is linked to diet and exercise
- explain why hormone replacement therapy may prevent the development of osteoporosis
- describe the typical features of osteoarthritis.

Fact of life

Each year, osteoporosis (brittle bone disease) affects over 1.7 million people worldwide. The number is expected to grow to over 6 million a year by 2050. For those who suffer complications such as hip fractures, the risk of death is high.

Osteoporosis

Osteoporosis (brittle bone disease) is characterised by a reduction in bone mass: more bone is broken down in the body (bone resorption) than is laid down (bone deposition). The disease is usually associated with loss of weight. Other symptoms include tiredness, bone pains, and acute back pain. However, the most harmful effect of the disease is that the bone becomes porous, brittle, and liable to fracture.

Osteoporosis is an age-related disease, primarily affecting post-menopausal women (figure 2). However, it can occur much earlier in life and can affect men whose bone densities are particularly low.

Bone is living tissue consisting of a calcified matrix and fibres of collagen (spread 11.2). As a bone grows, it responds to the mechanical stresses that it is subjected to. A femur of an active person who performs regular load-bearing exercises, for example, is generally thicker and denser than the femur of an inactive person. Bone elongation stops at maturity, but bones continue to change in density and strength throughout adulthood. There is generally some degree of bone weakening with age (figure 2), but insufficient exercise or a low calcium diet (especially in early life when bones are elongating) may accelerate the onset of osteoporosis. Excessive exercising and weight-loss dieting are also linked with development of the disease.

Some people may be genetically predisposed to osteoporosis. A gene affecting vitamin D metabolism has been found in Australians which may largely account for the genetic variation in their bone density (vitamin D enhances calcium and phosphorus absorption from the gut and, with parathyroid hormone, lays down these minerals in bones). If this gene occurs in a wider population, it may not be long before genetic screening can be used to identify those at risk of osteoporosis so that preventative measures can be taken.

Preventing osteoporosis

Moderate weight-bearing exercises (such as cycling, dancing, and walking) stimulate calcium deposition, strengthen bones, and reduce the risk of osteoporosis. Young people benefit most from exercise because strong and healthy bones are more easily established in adolescents and young adults.

Slightly overweight post-menopausal women are generally less likely to suffer from osteoporosis than slim women. The extra fat may act as a protective cushion for bones, and the bones of plump women

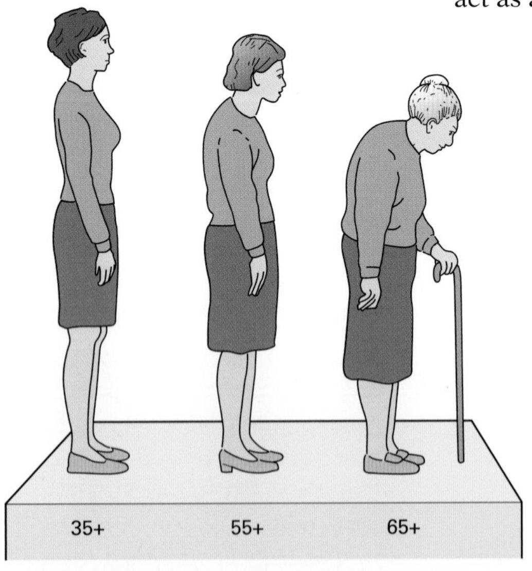

Figure 1 Osteoporosis images such as these have been used by the National Osteoporosis Society (UK) to heighten awareness of the disease and to encourage women to take preventative measures.

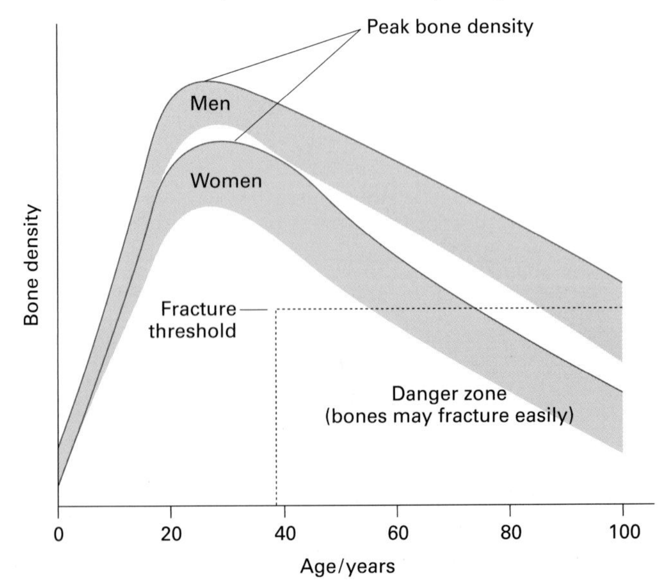

Figure 2 Bone density decreases with age, increasing the risk of osteoporosis.

may be heavier and stronger. However, the most likely reason for the lower incidence of osteoporosis in plump women is that the extra fat enables them to produce more oestrogen. This hormone is involved in calcium metabolism in both males and females. Its presence reduces the demineralisation of bones. Some doctors believe that the best way to prevent osteoporosis in post-menopausal women is by **hormone replacement therapy (HRT)** which gives low doses of oestrogen. Unfortunately, this treatment is not suitable for all women.

A diet that includes a good intake of calcium-rich foods (such as dairy products) may help prevent osteoporosis. It is especially important for pre-menopausal women to have adequate calcium intake so that they can build up high peak bone densities in early adult life (table 1).

By a careful mixture of exercise and a calcium-rich diet, most people should be able to prevent the development of osteoporosis. The prevention of another age-related bone disease, osteoarthritis, is much more problematic.

Table 1 *Daily calcium intakes recommended by the National Osteoporosis Society (UK).*

Population group	Calcium intake per day (mg)
Children 7–12 years	800
Teenagers/adults	1000
Women over 45 years old (not taking HRT)	1500
Women over 45 years old (taking HRT)	1000
Pregnant and lactating teenagers	1500
Pregnant and lactating women	1200
Men over 45 years old	1500

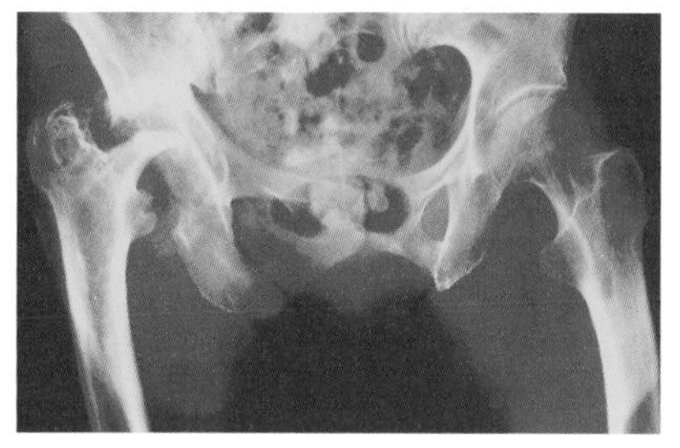

Figure 3 *Coloured X-ray of a hip joint showing destruction of the right hip (left on image) due to osteoarthritis.*

Osteoarthritis

Osteoarthritis (osteoarthrosis) is a degenerative disease of the cartilage that covers bones in joints. It is characterised by stiffness in heavily used joints such as in the knees, hips, and fingers. Osteoarthritis may progress into the bone itself, causing pain and stiffness. X-rays of people with osteoarthritis reveal that the joint space (the space between the bones) narrows because of loss of cartilage, and bony deposits and other irregularities develop at the edges of the bones (figure 3).

Although osteoarthritis affects some children, it is more common in those past middle life. It results in more than 50 000 hip and knee replacements each year in the UK. Its development is linked with obesity, low bone density, congenital abnormalities in the structure of joints, and repeated mechanical stress. However, a comparison of the incidence of osteoarthritis in pairs of identical and non-identical twins indicates that the disease is partly genetically determined.

The development of osteoarthritis is usually slow and irreversible, but it is rarely crippling. In most cases, the symptoms can be kept under control for many years by proper treatment which usually includes reduction of pressure across the joint (for example, by losing weight and using a walking stick), the use of analgesics (painkillers) and anti-inflammatories such as acetylsalicylic acid (aspirin), and corrective surgery.

Osteoarthritis is sometimes confused with another type of joint disease called **rheumatoid arthritis (RA)**. However, RA is a chronic inflammatory disorder rather than degeneration of the cartilage. It starts gradually and may occur at any age, but usually occurs between the ages of 30 and 40 years. Typically, many joints are affected at the same time (especially the fingers, wrists, and ankles). RA is an autoimmune disease in which the immune system attempts to destroy its own tissue (spread 15.5). Treatment includes anti-inflammatories (such as aspirin) and other drugs, gentle exercise to maintain joint mobility, and ice treatment to relieve swelling. In very severe cases, joints may be replaced.

Quick check

1 Explain why load-bearing exercise during adolescence can prevent the development of osteoporosis in later years.

2 Explain why hormone replacement therapy may prevent osteoporosis.

3 Describe the typical features seen on an X-ray of a joint of someone with osteoarthritis.

Food for thought

At one time, milk was supplied free of charge in schools to all children. Suggest why discontinuing free school milk may contribute to an increased incidence of osteoporosis.

16.5

Cancer

OBJECTIVES

By the end of this spread you should be able to:

- discuss the causes of cancer
- describe how a malignant cancer can spread throughout the body
- discuss methods of prevention and treatment of cancer.

Fact of life

In 2008 an estimated 12.7 million new cancer cases were diagnosed worldwide. In the UK, half of people diagnosed with cancer now survive their disease for at least five years.

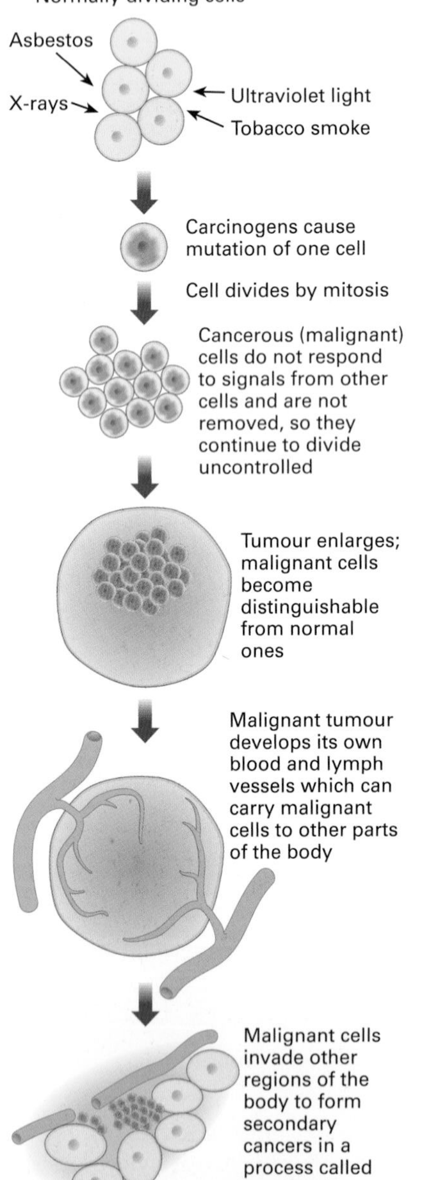

Figure 1 *The development of a malignant tumour.*

Cancers are among the most feared of diseases. There are more than 200 known types. They have many causes and treatments but all result from uncontrolled cell division.

Cells dividing out of control

In healthy, normal body tissue, cells divide by mitosis. The process is carefully controlled mainly by **cell signalling** pathways that switch genes, such as **proto-oncogenes** and **tumour supressor** genes, on and off. Proto-oncogenes code for proteins that stimulate cell division, whereas signals from tumour suppressor genes inhibit cell division. In a healthy cell, the activities of these two types of gene are in balance. Problems arise when the genes mutate or other control mechanisms break down so that cells divide uncontrollably (figure 1). Most cancers are probably caused not by a single factor but by a combination of genetic and environmental factors operating over several years.

Cancer-causing agents in the environment are called **carcinogens**. They probably trigger cancers by causing the proto-oncogenes to mutate into oncogenes (*onkos* means tumour). Most mutated cells are either destroyed by the body's immune system or die, causing no harm to the body. However, a single mutated cell may divide uncontrollably to form a clone of identical cells. Eventually a mass of abnormal cells called a **tumour** is formed. Most tumours, such as common warts, are benign and do not spread from their point of origin. However, some **benign tumours** (such as ovarian cysts) can compress surrounding tissues and displace them. Tumours that spread through the body are called **malignant tumours**. Malignant tumour cells can be carried by the bloodstream or lymphatic system to invade other tissues, causing **secondary cancers**. This process is called **metastasis**. It is the most dangerous characteristic of cancer, as it can be very difficult to find secondary cancers and remove them.

Causes of cancer

The causes of cancer are far from fully understood, but several specific triggers have been identified. At least 15 per cent of all cancers worldwide are a consequence of chronic infectious disease, the most important being hepatitis B and C viruses (linked to liver cancer), HIV (the human immunodeficiency virus, linked to cervical cancer), and HPV (the human **papillomavirus**, linked to genital and throat cancers).

About 5 per cent of human cancers tend to have a family history, indicating that there is a strong genetic predisposition towards these cancers. It is thought that these predispositions are due to the inheritance of particular oncogenes, or of proto-oncogenes that are readily transformed into oncogenes if exposed to particular carcinogens.

Carcinogens include radiation, ultraviolet light from the Sun, and X-rays, which can all damage DNA and cause mutations which may lead to cancers. Ultraviolet light from the Sun is the most common form of carcinogenic radiation. Exposure to certain wavelengths of ultraviolet light is linked to the development of skin cancers, including a highly malignant form called a **melanoma**.

Chemical carcinogens include inorganic arsenic compounds linked to skin cancers, tars in tobacco smoke (spread 16.6), and asbestos products which cause lung cancers, along with some dietary substances. Many common foods are believed to contain carcinogens, but the levels are usually too low to be harmful.

Several detailed investigations have revealed a link between diet and cancer. For example, a high-fibre diet consisting of plenty of fruit and vegetables reduces the risk of cancer of the colon.

Treating cancers

There is no single cure for all cancers, but many can be prevented or treated successfully. Prevention includes minimising exposure to known carcinogens such as asbestos dust, tobacco smoke, and ultraviolet light. Effective treatment often depends on early diagnosis of a cancer. Most cancers of the skin, colon, breast, and cervix can be cured if diagnosed early.

Diagnosis of cervical cancer and breast cancer involve **screening programmes** in which members of a population who show no symptoms are tested to find early signs of the disease. Screening for cervical cancer includes the cervical smear test. Cells are gently scraped from the cervix and examined microscopically for abnormalities (figure 2). Screening for breast cancer, which affects about one woman in twelve at some stage in their lives, involves self-examination for any unusual lumps; mammography, in which the breasts are X-rayed; and biopsy, in which sample cells are examined for malignancy. Advances in DNA technology have led to the use of **DNA probes** as diagnostic tools for a variety of cancers (spread 18.9). A DNA probe is a short segment of cloned DNA or RNA which binds to a target DNA with a complementary nucleotide sequence. The probe can be genetically engineered to match the mutant DNA in cancerous cells; it is labelled (for example, with a fluroescent protein) to make identification of the target cells easier.

Treatments for cancer include:

- surgical removal of the tumour
- radiotherapy: the application of radiation to destroy cancerous cells selectively
- chemotherapy: the use of chemicals to kill cancerous cells preferentially.

Monoclonal antibodies (spread 15.6) can be used to carry radiation and chemicals specifically to cancerous cells.

Cancer research is a very active field and new knowledge is gained daily, enabling doctors to establish the causes of the diseases and to design effective treatments. However, much remains unknown. Finding cures for cancers will continue to be one of our greatest challenges in the twenty-first century.

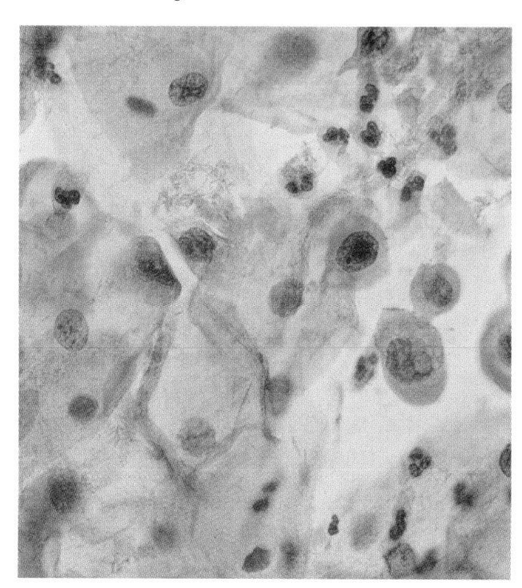
(a)

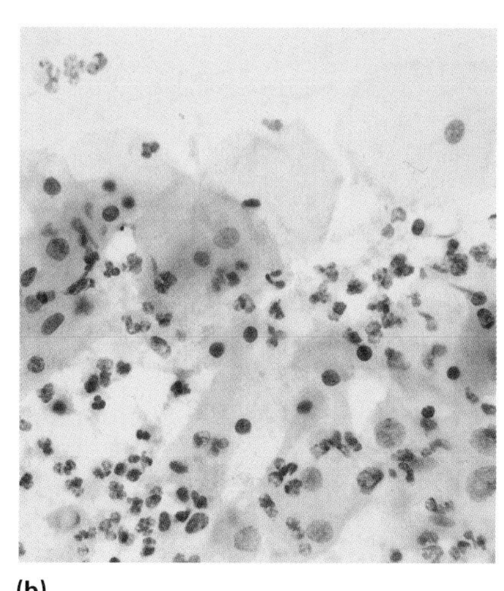
(b)

*Figure 2 Results of cervical smear tests. **(a)** Normal epithelial cells stained pink and blue. **(b)** Malignant epithelial cells. Indicators of malignancy include abnormal size and shape (e.g. tadpole cells), abnormal nuclear size, shape, content or mitotic division.*

Cancer screening techniques

In addition to radiography and smear tests, cancer screening techniques include **ultrasonography** and **endoscopy**. In ultrasonography, a controlled beam of ultrasound above 30 000 Hz is directed into the body to produce images of internal structures. In endoscopy, a tube is inserted into the body. A light at the end of the tube allows an optical system or miniature video camera to transmit an image to the examiner's eye. Both techniques can be used to detect unusual growths so that a biopsy can be performed. A **biopsy** involves removing a small piece of the unusual tissue for microscopic examination. Now that the human genome project (spread 19.6) has identified the location and nucleotide sequence of all human genes, it is likely that future cancer screening will include **DNA profiling** (spread 18.11) to identify people with a genetic predisposition to particular cancers. For example, BRCA1 and BRCA2 are tumour suppressor genes. Mutation of these genes has been linked to hereditary breast and ovarian cancer.

Quick check

1 What is a carcinogen?

2 What is the name given to the process by which a malignant tumour can spread through the body?

3 Explain why the risk of developing cervical cancer can be reduced by screening.

Food for thought

In economically developed countries, breast cancer is the most common form of cancer among women. The chance of treating the disease successfully improves dramatically if the cancer is detected and treated in its early rather than later stages. Many governments are promoting educational campaigns to encourage more woman to go for breast cancer screening so that they can reduce deaths. What information would you include in an educational leaflet about breast cancer screening procedures for the general public?

16.6

Smoking and disease

OBJECTIVES

By the end of this spread you should be able to:

- name the main harmful components of cigarette smoke
- explain how nicotine can contribute to cardiovascular disease and osteoporosis
- discuss the reasons why cigarette smoking increases the risk of lung diseases.

Fact of life

The link between lung cancer and tobacco was established in 1950 with a study that showed a 26-fold increased risk of lung cancer among smokers of 15–24 cigarettes a day, compared with non-smokers. Although this study demonstrated a clear **correlation** between smoking and lung cancer, it did not prove a **causal relationship** (i.e., that smoking causes lung cancer). Lawyers defending tobacco companies argued that it was not smoking that caused the lung cancers, but some other factor strongly associated with the two. Lung cancer is a complicated multifactorial disease. Some heavy smokers never develop lung cancer while some non-smokers do. Nevertheless, since the 1950s so much evidence has accumulated confirming a strong link between smoking and cancer that it is generally accepted that smoking is the main cause of lung cancers.

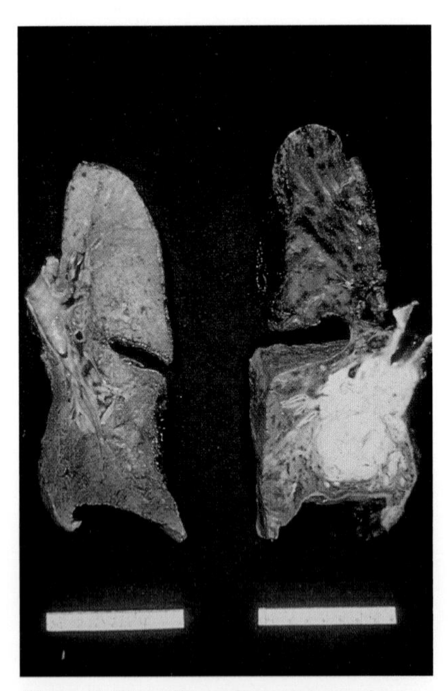

***Figure 1** The effect of cancer on a lung. The left-hand lung is free of cancer. The right-hand lung is riddled with a cancerous growth (the large white area in the lower part of the lung).*

Many smokers claim that their habit helps them to relax, but health statistics show that this relaxation comes at a high price. In the USA and some other economically developed countries, almost 20 per cent of deaths have been attributed to diseases associated with inhaling the products of combustion from the tobacco plant *Nicotiana tobacum*. These products include nicotine, carbon monoxide, and tars.

Nicotine

Cigarette smoking (or the passive inhalation of someone else's cigarette smoke) is the most common method of inhaling tobacco smoke, along with cigar and pipe smoking. The **nicotine** in the smoke gives cigarettes their psychological and addictive effects. It also contributes to some of the harmful effects of smoking on the cardiovascular system and the lungs.

Studies carried out in the 1980s revealed that nicotine mimics the actions of some neurotransmitters, especially dopamine in the brain and acetylcholine in the parasympathetic nervous system. The dopamine receptors which are stimulated by nicotine are known to be involved in addictions to other substances such as amphetamines and cocaine. The stimulatory action of nicotine on the parasympathetic nervous system has a number of effects: it constricts the finer bronchioles in the lungs, reducing air flow in and out of the lungs; it paralyses cilia which remove dirt and bacteria from the trachea; and it raises the blood pressure and heart rate, increasing the risk to smokers of **cardiovascular disease** (spread 16.7).

Nicotine also affects hormone production. For example, cigarette smoking lowers blood oestrogen levels and therefore reduces bone mineralisation, increasing the risk of **osteoporosis** (spread 16.4).

Carbon monoxide

Every time a smoker inhales tobacco smoke he or she is poisoning some red blood cells. Like car exhaust fumes, tobacco smoke contains **carbon monoxide**, which passes into the bloodstream where it can combine irreversibly with haemoglobin to form **carboxyhaemoglobin**. As haemoglobin has a much higher affinity for carbon monoxide than it does for oxygen, the formation of carboxyhaemoglobin reduces the oxygen-carrying capacity of the blood. Among heavy smokers, this reduction may be as much as 10 per cent, reducing their ability to take strenuous exercise, and accounting for some of the breathlessness experienced by smokers.

Tars

Cigarette smoke contains **tars**. These are organic substances which can stick on to cells in the lungs. The most important of these substances are polycyclic hydrocarbons, which can release carcinogenic free radicals into cells, increasing the risk of cancer, especially of the lungs (figure 1). Lung cancers are slow to develop and by the time they are diagnosed metastasis (spread 16.5, figure 1) has usually happened.

Free radicals are unstable, chemically incomplete substances that 'steal' electrons from other molecules. They are highly reactive chemicals. It appears that antioxidants in our diet (especially vitamins A, C, and E) mop up free radicals, enabling us to cope with small amounts of them, but large amounts of free radicals can damage chemicals such as enzymes, reducing their effectiveness. Free radicals can also damage DNA, disrupting the delicate store of information; this may lead to the development of cancers. Medical scientists believe that free radicals contribute to at least 50 other major diseases including atherosclerosis, heart disease, rheumatoid arthritis, and lung disease. They may even accelerate the ageing process. In addition to smoking, our exposure to free radicals can be increased by pollution, certain food additives, and ultraviolet light. The high content of free radicals in tar may explain why smokers break down vitamin C much faster than non-smokers. In the UK, it is estimated that the vitamin C requirements of smokers may be twice as much as for non-smokers and make conditions such as **pneumoconiosis** (see right) much worse.

Tars in cigarette smoke can irritate and damage lung tissue both mechanically and chemically. This can lead to a **chronic obstructive pulmonary disease** (COPD), a group of lung diseases characterised by chronic breathlessness due to an airway obstruction. COPDs include emphysema, chronic bronchitis, and pulmonary fibrosis. **Emphysema** results from the gradual breakdown of the alveoli, decreasing the total surface for gaseous exchange (figure 2). **Bronchitis** is an inflammation of the lining of the bronchioles. Acute bronchitis lasts a few days and is commonly associated with colds. Chronic bronchitis is longer lasting and usually much more serious. It is often caused by cigarette tars irritating the bronchiole walls and triggering an **inflammation response**. Coughing associated with bronchitis can lead to emphysema. **Pulmonary fibrosis** is the formation of excess fibrous connective tissue in the lungs, causing scar tissue which decreases the ability of the lungs to absorb oxygen by diffusion. Pulmonary fibrosis can occur without any known cause but cigarette smoking can increase the risk or aggravate the condition.

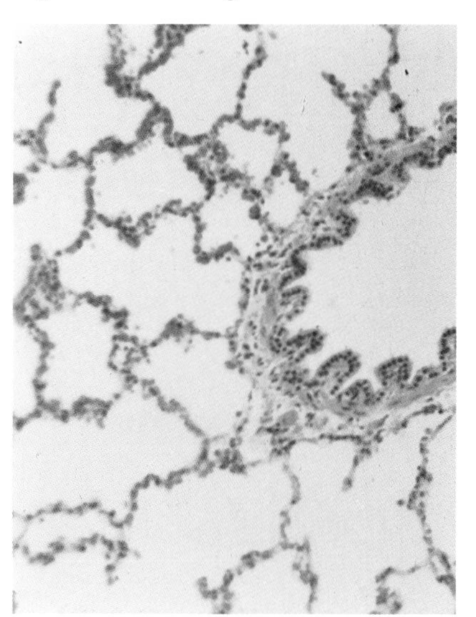

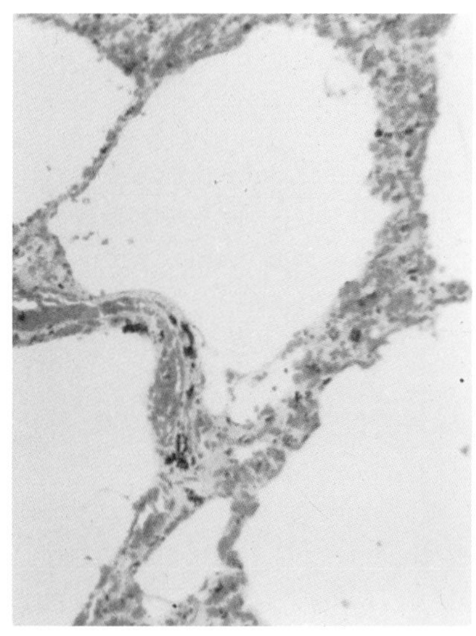

Figure 2 *Section of a lung from a patient who had emphysema (right) compared with a healthy lung (left). The air spaces (white areas) are much larger in the emphysemic lung.*

The problems of addiction

In the UK, each packet of cigarettes comes with a government health warning and everyone is aware of the health hazards. Some ignore them, but others take heed and try to stop smoking. However, the nicotine in cigarettes makes smoking highly addictive and it is difficult to 'kick the habit'. Smokers who try to give up commonly suffer withdrawal symptoms including a persistent craving for tobacco, irritability, poor concentration, and weight gain. The symptoms can be relieved by using nicotine patches or chewing gum containing nicotine. These devices often help smokers to reduce the number of cigarettes they smoke or to stop smoking, so removing their exposure to carcinogenic tarry substances (nicotine is not thought to be carcinogenic), but they do not remove the harmful effects of nicotine on the cardiovascular system.

Tobacco smoking is a major preventable factor leading to disease and death. There are other less lethal ways of relaxing than by the consumption of a cocktail of carcinogens. The easiest way to avoid the health risks associated with smoking is never to start smoking in the first place.

Pneumoconiosis

Pneumoconiosis refers to a group of lung diseases caused by inhaling dust particles (usually coal, silica, or asbestos) less than 0.5 μm in diameter. These particles are small enough to penetrate into the depths of the lung, where they can damage delicate alveoli. Over a long period, the damage may become visible as shadows on a chest X-ray, and breathlessness develops. Inhaling coal dust may lead to coal-worker's lung, asbestos to asbestosis, and silica to silicosis. The incidence of lung cancer is high in those suffering pneumoconiosis. Even if patients avoid cancer, the symptoms of pneumoconiosis are made worse by smoking, particularly if they smoke cigarettes.

Quick check

1. Name the gas in cigarette smoke that prevents oxygen from binding to haemoglobin.
2. Explain why cigarette smoking increases the risk of osteoporosis.
3. How does tar in cigarette smoke contribute to the development of emphysema?

Food for thought

The connection between cigarette smoking and lung cancer is irrefutable. Epidemiological studies have shown that cigarette smoking is a common feature in the majority of cases of lung cancer. Twenty-five per cent of smokers die of lung cancer. An association between smoking and lung cancer was first recognised by medical scientists in the 1950s. Suggest why it has been so difficult to convince everyone that smoking causes lung cancer.

16.7

Cardiovascular diseases

OBJECTIVES

By the end of this spread you should be able to:

- describe the main cause of coronary heart disease
- explain what happens if heart or brain tissue is deprived of oxygen
- list the main risk factors for coronary heart disease.

Fact of life

Cardiovascular diseases are a major killer. But in England between 2002 and 2010 fatalities from **myocardial infarctions** (heart attacks) fell by about 50% due to improved preventions and treatments.

Cardiovascular diseases are diseases of the heart and blood vessels. They are not confined to the middle aged and elderly: in Britain, electrocardiograms have revealed the beginnings of cardiovascular disease in children as young as seven years old. These heart abnormalities are thought to be linked to the children's lack of exercise and a poor diet. About a third of 10-year-olds walk continuously for less than 10 minutes a week. Many are transported to school by car and spend their leisure time in front of a television or computer screen, living off a diet of fatty foods.

Atherosclerosis and aneurysm: blocking the arteries

Atherosclerosis is a thickening of the inner layers of arterial walls (figures 1 and 2) with deposits of cholesterol, fibrous tissue, dead muscle cells, and blood platelets. These deposits are referred to as atheromatous plaques or **atheroma**. The fibrous tissue within an atheroma often becomes calcified and hardened, contributing to a loss of elasticity of the arteries, known as **arteriosclerosis**.

As an atheroma enlarges, it bulges into the lumen of the vessel, causing it to narrow and so reducing the rate of blood flow. If the atheroma breaks through the smooth endothelial wall of an artery and penetrates into the lumen, the rough surface of a plaque can trigger the formation of a blood clot. The clot may remain at its site of origin (a **thrombus**) or be carried away to another artery (an **embolus**). In either case, if the clot is large enough it can block an artery.

Arterial blockage may occur in another way. An arterial wall weakened by the development of an atheroma may form a balloon-like structure called an **aneurysm**. If the weakening is severe enough, the arterial wall may rupture, causing an internal haemorrhage which blocks the artery.

However it occurs, blockage of an artery can deprive tissue of oxygen, causing the tissue to become severely damaged or to die.

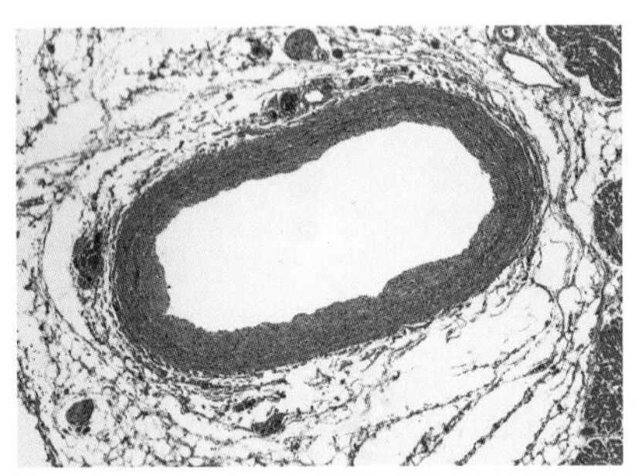

(a)

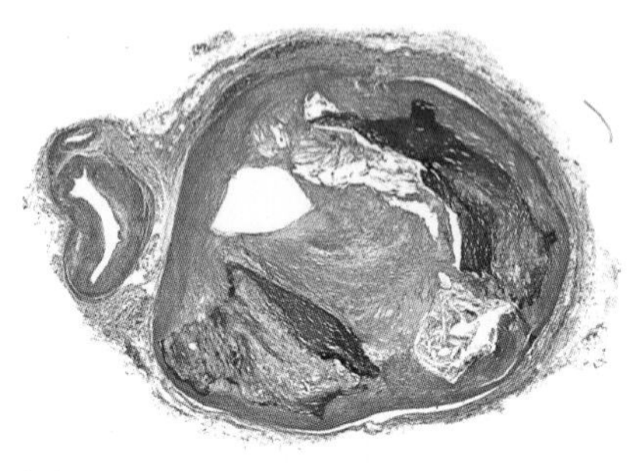

(b)

Figure 1 Transverse section through two coronary arteries. ***(a)*** *A normal artery.* ***(b)*** *An artery that is partly blocked by an atheroma. This type of blockage can lead to a heart attack.*

Stroke

An interruption of the blood supply to the brain may lead to a **stroke** (known medically as a **cerebrovascular accident**). A blood clot, head injury, or a burst blood vessel associated with an aneurysm can cause a stroke. Strokes are most commonly associated with atherosclerosis. A stroke deprives a portion of the brain of oxygen. Major strokes can result in severe paralysis or even death; minor strokes may occur without symptoms.

Preventing and treating heart attacks

The risk of having a heart attack is reduced by decreasing cigarette smoking, lowering blood cholesterol levels by eating a balanced diet low in saturated fats, and taking regular exercise. Drugs are administered both to prevent and treat heart attacks. **Anti-hypertensives** are a class of drug that reduce high blood pressure by different means. **Statins** reduce high blood cholesterol levels by inhibiting an enzyme involved in the production of cholesterol in the liver. **Warfarin** (a common constituent of rat poison) is an **anticoagulant** administered in low doses to make the blood flow more easily in vessels, reducing the risk of blood clots. **Platelet inhibitors** such as **aspirin** reduce the risk of blood clotting by inhibiting the clumping together of platelets (spread 7.8). Invasive treatments include **heart transplants** (spread 15.8), **coronary bypass operations** in which a segment of a diseased coronary artery is bypassed by a compatible section of healthy vein or artery, and **balloon angioplasty** in which a small balloon catheter is inserted via the groin or arm into a partly blocked coronary artery. Once in the coronary artery, the balloon is inflated to enlarge the artery, relieve chest pains, and reduce the risk of a heart attack.

Coronary heart disease

In economically developed countries, the most common form of cardiovascular disease is **coronary heart disease** (**CHD**). The main cause of CHD is atherosclerosis.

Blockage or severe narrowing of one of the coronary arteries that supply the heart (for example, by a coronary thrombosis) may lead to a 'heart attack', known medically as a **myocardial infarction** ('myocardial' refers to the heart muscle; 'infarction' means suffocation due to oxygen deprivation). The section of the heart deprived of oxygen will die. If the blockage is high up in a coronary artery, the heart attack may cause the heart to stop beating. Death will follow unless the heart is restarted, for example by cardiac massage or an electric shock. A blockage towards the end of a coronary artery may have such mild effects that the patient is not even aware of having a heart attack.

Angina pectoris is a less severe symptom of CHD, caused by a narrowing of coronary arteries. This is a gripping, vice-like pain in the chest which sometimes extends down the left arm. It is brought on by physical activity and is relieved by rest. The pain, which usually lasts about 15 minutes, is produced when the heart muscle does not receive enough oxygen to cope with its workload. Like a skeletal muscle deprived of oxygen, it goes into a cramp.

CHD risk factors

Extensive research has established a list of factors that increase the risk of developing atherosclerosis and having CHD (table 1). These CHD risk factors include **age**, **sex**, and **family history**, over which we have little control.

Although CHD may begin at an early age, most deaths are among the elderly. Some scientists believe that atherosclerosis is an inevitable consequence of ageing, but others believe that it can be delayed by taking regular exercise and eating a balanced diet, low in saturated fats but rich in fibre and vitamins.

Death rates from CHD are much lower among women than men. This is thought to be related to the protective effect of female sex hormones. Oestrogens, for example, reduce the proportion of low-density lipoproteins (LDLs) in the blood. LDLs carry large amounts of cholesterol (spread 16.8) which they may deposit in the walls of blood vessels, contributing to the development of atherosclerosis. Testosterone, the predominantly male sex hormone, is thought to increase the proportion of LDLs in the blood.

The risk of CHD is increased by having a close relative who has had the disease. If, for example, a man has one parent who died from CHD, his risk of suffering a similar fate is doubled. If both his parents died from the disease, the risk is increased five fold. These figures indicate that there is a strong genetic predisposition to the disease. Geneticists are in the process of identifying the genes responsible for CHD. It will not be long before genetic screening will be able to identify those at most risk so that they can take precautionary measures.

We cannot do much about our age, sex, or genes, but the next spread deals with the remaining CHD risk factors over which we do have some control.

(a) Artery from healthy person

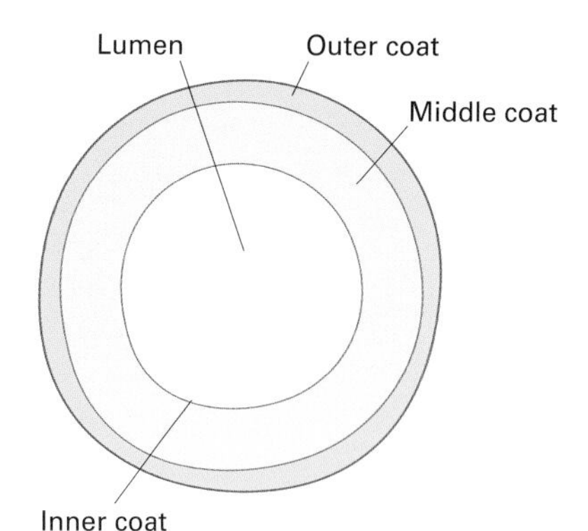

(b) Artery from a person with atherosclerosis

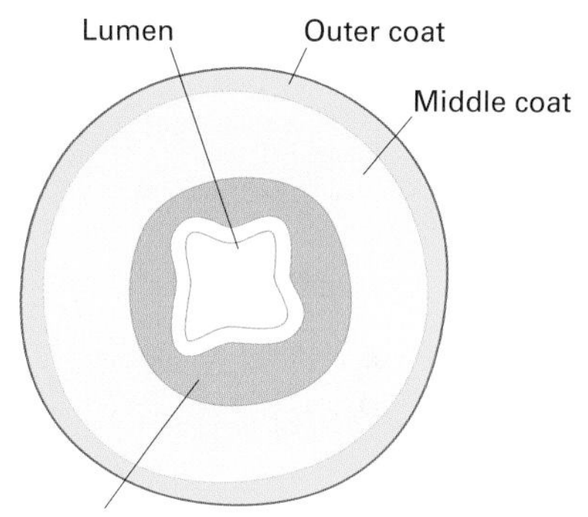

Figure 2 *The effects of atherosclerosis on the lumen of an artery.*

Table 1 *Coronary heart disease risk factors.*

- Old age
- Male sex
- Family history of heart disease
- High blood pressure
- High blood cholesterol levels
- High emotional stress
- Lack of exercise
- Obesity
- Tobacco smoking
- High alcohol intake

Quick check

1 Distinguish between the terms atherosclerosis and arteriosclerosis.

2 State the likely outcome of a blockage of a coronary artery.

3 List three coronary risk factors that are uncontrollable.

Food for thought

Heart disease is very rare in pre-menopausal women. Suggest why it increases after menopause, and what action could be taken to reduce the risk of CHD in some women.

16.8

Controlling risk factors for CHD

OBJECTIVES

By the end of this spread you should be able to:

- explain how high blood pressure can contribute to CHD
- explain why a high-fat diet can increase the risk of CHD
- discuss how diet and exercise can reduce the risk of CHD

Fact of life

According to the World Health Organization, the number of people with uncontrolled hypertension rose from 600 million in 1980 to nearly 1 billion in 2008. This was attributed mainly to the growth and ageing of the global population.

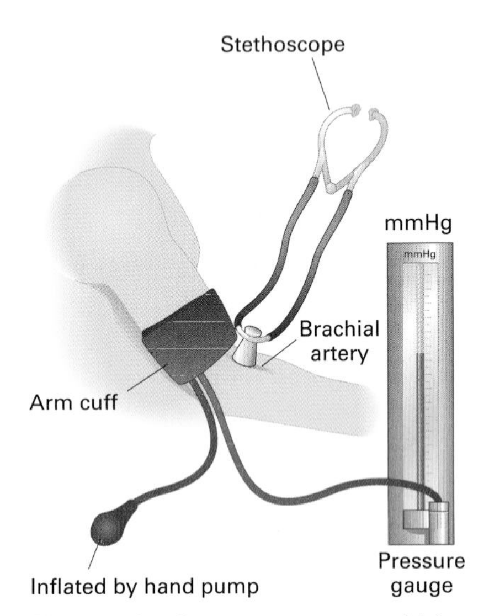

***Figure 1** A sphygmomanometer which measures blood pressure mechanically. An inflatable cuff connected to a pressure gauge is wrapped around the arm and pumped up sufficiently to stop the pulse as felt at the wrist or heard with a stethoscope placed on the artery at the bend of the elbow. As the pressure in the cuff is reduced, blood starts to flow again in the artery and the pulse re-starts. The pressure at this point represents systolic pressure (the highest blood pressure, due to contraction of the heart). The pressure in the cuff is reduced further and the pressure at which there is a full flow of blood, indicated by a marked change in the sound heard through the stethoscope, represents diastolic pressure (the lowest blood pressure, associated with the relaxation of the heart). More modern blood pressure meters also use a cuff around the arm, but the measurements of blood pressure are made automatically and recorded electronically.*

Although the first three of the risk factors for CHD in table 1 on spread 16.7 are out of our control, we can take measures against the others.

High blood pressure

Blood pressure refers to the pressure caused by blood pushing against the inside walls of the main arteries. The pressure results from the heart pumping blood around the body. It is highest during systole (spread 7.6) when the ventricles are contracting (systolic pressure), and lowest during diastole when they are relaxing and filling up with blood (diastolic pressure).

Traditionally, blood pressure is measured in millimetres of mercury (mmHg) using a **sphygmomanometer** (figure 1). A stethoscope is placed at the artery inside the elbow as this is an accessible place where the blood pressure is similar to that of the blood leaving the heart. The mean pressures for a healthy young person are about 120 mmHg (16 kPa) for systolic pressure, and 80 mmHg (10 kPa) for diastolic pressure. This is usually expressed as '120 over 80' or 120/80.

Blood pressure depends mainly on the amount of blood pumped out of the heart per beat (the stroke volume) and the resistance it meets as it passes through the blood vessels (peripheral resistance). If the arteries become narrower and hardened by atherosclerosis (spread 16.7), the blood pressure increases, causing chronic, persistent high blood pressure called **hypertension** (high blood pressure). It is difficult to define precisely what is 'normal' blood pressure, but it is generally agreed that a desirable blood pressure is less than 140/90.

A systolic blood pressure of 180 is not uncommon, and it may be as high as 280. This can put strain on the heart and blood vessels, leading to the rupture of a blood vessel or to a thrombosis. Men below the age of 50 years with a blood pressure of 170/100 are twice as likely to suffer CHD as similar men with a blood pressure of 120/80. Hypertension is sometimes called the 'silent killer' because it can develop without symptoms.

Hypertension is associated with a number of factors, including excessive salt intake, high-fat diet, lack of exercise, obesity, stress, smoking, and high alcohol consumption. All these factors can be controlled to some extent by changes in lifestyle. For example, regular vigorous aerobic exercise (exercise of relatively long duration using large muscle groups and depending on aerobic respiration) at a safe level can help prevent hypertension and reduce blood pressure. Moderating the intake of fat, salt, and alcohol also has beneficial effects. Having a healthy lifestyle which does not encourage the development of hypertension is especially important for people with a genetic predisposition to the disease.

Treatment of hypertension may include the use of drugs known as beta-blockers. Beta-blockers act as antagonists to adrenaline and noradrenaline, preventing adrenaline and noradrenaline from binding to receptors called beta-adrenoceptors on cells in the pacemaker region of the heart. Beta-blockers therefore inhibit the normal stimulatory actions of adrenaline and noradrenaline on the heart and cause the heart rate and blood pressure to be lowered.

High blood cholesterol levels

Cholesterol is a white waxy lipid which has several important functions in the body (spread 2.8). However, excess cholesterol accumulates in the blood, where it is transported by lipoproteins (see box). Low-density lipoproteins tend to unload cholesterol so that it can be deposited in the walls of blood vessels. Thus high blood cholesterol levels (more than 251–300 mg/100 cm^3) accelerate atherosclerosis and increase the risk of CHD. The most important factor in raising blood cholesterol levels is eating foods high in saturated fat. Diets rich in fibre and certain

Lipoproteins

Lipoproteins are a combination of lipid and protein. They transport fatty acids and cholesterol in the bloodstream and lymph. There are three main types: high-density lipoproteins (HDLs), low-density lipoproteins (LDLs), and very low-density lipoproteins (VLDLs). The density reflects the relative amount of protein and lipid; the lower the density, the higher the lipid and cholesterol contents. LDLs have been called 'bad cholesterol' because a high proportion of LDLs in the blood increases the risk of CHD. These lipoproteins tend to deposit their load of cholesterol in the walls of blood vessels. HDLs have been called 'good cholesterol' because they tend to remove cholesterol from the blood, reducing the likelihood of cholesterol becoming deposited in arterial walls. People with a high proportion of HDLs tend to have a low risk of CHD. Regular aerobic exercise can increase the proportion of HDLs, whereas a diet high in saturated fats can increase the proportion of LDLs.

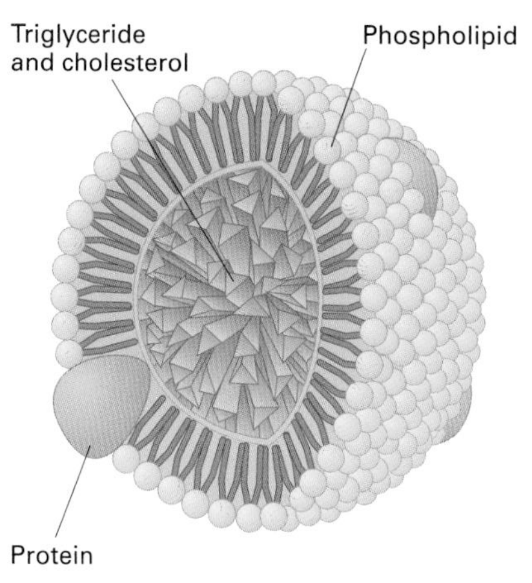

Figure 2 *A lipoprotein molecule.*

unsaturated fats, along with regular aerobic exercise, can reduce blood cholesterol levels.

High emotional stress

Although it is quite easy to recognise psychological and emotional stress, it is difficult to quantify it. Nevertheless, high stress is generally accepted as an important factor in triggering heart attacks and angina. There is a probable link with the secretion of high levels of adrenaline which increases the heart rate and constricts some blood vessels.

Lack of exercise

Many health experts in the UK and USA believe that each year tens of thousands of deaths from heart disease could be prevented if more people took regular exercise. As little as 8–16 km of walking or jogging each week can offer some protection against heart disease. Aerobic exercise strengthens the heart, enabling it to contract more forcefully so that, for any given workload, the stroke volume increases and the heart rate decreases. Exercise does not guarantee protection from CHD, but active people are less likely to have a heart attack than those who are inactive.

Obesity

Obesity is a condition in which excess fat has accumulated in the body. It usually results from chronic overeating. Obesity is a major health hazard: obese people are predisposed to a number of other diseases in addition to CHD. Some medical experts believe that life expectancy decreases by approximately one per cent for each half-kilogram of excess fat carried by an individual between the ages of 45 and 50 years. Obesity can be avoided by eating sensibly so that food consumption does not exceed the energy required for daily activities.

Tobacco smoking

Almost 40 per cent of deaths attributed to CHD are due to smoking. People under the age of 45 years who smoke more than 25 cigarettes a day are more than 15 times as likely to die of CHD than are non-smokers. Cigarette smoke weakens the lining of blood vessels and encourages platelets to stick to the lumen walls. Nicotine increases heart rate and blood pressure, and constricts blood vessels. Carbon monoxide in smoke reduces the oxygen-carrying capacity of blood. These factors combine to put stress on the heart, accelerate the development of atherosclerosis, and increase blood pressure.

High alcohol intake

High alcohol consumption can increase the risk of CHD by increasing blood pressure. There is some evidence that moderate levels of alcohol consumption may actually reduce the risk of CHD.

Quick check

1 Why is hypertension sometimes called the 'silent killer'?

2 Which type of fat in the diet is associated with high levels of cholesterol in the blood?

3 Name two forms of aerobic exercise which can reduce the risk of coronary heart disease.

Food for thought

Lipoprotein(a) is a unique lipoprotein that acts as an independent risk factor for developing vascular disease. It consists of a low-density lipoprotein (LDL) particle attached to a protein called apoprotein(a). Apoprotein(a) is thought to increase blood clot formation. Consequently, people with high lipoprotein(a) levels have an increased risk of CHD. High levels of lipoprotein(a) appear to be inherited, but other factors, such as kidney disease, also affect the levels. Unlike other forms of lipoprotein, diet does not seem to have a major effect on lipoprotein(a) levels. What advice would you give to a person who has a high level of lipoprotein(a)?

16.9

Eating disorders

OBJECTIVES

By the end of this spread you should be able to:

- discuss the causes and symptoms of anorexia
- describe the main features of bulimia
- define obesity and describe the link between obesity and disease.

Fact of life

In 2011 in the UK, more than a million people were affected by an eating disorder. Eating disorders, especially anorexia nervosa and bulimia, are responsible for more loss of life than any other psychological illness.

Figure 1 *A patient receiving counselling for an eating disorder such as anorexia nervosa or bulimia nervosa. The counsellor is attempting to correct any distorted beliefs that the patient may have about her body shape.*

The preoccupation of Western culture with slimness and the negative stereotyping of plump people have contributed to an upsurge in eating disorders, especially among young women. Women are continually bombarded with images from the media reinforcing the notion that they have to be slim to be beautiful, successful, healthy, and happy.

An **eating disorder** is a potentially dangerous disturbance in the pattern of eating. The term usually refers to two main groups of disorder: **anorexia nervosa** and **bulimia nervosa** (often abbreviated to anorexia and bulimia, respectively). In reality, there is a whole range of eating disorders and it is not always easy to put a particular disorder neatly into either of the two main groups. Patients with such a disorder are said to have an 'eating disorder not otherwise specified'.

Anorexia

Anorexia, sometimes referred to as **self-starvation syndrome**, is a potentially fatal eating disorder in which loss of appetite or loss of desire for food leads to a severe loss of body mass. The patient's **energy budget** becomes seriously imbalanced: less energy is consumed than expended. Anorexics are ten times more likely to be female than male. The condition is recognised as a serious psychological disturbance. It has been linked to dietary problems in early life, parental obsession with food, problems within the family, and rejection of adult sexuality. It is also seen as a constant attempt to please others.

Whatever the cause, the effects of anorexia are dramatic and potentially very dangerous. Symptoms include:

- loss of body mass to less than 85 per cent of ideal body mass
- muscle wasting (including heart muscle)
- intense fear of becoming obese, even when underweight
- disturbance of body image (that is, feeling fat even when very thin).

In women, the normal menstrual cycle ceases, resulting in infertility.

Anorexia is much more than 'dieting gone wrong'. It requires medical treatment and may respond to psychotherapy. The more long-standing the condition, the more difficult it is to treat. If treated early, most of the physical symptoms of anorexia can be corrected through nutrition and the gradual restoration of normal body mass. In severe, well developed cases, the heart may become so weakened by loss of cardiac muscle that the chances of recovery are small.

Bulimia

Bulimia (**binge–purge syndrome**) is characterised by sequences of excessive eating followed by purging (efforts to remove the food from the body). Self-induced vomiting is the most common method of purging, but fasting, excessive exercise, and taking laxatives (drugs that stimulate bowel emptying) or diuretics (drugs that stimulate the kidney to eliminate large volumes of urine) are all used to counteract the effects of the binge.

Like anorexia, chronic bulimia usually has underlying psychological causes. Binge eating may help release pent-up emotions and enable the bulimic to be distracted from problems. However, a typical bulimic then has to purge to remove feelings of guilt associated with the bingeing. Short episodes of bulimia are quite common, but bulimic behaviour may become chronic if a person develops a psychological need to binge. Repeated purging can cause an imbalance of mineral salts which can lead to cardiovascular problems.

Bulimia may occur as a phase of anorexia. Both disorders must be regarded as serious, and medical advice must be sought. However, it is not easy to identify bulimics. They may not show symptoms until late in the course of the illness and, unlike anorexics, they may have a normal body mass. Both bulimics and anorexics are often secretive about their eating behaviour and usually deny that they have a problem.

Obesity

In addition to eating disorders, most cases of **obesity** also reflect a disturbed pattern of eating. Most obesity results from eating too much and not exercising enough. As in anorexics, the energy budget becomes seriously imbalanced. But in obese persons, more energy is consumed than expended.

Although obesity is an excess of body fat, it is commonly defined in terms of **body mass index** (table 1). However, obesity is not necessarily the same as being overweight. A very muscular person may be heavy but still have less than 10% body fat. It is generally agreed that body fat should not exceed 20–25% of total body mass in men and 28–30% in women.

Obesity is a major health hazard. Obese people are predisposed to a number of diseases including diabetes mellitus (spread 8.3), certain cancers (spread 16.5), cardiovascular disease (spread 16.7), and hypertension (spread 16.8). It also increases the risk of developing a hernia, gallstones, and varicose veins.

Table 1 *Guideline used to define different body conditions.*

Body mass index (BMI)	Condition
<20	Underweight (may need to gain weight)
20–25	Advisable range
25–30	Overweight (some weight loss may be beneficial to health)
30–40	Obese (need to lose weight)
>35	Severely obese (urgent need to lose weight; advised to consult doctor)

Body mass index (BMI) *is calculated as:*

$$\text{BMI} = \frac{\text{body mass (kg)}}{(\text{person's height})^2\ (\text{m}^2)}$$

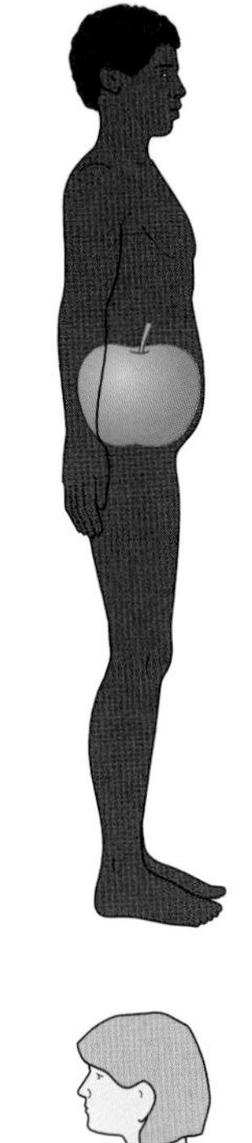

(a)

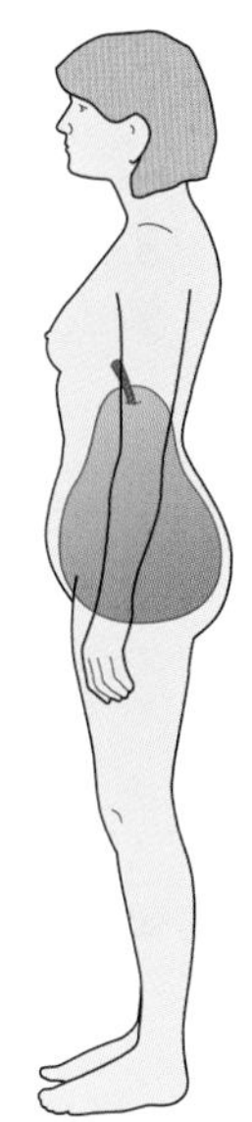

(b)

Figure 2 *Two main types of fat distribution:* ***(a)*** *android fat distribution causing an apple-shaped body;* ***(b)*** *gynoid fat distribution causing a pear-shaped body. The health risk associated with obesity varies according to the distribution of fat deposits. So-called 'pear-shaped' people (mainly women) have a gynoid fat distribution with fat deposited predominantly around the hips, thighs, and buttocks. They seem to be less prone to disease than 'apple-shaped' people with an android fat distribution (mainly men) whose excess fat is distributed predominantly around the abdomen and trunk. The difference in health risks may be because fat around the hips is less likely to be mobilised than fat around the waist. Fat therefore remains at the hips and does not enter the bloodstream where it can become deposited in the walls of blood vessels.*

Quick check

1 Describe the characteristic body mass of a patient with anorexia nervosa.

2 Explain why bulimia nervosa can be difficult to detect.

3 Define obesity.

Food for thought

Being short and having a low body mass can be a great advantage to female artistic gymnasts. Some young gymnasts who aspire to be champions are tempted to lose weight artificially by fasting, by inducing vomiting, or by taking diet pills. If you were a coach of a young female gymnast whom you suspected was using an extreme weight control method, what advice would you give her?

16.10

Deficiency diseases

OBJECTIVES

By the end of this spread you should be able to:

- interpret dietary reference values
- define a deficiency disease
- describe the symptoms of the diseases caused by a deficiency of vitamins A and D, and iron.

Fact of life

At the beginning of the twentieth century, Sir Frederick Gowland Hopkins performed a famous experiment showing that a diet consisting of pure carbohydrate, fat, protein, mineral salts, and water was not enough to maintain health and growth in young rats. He took two sets of eight young rats and fed both sets a diet consisting of pure starch, sucrose, lard, casein (a milk protein), mineral salts, and water. The first set of rats also received 3 cm^3 of milk every day for the first 18 days of the experiment. On day 18, the milk supplement was no longer fed to the first set but given to the second set instead. The results of the experiment (figure 1) indicated that certain 'factors' other than carbohydrate, fat, proteins, mineral salts, or water are essential for healthy growth and development. These factors are now known as vitamins.

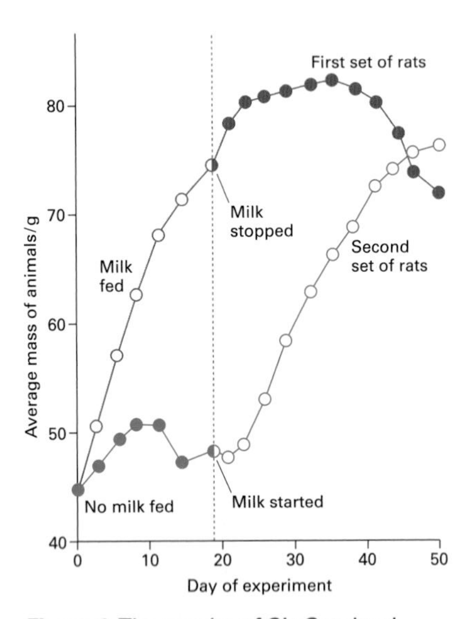

***Figure 1** The results of Sir Gowland Hopkins' experiment on the effects of feeding milk to young rats.*

Dietary reference values

Nutritionists have established that good health depends on eating a balanced diet (spreads 9.7 and 9.8) and that a deficiency of, for example, a vitamin can lead to disease. However, nutritionists have great difficulty in defining the precise daily requirement of each nutrient for particular individuals because:

- individual requirements differ greatly
- nutritional studies on human subjects tend to be difficult to carry out
- results from animal studies are not always applicable to humans.

In 1991, the British Government's Committee on Medical Aspects of Food Policy (COMA) looked at all the available data on nutrition and introduced a series of tables of **dietary reference values**. The values are based on the requirements of population groups and not of individuals. One of the most commonly cited values in these tables is the **reference nutrient intake** (**RNI**). The Appendix, page 571, shows the RNIs for protein and for some vitamins and minerals. Each RNI represents enough or more than enough of a nutrient to meet the needs of 97.5 per cent of a particular group in the population. In most cases, the RNI is nearly equivalent to the **recommended daily amounts** (**RDAs**) found on food labels in the UK and other European countries. However, RDAs differ from RNIs in that RDAs apply to adults and give large enough quantities for the population as a whole, rather than for particular groups of people. For example, the RNI for vitamin C for adults who are not pregnant or lactating is 40 mg, whereas the RDA is 60 mg.

Lack of an essential nutrient in the diet leads to a **deficiency disease** which has a well defined set of signs and symptoms (spread 15.1).

Vitamin A (retinol) deficiency

Vitamin A is essential for the formation of visual pigments in the eye, for the maintenance of a healthy skin, and for the growth and repair of mucous membranes. A deficiency leads at first to a condition known as 'night blindness'. This is poor adaptation to conditions of low light intensity due to a deficiency of retinal (the pigment used for rhodopsin formation; see spread 10.12). Prolonged vitamin A deficiency can cause rods (photoreceptors in the retina) to become permanently damaged, and the cornea and conjunctiva to become dry, a condition known as **xerophthalmia**. This may eventually lead to keratomalacia (blindness resulting from ulcers on the cornea). Vitamin A deficiency affects more than 200 million people worldwide. Young children are particularly susceptible: chronic vitamin A deficiency causes blindness in about 3 million children under the age of 10 years, and it can be fatal.

Excessively high vitamin A intakes can also be dangerous, especially for pregnant women. Intakes above 3300 µg per day taken by pregnant women can lead to birth defects, therefore pregnant women should not take vitamin A supplements unless advised by a doctor to do so.

Calcium deficiency

Calcium is required for the normal functioning of muscles and nerves, for the growth and repair of teeth and bones, and for normal blood clotting. In children, calcium deficiency can retard growth and cause rickets. In adults, calcium deficiency may lead to the development of soft, inadequately mineralised bones (osteomalacia) and osteoporosis (see spread 16.4).

Vitamin D (calciferol) deficiency

Vitamin D regulates the deposition of calcium in bones. Enzymes in the liver and kidneys convert vitamin D into an active form called 'active vitamin D'. This functions as a hormone, stimulating the absorption of calcium from the gut into the bloodstream. In children vitamin D deficiency may lead to **rickets**, characterised by poor bone growth. The bones become weakened by the lack of calcium and phosphate; weak bones may cause a permanent bowing of the legs (figure 2) and bending of the spine. In adults, vitamin D deficiency may lead to **osteomalacia**, a softening of the bones.

Vitamin D is sometimes called the 'sunshine vitamin' because the action of sunlight on the skin is by far the most important source. Children and adults living in the UK can usually synthesise enough vitamin D in the summer months to satisfy their requirements for the whole year (vitamin D can be stored in the liver and adipose tissues). An RNI of 10 μg is recommended for pregnant and lactating women (whose requirements are especially high) and people whose skin is not exposed sufficiently to sunlight.

Supplementation to 7 μg per day for infants and children up to 2 years is also recommended to overcome seasonal depletion during winter months. As with vitamin A, excessively high amounts of vitamin D supplements can be toxic if taken over a long period of time.

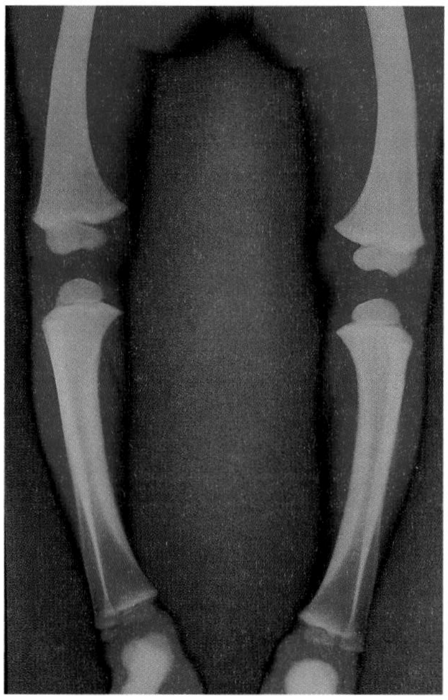

Figure 2 *An X-ray of the leg bones of a child with rickets.*

Iron deficiency

Iron is a component of haemoglobin, myoglobin, cytochromes, and other chemicals involved in respiration and other vital metabolic activities. Iron deficiency may lead to **anaemia**, a reduction in the haemoglobin concentration of blood. Anaemia results in tissues being starved of oxygen because of the decreased oxygen-carrying capacity of blood. Anaemia is characterised by loss of appetite, abdominal pains, tiredness, shortness of breath, and headaches. Iron deficiency may result from lack of iron in the diet, inadequate absorption from the gut, or losses, usually through bleeding. For example, iron deficiency anaemia affects 10–15 per cent of women of menstruating age because the iron they lose in menstrual blood exceeds the iron obtained from food. Therefore, women with high menstrual losses may need higher iron intakes than the RNIs shown on page 571. Too much iron is toxic. An excess can damage the heart, liver, and pancreas, and irritate the gut causing constipation or diarrhoea.

'Golden rice'

For many millions of people living in remote rural areas of Asia, rice is the staple food. Common rice grains are white because they are virtually devoid of carotenoids, orange–yellow pigments that include beta-carotene. When consumed, beta-carotene is converted into vitamin A. Lack of beta-carotene in the diet of rice-dependent people puts them at risk of vitamin A deficiency. In the 1990s, molecular biologists discovered that normal rice had a complex biosynthetic pathway which, by the genetic engineering of two genes, could be modified to make carotenoids. This genetic modification led to the creation of '**golden rice**', a new variety of rice so-named because its grains contain high concentrations of beta-carotene and so are golden in colour. Supporters of its use believe that making 'golden rice' generally available to those in poor, rural rice-dependent areas would significantly reduce the global incidence of vitamin A deficiency. Opponents argue that 'golden rice' would not solve other problems of malnutrition associated with a diet dependent on milled rice, such as iron deficiency. They also argue that the extensive cultivation of this genetically modified rice could have harmful socioeconomic and environmental consequences.

Quick check

1 Explain why the RNI for a particular nutrient may not fully satisfy the requirements of some individuals.

2 **a** Why does vitamin A deficiency lead to night blindness?

b Which childhood disease is caused by a deficiency of vitamin D?

c Why do women who suffer high blood losses during menstruation often have headaches, feel tired, and suffer from a shortness of breath?

Food for thought

Dietary reference values (usually RNIs) are used to plan diets for groups, individuals, and institutions. Suggest why one of the least suitable uses of dietary reference values is in assessing whether an individual's existing diet is a suitable one.

16.11

Drugs

OBJECTIVES

By the end of this spread you should be able to:

- explain what a drug is
- discuss how psychoactive drugs work
- describe the harmful effects of alcohol abuse.

Fact of life

L-DOPA (levodopamine) is a psychoactive substance that occurs naturally in some animals and in plants such as herbs. In humans, it is a precursor of **dopamine**, **noradrenaline**, and **adrenaline**. It is used as a medicine to treat Parkinson's disease because of its effects on parts of the brain controlling muscular movements (spread 10.6). Dopamine also affects the pleasure centres in the brain. Stimulants such as the highly addictive amphetamines and cocaine cause a build-up of dopamine in the brain, making abusers of these drugs feel intense pleasure. Over-use of these drugs can lead to desensitisation of dopamine receptors in the brain, reducing the ability to feel any pleasure and inducing feelings of anxiety and paranoia. Although the side-effects of L-DOPA may include anxiety, this substance is not addictive.

***Figure 1** Tommy Simpson was the first Briton to win the World Championship Road Race. During the 1967 Tour de France, he died of dehydration and exhaustion while cycling up the slopes of Mont Ventoux. A post mortem revealed that he had taken a cocktail of amphetamines and alcohol to boost his performance. The cocktail, combined with intense heat, intense effort, and a stomach complaint, proved fatal. **Ergogenic aids** (performance-enhancing drugs) such as amphetamines, narcotics, beta blockers, and blood doping or erythropoietin (EPO, used to boost red blood cell counts) are on the list of substances that are banned from use in sports.*

Are you a drug taker? A **drug** is any substance that alters the body's actions and its natural chemical environment. Therefore, if you eat chocolate, or drink coffee, tea, or alcohol, or if you take an aspirin or another painkiller for a headache, your answer should be 'yes'. According to the above definition of a drug, it is almost impossible not to be a drug taker.

Drugs are widely used to prevent, diagnose, and treat disease. The term **medicine** is often used to distinguish the use of therapeutic chemicals from drugs that are used to alter the state of a healthy body. Drugs that interfere with the nervous system and cause changes in the mental state and behaviour are called **psychoactive drugs**. Habitual use of psychoactive drugs such as morphine, heroin, and alcohol can lead to **addiction**, a state of physiological dependence on the drug.

Psychoactive drugs

Most psychoactive drugs bring about their effects by interfering with the transmission of nerve impulses across synapses in the brain (see spread 10.6 for an account of synaptic transmission). **Narcotics** such as **heroin** and **morphine** mimic (copy) the action of endorphins, peptide neurotransmitters produced naturally in the brain. By binding onto special receptors (called opiate receptors), endorphins and their mimics block the sensation of pain and induce a feeling of well-being. Unfortunately, narcotics are potent **depressants**: they reduce the normal action of the body. They can also reduce (or depress) the activity of the respiratory and cardiac centres in the brain. This can prove fatal if the diaphragm stops contracting or the heart stops beating.

Stimulants increase the activity of the nervous system. **Caffeine** stimulates the nervous system by increasing cell metabolism. This leads to the release of neurotransmitters which speed up the heart and breathing rate, making a person feel more alert and energetic. **Amphetamines**, such as MDMA (3,4-methylenedioxy-N-methylamphetamine, an ingredient of ecstasy pills), affect synaptic transmission in the brain by interfering with the storage and release of neurotransmitters including dopamine, noradrenaline, and serotonin. Users of MDMA often feel more aroused, less inhibited, and euphoric, but these feelings are often gained at the expense of good judgement and self-control. **Cocaine** delays the breakdown of noradrenaline, and has a stimulatory effect similar to that of amphetamines. Both amphetamines and cocaine can affect the heart adversely. Side-effects of cocaine abuse include heart irregularities, high blood pressure, blockage of coronary arteries, and mental seizures.

Nicotine produces its effects on the heart and blood vessels (spread 16.6) because it mimics the action of the neurotransmitter acetylcholine, one of the main neurotransmitters in the parasympathetic nervous system. Acetylcholine is also the neurotransmitter at neuromuscular junctions. Organophosphate insecticides and nerve gases act at these junctions by inhibiting cholinesterase, the enzyme that catalyses the breakdown of acetylcholine. The continued presence of acetylcholine causes muscles to contract continuously, in some cases with fatal results.

In contrast to organophosphates, **curare** blocks the action of acetylcholine and prevents the contraction of voluntary muscles. Curare is a poison found in the fruits of *Strychnos toxifera*, a tropical rainforest plant. American Indians put it on their arrowheads to immobilise prey and enemies. **Strychnine**, found naturally in another plant, *Strychnos nux-vomica*, has the opposite effect to that of curare: in high doses it causes severe muscular spasms which can be fatal; in low doses it can act as a stimulant, and was once included in 'nerve tonics'.

Benzodiazepine, such as Valium, act as **tranquillisers**: drugs that produce a calming effect. Benzodiazepines increase the effect of GABA, an inhibitory neurotransmitter which reduces the transfer of information between certain brain neurones. **Phencyclidine** (the active ingredient of 'angel dust'), on the other hand, reinforces the action of excitatory neurotransmitters in the brain. This causes the transmission of excessive and inappropriate information which may lead to hallucinations.

Alcohol

Alcohol (ethanol) is the most popular psychoactive drug in the world. Absorbed quickly into the bloodstream from the mouth cavity and stomach, it acts as a depressant on the central nervous system. This may reduce inhibition and feelings of fatigue, but it also reduces self-control and concentration. Reaction times and muscle coordination are impaired. The effect of alcohol depends on the amount taken as well as on individual metabolic differences. Moderate consumption of alcohol may cause no harm: in fact, it might have a beneficial effect on health by preventing platelets in the blood from sticking together, reducing the risk of coronary heart disease (spread 16.8). However, chronic, heavy drinking is a significant health risk. It causes the brain to shrink. It irritates the stomach and intestines, encouraging the development of ulcers and disturbing the uptake of minerals and vitamins (vitamin deficiency is one of the symptoms of alcoholism). Long-term heavy drinking can contribute to cardiovascular disease. It can also damage the liver, causing liver cells to be gradually replaced by fibrous scar tissue. This condition is called **liver cirrhosis**.

Attitudes to drugs and society

The acceptability of a drug depends on cultural attitudes as well as the relative benefits and harm it produces. In some societies and among certain religious and ethnic groups alcohol is disapproved of, forbidden, or even illegal. In the UK, the sale, possession, and consumption of alcohol is legal, but not of cannabis.

Cannabis, like alcohol, is an intoxicant. Smoked or swallowed, it produces euphoria and hallucinations, and has sedative effects inducing relaxation. Although chronic over-use may cause brain damage, many people regard cannabis as no more harmful than alcohol. On the contrary, for people suffering degenerative diseases such as multiple sclerosis, cannabis may be their only source of relief from unpleasant symptoms.

Some highly addictive and potentially legal drugs, such as morphine, have valuable medical uses. However, they should be used only under prescription with the doctor making an objective assessment of their relative benefits. These drugs are not suitable for recreational use and should be avoided by anyone who wishes to remain healthy and free from drug dependence.

Most modern medicines are created artificially in the laboratory where it is possible to vary their precise ingredients and potency. Illegal drug manufacturers take advantage of this ability to vary the composition of medicines to create '**designer drugs**' that are difficult to detect. With advances in DNA technology and the ability to map individual genomes, medical scientists are working towards creating new **personalised medicines** that match individual genetic profiles. This should make medicines more effective and reduce their side-effects.

All drugs interfere with the normal functioning of the body. Even medicines such as aspirin have harmful as well as beneficial effects. In 1775, William Withering tested the effectiveness of digitalis (a drug obtained from foxgloves) as a treatment for congestive heart failure. He gave 163 patients extracts of foxglove as a digitalis soup. He increased the dosage gradually until the patient showed signs of side-effects, such as vomiting and yellow/green vision. Overdoses of foxglove extract can be fatal. Some of Withering's patients nearly died. He then reduced the dosage slightly to obtain a dose that improved heart function. Although Withering's drug tests, published in 1785 in *An Account of the Foxglove and some of its Medical Uses*, led to the successful use of digitalis as a medicine, his tests would not be regarded as ethical today. Modern **drug-tests protocols** are designed to test the effectiveness and safety of drugs in an ethical way (see box opposite).

Modern drug testing

In the UK, modern drug testing usually takes place as preclinical testing followed by three phases.

Preclinical testing is laboratory testing of the drug on cell cultures and whole animals to find out its effects on target cells and to see if there are any side-effects on non-target cells.

Phase 1 clinical trials are carried out on a small group of healthy volunteers who are told what the drug does (including its known side-effects). The effects of different doses are assessed, and the distribution, absorbance, and metabolism of the drug are monitored, to find an optimum dose. An independent organisation decides whether or not the trials proceed to phase 2.

Phase 2 clinical trials are similar to phase 1 trials, but instead of healthy volunteers, small groups of people with the disease are given the drug and the optimum dose is determined.

In **phase 3 clinical trials** a large group of people with the disease are given optimum doses of the drug or a **placebo** (an inactive substance). The large numbers reduce the effects of chance and enable the significance of the results to be analysed statistically. To avoid bias, the trials are **double blind**: neither the patients nor the doctors know who has the drug or placebo. The drug is put forward to the licensing authority only if it has had a significant positive and safe effect in the treatment of the disease.

Quick check

1 Distinguish between a drug and a food.

2 Explain why alcohol is regarded as a psychoactive drug.

3 Explain why cirrhosis of the liver is so harmful.

Food for thought

Research has shown that alcohol inhibits the release of antidiuretic hormone from the posterior lobe of the pituitary gland. Suggest what effect this has on the body and how it might contribute to the unpleasant 'morning after' feelings commonly described as a 'hangover'.

Summary

Non-infectious diseases include **congenital malformations**, **inherited diseases**, **allergies** (including **hay fever** and **asthma**), **age-related diseases** (e.g. **Alzheimer's disease**, **osteoporosis**, and **osteoarthritis**), most **cancers**, **coronary heart disease**, **eating disorders**, and **deficiency diseases**. **Genetic predisposition** is one of the most important risk factors in many non-infectious diseases.

Sickle-cell anaemia is an inherited disease which is most common where **malaria** occurs, probably because heterozygous individuals with **sickle-cell trait** have some resistance to malaria.

Allergies include **hay fever** and **asthma**. **Smoking** tobacco, exposure to radiation and certain viruses, and a poor diet are associated with a higher risk of cancer. **Screening programmes**, such as **cervical smear tests**, are important in the diagnosis and successful treatment of cancers. Smoking also increases the risk of other diseases, especially **cardiovascular** and **coronary heart diseases (CHDs)**. Other factors that increase the risk of CHDs include lack of regular exercise, fatty foods, and high alcohol consumption. Eating disorders include **anorexia nervosa** (**self-starvation syndrome**), **bulimia** (**binge–purge syndrome**), and overeating. Overeating may lead to **obesity**, CHDs, and some other diseases. **Deficiency diseases** usually result from an inadequate dietary intake of **minerals** and **vitamins**.

Drugs are widely used to prevent, diagnose, and treat diseases. They are also abused because of their psychological effects or physiological effects, for example. Drugs include **caffeine**, **alcohol**, and **nicotine**. High **alcohol** consumption can lead to **cirrhosis** of the liver and **alcoholic liver disease**.

PRACTICE EXAM QUESTIONS

1 The diagram shows a section through a human hip joint.

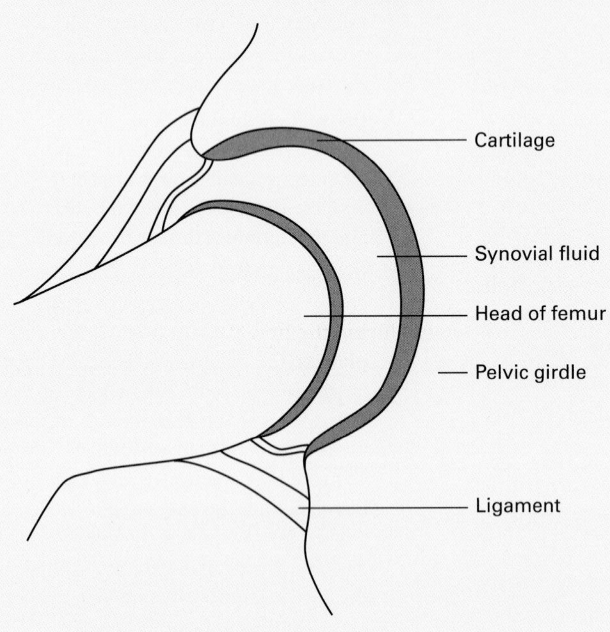

a Describe how the structure of this joint would be affected by osteoarthritis. [2]

b The table shows the percentage of women of different age groups reporting to the accident and emergency unit of a large hospital with fracture of the femur.

Age group/years	Percentage of woman in age group reporting with fracture of the femur
20–29	0
30–39	0
40–49	0.4
50–59	1.1
60–69	2.6
70–79	7.4

Apart from an increased likelihood of falling with age, suggest an explanation for the trend shown by the figures in this table. [3]

[Total 5 marks]

2 The diagram shows a section through a healthy artery and a section through an artery with atheroma.

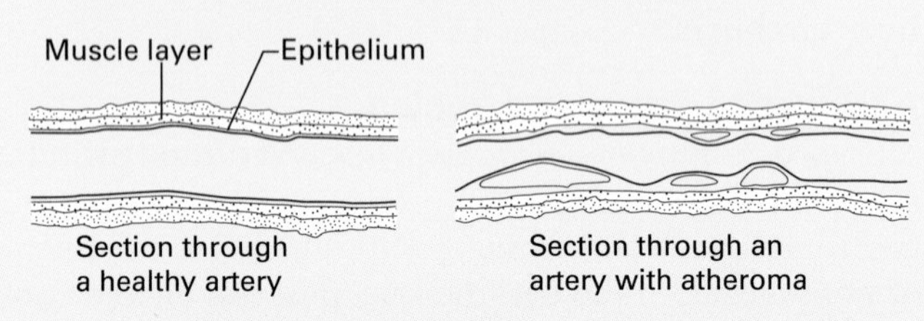

a Give **one** way in which the atheroma has changed the structure of the artery wall. [1]

b **i** Atheroma may lead to damage to red blood cells. Describe how this could lead to the formation of a blood clot. [2]

ii Explain the effects of a blood clot in the coronary artery. [2]

c Give **two** ways in which changes in diet could reduce the risk of cardiovascular disease. [2]

[Total 7 marks]

3 **a** The table shows the results of an investigation into the effects of five minutes standard exercise on the pulse rate of four men of the same age and mass.

Individual	Resting pulse rate/ beats per minute	Pulse rate immediately after exercise/beat per minute
John	64	82
David	70	105
Anthony	78	135
Neil	68	98

i Calculate the mean percentage increase in pulse rate for these men. Show your working. [2]

ii Explain why it is an advantage for the pulse rate to increase during exercise. [2]

b John plays football regularly. His resting pulse is lower than that of the other men and increases only by 28% after five minutes standard exercise. His blood pressure is lower than that of the other men.

i Explain how regular exercise has brought about these effects. [3]

ii Explain why John has a lower risk of developing cardiovascular disease. [1]

[Total 8 marks]

4 People who drink large quantities of alcohol are at risk of developing cirrhosis of the liver.

a State **three** ways in which the liver of a person with cirrhosis differs from the liver of a healthy person. [3]

Various forms of alcohol-related harm are associated with heavy drinking. One of the best indicators of the amount of harm caused by excessive consumption of alcohol is the incidence of cirrhosis in a population.

b Suggest why cirthosis is one of the best indicators of the amount of alcohol-related harm in a population. [3]

c **i** Explain why people who drink excessive quantities of alcohol are often malnourished. [2]

ii Describe **one** consequence of alcohol-related malnutrition on the nervous system. [2]

It is claimed that reducing the mean alcohol consumption in a population will reduce the percentage of people who drink excessive quantities. A survey of alcohol consumption was carried out on over 10 000 people in 52 different locations across the world. The mean weekly consumption of alcohol per person in each location was calculated. The people who drank in excess of 240 g of alcohol a week at each location were classified as heavy drinkers. The mean weekly consumption of alcohol per person was plotted against the percentage of heavy drinkers at each location. The results are shown in the figure below.

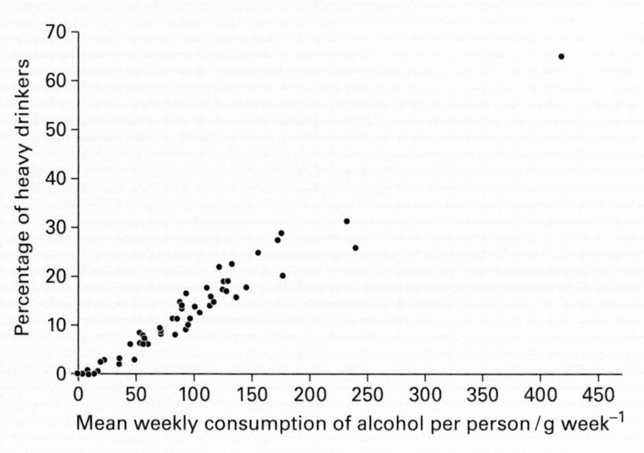

d Comment on the claim that the results of this survey show that reducing the mean alcohol consumption of a population will reduce the harm that it causes. [4]

e State **two** social benefits that may be derived from a reduction in alcohol consumption. [2]

[Total 16 marks]

5 **a** State **two** ways in which cancer cells differ from normal body cells. [2]

b State **three** factors which increase the chances of cancerous growth. [3]

c Suggest why cancerous growths are most likely to occur in tissues where cells normally divide at a higher rate, such as in epithelial tissues which line organs. [2]

Chemotherapy is one of the methods used to treat cancerous growths. Drugs are administered which kill dividing cells. A typical chemotherapy drug programme is illustrated in the figure below. The drug is administered at 21-day intervals, as indicated on the graph by arrows. The effect of the treatment on the numbers of normal and tumour cells per unit volume is recorded.

d With reference to the figure below:

i describe the way in which **both** types of cell respond during the first 30 days of treatment; [3]

ii explain the difference between the responses of the normal and tumour cells over the period of the drug programme. [3]

The tumour cells could be destroyed more effectively if the amount of drug given each time was increased and if it was given at shorter time intervals.

e Suggest why:

i the amount of drug given on each occasion is not increased; [1]

ii the drug is not given at shorter intervals (e.g. weekly). [1]

[Total 15 marks]

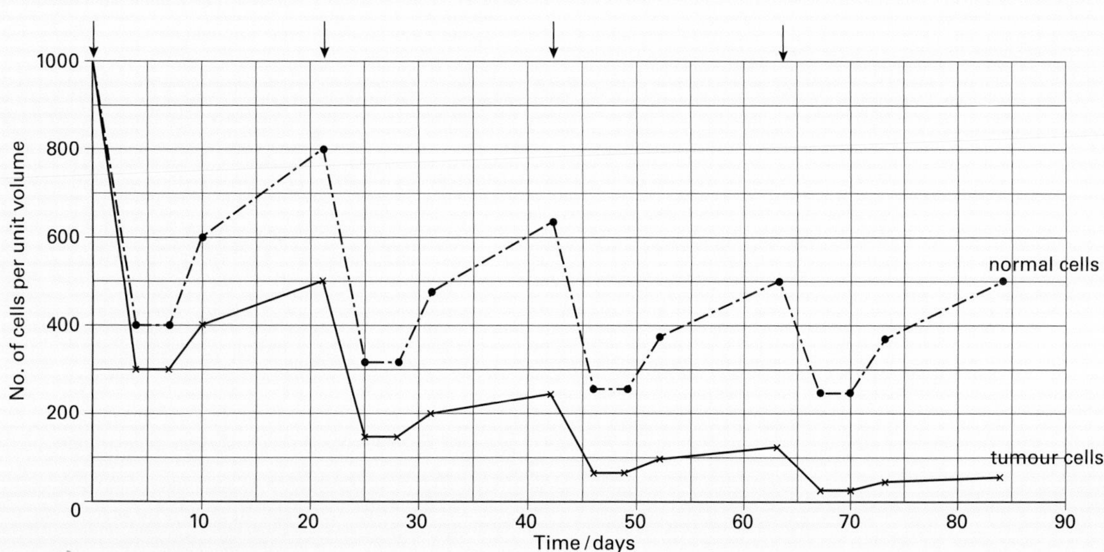

17 Microorganisms and their applications

17.1 Viruses

OBJECTIVES

By the end of this spread you should be able to:

- describe the basic structure of viruses
- explain how a retrovirus reproduces
- describe how plant viruses can be transmitted
- discuss how viruses are different from living cells.

Fact of life

Until recently, it was assumed that the oceans did not contain many viruses. Recent studies have changed that view radically. By centrifuging sea water at very high speeds and examining the sediment under an electron microscope, scientists have found that the oceans are teeming with viruses. It has been estimated that the top 1 mm depth of the world's oceans could contain a total of over 3×10^{30} virus particles.

Microbiology is the study of organisms and biologically important agents which are too small to be seen clearly by the unaided eye (less than about 1 mm in diameter). Its subjects include viruses, bacteria, many fungi, and single-celled organisms.

Viruses are a group of unique particles that infect every type of living organism, from bacteria to flowering plants, and mammals. They have had an enormous impact on the lives of humans, causing death and suffering to countless individuals. They may even have influenced the course of history; for example, the colonisation of North America by Europeans led to smallpox and measles epidemics which reduced the armed resistance of hostile native Americans. Although the threat of smallpox has been removed, viruses, including new ones such as HIV (the human immunodeficiency virus responsible for AIDs, spread 15.2), continue to reap their harvest of death.

The structure of viruses

Viruses are small infectious agents with a wide variety of shapes and structures (figure 1). A complete virus particle (called a **virion**) ranges in size from about 10 to 400 nm in diameter. The smallest particles are little larger than ribosomes and can be viewed only with the aid of an electron microscope. Pox viruses, such as vaccinia which causes cowpox, are among the largest. They are about the same size as bacteria and can be seen with a powerful light microscope. Typically, each virion consists of one or two molecules of DNA or RNA which form the **core**, enclosed in a protein coat called the **capsid**. This coat protects the viral genetic material and helps to transfer this material from one host to another. Some viruses, such as HIV (figure 2) and the influenza virus, have an additional outer layer or **envelope** surrounding the capsid. This envelope can be a very complex mix of carbohydrates, lipids, and different proteins.

Viral life cycles

Viruses occur in two phases: an **extracellular phase** and an **intracellular phase**. The extracellular phase is non-reproducing and contains few if any enzymes. During the intracellular phase, the virus exists mainly as replicating nucleic acids which take over the metabolic machinery of the host to synthesise new viruses.

Viruses can reproduce only inside living cells. They use a variety of strategies to enter a cell. Some bacteriophages (viruses which infect bacteria) insert their DNA into the host cell, leaving their protein coat on the cell surface (spread 18.4, figure 1); some animal viruses enter cells by endocytosis; many plant viruses are injected into the phloem by sap-sucking insects such as aphids.

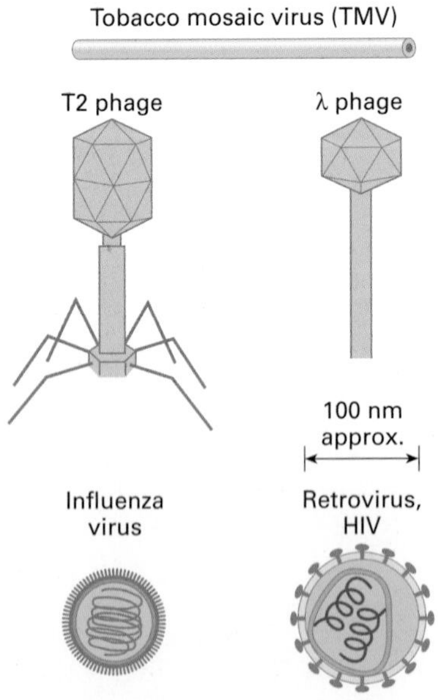

Tobacco mosaic virus (TMV)
T2 phage
λ phage
100 nm approx.
Influenza virus
Retrovirus, HIV

Figure 1 Simplified diagrams of a few representative viruses showing their different sizes and shapes.

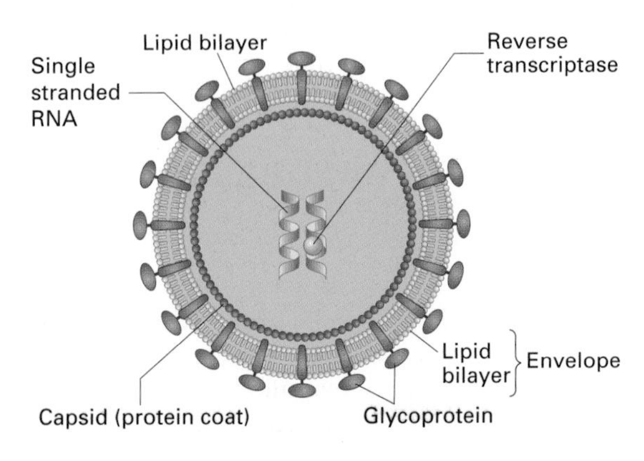

Figure 2 Human immunodeficiency virus (HIV). The virus is cut open to show its two copies of RNA and reverse transcriptase. The protein coat is surrounded by an outer envelope consisting of a lipid bilayer (taken from the cell surface membrane of the previous host cell) and protein. The envelope contains glycoproteins which bind the virus to specific receptors on the surface of certain white blood cells called helper T cells (spread 15.5).

(a) Tobacco mosaic disease (mottling of leaves)

(b) Electron micrograph of tobacco mosaic virions (× 25 000)

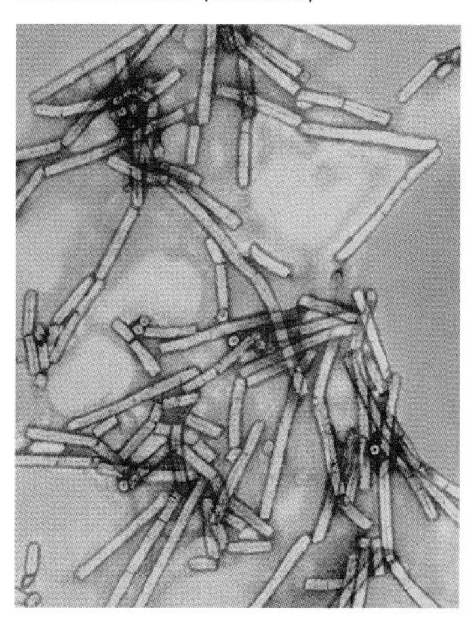

(c) Structure of the virus

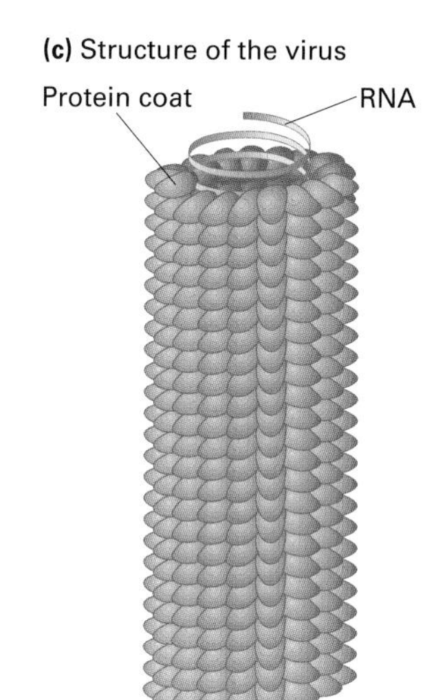

Figure 3 *Tobacco mosaic virus (TMV).* ***(a)*** *Inhibition of chloroplast synthesis by TMV causes mottling of the leaves of tobacco plants.* ***(b)*** *A TMV virion consists of a protein coat and a single strand of RNA.* ***(c)*** *The coat is made of 158 amino acids arranged helically around a central RNA molecule. TMV can be transmitted in smoke blown from cigarettes made from infected leaves, and by animals such as sap-sucking aphids. Mechanically damaged leaves are most easily infected. Once inside the host, TMV spreads from cell to cell through plasmodesmata. The virus takes over the host cell, directing it to make new viral RNA and protein coats which assemble spontaneously into complete virions.*

Virus reproduction varies from virus to virus. **Retroviruses**, such as HIV, contain the enzyme **reverse transcriptase**. This enzyme causes the host cell to make viral DNA on a template of viral RNA. The process is called **reverse transcription**; it is the reverse of the normal cell process by which RNA is made on a template of DNA during protein synthesis (see spread 18.7 for a description of transcription). The viral DNA is then incorporated into the host DNA where it acts as a gene. It may continue to direct the production of virus particles, or it may remain inactive for many cell generations. The integrated viral DNA (called a **provirus**) is passed on to all the host daughter cells. When the viral DNA becomes active, it can cause many thousands of new viruses to be produced, which eventually burst out of the host cells.

Bacteriophages called T2 phages, and also lambda (λ) bacteriophages that infect *Escherichia coli*, reproduce by inserting their DNA into, and taking over, the protein-synthesising machinery of their bacterial host (figure 1, spread 18.4). Bacteriophage DNA may direct the production of new viruses immediately, or it may remain dormant until activated. In either case, the build-up of new bacteriophages eventually causes the bacterial cell to burst (a process called lysis). Each bacteriophage is then able to infect a new host.

Living or non-living?

Viruses do not belong to any of the five kingdoms into which life is classified (see spread 21.1). In fact, it is difficult to decide whether they are living or non-living. They differ from living cells in at least three ways. First, they have a relatively simple acellular organisation (they are not made of cells). Secondly, they have only one type of nucleic acid, either DNA or RNA (living cells contain both types of nucleic acid). Thirdly, they are unable to reproduce independently of cells.

The origin of viruses is shrouded in mystery. Some scientists think that large viruses such as the vaccinia that causes cowpox evolved from bacterial parasites which became increasingly dependent on their host. Consequently, they lost their complex organelles, leaving only the bare essentials: a nucleic acid for reproduction and a coat for protection. Other scientists believe that viruses originated from fragments of cellular nucleic acid that escaped from host cells. A few mutations could have converted this nucleic acid into infectious agents whose reproduction could not be controlled by the host cell. This idea is supported by the observation that the nucleic acids of many viruses contain DNA sequences similar to those of their host cells.

Whatever their origins, viruses continue to evolve into new forms. Some, such as HIV and the Ebola virus, have emerged as major threats to human health. Joshua Lederberg, the Nobel prizewinning geneticist, warned that we live in evolutionary competition with viruses such as these, and that there is no guarantee that we will be the victors.

Quick check

1. What are the two main components of a typical complete virus particle?
2. What is the function of reverse transcriptase in HIV?
3. How can aphids transmit plant viruses?
4. List three differences between viruses and living cells.

Food for thought

Viruses can be produced only in living host cells. Bacteriophages can be grown relatively easily in cultures of bacteria, in either a liquid or a semi-solid medium. Plant viruses are grown in cultures of plant plastids (plant cells from which the cell walls have been removed). In the past, animal viruses were produced mainly in living animals such as rabbits, mice, and guinea pigs, or in fertilised eggs. Suggest why this method of virus production has been largely replaced by growing animal viruses in animal cell cultures.

17.2

OBJECTIVES

By the end of this spread you should be able to:

- explain why archaebacteria are thought to have been the first forms of life
- describe the main structural forms of eubacteria
- discuss the nutrition of eubacteria.

Fact of life

Most bacteria are very small, less than 10 μm in diameter. However, there are exceptions. In 1993, *Epulopiscium fisheloni*, a bacterium measuring 600 by 80 μm, was extracted from the intestines of a brown sturgeonfish (*Acantharus nigrofuscus*). When discovered, this bacterium was called a 'giant among prokaryotes', but it is dwarfed by a species found in sediments on the sea floor off the coast of Namibia. Called *Thiomargarita namibiensis* ('Sulphur Pearl of Namibia') because it can accumulate pearly-white compounds of sulphur, the width of this bacterium can reach up to 750 μm, clearly visible to the naked eye.

BACTERIA

Bacteria are an ancient group of small cellular organisms ranging in size from 0.1 to 10 μm. They have inhabited the Earth for more than 3500 million years. During that time, they have diversified into a bewildering variety of forms which represent almost every shape, physiology, and lifestyle possible for such small organisms.

Today, they occupy every region of the biosphere (the living world) from cracks in deep oceanic trenches to mountaintops, and from the frozen poles to hot volcanic springs. They play a vital role in the food chains of every ecosystem, and also in recycling nutrients.

Bacteriology (the study of bacteria) is in a state of great flux. About 10 000 species of bacteria have been identified and new ones are being discovered every day, but there is no generally accepted system of classification. New discoveries are even making taxonomists (scientists who study biological classification) question the position of bacteria in the five-kingdom classification system of life (see spread 21.1). However, there is general agreement that there are two very different types of bacteria: the archaebacteria and the eubacteria. Bacteria are **prokaryotes**, differing from eukaryotes (animals, plants, fungi, and protoctists) in having no membrane-bound nucleus or double-membraned organelles, and possessing circular DNA (see spread 21.2).

Archaebacteria

The discovery of **archaebacteria** (from the Greek *archaios*, ancient, and *bakterion*, rod) in the late 1970s left many biologists incredulous. These bacteria were found thriving in extreme environments where few other organisms survive:

- **Methanogenic archaebacteria** inhabit anaerobic habitats and give off methane as a product of their metabolism. They live in the guts of cattle and are responsible for the intestinal gases that cattle produce. A cow can belch out 200–400 dm^3 of methane each day.
- Extremely **halophilic** ('salt-loving') **archaebacteria** live only in very salty conditions such as salt flats, where heat from the Sun has evaporated sea water. They will grow in salt concentrations approaching saturation, about ten times as salty as oceanic sea water.
- **Thermoacidophilic** ('heat- and acid-loving') **archaebacteria** can live in hot acidic springs where temperatures may exceed 100 °C and the pH may be as low as 2 (figure 1).

Because they tolerate extreme conditions similar to those that are thought to have existed at the dawn of life, archaebacteria are believed to have been the first forms of life on Earth. Also, some features of their ultrastructure and metabolism suggest that they are the ancestors of both eubacteria and eukaryotes.

Figure 1 *Archaebacteria live in extreme environments such as this sulphur cauldron in Yellowstone National Park, growing well in the boiling sulphur-rich water.*

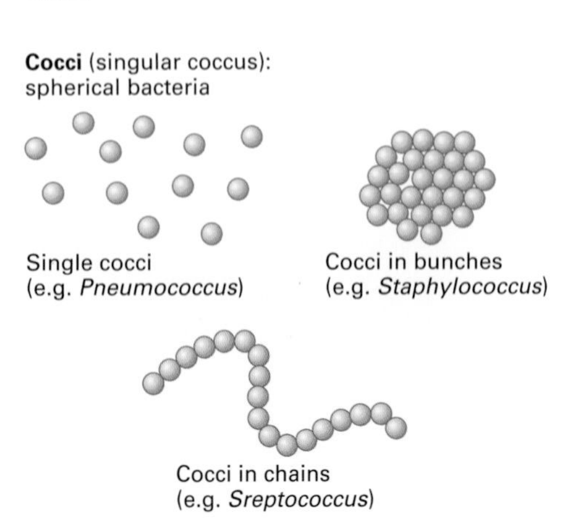

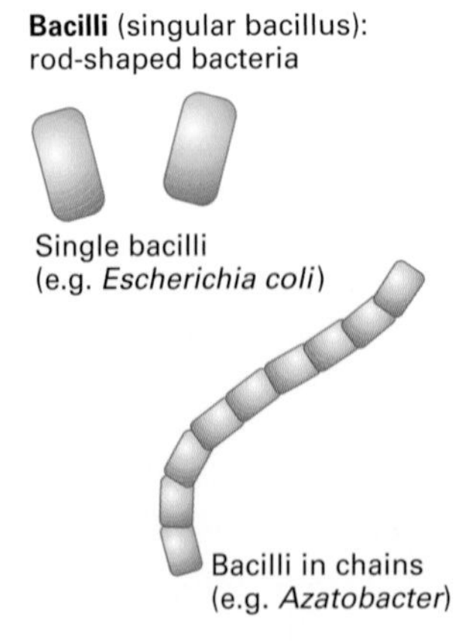

Spirilla (spirochaete): spiral-shaped bacteria

(e.g. *Spirillum*)

Vibrios: comma-shaped bacteria

(e.g. *Vibrio cholerae*, causes cholera)

10 μm approx.

Figure 2 *A few of the shapes of bacterial cells and groups of cells.*

Eubacteria

Eubacteria (from the Greek *eu*, good or true) make up the remaining bacteria. They are found in all but the most extreme environments: in soils and sediments, floating free in air or water, and both in and on other organisms. They occur mainly as single cells, but some bacteria form filaments (rows of cells) or group together in chains or grape-shaped clusters (figure 2). The branching filaments of one group, the **actinomycetes**, were once mistaken for fungal moulds which have a similar growth pattern. Actinomycetes are extremely important medically because they produce most of our antibiotics.

Every type of nutrition and metabolism is represented among the thousands of known species. **Photoautotrophs** ('light self-feeders') include the **cyanobacteria** (also called blue-green bacteria) which photosynthesise in a similar way to plants and algae:

$$\text{Carbon dioxide} + \text{water} \xrightarrow{\text{light}} \text{carbohydrate} + \text{oxygen}$$
$$CO_2 + H_2O \longrightarrow (CH_2O)_n + O_2$$

In addition to having photosynthesising cells, filamentous blue-green bacteria such as *Anabaena* (figure 3) contain special cells which have the ability to fix nitrogen gas from the air to form ammonia, which is used to synthesise protein. These special cells (called heterocysts) lack photosynthetic pigments and have thick walls which prevent oxygen gas from interfering with nitrogen fixation (spread 22.12).

Certain photoautotrophic purple and green bacteria live in anaerobic muds. Instead of using water as their source of hydrogen and electrons, they use other compounds such as hydrogen sulphide, and do not produce oxygen as a waste gas:

$$\text{Carbon dioxide} + \text{hydrogen sulphide} \xrightarrow{\text{light}} \text{carbohydrate} + \text{water} + \text{sulphur}$$
$$CO_2 + 2H_2S \longrightarrow (CH_2O)_n + H_2O + 2S$$

Chemoautotrophs ('chemical self-feeders') use carbon dioxide as a raw material for making organic compounds, but obtain their energy by oxidising inorganic chemicals such as ammonia and nitrite. Some play a key role in the nitrogen cycle (spread 22.12).

Most bacteria are **chemoheterotrophs** ('chemical other feeders'), obtaining their carbon atoms and energy from organic compounds. Heterotrophic bacteria may be:

- predators, eating other microorganisms
- decomposers, breaking down dead organisms
- parasites, infecting other organisms and causing disease
- mutualistic organisms, living in mutual harmony with their host.

Quick check

1. What evidence is there that archaebacteria were the first forms of life on Earth?
2. Describe the shape of bacilli.
3. Explain briefly the difference between photoautotrophic bacteria and chemoautotrophic bacteria.

Food for thought

The first bacteria are thought to have been heterotrophic, feeding on organic molecules formed by the action of electrical storms on carbon dioxide, methane, ammonia, and water in the atmosphere and in the primeval seas. Suggest why David Attenborough in his book *Life on Earth* says that the arrival of blue-green bacteria 'marked a point of no return in the history of life'.

Electron micrograph of *Anabaena* (×1000)

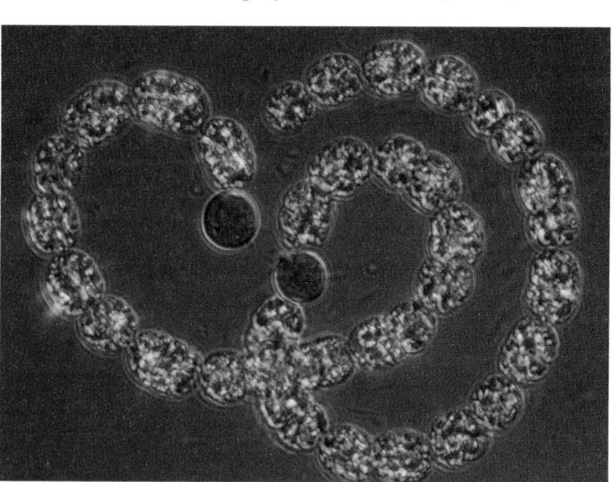

Diagram of a few photosynthesising cells

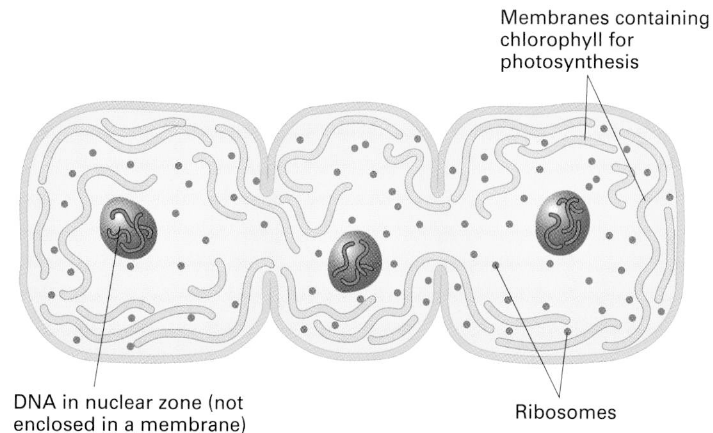

Figure 3 Anabaena, *a blue-green bacterium (cyanobacterium). The oxygen produced by primitive blue-green bacteria is thought to have made a major contribution to our oxygen-rich atmosphere.*

17.3 *E. coli* and food poisoning

OBJECTIVES

By the end of this spread you should be able to:

- describe the structure and life cycle of *Escherichia coli*
- explain how a harmless bacterium can be changed into a potentially lethal one
- discuss methods of reducing the risk of food poisoning by pathogenic bacteria.

Fact of life

In August 1997, Hudson Foods of Nebraska recalled 25 million pounds of beef to be destroyed because it contained a harmful strain of *Escherichia coli*. In 2011 an *E coli* outbreak killed 23 people across Europe.

In a group as diverse as bacteria, there is no such thing as a 'typical' species. However, *Escherichia coli* is often given that title because we know more about this bacterium than any other. It is one of the few organisms whose entire genetic code has been determined. Research teams worldwide gather around this tiny organism because it has become the bacterial equivalent of the laboratory rat: it is used to study many physiological and structural characteristics of bacteria. More importantly, it is the organism most commonly genetically engineered to produce desirable products such as human insulin (spread 18.10). Biotechnologists know a great deal about its ecological requirements because such knowledge makes the difference between profit and loss.

Structure and life cycle of *E. coli*

E. coli is a rod-shaped bacterium measuring about 2.5 µm by 0.5 µm. If suitably stained, it is large enough to be seen under a high-power light microscope (figure 1).

E. coli is found in the guts of vertebrates. It is a chemoheterotroph, capable of thriving on a variety of organic molecules. Its presence in water indicates contamination by faeces. As much as 40 per cent of mammalian faeces is made up of intestinal bacteria.

On reaching a certain size, *E. coli* reproduces asexually by binary fission (splitting into two). This form of reproduction gives *E. coli* populations a tremendous capacity for growth; under ideal conditions, they can double in number every 20 minutes. *E. coli* can also take part in a primitive form of sexual activity called **conjugation**. Cells make direct contact via cellular extensions called **pili** (figure 2). Genetic material is passed in one direction from one bacterium to another, probably through a pilus. Although conjugation does not in itself produce new offspring, after the process has finished the bacteria reproduce asexually, passing on their new genetic make-up to their offspring.

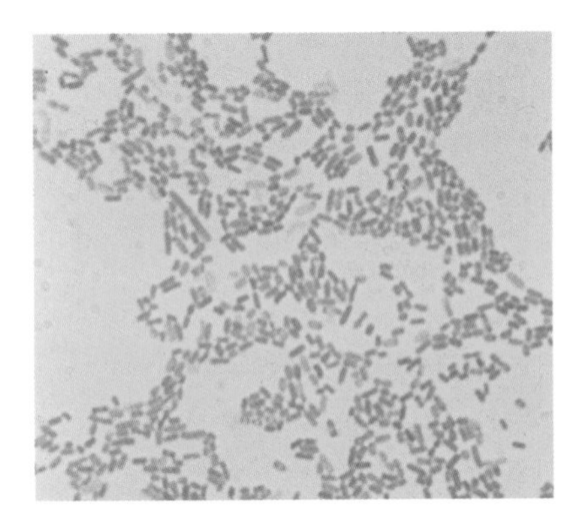

Light micrograph of *E. coli* (×1000)

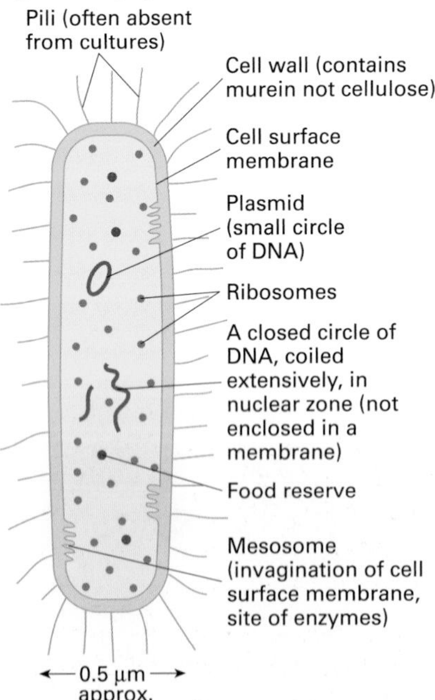

Figure 1 Escherichia coli, *a rod-shaped bacterium.*

The evolution of a harmful strain

Until recently, *E. coli* was thought to be a relatively harmless resident of the human gut which might be linked to the occasional upset stomach and mild diarrhoea. In fact, the presence of massive colonies of mutualistic bacteria in the gut, including most strains of *E. coli*, keep other more harmful bacteria away by starving them of food. They also help make vitamin K (a vitamin important for blood clotting). But in 1982, it became clear that a new strain of *E. coli* had evolved into a much more troublesome organism. The strain had acquired a gene that enabled it to produce a powerful toxin which damages the intestinal wall, causing severe diarrhoea and internal bleeding. This may lead to serious dehydration and, in small children and frail elderly people, death. In some cases, an *E. coli* infection can cause kidney failure and haemorrhaging (bleeding) in the gut. In the majority of cases, infections of the pathogenic strain of *E. coli* are not fatal and the disease clears without treatment.

E. coli probably obtained its toxin-producing gene from a bacteriophage (a virus which infects bacteria, spread 17.1). The virus can insert its own DNA into the bacterial DNA. The viral DNA remains incorporated with the bacterial DNA, so every time the bacterium reproduces it passes on the viral DNA to its daughter cells. These cells make up the harmful *E. coli* strain.

Sources of infection

E. coli is everywhere in the environment, but the deadly strain is rare. Nevertheless, it takes only ten or so of the pathogenic *E. coli* (the **infectious dose**) to infect a person. Therefore touching a source of contamination and not washing hands before handling food may be sufficient to cause infection.

Many outbreaks of food poisoning can be traced back to meat contaminated with *E. coli*. This meat must have been either cooked incompletely or handled by a contaminated person after cooking, because the bacterium is killed by thorough cooking (that is, cooking until the juices run absolutely clear and the internal temperature of the meat is higher than 72 °C).

Meat is not the only source of infection. A contaminated person can pass the bacteria on to vegetables and other foods. It is therefore important to practise good hygiene even if not handling raw meat. In November 1996, an outbreak of *E. coli* food poisoning was traced to drinking fresh apple juice. If similar outbreaks recur, it might be necessary to pasteurise all fresh fruit juices, just as milk is required to be pasteurised (see box).

It is easy to panic after reading alarming reports about the upsurge in food poisoning and about the details of an infection. However, although it is important to realise that there is always a risk, taking common-sense precautions when dealing with food will minimise that risk. It is important, for example, to store and package food correctly (see box).

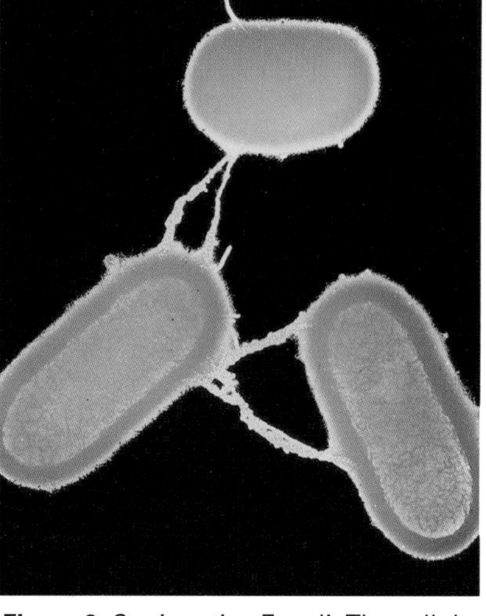

Figure 2 *Conjugating* E. coli. *The cellular extensions are pili, through which the genetic material is thought to pass.*

Storage and packaging

Adding salt, sugar, or vinegar to food can increase its storage life by creating an environment unsuitable for microbial growth. Salt and sugar make solutions hypertonic and so dehydrate microbes. Pickling in vinegar makes solutions too acidic for microbial growth. Irradiation and heat treatment (for example, in drying) increase storage times by killing any microbes in the food. Refrigeration reduces the growth of any microbes that are in or on the food. Optimum storage conditions differ: raw meat is best kept at around 0 °C; meat products at 1–4 °C; and canned or dried foods in airtight containers at 10–15 °C. For long-term storage, meats and fish can be frozen and kept in a freezer at –18 °C, or heat sealed and then vacuum packed in laminated plastic or nylon pouches. Packing conditions must be hygienic to avoid contamination with microbes such as *Clostridium botulinum* (botulism-causing bacterium) which grows well in the oxygen-free conditions of the packs. Canning provides another means of long-term storage. Food is heated to destroy bacteria prior to putting the food in a can. The can is usually lined with protective coatings and is sealed so that its headspace (the space between the top of the can and the food inside) is a vacuum. Food additives such as antioxidants may also be used to preserve foods and increase their shelf life.

Pasteurisation and sterilisation

In the nineteenth century, Louis Pasteur developed a process that has since been used to kill harmful bacteria in milk and other foods. In the process, called **pasteurisation**, food and drinks are heated to a temperature that kills disease-causing microorganisms such as *Mycobacterium tuberculosis* (spread 15.3). In conventional low-temperature-holding pasteurisation, the liquid is maintained at 62.8 °C for 39 minutes. Foods and drinks can also be held at 71 °C for 15 seconds, a process called flash pasteurisation or high-temperature short-time pasteurisation. Milk can be treated at 141 °C for 2 seconds (ultrahigh-temperature or UHT process). Such heat treatment reduces the number of harmful microorganisms to below the infectious dose. However, it does not kill all bacteria; only sterilisation will do this.

Quick check

1 By what processes do *E. coli* reproduce?

2 What is the probable source of the gene that transforms harmless *E. coli* into pathogenic *E. coli*?

3 At what temperature is *E. coli* in meat killed?

Food for thought

After chairing inquiries into *E. coli* outbreaks in Scotland and Wales, Professor Hugh Pennington, an expert on microbiology and food safety, concluded that most cases of food poisoning were preventable – but they were not being prevented. He regarded food poisoning as 'an unnecessary problem'.

Do you think that *E. coli* food poisoning outbreaks are 'an unnecessary problem'? Suggest how the risk of food poisoning could be reduced in your home.

17.4

Studying bacteria

OBJECTIVES

By the end of this spread you should be able to:

- describe the main features of the aseptic technique
- explain how pure cultures of bacteria can be obtained
- distinguish between Gram-negative bacteria and Gram-positive bacteria.

Fact of life

Joseph Lister (1827–1912) was the first person to develop a system of antiseptic surgery designed to prevent infections of wounds. He heat-sterilised instruments and sprayed the surgical dressings and surgical area with phenol. The technique transformed surgery after Lister published his findings in 1867.

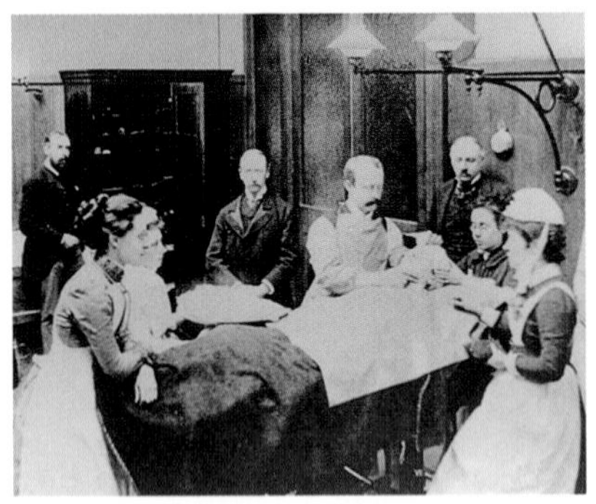

***Figure 1** A group of medical students in an operating theatre of the Royal Free Hospital, London, in the 1880s. Joseph Lister pioneered the use of antiseptics during surgery: the operating area was sprayed with phenol.*

Although new molecular techniques are revolutionising the study of bacteria (bacteriology), traditional methods continue to be used extensively. These traditional methods include the aseptic technique, growing bacteria in culture media, and staining bacteria using Gram staining.

Asepsis

Asepsis (the state of being free from disease-causing microorganisms) is central to all bacteriological investigations. Aseptic techniques include sterilising equipment and culture media before an investigation, confining the investigated bacteria within culture containers during the investigation, and destroying the bacteria afterwards.

Sterilisation is the removal or destruction of all living organisms, including spores (inactive structures that enable some bacteria, algae, fungi, and plants to survive through unfavourable periods). This can be done in a pressure cooker or an autoclave (a heavy-duty laboratory version of a pressure cooker) in which water is heated under pressure so that it boils at 121°C or above (boiling at 100°C does not kill bacterial spores). Autoclaves are used to sterilise equipment and culture media before experiments, and also to sterilise equipment and specimens afterwards before disposal. In addition to sterilising equipment and media, great care is taken by experimenters to ensure that they are not themselves a source of contamination and that they do not become infected by the bacteria under investigation (for example, experimenters wear clean laboratory coats and wear gloves, and do not breathe on cultures). In schools and colleges, basic precautions are adequate because the bacteria studied should be non-pathogenic and are cultured at temperatures well below human body temperature. However, special laboratories are needed for routine bacteriological work in hospitals and research laboratories where pathogenic bacteria may be cultured. These laboratories have easily cleaned surfaces and special enclosed benches or isolated chambers which receive sterile air.

Growing bacteria

Bacteria may be cultured in a liquid broth or on solid agar (agar is a gelatinous substance extracted from red algae). Broth is usually contained in a small specially designed bottle which can be sterilised easily. In **plate cultures**, bacteria are grown on a solid agar medium in a Petri dish (a shallow circular dish made of glass or plastic, with a lid to prevent contamination). Liquefied agar is added to a Petri dish, a process called **plating**, after which the agar sets to a gel. In both broth cultures and plate cultures, the medium is sterilised and contains specific nutrients for the growth of particular types of bacteria. A bacterial sample is introduced into the culture medium; this is called **inoculation**. It can be

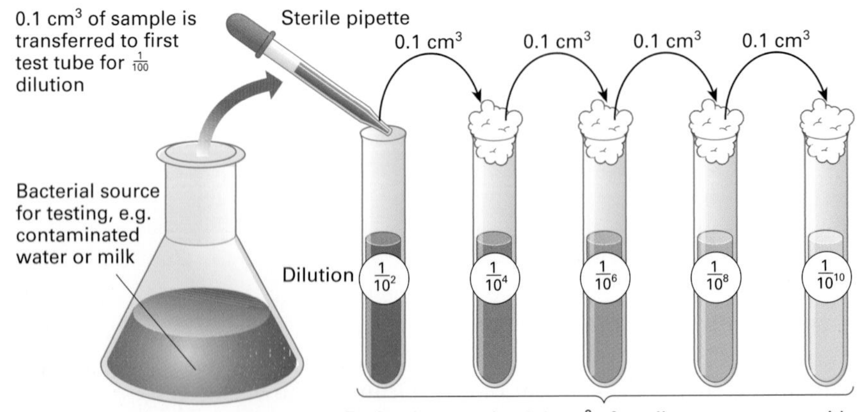

***Figure 2** Making a serial dilution.*

carried out by dipping a small sterile wire loop first in the bacteria source and then into or onto the culture medium. If the source has a high concentration of bacteria, it is often necessary to dilute the bacteria with sterile water (figure 2). After inoculation, the medium is incubated at a specific temperature.

Most broths turn cloudy when the population of bacteria reaches a certain density. Some broths contain indicators which change colour due to the activity of particular types of bacteria. For example, bacteria that break down milk sugar (lactose) to acids cause a milk broth containing the pH indicator neutral red to appear red.

Figure 3 shows the results of growing bacteria on an agar plate using a technique called **streaking**: lines are stroked on the agar with a sterile wire loop which has been dipped into a source of bacteria. Each small circular growth represents a colony of identical individuals (a clone) formed from an individual bacterium. A sample colony can be removed from this plate and grown separately on a different medium to obtain a pure culture. The bacteria within the pure culture can be identified using a combination of methods:

- investigating their growth requirements (media providing specific nutrients allow some bacteria to grow while others from the same source will die)
- size and shape of the colony (figure 4) or of individual bacteria (spread 17.2, figure 2) by looking at the bacteria under a microscope
- biochemical reactions, including sensitivity to specific antibiotics
- reactions to specific stains, such as Gram stains.

Gram staining

Gram staining was devised by the Danish physician H.C.J. Gram. During Gram staining, heat-fixed smears of bacteria are stained with crystal violet solution and then washed in ethanol. This decolorises Gram-negative bacteria, whereas Gram-positive bacteria remain violet. The smears are then flooded with a counterstain, safranin, which stains any Gram-negative bacteria pink and gives the Gram-positive bacteria a purple colour (figure 5). Gram-positive bacteria, such as *Staphylococcus* and *Bacillus*, have thick cell walls made of a network of fibres of a substance called murein, along with other components (including polysaccharides and proteins). These cell walls bind with the crystal violet stain. Gram-negative bacteria, such as *Salmonella* and *E. coli*, have thin cell walls which contain lipid and polysaccharides. The lipid is dissolved by the ethanol so that Gram-negative bacteria do not retain the crystal violet.

The Gram reaction is used in the classification of bacteria, and it also has medical applications. Many Gram-positive bacteria can be controlled quickly by penicillin or penicillin-like antibiotics. Gram-negative bacteria are usually controlled more quickly by a different type of antibiotic (for example, streptomycin).

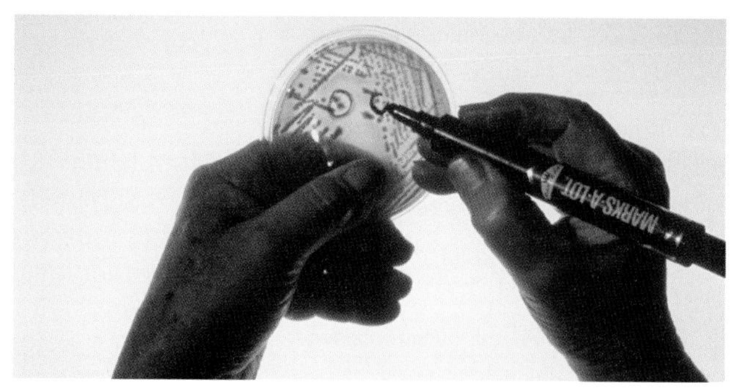

Figure 3 Escherichia coli *growing on an agar plate.*

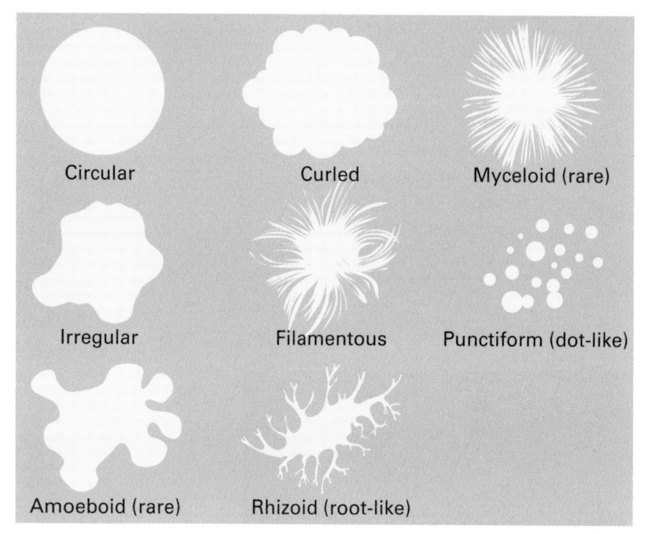

Figure 4 *Some shapes of colonies used in the identification of bacteria.*

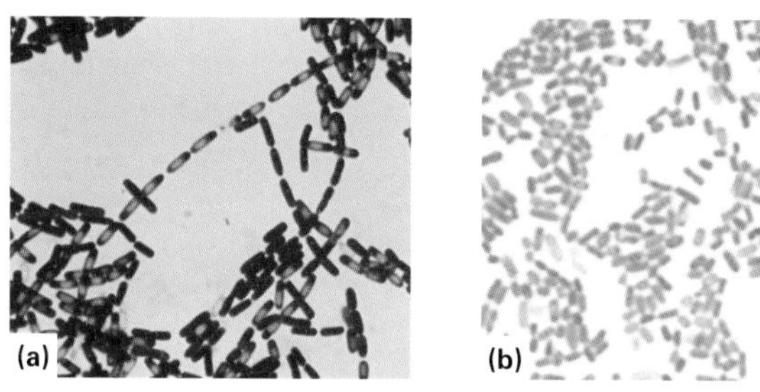

Figure 5 ***(a)*** *Gram-positive bacteria.* ***(b)*** *Gram-negative bacteria. The Gram-negative bacteria appear pink because of the counterstain safranin, whereas the Gram-positive bacteria appear purple.*

Quick check

1. Explain why boiling at 100 °C does not sterilise laboratory equipment.
2. Why is a single relatively large area of growth on agar assumed to be a pure culture of a bacterium?
3. Discounting the possibility of an adverse reaction to antibiotics, do you think a doctor would use a penicillin-like antibiotic or streptomycin for controlling an infection of bacteria which appear pink after Gram staining?

Food for thought

Traditionally, bacteria have been identified only after growing cultures in nutrient media. Suggest why this has resulted in only about 10 000 species being discovered, whereas modern techniques are adding many new bacteria every day to the list of species.

17.5

MOULDS: *RHIZOPUS*, *MUCOR*, AND NON-FUNGAL MOULDS

OBJECTIVES

By the end of this spread you should be able to:

- describe the main features of moulds
- describe the structure of a *Mucor* hypha
- explain how *Mucor* and *Rhizopus* feed
- describe how *Mucor* and *Rhizopus* reproduce.

Fact of life

In 1845, Irish potatoes were infected with potato blight (the mould *Phytophthora infestans*). In 1846 the entire crop, including seed potatoes, was lost. Between 1845 and 1851, the population of Ireland declined by about 2.5 million. 1.5 million had died of starvation and disease and a further million emigrated.

Moulds pervade our world, living wherever moisture is present. Some are of great benefit to humans, providing antibiotics, acting as decomposers so that nutrients can be recycled, or taking part in industrial processes. Other moulds produce diseases which cause immense suffering and damage.

Moulds have cells arranged in long thread-like filaments, the **hyphae**, that form a tangled woolly mass called a **mycelium**. Moulds are usually thought of as fungi, but moulds may also be formed by filamentous bacteria, slime moulds, and water moulds.

Non-fungal moulds

Bacterial moulds include those of *Streptomyces griseus*, which secretes the antibiotic streptomycin.

Slime moulds are a peculiar group of organisms that resemble fungi in appearance and lifestyle, but are related more closely to protoctists such as *Amoeba* in their cellular organisation, reproduction, and life cycles. There are two main groups: **plasmodial slime moulds** (also called Myxomycotae) and cellular slime moulds (Acrasiomycotae). For part of their lives, plasmodial slime moulds (these have no connection with the parastic protoctists belonging to the genus *Plasmodium*, which cause malaria) exist as thin, streaming masses of colourful protoplasm that creep along moist, rotting logs and leaves (figure 1a). They move in an amoeboid fashion, engulfing food particles by phagocytosis. A single mould may extend for several centimetres, but it is not multicellular. It is made up of a continuous mass of cytoplasm with many nuclei called a **coenocytic mass**.

Cellular slime moulds have a unicellular feeding stage resembling an amoeba, with each cell functioning individually. When food is scarce, the individual cells group into a mass resembling that of plasmodial slime moulds. However, the individual cells of cellular slime moulds retain their identity and have separate cell surface membranes.

Water moulds (Oomycota) are sometimes included with the fungi. Although water moulds and fungi are closely related and have a similar structure, water moulds are generally regarded as a separate and more ancient group belonging to the protoctists. Water moulds include rusts and mildews which, like plasmodial slime moulds, consist of coenocytic masses of hyphae similar to fungi. However, most water moulds have cell walls made of cellulose, while the walls of true fungi are made of chitin, a different polysaccharide. Some of the most devastating plant diseases are caused by water moulds. Downy mildew threatened the French vineyards in the 1870s, *Phytophthora infestans* (figure 1b) causes potato blight which contributed to the Irish famine in the nineteenth century, and *Pythium*, a relatively unspecialised parasite, attacks a great variety of plants causing soft rot. Water moulds reproduce asexually by structures called conidia, and by moving spores with flagella, called zoospores. They reproduce sexually by producing moving male gametes that fertilise large immobile egg cells. These eggs gives the group its name Oomycotae (the prefix 'oo-' refers to eggs).

***Figure 1** Non-fungal moulds.*
***(a)** Plasmodial slime mould.*
***(b)** False-coloured scanning electron micrograph of spore-bearing stalks of potato blight (dark green) emerging from lip-like stomata on a leaf.*
***(c)** A water mould growing on a dead water beetle larva.*

Fungal moulds: *Rhizopus* and *Mucor*

All fungi that produce mycelia can be called moulds, but the term is usually used for an organism in which the mycelium forms the main body of the fungus.

In the black bread mould *Rhizopus* and the pin mould *Mucor*, the mycelium consists of a tangled mass of hyphae with many nuclei. These hyphae are called **coenocytic** because the fungal tissue is not separated by cell walls. Consequently, there is nothing to stop cytoplasm flowing from one part of a hypha to another. Fungal hyphae have an outer cell wall made of chitin (not cellulose) and an inner lumen which contains the cytoplasm and organelles. A cell surface membrane surrounds the cytoplasm and sticks tightly to the cell wall.

Rhizopus and *Mucor* are **saprotrophic**, obtaining their nutrients from dead organic material. *Rhizopus nigricans* and *Mucor mucedo* can live on bread, but some species of *Rhizopus* feed on living plants, and *Mucor* commonly grows on rotting fruit and vegetables, in the soil, or on dung. *Rhizopus* and *Mucor* secrete hydrolytic enzymes onto their food source and digest the food extracellularly (outside the organism). They then absorb the soluble digestion products and assimilate them (figure 2).

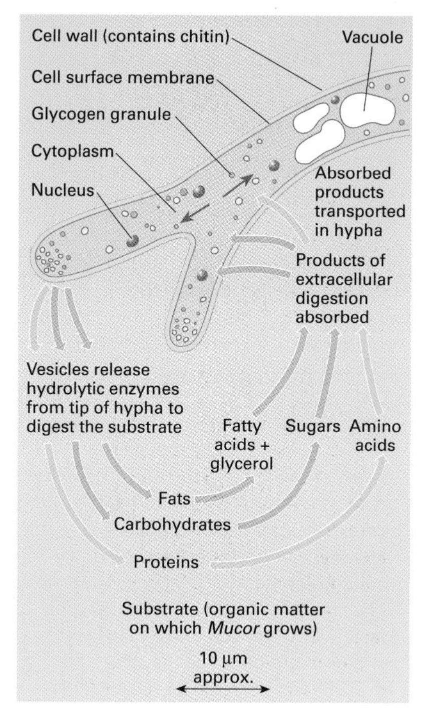

Figure 2 *Structure of a* Mucor *hypha and nutrition in* Mucor.

Life cycle of *Rhizopus*

Rhizopus and *Mucor* belong to the fungal phylum **Zygomycota**. The phylum acquired its name because its members produce two kinds of spores: sexual **zygospores** as well as asexual **sporangiospores**. The life cycle of *Rhizopus* is shown in figure 3. Asexual sporangiospores, formed by mitosis, develop in a sac (a **sporangium**) at the tip of a hypha. When the sac bursts open, the spores are released. In most species of *Mucor*, the sporangium wall dissolves, water enters the spore mass, and the spores are dispersed by a raindrop or are transported by insects. In most *Rhizopus* species, the sporangium wall fractures and dry spores are dispersed by the wind.

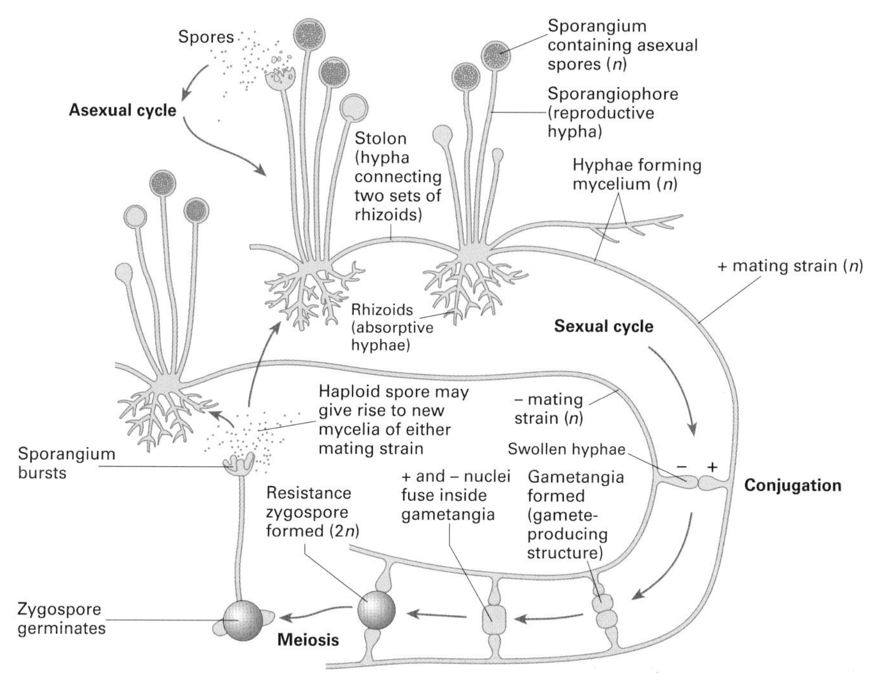

Figure 3 *Life cycle of* Rhizopus. *The mycelium and spores are haploid: the only diploid part of the life cycle is the zygospore, which undergoes meiosis on germination.*

Sexual reproduction involves **conjugation**. Usually, hyphae from mycelia of different mating types meet and interconnect via outgrowths. However, *Rhizopus sexualis* is exceptional in that conjugation can occur between hyphae of the same strain. The interconnecting walls break down and their cytoplasm (containing haploid nuclei) mix. The zygote formed by the fusion of two nuclei develops a thick, rough, black coat and becomes a dormant **zygospore**. Meiosis probably occurs at the time of germination; the zygospore cracks open to liberate numerous haploid spores which can give rise to asexual sporangia and mycelia of either mating strain.

Use of moulds

Although species of *Mucor* and *Rhizopus* are responsible for the spoilage of food, they are also used to make human foods. In eastern Asia, *Mucor* is used with soya beans to make a cheese called sufu. In Indonesia, *R. oligosporus* and *R. oryzae* are used to produce a food called tempeh from boiled skinless soya beans. Other fungal moulds belonging to the Zygomycota are used to make anaesthetics, birth control pills, meat tenderisers, and the yellow colouring agent used in margarines and butter substitutes.

Quick check

1 What is the main feature of moulds?

2 Why is the hypha of *Mucor* called coenocytic?

3 Explain what is meant by extracellular digestion.

4 Distinguish between a sporangiospore and a zygospore.

Food for thought

Lung infections of *Mucor* are generally very rare. Suggest why the lungs of an AIDS patient can more easily become infected with *Mucor* than the lungs of a healthy person.

17.6

Moulds: *Penicillium* and *Saccharomyces*

OBJECTIVES

By the end of this spread you should be able to:

- distinguish between the structure of *Penicillium* and that of *Mucor*
- describe the structure of a yeast cell
- explain how *Saccharomyces* reproduces.

Fact of life

In April 1996, an international team of biologists from 92 laboratories finished spelling out the entire genetic code of the yeast *Saccharomyces cerevisiae.* The work included identifying 12 057 500 chemical subunits contained in the yeast's DNA. The incentive for the research was the fact that the genetic make-up of yeast shares a number of similarities with that of humans. Yeast has about 6000 genes located in 16 chromosomes. It is estimated that about one-third of these genes are related to those of humans. These genes play a critical role in cell function in both species. The genetic information provided from studies of yeast have already increased our understanding of the function of human genes involved in medical problems such as cancer, neurological disorders, and skeletal disorders.

Although most fungi are something of a mixed blessing to humans, *Penicillium* and *Saccharomyces* have been of overwhelming benefit. The genus *Penicillium* contains species of mould which synthesise the antibiotics penicillin and griseolfulvin. Other species give the Gorgonzola, Camembert, and Roquefort cheeses their distinctive tastes (see spread 17.8). The genus *Saccharomyces* includes species of yeast essential to many industrial fermentation processes. These organisms enable bread to rise and make beers and wines alcoholic (see spread 17.10).

Penicillium and antibiotics

Penicillium is famous for producing penicillin, the first antibiotic to be discovered. In 1928 Sir Alexander Fleming was culturing some *Staphylococci* bacteria as part of his medical research. After leaving some Petri dishes for several days, he found a mouldy growth of *Penicillium notatum* contaminating a corner of one of the dishes, in addition to the expected *Staphylococci*. Fleming noticed that the *Staphylococci* next to the mould had been destroyed. Instead of discarding the dish as a mistake, Fleming studied it closely. He deduced that the *Penicillium* mould was producing a substance that killed the *Staphylococci*. He then went on to find that a broth of the *Penicillium* mould contained penicillin, which could destroy pathogenic bacteria. Fleming was not convinced that penicillin would be effective in the body, and in 1931 he dropped his research. Howard Florey and Ernst Chain went on to produce purified penicillin. Successful trials were reported in 1940, and the treatment of wounded soldiers with penicillin saved many thousands of lives in the Second World War. In 1945 Fleming, Florey, and Chain received a Nobel prize for the discovery and production of penicillin.

Although penicillins are very effective when they work, they have a narrow antibacterial spectrum (that is, the range of bacterial species against which they act is small). Other antibiotics are often required to fight bacteria that are resistant to penicillins.

Griseofulvin is produced by another *Penicillium* species, *P. griseofulvum*. It is active against plant pathogens, particularly some rusts and mildew.

Figure 1 Penicillium.

Circular *Penicillium* mould growing in a Petri dish

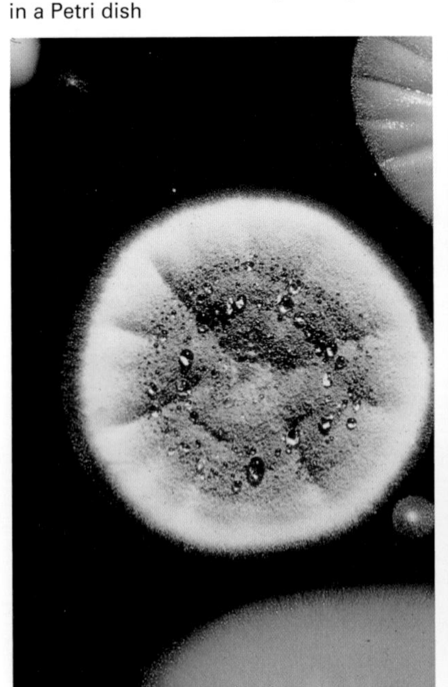

Penicillium notatum (×1200)

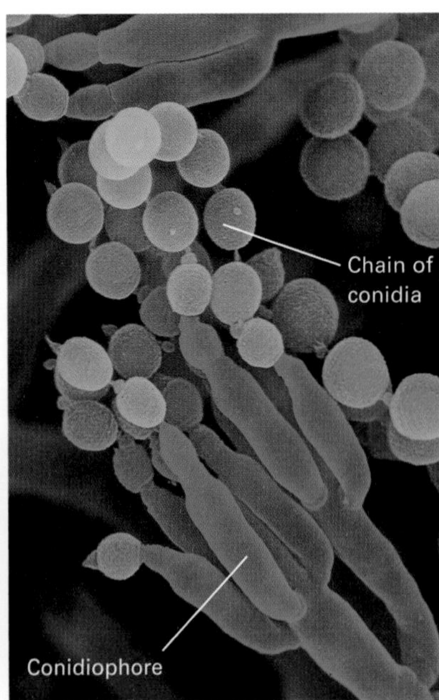

The structure of *Penicillium*

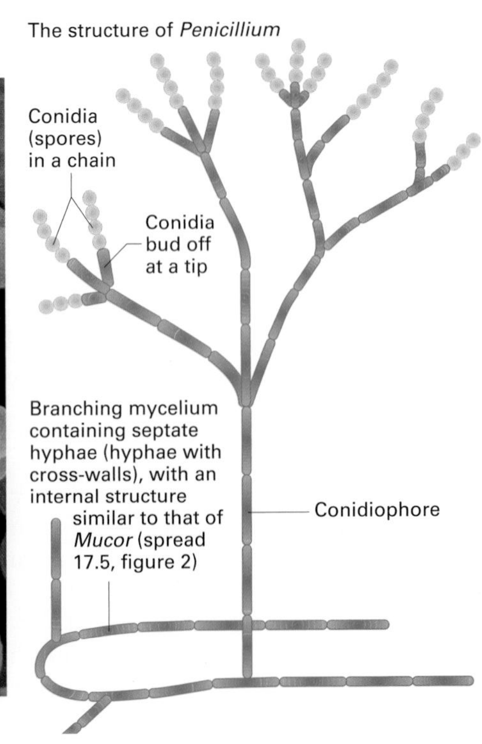

The structure of *Penicillium*

Penicillium is **septate**; that is, its hyphae have cross-walls called **septa**. However, the septa are not formed by cell division, and there is usually a pore at the centre of a septum which allows cytoplasm to flow from one compartment to another. Each compartment may contain one or more nuclei. So despite having septa, *Penicillium*, like the non-septate moulds *Mucor* and *Rhizopus*, is a coenocyte (spread 17.5).

Penicillium is a saprotroph, feeding on organic matter in damp soils, leather, bread, and decaying fruit. Typically, the mycelia of *Penicillium* species form circular green, yellow, or blue moulds (depending on the species). *Penicillium* reproduces asexually by means of spores called **conidia** formed at the tip of special hyphae called **conidiophores** (figure 1). Whereas the spores of *Mucor* are enclosed in sporangia, those of *Penicillium* are exposed and free to be dispersed as they mature.

Saccharomyces

In a general sense, the **yeasts** include all unicellular fungi that reproduce asexually by budding. They occur commonly on faeces, in the soil, and on the surfaces of plants and animals.

The most familiar and industrially important yeast is *Saccharomyces cerevisiae*. The tiny cells of this yeast are very active metabolically. They usually respire aerobically, but when deprived of oxygen they switch to anaerobic metabolism, producing carbon dioxide and ethanol (alcohol) as waste products. It is these waste products that are important to us industrially. Each cell of *S. cerevisiae* has a single nucleus and is usually egg shaped. Individual cells vary in size; some are not much bigger than bacteria. Nevertheless, they contain most of the organelles of a typical eukaryote (figure 2). *S. cerevisiae* can reproduce either asexually or sexually. In asexual reproduction, a single cell divides by budding and separates into two cells. Some buds group together to form colonies; others separate to grow individually into a new yeast. In sexual reproduction, two cells fuse to form a diploid cell which then forms haploid spores by meiosis (figure 3).

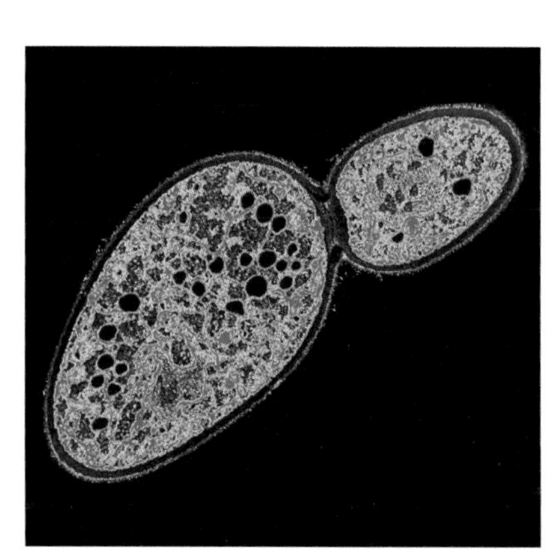

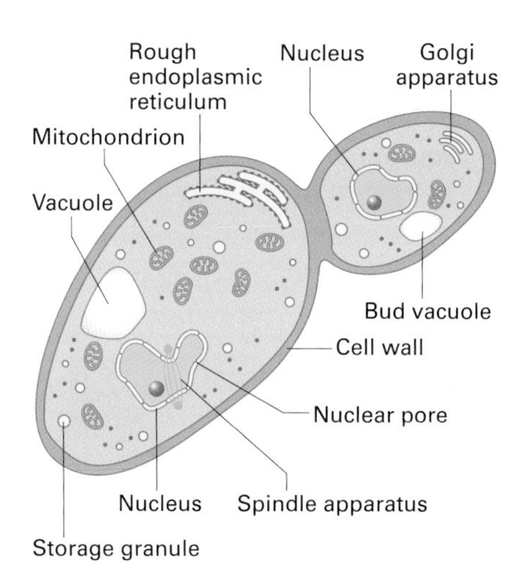

Figure 2 Saccharomyces cerevisiae.

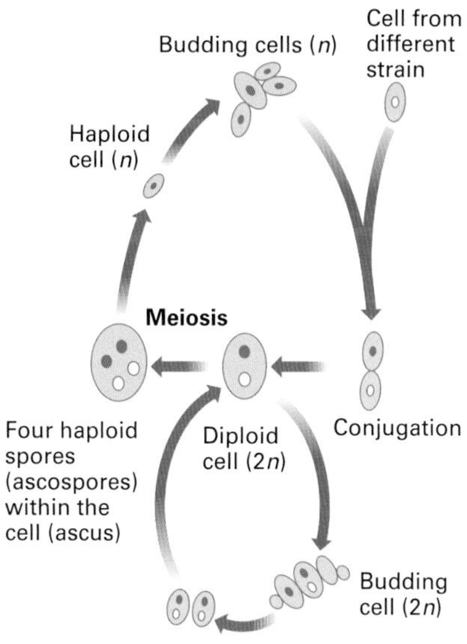

Figure 3 *Life cycle of a yeast such as* S. cerevisiae.

QUICK CHECK

1 How do the hyphae of *Penicillium* differ from those of *Mucor*?

2 Which feature do all yeasts have in common?

3 Explain what is meant by budding.

Food for thought

People have fermented grapes to produce wine for thousands of years. This was one of the first microbiological processes to be harnessed for domestic purposes. However, yeast, the organism responsible for fermentation, was shown to be a living organism only in 1837. Suggest how fermentation might be shown to be caused by the activities of a living organism.

17.7

FERMENTATION

OBJECTIVES

By the end of this spread you should be able to:

- discuss the problems of scaling up
- distinguish between batch cultures and continuous cultures
- describe the main features of downstream processing.

Fact of life

Fermenters 40 metres high are used to culture the filamentous fungus *Fusarium venenatum* from which mycoprotein is extracted to make meat substitutes that can be eaten by vegetarians.

The metabolic activities of microorganisms have been used by humans for thousands of years to produce a variety of foods and drinks. Genetic engineering has given us the potential to extend the range of microbial products enormously. We can now harness the biochemical machinery of microorganisms to produce human substances. This is an achievement which would have been thought unbelievable only 50 years ago.

Industrial fermentation

In the laboratory, it is relatively easy to grow microbes on a small scale in Petri dishes, test tubes, and flasks, given a suitable nutrient medium and the right environmental conditions. Producing substances such as penicillin from microbes on an industrial scale causes more problems because massive numbers of organisms have to be grown for the venture to be commercially viable.

Modifying a laboratory procedure so that it can be used on an industrial scale is called **scaling up**. The microorganisms are grown in very large vessels called fermenters. The term **fermentation** refers specifically to the anaerobic breakdown of organic compounds by living cells, especially microorganisms, often with the production of heat and a wide variety of end products (such as ethanol and carbon dioxide or lactate). The term has been extended to include the process of growing microorganisms in liquid under any conditions, including the large-scale culturing of cells in either aerobic or anaerobic conditions. The vessels in which such cultures grow are therefore called **fermenters**, or **bioreactors** (figure 1).

Penicillin production

Penicillin is produced commercially by growing the fungus *Penicillium chrysogenum* in large stirred fermenters (figure 1). A solution of essential salts and a nitrogen source are put into the fermenter together with an inoculum of the fungus. All procedures are performed aseptically. The pH of the medium is regulated with ammonium salts at 6.5 to 7.0. Lactose (a slowly hydrolysed disaccharide) or a slow continuous feed of glucose (in 70% solution) is added to promote cell growth and reproduction and maximise penicillin production. On completion of fermentation (usually after 6–7 days) the broth is separated from the fungal mycelium and penicillin extracted. This penicillin can then be modified by chemical procedures to yield a variety of semisynthetic penicillins.

Scaling up

Effective scaling up requires biologists and engineers to solve several problems:

- Large-scale fermenters have to be built to very strict design and building specifications.
- There must be no risk of contamination. Only the desired organism must be allowed to grow in the vessel; all others must be excluded.
- The organisms must be kept in conditions that optimise the production of the required substance. This means installing highly sensitive equipment that maintains pH, temperature, and fluid volume within very strict limits.

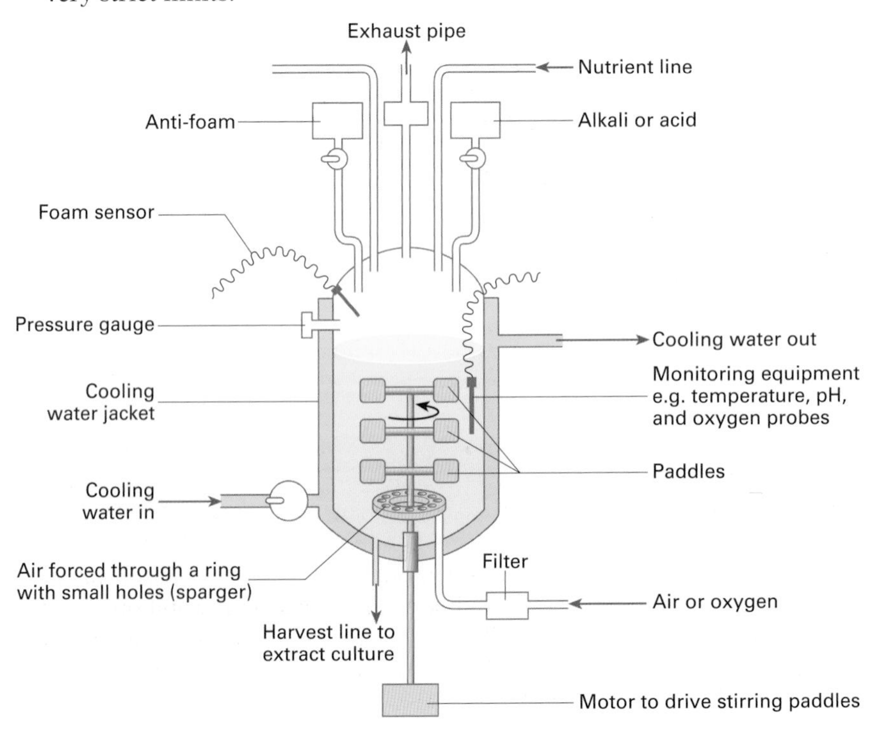

***Figure 1** A fermenter: a large vessels used to culture microorganisms in an industrial process.*

- High levels of microbial activity generate large amounts of heat which must be removed via a heat exchanger so that a constant temperature can be maintained.
- Nutrient levels must be kept at optimal levels as the microbial population grows.
- The build-up of end-products or inhibitors which impair production must be minimised, or the microorganism's resistance to them maximised.
- The formation of foam (an unavoidable consequence of carbon dioxide production in a nutrient-rich solution) must be monitored and controlled.
- The motors that mix the microbial suspension must be powerful enough to work even when the growth of the microbial population gives the suspension a porridge-like consistency.
- Cultures of aerobic organisms have to be given an adequate oxygen supply. This is frequently the major limiting factor for large cultures. Often the only way to provide sufficient oxygen is by aeration with small bubbles of sterile air which have a large surface area to volume ratio.

Batch cultures and continuous cultures

There are two main types of culture used in industrial processes: batch cultures and continuous cultures. In **batch cultures**, cells are grown in a fixed volume of liquid medium in a closed vessel. No microorganisms, fluid, or nutrients are added or removed from the culture during the incubation period. Batch cultivation is used for producing secondary metabolites, such as penicillin and other antibiotics, which are relatively unstable and not essential for the growth of the culture. These secondary metabolites can be extracted economically only when they reach a high concentration in the culture suspension.

In **continuous cultures**, nutrients are added and cells harvested at a constant rate, so that the volume of suspension is also kept constant. This means that the fermenter does not have to be emptied, cleaned, and refilled very often, and also means that production is almost continuous. However, continuous cultivation is expensive because it needs sophisticated equipment to maintain constant conditions, and highly skilled staff to operate the equipment.

Downstream processing

A glance at a large stainless steel fermenter with its complicated control systems might suggest that this hardware forms the most expensive part of biotechnological processes. However, almost 80 per cent of the cost of many biotechnological processes is accounted for by **downstream processing**: the extraction and purification of the product at the end of the culture process.

Downstream processing uses a variety of techniques. The first stage often involves separating the cells from the liquid part of the suspension. This might be done by sedimentation, centrifugation, or filtration. If the cells themselves are the desired product, these need only be washed, dried, and packaged after separation from the fluid. However, if the desired product is a chemical within the cells, the cells will have to be broken apart to release the chemical and the cellular components removed, again by sedimentation, centrifugation, or filtration. The desired chemical might then be extracted and purified by any of a number of techniques, including precipitation and chromatography (see appendix). Similar extraction and purification techniques might be used if the desired product is a chemical produced by the microbes but secreted out of the cells, so that it is present in the fluid part of the suspension. For example, extracting and purifying the secondary metabolite penicillin dissolved in the nutrient broth involves filtration, concentration (by partial drying), and precipitation of penicillin crystals by the addition of potassium compounds.

Primary and secondary metabolites

Metabolites are any substances produced during the metabolic activities of an organism. Whereas **primary metabolites** are involved directly in growth, development, or reproduction, **secondary metabolites** are not (although they may increase the survival of their producers). Metabolites fermented on an industrial scale include primary metabolites (such as alcohol, vitamin B, mycoprotein, and proteases) and secondary metabolites (such as penicillin and morphine). **Proteases** are protein-digesting enzymes that are essential to all forms of life on Earth. They are also the most highly exploited enzymes; they are used in waste management and the food, detergent, and pharmaceuticals industries.

QUICK CHECK

1 Explain the significance of the small size of the air bubbles used to aerate large cultures of microorganisms.

2 Which type of cultivation is better for producing an antibiotic such as penicillin: batch cultivation or continuous cultivation?

3 What is downstream processing?

Food for thought

Contamination of an industrial fermenter with pathogenic microorganisms can have disastrous consequences. Suggest how the risk of contamination can be minimised for a fermenter such as the one in figure 2.

Figure 2 A commercial fermenter used to make products from genetically engineered organisms.

17.8

METAL EXTRACTION, CHEESE, AND YOGHURT

OBJECTIVES

By the end of this spread you should be able to:

- outline the use of microorganisms in the extraction of heavy metals from low grade ores
- explain how fermentation preserves milk
- describe how true curd cheese is made
- discuss the role of primary microorganisms and secondary microorganisms in cheese-making
- describe how yoghurt is made.

Fact of life

Freshly prepared yoghurt contains 10^9 bacteria per gram.

Bioleaching

Bioleaching is a process by which sulphur-eating bacteria are used to extract precious metals such as copper, gold, silver, and uranium from low-grade ore by oxidation. The process depends on the availability of Fe^{3+} ions. It occurs spontaneously in the ore without microorganisms, but the bacteria accelerate it by regenerating Fe^{3+} ions from Fe^{2+} ions as they use the sulphur-containing ores as a source of energy. They oxidise sulphides into sulphates and pure metal, using electrons from the Fe^{2+} ions and obtaining energy from the process. Eventually, sufficient pure metal accumulates to be filtered from the unprocessed material and the waste solution. Although slow, bioleaching can be more efficient, more environmentally friendly, and cheaper than other processes such as smelting.

Many people associate the word 'bacteria' with disease and decay. However, in nature bacteria play a vital role in many processes, including recycling nitrogen and other nutrients (spread 22.12), breaking down cellulose eaten by sheep and cattle (spread 9.10), and releasing vitamins in our guts (spread 9.5). They also have their industrial uses. These include extracting metals from low grade ore (see box 'Bioleaching'), and fermenting milk into cheese and yoghurt.

Cheese

Milk has a well deserved reputation for being a highly nutritious food. Unfortunately, its high nutrient content is not only attractive to humans. If left standing for any length of time, the nutrients enable microorganisms to grow which make the milk unfit for human consumption. In ancient times, the main way of preserving milk was to convert it into cheese.

Historians believe that cheese became part of the human diet about 8000 years ago, making it the first fermented food. It was probably produced accidentally through the practice of carrying milk in pouches made from animal stomachs. Enzymes in the digestive juices from the stomach and bacteria in the milk worked together to form a curd and then a crude cheese. Cheese-making artefacts dating from 2000 BC have been found.

True curd cheese is a solidified mass of fermented milk. The milk is usually from cows, but goat's milk and sheep's milk are also used. The fermentation is carried out by bacteria which produce lactate by fermenting lactose (milk sugar). The lactate discourages the growth of other organisms that would spoil the food or cause disease.

Modern cheese-making involves four main processes: milk pretreatment; curdling; curd extraction; and salting, pressing, and moulding.

Pretreatment

Pretreatment begins with the pasteurisation of raw milk and testing the milk to make sure it contains no unwanted antibiotics. After pretreatment, the milk is cooled to 30 °C and a starter culture of bacteria added. The two most commonly used species of bacteria are *Lactobacillus casei* and *Streptococcus lactis*. Modern cheese-making uses bacteria-free milk to which purified bacterial cultures are added so that the population of bacteria in the cheese is predictable and safe to eat.

Curdling

Curdling is the process by which milk is separated into a solid **curd** and a liquid **whey**. The process happens in the acidic conditions caused by the lactate (lactic acid) excreted by the bacteria. Enzymes are usually added to help coagulate the milk. Until recently, the main source of these

Figure 1 Cheddar cheese, the most common type of cheese in the world, is produced by 'cheddaring'. The curd is turned and piled to press out the whey and develop the desired texture.

enzymes was **chymosin** (**rennin**), sold as rennet. Chymosin is produced in the stomachs of mammals to coagulate casein (a milk protein) and delay the passage of milk from the stomach to the small intestine. Originally extracted from the stomachs of calves slaughtered for food, bovine chymosin has now been largely replaced by edible plant enzymes or chymosin produced by genetic engineering. The chymosin gene in calf cells is transferred to bacteria or yeast cells, which then produce the enzyme on demand (spread 18.9 explains how gene transfer is accomplished).

Curd extraction

Curd extraction (figure 1) involves draining off the whey (this may be used in animal feedstuffs) and cutting the solid curd into small pieces and scalding it at a temperature of 32–42 °C. This changes the texture of the curd.

Salting, pressing and ripening

After allowing the curd to settle and knit together, blocks of curd are cut, salted, pressed into moulds, and allowed to ripen for a period of time which varies according to the cheese type. Soft cheeses have a short ripening time (1–5 months); hard cheeses may require many months of ripening (Parmesan requires 12–16 months). In some cheeses, the ripening is aided by fungi such as *Oidium* which grow on the surface of the curd. During ripening, the bacteria in the curd die and are digested by their own enzymes (a process called autolysis). This releases substances which flavour the cheese.

Secondary microorganisms

The unique characters of different cheeses are brought about by differences in the physical composition (water, protein, and fat content) and also in the **secondary microorganisms** that grow in or on cheese (as opposed to the **primary microorganisms** used to ferment the milk to curd). In modern cheese-making, secondary microorganisms are introduced by the cheese-maker for a particular purpose. For example, needles are used to press the blue spores of *Penicillium roqueforti* deep into Roquefort or Stilton cheeses (figure 2); and *Penicillium camemberti* is sprayed over the surface of raw Camembert cheese. The fungal hyphae of these moulds give the cheeses their distinctive flavours. In Swiss cheeses, such as Gruyère and Emmenthal, *Propionibacterium* grows in the curd. The bacteria produce propionic acid, responsible for the characteristic flavour, and carbon dioxide, responsible for the 'holes'.

Yoghurt

Like cheese, yoghurt is produced from milk by the action of lactate-producing bacteria, especially *Lactobacillus bulgaricus* and *Streptococcus thermophilus*. These bacteria are commonly used in yoghurt starter cultures. They are added to milk in approximately equal proportions. Most commercial producers pasteurise the milk (at 90 °C for 20 minutes) before adding the bacteria. The mixture is incubated at around 45 °C for 5 hours during which time the bacteria break down milk proteins into peptides. The fermentation produces lactate which brings the pH down to about 4.0. Fermentation by-products, including ethanal and methanoic acid, give yoghurt its characteristic flavour. The product is cooled to prevent further fermentation. Sometimes fruit pulp, colouring, and flavours are added before packaging. Some yoghurts are heat-treated before or after packaging to kill any bacteria, but most yoghurts contain live bacteria.

Fermented milks

Acidophilous milk, a special type of fermented milk containing *Lactobacillus acidophilus*, is thought to have special health-giving properties. Although the exact nature and extent of the effects of *L. acidophilus* are unclear, it may modify the population of microorganisms in the large intestine, lowering blood cholesterol levels, and possibly lowering the risk of colon cancer. *L. acidophilus* is available in some health food shops in tablet form.

Figure 2 *The dark areas in this Roquefort cheese are the result of an extensive growth of* Penicillium roqueforti.

Quick check

1. How does fermentation preserve milk?
2. Explain the function of chymosin in cheese-making.
3. Name two types of mould commonly used as secondary microorganisms in cheese-making.
4. Name a species of bacteria commonly used in the starter cultures of yoghurt.

Food for thought

Cultures of lactate-producing bacteria, called starter cultures, are added to milk during the preparation of cheese. *Streptococcus lactis* and *S. cremoris* are commonly used. One of the greatest problems for the cheese-making industry has been the unreliability of the starter cultures. Under certain circumstances, lactate production is brought to a halt within 30 minutes. Examination of the bacteria reveals that they have undergone lysis (rupture and physical disintegration of the cell). Suggest what might cause this problem and how it might be controlled.

17.9

Single cell protein and mycoprotein

OBJECTIVES

By the end of this spread you should be able to:

- explain what single cell protein (SCP) and mycoprotein are
- describe how a bacterium, a fungus, and an alga can be used to make fermented foods
- discuss the reasons for making fermented foods.

Fact of life

Spirulina, a cyanobacterium (blue-green bacterium) inhabiting freshwater lakes, was used as a high-protein staple food by the Aztecs of Mexico and continues to be used by people living around Lake Chad in Africa. *Spirulina* is sold in healthfood shops as a rich source of vitamin B_{12}. Vitamin B_{12} activity is assessed by seeing whether the substance causes growth of *Lactobacillus*, and *Spirulina* contains B_{12}-related compounds that do cause this growth. However, some scientists doubt whether these compounds are active in humans. They suggest that the B_{12} activity of some *Spirulina* extracts may come from faecal bacteria contaminating the extracts.

In the late 1960s, against a background of increasing fears about the growth of the world's human population outstripping food supplies, biologists searched for new food sources. Of particular concern was the possible shortage of protein, so there was considerable excitement when biologists discovered that certain microorganisms have an unusually high protein content. The excitement grew when biotechnologists demonstrated that these microorganisms (species of bacteria, algae, and fungi) could be cultured on a large scale for human consumption or animal feed. The foods grown from unicellular microorganisms are called **single cell protein (SCP)**.

Microorganisms used to produce SCP should have the following characteristics: they should be non-pathogenic to humans and other animals; they should grow quickly on cheap sources of carbon, nitrogen, and energy; they should be able to tolerate the heat generated by the metabolic activity of large cultures; and they should produce a food which has a high protein content and is safe to eat.

Pruteen

In the 1980s, Pruteen was one of the first SCPs produced on a large scale. It was made by the bacterium *Methylophilus methylotrophus* cultured in a huge fermenter at Billingham in north-east England. This bacterium doubles its numbers in 2–5 hours; it is non-pathogenic; and it is able to convert methanol (obtained from oil) aerobically to SCP:

Methanol + nitrogen + mineral salts + oxygen → SCP + carbon dioxide + water

Two tonnes of methanol produce one tonne of Pruteen, which is 72% protein and 8% moisture. If the fermenter temperature was kept between 30 °C and 40 °C, and the pH at 6.7, the SCP could be cultured continuously for several months. The potential output of the fermenter was 70 000 tonnes of Pruteen each year. The carbon dioxide was not wasted; it was bottled and sold.

Pruteen was used as an animal feed. Before packaging, the cells were separated, dried, and ground to make them more digestible, and the pH and mineral content were adjusted. Commercially Pruteen was not very successful, partly because methanol increased in price while animal feeds became cheaper. However, the lessons learned in mass culturing *M. methylotrophus* were invaluable in producing other SCPs.

Chlorella

Chlorella is a non-moving, single-celled alga (figure 1). It has a high reproductive capacity. This makes it a potential problem to owners of swimming pools and managers of ponds, lakes, and canals as blooms of the alga can cause the water to resemble pea soup.

However, its reproductive capacity is an advantage to food producers. *Chlorella* is highly nutritious. The dry weight of each cell is about 50% protein, 20% carbohydrate, and 20% fat, with significant amounts of vitamins and minerals. Toxicity tests show that it is safe to eat. The cell walls are cracked open so that the contents can be digested easily.

Although reproductive capacity is high, production efficiency is low unless the cultures are supplied with intensive artificial light as an energy source and highly carbonated water as a source of carbon dioxide. This makes mass production expensive. Consequently, *Chlorella* is cultured commercially mainly for the health food market. Advertisements claim that it has a higher percentage of chlorophyll than any edible plant (though *Chlorella* is a protoctist, not a plant; see spread 21.3) and that it is a complete protein with all the essential amino acids, vitamins, and minerals. Since the 1950s, Japan and Taiwan have been the main producers of *Chlorella* products.

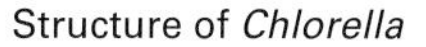

Mycoprotein

The experience gained in fermenting *M. methylotrophus* to make Pruteen has been used to produce a mycoprotein (fungal protein) from a mould called *Fusarium venenatum*. One mycoprotein, Quorn®, is used for human foods. To make the mycoprotein, *F. venenatum* is fermented by a continuous process at 30 °C for about six weeks. The nutrient medium in which the mould is fermented includes glucose as a source of carbon and energy, and ammonium salts as a source of nitrogen. Although aseptic procedures are used, the mould is grown at a low pH to inhibit bacterial growth and reduce the risk of contamination. The doubling time of the mould is about 5.5 hours. About 0.5 kg of mycoprotein is produced for each kilogram of glucose used. After fermentation, the liquid in the culture medium is filtered off and the mycelium (mass of fungal threads) is extracted.

Unless specially treated, mycoprotein has a dangerously high nucleic acid content (about 5–15 per cent dry weight), mainly RNA. The consumption of more than 2 g of nucleic acid per day can lead to kidney stones and gout. The RNA content of mycoprotein used to make Quorn is reduced by heating the culture to 60 °C for 20–30 minutes. This breaks down the RNA by the action of natural RNAase enzymes without affecting the protein content. The RNA content of Quorn is less than one per cent, well below the World Health Organization's recommended maximum of two per cent.

After the mycoprotein has been extracted, flavourings and colouring are added, along with a little egg albumen. Quorn is marketed as a meat substitute, low in fat and cholesterol and high in protein, fibre, zinc, and thiamin.

The future of fermented foods

The production of SCP and mycoprotein has been boosted by increasing public concern about the ethics of intensive farming and the safety of animal foods (see spread 15.10 for an account of CJD and beef). It is likely that these novel sources of food will become more important in the future.

Quick check

1 Explain what single cell protein is.

2 Give the carbon source used in cultures of:

a *Methylophilus methylotrophus*

b *Chlorella*

c *Fusarium graminearum.*

3 What is the main use of SCP derived from *Chlorella?*

Food for thought

Many astrobiologists believe that organisms such as *Chlorella* will play a vital part in space travel in the future. What benefits could *Chlorella* provide for astronauts on long journeys in space?

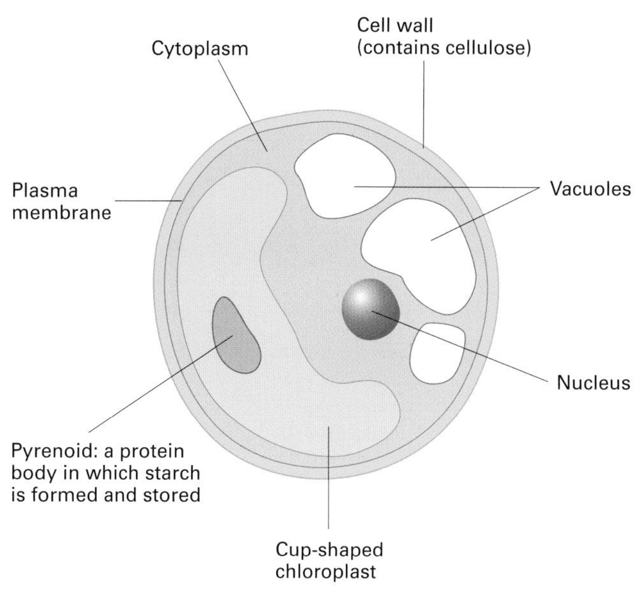

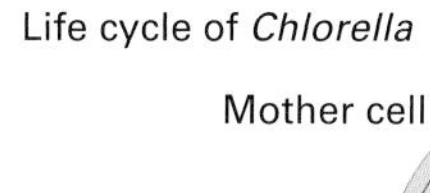

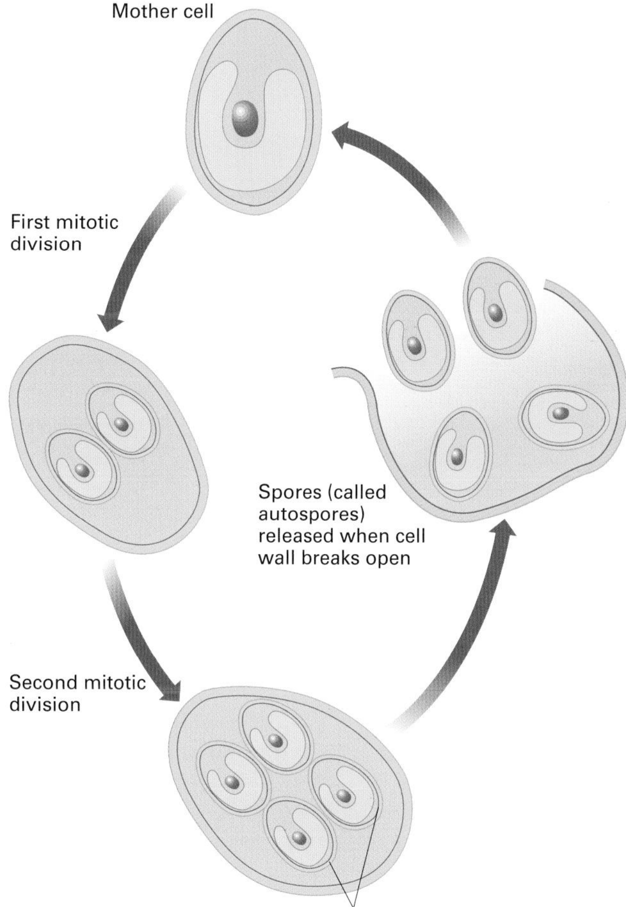

***Figure 1** The structure and life cycle of* Chlorella. *The organism is widespread in both fresh water and sea water, and also in soil. It reproduces only asexually.*

17.10 BREAD, BEER, AND WINE

OBJECTIVES

By the end of this spread you should be able to:

- describe how bread is made
- explain what happens during the main stages of beer-making
- discuss how different wines are made.

Fact of life

The Ancient Egyptians were known as the 'bread eaters'. Wall paintings show them using yeast to make leavened bread (that is, bread that has risen) and samples of bread dating from 2100 BC are on display in the British Museum. Pictures on tomb walls (figure 1) also show that beer drinking was popular in Egypt 4000–5000 years ago.

Figure 1 *A scene from a tomb wall showing that beer drinking occurred in Egypt between 4000 and 5000 years ago.*

Fermented plant products

In additon to bread, beer, and wine, fermented plant products include sauerkraut, tofu, and soy sauce. Sauerkraut or sour cabbage is produced from wilted shredded cabbage by lactic acid fermentation. Addition of 2.2 to 2.8% sodium chloride restricts the growth of undesirable gram-negative bacteria while flavouring the development of the desirable lactic acid bacteria. Fermentation takes 20–30 days, during which the cabbage is dehydrated and the lactic acid produced by a succession of microorganisms. Tofu is chemically coagulated soybean milk product. Soy sauce is fermented from soy beans by the action of fungi.

Leavened bread, beer, and wine are among the best known products made by yeast fermenting sugars to ethanol (alcohol) and carbon dioxide. The most commonly used species of yeast is *Saccharomyces cerevisiae* (spread 17.6), but in practice highly selected strains are used for different purposes.

Bread

The main ingredients of leavened bread are yeast, flour (obtained from grinding cereal grain, usually wheat), and water. Most breads also include salt, sugar, and ascorbic acid. The sugar speeds up the fermentation and the ascorbic acid (vitamin C) makes gluten, a protein in the dough, more springy.

Modern bread-making involves several steps. First, amylases in the moistened dough produce maltose and sucrose from the starch in the flour. Then a baker's strain of *S. cerevisiae* is added. The yeast produces a mixture of enzymes (including maltase) which break down the starch to glucose and other sugars. These sugars provide the yeast with a source of energy and carbon. Fermentation by the yeast produces carbon dioxide gas, which helps the dough to rise and swell. This stage of bread-making is called **proving**. Traces of other fermentation products contribute to the flavour of the bread. The fermentation is carried out mainly under aerobic conditions to maximise carbon dioxide and minimise ethanol production. Any ethanol produced during fermentation evaporates during the baking process, which follows proving.

Coeliac disease (gluten intolerance)

Gluten is a mixture of two proteins, gliadin and glutenin, found in rye and wheat flour but absent from flour made from oats, barley, or maize. The gluten gives flour a springy quality. People with the digestive disorder **coeliac disease** are intolerant of gluten. If they eat bread containing gluten, the intestinal wall swells and the microvilli of the cells lining the wall become damaged. This causes general malabsorption (the absorption of all nutrients is impaired). Coeliac disease is controlled by having a strict, lifelong gluten-free diet. People with coeliac disease can eat bread that contains no gluten.

Beer

Beer is brewed from barley grain. The process (figure 2) involves six stages:

1. During **malting**, dry grain is soaked (steeped) in water and then layered in a malting tower or spread on a malting floor. This allows the grain to germinate partially so that its store of starch is converted into sugars (including maltose). Malting also converts proteins to amino acids. These breakdown products are used by yeast for fermentation. Germination is usually accelerated and controlled by adding gibberellins (plant growth substances; spread 14.10) and amylases. The process is stopped by slowly heating the grains to 80 °C.
2. **Milling** breaks down the grains produced by malting (called malted grain) into two or three small pieces. The crushed grain or **grist** is then mashed.
3. **Mashing** takes place in a 'mash tun' in which hot water trickles through the grains, softening them and releasing **wort**, a nutrient-rich liquid.
4. Hops (dried ripe flowers from the hop plant) are added to the wort to give flavour and for their antiseptic properties. **Boiling** involves heating the mixture of wort and hops to a high temperature. The resulting concentrated, sterile liquid (called **mash**) is strained to remove the hops and then allowed to cool before fermentation.

5 In **fermentation**, brewer's yeast is added to the mash. Two commonly used yeast species are *S. cerevisiae* and *S. carlsbergensis*; the latter is used to make lager. Fermentation usually takes place in large conical batch fermenters (spread 17.7). During this process, the yeast converts sugars into ethanol and carbon dioxide as in bread-making. However, in beer-making, the fermentation is mainly anaerobic and the ethanol is kept in the brew. Fermentation is generally completed after about five days, when the ethanol content has reached 4–6 per cent, but beer with a higher or lower ethanol content can be made by varying the brewing process.

6 In the final stage, **finishing**, the beer is separated from the yeast and packaged for sale. Bottled and canned beers are usually pasteurised at 60 °C or higher to kill off any microorganisms in the brew. Aqueous extracts of brewer's yeast are used as a food, rich in vitamin B.

Wines

Wines are made by the fermentation of fruit juices. Grapes are the fruit used in most commercial wine-making. After being picked, the grapes are crushed, usually mechanically although some wine-makers still 'tread' the grapes by foot. The resultant liquid, called **must**, contains the sugar used in fermentation.

Fermentation usually takes place in large fermentation vats. Natural grape-skin microorganisms, a mix of different species of bacteria and yeast, may be used for this fermentation. This gives unpredictable results, producing wines of a variable quality: some excellent, others less palatable. In order to make a wine of consistent quality, most commercial wine-makers now use cultivated strains of yeast belonging to the species *S. cerevisiae* or *S. ellipsoideus*. Fresh must is first treated with sulphur dioxide to kill unwanted microorganisms, and the desired strain of yeast then added. The must is fermented for 3–5 days at temperatures varying between 20 and 28 °C. During fermentation, ethanol and carbon dioxide are released. The processes of filtering, heating, and ageing complete the wine-making. The details of the final processes vary with the type of wine.

One of the great attractions of wines, even those made under carefully controlled conditions, is that they are extremely variable. Wine colour depends on how long the must is allowed to remain in contact with grape skins (it does not depend on the grape colour; all grapes release a white liquid). Red wines are produced by allowing the must to remain in contact with grape skins longer than is the case with white wines. The ethanol content of wine depends on the ethanol tolerance of the yeast strain: when the ethanol reaches a certain concentration in the fermentation process, the yeast is killed by it. Most table wines contain 8–14 per cent ethanol. Sweetness is varied by regulating the initial sugar concentrations of the must. With higher levels of sugar, the fermentation stops before all the sugar is used. Very dry wines have no sugar remaining at the end of fermentation.

Quick check

1 What happens during the proving stage of bread-making?

2 Explain the main purpose of malting.

3 How are red wines produced when the juice of all grapes is white?

Food for thought

Champagnes and other expensive sparkling wines are produced by allowing fermentation to continue in bottles. Suggest why the process is much more difficult than making ordinary wines.

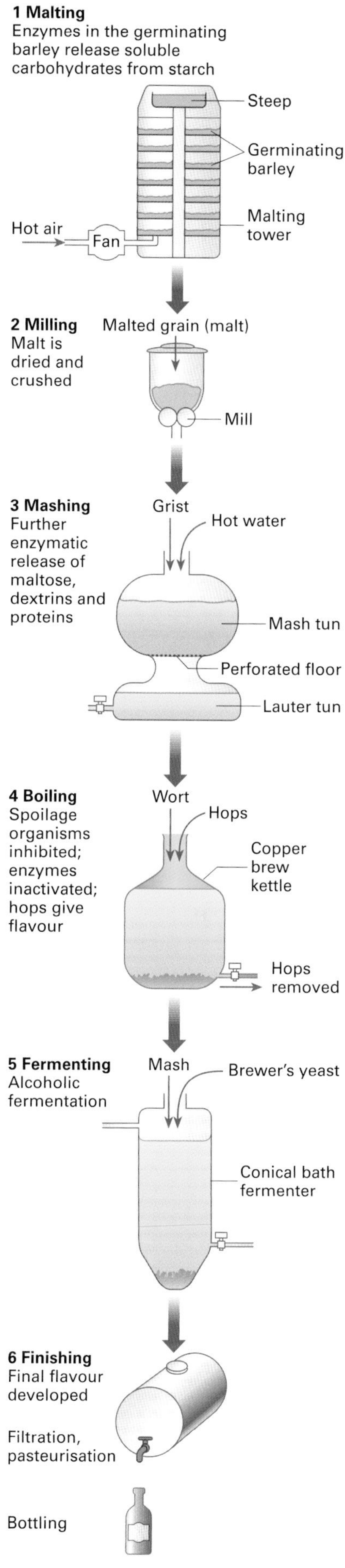

Figure 2 *Commercial beer-making.*

17.11

Biofuels

OBJECTIVES

By the end of this spread you should be able to:

- describe the three stages of biogas production
- discuss the role of bioreactors in economically poor, rural communities
- describe how gasohol is produced.

Fact of life

In 2010, the worldwide production of biofuel was more than 100 billion litres and provided fuel for 2.7% of the world's road transport. According to the International Energy Agency, biofuels have the potential to meet more than a quarter of the world demand for transportation fuel by 2050.

Biofuels are energy sources derived from the conversion of **biomass** to a combustible product. They are becoming increasingly important as alternative sources of power as non-renewable fossil fuels begin to run out. The biomass of biofuels is derived from organisms that, unlike fossil fuels, are still actively involved in the carbon cycle. Therefore, biofuels are renewable while fossil fuels are not. Two biofuels that harness the activity of microorganisms are biogas and gasohol.

Biogas

Biogas is a gas produced by the anaerobic fermentation of organic waste. The gas contains 40–70% methane and is similar to natural gas, which is about 80% methane. Biogas production depends on methanogens (methane-producing archaebacteria), the same type of microorganisms that produce methane naturally (spread 17.2).

More than 30 years ago, specially designed fermenters called **bioreactors** (figure 1) started to be constructed to harness the methane-producing ability of methanogens for human use. The production of biogas involves three stages and three communities of microorganisms:

1. anaerobic fermentation by eubacteria including *Lactobacillus*, which convert the organic waste into a mixture of organic acids and alcohol, with some hydrogen, carbon dioxide, and acetate
2. acetogenic (acetate-producing) reactions by bacteria such as *Acetobacterium* which, in addition to acetate, produce hydrogen and carbon dioxide from the organic acids and alcohol
3. methanogenic (methane-producing) reactions by archaebacteria, including *Methanobacterium*, *Methanococcus*, and *Methanospirillum*.

The archaebacteria generate methane either by reducing the carbon dioxide:

$$\text{Carbon dioxide} + \text{hydrogen} \longrightarrow \text{methane} + \text{water}$$
$$CO_2 + 4H_2 \longrightarrow CH_4 + 2H_2O$$

or by converting acetate:

$$\text{Acetate} \longrightarrow \text{methane} + \text{carbon dioxide}$$
$$CH_3COOH \longrightarrow CH_4 + CO_2$$

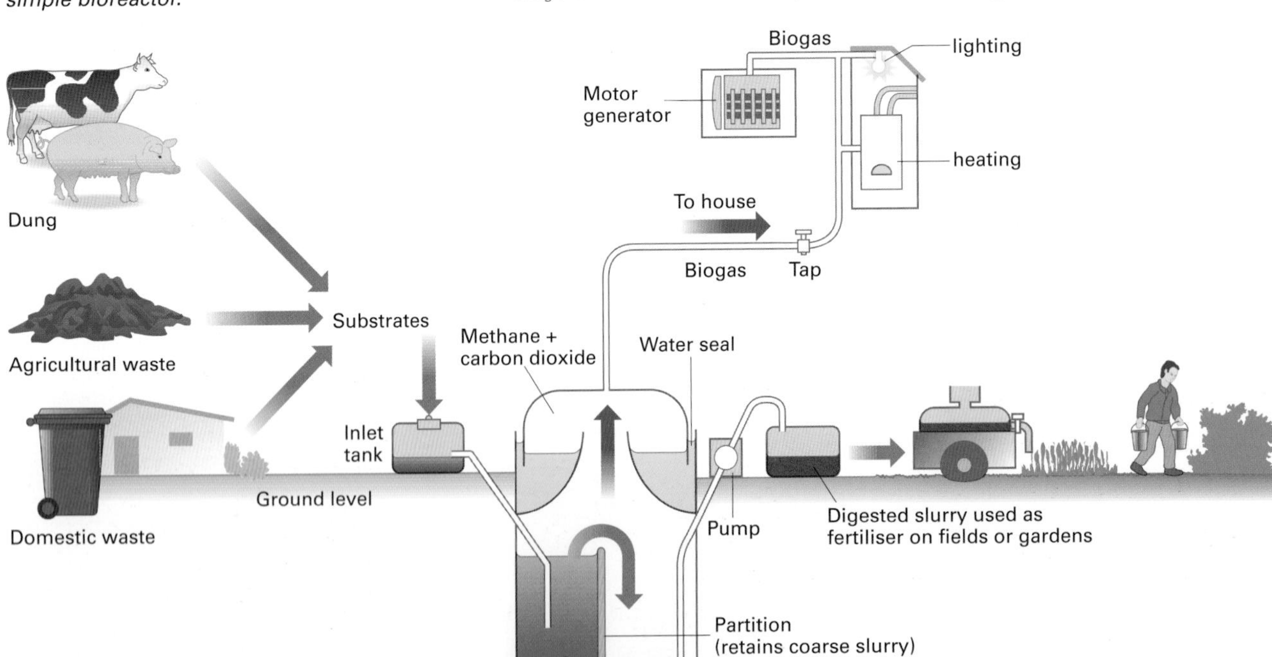

***Figure 1** Biogas production using a simple bioreactor.*

The biogas accumulates at the top of a bioreactor and can be collected and used as a domestic fuel, or compressed in cylinders to be used in cars and tractors. The solid waste, known as **digested slurry** or **sludge**, can be used as a fertiliser on fields or in gardens. A single tonne of cow manure, for example, provides enough fertiliser for 50 kg of grain.

In principle, all organic waste can be fermented. In the USA water hyacinth, a weed that blocks canals, has been used. However, in small, simple bioreactors, only more liquid substances (human sewage along with faeces from cattle, pigs, and poultry) are used. The organic waste is diluted with an equal amount of water (or urine) and put into the bioreactor.

Today millions of small bioreactors are being used in rural areas throughout the world to make biogas. In China, for example, over 18 million small-scale bioreactors provide families with fuel for cooking, lighting, and vehicles. In addition to producing a clean fuel, bioreactors are an efficient way of dealing with organic wastes and produce pathogen-free nutrient-rich wastes which can be used as fertilisers.

Gasohol

Gasohol is a blend of 10% ethanol and 90% petrol. It is used successfully as a fuel in Brazil (figure 2) and the USA. In Brazil, all cars have been converted to run on gasohol or on pure ethanol (the authorities insist that some petrol is added to all fuels to dissuade people from drinking them). In 1990, the USA produced about 875 million gallons of ethanol.

Ethanol production involves three main steps:

1 extracting sugars from a suitable source
2 fermenting the sugar to produce ethanol
3 distilling the ethanol from water by boiling.

Sugar cane is the source of ethanol in Brazil. The cane is crushed and sucrose extracted from it, leaving sugar molasses and fibrous material. The molasses is a syrup containing glucose and frutose, which are used as the substances for fermentation. In the USA, ethanol is made from maize syrup. In both countries, fermentation is carried out by *Saccharomyces cerevisiae*, the same species of yeast used in brewing and baking. Alcoholic fermentation by yeast produces at best a 16 per cent ethanol solution. This needs to be distilled as cars will run only on a pure ethanol/petrol mix, or on a solution with 95 per cent or more ethanol. The heat required for the distillation is supplied by burning the dried fibrous sugar cane waste, making ethanol a cheap fuel in Brazil.

***Figure 2** A field of rapeseed. Enormous tracts of land are being used for biofuel production.*

Land clearing and biofuels

Declining oil stocks, the demand for more energy by emergent economies, and the greenhouse effects of carbon dioxide make the search for alternative, environmentally friendly sources of energy increasingly urgent. Biofuels appear to fit the bill: they are renewable and burn cleanly, but whether they are truly environmentally friendly depends on how they are produced. In USA, Brazil, and south-east Asia, the destruction of vast tracts of rainforests, peat land, savannas, or grasslands to grow crops for biofuels has resulted in the release of much more carbon dioxide than is saved by burning biofuels rather than fossil fuels. Consequently, there is a net increase in global greenhouse gas (GHG) when crop-based biofuels are used. Also, using good agricultural land for biofuel crops means that less land is available for food production, pushing food prices up. However, biofuels made from biomass obtained from degraded or abandoned agricultural land produces little or no net increase in GHG and does not diminish the area of land used for food crops.

Quick check

1 What part do acetogenic reactions play in the production of biogas?
2 Family-sized bioreactors are used in China to produce biogas. Give two other functions of the bioreactors.
3 What is gasohol?

Food for thought

Gasohol has been portrayed as 'environmentally friendly' by its supporters. Unlike fossil fuels, ethanol is a renewable fuel which burns cleanly. It is used as a substitute for lead in petrol and reduces sulphur emissions that contribute to acid rain. However, many environmentalists believe that dependence on gasohol damages the environment and could lead to an increase in the price of food crops. Suggest reasons for their concern.

Summary

A typical **virion** (complete **virus** particle) consists of a **protein capsule** enclosing **nucleic acid**, either DNA or RNA. Viruses can only reproduce and function inside living cells so it is difficult to decide whether they are living or non-living. Viruses infect a wide range of organisms and cause many diseases; for example, **tobacco mosaic virus** causes mottling of the leaves of tobacco plants, and **HIV** causes AIDS.
Bacteria are small cellular organisms forming the **prokaryotes**. They are divided into two main groups: **archaebactertia** and **eubacteria**. *E. coli* is a eubacterium that is usually harmless to humans, but which may cause food poisoning. **Pasteurisation** reduces the number of bacteria in foods and drinks but only **sterilisation** kills all bacteria. Bacteria are studied using **aseptic techniques**. The bacteria can be cultured in **broths** or on **plates** covered with **agar**. **Gram staining** distinguishes two groups of bacteria: **Gram-negative** bacteria, such as *E. coli*, and **Gram-positive bacteria**, such as *Staphylococcus sp*.
Moulds may be formed from bacteria, fungi (e.g. *Rhizopus*, *Mucor*, and *Penicillium*), and protoctistans (e.g. **slime moulds**). *Rhizopus* and *Mucor* are saprotrophic (saprobiontic) and have life cycles including sexual and asexual phases. *Penicillium* moulds produce antibiotics that have been commercially synthesised.
The products of microorganisms are produced commercially in **fermenters** or **bioreactors** either as **batch cultures** or **continuous cultures**. **Downstream processing** is usually the most expensive part of the process. **Cheese** and **yoghurt** are made using the activities of bacteria and moulds.
Single cell protein can be made from bacteria. *Chlorella*, a photosynthesising single-celled alga, is cultured to make a nutritious food. **Mycoprotein** is produced from a mould called *Fusarium graminearum*. Yeasts belonging to the genus *Saccharomyces* are used in **fermentation** to make **bread**, **beer**, and **wine**. Microrganisms can also be used to make fuels: **biogas** is made from organic waste; **gasohol** is made from the fermentation of sugar by yeast.

PRACTICE EXAM QUESTIONS

1 The diagrams show three types of virus: Herpes, T_2 bacteriophage and Tobacco Mosaic Virus.

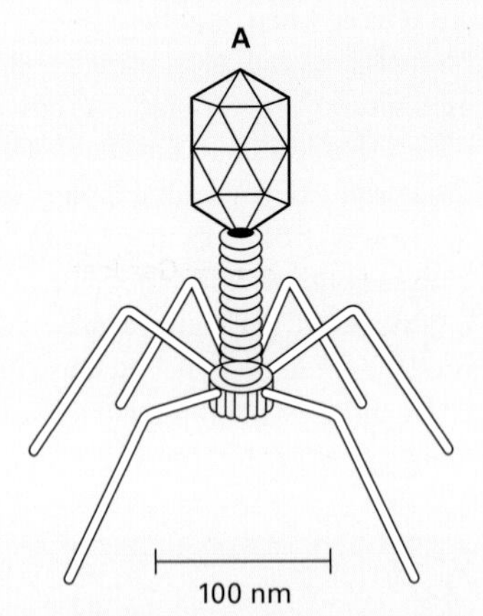

(Reproduced from *Microorganisms and Biotechnology*, Lowrie and Wells, by permission from Cambridge University Press.)

a **i** Which of the viruses, **A**, **B** or **C**, is a bacteriophage? [1]

ii Label the nucleic acid on a sketch of the diagram of virus **B**. [1]

b Name the type of nucleic acid found in HIV. [1]

c Copy and complete the table comparing bacteria and viruses.
Use a (✓) if the feature is present or a (✗) if the feature is absent.

	protein	ribosomes	mitochondria
Bacteria			
Viruses			

[2]

d Describe **one** method by which viruses can be cultured. [3]

[Total 8 marks]

2 The diagram right shows the structure of a Tobacco Mosaic Virus.

a **i** Name the features **A** and **B** shown on the diagram. [2]

(Reproduced from *Introduction to Molecular Biology*, Haggis, by permission of Addison, Wesley Longman Ltd.)

ii Give **two** ways in which the structure of a Tobacco Mosaic Virus differs from that of a T_2 bacteriophage. [2]

b Tobacco Mosaic Virus infects the leaves of tobacco and tomatoes. The symptoms include bright yellow and light green patches on the dark green leaf. Explain how infection by this virus would lead to a decrease in crop yield. [2]

[Total 6 marks]

3 The table below refers to features of two viruses, tobacco mosaic virus (TMV) and human immunodeficiency virus (HIV). Copy and complete the table by writing the appropriate word or words in the boxes.

Feature	Tobacco mosaic virus (TMV)	Human immunodeficiency virus (HIV)
Type of nucleic acid		
Means of entering host	Transmission by aphids	
One sign of infection		Swollen glands

[Total 4 marks]

4 The drawings have been made from scanning electron micrographs of three types of bacteria.

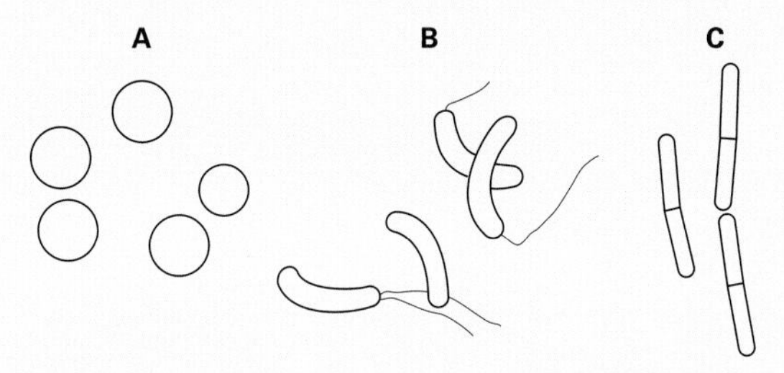

a Identify each type of bacterium from its shape. [3]

b Which structure present in bacteria is the site for

i translation; [1]

ii respiration? [1]

c **i** Give **one** way in which a bacterial cell wall differs from a plant cell wall. [1]

ii Name the test carried out to distinguish between different types of bacterial cell wall. [1]

[Total 7 marks]

5 The diagram below shows an organism of the genus *Rhizopus*.

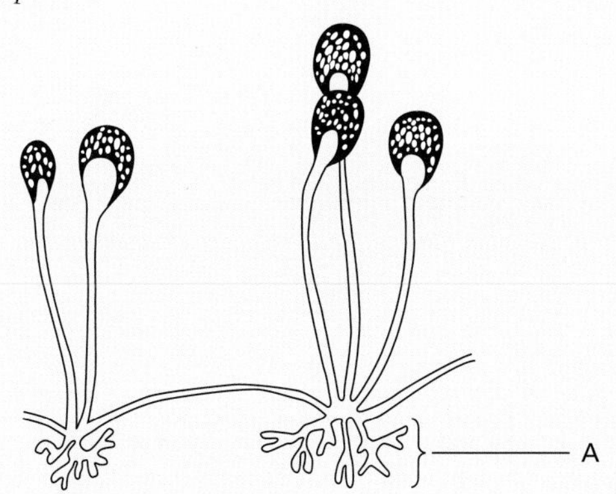

a Name the major taxonomic group to which this organism belongs and give *one* external feature characteristic of this group. [2]

b Describe the role of part A in the nutrition of the organism. [2]

c Explain how parasitic nutrition differs from the nutrition of *Rhizopus*. [2]

[Total 6 marks]

6 **a** **i** To which group of microorganisms do *Penicillium* and the yeast, *Saccharomyces*, belong? [1]

ii State **one** difference in the structure of these two microorganisms. [1]

iii List **three** differences between the structure of these microorganisms and bacteria. [3]

The antibiotic penicillin is produced by *Penicillium* as a secondary metabolite.

b **i** Explain what is meant by the term *secondary metabolite*. [1]

ii Suggest another possible function of a secondary metabolite, in addition to acting as an antibiotic. [1]

c Suggest **one similarity** and **one difference** between the production of penicillin and the production of alcohol by fermentation. [2]

Penicillin is produced in batch fermenters.

d Explain how a batch fermentation differs from a continuous fermentation. [4]

[Total 13 marks]

7 The diagram shows a fermenter developed for the production of protein from bacteria.

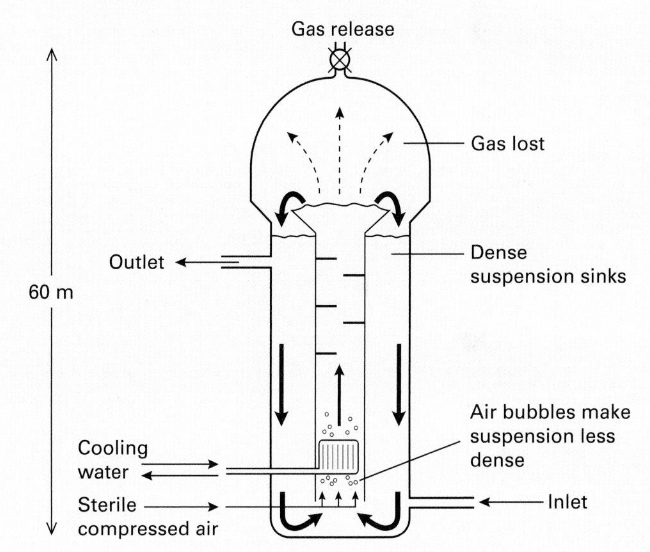

a **i** Suggest **two** functions of the compressed air. [2]

ii Explain why the contents of the fementer are continually circulated. [2]

b Some fermenters are designed for continuous production. [2]

i Give **two** differences between continuous and batch production. [2]

ii Give **two** advantages of continuous production. [2]

[Total 8 marks]

8 Copy the account of the production of gasohol then write on the dotted lines the most appropriate word or words to complete the account.

Gasohol is a mixture of petrol and .... which is produced by microorganisms such as ....
during the .... respiration of glucose.
The glucose can be obtained from sugar cane or from any .... source, such as ....

[Total 5 marks]

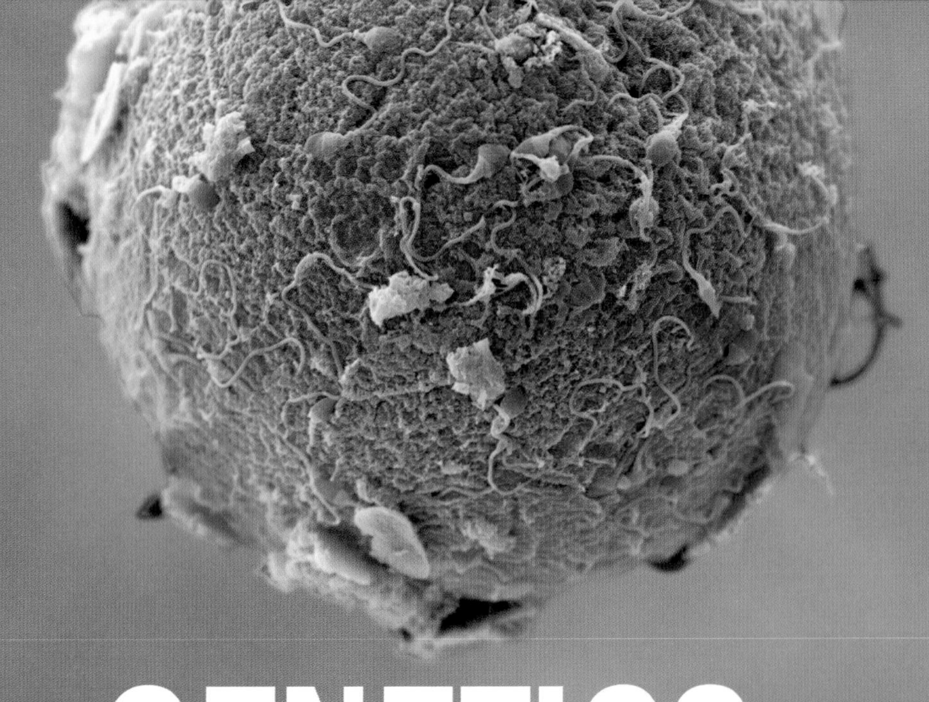

GENETICS *and* EVOLUTION

The whole world of viruses and related particles appears as an intricate fleet of gene carriers, which probably originated in the first place through the evolutionary emancipation of mobile genes from complex prokaryotic and eukaryotic genomes. Once these facts became appreciated and scientists began to understand the mechanisms involved, the idea of using the fleet for the transport of certain specifically chosen passenger genes naturally presented itself. Nature provided the tools, and so genetic engineering was born, perhaps to become one of the most powerful techniques ever developed by mankind.

Christian de Duve

In this section of the book we learn about DNA, how it is used to pass on information from one generation to the next, and how it determines the structure and function of individuals. We consider how our knowledge of genetic material has revolutionised our ability to change the characteristics of organisms, including ourselves. We examine genes and their role in inheritance. Then we look at evolution, the process by which organisms have changed over time and new species have arisen. On the way, we discuss the major steps in human evolution. The final chapter in this section is devoted to a description of the products of evolution, the main groups of organisms living on Earth today.

An orang-utan (Pomo pygmaeus) *has many genes in common with humans, indicating a similar evolutionary history.*

18 Molecular biology of the gene

DNA STRUCTURE

18.1

OBJECTIVES

By the end of this spread you should be able to:

- distinguish between a nucleoside, a nucleotide, and a polynucleotide
- explain how a phosphodiester bond forms
- discuss the significance of complementary base pairing in DNA.

Fact of life

In April 1953, the biologist James Watson and the physicist Francis Crick published the first description of the structure of DNA, in a letter to the journal *Nature*. They based their description on a model they had constructed, but they did little experimental work themselves. The information they used for their model came from work carried out by Erwin Chargaff on the base composition of DNA (table 1), and X-ray data obtained by Rosalind Franklin, working with Maurice Wilkins at King's College, London. One particularly good X-ray diffraction photograph obtained by Franklin in the winter of 1952–3 gave crucial support to the idea that DNA has a helical structure (figure 1). Other data from Franklin showed that DNA has two strands, not three or more as some scientists had proposed. In 1962, Watson, Crick, and Wilkins received a Nobel Prize for their discoveries. Tragically, Rosalind Franklin died of cancer in 1958 at the age of 37. Nobel prizes cannot be given posthumously.

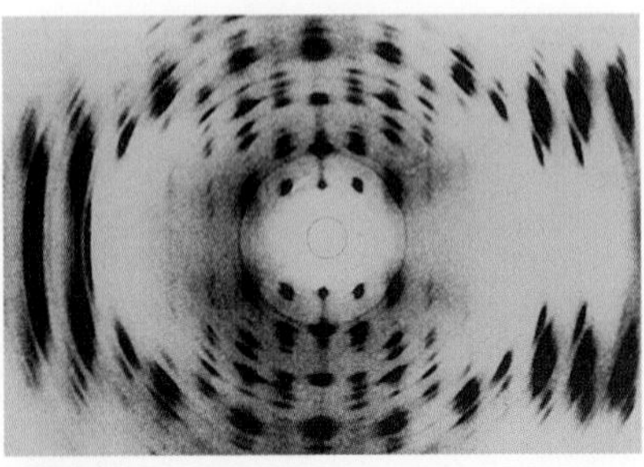

Figure 1 X-ray diffraction photograph of DNA.

The description of the double helical structure of DNA (deoxyribonucleic acid) by Watson and Crick in 1953 (see Fact of life) was a landmark in science history. Their discovery sparked off a new era in scientific research which has had, and will continue to have, far-reaching consequences.

A polymer of nucleotides

- Each DNA strand is a polymer made up of **nucleotide** subunits. The nucleotides join together to form long unbranched polynucleotide chains (spread 2.11).
- Each **nucleotide** consists of **deoxyribose** (a five-carbon or pentose sugar), an organic nitrogen-containing **base** (of which there are four different types), and **phosphoric acid**.
- The sugar and the organic base join together by a **condensation reaction** to form a **nucleoside**. (A condensation reaction results in the removal of a water molecule.)
- Another condensation reaction joins the nucleoside with phosphoric acid to form the nucleotide. This bond forms between carbon 5 of the sugar and the phosphate, and is called a **phosphoester bond** (figure 2).
- The organic bases present in DNA are either **purines** (guanine, G and adenine, A) or **pyrimidines** (cytosine, C and thymine, T). Purines have a double ring structure; pyrimidines have a single ring structure (figure 4a).
- Two nucleotides can join together by a condensation reaction between the phosphate group of one nucleotide and the hydroxyl group on carbon 3 of the sugar of the other nucleotide. The bonds linking the nucleotides together are strong, covalent **phosphodiester bonds** (figure 3).
- The process can be repeated so that a polynucleotide chain builds up (figure 4b). The chain has a sugar–phosphate backbone with the organic bases projecting outwards.
- Each chain has two distinct ends: a 3′ ('three prime') end and a 5′ ('five prime') end. At the 3′ end, the carbon 3 of the deoxyribose is closest to the end; at the 5′ end, the carbon 5 of the deoxyribose is closest to the end.

Phosphoric acid
Nucleoside
H_2O Condensation
H_2O Condensation
Pentose sugar (deoxyribose)
Cytosine (base)
Phosphoester bond
Nucleotide + $2H_2O$

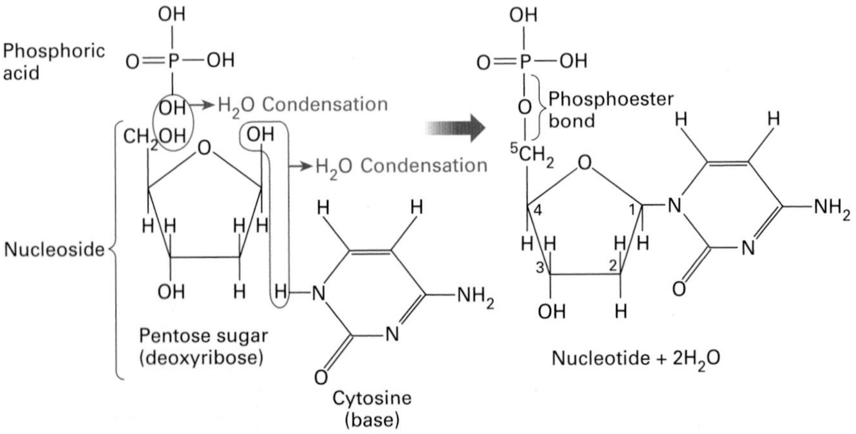

Figure 2 Formation of a nucleotide by condensation reactions. (Note the phosphoester bond forming between the sugar and phosphate group.)

The double helix

DNA consists of two polynucleotide chains coiled around each other to form a **double helix** (figure 4b). The double helix is held together by hydrogen bonds between pairs of bases in the two chains. The pairings depend on the shapes of the bases (a purine can only bond with a pyrimidine) and on their ability to form hydrogen bonds:

- Adenine (a purine) pairs with thymine (a pyrimidine), forming two hydrogen bonds (A=T).
- Guanine (a purine) pairs with cytosine (a pyrimidine), forming three hydrogen bonds (G≡C).

Complementary base pairing

These **complementary base pairs** (figure 4a) are the only ways the bases can bond and join the two nucleotide chains. Thus, the sequence of bases along one polynucleotide chain determines the sequence along the other: an adenine on one chain means there must be a thymine on the other chain at that point, and so on. Complementary base pairing forms the basis of DNA replication (spread 18.3) and its ability to form messenger RNA during protein synthesis (spread 18.7).

Complementary base pairing can happen only if the two polynucleotide chains are **antiparallel**. Antiparallel chains run in opposite directions; one chain runs from 3′ to 5′, and the other from 5′ to 3′.

Watson and Crick's model of DNA showed that the base pairs are 0.34 nm apart, and that each complete turn of the helix has ten base pairs.

In summary

- DNA is a double helix made of two polynucleotide chains.
- Each chain has a sugar–phosphate backbone on the outside with organic bases on the inside.
- The two chains are held together by complementary base pairing.
- The chains are antiparallel (the 5′ end of one chain lies next to the 3′ end of the other chain).

Quick check

1 Distinguish between a nucleoside and a nucleotide.

2 By what type of chemical reaction does a phosphodiester bond form?

3 If one strand of DNA has the base sequence AATCCG, what will be the corresponding base sequence of its complementary strand?

Food for thought

In 1948 the chemist Erwin Chargaff began using paper chromatography to analyse the base composition of DNA from a number of species. Table 1 shows the types of results he obtained. Suggest how these results may be interpreted to support Watson and Crick's double-helix hypothesis. What other interpretations could be given?

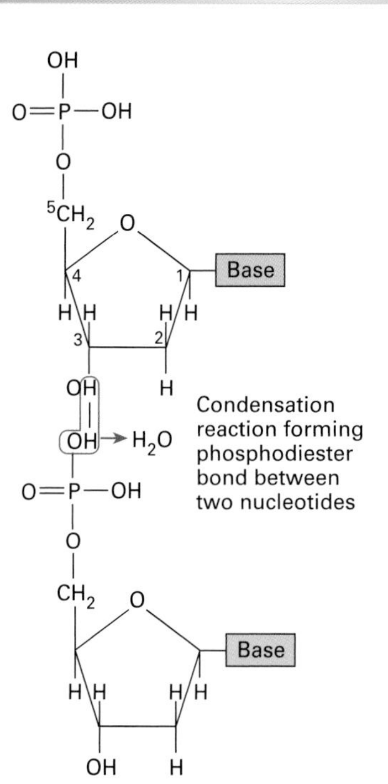

Figure 3 *Two nucleotides join together to form the start of a polynucleotide chain (note the phosphodiester bond linking the nucleotides.)*

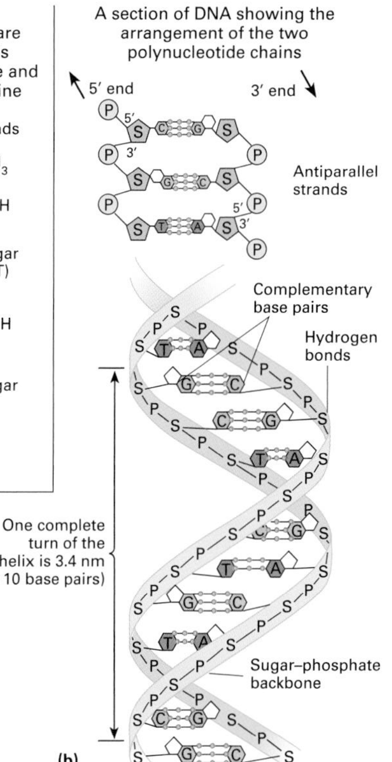

Figure 4 *The structure of DNA. A model of the molecule is shown on spread 18.2, figure 2.*

Table 1 *Percentage base composition of DNA from different species.*

DNA source	A	G	C	T
Human	30.9	19.9	19.8	29.4
Sheep	29.3	21.4	21.0	28.3
Hen	28.8	20.5	21.5	29.2
Turtle	29.7	22.0	21.3	27.9
Salmon	29.7	20.8	20.4	29.1
Locust	29.3	20.5	20.7	29.3
Wheat	27.3	22.7	22.8	27.1
Yeast	31.3	18.7	17.1	32.9
Escherichia coli	24.7	26.0	25.7	23.6
Staphylococcus aureus	30.8	21.0	19.0	29.2

18.2

Chromosomes

OBJECTIVES

By the end of this spread you should be able to:

- explain how DNA is folded in a chromosome
- describe the structure and function of centromeres
- discuss the role of telomeres.

Fact of life

The largest human chromosome is chromosome 1 (spread 4.10). The length of a chromosome depends mainly on the number of nucleotide bases in its DNA. Because DNA is usually double stranded, its length is measured in units of nucleotide base pairs (bp). One base pair corresponds to a length of about 0.34 nm. The following abbreviations are commonly used to describe the length of DNA:

- kb (1 kilobase pair = 1000 bp)
- **Mb** (1 **megabase** pair = 1 000 000 bp) and
- Gb (1 gigabase pair = 1 000 000 000 bp).

Chromosome 1 is 247 Mb long. The total length of all the chromosomes in a human haploid cell of 23 chromosomes is estimated to be about 3.2 Gb.

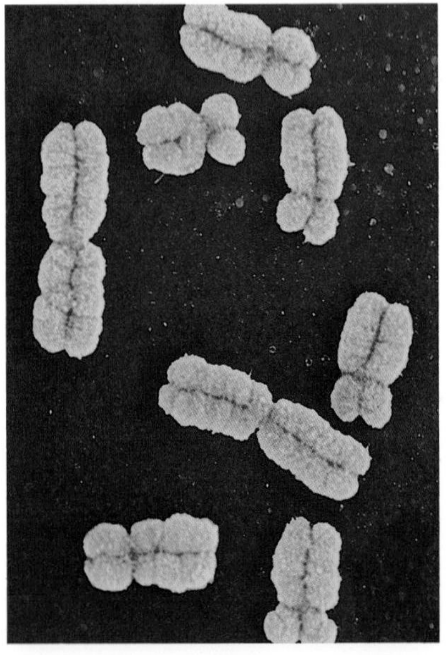

Figure 1 Scanning electron micrograph of chromosomes. They have replicated their DNA, hence their X-shape.

Prokaryotic DNA

Prokaryotic cells have a single circular DNA molecule which lies free in the cytoplasm. This structure is also often called a chromosome; however, in this spread the term is restricted to eukaryotic chromosomes.

What's in a chromosome?

A chromosome consists of hundreds or thousands of genes (a gene is the basic unit of inheritance, see spread 18.4), and specialised parts that are thought to be important to the chromosome's stability and function. The deoxyribonucleic acid (DNA, figure 2) that makes up the genes is packaged with the aid of proteins to form a complex structure. Chromosomes also contain small amounts of ribonucleic acid (RNA).

DNA is packaged in chromosomes

Each human chromosome contains one very long DNA molecule which unravelled would measure about 4.8 cm in length. The total length of DNA in the nucleus of a human cell has been estimated to be about 2.2 m. This poses a packaging problem: how does a chromosome measuring on average 6 μm long contain about 8000 times its length of DNA? The answer is that chromosomal DNA is intricately folded and is tightly bound to protein molecules called **histones**. Histones are small proteins that are rich in the amino acids lysine and/or arginine.

The complex formed between DNA and histones is called **chromatin**. Chromatin takes up stain and is visible in non-dividing nuclei. Individual chromosomes can be seen under the light microscope only during cell division (mitosis or meiosis – spreads 4.10–4.12) (figure 1).

Nucleosomes – the basic structural unit

- Each DNA molecule is wound around histones arranged in groups of eight known as **octamers**.
- The DNA and octamers form bead-like structures known as **nucleosomes** (figure 3). Positively charged groups on the side-chains of the histones form strong ionic bonds with negatively charged phosphate groups in the backbone of the DNA.
- In each nucleosome, a length of DNA containing about 150 base pairs is wrapped around the octamer.
- Another histone molecule attached to the outside of the nucleosome binds DNA to the octamer.
- The nucleosome is regarded as the basic unit of the structure. The **linker region**, the stretch of DNA between the nucleosomes, varies in length from 14 to over 100 base pairs.

Nucleosomes fold to form solenoid fibres

More histones in the linker region help to fold the thread of DNA and nucleosomes (the nucleosome fibre) into a tightly coiled structure called a **solenoid**. The solenoids are thought to be further looped and coiled around non-histone proteins called **scaffolding proteins**. The precise details of this higher level of folding are not known.

The centromere

Each chromosome has a **centromere** which usually appears as a constriction when the chromosomes condense during mitosis and meiosis. The position of the centromere can be used to distinguish between different chromosomes.

Centromeres do not contain any genes. However they do contain large segments of highly repetitive DNA, called **alpha satellite DNA**. This is thought to play a significant role in centromere function. The centromere contains the **kinetochore**. This is a densely staining structure that attaches the chromosome to the spindle apparatus during nuclear division. Centromeres control the distribution of chromosomes during cell division. Chromosomes that do not have centromeres cannot divide.

Telomeres

Telomeres consist of DNA and protein. They 'seal' the ends of linear DNA and have been likened to the tips of shoe laces. They have a similar function: they stop DNA fraying. Experimental removal of telomeres causes chromosomes to disintegrate. In nature, telomeres play an important role in cell division and apoptosis (programmed cell death). Telomeres contain repeating sequences of nucleotide bases which are synthesised with the help of **telomerase**. The enzyme consists of an RNA component that serves as a template for the repeating sequences of bases, and a protein called **telomerase reverse transcriptase** (**TERT**) that transcribes the single-stranded RNA into single-stranded DNA.

In the absence of telomerase, telomeres shorten with each cell division. When shortened to a critical length, the cell stops dividing and dies. In normal somatic cells, telomerase activity is suppressed. In **stem cells** telomerase activity is maintained so that telomeres do not shorten when cells divide. In some cancer cells, telomerase is 10–20 times more active than in normal body cells. This gives a selective growth advantage to some types of tumour. If telomerase activity could be turned off, then telomeres in these cancer cells would shorten. This would prevent the cancer cells from dividing uncontrollably.

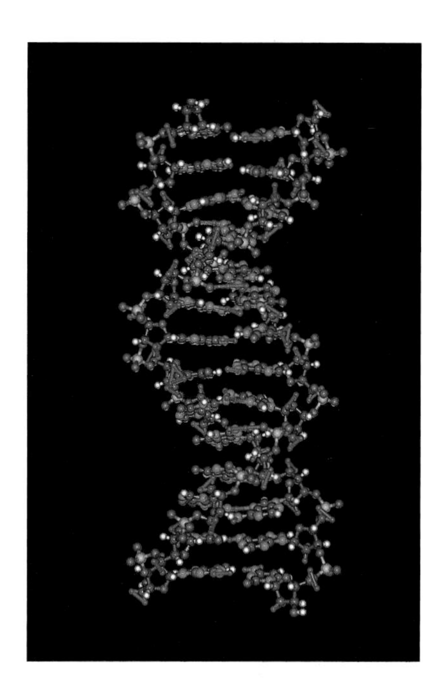

Figure 2 *Model of one molecule of DNA. The molecule contains two strands of DNA held together by base pairing. Before cell division, when the chromosomes become visible as two chromatids joined at the centromere, the DNA replicates. Each chromatid contains a double-stranded molecule of DNA.*

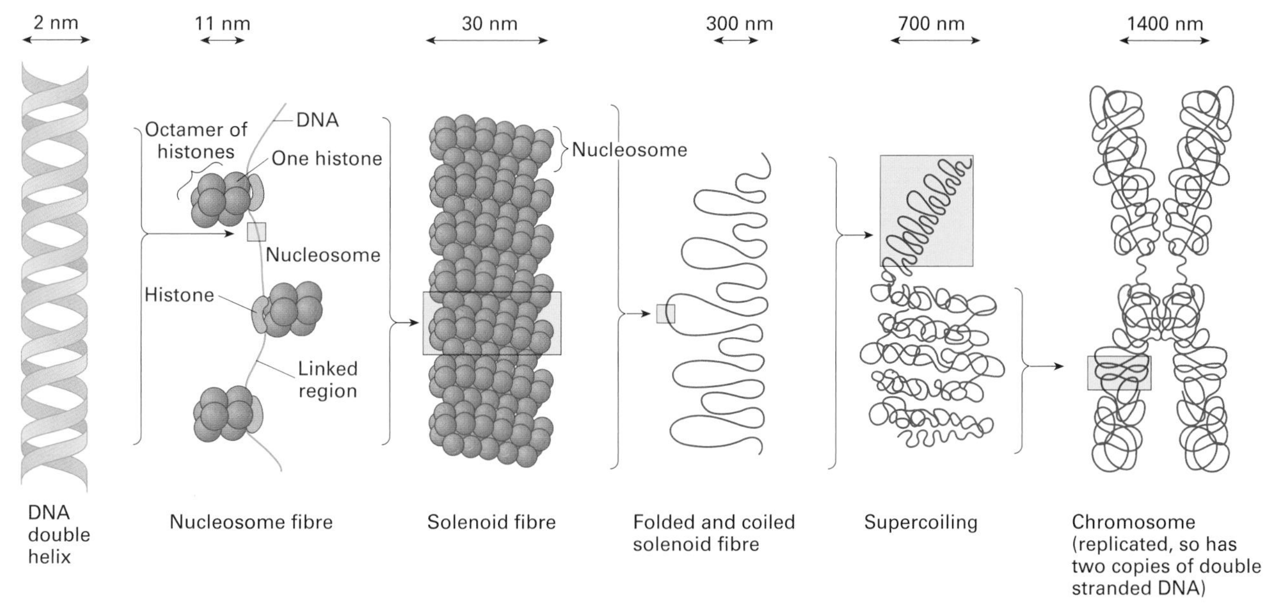

Figure 3 *The structure of a chromosome showing the basic structural unit: the nucleosome.*

QUICK CHECK

1 What is a nucleosome?

2 Centromeres contain no genes. What is their main function?

3 Why have telomeres been compared with the tips of shoelaces?

Food for thought

Telomeres can be lengthened by introducing a gene for an enzyme called telomerase reverse transcriptase (hTRT). This enzyme causes cells to produce active telomerase, the enzyme that repairs telomeres. Suggest how techniques for manipulating telomere length might be used to treat age-related diseases and cancers. What risks might be associated with deliberately changing the length of telomeres in body cells?

18.3

DNA REPLICATION

OBJECTIVES

By the end of this spread you should be able to:

- describe how DNA can be made in the laboratory
- interpret Meselsohn and Stahl's experiment on semiconservative replication
- describe how semiconservative replication takes place.

Fact of life

DNA replication is a very complex process during which mistakes happen. Uncorrected mistakes may lead to harmful mutations. In the living cell, errors are kept to a very low frequency (typically about 1 in 10^9) by a number of repair mechanisms. One such mechanism is **mismatch repair**. This is carried out by the enzyme **DNA polymerase** which 'proofreads' newly formed DNA against its template as soon as it is added to the strand. If it finds an incorrectly paired nucleotide, the polymerase reverses its direction of movement, removes the incorrect nucleotide, and replaces it before replication continues. The process is similar to correcting a typing error by going back a space, deleting the error, and typing in the correct letter before continuing.

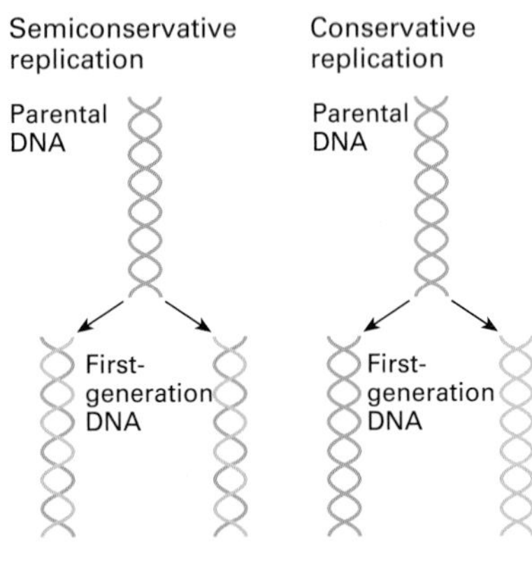

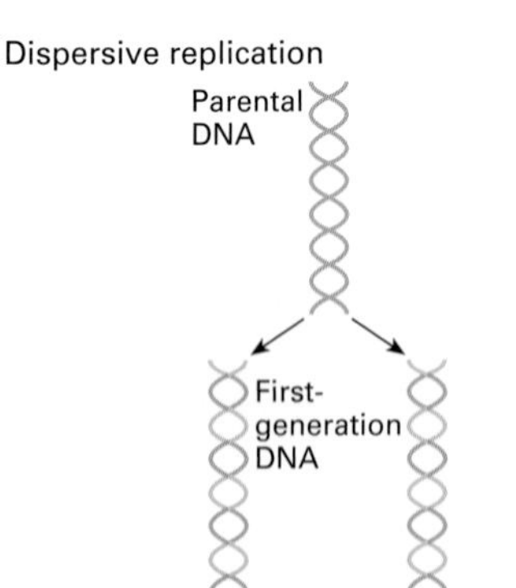

***Figure 1** Three possible mechanisms of replication.*

A possible mechanism for replication

A chemical that carries inherited information must be able to copy itself exactly. Complementary base pairing between adenine and thymine and between cytosine and guanine (spread 18.1) makes this possible.

Watson and Crick's description of DNA suggested that, during replication, the hydrogen bonds connecting base pairs are disrupted, allowing the two polynucleotide chains to unwind from one another. Each chain then acts as a template for the synthesis of a new complementary polynucleotide chain. It was suggested that the DNA molecule 'unzips' from one end and new nucleotides already present in the nucleus bind with their complementary bases in each exposed chain. This therefore forms two identical molecules of DNA from the single parent molecule.

Experimental evidence

Arthur Kornberg and his colleagues were the first to successfully replicate DNA in a test tube. They used the following ingredients:

- intact DNA (to act as a template)
- a mixture containing all four nucleotides
- DNA polymerase (an enzyme which catalyses the synthesis of DNA)
- ATP (as a source of energy).

New DNA molecules were formed, which contained the same proportions of the four bases as the original parent DNA. This was a strong indication that DNA can copy itself by complementary base pairing.

Semiconservative replication

The idea that DNA unzips before replication is an attractively simple one. This mechanism is called **semiconservative replication**, because each new molecule of DNA (daughter DNA) contains one intact strand from the original DNA (parental DNA) and one newly synthesised strand. However, semiconservative replication is not the only means by which DNA might replicate by complementary base pairing (figure 1).

Meselsohn and Stahl

In 1958, two American biochemists, Matthew Meselsohn and Franklin Stahl, conducted a neat experiment which gave strong support for the theory of semiconservative replication (figure 2).

- First, they grew *Escherichia coli* bacteria for many generations in a medium containing ^{15}N, a heavy isotope of nitrogen. The bacteria incorporated the ^{15}N into their DNA. This made the DNA denser than normal ('heavy' DNA).
- A control culture of bacteria was grown in a medium with ^{14}N, the normal, lighter isotope of nitrogen. These bacteria had normal 'light' DNA.
- The bacteria grown in ^{15}N were then transferred to a ^{14}N medium and left for periods of time that corresponded to the generation time of *E. coli* (about 50 minutes at 36 °C).
- Samples of bacteria were taken at intervals to analyse the parental, first-generation, and second-generation DNA.
- The composition of the DNA was analysed using **density gradient centrifugation**. The mixture of the three DNA types was suspended in a solution of caesium chloride and spun at high speed in a centrifuge. The DNA separated according to its density: heavy DNA (which contained ^{15}N) formed a band lower down the tube than the light DNA (which contained ^{14}N). The bands became visible when the tubes were exposed to ultraviolet light.

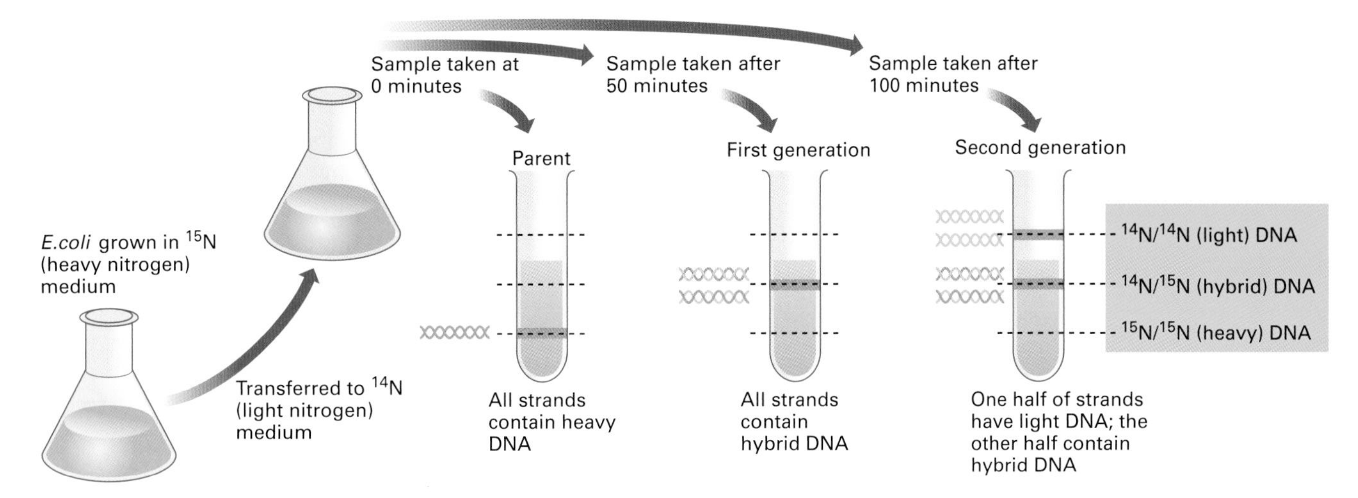

***Figure 2** Meselsohn and Stahl's experiment to test the semiconservative hypothesis of replication.*

The results gave overwhelming support to the semiconservative hypothesis (figure 3):

- In the first generation, all the DNA had a density midway between that of heavy DNA and light DNA. Thus it contained equal amounts of each.
- In the second generation, two sorts of DNA were detected: one was light DNA; the other containing equal amounts of ^{14}N and ^{15}N (i.e. it was like the DNA in the first-generation bacteria).
- Throughout the investigation, DNA from the control culture produced only light bands, indicating that it contained only ^{14}N.

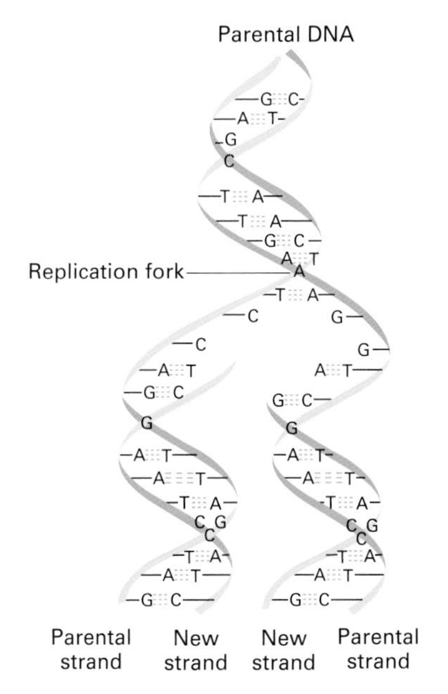

***Figure 3** Semiconservative replication. Each parental strand acts as a template for the synthesis of a new strand, which forms by complementary base pairing.*

The enzymes involved in replication

DNA replication is a complex process involving several different enzymes:

- **Helicases** separate the two DNA strands. Their action uses energy from ATP.
- **DNA binding proteins** keep the strands separate during replication.
- **DNA polymerases** catalyse the polymerisation of nucleotides to form a polynucleotide chain in the 5′ to 3′ direction. This allows one strand to be replicated continuously.
- The other strand is not replicated continuously but in small sections. The pieces of polynucleotide chain are joined together by an enzyme called **DNA ligase**.

DNA is a long molecule. DNA replication would take a long time if it started at one end and proceeded nucleotide by nucleotide along the entire length of the molecule. In fact, the double helix opens up and replicates simultaneously at a number of different sites, known as **replication forks**. DNA ligases then join the segments of DNA together, completing the synthesis of new DNA strands.

Quick check

1 List the ingredients Kornberg used to make DNA in the test tube.

2 Suppose DNA replication were conservative. What results would Meselsohn and Stahl have obtained in the first generation?

3 During DNA replication, what is the function of:

a helicases

b DNA binding proteins

c DNA polymerase

d DNA ligase?

Food for thought

Bacterial DNA can replicate at the rate of one million base pairs per minute. Eukaryotic DNA replicates much more slowly, at less than 5000 base pairs per minute. A human DNA molecule may have as many as 6×10^9 base pairs. If replication started at one end of the molecule and took place continuously until the other end was reached, how long would it take for this DNA molecule to be replicated? Why does it actually take only a few hours?

18.4 THE CHEMICAL NATURE OF GENES

OBJECTIVES

By the end of this spread you should be able to:

- explain the significance of Griffith's work on *Pneumococcus*
- describe how Avery and other workers analysed the transforming factor
- describe Hershey and Chase's experiment.

Fact of life

Some bacterial cells have a well organised layer outside the cell wall called the **capsule**. Unlike a slime layer, it is not easily washed off. Although capsules are not essential for bacterial growth and reproduction in laboratory conditions, they can make the difference between life and death in natural situations. For example, *Streptococcus pneumoniae* (a member of the pneumococci, the group of pneumonia-causing bacteria used in Griffith's experiment; see text) has non-capsulated and capsulated strains. Those lacking a capsule are easily destroyed by the host and do not cause disease. However, the capsulated strain kills mice quickly. The capsule helps the bacterium resist phagocytosis by host cells. It contains a great deal of water, protecting the bacterium from desiccation; it keeps out detergents which could destroy the cell surface membrane; and it helps bacteria attach to host cells.

Table 1 *Results of Griffith's experiment.*

Bacteria (*S. pneumoniae*) injected into mice	Reaction of mice
Live strain R (no capsule)	Survived
Live strain S (capsule)	Died
Dead strain S	Survived
Live strain R + dead strain S	Died

We know today that DNA is the chemical in which information is passed from parent to offspring. This spread looks at how researchers established this link between DNA and inheritance.

In the 1860s, nearly 100 years before Watson and Crick's work on the structure of DNA, Gregor Mendel established that inheritance depends on factors that are transmitted from parents to offspring (spread 19.2). In 1909 it was found that patterns of inheritance were reflected in the behaviour of chromosomes. Wilhelm Johannsen referred to these factors as **genes**. Genes were assumed to be located on the chromosomes because genes that are inherited together (linked genes) were found to be carried on the same chromosome. However, the chemical composition of genes was not known.

Protein or DNA: which is the genetic material?

Chromosomes were known to contain both protein and DNA. Most biologists assumed that proteins, with their highly complex and infinitely variable structure, were the inherited material. The nucleic acids were thought to be too simple to carry complex genetic information. This view was reinforced by the work of Phoebus Aaron Levene. Levene made major contributions to the chemistry of nucleic acids but believed, mistakenly, that DNA was a very small molecule, probably only four nucleotides long.

In 1928 Fred Griffith, an English medical bacteriologist, published a paper describing experiments on pneumococci. His results set the stage for the research that finally showed that DNA is the genetic material.

Griffith's experiment: transformation of pneumococci

Pneumococci are bacteria that cause pneumonia. They occur in two strains: a disease-causing smooth strain (strain **S**), and a harmless rough strain (strain **R**). Strain S has a capsule on its cell surface; this capsule is absent from the harmless strain R (see Fact of life).

Griffith found that mice injected with live strain S soon died, but those injected with live strain R survived. Mice injected with dead strain S bacteria (killed by heat) all survived. The results of this series of experiments were as expected. However, the results of Griffith's next series of experiments were thoroughly baffling: mice injected with a mixture of heat-killed strain S and live strain R died (table 1). Moreover, Griffith recovered live strain-S bacteria from the dead mice.

After many careful experiments, Griffith concluded that hereditary material had passed from the dead bacteria to the live bacteria. This changed harmless strain R bacteria into virulent strain S pathogens. This process is called **transformation**.

Avery's experiment: DNA was the transforming agent

In the 1940s, Oswald T. Avery, Colin MacLeod, and Maclyn McCarty showed that DNA was responsible for transformation.

- They used enzymes that hydrolysed polysaccharide, DNA, RNA, and protein on samples of the disease-causing strain-S pneumococci.
- Different samples had different parts of their cells destroyed by these enzymes.
- The researchers then exposed strain-R pneumococci to the treated samples of strain S.
- The transformation of strain R to strain S was blocked only when the DNA in the sample was destroyed.

These results provided strong evidence that DNA carried genetic information for transformation. However, many scientists remained unconvinced.

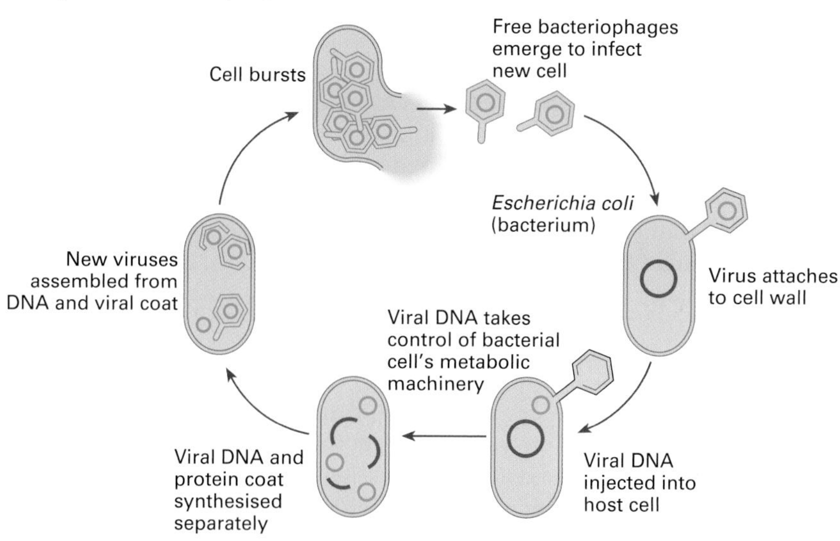

Figure 1 The structure and life cycle of T2 bacteriophage (a virus that infects bacteria).

Hershey and Chase: the role of DNA in the T2 phage life cycle

In 1952, Alfred D. Hershey and Martha Chase performed several experiments with T2 bacteriophage, a virus that infects bacteria. Their results convinced even the sceptics that DNA, and not protein, was the genetic material.

Electron micrographs indicate that T2 bacteriophage infects *Escherichia coli* by injecting its DNA into the bacterium while leaving its protein coat on the outside. The phage takes over the genetic machinery of the host cell to make new phages. Eventually, the bacterial cell bursts (a process called lysis), releasing new phages to infect other bacteria (figure 1).

Hershey and Chase wanted to test the hypothesis that only the viral DNA entered the bacterium. They made use of the fact that DNA contains phosphorus but not sulphur, whereas protein contains sulphur but not phosphorus.

- With some T2 phages, they labelled the viral DNA with a radioactive isotope of phosphorus (^{32}P). With other T2 phages, they labelled the viral protein coat with a radioactive isotope of sulphur (^{35}S).
- They added the viruses to a culture of *E. coli* and gave them enough time to infect their host cells (but not enough time to reproduce).
- The viral coats were then separated from the infected bacteria by shaking the mixture vigorously in a blender.
- When *E. coli* was infected with a T2 phage containing ^{35}S (labelled protein), little radioactivity occurred within the bacterial cells.
- With a T2 phage containing ^{32}P (labelled DNA), the bacterial cells were radioactive. Moreover, when the bacterial cells burst open, the new viruses that emerged were radioactively labelled with ^{32}P. When the protein was labelled, new viruses were only slightly radioactive.

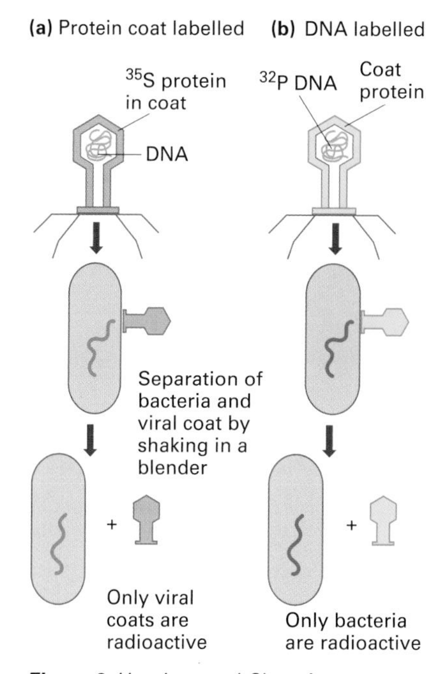

*Figure 2 Hershey and Chase's experiment using: **(a)** labelled protein **(b)** labelled DNA.*

Quick check

1 How can the harmless rough strain of pneumococcus be transformed into the pathogenic smooth strain?

2 How can the DNA in the disease-causing smooth strain of bacteria be extracted from RNA and proteins?

3 **a** Describe the distribution of protein and DNA in T2 bacteriophage.

b Explain how they can each be labelled.

Food for thought

Following the publication of the work of Avery and his colleagues on the transforming principle in pneumococci, more than 30 studies were conducted on transformations in other bacteria. In each case, the results were similar. Suggest why many scientists remained unconvinced that DNA was the inherited material until Hershey and Chase reported the results of their experiments.

18.5

OBJECTIVES

By the end of this spread you should be able to:

- explain how certain inherited metabolic disorders indicate that genes exert their effects through enzymes
- describe Beadle and Tatum's experiment on *Neurospora* which led to the one gene–one enzyme hypothesis
- discuss why the one gene–one enzyme hypothesis had to be modified.

Fact of life

Phenylketonuria (PKU) occurs in about one in 10 000 live births among white Europeans. If untreated, a patient may have an IQ (intelligence quotient) of less than 20 (the average IQ is 100).

The disorder is treated by reducing the intake of phenylalanine in the diet to an absolute minimum. A child with PKU must avoid products that are rich in phenylalanine such as drinks and confectionery that are sweetened with aspartame. (Aspartame contains a mixture of two amino acids: aspartic acid and phenylalanine.)

High blood levels of phenylalanine are not damaging in adulthood (probably because brain growth is complete). Therefore, with the exception of women who are pregnant or breast-feeding, adults with PKU can have a normal diet.

THE ONE GENE – ONE POLYPEPTIDE HYPOTHESIS

In the 1940s and early 1950s, researchers established that genes are made of DNA (spread 18.4). At the same time, other researchers wanted to know how genes determine inherited characteristics. Clues came from research carried out in the early 1900s by Sir Archibald Garrod. He observed that two human inherited diseases – **alkaptonuria** and **phenylketonuria (PKU)** – were each caused by absence of a specific enzyme. (He called these diseases 'inborn errors of metabolism'.)

Alkaptonuria

People suffering from alkaptonuria lack an enzyme called homogenistic acid oxidase. This enzyme breaks down the amino acids tyrosine and phenylalanine. When the enzyme is absent, an intermediate product known as homogenistic acid accumulates. This causes a dark brown discoloration of the skin and eyes, and progressive damage to the joints, especially the spine.

Phenylketonuria

Normally phenylalanine is converted into another amino acid by the enzyme **phenylalanine hydroxylase**. This enzyme is absent from people with PKU, causing phenylalanine to accumulate in the blood. High concentrations of phenylalanine damage the nervous system, leading to severe mental retardation. Nowadays, routine postnatal screening detects the condition early enough that the diet can be modified to prevent brain damage (see Fact of life).

Garrod's observations indicated that genes probably exert their effects through enzymes, but the evidence was only circumstantial. Scientists wanted more direct proof that genes brought about their effects by determining which enzymes were made in cells. This proof came with the work of George Beadle and Edward Tatum on *Neurospora crassa* (figure 1).

Beadle and Tatum: the one gene – one enzyme hypothesis

Neurospora crassa is a common pink mould (a fungus) which is a particularly damaging pest in bakeries because it can turn bread mouldy. It reproduces by spores and grows in the bread as a mycelium (a mass of threads). It has several features which make it suitable for genetic research (table 1). One of the most important is its ability to produce haploid spores asexually. These spores are identical, and have only one set of chromosomes. They therefore have only one allele for each characteristic (spread 19.3). This means that a recessive mutation is not masked by a dominant allele; it is always expressed in the haploid organism.

Neurospora can grow on a culture medium called **minimal medium**. This contains sugar, a source of nitrogen, mineral ions, and the vitamin biotin. The fungus can synthesise all the other carbohydrates, fats, proteins, and nucleic acids it needs using enzymes produced by its cells.

- Beadle and Tatum grew *Neurospora* on minimal medium and exposed the culture to a dosage of X-rays that caused the formation of mutations.
- Occasionally a mutant spore was produced that was unable to grow on minimal medium. However, it would grow and reproduce if provided with all 20 amino acids.
- After isolating a mutant *Neurospora*, Beadle and Tatum attempted to grow it on 20 different minimal media, each of which was supplemented with a different single amino acid.
- They discovered that the mutant that could not grow on the minimal medium needed only one particular amino acid in order to grow and reproduce normally.

Table 1 *Some features of* Neurospora crassa *which make it suitable for genetic research.*

- Can be grown on minimal medium (easy to grow)
- Small size (can be bred in very large numbers in the confined space of a laboratory)
- Short life cycle (10 days)
- Diploid and haploid stages (allows effects of different crosses to be studied)
- Diploid organism has only seven pairs of chromosomes on which gene positions are easily mapped
- Produces spores which can be easily separated from the parent mould and grown individually
- Relatively long haploid, asexually reproducing stage

- They concluded that the mutant lacked the enzyme required to synthesise that particular amino acid.

Further experiments indicated that other mutants lacked different enzymes, each of which was dictated by a particular gene. In each case, Beadle and Tatum found that the inability to synthesise a specific enzyme was inherited in a normal Mendelian manner (see spread 19.2). They concluded that each gene in an organism coded for the production of one enzyme. This became known as the **one gene–one enzyme hypothesis.** The hypothesis was soon extended to a **one gene–one protein hypothesis** when it was shown that proteins other than enzymes could also be determined by specific genes.

Refining the theory

The hypothesis was modified into the **one gene–one polypeptide hypothesis** when it was realised that proteins could consist of more than one polypeptide chain, each determined by the action of a different gene.

For example, haemoglobin has four polypeptide chains, two identical alpha and two identical beta chains. These two different types of polypeptide are determined by two separate genes. Sickle-cell anaemia (spread 16.1) is caused by a mutation in a single gene which results in just one amino acid being changed in the beta chain of haemoglobin.

Figure 1 *Beadle and Tatum's experiment tested the 'one gene–one enzyme' hypothesis by exposing the mould* Neurospora crassa *to X-rays to induce mutations.*

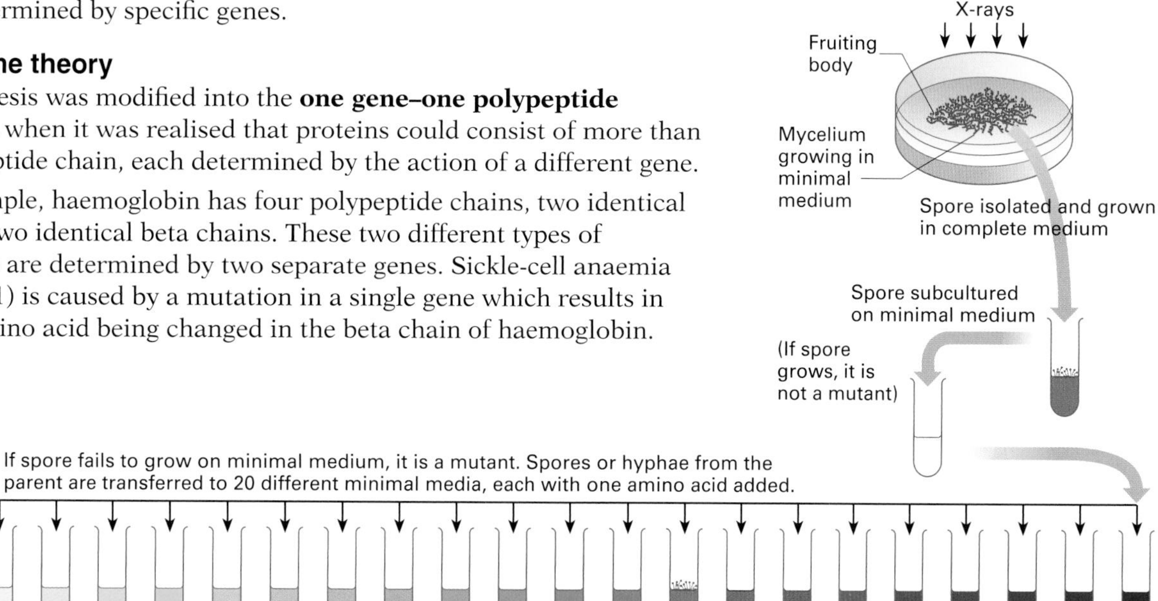

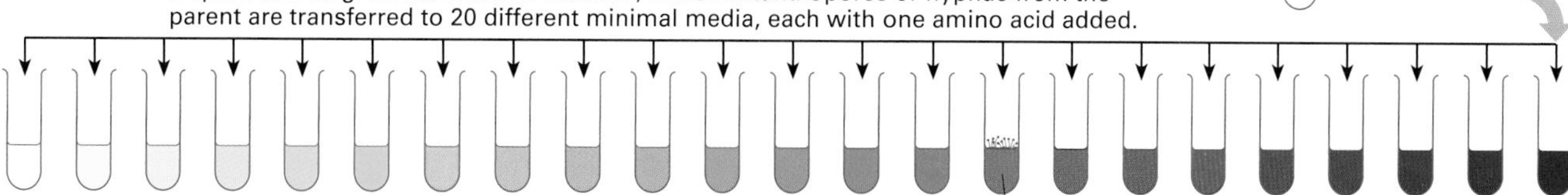

Table 2 *Results of irradiating* Neurospora *with ultraviolet light. (mm = minimal medium; a = arginine; c = citrulline; o = ornithine. Growth is indicated by + and no growth by –.)*

Nutrient medium	'Wild type'	Mutant 1	Mutant 2	Mutant 3
mm	+	–	–	–
mm + a	+	–	+	–
mm + a + c	+	+	+	–
mm + a + c + o	+	+	+	+

QUICK CHECK

1 How does phenylketonuria indicate that genes exert their effects through the production of specific enzymes?

2 How does Beadle and Tatum's experiment on *Neurospora* support the one gene–one protein hypothesis?

3 Explain why the one gene–one protein hypothesis needed to be modified in the light of conditions such as sickle-cell anaemia.

Food for thought

'Wild type' *Neurospora* can convert the amino acid glutamic acid to another amino acid, arginine, in the following steps (ornithine and citrulline are also amino acids):

$$\text{Glutamic acid} \xrightarrow{e1} \text{ornithine} \xrightarrow{e2} \text{citrulline} \xrightarrow{e3} \text{arginine}$$

Each step requires a specific enzyme (e1, e2, and e3). Exposure of *Neurospora* to ultraviolet light results in mutant strains that are unable to grow on minimal medium. However, the mutants can grow on media supplemented with one or more amino acids. From the data in table 2, suggest which enzyme (e1, e2, or e3) was no longer formed by mutant 1, mutant 2, and mutant 3.

Another mutant failed to grow on any of the media shown in table 2. What does this indicate about the amino acid metabolism in this mutant? How could you further investigate the amino acid metabolism of this mutant?

18.6

The genetic code

OBJECTIVES

By the end of this spread you should be able to:

- describe the relationship between DNA, messenger RNA, and proteins
- explain how frame-shift experiments support the triplet code hypothesis
- discuss the main features of the genetic code.

Fact of life

Prokaryotes and eukaryotes share the same 'language of life'. Comparisons of DNA sequences with the corresponding protein sequences reveal that (with a few exceptions) an identical genetic code is used in both prokaryotes and eukaryotes. This means that bacteria can be genetically engineered to make human proteins. The universal nature of the code suggests that all living things are descended from a single pool of primitive cells which first evolved this code.

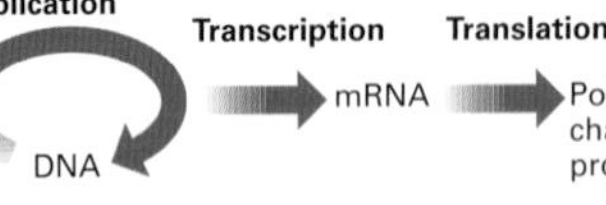

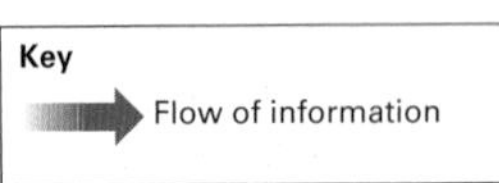

***Figure 1** The one-way flow of information between DNA and protein is often called the **central dogma** of biology. This is the concept that genetic information is transferred irreversibly from DNA to mRNA by a process called transcription, and from mRNA to protein by translation. The concept has been modified to take into account the transfer of information in the other direction from RNA to DNA by reverse transcription, carried out by some viruses.*

One of the most remarkable facts of life is that each cell in an organism contains all the information required to determine all the characteristics of that whole organism. This information is stored in DNA, and is known as the **genetic code**. Deciphering that code has been one of the major scientific breakthroughs of the twentieth century. It has given us an understanding of how genes function, and it has opened the way for most of the recent developments in genetic engineering and biotechnology.

Transcribing the genetic code from DNA to mRNA

The genetic code is held in the order of bases along the DNA molecule. Sections of DNA called **cistrons** (commonly referred to as **genes**) contain the information needed to make a particular polypeptide. However, DNA does not carry out polypeptide synthesis directly. When the DNA in a cistron is activated, the information is transferred to a molecule of ribonucleic acid (RNA) called **messenger RNA** (**mRNA**), which acts as a template for the synthesis of the polypeptide.

The central dogma of biology

The relationship between DNA, mRNA, and polypeptides in a eukaryotic cell is often called the **central dogma** of biology (figure 1).

- mRNA is made on a DNA template in the nucleus, in a process called **transcription** (spread 18.7).
- The mRNA then moves into the cytoplasm, where it combines with ribosomes to direct protein synthesis by a process called **translation** (spread 18.8).
- When the information in a cistron is used to make a functional polypeptide chain by transcription and translation, **gene expression** is said to have taken place.

mRNA is made from the DNA template

mRNA is a large polynucleotide polymer, chemically similar to DNA but differing in that:

- mRNA consists of only one chain of nucleotides, not two
- mRNA contains the sugar ribose instead of deoxyribose
- mRNA contains the base uracil instead of thymine.

During transcription, DNA acts as a template for making mRNA by complementary base pairing. Thus a particular short sequence of DNA may be transcribed as follows:

DNA base sequence: T A G G C T T G A T C G
mRNA base sequence: A U C C G A A C U A G C

The triplet code: frame-shift experiments

Twenty amino acids make all the proteins in living organisms.

- If a code consisted of one base for one amino acid, only four combinations would be provided (there are four bases).
- If two bases coded for one amino acid there would be 16 (4^2) possible combinations.
- A three-base (**triplet**) code provides 64 (4^3) possible combinations, more than enough for all 20 amino acids.

Francis Crick and his co-workers confirmed that the genetic code is a **triplet code**. Using enzymes, they added or deleted nucleotide bases in the DNA of a virus that infects bacteria. They found that when one or two bases were added or deleted, the viruses were unable to infect the bacteria. But when three bases were added or deleted, the virus was able to infect the bacteria. They concluded that adding or removing one or two bases caused a **frame shift** which inactivated the gene. However, adding or removing three bases only partially affected the gene. Thus the sequence of bases shown above would contain the following sequences of DNA base triplets and mRNA **codons**:

DNA base triplet sequence: TAG GCT TGA TCG

mRNA codon sequence: AUC CGA ACU AGC

If one base (for example, guanine) is added to the DNA the frame shifts and the sequence of triplets and codons is changed:

DNA base triplet sequence: **G**TA GGT TTG ATC G

mRNA codon sequence: **C**AU CCG AAC UAG C

The results of the frame-shift experiments also showed that the code is **non-overlapping**:

- Each triplet in DNA specifies one amino acid.
- Each base is part of only one triplet, and is therefore involved in specifying only one amino acid.

A non-overlapping code requires a longer sequence of bases than an overlapping code (see box): however, replacing one base for another has a small or no effect.

Overlapping codes

If the genetic code was an **overlapping code**, a much shorter length of DNA would code for a polypeptide. For example, the base sequence TGACT would provide three triplets: TGA, GAC, and ACT.

However, if one central base was replaced with another, all three codons in the mRNA would be different. For example, if the sequence above was changed to TGGCT, there would be three new triplets: TGG, GGC, GCT, and therefore three new codons. This could have potentially serious consequences.

Cracking the genetic code

To crack the genetic code, scientists had to work out which of the 64 codons determined each amino acid. To do this, they made mRNA molecules with a known sequence of bases. This mRNA was added to a cell-free system that contained isolated ribosomes, radioactively labelled amino acids, and all the enzymes needed for polypeptide synthesis. The polypeptides that were synthesised were then analysed to determine their amino acid sequence.

The first synthetic mRNA molecule made was a chain of uracil bases and was called poly-U. The polypeptide chain synthesised from it contained only phenylalanine. It was therefore concluded that the codon UUU codes for phenylalanine.

The complete genetic code was finally deciphered in 1966 (see appendix). Its features are summarised in table 1.

***Table 1** Main features of the genetic code.*

Features of code	Comments
Triplet code of three nucleotides	Each of the 20 amino acids used to make proteins is represented by a three-letter abbreviation (a **base triplet** in DNA or a **codon** in mRNA)
Linear code reads from a starting point to a finishing point	The codon is always read in the 5′ → 3′ direction
Degenerate code	The code is **degenerate** (there are more codons than amino acids; most amino acids are coded for by more than one codon)
Punctuation codons	The start and end of a coding sequence in a cistron is determined by specific codons: the 'start' signal is given by AUG, which codes for methionine; there are three 'stop' signals (UAA, UAG, and UGA)
Almost universal	Most organisms share the same code; chloroplast and mitochondrial DNA have a slightly modified code; other exceptions to the universal genetic code are rare

Quick check

1. What is the 'central dogma' of biology?
2. What is the name given to the result of adding one or two nucleotide bases to a DNA sequence?
3. Use the genetic code in the appendix to work out which amino acid has the codon CAU.

Food for thought

Francis Crick and his co-workers reasoned that if the genetic code was a triplet code, adding or deleting one or two nucleotides would make the genetic information in the DNA meaningless. A sentence consisting of three-letter words can be used as a simple analogy of the triplet code. Each letter is analagous to a nucleotide base and each word to a triplet code for an amino acid. The sentence is equivalent to the information in DNA which determines the sequence of amino acids in a polypeptide chain. Devise a sentence of different three-letter words and show what happens to the message when one, two, or three letters in a word are deleted.

18.7

PROTEIN SYNTHESIS: TRANSCRIPTION

OBJECTIVES

By the end of this spread you should be able to:

- outline the process of protein synthesis in eukaryotes
- describe the process of transcription
- distinguish between mRNA, tRNA, and rRNA in terms of their structures and functions.

Fact of life

In mammals, insects, and birds, mRNA (which carries the information from DNA to the cytoplasm for protein synthesis) is only one-fifth the length of the 'average' gene. This is because as well as coding for proteins, eukaryotic genes also contain nucleotide base sequences called introns that do not code for functional proteins. When first discovered in 1977, introns were loosely called 'junk DNA' because they are discarded during the formation of mRNA, and appeared to serve no function. However, it is now known that some introns bring about certain chemical reactions in the nucleus. Introns may, for example, form enzymatic RNA molecules known as **ribozymes**. One role of ribozymes is to act as self-splicing catalysts: they cut themselves out of nuclear RNA during the formation of the final mRNA (figure 3).

Proteins are essential components of cells. DNA carries the instructions (the 'blueprint') for making them. But DNA needs to be kept intact as the permanent store of genetic information; therefore it is not used directly for protein synthesis. Instead, the information encoded in DNA in the nucleus is transferred to smaller molecules of another nucleic acid called **messenger RNA** (**mRNA**). mRNA carries this information to ribosomes in the cytoplasm, where it is used to make the polypeptides that make up a protein.

- The transfer of information from DNA to mRNA is called **transcription**.
- The conversion of the information in mRNA to make polypeptides is called **translation** (figure 1).

The blueprint for protein synthesis

The section of DNA that holds the information for one polypeptide chain is called a **cistron** or **gene**. Before transcription can take place, the double helix in the cistron has to unwind and the two polynucleotide chains have to separate ('unzip') in order to expose the nucleotide bases (figure 2). This is done with the help of enzymes. Only one of the strands is used as a template for the synthesis of mRNA. This transcribing or coding strand is known as **antisense DNA**. The non-transcribing strand is known as the **sense** strand (because its sequence of nucleotides looks like the protein codon sequence).

Starting transcription: making mRNA

- The hydrogen bonds holding the double helix of DNA together are broken by an enzyme called **helicase**, exposing the bases in the antisense DNA strand.
- For transcription to take place, an enzyme called **polymerase** must attach onto the DNA at a particular base sequence called the **promoter** site. Proteins called **transcription factors**, which have activator and repressor sequences that regulate gene expression, are also attached to promoters.

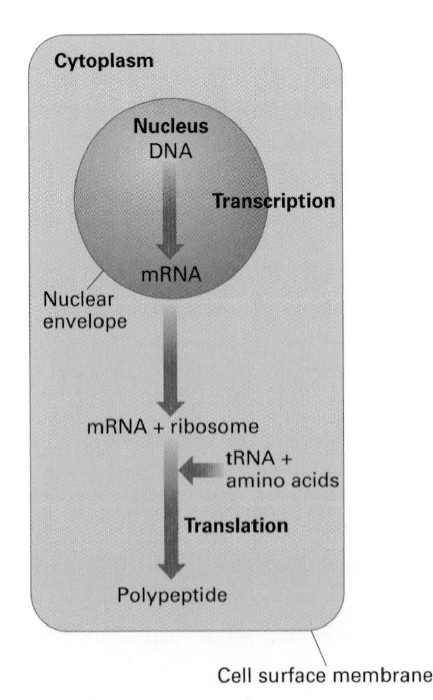

Figure 1 Summary of protein synthesis in a eukaryotic cell.

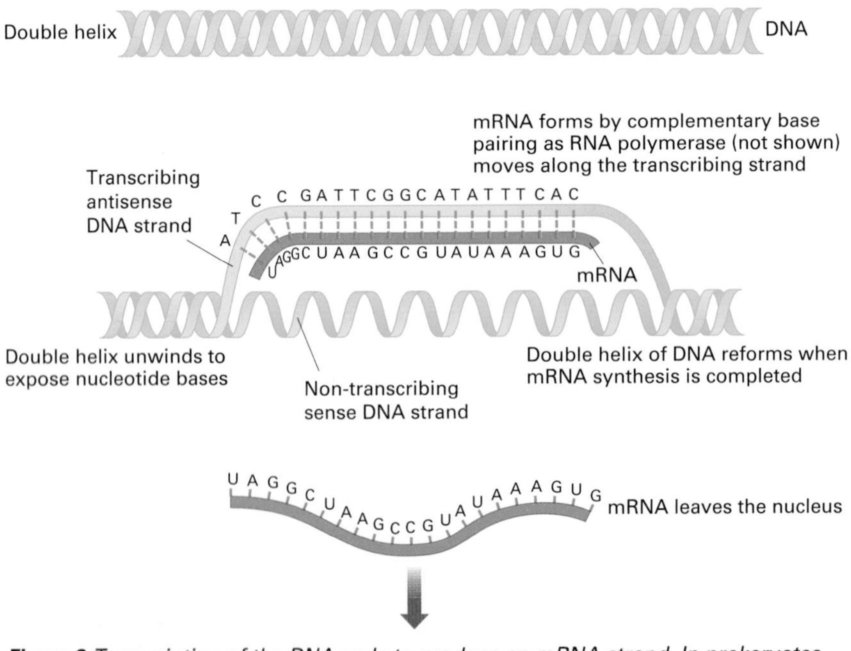

Figure 2 Transcription of the DNA code to produce an mRNA strand. In prokaryotes and some eukaryotes, mRNA passes to the cytoplasm immediately to direct the process of translation. In most eukaryotes, mRNA is processed before leaving the nucleus (figure 3).

- During transcription, polymerase moves along the cistron in the 5'–3' direction. It passes over the nucleotides in the transcribing DNA strand one at a time and builds up mRNA by adding complementary nucleotides (U to A, C to G, etc.) as it goes.
- When the enzyme moves on to another region of the antisense strand, the double helix of the DNA reforms behind it (the 'zip' closes).
- On reaching a special 'stop' sequence called a **terminator**, the enzyme detaches and the mRNA molecule peels away from the DNA.

Post-transcription processing of mRNA

In multicellular eukaryotic cells, **pre-mRNA** (precursor mRNA transcribed from the antisense DNA) has to be processed to remove **introns** before it leaves the nucleus (figure 3). Introns are sequences of nucleotides that lie between the cistrons that code for polypeptides. (They are called 'introns' because they 'intrude' into the cistron but are not expressed in the final protein.) After the introns have been cut out and removed from the pre-mRNA, the remaining sections of RNA (called **exons** because they are expressed) are spliced together to form the final mRNA. Final mRNA molecules are mobile templates of the protein-coding part of the cistron. They are dispatched out of the nucleus through the nuclear pores into the cytoplasm where they associate with ribosomes to make polypeptides. Variations in the splicing can lead to more than one type of polypeptide being synthesised.

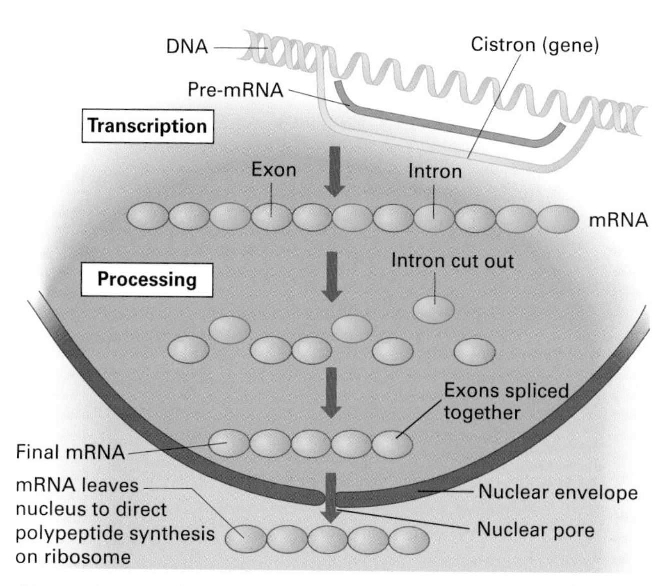

***Figure 3** Transcription and processing in a typical eukaryotic cell. Prokaryotes and some eukaryotes (such as single-celled organisms) have no introns so their mRNA does not need to be processed. After transcription, prokaryotic mRNA goes directly to the ribosomes to be translated.*

Different types of RNA

Three kinds of RNA involved in protein synthesis are **transfer RNA** (tRNA), **ribosomal RNA** (rRNA), and **small interfering RNA** (siRNA). All are made by transcription of special genes.

tRNA transports amino acids to the ribosomes during protein synthesis. There are about 20 groups of tRNA. Each tRNA is specific for one kind of amino acid and consists of a single polynucleotide strand of RNA, about 80 bases long (figure 4). The strand is folded to form a clover-leaf arrangement held in place by hydrogen bonds between complementary bases. One end of tRNA acts as an attachment site for a specific amino acid. A region called the **anticodon** contains three bases which are complementary to the codon for the amino acid it carries.

rRNA combines with proteins to make ribosomes. In eukaryotes it is synthesised in a region of the nucleolus known as the **nucleolar organiser** and then exported to the cytoplasm.

siRNA consists of double-stranded RNA usually between 19 and 21 nucleotides long. It plays a key role in **RNA interference**. By unwinding and then combining one of its strands (the **guide strand**) by complementary base pairing with its target mRNA, siRNA interferes with the translation of the mRNA into a polypeptide, thereby stopping expression of the gene. Targets for natural siRNA include the RNA of viruses. Synthetic siRNA is used as a valuable tool in research and medicine.

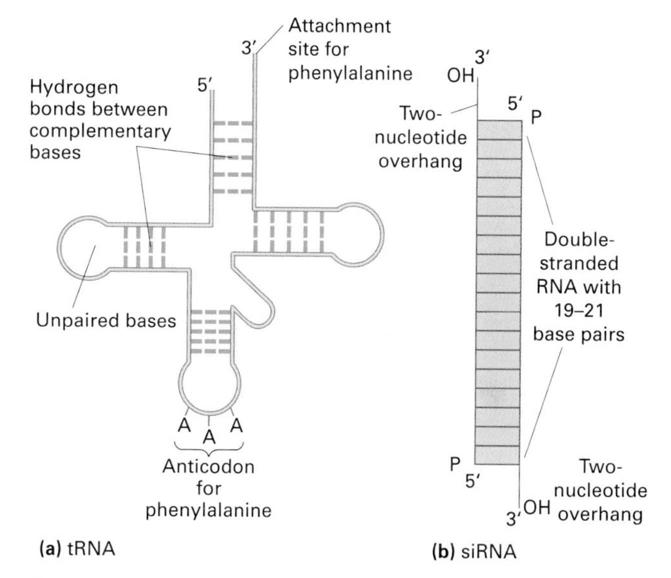

***Figure 4** Schematic diagrams of RNA:*
***(a)** tRNA for phenylalanine;*
***(b)** siRNA. P = phosphate; OH = hydroxyl group.*

Quick check

1 Distinguish between transcription and translation.

2 A transcribing DNA strand contains the base sequence CGGAATCGT. What will be the base sequence in the mRNA transcribed from it?

3 What is the role of tRNA?

Food for thought

Suggest why the genes for rRNA are repeated many times in chromosomes. What do you think is the significance of the fact that the base sequence of rRNA is similar in all organisms?

18.8

OBJECTIVES

By the end of this spread you should be able to:

- describe how a new amino acid is added to a polypeptide chain during translation
- describe the functions of tRNA, mRNA, and ribosomes during translation
- explain how polysomes speed up protein synthesis.

Fact of life

Even after an organism is fully grown, protein synthesis is essential for survival. Proteins such as digestive enzymes, haemoglobin, and collagen are constantly being broken down and resynthesised. Under normal conditions, the average cell is probably synthesising several thousand new protein molecules every minute.

Protein synthesis: translation

Translation is the process by which information encoded within mRNA is used to make a specific polypeptide chain. It is a complex process which happens on ribosomes in the cytoplasm. Each ribosome consists of two subunits made up of proteins and ribosomal RNA (rRNA).

- The **small subunit** has a binding site for mRNA.
- The **large subunit** has binding sites for transfer RNA (tRNA).

The process of translation

mRNA carries information in the form of codons which correspond to the amino acids to be used. A ribosome moves along a strand of mRNA in the 5' → 3' direction and the codons are 'read' sequentially. tRNA molecules bring amino acids to the ribosome, and these are added one by one to the growing polypeptide chain. The ribosome allows two molecules of tRNA to combine with the mRNA at any one time. One tRNA molecule holds the growing polypeptide chain; the other carries the next amino acid to be added to the chain.

Before a tRNA molecule moves to a ribosome, a specific enzyme makes sure that it is carrying the correct amino acid specified by its anticodon. The attachment of amino acids to tRNA requires energy; this is supplied by ATP.

Figure 1 shows how a new amino acid is added to a growing polypeptide chain.

- The ribosome has reached a part of the mRNA strand containing the codon for proline (CCU).
- The tRNA molecule with the anticodon GGA and carrying proline has attached onto the mRNA.
- An enzyme (peptidyl transferase) catalyses the formation of a peptide bond between the proline and the histidine at the end of the growing polypeptide chain. The polypeptide chain is therefore transferred to proline.
- The tRNA molecule that was holding the histidine is now released from the ribosome, and is free to carry another histidine molecule.
- Meanwhile, the ribosome moves one codon further along the mRNA strand, exposing the next codon (GGU) so that glycine can be added to the chain.

Complementary base pairing between anticodons on tRNA and codons on mRNA ensures that the correct sequence of amino acids is built up in the polypeptide chain, according to the information on the mRNA molecule.

__Figure 1__ The process of translation: the building up of a polypeptide chain according to the code specified on DNA in the nucleus.

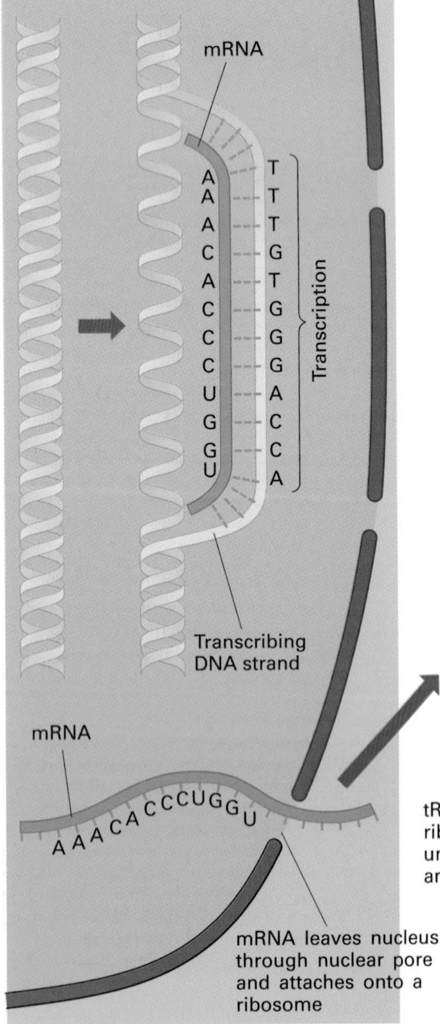

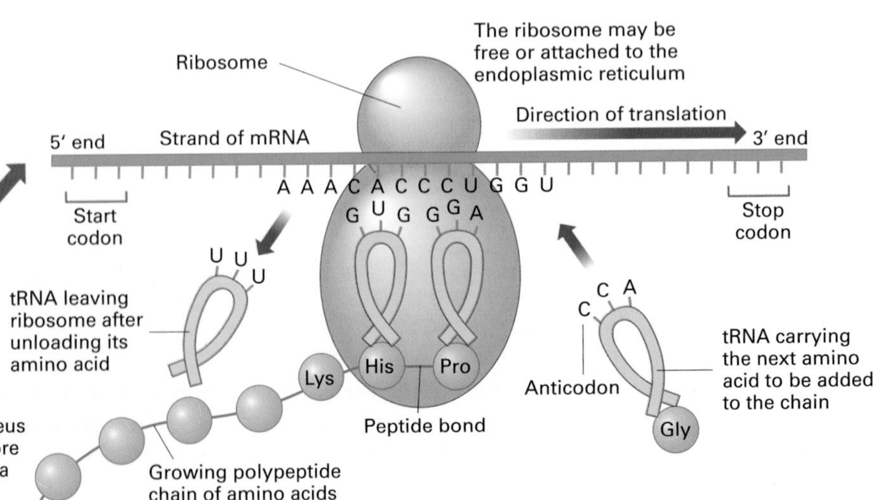

Starting and stopping

Polypeptide synthesis is usually initiated by the codon AUG. This is also the codon for methionine so methionine is the first amino acid in the chain. If the methionine is not needed as a constituent of the final polypeptide chain, it is removed when synthesis is completed. Synthesis ends when the ribosome reaches a 'stop' codon (UAA, UGA, or UAG). The mRNA, ribosome, and tRNA molecules separate, and the polypeptide chain is released.

Processing the polypeptide

On leaving the ribosome, the polypeptide chain is processed according to the final destination of the protein. This may include folding to form the secondary and tertiary structure and, in some proteins, adding chains to form a quaternary structure.

- Proteins used inside the cell, such as haemoglobin, are usually made on free ribosomes and released into the cytoplasm.
- Polypeptides of proteins that are to be exported from the cell, such as digestive enzymes, are made on the rough endoplasmic reticulum. As the polypeptide is made, it is threaded through pores in the endoplasmic reticulum and it builds up in the cisternae. It is transported in vesicles to the Golgi apparatus, where it is modified and packaged. Golgi vesicles then transport the protein to the cell surface membrane from which it is secreted by exocytosis.
- Polypeptides that form membrane proteins follow the same route, but they remain on the cell surface membrane rather than being exported.

Quick check

1 What type of bond attaches a new amino acid to a polypeptide chain?

2 By what process does a tRNA molecule attach onto mRNA during translation?

3 What is a polysome?

Food for thought

In the lac operon the structural gene for beta-galactosidase is not always switched on in the presence of lactose. For example, when glucose and lactose are available together, *E coli* uses glucose preferentially until this sugar is exhausted. Then, after a short lag, growth resumes with lactose as the carbon source. This two-phase growth pattern called **diauxic growth** suggests that glucose in some way 'switches off' the structural gene. Suggest the advantage to the bacterium of having this pattern of growth.

Polysomes: speeding up the process

A single polypeptide chain grows at the rate of about 15 amino acids per second. This is not fast enough for cells that have to produce a great deal of protein quickly. Protein synthesis in these cells is accelerated by the formation of polysomes.

A **polysome** consists of 5–50 ribosomes on the same mRNA strand. Each ribosome translates a different part of the mRNA at the same time (figure 2). When a ribosome reaches the end of the strand, it releases its polypeptide chain and returns to the beginning. Ribosomes can carry out polypeptide synthesis repeatedly on the same mRNA strand, so that a large number of polypeptide chains can be synthesised from one mRNA molecule.

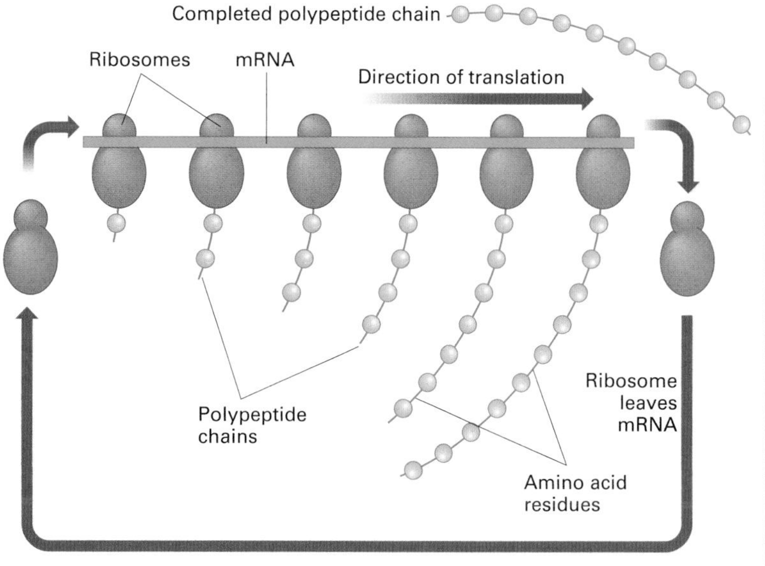

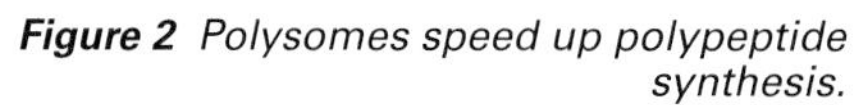

***Figure 2** Polysomes speed up polypeptide synthesis.*

Regulation of gene expression

Every cell has many genes, of which only a few are expressed at any one time. Protein synthesis has to be highly regulated so that the right protein is made in the correct amounts at the appropriate time. Gene expression can be regulated at any stage of protein synthesis, but the best known mechanisms occur during transcription. One mechanism involves **transcription factors**, regulatory proteins that bind to DNA and stimulate the transcription of specific genes. For example, when oestrogen is attached to its receptor proteins it acts as a transcription factor for a number of genes (spread 10.3). *Escherichia coli* has a **gene switch** for three enzymes that catalyse the absorption and breakdown of lactose. The switch consists of structural genes, a regulator gene, and an operator gene. The **structural genes** code for the three enzymes. The **regulator gene** produces a repressor molecule which, in the absence of lactose, inhibits the structural genes. The repressor molecule does not affect the structural genes directly but acts through the **operator gene** next to the structural gene. The operator and structural genes are known collectively as the **lac operon**. This gene switch ensures that the enzymes are produced only when lactose is present.

18.9

GENETIC ENGINEERING TECHNIQUES

OBJECTIVES

By the end of this spread you should be able to:

- explain how genes can be isolated
- name the enzymes that cut and splice DNA
- describe how genes can be transferred from one organism into another.

Fact of life

Plasmids are self-replicating fragments of DNA which make up about 5 per cent of the total DNA in many bacteria. They are independent of the main circular bacterial DNA. When a bacterium dies and bursts apart, its contents pass into the environment. Some plasmids are transferred quite easily from one species of bacteria to another. This is unfortunate for us, because these plasmids may carry genes for antibiotic resistance.

Genetic engineering is the alteration of the genetic make-up of an organism by artificial means, such as **recombinant DNA (rDNA) technology** which has the potential to transfer DNA from one species to another. rDNA technology enables the genetic engineer to copy a piece of DNA from one organism (the donor) and join or recombine this copy with DNA in another organism (the host).

To create new gene combinations, genetic engineers must be able to:

- identify and locate a specific gene in the donor cell
- isolate this gene in a piece of donor DNA
- modify the donor DNA in a highly selective way
- transfer the modified donor DNA into host cells in such a way that the gene will be expressed strongly enough to be of practical use.

Locating and isolating genes

A variety of techniques are used to locate and isolate genes, including DNA or gene probes and DNA hybridisation, reverse transcription, artificial DNA synthesis, and restriction mapping using restriction endonucleases.

DNA probe and DNA hybridisation

A **DNA probe** (also known as a **gene probe**) can be used if at least part of the DNA base sequence in the sought-after gene is known. The probe consists of a single strand of DNA that contains bases complementary to the known sequence (e.g., if the known base sequence is CTA AGT CCA, the probe will have the bases GAT TCA GGT). The probe is labelled with a radioactive or fluorescent marker. Given suitable conditions, the bases in the probe hybridise with the bases on the DNA in the test material. The marker enables the DNA probe to be tracked and the sought-after gene to be located. Once located, the gene can be extracted.

Reverse transcription

Cells that produce large amounts of a particular polypeptide will have large amounts of mRNA for that polypeptide. If this mRNA can be isolated, its **complementary DNA** (cDNA) can be synthesised from it by a process called **reverse transcription**. This process is the reverse of normal transcription because the mRNA acts as a template for DNA synthesis. Reverse transcription requires DNA nucleotides and enzymes called **reverse transcriptases**. After the mRNA strand has been copied into cDNA, the mRNA is removed and a second strand of DNA is made by adding the enzyme **DNA polymerase** and more DNA nucleotides. The result is a double-stranded DNA molecule identical to the original DNA molecule.

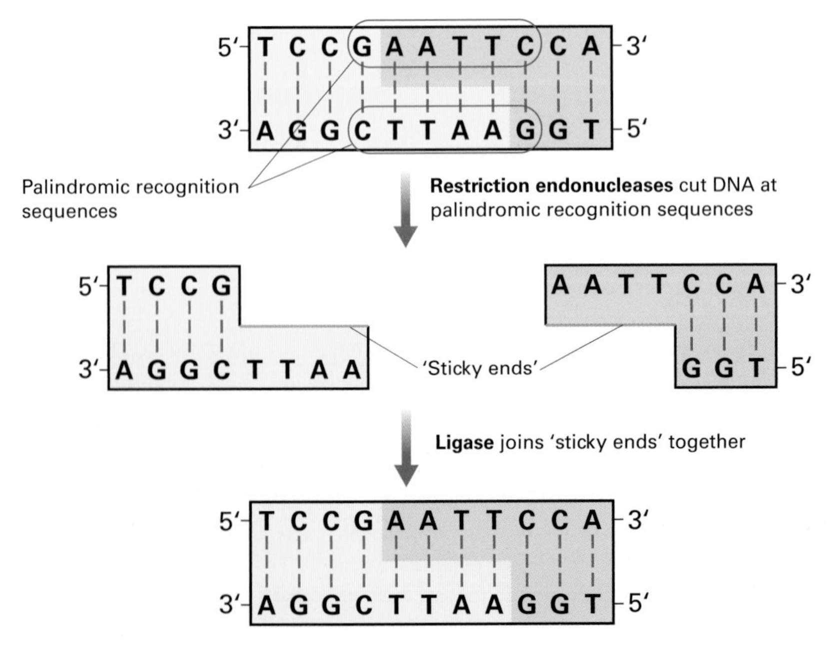

Figure 1 A piece of double-stranded DNA is cut by a restriction endonuclease which recognises a palindromic sequence of bases. Two fragments are created with overlapping 'sticky ends' that can be joined together using ligase.

Reverse transcriptases were first obtained from retroviruses, which contain RNA (spread 17.1). These viruses use reverse transcriptases to copy their RNA into DNA in the host cell.

Artificial DNA synthesis

A piece of DNA coding for a particular protein can be made artificially if the base sequence is known. This base sequence can be deduced from the amino acid sequence of the protein. The process is most useful for small proteins, which have a correspondingly short DNA base sequence.

Restriction endonucleases

A restriction endonuclease is an enzyme that cuts DNA at a **restriction site** near or overlapping a specific base sequence called a **recognition sequence**. The two DNA strands may be cleaved directly opposite each other to create 'blunt' ends, or the cleavage sites may be offset slightly to create 'sticky ends' that can be joined together by using **DNA ligase**. In nature, DNA ligase enzymes stick lengths of DNA together during replication and repair.

Genetic engineers have a wide choice of specific restriction endonucleases and ligases which they can use as molecular tools to 'cut and splice' DNA. Joining together 'blunt' ends requires higher enzyme concentrations and more demanding reaction conditions than joining 'sticky' ends.

One widely used restriction enzyme has the recognition sequence GAA TTC. This sequence is termed a **palindrome** because it is the same as the complement of bases (CTT AAG) read backwards. By cleaving the G—A bond on both strands, the restriction enzyme creates fragments with 'sticky ends' in which the exposed bases form hydrogen bonds with complementary 'sticky ends' from other DNA molecules cut by the same restriction endonuclease (figure 1).

Different restriction endonucleases can be used to compile a **restriction map** showing the location of specific restriction sites on test DNA. Restriction mapping involves marking the ends of the test DNA with, for example, a fluorescent dye, cutting the DNA with two or more different restriction endonucleases, and separating the resulting fragments by gel electrophoresis (see appendix).

In nature, bacteria use restriction endonucleases to defend themselves against bacteriophages, chopping viral DNA into smaller non-infectious fragments.

Transferring DNA from donor to host: DNA vectors

A **DNA vector** transfers DNA from one cell to another. The most commonly used vectors are **plasmids** (small rings of DNA contained within some bacteria). Plasmids are quite separate from the main bacterial DNA. They replicate independently and can be transferred from one bacterial cell to another, and even from bacteria to plant cells. This process occurs naturally, but it can also be encouraged artificially.

Using restriction endonucleases and DNA ligases, a plasmid can be cut open so that short stretches of donor DNA obtained from another organism can be inserted. The donor DNA may be from any other living organism. After inserting the donor DNA, the modified plasmid can be transferred to another living cell (the host). The universality of the genetic code (spread 18.6) means that the host cell can use the donor DNA even though it comes from another organism or species. If appropriate fragments of DNA (called **control sequences**) are also incorporated into the plasmid, the host cell will produce the polypeptide coded for by the donor DNA (figure 2).

Plasmids can carry only relatively short DNA sequences; consequently their use is limited. Other biological vectors include viruses, which transfer DNA into cells naturally. Non-biological vectors include **ballistic impregnation** (the use of a gun to fire gold or tungsten coated with DNA into the target cell), **microinjection** (the injection of a gene, such as that for beta-carotene in 'golden rice', through a very fine pipette into the target cell), and **liposome transfer** (the carriage of DNA in a bubble of fatty substance called a liposome through the cell surface membrane and into the host cell).

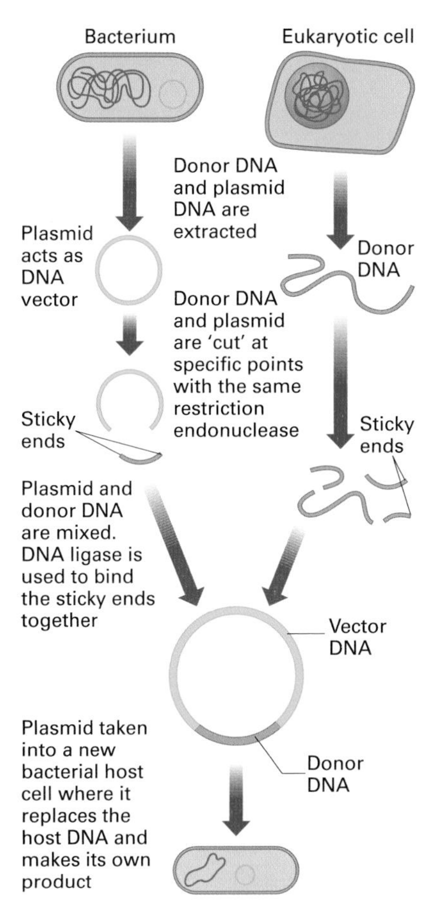

Figure 2 *Recombinant DNA: transfer of DNA from a eukaryotic cell to a bacterial cell using a plasmid. Note that by using the same restriction endonuclease on the plasmid DNA and donor DNA, the sticky ends will match (they will have complementary bases).*

QUICK CHECK

1 What does a genetic probe consist of?

2 Which enzymes:

   a cut DNA at specific points

   b join together cut ends of DNA?

3 What are plasmids?

Food for thought

Humans have been changing the genetic make-up of organisms since the beginning of agriculture: modern domestic plants and animals are the results of thousands of years of selective breeding. How does traditional selective breeding differ from modern methods of changing the DNA of animals and plants?

18.10

OBJECTIVES

By the end of this spread you should be able to:

- explain what gene cloning is
- describe how human insulin can be made by recombinant DNA technology
- discuss cloning of mammals.

Fact of life

Jerry Hall of George Washington University was the first to clone a human embryo artificially. He divided an early embryo into separate cells which were then grown *in vitro* into separate embryos. They survived long enough (6 days) to become implanted into a uterus and had the potential to develop into reproductive adults. However, this was never done.

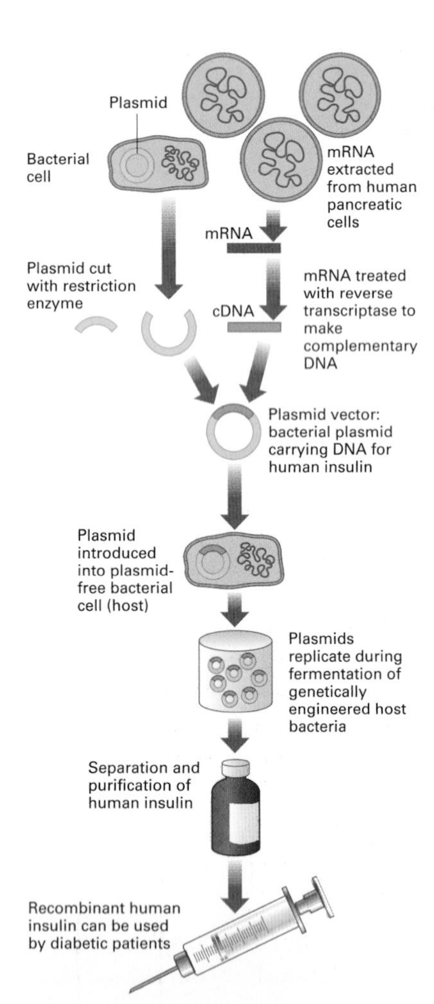

***Figure 1** The manufacture of recombinant human insulin.*

Cloning

Genetic engineers use rDNA technology to identify, isolate, and transfer DNA from one cell to another (spread 18.9). The host cell can be used to make the product coded for by the donor DNA. But genetic engineers must be able to make multiple copies of the gene to produce useful quantities of the product.

Gene cloning

The replication of donor genes in bacterial or other living host cells is called *in vivo* **gene cloning** or **DNA cloning**. A clone is a group of cells, organisms, or genes that are exact copies of each other. Clones occur naturally by asexual reproduction (spread 12.1) as well as artificially by genetic engineering. A donor gene inserted into a bacterium is copied every time the plasmid containing it replicates. A newly inserted plasmid replicates several times inside a bacterium both before and after the bacterium divides. Genes can therefore be cloned by growing bacteria in industrial fermenters (spread 17.7).

Genetic markers

Genetic engineers use a **genetic marker** to ensure that only bacteria carrying donor genes grow. Until recently, the most commonly used genetic marker was a gene for antibiotic resistance attached to the donor gene before it was inserted into the host's plasmids. Plasmids with antibiotic resistant genes are called **r-plasmids**. By a process called **r-plasmid identification**, the antibiotic to which there is resistance is then used to kill all the unwanted bacteria that do not have the antibiotic resistance gene and therefore do not have the donor gene. Now, because of the possibility of antibiotic resistance genes being transferred unintentionally to pathogenic bacteria, genes for enzymes that produce fluorescent or easily stained substances are replacing antibiotic resistance genes as genetic markers (spread 21.9).

Industrial processes

One problem with using bacteria to manufacture a useful product is that the product often inhibits bacterial growth at concentrations that are too low for extraction and purification of the product to be economical. In some cases, this problem can be overcome by separating the growth phase from the production phase.

- The bacteria are allowed to multiply rapidly at first, without producing the required product (the **growth phase**).
- Once there are enough bacteria, the donor gene can be 'switched on' by activating a key enzyme in the synthetic pathway.
- The product is then made by the bacteria (the **production phase**).

Another problem with industrial-scale fermentation is that the microorganisms may be damaged by the heat they generate when in large populations. Cooling the fermenter requires energy and costs money. Genetic engineers sometimes overcome this problem by using a thermophilic organism to synthesise the desired product, or by inserting the gene for temperature tolerance into the host DNA.

After the required product has been synthesised, it must be separated from the host cells and then purified by downstream processing (spread 17.7).

Human insulin

Human insulin was the first human protein to be manufactured by gene cloning and used by humans. It is now produced on a very large scale (figure 1) and is used by many thousands of people with diabetes. Human insulin made by recombinant DNA technology produces fewer side-effects than insulin prepared from cow or pig pancreatic extracts, previously the main source of insulin. Other uses of gene cloning include the manufacture of human growth hormone, vaccines, silk, and adhesives.

Transgenic animals: human proteins in milk

Modified DNA can also be introduced into animals such as cows or sheep so that they produce human proteins. These **genetically modified organisms** are called **transgenic animals**. Females are particularly useful if the protein can be produced in the milk. The product can then be obtained continuously at a predictable quality without distressing the animal. Medically important substances produced in transgenic mammals include:

- blood clotting Factor VIII, to treat haemophilia
- alpha-1-antitrypsin, to help protect the lungs from damage during infections.

Both of these products were previously obtained from human blood. However, they are still very expensive to produce by genetic engineering, and demand exceeds supply.

Manufacture of alpha-1-antitrypsin

- The donor gene for the protein is cloned from cultured human cells.
- It is introduced into the fertilised eggs of sheep by microinjection.
- The gene is pre-programmed to work in the mammary gland when the sheep lactates. The protein can be separated and purified from the other milk constituents after milking.
- The gene for alpha-1-antitrypsin can be inherited by the sheep's offspring, so they will also produce this valuable protein.

Cloning of organisms

Cloning is not confined to genes, and sometimes happens naturally. Clones of whole organisms result from asexual reproduction, and also from multiple births originating from the same fertilised egg. Identical twins, for example, are clones. Whole organisms can also be cloned artificially. Since 1979, sheep, rabbits, toads, and other sexually reproducing animals have been cloned by dividing up an embryo and transplanting them into surrogate mothers. The number of clones is limited because if you repeatedly divide an embryo to make more embryos, they become less vigorous and will not survive.

In February 1997, scientists at the Roslin Institute near Edinburgh cloned a sheep from a single cell taken from the udder of a six-year-old ewe (figure 2). The sheep, known as Dolly, was the first mammal to be created from the non-reproductive tissue of an adult mammal.

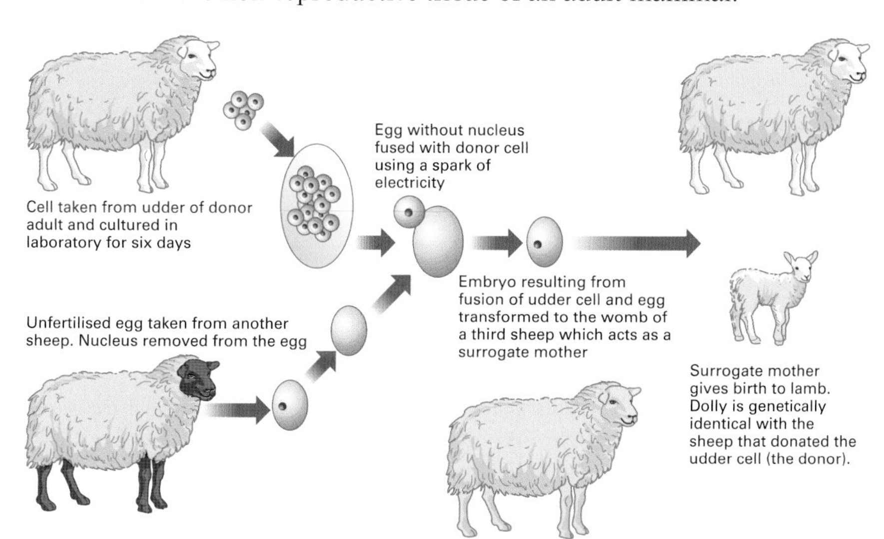

Figure 2 *Dolly the lamb was cloned from a non-productive cell of an adult ewe. Cloning is of interest to farmers because it produces uniform offspring. High-quality sheep produced by selective breeding could be cloned to produce genetically identical lambs instead of the variable offspring produced by sexual reproduction.*

Reproductive and therapeutic cloning

Reproductive cloning creates whole animals that have the same nuclear DNA as the donor, and which can develop into adults that can produce offspring. Dolly the sheep, the first mammal created by reproductive cloning, was a mother to six lambs bred in the traditional way. Reproductive cloning is expensive and inefficient; less than 10 per cent of cloning attempts produce viable offspring. **Therapeutic cloning** is a form of **non-reproductive cloning**. The aim is to produce genetically identical embryos (not whole animals) from which **stem cells** can be harvested for research and medical use. Cloned animals and embryos have the same nuclear DNA, but they may not be phenotypically identical because the genes may not all be expressed in the same way. Each type of cloning has its potential benefits and risks, and has its own ethical dilemmas.

Quick check

1 What is a clone?

2 How is the human insulin gene cloned in bacteria?

3 What made the cloning of Dolly the sheep such an important scientific breakthrough?

Food for thought

The reproductive cloning of humans is banned in Europe. However, some people think that, because we have the technology, the creation of artificial human clones is inevitable. What do you think? Suggest:

a an argument for human cloning

b an argument against human cloning.

18.11

DNA PROFILING

OBJECTIVES

By the end of this spread you should be able to:

- explain what a hypervariable region is
- explain what a DNA profile is and why it is commonly called a DNA fingerprint
- describe the main stages in making a DNA profile.

Fact of life

In 1987, Robert Melias was the first person to be convicted of a crime (rape) on the basis of evidence including DNA profiling. DNA profiling is now used in many thousands of criminal investigations each year.

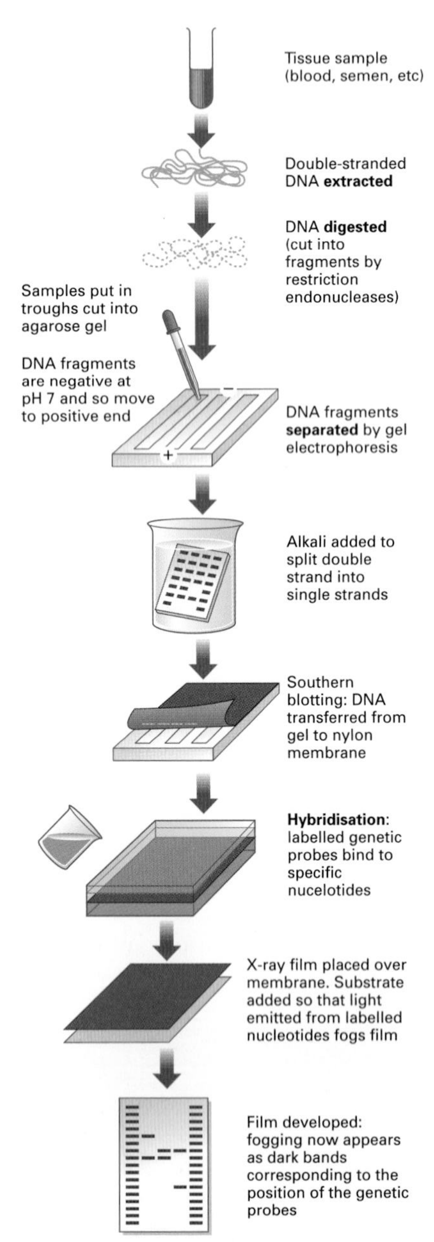

Figure 1 *The main stages of DNA profiling.*

A human chromosome contains a single molecule of DNA, elaborately coiled and intertwined with proteins (spread 18.2). One DNA molecule may contain up to 4000 genes at different **loci** (positions) along the chromosome. However, about 95 per cent of human DNA does not code for polypeptides. Stretches of non-coding DNA may lie between different genes, and the genes themselves contain non-coding portions called **introns** (spread 18.7).

DNA profiling

In 1984, Alec Jeffreys at Leicester University devised a method of identifying individuals using the non-coding regions of DNA. The method is called **DNA profiling**, popularly known as **DNA fingerprinting** because of its use in forensic investigations.

The principle

Non-coding DNA contains **hypervariable regions**. These regions have repeating nucleotide sequences called **core nucleotide sequences**. The number and length of the repeats vary between individuals, but are similar in related individuals; the closer the relationship, the greater the similarities. Figure 1 shows the main steps involved in DNA profiling.

Extraction and digestion

Extraction is normally carried out by mixing a tissue sample with a solvent, such as a mixture of water-saturated phenol and chloroform, which dissolves DNA but precipitates protein. Restriction endonucleases are then used to digest the DNA into fragments. The enzymes cut the DNA at specific points and produce fragments of varying size, some of which contain hypervariable regions.

Separation and analysis

DNA fragments can be separated by **gel electrophoresis** (see appendix). The mixture of fragments is placed in a trough cut into agarose gel which is immersed in buffer at pH 7 and a current applied to the gel. All the DNA fragments move towards the positive electrode because they are negatively charged at pH 7. However, the smaller fragments move faster than the larger ones. Therefore, gel electrophoresis separates fragments according to their size. **Markers** (DNA fragments of known size) are also separated in the gel. Up until this point, the DNA is double stranded. After electrophoresis, it is converted into single strands by immersing the gel in an alkali. The DNA fragments are blotted onto a nylon membrane using a technique called **Southern blotting**. A thin nylon membrane is laid over the gel and several sheets of blotting paper or filter paper are laid on top. The buffer containing the DNA is drawn up through the filter paper by capillarity and some of the DNA is deposited on the nylon membrane; the positions of the DNA fragments on the membrane correspond to their positions on the gel. The DNA is fixed on the membrane by exposing it to short-wavelength ultraviolet light.

Hybridisation

The separated single-stranded DNA is mixed with DNA probes. These contain nucleotide sequences complementary to the core nucleotide sequences known to occur in hypervariable regions. The probes are labelled with the enzyme **alkaline phosphatase**. When the nylon membrane is incubated at the correct temperature, pH, and ionic strength, complementary base pairing occurs between the probes and the DNA from the sample.

After hybridisation, the membrane is covered with a phosphate-containing substrate and placed on an X-ray film in the dark. The alkaline phosphatase removes the phosphate. This causes the substrate to fluoresce, fogging the X-ray film. When developed, the film shows dark bands where the probes were bound to the DNA in the hypervariable regions. The positions of these bands (the genetic profile or fingerprint) can be used to identify an individual (figure 3). With the exception of identical twins, the probability of two people having the same banding is very small. However, forensic scientists sometimes have to work with poor quality tissue samples containing degraded DNA which may give only a few bands. The fewer the bands, the less reliable is the fingerprint.

Uses of DNA profiling

DNA profiling can be used to compare DNA from sources found at the scene of a crime with that of suspects, to detect inherited diseases in embryonic cells, to resolve paternity disputes, and to confirm evolutionary relationships between organisms. A method called the **polymerase chain reaction** (PCR) enables DNA profiling of very small samples of even partly degraded DNA from ancient, long-dead sources, such as human remains in archaeological sites.

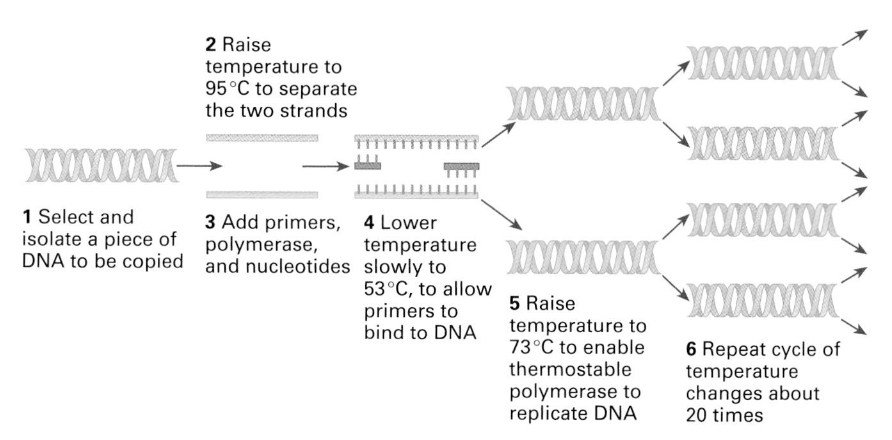

Figure 2 Polymerase chain reaction (PCR): a method of gene cloning without bacteria. The temperature is controlled very carefully using a device called a thermocycler.

Polymerase chain reaction: *in vitro* cloning

In 1983, Kary Mullis devised an *in vitro* method of cloning DNA called the **polymerase chain reaction** (PCR) in which a piece of DNA to be cloned is mixed with DNA polymerase in a solution containing nucleotides. Special **DNA polymerases** are obtained from thermophilic bacteria so that they can function at high temperatures. Those obtained from *Thermus aquaticus* are called **Taq polymerases**. By a very precisely controlled cycle of temperature changes, the two strands of DNA are separated, specific **primers** are added to each strand, and the strands are replicated (figure 2). Primers are essential ingredients of PCR. They are short nucleotide sequences complementary to the ends of the target DNA. They act as 'start' signals for replication and flag the beginning and end of the stretch of DNA to be copied. Each time the cycle of temperature changes is repeated, the DNA is replicated. Typically the process is repeated 20 times, so that millions of copies of the DNA can be produced in about an hour. Originally, replicated DNA was digested and then analysed by gel electrophoresis. However, a new technique called **quantitative PCR** (Q-PCR) allows DNA fragments to be analysed as they are produced. Special fluorescent dyes react with specific base sequences in the DNA so that the nucleotide sequences present in the DNA can be identified and their amounts measured. *In vitro* cloning by PCR requires only a very small sample of DNA and is quicker than *in vivo* cloning (spread 18.10), but specific primers are needed which require prior knowledge of at least some of the nucleotide sequence in the target DNA. There is also a gradual loss of cloning efficiency which limits the DNA yield and, because cloning becomes more difficult as the size of the sequence increases, only relatively short sequences can be cloned. *In vivo* cloning is possible with sequences approaching 2 megabase pairs and is the best way to produce large amounts of a particular gene.

QUICK CHECK

1 What are core nucleotide sequences, and in which region of DNA are they found?

2 Why is DNA profiling commonly referred to as DNA fingerprinting?

3 In DNA profiling, what is the function of alkaline phosphatase?

Food for thought

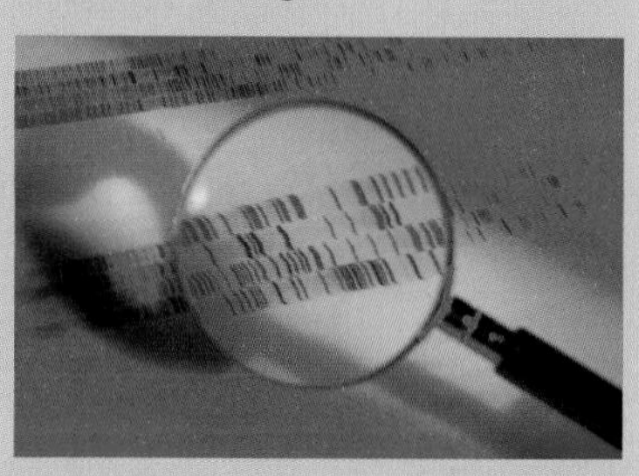

Figure 3 A genetic fingerprint.

Figure 4 shows the genetic fingerprints of a victim's blood, a specimen of semen taken from the victim, and the blood of three suspects. On the basis of the DNA fingerprints, which suspect was the rapist? How reliable do you think this evidence would be? Suggest how the reliability could be improved.

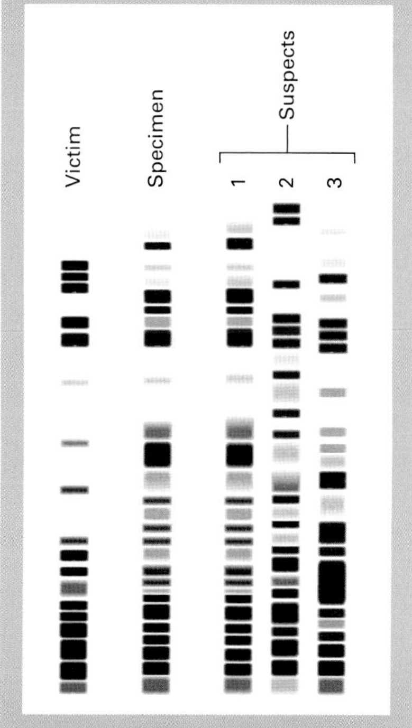

Figure 4 Diagram of genetic fingerprints of a victim's blood.

Summary

DNA structure is described as a **double helix** consisting of two **antiparallel polynucleotide chains** linked by **hydrogen bonds** between **complementary base pairs** (adenine to thymine, and cytosine to guanine). DNA occurs mainly with proteins and small amounts of RNA in **chromosomes**. Chromosomal DNA is packaged in bead-like structures called **nucleosomes** connected by **linker regions** and tightly coiled into **solenoids**. Other chromosomal structures include **centromeres** and **telomeres** which occur at the ends of chromosomes.
Meselsohn and Stahl showed that **DNA replication** is **semiconservative**: each daughter molecule of DNA contains one parental strand and one new strand. Replication requires **binding proteins**, ATP, and a number of enzymes, including **DNA polymerases** and **DNA ligases**.
The chemical nature of **genes**, the units of inheritance, was determined from **Griffith's** discovery that harmless strains of *Pneumococcus* can be transformed into disease-causing strains and **Avery** and his colleagues' demonstration that DNA caused the **transformation**. The fact that genes are made of DNA was confirmed by **Hershey and Chase's experiments** on bacteriophage reproduction. **Beadle and Tatum** proposed the **one gene–one enzyme** hypothesis, based on experiments with *Neurospora*. This has been refined into the **one gene–one polypeptide hypothesis**. Each gene carries information in the form of a **genetic code**. It is a **triplet, non-overlapping, degenerate code**. Information in the code is transcribed into information in mRNA which is then translated into the amino acid sequence that produces a specific polypeptide.
Transcription takes place in the nucleus; **translation** occurs on ribosomes and involves tRNA, in addition to mRNA.
The cracking of the genetic code and the discovery that it is almost **universal** has led to **genetic engineering** by **recombinant DNA technology**. Molecular tools used in this technology include **genetic probes, restriction endonucleases** (enzymes which cut DNA at specific points), **ligases** (enzymes that join bits of DNA together), and **vectors** (e.g. **plasmids**) that carry DNA from a donor to a host cell. Large quantities of genetically engineered product can be made by **cloning** bacterial host cells and other **transgenic organisms**.
DNA profiling, popularly known as **DNA fingerprinting**, is a method of identifying individuals by their DNA. Non-coding DNA in **hypervariable regions** is extracted and digested into fragments. Multiple copies of the DNA may be made by **polymerase chain reactions**. Different DNA fragments are separated by **electrophoresis**.

PRACTICE EXAM QUESTIONS

1 The drawing shows a section of a DNA molecule.

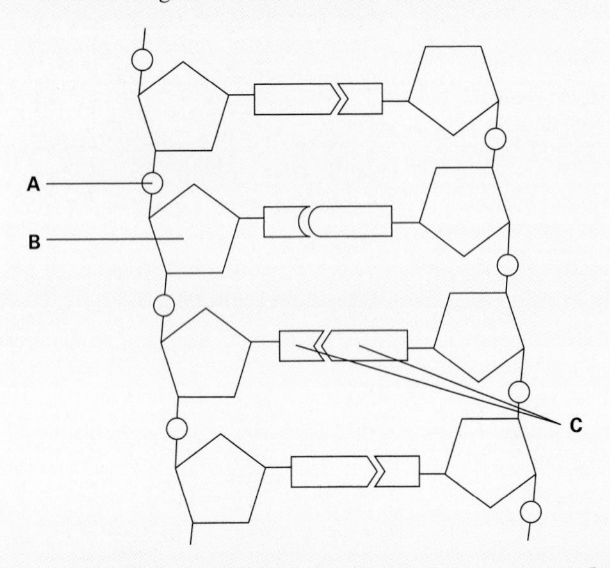

a Name the parts labelled A, B, C. [3]

b The mRNA code for the amino acid serine is UCA.

i Give the DNA code for serine.

ii Give the tRNA code for serine. [2]

c **i** What type of molecule is the end product of translation? [1]

ii Describe the role of tRNA in the translation process. [2]

[Total 8 marks]

2 **a** Name the type of bond that holds together the two strands of nucleotides in a DNA molecule. [1]

Genetic drugs are short sequences of nucleotides. They act by binding to selected sites on DNA or mRNA molecules and preventing the synthesis of disease-related proteins. There are two types.
Triplex drugs are made from DNA nucleotides and bind to the DNA forming a three-stranded helix.
Antisense drugs are made from RNA nucleotides and bind to mRNA.

b Name the process in protein synthesis that will be inhibited by:

i triplex drugs; [1]

ii antisense drugs. [1]

c The table shows the sequence of bases on part of a molecule of mRNA.

Base sequence on coding strand of DNA									
Base sequence on mRNA	A	C	G	U	U	A	G	C	U
Base sequence on antisense drug									

Copy and complete the table to show:

i the base sequence on the corresponding part of the coding strand of a molecule of DNA; [1]

ii the base sequence on the antisense drug that binds to this mRNA. [1]

[Total 5 marks]

3 The diagram below shows the process of DNA replication.

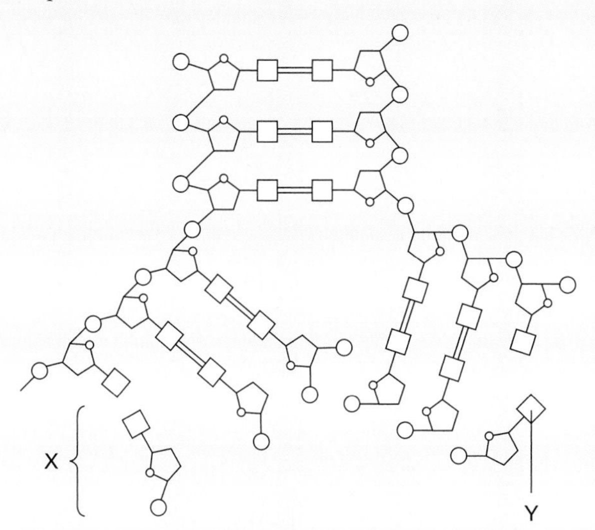

a Name the parts labelled X and Y. [2]

b Name *one* enzyme involved in DNA replication and state the type of reaction it catalyses. [2]

c Suggest why DNA replication is described as *semi-conservative*. [1]

d Name the stage of the cell cycle during which DNA replication occurs. [1]

[Total 6 marks]

4 The polymerase chain reaction is a process which can be carried out in a laboratory to make large quantities of identical DNA from very small samples. The process is summarised in the flowchart.

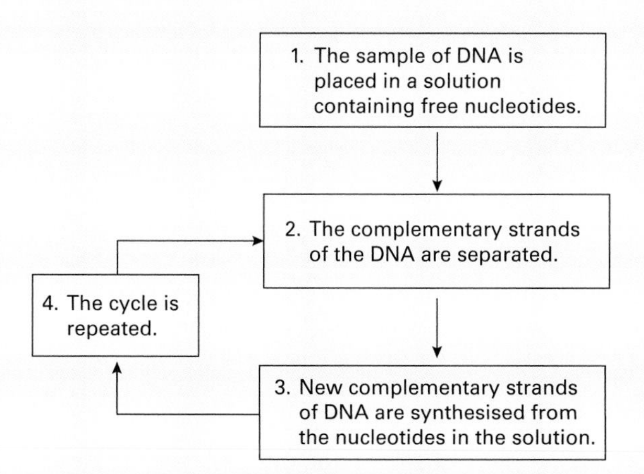

a **i** At the end of one cycle, two molecules of DNA have been produced from each original molecule. How many DNA molecules will have been produced from one molecule of DNA after 5 complete cycles? [1]

ii Suggest **one** practical use to which this technique might be put. [1]

b Give **two** ways in which polymerase chain reaction differs from the process of transcription. [2]

c The polymerase chain reaction involves semi-conservative replication. Explain what is meant by *semi-conservative* replication. [2]

[Total 6 marks]

5 Scientists have perfected the technique of cloning sheep in recent years. The process begins with the fertilisation of eggs by sperm. Nine days later, when the zygote has divided to form about 400 cells, the cells of the embryo can be separated. When separated each cell can grow into a new embryo that can be implanted into an adult female sheep.

a Explain why an embryo cell can divide to form an embryo, but an egg cell will not. [1]

b Explain why the sheep formed from the embryo cells are all genetically identical. [2]

c Give **one** advantage and **one** disadvantage of producing sheep in this way. [2]

[Total 5 marks]

6 **a** Define the terms 'transcription' and 'translation', indicating why these processes are necessary for an organism's survival. [5]

b Explain fully how the cell carries out these processes. [15]

[Total 20 marks]

Inheritance

Variation

19.1

OBJECTIVES

By the end of this spread you should be able to:

- define the following genetic terms: allele; homozygous; heterozygous; dominant; recessive; polygenic
- distinguish between genotype and phenotype
- distinguish between continuous variation and discontinuous variation
- explain how mutations contribute to variation.

Fact of life

Mutations (changes in DNA) are the ultimate source of inherited variation. They can either arise spontaneously or be induced by agents called mutagens (such as X-rays, mustard gas, or ultraviolet radiation). The rate of spontaneous mutations varies for different genes and in different organisms. Each human gene has about a one in 100 000 chance of mutating. Mutations are usually thought of as harmful, and they often are. However, because we have so many genes, even the healthiest of us probably have at least a few spontaneously mutated genes hidden in the recessive form which do not affect us. X-rays and other mutagens increase the mutation rate, and the higher the dosage of radiation, the higher the rate of mutation.

Every organism on Earth is unique. Even clones have some subtle, environmentally induced differences. Differences between individuals of different species are called **interspecific variation**; those between members of the same species are called **intraspecific variation**. Both types of variation result from an interaction between genetic and environmental factors.

Genetic factors and genotype

Genetic differences reflect the **genotype** (the genetic make-up) of an organism. A **diploid** organism has two sets of chromosomes and two forms (**alleles**) of each particular gene. When the two alleles are the same, the organism is said to be **homozygous** for that gene; when they are different, the organism is said to be **heterozygous** for that gene. In the heterozygous condition, one of the pair of alleles (the **dominant allele**) may mask the other allele (the **recessive allele**). The dominant allele is therefore expressed in either the heterozygous or the homozygous condition, whereas the recessive allele is expressed only in the homozygous condition. If an organism is haploid (that is, it has only one set of chromosomes), recessive alleles are not masked by dominant alleles.

Environmental factors and phenotype

The measurable physical and biochemical characteristics of an organism, whether observable or not, make up its **phenotype**. The phenotype results from the interaction of the genotype and the environment. The genotype determines the potential of an organism, whereas the environmental factors to which it is exposed determine to what extent this potential is fulfilled. In humans, potential hair colour is genetically determined, but it can be modified by dyes; potential height is genetically determined, but actual height depends on environmental factors, especially diet. Genetic and environmental factors also interact to bring about diseases such as cancer (see spreads 16.5 and 16.6) and disorders such as forms of depression linked to the gene for the enzyme **monoamine oxidase**. However, the relative contributions of genes and the environment are often difficult to determine because of lack of data.

Phenotypic variation may be continuous or discontinuous. In **continuous variation**, differences are slight and grade into each other. Human height is a continuous variation determined by a large number of genes (it is **polygenic**) and influenced greatly by environmental factors. In **discontinuous variation**, differences are discrete (separate) and clear cut: they do not merge into each other. Discontinuous variations are generally caused by one, two, or only a few genes. Continuous variations are usually quantitative (they can be measured) whereas discontinuous variations are qualitative (they tend to be defined subjectively in descriptive terms). Thus height in humans is a continuous variation

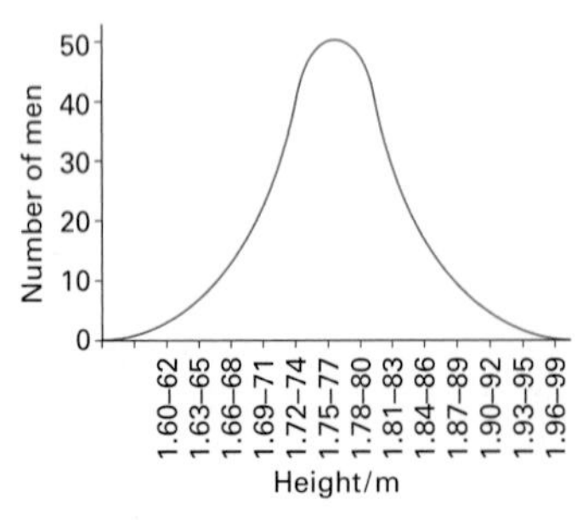

Continuous variation of height among adult males in a human population. There is a normal distribution about a mean and a single mode. Height in humans is determined by the interaction of several genes and is also influenced by environmental factors.

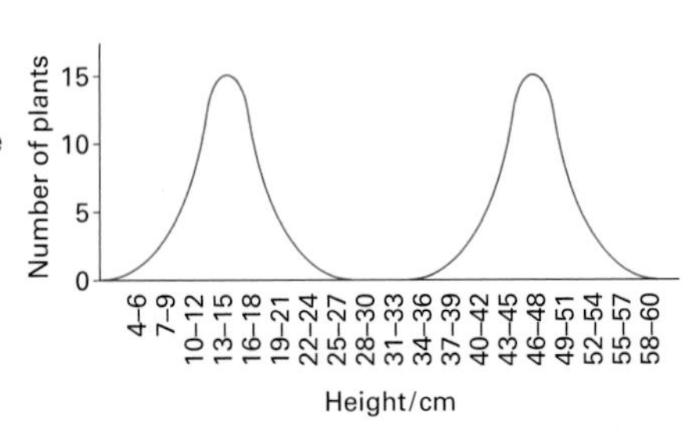

Discontinuous variation of height among sweet pea plants. There is a bimodal distribution (two groups) with no overlap between the 'dwarf' group and the 'tall' group. The 'tall' and 'dwarf' traits are determined by a single gene. The variations within each group are due mainly to environmental factors.

***Figure 1** Variation: continuous and discontinuous.*

measured in metres, whereas height in sweet pea is a discontinuous variation described as 'tall' or 'dwarf' (figure 1).

Mutations: more variation

Genetic variation arises partly from sexual reproduction by a combination of independent assortment, crossing over, and random fertilisation (spreads 4.12 and 12.1). However, these processes merely shuffle the existing pack of genes so that new combinations are made. The ultimate source of inherited variations is mutations.

A **mutation** is a change in the amount or the chemical structure of DNA. If the information contained within the mutated DNA is expressed (that is, transcribed into mRNA and translated into a specific polypeptide chain; see spreads 18.7 and 18.8) it can cause a change in the characteristics of an individual cell or an organism. Mutations in the gametes of multicellular organisms can be inherited by offspring. Mutations of the body cells of multicellular organisms (somatic mutations) are confined to the body cells derived from the mutated cell; they are not inherited.

Mutations can happen spontaneously as a result of errors in DNA replication or errors during cell division, or they can be induced by various environmental factors (such as certain chemicals, X-rays, and viral infection). Factors that induce mutations are called **mutagens**.

Chromosome mutations and gene mutations

Alterations in the number or structure of chromosomes are called **chromosome mutations**. Chromosome mutations can happen during mitosis and meiosis when chromosomes are being condensed and pulled apart. Homologous chromosomes may fail to separate, resulting in **non-disjunction** (see spread 19.7). Chromosome mutations also occur during interphase when DNA replicates, and during crossing over (see spread 4.12) when sections of chromosomes are exchanged (figure 2).

Gene mutations are changes in the nucleotide base sequence in a cistron (the stretch of DNA coding for an allele of a single gene). This can result in non-functional proteins, including non-functional enzymes. Point mutations include:

- **substitution** – the replacement of one nucleotide with another containing a different base
- **deletion** – the loss of a nucleotide
- **insertion** or **addition** – addition of an extra nucleotide.

Sickle-cell anaemia is an example of an inherited condition that results from a substitution (spread 16.1). Gene mutations may also result from **duplication** (repetition of a portion of a nucleotide sequence within a cistron) and **inversion** (reversal of the portion of the nucleotide sequence in the cistron).

Most mutations, if expressed, are harmful. Note, however, that in diploid organisms such as ourselves, mutations usually result in recessive alleles. These are expressed only in the homozygous condition unless the mutation is on the X chromosome (see spread 19.8.) Many mutations result in a change in the shape of a protein so that the protein cannot function properly (for example, the mutation that causes sickle-cell anaemia). Mutations that affect large sections of a gene and chromosome mutations are often lethal. However, some mutations have no effect: a mutation may occur in a non-coding part of DNA; it may produce a different codon for the same amino acid; or the altered amino acid sequence may not affect the protein's shape or function. Occasionally, a mutation is beneficial, changing the phenotype so that an organism has a better chance of surviving and reproducing. Although beneficial mutations are very rare events, they are bound to happen sooner or later if there is a large number of individuals in a population. These mutations are of immense importance because they are the ultimate source of all variation: the raw material for the evolution of new species by natural selection (spread 20.4).

Deletion: loss of a chromosome fragment. In this example, several genes are lost. This type of mutation will probably be lethal.

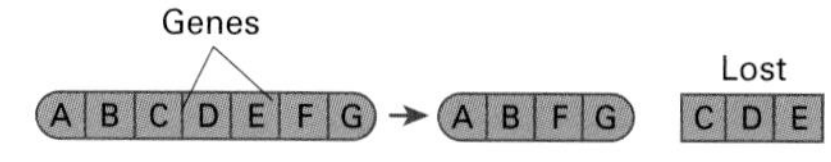

Duplication of genes in one chromosome. This may happen along with deletion of genes in a homologous chromosome, due to a mistake during crossing over.

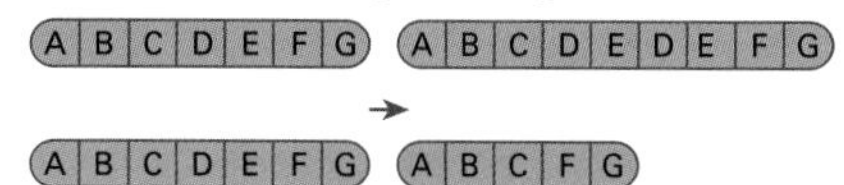

Inversion: a broken chromosome segment is reinserted in the wrong direction, changing the order of the genes, which will be transcribed and translated backwards.

Reciprocal translocation between non-homologous chromosomes

Figure 2 *The main types of chromosome mutation.*

Quick check

1. If a diploid organism has two different alleles for the same gene, is it homozygous or heterozygous?
2. Distinguish between the genotype and the phenotype of an organism.
3. Is weight in humans an example of continuous variation or discontinuous variation?
4. What is a mutagen? Give one example.

Food for thought

Twins (pairs of children born at the same time) may be **dizygotic** or **monozygotic**. Each dizygotic or **non-identical twin** develops from a different egg and may be of a different sex. Monozygotic twins or **identical twins** develop from one egg and contain identical genetic information; they are always of the same sex. Suggest how the study of twins may be used to distinguish between the effects of inheritance and environmental factors on the variations of an individual character.

19.2

Mendelian inheritance

OBJECTIVES

By the end of this spread you should be able to:

- explain how Mendel's work on the inheritance of single characteristics led to his law of segregation
- explain how Mendel's work on the inheritance of two characteristics led to his law of independent assortment
- discuss how Mendel's work laid the foundations for modern genetics.

Fact of life

Some scientists believe that Mendel's results were too good to be true. Taking the results of all his experiments together and assuming Mendel's laws are true, the great statistician R. A. Fisher calculated that the chance of obtaining results so close to the ideal is only 0.007 per cent, or one in 14 000.

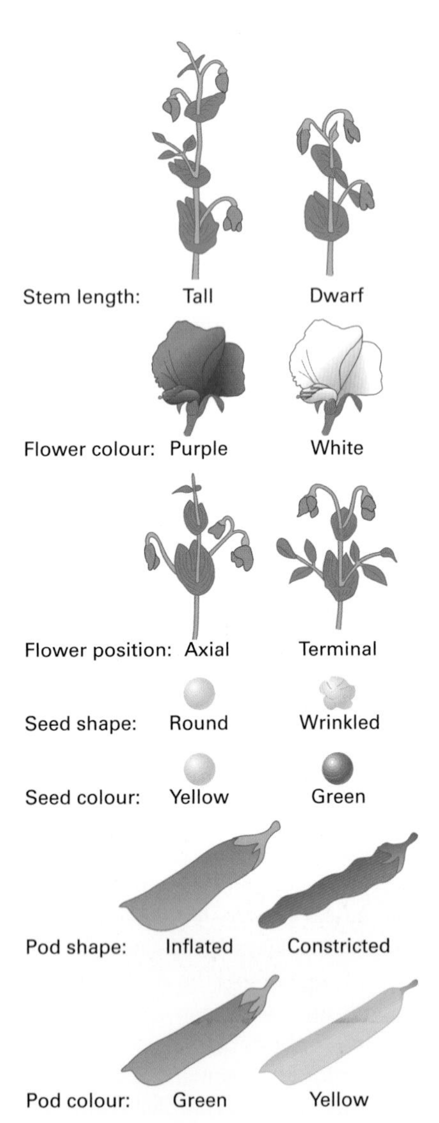

***Figure 1** The seven characteristics of peas studied by Mendel.*

Early ideas of inheritance

Even the earliest humans must have noticed the remarkable similarities that often occur between parents and their offspring. They must also have noticed that offspring differ from each other and from their parents in many respects. They probably realised that some characteristics are passed on to the next generation while others appear to be lost. However, it was not until the 1850s that **genetics** (the scientific study of inheritance) really began. Starting in about 1856, Gregor Mendel carried out a large number of experiments which laid the foundation for genetics.

Before Mendel's experiments, most people believed that parental features were blended and transmitted to offspring as though they were fluids: for example, marriage between a short person and a tall person would produce children who would grow to an intermediate height. Even though Mendel had no knowledge of genes or chromosomes, his work showed that inheritance is particulate: it depends on the transfer of discrete (separate) factors from parents to offspring.

Mendel's experiments

Mendel was an Austrian monk who studied the inheritance of characteristics in garden peas (*Pisum sativum*), which he grew in the vegetable garden of his monastery. He chose peas because they were easy to grow, they had a short life cycle, their pollination could be controlled, and they had easily observable characteristics. He studied seven characteristics, each of which has two contrasting alternatives (figure 1).

Mendel first had to establish **pure-breeding** plants, that is, plants which when self fertilised produce identical offspring generation after generation. Mendel then selected for his experiments two pure-breeding plants with alternative expressions of a particular characteristic (for example, a tall plant and a dwarf plant). He crossed the plants by transferring pollen from one plant (the 'male' parent) to the stigma of a second plant (the 'female' parent). Mendel made sure that the 'female' plant could not be pollinated either by its own pollen or by any other by removing its anthers and covering its flowers with fine muslin bags. Mendel collected the seeds produced by the 'female' parent and grew them the next year to give the first-generation offspring (the offspring of pure-breeding parents are often called the **first filial generation** or the $\mathbf{F_1}$). He carefully recorded the characteristics of these plants and then crossed two plants from this generation. This type of cross involving plants of the same generation is called a **self-cross**. Again, the seeds produced were collected and grown the following year to give the second-generation offspring (the offspring produced by a cross between F_1 parents is often called the **second filial generation** or the $\mathbf{F_2}$). The characteristics of each plant were again recorded.

Mendel's results: single characteristics and the law of segregation

Figure 2 gives the results of one of Mendel's experiments involving a single characteristic. The F_2 plants were a mixture of tall plants and dwarf plants in the approximate ratio of 3:1. Similar crosses involving other single characteristics produced similar results, all giving an approximate ratio of 3:1 in the F_2. The most striking features of these results were that:

1. there were no plants of intermediate height
2. there were no dwarf plants in the F_1 generation, though they reappeared in the F_2.

From the first observation, Mendel concluded that characteristics are not blended together like different colours of paint, but that they are determined by definite, discrete particles which he called 'factors'. From the second observation, Mendel concluded that the factor for dwarfness must be carried in the F_1 plants, but that it is 'hidden' by the factor for tallness. The dwarf factor is expressed only in the absence of the tall factor. Therefore the plants must carry two factors for each characteristic, one factor coming from each parent. As all F_1 plants are tall, the factor for tallness must be dominant to the factor for dwarfness, which is the recessive factor. Mendel combined these conclusions in his first law of inheritance, the **law of segregation**. In modern terms, the law can be stated as: *The characteristics of a diploid organism are determined by alleles which occur in pairs. Of a pair of such alleles, only one can be carried in a single gamete.*

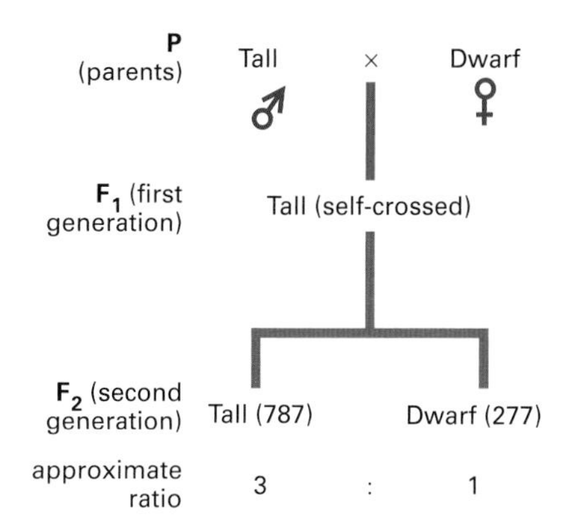

Figure 2 *Inheritance of one characteristic. A pure-breeding tall 'male' (♂) plant was crossed with a pure-breeding dwarf 'female' (♀) plant.*

Two characteristics and the law of independent assortment

Figure 3 gives the results of one of Mendel's experiments involving two characteristics: height and flower colour. From these results, Mendel concluded that, as well as tall being dominant to dwarf, purple flower colour is dominant to white. The four different kinds of plants in the F_2, in an approximate ratio of 9:3:3:1, suggested that the two pairs of factors are transmitted from parents to offspring independently of each other and assort freely (tall plants may have either purple or white flowers, as may dwarf plants; height and flower colour are inherited independently). Mendel described this in his second law of inheritance, the **law of independent assortment**. In modern terms, this says: *Each of a pair of alleles for a particular gene can combine randomly with either of another pair of alleles for a different gene.*

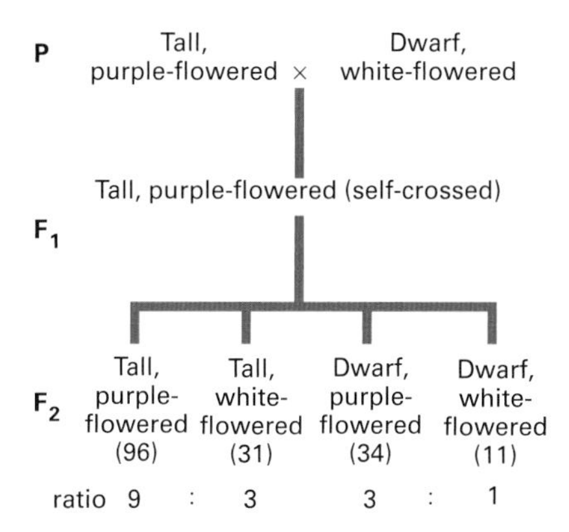

Figure 3 *Inheritance of two characteristics.*

Mendel's success

Mendel was fortunate in choosing simple features controlled by single genes. However, his good fortune would have amounted to nothing without his hard work, good scientific method, and insight in interpreting the results of his breeding experiments. Mendel established the following important facts which formed the foundation of modern genetics:

1. Inheritance is particulate; it depends on factors or particles.
2. The factors are passed from generation to generation unaltered.
3. The factors may be dominant or recessive, so that some factors are not expressed in every generation.
4. Each organism has a pair of factors controlling a given characteristic.
5. Factors segregate during gamete formation so that each gamete contains only one of a pair of factors for a given character.
6. Different pairs of factors are assorted independently of each other during the formation of gametes.

Quick check

1. State the law of segregation.
2. State the law of independent assortment.
3. Explain how Mendel's experiments showed that inheritance depends on particles or factors, and is not the result of blending.

Food for thought

Mendel is now one of the most famous of scientists, hailed as the 'father of genetics' (figure 4). Mendel presented his results in a paper published by the Natural History Society of Brünn in 1866. It was a brilliant paper which reflected Mendel's genius for experimentation and interpretation. However, the paper was ignored by the scientific community until 1900, 16 years after his death. Suggest why Mendel's ideas remained in obscurity for so long.

Figure 4 *Gregor Mendel, 1822–84.*

19.3

Monohybrid inheritance

OBJECTIVES

By the end of this spread you should be able to:

- use a complete and accurate format to show a genetic cross
- describe the results of a simple monohybrid cross
- explain how to conduct a test cross.

Fact of life

It is now known that the height of a pea plant is determined by a single gene with two alleles which affect the production of gibberellins. The dominant *Le* allele enables the plant to produce an enzyme required for the synthesis of gibberellins that promote stem elongation. Pea plants with the recessive *le* allele lack this ability.

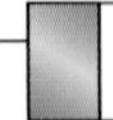

Monohybrid inheritance in humans

Examples of monohybrid inheritance in humans include:

- albinism
- cystic fibrosis
- haemophilia
- Huntington's disease
- rhesus blood groups
- sickle cell anaemia.

Huntington's disease is caused by a dominant allele; the others are recessive traits. All except haemophilia are inherited in a simple Mendelian manner. Haemophilia is sex linked (see spread 19.7).

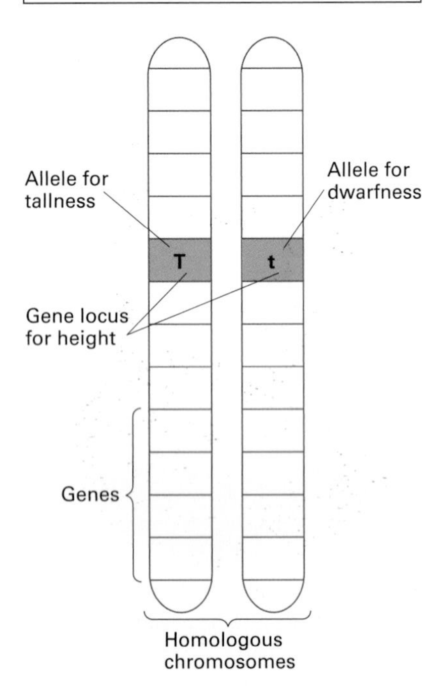

Figure 1 Outline of a pair of homologous chromosomes (see spread 4.10) to show the distinction between genes and alleles.

When Mendel interpreted his historic breeding experiments on garden peas, he did not have the advantage of our knowledge of chromosomes and genes. In this spread, we are going to use this knowledge to explain in modern terms the results of some of his experiments.

Mendel's breeding experiment with tall and dwarf plants (spread 19.2, figure 2) is an example of **monohybrid inheritance**: that is, inheritance involving a single characteristic determined by one gene.

Genes and alleles

In genetics, a **gene** can be regarded as a specific length of DNA which occupies a position on a chromosome called a **locus**. The gene determines a particular characteristic of an organism, in this case, the height of the plant. Mendel recognised that the two ways in which height could be expressed (tall and dwarf) were determined by factors. We call these factors **alleles**. Alleles are different forms of the same gene which may alter the way in which a particular characteristic (such as the height of a plant) is expressed. More accurately, an allele is a particular sequence of nucleotide bases making up a gene. The **phenotype** (the observable or measurable features of an organism) may be changed depending on which allele is expressed. An allele may or may not be expressed in the phenotype (for example, a recessive allele will not be expressed if a dominant allele is also present). Thus, in adult peas the gene for height is always found at the same locus on a particular pair of homologous chromosomes, but either the allele for tallness or the allele for dwarfness will be present at each locus (figure 1).

Representing genetic crosses

By convention, a gene is represented by a letter or letters. If the letters are to be written by hand, they are chosen with care so that the upper and lower case versions can be distinguished easily. The upper case (capital) letter represents the dominant allele and the lower case letter represents the recessive allele. In our example of height in pea plants:

T represents the allele for tallness

t represents the allele for dwarfness.

Mendel recognised that diploid organisms have two factors (alleles) for each characteristic (gene), but only one of the factors is carried by each gamete. In the cross between pure-breeding tall and dwarf plants, all the offspring were tall, therefore the allele for tallness must be dominant to the allele for dwarfness. As the parent plants were pure breeding, they must have been homozygous, therefore the genotypes of the parents can be represented as **TT** (tall) and **tt** (dwarf). The male tall parent plant produced gametes which each carried one dominant allele; the female gametes each carried one recessive allele. The offspring were all heterozygous tall plants, **Tt** (they carried two different alleles, one for tallness, the other for dwarfness). We can summarise these points in a genetic diagram (figure 2a). In the diagram, a circle is drawn around the genotype of the gametes to reinforce the idea that each allele is discrete and can potentially join with alleles in any other gamete from the sexual partner.

The genetic explanation of the subsequent self-cross between heterozygous tall plants is summarised in figure 2b. The diagram shows that there are only three different possible genotypes in the offspring: **TT**, **Tt**, and **tt**. Assuming random fusion of gametes, the diagram also shows us the probability of each genotype occurring in the offspring:

- there is a 1 in 4 (or 0.25) chance that a pea plant will have the genotype **TT**
- there is a 2 in 4 (or 0.5) chance that it will be **Tt**
- there is a 1 in 4 (or 0.25) chance that it will be **tt**.

As all plants carrying at least one **T** allele are tall, 3 out of every 4 plants would be expected to be tall. That is, we would expect a ratio of 3 tall offspring to 1 dwarf offspring. Thus, in Mendel's experiment, out of 1064 offspring, 798 would be expected to be tall, and 266 dwarf. However, because these figures are based on chance and probabilities, it is unlikely that the actual number of offspring would match the expected value precisely. Mendel obtained 787 tall offspring and 277 dwarf offspring. A chi-squared test (a statistical test, described in the appendix, for testing the significance of the differences between observed numbers and expected numbers) would show that these results do not differ significantly from the 3:1 ratio predicted from Mendel's law of segregation. A 3:1 ratio is typical of the results of a monohybrid cross between two heterozygous organisms, where one of the alleles of a gene is dominant and the other is recessive.

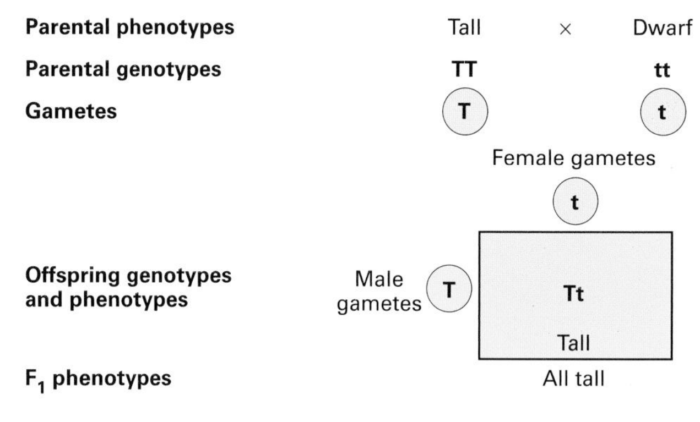

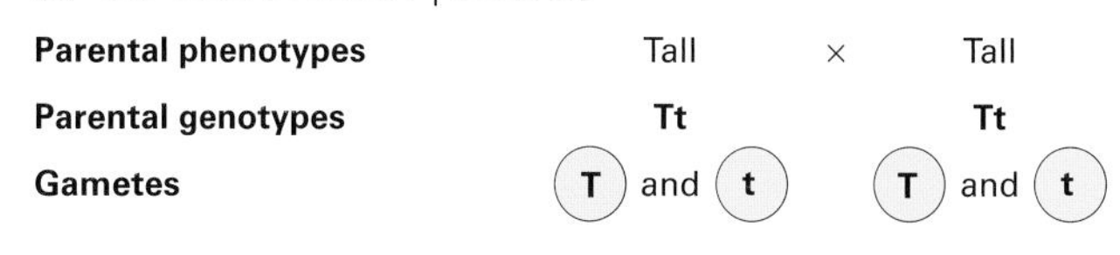

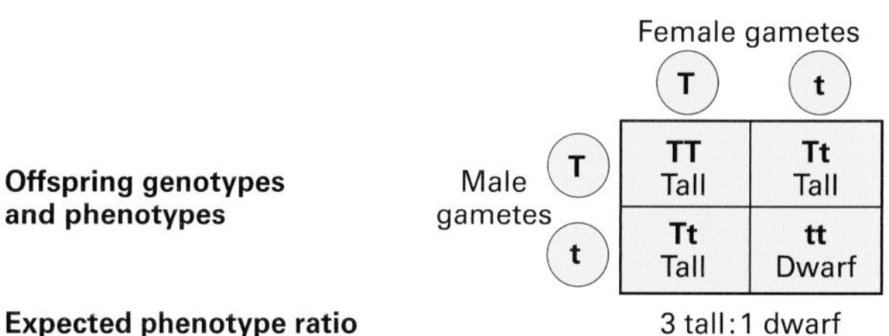

Figure 2 *Diagrams showing a genetic explanation of Mendel's monohybrid cross. The circles show the alleles carried in the gametes, and the boxes (Punnett squares, see spread 19.5, figure 2) show the various outcomes resulting from random fusion of the gametes.*

Test crosses

Organisms with the recessive characteristic must be homozygous for that condition, but organisms showing the dominant characteristic may be either homozygous or heterozygous. If you have a tall pea plant, how can you tell whether it is heterozygous (**Tt**) or homozygous (**TT**)? It is possible to use gene probes to show whether a particular allele is present or not, but test crosses are used by most breeders. In a **test cross**, an organism showing the dominant characteristic is crossed with another organism that is homozygous recessive. For example, to find out whether a tall pea plant is homozygous (**TT**) or heterozygous (**Tt**), it is crossed with a dwarf plant (**tt**). If homozygous, it will not produce any dwarf offspring; if heterozygous, it will produce dwarf and tall offspring in the expected ratio of 1:1 (figure 3). These are the results expected by probability theory, and the actual results must be interpreted with caution. An absence of dwarf plants may not show conclusively that the tall plant was homozygous unless the number of offspring is large. However, if only one dwarf plant is produced, then the unknown genotype *must* be heterozygous.

Parental phenotypes: Tall × Dwarf
Parental genotypes: Tt, tt
Gametes: T and t; t and t
Female gametes: T, t
Offspring genotypes and phenotypes: Male gametes t — Tt Tall, tt Dwarf
Expected phenotype ratio: 1 tall:1 dwarf

Figure 3 *Results of a test cross in which the unknown plant turns out to be heterozygous.*

Quick check

1. What symbols are used to describe the genotype of a heterozygous tall pea plant?
2. Show the genotypes of parents that produce offspring in the ratio of 3 tall plants to 3 dwarf plants.
3. Draw a genetic diagram to show the results of a test cross on a tall plant which turned out to be homozygous for height.

Food for thought

Cystic fibrosis is an inherited condition caused by a single recessive allele inherited in a simple Mendelian manner (see spread 19.10). In the UK population, about one person in 25 is heterozygous for the condition. These people are healthy because the normal allele is dominant and masks the recessive allele. In a marriage between two healthy people, what is the probability of both partners being heterozygous for cystic fibrosis? What is the probability of two healthy people being parents to a child born with cystic fibrosis? (Hint: according to the **multiplication law of probabilities**, the probability of two events A and B happening together is equal to the product of the probability of event A happening and the probability of event B happening, providing the events are independent.)

19.4

OBJECTIVES

By the end of this spread you should be able to:

- explain why monohybrid ratios of 1:2:1 occur
- describe an example of inheritance involving multiple alleles
- discuss the effect of lethal genes on phenotype ratios.

Codominance, multiple alleles, and lethal alleles

Mendel only studied characteristics determined by single genes that had two alleles, one of which was dominant and the other recessive. In this spread, we are going to look at three phenomena not considered by Mendel: codominance, multiple alleles, and lethal alleles.

Codominance

Codominance is shown by some alleles such that the heterozygous condition produces a different phenotype from the homozygous state of either allele. For example, in *Antirrhinum* (snapdragon), two alleles for flower colour are codominant. Homozygous individuals may be either red or white, whereas heterozygous individuals are pink. This is because in the heterozygous individual, both alleles of the pair are expressed in the phenotype. The heterozygous *Antirrhinum* is part-way between the phenotypes of the two types of homozygous individuals, and this is often the case with codominant alleles.

As neither allele is dominant, the different alleles are not represented by upper case and lower case letters. Instead, the gene is given an upper case letter and each allele is represented by a superscript upper case letter. For example, $\mathbf{C^R}$ could represent the allele for red flowers and $\mathbf{C^W}$ the allele for white flowers. The possible genotypes and phenotypes of snapdragon flowers are therefore as shown below:

$\mathbf{C^RC^R}$ – red flower

$\mathbf{C^RC^W}$ – pink flower

$\mathbf{C^WC^W}$ – white flower.

If pollen from a white-flowered plant pollinates a red-flowered plant, the offspring will all have pink flowers. If these flowers are self-crossed, there is a 1 in 4 chance (0.25 probability) that offspring will have red flowers, a 2 in 4 chance (0.5 probability) that they will have pink flowers, and a 1 in 4 chance (0.25 probability) that they will have white flowers (figure 1).

The alleles involved in sickle-cell conditions (spread 16.1), the human ABO blood group system, and coat colour in some breeds of cattle are all examples of characteristics involving codominance.

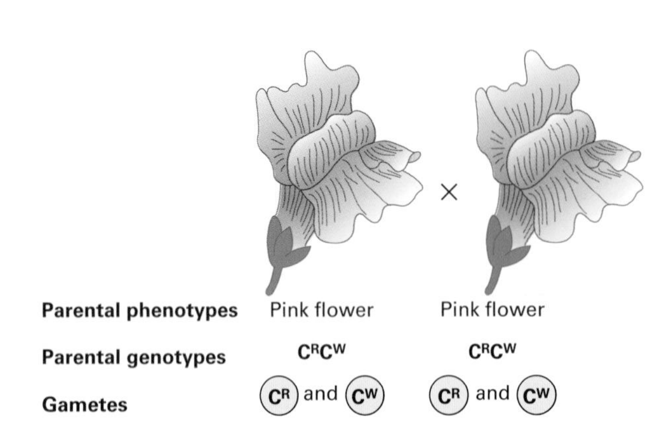

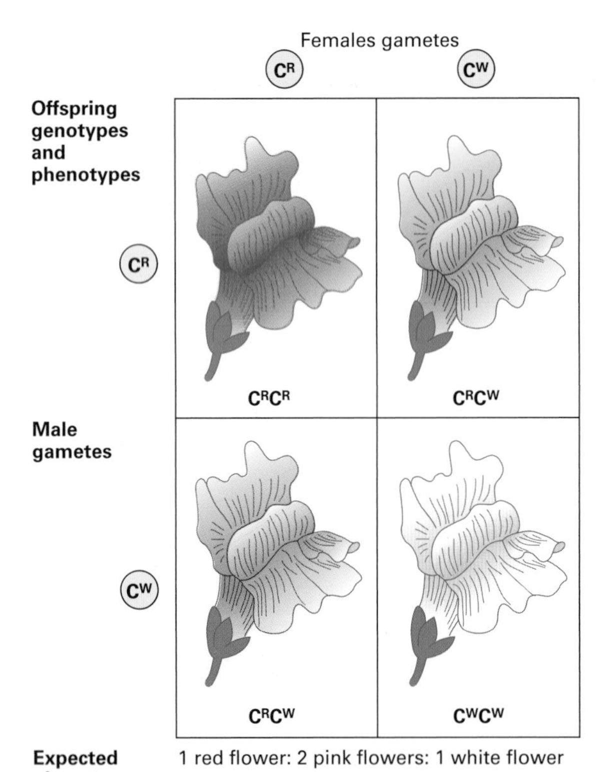

Figure 1 *Codominance of flower colour in snapdragons.*

Fact of life

In addition to the familiar ABO and rhesus blood groups, there are many other blood group systems that have multiple alleles. In the 1970s, more than 303 264 phenotypic combinations could be identified in English people. Before the advent of DNA fingerprinting, identifying the blood groups of a sample of blood was an important way of deciding paternity disputes and eliminating people from police enquiries.

Multiple alleles

The human ABO blood group system is not only an example of codominance, but it is also an example of multiple alleles. **Multiple alleles** are more than two alternative forms of a particular gene that occupy a gene locus. In some cases there are as many as 100 possible alleles for a particular gene, resulting in many slightly different phenotypes for the same trait. Of course, only two alleles can occur in any one diploid organism. By convention, in genetic diagrams the gene is given an upper case letter and each allele a superscript. For example, in the ABO blood group system:

I (for immunoglobulin) represents the gene locus

A, **B**, and **O** represent alleles

thus the alleles are indicated as $\mathbf{I^A}$, $\mathbf{I^B}$, and $\mathbf{I^O}$.

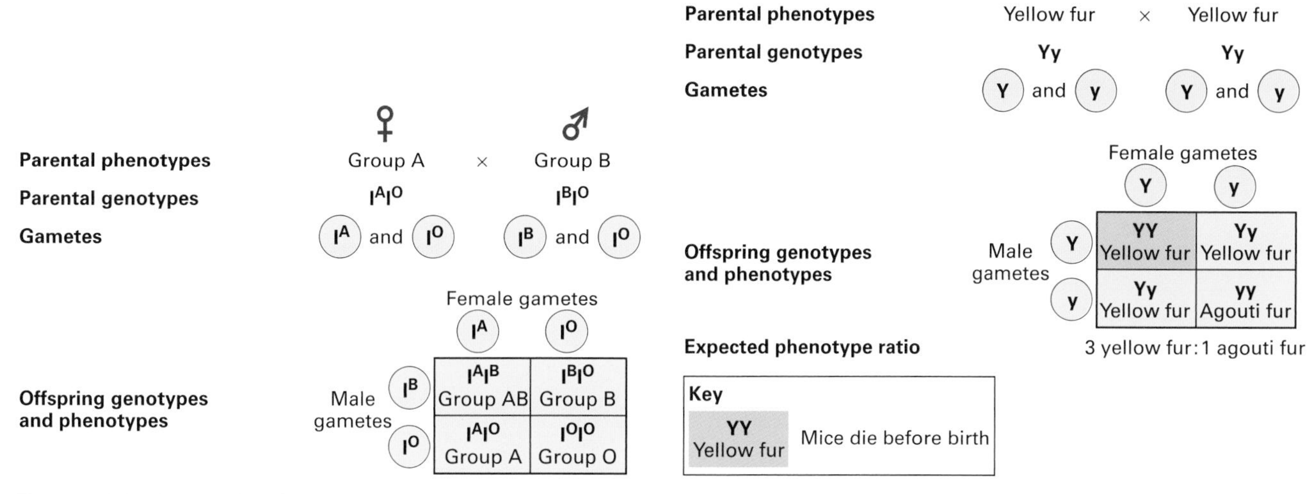

Figure 2 *Inheritance of ABO blood groups in humans: possible offspring of a heterozygous blood-group A mother and a heterozygous blood-group B father.*

Figure 3 *Genetic explanation of the 2:1 ratio of offspring from a monohybrid cross between two yellow mice.*

The alleles $\mathbf{I^A}$ and $\mathbf{I^B}$ are codominant, whereas both these alleles are dominant to $\mathbf{I^O}$. Therefore these three alleles give six possible genotypes and four phenotypes:

- blood group A (genotypes $\mathbf{I^AI^A}$ and $\mathbf{I^AI^O}$)
- blood group AB (genotype $\mathbf{I^AI^B}$)
- blood group B (genotypes $\mathbf{I^BI^B}$ and $\mathbf{I^BI^O}$)
- blood group O (genotype $\mathbf{I^OI^O}$).

Figure 2 shows how a child of a heterozygous mother with blood group A and a heterozygous father with blood group B could have any one of the four blood groups. The significance of the ABO blood group system in blood transfusions is discussed in spread 15.8.

Lethal alleles

There are several conditions where a single gene may affect more than one characteristic, including mortality. One classic example is the gene for fur colour in mice. Wild mice have grey-coloured fur, a condition known as agouti. Occasionally, a mouse with yellow fur is born. A self-cross between mice with yellow fur produces offspring in the ratio of 2 yellow to 1 agouti. These results suggest that the allele for yellow (**Y**) is dominant to the allele for agouti (**y**). However, the ratio of 2:1 is not a typical Mendelian ratio. The simplest explanation is that all live yellow mice are heterozygous, and that homozygotes are either not conceived or die in the uterus. Examination of the uteri of female yellow mice made pregnant by a yellow male revealed dead yellow fetuses. No dead fetuses were found in yellow mice impregnated by agouti mice. These results are consistent with the conclusion that the allele for yellow fur is **lethal** in the homozygous (**YY**) condition (figure 3). So, although the **Y** allele is dominant for fur colour, it is recessive for the lethal characteristic.

Like the **Y** allele in yellow mice, most lethal alleles are recessive for mortality. This means they persist in the population by being passed on in the heterozygous form, only causing problems when two heterozygotes mate and produce a fetus with two lethal alleles. However, dominant lethal alleles can also persist in the population if the individuals carrying the allele survive long enough to reproduce. This is the case with Huntington's disease. Premature death is the ultimate result of the disease, but the disease does not show itself until after the age of 40 or so, allowing plenty of time for the allele to be passed to offspring.

Figure 4 *A Manx cat.*

Quick check

1. Two heterozygotes for a particular condition produce offspring in the expected ratio of 1:2:1. What sort of inheritance does this gene show?
2. Give the possible genotypes of a person belonging to blood group B.
3. Why are most lethal alleles recessive?

Food for thought

Manx cats (figure 4) are noted for their lack of tail. Suggest an explanation for the following observations:

- When a Manx cat is mated with a normal long-tailed cat, approximately half the offspring will be long-tailed and approximately half will be Manx.
- When two Manx cats are mated, the ratio of offspring is 2 Manx cats to 1 long-tailed cat.

19.5

Dihybrid inheritance

OBJECTIVES

By the end of this spread you should be able to:

- give a genetic explanation of Mendelian dihybrid inheritance
- explain the use of test crosses to determine unknown genotypes in studies of dihybrid inheritance
- explain the significance of recombination.

Fact of life

Mendel would have had a problem interpreting the genetics of coat colour in Labrador retrievers (figure 1) because this is controlled by two genes, not one. Labrador retrievers have three possible coat colours (black, yellow, and chocolate or liver) controlled by two sets of genes. One set determines whether the retriever will be dark (either black or chocolate) or light (yellow). Dark is dominant to light. The second set comes into play only if the dog is dark. This set determines whether the dog is black (the dominant trait) or chocolate (the recessive trait). Labrador retrievers were bred originally as water dogs, trained to haul cod nets ashore and retrieve items lost overboard.

Figure 1 *Labrador retriever pups: the coat colour is determined by the interaction of two genes.*

Table 1 *Phenotypes and genotypes in Mendel's dihybrid cross.*

Phenotype	Genotypes
Tall, purple-flowered	**TTPP TTPp TtPP TtPp**
Tall, white-flowered	**TTpp Ttpp**
Dwarf, purple-flowered	**ttPP ttPp**
Dwarf, white-flowered	**ttpp**

The inheritance of two characteristics

Dihybrid inheritance is the inheritance of two characteristics, each controlled by a different gene at a different locus. In one experiment, summarised in spread 19.2, figure 3, Mendel studied dihybrid inheritance by crossing plants from two pure-breeding strains: one tall with purple flowers, the other dwarf with white flowers. All the offspring in the F_1 generation were tall with purple flowers, these being the dominant characteristics. The F_1 generation were self-crossed, producing the following phenotypes and ratios in the F_2 generation:

- 9 tall purple-flowered
- 3 tall white-flowered
- 3 dwarf purple-flowered
- 1 dwarf white-flowered.

Mendel observed that two phenotypes resembled one or other of the parents, and two phenotypes had combined the characteristics of both parents. He also observed that the ratio of tall plants to dwarf plants was 3:1, and that the ratio of purple-flowered plants to white-flowered plants was 3:1. This was the same ratio that occurred in the monohybrid crosses. He concluded from these results that the two pairs of characteristics behave quite independently of each other. This led him to formulate his law of independent assortment, which states that any one of a pair of characteristics may combine with any one of another pair.

Interpreting the results of a dihybrid cross

Mendel's results can be explained in terms of alleles and the behaviour of chromosomes during meiosis (figure 2). Notice that the two alleles for one gene are always written together (for example, **TtPp**, *not* **TPtp**). This makes it easier to interpret the crosses. The pure-breeding adult plants, being diploid, have two alleles for each gene. The genes for height and flower colour are carried on separate chromosomes (we shall see in spread 19.6 what happens when two genes are carried on the same chromosome). During gamete formation, meiosis occurs, producing gametes containing one allele for each gene (figure 3). In the F_1 generation, the only possible genotype is **TtPp**. When these plants are self-crossed, there are four possible combinations of alleles in both the female and male gametes: **TP**, **Tp**, **tP**, and **tp**. Assuming fertilisation is random, any male gamete can fuse with any female gamete, so there are 16 possible combinations for the offspring, as shown in the Punnett square. These combinations can produce four different phenotypes from nine genotypes (table 1).

The only genotype that can be worked out simply by looking at the plants is that of the dwarf white-flowered plants. Genotypes of the other plants can be established by test crosses. The result of one such test cross is shown in figure 4.

Recombination

As already stated, two of the phenotypes in the F_2 resemble the original parents (tall purple-flowered and short white-flowered), and two show new combinations of characteristics (tall white-flowered and dwarf purple-flowered). This process that results in new combinations of characteristics is called **recombination**, and the individuals that have the new combinations are known as **recombinants**. Recombination is an important source of genetic variation, contributing to the differences between individuals in a natural population.

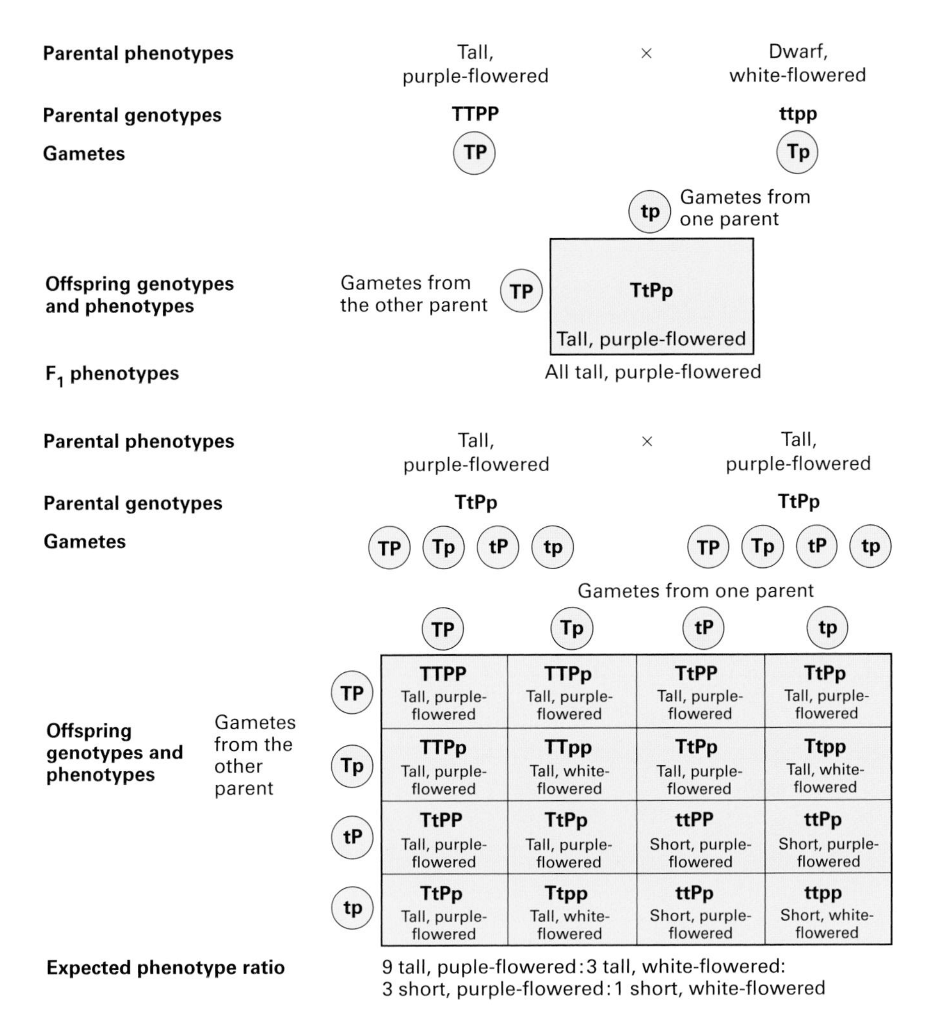

	TP	Tp	tP	tp
TP	TTPP Tall, purple-flowered	TTPp Tall, purple-flowered	TtPP Tall, purple-flowered	TtPp Tall, purple-flowered
Tp	TTPp Tall, purple-flowered	TTpp Tall, white-flowered	TtPp Tall, purple-flowered	Ttpp Tall, white-flowered
tP	TtPP Tall, purple-flowered	TtPp Tall, purple-flowered	ttPP Short, purple-flowered	ttPp Short, purple-flowered
tp	TtPp Tall, purple-flowered	Ttpp Tall, white-flowered	ttPp Short, purple-flowered	ttpp Short, white-flowered

Grids or boxes such as this which show all the possible genotypes that can result from the random fusion of gametes are called **Punnett squares** after R. C. Punnett, the professor of genetics who devised them. Note that the genotypes indicated are only *possible* combinations. The likelihood of obtaining all possible genotypes depends on the number of offspring produced. Also, the actual ratio of genotypes is likely to match the predicted ratio only where the number of offspring produced is high.

Figure 2 *Genetic explanation of Mendel's cross between pure-breeding tall pea plants with purple flowers and dwarf pea plants with white flowers. After self-crossing the F_1 plants, the ratio of phenotypes in the F_2 generation is 9:3:3:1.*

Quick check

1 Show the types of gametes produced by a plant with round seeds and yellow cotyledons, and the genotype **RrYy**.

2 In figure 4, why is a dwarf white-flowered plant used in the test cross?

3 What are recombinants?

Food for thought

Two different genes may affect the same characteristic. In some cases, the effect of one gene is suppressed or masked by another **non-allelic** gene. This phenomenon is called **epistasis.** Coat colour in some mice is an example of epistasis. The colours of agouti and black mice are determined by a gene with alleles **A** and **a**. This gene codes for the distribution of melanin pigment in the hairs. The coat of agouti mice is made up of banded hairs, producing a grey-brown coloured coat. The dominant allele **A** codes for the presence of this banding. The recessive allele **a** codes for the uniform black colour of a black mouse.

A second gene with alleles **C** and **c** determines the production of melanin. The dominant allele **C** is required for colour of any type to develop in the coat. A mouse with the genotype **cc** does not make melanin and is therefore albino. What would be the expected result of a dihybrid cross between two agouti mice with the genotypes **AaCc**?

Diploid cell undergoing meiosis
Homologous chromosomes
Segregation of homologous chromosomes during meiosis
Independent assortment of alleles to produce gametes with four different genotypes

Figure 3 *Homologous chromosomes separate during meiosis, leading to the independent assortment of alleles carried on separate chromosomes.*

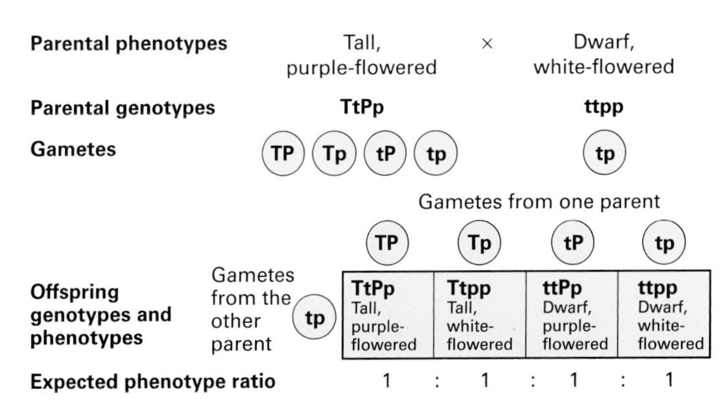

	TP	Tp	tP	tp
tp	TtPp Tall, purple-flowered	Ttpp Tall, white-flowered	ttPp Dwarf, purple-flowered	ttpp Dwarf, white-flowered

Figure 4 *Test cross in which the unknown plant turns out to be a dihybrid (that is, it is heterozygous for both genes).*

19.6

LINKAGE AND CROSSING OVER

OBJECTIVES

By the end of this spread you should be able to:

- explain why linked genes do not show independent assortment
- explain how recombinants form when linkage is partial
- explain how cross over values can be used to make a chromosome map.

Fact of life

The multinational **Human Genome Project** (HGP) was the largest biological investigation ever undertaken. Begun in 1990, in 2003 it fulfilled its main aim to determine the sequences of the 3 billion nucleotide base pairs that make up human DNA and to store this information in databases. The sequencing required the use of many high-powered computers programmed with sophisticated software and operated by specially trained computer analysts. Work continues on the Project's other aims: to identify all the 20 000–25 000 genes on the 24 different chromosomes (surprisingly, researchers found that our genome has far fewer genes than expected at the start of the HGP); to improve tools for data analysis; to transfer related DNA technologies to the private sector (e.g. pharmaceutical companies); and to address the ethical, legal, and social issues that may arise from the project. Nevertheless, the results of the HGP are already affecting us. As well as providing us with valuable insights into embryonic development and human evolution (spread 20.12), they are also helping us understand, diagnose, treat, and prevent diseases with a significant genetic component. For example, defects on chromosome 1, the largest human chromosome, are probably involved in more than 350 diseases, including Parkinson's disease and Alzheimer's disease, as well as some cancers.

Mendel's law of independent assortment, describing how two different pairs of characteristics are inherited independently of each other, was made on the basis of experiments involving genes carried on different chromosomes. These genes separate at meiosis, allowing their alleles to undergo independent assortment (spread 19.5, figure 3). However, the alleles of different genes cannot assort independently if the genes are on the same chromosome. Such genes are said to be **linked** because they tend to be inherited together. All the genes on the same chromosome are said to form a **linkage group**.

Bateson and Punnett's experiments with sweet peas

The fact that some characteristics are not inherited in a simple Mendelian manner was realised soon after the rediscovery of Mendel's work in the early 1900s. In an effort to check Mendel's results, geneticists carried out many different dihybrid crosses. Some gave the 9:3:3:1 pattern predicted by the law of independent assortment, but others did not give any simple ratio. One such experiment was conducted by William Bateson and Reginald Punnett (the professor of genetics who devised the Punnett square). They used the sweet pea (*Lathyrus odoratus*), a different species from the garden pea used by Mendel. They crossed pure-breeding sweet pea plants which had purple flowers and long pollen grains with plants having red flowers and short pollen grains. All the F_1 plants were purple-flowered with long pollen, indicating that the allele for purple flower was dominant to the allele for red flower, and the allele for long pollen was dominant to that for short pollen. So far, these results were as expected from simple Mendelian inheritance. However, when the F_1 plants were self-crossed, most of the F_2 offspring resembled the parental phenotypes, with a small number of recombinants. The results of Bateson and Punnett's experiment are shown in table 1.

Although there is no simple Mendelian ratio between the four different genotypes, Bateson and Punnett found that the ratio of the total number of purple-flowered (315) to red-flowered plants (112) was approximately 3:1, and the ratio of plants with long pollen (323) to plants with short pollen (104) was also approximately 3:1. From this and similar results, geneticists concluded that certain characteristics are inherited together.

Recombinants and crossing over

We now know that Bateson and Punnett's results are due to the genes for flower colour and pollen shape in sweet peas being carried on the same chromosome. The genes are therefore linked, and tend not to segregate

Table 1 *Phenotypes of Bateson and Punnett's F_2 sweet peas. (The results expected from a Mendelian ratio of 9:3:3:1 are given to the nearest whole figure.)*

Phenotype	Actual results	Expected results (with ratios)
Purple flower, long pollen	296	240 (9)
Purple flower, short pollen	19	80 (3)
Red flower, long pollen	27	80 (3)
Red flower, short pollen	85	27 (1)

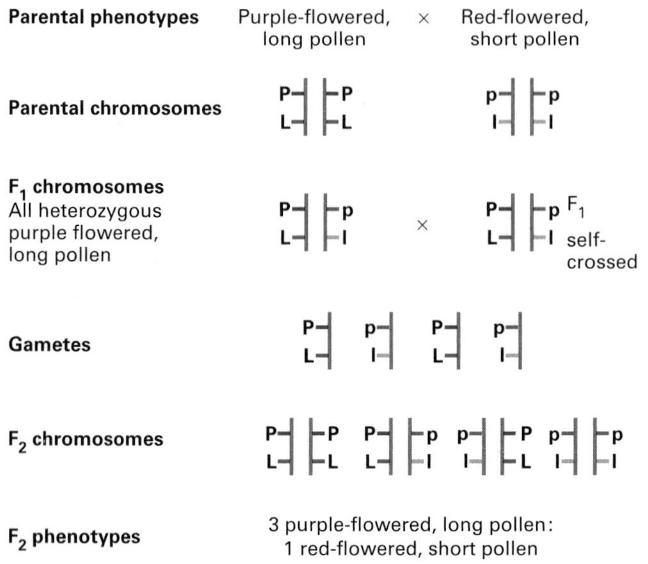

Figure 1 *The gene for flower colour and the gene for pollen shape are linked, which explains why most of the offspring in Bateson and Punnett's experiment were parental types.*

Key
P allele for purple flower colour
p allele for red flower colour
L allele for long pollen
l allele for short pollen

Parental phenotypes Purple-flowered, long pollen × Red-flowered, short pollen

Parental chromosomes P L P L   p l p l

F_1 chromosomes All heterozygous purple flowered, long pollen: P L p l × P L p l F_1 self-crossed

Gametes P L, p l, P L, p l

F_2 chromosomes P L P L, P L p l, p l P L, p l p l

F_2 phenotypes 3 purple-flowered, long pollen: 1 red-flowered, short pollen

during meiosis. Figure 1 shows what would happen if linkage were complete: the F_2 generation would contain only parental types (purple-flowered plants with long pollen, and red-flowered plants with short pollen). The occurrence of recombinants shows that linkage was not complete. These recombinants can be accounted for by the crossing over that can occur during prophase I of meiosis (figure 2).

During meiosis, at least one chiasma forms between two homologous chromosomes. However, it may not form between a particular pair of genes (figure 3). The further apart the genes are, the more likely it is that crossing over will result in the formation of recombinants.

The proportion of recombinants resulting from a dihybrid test cross (see spread 19.5, figure 4) is used to calculate the **cross over value** (**COV**), a measure of linkage, and, if linkage occurs, the distance between genes.

$$\text{COV} = \frac{\text{total number of recombinants}}{\text{total number of offspring}} \times 100$$

If the genes are not linked, the expected phenotype ratio of such a cross is 1:1:1:1 and, as there is a 50 per cent chance that alleles on separate chromosomes will be inherited together, the expected COV is 50 per cent. Linkage results in the COV being significantly less than 50 per cent; the lower the value, the closer the genes. Thus the COV can be used to locate the relative positions of genes on chromosomes, a process called **chromosome mapping** or genetic mapping. By convention, one per cent COV is equivalent to one map unit (figure 4).

QUICK CHECK

1. What is a linkage group?
2. By what process do recombinants form when genes are partially linked?
3. Calculate the cross over value between the genes for flower colour and pollen shape in sweet peas using the results given in table 1.

Food for thought

About 5 per cent of the budget of the HGP was spent on addressing its ethical, legal, and social issues (ELSI). This represents the world's largest bioethics programme. The overarching aim is to maximise the benefits of the information gained from the HGP while minimising the potential for social, environmental, and personal harm. One of the ethical concerns being addressed is fairness in the use of the genetic information. Suggest how the information about a person's genome might affect his or her ability to obtain life insurance. Who should have access to personal genetic information and how should it be used?

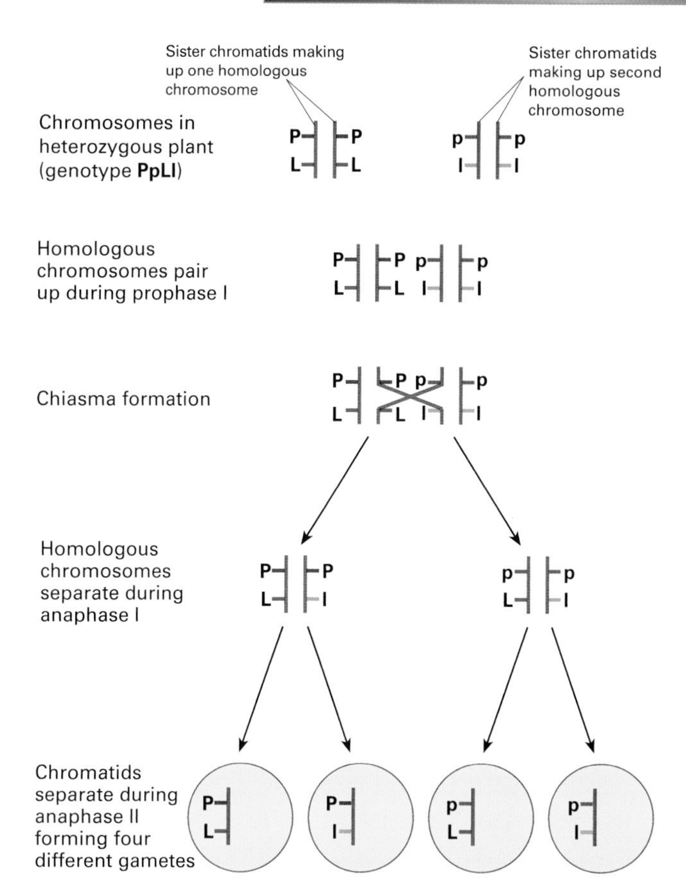

Figure 2 *Formation of recombinants by crossing over during meiosis.*

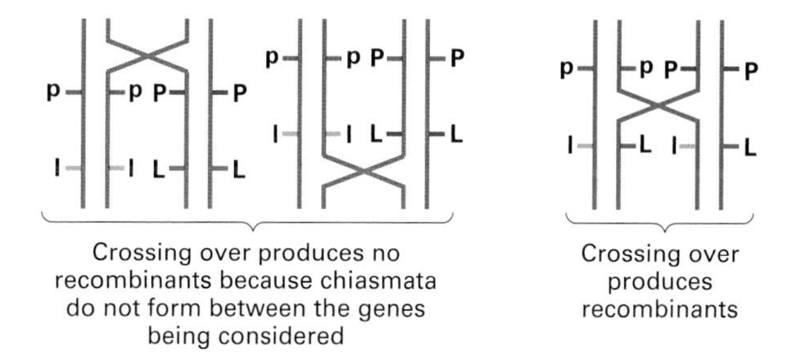

Figure 3 *The position of chiasma formation determines whether recombinants occur.*

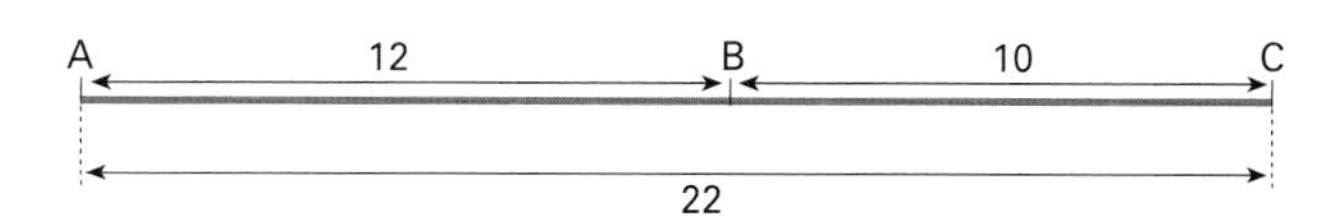

Figure 4 *The positions of three genes on a chromosome. In the diagram, the results of breeding experiments showed that A and B have a COV of 12% and are 12 units apart; A and C have a COV of 22% and are 22 units apart; and B and C have a COV of 10% and are 10 units apart. This indicates that the genes are arranged in the order A, B, C. (If the order were A, C, B, then A and B would be 32 units apart.) The numbers refer to map units.*

19.7

Sex determination

OBJECTIVES

By the end of this spread you should be able to:

- explain how sex is determined in humans
- discuss the role of the sex related Y gene in determining sex
- describe how non-disjunction can affect the distribution of sex chromosomes in gametes and offspring.

Fact of life

One of the most fascinating marine animals is the slipper limpet, a mollusc with the intriguing scientific name of *Crepidula fornicata* (figure 1). It was given this name because it has the surprising ability to change its sex. The limpets are immobile for most of their lives, growing in chains. The sex of each limpet depends on its size and its position in the chain. The young, small individuals are males, with long tapering penises which fertilise females lower in the chain. In due course, when a male has grown and has been settled on by another smaller limpet, the male loses its penis and grows into a female. Thus large females occur at the base of the chain, with animals changing sex above them, and males at the apex. In this way, the limpets have been able to combine immobility with internal fertilisation.

Figure 1 *A chain of slipper limpets,* Crepidula fornicata.

Sex chromosomes

There have been many weird and wonderful ideas about sex determination in humans. Some Ancient Greeks thought that the sex of a baby was determined by which testicle the sperm came from. Apparently, this belief was adopted by some European kings who tied off or removed their left testicle to ensure a male heir to the throne. Other people believed that the sex of a baby could be controlled by conceiving when the Moon was in a particular phase, when the wind was blowing in a certain direction, or whilst speaking certain words. We now know that human sex is determined by a pair of **sex chromosomes** called **X** and **Y**. Because these chromosomes do not look alike, they are sometimes called **heterosomes**. All other chromosomes are called **autosomes**. Females have two X chromosomes (XX). Males have one X and one Y chromosome (XY). Although the sex chromosomes determine the sex of an individual, it is important to realise that they do not carry all the genes responsible for the development of sexual characteristics.

During meiosis, the sex chromosomes pair up and segregate into the daughter cells. Males are called the **heterogametic sex** because they produce different sperm: approximately 50% contain an X chromosome and 50% have a Y chromosome. Females are called the **homogametic sex** because (usually) all of their eggs contain an X chromosome. This arrangement applies to all mammals and some insects (including *Drosophila*, the fruit fly commonly used in genetic experiments). However, in birds, moths, and butterflies, females are the heterogametic sex with the XY genotype (or XO, meaning the second sex chromosome may be absent). In some species, sex determination depends on a complex interaction between sex chromosomes and autosomes, or between inherited factors and environmental ones. The sex of some turtles, for example, depends on the temperature of the sand in which eggs are laid: those laid in sand warmed by the Sun develop into females; those laid in cool sand in the shade develop into males.

In humans, the father's sperm determines the sex of the baby: if a baby inherits a Y chromosome from its father it will be a boy; if it inherits an X chromosome from its father it will be a girl. So the sex of a baby depends on which sperm fertilises the egg cell: a sperm with an X chromosome or one with a Y chromosome (figure 2). However, there are cases where having a Y chromosome does not necessarily mean that an embryo will become a boy.

The SRY gene

In the early stages of development, human embryos have no external genitalia. Whether they develop testes or ovaries depends on the presence and activity of a particular gene on the Y chromosome. This gene, called the **sex related Y gene** (**SRY gene**), was discovered when geneticists were studying some interesting people: men who had two X chromosomes and women who had one X and one Y chromosome. Microscopical examination of the sex chromosomes of these people revealed that the XX males had a very small piece of Y chromosome in their X chromosomes, whereas this piece was missing from the Y chromosome of the XY females. The geneticists found the SRY gene within this small piece of Y chromosome (figure 3).

The SRY gene codes for a protein called **testis determining factor**. This switches on other genes, causing the embryo to develop male structures. The testes develop and androgens (hormones which promote the development of male sexual organs and secondary sexual characteristics) are secreted. At about 16 weeks, an embryo with the SRY gene begins to produce immature sperm. In addition to stimulating male structures to grow, SRY suppresses the development of female structures by activating

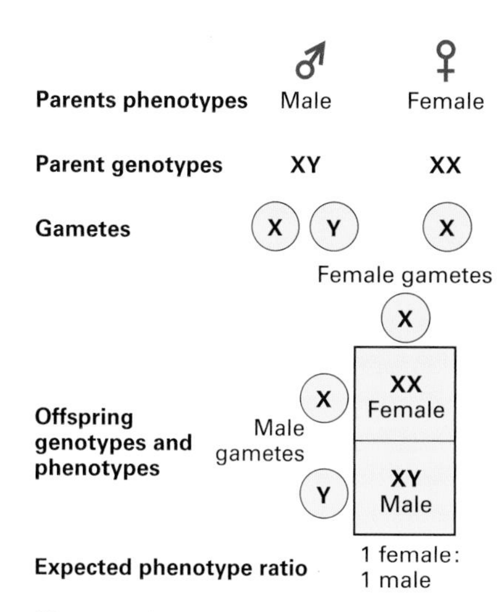

Figure 2 Sex determination in humans. A normal female has two X chromosomes, and males have one X and one Y chromosome.

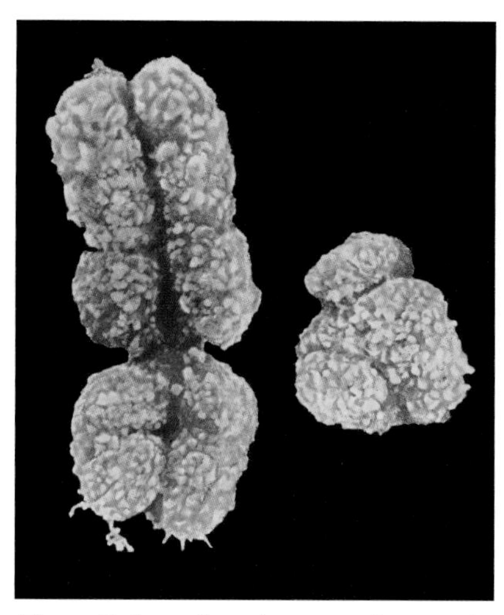

Figure 3 Scanning electron micrograph of human X and Y chromosomes (×10 000). The sex related Y gene occurs in the top of the short arm of the Y chromosome.

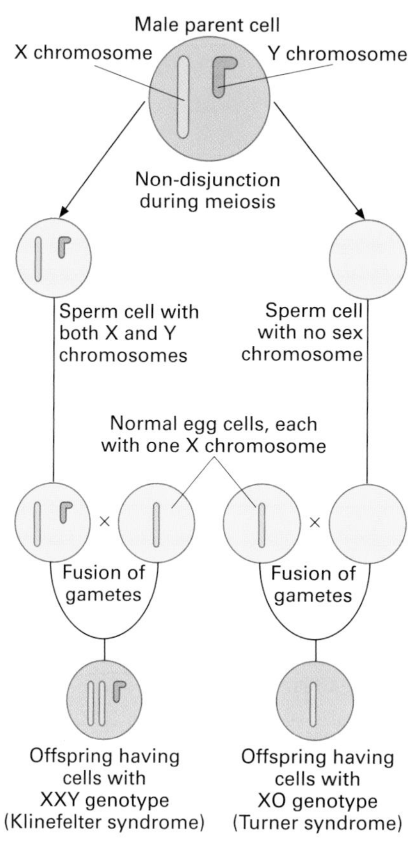

Figure 4 How Klinefelter syndrome and Turner syndrome may arise from non-disjunction of the X and Y chromosomes during sperm formation.

a gene on chromosome 19. This activation leads to the production of a protein called **Müllerian-inhibiting substance**, which destroys female structures early in their development. Lack of testis determining factor results in the development of female genital organs. Therefore, all embryos are female unless active testis determining factor makes them male.

Sex testing

The governing bodies of all-female sports sometimes use sex tests to make sure participants in their sports are female. The first attempts at gender verification were by the International Amateur Athletic Federation, whose sex test included parading naked female athletes before a panel of male doctors. In 1968 this rather dubious procedure was dropped, and the International Olympic Committee adopted the **Barr test**. This test uses the presence of stainable particles called **Barr bodies** as sex indicators. Barr bodies occur in epithelial cells in the mouth (buccal epithelial cells), and are thought to be derived from inactive X chromosomes. Females therefore usually have one Barr body in their buccal epithelial cells and males usually have none. At the 1992 Barcelona Olympics, the Barr test was replaced by the polymerase reaction test. In this test, the **polymerase** chain reaction (spread 18.11) replicates the SRY gene, so that if it is present it can be detected.

Sex testing is complicated by the fact that, on rare occasions, sex chromosomes fail to segregate at meiosis. This phenomenon, known as **non-disjunction**, can result in a sperm cell either having both an X and a Y chromosome or having no sex chromosome (figure 4), and an egg cell either having two X chromosomes or having no sex chromosome. Non-disjunction can lead to unusual genotypes (table 1). Sex testing is confused even further by the occurrence of **chimaeras**. A chimaera is any animal or plant consisting of some cells with one genetic constitution and some with another. Very rarely, chimaera formation can occur during the early stages of embryonic development when chromosomes in mitotically dividing cells fail to segregate properly (for example, some cells can have the genotype XXX, others XO, while the majority are XX!).

Table 1 *Some sex chromosome abnormalities.*

Genotype	Syndrome	Some typical characteristics
XO	Turner	Females: lack of ovaries; no sexual maturity
XXY	Klinefelter	Males: no secondary sexual characteristics
XXX	Triple-X	Females: tall; menstrual irregularities
XYY	Jacob	Males: 96% apparently normal

Quick check

1. Why are males called the heterogametic sex?
2. Explain why an embryo with an XY genotype may develop female sexual organs.
3. Explain why a person may have buccal epithelial cells with two Barr bodies.

Food for thought

Human sperm that contain a Y chromosome usually give rise to male babies, and sperm carrying an X chromosome usually give rise to female babies. Assuming random fusion of gametes, 50% of babies should be boys, and 50% girls. However, this is not the case. Suggest why the proportion of males to females varies in the following manner: there are about 114 boys conceived for every 100 girls; at birth there are 106 boy babies for every 100 girl babies; at puberty the proportions of males and females are equal; and in old age, females outnumber males.

19.8

Sex linkage

OBJECTIVES

By the end of this spread you should be able to:

- define sex linkage
- explain how haemophilia and colour blindness are inherited
- distinguish between sex-linked characters and sex-limited characters.

Fact of life

Haemophiliacs in the UK have until recently been treated with Factor VIII made from pooled blood donations. Many of them have become infected with hepatitis C and HIV as a result. Genetically engineered Factor VIII is now to be used instead, which is more expensive. To reduce costs, most haemophiliacs are given Factor VIII only when they bleed.

Conventions used to describe the genetics of sex linkage

In sex linkage studies, the sex chromosomes XX and XY are always shown and the dominant or recessive alleles of particular genes are represented by superscript upper and lower case letters. Thus in the transmission of haemophilia, $\mathbf{X^HX^h}$ would represent a female carrier. A cross between a female carrier of haemophilia and a 'normal' male would be written as shown in figure 1.

Genes that are located on one or other of the sex chromosomes are said to be **sex linked**. Sex linkage results in an inheritance pattern different from that shown by genes carried on autosomes (chromosomes other than the sex chromosomes). Sex-linked genes may be on the X chromosome (X linkage) or the Y chromosome (Y linkage).

X linkage

Because the X chromosome is much larger than the Y chromosome, most sex-linked genes are located on the X chromosome. In addition to carrying genes that are concerned with the determination of female sex characteristics, the X chromosome also carries genes for non-sexual characteristics such as the ability to see particular colours and the ability to clot blood efficiently.

In X linkage, a recessive allele is expressed more often in males than in females because males carry only one allele for an X-linked gene whereas females have two alleles for the gene. Thus, a female with a recessive allele on one of her X chromosomes will not be affected by it, just as someone who is heterozygous for a recessive allele on one of the autosomes is unaffected by it. However, a male with a recessive allele on his X chromosome has no dominant allele to counteract it, because his Y chromosome lacks the corresponding gene. The phenotype of the male will therefore show the presence of a single recessive allele.

In animals such as *Drosophila* which are used in genetics experiments, X linkage is distinguished from autosomal linkage by making reciprocal crosses. In a reciprocal cross, parents are chosen such that the new male parent has the same phenotype as the female parent in the original cross, and the new female parent has the same phenotype as the original male parent. In autosomal linkage, reciprocal crosses produce similar results to the original cross. By contrast, in X linkage, reciprocal crosses produce different results. Thus in the inheritance of X-linked eye colour in which the allele for red eye is dominant to the allele for white eye, a cross between a male with red eyes and a female with white eyes produces 50% red-eyed males and 50% white-eyed females, whereas in a cross between a male with white eyes and a female with red eyes, all the offspring (male and female) are red eyed.

Haemophilia: an X-linked character

Haemophilia is a condition in which the blood does not clot normally. Haemophilia often results in excessive bleeding, both internally and externally. The condition is due to the lack of one or more clotting factors, which play an important role in the complex chain reaction of blood clotting (spread 7.8). Haemophilia A individuals (the most common type of haemophilia) lack **Factor VIII** (also called **antihaemophiliac globulin**, AHG). Haemophilia A individuals can lead a normal and active life when treated with regular injections of Factor VIII.

Haemophilia A is a sex-linked character caused by a recessive allele carried on the X chromosome. Females have a pair of alleles for the gene that controls the production of Factor VIII, but males have only one. Therefore, if a male inherits one allele for haemophilia A, he has the disease because he cannot possess another allele which might mask its effects. On the other hand, a female has the disease only if she inherits two recessive alleles, one from each parent. Heterozygous females are **carriers** of the disease: although they have normal blood, one of

Parental phenotypes   Female carrier × Normal male

Parental genotypes X^HX^h X^HY

Gametes (X^H) and (X^h) (X^H) and (Y)

Offspring genotypes and phenotypes

Male gametes \ Female gametes	X^H	X^h
X^H	X^HX^H Normal female	X^HX^h Female carrier
Y	X^HY Normal male	X^hY Haemophiliac male

***Figure 1** The inheritance of haemophilia.*

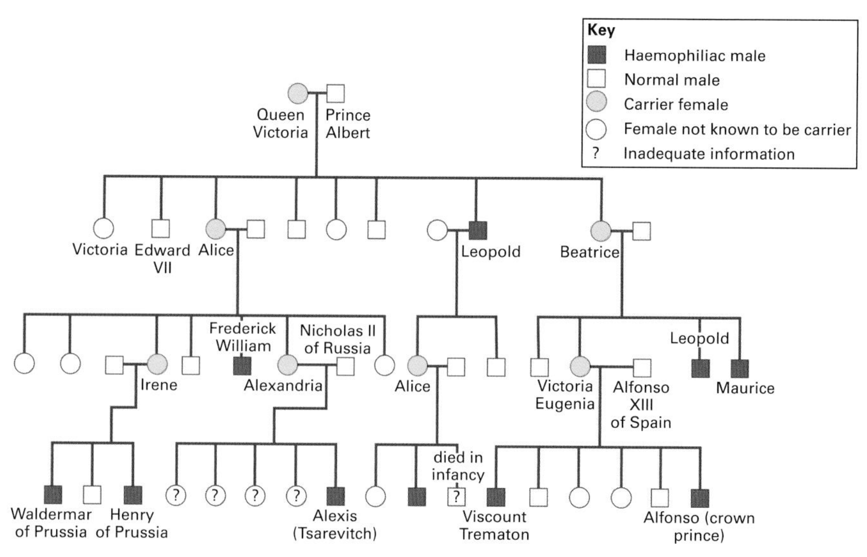

Figure 2 A pedigree showing the inheritance of haemophilia in Queen Victoria's family.

their X chromosomes carries the harmful recessive allele. Therefore, there is a 50 per cent chance of a carrier mother transmitting the recessive allele to any one of her sons (figure 1).

The inheritance of haemophilia within a family can be shown in a **pedigree** (a chart of the ancestral history of a group of related individuals). One of the best established pedigrees for haemophilia is that of Queen Victoria's family (figure 2). By convention:

- circles represent females; squares represent males
- shaded circles or squares represent affected individuals; unshaded circles or squares represent unaffected individuals
- two parents are linked by a horizontal line joining a circle and a square
- vertical lines run down from parents to children
- the children in one family are linked by a horizontal line above them.

Sex-limited characters

Sex-limited characters should not be confused with sex-linked characters. Usually, sex-limited characters are expressed in only one or other of the sexes, even though both sexes may carry genes for such characters. Sex-linked characters such as haemophilia which is X linked may occur in both males and females, though far more males show the condition than females.

Sex-limited characters include lactation in female mammals and facial hair in male humans. Sex-limited genes may be carried on autosomes and are strongly affected by the level and types of sex hormones present in the body (see spread 19.7).

Quick check

1 Why is X linkage more common than Y linkage in humans?

2 What is the probability of a haemophiliac father and a mother carrying the allele for haemophilia having a haemophiliac daughter?

3 Distinguish between sex linkage and sex limitation.

Food for thought

One much-quoted example of Y linkage is very hairy ears, a condition found among a high percentage of men from certain parts of Israel, India, and Pakistan. This not a clear example of Y linkage because sons do not *always* inherit the condition from their fathers. However, some geneticists argue that this might be due to **incomplete penetrance**; an allele is not always expressed, possibly because of the inhibitory effects of other genes. Suggest any alternative genetic explanations for hairy ears.

Y linkage

In humans, the Y chromosome is very short and does not carry many genes. In theory, any unusual character determined by a gene on the Y chromosome which has no corresponding gene on the X chromosome would give rise to a very distinctive family tree. Because a man always passes on his Y chromosome to his sons, all the males should be affected by an abnormal allele on the Y chromosome.

Early textbooks often cited 'porcupine men' as examples of Y linkage. These men had strikingly thick, spiny, and dark skin. The first affected man was born in Suffolk in 1716. He was reported to have six sons all affected with the condition, and for four generations all his male descendants, but no females, were similarly affected. However, a recent detailed examination of the parish records indicate that there were almost certainly unaffected male descendants, and even some affected females. This condition was probably due to a dominant gene carried on an autosomal chromosome. There are no undisputed examples of Y-linked inheritance in humans, although some species of fish and mice show examples of Y linkage.

Colour blindness

Colour blindness refers to any condition in which colours cannot be distinguished. The most common form is red–green colour blindness, in which a person cannot distinguish between red and green. This is caused by a recessive allele carried on the X chromosome. The pattern of inheritance of red–green colour blindness is therefore similar to that of haemophilia: men are more likely to show the defect (6–8% of Caucasian males are red–green colour blind, but less than 0.5% of women have this condition).

19.9

Down's syndrome, genetic screening, and diagnostic testing

OBJECTIVES

By the end of this spread you should be able to:

- explain how Down's syndrome arises
- compare the main features of amniocentesis and chorionic villus sampling
- discuss the role of a genetic counsellor.

Fact of life

Mice are relatively small, easy to breed in the laboratory, and have many genes similar to our own. Therefore, laboratory mice are being used as animal models for studying the physiological roles of genes that have their nucleotides sequenced but whose functions are not known. **'Knockout' mice** have been genetically engineered to study the effects of inactivating a gene or replacing it with an artificial piece of DNA. **'Knockin' mice** have been genetically engineered to study the effects of adding a gene.

Advances in DNA technology have brought a new era in preventative medicine. We can now detect a large range of inherited diseases before birth, one of the most common of which is Down's syndrome.

Down's syndrome: trisomy 21

Down's syndrome is the most common single cause of learning disability in children of school age. Children with the syndrome typically have a round, flat face, and eyelids that appear to slant upwards (figure 1). In addition to some learning disability, they also have an increased risk of infection (particularly respiratory and ear infections), and heart defects occur in about one-quarter of those with the syndrome.

Figure 1 A young boy with Down's syndrome.

The syndrome is named after John Langdon Down, a nineteenth century doctor who first described the condition in 1866. In 1959, the French physician Lejeune used chromosome-staining techniques to show that Down's syndrome is caused by an extra chromosome 21 (figure 2). Having one extra chromosome is known as **trisomy**, hence Down's syndrome is also known as **trisomy 21**. The extra chromosome usually comes from the egg cell due to non-disjunction of chromosome 21. About 70% of the non-disjunctions occur during meiosis I, when homologous chromosomes fail to separate; 30% occur during meiosis II, when sister chromatids fail to separate. Whether it occurs during meiosis I or meiosis II, non-disjunction leads to trisomy (figure 3). In a few cases, the extra chromosome comes from the sperm of the father (that is, a sperm that contains two chromosomes 21).

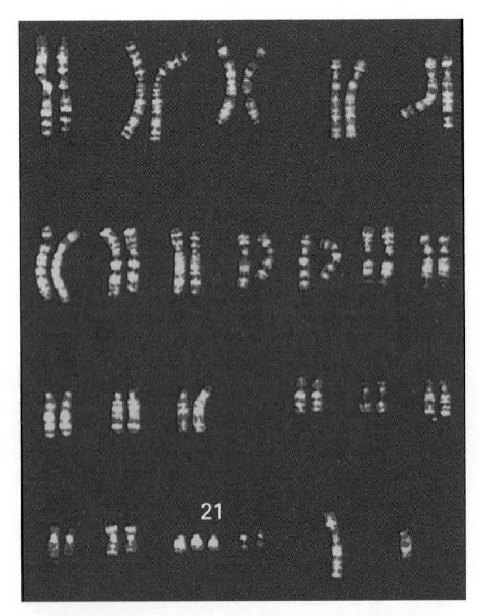

Figure 2 The chromosomes of a boy with Down's syndrome. Non-disjunction has resulted in three chromosomes 21.

In about 3% of cases, Down's syndrome results from translocation of an extra chromosome 21. A region of the chromosome breaks off and rejoins with either the end of the other chromosome 21 or with another non-homologous chromosome (commonly chromosome 15). In these cases, a person may have the normal number of chromosomes, but one of the chromosomes will be abnormally long.

Screening and diagnostic testing

Until relatively recently, a mother's age was the only factor available to assess the risk of a baby having Down's syndrome. Now testing blood using ultrasound scans and biochemical markers has become an important part of screening. For example, a pregnant woman with a high chance of having a baby with Down's syndrome tends to have about twice as much chorionic gonadotrophin (a sex hormone produced by the placenta) in her blood serum as women with normal pregnancies. Screening tests do not show with certainty the presence or absence of a Down's baby, but they can be used in conjunction with the mother's age and medical history to predict the probability of her baby having Down's syndrome. The results are given as a chance, for example one in 1000. This represents a relatively low risk; it means that one in 1000 pregnant women with that result will have a baby with Down's syndrome. If the risk is high, the expectant mother can then decide whether or not to have a diagnostic test. Two of the main diagnostic tests for Down's syndrome are amniocentesis and chorionic villus sampling.

Amniocentesis is usually carried out at 15–16 weeks of pregnancy. It involves passing a very fine needle into the uterus, observed with an ultrasound image, and withdrawing a sample of amniotic fluid containing fetal cells (figure 4). The karyotype (the chromosomes cut out from a picture and arranged in pairs, spread 4.10) of the fetal cells is then analysed to test for Down's syndrome. The fetal cells can also be cultured in a suitable medium in a laboratory so that further tests, such as DNA analysis, can be carried out.

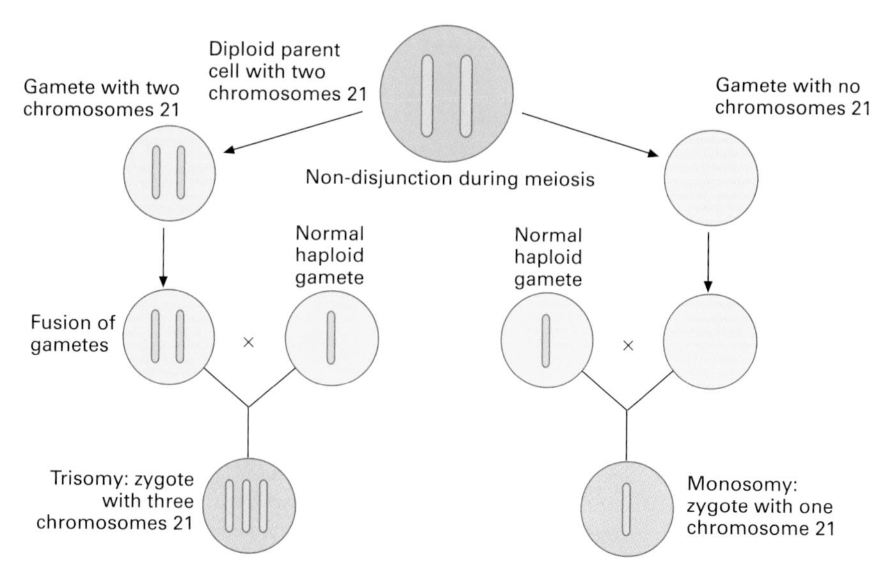

Figure 3 *Simplified diagram to show how non-disjunction of chromosome 21 leads to polysomy and Down's syndrome.* ***Polysomy*** *is the presence of more than two copies of any particular chromosome in diploid cells; in this case, there are three chromosomes 21 so this type of polysomy is called* ***trisomy****. There is an equal probability that non-disjunction will result in an egg cell with no chromosome 21 and, therefore, monosomy (one chromosome) in the fertilised egg. However, a fetus with this condition dies at an early stage.*

Amniocentesis is performed under local anaesthetic and most women do not find it too uncomfortable. However, there is a 0.5–1 per cent risk of spontaneous miscarriage after the procedure. Therefore, amniocentesis is usually recommended only for those at high risk of carrying a Down's baby.

Chorionic villus sampling (**CVS**) is an alternative test that can be carried out earlier than amniocentesis. In CVS, a sample of cells is taken from the chorionic villus (small finger-like processes which grow from the embryo into the mother's uterus). The sample is obtained either by inserting a needle through the abdomen, or inserting a catheter (a flexible tube) through the vagina and cervix. The fetal cells in the sample can then be analysed in the same way as for amniocentesis.

CVS is carried out between weeks 10 and 13 of pregnancy. If the test shows the fetus has Down's syndrome, a decision about abortion can be made earlier than with amniocentesis. Early abortions are usually less difficult, both physically and mentally, than later abortions. However, a higher risk of miscarriage is associated with CVS than with amniocentesis.

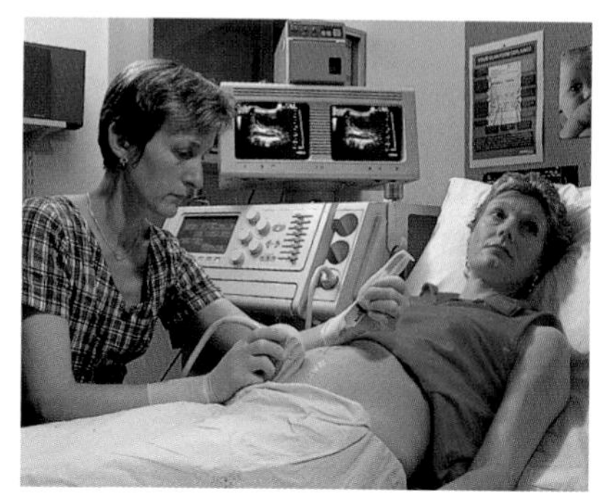

Figure 4 *Amniocentesis: amniotic fluid being withdrawn from the abdomen of a pregnant woman, using an ultrasound image to guide the positioning of the needle.*

Genetic counselling

Screening and diagnostic testing for Down's syndrome should be followed by **genetic counselling**, the giving of advice and information about the risks of a genetic disease and its outcome. Counselling is a very challenging task. Counsellors must have a good understanding of medical genetics and need to be well trained in sympathetic counselling techniques. They must give information which helps clients come to their own decision rather than imposing their own views on the clients. Clients should be made aware that the features of Down's syndrome vary widely. The condition often results in individuals with severe mental disability who require a great deal of support, but many people with Down's syndrome lead independent, long, and fulfilling lives, and they are often very loving individuals. It should not be assumed that mothers carrying a fetus with Down's syndrome would automatically opt for termination of pregnancy.

Quick check

1 What is non-disjunction?

2 Compare amniocentesis and chorionic villus sampling with respect to:

a when they can be carried out

b the risk of inducing a miscarriage.

3 Name one biochemical marker which can help a genetic counsellor assess the risk of Down's syndrome for a client.

Food for thought

Modern genetics is making it much easier to detect genetic disorders and to screen potential parents, fetuses, and babies. Suggest what benefits and problems might be associated with large-scale genetic screening. How do you think society should respond to parents who choose to proceed with a pregnancy likely to bring into the world a child who has a genetic disorder?

19.10

Cystic fibrosis and gene therapy

OBJECTIVES

By the end of this spread you should be able to:

- describe the cause of cystic fibrosis
- describe traditional methods of treatment
- discuss how gene therapy may be used to treat cystic fibrosis.

Fact of life

In 2012, cystic fibrosis affected more than 9000 people in the UK and over 2 million people carried the faulty allele that causes the disease. Genetic screening could significantly reduce the number born with the disease (at present, over 250 per year), and gene therapy has the potential to correct the underlying cause. When trials started, it was thought that the airways would be a simple and easy-to-reach target for a gene therapy product. But this was found not to be the case: the airways (figure 2) actually have more barriers to successful gene therapy than most other organs.

Severe combined immunodeficiency disease

Severe combined immunodeficiency disease (SCID) is a group of genetic disorders. The most common is linked to the X chromosome. Those with SCID have extremely compromised immune systems and are highly vulnerable to infections. The most effective treatment is bone marrow transplants containing blood-forming **stem cells** of a healthy person. In 1990, **gene therapy** successfully restored the immune systems of two girls with one form of SCID. Treatment consisted of removing some of the girls' own T white blood cells, inserting a normal gene into the cells, culturing the genetically modified cells and returning them to the girls' bodies through a vein. Gene therapy trials were halted worldwide for several years when the treatment activated cancer-causing genes (oncogenes) in some patients. Research is now taking place to find ways of correcting the faulty genes in blood-forming stem cells without activating oncogenes.

The symptoms of cystic fibrosis

Cystic fibrosis (CF) is caused by a recessive allele of the **CFTR (cystic fibrosis transmembrane regulator)** gene located on chromosome 7. The normal allele controls the production of a membrane protein called **cystic fibrosis transmembrane protein (CFTP)**. CFTP is essential for the healthy functioning of epithelial cells. It is a carrier protein that transports chloride ions out of cells and into the mucus. As chloride ions leave a cell, water follows by osmosis, making normal epithelial mucus watery. The mucus of normal epithelial cells lining the trachea helps prevent lung infections by trapping dirt and microorganisms. The mucus is carried up the throat by the beating of cilia on the epithelial cells. Eventually, it reaches the buccal cavity and is swallowed or blown out of the mouth or nose.

In CF patients, mucus becomes abnormally thick, sticky, and difficult to clear. It tends to clog up the airways, and it becomes a rich breeding ground for bacteria. In the pancreas, lack of CFTP leads to the formation of fibrous cysts, which gave the disease its name. The cysts can lead to **pancreatitis**, inflammation of the pancreas characterised by severe abdominal and back pain. They may block the pancreatic duct, interfering with the digestion of food which can cause a CF patient to be undernourished. Another feature of CF patients is that the sweat is abnormally salty.

Cloning the CF gene: identifying causes and genetic screening

In 1989, the gene for CF was cloned, enabling geneticists to identify the many different mutations that cause CF. Some mutations give rise to mild forms of the disease. In 70 per cent of cases the disease is caused by a mutation that deletes three bases. This leads to the loss of one phenylalanine residue from CFTP.

Cloning the gene has also allowed geneticists to develop genetic screening methods. A simple mouthwash technique is used to obtain cells for DNA analysis. The test can identify some but not all adult carriers of the disease. The child of two symptomless heterozygous carriers has a 25 per cent chance of having CF (CF is inherited in a Mendelian manner; see spread 19.3). Genetic screening using a DNA test for CF gives couples expecting a child (or planning to conceive) the chance to decide not to have children if the tests show that they are carriers, or to decide to have fetuses screened and to consider aborting them if affected.

Treatment of CF

Ninety-five per cent of CF patients die as a result of lung complications. Traditional treatment is aimed at keeping the airways free of blockage. As often as five times a day, a CF patient will have physiotherapy which includes slapping the back to dislodge mucus from the lungs.

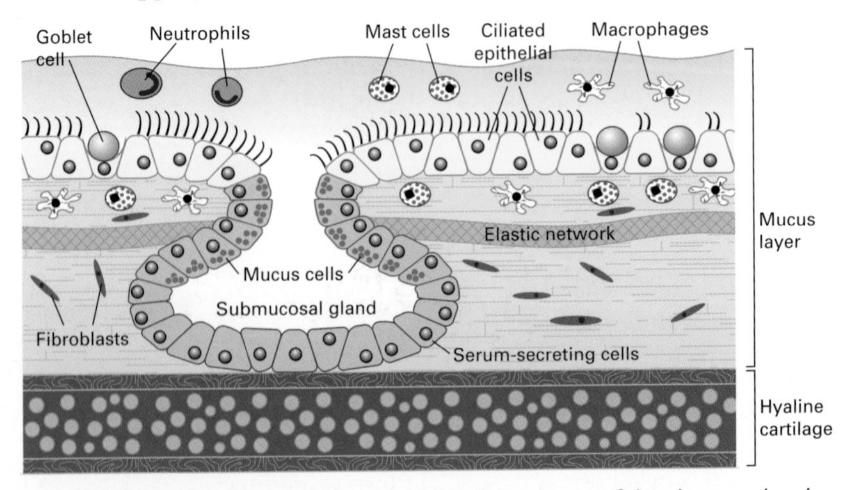

Figure 1 *Schematic diagram of a cross-section through part of the airways, showing that the mucus and serum-secreting cells, the main targets for gene therapy, are not easy to reach because they are surrounded by mucus and other cells.*

In addition, antibiotics are given to reduce infections, and enzyme treatments used to remove accumulations of DNA that come from dead bacteria and dead white blood cells.

Physiotherapy and drugs can extend the life expectancy of CF patients. Genetic screening can help reduce the incidence of the disease, but it does not cure it. However, isolation and cloning of the CFTR gene has led to the exciting possibility of using gene therapy to correct the disease.

Gene therapy may involve gene replacement (replacing a defective allele with a normal one) or gene supplementation (adding one or more copies of a normal allele to a cell without removing any of the pre-existing ones). In gene supplementation, the added alleles are dominant and mask the effects of recessive alleles.

Gene therapy may be either **germ-line gene therapy** or **somatic-cell gene therapy**, depending on which cells are treated. In germ-line gene therapy, sperm or egg cells are treated so that every cell of the offspring functions normally. In somatic-cell gene therapy, the body cells are altered, but not the sperm or eggs. The alteration is therefore not passed on to future generations. In Britain, germ-line gene therapy is regarded as unethical so that only somatic-cell gene therapy is being attempted.

Although every cell in the body of a CF patient contains a defective allele, the most harmful effects are in the airways. Gene therapy therefore targets airway tissue. The aim is to get normal alleles into epithelial cells so that they can make normal CFTP. A variety of vectors could be used, including a harmless virus which would insert the healthy gene into cells (as viruses do naturally), and liposomes (see spread 18.9). Liposomes are spheres of lipid which, because of their lipid solubility, can pass easily through cell membranes. Those used in gene therapy would contain plasmids carrying recombinant DNA coding for the healthy CFTP. They would be inhaled into the airways as a nasal spray and, if all goes well, the healthy CFTR gene would be deposited into the cells of the airway where it could correct the defects causing CF (figure 2).

These methods of gene therapy would not provide a permanent cure since epithelial cells have a short life and are constantly being renewed. The treatment would therefore have to be repeated regularly. Also, as CF affects the entire body, treating the lungs would not cure chronic pancreatitis associated with the disease and the patient would still need to take enzyme supplements.

An alternative to gene therapy would be to use **stem cells** to introduce a healthy CFTR gene inside a cell that can integrate into a patient's tissue and correct the problems. However, although research into using stem cells from cord blood to treat cystic fibrosis is taking place, this is only at its early stages and it will probably be a long time before stem cell therapy of CF becomes generally available to patients.

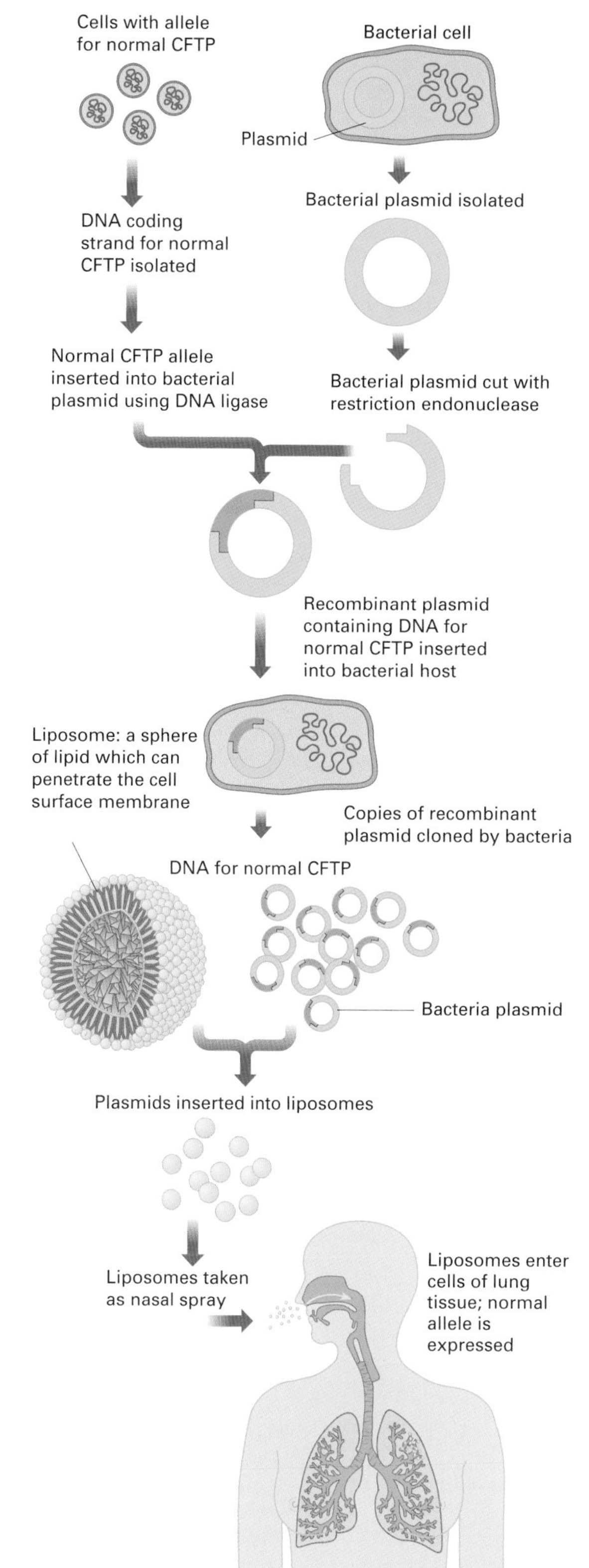

Figure 2 *Outline of gene therapy for cystic fibrosis using liposomes as vectors.*

Quick check

1. What is the function of cystic fibrosis transmembrane protein?
2. Why do CF patients have to take tablets containing pancreatic enzymes?
3. What are liposomes?

Food for thought

In Britain the Warnock Committee concluded that germ-line gene therapy is unethical. Suggest reasons for and against germ-line gene therapy.

Summary

Variation among organisms is genetic and phenotypic. Genetic variation is caused by differences in **genotype** (genetic make-up), whereas **phenotype variation** refers to the observable characteristics of an organism. Phenotype variation may be **continuous** or **discontinuous**. Genotype variation results from **mutations**, either chromosome mutations or **gene (point) mutations**.

Genetics is the study of inheritance. Mendelian inheritance refers to inheritance of characteristics that obey Mendel's laws of inheritance, the **law of segregation** and the **law of independent assortment**. **Monohybrid inheritance** involves a single characteristic determined by one gene whereas **dihybrid inheritance** involves two characteristics determined by two genes. Each **gene** occupies a specific **locus** on a chromosome and may take different forms called **alleles**. Non-Mendelian inheritance may involve **codominance**, **multiple alleles**, **lethal alleles**, **linkage**, and **cross over**. **Cross-over values** have been used to locate genes on chromosomes, a process called **chromosome mapping**.

In humans, **sex** is determined by X and Y chromosomes. Development of external genitalia depends on the presence and activity of the **sex related Y gene**. **Sex linkage** (or **X linkage**) for characteristics such as haemophilia results from the Y chromosome being much shorter than the X chromosome and carrying fewer genes. Inherited diseases include **Down's syndrome** and **cystic fibrosis**. These conditions can be detected before birth by **genetic screening**, for example by **amniocentesis** or **chorionic villus sampling**. **Gene therapy** is being developed to treat inherited diseases.

PRACTICE EXAM QUESTIONS

1 The histogram shows the heights of wheat plants in an experimental plot.

a What evidence from the data suggests that there were two strains of wheat growing in the experimental plot? [1]

b i Which type of variation is shown by the height of each of the strains of wheat plants? Give the reason for your answer. [2]

ii Explain why the height of the wheat plants varies between 45 cm and 120 cm. [1]

[Total 4 marks]

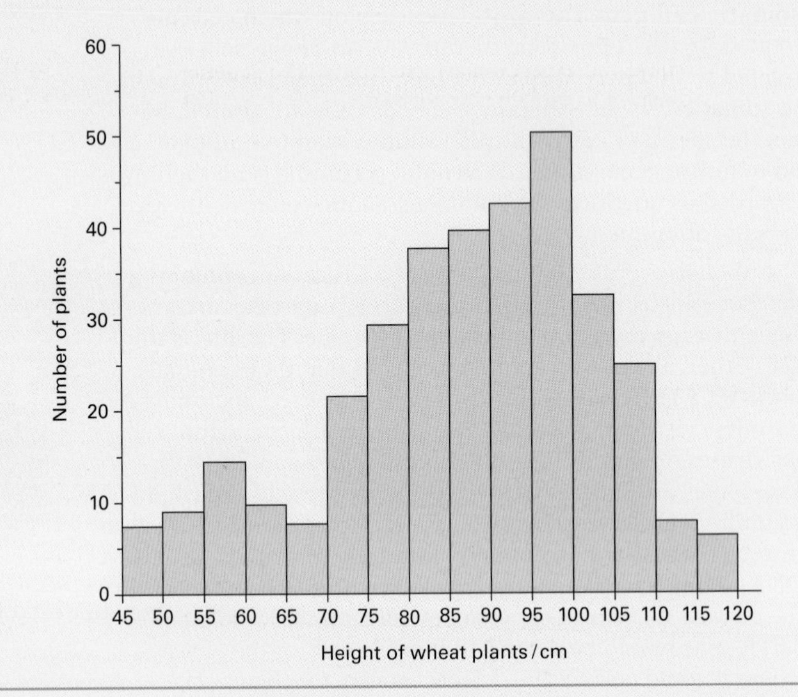

2 Night blindness is a condition in which affected people have difficulty seeing in dim light. The allele for night blindness, **N**, is dominant to the allele for normal vision, **n**. (These alleles are not on the sex chromosomes.)

The diagram below shows part of a family tree showing the inheritance of night blindness.

a Individual 12 is a boy. What is his phenotype? [1]

b What is the genotype of individual 1? Explain the evidence for your answer. [2]

c What is the probability that the next child born to individuals 10 and 11 will be a girl with night blindness? Show your working. [2]

[Total 5 marks]

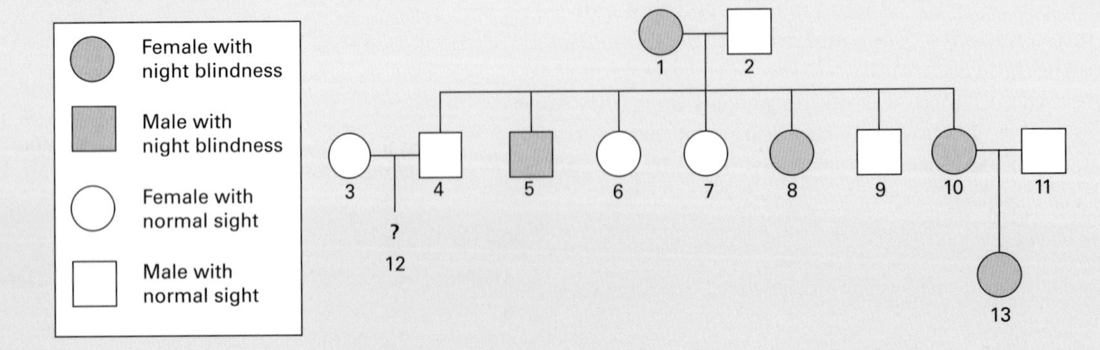

3 In tomatoes, the allele for red fruit, **R**, is dominant to that for yellow fruit, **r**. The allele for tall plant, **T**, is dominant to that for short plant, **t**. The two genes concerned are on different chromosomes.

a A tomato plant is homozygous for allele **R**. Giving a reason for your answer in each case, how many copies of this allele would be found in:

i a male gamete produced by this plant,

ii a leaf cell from this plant? [2]

b A cross was made between two tomato plants.

i The possible genotypes of the gametes of the plant chosen as the male parent were **RT**, **Rt**, **rT** and **rt**. What was the genotype of this plant? [1]

ii The possible genotypes of the gametes of the plant chosen as the female parent were **rt** and **rT**. What was the phenotype of this plant? [1]

iii What proportion of the offspring of this cross would you expect to have red fruit? Use a genetic diagram to explain your answer. [3]

[Total 7 marks]

4 **a** Distinguish between the terms *gene* and *allele*. [3]

b The diagram below shows a family tree in which the blood group phenotypes are shown for some individuals.

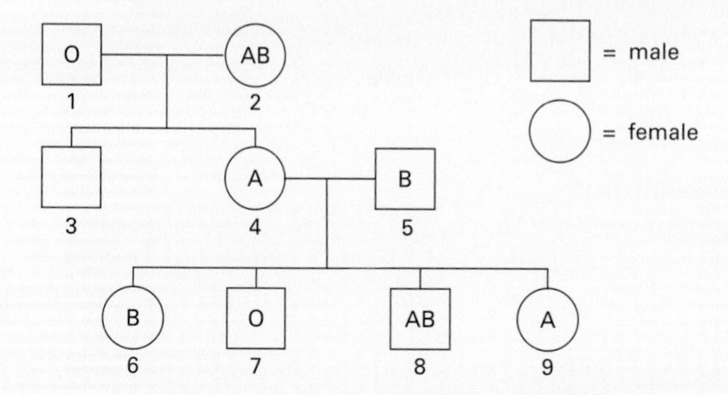

i Using the symbols I^A, I^B and I^C to represent the alleles, write the genotypes of the following people.
1 2 4 5 6 [5]

ii State the possible blood groups of person 3. Explain your answer. [3]

[Total 11 marks]

5 In cats, one of the genes for coat colour is present only on the X chromosome. This gene has two alleles. The allele for ginger fur, $\mathbf{X^B}$, is dominant to that for black fur, $\mathbf{X^b}$.

a All the cells in the body of a female mammal carry two X chromosomes. During an early stage of development one of these becomes inactive and is not expressed. Therefore female mammals have patches of cells with one X chromosome expressed and patches of cells with the other X chromosome expressed. Tortoiseshell cats have coats with patches of ginger and patches of black fur.

i What is the genotype of a tortoiseshell cat? [1]

ii Explain why there are no male tortoiseshell cats. [1]

b A cat breeder who wished to produce tortoiseshell cats crossed a black female cat with a ginger male. Copy and complete the genetic diagram and predict the percentage of tortoiseshell kittens expected from this cross.

Parental phenotype:	black female	ginger male
Parental genotypes:	....	....
Gamete genotypes:	....	....
Offspring genotypes:	....	

....

Percentage of tortoiseshell kittens: ....

.... [3]

[Total 5 marks]

6 **a** In the inheritance of sickle-cell anaemia, the normal allele is represented by $\mathbf{Hb^A}$ and the sickle-cell allele by $\mathbf{Hb^S}$.

i Copy and complete the genetic diagram to show the possible phenotypes and genotypes of the offspring of a couple heterozygous for these alleles.

Parental phenotypes

Parental genotypes

Genotypes of gametes

Genotypes of offspring

Phenotypes of offspring

ii The first child born to this couple had sickle-cell anaemia. What is the probability that their second child will have sickle-cell anaemia? [4]

b Of 12 387 adults examined in Nigeria, 29 were $\mathbf{Hb^S}$ $\mathbf{Hb^S}$, 2993 were $\mathbf{Hb^A}$ $\mathbf{Hb^S}$, and 9365 were $\mathbf{Hb^A}$ $\mathbf{Hb^A}$. Explain why there were so many heterozygotes in this population compared to the low number of individuals homozygous for sickle-cell anaemia. [3]

[Total 7 marks]

7 In humans, cystic fibrosis is caused by a recessive allele (**f**) of a particular gene. The albino condition, in which no melanin is produced in the skin, is produced by a recessive allele (**a**) of another gene.

These two genes are found on different chromosomes.

a What is meant by the term *allele*? [1]

b **i** An unaffected individual, who is heterozygous for both albinism and cystic fibrosis, can produce a range of gametes with respect to these two genes. With reference to meiosis, describe how the differences in the gametes can arise. [2]

ii Two individuals, heterozygous for cystic fibrosis, produce a child. Calculate the probability that this child has cystic fibrosis. [2]

[Total 5 marks]

20 Evolution

Theories of evolution

20.1

OBJECTIVES

By the end of this spread you should be able to:

- explain the biological meaning of evolution
- distinguish between neo-Darwinism and Darwinism.

Fact of life

Highly sensitive dating techniques tell us that the Earth is between 4.5 and 5.0 thousand million years old. It is generally agreed by scientists that the Earth was originally devoid of life, and that the first living organisms arose by biochemical evolution from complex organic chemicals formed in the atmosphere and seas of early Earth (see spread 9.1, figure 1). These first forms of life gave rise to countless millions of species. Most have become extinct, but some have evolved into organisms found today. According to the latest estimates, 20–30 million species share our planet.

One of the most fundamental questions in biology is: 'where do all living things come from?'. According to most biologists, the millions of species living on Earth today (including humans) are descended from other species that inhabited the world in the past. This change has come about by a process called **evolution**. Evolution happens when the genetic composition (allele frequency) of a population changes over successive generations. When the changes are sufficiently great, a new species may be formed. (A **species** is a group of closely related organisms potentially capable of interbreeding to produce fertile offspring; see spread 20.8 for a fuller description.) Although a few biologists believe in creationism (see box), most accept that the scientific evidence for evolution, examined in the next two spreads, is overwhelming.

Creationism

Creationism is the belief that God created all species in six days as described in Genesis, the first book of the Judaeo-Christian Bible. According to creationism, all species present on Earth today have remained unchanged since they were created by God. Darwin's theory of evolution contradicts this belief. When Darwin's book was first published, it received a very hostile reception from influential church leaders who felt that it was promoting atheistic ideas. Today, many Christians feel that a Darwinian theory of evolution does not exclude God: they believe that evolution is God's way of creating things. Nevertheless, there are still some people who believe in creationism and argue that evolution has never taken place.

The mechanism of evolution

Evolution is not a modern concept. Since ancient times, a number of philosophers and naturalists (including Confucius in China and Aristotle in Greece) have suggested that complex species evolve from simpler pre-existing ones by a process of continuous and gradual change. However, it was not until the nineteenth century that scientists came up with plausible mechanisms for evolution. The mechanism that is widely accepted among biologists today is called neo-Darwinism. It is a modern theory based on the work of the nineteenth-century naturalist Charles Darwin.

Between 1831 and 1836, Darwin was the naturalist on board HMS *Beagle*, a research vessel engaged in mapping different parts of the world. After spending over three years surveying the coast of South America, the *Beagle* landed on the Galapagos Islands in the Pacific Ocean. Darwin compared the organisms on these islands with those on the South American mainland (figure 1), and this led him to develop his theory of evolution. He came to the conclusion that, over successive generations, a new species comes into being by slow and gradual changes from a pre-existing one. He believed that these changes are brought about by a process which he called **natural selection**.

Figure 1 Of all the places he visited, the Galapagos Islands made the greatest contribution to the development of Darwin's theory of evolution. Darwin observed that animals of the same species, such as the giant tortoise, were slightly different on the different islands, and suggested that these different branches evolved from a common ancestor.

Darwin's theory was based on three main observations:

1. Within a population are organisms with varying characteristics, and these variations are inherited (at least in part) by their offspring.
2. Organisms produce more offspring than are required to replace their parents.
3. On average, population numbers remain relatively constant and no population gets bigger indefinitely.

From these observations, Darwin came to the conclusion that within a population many individuals do not survive, or fail to reproduce. There is a 'struggle for existence'. For example, members of the same population compete to obtain limited resources, and there is a struggle to avoid predation and disease, or to tolerate changes in environmental conditions such as temperature. In this struggle for existence those individuals that are best adapted to their environment will have a **selective advantage**: they will be more likely to survive and produce offspring than less well-adapted organisms. In Darwin's own words:

> *As many more individuals of each species are born than can possibly survive, and as, consequently, there is a frequently recurring struggle for existence, it follows that any being, if it vary however slightly in any manner profitable to itself, under the complex and sometimes varying conditions of life, will have a better chance of surviving and thus be naturally selected. From the strong principle of inheritance, any selected variety will tend to propagate its new and modified form.*

The Origin of Species

For more than 20 years, Darwin collected evidence to support his theory and refined his ideas. He delayed publishing his ideas until 1858, when Alfred Russel Wallace sent him a letter describing a theory of evolution identical to Darwin's own. Wallace was a British naturalist who had worked in the Malay Archipelago for eight years. He concluded from his research that some organisms live while others die because of differences in their characteristics, such as their ability to resist disease or escape predation. Darwin and Wallace published a paper jointly describing their theory of evolution by natural selection. However, Darwin's name has become more strongly linked with the theory because of a book he published on 24 November 1859. The book, entitled *The Origin of Species by Means of Natural Selection or the Preservation of Favoured Races in the Struggle for Life*, has been called the most important biology book ever written. It not only gives a full description of the theory of evolution by natural selection, but also contains a huge mass of evidence to support the theory.

The reaction to Darwin

Many people found it difficult to accept Darwin's ideas, especially the idea that modern humans and apes are probably descended from a common ancestor. However, his theory is supported by so much evidence that the majority of biologists accept it. Evolution by natural selection has become a central theme which underpins much of modern biology. The modern theory of evolution is called **neo-Darwinism** (neo = new) because it incorporates new scientific evidence, particularly from genetics and molecular biology. For example, we now know that the variations that are so important in natural selection come about by random and spontaneous changes in genes, particularly from mutations in reproductive cells. Despite modifications to Darwin's theory in neo-Darwinism, natural selection is still the driving force behind evolution, or *the theory of evolution by the natural selection of inherited characteristics.*

Quick check

1 Give the biological meaning of evolution.

2 How does neo-Darwinism differ from Darwin's original theory of evolution?

Food for thought

In 1809 Jean-Baptiste de Lamarck suggested that the driving force behind evolution was the need for organisms to adapt to changing environmental conditions. His theory became known as *the theory of evolution by the inheritance of acquired characteristics.* He believed that adaptations developed by an organism during its lifetime could be passed on to its offspring. According to Lamarck, modern giraffes might have evolved from a short-necked ancestor in the following way. Giraffes feed on leaves ripped off the branches of trees. When leaves on the lower branches were removed, or when the trees became taller, the ancestral giraffe needed to stretch to reach leaves on higher branches. By continually stretching, their necks lengthened and the ability to grow a slightly longer neck was inherited by the next generation which carried on stretching, and so on.

Figure 2 *A giraffe's long neck allows it to reach leaves that other animals cannot reach.*

We know that this explanation of the evolution of the giraffe's long neck is untrue because activities such as stretching to feed do not affect the gametes. Therefore, this type of characteristic acquired during the life of an organism is not inherited by its offspring. Expressed in modern terms, Lamarckism would mean that changes in phenotype could determine the genotype of future generations. This does not agree with modern genetics, and there are no generally accepted examples of acquired characteristics being inherited. Suggest a neo-Darwinian explanation for the evolution of the modern long-necked giraffe (figure 2) from a short-necked ancestor.

20.2

Evidence for evolution (1)

OBJECTIVES

By the end of this spread you should be able to:

- explain the significance of fossils to evolutionary theory
- discuss how comparative biochemistry can be used to show evolutionary relationships.

Fact of life

The **Burgess Shale**, a rock formation in British Columbia, contains the most celebrated fossil field in the world. Charles Walcott discovered the field in 1909. By 1924, he had amassed over 65 000 specimens of extraordinarily diverse fossils preserved from the Cambrian period, over 500 million years ago. Many have their soft body parts exceptionally well preserved and are early ancestors of modern species. Others, like *Hallucigenia* (figure 1), are spectacularly peculiar and seem to be unrelated to any living forms. Their later disappearance from our fauna is an intriguing mystery.

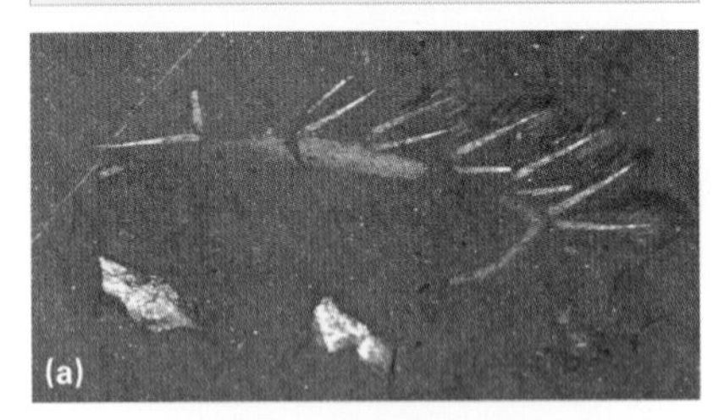

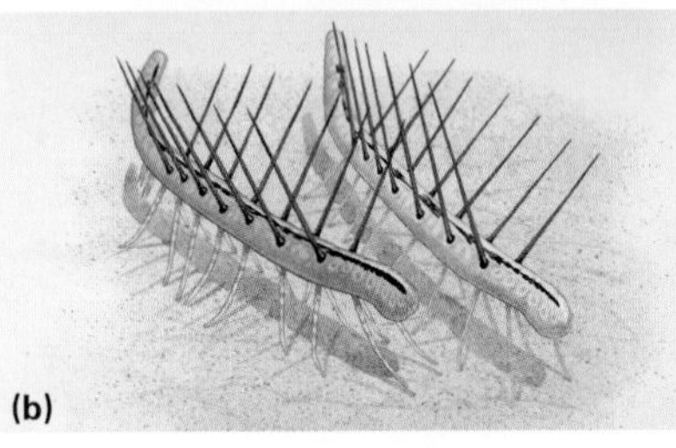

Figure 1 ***(a)*** *A fossil and* ***(b)*** *an artist's impression of* Hallucigenia, *so named by Simon Conway Morris because of the animal's 'bizarre and dream-like quality'.*

Comparative embryology

Closely related species have a similar embryonic development, even if their adult stages are very different. For example, in echinoderms and chordates the blastopore (the first opening in the blastula, the hollow fluid-filled ball of cells that develops from the zygote, spread 12.7) becomes the anus, whereas in annelids, arthropods, and molluscs it becomes the mouth. Because of this and other aspects of their embryological development, echinoderms and chordates have been placed in the superphylum Deuterostomia whereas annelids, arthropods, and molluscs have been placed in the superphylum Protostomia.

Darwin's theory of evolution had a great impact because it was supported by a wealth of evidence. Since Darwin first published his book on the origin of species, his theory has been further supported by evidence from many different branches of science. In this spread we shall consider briefly some of the evidence which, in the minds of most scientists, has confirmed evolution as the central process in biology.

According to Darwin's theory:

1 Each species living today arose from a pre-existing species.
2 All species have evolved from one ancestral type.
3 Natural selection provides the mechanism for one species to change into another.

The main evidence for his first suggestion, which has been called **descent with modification**, comes from fossils.

Palaeontology: the study of fossils

A **fossil** is the remains of an organism that lived in the past, preserved by a natural process (for example, in rock, peat, or ice). Fossils include bones, shells, footprints, and faeces. Most fossils are found in sedimentary rocks formed by layers of silt. The sediment preserves the bodies of dead organisms before they have a chance to decay. In some places, such as the Grand Canyon in Arizona, USA, sedimentary rock has been laid down in layers called **strata**. The depth of each stratum gives the relative age of any fossils it contains: the deeper the stratum, the older the rocks and fossils in it.

Rocks and their fossils can be dated approximately on the basis of how long it takes for sedimentary rocks to be laid down. However, these estimates are very rough. More accurate estimates come from measuring the radioactivity of crystals of igneous rock in the strata. Igneous rock is formed from molten material from beneath the Earth's crust which is brought up to the surface by volcanic activity. Crystals of igneous rock contain radioactive isotopes, such as isotopes of uranium and potassium. The level of radioactivity is greatest when the crystals first form. As they age, the isotopes decay: uranium to lead, and potassium to argon. The older the rock, the less original radioactive material remains. Fossils can therefore be dated by analysing the amounts of uranium and lead, or potassium and argon, they contain.

Uranium–lead dating depends on two separate decay series: the uranium series from ^{238}U to ^{206}Pb, with a half-life of 4.47 billion years; and the actinium series from ^{235}U to ^{207}Pb, with a half-life of 704 million years. Using U–Pb dating, zircon-containing rock from Western Australia is estimated to be about 4.4 billion years old. This supports the hypothesis that the Earth is about 4.6 billion years old. Radioactive **carbon dating** is used to date younger fossils. The atmosphere contains carbon of two types, ordinary carbon-12 and radioactive carbon-14. During its life, an organism takes in both isotopes. However, when it dies, the radioactive carbon-14 decays over time. The proportion of radioactive carbon present in, for example, peat, gives an idea of its age (assuming no more carbon-14 has been incorporated into the organism since it was alive, and also that the amount of carbon-14 in the atmosphere has remained constant). As the decay of carbon-14 is quite rapid, this method of dating cannot be used on fossils more than about 50 000 years old.

After being dated, fossils can be placed into chronological order, giving an idea of how one group of organisms may have evolved into another. Most fossilised organisms have become extinct, but some (mainly the more recent ones) belong to species living today.

Fossils of extinct organisms which lie in between two present-day types provide strong evidence in support of evolution. Figure 1 shows a fossil found in rocks some 150 million years old. Although fossils give such strong support for evolution, fossil evidence is far from perfect. Dating is often only approximate and there are no fossils of the majority of early or soft-bodied organisms. Fossils of most organisms are so rare that it is not possible to trace their evolutionary pathway. Even making an evolutionary pathway for the modern horse, which seems to have many fossilised ancestors, has produced hot debate. There is much evidence, but it comes from different geographical regions, and is open to various interpretations.

Comparative biochemistry and cell biology

The most persuasive evidence that all organisms have evolved from a common ancestor comes from studies comparing the biochemistry and molecular biology of different organisms. Such studies reveal that:

- the genetic code contained within nucleic acids is almost universal (see spread 18.6)
- physiological processes, such as respiration, vital to all organisms, follow very similar metabolic pathways and use closely related molecules
- ATP is the universal energy currency.

Differences in the biochemistry and cell biology of organisms can reveal their evolutionary relationships. Closely related species would be expected to differ less than distantly related species. Detailed comparisons of DNA, metabolic pathways, key proteins, and organelles from different organisms have been used to work out their evolutionary relationships. For example, **DNA sequencing** in flowering plants has shown that they can no longer be divided simply into monocots and dicots (see spread 21.7). **Cytochrome c**, a key component in all electron transport systems (spread 6.4), has the same amino acid sequence in closely related organisms but differs in more distantly related organisms (for example, cytochrome c in humans and chimpanzees is identical but differs from that in rhesus monkeys). **Ribosomes** inside mitochondria and chloroplasts are similar to those in bacteria, suggesting that these organelles may have evolved from bacteria (see spread 4.4). Non-human primate **blood proteins** have been analysed to see how similar they are to human blood proteins: blood serum from the primate in question is added to rabbit serum containing anti-human antibodies. A precipitate forms due to an antibody–antigen reaction (see spread 15.5). The degree of precipitation is compared with that caused by human serum (taken as 100%): chimpanzee serum causes 97% precipitation, gorilla serum 97%, and gibbon serum 79%. Comparing **haemoglobin** in the four species gives similar results: the amino acid sequence in chimpanzee haemoglobin is identical to that in humans (although the amino acids in other polypeptides differ); gorilla haemoglobin differs from human haemoglobin by 3 amino acids; and gibbon haemoglobin differs by 6 amino acids. In contrast, the haemoglobin of lamprey (a jawless fish-like parasite) differs from human haemoglobin by 125 amino acids.

DNA–DNA hybridisation

A single DNA molecule consists of two nucleotide strands which can be separated by heating. On cooling, the strands bind together again by complementary base pairing. When individual nucleotide strands from two species are mixed together, a double-stranded DNA hybrid forms. The strength of the binding between the strands depends on how closely the two sources of DNA are matched. When DNA–DNA hybridisation is used to study the evolutionary relationship between two species, DNA from each species is cut into small segments. The segments are heated to produce single nucleotide strands. The strands of one species are labelled to distinguish them from the unlabelled strands of the other species. When mixed and allowed to cool, strands from the two species bind together to form hybrid DNA. On heating, the double-stranded hybrid DNA 'melts' (separates into single strands). Hybrid DNA made from closely related species requires more heat to melt than hybrid DNA from more dissimilar species. In practice, hybrid DNA is heated gently in small increments. The temperature at which 50 per cent of the hybrid DNA melts is compared with the temperature at which 50 per cent of self-hybridised DNA (i.e., DNA reconstituted from two nucleotide strands of the same species) melts. When used to study primate relationships, DNA–DNA hybridisation supported other evidence indicating that human and chimpanzee DNAs are more similar to each other than to the DNA of orangutans or gorillas.

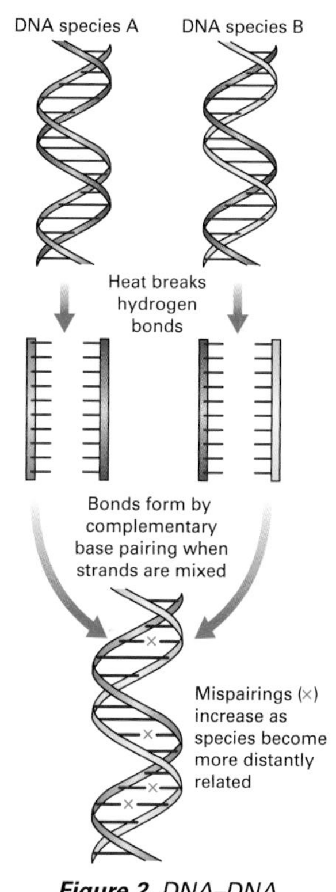

Figure 2 *DNA–DNA hybridisation.*

Quick check

1 Why are fossils like those of *Archaeopteryx* so important?

2 Give two pieces of evidence from comparative biochemistry that support the theory that all species living today are decended from a common ancestor.

Food for thought

One of the main criticisms of the fossil record is that it is incomplete. There is no fossil record of the majority of early organisms or of most of those that followed. There is no complete sequence of fossils which shows how a group of related species have evolved from a common ancestor. Suggest why the fossil record is incomplete.

20.3

EVIDENCE FOR EVOLUTION (2)

OBJECTIVES

By the end of this spread you should be able to:

- distinguish between homologous organs and analogous organs
- distinguish between divergent evolution, parallel evolution, and convergent evolution
- describe two examples of adaptive radiation.

Fact of life

Henry Bates was a distinguished nineteenth-century naturalist who collected animals in the Amazon rainforests for eleven years. He found that humming-bird moths and humming-birds (figure 1) looked so alike in flight that when he tried to collect a humming-bird, he often got a moth by mistake. These unrelated animals have independently evolved a similar body form as an adaptation to obtaining nectar from flowers.

Figure 1 *A striking example of convergent evolution shown by* ***(a)*** *a humming-bird moth and* ***(b)*** *a humming-bird feeding on nectar.*

Geographical isolation and adaptive radiation

Two populations that are geographically isolated from each other by physical barriers such as seas or mountains may experience different environmental conditions. If the theory of evolution by natural selection is correct, you would expect that the separate populations in different environments would become more and more different from each other as time goes on. This is precisely what Darwin concluded from his observations of animals and plants on the Galapagos Islands, a group of volcanic islands in the Pacific Ocean, about 580 miles from the South American mainland. Although he found only a few species here, a large proportion of them were **endemic species** (that is, they were found nowhere else). The species on the various islands resembled those on the mainland but differed from each other and from the mainland species in subtle but significant ways. Darwin thought that the islands must have been colonised from the mainland by a few individuals belonging to the same species, and the species then evolved independently to fill the ecological niches on each island. (The **ecological niche** refers to the role of an organism in the community, including the food it eats and the precise habitat it lives in; see spread 22.5.) This type of evolution is shown by a group of birds now known as **Darwin's finches**. There are 14 separate species of finch in the group, all of which probably evolved from individuals belonging to one mainland species. The islands have few other bird species. In the absence of competition, the finches became adapted to fill all the available niches. In particular, they evolved a wide range of beak sizes and shapes so that they could take advantage of the food sources on the different islands (figure 2). The evolution of an ancestral species into different species to fill different niches is called **adaptive radiation**.

Continental drift and independent evolution

The continents which now exist have not always appeared as they do today. At one time, the Earth had a single large land mass called Pangaea. This is thought to have broken up into two parts, a northern Laurasia and a southern Gondwanaland. Over millions of years, the two great land masses split up and moved by a process called **continental drift** to form our present continents. The theory that these land masses were once joined is supported by the discovery in Australia, South Africa, South America, and Antarctica of fossils belonging to the same extinct species. Fossils in North and South America show differences between the species, suggesting that these two continents have only joined together relatively recently. Before this, their fauna (animals) and flora (plants) were geographically isolated and evolved independently.

Australia shows many excellent examples of species that evolved independently following its geographical isolation. It is thought that Australia became isolated about 120 million years ago, when marsupials (mammals without a placenta but with a pouch in which the young develop) and eutherian mammals (mammals with a true placenta) diverged

Figure 2 *Five of Darwin's finches.*

Large ground finch has strong beak with which it can crack nuts.

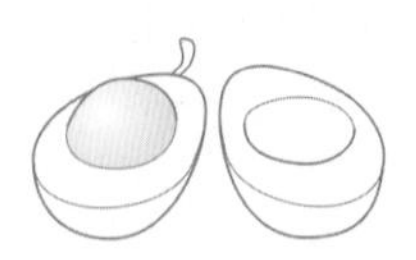

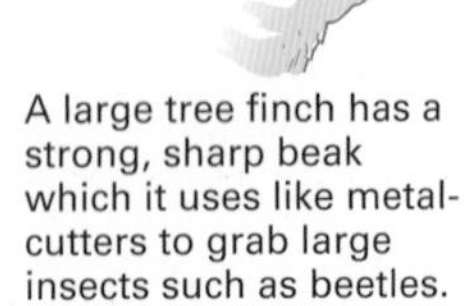

A large tree finch has a strong, sharp beak which it uses like metal-cutters to grab large insects such as beetles.

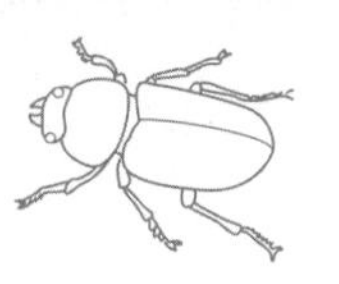

A warbler finch has a small pointed beak with which it catches flying insects

The cactus finch has a long, tough beak with which it can probe cactus flowers for nectar. It also eats cactus seeds and nectar.

The woodpecker finch has a hard beak with which it hammers wood. It is very unusual in being a tool-using bird: it uses a cactus spine to probe for insect larvae in the wood.

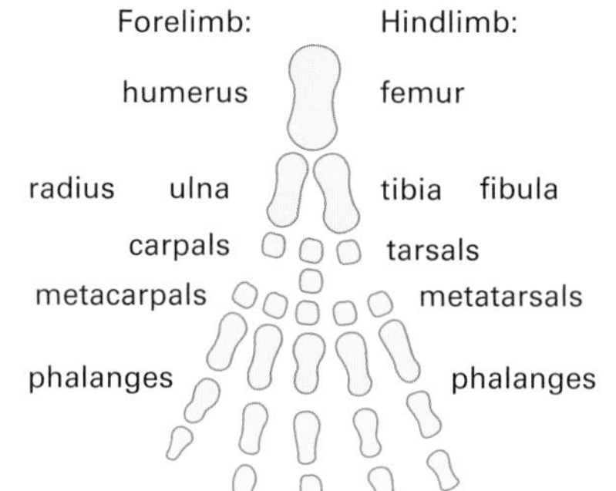

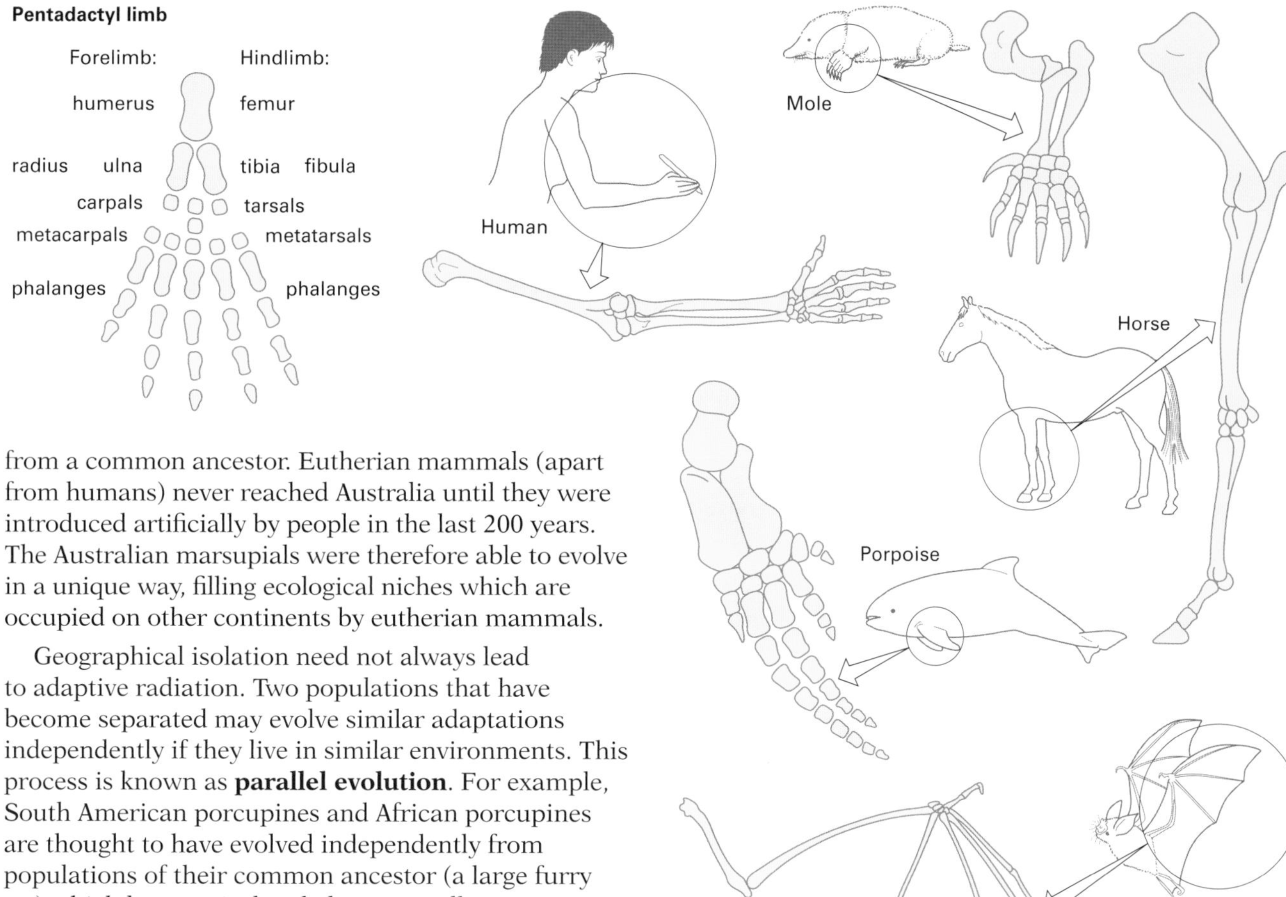

from a common ancestor. Eutherian mammals (apart from humans) never reached Australia until they were introduced artificially by people in the last 200 years. The Australian marsupials were therefore able to evolve in a unique way, filling ecological niches which are occupied on other continents by eutherian mammals.

Geographical isolation need not always lead to adaptive radiation. Two populations that have become separated may evolve similar adaptations independently if they live in similar environments. This process is known as **parallel evolution**. For example, South American porcupines and African porcupines are thought to have evolved independently from populations of their common ancestor (a large furry rat) which became isolated about 70 million years ago. Today, the two groups of porcupine look very alike, and have evolved similar sharp spines for defence.

Comparative anatomy

Comparative anatomy is the study of biological structures in different organisms. Structures from different species that have a similar internal frame work, position, and embryonic development are said to be **homologous**. Homologous structures can become modified to perform different functions in species that live in different environments. For example, the basic plan of the bones found in the limbs of all mammals is the same, and is called the **pentadactyl limb** (five-fingered limb). As figure 3 shows, the pentadactyl limb has become adapted to different environmental conditions and modes of life. The fact that different mammalian species have pentadactyl limbs is a strong indication that they have evolved from a common ancestor. This type of evolution is called **divergent evolution**, and it clearly results in adaptive radiation.

The fact that two different organisms look alike does not necessarily suggest a close evolutionary relation-ship. Structures of unrelated species can evolve to look alike because the structures are adapted to a similar function. These are called **analogous structures**. Examples include the wings of birds, bats, and insects; the streamlined shapes of marine fish, birds (such as penguins), and mammals (such as dolphins); and the eyes of an octopus and a human. Analogous structures differ from each other in their microscopic details and their embryonic development. The process by which they evolve to resemble each other is called **convergent evolution**.

Figure 3 *The pentadactyl limb in five different species of mammal. Darwin wrote:* What can be more curious than that the hand of man, formed for grasping, that of a mole for digging, the leg of a horse, the paddle of a porpoise, and the wing of a bat should all be constructed on the same pattern, and should include similar bones in the same relative positions?

Quick check

1 State whether the wings of a humming-bird moth and a humming-bird are homologous or analogous structures.

2 By what process do:

a analogous structures evolve so that they look alike

b two related but geographically separate groups evolve similar adaptations independently?

3 Name two examples of adaptive radiation.

Food for thought

A genus (plural genera) is a group of closely related species. Suggest why dinosaur fossils belonging to one genus have been found in North America and East Africa, and dinosaur fossils belonging to another genus have been found in South America and India, whereas the mammals found naturally in all these regions are different.

20.4

Natural selection

OBJECTIVES

By the end of this spread you should be able to:

- explain what is meant by 'survival of the fittest'
- distinguish between directional selection, stabilising selection, and disruptive selection.

Fact of life

You may think that natural selection results in change and diversification. This is not always the case. For example, natural selection helps to keep the average birth mass for human babies around 3.3 kg. Not suprisingly, extremely small or large babies have low rates of survival under natural conditions (figure 1).

Survival of the fittest

Darwin had the idea that natural selection is the mechanism that drives evolution after reading *An Essay on the Principle of Population* by Thomas Malthus, a clergyman and political economist. Malthus argued that, in time, the growth of human populations will outstrip the food supply, and that this will lead to 'famine, pestilence, and war'. Darwin applied this idea to populations of other animals and of plants. In his book on the origin of species, Darwin wrote: *There is no exception to the rule that every organic being naturally increases at so high a rate that, if not destroyed, the Earth would soon be covered by the progeny of a single pair.* In spite of reproducing quickly, no single species has completely over-run the planet, although the populations of some species may be increasing at any one particular time. Darwin concluded that populations are kept in check by a 'struggle for existence' as they compete for limited resources and are exposed to disease. Environmental factors that keep populations in check are called **selection pressures** or **environmental resistances**. These include:

- disease
- competition for resources such as food and a place in which to live
- predation
- lack of light, water, or oxygen
- changes in temperature.

Those organisms best suited to the environmental conditions, with characteristics that give them an advantage in the 'struggle for existence', will have the best chance of surviving and producing offspring. Their high **natality** (birth rate) gives them a **selective advantage**. On the other hand, those with unfavourable characteristics are more likely to die. Their high **mortality** (death rate) gives them a **selective disadvantage**. Darwin argued that this difference in natality and mortality results in natural selection. As environmental conditions change, certain characteristics within a randomly varying population are favoured, and natural selection occurs. This has become known as the **'survival of the fittest'**.

In evolution, **fitness** is defined as the ability of an organism to pass on its alleles to subsequent generations, compared with other individuals of the same species. The 'fittest' individual in a population is the one

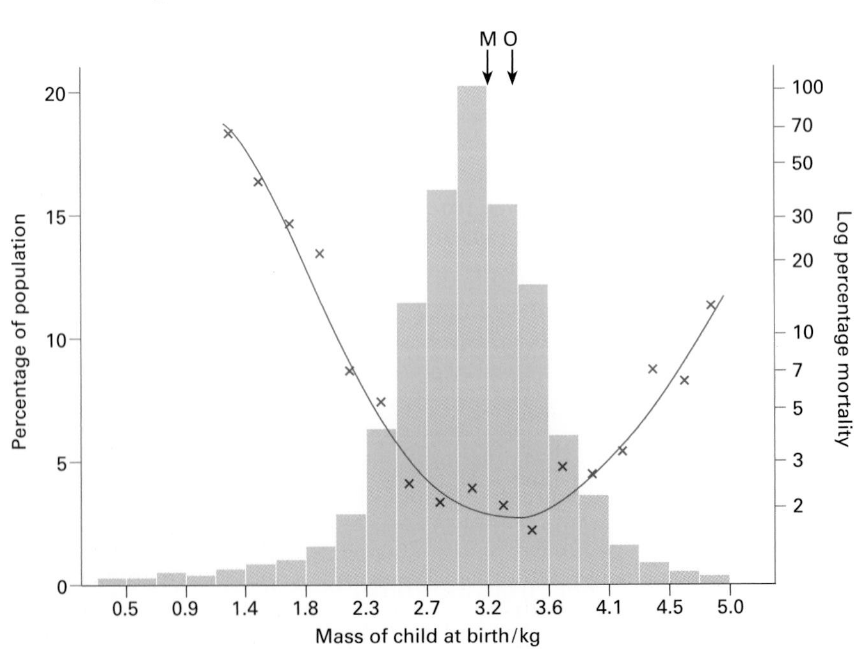

***Figure 1** A histogram (blue) showing the distribution of mass at birth of children born in University College Hospital over a 12-year period, and a line graph (red) showing the percentage perinatal mortality (those failing to survive for four weeks) on a logarithmic scale. At the optimal birth mass (O), the percentage mortality is lowest. M is the mean birth mass.*

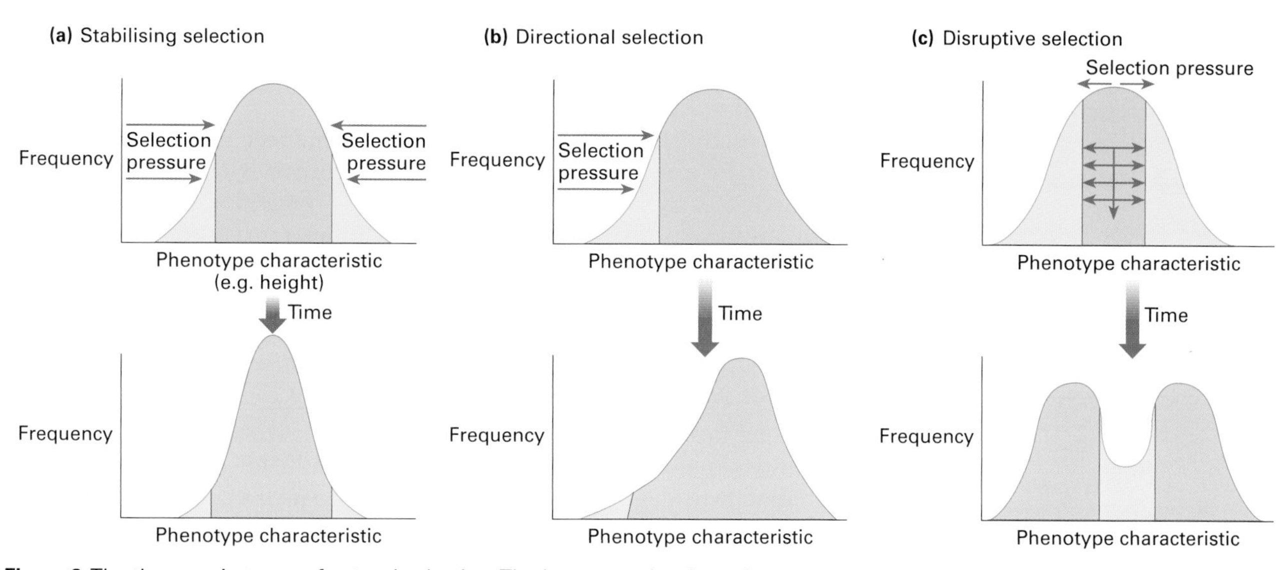

***Figure 2** The three main types of natural selection. The lower graphs show the population after natural selection has removed individuals with a selective disadvantage (shaded yellow) from the populations in the top graphs.*

that produces the largest number of offspring that survive to reproduce themselves. Natural selection by 'survival of the fittest' means that the genetic characteristics of a population gradually change from generation to generation in response to changes in the environment. As we shall see in the following spreads, natural selection affects a gene pool by increasing the frequency of alleles that give an advantage, and reducing the frequency of alleles that give a disadvantage. (A **gene pool** is all the genes and their different alleles present in an interbreeding population.)

Three types of natural selection

Natural selection is not always a mechanism for change. There are three different types: stabilising selection, directional selection, and disruptive selection. These are three different ways in which natural selection acts on the phenotypes in a population (the observable characteristics such as height or colour). Typically, the frequency in the population of each phenotype has a normal distribution, described by a bell-shaped curve.

Stabilising selection (figure 2a) happens in an unchanging environment. Extremes of the phenotype range are selected against, leading to a reduction in variation (more individuals tend to conform to the mean). Stabilising selection occurs in the natural selection of birth mass in humans (figure 1).

Directional selection (figure 2b) favours one extreme of the phenotype range and results in a shift of the mean either to the right or to the left. This type of selection usually follows some kind of environmental change. The long neck of the giraffe (spread 20.1, figure 2) is thought to have evolved in this way. Probably, when food was in short supply, only the tallest individuals could reach enough food to survive. They passed on their genes to the next generation.

Disruptive selection (figure 2c) selects against intermediate phenotypes and favours those at the extremes. This leads to a **bimodal distribution** (the distribution curve has two peaks or modes) and two overlapping groups of phenotypes. If the two groups become unable to interbreed, then each population may give rise to a new species. Disruptive selection may have contributed to the evolution of Darwin's finches (spread 20.3, figure 2). Because there were few other birds to compete, finches with short strong beaks had exclusive use of nuts as a food source, while those with long slender beaks had almost exclusive use of insects. Those finches with an average, unspecialised beak were more likely to have been in competition with other species of bird and would have reproduced less successfully.

Quick check

1 What is meant by fitness in evolutionary terms?

2 Some individuals of the European swallowtail butterfly (*Papilio machaon*) pupate on brown stems or leaves; others pupate on green stems or leaves. Two distinct colour forms of the pupae are found, namely brown and green, with very few intermediates.

  a What type of natural selection does this example show?

  b Explain why the intermediate colour forms would be at a selective disadvantage.

Food for thought

The **extinction** of animal and plant species is of great concern today because it is accelerated by direct and indirect results of human activities. However, extinction is a natural process that has occurred since the dawn of life. The 20–30 million species that inhabit the Earth today represent only a minute proportion of all the species that have ever existed. Suggest why more than 99.9 per cent of all species that ever evolved have become extinct by natural processes. Explain why the highest rates of extinction in recent times have occurred among species that live only on small oceanic islands.

20.5

Evolution in action

OBJECTIVES

By the end of this spread you should be able to:

- explain how antibiotic resistance, insecticide resistance, and heavy-metal tolerance can evolve
- discuss the evolution of industrial melanism in the peppered moth.

Fact of life

DDT was introduced as a pesticide in the 1940s. Within a few years, many insects (including *Anopheles gambiae*, a mosquito vector for malaria) had become resistant to it. Several mechanisms of resistance determined by different genes evolved independently, enabling DDT to be detoxified by the pests, or preventing it from penetrating specific tissues. *Anopheles gambiae* was one insect that evolved DDT resistance. In villages that had never been sprayed before, almost no mosquitoes became resistant. However, in villages with a history of regular spraying, about 90 per cent of mosquitoes became resistant.

Figure 1 *Spring sandwort (*Minuartia verna*) growing on a lead spoil heap in Derbyshire. The heaps are largely devoid of vegetation because of toxic heavy metals (including copper, lead, and zinc) in the soil. However, scattered in the heaps are a few clumps of spring sandwort. In unpolluted areas, random mutations result in a few individuals with greater tolerance to heavy metals than others. The tolerant plants may be able to trap heavy metals on organic molecules in the cellulose cell wall, confine the metals to the vacuoles, or excrete the metals back into the environment. In all cases, tolerance to heavy metals is inherited and appears to have evolved by directional selection. Heavy-metal-tolerant plants are less competitive (have a selective disadvantage) in unpolluted areas and rarely survive there. However, tolerant individuals flourish in polluted areas as the heavy metals kill their rivals, and they pass on their tolerance to their offspring.*

Inherited variations

Darwin established that natural selection is the mechanism that drives evolution by causing differential mortality (unequal chances of dying), or 'survival of the fittest'. Within a population, individuals vary and those with favourable characteristics are more likely to survive than those with unfavourable characteristics. Darwin recognised that for evolution to occur, the characteristics of survivors have to be inherited by their offspring. However, Darwin had no knowledge of genes. We now know that inherited variations are determined by genes, and the ultimate source of these variations is mutation (spread 19.1).

A **gene mutation** is a random event resulting in a new allele. Most mutations that affect the phenotype are harmful. However, occasionally a mutation provides a new phenotype such that the mutant has a selective advantage over other individuals in a population. Over many generations, populations may gradually change so that individuals with the mutant allele become more frequent, a process called **microevolution**. Examples of such a change include the microevolution of antibiotic resistance in bacteria (spread 15.9), heavy-metal tolerance in grasses (figure 1), and pesticide resistance in rats (see below). In contrast, **macroevolution** refers to evolutionary changes at a higher level than the species level that result in the formation of a higher taxonomic group such as a new genus or class (see spread 21.1). Some biologists believe that macroevolution results from a build-up of small changes by microevolution; others believe that macroevolution is quite distinct from microevolution.

Pesticide resistance

Attempts to eradicate pests with deadly chemicals is continually frustrated by organisms evolving ways to resist them. **Warfarin** is an anti-coagulant used as a poison against rats and mice. It binds with and inhibits vitamin K epoxide reductase, an enzyme that catalyses the production of clotting factors. It is used in low concentrations as a medicine for people at risk of strokes and heart attack. In high concentrations, it causes potentially lethal internal bleeding, hence its use as a rat poison.

After repeated exposures over several generations, some rodent populations have become resistant to warfarin by evolving enzymes onto which warfarin cannot bind so readily, or by evolving other enzymes that can break down the poison. Resistance is due to random mutation of the genes responsible for the enzymes; it is not caused directly by the warfarin. The allele for resistance would have been present in a few individuals in a population before warfarin was ever used as a rat poison. However, repeated exposures to the poison kills most rats and mice, leaving mainly warfarin-resistant ones to survive and reproduce, resulting in a gradual increase in the frequency of the resistance allele (and resistant rodents) in the population. With the increase in warfarin resistance, and the development of more potent alternatives, the use of warfarin as a rodent poison has declined.

Antibiotic resistance, pesticide resistance, and heavy-metal tolerance show how natural selection can affect a gene pool by increasing the frequency of alleles that can give an advantage and reducing the frequency of alleles that give a disadvantage.

Industrial melanism

Many species, especially insect species, have two or more adult body forms called **morphs** that are genetically distinct from one another, but that are contained within the same interbreeding population. This condition is known as **polymorphism**. In the UK, the peppered moth (*Biston betularia*), for example, has three morphs for wing colour: one morph called *typica* has pale wings speckled with dark markings; another morph called *carbonaria* is melanic because its wings contain large amounts of the black pigment melanin; the third, known as *insularia*, is an intermediate partly melanic morph. Until 1849, Manchester moth collections contained predominantly the pale form, but by 1895, 98 per cent of moths caught in Manchester were melanic. This increase in the frequency of melanic forms was linked to air pollution and is called **industrial melanism**. Between 1849 and 1895, Manchester had become so highly industrialised that soot from factory chimneys had blackened trees and fences in and around the city.

In the 1950s, H. B. D. Kettlewell demonstrated experimentally that industrial melanism was probably due to selective predation by birds. He bred pale and dark moths, marked them, released both forms in two separate areas, one in Dorset (unpolluted) and the other in Birmingham (polluted), and recorded the percentage recaptured (table 1). Using binoculars and cine-film, he observed that the peppered moth flies by night and rests by day on tree trunks and wooden fences. During the day, it is hunted by birds that rely on sight to catch their prey. He concluded that these birds produced the main selection pressure by feeding differentially on moths according to their background. In unpolluted areas with lichen-covered trees, the pale form had the selective advantage of not being clearly visible, whereas in polluted areas with soot-blackened trees and no lichens (lichens are generally very intolerant of air pollution), the melanic form had the selective advantage (figure 2).

Table 1 *Data from Kettlewell's mark–release–recapture experiment.*

Polluted region	Pale form	Melanic form
Marked and released	137	447
Recaptured later	18	123
Percentage recaptured	13.1	27.5
Relative recaptures (pale: melanic)	0.48	1
Unpolluted region		
Marked and released	496	473
Recaptured later	62	30
Percentage recaptured	12.5	6.3
Relative recaptures (pale: melanic)	1	0.51

Whether a peppered moth is pale, intermediate, or fully melanic is determined genetically, with the allele for melanism producing the *carbonaria* morph being controlled by a single gene locus. Figure 3 shows that in 1958 the frequency of the melanic moths was high in industrialised areas or downwind of them. Since the 1970s, air pollution has decreased sufficiently for trees and fences to have become cleaner so that lichens grow on them. Consequently, the frequency of pale forms has increased.

The mutation that produces the allele for melanism occurs spontaneously; it is not produced by pollution. A change in an environmental factor, such as air pollution, changes the frequency of an allele in a population, but it does not affect the rate at which such an allele arises by mutation. Before 1849, melanic moths probably occurred but were too rare to have been collected; they would almost certainly have been eaten by birds before the moths could reproduce.

Although industrial melanism of the peppered moth demonstrates how natural selection can drive the evolutionary process, the different morphs have remained members of the same species because they can interbreed. Nevertheless, if the moths in polluted and unpolluted areas were prevented from interbreeding for long enough, they would probably become sufficiently different to evolve into two distinct species (see speciation, spread 20.8).

Figure 2 *Pale and melanic forms of the peppered moth on bark from:*
(a) *an unpolluted area, and*
(b) *a polluted area.*

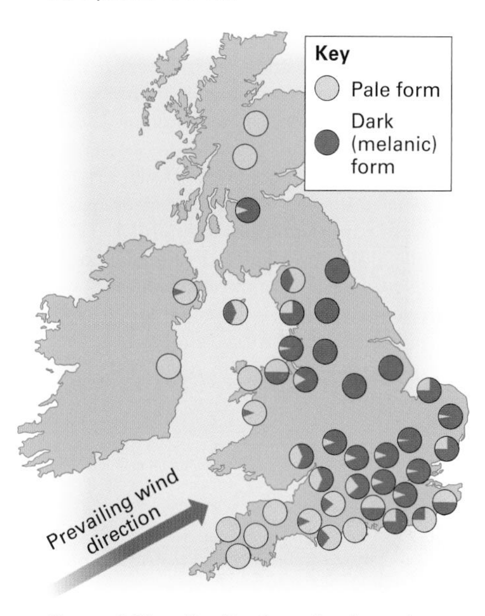

Figure 3 *The distribution of pale and melanic forms of the peppered moth in the British Isles in 1958.*

Quick check

1 Why are heavy-metal tolerant plants rare in unpolluted regions?

2 What effect did industrial pollution have on:

a the frequency of the alleles determining melanism within a population of peppered moths

b the rate of mutation of the alleles determining melanism?

Food for thought

The two colour morphs of the peppered moths studied by Kettlewell are determined by a single gene with two alleles. The allele for melanism is dominant. Suggest how the rate of evolution might have differed if the allele for melanism were recessive.

20.6

OBJECTIVES

By the end of this spread you should be able to:

- discuss why the allele for sickle-cell anaemia is retained in a population
- state the conditions necessary for allele frequency to remain constant in a population
- explain how a change in allele frequency in a population can be used to measure evolution.

Fact of life

If people with sickle-cell trait contract malaria, they have only about one-third the number of *Plasmodium* parasites in their blood as do people with normal blood. In one study, 99% of children who died from malaria were homozygous for the normal allele, although within the population as a whole 20% of the children were heterozygous. In some malaria-infested parts of Africa, almost 14% of babies are born with sickle-cell anaemia.

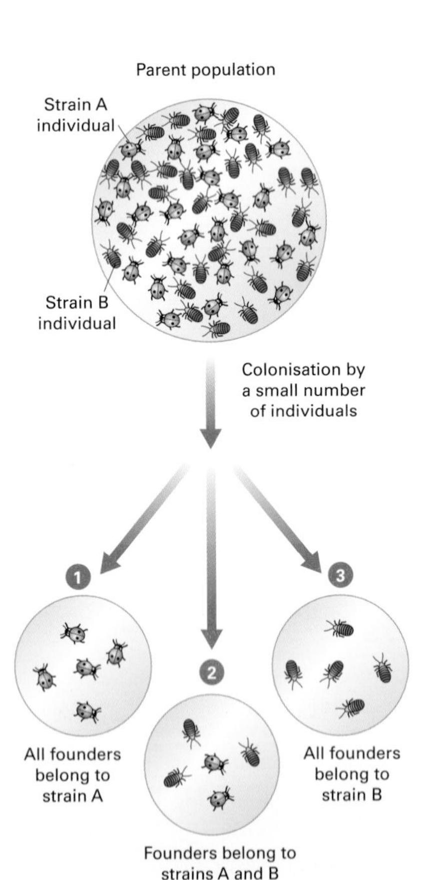

***Figure 1** The founder effect: in this example, the colonisation of three islands by a small number of individuals from a parent population containing two distinct genetic strains results in three founder populations.*

Natural selection and allele frequency

Natural selection can change the frequency of particular alleles in a **gene pool** (the sum of all alleles within a population). For example, during the nineteenth century, industrialisation acted as a selection factor, increasing the frequency of the melanin allele in peppered moth populations around Manchester (spread 20.5). However, natural selection does not always result in change; it can also be an agent for stability (see spread 20.4, figure 2a). This is illustrated by the fact that the sickle-cell allele is retained in human populations.

Sickle-cell anaemia

Sickle-cell anaemia is caused by a single gene with two alleles $\mathbf{Hb^A}$ and $\mathbf{Hb^S}$. People with the genotype $\mathbf{Hb^AHb^A}$ have normal blood. Those with the genotype $\mathbf{Hb^AHb^S}$ have sickle-cell trait, and they may have problems when oxygen demands are high such as during strenuous exercise, or when oxygen partial pressures are low, such as at high altitude. Those with the genotype $\mathbf{Hb^SHb^S}$ have abnormal blood that causes sickle-cell anaemia, a very serious disorder which can be fatal (spread 16.1). You might expect such a disadvantageous allele as $\mathbf{Hb^S}$ to be selected against and eventually eliminated from human populations. However, in some parts of the world this allele is common and many babies are born with sickle-cell anaemia.

Why has the sickle-cell allele survived?

The distribution of the sickle-cell allele corresponds closely with that of malaria in Africa and parts of Asia. Malaria is caused by the single-cell proctotist parasite *Plasmodium* which can be introduced into the bloodstream by the bite of a mosquito (see spread 15.4). Malaria is a major disease in many parts of the world, killing or debilitating millions of people. Research has shown that people who are heterozygous for the sickle-cell allele are much less likely to suffer from a serious attack of malaria than people who are homozygous for the normal allele. So there are two conflicting selection pressures acting on the sickle-cell allele: there is a strong selection pressure against the allele in the homozygous condition ($\mathbf{Hb^SHb^S}$), whereas in malaria-infested areas there is a strong selection pressure for the allele in the heterozygous condition ($\mathbf{Hb^AHb^S}$). In addition, people who are homozygous for the normal allele ($\mathbf{Hb^AHb^A}$) will have a selective disadvantage in malaria-infested regions. The situation in which the heterozygote is fitter than either of the homozygotes is known as **heterozygous advantage**. The net result of the opposing selection pressures is to retain the sickle-cell allele within populations in places where malaria is an important environmental factor. The actual frequency of the allele in the population varies according to the amount of malaria. As expected, where malaria has never been present, the sickle-cell allele is eliminated quickly and has never become established among the natural population. Of course, people migrating from a malaria-infested region to an uninfested region may carry the sickle-cell allele and pass it on to their children. Hence about one in 400 black people in Britain has sickle-cell anaemia, and one in ten has sickle-cell trait.

Changes in allele frequency

In 1908, G. H. Hardy and W. Weinberg independently devised an equation to calculate allele and genotype frequencies in a population. The equation is based on the fact that the total frequency of alleles of a gene must equal 100% (expressed as a probability of 1.0). In a population that has a gene with two alleles, **A** dominant and **a** recessive, the total allele frequency must therefore equal 1.0.

They expressed this in an equation as:

$$p + q = 1.0$$

where p represents the dominant allele (**A**) and q represents the recessive allele (**a**).

To calculate genotype frequencies in a diploid organism, they derived from this equation another equation, now known as the **Hardy–Weinberg equation**:

$$p^2 + 2pq + q^2 = 1.0$$

where p^2 represents the frequency of the **AA** genotype, $2pq$ represents the frequency of the **Aa** genotype, and q^2 represents the frequency of the **aa** genotype (see appendix page 577 for an example of how this equation can be used).

Using this equation they identified a mathematical relationship between the alleles and genes in populations. The relationship, called the **Hardy–Weinberg equilibrium**, states that, provided certain conditions are met:

The frequency of dominant and recessive alleles in a population will remain constant from generation to generation.

The conditions that need to be met are that:

1 the population is large
2 mating is random
3 no mutations occur
4 there is no immigration or emigration from the population
5 all genotypes are equally fertile so that no natural selection occurs.

If the first four conditions are met, a change in allele frequency provides a means of measuring the rate of evolutionary change.

Hardy and Weinberg recognised that allele frequencies are likely to remain constant only if populations are large. In small populations, chance factors play a significant part in determining which alleles are passed on to the next generation. The smaller the population, the greater the probability that the allele frequency will differ between one generation and the next. The name given to the change in allele frequency due to chance is **genetic drift**. It may occur when a few individuals from a large population colonise isolated areas such as an island (the so-called **founder effect**, figure 1). Or it may occur when a catastrophic event creates a **genetic bottleneck** (figure 2): the population declines dramatically with only a few individuals acting as the source of genes for future generations, and inbreeding is increased due to the reduced pool of possible mates. In either case, there will be less genetic variation within the small population than in the original population. In addition, some alleles from the large, original population may become absent in the small population and others may be disproportionately represented.

***Figure 2** The genetic bottleneck analogy. Shaking just a few beads through the narrow neck of a bottle is like a catastrophic event allowing only a few individuals of a population to survive. By chance, shaking the bottle results in the black beads being over-represented in the surviving population and the yellow beads being absent. Similarly, genetic bottlenecks tend to result in populations with reduced genetic diversity.*

Quick check

1 Explain what is meant by heterozygous advantage, using the sickle-cell allele as an example.

2 Under what conditions can changes in allele frequency be used to measure evolution by natural selection?

3 If one in 2200 people in a population has cystic fibrosis, an inherited disorder caused by a single recessive allele, what is the frequency of carriers?

Food for thought

Suggest how, despite reducing genetic variation within a population, genetic drift can increase variation in a species as a whole and contribute to the evolution of a new species.

20.7

Artificial selection

OBJECTIVES

By the end of this spread you should be able to:

- describe one example of artificial selection
- distinguish between inbreeding and outbreeding
- explain the meaning of hybrid vigour.

Fact of life

With the advent of genetic engineering, artificial selection has entered a new phase. It is now possible to breed clones of cattle and sheep which have genes for producing specific human proteins (see spread 18.10). What is more, nuclei of two different species can be combined to form a completely new type of animal. In this way, a hybrid that combines the characters of a sheep and a goat has been formed: this new species has been dubbed a 'geep' by the popular press. Plants can also be genetically engineered to incorporate characters of a number of different species; for example, potatoes with a high starch content and high productivity can be genetically engineered to produce the beta-carotene of green vegetables and the vitamins of citrus fruits. One day it might be possible to design foods on a computer by choosing characteristics from a palette of tastes, colours, textures, and nutrients.

Figure 1 *Some modern wheat varieties are less than half as tall and yield three times as much grain as nineteenth-century varieties such as that shown here.*

The cultivation of wheat

Ever since farming began in the Middle East about 10 000 years ago, humans have been breeding animals and plants selectively to produce specific desirable qualities. Wheat was probably among the first crops to be cultivated. By selective breeding over thousands of generations, wild wheat has been converted into the modern types which produce much higher yields. In **selective breeding**, particular individuals are chosen and allowed to breed, whereas others are prevented from breeding. This means that alleles that give characteristics favoured by humans are retained, while those that give undesirable characteristics are eliminated. Artificial selection is therefore similar to directional selection (spread 20.4), in that selection pressure brings about a gradual change in the genotype of a group of organisms. However, in artificial selection it is humans, not environmental factors, that act as the selection pressure, gradually bringing about changes in allele frequencies.

We can only speculate as to how wheat cultivation began. Perhaps people who gathered wild seeds for food observed that seeds spilled accidentally sprouted new plants from which more seeds could be harvested. This might have encouraged them to save some seeds to sow for the following season's crop.

Wild wheat sheds its grains as soon as they are ripe. This makes harvesting difficult. Therefore, grains were most likely to be gathered from plants that by chance retained their grains a little longer. By using this grain for the next crop, farmers would inadvertently have started the process of selective breeding.

The next stage in the cultivation of wheat would have been the deliberate selection of varieties with desirable qualities. Early farmers appear to have selected grains from plants which gave the greatest yield, and produced grain which was easy to separate from its husk. Eventually, over many generations, the variety of cultivated wheat changed. This led to the ancestor of our modern wheat, in which the grains are held so firmly that they must be removed by a separate operation (threshing) after harvest. Selective breeding of wheat continues today by a combination of inbreeding and outbreeding.

Inbreeding

Inbreeding involves breeding between closely related individuals which, by chance, possess some desirable character. In wheat, desirable characters include:

- high yield
- short stem length (allowing the plant to devote more energy to the production of seeds, which have a much higher value than straw from stems)
- pest resistance (for example, to fungal moulds and rusts)
- high protein content of the grain.

Inbreeding is carried out to try and retain the desirable characters in future generations. Wheat plants are particularly suitable for selective breeding because they self pollinate naturally. They are unlikely to cross fertilise without the intervention of the plant breeder.

Inbreeding allows a farmer to produce a uniform crop which is easy to harvest and has, given certain conditions, predictable characters. However, this uniformity of characters is at the expense of genetic diversity. In wheat, genetic diversity may be reduced to such an extent that every individual has identical alleles (a condition known as **complete homozygosity**). Such a wheat strain cannot be changed because there are no other alleles present that could produce genetically different plants.

Another problem is that if genetically identical plants are exposed to new diseases to which the plants have no resistance, all the plants may be killed. (This is precisely what happened in Ireland in 1845 when potato blight caused by *Phytophthora infestans* destroyed the whole potato crop and led to famine. All the potatoes belonged to the same variety, which had no resistance to the disease.)

Similar techniques of selective breeding have been used to develop domestic and farm animals. Although complete homozygosity has not been reached in any animals, inbreeding increases the risk of a harmful recessive allele occurring in the homozygous condition and being expressed. Because of these disadvantages, inbreeding is not carried out indefinitely. New alleles are introduced by outbreeding with other stock.

Outbreeding

Outbreeding involves crossing individuals from genetically distinct strains. The offspring from such a cross are called **hybrids**. If the parental stocks are pure breeding, the offspring are called F_1 hybrids. F_1 hybrids often have characters, such as grain yield in wheat, which are superior to the characters in either parent. This phenomenon is called **hybrid vigour** or **heterosis**. Hybrid vigour probably results from an increased heterozygosity arising from the mixing of alleles. Harmful recessive alleles are less likely to be present in the homozygous condition. Hybrid vigour is also thought to result from some form of interaction between particular combinations of alleles in the hybrid. Whatever the explanation of hybrid vigour, if the descendants of F_1 hybrids are continually inbred, the vigour decreases as the plants become more homozygous again.

Outbreeding depends on the availability of genetically distinct animals and plants. It is therefore important to maintain sources of genetic diversity. This may be done by maintaining seed banks of old or wild varieties of plants (the genetic diversity of wheat, rice, cabbages, and carrots is maintained in this way). Also, adults of old varieties of animals and plants with little or no commercial value may be maintained as a source of new alleles for future breeding programmes.

Quick check

1 Which type of natural selection does artificial selection resemble?

2 What effect does:

  a inbreeding

  b outbreeding

  have on the genetic diversity of a population?

3 Give two possible explanations of hybrid vigour in plants produced by a cross between two different strains of pure-breeding plants.

Food for thought

The dog is thought to have been the first domesticated animal. For at least 12 000 years, it has been subjected to artificial selection. Dogs have been bred to do specific types of work (for example, Labrador retrievers for retrieving fishing gear, Old English sheepdogs for rounding up sheep, and poodles for retrieving ducks) or for show. Suggest why pedigree dogs bred for show tend to have more genetic disorders than mongrels and cross-breeds (for example, highly inbred pedigree Labradors often have hip problems, St Bernards suffer eye problems, and Pekineses, figure 2, often have respiratory problems).

Cattle breeding

Cattle are among the most intensively bred farm animals. Artificial selection techniques used to improve milk production include:

- **progeny testing**, in which the genetic capabilities of an organism are established by examining the performance of its offspring. In this case, the 'milk-yield' genes of a bull can be estimated by measuring the milk productivity of its daughters.
- **artificial insemination** (**AI**), in which sperm from a suitable bull are collected in an artificial vagina, tested for motility, stored in a suitable medium (e.g. milk and egg yolk mixed with a citrate buffer) and, at an appropriate time, inserted into the uterus of selected cows using a tube called a **catheter**. AI allows the sperm of one superior male to be used to fertilise many different females.
- **embryo transplantation**, in which an embryo is removed from a real mother and transferred to a surrogate mother. Embryo transplantation can be used to increase the reproductive rate of a good animal; a particularly good milk-producing cow can be treated with hormones to produce large numbers of eggs which, after fertilisation, can be transferred to a number of surrogates. It is also used in **cloning**; embryos are split into two separate cells after first division and implanted into surrogates (see spread 18.10).

Figure 2 *Pekinese dogs were bred in China, and used to be carried in the sleeves of Chinese robes by courtiers.*

20.8

Speciation (1)

OBJECTIVES

By the end of this spread you should be able to:

- explain the meaning of the term species
- list the main types of isolating mechanism
- describe one example of allopatric speciation.

Fact of life

African cichlids are freshwater fish belonging to the same group as the wrasses. There are almost 2000 species in the Great Lakes of East Africa that have evolved over the last 10 million years. In each of the major lakes, one or a few ancestral species have undergone adaptive radiation that has resulted in several hundred species. These are closely related genetically but are phenotypically very diverse. Lake Victoria alone has hundreds of species. Many are endemic and can be found nowhere else in the world. A species adapted to a particular niche in one lake may look similar to a species adapted to the same niche in another lake by a process called convergent evolution (see spread 20.3), but the two species do not usually interbreed because they are geographically isolated. Different species living in the same lake do not usually interbreed because of reproductive isolating mechanisms. Most mechanisms involve distinct courtship rituals that result in mates being chosen only from members of the same species.

(a)

(b)

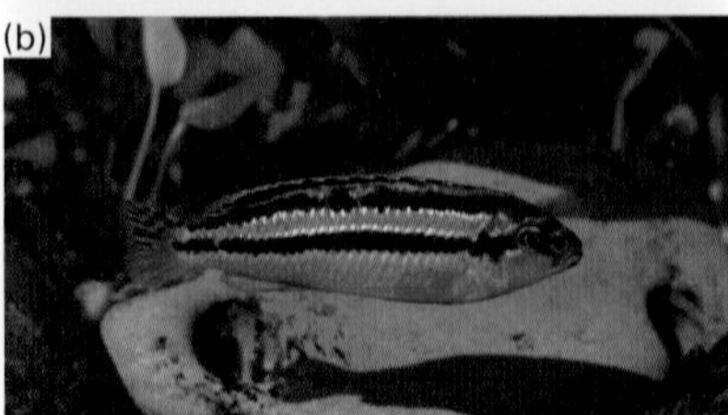

Figure 1** Two species of cichlid: **(a) Julidochromis ornatus *from Lake Victoria and **(b)*** Melanochromis auratus *from Lake Malawi; they look similar because of convergent evolution, but they do not normally interbreed because they are geographically isolated.*

Evolution occurs whenever the inherited characteristics of a population or of a species change over a period of time. When these changes lead to the formation of one or more new species, **speciation** has taken place.

A definition of species

A species can be defined as: *a group of organisms with similar features which can interbreed to produce fertile offspring, and which are reproductively isolated from other species.*

The central part of this and most other definitions of species is that members of the same species can interbreed to produce fertile offspring. Thus, although donkeys can interbreed with horses to produce offspring called mules, donkeys and horses are regarded as separate species because mules are infertile.

Organisms which do not interbreed to produce fertile offspring under normal circumstances are regarded as **reproductively isolated**, and they belong to separate species. In nature, the coyote (*Canis latrans*) and the wolf (*Canis lupus*) do not normally interbreed and are therefore classed as different species even though they are capable of interbreeding and producing fertile offspring in captivity (figure 1).

Reproductive isolation: becoming unable to interbreed

Rarely do all the members of a species exist in a single large population. They usually live in a number of local interbreeding populations called **demes**. Each deme has its own distinct **gene pool**, but two or more demes will belong to the same species as long as some individuals from the different demes continue to interbreed and produce fertile, hybrid offspring. On the other hand, if **reproductive isolation** occurs between the demes (the members of different demes are unable to interbreed at all), the demes may over a long period of time evolve by natural selection into new species.

Problems in defining species

Whether species interbreed or not can only be observed in sexually reproducing organisms. This definition of a species cannot therefore be used where reproductive behaviour has not been observed, for example in extinct organisms.

In practice, organisms are often placed in a particular species on the basis of having 'similar features', particularly **morphological features** (structural features which determine the organism's external appearance). Sometimes these are the only clues available (for example, with most fossils). However, biologists often have difficulty deciding which morphological features to use, and also how similar two specimens must be to belong to the same species. In addition, many species, such as the peppered moth, are polymorphic (they have two or more different structural forms), yet the different forms can interbreed and obviously belong to the same species.

Wherever possible, several features are used to define a species, including:

- physiological features (the way the organism carries out functions such as respiration and excretion)
- behavioural features (for example, the way organisms communicate with each other, such as the different songs of birds)
- ecological features (generally, members of the same species share the same ecological niche; see spread 22.5)
- biochemical features (for example, comparison of amino acid sequences in haemoglobin)
- genetic features (members of the same species will have a similar karyotype, that is, similar chromosomes).

Even when several features are used, there is no common agreement as to what precisely makes up a species. This can lead to great confusion because some biologists lump together organisms with slight variations into one species, whereas other biologists split them up into different species.

Mechanisms that prevent breeding between populations and which can eventually lead to speciation are called **isolating mechanisms** (table 1). Mechanisms that prevent the formation of hybrids are called **prezygotic isolating mechanisms**; mechanisms that affect the ability of hybrids to produce fertile offspring are called **postzygotic isolating mechanisms**. The most important isolating mechanism is thought to be **geographical isolation**, in which two populations originally of the same species are separated from each other by a physical barrier such as a mountain, river, or ocean.

When geographical isolation leads to new species being formed, **allopatric speciation** is said to have occurred. (Allopatric means literally 'different countries'.) Any physical barrier that prevents members of different populations from meeting must inevitably prevent them from interbreeding. Figure 2 gives a hypothetical example showing how allopatric speciation might take place. Note that although geographical isolation is the original cause of allopatric speciation, the two isolated populations diverge so much from each other that when reunited they are unable to interbreed. Other isolating mechanisms now keep the two species from breeding together.

When speciation occurs in two or more demes living in the same geographical location, **sympatric speciation** is said to occur. (Sympatric means 'same country'.) Clearly, sympatric speciation results from mechanisms other than geographical isolation. We shall consider these mechanisms in the next spread.

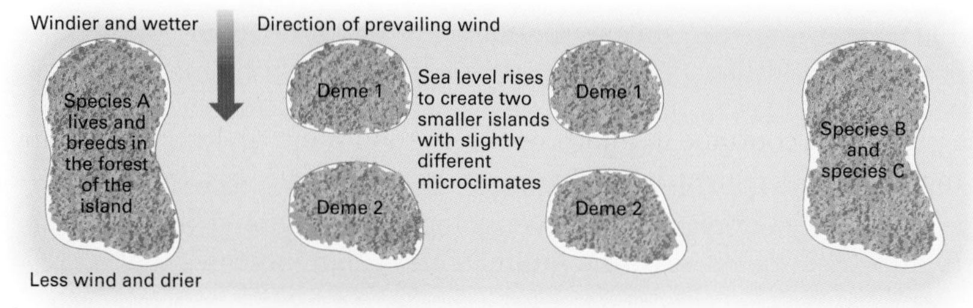

Figure 2 *A hypothetical example of allopatric speciation.*

Table 1 *Isolating mechanisms.*

Type of isolation	Reason for the isolation
Geographical isolation	Organisms isolated by a physical barrier, such as a mountain, river, or ocean
Temporal isolation	Organisms breed at different times of year
Ecological isolation	Organisms live in different habitats within the same area
Behavioural isolation	Organisms have different behaviour patterns, e.g. they use different behaviour to attract a mate. In the fruit fly *Drosophila*, for example, normal mating involves males performing a ritualised 'dance' that has a definite sequence of wing and body movements. Closely related species will not normally mate because the courtship dances of the males are different. But experiments have shown that, in some cases, mating will occur if the antennae of the female are removed. Presumably, the female is unable to detect the wrong courtship dance and permits mating.
Mechanical isolation	Organisms cannot mate because of anatomical differences which make it impossible for gametes to come together
Gametic isolation	Genetic or physiological incompatibility between different organisms prevents hydrids forming, e.g. pollen may fail to grow on a particular stigma with incompatible genes
Hybrid isolation	Different organisms interbreed but offspring do not survive or are infertile

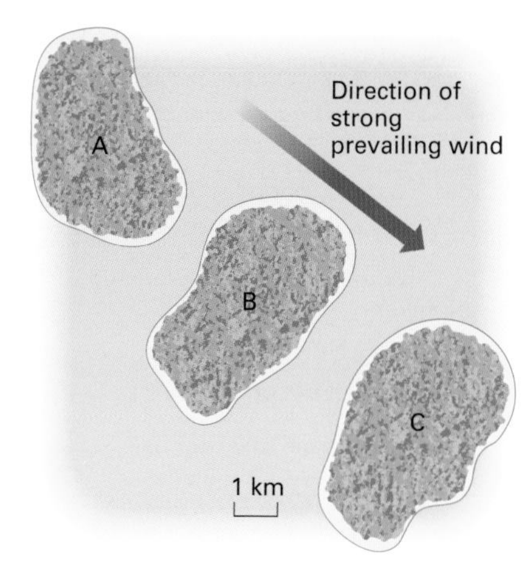

Figure 3 *Three populations of butterflies living on three islands.*

Quick check

1 Briefly explain why two types of organism may be regarded as separate species even though they can interbreed to produce fertile offspring.

2 What type of isolating mechanism would occur if the only dogs left in the world were St Bernards and chihuahuas?

3 Distinguish between allopatric speciation and sympatric speciation.

Food for thought

Three populations of butterflies, A, B, and C, live separately on three oceanic islands (figure 3). The butterflies all feed on nectar, but have slightly different wing colouring. The islands are swept through the year by strong prevailing winds from the north-west. Populations A and B can interbreed and produce fertile offspring. Population B can mate with population C, but the offspring are infertile. Matings do not occur at all, even in captivity, between populations A and C. Suggest how this situation may have arisen.

20.9

OBJECTIVES

By the end of this spread you should be able to:

- explain how polyploidy occurs
- describe one example of sympatric speciation.

Fact of life

Botanists estimate that over half of the world's flowering plants are polyploid (they have more than two sets of chromosomes). Many of our food plants are polyploid: for example, bread wheat is hexaploid (6*n*), and cultivated blackberries and strawberries may be octoploid (8*n*), whereas the wild varieties of these plants are usually diploid (2*n*).

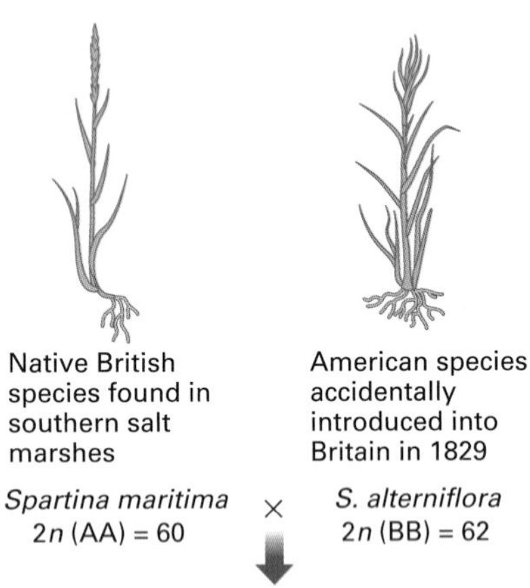

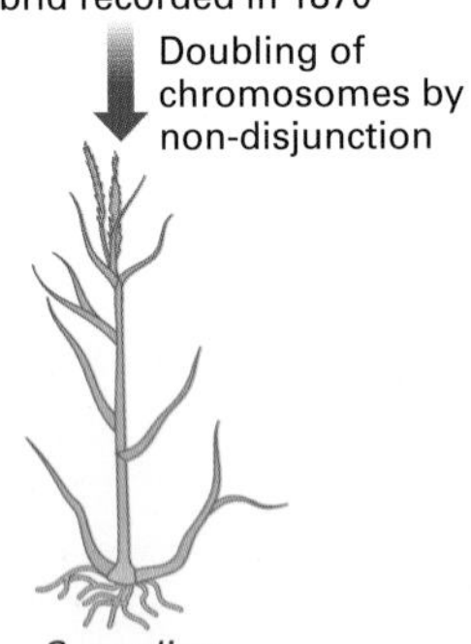

***Figure 1** Polyploidy in* Spartina. *(n = number of chromosomes; each set is represented by a letter.)*

Speciation (2)

New animal species usually arise in geographically isolated areas by **allopatric speciation** (spread 20.8). New plant species, however, are more likely to evolve from individuals living in the same area; a process called **sympatric speciation**. The most common method of sympatric speciation involves polyploidy. A **polyploid** organism is one that has more than two sets of chromosomes. One of the best understood examples of polyploidy leading to the formation of new species occurred in the genus of cord-grass called *Spartina* (figure 1); its speciation is an example of evolution that has been observed in action.

The story of *Spartina*

Cord-grasses are a very distinct genus of coastal plants living on mudflats and salt marshes. They are stout perennials with thick tough leaves and an extensive system of rhizomes (underground stems). Until the nineteenth century, there was one species of *Spartina* in the British Isles, *S. maritima*. This is a small cord-grass that grows in the fringes of gullies and the shallows of pools on salt marshes. In 1829, a larger cord-grass with longer leaves was noticed growing in the River Itchen near Southampton. It turned out to be a North American species *S. alterniflora*, introduced from the United States, probably via a ship.

For the next 40 years, *S. alterniflora* spread along the coast. Then, in 1870, it interbred with *S. maritima* to form hybrids. These hybrids were similar to the American species, but displayed hybrid vigour, being more vigorous and healthier than either parent species. As with many grass hybrids, the anthers of the new plants were devoid of pollen, making the plants sterile. This is because the hybrid has two different sets of chromosomes; there are no homologous chromosomes to pair up so meiosis cannot take place. The new cord-grass was named Townsend's cord-grass, *S.* × *townsendii* (the multiplication sign indicates that it is sterile).

Despite its sterility, the ability of Townsend's cord-grass to reproduce vegetatively (for example, by fragments of rhizomes being uprooted during storms and carried along the coast by tidal currents), its fast growth, and its robust nature ensured that it thrived and spread. One of its greatest strengths is its ability to colonise tidal mudflats, often to the exclusion of other species. This has made Townsend's cord-grass ideal for planting in areas that need to be established for land reclamation schemes.

In 1892, another type of cord-grass was discovered. It resembled Townsend's cord-grass, but had broader leaves and was remarkable in having perfectly formed anthers which produced pollen. After examining its chromosome number and comparing it with related species, botanists concluded that it is a version of Townsend's cord-grass which, by chance, has doubled its chromosome number. This came about by non-disjunction of all the chromatids (that is, all the pairs of chromatids or homologous chromosomes failed to separate and go to opposite poles at mitosis or meiosis). This made the plant fertile because each chromosome can now pair up with its homologous partner during meiosis. Because of its fertility, this new plant is regarded as a new species and is called *S. anglica*.

S. anglica is an even more successful coloniser than Townsend's cord-grass. Today, it can be found in all suitable habitats in England and Wales. It has also been introduced into Scotland and Ireland to stabilise mudflats.

Polyploidy in other organisms

The species of wheat that we use today in bread is thought to have evolved in a similar way to *S.anglica* (figure 2). Modern bread wheat, *Triticum aestivum*, is a hexaploid (6*n*). Its chromosomes can pair up during meiosis, therefore it is fertile.

Allopatric speciation is a slow process that may be barely detectable even after many generations, but sympatric speciation involving polyploidy can occur in a very short time. In fact, in the laboratory, polyploidy can be induced in individual cells almost instantaneously by the application of colchicine (a chemical extracted from crocus, *Colchicum*). In the correct concentration, colchicine prevents cell division but not chromosome replication. The polyploid cells that are formed may then be cultured to produce polyploid organisms with sex organs that make gametes by meiosis in the normal way. Colchicine is used to induce polyploidy in various crops, including wheat, black-berries, and strawberries because it increases hardiness and vigour. Polyploidy that involves duplication of chromosomes derived from only one species is called **autoploidy** (all chromosomes come from the same original species). Polyploidy involving duplication of chromosomes derived from two or more species is called **allopolyploidy** (figure 1 shows that S. anglica is an allopolyploid formed by the doubling of chromosomes in *S.* × *townsendii*, which is a hybrid between two species, *S. maritima*, and *S. alterniflora*).

Polyploidy is rare in animals, probably because there are few instances of cross-breeding between species. Also, many animals have sex chromosomes, the numbers and types of which are essential for normal development.

Polyploidy is not the only means by which sympatric speciation can happen. Any of the isolating mechanisms shown in spread 20.8, table 1 (other than geographical isolation) can lead to the failure of different populations living in the same area to interbreed. For example, the maggot fly is a pest of the North American hawthorn tree. When apples were introduced into the USA in the nineteenth century, the maggots attacked them too. The flies have become specialised into two distinct populations: one infecting hawthorn trees, and the other attacking apple trees. Although the two populations can still interbreed, they have developed a number of differences in addition to which fruit they prefer. For example, in controlled laboratory conditions, the flies from hawthorns mature in about 54–61 days, whereas those from apples mature in 40 days. It is likely that these flies will continue to diverge so that eventually they will not be able to interbreed. Then they will become two different species.

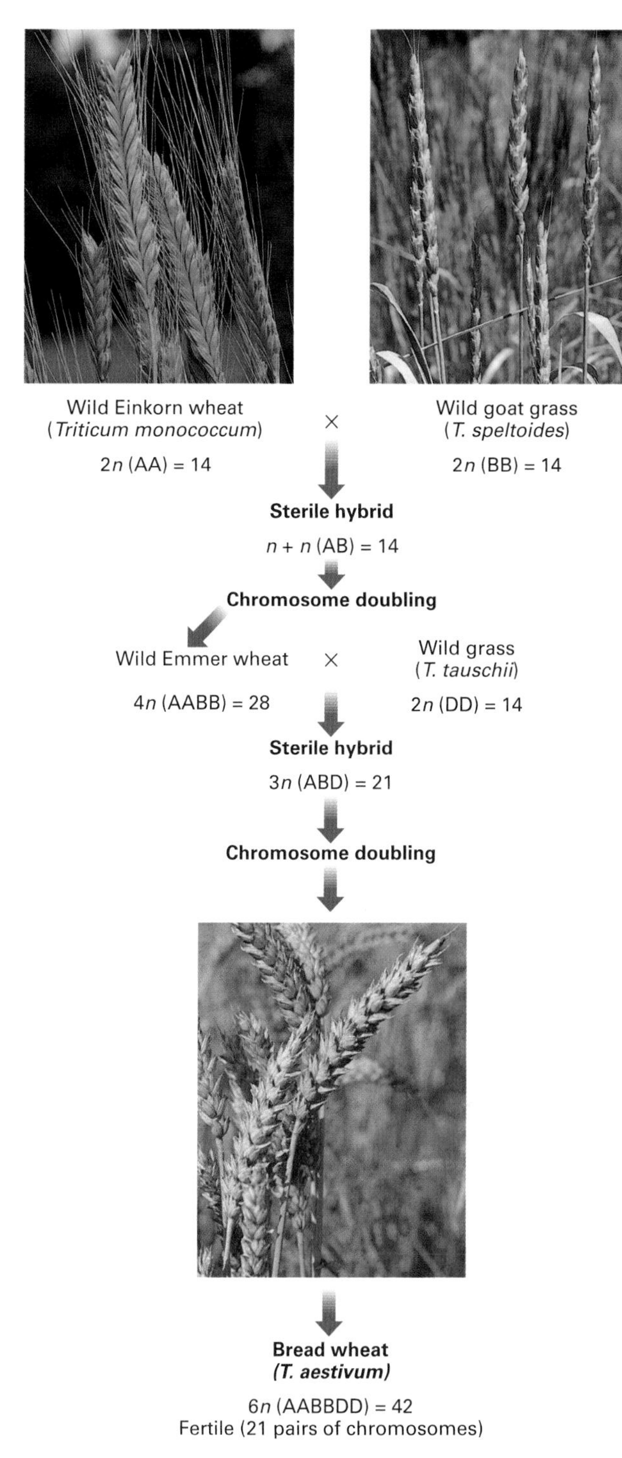

Figure 2 *Polyploidy in wheat. The capital letters represent sets of chromosomes that can be traced to a particular species. Bread wheat has chromosome sets derived from three ancestral species.*

Quick check

1 Polyploids are usually bigger and more vigorous than their diploid counterparts. They are important to plant breeders in the search for bigger and better crops.

a Name a chemical that can induce polyploidy in crop plants.

b Briefly explain how this chemical causes polyploidy.

2 Why are the hybrids formed by a cross between *Spartina maritima* and *S. alterniflora* sterile?

Food for thought

Polyploidy in *Spartina anglica* arose spontaneously by chance. Suggest the part natural selection might have played in the evolution of this species.

20.10

OBJECTIVES

By the end of this spread you should be able to:

- explain the significance of the adaptions of primates to an arboreal mode of life.

Fact of life

Lemurs are cat-like primates that live exclusively in the tropical rainforests of Madagascar. It is thought that ancestral lemurs became isolated on the island about 50 million years ago and gradually diversified into 40 species (an example of adaptive radiation). Lemurs have retained numerous primitive characteristics while at the same time developing many features in parallel with the monkeys and apes that evolved on the mainland. During this evolution, body mass gradually increased (the ancestral species was very small) which corresponds with a shift away from mainly nocturnal (night-time) activity to diurnal (day-time) activity. This evolutionary trend is also seen among the monkeys and apes. Primitive lemur species are small nocturnal animals that spend nearly all their time climbing and leaping in trees, living mainly on insects. Several other species of lemur (including *Lemur catta*, figure 1) live on the ground. These more advanced lemurs evolved to live in social groups, associated with their becoming diurnal. The young grow up within a troop and much time is spent learning the skills of life. Individuals cooperate within the group to gather food (fruit and leaves as well as insects) and avoid predators. However, none of the lemurs have the manual dexterity or intelligence of apes and monkeys.

Figure 1 *The ring-tailed lemur,* Lemur catta.

HUMAN EVOLUTION: PRIMATE ANCESTORS

The theory of evolution applies just as much to humans as to other organisms. All humans are in some way related and, in the words of Darwin, are 'descended with modification' from a common ancestor. Although our social and technological developments have freed us from many of the effects of natural selection, our present-day physical and behavioural characteristics are rooted in the adaptations of our ancestors. By finding out more about our ancestors, we can learn more about ourselves.

Adaptations of primates

The classification of humans (table 1) reflects our evolutionary relationships. About 150–170 million years ago, all mammals were small insectivores rather like the shrews of today. About 75 million years ago some of these insectivores adopted an arboreal (tree-dwelling) mode of life and evolved into lemur-like primates. The adaptations of these ancestral primates to their new tree-living mode of life are thought to have included a short nose, large eyes and prominent ears, long flexible fingers with nail-like claws, and teeth well adapted for eating insects. These features are found in tarsiers (lemur-like primates) living today in Indonesia. Many other features that evolved in ancestral primates as adaptations to an arboreal life have been retained by modern primates. These features include:

- **A prehensile (grasping) limb**: the hands (and often the feet) of primates have long and highly mobile digits so that they can grasp the branches of trees. The first digit can oppose the remaining four digits, giving primates a powerful grip. (This opposing movement can be seen if you hold your hand palm upwards and move your thumb across your palm to point towards your little finger.) Primates have flattened nails that support pads of sensitive skin on the fingers or toes.
- **A mobile forearm**: the clavicle (collar bone) and scapula (shoulder blade) are adapted to allow a wide range of movements. Mobile forearms are essential for moving from tree to tree, and for manipulating objects in the hand; for example, to transfer food to the mouth or to bring an object to the eyes for closer examination.
- **Well developed stereoscopic vision**: the ability to judge distances is essential for leaping from branch to branch. Primates have large, well developed, forward-looking eyes with overlapping fields of view. The development of stereoscopic vision has been associated with a flattening of the face.
- **A reduced sense of smell**: it is not easy to locate scents through the canopies of trees and primates have a reduced sense of smell and a relatively small nose. Combined with the flattening of the face, a shorter nose is associated with the development of stereoscopic vision, and has allowed the development of facial muscles which play an important part in non-verbal communication.
- **An unspecialised digestive system**: primates have relatively unspecialised teeth and guts and they can exploit a wide range of food sources. Although some primates have a specialised herbivorous diet, all primate families have some omnivorous members that have a mixed diet.
- **A skull modified for upright posture**: primates have an upright posture associated with having a forward-looking face. The skull rests on top of the vertebra and has a large opening, the foramen magnum, through which the medulla of the brain emerges and extends downwards as the spinal cord.

- **Reduced number of offspring**: life in the trees is difficult and dangerous, especially for young animals. Some arboreal animals, such as birds and squirrels, build nests in which the young can be protected until they are old enough to fend for themselves. Primates have adopted another strategy: from birth, the young cling to the mother's body and only slowly gain independence. Primates produce few young but look after them for a long time: they have a long gestation period and a prolonged period of dependency after birth.
- **A large brain**: an active life in the trees requires precise movements and therefore good muscular coordination, vision, tactile senses, memory, thought, and learning. These processes depend on a large and highly developed brain.
- **Social groupings**: all primates live to some degree in social groups in which members cooperate with each other. Complex social behaviour probably stems from the strong pair bond which enables a mother and her young to remain closely together for a long time. Lengthy rearing of a small number of young is most successful when the mother has support from other adults. The continued success of a group of animals depends on the recruitment of young helpers, and so evolves a social interdependency which is the basis of our own human society.

The groups of modern primates

At about the same time as the dinosaurs became extinct, about 65 million years ago, the primitive primates diverged quickly to give rise to two main suborders: the **prosimians** (meaning 'before apes') and **anthropoids** (meaning 'ape form'). The prosimians are represented today by lemurs, lorises, and tarsiers, and the anthropoids by monkeys, apes, and humans (figure 2).

Monkeys are distinguished from apes in having long tails, and the forelimbs are not usually longer than the hindlimbs. They are believed to have evolved from two different groups of lemur-like animals which became isolated when continental drift separated Eurasia from North America. The North American group evolved into New World monkeys which died out in North America but somehow colonised South America. The Eurasian group gave rise to Old World monkeys, from which apes and humans evolved. There are several differences between Old World monkeys and New World monkeys which show their separate evolution. For example, the nostrils of monkeys from South America are wide open and far apart, and New World monkeys have a long tail that is prehensile (adapted for grasping branches); the nostrils of monkeys from Africa and Asia are narrow and close together and no Old World monkey has a prehensile tail.

Table 1 *Classification of humans. The table shows the hierarchical relationship between the taxa to which humans belong and which are mentioned in this and the following two spreads.*

kingdom	Animalia
phylum	Chordata
class	Mammalia
order	Primate
suborder	Anthropoid
superfamily	Hominoid
family	hominid
genus	*Homo*
species	*sapiens*

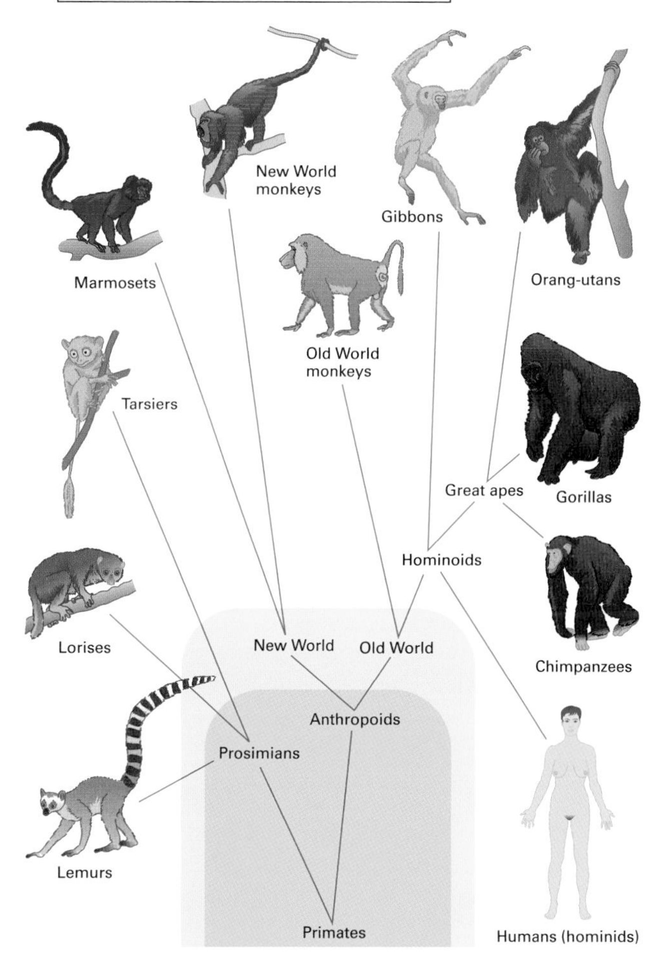

Figure 2 *Modern primates and their evolutionary relationships.*

Quick check

1 Primates evolved as a group adapted to an arboreal mode of life. Briefly explain the importance of the following adaptations:
 a reduced sense of smell
 b opposable thumb
 c small, single uterus
 d flexible pectoral girdle.

Food for thought

Suggest how stereoscopic vision evolved by natural selection in arboreal primates.

20.11

OBJECTIVES

By the end of this spread you should be able to:

- define brachiation
- discuss the significance of bipedalism.

Fact of life

Fossil evidence and DNA analysis suggest that chimpanzees and humans have a common ancestor that lived in Africa between 5 and 7 million years ago (equivalent to about half a million generations ago). To appreciate the unbroken chain of parent–child bonds between the ancestor and ourselves, Professor Richard Dawkins, an expert in evolutionary biology, suggests the following mind game. Imagine getting all your ancestors together and holding hands to form a human chain. The chain would start with you holding your mother's hand; your mother also holds the hand of her mother, your grandmother; your grandmother also holds her mother's hand; and so on until the chain reaches our common ancestor with the chimpanzees. Dawkins estimates that this entire chain would be less than 480 km long (allowing about 1 m for each person).

Apes and other hominoids

Modern apes (gibbons, orang-utans, gorillas, and chimpanzees) are our closest living primate relatives, according to all the available anatomical, biochemical, and genetic evidence. DNA profiling, for example, shows that humans and chimpanzees are very similar. We did not evolve from them, but we share a common ape-like ancestor.

Brachiation: a step on the way to walking

Modern apes are fairly large animals that have no tail, a relatively large skull and brain, and very long and powerful arms. The long arms of gibbons and orang-utans are an adaptation to **brachiation** (figure 1). This is a way of moving through trees in which the ape swings hand over hand from branch to branch. Brachiation does not occur fully in monkeys. Its development in primitive apes depended on two things: being able to rotate the shoulder joint through 360°, and a high level of flexibility so that the hand and lower arm have a wide range of movement. (This high mobility of the forearm is possible in apes because the scapula is at the back of the body.) Brachiation, like knuckle-walking (see below) and the absence of a tail, is thought to have been a preadaptation to **bipedalism** (walking on two legs). (A **preadaptation** is any anatomical structure, physiological process, or behavioural pattern evolved in one environmental situation which, quite by chance, proves advantageous in another situation, allowing the organism to radiate into the new situation and exploit a new ecological niche or habitat.)

Hominids walk on two legs

Bipedalism and a completely upright posture are the main distinguishing features of hominids, the family to which humans belong, and of which we are the only living representative. Many of the differences between apes and humans are linked to the structural adaptations of humans for bipedalism. For example, in humans the spine and bones are aligned so that, when a person stands upright, the centre of gravity falls between the feet, whereas in apes it falls between the four limbs (figure 2). Bipedalism is also associated with changes to the bones and muscles of the back, hips, and legs. The knee can be locked into position when a person is standing upright, and the knee joint is not at right angles to the femur shaft (figure 3).

Figure 1 *Brachiation: a form of locomotion which is thought to be a preadaptation to bipedalism. On broad branches and on the ground, brachiators such as gibbons tend to move by bipedal walking.*

The human foot takes a tremendous pounding during walking and running. It has a rigid ankle joint and a large calcaneum (heel bone) to which the gastrocnemius, a powerful calf muscle that extends the foot, is attached via the Achilles tendon. In order to support the body weight and transmit forces efficiently, the human foot is arched and the big toe is not opposable but aligned parallel with the others. Our bipedal gait has freed our arms and hands so that we can use tools and manipulate objects. This has led to our ability to control our environment to a greater degree than any other animal.

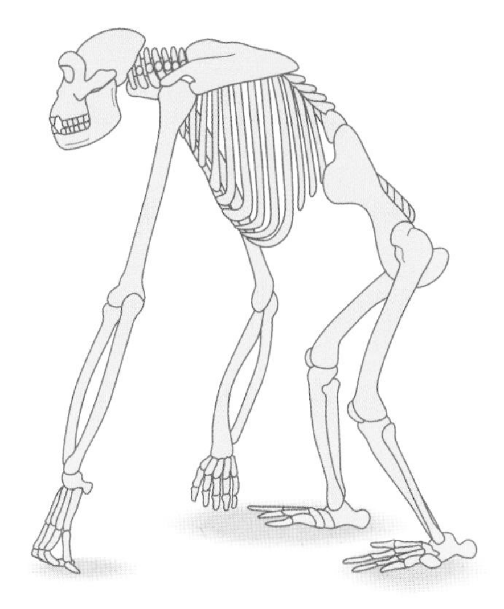

Figure 2 *In an ape such as a gorilla, which moves by knuckle-walking, its centre of gravity falls between the four limbs, whereas in an upright human the centre of gravity falls between the two feet.*

Ancestral human–apes and the early evolution of humans

Fossils of animals which have both human- and ape-like features, dating from about 22 million years ago until 10 million years ago, have been found in Europe, Asia, and Africa. These fossils have been named 'dryopithecine apes'; we shall refer to them as human–apes. Their chimpanzee-like nature was recognised in the nineteenth century (a genus from East Africa was called *Proconsul*, in honour of Consul, a famous chimp in Manchester zoo). These ancient primates are thought to have had light bones (more like those of humans than apes), and they were probably knuckle-walkers. **Knuckle-walking** is a method of locomotion typical of modern gorillas and chimps in which the knuckles of the ape's hands make contact with the ground. Knuckle-walking allows the ape to be fairly upright. Changes in the pelvis and pelvic muscles to support soft abdominal tissues appear to be associated with this posture and type of walking; quadruped apes (apes that walk on all four legs) use their abdominal muscles to support these tissues.

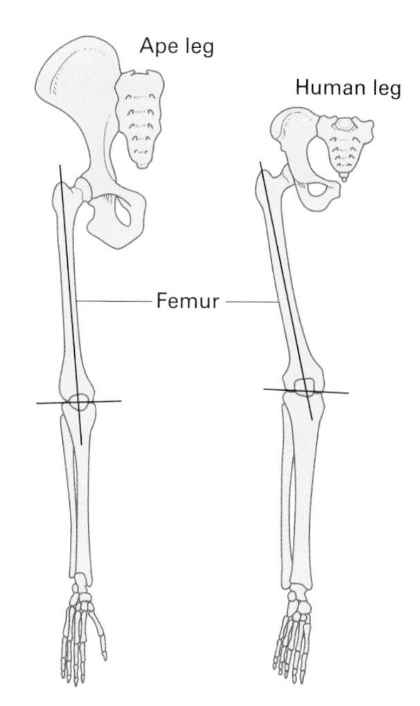

Figure 3 *The angle between the femur and the lower leg at the knee (the valgus angle) differs between apes and humans. The wider angle in humans allows the foot to be placed underneath the centre of gravity when walking; the straighter alignment of upper and lower bones in the ape's limb means that the ape 'waddles' during bipedal locomotion.*

The generally accepted view of the evolution of modern humans is that it happened exclusively in Africa. At about the time the human–apes were evolving in Africa, the global climate changed dramatically, probably as a result of continental drift. At times, the climate became significantly drier and the African rainforests were reduced to a few isolated areas. A more open type of woodland and savannah grassland replaced the forests. It is thought that human–apes with a preadaptation to bipedalism had a distinct selective advantage in these new habitats. Once they had gained an upright posture, they could see above the vegetation, and their arms and hands were free to reach up and gather more food.

Interestingly, modern chimps stand to reach their food most of the time. The larger and more highly ranked a chimp is in its social group, the more time it spends in an upright position.

Whereas apes have canines and molars with relatively small biting surfaces adapted to eating the soft fruits and leaves of the forests, our human–ape ancestors had dentition adapted to the tougher vegetable diet of open woodland. Like us, they had heavily enamelled premolars and molars with large biting surfaces to cope with chewing hard seeds and small nuts in fruit. They also had reduced canines which made sideways chewing easier.

Human–apes which have adopted a completely upright posture and move almost exclusively on two legs are regarded as hominids, the subject of our next spread.

Quick check

1 Why are monkeys unable to perform brachiation?

2 Why is the big toe (the hallux) of an ape not suitable for complete bipedalism?

Food for thought

In addition to freeing the arms and hands to manipulate objects, bipedalism might have provided a significant selective advantage to hominids living in a sunny, hot, and open environment. Suggest what this advantage might be.

The evolution of bipedalism involved disadvantages as well as advantages. For example, the bipedal gait requires a smaller pelvis. Suggest how this is a disadvantage to women. What other disadvantages might be associated with bipedalism?

20.12

Hominids

OBJECTIVES

By the end of this spread you should be able to:

- discuss methods of studying hominid evolution.

Fact of life

Fossil bones are not the only source of evidence of human evolution. Preserved in the volcanic ash in the Afar region of Ethiopia, fossil footprints (figure 1) made over 3.5 million years ago give us important clues about our evolution. Their size and stride length indicate that they were made by two bipedal hominids about 140 cm and 120 cm tall. Many scientists claim that the footprints are effectively identical to those of modern humans, but others believe that they are more ape-like.

***Figure 1** The hominid footprints preserved in volcanic ash might have been made by members of* Australopithecus afarensis, *the same species to which 'Lucy', the oldest hominid fossil, belongs.*

***Figure 2** Neanderthal (*Homo sapiens neanderthalensis*): photograph of a reconstruction in the Neanderthal Museum, Mettmann, Germany.*

Hominids are distinguished from apes by being able to walk completely upright on two legs, and by having a larger brain. The frontal and occipital lobes (areas of the brain concerned with complex behaviour including communication by speech) are particularly well developed. Hominids also have a slow postnatal development which has favoured the evolution of complex social behaviour. The hominid family includes the genera *Australopithecus* and *Homo*. We belong to the species *Homo sapiens* and are the only living hominids.

Studying hominid evolution

Until recently, the study of the evolution of modern humans from hominid ancestors was based mainly on fossil evidence and was very speculative. Only a few thousand fossils, and most of these only fragments of whole skeletons, have been discovered. Sometimes anatomically similar bones collected over a wide area have been assumed to belong to the same individual when this may not be the case. Recent advances in DNA technology, such as **DNA–DNA hybridisation** (see spread 20.2), molecular clocks, and DNA sequencing, have provided fresh ways of investigating human evolution. The DNA evidence collected so far is limited, but is already making us revise some our fossil-based ideas about human evolution.

Molecular clocks are based on the hypothesis that evolutionary changes occur in a clock-like manner, with mutations accumulating in a stretch of DNA at a statistically predictable rate. For example, if the mutation rate for a particular gene is 5 mutations every million years, and the molecular biologist finds a difference between two DNA sequences of 25 bases, then the sequences diverged 5 million years ago.

It has been estimated that the genes coding for the alpha and beta chains of haemoglobin (spread 2.10) have been changing at a constant rate since they first appeared about 500 million years ago. If this rate is reliable, it could be used to investigate human evolution, including the date of the chimpanzee/human divergence. However, the assumption that such molecular clocks are reliable has been questioned because the rate can change, for example in response to environmental stresses such as increased exposure to ultraviolet light.

DNA sequencing is the process of determining the exact sequence of nucleotide bases (adenine, cytosine, guanine, and thymine) in DNA. Early methods, based on traditional techniques such as **two-way chromatography** (see appendix), were very laborious and expensive. A major breakthrough came in the 1970s when Frederick Sanger and his team at the university of Cambridge devised the **chain-termination method**. This starts with a sample of DNA which is used as a template to make a series of DNA fragments differing in length by only one nucleotide. The synthesis of the fragments requires a supply of the four types of nucleotides, the enzyme **DNA polymerase**, and a **primer** (a short sequence of DNA attached to the template which initiates DNA replication). The base sequence in the newly synthesised fragment is complementary to that in the template from which it is replicated. The last nucleotide at one end (the 3' end) of each fragment has its base modified and uniquely labelled with a fluorescent dye so that it can be identified. Each DNA fragment therefore contains a sequence of natural bases with an artificial nucleotide (called an **analogue nucleotide**) at the end. Sanger incorporated modified bases called **dideoxy bases** in the replication of each newly synthesised fragment. Four types of dideoxy base were used, one specific for each type of the natural bases (adenine, cytosine, guanine, and thymine). Once incorporated into the new replicated DNA, the dideoxy bases stop replication, hence they are called **chain terminators**.

Gel electrophoresis (a technique that can separate particles according to their size, shape, and electrical charge) is used to separate different DNA fragments according to their size only (all the fragments have a similar shape and are pre-treated so that they have the same charge). Once the fragments are separated, the bases at the 3' carbon end of each chain is identified. From this identification, the original sequence of bases in the parent DNA can be determined (figure 3).

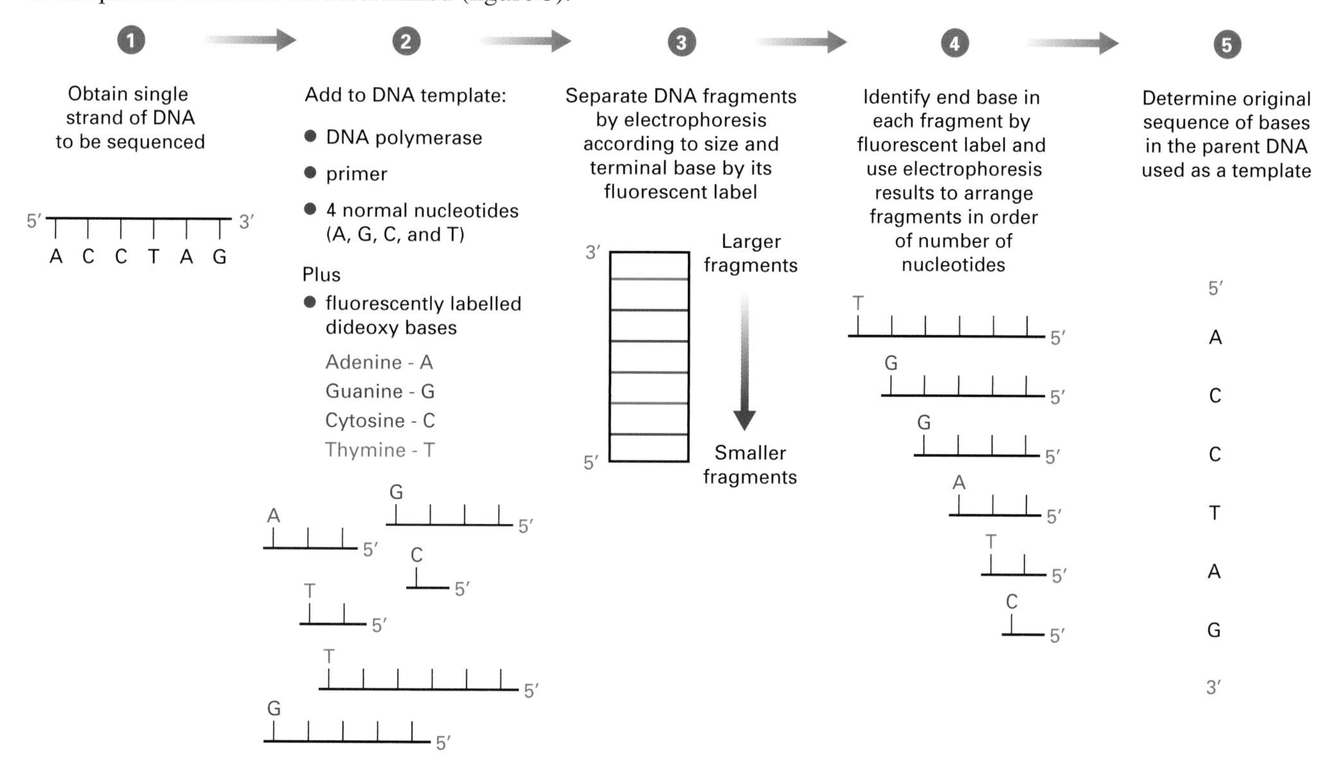

Figure 3 An outline of DNA sequencing using the chain-termination method.

Sanger's chain-termination method has given rise to many DNA sequencing methods that vary in the technologies they incorporate. For example, in **dye-terminator sequencing** each of the four dideoxy-base-chain terminators is labelled with a separate fluorescent dye, which fluoresces at a different wavelength, enabling sequencing in a single reaction rather than four as in the original method. Whatever the precise technique used, the general trend has been to automate the processes so that DNA sequencing can become cheaper, faster, and more accurate. Central to this automation has been the use of sophisticated computer technology.

It is estimated that the first sequencing of human DNA cost $3 million. By 2012, equivalent sequencing cost about $50 000, with biotechnology aiming to reduce the cost per sequence to under $1000 so that it can be used routinely in medical testing. One result of DNA sequencing has been to improve our understanding of the relationship between Neanderthals and modern humans.

Fossils of Neanderthals (*Homo sapiens neanderthalensis*) were first discovered in the Neander Valley, Germany, in 1857. These showed that Neanderthals were short, burly hunter–gatherers characterised by a massive jaw, large brow ridges, and a sharply receding forehead. They had brains larger than those of modern humans (averaging about 1530 cm^3 compared with ours averaging about 1400 cm^3). Other archaeological evidence shows that Neanderthals made a variety of tools, took care of their sick and young, and probably performed ritual burials of their dead.

Neanderthals are the extinct hominid group most closely related to modern humans. Although Neanderthals and modern humans had a common ancestor, the evolution of the two species probably diverged about 500 000 years ago.

Quick check

1 How does the brain of a Neanderthal compare with that of a modern human?

Food for thought

Evolutionary relationships between primates have been studied by analysing the DNA nucleotide base sequences of particular alleles. The analysis involves a technique known as DNA–DNA hybridisation (see spread 20.2). A geneticist analysing the alleles of particular genes from humans and from chimpanzees found that they had a greater similarity than alleles of other genes from two different people. Suggest why the results do not necessarily imply that chimpanzees are humans, or that people with different alleles are from different species.

20.13

Major trends in hominid evolution

OBJECTIVES

By the end of this spread you should be able to:

- describe the main trends in human evolution
- discuss the significance of farming, writing, and genetic engineering to the evolution of human culture.

Fact of life

The brain has steadily become larger in the evolution of hominoids. The human brain is over three times as big as an ape's would be if it had the same body size. Each increase in brain capacity has been accompanied by an even greater learning capacity.

Table 1 *Cranial capacities of different hominoids*

Species	Cranial capacity (cm^3)
Chimpanzee	400
Gorilla	550
Australopithecus sp.	500
Homo habilis	650
Homo erectus	1000
Homo sapiens	
a) Neanderthals	1500
b) Fully modern	1400

The brain

Perhaps the most significant trend in the evolution of hominid species has been the enlargement of the brain. This is reflected by changes in cranial capacities (table 1). The **cranial capacity** is the volume of space within the cranium, the part of the skull that encloses the brain. Assuming the brain fills this space, cranial capacity is equivalent to brain size.

Although an increase in the size of the brain is usually associated with an increase in intelligence and complexity of behaviour, the relative size of an animal also needs to be taken into account. Hence a gorilla has a larger brain than a much smaller chimpanzee, although the chimp scores more highly on most tests of intelligence. As important as the size of the brain is its shape. Irregularities on the internal surface of a cranium can be used to work out the shape of a brain (figure 1).

The skeleton

As well as changes to the cranium, other changes have occurred in the skeleton during the evolution of hominids. These changes are summarised in figure 2.

Culture

Culture has been defined as: *a store of information and set of behaviour patterns that can be transmitted from one generation to another not by genetic inheritance, but by learning, by imitation, by instruction, or by example.* Many animals other than hominids have a culture (for example, African wild dogs; see spread 11.11), but humans have the most advanced and sophisticated cultures of all. The evolution of modern human culture has involved a process called **hominisation**, the development of a number of different features that give an increasingly human quality. These include sharing food, the development of speech and language, a longer period of childhood, and the development of manipulative skills to make tools.

A **tool** is an object that is held, carried, or otherwise manipulated by an animal in order to achieve an objective. Tool-making is the characteristic traditionally used to distinguish humans from other hominids. Although other animals (including one of Darwin's finches; spread 20.3) used tools, none use them to the extent of humans.

In the past, two major cultural developments have had a profound influence on human evolution. They are farming and writing.

Farming comprises four main activities: the propagation of seeds; harvesting; caring for animals; and the artificial selection of animals and seeds from which to breed (see spread 20.7). Farming is thought to have started about 10 000 years ago. It enables humans to settle in one place, and has freed them from the need to spend most of their time and energy searching for food.

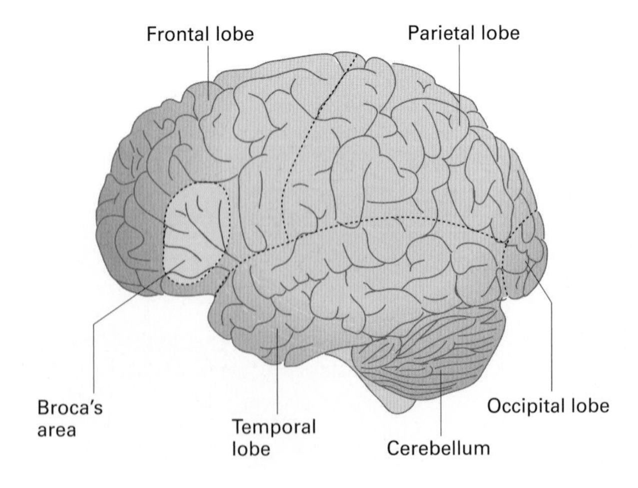

Figure 1 *In modern humans, one brain region (called **Broca's area**) towards the front of the left hemisphere controls the muscles of the mouth, tongue, and throat when we speak. Because it contains Broca's area, the left hemisphere is rather larger than the right one, and there is a detectable lump over the area; this produces an imprint on the cranium. The imprint is more pronounced in the more recent hominids: it is small in australopithecines, becomes more distinct in* Homo habilis, *is even more strongly marked in* H. erectus, *but is most definite in modern humans. Some scientists believe that the imprint in the skull produced by Broca's area indicates the ability to use language to communicate. Modern chimpanzees also have a swelling which might correspond to Broca's area. Other regions of the brain that have become increasingly dominant during the evolution of humans are the parietal lobe, which integrates information from the senses, and the temporal lobe, which is concerned with memory.*

Although carved and painted figures in caves are probably more than 25 000 years old, the oldest known writing was produced by the Sumerians about 3500 BC. Cave painting and writing provide a permanent means of recording information and transmitting it from one generation to another, enabling each generation to examine past ideas and add new insights of their own.

A recent cultural development is likely to be even more significant than either farming or writing. **Genetic engineering**, our ability to manipulate genes (including our own), is taking us into a new stage of evolution. Let us hope that we live up to our name *Homo sapiens* ('wise man'), and use our new-found power wisely.

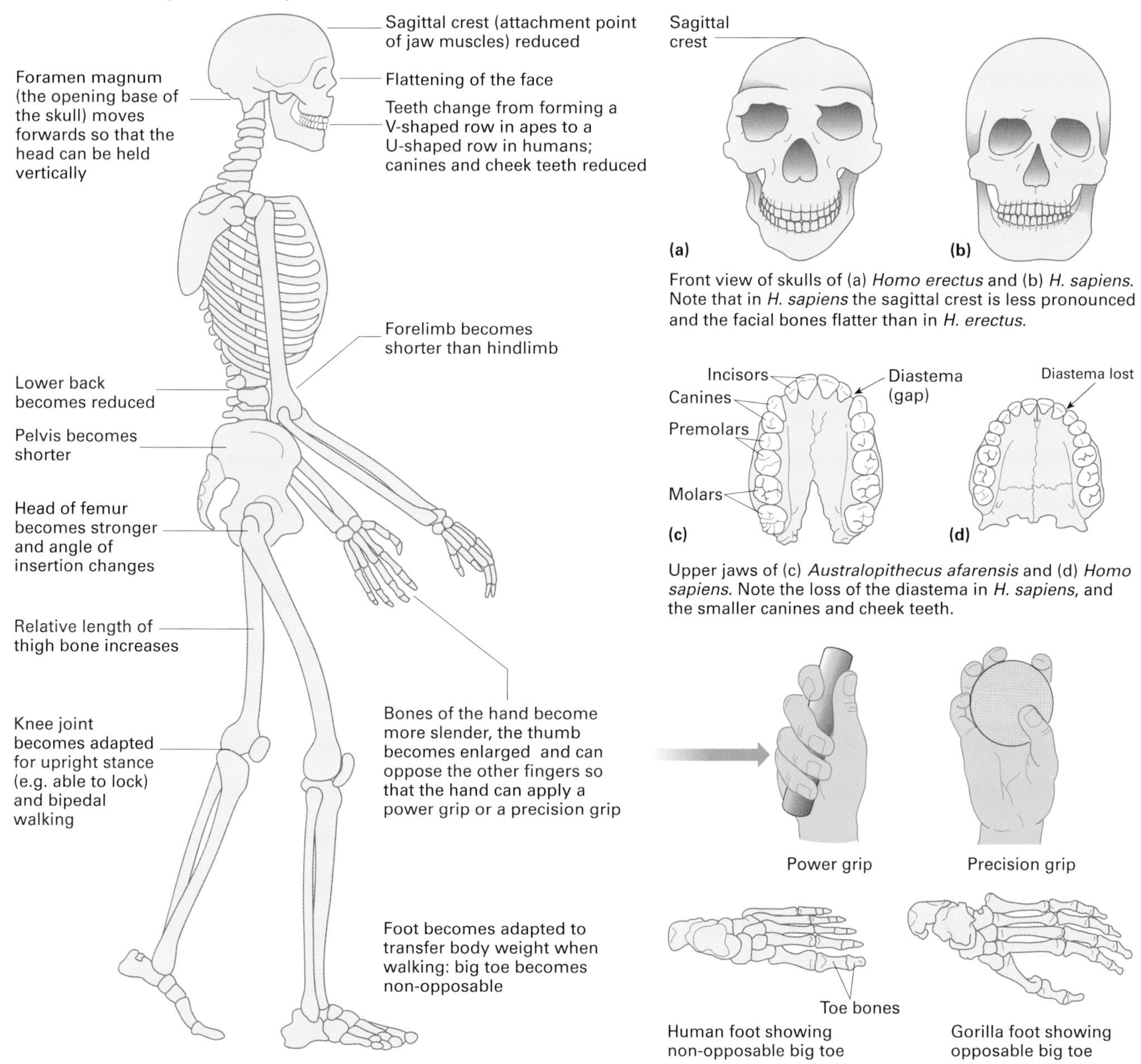

Figure 2 *Some trends in the evolution of the hominid skeleton.*

Quick check

1 Explain the significance of each of the following trends in human evolution:

a an increase in the size of Broca's area

b narrowing of the pelvic girdle

c the ability of the thumb to oppose any other finger

d a flattening of the facial bones.

Food for thought

Has human evolution reached an end? We have control over most of the factors involved in natural selection and we can potentially regulate our own genome. Suggest what this means for our future.

Summary

Theories of evolution include those of **Darwin** and **Lamarck**. Darwin proposed that new **species** evolve by **natural selection**; Lamarck suggested that new species evolve by the **inheritance of acquired characteristics**. Evidence for evolution comes from the study of **fossils**, **comparative biochemistry**, **comparative anatomy**, **cell biology**, and **adaptive radiation** of geographically isolated organisms. Natural selection results from **selection pressures** that keep populations in check; the **fittest** individuals survive to produce the most offspring. There are three types of natural selection: **stabilising selection**, **directional selection**, and **disruptive selection**. **Industrial melanism**, **antibiotic resistance** in strains of bacteria, and **pesticide resistance** in insects are examples of **evolution in action**. Natural selection can lead to a change in **allele frequency**; it can also result in alleles, such as that for **sickle-cell anaemia**, being retained. Allele frequency can be calculated using the **Hardy–Weinberg equation**. **Artificial selection** by **selective breeding** has been carried out by farmers for thousands of years. **Inbreeding** can lead to an increase of inherited diseases whereas **outbreeding** by crossing genetically distinct individuals can produce **hybrid vigour**, reducing the incidence of inherited diseases.
Speciation is the formation of new **species**. **Isolation mechanisms** which can lead to **reproductive isolation** and **speciation** include **geographical isolation**, **temporal isolation**, **ecological isolation**, **behavioural isolation**, **mechanical isolation**, **gametic isolation**, and **hybrid isolation**. **Allopatric speciation** occurs in geographically isolated populations; **sympatric speciation** (for example, in the cord grass *Spartina*) occurs among individuals living in the same area.
Studies of **human evolution** indicate that we had a lemur-like ancestor adapted to an **arboreal existence**. Adaptations occurring in ancestral primates and retained by modern ones include a **prehensile limb**, a mobile forearm, well-developed **stereoscopic vision**, a reduced sense of smell, an unspecialised digestive system, a skull adapted to an upright posture, reduced number of offspring, a large brain, and social groupings. **Modern apes** are our closest living primate relatives. They have long, powerful arms for **brachiation**.
Brachiation and **knuckle-walking** are thought to have major preadaptations for **bipedalism**, a characteristic feature of hominids. **Hominids** include the genera ***Australopithecines*** and ***Homo***. Extinct *Homo* species include *H. habilis* and *H. erectus*; *H. sapiens* is the only living species. Major trends in hominid evolution include an increase in **brain size**, adaptations of the skeleton to bipedalism and a **precision grip** of the hand, and the development of culture, including **tool-making**, **farming**, and **writing**.

PRACTICE EXAM QUESTIONS

1 The peppered moth, *Biston betularia*, produces a black variety from time to time. The mutation causing this black variety results in a dominant allele, **B**. The black variety was first observed in 1848 in Manchester, but by 1895 it had increased to 95% of the population in the city.

a What was the frequency of the dominant allele, B, in the 1895 population of the moth? Show your calculations. [3]

b Explain why there were always some light-coloured forms of the moth present in urban populations after 1895. [2]

c Explain why, in rural populations, the black form of the moth remains very rare. [3]

[Total 7 marks]

2 a Describe, with the use of examples, the genetic basis of resistance. [9]

b Discuss the development of resistance in a named organism. [7]

[Total 16 marks]

3 The coyote, jackal and dingo are closely related species of the dog family. Their distribution is shown on the map.

Suggest and explain how these three distinct species evolved from a common ancestor. [4]

[Total 4 marks]

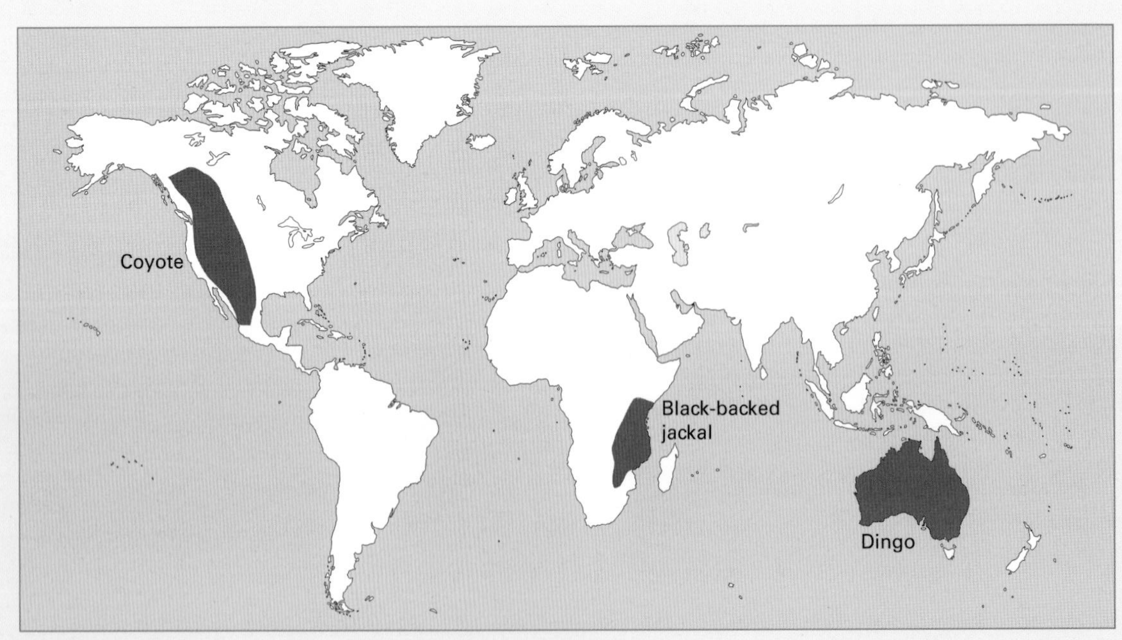

4 *Spartina anglica* is a species of grass which has originated as a result of the formation of a hybrid between two related species, *S. maritima* and *S. alterniflora*, as shown in the diagram below. The diploid numbers of chromosomes for *S. maritima* and *S. alterniflora* are given in the boxes.

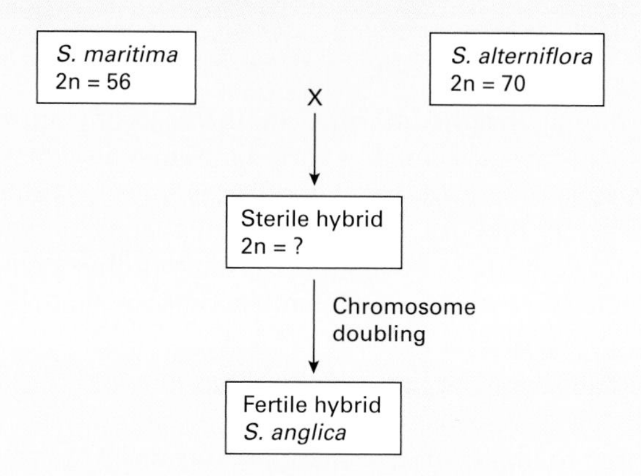

a **i** Give the expected diploid number (2n) of chromosomes for the sterile hybrid. [2]

ii Explain why this hybrid is sterile. [2]

b Suggest how doubling of the chromosomes may have occurred to produce *S. anglica*. [3]

[Total 7 marks]

5 Snails in the genus *Discula*, found on one of the Canary Islands, have originated from founder species which were carried to the island on floating logs or by birds. The genus *Discula* is now split into two subgenera as shown in the diagram.

Species **C** and **D** have been found to inhabit separate mountains on the island.

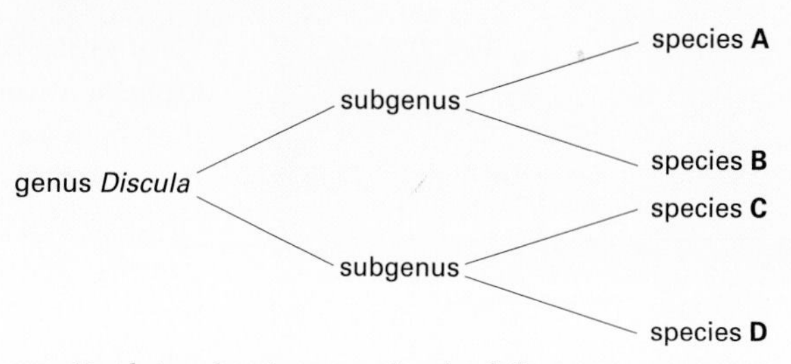

a Explain what is meant by the following terms:

i *Genus* [2]

ii *Species* [2]

b **i** Suggest why species **C** and **D** may eventually show changes in gene frequency. [1]

ii Explain briefly how these eventual changes in gene frequency might occur. [3]

c Explain why different species of *Discula* found in the same habitat could be expected to have different food requirements. [2]

d Suggest why competition might be greater between species **A** and **B** rather than between species **A** and **D**. [2]

[Total 12 marks]

6 Fossil bones provide important evidence for human evolution.

a Name two techniques that could be used to determine the age of a fossil bone found in sedimentary rock between layers of volcanic rock. [2]

b Explain how a palaeontologist might reconstruct the appearance of a human ancestor from fossil bones. Give three difficulties that might be encountered in this. [6]

[Total 8 marks]

21 The variety of living things

Classification

21.1

OBJECTIVES

By the end of this spread you should be able to:

- distinguish between artificial and natural classifications
- explain the binomial system of nomenclature
- describe the main distinguishing features of each of the five main kingdoms.

Fact of life

Carl Linnaeus (1707–78) named about 12 000 different species.

Cladistics and phenetics

Cladistics and phenetics are two different methods of classifying organisms. **Cladistics** aims to group organisms according to their evolutionary relationships. It is constructed on the basis of **derived characteristics**. A derived characteristic is a new characteristic shared only by a group and its hypothetical ancestor (this group is called a **clade**). Hypothesised evolutionary relationships between organisms are hierarchically depicted in an evolutionary tree called a **cladogram** (figure 2). Different cladograms are constructed for competing hypotheses.

Phenetics classifies organisms into groups based on the overall similarity of their observable characteristics, regardless of their evolutionary relationship. The characteristics most commonly used are measurable and morphological. It can sometimes result in different organisms being wrongly placed in the same group on the basis of **analogous characteristics** that look similar but that have evolved separately by **convergent evolution** (see spread 20.3).

Why classify?

Classification is the grouping together of things on the basis of features they have in common. This process emerges early in human development. Leave a child to play with a mixture of football cards, stamps, or coins, and it will not be long before they are sorted into groups. The ability to classify things is probably essential to our survival. Without this ability, we would be overwhelmed by the chaos and confusion of having to deal with an endless number of individual things.

One of the most potentially confusing things in our world is the enormous variety of living organisms. As far back as we can trace, humans have been trying to make sense of these organisms by classifying them.

Biological classification systems

Classification systems are either artificial or natural. **Artificial classifications** group organisms for convenience. The groupings are based on observed features, such as where the organisms live, how they move, their colour, or their size. Most modern classification systems are natural. **Natural classifications** attempt to group organisms according to their evolutionary relationship or **phylogeny**. The groupings are based on many features, internal as well as external, and use information from many branches of biology, especially molecular biology. **DNA sequencing** and comparison of **genomes** is becoming increasingly important in natural classifications.

Modern biology places organisms into **taxa**: a series of groups arranged in a hierarchy. Each group is called a **taxon** and contains organisms sharing key features called **diagnostic characteristics** which indicate that they have a common ancestry. Table 1 shows a classification system with seven main taxa arranged in descending order of size: **kingdom** is the largest taxon and contains the most organisms with the fewest features in common; species is the most exclusive group and contains the fewest organisms with the most features in common. Members of a species are so similar that they can interbreed.

Table 1 *Classification of the grey wolf and maize, illustrating the hierarchical system of classification.*

Taxon	Grey wolf	Maize
kingdom	Animalia	Plantae
phylum	Chordata	Angiospermophyta
class	Mammalia	Monocotyledoneae
order	Carnivora	Commelinales
family	Canidae	Poaceas
genus	*Canis*	*Zea*
species	*Canis lupus*	*Zea mays*

The study of the classification of life is called **taxonomy**. It includes the principles, practice, and rules of classification. Taxonomists are concerned not only with deciding how organisms are placed into groups, but also with **nomenclature**, or what names to give the groups.

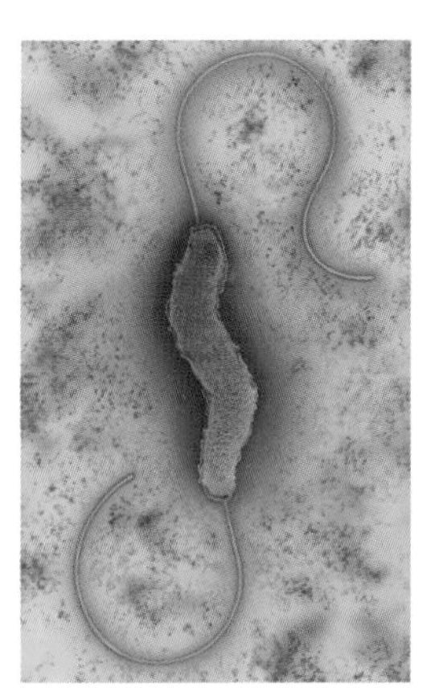

Prokaryotes
No nucleus; no double-membraned organelles; includes all bacteria

Protoctists
Eukaryotic (defined nucleus and double-membraned organelles); mainly single-celled, or organisms with cellular level of organisation; includes protozoa and algae

Plants
Multicellular, eukaryotic and photosynthetic organisms; cell walls contain cellulose; includes mosses, ferns conifers, and flowering plants

Fungi
Multicellular, eukaryotic organisms; cell wall not of cellulose; non-photosynthetic with absorptive methods of nutrition; includes yeast, moulds, mushrooms, and toadstools

Animals
Eukaryotic multicellular, heterotrophic organisms with nervous system; includes worms, insects, fish, birds, and mammals

Figure 1 *The five-kingdom classification.*

Biological nomenclature

Biological nomenclature is based on the **binomial system** devised by the Swedish naturalist Carl Linnaeus (1707–78). In this system, each type of organism is given a name consisting of two words: the first is the generic name (the name of the genus) and begins with an upper case letter; the second is the specific name (the name of the species) and begins with a lower case letter. The two words are printed in italics (or underlined when handwritten) to show that this is a biological name accepted by scientists throughout the world, rather than a common name which may be known only locally. Humans belong to the genus *Homo* and the species *sapiens* and are given the name *Homo sapiens*. By convention, the species name is written in full when first mentioned but thereafter may be abbreviated using only the first letter of the genus: *H. sapiens*. The species name is never used on its own.

Generic and specific names are laid down by International Codes of Nomenclature. No such codes exist for higher taxa. Consequently, there are several different classification systems. In the **five-kingdom classification system** (figure 1) all single-celled organisms with no nuclear membrane or double-membraned organelles are included in the kingdom Prokaryotae (**prokaryotes**). Another classification system adds a taxon, the domain, above that of the kingdom. It divides prokaryotes into the **archaea** and **bacteria**; all other organisms are placed in the **eukaryotes**. This **three-domain system** of classification emphasises that archaea and bacteria are two distinct groups that arose separately from a common ancestor.

Evolutionary relationships between the various taxa are commonly depicted as branching diagrams called **phylogenetic trees** or trees of life. Figure 2 shows a special phylogenetic tree called a **cladogram** used in cladistics (see box).

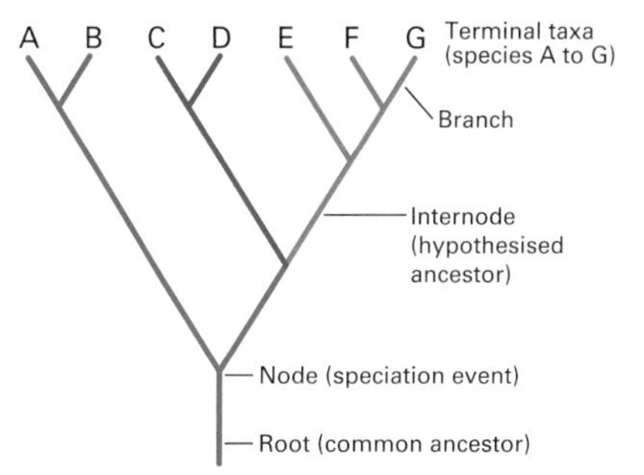

Figure 2 *A vertical cladogram. It starts at the root which splits several times at nodes into two branches or sub-branches. Each **node** represents the point at which a new species evolved (a **speciation event**). The **internode** (the line between two nodes) represents a hypothetical ancestor. The species at the end of the branches are called **terminal taxa**. The closer the species are on the diagram, the closer the hypothesised relationship between the species.*

Viruses

Viruses are generally considered to be non-living because they are neither self-replicating nor capable of metabolism. Although viruses can reproduce, they do so only inside the cell of a living organism. Electron microscopy has revealed that viruses are nothing more than complex associations between macromolecules (mainly protein and a nucleic acid) which can be crystallised and kept in a test tube for long periods of time. As they are non-living, viruses are not put into a kingdom or classified in the same way that organisms are.

Quick check

1 Is the five-kingdom classification natural or artificial?

2 Rewrite correctly the following scientific name for a tiger: panthera tigris.

3 **a** Which of the five kingdoms are eukaryotic?
b Which of the five kingdoms have a structure that includes cells with a cell wall?

Food for thought

Suggest why the binomial system of nomenclature is used by biologists in scientific publications in preference to common names of organisms.

21.2

OBJECTIVES

By the end of this spread you should be able to:

- describe the main features of prokaryotes
- distinguish between a prokaryotic cell and a eukaryotic cell.

Fact of life

Prokaryotes can be distinguished from eukaryotes by the size of their ribosomes, expressed in units called Svedberg (S) units. A Svedberg unit is a measure of how quickly particles sediment in an ultracentrifuge, which gives an indirect measurement of size and molecular weight. The 70S ribosomes of prokaryotes are significantly smaller than the 80S ribosomes found in the cytosol (the fluid part of the cytoplasm) of eukaryotes. Intriguingly, mitochondria and chloroplasts of eukaryotes contain 70S ribosomes, supporting the idea that these organelles may have evolved from symbiotic bacteria.

Prokaryotes

The kingdom **Prokaryotae** (prokaryotes) is made up of relatively small and simple organisms commonly known as bacteria. They lack nuclei (the term prokaryote means 'before nucleus', whereas eukaryote means 'true nucleus') and their genetic material consists of DNA lying free in the cell. Members of the other four kingdoms are **eukaryotes**. Most of their genetic material is contained in a well defined nucleus bounded by a double-membraned nuclear envelope, and their nuclear DNA is associated with RNA and proteins to form chromosomes. These and other differences between prokaryotes and eukaryotes are summarised in table 1.

The prokaryotes can be divided into two main groups: **archaebacteria**, which are thought to have been the first organisms to have evolved on Earth, and **eubacteria**, which include cyanobacteria (see also spread 17.2). Radioactive dating of fossil stromatolites (figure 2) and rocks containing the compounds **hopane** and acyclic isoprenoid (molecules derived from steroids and membrane lipids unique to prokaryotes) suggest that prokaryotes originated at least 3.5 billion years ago. The existence of complex sterane molecules in isotope-dated rocks suggests that eukaryotes originated about 2.7 billion years ago (sterane molecules are derived from cholesterol, found only in the cell membranes of eukaryotes).

Although *Escherichia coli* has often been referred to as a 'typical bacterium' (spread 17.3), there is really no such typical organism. Bacteria form a diverse group with members that range widely in size and shape. Most occur as single cells, but some stick together to form chains or clusters (spread 17.2, figure 2).

Figure 1 shows a generalised bacterium which contains all the main features of prokaryotes. Note that even this theoretical cell contains little structure compared with a typical eukaryotic cell.

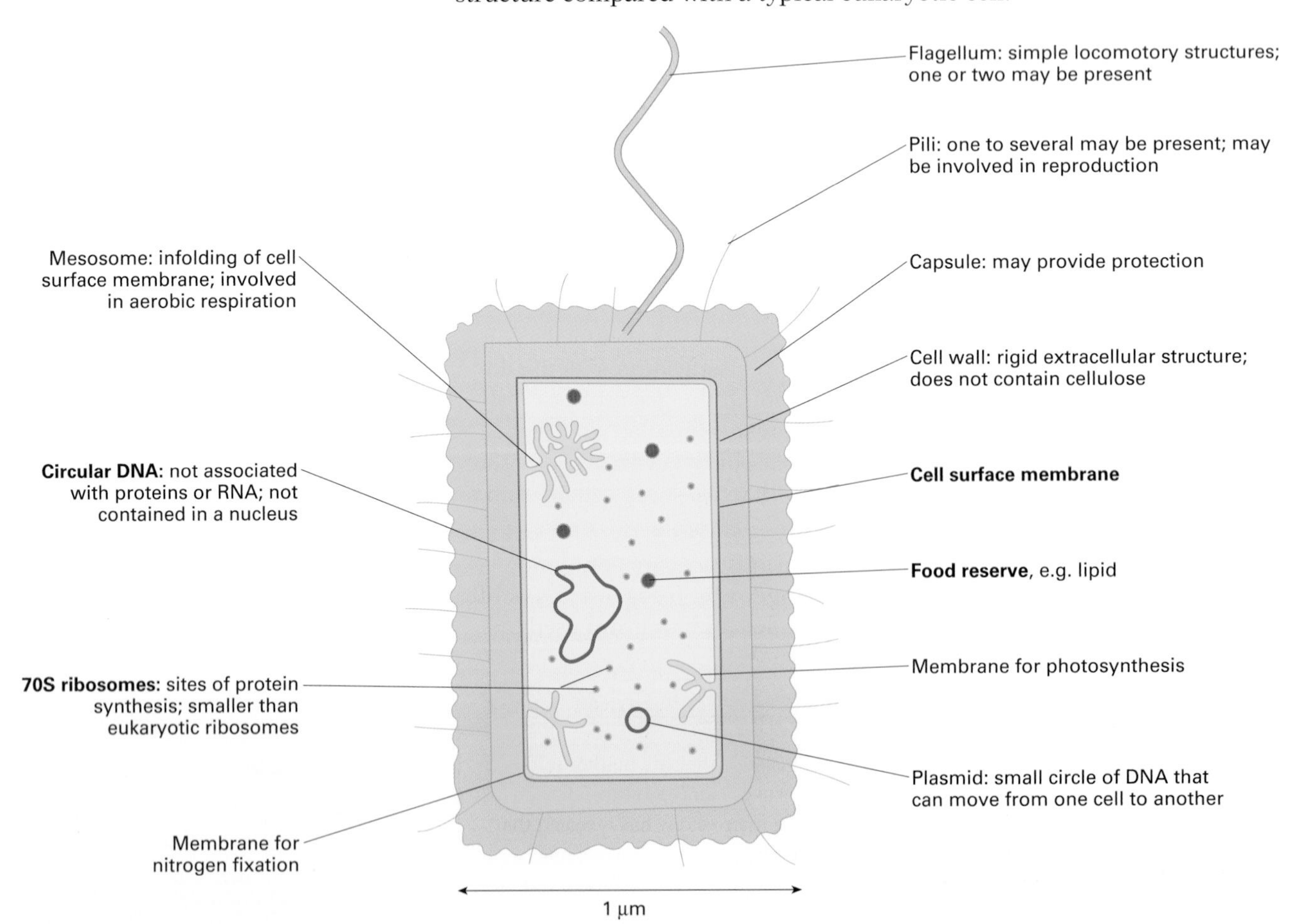

***Figure 1** 'Typical' bacterial cell showing the features of prokaryotes. Those shown in bold type are present in all bacteria; other features are present in some but not all bacteria.*

***Table 1** Comparison of the main features of prokaryotes and eukaryotes.*

Feature	Prokaryotic cell	Eukaryotic cell
Kingdom	Prokaryotae	Protoctista, Plantae, Fungi, Animalia
Size	Most 0.5–10 μm in diameter	Commonly 10–100 μm in diameter
Genetic material	Most DNA is 'naked', i.e. it is not incorporated in chromosomes but is a single, circular strand lying free in the cytoplasm; there is no nucleus. Some bacteria have additional small circular pieces of DNA called plasmids.	Most DNA is linear and incorporated with proteins (including histones) and RNA in chromosomes; there is a nucleus bound by a double-membraned nuclear envelope. Circular DNA occurs in mitochondria and, when present, chloroplasts.
Cell division	Usually by binary fission but sometimes by conjugation (see spread 17.3, figure 2); no spindle formation	Mitosis, meiosis, or both; spindle formation occurs
Ribosomes	70S ribosomes that only occur free in the cytoplasm	80S ribosomes that may occur either free in the cytoplasm or bound to the endoplasmic reticulum to form the rough endoplasmic reticulum
Endoplasmic reticulum	No endoplasmic reticulum	Smooth and rough endoplasmic reticulum
Organelles	Few organelles; no organelles are bounded by a double membrane	Many organelles; organelles bounded by a double membrane include the nucleus, mitochondria, and, in plants and some protoctists such as algae, chloroplasts
Hair-like structures	Flagella and pili are simple: they are extracellular (i.e. not extensions of the cell surface membrane) and they lack an internal system of microtubules	Flagella and cilia are complex extensions of the cell surface membrane and contain an intracelluar arrangement of microtubules, typically with a 9 + 2 arrangement (see spread 21.4, figure 5)
Cell wall	Rigid cell wall made of **peptidoglycan** (a polymer consisting of sugars and amino acids). Murein is the main strengthening compound, not cellulose.	Rigid cell walls of plants and algae contain cellulose, whereas those of fungi contain chitin. Animal cells have no cell walls.
Respiratory membranes	Mesosomes (infoldings of the cell surface membrane) act as a respiratory surface in some bacteria; no mitochondria	Cristae, infoldings of the inner membrane of mitochondria, act as the main surface for aerobic respiration
Photosynthetic membranes	Intracelluar membranes not organised into stacks (grana); no chloroplasts	Membranes contained within chloroplasts; in plants and some algae the membranes form thylakoids stacked into grana

***Figure 2** Stromatolites, colonial structures formed by reef-building cyanobacteria, the dominant life-form on Earth for about 2 billion years. Today, stromatolites are nearly extinct, existing in only a few places worldwide, the most famous being Shark Bay, western Australia. In 2011, small patches of living stromatolites were recorded from the Giant's Causeway in Northern Ireland.*

Quick check

1 Explain what is meant by 'naked' DNA.

2 Outline the differences shown in the following structures between prokaryotes and eukaryotes:

 a flagella

 b ribosomes.

Food for thought

Suggest why all prokaryotes are smaller than 1 mm in diameter.

21.3

Protoctists: general features

OBJECTIVES

By the end of this spread you should be able to:

- give the main distinguishing features of organisms in the kingdom Protoctista.

Fact of life

Many prokaryotes and protoctists consisting of colonies or filaments of cells occurred very early in the evolution of life. But it is difficult to be certain when truly integrated **multicellular organisms** with differentiated tissues and organs evolved, because the evidence is based mainly on fossils which lack soft body parts. For example, fossil vendobionts are estimated to be 600 million years old. We can deduce from their remains that they were peculiar flat-lying organisms that sprawled on the ocean floor and that their bodies were made of many cells. However, it is not clear if these were differentiated into true organs and tissues. Known only in fossil form, and having no closely related modern species, scientists are not sure whether vendobionts were protoctistan, fungal, or animal.

The kingdom **Protoctista** is a diverse collection of eukaryotic organisms. It contains all unicellular eukaryotes and their direct multicellular descendants. All protoctists therefore have a nucleus and mitochondria bound by a double membrane. Table 1 shows the main protoctistan phyla. Chlorophyta, Rhodophyta, and Phaeophyta are commonly referred to as **algae**, and flagellates, sporozoans, and ciliates are commonly referred to as **protozoa**, but neither algae nor protozoa are taxonomic names.

Cellular structure and organisation

As most protoctists are unicellular, they are regarded as the simplest eukaryotic organisms. However, the cells of some unicellular protoctists, such as *Paramecium* (figure 1), are among the most complex in the living world. This is not really surprising, because each cell is a complete eukaryotic organism equivalent to an entire plant, animal, or fungus.

Many protoctists consist of colonies or filaments of cells. Some are very simple, with the cells being almost identical to each other. In other protoctists, certain cells and regions have become specialised and interdependent. For example, *Fucus vesiculosus*, commonly known as bladderwrack, is a relatively large and complex brown alga. Its body is differentiated into a holdfast, stipe, and fronds (figure 2). Although these regions carry out different functions, they are not as specialised as the true roots, stems, or leaves of plants; for example, *F. vesiculosus* does not have a vascular system of phloem and xylem. Seaweeds are included in the Protoctista because they are direct descendants of unicellular algae.

Importance of Protoctista

Protoctists are thought to hold a pivotal position in the history of life on Earth. The first ones arose from prokaryotes, and their descendants evolved into plants, fungi, and animals. Recent molecular studies suggest that the different protoctistan phyla probably evolved from different prokaryotic ancestors. So apart from having a well defined nucleus, protoctists have little else in common except that they do not fit neatly into the three eukaryotic kingdoms. Some biologists regard the Protoctists as the least secure of the five kingdoms, devised mainly for the classification of awkward organisms which do not fit easily into any other kingdom.

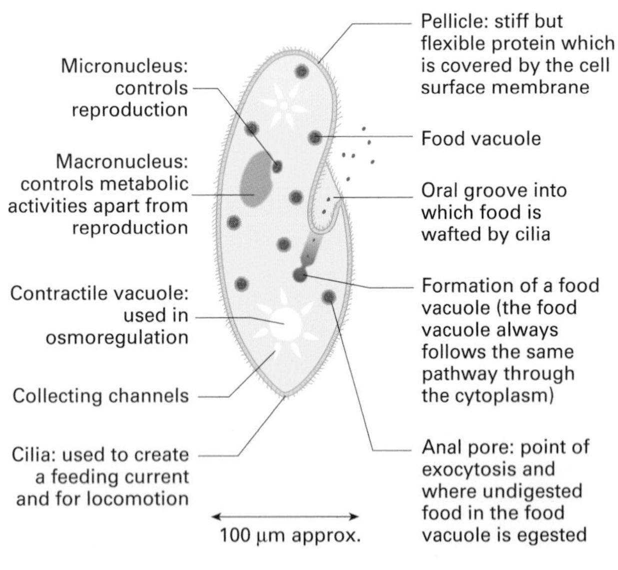

Figure 1 Paramecium caudatum, *ciliate protozoan.*

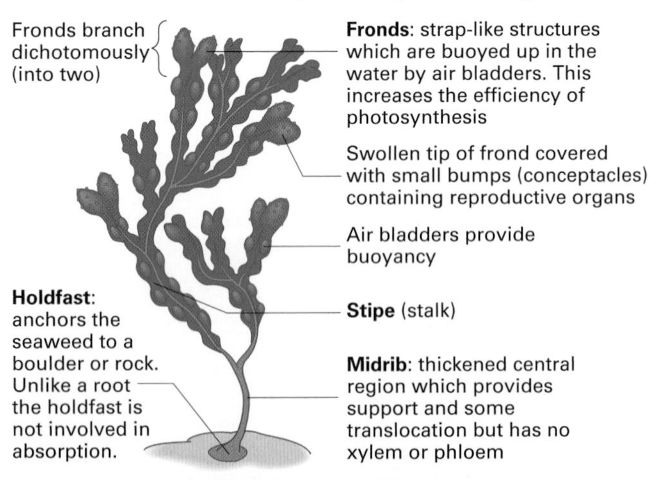

Figure 2 Fucus vesiculosus, *bladderwrack, is a large brown seaweed common on rocky shores. It is covered with slimy mucilage to prevent it drying out at low tide. The relatively simple plant-like body consists of fronds (sometimes called lamina or blades), a stipe and a holdfast; the entire body is referred to as a thallus.*

Table 1 *The main protoctistan phyla. This table is provided for reference purposes only; A-level students are not expected to learn details of the phyla in this kingdom.*

Phylum	Characteristic features	Example	Comments
Rhizopoda (rhizopods)	Organisms have **pseudopodia** for locomotion	*Amoeba* (spread 4.8, figure 5)	Pseudopodia (literally 'false feet') are cellular extensions which, as well as being involved in locomotion, may also be used for feeding by phagocytosis (spread 4.8)
Zoomastigina (flagellates)	Heterotrophic organisms which have at least one **flagellum** for locomotion	*Trypanosoma* (spread 15.4, figure 4)	Protoctistan flagella are quite different from those of prokaryotes because they are cellular extensions which contain microtubules
Apicomplexa (sporozoans)	Mainly parasitic organisms which reproduce by multiple fission	*Plasmodium* (spread 15.4, figure 1)	Sporozoans usually have no feeding or locomotory organelles
Ciliophora (ciliates)	Organisms with **cilia**	*Paramecium* (figure 2)	Cilia (which are less than 5 μm long) have a structure similar to that of flagella (which are about 100 μm long) and are collectively called **undulipodia** (spread 21.4, figure 5). Cilia may be used to attach the ciliate to a substrate, or for feeding.
Euglenophyta (euglenoid flagellates)	Organisms with **flagella** but with a biochemistry quite distinct from that of flagellates	*Euglena* (spread 21.4, figure 4)	The phylum contains both photosynthetic and non-photosynthetic members. Some species of euglenoid flagellates can switch from one type of nutrition to another.
Oomycota (water moulds; oomycetes)	Similar to fungi except that they have undulipodia and cell walls which contain cellulose	*Phytophthora* (spread 17.5, figure 1b)	These moulds appear very similar to fungal moulds but they are placed in the kingdom Protoctista because of their non-fungal features
Chlorophyta (green algae)	Photosynthetic organisms with chlorophyll pigments similar to those of plants	*Chlorella* (spread 17.9, figure 1)	This phylum includes unicellular and multicellular forms; the latter are merely groups or colonies of cells which show little differentiation
Rhodophyta (red algae)	Photosynthetic organisms with organelles which contain red pigments as well as chlorophyll	*Chondrus*	As above
Phaeophyta (brown algae)	Photosynthetic organisms with organelles which contain brown pigments as well as chlorophyll	*Fucus* (figure 3)	As above

Quick check

1 a How do euglenoid flagellates differ from other flagellates?

b Why are oomycetes not included in the kingdom Fungi?

c What is the main difference between members of the Chlorophyta and the Rhodophyta?

Food for thought

Pelomyxa palustris, a rhizopod (figure 3), is thought to be one of the most primitive living eukaryotes. It has a membrane-bound nucleus, and its DNA is naked and not associated with proteins. In addition, it reproduces by binary fission, apparently without the involvement of a spindle apparatus. Although its cytoplasm does not have any endoplasmic reticulum or mitochondria, it contains symbiotic bacteria which can respire aerobically. Suggest why some biologists think that this organism provides strong support for the evolution of eukaryotes by endosymbiosis (see spread 4.4). Why is *P. palustris* placed in the kingdom Protoctista rather than the kingdom Prokaryotae?

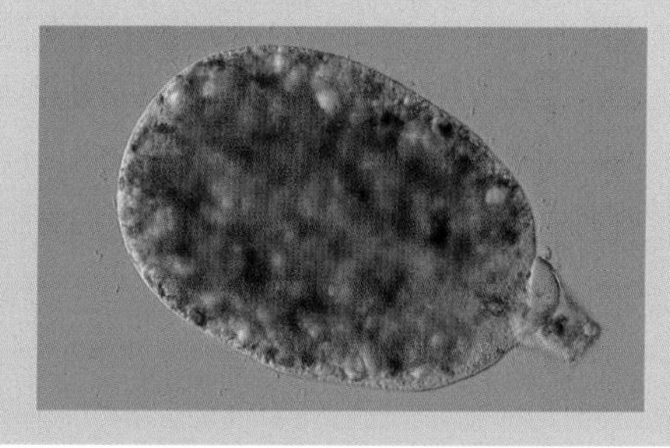

Figure 3 Pelomyxa.

21.4

Protoctists: *Amoeba* and *Euglena*

OBJECTIVES

By the end of this spread you should be able to:

- compare the nutrition and locomotion of *Amoeba* and *Euglena*
- describe the structure and function of a flagellum.

Fact of life

The largest amoeba is about 0.5 mm long; large enough to be seen without the aid of a microscope.

Amoeba (kingdom Protoctista; phylum Rhizopoda)

If you look at a drop of water containing amoebae you can just about see the organisms with the naked eye. They appear no bigger than dust particles. Under the low power of a light microscope, amoebae are like animated blobs of jelly. It seems incredible that such small and apparently simple creatures can lead an independent existence. However, a more detailed examination using high-power microscopes and biochemical analysis gives an insight into their complex lives. You will discover that though multicellular organisms need a whole battery of organ systems to carry out their vital physiological functions, *Amoeba* performs the equivalent functions within a single cell. Contrary to popular misconception, *Amoeba* is not a primitive organism. It has evolved over millions of years to become highly adapted to its mode of life. Most species of *Amoeba* are free living and are found in freshwater and marine environments, as well as in soil water. Some species are parasitic, living, for example, in the human gut and causing amoebic dysentery.

All amoebae move by putting out cytoplasmic extensions called **pseudopodia**. Only the tips of pseudopodia are in contact with the ground, so they act like tiny temporary legs, new ones forming as old ones are reincorporated into the main body. This form of locomotion is so characteristic of *Amoeba* that it has been called **amoeboid movement**; white blood cells also use this method of locomotion. The exact mechanism of amoeboid movement is not understood, but it appears to involve a movement of cellular contents (a process called cytoplasmic streaming) and an interconversion between the gel-like outer part of the cytoplasm (ectoplasm) and the more fluid inner part of the cytoplasm (endoplasm). Amoeboid movement requires energy (metabolic poisons inhibit it) and calcium ions. It may also involve protein molecules similar to the actin that occurs in animal muscles.

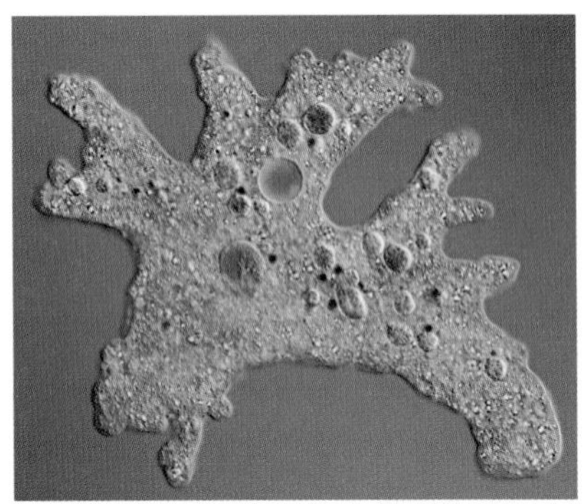

Figure 1 Amoeba proteus *(×400).*

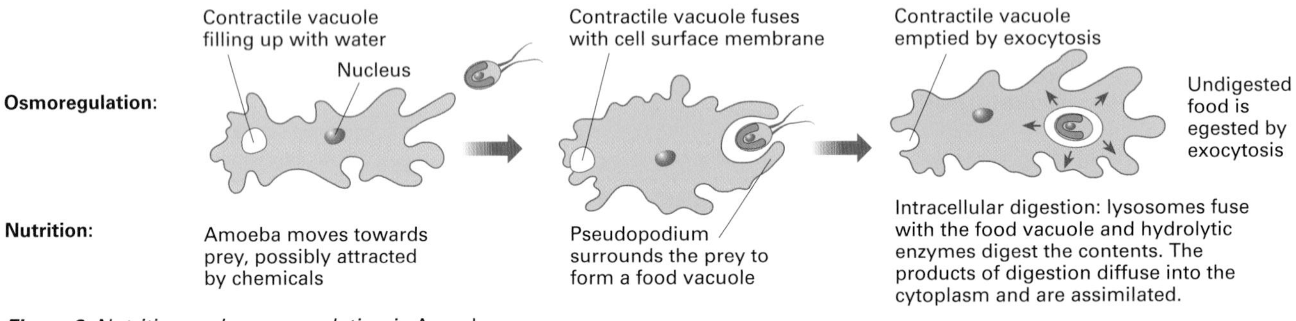

Figure 2 *Nutrition and osmoregulation in* Amoeba.

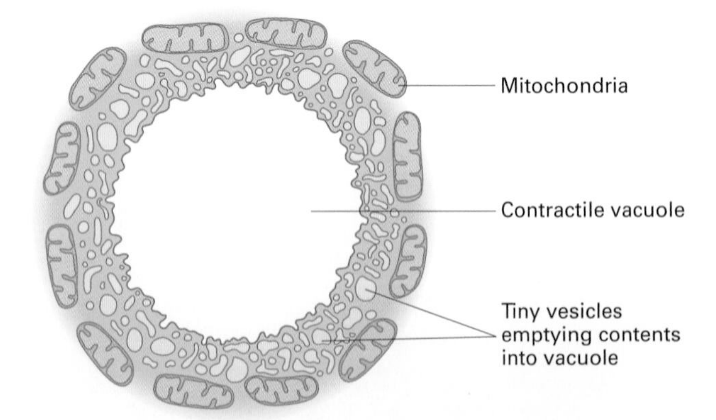

Figure 3 *The contractile vacuole of* Amoeba proteus *is surrounded by small vesicles filled with fluid that appears to empty into the vacuole. Mitochondria form a ring around the vesicles. An osmotic gradient is established which draws water by osmosis from the surrounding cytoplasm into the vesicles. When full, the contractile vacuole expels its contents by exocytosis. Metabolic inhibitors reduce the activity of contractile vacuoles.*

Amoebae are constantly changing shape as they meet obstacles and change direction, or when they respond to environmental stimuli. Adverse stimuli, such as acidic water or very bright light, cause them to retreat, while favourable stimuli, such as contact with food, cause them to advance. Amoebae use their pseudopodia to capture the food (figure 2). They are heterotrophic organisms which feed by phagocytosis (spread 4.8).

Amoebae are less tham 1 mm in diameter and respiratory gases are exchanged by diffusion over the whole cell surface (see Fick's law, spread 4.7). Freshwater amoebae live in an external environment that has a higher water potential than the cytoplasm, so water tends to move into the cell. If this osmotic influx of water continued, the cell would eventually burst. Freshwater amoebae have a **contractile vacuole**: an organelle that fills with water and expels its contents from time to time, allowing the cell to maintain an almost constant volume (figure 3).

Euglena (kingdom Protoctista; phylum Euglenophyta)

Euglena is a widespread unicellular protoctist found in marine and freshwater habitats. It is often so plentiful that it produces a green scum on the surface of the water. There are a number of species of *Euglena*. Their nutrition varies: about a third photosynthesise, the remainder lack chloroplasts and are heterotrophs. Several species are photosynthetic in the light and heterotrophic in the dark, feeding saprophytically on decaying matter. Euglenophyta or euglenoid flagellates which lack chloroplasts are very similar to flagellate protozoans. This has led some biologists to believe that they are actually protozoans that have acquired chloroplasts by endosymbiosis (spread 4.4, figure 1). They suggest that ancestral euglenoid flagellates were colourless, as are some present-day ones.

One of the most studied members of the genus is *Euglena gracilis*. It has both protozoan and algal features (figure 4). Its flask-shaped body is covered with a tough, flexible layer called a pellicle which, unlike algal cell walls, is made of a protein rather than cellulose, and lies inside the cell surface membrane. *E. gracilis* has two flagella. One flagellum is very short and its function is unknown; the other is long and is involved in locomotion (figure 5a, b). Flagella and cilia of protoctists have a similar structure and are called undulipodia (figure 5c).

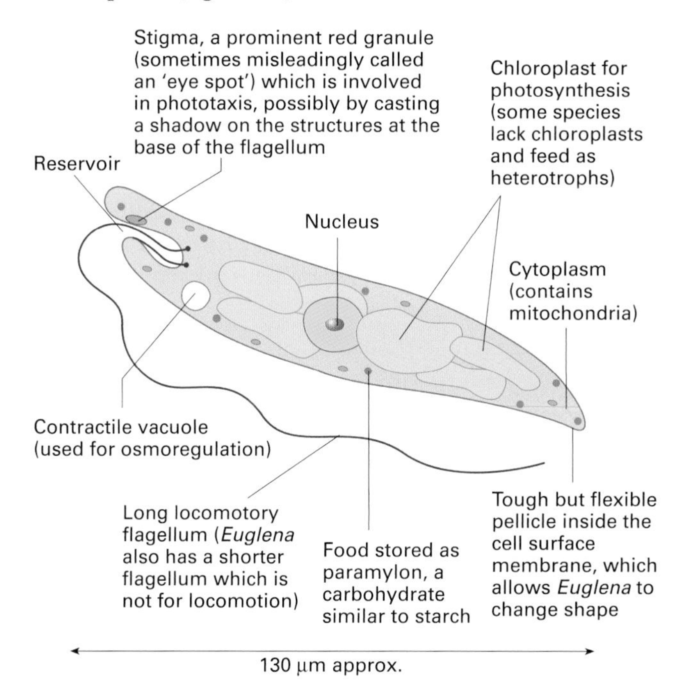

Figure 4 Euglena gracilis.

(a) *Euglena* moves quickly in the water by whip-like movements of its long flagellum

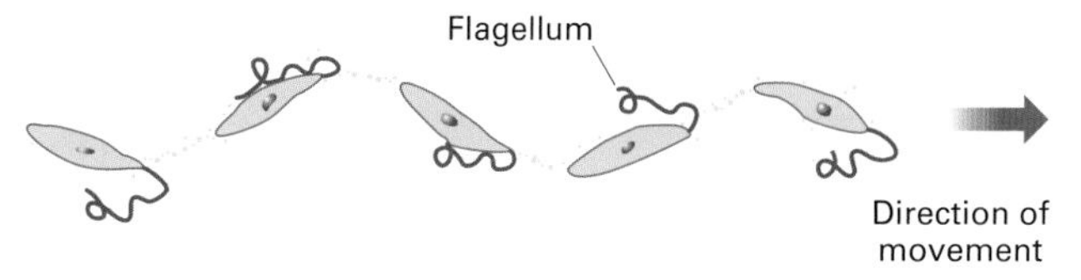

(b) Beating pattern of a flagellum

Direction of movement

Waves moving down the flagellum

Thrust of flagellum

Outer ring of 9 pairs (doublets) of microtubules

Cell surface membrane

Dynein arms

Central microtubules

(c) Transverse section through a flagellum or cilium (×240 000). Flagella and cilia are extensions of the cell and are covered by a cell surface membrane; both have a similar 9+2 pattern of microtubules. However, flagella are about 100 μm long whereas cilia are usually less than 5 μm long, and the beating pattern of flagella and cilia differ. Protoctistan flagella and cilia are called undulipodia. Locomotion depends on the action of a protein called dynein which forms small arm-like structures attached to each doublet microtubule. The 'arms' latch onto a neighbouring microtubule so that one microtubule can slide along the next one, causing an undulipodium to bend. Dynein has ATPase, and ATP provides the energy for ciliary and flagellar movement.

Figure 5 *Locomotion in* Euglena.

Quick check

1 a What is the main function of a contractile vacuole?

b State two functions of pseudopodia.

2 a What is the significance of dynein having ATPase activity?

b How does the flexible pellicle of *Euglena* differ from an algal cell wall?

Food for thought

At one time all organisms were classified into two kingdoms: the animal kingdom and the plant kingdom. Suggest why euglenoid flagellates were once claimed by zoologists to be animals and by botanists to be plants.

21.5

FUNGI AND PLANTS

OBJECTIVES

By the end of this spread you should be able to:

- explain why fungi have a kingdom of their own
- describe the main features of fungi
- describe the main features of plants.

Fact of life

The visible part of most fungi, such as mushrooms, is the spore-producing fruiting body, which is usually only a small part of the whole organism. The most extensive part consists of hyphae (fine threads) which remain hidden in the substrate on which the fungus is growing. Hyphae of some rainforest fungi can spread for up to 500 m. Hyphae can grow very quickly; a single fungal spore can produce more than 1 km of hyphae in 24 hours.

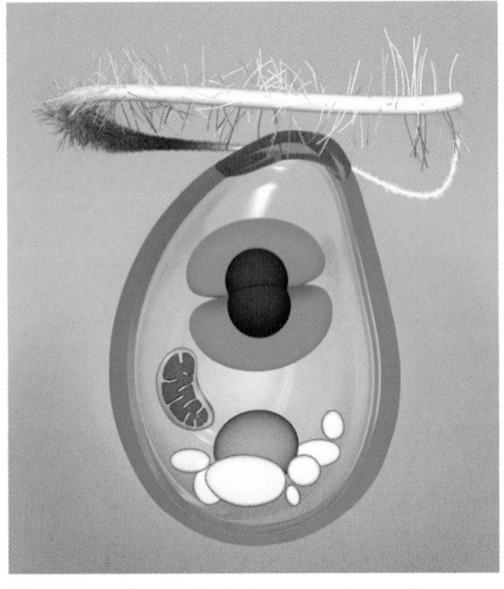

***Figure 1** An artist's impression of* Cyanophora paradoxa, *a member of the algal phylum Glaucophyta which in 2012 had its nuclear genome published. It measures only about 7 μm wide, has two flagella and two plastids. It is regarded as a 'living fossil' because it retains several features found in cyanobacteria: its blue–green plastids can photosynthesise and fix nitrogen; it has a peptidoglycan wall similar to that of bacteria; and it has some genes similar to those of cyanobacteria and other ancient bacteria.*

When all of life was classified into either the plant kingdom or the animal kingdom, fungi presented a particular problem. They were usually included in the plant kingdom because, like plants, they are mainly sessile and they have cell walls. But unlike plants, fungi have no chlorophyll, no cellulose in their cell walls, and are unable to photosynthesise. Nutritionally, they are more like animals; they are heterotrophic organisms which obtain complex nutrients from their environment. However, unlike most animals, they cannot ingest large particles of food and digest them inside the body; they absorb their nutrients from outside the body. They must therefore live in or on their food source, which is either a dead or a living organism, and the food has to be made soluble before it can be absorbed. Fungi differ from both plants and animals in so many ways that the five-kingdom classification gives them a kingdom of their own.

Kingdom Fungi

Fungi share the following characteristic features:

- They have eukaryotic cells with a rigid, protective cell wall. The wall is made of chitin; it does not contain cellulose.
- Their body is usually organised into multinucleate hyphae (thread-like structures).
- They cannot photsynthesise: they are decomposers or **lysotrophs** (from lysis = breakdown, troph = feeder), a type of heterotroph that feeds saprotrophically, digesting their food extracellularly by secreting powerful hydrolytic enzymes. They then absorb the soluble products of digestion through their cell walls and into their hyphae.
- Carbohydrate stores, when present, consist of glycogen, not starch.
- They produce very large numbers of tiny reproductive spores which have no flagella (they lack undulipodia). Sometimes the spores are produced asexually, sometimes sexually.

(Note that although *Pythium*, *Phytophthora*, and other oomycotes resemble fungi in their appearance and nutrition, they have cellulose cell walls and flagellated spores and are generally regarded as protoctists; see spread 21.3, table 1.)

There are three main phyla of fungi: Zygomycota, which includes *Mucor* and *Rhizopus*, characterised by the formation of zygospores during sexual reproduction (spread 17.5); Ascomycota, which includes *Neurospora*, *Saccharomyces* (yeast), and *Penicillium*, characterised by the production of spores in a sac-like structure called an ascus (spread 17.6); and Basidiomycota, which includes the field mushroom *Agaricus* (figure 2). Its characteristic features are described below.

Basidiomycota (basidiomycetes) take their name from the **basidium**, a small club-shaped structure which carries microscopic spores. The familiar cap (pileus) of a mushroom is its fruiting body. The underside of the cap consists of gills (lamellae) on which the basidia are formed. Each mushroom can release up to 100 million spores each hour, and a total of more than 15 000 million during its relatively brief existence.

The main body of a mushroom consists of septate hyphae (hyphae with cross-walls) which lie on or in the substrate (for example, cow dung or nutrient-rich soil). Collectively, the mass of hyphae is called a **mycelium**.

Kingdom Plantae

Plants have the following characteristics:

- They are multicellular.
- They have eukaryotic cells with cellulose cell walls.
- They feed by photosynthesis: some cells contain chloroplasts. (A few parasitic plants have lost the ability to photosynthesise and have no chloroplasts, but they are so clearly related to plants that photosynthesise that they are placed in this kingdom.)

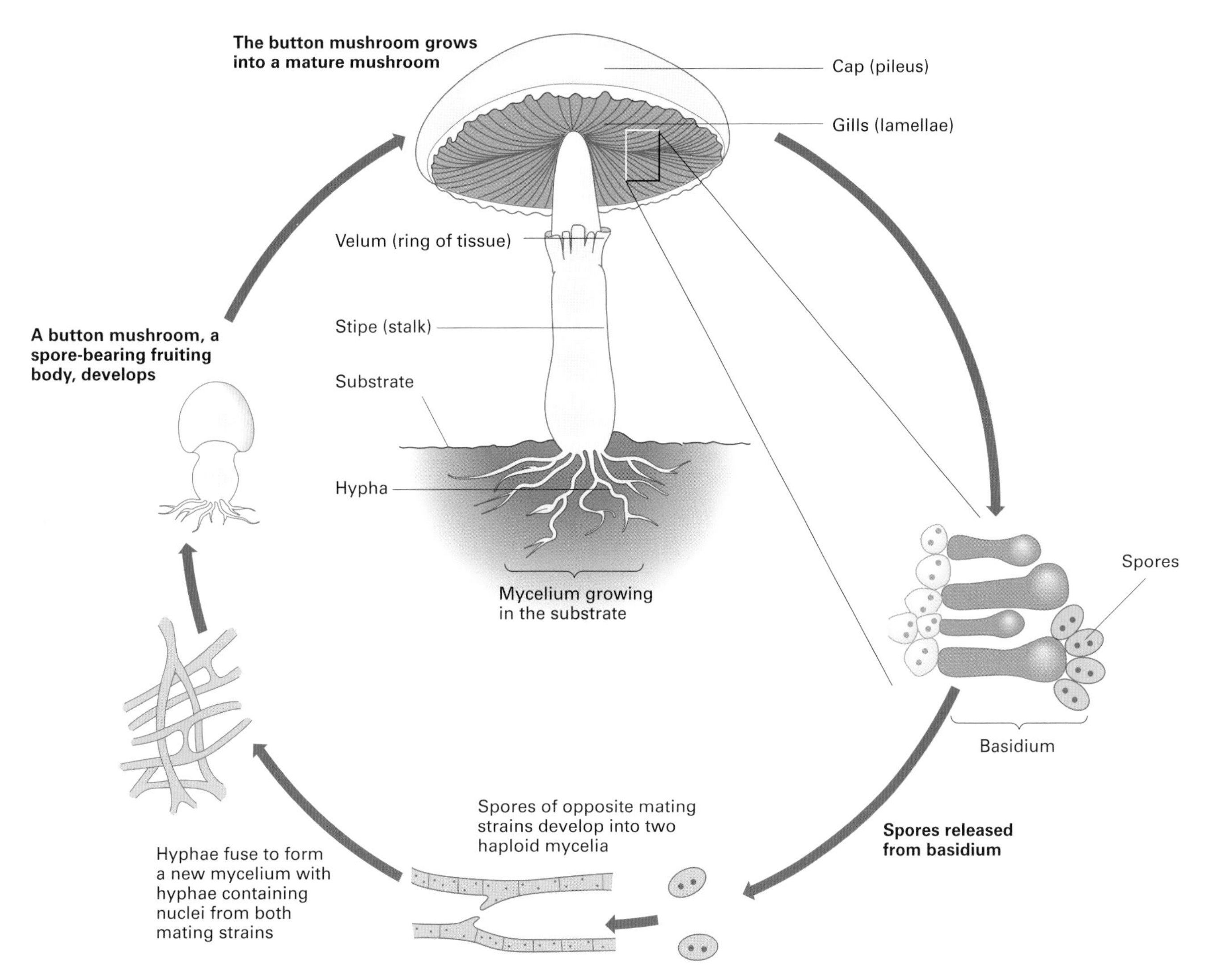

Figure 2 *The field mushroom (*Agaricus campestris*).*

- Most plants store carbohydrates as starch or sucrose.

It is generally accepted that all plants and the algae from which they evolved had a single ancestor that arose by endosymbiosis (see spread 4.4). In 2012, this theory was supported by DNA analyses of *Cyanophora paradoxa* (figure 1) which showed that these glaucophytes have genes similar to those of ancient bacteria and blue–green algae.

During the course of their evolution, plants have become increasingly adapted to life on land. This is reflected in the lives of the four plant phyla that we shall examine in the next two spreads. The phyla are Bryophyta (liverworts and mosses), Filicinophyta (ferns), Coniferophyta (conifers), and Angiospermophyta (flowering plants or angiosperms).

Figure 3 *The plants growing in this garden show a range of evolutionary features adapting them to life in the water, in damp conditions, or on land.*

Quick check

1 How do fungi obtain their nutrients?

2 How do the cell walls of fungi and plants differ?

Food for thought

Suggest why the feeding part of a fungus is usually hidden from sight, whereas the fruiting body is often clearly visible.

21.6

Bryophytes and ferns

OBJECTIVES

By the end of this spread you should be able to:

- explain the meaning of alternation of generations
- describe the characteristic features of bryophytes
- describe the characteristic features of ferns.

Fact of life

Ferns vary in size from tiny water ferns, only a few centimetres across, to tree ferns as much as 25 m high.

Plants have life cycles very different from ours. We exist as a diploid organism for most of our lives; the only haploid stage in our life cycle is a sperm or egg cell. The life cycle of sexually reproducing land plants has two distinct stages or generations: a haploid stage which produces gametes, called the **gametophyte generation**; and a diploid stage which produces spores, called the **sporophyte generation**. Gametes, like their parent cells, are haploid and are produced by mitosis (in contrast with gamete production in animals which is by meiosis). When a female gamete is fertilised by a male gamete, a diploid zygote is formed. This grows up into a diploid sporophyte which produces haploid spores by meiosis. In appropriate conditions, these spores germinate into the gametophyte generation, thus completing the life cycle (figure 1). This **alternation of generations** provides plants with two opportunities to produce large numbers of offspring:

- The gametophyte undergoes sexual reproduction to produce gametes which fuse to form zygotes.
- The sporophyte undergoes asexual reproduction to produce spores.

The gametophyte and sporophyte generations are always very different, and one is always larger and more conspicuous than the other. In seed plants (conifers and flowering plants), alternation of generations is not obvious, but it still happens (spread 21.7). In bryophytes and ferns, both stages are clearly visible.

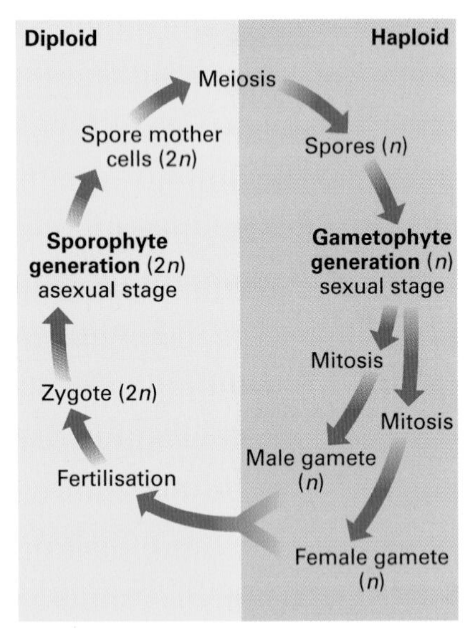

***Figure 1** The alternation of generations seen in the life cycle of plants.*

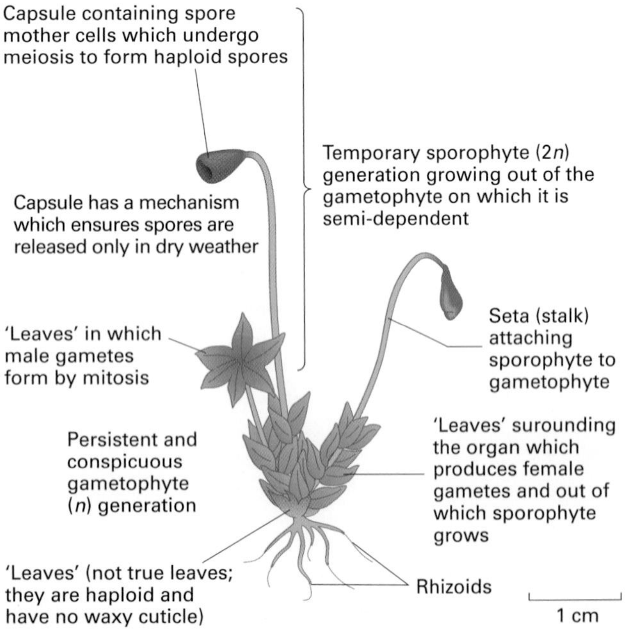

Kingdom Plantae; phylum Bryophyta (bryophytes)

Bryophytes (liverworts, hornworts, and mosses) have the following characteristic features:

- The gametophyte generation is much more conspicuous and lasts longer than the sporophyte generation (this can be seen in the life cycle of the moss *Funaria hygrometrica*, figure 2).
- The body is anchored in the substrate by **rhizoids**, not roots (rhizoids are haploid, filamentous outgrowths of the gametophyte; roots are diploid).
- They do not have true stems or leaves (like the rhizoids, the so-called 'leaves' of mosses are haploid and are not homologous to true leaves, which are diploid).
- They have no true vascular system (they lack xylem and phloem).

Bryophytes often live in habitats which other plants do not fill, such as in the cracks of rocks and on the bark of trees. Among all plants living today, however, they are anatomically the least well adapted to life on land. Many species are confined to damp places for most of their active lives, surviving during dry periods by adaptations that help them tolerate desiccation and by producing resistant spores. During sexual reproduction, all bryophytes need water because they have motile male gametes that can reach the female gametes only by swimming.

Capsule containing spore mother cells which undergo meiosis to form haploid spores

Temporary sporophyte (2*n*) generation growing out of the gametophyte on which it is semi-dependent

Capsule has a mechanism which ensures spores are released only in dry weather

Seta (stalk) attaching sporophyte to gametophyte

'Leaves' in which male gametes form by mitosis

'Leaves' surounding the organ which produces female gametes and out of which sporophyte grows

Persistent and conspicuous gametophyte (*n*) generation

'Leaves' (not true leaves; they are haploid and have no waxy cuticle)

Rhizoids

1 cm

***Figure 2** Funaria hygrometrica, a common moss showing both stages of its life cycle: the conspicuous and longer lasting gametophyte, and the dependent and short-lived sporophyte.*

Kingdom Plantae; phylum Filicinophyta (ferns)

The characteristic features of the phylum Filicinophyta are shown in figure 3 and include the following:

- The diploid sporophyte is the most conspicuous and long-lived generation.
- The gametophyte is reduced to a small structure called a prothallus.
- The body has true leaves, stems, and roots with an elaborate vascular system of xylem and phloem.

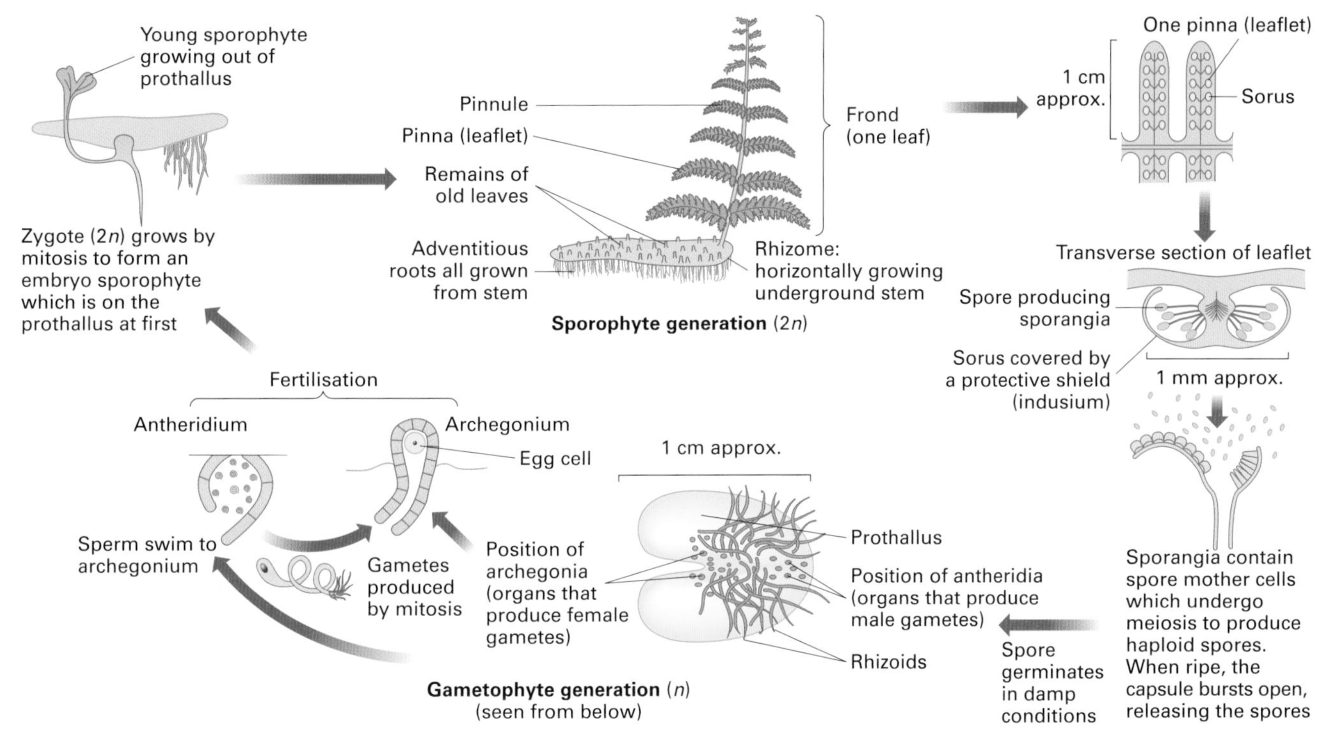

Figure 3 The life cycle of the male fern, Dryopteris felix-mas.

- The young leaves of the sporophyte are coiled and are known as fiddleheads.
- **Sporangia** (structures on the undersides of leaves in which asexual non-motile spores are produced) occur in clusters called **sori**.

Ferns are the most primitive group of plants to evolve a true vascular system. The system consists of xylem and phloem, supported by some lignified tissue. Most ferns also have complex leaves (commonly called fronds) with a waxy cuticle. All plants which share these features are called **vascular plants** or **tracheophytes**. Vascular plants include the conifers and flowering plants as well as the ferns. Ferns are much better adapted to life on land than are bryophytes. The waxy cuticle of fern fronds reduces water losses, and the partly lignified vascular system provides mechanical support and transports materials efficiently between roots and leaves. This means that ferns can develop larger, more complex bodies, grow taller (some tree ferns are more than 20 m high), and live in drier habitats than bryophytes. However, these adaptations to life on land are found only in the sporophyte generation. In ferns, the gametophyte remains dependent on wet conditions because it has no waxy cuticle; it has no true vascular system; and, most crucially, it produces flagellated male gametes which need water for fertilisation. The gametophyte is reduced to a short stage in the life cycle of ferns, but it is a significant stage, and this confines ferns to habitats that are damp, at least for the period during which the gametophyte exists.

Figure 4 Tree fern in the understorey of a forest in Tasmania.

Quick check

1. Copy and complete the following sentence, using the words **asexually**, **gametophyte**, **sporophyte**, and **sexually**.

   In all land plants the ______________ generation is haploid and reproduces ______________, whereas the ______________ generation is diploid and reproduces ______________.
2. How do the 'leaves' of bryophytes differ from the true leaves of ferns?
3. Which part of the life cycle of ferns is most dependent on water?

Food for thought

Suggest why tree ferns are found mainly in the understorey of tropical rainforests, below the tree canopy (figure 4).

21.7 Conifers and flowering plants

OBJECTIVES

By the end of this spread you should be able to:

- state the main features of conifers and discuss their significance
- state the main features of angiosperms and discuss their significance
- distinguish between monocots and dicots.

Fact of life

Conifers reproduce by producing seeds which can be remarkably tolerant of extremes of heat, cold, and drought. Unlike free-living gametophytes, seeds can remain dormant until conditions are suitable for germination. In unmanaged conifer forests of North America, for example, fires started by lightning are relatively frequent. The jack pine, *Pinus banksia*, keeps its seeds inside cones sealed by resin. During a fire, heat melts the resin and releases the seeds. They germinate and take advantage of the space produced by the fire. The life cycle of the jack pine has become so dependent on forest fires that it will stop reproducing in the absence of fires.

Kingdom Plantae; phylum Coniferophyta (conifers)

Conifers have the following characteristic features:

- They are cone-bearing plants: female cones produce ovules and male cones produce pollen.
- The fertilised ovule develops into a seed on the surface of a cone; it is not enclosed in an ovary.
- They have no flowers, and no fruits.

Conifer trees have a well developed vascular system and most have narrow needle-like leaves protected by a waxy cuticle. Like other land plants, conifers have an alternation of generations, as shown by the Scots pine (*Pinus sylvestris*, figure 1). The most conspicuous stage of its life cycle is the diploid sporophyte generation, the tree. Because conifers reproduce by seeds, the gametophyte generation is much reduced and is never free living.

Conifers are better adapted to life on land than ferns. Conifers (and angiosperms) do not need water as a medium for fertilisation because they produce pollen. A pollen tube grows down from the pollen grain to an egg cell, and fertilisation can take place without the male gamete having to swim through water. Seeds reduce the role of the vulnerable gametophyte stage, which becomes totally dependent on the sporophyte.

Conifers are more widespread in cold and dry environments than either ferns or flowering plants, and they often dominate the landscapes of mountainous regions. Their leaves and stems are well adapted to periods of drought (for example, when the ground is frozen). Many conifers have evolved above-ground structures that can persist even in harsh environments. This has enabled bristlecone pines (spread 14.9, figure 1), for example, to grow on bitterly cold mountain ridges in California.

The hardiness of conifers makes them especially important economically as a tree crop for pulp and timber in areas unsuitable for other crop production. In addition, as most conifers are evergreen they are valued as landscape plants. However, the evergreen habit has disadvantages. Premature leaf death caused by pollution, disease, or insect attack can be more damaging to conifers than to deciduous plants which produce a new flush of leaves each spring.

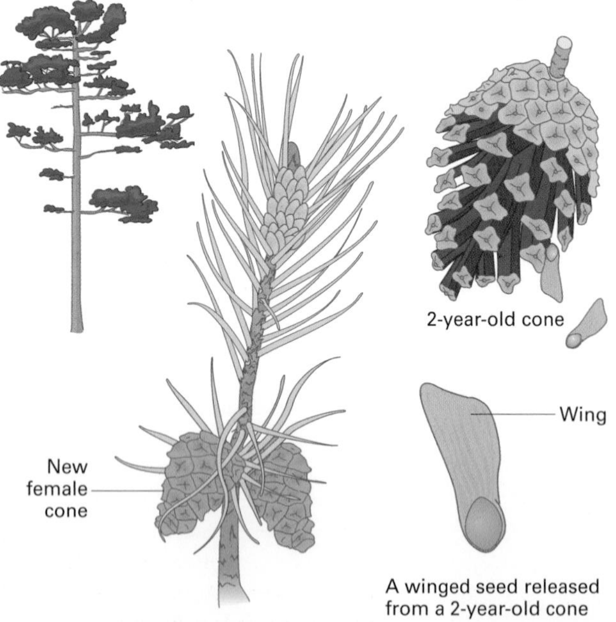

Figure 1 *Scots pine,* Pinus sylvestris *showing male and female cones. Pollen is released from male cones, and fertilisation occurs in the female cone in which the seeds develop. When it is two years old, the female cone ripens and its scales open, releasing seeds which are carried by the wind.*

The life cycle of conifers

Conifer trees, the main part of the sporophyte generation, bear female and male cones which produce two different types of spore, a condition known as **heterospory**. Female cones contain ovules, each of which produces a single diploid **megaspore** (large spore). Megaspores divide meiotically to form a tiny haploid, egg-containing female gametophyte. This remains attached to the cone on the parent plant. Male cones contain microspore mother cells. Each cell divides meiotically to produce four haploid **microspores** (small spores) which are released as pollen grains. The total output of pollen from a cone can be enormous. If pollen grains reach a compatible female cone (usually after being dispersed by the wind), a pollen tube develops and the haploid microspore divides mitotically to form the male gametophyte. A male gamete travels down the tube towards the female gametophyte in the ovule. After fertilisation of the egg cell by one male gamete, a new diploid sporophyte embryo develops in the ovule and is released as a winged seed. The seed consists of an embryo plant and food reserve enclosed in a protective testa (seed coat).

Kingdom Plantae; phylum Angiospermophyta (angiosperms, flowering plants)

Angiosperms have the following characteristic features:

- They have true flowers.
- They are seed-bearing plants.
- The seed is enclosed in a fruit formed from an ovary.

Flowering plants or angiosperms are the largest phylum of plants, outnumbering all the others put together. They include most of our crop and ornamental plants. They vary in size from tiny duckweed which lives on the surface of ponds to giant trees which can grow over 90 m tall. Although not as hardy as conifers, angiosperms occupy a wide variety of habitats from equatorial deserts and rainforests to arctic tundra.

The success of angiosperms comes from their evolution of flowers in which the reproductive structures are carried. Conifers must produce enormous quantities of pollen because they rely on wind to distribute it. In contrast, ancestral angiosperms evolved flowers with features that encouraged insects to carry pollen from one plant to another and place it exactly where it was needed for fertilisation. The magnolia (figure 2) was one of the first plants to adopt this efficient strategy.

After fertilisation, the ovary develops into a fruit which encloses one or more seeds ('angiosperm' means 'container of seed'). The fruit protects the seeds and helps to disperse them efficiently. You can read more about the life cycle of angiosperms and their reproductive strategies in spreads 14.1 to 14.6.

Traditionally, flowering plants have been classified by morphological features into one of two classes: the **Monocotyledonae** (monocots) and **Dicotyledonae** (dicots). Table 1 shows their main distinguishing features.

Figure 2 *A magnolia flower being pollinated by a bee.*

Table 1 *Main distinguishing features of monocots and dicots.*

Feature	Class Monocotyledonae	Class Dicotyledonae
Embryo	One cotyledon (seed leaf)	Two cotyledons (seed leaves)
Leaf veins	Usually parallel	Net-like or branching
Flower parts	Often in threes	Often four, five, or many
Secondary growth	Never woody	Woody or herbaceous
Example	*Titicum* (wheat); *Zea* (maize)	*Ranunculus* (buttercup)

New classifications of flowering plants

DNA sequencing techniques have enabled plant taxonomists to make detailed comparisons of the **genomes** of flowering plants and these are leading to a major revision of their classification. The traditional division of the flowering plants into the monocots and dicots is disappearing. Although the monocots grouping is recognised as a **clade** (a major grouping of organisms that have a single common ancestor), the dicots grouping is not. A number of former dicots are being placed in separate groups that are basal to monocots, and the remaining dicots are classified as the **eudicots** or 'true dicots'. (A basal clade is the earliest clade to branch in a larger clade; it appears at the base of an evolutionary diagram called a **cladogram**.)

Quick check

1 What is the function of the cuticle in a conifer?

2 Why do many angiosperms produce less pollen than conifers?

3 How does the leaf of a monocot differ from that of a dicot?

Food for thought

Giving reasons, suggest why both angiosperms and conifers can be regarded as successful land plants. Which do you think is the more successful group?

21.8

Animals

OBJECTIVES

By the end of this spread you should be able to:

- list the main characteristic features of animals
- distinguish between bilateral symmetry and radial symmetry.

Fact of life

Sponges are sessile, mainly marine organisms which filter feed by passing a current of water through pores in their bodies. Although classified as animals, they have no tissues or organs, no mouth and no digestive system; their entire body structure is built around a unique water canal system which they use for their filter feeding. Unlike any other animal, they have no nervous coordination. They are commonly shaped liked a vase and may have an internal skeleton consisting of glass-like spicules or rods (figure 1). However, their body shape varies with the direction and force of water currents.

Sponges reproduce asexually. They have an amazing capacity for regeneration. If living sponge tissue is forced through a silk mesh, the separated cells reorganise after a short time and reform themselves into several new sponges. This regeneration has been used to produce commercial sponges in the overfished waters of Florida. Pieces of sponge called 'cuttings' are attached to cement blocks and dumped into the water. After a few years, each cutting grows into an individual sponge of marketable size. Because of their sessile nature and regenerative ability, Aristotle, Pliny, and other ancient naturalists regarded sponges as plants. In fact, it was not until 1765, when their filter feeding currents were observed, that their animal nature was clearly established.

Although there are over 10 000 species of sponges living today, they appear to be an evolutionary dead end, giving rise to no other members of the animal kingdom.

Kingdom Animalia

Members of the animal kingdom have the following characteristic features:

- They are multicellular.
- They have eukaryotic cells with no cell walls.
- They are non-photosynthetic, feeding heterotrophically.
- Except for sponges, they all have nervous coordination.

The animal kingdom therefore contains all multicellular eukaryotes that do not have cell walls enclosing their cells. They are all heterotrophic and all, except sponges, have nervous coordination and a muscle system so that they can respond quickly to environmental stimuli. No other kingdom has members with either nerve cells or muscle cells. Animals form the majority of known multicellular organisms and are extraordinarily diverse in their body shapes and functions.

***Figure 1** The skeleton of the sponge Venus's flower basket (*Euplectella*) is made of glass-like spicules fused together.*

Animal symmetry

A typical animal is a **motile** creature; it can move spontaneously from place to place, for example, to feed. Usually, one part of the body travels forwards first, which means an animal has:

- a definite **anterior** end
- a **posterior** end
- a **ventral** surface (the underside of an animal, or the side normally directed downwards; in bipedal primates such as humans, it is the 'front')
- a **dorsal** surface (the 'back' or that part which is normally upwards; in bipedal primates it is directed backwards)
- two side or **lateral** surfaces
- **bilateral symmetry** (only a single line or plane can divide the body into equal halves; figure 2).

The anterior end is likely to meet new stimuli before the rest of the body, and a typical animal has a distinct head at the anterior end which is well supplied with sense organs.

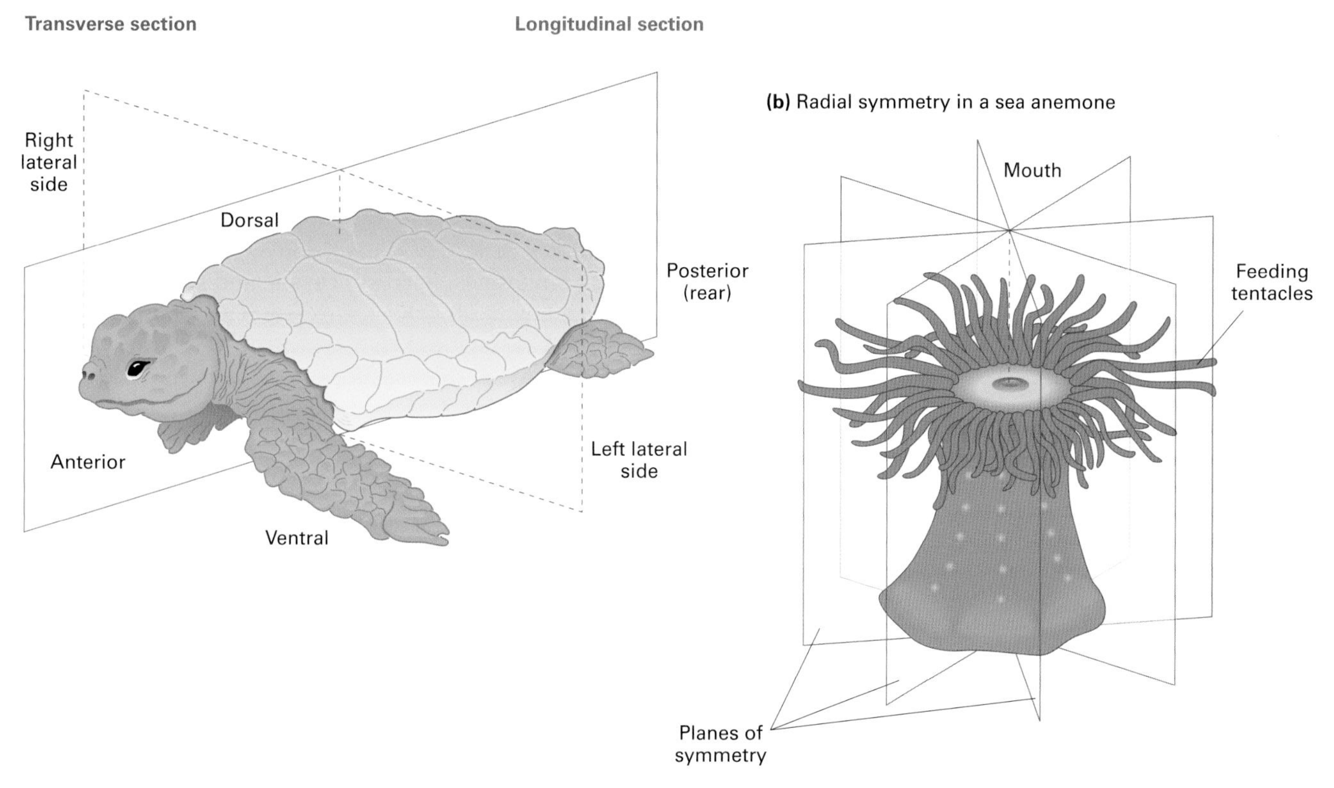

Figure 2 *Most animals are bilaterally symmetrical; some are radially symmetrical.*

Many animals, however, do not conform to the typical body plan. For example, adult barnacles (see spread 22.5, figure 3) are **sessile** (they are fixed permanently in one place) and have no head. They glue themselves to their substrate, commonly a rock, and wait for the sea to bring their food to them. Sea anemones and *Hydra* are **sedentary** (slow-moving) creatures that also wait for their food to come to them. They have no definite head end but have a ring of tentacles armed with special stinging cells with which to capture their prey (see spread 21.9). These sessile and sedentary creatures are often (at least superficially) **radially symmetrical** (the body can be divided into roughly equal halves by more than one straight line or plane which passes through the central point of the body). As sea anemones have no anterior end, any part of the body may be the first to move forward during locomotion. The radial arrangement of their feeding, sensory, and defensive structures means that they are aware of the environment all around them and can respond to stimuli from any direction. Some animals, such as certain types of coral, have no regular body shape and are said to be **asymmetrical**.

Classification of animals

Most zoologists (scientists who study animals) recognise about 30 animal phyla; the following spreads deal with eight of the more familiar ones:

- Cnidaria (for example, sea anemones and jellyfish)
- Mollusca (for example, limpets, mussels, and octopus)
- Nematoda (roundworms)
- Platyhelminthes (flatworms)
- Annelida (truly segmented worms)
- Arthropoda (for example, crustaceans, insects, spiders, and mites)
- Echinodermata (for example, sea urchins and starfish)
- Chordata (including fish, amphibians, reptiles, birds, and mammals).

Quick check

1 Which feature do all animals (except sponges) have which distinguishes them from plants and fungi?

2 Distinguish between bilateral and radial symmetry in terms of the number of planes of symmetry.

Food for thought

Suggest why most animals that show radial symmetry are aquatic, whereas most terrestrial animals have bilateral symmetry.

21.9

OBJECTIVES

By the end of this spread you should be able to:

- list the main characteristic features of the phylum Cnidaria and discuss their significance.

Fact of life

The Great Barrier Reef off the coast of Australia is formed mainly from the skeletons of cnidarians called corals. It is over 2000 miles long and is home to over 3000 species of marine organism.

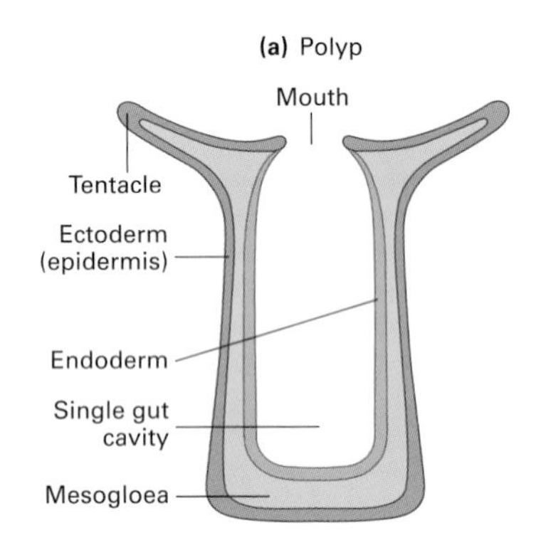

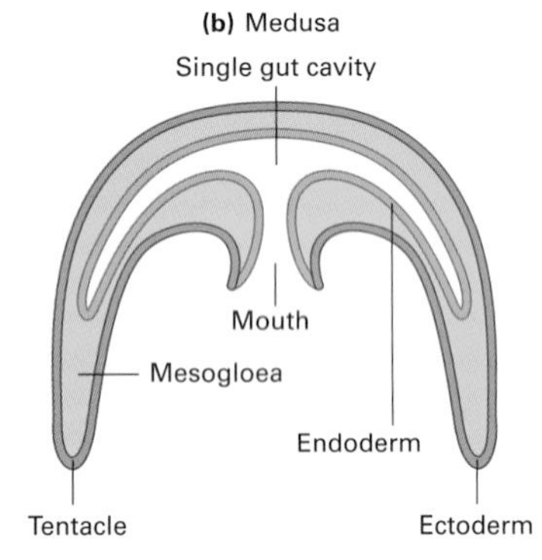

***Figure 1** Longitudinal sections showing alternative body plans of cnidarians: **(a)** a polyp **(b)** a medusa.*

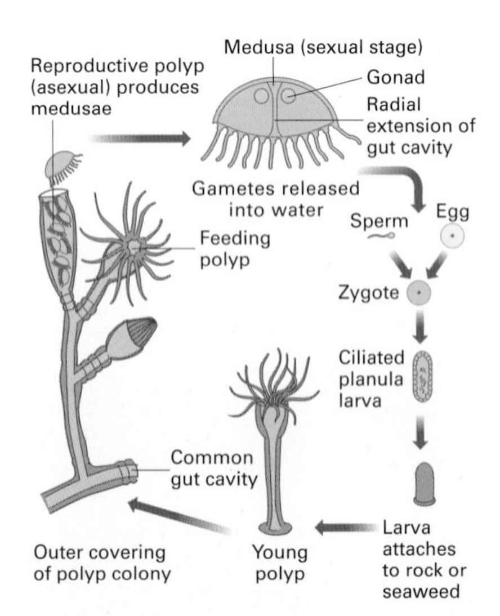

***Figure 2** The life cycle of* Obelia *alternates a polyp stage with a medusa stage.*

CNIDARIANS

Kingdom Animalia; phylum Cnidaria (cnidarians)

Cnidarians (the 'c' is silent) have the following characteristic features:

- They are **diploblastic** animals: they have two cell layers separated by mesogloea (a jelly-like, non-cellular layer).
- They have nematoblasts (stinging cells).
- They are radially symmetrical.
- They have tentacles.

Cnidarians include sea anemones, jellyfish, corals, and hydras. Most are marine, but there are a few freshwater forms (such as the hydras). Typically, cnidarians are radially symmetrical (spread 21.8, figure 2) and many are brightly coloured, giving them a plant-like appearance. This belies their true nature for cnidarians are animals and some, such as the Portuguese man-of-war, are among the most deadly predators in the seas.

Cnidarian body plan

Cnidarians are soft-bodied animals that can exist in two forms: as a **polyp** or as a **medusa** (figure 1). Typically, a polyp spends most of its time attached to the substrate (such as a rock). It has a tubular body, usually with a mouth and a ring of tentacles on top. In contrast, a medusa is a free-swimming bell- or saucer-shaped individual with a tube hanging down in the centre; the tube ends in a mouth and tentacles usually line the edge of the bell. Sea anemones are polyps while jellyfish are medusae. Some cnidarians have both polyp and medusa forms in their life cycles (figure 2). This condition in which a species exists in two or more different forms is called **polymorphism**.

All cnidarians have a relatively simple body plan with no special respiratory, circulatory, or excretory systems. Exchange with the surrounding water and between one cell and another takes place by diffusion: cnidarians are generally small with a large surface area to volume ratio. They also have water circulating through their gut cavity.

In addition to being radially symmetrical, cnidarians are **diploblastic**. This means that they have a body wall made of two layers of cells: an outer layer or **ectoderm** which forms the epidermis, and an inner layer or **endoderm** which forms the internal structures. The **mesogloea** (a non-cellular jelly-like layer through which cells are able to migrate) separates the ectoderm from the endoderm. In jellyfish, the watery mesogloea forms a large part of the body and acts as a hydrostatic skeleton (see spread 11.1). The body wall of cnidarians encloses a mouth, usually surrounded by tentacles, which leads into a gut cavity or **enteron**. There is no anus; the mouth acts as both the entry point for ingested food and as the exit point for egested material.

Coordination in cnidarians

Unlike sponge cells which are only loosely associated with each other (spread 21.8), the cells of cnidarians are organised into tissues. This **tissue level of organisation** (figure 3) enables cells to act together in a relatively coordinated manner, carrying out various functions much more efficiently. Stinging cells known as **nematoblasts** occur in the ectoderm of tentacles. When touched or stimulated with a chemical, these can discharge different types of threads which are used to capture prey and to defend against predators. One type of nematoblast can inject toxin into a prey or predator. A single nematoblast would probably have little effect, but a battery of them can deliver enough toxin to paralyse or even kill a small animal. Toxins produced by some jellyfish and corals produce severe pain in humans.

There are contractile muscle-like fibrils in the ectoderm and endoderm of cnidarians, and the gut cavity acts as a hydrostatic skeleton against which the muscle fibrils act. On its own, a single fibril would produce little force, but the combined action of many fibrils can produce strong locomotory movements. In *Hydra*, fibrils in the ectoderm are arranged

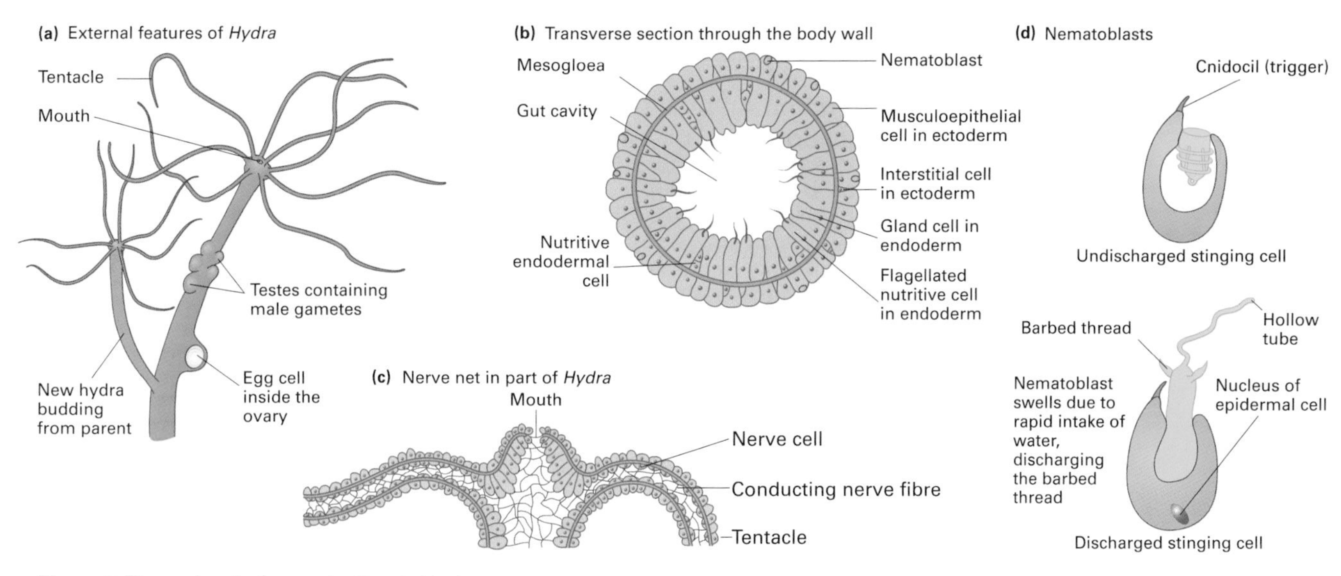

Figure 3 *Tissue level of organisation in* Hydra.

longitudinally so that they shorten the body when they contract. Fibrils in the endoderm lie laterally and lengthen the body when they contract.

A simple **nerve net** coordinates movement. Nerve cells in the body wall connect sense organs with muscle fibrils. A weak stimulus can bring about a localised response, but a very strong stimulus may cause the entire animal to contract. Although there may be a slight concentration of nerve cells around the mouth, there is no head and no centralisation of nervous tissue. Nevertheless, cnidarians show a surprisingly high degree of coordination. During feeding, tentacles appear to work together to grasp prey and carry it towards the central mouth.

Cnidarians and green fluorescent protein

The snakelocks anemone (figure 4) is one of the most colourful cnidarians on our rocky shores. Its bright green coloration is due to a green fluorescent protein (GFP), homologous to that originally isolated from the jellyfish *Aequorea victoria*. In genetic engineering, the GFP gene is frequently used as a genetic marker (see spread 18.10). Because GFPs are relatively harmless and can be attached to virtually any protein, they are used extensively in cell and molecular biology. With the aid of fluorescence microscopy and ultrafast digital cameras, glowing GFP-labelled genes and proteins can easily be tracked in living systems to monitor cellular processes. In 2008, Martin Chalfie, Osamu Shimomura, and Roger Y. Tsien were awarded the Nobel Prize in Chemistry for their discovery and development of the green fluorescent protein. Since their discovery in cnidarians, GFPs have been found in many other organisms. Although the natural function of GFPs is not yet known with certainty, they can act as light-induced electron donors in photochemical reactions with various electron acceptors, similar to the way that chlorophyll donates electrons in photosynthesis. The snakelocks anemone is host to symbiotic unicellular algae. Perhaps the GFP acts as a **light-harvesting** accessory pigment and donates electrons to the photosynthesising 'guests'.

Figure 4 *A snakelocks anemone,* Anemonia viridis.

Quick check

1 **a** What is a nematoblast?
b Distinguish between a polyp and a medusa.
c List all the layers of a cnidarian.
d Which is the non-cellular layer?

Food for thought

Obelia is a cnidarian that has both polyp and medusoid stages in its life cycle. The polyp stage consists of hydra-like bodies living in colonies, which reproduce asexually, whereas the medusoid stage produces gametes and reproduces sexually (figure 2). Both stages are diploid. Suggest the advantages of having two distinctly different stages in the life cycle. How does this type of life cycle differ from the true alternation of generations shown by plants (spread 21.6, figure 1)?

21.10

OBJECTIVES

By the end of this spread you should be able to:

- list the main characteristic features of the phylum Platyhelminthes
- discuss the biological significance of having a triploblastic body plan.

Fact of life

Most flatworms are tiny aquatic organisms, but some species live on land in humid tropical forests. One of these grows to a length of 60 cm. Parasitic flatworms can be even longer; for example, tapeworms living in the human gut may be as much as 10 m long.

PLATYHELMINTHS: GENERAL FEATURES

Kingdom Animalia; phylum Platyhelminthes (platyhelminths, flatworms)

Platyhelminths have the following characteristic features:

- They are flat, unsegmented, bilaterally symmetrical animals.
- They have three layers of cells (they are **triploblastic**).
- They have a mouth (except cestodes), but no anus.
- They are usually hermaphrodites with a complex reproductive system.

All animals except sponges and cnidarians develop from a three-layered embryo and are called **triploblastic**. The three body layers are the ectoderm, endoderm, and mesoderm. The **mesoderm**, found between the ectoderm on the outside and the endoderm on the inside, forms the bulk of the body and gives rise to many important body structures including true muscles. Whereas cnidarians show a tissue level of organisation (spread 21.9), flatworms and all other animals (except sponges) have an **organ level of organisation**. Each organ is made up of different types of tissue which work together to carry out a specific role.

Flatworms have only one body cavity, the gut or enteron. Most other triploblastic animals have a second body cavity called the **coelom**. This is a fluid-filled space in which the organs are suspended (spread 21.12). Being **acoelomate** (without a coelom), flatworms have their organs simply crammed in the body, giving it a fairly dense and solid appearance (figure 1).

Blood vessels evolved from the mesodermal lining of the coelom and lie within the coelom, so flatworms have not evolved a blood vascular system. Their flattened shape, however, means that they have a large enough surface area to volume ratio for gases to be exchanged with the environment and transported in the body by simple diffusion alone. In larger flatworms, the gut may be highly branched and spread into tissues so that digested food molecules can diffuse easily to each cell requiring them.

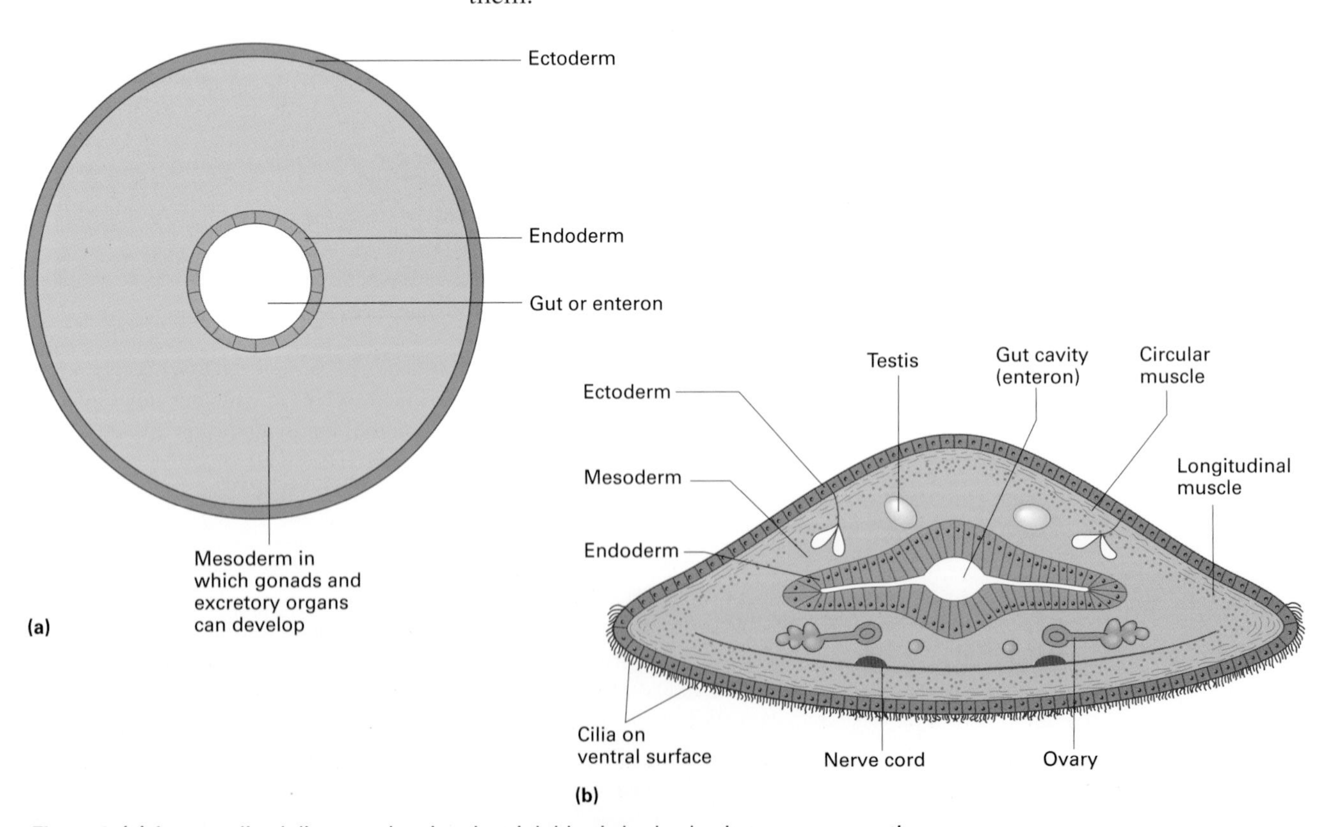

***Figure 1** **(a)** A generalised diagram showing the triploblastic body plan in transverse section.*
***(b)** Transverse section through a flatworm showing its dorsoventral flattening and its triploblastic body plan.*

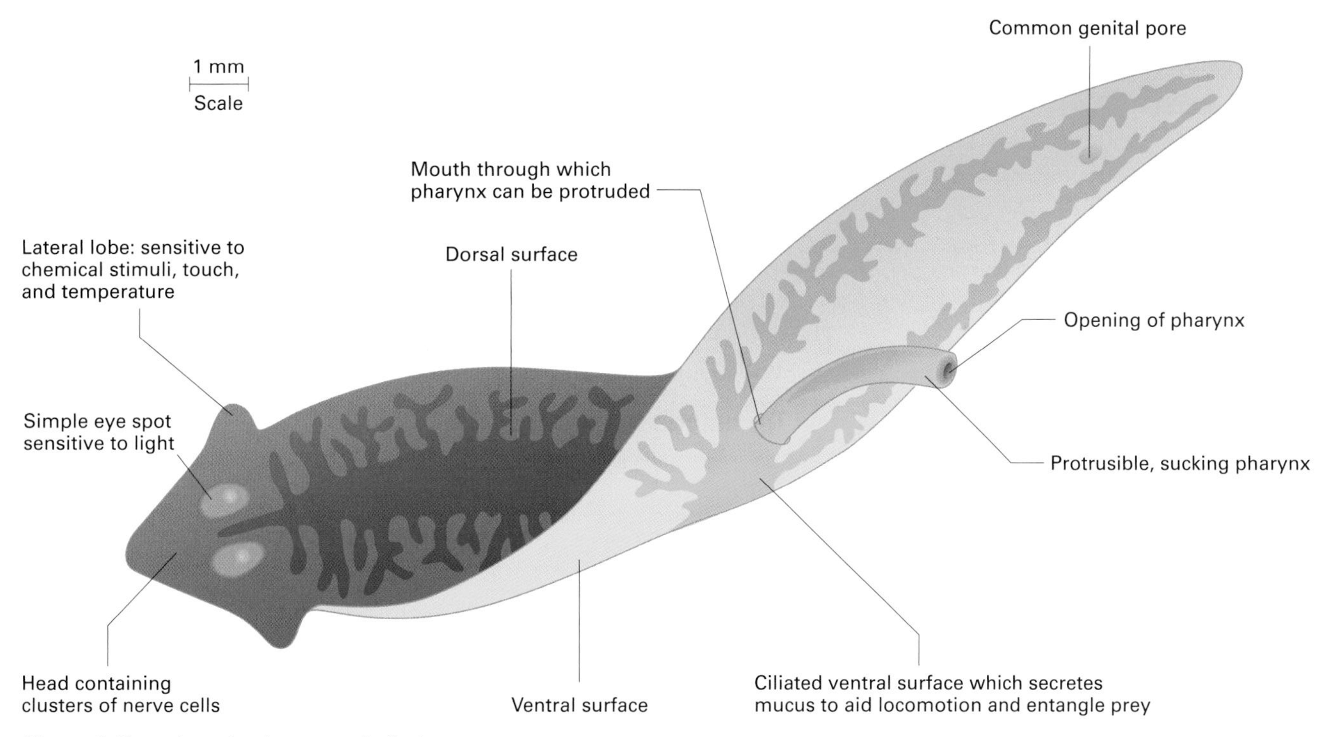

Figure 2 Planaria, *a freshwater turbellarian.*

Free-living flatworms

If you drop a piece of meat into a pond with thick underwater vegetation, you are likely to find many flatworms gliding out to settle on the bait. These free-living, predominantly aquatic animals belong to the class of flatworms called **Turbellaria** (turbellarians).

Turbellarians are characterised by having a ciliated outer surface. The beating of the cilia propels the animal along and in so doing creates turbulence in the water, hence the name of the class. Turbellarians can also move around by wave-like undulations caused by muscle contractions.

Planaria (figure 2) is a typical free-living, carnivorous flatworm commonly found in freshwater streams and ponds. Its bilateral symmetry is associated with its active mode of life (spread 21.8). It has a head at its anterior end with a pair of simple eyes, and chemoreceptors on the lateral lobes. *Planaria* has a clearly defined central nervous system. Dense clusters of nerve cells form a simple brain in the head, and a pair of nerve cords connect the brain with small nerves throughout the body. Together, the nerves and the sense organs make up an organ system, the nervous system.

The behaviour of free-living flatworms is more varied than that of animals such as cnidarians with simpler nervous systems. Flatworms are capable of learning. When placed in a T-maze, *Planaria* can, for example, learn to move to one side rather than the other to obtain food.

Planaria is a nocturnal animal, usually hiding under stones during the day and emerging at night to feed on small worms, crustaceans, and on the dead bodies of larger organisms. Its chemoreceptors help it locate food. Gland cells on its ventral surface secrete mucus which entangles prey. Once the prey has been immobilised, the pharynx can be protruded through the mouth and enzymes poured over the prey. When partially digested, the food is sucked into the body by the pumping action of the pharynx. The mouth leads into a well developed digestive tract where digestion is completed by amoeboid cells. *Planaria* has no anus; undigested food leaves the body by the same route it entered, through the mouth.

Like most flatworms, *Planaria* is hermaphrodite (individuals have both male and female sex organs). However, cross fertilisation usually occurs. *Planaria* can also reproduce asexually by splitting into two and regenerating the missing half.

Quick check

1. Briefly state how the body plan of platyhelminths differs from that of cnidarians (spread 21.9).
2. In which body layer of platyhelminths do most organs develop?

Food for thought

Although there are a few semi-terrestrial species of free-living flatworms, they are mainly aquatic creatures. Suggest why flatworms such as *Planaria* are poorly adapted for life on land.

21.11

OBJECTIVES

By the end of this spread you should be able to:

- distinguish between trematodes and cestodes
- discuss the adaptations of liver flukes and tapeworms to their parasitic mode of life.

Fact of life

Parasitic flatworms can have a very high reproductive output. The rat tapeworm, for example, may produce 250 000 eggs per day for as long as its host lives, a potential production of 100 million individuals. If all these offspring survived to maturity, they would amount to more than 20 tonnes of tapeworm. However, this high reproductive output is an adaptation to the hazards of the parasitic mode of life, and most of the offspring perish before they reach maturity.

PLATYHELMINTHS: PARASITIC FLATWORMS

Whereas *Planaria* has adopted a free-living predatory mode of life (spread 21.10), many platyhelminths are parasitic. Most are endoparasites, living unseen within the bodies of animals belonging to different species. By definition, parasites benefit from the relationship while the other animal (the host) is harmed in some way. However, unlike *Planaria*, parasitic flatworms do not kill their host in order to obtain nutrients. Instead, they feed off living tissue or absorb nutrients within the host's body.

There are two main classes of parasitic flatworms: the Trematoda and the Cestoda.

Class Trematoda (trematodes, flukes)

Trematodes have the following characteristic features:

- They are endoparasites.
- They do not have cilia on their outer surface.
- They possess one or more suckers.
- They have a gut cavity (enteron) which is usually highly branched.

The liver fluke *Fasciola hepatica* is a typical trematode. The adult (figure 1a) lives in the bile ducts of sheep, goats, cattle, and occasionally humans. It retains the typical flatworm body shape, but has become specially adapted for its parasitic mode of life. It has a complex life cycle involving both a **primary host** (a host in which the parasite is sexually mature) and a **secondary** or intermediate host (a host containing sexually immature stages; figure 1b).

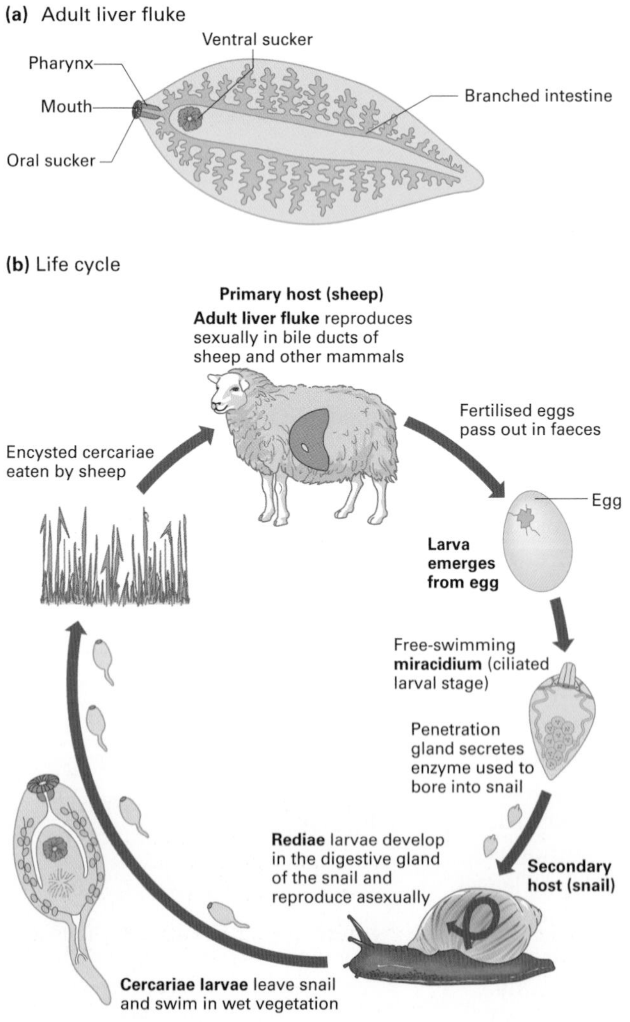

Class Cestoda (cestodes, tapeworms)

Cestodes have the following characteristic features:

- They are endoparasites.
- They have no gut cavity; they absorb their food, predigested by the host, through the cuticle.
- They have a flat or elongated body with no cilia.
- The body is usually divided into sexually reproducing sections called **proglottids**.
- They have a region at the anterior end (a **scolex**) bearing suckers and/or hooks.

Figure 2 shows the life cycle of the tapeworm *Taenia solium*, a species of tapeworm that infects pigs and humans.

Tapeworms look very different from turbellarians or trematodes. As adults, they bury their heads in the wall of their host's gut, where they are bathed in ready-digested food. Tapeworms therefore do not need a gut or digestive system of their own. However, the body wall of *Taenia* is highly folded, providing a large surface area to volume ratio, and it is metabolically very active; large numbers of mitochondria provide the energy required for active transport of nutrients. *Taenia* also secretes some enzymes close to the body wall to increase the availability of digested food.

Figure 1 *The liver fluke* Fasciola hepatica *has a dorsoventrally flattened body covered with a thick cuticle which protects against the host's enzymes. Backward-pointing spines help hold it in position in the host.*

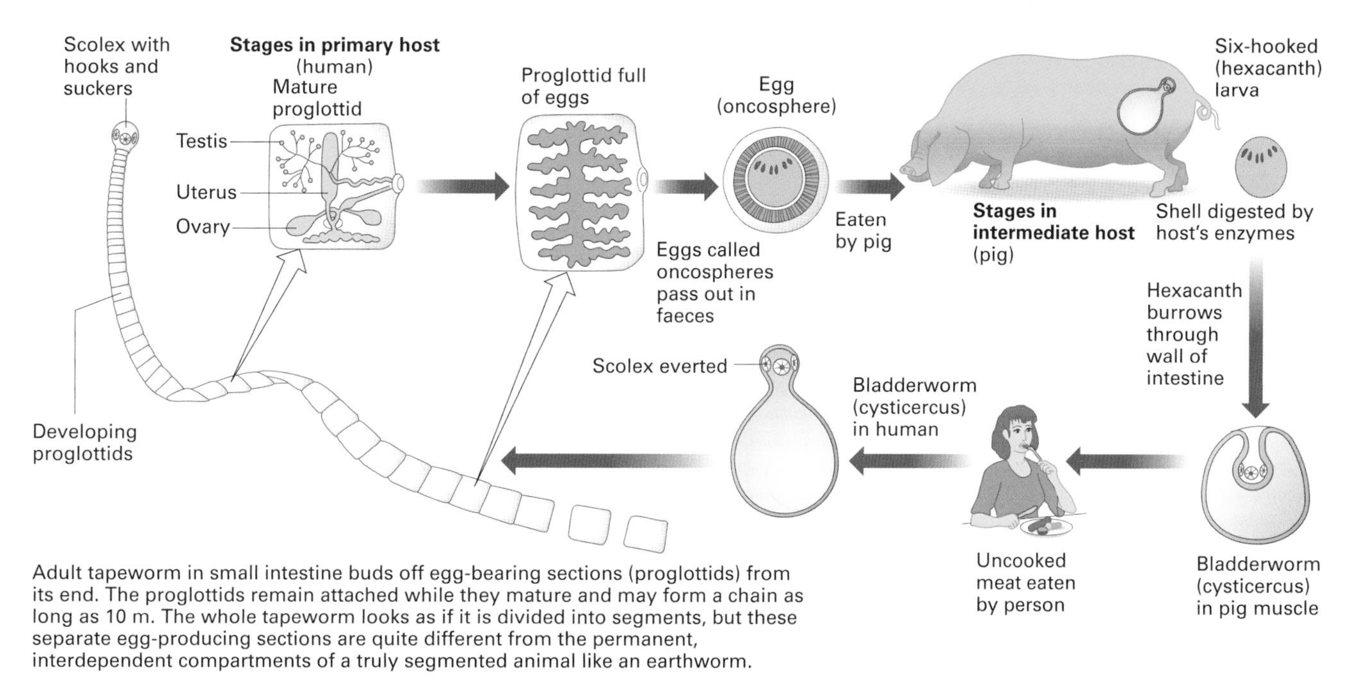

***Figure 2** Life cycle of* Taenia solium. T. saginatum, *another species that infects humans, is found in uncooked beef. Both species can be eradicated from an infected human by appropriate drugs.*

Adaptations of parasitic flatworms

Parasitic flatworms have some general adaptations in common with most parasites:

- They have a special way of gaining entry into the body of the host, but locomotory structures are generally reduced or absent.
- They have structures which anchor them onto their host. Liver flukes have suckers; tapeworms have both hooks and suckers.
- They have lost organ systems and functions that they no longer need (a phenomenon called parasitic degeneration). Liver flukes have lost their external sensory organs (for example, they have no eye spots), and tapeworms have no digestive system. Both liver flukes and tapeworms have lost the function of locomotion.
- They protect themselves against the internal defences of their host. The tough cuticle of parasitic flatworms is resistant to host chemicals, and parasitic flatworms produce inhibitory substances to prevent their being digested by host enzymes.
- They have complex life cycles; *Fasciola* and *Taenia* have a secondary host which transfers the parasite from one primary host to the next.
- They have a very high reproductive output (high levels of fecundity). Adult *Fasciola* devote much of their energy and body space to sexual reproduction.

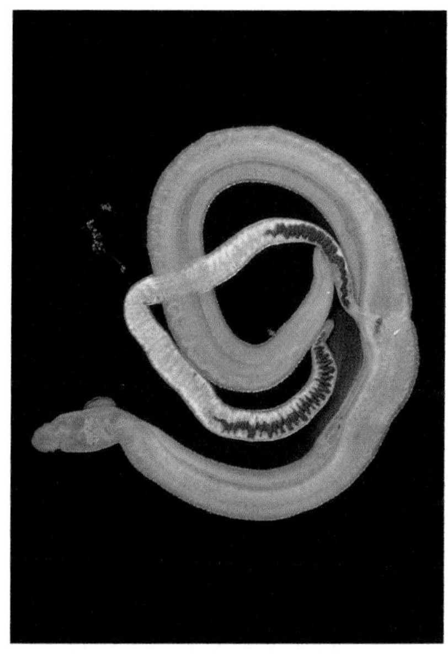

Figure 3 Schistosoma. *The larger male schistosome is attached permanently to the female, fertilising eggs as they are produced.*

Quick check

1 Give two features which distinguish tapeworms from liver flukes.

2 Explain what is meant by parasitic degeneration.

Food for thought

Schistosoma (figure 3) is a trematode common in tropical areas. Its life cycle involves aquatic snails and humans. Infected snails release a larval stage into fresh water. These larvae can bore their way through the skin of a person bathing in infested waters. Once in the bloodstream, the larvae develop into sexually reproducing adults. They produce spiked eggs which can pass out of the body in the urine or faeces. In fresh water, the eggs hatch and develop into different larval forms which infect snails so that the cycle can be repeated. Some eggs are retained in the human body and become lodged in organs such as the kidneys, liver, spleen, and bladder, where they cause a debilitating disease called schistosomiasis or bilharzia. The disease affects millions of people worldwide and kills over 200 000 each year. Using the information above, suggest ways of controlling this disease.

21.12

Annelids

OBJECTIVES

By the end of this spread you should be able to:

- list the characteristic features of the phylum Annelida
- discuss the biological significance of metameric segmentation and the coelom
- explain how earthworms exchange respiratory gases.

Fact of life

The possession of a tubular gut with different regions through which food passes enables earthworms to deal efficiently with huge amounts of ingested soil. Soil is taken in through the mouth at the anterior end and passes via the **pharynx** and **oesophagus** into a **crop** (which moistens and stores food temporarily), then into a **gizzard** (which churns the food, usually aided by grit or gravel taken in with the food), after which it passes into the **intestine**. Most chemical digestion and nutrient absorption takes place in the intestine; this is inwardly folded to form a structure called the **typhlosole** which increases its surface area. Finally, undigested food exits through the anus at the posterior.

Kingdom Animalia; phylum Annelida (annelids, segmented worms)

Annelids have the following characteristic feature:

- They are worm-like animals that are clearly segmented.

Annelids include marine ragworms, earthworms, and leeches. These are representatives of the three classes of the phylum, Polychaeta, Oligochaeta, and Hirudinea respectively (figure 1). Like platyhelminths, annelids are triploblastic and bilaterally symmetrical, but unlike platyhelminths they are segmented, they have a second body cavity (the **coelom**), and they have a through gut (a gut with both a mouth and an anus). This has allowed different gut regions to carry out different digestive functions, improving efficiency.

Segmentation

The word 'Annelida' means 'ringed' and derives from the fact that the annelid body is divided into many ring-like **segments**, a condition known as **metameric segmentation**. Unlike segments of tapeworms, annelid segments are fixed in number and are all of the same age.

Segmentation may have evolved as an adaptation to burrowing which enabled worms to escape from predators, avoid adverse environmental conditions, and exploit a new environment. Segments of ancestral annelids were separated by partitions called **septa**. Each segment contained a similar pattern of structures such as nerves, blood vessels, and muscles interconnected through the septa so that they could function together for locomotion. This repetition of body parts is still seen in polychaetes such as ragworms, but in many other worms segments in different parts of the body have become specialised for different functions. For example, some polychaetes have anterior segments with ciliated tentacles which can move independently of the rest of the body and are specially adapted for filter feeding. The ability to use separate parts of the body for different functions is probably one reason for the success of segmentation.

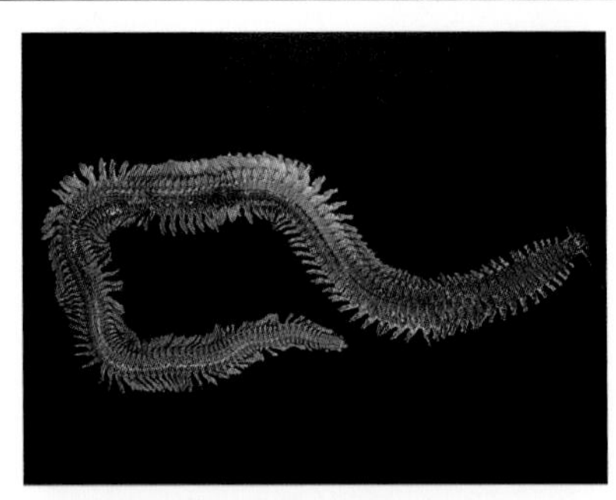

Figure 1 (a) Nereis *(ragworm).*

Class Polychaeta (polychaetes; marine worms)

Polychaetes have the following characteristic features:

- They have numerous bristles (**chaetae**) on lateral projections from the body (**parapodia**).
- Most are marine.
- Most have a distinct head.

An example is *Nereis* (ragworm).

(b) Lumbricus *(earthworm).*

Class Oligochaeta (oligochaetes; earthworms)

Oligochaetes have the following characteristic features:

- They have few chaetae and no parapodia.
- They do not have a distinct head.
- They live in fresh water or soil.

An example is *Lumbricus* (earthworm).

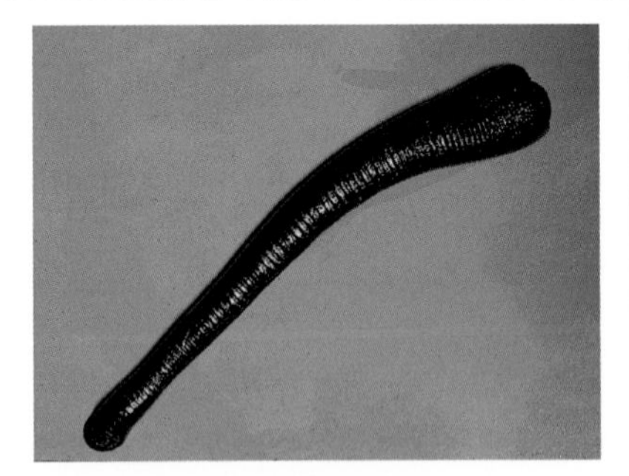

(c) Hirudo *(medicinal leech).*

Class Hirudinea (leeches)

Leeches have the following characteristic features:

- They have no chaetae or parapodia.
- They do not have a distinct head.
- They are usually ectoparasites or predators.

An example is *Hirudo* (medicinal leech).

The coelom

Animals with a coelom are called **coelomates** (as opposed to animals without a coelom, which are called **acoelomates**). A coelom is a body cavity lined with mesoderm (figure 2). The coelom:

- provides a space in which internal organs can grow, develop, and function independently of each other
- contains a fluid (coelomic fluid) that bathes the organs and which can act as a hydrostatic skeleton (see spread 11.1)
- allows the animal's internal organs to move independently of each other and of the body wall. For example, the gut can perform peristalsis without causing the body wall to go into waves of contraction.

The coelom separates the body wall from the gut. All coelomates have a blood vascular system which transports materials between the regions of the body. The importance of this system in the transport of respiratory gases is illustrated by its role in earthworms.

Gaseous exchange in earthworms

In earthworms, gaseous exchange takes place by diffusion over the whole surface of the body. No special organs are required for gaseous exchange because the cylindrical bodies of earthworms give them a high surface area to volume ratio, and their relatively low levels of activity mean that they have low rates of oxygen consumption. However, the blood vascular system plays a vital role in transporting respiratory gases inside the body.

According to Fick's law (spread 4.7), the rate of diffusion is directly proportional to the surface area and the concentration gradient, and inversely proportional to the diffusion distance. These factors are optimised in earthworms by the following:

- Earthworms have a large number of looped blood capillaries in the epidermis, immediately beneath the cuticle, providing a large surface area for diffusion (figure 3).
- The pumping action of the pseudohearts (see box) circulates the blood with its dissolved gases to and from the body wall and maintains a steep concentration gradient.
- The distance between the body surface and the looped blood vessels is short.

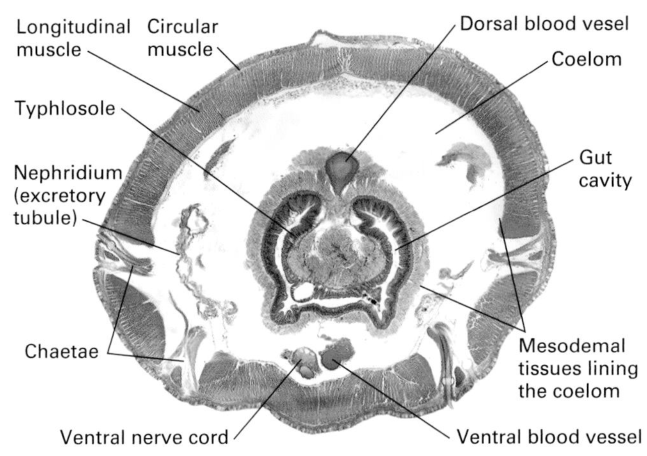

Figure 2 *Transverse section through an earthworm,* Lumbricus.

The closed circulatory system of earthworms

Earthworms have a closed circulatory system consisting of dorsal blood vessels, in which blood flows from tail to head, and ventral blood vessels, in which blood flows in the opposite direction. Connecting these two sets of blood vessels are **pseudohearts**, contractile vessels with valves which maintain blood flow. The blood contains the respiratory pigment haemoglobin in solution, which transports most of the oxygen from the body wall to respiring tissues.

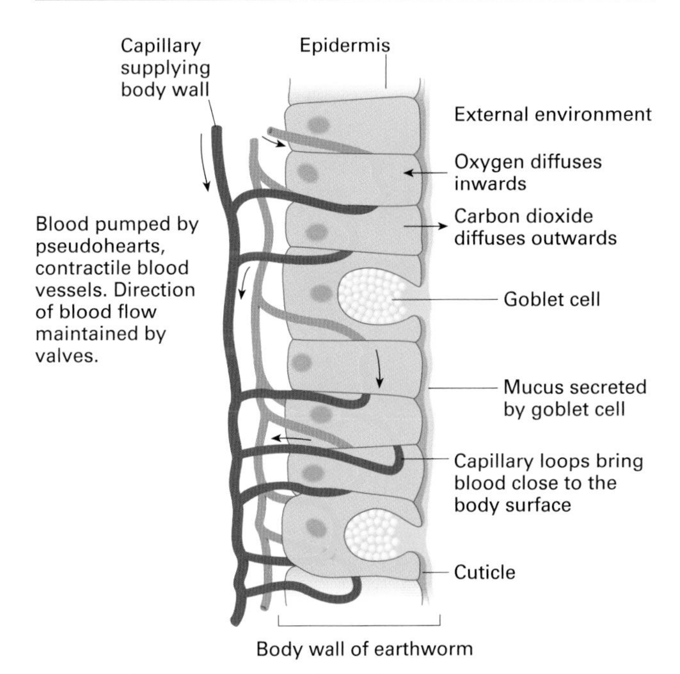

Figure 3 *Gaseous exchange in earthworms.*

QUICK CHECK

1. Distinguish between the segments in earthworms and the proglottids in tapeworms.
2. Explain briefly why gut peristalsis in annelid worms does not interfere with the contractions of muscles for locomotion in the body wall.
3. Why do earthworms not require special organs for the exchange of respiratory gases?

Food for thought

Earthworms live in soil, and they secrete mucus from epidermal glands to keep their body surface moist. However, they are not very well adapted to life on land and quickly die in dry conditions. Suggest why the adaptations that maximise the exchange of respiratory gases confine the earthworms to damp conditions.

21.13

Arthropods

OBJECTIVES

By the end of this spread you should be able to:

- describe the characteristic features of the phylum Arthropoda
- distinguish between the main groups of arthropods.

Fact of life

In terms of their relatively large size, number of individuals, and species diversity, arthropods dominate life on Earth. The world population of crustaceans, centipedes, millipedes, arachnids, and insects is estimated to total one billion billion (10^{18}) individuals. More than three-quarters of all known species are arthropods; insects alone account for more than half the total known species.

Kingdom Animalia; phylum Arthropoda (arthropods)

Arthropods have the following characteristic features:

- They are triploblastic coelomates.
- They are metamerically segmented and bilaterally symmetrical.
- They have an exoskeleton which may be rigid, stiff, or flexible.
- Typically, each segment bears a pair of jointed appendages.

The phylum Arthropoda includes crustaceans, centipedes, millipedes, arachnids, and insects (see opposite). Arthropods have more individual animals and more different species than any other phylum on Earth. They owe their success to a very adaptable body design that has enabled them to exploit every type of environment on land, in fresh water, and in the sea.

The arthropod body plan has evolved from that of annelids. Members of both phyla are bilaterally symmetrical, triploblastic coelomates that are clearly segmented. Arthropods, however, have two additional and unique features: a hard exoskeleton and jointed limbs.

Exoskeleton

The structure of the arthropod exoskeleton and its function as an attachment point for muscles is discussed in spread 11.1. Having an exoskeleton which is both strong and light has enabled arthropods to adopt many different types of locomotion, including burrowing, swimming, running, and walking. In insects, the presence of a highly elastic protein in the exoskeleton also enables them to jump and fly.

The variable hardness and flexibility of the exoskeleton is a key feature to its usefulness. Joints are relatively soft and flexible, while mouthparts and feeding appendages (such as crab claws) can be very hard and rigid. Also, the skeleton can be transparent in places, allowing, for example, light to pass into the eyes and providing camouflage in water.

As mentioned in spread 11.1, having an exoskeleton brings not only advantages but at least two serious disadvantages: its weight to strength ratio decreases with the size of an animal, making it less efficient as animals become larger; and it restricts growth to periods when the exoskeleton is shed. During the period of shedding, the body of an arthropod is soft and vulnerable.

Jointed appendages

The jointed limbs are the most obvious feature of arthropods (the word arthropod means 'jointed foot'), but other appendages (structures attached to the main body) are also jointed. The jointed appendages of arthropods have evolved into many different forms. Some have a specific locomotory function, while others serve as sense organs, jaws, mating structures, or respiratory organs. In contrast to the multi-purpose limbs of humans, arthropod appendages are often specialised for a single function. Thus looking at an arthropod's limbs reveals almost everything about its lifestyle: how it moves, where it lives, and how it feeds.

Quick check

1 a List four features that annelids and arthropods have in common.

b List two features found in arthropods but not in annelids.

2 How do the legs of:

a millipedes differ from centipedes

b insects differ from arachnids?

Food for thought

We tend to regard ourselves and other mammals as the pinnacle of evolution and the most successful animals on Earth. However, in terms of numbers, arthropods are much more successful than mammals. In what other ways could arthropods be regarded as more successful than mammals? In what ways are mammals more successful than arthropods?

Class Crustacea (crustaceans)

Crustaceans have the following characteristic features:

- They are aquatic or live in damp habitats.
- They have a cephalothorax (the head is not distinct from the thorax).
- They have two pairs of antennae and at least three pairs of mouthparts (jaws).
- They lack a waterproof exoskeleton.

Examples include woodlice (figure 1), *Daphnia* (water flea), crayfish, crabs, lobsters, and barnacles.

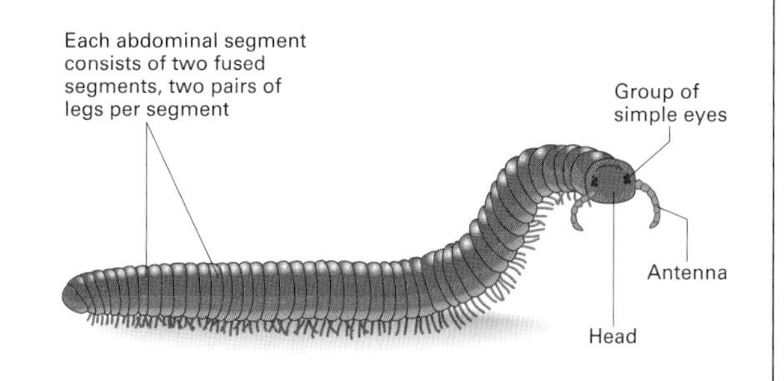

Figure 1 *Dorsal view of a woodlouse.*

Class Diplopoda (millipedes)

Millipedes have the following characteristic features:

- They are mainly terrestrial and mainly herbivorous.
- They have a distinct head.
- They have one pair of antennae and one pair of mouthparts (jaws).
- They have numerous similar limbs along the length of the body; two pairs on each segment.

Examples include *Iulus* (millipede, figure 2) and wireworms.

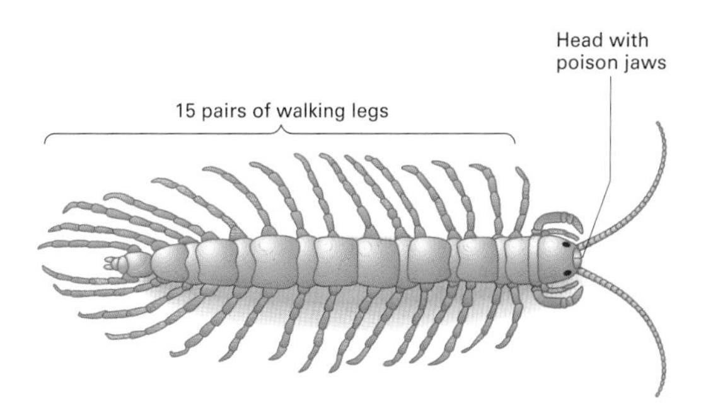

Figure 2 *Lateral view of* Iulus, *a millipede.*

Class Chilopoda (centipedes)

Centipedes have the following characteristic features:

- They are terrestrial and mainly carnivorous.
- They have a distinct head.
- They have one pair of antennae and one pair of mouthparts (jaws).
- They have similar legs along the length of the body; one pair on each segment.

Examples include *Lithobius* (centipede, figure 3).

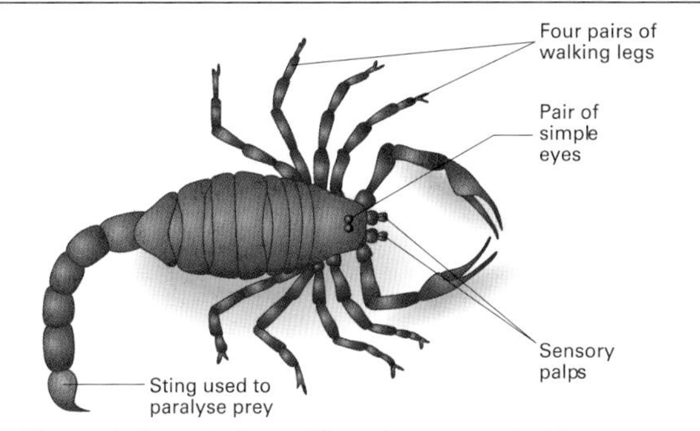

Figure 3 *Dorsal view of* Lithobius, *a centipede.*

Class Arachnida (arachnids)

Arachnids have the following characteristic features:

- They are mainly terrestrial and mainly carnivorous.
- The body is divided into two parts: the cephalothorax and abdomen.
- They have no antennae and no true mouthparts.
- They have four pairs of walking legs.
- They have one pair of appendages adapted to capture prey and one pair of sensory appendages (palps).

Examples include *Scorpio* (scorpion, figure 4), spiders, mites and ticks.

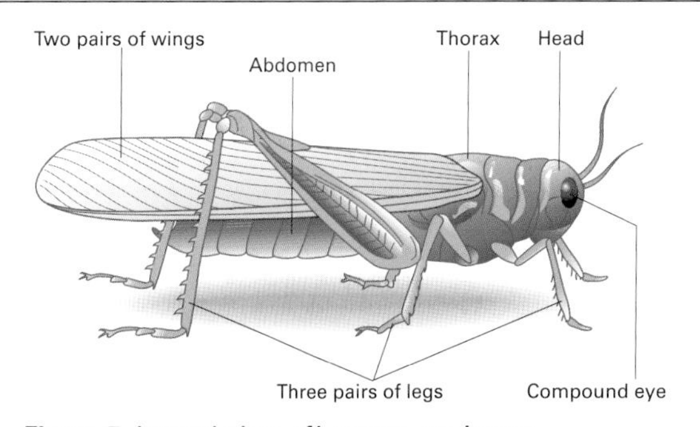

Figure 4 *Dorsal view of* Scorpio, *an arachnid.*

Class Insecta (insects)

Insects have the following characteristic features:

- They are mainly terrestrial.
- The have a distinct head, thorax, and abdomen.
- They usually have three pairs of mouthparts and one pair of antennae.
- They have three pairs of walking legs.
- Adults usually have wings.

Examples include *Locusta* (locust, figure 5), bees, ants, beetles, aphids and fleas.

Figure 5 *Lateral view of* Locusta, *an insect.*

21.14

INSECTS: LOCOMOTION

OBJECTIVES

By the end of this spread you should be able to:

- list the main adaptations of insects for life on land
- describe how insects walk
- explain how insects fly.

Fact of life

Locusts have a relatively primitive insect body plan. The wings are flapped by muscles attached directly to the wing base. The two pairs of wings beat out of phase with each other at about 20 beats per second. Flies, on the other hand, have evolved an advanced aeronautical system. Their single pair of wings is flapped indirectly by vibrations of the thorax at an incredible 1000 beats per second.

***Figure 1** A fly in flight.*

Although there are many millions of insect species, they are all built on the same three-part body plan (spread 21.13, figure 5):

- a prominent head bearing the mouth and most of the sense organs
- a thorax containing powerful muscles which operate three pairs of legs below and, usually, one or two pairs of wings above
- an abdomen carrying the organs needed for digestion, excretion, and reproduction.

As with other arthropods, these body sections are enclosed within an exoskeleton. The insect body plan is highly versatile and has enabled insects to exploit a diverse range of habitats from the equator to the poles. Insects can be found swarming in deserts, crawling among the undergrowth of forests, swimming in ponds, and flying over mountains.

By any standards, insects are the best adapted of all animals for life on land. Insects such as locusts for example, have a waxy exoskeleton and can produce semi-solid uric acid as their main nitrogenous waste product. These adaptations enable insects to conserve water and survive in the driest of terrestrial habitats. Other adaptations which make them successful colonisers of land relate to their locomotion, gaseous exchange, and reproduction. In the remainder of this spread, we shall discuss insect locomotion; in the next spread we deal with gaseous exchange and reproduction.

Dipterans: masters of the air

The order Diptera includes flies, midges, and mosquitoes. They are characterised by having only a single pair of wings, the fore-wings, and they are the most aerobatic of all insects. Their hind-wings have been reduced to small club-shaped structures called **halteres**. The halteres are attached to the thorax and act partly as stabilisers, like gyroscopes, and partly as sense organs, presumably providing information about the angle of the body and the direction of movement. Information about flight speed comes from the antennae which vibrate as the air flows through them. By using indirect muscles (figure 4) to vibrate the thorax, dipterans are capable of beating their wings at incredible rates. The thorax and wings of insects contain pads of a rubber-like protein called resilin which stores the kinetic energy at the end of each beat. This means that a rapid beating rate can be maintained without the insect using too much energy (or producing too much heat).

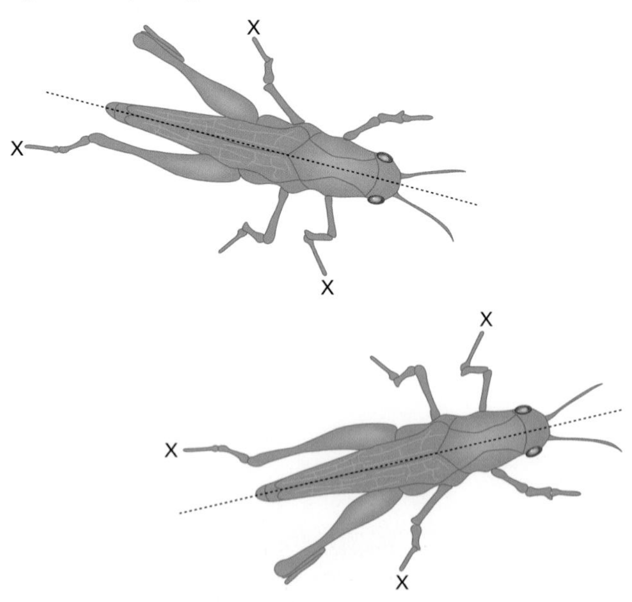

***Figure 2** When walking, a locust always keeps three of its legs (marked X) in contact with the ground.*

Locomotion on the ground

An insect's limb, with its multiple joints and antagonistic muscles (spread 11.1, figure 4), is strong enough to lift the insect's small body off the ground, and mobile enough to allow a wide range of movements. The six legs are attached in pairs to the three segments of the thorax. During walking, the body is always supported on a tripod of three legs while the other legs pull or push the body forwards (figure 2).

Many insects have claws or adhesive pads to reinforce their grip on the substrate. Flies have pads which are so efficient that flies can even walk upside down. Some insects, such as locusts and fleas, have back legs specially adapted for jumping. Before the jump, the locust flexes the leg fully. Energy is stored in highly elastic tendons and released when the flexor muscles are relaxed, propelling the locust forward in the air.

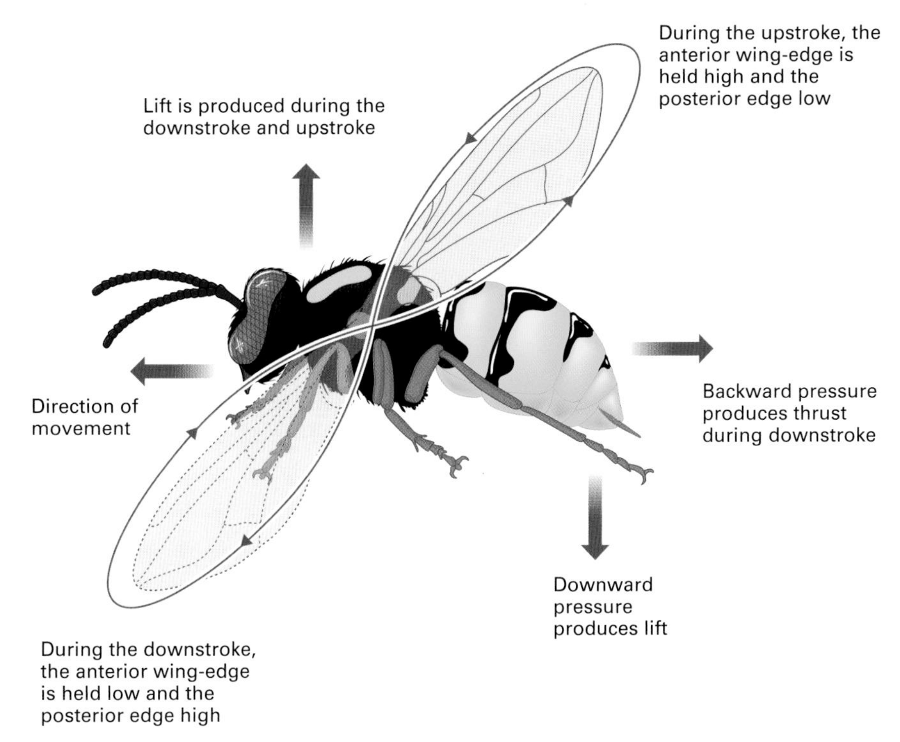

***Figure 3** The figure-of-eight pattern of wing movements of an insect in flight.*

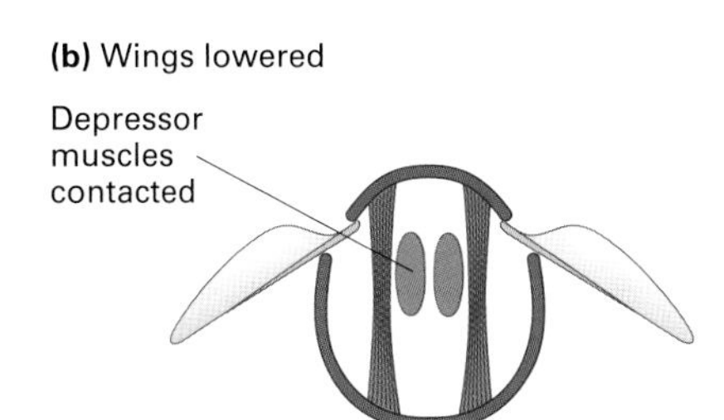

***Figure 4** Transverse section of the thorax showing the indirect flight muscles of most insects.*

Flying

Insects are the only animals without a backbone that can fly. This has undoubtedly contributed to their great success, giving them an efficient means of dispersal, enabling them to escape from predators or adverse conditions, and, probably most important, allowing them to exploit new food sources and habitats, especially those associated with tall flowering plants.

Flying insects usually have two pairs of wings. Dragonflies and locusts beat their fore-wings and hind-wings just out of phase with each other. In bees and wasps the fore- and hind-wings are hitched together so that they function as a single pair of wings. In flies, midges, and mosquitoes the hind-wings are reduced and act as balancing organs, while in beetles the fore-wings are modified into hardened protective structures called elytra (singular elytron) so that only the hind-wings flap in flight.

The wings are flattened extensions of the exoskeleton, reinforced by ridges known as veins. In cross-section, the wing has the shape of an aerofoil. Air flows over it in the same way as it does over the wing of a bird (spread 11.6, figure 3), generating lift on the undersurface. During flight, an insect's wing tip moves in a path shaped like a figure of eight to provide both thrust and lift (figure 3).

In most insects, wing movements are brought about mainly by two antagonistic pairs of flight muscles which change the shape of the thorax (figure 4). A pair of **elevator muscles** runs from the roof of the thorax to its floor. When these muscles contract, they pull the roof of the thorax downwards and force the wings up. A pair of **depressor muscles** runs along the thorax from its anterior surface to its posterior surface. When they contract, they cause the roof of the thorax to bow upwards and the wings to be forced downwards. As these two pairs of muscles are not attached to the wings, they are called **indirect flight muscles**.

Other muscles attached to the bases of the wings, called **direct flight muscles**, are usually used to alter the angle of the wings during flight and to fold the wings at rest. In flight, the wing angle is continually and automatically adjusted so that the wings provide lift in up to 85 per cent of wing movements. Some insects, such as locusts, use the direct flight muscles to provide the main power for flying. They tend to have a much slower wing beat than insects that use indirect muscles for flight.

Quick check

1 Urine is the main nitrogenous excretory product of mammals.
   a What is the main nitrogenous excretory product of terrestrial insects?
   b How does this make insects better adapted than mammals for life on land?

2 During walking, how many of an insect's legs remain on the ground at any one time?

3 A bee can be described as functionally having a single pair of wings. Describe how this comes about.

Food for thought

Insect wings are analogous to those of a bird (spread 11.6). Suggest how they differ in their structure and function.

21.15

Insects: gaseous exchange and life cycles

OBJECTIVES

By the end of this spread you should be able to:

- describe the tracheal system for gaseous exchange
- distinguish between hemimetabolous and holometabolous life cycles
- discuss the advantages to insects of undergoing metamorphosis.

Fact of life

The tracheal system of insects is a network of tubes that carries air directly to respiring tissues. The finest tubes (tracheoles) are as little as 0.5 μm in diameter and they can extend into individual cells such as muscle fibres. In very active tissues, such as locust flight muscles, tracheoles may lie as close as 0.07 μm to a mitochondrion.

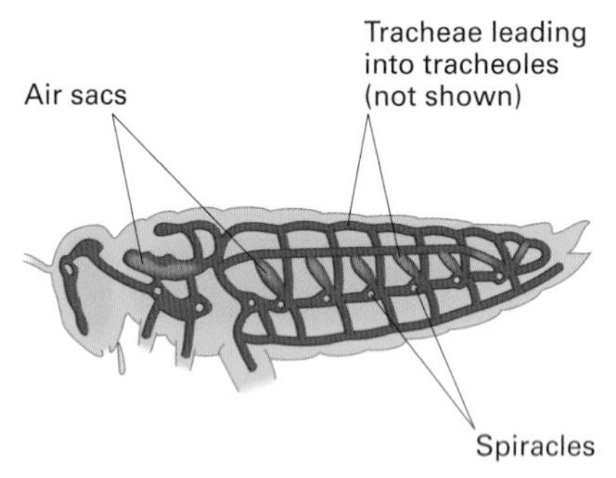

Figure 1 Longitudinal section of an insect showing the tracheal system.

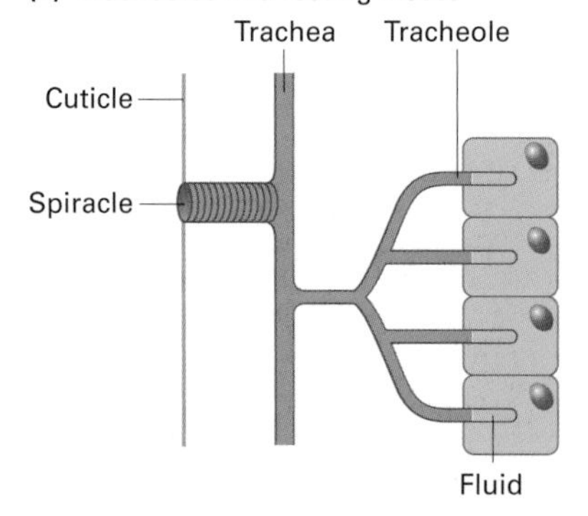

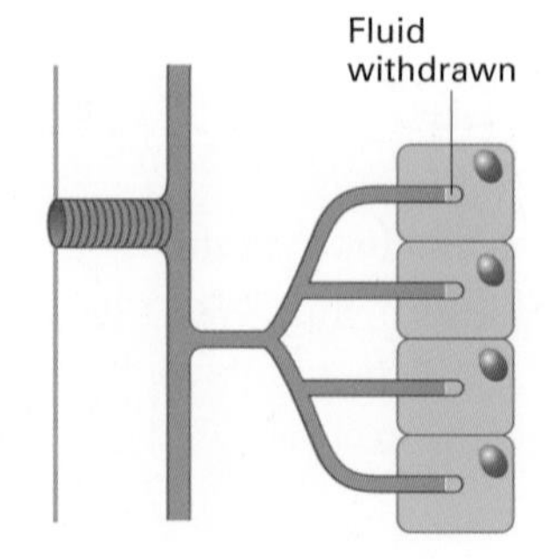

Gaseous exchange and conserving water

Life on land poses a continual conflict between the need for oxygen and the need for water. Conditions which favour gaseous exchange from the body of a terrestrial animal also favour water loss. It is vital for terrestrial insects, especially those such as locusts that live in desert conditions, to be able to breathe air without losing too much water.

Water loss through the body surface is minimised by insects having a hard cuticle covered with wax which is impermeable to water. However, this also makes the body surface virtually impermeable to gases as well. In order to exchange gases through their body surface, insects have evolved the **tracheal system**. This consists of a series of small holes in the cuticle which leads to a system of internal air-filled tubes (figure 1). The holes are called **spiracles**. In many insects (for example, locusts) the spiracles have a system of valves which control the size of the hole in response to carbon dioxide levels (high carbon dioxide levels, occurring when respiratory demands are high, stimulate the spiracles to open). This enables insects to conserve water by keeping the spiracles closed as much as possible. When the insect is inactive, a few spiracles may open briefly, but when the demand for oxygen is high or carbon dioxide levels build up, more spiracles are open for longer periods.

The first tubes into which the spiracles lead are called **tracheae**. These tubes are lined with chitin which supports them and also makes them impermeable to gases. The tracheae branch off into **tracheoles**, small unlined tubes in which respiratory gases are exchanged (figure 2). The vast number of tracheoles in an insect provide a very large surface area for gaseous exchange and bring air very close to the respiring tissues, minimising the diffusion distance and maximising the rate of diffusion (see Fick's law, spread 4.7). Tracheae allow gaseous exchange to take place independently of the circulatory system; insect blood has no direct role in transporting respiratory gases.

In small insects and those that are relatively inactive, air moves along the tracheal system by diffusion alone. However, in some large and active insects such as locusts, air can be actively pumped by movements of the thorax or abdomen. These insects have collapsible tracheae called **air sacs** which are inflated and deflated by these ventilatory movements, moving more air through the tracheal system. Ventilatory movements become more frequent and more forceful as activity increases.

Insect life cycles

Eggs and sperm cannot survive unprotected on land. Like many sexually reproducing terrestrial animals, insects overcome this problem by carrying out internal fertilisation: the male introduces sperm directly into the female reproductive tract. Many male insects use an extensible penis to introduce the sperm; others use another special appendage to transfer the sperm, and some present sperm to the female enclosed in a sac. Sometimes mating involves elaborate manoeuvres and may take a considerable time; some male butterflies cement themselves to the female with a special secretion during mating.

Insect life histories are variable and often complex. Most insects, however, change form or structure during the course of their life cycle (a process called **metamorphosis**). They do this in one of two main ways: by **complete metamorphosis** (also called **holometabolous**

Figure 2 Gaseous exchange in tracheoles. At rest, the ends of the tracheoles contain a relatively large volume of watery fluid. Diffusion of oxygen in water is about 300 000 times slower than in air, so the presence of this fluid slows the uptake of oxygen. However, during activity, water is withdrawn from the ends of tracheoles, so that gases can diffuse more quickly.

Aquatic larvae: surviving the cold

Insects are primarily terrestrial creatures, but many species have aquatic larvae (figure 3). In seasonal environments, these larvae allow the insects to survive periods of extreme cold. Living in the bottom of a deep pond, the larvae avoid the freezing conditions on land; although ice may form on the top of the pond, it rarely fills a deep pond. All aquatic larvae have a tracheal system for gaseous exchange, but it has become modified in a number of different ways. Mayfly larvae have extensive external gills on their abdomens; they have no functional spiracles. The gills are thin and provide a large surface area for gaseous exchange between the water and air within the tracheal system. Movements of the external gills ensure a continual flow of water across their surface; the rate of movement increases as oxygen concentration in the water decreases.

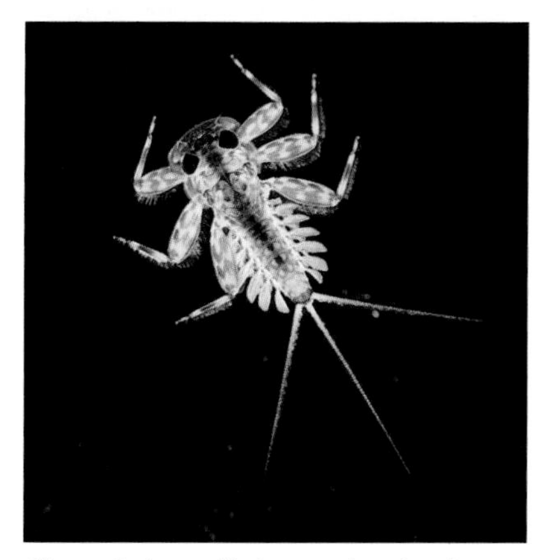

Figure 3 *A mayfly larva, showing its external gills.*

metamorphosis, figure 4a); or by **incomplete metamorphosis** (also called **hemimetabolous metamorphosis**, figure 4b). Metamorphosis enables juvenile and adult forms to live in different habitats and exploit different resources. This reduces competition between the different developmental stages. Metamorphosis also allows the larval and adult stages to become highly specialised for particular functions; usually, the larval stage is specially adapted for feeding and the adult for reproduction.

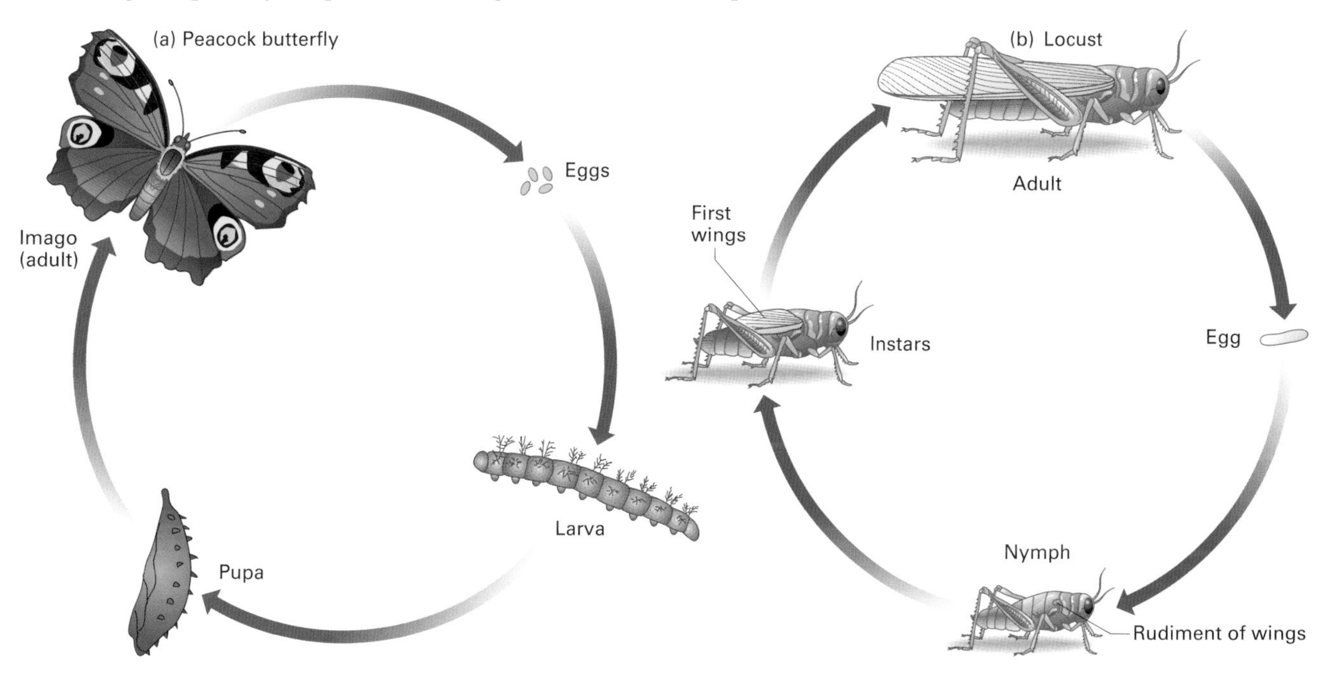

Figure 4 *Metamorphosis.* ***(a)*** *Complete metamorphosis: the embryo of an insect (for example, a moth, butterfly, beetle, or fly) which undergoes complete metamorphosis develops into a young form called a larva which appears very different from the adult. Larvae often lack many of the structures of the adult. Butterfly larvae have no wings and lack compound eyes and jointed legs; they have become little more than feeding machines whose primary function is to find and consume food. Once it has reached a certain size, the larva stops feeding and becomes a pupa by enclosing itself in a protein case. Within the case, tissues are broken down and reorganised so that they undergo a remarkable transformation to the adult form (the* ***imago****). Once the adult has emerged from the case with fully developed wings, it can no longer moult. This restricts growth.*

(b) *Incomplete metamorphosis: the embryo of an insect such as a grasshopper, cockroach, or locust undergoes incomplete metamorphosis. It develops into a nymph which closely resembles the adult form, but which has a number of adaptive features that enable it to live in a different habitat and eat different food from the adult. In order to grow, the nymph moults several times and goes through a number of developmental stages called* ***instars****. The final instar emerges as an adult, with all the adult organs.*

Quick check

1. In which structures in the tracheal system does most gaseous exchange take place?
2. Distinguish between a holometabolous life cycle and a hemimetabolous life cycle.
3. Give two advantages of a life cycle that incorporates metamorphosis.

Food for thought

A number of changes take place in respiring tissues which cause water to be withdrawn from the ends of tracheoles (figure 2). Suggest what these changes might be, and by what process water is withdrawn.

21.16

OBJECTIVES

By the end of this spread you should be able to:

- list the main characteristic features of nematodes, molluscs, and echinoderms.

Fact of life

The phylum Nematoda (roundworms) contains some of the most widespread and numerous of all animals.

Free-living roundworms are found in all types of environments including deserts, hot springs, high mountains, and great ocean depths. They are thought to be the most numerous animals in the world. It has been estimated that the surface layer of 1 m^2 of bottom mud off the Dutch coast contains over four million nematodes; a single decomposing apple on the ground of an orchard has yielded more than 90 000 roundworms. Parasitic species are equally widespread and numerous, attacking all groups of plants and animals, giving them great economic significance. The large roundworm (*Ascaris*) infects about a quarter of the world's human population. Its success may be due to the fact that a single female can lay as many as 200 000 eggs a day, and these eggs can remain viable in soil for years.

Nematodes, molluscs, and echinoderms

In addition to the cnidarians, platyhelminths, annelids, and arthropods dealt with in the previous spreads, there are several other phyla of non-chordate animals. Some are rarely encountered and are beyond the scope of this book. This spread deals with three common non-chordate phyla: the Nematoda, Mollusca, and Echinodermata.

Kingdom Animalia; phylum Nematoda (nematodes, roundworms)

Nematodes are triploblastic, unsegmented worms which lack a true coelom. They have a slender, cylindrical body which tapers at both ends and is covered with a thick elastic cuticle. Unlike platyhelminths, nematodes have a through gut which allows food to be continuously processed; it also allows different parts of the gut to be specialised for different functions.

Nematodes have the following characteristic features:

- They are animals with unsegmented cylindrical bodies.
- They have a mouth and an anus.
- The sexes are separate.

An example is *Ascaris* (figure 1).

Figure 1 *The roundworm* Ascaris lumbricoides *can infect pigs and humans. It can be longer than 30 cm.*

Kingdom Animalia; phylum Mollusca (molluscs)

Molluscs are coelomate triploblastic animals which have little trace of segmentation. They are the second largest animal phylum. They include snails, slugs, cockles, mussels, octopus, and squid. All members of this phylum have a soft body which is often covered by a shell, although this may be reduced or lost. The shell is secreted by special epidermal tissue called the **mantle**.

Molluscs have the following characteristic features:

- Their bodies are outwardly unsegmented, usually with a **head**, **foot**, and **visceral hump** (a central mass of internal organs).
- Many species have a shell containing calcium carbonate (a **calcareous shell**).

There are three main classes of molluscs: **Gastropoda** (gastropods), **Pelecypoda** (formerly called Bivalvia, bivalves), and **Cephalopoda** (cephalopods).

Class Gastropoda

Gastropods have the following characteristic features:

- They have a distinct head with eyes and sensory tentacles.
- The shell, when present, is single and often coiled.
- They have a radula (a broad tongue-like organ armed with rows of teeth) with which they feed.

An example is *Helix*, the garden snail (figure 2).

Figure 2 Helix, *a common gastropod.*

Class Pelecypoda

Bivalves have the following characteristic features:

- Their head is reduced and has no tentacles.
- The shell has two hinged valves.
- They are usually filter feeders.

An example is *Pecten*, the scallop (figure 3).

Other bivalves include oysters, clams, cockles, and mussels. Most of them are immobile filter feeders.

Figure 3 Pecten, *an example of a pelecypod (bivalve).*

Class Cephalopoda

Cephalopods have the following characteristic features:

- They have a conspicuous head with well developed eyes and tentacles.
- They have a beak and a radula used for feeding.

An example is *Octopus* (figure 4).

Cephalopods are active, fast-moving animals, and are regarded as the most intelligent invertebrates. In the laboratory, they are capable of learning quite complex tasks. Members of the class Cephalopoda have reached the largest size of any animal without a backbone. The North Atlantic giant squid (*Architeuthis*) is up to 18 m long and has a body circumference of 4 m.

Figure 4 Octopus, *a cephalopod.*

Kingdom Animalia; phylum Echinodermata (echinoderms)

Echinoderms have the following characteristic features:

- They are marine mammals with five-way radial symmetry.
- They have a water vascular system (a system of internal tubes containing a watery fluid) which enables tube feet to move.

An example is *Asterias*, starfish (figure 5).

Echinoderms are all marine animals with spiny skins (the word 'echinoderm' means 'spiny skin'). Adults are relatively slow moving, depending on tube feet for locomotion. The phylum includes starfish, brittle stars, sea cucumbers, and sea urchins.

Figure 5 Marthasterias gracilis *species, a typical echinoderm.*

Quick check

1 a How does the gut of a nematode differ from that of a platyhelminth?

b How does the shell of a gastropod differ from that of a pelecypod?

c What feature do echinoderms and cnidarians (spread 21.9) have in common?

Food for thought

Suggest why adult starfish have a five-way radial symmetry (spread 21.8) whereas their planktonic larvae are bilaterally symmetrical.

21.17

Chordates: fish

OBJECTIVES

By the end of this spread you should be able to:

- list the characteristic features of chordates
- distinguish between cartilaginous fish and bony fish
- describe the gaseous exchange system of a bony fish.

Fact of life

Bony fish are the most varied of all the chordate classes. They show a tremendous diversity of form, ranging in size from the tiny pygmy goby, about 1 cm long, to the beluga (*Huso huso*), a member of the sturgeon family which can exceed 5 m in length and weigh up to 1200 kg. A single female beluga may contain up to 7 million sticky black eggs, which are collected, salted, and eaten as caviar.

Kingdom Animalia; phylum Chordata (chordates)

Chordates have the following characteristic features:

- They are animals with a **notochord** (a slender, stiff, but flexible rod running along the back) at some stage in their life cycle.
- They have a dorsal hollow nerve cord (non-chordates such as worms and insects have a ventral solid nerve cord).
- They have **visceral clefts**. These are perforations on each side of the pharynx which occur in all chordate embryos. They are retained in primitive chordates, become gills in fish, but are reduced or modified in other chordates.
- They have a post-anal tail at some stage in their life cycle, which may be reduced or lost in some adult chordates.

Chordates are bilaterally symmetrical, metamerically segmented, triploblastic coelomates. The name chordate reflects the fact that they all have a **notochord**. This is a slender rod of cells originating from the mesoderm, that runs along the back in the embryo of all chordates. It persists in primitive chordates, but in cartilaginous fish, bony fish, amphibians, reptiles, birds, and mammals it is replaced by the backbone. In this spread we shall examine two chordate classes: class **Chondrichthyes** (the cartilaginous fish) and class **Osteichthyes** (the bony fish). In the next spread we shall deal with amphibians, reptiles, birds, and mammals.

Class Chondrichthyes

Cartilaginous fish have the following characteristic features:

- They have a skeleton made of cartilage (a cartilaginous skeleton is strong and rigid, but also flexible and light).
- The mouth is ventral.
- Their fins are fleshy.
- They have gills with separate openings.
- The skin contains **dermal denticles** (tooth-like structures with a central pulp cavity surrounded by an outer covering of enamel).
- They typically have a **heterocercal tail** (the dorsal lobe of the tail fin is usually much larger than the ventral lobe).

An example is *Scyliorhinus*, dogfish (figure 1).

Figure 1 Scyliorhinus *(dogfish), an example of a cartilaginous fish (20–30 cm long).*

Class Osteichthyes

Bony fish have the following characteristic features:

- They have a skeleton made of bone.
- They have a terminal mouth which can often be protruded.
- They have fins supported by rays.
- They have gills covered with a bony flap (the **operculum**).
- The skin usually contains **scales** (bony plates covered with skin).
- They typically have a **homocercal tail** (the dorsal and ventral lobes of the tail fin are usually the same size).

An example is *Clupea*, herring (figure 2).

Figure 2 Clupea *(herring), an example of a bony fish (10–20 cm long).*

Gaseous exchange in bony fish

Fish are aquatic animals adapted to extracting oxygen from water. Water has a highly variable oxygen content, rarely exceeding 0.8% compared with the 20.9% in air. In order to extract enough oxygen for respiration, aquatic organisms have to pass a larger volume of water over their gaseous exchange surface than the volume of air moved by terrestrial animals. Aquatic animals also have to work harder than terrestrial animals to ventilate their gaseous exchange surfaces because water is much denser and more viscous than air. However, when exchanging gases, aquatic organisms do not have the problem of conserving body water. Also, thin respiratory structures such as gills are supported by water but would collapse in air.

Bony fish have a highly efficient gaseous exchange system, extracting as much as 80 per cent of the available oxygen passing through it. The efficiency is due to:

- the gills providing a large surface area to volume ratio
- the gills being very thin
- a high concentration gradient being maintained between the blood in the gills and the water passing over them (see Fick's law, spread 4.7).

A bony fish has four pairs of **gill arches** which support **gill filaments** on which are **gill lamellae** (gill plates, figure 3). Gill lamellae provide a large surface area for gaseous exchange, and contain blood vessels which transport respiratory gases to and from the gills. The lamellae are very thin so that the blood is only a short distance from sea water (about 5 μm in active fish such as mackerel). Water moves across the gill lamellae in the opposite direction to the blood flowing through them. This provides a **countercurrent exchange mechanism** (spread 8.7) so that the water always has a higher oxygen concentration than the blood it is flowing past. Diffusion can therefore occur over the whole surface of the lamellae. This is much more efficient than the parallel flow system of dogfish, where blood and water move in the same direction and only about 50 per cent of the oxygen is absorbed.

Fish gills are internal, covered by a bony flap called the operculum. This protects the delicate gills, and also provides bony fish with two ventilatory pumps (an **opercular suction pump** and a **buccal pressure pump**) which can maintain an almost continuous flow of water across the gills (figure 4). The efficient gaseous exchange system of fish enables them to live very active lives and colonise all types of water.

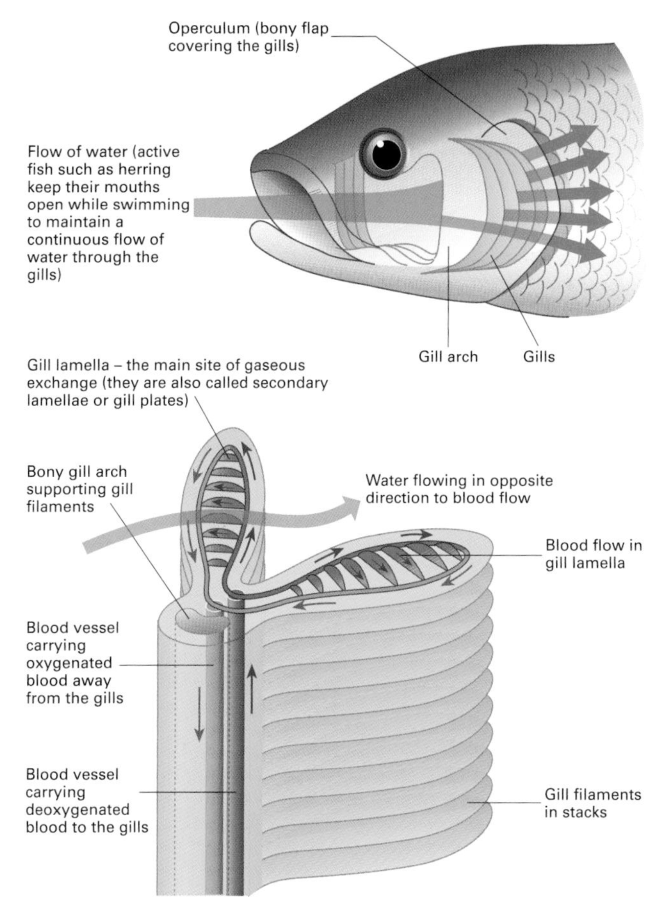

Figure 3 Gaseous exchange in bony fish.

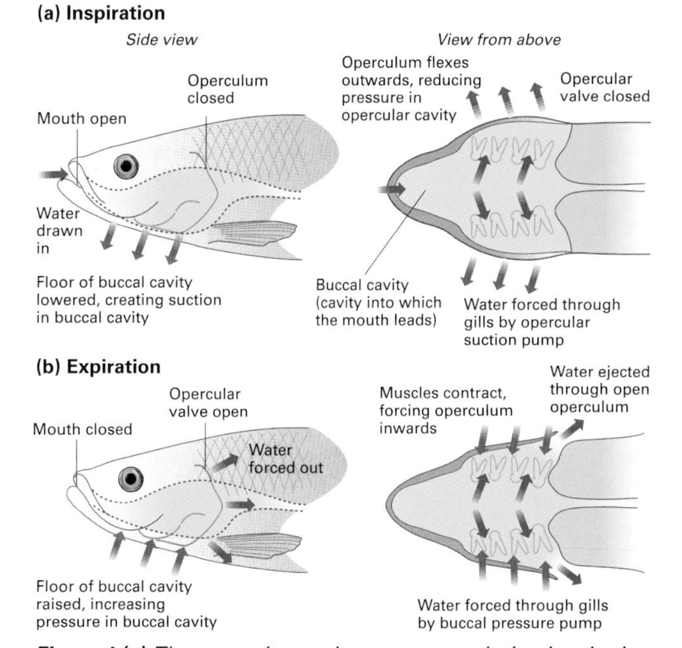

***Figure 4 (a)** The opercular suction pump acts during inspiration and **(b)** the buccal pressure pump acts during expiration to maintain an almost continuous flow of water through the gills.*

Quick check

1 What is a notochord?

2 How do the fins of cartilaginous fish differ from those of bony fish?

3 Explain the term countercurrent exchange with reference to gaseous exchange in fish.

Food for thought

Suggest why gills are not suitable structures for gaseous exchange on land.

21.18

Chordates: adaptations to life on land

OBJECTIVES

By the end of this spread you should be able to:

- discuss the adaptations of tetrapods to life on land
- list the main characteristic features of amphibians, reptiles, birds, and mammals.

Fact of life

Frogs are renowned for their ability to leap. They have become specialised for jumping by developing a shortened backbone and long, highly muscular legs. They jump not only to get from one point on the ground to another, but also to escape from predators or to obtain food (figure 1). A few tree-living species can travel 15 m or so through the air, about 100 times their body length. They glide through the air using the membranes between their toes as parachutes.

Figure 1 *A frog leaping to attack a damselfly. When within striking distance, it will shoot out its sticky tongue to grasp its prey and carry it back to the mouth. The tongue, unlike ours, is attached to the front of the mouth. Frogs can therefore stick their tongues out much further than we can.*

One of the most important episodes in the evolution of animal life on Earth was the move from living in water to living on land. This momentous change presents animals with four major problems:

1. how to move about in a low-density medium which gives little support
2. how to breathe air without dying from desiccation
3. how to reproduce out of water
4. how to cope with the variable environmental conditions, especially wide fluctuations in temperature, experienced on land.

Locomotion on land

Amphibians, reptiles, birds, and mammals are chordates which have (to varying degrees) solved these problems. They have overcome the first problem by having a bony endoskeleton and by being **tetrapods** (four-limbed animals) with limbs built on the pentadactyl plan (spread 20.3, figure 3). The tetrapod limb evolved from the fish fin and, together with modifications of muscles and structures around the girdles, it can be used to raise the body off the ground and propel the animal forwards.

Gaseous exchange

All land-living tetrapods have internal organs of gaseous exchange, the lungs, adapted for breathing air. However, amphibians also breathe through their skin which is therefore soft, thin, and permeable to water. This means that to prevent desiccation, amphibians are generally confined to damp habitats. Reptiles, birds, and mammals, like insects, have evolved a water-tight skin which enables them to live in dry habitats. Reptiles and birds have the added advantage of being able to produce a semi-solid nitrogenous waste containing uric acid. This allows them to conserve water more effectively than amphibians or mammals which produce urine, a liquid nitrogenous waste containing mainly urea.

Reproduction

Amphibians have also failed to overcome the problem of reproducing on land. The membrane covering their eggs is permeable to water and would dry out quickly if exposed to air. Amphibians (with a few exceptions) therefore breed where water is plentiful. The amphibian egg does not develop into a miniature version of the adult, but into a larval stage (such as the tadpole of frogs, figure 2) which metamorphoses into the adult stage. In most species, the larvae have external gills and live in ponds or streams. Reptiles, birds, and mammals have evolved an egg which is surrounded by protective membranes, enabling the embryo to develop within its own aqueous environment. These chordates have therefore become fully terrestrial throughout their life cycle.

Environmental variations

The fourth problem, that of living in an environment of fluctuating temperatures, has been overcome completely only by birds and mammals. These two classes of chordates have evolved **endothermy**, the ability to maintain a constant body temperature by physiological mechanisms. Endothermy provides a very stable internal environment which has allowed birds and mammals to colonise all terrestrial habitats, from the equator to the poles. It is also thought to have been essential for the extraordinary development of the brain shown by these two groups of animals.

Class Amphibia (amphibians)

Amphibians have the following characteristic features:

- They have four limbs.
- They have a soft skin, permeable to water.
- Their life cycle usually includes an aquatic larval stage with gills.
- Land-living adults have lungs.

Examples include *Rana*, frog (figure 2), newts, and salamanders.

Figure 2 *A tadpole of* Rana *(frog) showing the external gills.*

Class Reptilia (reptiles)

Reptiles have the following characteristic features:

- They usually have four limbs (but these may be reduced or lost).
- They have a scaly, water-tight skin.
- They produce eggs with shells.
- They have lungs.

Examples include *Crocodilus*, crocodile (figure 3), lizards, and snakes.

Figure 3 *Crocodiles are some of the largest reptiles.*

Class Aves (birds)

Birds have the following characteristic features:

- One pair of limbs modified to form wings.
- Their skin has feathers.
- They produce eggs with shells.
- They have lungs.
- They are endothermic.

Examples include *Hirundo*, swallow (figure 4), penguins, eagles, and ostriches.

Figure 4 Hirundo rustica *is a familiar summer visitor*

Class Mammalia (mammals)

Mammals have the following characteristic features:

- They have four limbs (reduced in some marine mammals).
- Their skin has hair in follicles.
- They are mostly **viviparous** (they give birth to active young, rather than laying eggs) and feed the young on milk.
- They have lungs.
- They are endothermic.

Examples include marsupials (pouched mammals such as kangaroos), monotremes (egg-laying mammals such as the duck-billed platypus), and eutherians (placental mammals such as humans and chimpanzees (*Pan*; figure 5).

Figure 5 Pan *(chimpanzee), a typical mammal.*

Quick check

1 a What is the main nitrogenous excretory product of most reptiles and birds?

b Briefly explain how this product makes reptiles and birds better adapted for life on land than mammals.

2 a Which chordate classes produce an egg with a shell?

b Which chordate classes are endothermic?

Food for thought

The production of heat by metabolism is a function of an animal's body mass and is directly proportional to its volume. On the other hand, heat loss from the body to the environment is a function of surface area. With reference to metabolic rates and heart rates, the relative amount of food eaten, and adaptations to retain or lose heat, suggest the significance of these facts to: **a** a small mammal, such as a pygmy shrew **b** a large mammal, such as an elephant.

Summary

Classification of organisms may be artificial or natural. **Natural classifications** aim to group organisms according to their evolutionary relationships whereas **artificial classifications** do not. Modern biology places organisms into a hierarchical system of taxa. The seven main **taxa** are **kingdom**, **phylum**, **class**, **order**, **family**, **genus**, and **species**. Species are named according to the **binomial system of nomenclature**; each species has a generic and specific name.
Prokaryotes include all the **bacteria**. The lack of a nucleus and double-membrane organelles are two of several features that distinguish them from **eukaryotes**. **Protoctists** are eukaryotes with a cellular structure and organisation. They include **algae** and **protozoa**. *Amoeba* is a protoctistan belonging to the phylum **Rhizopoda**. It is characterised by having **pseudopodia** and **amoeboid movement**. *Euglena* is a protoctistan belonging to the phylum **Euglenophyta**. It is characterised by the possession of **flagella** and **euglenoid movement**. The phylum has both photosynthetic and non-photosynthetic species. **Fungi** are heterotrophic and sessile. **Plants** are photoautotropic, multicellular eukaryotes that include **bryophytes**, **ferns**, **conifers**, and **flowering plants**. All plant phyla exhibit **alternation of generations** but there is an evolutionary trend for a reduction in the **gametophyte generation**, with the **sporophyte generation** becoming the most persistent stage.

Animals are multicellular, heterotrophic organsims which are typically motile. **Cnidarians** have a **diploblastic**, **acoelomate**, **radially symmetrical** body form and simple nervous system; **Platyhelminthes** are **acoelomate triploblastic** animals. **Turbellaria** are free-living whereas **trematodes** and **cestodes** are **parasitic**. **Annelids** are **triploblastic coelomates** with **metameric segmentation**. They include **polychaetes** (marine worms), **oligochaetes** (earthworms), and **leeches**. Gaseous exchange in earthworms is by simple diffusion across the body surface. **Arthropods** are triploblastic, segmented coelomates with **jointed limbs** and an **exoskeleton**. They, include **crustaceans**, **millipedes**, **centipedes**, **arachnids**, and **insects**. Forms of **insect locomotion** include walking and flying. The dipterans are best adapted for **flight**. In insects, gaseous exchange takes place in the **tracheal system** which includes **spiracles**, **tracheae**, and **tracheoles**. Three common non-chordate phyla are **nematodes**, **molluscs**, and **echinoderms**. **Chordates**, animals with a notochord at some stage in their life, include **cartilaginous fish**, **bony fish**, **amphibians**, **reptiles**, **birds**, and **mammals**. Fish have **gills** for gaseous exchange; bony fish have a **countercurrent exchange mechanism** for maximising the rate of diffusion across the gills. Chordate **adaptations to life on land** include the evolution of a **tetrapod limb** for locomotion, **lungs** for gaseous exchange, and independence of a watery external environment for reproduction.

PRACTICE EXAM QUESTIONS

1 The classification system for living organisms is a hierarchy of phylogenetic groupings.

a Explain what is meant, in this context, by:

i a hierarchy; [2]

ii phylogenetic. [2]

b Copy and complete the table to show the classification of the ocelot. [2]

Kingdom	Animalia
	Chordata
Class	Mammalia
	Carnivora
Family	Felidae
	Leopardus
	pardalis

[Total 6 marks]

2 The diagram below shows how organisms may be separated into five kingdoms.

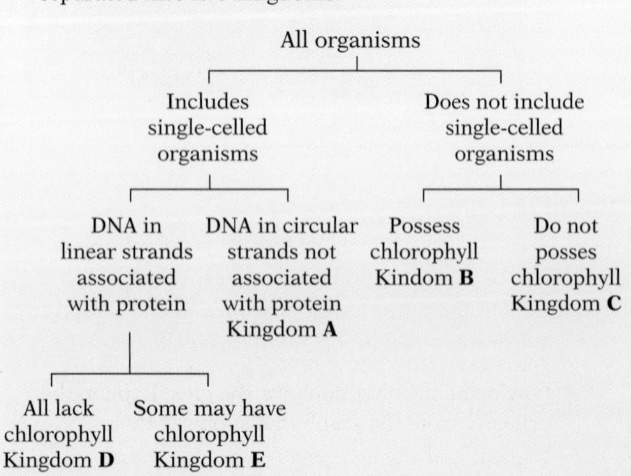

a **i** Name Kingdom **B**. [1]

ii Give one characteristic, other than the possession of chloroplasts, which could be used to distinguish cells of organisms in Kingdom **B** from those of organisms in Kingdom **C**. [1]

b Which of Kingdoms **A**, **B**, **C**, **D** or **E** represents the Fungi? [1]

c *Microactinium* is a single-celled eukaryotic organism. It is an autotroph. Which of kingdoms **A**, **B**, **C**, **D** or **E** includes *Microactinium?* [1]

[Total 4 marks]

3 The drawing shows a section through part of a filamentous fungus.

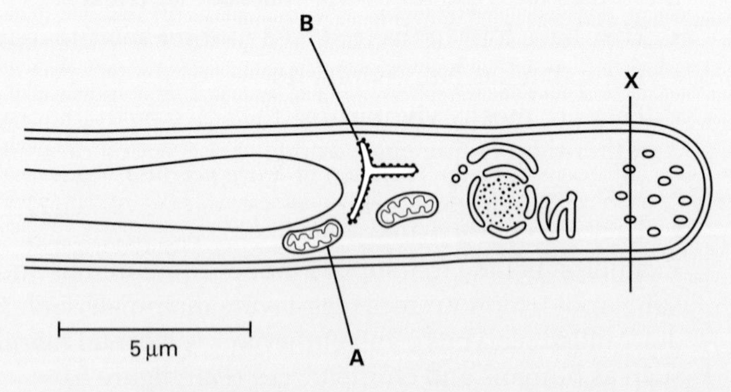

a **i** Name the features **A** and **B** shown on the drawing. [2]

ii Give **two** features other than those labelled **A** and **B** on the drawing which show that this fungus is a eukaryote. [2]

b Calculate the diameter of the part of the fungus shown at position **X** in the drawing. Show your working. [2]

[Total 6 marks]

4 Three different types of organisation of the body layers of animals are shown in the diagrams.

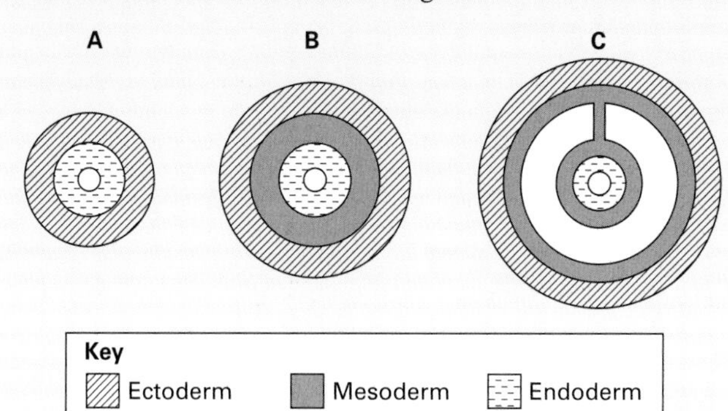

a Give the name of **one** phylum whose members:

i are radially symmetrical and have the type of organisation shown in diagram **A**; [1]

ii have the type of organisation shown in diagram **B**. [1]

b i Describe the type of organisation shown in diagram **C**. [1]

ii Explain how the type of organisation shown in diagram **C** allows for greater complexity of structure than the type of organisation shown in diagram **A**. [2]

[Total 5 marks]

5 The diagram below shows a part of the beef tapeworm *Taenia saginata*.

a Explain the importance of the part labelled A in the life of the tapeworm. [2]

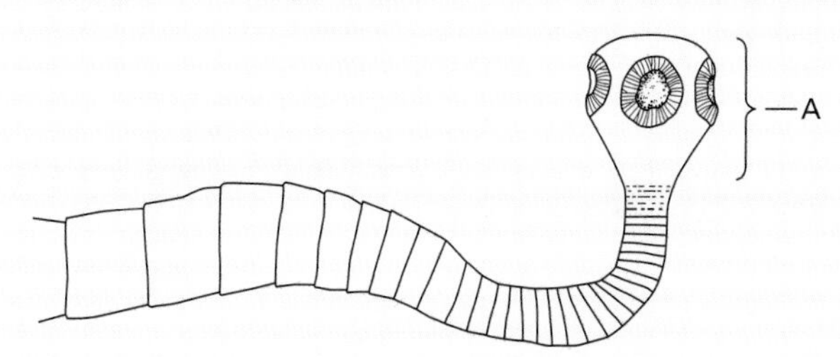

b i Describe how the tapeworm obtains its nutrition. [2]

ii How does the nutrition of *Rhizopus* differ from that of the tapeworm? [2]

[Total 6 marks]

6 The diagrams below show *three* organisms. Each belongs to a different phylum (major group).

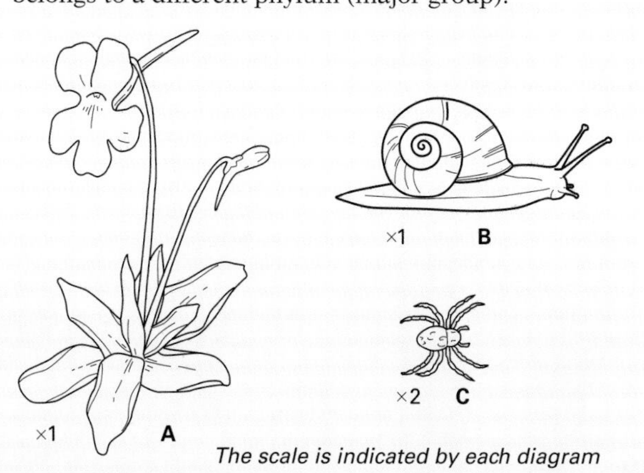

The scale is indicated by each diagram

Copy the table below. Fill in the name of the phylum to which each organism belongs and give *one* external feature, shown in the diagram, which is characteristic of this phylum.

Organism	Phylum	*One* visible external feature
A		
B		
C		

[Total 6 marks]

7 The drawing shows some of the main flight muscles of a locust.

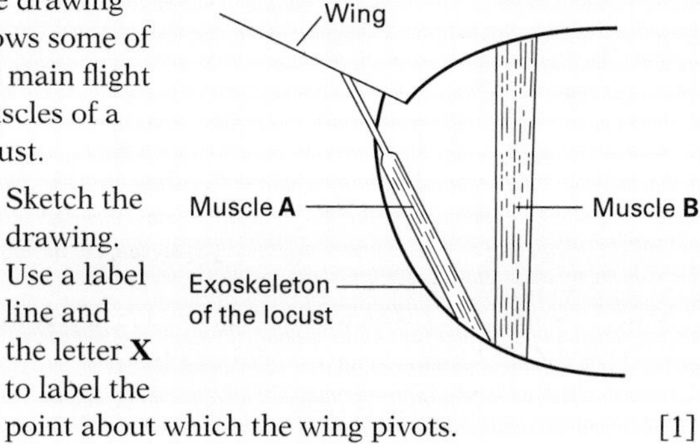

a Sketch the drawing. Use a label line and the letter **X** to label the point about which the wing pivots. [1]

b i What name is used to describe the action of a pair of muscles such as **A** and **B**?

ii Explain how the action of muscles **A** and **B** brings about movement of the wing. [3]

c Suggest **one** function of the exoskeleton in the movement of the wing. [1]

[Total 5 marks]

8 a i Give **one** similarity between the way in which oxygen from the atmosphere reaches a muscle in an insect and the way it reaches a mesophyll cell in a leaf. [1]

ii Give **one** difference in the way in which carbon dioxide is removed from a muscle in an insect and the way in which it is removed from a muscle in a fish. [1]

The diagram shows the way in which water flows over the gills of a fish. The graph shows the changes in pressure in the buccal cavity and in the opercular cavity during a ventilation cycle.

b Use the graph to calculate the rate of ventilation in cycles per minute. [1]

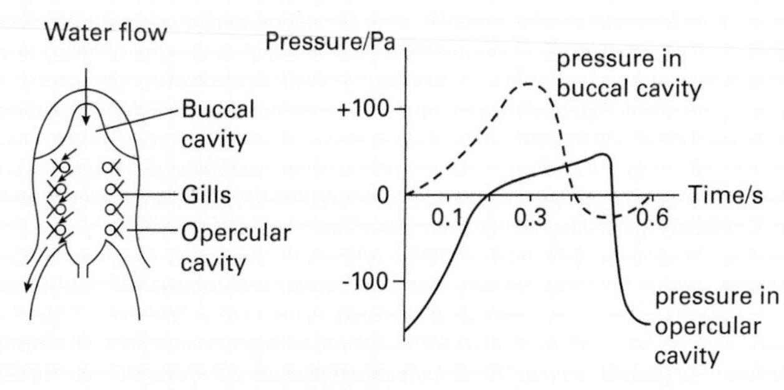

c For most of this ventilation cycle, water will be flowing in one direction over the gills. Explain the evidence from the graph that supports this. [2]

d Explain how the fish increases pressure in the buccal cavity. [2]

[Total 7 marks]

ECOLOGY *and* CONSERVATION

Ecosystems are intricate and vulnerable; once misused, disfigured and greedily exploited they will vanish to our detriment. Used wisely they provide boundless treasure. Used unwisely they create misery, starvation and death to the human race and to a myriad other lifeforms.

Gerald Durrell

Ecology is the study of how organisms interact with each other and with other parts of their environment. It touches on all aspects of biology, so it is fitting that ecology is included in the last section of the book. In the final chapter, we consider the impact humans have had on the environment, through agriculture, deforestation, fisheries, and pollution. We conclude the book with a discussion of biological conservation, probably the greatest challenge to face humans in the twenty-first century.

Elephants grazing in the plains below Mount Kilimanjaro. Maintaining the habitat in which these magnificent beasts can thrive will be a great challenge.

22 Environmental biology

Biosphere and biomes

22.1

OBJECTIVES

By the end of this spread you should be able to:

- explain the terms biosphere and biome
- discuss the factors that determine the type of biome occurring in a particular region.

Fact of life

The radius of the Earth is 8371 km, on average. However, most of the biosphere (the region occupied by living organisms) is contained within a thin layer no more than 28 km deep. Deep-sea organisms have been found in the Marianas Trench, the deepest part of the oceans, which plunges about 11 km below sea level. These organisms live on the organic material falling down on them, like rain, from above. Active organisms are also found in the atmosphere, at heights up to about 6.5 km above sea level. Higher than this, the air density, oxygen levels, and temperature become too low to support powered flight. However, dormant bacterial and fungal spores can sometimes reach altitudes of 17 km at the equator.

Figure 1 *Savannah grassland, one of about 40 biomes in the biosphere.*

Where organisms live

The part of the Earth occupied by living organisms is called the **biosphere**. It includes streams, rivers, lakes, and seas, land down to a soil depth of a few metres, and the atmosphere up to an altitude of a few kilometres. Although the biosphere is thin, the organisms in it are very diverse and vary from one geographical area to another.

The largest subunits of the biosphere are called **biomes**. Each biome is defined in terms of its living organisms and their interactions with the environment. There are about 40 biomes; some are aquatic (occur in water) whereas others are terrestrial (occur on land). The locations of the major terrestrial biomes, shown in figure 2, correspond broadly with climatic regions (temperature and precipitation being the most important climatic factors). Other environmental factors such as soil type and surface features such as mountains, however, may also play an important part in determining the location of a biome. Although a particular biome, such as a tropical rainforest, supports plants and animals of a certain type, the actual species differ from one rainforest to another: the various species in an Amazonian rainforest may be quite different from those in a rainforest in Malaysia, for example.

Gaia

James E. Lovelock has been a pioneer in the ecological study of the biosphere, investigating how, for example, the activities of living organisms may affect the climate of the Earth and its chemical composition. With Lynn Margulis, he formulated the **Gaia hypothesis**, named after Gaia, the Greek mother-goddess representing Earth. This hypothesis says that on any planet supporting life, the physical and chemical conditions necessary for life are maintained through the organisms' own feedback responses. The planet can be viewed as a single self-regulating living organism. Deviations from optimum conditions trigger biological processes which, within certain limits, bring conditions back to the optimum. Lovelock has illustrated the hypothesis using a computer model: 'Daisyworld'.

In 'Daisyworld', the only living things are daisies, either white or black. When there is less solar radiation and the environmental temperature falls, the black daisies are at a selective advantage because they can absorb more heat and remain warm. When solar radiation increases and the temperature rises, the white daisies are at a selective advantage, able to reflect more light and keep cool. During cold periods of low solar radiation, the black daisies cover most of 'Daisyworld' whereas during hot periods of high solar radiation, white daisies are more common. The amount of solar radiation absorbed or reflected by the ground depends on which daisies are covering the surface. The daisies therefore regulate the surface temperature of 'Daisyworld', preventing the climate from becoming too hot or too cold for either type of daisy to survive.

Since being formulated in the 1970s, the Gaia hypothesis has continued to be highly controversial. Many question the scientific validity of regarding the Earth as a single, giant, living organism. Nevertheless, the idea that the Earth behaves *like* a living organism in regulating atmospheric and oceanic conditions has been a very powerful metaphor. It has inspired huge amounts of research across a wide range of disciplines from the sciences to politics and philosophy. It has also highlighted the vital role of living organisms in driving geochemical cycles that affect the whole of life on Earth.

Studying interactions between organisms

Biological systems are dynamic: organisms interact with each other and with their surroundings, the environment. The scientific study of these interactions is called **environmental biology** or **ecology**. Ecologists may study environmental interactions at any level, from the biosphere to individual organisms.

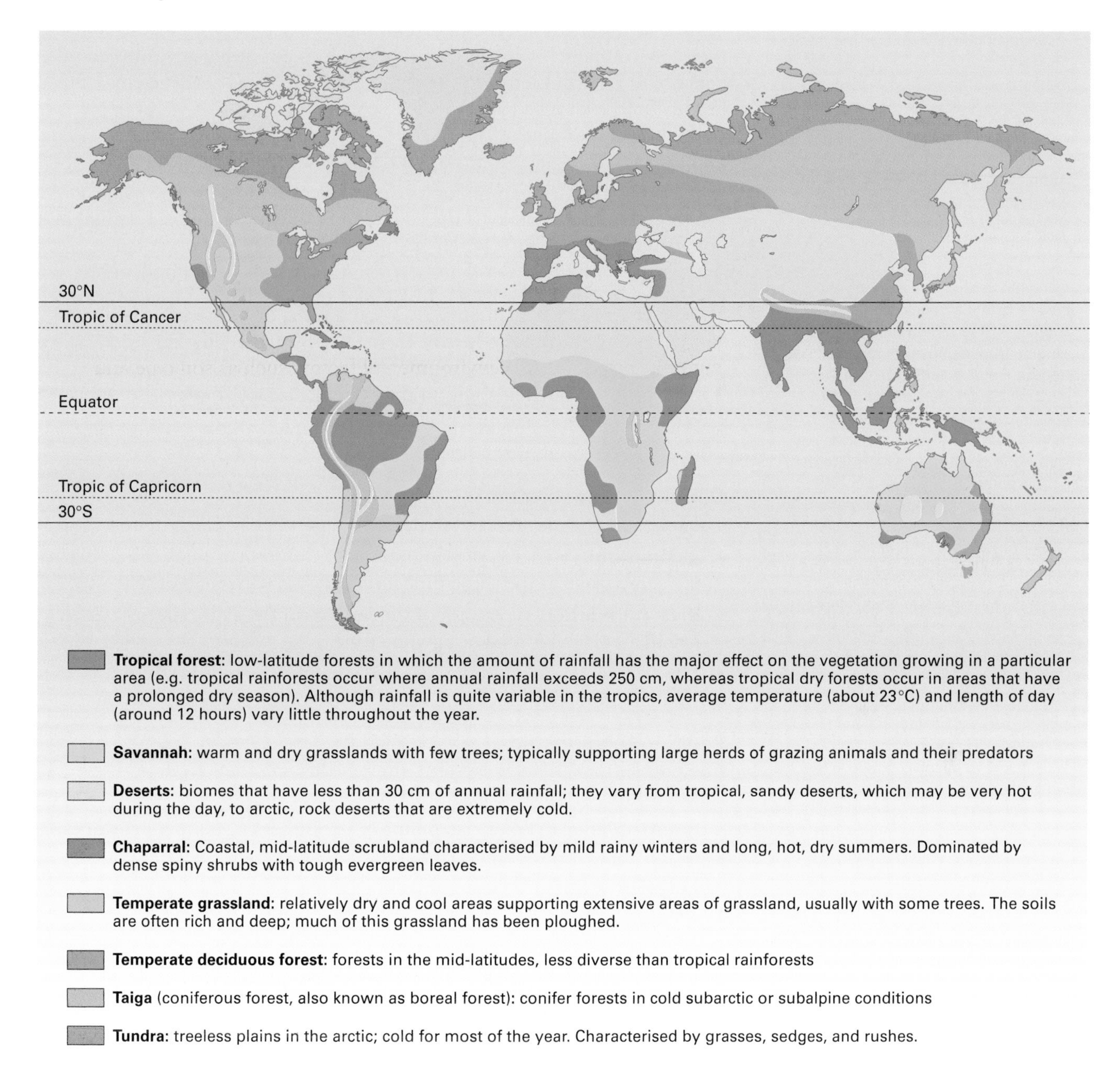

Figure 2 *The major terrestrial biomes. The map shows sharp boundaries between the biomes, but they actually merge into each other.*

Quick check

1 What sort of organisms occur in the highest parts of the atmosphere?

2 Which two climatic factors are the most important in determining the type of terrestrial biome?

Food for thought

All scientists agree that the activities of living organisms play an important role in driving biogeochemical cycles, and that organisms shape their environment to a considerable extent. Suggest how, for example, herbivores affect their grassland environment. What might happen if herbivores were removed from grassland in Britain? What might happen to grassland in a tropical region if overgrazing occurred?

22.2 ECOSYSTEMS

OBJECTIVES

By the end of this spread you should be able to:

- describe the main components of an ecosystem
- distinguish between abiotic and biotic factors.

Fact of life

In 1991, scientists tried to create a completely self-contained ecosystem that mimicked conditions on Earth, and which could support a small group of people. The artificial ecosystem, called Biosphere 2 (Biosphere 1 being the Earth itself), consists of an area of 15 000 m^2 enclosed in glass, containing five wilderness areas ranging from a miniature rainforest to desert, and including lakes and streams. Biosphere 2 (figure 1) was built in Oracle, Arizona, cost $200 million, and was seen as the first step towards colonising Mars. Eight people entered Biosphere 2 in 1991, and the plan was for them to remain totally isolated from the rest of the world for two years, breathing air recirculated by plants, and eating food grown on soils containing organic matter. However, within 18 months oxygen levels became so low that the people began to suffocate and extra oxygen had to be pumped in. Oxygen consumption by bacteria was higher than anticipated, and some of the oxygen produced by plants was absorbed by the concrete in the walls of Biosphere 2. Food production was reported to be so low that the inhabitants somehow smuggled in extra supplies. Scientists involved in the project blamed the failure of the project on a lack of understanding about the correct proportions of different organisms required to maintain a self-sustaining system. In 2011, the University of Arizona assumed ownership of Biosphere 2. It is now being used by the university's scientists as a laboratory for large-scale experiments designed to quantify some of the effects of global climate change.

What is ecology?

To many non-scientists, the term 'ecology' conjures up images of activists campaigning to preserve environmental quality and promoting ways of living in harmony with nature. In its biological sense, however, **ecology** is a science that studies the interrelationships between organisms and their environment. In this sense, ecology does not imply environmental concern or environmental conservation, though these activities may draw heavily on the knowledge gained by ecologists.

Ecology is a relatively new science. The term comes from two Greek words (*oikos*, meaning 'house', and *logos*, meaning 'discourse') and was first used by the German biologist Ernst Haeckel in 1866. The basic functional unit of ecology is the ecosystem. An **ecosystem** is a definable area containing a relatively self-sustained community of organisms interacting with their non-living surroundings. A good example of an ecosystem is a pond (figure 2). Two major processes take place within an ecosystem:

1 the flow of energy through the system
2 the cycling of nutrients within the system.

Energy flow is dealt with in spreads 22.3 and 22.4; the cycling of nutrients is covered in spreads 22.12 and 22.13.

Defining an ecosystem

The following terms are used to define the parts of an ecosystem:

- A **community** consists of populations of different species which live in the same place at the same time, and interact with each other.
- Each **population** is a group of individuals of the same species which live in a particular area at any one time.
- The area in which an individual lives is called its **habitat** or, if it is very small, its **microhabitat**. Most ecosystems contain several habitats.

Feeding relationships

The populations in an ecosystem interact through their feeding relationships. A typical community has:

- **autotrophs** (also known as **primary producers** or simply **producers**, for example green plants) which manufacture their own food
- **herbivores** (**primary consumers**) which eat the autotrophs
- **secondary consumers** which eat the herbivores
- **tertiary consumers** which eat the secondary consumers
- **decomposers**, mainly bacteria and fungi, which obtain nutrients by breaking down the remains of dead organisms.

Figure 1 *Biosphere 2.*

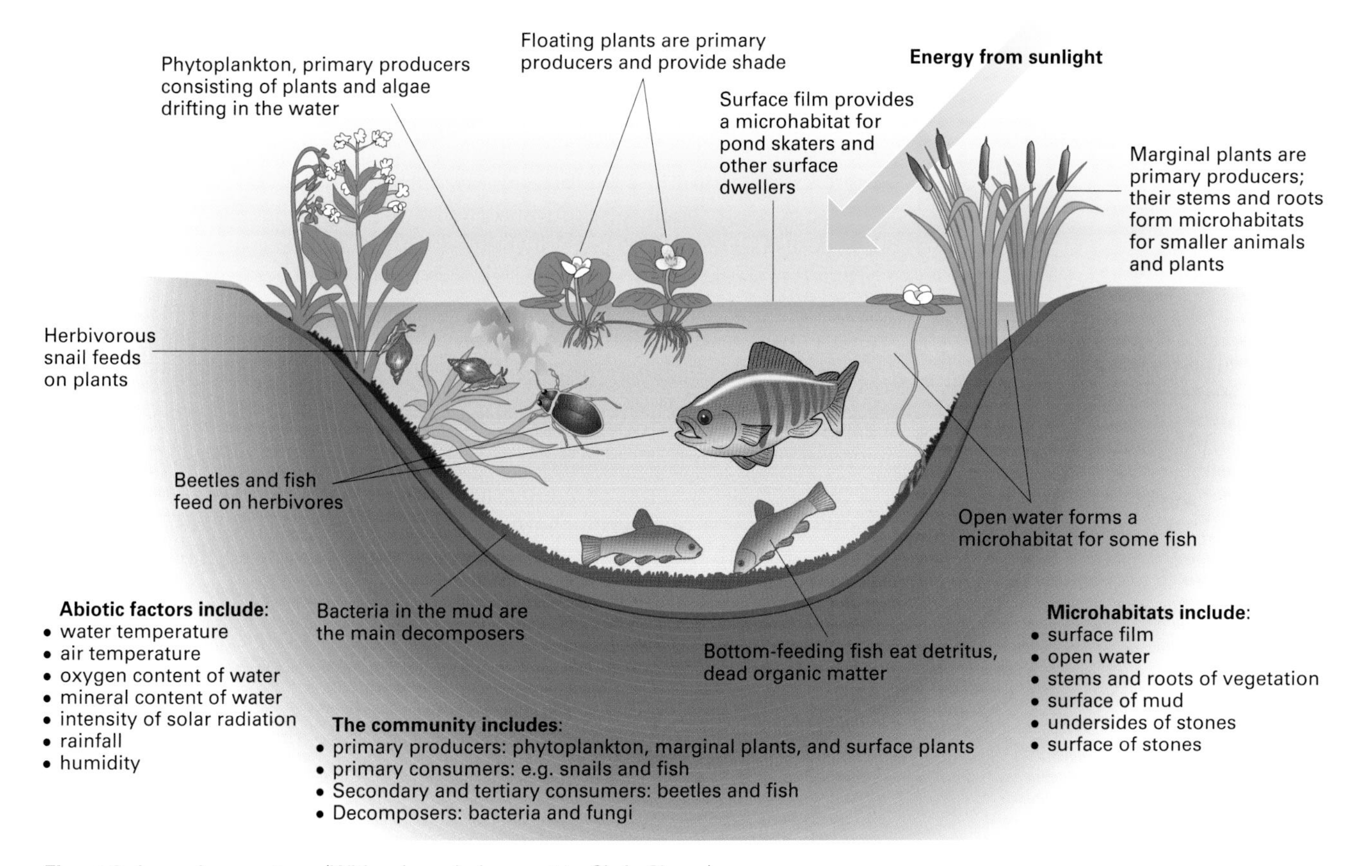

Figure 2 *A pond ecosystem. (With acknowledgement to Chris Clegg.)*

The feeding relationships (also called **trophic relationships**) within a community are discussed more fully in the next spread. **Ecological niche**, the role of each species within an ecosystem, is discussed in spread 22.5.

Biotic and abiotic factors

The complete range of external conditions in which an organism lives is called its **environment**. Within an ecosystem, the community forms the living or **biotic environment**. Factors such as predation and disease which result from the activities of living organisms are called **biotic factors**. The non-living part of an ecosystem forms the **abiotic environment**. **Abiotic factors** include:

- **climatic factors** such as light, temperature, water availability, and wind (in a microhabitat these form the **microclimate**)
- **edaphic factors**: factors associated with the soil, such as its texture, pH, temperature, and organic content (see spread 23.1)
- **topographic factors** such as the angle and aspect of a slope.

All these abiotic factors interact with each other and with the biotic component, the organisms living in the ecosystem.

Applying the idea of an ecosystem

The ecosystem concept can be applied to any scale from a small pool of water to a whole ocean, or even to the biosphere (spread 22.1). Whatever its size, an ecosystem is relatively self-contained and tends to maintain itself by recycling minerals. Thus in a pond, organisms die and decompose and the nutrients in them are returned to the water. In turn, the autotrophs remove these nutrients from the water and use them in growth.

Ecosystems are often thought of as closed systems, but they are not really closed. No ecosystem is completely self-contained; they all need an outside source of energy, usually the Sun. Although ecosystems must be definable, they do not have sharp boundaries. Even in a pond, energy, materials, and organisms are imported and exported.

Quick check

1 Distinguish between a community and a population.

2 Distinguish between abiotic factors and biotic factors.

Food for thought

The microclimate, the precise atmospheric conditions in which an organism lives, often varies substantially within a single habitat.

a Suggest why plants in the arctic tundra tend to be short.

b How will the microclimate at ground level differ from that 1 m above the ground?

c In terms of microclimate, what advantage may plants have in growing closely packed together?

22.3

Energy flow (1)

OBJECTIVES

By the end of this spread you should be able to:

- describe a food chain and a food web
- discuss the relative merits of pyramids of numbers, biomass, and energy.

Fact of life

The osprey is the top predator in one North American aquatic food chain. In a study, ospreys were found to contain over 8 million times more DDT than the water taken in by primary producers at the bottom of the food chain (figure 1). These high concentrations were not enough to kill the birds, but they caused their egg shells to be weakened so that few offspring survived.

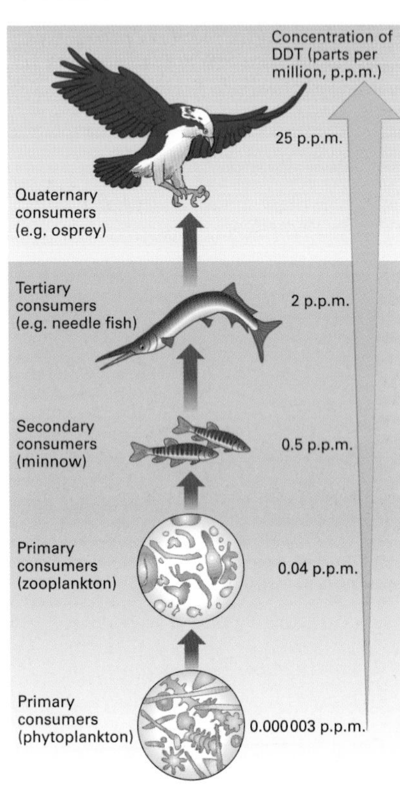

***Figure 1** DDT (dichlorodiphenyl-trichloroethane) was the first major synthetic pesticide. Introduced in the mid-1940s to control insect-transmitted diseases in American troops, DDT has been remarkably effective in killing insect pests such as malarial mosquitos and head lice which transmit typhus. DDT is cheap and effective, but is not easily broken down by natural processes and can build up in the environment. When consumed by organisms, DDT is stored in fatty tissue, so it tends to become more concentrated in animals higher up a food chain. This process is called **bioaccumulation** or **biological magnification**. It can result in top predators having body fat with levels of DDT high enough to kill them or to prevent them from breeding. Since 1972, the use of DDT has been banned in many countries.*

Life cannot exist without energy. Energy from the Sun sustains most terrestrial and aquatic ecosystems. The energy flows in one direction through the ecosystem; green plants and other photoautotrophs harness light energy from the Sun for photosynthesis, and animals and other consumers eat the autotrophs and each other. This one-way flow of energy is a consequence of the laws of thermodynamics: energy can be neither created nor destroyed, and when energy is transformed from one type to another, some is converted into heat (see spread 3.1). The transfer of food energy in an ecosystem can be shown in a number of ways including food chains, food webs, and pyramids of numbers, biomass, and energy.

Food chains and food webs

A **food chain** shows a sequence of organisms, usually starting with a producer and ending with a top consumer, in which each organism is the food of the next one in the chain. The arrows show the direction of flow of energy, with the arrow head always pointing away from the organism being eaten. Each organism within the chain represents a **trophic level**. The following food chain has five trophic levels:

Euglena → water flea → minnow → needle fish → osprey

Primary producer → primary consumer → secondary consumer → tertiary consumer → quaternary consumer

The number of links in a food chain is usually limited to four or five. This is because at each trophic level, a high percentage of the useful energy consumed as food is converted to heat, and lost from the food chain.

Food chains help to show how substances such as insecticides are transferred in an ecosystem (figure 1). However, they do not reflect the complex feeding relationships that exist in most ecosystems, and they give the false impression that each organism feeds on only one other type of organism. Most organisms in fact have a wide range of food items, and belong to more than one trophic level. The food chains in an ecosystem are usually interconnected, and this can be shown as a **food web** (figure 2).

Even food webs have their limitations. In large ecosystems, the relationships are often so complex that related species with similar diets are grouped together on the food web. Only a part of the ecosystem is usually shown on a food web (most, for example, omit decomposers). A food web showing all the feeding relationships of every species would have so many lines that it would be almost incomprehensible. Also, food webs give no information about *how much* energy flows from one part of an ecosystem to another.

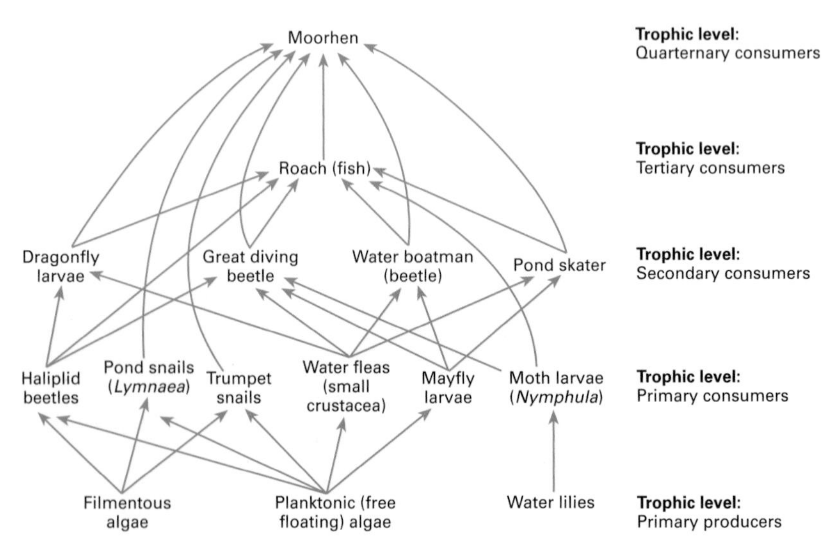

***Figure 2** A pond food web.*

Pyramids of numbers, biomass, and energy

A **pyramid of numbers** is one graphical method that shows the quantitative relationships between organisms at each trophic level. Each trophic level is represented by a rectangle, the length of which is proportional to the number of organisms at that level. Starting with the rectangle for the first trophic level (primary producers) as the base, the next rectangle is drawn on the centre of the first, and so on. Figure 3a shows a simple ecosystem in which there are many more organisms at the lower trophic levels than at higher ones. However, pyramids of numbers do not always have this shape. A single large tree, for example, can form an ecosystem which supports many thousands of insects (figure 3b), and when parasites are included in a pyramid of numbers the upper level can be much larger than the one below it (figure 3c). These are called **inverted pyramids**. They emphasise the fact that numbers of organisms do not always reflect the amount of energy: the one large tree in figure 2b, for example, contains much more energy than the organisms that feed on it.

Pyramids of biomass (figure 4) take account of this difference in size between organisms. In such pyramids, the horizontal axis is the total dry mass (biomass) of organisms at a particular trophic level, not their number. Unfortunately, even pyramids of biomass do not give a meaningful comparison between masses of different trophic levels in the same ecosystem, or between masses of the same trophic level in different ecosystems. This is because the biomass usually refers to the **standing crop**, the amount of dry mass measured at any one time. Whereas the standing crop of trees, for example, may result from the accumulation of biomass over 100 years or so, that of algae may represent only a few days' growth. Moreover, the biomass of populations of small organisms may fluctuate considerably throughout the year. The shape of the 'pyramid' therefore depends on the time of the year the sampling takes place. In aquatic ecosystems, this can result in an inverted pyramid of biomass (figure 5a). If the total biomass accumulated by each trophic level over the whole year was measured, we would usually obtain a pyramid of biomass that has a pyramidal shape. But this still may not be an accurate reflection of energy relationships in an ecosystem because two organisms of the same mass do not necessarily have the same energy content.

The type of pyramid that best reflects energy relationships in an ecosystem is a **pyramid of energy**, in which the horizontal axis shows the energy entering each trophic level in an ecosystem over a whole year. It shows the energy taken in by photosynthesis in the primary producer level and the energy entering the mouths of organisms in each consumer level over the whole year (figure 5b). The energy content of food eaten by consumers is measured using a bomb calorimeter (spread 9.6).

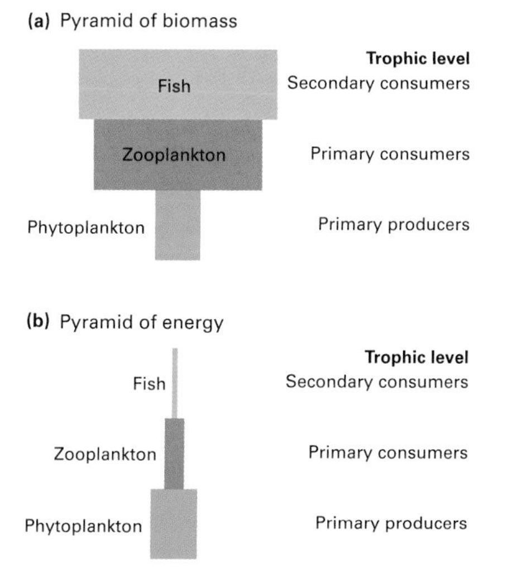

***Figure 5** Pyramids of biomass and energy for a marine ecosystem. **(a)** The pyramid of biomass shows that at any one particular time, the biomass of zooplankton (small animals drifting in the sea) is much greater than the mass of phytoplankton (primary producers drifting in the sea). This inverted pyramid results from the fact that most members of the phytoplankton have a life expectancy of a few days whereas zooplankton might live for several weeks and fish might live for years. It does not take into account the rapid turnover rate of phytoplankton (i.e. their rapid replacement by reproduction). **(b)** The pyramid of energy for the same ecosystem shows that the rate of energy flow decreases at each higher trophic level.*

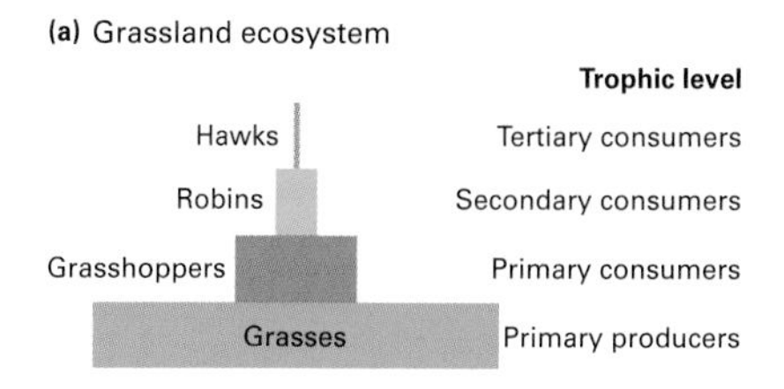

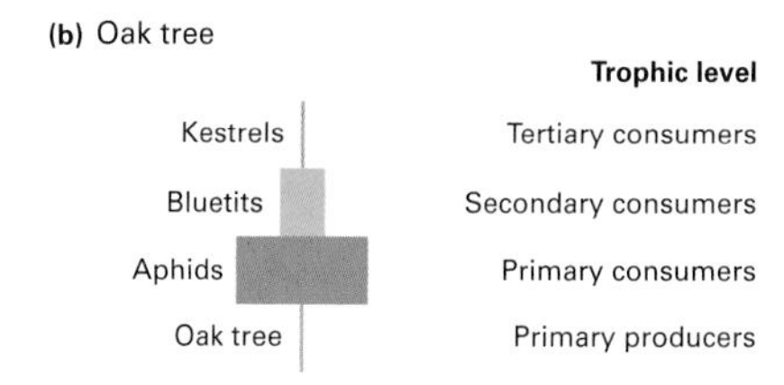

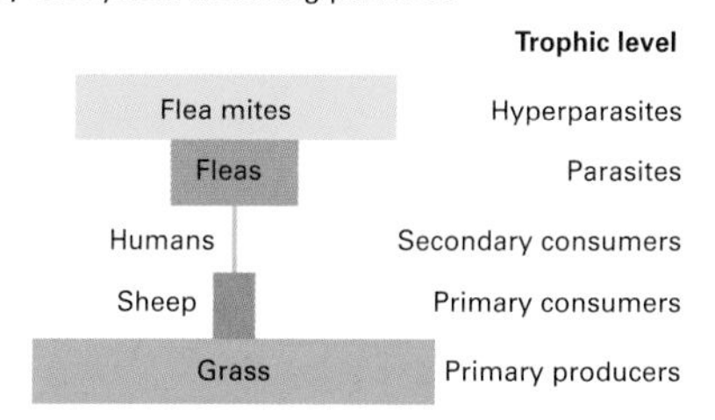

***Figure 3** Pyramids of numbers for: **(a)** a grassland ecosystem **(b)** an oak tree ecosystem **(c)** an ecosystem that includes parasites.*

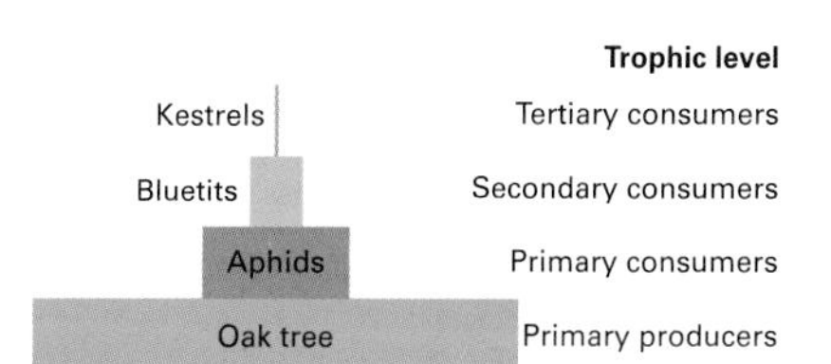

***Figure 4** Pyramid of biomass for an oak tree ecosystem.*

QUICK CHECK

1 In figure 2:
   - **a** What do the arrows represent?
   - **b** List the primary producers.
   - **c** Which major group of organisms within the ecosystem is not shown?

2 Why do the rectangles in a pyramid of energy get smaller at each higher trophic level?

Food for thought

Suggest how the community shown in figure 2 might be altered if the population of great diving beetles was wiped out.

22.4

ENERGY FLOW (2)

OBJECTIVES

By the end of this spread you should be able to:

- interpret energy flow diagrams
- distinguish between gross primary production, net primary production, and secondary production
- explain what is meant by trophic efficiency.

Fact of life

Tropical rainforests are the most productive biomes on land. They have a mean net primary production (NPP) of about 40 000 kJ m^{-2} y^{-1}, whereas temperate deciduous forests have a mean NPP of about 26 000 kJ m^{-2} y^{-1}. In extreme deserts, the mean NPP is only 260 kJ m^{-2} y^{-1}.

Energy flow diagrams

Pyramids of energy give a clear visual image of energy relationships within an ecosystem (spread 22.3, figure 5). However, most are incomplete because they do not show detritivores and decomposers. These organisms play a vital role in the energetics of ecosystems by eating and breaking down dead organic matter. Up to 80 per cent of all the energy incorporated into primary producers may not be eaten by consumers, but is transferred directly to detritivores or decomposers. Pyramids of energy also do not show how much energy passes to each part of the ecosystem. An alternative way of illustrating the transfer of energy through an ecosystem is by an **energy flow diagram** (figure 1). This type of diagram shows the amounts of energy (usually expressed as kJ m^{-2} y^{-1}) entering different parts of an ecosystem, and what happens to that energy. Using diagrams like this, it is possible to compare the efficiencies of energy transfer at different trophic levels or in different ecosystems.

Efficiency of primary production

The amount of solar energy reaching the Earth's surface varies with latitude, season, weather conditions, and details of location (for example, aspect). Each square metre of Earth receives up to 5 kJ of solar energy per minute. However, the proportion of solar energy actually converted into the chemical energy of plant tissues is astonishingly low, generally about 2%, and even in rapidly growing young plants under ideal conditions, not more than 8%. Figure 2 shows why only a small percentage of incident solar energy actually ends up as new plant tissue.

The total rate at which plants and other autotrophs synthesise organic material (such as carbohydrate) is called **gross primary production (GPP)**. It includes organic material used in respiration during the measurement period as well as organic material stored as new plant tissue. The rate at which autotrophs store organic material as new tissue is called **net primary production (NPP)**. So:

Net primary production = gross primary production – respiration

Only the energy that is stored in plant tissues is available to consumers, so NPP values are often used to compare the productivity of different ecosystems.

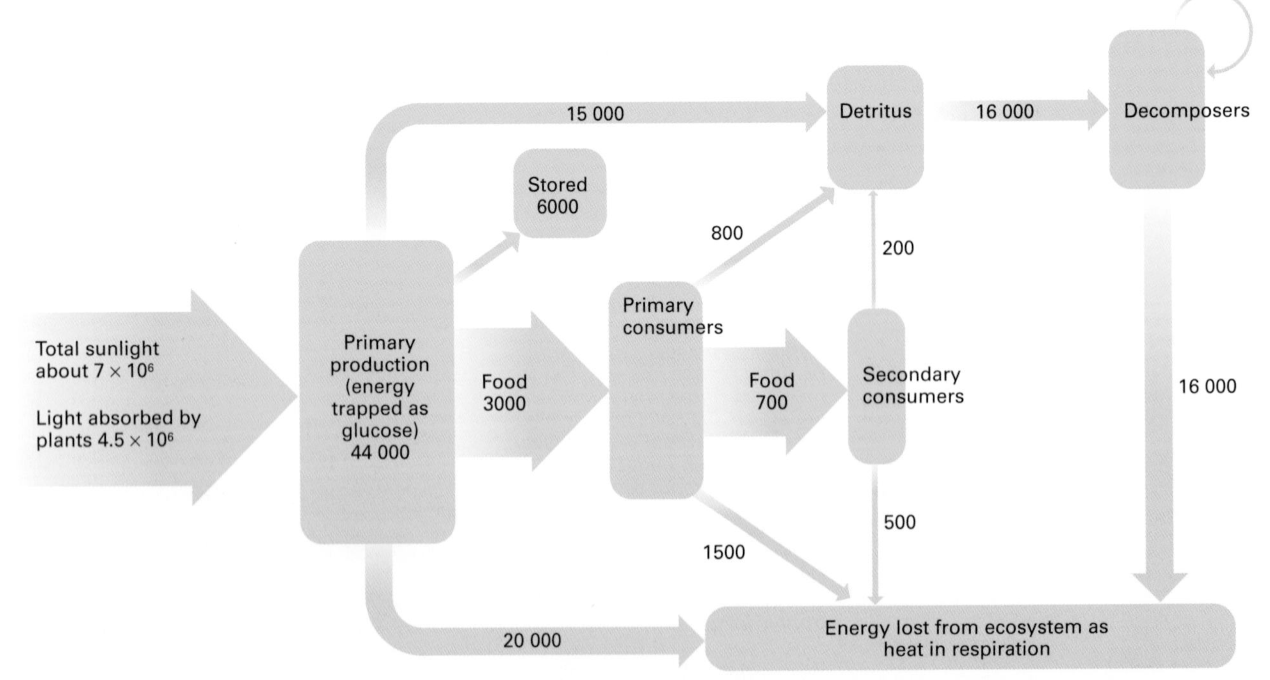

Figure 1 *Energy flow diagram for a forest ecosystem. Note that the transfer of energy obeys the first law of thermodynamics; energy is neither created nor destroyed as it passes from one trophic level to another. The figures show energy in kilojoules per square metre per year, kJ m^{-2} y^{-1}.*

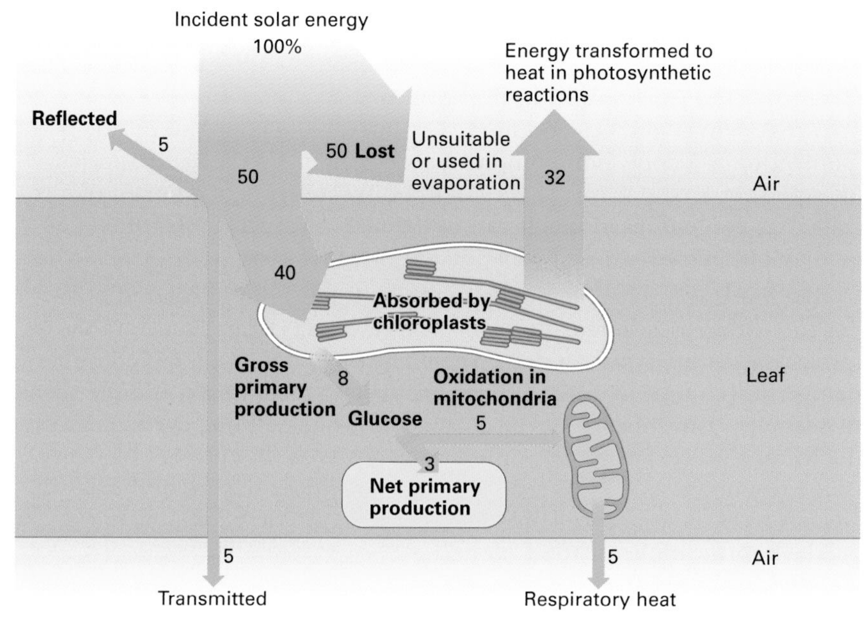

Figure 2 *What happens to solar energy falling on a leaf. In this example, about 60% of the solar energy is immediately lost from the plant to the environment because it is reflected, transmitted, not absorbed by chlorophyll, or used to evaporate water from leaves. Once light energy of a suitable wavelength enters the chloroplast, about 20% of absorbed light energy is converted into organic molecules; the rest is converted into heat in the environment.*

Efficiency of secondary production

Usually only a small proportion of the energy stored in plant tissue is transferred to the next trophic level; as we have seen, most plants die and decompose without being used by animals. Figure 3 shows the components of an **energy budget** (energy consumption against energy expenditure) of a herbivore. A similar diagram could be drawn for an omnivore or carnivore.

Some of the plant material remains undigested and is egested in the faeces. Digested plant material is absorbed by the herbivore, and some energy in this material will be used in respiration and transferred to the environment as heat; some will be transferred to the environment in urine and other secretions; and some will be used for growth of new tissue and reproduction. The rate at which energy is used to make new consumer tissue is called **secondary production**.

It is generally stated that about 10% of the energy in one trophic level ends up in the next trophic level. However, this figure (called **trophic efficiency**) is only a rough average: some herbivores eating large amounts of indigestible plant material have a trophic efficiency of less than 1%; zooplankton feeding on phytoplankton, on the other hand, may have a trophic efficiency of up to 40%. Nevertheless, increasingly small amounts of energy are passed from one trophic level to the next. This is why the number of trophic levels in an ecosystem is limited to four or five.

Farming efficiencies

Secondary production can be greatly increased by intensive farming techniques, for example by keeping animals indoors in a confined space under controlled environmental conditions and feeding them highly digestible and easily assimilated diets supplemented with growth stimulants. In addition, animal movements can be restricted to maximise the conversion of food consumed into useful products such as eggs, milk, and meat. Similarly, primary production can be increased enormously by growing plants in glasshouses to maximise the conversion of light energy into the plant product desired. The efficiency of animal and plant production can also be increased by using varieties and strains that have been bred artificially to maximise production.

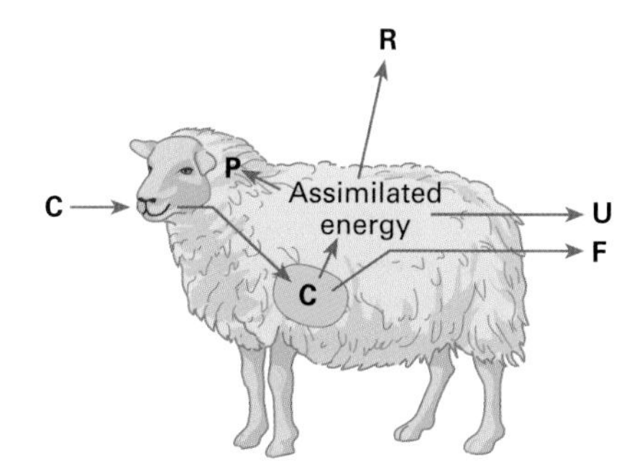

Figure 3 *The flow of energy through a consumer.*
C *= energy consumption (food ingested)*
P *= secondary production (energy stored in new body tissues or used to make reproductive products)*
R *= respiration (energy transferred to the environment as heat as a result of the body's metabolic activities)*
U *= energy in urine*
F *= energy in faeces*
Note that, according to the first law of thermodynamics, energy is conserved so energy input must equal energy output:
$C = P + R + U + F$.

Quick check

1. What is the relationship between gross primary production, net primary production, and respiration?
2. Why is net primary production often used to compare the efficiencies of different ecosystems?
3. What is trophic efficiency?

Food for thought

Suggest why:

a some people believe that above some critical density, human populations must become vegetarian to survive

b large, carnivorous animals are rare

c the conversion rate of food into tissue is generally less in a small mammal than in a fish of comparable size and given the same type of food.

22.5

ECOLOGICAL NICHE

OBJECTIVES

By the end of this spread you should be able to:

- define ecological niche
- distinguish between fundamental niche and realised niche
- explain the competitive exclusion principle.

Fact of life

Until the end of the nineteenth century, the red squirrel (*Sciurus vulgaris*) was found in all types of woodland throughout the British Isles. But since then its population has gradually declined. Today, it is found only in coniferous woods, mixed woods dominated by conifers, and in deciduous woods on some islands. Although no one knows precisely the reason for the decline, competition for resources with the grey squirrel (*Sciurus carolinensis*) has been blamed. This species was introduced from North America in the nineteenth century and became established in Britain by the beginning of the twentieth century. Today, grey squirrels have become the most common type of squirrel in Britain. Although bigger than reds, grey squirrels do not drive out red squirrels. In some places, reds and greys appear to live quite amicably together, feeding side by side. The greater success of grey squirrels may be due to their ability to digest acorns more efficiently than can red squirrels, allowing them to take better advantage of this vast food resource. Oak trees produce toxins to protect their leaves and acorns from herbivores: greys can detoxify these chemicals. Many grey squirrels also carry a viral disease that seems to do them no harm but which can be lethal to red squirrels. In 2008, an 80 per cent reduction in the red squirrels at Formby was attributed to this disease.

The role of a population in its habitat

An ecosystem contains a community of interacting populations. Each population is a collection of individuals belonging to the same species that play a particular role in the community. This role of a population is called its **ecological niche**. It can be defined as the sum total of biotic and abiotic resources used by a population in its habitat.

The **habitat** is where a population lives. It has been likened to a person's address, whereas the ecological niche has been likened to the profession or job a person does. The niche of a population includes the temperature range in which it lives, the type of food it eats, and the space it occupies. Time also has to be considered when defining the ecological niche: two species may have different niches even though they use the same resources, as long as they use them at different times of the day or year. A niche is sometimes broken down into different components (such as the **habitat niche** or **food niche**.)

Competition

Communities consist of individuals living together. **Competition** occurs whenever two or more individuals have to share resources in short supply. Competition between members of two different species, such as that between red squirrels and grey squirrels, is called **interspecific competition**, and that between members of the same species is called **intraspecific competition**. Intraspecific competition may be for nutritional needs, for mates, or for breeding sites; it becomes more intense with an increase in population density. In many animals, intraspecific competition for limited resources has led to the evolution of special forms of behaviour, such as ritualised fighting, territoriality, and dominance hierarchies (see spread 11.13). In 1934 G. F. Gause, a Russian ecologist, showed experimentally that when the ecological niches of two species are so similar that they compete for the same limited resources, they cannot coexist together indefinitely. This has become known as the **competitive exclusion principle**. His classic experiment involved two closely related species of protoctists: *Paramecium aurelia* and *P. caudatum* (figure 1). Gause cultured the two species separately under controlled conditions, supplying a constant amount of food each day. Both populations thrived, growing rapidly until reaching a maximum, after which they levelled off. However, when Gause cultured the two species together, the population of *P. aurelia* grew more slowly than when on its own, but the population of *P. caudatum* was eliminated. He concluded that the two species were competing for the limited amount of food and that *P. aurelia* had the competitive edge, feeding more efficiently and reproducing more rapidly than *P. caudatum*. This led to the death of the inferior competitor.

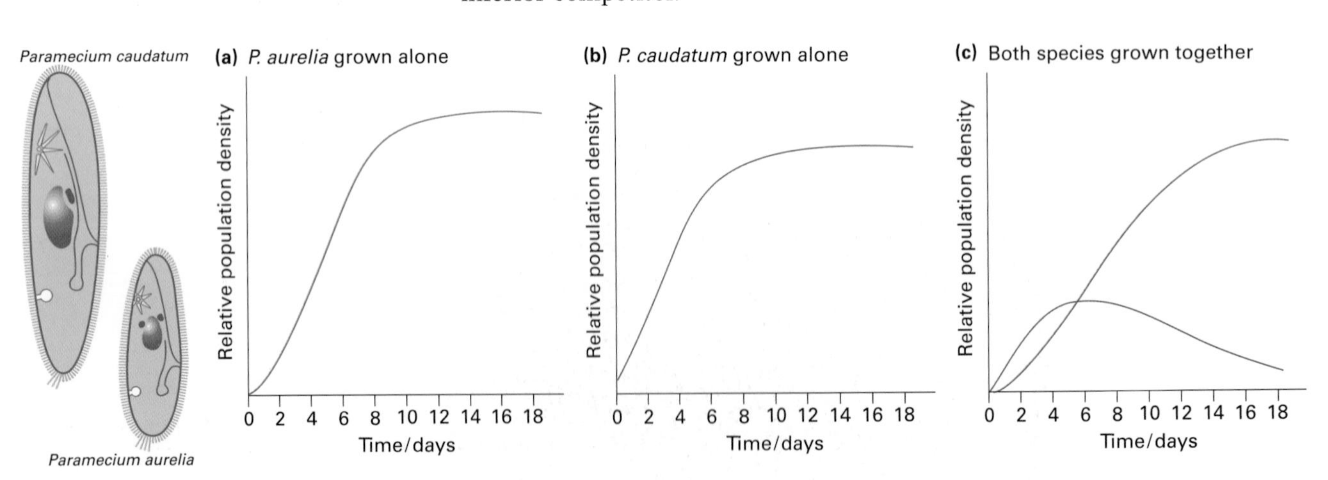

Figure 1 *Competitive exclusion principle: Gause's experiment.*

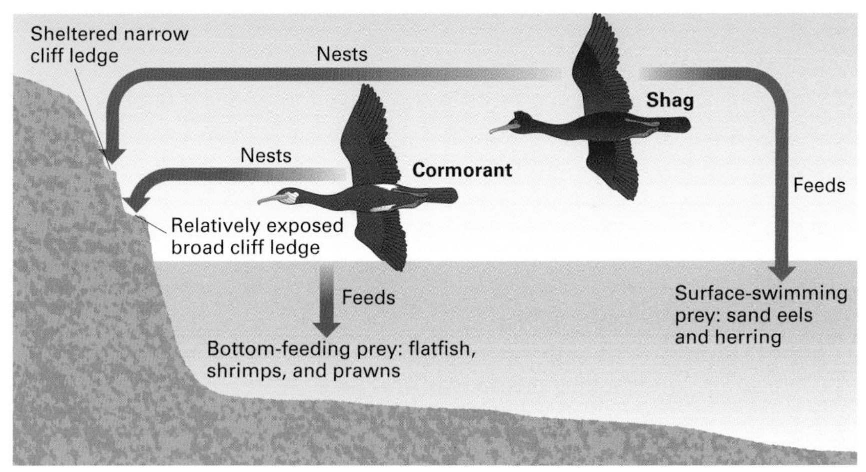

Figure 2 *The niches of shags and cormorants.*

Living together: realised niche and fundamental niche

Some species appear to exist together as exceptions to the competitive exclusion principle. However, on closer examination they are found to support Gause's conclusions after all. For example, two of our largest sea birds, cormorants (*Phalacrocorax carbo*) and shags (*P. aristotelis*), appear to have similar niches: they can often be seen in the same coastal areas, bobbing about in the water in their continual search for food. The birds look so similar that even experienced ornithologists sometimes have difficulty telling them apart. However, a detailed study shows that they have different ecological niches (figure 2). Shags feed mainly on surface-swimming prey (sand eels and herring) and choose sheltered coastal sites for breeding, such as crevices in rock gullies, ledges in the roofs of caves, or among boulders on steep slopes. They can nest in small sites, for example on a ledge only 30 cm square. In contrast, cormorants feed mainly on bottom-feeding prey (flatfish, shrimps, and prawns) and prefer to breed on small rocky islands and broad cliff edges which are often more exposed to the elements than the nest sites of shags. Cormorants also need more space as their nests may be up to a metre wide and 80 cm high. By dividing environmental resources between them, a process called **resource partitioning**, shags and cormorants can coexist in the same place. But competition between the two species probably means that each has a more limited ecological niche than if it was living alone. This niche, restricted by the presence of a competitor, is called the **realised niche**; the larger, potential niche that would occur without the competitor is called the **fundamental niche** (figure 3).

Two species cannot share the same ecological niche, but individuals belonging to the same species may have different niches. For example, animals that include metamorphosis in their life cycle occupy different niches as larvae and as adults: young frog larvae live entirely in water and use different resources from the adults which are mainly terrestrial creatures.

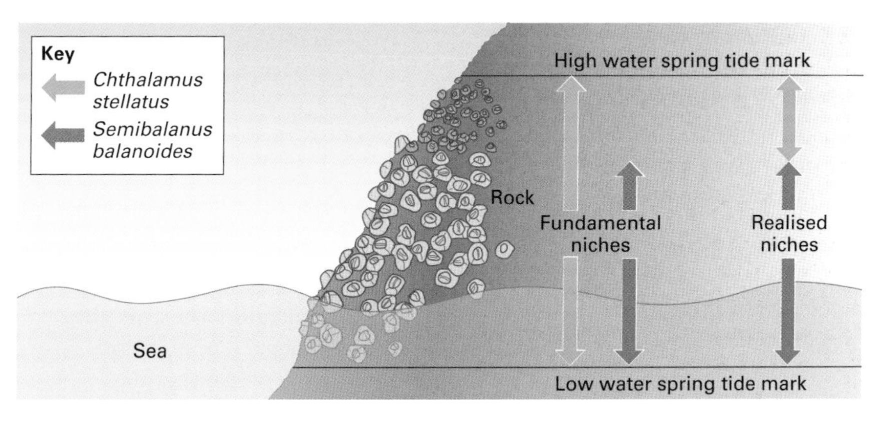

Figure 3 *The realised and fundamental niches of two species of barnacle.*

Quick check

1 The habitat of an organism has been likened to a person's address. To what has the ecological niche been likened?

2 Which is usually the larger, the realised niche or the fundamental niche?

3 Why is it not possible for two different species to have identical ecological niches?

Food for thought

Chthalamus stellatus and *Semibalanus balanoides* are two species of barnacle that coexist on tidal rocky shores along western Scottish coasts. Figure 3 shows the realised and fundamental niches of the two species. When together, they compete for space. Faster-growing barnacles with heavier shells can edge their shells underneath slower-growing and lighter shells and literally lever them off the rock. Barnacle species also vary in their tolerance to changes in abiotic factors.

a Which species do you think has the heavier shell and is faster growing?

b Which species has the greater tolerance of variations in temperature?

c Suggest what would happen to the distribution of *C. stellatus* if *S. balanoides* were removed from the lower part of the shore.

d What would you expect to happen to *S. balanoides* if they were transferred on rocks to the upper zone in which *C. stellatus* live?

22.6

Population growth

OBJECTIVES

By the end of this spread you should be able to:

- describe a sigmoid population growth curve
- explain the carrying capacity of an environment
- distinguish between density-dependent and density-independent factors
- discuss how natural populations grow.

Fact of life

Reproductive strategies involve a trade-off between quantity (number of offspring produced) and quality (amount of energy invested per offspring). Generally, as parents produce more offspring, they invest less energy in each one. In the 1960s, R. MacArthur and E. O. Wilson, two American ecologists, proposed that environmental and density-dependent selection pressures drive the evolution of reproductive strategies in one of two directions: r- or k-selection. In under-exploited and unstable environments, **r-selection** predominates, resulting in species with a high **biotic potential** (*r*), fast growth rate, and short lifespan; parents produce a large number of offspring but invest little energy on each. In crowded, stable environments **k-selection** predominates. This results in populations living at densities close to the carrying capacity (*k*) of the environment and species having a low biotic potential (parents produce few offspring but invest a lot of energy in each), slow growth rate, and long lifespan. Population growth curves of r-selected species are typically J shaped (figure 2), whereas population growth curves of k-selected species are typically S shaped (figure 1). In recent years, the r/k selection theory has been largely replaced by life-history theories that recognise the importance of factors such as age-related mortality in driving the evolution of reproductive strategies.

Populations are dynamic, constantly changing components of ecosystems. They are commonly described using the following terms:

- **population size**: the number of individuals in a population
- **population density**: the number of individuals per unit area
- **population growth**: a change in the number of individuals; this is referred to as positive growth when the numbers increase and negative growth when they decrease
- **population growth rate**: the change in number of individuals per unit time; again, this may be positive or negative.

A population under ideal conditions: sigmoid growth curves

Figure 1 shows the growth curve of a population colonising a new habitat in which conditions are initially ideal: there is an abundant food supply; no competitors, predators, or disease; optimal temperature; and so on. Populations of bacteria growing on agar, or yeast growing in broth, typically have this type of growth curve, called a **sigmoid growth curve**.

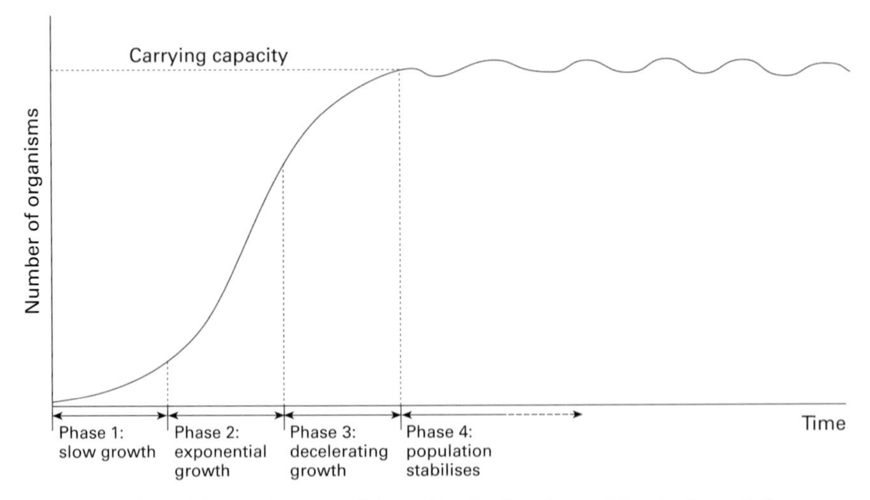

Figure 1 *A sigmoid growth curve. ('sigma' is the Greek word for the letter 's').*

Initially, the growth curve rises gently, indicating that the population growth rate is slow (phase 1, sometimes called the **lag phase**). Growth is slow because there are few reproducing individuals; those that reproduce sexually may not be able to find a mate when the population density is low. In phase 2 of the growth curve, the population grows at its **biotic potential**, that is, its maximum rate. Birth rate exceeds death rate and the population size doubles at regular intervals. This phase is called the **exponential growth phase** (or the **log phase**, because the curve is linear when plotted on a logarithmic scale). Exponential growth cannot go on forever. Eventually the population is prevented from increasing further by various factors in the environment, such as lack of food or space, increased competition for resources, a build-up of toxic chemicals produced by the organisms, or an increase in the incidence of parasitic infections and disease. The population growth rate slows down (phase 3, during which the growth curve becomes less steep), and stops (phase 4, during which the growth curve levels off). Collectively, the factors that limit the size of a population are referred to as the **environmental resistance**.

Boom and bust

The maximum population size that can be sustained over a relatively long period of time by a particular environment is called the **carrying capacity** of the environment. In some circumstances, a population increases so rapidly during the exponential growth phase that it overshoots the carrying capacity. As the environment cannot support the population, a population crash usually follows. Populations that show this type of growth curve are sometimes referred to as '**boom-and-bust**' populations (figure 2). Overpopulation can damage the environment, leading to a new, lower carrying capacity. After the crash, the population fluctuates around the new carrying capacity.

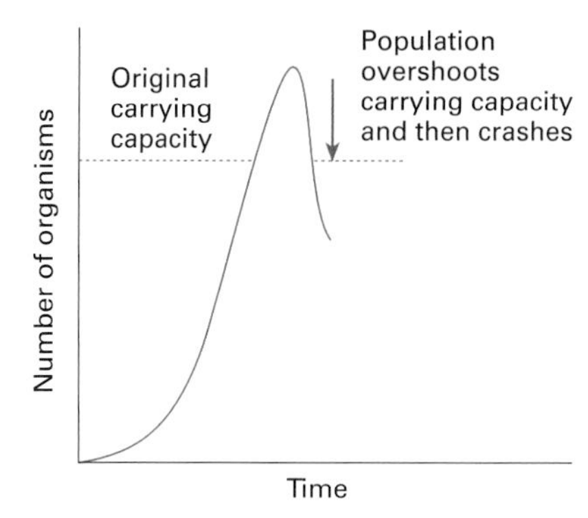

Figure 2 *A 'boom-and-bust' population growth curve, for example, of a population of* Daphnia *(water fleas) in a culture. The population increases so much it overshoots the carrying capacity and crashes.*

Environmental factors that limit population size

Exponential population growth is not often seen, except under controlled laboratory conditions or when a species colonises a new habitat. In natural populations a complex of environmental factors interact, some tending to decrease the population size, others tending to increase it (table 1).

Table 1 *Some of the factors affecting population size.*

Factors that increase population size	Factors that decrease population size
Plenty of suitable space available	Suitable space unavailable or limited
Good food supply	Inadequate food supply
Good water supply	Inadequate water supply
Ability to resist disease	Inability to resist disease
Small number of predators or ability to avoid predation	Inability to avoid predation
High reproductive rate	Low reproductive rate
Favourable light	Too much or too little light
Stable abiotic conditions (e.g. temperature and chemical conditions within the optimal range)	Unstable abiotic conditions (e.g. temperature and chemical conditions outside the optimal range)

These factors may be classified as:

- **density-dependent factors –** as the population density (number of organisms in a given space) increases, these factors have stronger effects, affecting a larger proportion of the population. They are usually biotic and include availability of food, disease, and predation.
- **density-independent factors –** these factors affect the same proportion of the population no matter what its density. They are usually abiotic and the most important are weather (short-term changes in atmospheric conditions, such as rainfall and temperature) and climate (long-term changes in atmospheric conditions). A sudden freeze can have a dramatic effect on populations, killing a fixed percentage of organisms irrespective of the population density.

Environmental factors determine the size of a population by affecting:

- the **birth rate** (number of births/number of adults in the population)
- the **death rate** (number of deaths/number of adults in the population)
- the amount of movement into the population (**immigration**)
- the amount of movement out of the population (**emigration**).

In favourable conditions, the population will grow if:

Number of births + number of immigrants
is greater than
number of deaths + number of emigrants

In unfavourable conditions, the population will decrease if:

Number of births + number of immigrants
is less than
number of deaths + number of emigrants

Quick check

1 What happens to the size of a population during an exponential growth phase?

2 Define carrying capacity.

3 Why is temperature a density-independent factor?

Food for thought

Suggest why laboratory cultures of yeast often have a **decline phase** in which the population decreases and may become extinct.

22.7

OBJECTIVES

By the end of this spread you should be able to:

- describe the major trends in world population for the last 7000 years
- describe the demographic pattern that has occurred in Britain since the Industrial Revolution
- discuss the significance of population pyramids.

Fact of life

Every second, five people are born and two people die: a net gain of three people to the world population.

Growth of the human population

There are more than 6 billion people inhabiting the Earth, double the population of the 1950s. When humans were primitive hunter-gatherers, the world population was probably less than a quarter of a million. The development of agriculture about 10 000 years ago brought an upswing in population growth rate so that the human population reached about 100 million around 5000 BC. It continued to grow steadily until the start of the Industrial Revolution. The J-shaped part of the curve in figure 1 shows that since the Industrial Revolution the human population has been going through a phase of exponential growth. Social and economic changes including increased mechanisation and advances in technology stimulated an increase in the birth rate; advances in medicine and hygiene reduced the death rate, particularly of newborn babies. The world population therefore increased dramatically as countries became industrialised. In 1650, the world population was about 500 million; by 1850 it had doubled to 1 billion; 80 years later it had again doubled. The doubling time is now about 40 years. This means that if the population continues to rise unabated it will reach 12 billion in 40 years, 24 billion in 80 years, and so on. It is difficult to predict future trends, but it is self-evident that exponential growth cannot be sustained indefinitely.

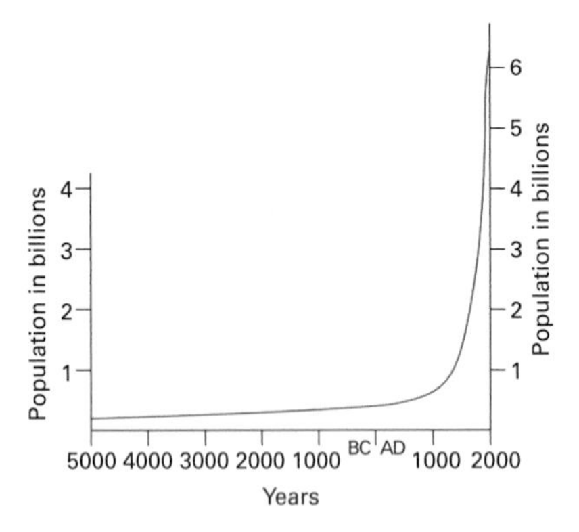

Figure 1 *Human population growth curve.*

Carrying capacity

The effect of human population growth will depend on the **carrying capacity** of the Earth: that is, the maximum number of people that the world can sustain for a long period of time. Unfortunately, the ideas of carrying capacity gained from animal studies do not apply to human populations, so no one can say precisely what our population limit is. New discoveries and inventions can change the carrying capacity of the land dramatically. This is precisely what happened during the industrial revolution of the eighteenth century, and continued with the green revolution in the twentieth century, when farming techniques improved agricultural outputs enormously. Today, world food production is growing at a faster rate than the world's population, and developments in genetic engineering are likely to increase food productivity even more.

Pessimists believe that the world population already exceeds the carrying capacity of the Earth, and that unless we do something very soon to curb our population growth and (just as important) to stop overexploiting natural resources, the result will be war, famine, and disease on a global scale. If this happens, our population growth curve would probably turn out to be similar to the boom-and-bust curve shown in spread 22.6, figure 2. Optimists, on the other hand, believe that our population growth curve (figure 1) will eventually become sigmoid shaped: the growth rate will slow down and the world population will stabilise below the carrying capacity. United Nations analysts are among the optimists: they predict that the world population will stabilise at about 12 billion in 120 years' time. There are already some signs that the population growth rate is slowing down (figure 2), but the 1994 figure of 1.6 per cent still implies a doubling time of 44 years.

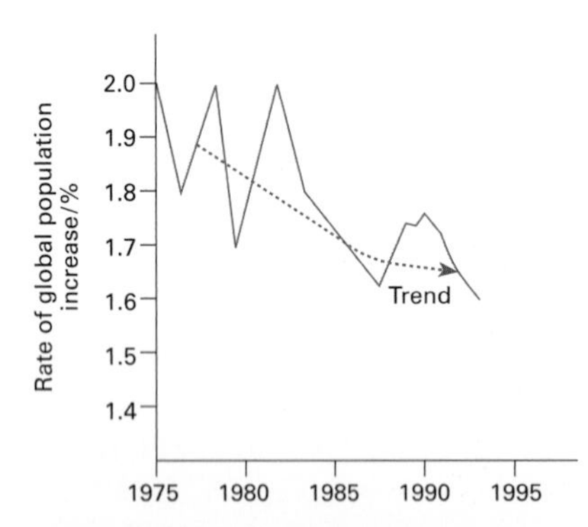

Figure 2 *Estimates of the rate of global population increase between 1975 and 1994. The figures are based on birth rates as a percentage of the population. They are estimates because some countries do not have reliable data.*

Demographic transition: a change in the population

Most of the population growth in recent years has taken place in developing countries. Some people believe that these countries will go through the same stages of population growth as industrially developed European countries. Before the industrial revolution, the population of Britain, for example, was relatively small and stable; birth rates and death rates were high and roughly in balance. The industrial revolution brought improved agriculture, better nutrition, and better medical knowledge. Initially birth rates remained high but infant mortality dropped rapidly (similar to the present situation in the developing countries). This allowed the population to grow very rapidly. As the

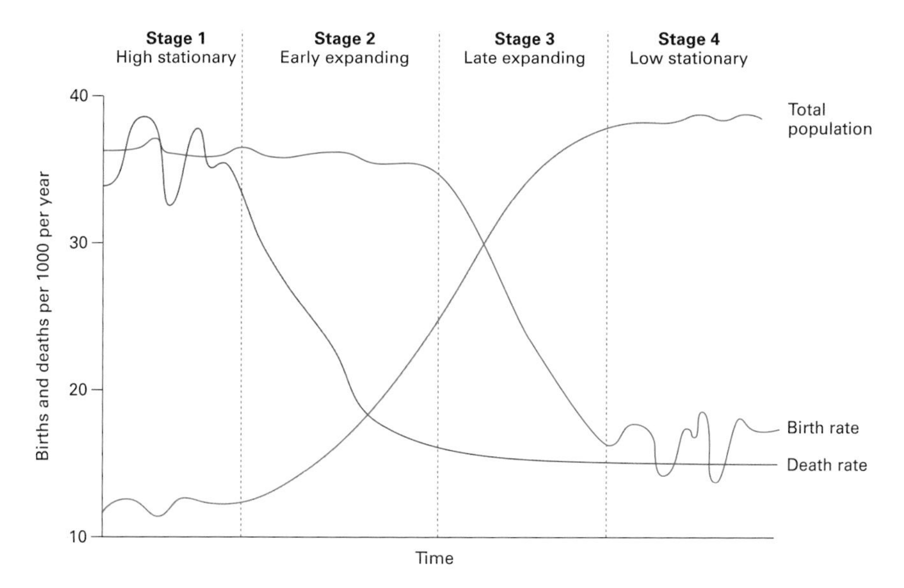

Figure 3 *Demographic transition model of population growth. Curves showing changes in total population, birth rate, and death rate in England and Wales since 1700 approximate to this demographic transition model.*

*Stage 1 (**high stationary stage**): birth and death rates are both high but because they are approximately equal, the population remains relatively stable at a low level.*

*Stage 2 (**early expanding stage**): the population starts to expand rapidly, mainly because of a decrease in death rate.*

*Stage 3 (**late expanding stage**): the birth rate declines but the population continues to expand as the birth rate still exceeds the death rate.*

*Stage 4 (**low stationary stage**): birth rates and death rates are approximately equal.*

economy improved, having a large family became less important, so the birth rate fell. Finally, the population size stabilised at all ages. This change in the human population from having high birth rates and high death rates to having low birth rates and low death rates is called **demographic transition** (figure 3). Although this change has happened in economically well developed nations and in some developing countries, it will not necessarily happen in all nations that become economically well developed.

Population pyramids

These changes in birth rates mean that the population structures may be very different in different countries. The statistical study of the size and structure of populations (for example, the age or sex distribution in the populations) and of the changes within them is called **demography**. Demographic trends can be shown in special diagrams called **population pyramids** or **population profiles** (figure 4). As you can see, the age distributions are completely different. In the developing countries, a much larger percentage of the population is younger than 15 years. This means that even if each has fewer children, the number of women reaching child-bearing age (generally taken to be between 15 and 49 years) will continue to rise steadily, so that the population size will continue to soar for many years.

(a) Economically well developed countries of Europe and North America

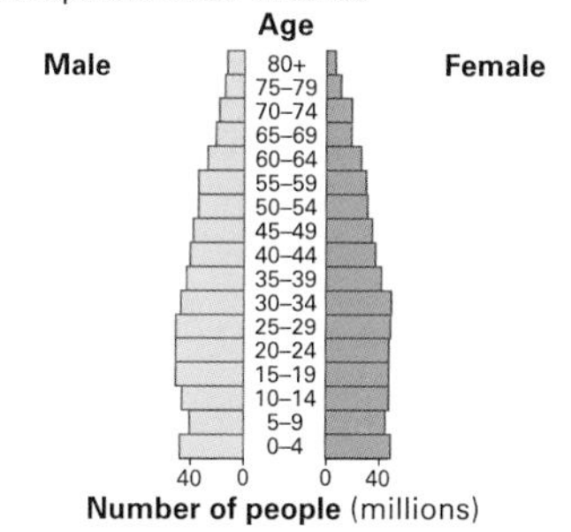

(b) Developing countries in Africa, Asia, Central and South America

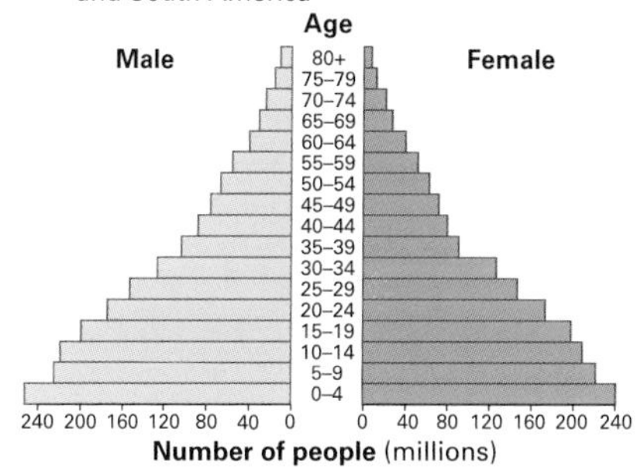

Figure 4 *Population pyramids showing the age and sex distributions in two contrasting types of population.*

Quick check

1 What brought about the first major upswing in world population about 6000–7000 years ago?

2 What is demographic transition?

3 Explain briefly why populations are likely to continue to increase in less well-developed countries even if the population growth rate decreases.

Food for thought

Thomas Malthus (1766–1834) was very anxious about the population explosion occurring in England and Wales in the eighteenth and nineteenth centuries: the population had grown from about 5.2 million in 1695 to 9.2 million in 1801. In 1798 he published a political essay in which he argued that population increases geometrically (that is, exponentially, for example, $2 \rightarrow 4 \rightarrow 8 \rightarrow 16 \rightarrow 32 \ldots$) whereas food production increases arithmetically (linearly, for example, $2 \rightarrow 4 \rightarrow 6 \rightarrow 8 \rightarrow 10 \ldots$) so that the population will eventually outstrip food production. The population will then be held in check by war, hunger, disease, or by choosing to produce fewer children. Suggest why Malthus' prediction has not been realised, despite the fact that the population in Britain has continued to rise (it was 17.9 million in 1851, 32.5 million in 1901, and about 63 million in 2011).

22.8

OBJECTIVES

By the end of this spread you should be able to:

- define interspecific relationships in terms of the effects of organisms on each other
- describe the benefits of mycorrhizal relationships
- distinguish between predation and parasitism.

Fact of life

Many interspecific relationships involve food, but some do not. One of the most extraordinary mutually beneficial relationships is that between male euglossid bees (a group of bees living in the tropics of Central and South America) and the orchids they pollinate. The bees do not collect pollen or nectar, but are rewarded instead with floral fragrances, which they modify to attract female bees.

Interspecific relationships

Individual organisms do not live in isolation in a community; they are continually interacting with each other. The interactions may be between members of the same species (**intraspecific interactions**) or between members of different species (**interspecific interactions**). Interspecific interactions are as many and varied as the organisms themselves, and there are many ways of classifying them. One way is according to the effects that individuals of interacting species have on each other

(+ = positive effects; 0 = no detectable effect; – = negative effects):

++ interactions are beneficial to individuals of both species.

+– interactions benefit individuals of one species but are harmful to those of the other species.

–– interactions are harmful to individuals of both species.

+0 interactions benefit individuals of one species and have no significant effect on individuals of the other species.

These interactions are often referred to as relationships. Table 1 shows six of the most common types of interspecific relationships and the effects they have on the individuals involved.

Relationships between organisms may be just casual and temporary, or more intimate and permanent. For example, the mutualistic relationship between a hermit crab and the sea anemone it carries on its shell is not a permanent one and the animals can live independently of each other. In contrast, some mutualistic relationships are so close that neither partner can exist without the other. This is the case for many lichens. These are combinations of algae (which provide photosynthetic products) and fungi (which conserve water and provide their partners with carbon dioxide and mineral salts). Organisms that do not have to adopt a certain mode of life are described as **facultative** (for example, facultative mutualists, parasites, anaerobes, etc.) whereas those compelled to follow a particular mode of life are described as **obligate**. Obligate relationships occur between cattle and rumen bacteria (spread 9.10), and between leguminous plants and nitrogen-fixing root-nodule bacteria (spread 22.12).

Mycorrhiza: a mutualistic relationship

A **mycorrhiza** (the term means 'fungus-root') is a mutualistic relationship between the roots of a plant and a fungus. An example is the relationship between the silver birch (*Betula pendula*) and the fly agaric fungus (*Amanita muscaria*) (figure 1). Fungal hyphae may grow just on the surface of the root, or some hyphae may penetrate into the plant tissue. In either case, masses of fungal hyphae in the soil help to break down organic matter, releasing soluble mineral nutrients (including phosphates, calcium salts, and potassium ions). Some of these are absorbed by the

Table 1 *Some interspecific relationships.*

Relationship	Type of interaction	Comments
Mutualism	+ +	Both partners benefit
Predation	+ –	One organism kills another in order to obtain food
Parasitism	+ –	The parasite obtains nourishment from living tissue (it does not kill its host in order to obtain food)
Competition	– –	Two (or more) species try to obtain the same limited resource. The relationship is harmful to both species because they are unable to exploit the resource as fully as they would in the absence of competition (see spread 22.5).
Commensalism	+ 0	One organism benefits but there is no apparent effect on the other organism
Allelopathy	+ –	One organism produces a chemical substance which has a harmful effect on another organism

Figure 1 *Mycorrhizal relationship between a birch tree and the fly agaric fungus.*

fungus and are passed on to the plant. In addition to breaking down the organic material, the extensive network of hyphae provides a much larger surface area for absorption than the plant roots alone. The plant in turn provides food for the fungus; in some cases as much as 40 per cent of the glucose synthesised by a tree may go to its fungal partner. It is estimated that more than three-quarters of all plants have mycorrhizal relationships. In many habitats, including beech woods, hyphae from the same fungus can form interconnecting mycorrhizal relationships with several species of plants. Although most mycorrhizal relationships are facultative, many plant species (including commercial crops such as citrus trees) grow much better with mycorrhiza and are more resistant to some pests than those without a fungal partner.

Figure 2 *The hydroid* Hydractinia *growing on a shell occupied by a hermit crab.*

Predators and parasites

We generally think of predators as carnivores that eat other animals, their prey. But in ecology a **predator** is any organism that uses other live organisms as an energy source and, in doing so, removes them from the population (that is, the predator eventually kills the prey). This definition of predators includes both herbivores and **parasitoids**. (Parasitoids are parasitic organisms that kill their host; an example is the ichneumon fly wasp which lays its eggs in insect larvae, such as beetle grubs. When the ichneumon fly larva emerges from the egg, it eats the grub alive.) We shall look at predator–prey relationships in more detail in the next spread.

Parasites differ from predators in that they do not normally kill their host to obtain food. In most parasitic relationships (such as platyhelminth parasites living in their mammalian hosts; spread 21.11), the parasite gains food and shelter from the host while inflicting some degree of harm. However, with the exception of parasitoids, death of the host is not usually caused directly by the parasite.

Commensalism and allelopathy

Commensalism, as shown in table 1, is a +0 interaction; one partner clearly benefits from the relationship and the other is not harmed. *Hydractinia*, for example, is a small marine animal living in colonies on shells occupied by hermit crabs (figure 2). It benefits from the relationship by being transported to new sources of food but, as far as we know, the hermit crab neither gains nor loses anything in the relationship. Although *Hydractinia* belongs to the phylum Cnidaria and has stinging cells, it is probably too small to offer the crab significant protection.

Allelopathy is a +– interaction, caused by one organism releasing a chemical substance into the environment that has a negative effect on another organism. For example, sunflower plants release chemicals from their roots and fallen leaves into the soil. The chemicals inhibit seed germination in some other plant species, but do not affect the germination of sunflower seeds, so giving the sunflower a competitive advantage. Allelopathy probably has a strong effect on the structure of plant communities.

Quick check

1 Define competition in terms of its effects on the populations involved.

2 Parasitism and predation are both +– interactions. How do they differ?

3 In a mycorrhizal relationship, what are the benefits to:

a the plant **b** the fungus?

Food for thought

The bird's nest plant (*Monotropa hypopitys*) lacks chlorophyll and grows in the dense shade of beech woods. It appears to obtain its organic nutrients from beech trees, and its mineral salts from soil fungi, with which it has a mycorrhizal relationship. When radiolabelled glucose is applied to the leaves of the beech, it rapidly finds its way into the bird's nest plant. Suggest how this happens.

22.9

OBJECTIVES

By the end of this spread you should be able to:

- discuss the relationship between predator and prey populations.

Fact of life

The intrepid ecologist Adolph Murie (1899–1974) carried out a long-term study of the predator–prey relationship between wolves and sheep on Mount McKinley, Alaska. In his work he collected 608 sheep skulls. After years of careful observation, Murie concluded that most of the skulls were from sheep killed and eaten by wolves (human hunters, the only other significant cause of death, would take the head). By analysing the growth rings in their horns, Murie was able to age each sheep. He found only two age groups: the very old and the very young. The wolves did not kill sheep in their prime; only ageing sheep and weak young ones. Similar observations have been made on the relationship between wolves and moose, between mountain lions and deer, and between some other big cats and their prey.

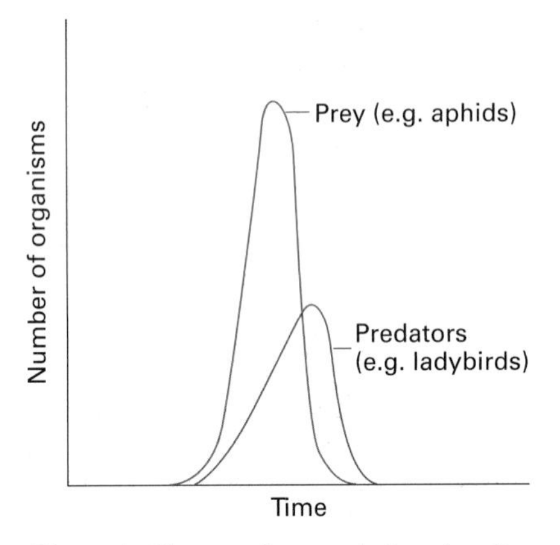

***Figure 1** Changes in population density of ladybird (predator) and aphids (prey).*

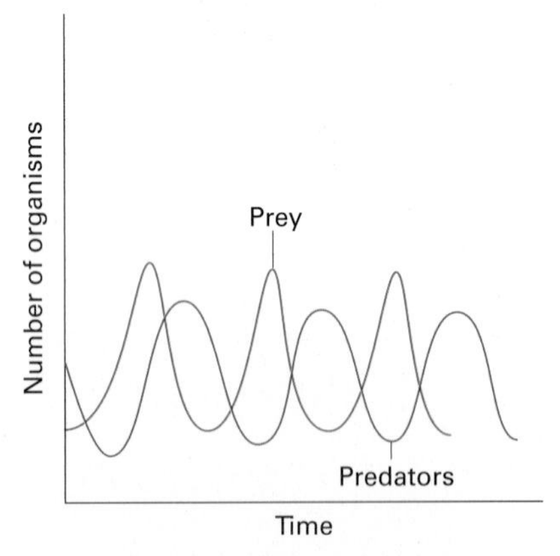

***Figure 2** Graph based on mathematical model of the relationship between predators and prey.*

PREDATION

As we saw in spread 22.8, predation refers to +– interactions between different species. Some ecologists regard parasites and herbivores as predators, but **true predators** are carnivores that gain their nourishment from killing and then eating other animals, their **prey**.

Coevolution of predators and prey

The relationship between predator and prey has been likened to an arms race. Among a prey species, individuals that have the best defensive strategies (such as being able to run fast, being camouflaged, and having deterrents such as horns or foul-tasting chemicals) will be more likely to escape capture. Among the predator species, individuals with the best hunting strategies (such as being able to run fast, having good eyesight so that they can see camouflaged prey, and having offensive weapons such as teeth and claws) will be more likely to catch prey. Through natural selection, this has led to a coevolution of predator and prey: as predators evolved more efficient hunting strategies, prey evolved more efficient defence mechanisms to combat them. During their evolutionary history, both predators and prey have therefore become larger, faster, and more intelligent, but the proportion of prey killed by predators has remained relatively constant.

Studying predator–prey relationships

A graph can be plotted of the population densities of a predator and its prey against time. Two patterns may be seen. The first (figure 1) usually applies to small predators and prey which have seasonal populations that peak in the summer. As conditions become favourable (for example, food plants become more abundant and the temperature reaches an optimal range), increases in the prey population are followed by increases in the predator population. At the end of the summer, both populations are low as they enter an overwintering stage in their life cycles. The second pattern (figure 2) shows wave-like oscillations in the population densities of both predator and prey, which tend to be out of phase with one another, the predator oscillations lagging behind those of the prey. The kind of pattern shown in figure 2 is a mathematical prediction for what happens when a population of efficient predators meets a population of its prey in a confined space. The predator population is limited by its food supply. In turn, the prey population is determined by the number of individuals killed by predators.

The Canada lynx and the snowshoe hare

One of the most studied predator–prey relationship is that of the Canada lynx (*Lynx canadensis*) and the snowshoe hare (*Lepus americanus*). Charles Elton, one of the pioneers of animal ecology, analysed the records of lynx and hare furs traded with the Hudson's Bay Company in Canada over a 200-year period. He showed that peaks and troughs in the population density of hare were followed by corresponding changes in lynx density, and that this pattern was shown for as long as records have been kept (figure 3). The graphs appear to support the mathematical model shown in figure 2, and have been interpreted as showing a stable relationship in which prey and predator regulate each other in a cyclical manner. The interpretation usually goes something like this:

> *When predators are scarce but prey are numerous, the predator population grows quickly. Inevitably, as predation increases, the size of the prey population falls. When the prey population falls, competition between predators for food is increased and predator numbers fall as some die of starvation. With fewer predators, the prey population starts to rise again, thus completing the cycle.*

Predator and prey populations are therefore regulated by a **negative feedback** mechanism that keeps the populations balanced at levels

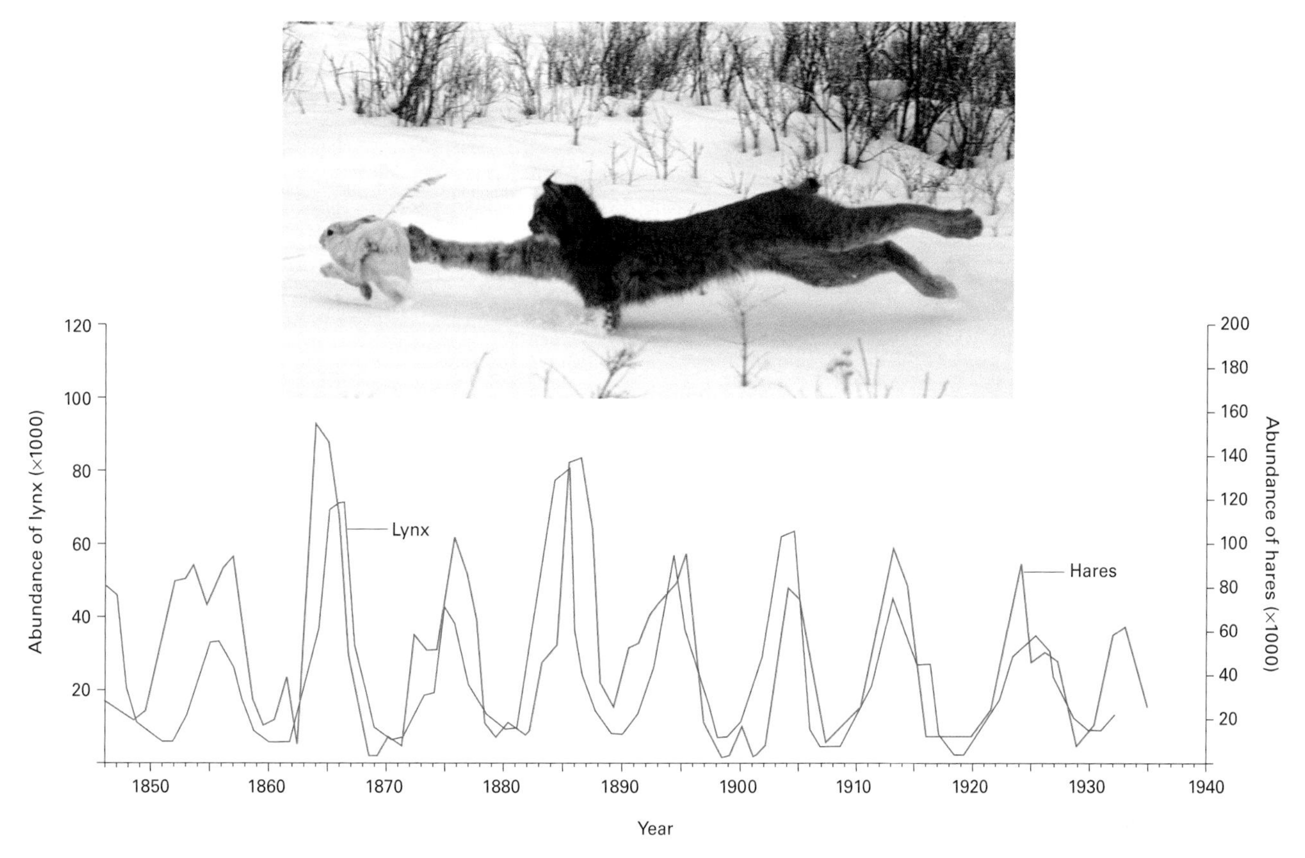

Figure 3 *Population cycles in snowshoe hare and lynx. The abundance figures were determined by the number of fur skins lodged with the Hudson's Bay Company in Canada.*

that the environment can support (figure 4). However, the description just given is an oversimplification. On islands where lynx are absent, snowshoe hare populations show similar cycles to those on the mainland. An alternative explanation for the oscillations is that high hare densities affect the grasses they eat. When heavily grazed, the grasses have a lower nutrient content, and they produce shoots with high levels of distasteful chemicals. This leads to increased hare mortality. On the mainland, therefore, although the lynx population probably oscillates in response to the hare population, the hare population is probably regulated by cyclic changes in the plants the hares eat.

It is likely that most cyclic patterns of population growth are caused by the interaction of many factors. So far, no predator–prey oscillation has been proven to be driven solely by predator and prey interactions.

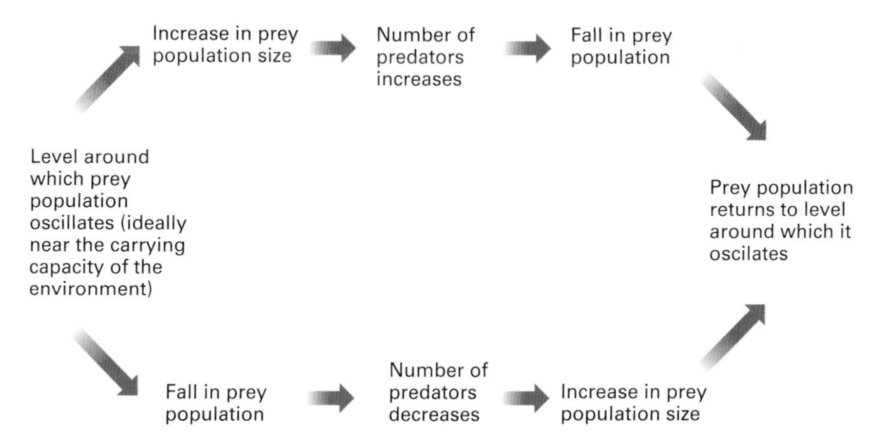

Figure 4 *Hypothetical negative feedback mechanism controlling prey and predator populations.*

Quick check

1 a In what way is the coevolution of predators and prey like an arms race?

b How did Elton estimate the population size of snowshoe hares and lynx in Canada?

c Which piece of evidence indicates that oscillations of snowshoe hare populations are not regulated entirely by the size of their predator population?

Food for thought

In a classic set of laboratory experiments, the introduction of a few *Didinium* (protoctistan predators) into test-tube cultures of *Paramecium caudatum* always resulted in the annihilation of the prey followed by death by starvation for the predator. In the real world, predators rarely wipe out their prey completely, though local devastations are common. Suggest differences between the laboratory situation and the real situations which account for the different outcomes.

22.10

Ecological succession

OBJECTIVES

By the end of this spread you should be able to:

- distinguish between primary, secondary, and deflected succession
- discuss the possible causes of succession.

Fact of life

In 1963, a new volcanic island called Surtsey emerged off Iceland's southern coast, rising to 150 m above the sea. As soon as the molten lava cooled, life began to colonise the bare rock. These **pioneer species** include lichens and mosses: hardy, drought-resistant organisms that can establish themselves and reproduce in areas where other organisms cannot live. These organisms act on the rock to produce soil (lichens, for example, secrete acids which break down rock) and when dead they add humus to the soil. On Surtsey, the soil is already deep and rich enough for plants to take root. The first plant was the sea rocket, a quick-growing annual. By 1996 there were 44 other plants flourishing on Surtsey. These plants produce large numbers of tiny seeds which are easily dispersed to colonise new areas. When fully established, annual plants will outcompete the pioneer species for nutrients and light and replace them. Some of the more recent plant arrivals are herbaceous perennials. If the climate allowed it, we would expect these to change the soil further and make it suitable for woody perennials (shrubs and trees) to colonise the area. If Surtsey were a tropical volcanic island, a woodland or forest might become established. Animal colonisation follows plant colonisation. At least five types of sea bird are now known to breed on Surtsey.

Ecosystems are dynamic. They are constantly changing, not only in response to physical factors such as climate, but also in response to biological factors resulting from the activities of organisms within the community. Communities gradually change from one type to another. The sequence of change, from the initial colonisation of a new area to establishing a relatively stable community, is called **ecological succession**. There are two main types: **primary succession** and **secondary succession**. Ecological succession happens in stages called **seres** or **seral** stages; each stage has its own community of organisms, and can usually be identified by the plants it contains.

Primary succession: colonising a new area

Primary succession happens in any newly formed area where no life previously existed, for example, volcanic islands such as Surtsey Island (figure 1), glacial deposits of rock, new ponds or lakes, and sand dunes (figure 2). The following types of plant communities usually occur on new rocky surfaces in favourable climates:

Pioneer species → annuals → herbaceous perennials → shrubs → trees

This succesion may be a very slow process, taking hundreds of years.

In ponds and lakes, organic matter builds up from the dead remains of plants and animals and sediment is brought in by water running off the land. The water becomes shallower and richer in nutrients, allowing rooted plants to crowd the shores and extend further and further into the lake or pond. In doing so, they tend to trap more sediment and make the water even shallower. Ponds and small, shallow lakes might develop into marshes, or perhaps tree-dominated swamps. Given the right climate and ecological conditions, pond and lake succession would ultimately lead to woodland, like succession on bare rock. Many ecologists, however, believe that this rarely, if ever, happens.

Secondary succession

Secondary succession takes place in areas where life is already present but has been altered in some way, for example by a natural disaster or human interference such as ploughing a field (figure 3). The changes are similar to those in primary succession on a rock but the process is usually faster as soil, often containing seeds, is present.

The process of succession

Succession is a complex process driven by many factors acting simultaneously. In the early stages, abiotic factors are the most important in determining which organisms colonise a habitat. Once pioneer species become established, biotic factors become increasingly important.

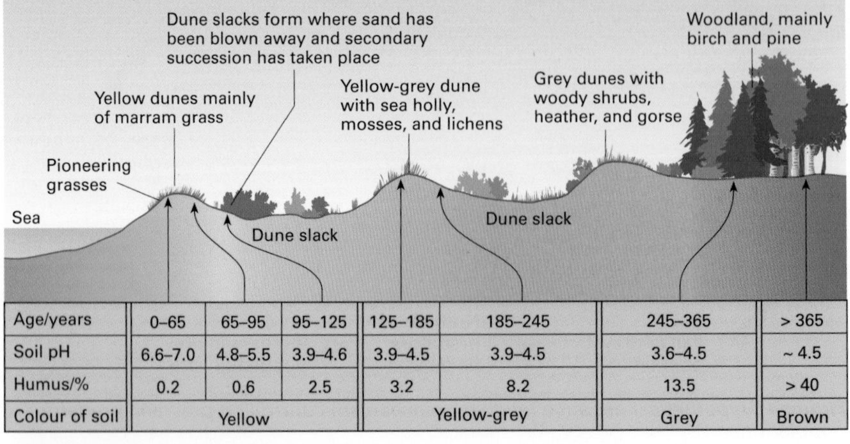

Age/years	0–65	65–95	95–125	125–185	185–245	245–365	> 365
Soil pH	6.6–7.0	4.8–5.5	3.9–4.6	3.9–4.5	3.9–4.5	3.6–4.5	~ 4.5
Humus/%	0.2	0.6	2.5	3.2	8.2	13.5	> 40
Colour of soil	Yellow			Yellow-grey		Grey	Brown

Figure 2 *Sand dune succession at Studland Bay, Dorset. The dunes furthest from the sea were formed first, and those nearest the sea are the most recent.*

Figure 1 *The formation of Surtsey Island in 1963.*

For example, early colonisers of rocky surfaces can often fix nitrogen, compensating for the poor nutrient content of the soils. When the colonisers die, they enrich the soil, making the habitat more suitable for other species.

According to the **climax theory**, ecological succession leads to a relatively stable community which is in equilibrium with its environment. This has been called the **climax community**. Some ecologists believe that there is a single type of climax community, called the **climatic climax**, for each climatic region. Oak woodland, for example, is often stated to be the climatic climax of much of lowland Britain. Grassland and heathland are **plagioclimaxes**: relatively stable plant communities arising from succession that has been deflected or arrested, either directly or indirectly, as a result of human activity. They occur in Britain mainly because controlled burning, grazing or some other factor prevents them from reaching their climatic climax. If the factor causing the plagioclimax is removed, these ecosystems may develop towards their climatic climax.

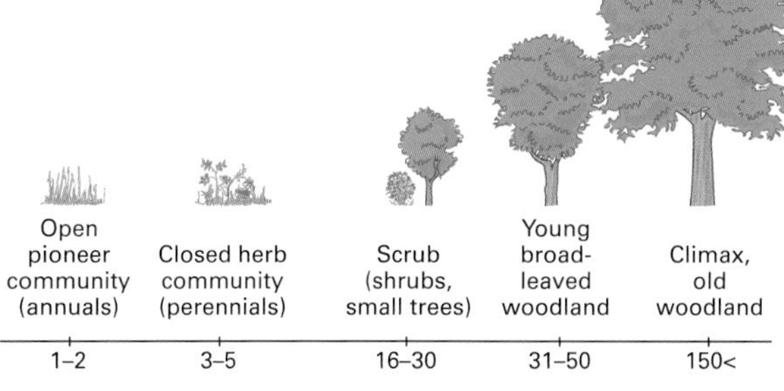

***Figure 3** Secondary succession: the succession from ploughed field to oak woodland is similar to primary succession on a rock, but the process is much more rapid because soil, often containing seeds, is already present.*

Other ecologists believe that an equilibrium between a community and climate can never be reached because, over long periods of time, the climate changes. They point out that, despite the impression often given by simplified accounts and stylised diagrams, ecological succession is not a completely predictable process; the sequence of events in one place may not be repeated exactly in another.

Pioneer species arrive haphazardly: those tolerant of the conditions survive. In surviving, they may alter the conditions in ways that favour other species more strongly than themselves. However, the pioneering stage does not always lead to further stages of succession. There appears to be a large element of chance in real successions. Alternative 'climax' communities can develop, and the precise structure of each seral stage depends on local circumstances (especially the soil condition). Sometimes stages may be omitted altogether.

In recent years many ecologists have abandoned the climax theory. The community of each ecosystem is complex and highly variable. So-called mature climax communities are dynamic and continually changing, even though the changes may appear very slow from a human point of view.

Zonation

Succession is sometimes confused with zonation. In succession, the species composition varies over a period of time. However, in **zonation**, the species composition varies with position. Organisms occur in distinct zones where there are different environmental conditions. On a large scale, plants may occur in broad zones, determined by climatic factors. There are also more local zones within a single ecosystem, for example, on a rocky shore where different species live in distinct horizontal strips, roughly parallel to the water's edge, from the low tide mark to the high tide mark (figure 4).

***Figure 4** Zonation on a rocky shore.*

Succession and determining time of death

Forensic investigators establish the time of death by looking at the extent of decomposition of the body, measuring body temperature (after death, core body temperature changes with time in a predictable manner in different environments), and assessing the degree of muscle contraction (muscles stiffen after death, a condition known as rigor mortis – see spread 11.5). The investigators include **forensic entomologists**. An ecological succession of different insects, other arthropods and their stages (larva, pupa, and adult) feed and lay eggs on a decomposing corpse. Forensic entomologists can identify these to determine whether a body has been dead for just a day or up to 3 or 4 weeks.

Quick check

1 Why is secondary succession usually much faster than primary succession?

2 How might pioneer species make the next stage of succession possible?

Food for thought

Pioneer species are usually opportunistic species. Suggest the effect this has on the precise composition of pioneer species in ecological successions.

22.11

Species diversity

OBJECTIVES

By the end of this spread you should be able to:

- explain how species diversity can be measured
- discuss the relationship between species diversity and ecosystem stability.

Fact of life

Since 1970 time-series data from more than 9000 populations of more than 2600 vertebrate species have been used to calculate an indicator of global biodiversity called the **living planet index (LPI)**. As 1970 was the baseline year, it was given an LPI of 1.0. Between 1970 and 2008 the index dropped by 28 per cent to 0.72, making this the period of greatest destruction of wildlife since the extinction of dinosaurs about 65 million years ago.

The richness of the natural world, generally referred to as **biodiversity**, is in dramatic decline. Biodiversity is an umbrella term that includes different types of biological diversities from local to global, from genetic and biochemical (see table 1) to whole organisms, and from local habitats to biomes.

Table 1 *Some secondary compounds of plants. A secondary compound is a metabolite which has no direct involvement in growth or reproduction. A decrease in biodiversity leads to reduced biochemical diversity, so the range of secondary compounds of plants, many of which have important human uses, is becoming smaller.*

Secondary compound	Source	Use
Alkaloids, e.g. quinine, morphine	Bark of the *Cinchona* tree Poppies (Papaveraceae)	Anti-malarial drug Potent painkiller
Oils, e.g. palm oil	Palm trees (e.g. *Cocos*)	Used as a biofuel
Pigments, e.g. carotene	Many vegetables and fruits	Food colouring
Polyisoprenes, e.g. rubber	Rubber trees, mainly *Hevea* spp.	Manufacture of tyres and other products

Measurement of biodiversity

Species richness refers to the number of species present in a habitat. On its own, this is not usually a very clear indicator of the richnesss of a habitat. For example, a habitat containing six species each with five individuals has a greater species richness than a habitat containing five species each with 100 individuals, but common sense would tell us that the latter is a 'richer' habitat.

A **species evenness** index refers to the relative abundance with which each species is represented in an area. For example, a rock pool community containing 3 fish, 4 sea anemones, and 4 brittlestars has a higher species evenness than a rock pool with 1 fish, 9 sea anemones, and 20 brittlestars.

The most commonly used measure of biodiversity is a **species diversity index**, such as the **Simpson species diversity index** (see box). This expresses species diversity as a single number, calculated using the number of species present (species richness) and the abundance of each species. Like other formulae for calculating species diversity, the Simpson species diversity index depends on **random sampling** (see appendix, pages 571–2) to obtain an accurate estimate of the numbers used in the equation.

The index can be related to the abiotic harshness of an environment (for example, during different stages of ecological succession) or the level of pollution in an ecosystem. Generally, the species diversity is greater in habitats in which abiotic conditions are less demanding or in which pollution levels are lower.

Simpson species diversity index

The formula used to calculate the Simpson species diversity index D is:

$$D = \frac{N(N-1)}{\Sigma\, n(n-1)}$$

where D is the diversity index, N is the total number of individuals of all species found, n is the total number of individuals belonging to a particular species, and Σ is 'the sum of'.

Species	Number of individuals
A	10
B	6
C	3
D	4
E	1
Total individuals = 24	

In the table each species is represented by a letter; it is not essential to be able to name the species in order to calculate the diversity index.

In this example, the species diversity index D is 4.0, calculated as follows:

$$D = \frac{24 \times 23}{(10{\times}9) + (6{\times}5) + (3{\times}2) + (4{\times}3) + (1{\times}0)} = \frac{552}{138} = 4.0$$

The Ecotron

The Ecotron, in Berkshire, England, consists of 16 small communities of microorganisms, plants, and animals housed in separate walk-in environmental chambers (figure 1). The climatic conditions in each chamber are computer controlled. The Ecotron has been used to make model communities that have a food-web structure typical of a temperate weedy field, albeit slightly simplified. During 1993, scientists studied the effect of species diversity on the functioning of the whole ecosystem. They found that reduced biodiversity leads to reduced community productivity and other changes. This was one of the first experimental demonstrations that declining species diversity can impair ecosystem performance.

Since March 2005 the Ecotron has operated as one of the Natural Environment Research Council's services and facilities. As such its use is open to the whole of the UK's scientific community.

Figure 1 *One walk-in chamber in the Ecotron.*

One weakness of the Simpson index is that it makes no allowance for differences in size of individuals; one elephant is treated as equal to one snail or even one bacterium. Also, it may be difficult to calculate the index for plants because it is not always easy to decide what constitutes an individual plant. A grove of birch trees, for example, may be genetically a single plant that pushes up stems from a single root system. Some species diversity indices are based on biomass rather than identifying individuals.

Species diversity and stability

The Simpson species diversity index is sometimes used as an indicator of ecological stability; the assumption being that the greater the diversity, the greater the stability. However, this use of species diversity is not very reliable because stability has several meanings:

- **resistance** – the ability of an ecosystem to resist a change following a disturbance
- **resilience** – the ability of the ecosystem to return to its original state after being changed
- **local stability** – the tendency of a community to return to its original state after a small disturbance
- **global stability** – the tendency of a community to return to its original state after a large disturbance.

Each of these types of stability is sometimes, but not always, related to species diversity.

Until recently, the accepted view in ecology was that a more complex community leads to a more stable ecosystem. Complex ecosystems with a high species diversity were assumed to be more stable because they have a large number of alternative links between different species. A change in the population density of one species would therefore be less likely to affect the density of others. Several species might be able to carry out the same function (such as pollination, decomposition, or photosynthesis) and if something happened to one species, another could take over.

Several experiments in both simulated environments (such as the Ecotron, figure 2) and natural environments support the idea that low species diversity leads to instability. However, there is also evidence that low diversity need not result in low stability. For example, natural monocultures (such as some salt marshes and mangrove swamps) with low species diversity appear to be stable. The relationship between species diversity and ecosystem stability probably varies with the environment. Robust and simple communities in environments with demanding abiotic conditions (for example, a mud-flat community in a temperate estuary) are likely to be more stable and less prone to human disturbance than fragile, complex communities in environments which are abiotically less demanding (for example, a coral reef community in the tropics). On the other hand, a tropical rainforest with a high species diversity and many interactions between species may well be more stable than another tropical rainforest with a low species diversity and few interactions.

Quick check

1 Given that there are five species in a one-metre square of land with 6, 10, 3, 4, and 5 individuals respectively, calculate the species diversity using the Simpson index.

2 Which type of ecosystem is likely to to be more stable:

  a an estuarine mudflat with low species diversity or a coral reef with a high species diversity

  b a woodland with a high species diversity or a woodland with a low species diversity?

Food for thought

Suggest why laboratory communities may be less stable than a natural community with the same level of species diversity.

22.12

OBJECTIVES

By the end of this spread you should be able to:

- describe the different components of the nitrogen cycle.

Fact of life

Human activities such as agriculture, transport, and manufacturing industries play a significant part in the cycling of nutrients in the biosphere. Almost as much nitrogen is fixed in the manufacture of nitrogen fertiliser (more than 50×10^6 tonnes per year) as is fixed biologically (about 54×10^6 tonnes per year). Human activities contribute about two-thirds of the phosphates carried in rivers to the sea.

Nutrient cycles (1): nitrogen and phosphorus

When an organism uses energy to carry out its activities, some of the energy is transferred to heat in the environment and is not available to be used again by another organism. Hence energy flows in one direction through an ecosystem. On the other hand, when an organism takes a nutrient into its body, it does so only temporarily. Eventually, the nutrient will be released into the external environment and can be used again. Nutrients are recycled in an ecosystem.

Many elements are involved in biogeochemical cycles. These are cycles in which elements pass through living organisms, as well as through soil, air, water, and rock. At any one time, an element may be in one of four main compartments:

- in living tissue or in the products of a living organism (such as faeces or shells); an element may pass from one organism to another through the food chain
- in solution (in aquatic environments or in soil water), in a form in which it can be taken up easily by plants and microorganisms
- in the atmosphere, as a component of the gases in the air
- in soils and rocks, in a form temporarily unavailable to organisms.

Biogeochemical cycles involving elements essential to life are called nutrient cycles. They include the nitrogen and phosphorus cycles, dealt with in this spread, and the carbon cycle, covered in the following spread.

The nitrogen cycle

Nitrogen is a vital component of every living organism. It forms part of all proteins, nucleic acids, and their products. Figure 1 summarises the nitrogen cycle. It consists of four main processes: nitrogen fixation, ammonification, nitrification, and denitrification.

Nitrogen fixation: nitrogen to nitrogen compounds

Nitrogen occurs in huge quantities in the air (the gas makes up 79 per cent of the atmosphere). Most organisms cannot use atmospheric nitrogen, or nitrogen dissolved in water, because nitrogen gas is chemically unreactive (each molecule consists of two nitrogen atoms linked by a triple bond). Before plants and animals can use nitrogen,

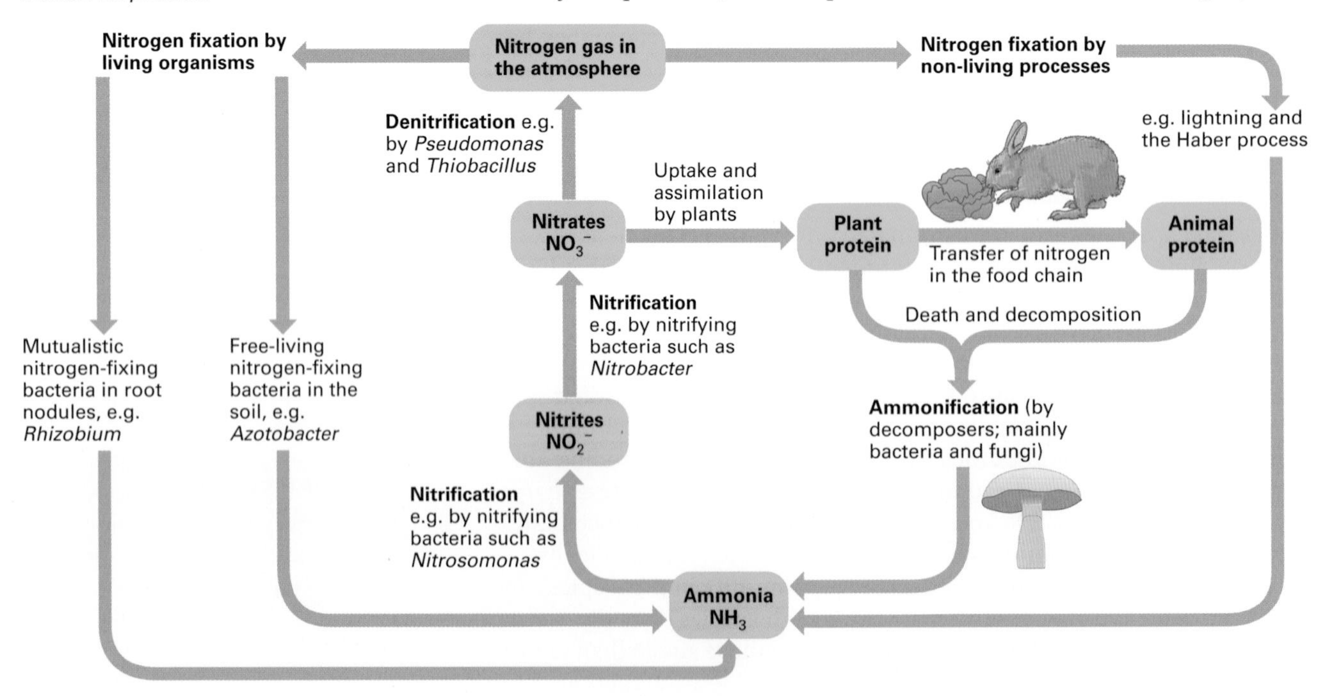

Figure 1 The nitrogen cycle. Note that without nitrogen fixation, most of the nitrogen in the atmosphere would not be available to living organisms. Most nitrogen fixation occurs in nitrogen-fixing bacteria which convert nitrogen to ammonia. The ammonia is then converted to nitrites and nitrates, either within root nodules of leguminous plants or by nitrifying bacteria in the soil. The plants use the nitrogen from nitrates to synthesise amino acids from other organic acids produced, for example, during the Krebs cycle of aerobic respiration.

it must first be converted to absorbable nitrogen compounds. This conversion is called **nitrogen fixation**.

In nature, a little nitrogen fixation occurs during thunderstorms. Lightning provides the energy to oxidise nitrogen to nitrogen oxides. These gases dissolve in rain droplets to form dilute nitric acid. However, more than 95 per cent of nitrogen fixation takes place not by lightning but biologically. The fixation is carried out by **nitrogen-fixing bacteria** such as *Azotobacter* and *Clostridium* which live free in the soil; *Rhizobium*, which lives mutualistically in root nodules on leguminous plants such as clover, beans, gorse, and sweet peas belonging to the family Papilonatae (figure 2); and by some aquatic blue-green bacteria such as *Anabaena* (see spread 17.2, figure 3). Nitrogen-fixing bacteria possess **nitrogenase**, an enzyme that enables them to reduce nitrogen to ammonia or ammonium compounds:

$$\text{Nitrogen} + \text{hydrogen} \xrightarrow{\text{nitrogenase}} \text{ammonia} \qquad N_2 + 3H_2 \longrightarrow 2NH_3$$

Nitrogen fixation is also carried out artificially by industry in the production of nitrogen fertilisers. One major industrial process, the Haber process, synthesises ammonia from atmospheric nitrogen and hydrogen by passing the gases at high temperature and pressure through an inorganic catalyst. The process requires large amounts of energy, usually from burning fossil fuels.

Figure 2 *Root nodules on a leguminous plant. The nodules contain nitrogen-fixing bacteria such as* Rhizobium. *The relationship between bacterium and plant is an example of mutualism: the plant obtains a source of nitrogen while the bacterium obtains a home and a constant supply of sugars from the plant. Leguminous plants also produce a red pigment called leghaemoglobin in their root nodules. This combines with oxygen to produce an oxygen-free environment around* Rhizobium. *Nitrogenase does not function in the presence of oxygen.*

Ammonification: ammonia from organic compounds

Ammonia (usually as ammonium ions) is also produced in ecosystems from organic nitrogen-containing compounds. These compounds occur in faeces, nitrogenous excretory products such as urea, and the dead bodies of organisms. The process by which these compounds are converted to ammonia is called **ammonification**. It is carried out mainly by bacteria and fungi which act as decomposers in ecosystems, breaking down the bodies of dead organisms.

Nitrification: ammonium ions to nitrites and nitrates

Ammonium ions in the soil or water are oxidised to nitrites and nitrates by **nitrifying bacteria**. For example, *Nitrosomonas* oxidises ammonia to nitrite (NO_2^-) while *Nitrobacter* oxidises nitrite to nitrate (NO_3^-). These bacteria are chemoautotrophs, obtaining energy from the redox reactions involved in nitrification. Nitrification requires oxygen, so it happens most rapidly in well aerated soils or well oxygenated bodies of water. The nitrate ions produced by nitrification can be taken up by plants and used to make proteins. Consumers obtain their nitrogen in the form of proteins when they eat plants or other animals.

Denitrification: nitrates back to nitrogen

The nitrogen cycle is completed by denitrifying bacteria such as *Pseudomonas denitrificans* and *Thiobacillus denitrificans*. These bacteria live in conditions of low oxygen and reverse the nitrifying process, converting nitrates to nitrites, and nitrites to nitrogen gas. This **denitrification** leads to the loss of nitrogen from the biotic component of an ecosystem to the atmosphere.

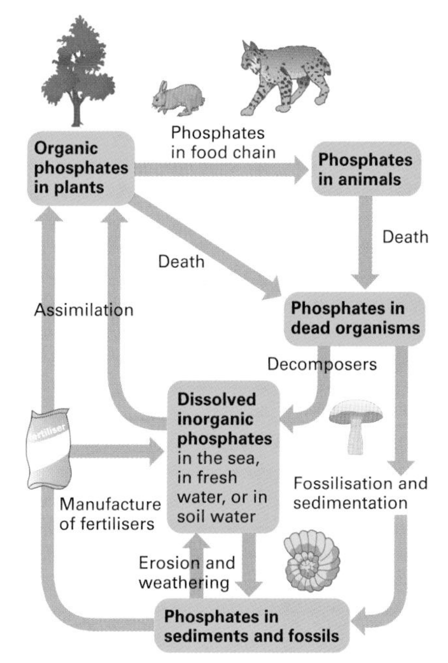

Figure 3 *The phosphorus cycle. Phosphorus is essential to life: it is part of adenosine triphosphate (ATP) and nucleic acids. Whereas nitrogen in the atmosphere is readily accessible to the biotic component of an ecosystem, most phosphorus is locked up in sediment and is made available to organisms mainly by slow physical processes.*

Quick check

1 By what process do:

a decomposers convert organic matter into ammonia

b bacteria convert gaseous nitrogen into ammonia

c *Nitrosomonas* convert ammonia into nitrites

d *Pseudomonas* convert nitrates into gaseous nitrogen?

Food for thought

Nitrogenase, the enzyme used for nitrogen fixation, will not function in the presence of oxygen even in aerobic bacteria or photosynthetic bacteria. *Anabaena* has nitrogenase and carries out nitrogen fixation in special cells called heterocysts. They have very thick walls and contain special photosynthetic membranes which generate ATP by cyclic photophosphorylation. Suggest why these cells do not carry out non-cyclic photophosphorylation and therefore do not photolyse water.

22.13

Nutrient cycles (2): carbon

OBJECTIVES

By the end of this spread you should be able to:

- describe the main components of the carbon cycle
- discuss the effect of human activities on the carbon cycle.e

Fact of life

Carbon dioxide levels in the atmosphere have increased steadily from about 280 parts per million (p.p.m.) in 1750 to over 380 p.p.m. in 2012 (figure 3), the highest levels for 650 000 years. The recent surge in atmospheric carbon dioxide levels has been attributed to three processes: industrial growth in emergent economies; heavy use of coal in China; and damage to forests, seas, and soils that act as 'sinks' absorbing carbon dioxide. Carbon dioxide concentration has been measured in air bubbles trapped in Greenland ice cores, indicating that carbon dioxide levels there have fluctuated from about 180 p.p.m. to 300 p.p.m. over the last 250 000 years. Figure 1 plots these long-term changes in atmospheric carbon dioxide on the same graph as temperature changes. There is a very strong correlation between the two variables, but no one can be sure if a high carbon dioxide level results from a warmer atmosphere or is the cause of it.

Carbon is central to life on Earth. It forms a part of all major biological molecules (carbohydrates, lipids, proteins, and nucleic acids). Figure 2 shows the **carbon cycle**, the pathway by which carbon atoms are passed from one organism to another, and between organisms and their environment.

Carbon in the atmosphere and the oceans

Although carbon dioxide gas makes up only a small proportion of the atmosphere (less than 0.04 per cent), it is the main source of carbon for organisms living on land. For aquatic organisms, the main source is hydrogencarbonate ions (HCO_3^-) formed from dissolved carbon dioxide (and carbonate rock). The store of carbon in the oceans is 50 times that in the atmosphere. Links in the carbon cycle between the atmosphere and the oceans ensure that, in the long term, atmospheric carbon dioxide levels stay relatively constant: as carbon dioxide concentration increases in the atmosphere, more becomes dissolved in water; this carbon dioxide is then converted to hydrogencarbonate ions. The reverse happens when atmospheric carbon dioxide concentrations are low.

Fixing carbon dioxide

The carbon in the atmosphere and oceans is made available to organisms by photosynthesis. Green plants and other photoautotrophs fix carbon dioxide and convert it to carbohydrates and other organic molecules. The carbon is then passed along the food chain as an animal eats a plant, or as one animal eats another. Carbon is returned to the atmosphere or oceans mainly as carbon dioxide, by the respiration carried out by all organisms. Unlike the nitrogen cycle, the assimilation of carbon into autotrophs does not depend on the action of decomposers. Nevertheless, decomposers play an important part in the cycle by releasing carbon when they break down the bodies of dead organisms. If the bodies of dead organisms are not broken down, they will eventually be incorporated in sediments. Shells of dead protoctists and molluscs, and vertebrate bones, may add to the sediments. Coal, oil, chalk, or limestone may eventually be formed and the return of carbon to the atmosphere and surface waters may be delayed by millions of years.

Releasing carbon dioxide into the atmosphere

Carbon in the lithosphere (the Earth's crust, including the mineral fraction of soil) may be made available to organisms by volcanic activity, or when sedimentary rocks are broken down by weathering. However, in the short term, these processes usually play only a minor role in the carbon cycle. In recent years, human activities have had a much more dramatic impact in releasing carbon trapped in sediments. The burning of fossil fuels produces more than 20 billion tonnes of carbon dioxide each year. A significant but smaller amount of carbon dioxide is also released into the atmosphere during deforestation and the production of cement from limestone. Some of the extra carbon dioxide is known to dissolve in surface waters, some is assimilated into extra biomass, some is building up in the atmosphere causing an increase in atmospheric carbon dioxide levels, but a significant proportion is unaccounted for. It may be assimilated into boreal forests (cold temperate forests in the northern hemisphere), but no one really knows.

The most spectacular evidence for rising carbon dioxide levels comes from measurements of air samples taken at an observatory at Mauna Loa in Hawaii (figure 3). The observatory is in an ideal position as a sampling point for atmospheric gases because it is in the middle of the Pacific Ocean, remote from the local effects of continental industries. The upward trend of the graph is direct evidence that the carbon dioxide concentration in the air is rising. This rise is thought to be due to the

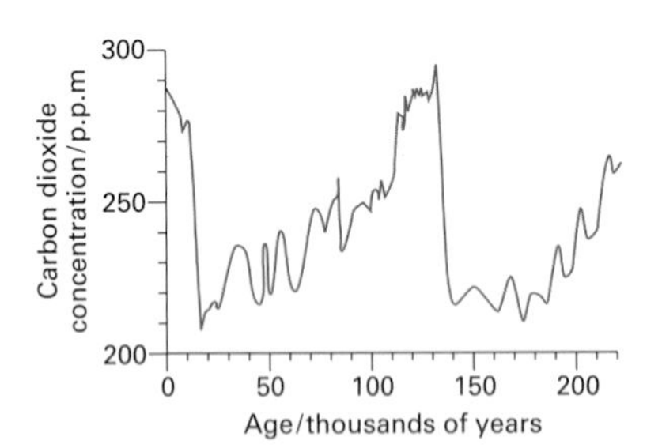

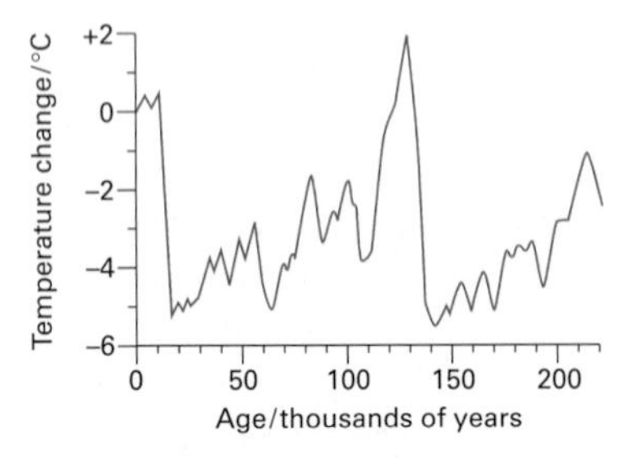

Figure 1 Long-term changes in atmospheric carbon dioxide concentration and temperature.

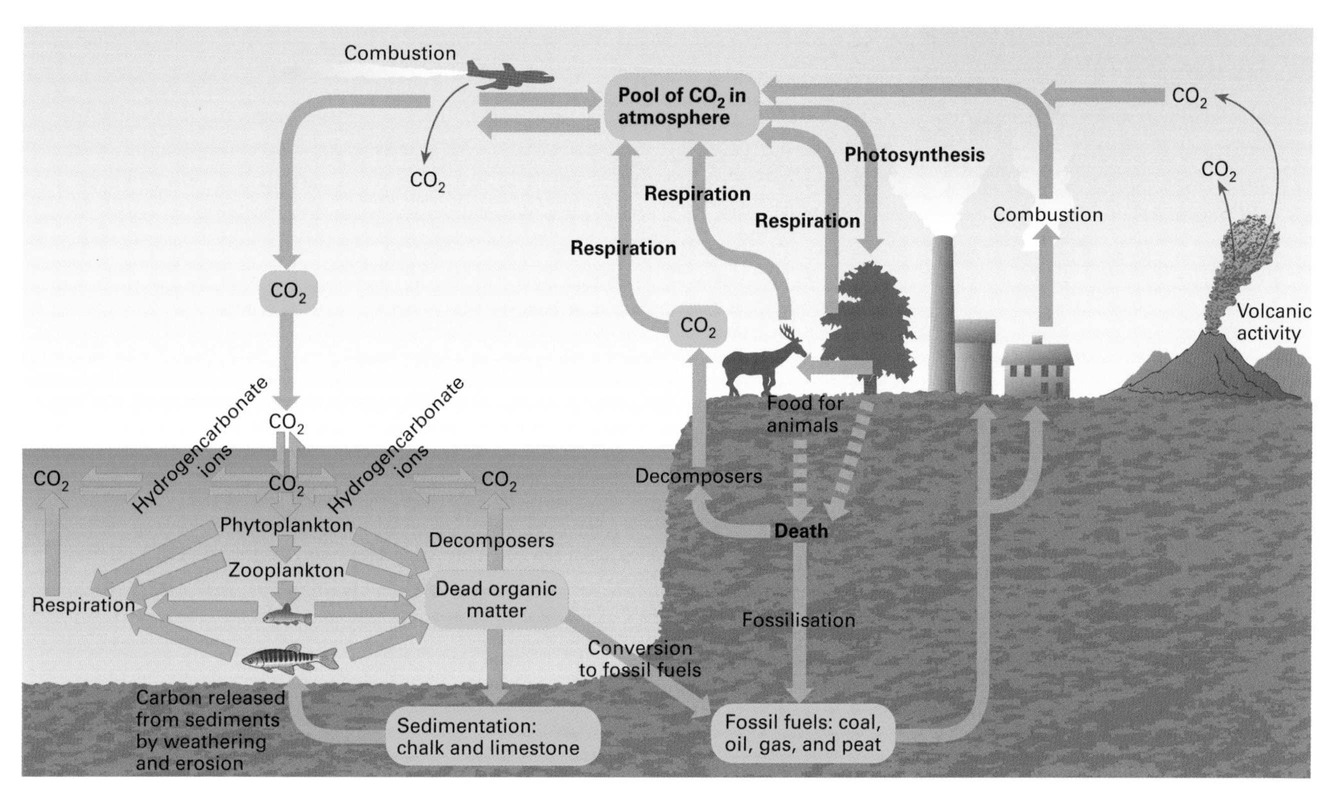

Figure 2 *The carbon cycle.*

burning of fossil fuels and deforestation. The peaks and troughs show that carbon dioxide is pumped in and out of the atmosphere with a yearly rhythm. The variations have been explained in terms of changes in the relative rates of photosynthesis and respiration of terrestrial ecosystems in the northern hemisphere (these cover a much larger surface area than those in the southern hemisphere). During the northern summer, the plants take in more carbon dioxide than they release, causing the levels to fall; during the northern winter, the reverse happens.

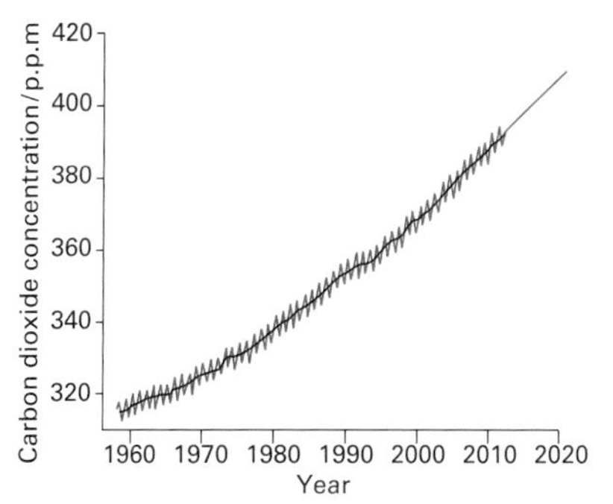

Figure 3 *Monthly mean atmospheric carbon dioxide (red curve) at Mauna Loa Observatory Hawaii based on data collected from 1958 to 2012 by the Scripps Institute of Oceanography and the National Oceanic and Atmospheric Administration of the USA. The black curve represents the seasonally corrected data. The green curve is an extrapolation of the graph.*

The greenhouse effect

Carbon dioxide acts as a **greenhouse gas**, preventing heat from escaping from the atmosphere. This effect, called the **greenhouse effect**, is a natural phenomenon that has been essential for the evolution of life on Earth; it allows surface temperatures to be much higher than they would otherwise be. Many scientists believe that an increase in the atmospheric concentration of carbon dioxide and other greenhouse gases is linked to global warming. It is predicted that if present carbon dioxide levels double, average global temperatures could increase by 1.5–4.5 °C, with a best estimate of 2.5 °C. The link between greenhouse gases and global warming is discussed further in spread 23.6.

Predicting future trends

Future trends in global carbon dioxide levels can be predicted by the **extrapolation** of graphs such as figure 3. Extrapolation means obtaining a value from a chart or graph that extends beyond the given data. The extrapolation in figure 3 is a linear extrapolation created by extending a line beyond and tangential to the black curve. Extrapolations can be useful in predicting future trends, but they are subject to great uncertainty. Trends predicted by extrapolations assume circumstances will be the same in the future as they are now; this may or may not be the case.

Quick check

1 By what process does carbon in the atmosphere become available to organisms in a community?

2 State three ways human activities add carbon to the atmosphere.

Food for thought

Emiliania huxleyi is a marine alga with a global distribution. It belongs to a group called coccolithophores which produce intricate calcareous plates, called coccoliths, around their cells. Suggest how increased levels of carbon dioxide in sea water may affect *E. huxleyi*.

22.14

Adaptations to environments (1)

OBJECTIVES

By the end of this spread you should be able to:

- describe the adaptations of camels to hot deserts
- describe the adaptations of desert rats to hot deserts.

Fact of life

Stretching from the Mediterranean coast to the scrublands of the northern Sudan and Mali, the Sahara is the greatest desert on Earth. No river crosses this vast expanse of sand, wind-polished gravel, and boulders; rain may not fall on parts of it for years on end. Shade temperatures can reach 58 °C. The absence of clouds means that while daytime temperatures soar, the nights may be freezing cold. Sand-laden winds blast rocks into grotesque shapes and make living in this inhospitable environment even more difficult. Nevertheless, in some parts of the Sahara and other deserts (figure 1), torrential storms happen regularly enough for rich communities of flowering plants to develop. They concentrate their lives into a brief period of frenetic activity when water is available. During this period, they grow and reproduce. The record for the shortest breeding season is held by *Boerhaavia repens*, which is said to take 8–10 days between sprouting and seed production. Most short-lived desert plants are annuals, surviving the intervening dry periods hidden and dormant in the desert sands as seeds. Many arthropods have the same strategy as plants, living through unfavourable conditions in a relatively inactive state called diapause.

Figure 1 The Arizona desert in bloom after a heavy fall of rain.

Every species is uniquely adapted to its own environment; if it were not, it would soon become extinct. One of the jobs of an ecologist is to find out the precise relationship between an organism and its physical environment. In this spread, we are going to examine how two different animals, the camel and the desert rat, cope with living in tropical deserts, one of the most adverse environments on Earth. In the next spread, we shall look at the way some plants are adapted to different conditions of water availability.

Camels

There are two species of camel: the Bactrian camel (*Camelus bactrianus*) of Central Asia, with two humps; and the Arabian camel (*C. dromedarius*), with one. Camels are widespread in the deserts of the world. Their ability to go for long periods without water and to survive on prickly desert vegetation has made them invaluable to desert-living people.

The dromedary (figure 2), a breed of Arabian camel, has a number of adaptations which enable it to survive the extremely harsh conditions

Heavy insulation of fur on the back

Sweating reduced to a minimum

Kidney has long loops of Henlé able to produce a concentrated urine

Legs long and not fatty

Faeces are so dry that they can be burned immediately

Tend to face the sun when at rest during the heat of the day

Specially long eyelashes protect eyes from sand

Walks on two toes protected from heat by pads of tissue

Undersurface bare of fur to allow heat to be lost readily by radiation and convection

Counter-current exchange mechanisim in nasal passages cools inhaled air and dries exhaled air

Tolerates high levels of dehydration of body tissues

Able to drink water rapidly to replenish losses but does not store water in the body

Tolerates wide range of body core temperature so that heat can be stored during the day and lost at night when the environment is cooler

Figure 2 Adaptations of camels to desert life.

Adaptations of the dromedary

Dromedaries can survive a loss of up to 25% of their ideal water content (a dehydration of about 10–12% is fatal for humans). A camel survives dehydration partly by maintaining the volume of its blood (figure 3). It can go without drinking water for up to eight days in desert conditions that would kill a human in one day. When water does become available, dromedaries may drink their loss of 25% body mass in about 10 minutes. (There is no evidence that they overdrink and can store water for future needs.) Dromedaries are exceptional in that they can tolerate wide fluctuations of body core temperature. In a well watered camel, body core temperature usually varies little, but in a dehydrated camel it may become as high as 41 °C during the day and fall as low as 34 °C at night. Only when the temperature exceeds 41 °C does the camel sweat freely.

Contrary to popular belief, the hump contains fat, not water. Although water is produced when the fat is metabolised, its main function is to insulate the underlying tissue from the heat of the sun, and to act as an energy store. This concentration of fat in one part of the body allows the legs to be skinny so that heat generated during locomotion can easily be lost to the surroundings. The thick fur on a camel's back also acts as insulation. Together with the hump, the fur reduces heat gain from the environment, reducing the need to lose precious water in cooling the body by evaporation (a sheared camel loses 50% more water than an unsheared camel).

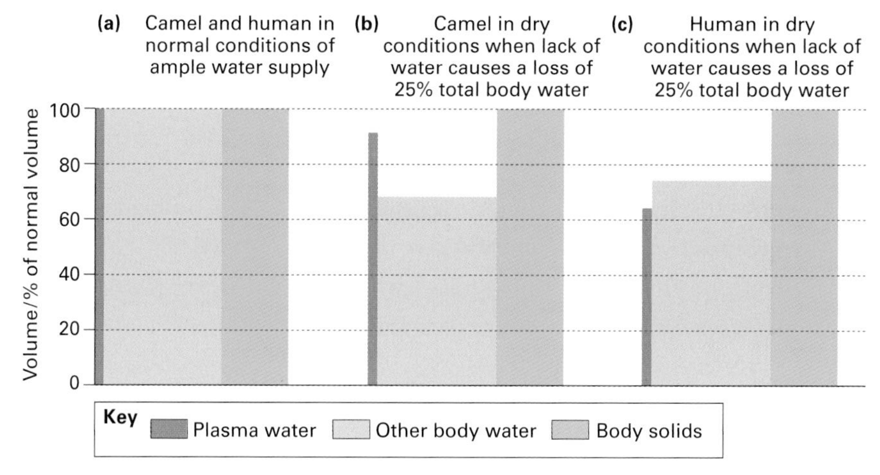

Figure 3 Effect of dehydration on the fluid compartments of camels and humans. (a) Under normal conditions, plasma water accounts for about 8% of the total body water in both human and camel. (b) When a camel has lost about a quarter of its body water, the blood volume drops by less than 10%. (c) When a human has lost a quarter of its body water, the blood volume drops by more than 30%. Under conditions of dehydration, human blood therefore becomes more viscous, circulates more slowly, and is less effective at carrying heat outwards to the skin than is the blood of the camel.

of hot deserts. Like all animals living in such conditions, it needs to be able survive long periods without water in an environment which, in the summer, is typically very hot during the day but cold at night.

Desert rats

Several species of rodent, commonly called desert rats or kangaroo rats, live in hot, very dry deserts such as the one in Death Valley, USA. These deserts often have no succulent vegetation and little if any water to drink, not even dew. Desert rats have adopted a different survival strategy to that of the camel; instead of tolerating the harsh conditions, they avoid them.

Unlike camels, desert rats cannot tolerate large fluctuations of body temperature. They cannot cope well with dehydration, nor can they drink vast volumes of water. They survive in hot deserts by their extraordinary capacity to conserve water and by their habit of remaining underground during the day.

Adaptations of desert rats

A desert rat lives in a burrow in the ground, coming out to feed only at night. It does not store water in its body, but obtains all the water it needs from the seeds it eats. Whereas many desert animals (including camels) supplement their water supplies by eating juicy plants, the desert rat never eats succulent vegetation and rarely if ever drinks water. Seeds have only a small water content but are rich in carbohydrate, which produces water when broken down by aerobic respiration. This water, called oxidation water or metabolic water, amounts to a very small volume (in one study, between 54 and 67 cm^3 during a five-week period). The desert rat can survive on this meagre water supply by keeping water losses to an absolute minimum.

Desert rats have kidneys with long loops of Henlé, giving them an efficient countercurrent exchange mechanism (see spread 8.7). They can produce urine so concentrated that it solidifies almost as soon as it is excreted: the concentration of urea in the urine can be as high as 24% whereas in humans the maximum is about 6%. Under experimental conditions, its extraordinarily efficient kidney enables a desert rat to drink sea water without problems.

Very little water escapes from the skin of a desert rat. Like all rodents, it has no sweat glands except on the toe pads, and the desert rat has fewer sweat glands there than most other rodents. It does, however, lose substantial amounts of water by evaporation during breathing. These losses are reduced by spending the day in the burrow where the air is always a little more humid than outside. A significant proportion of exhaled water is absorbed into the dry seeds stored in the burrow. Thus the rat can regain some of the exhaled water when it eats the seeds.

During the day, the animal stays in a cool burrow which is more humid than the outside environment

Respiratory moisture from the rat is absorbed by stored dry seeds which the rat eats

Food (mainly seeds) rich in carbohydrates which produce water when broken down in respiration

Few sweat glands confined to pads on toes

Faeces are very dry

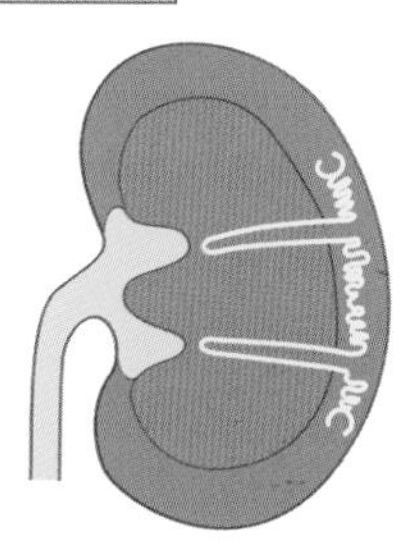

The kidney has long loops of Henlé enabling the desert rat to produce a concentrated urine

Figure 4 Adaptations of desert rats to hot dry deserts.

Quick check

1 What are the functions of the camel's hump?

2 a What is the main source of water for a desert rat?

b Why are the loops of Henlé of a desert rat about twice as long as those of a beaver?

c How does a desert rat regain the water it exhales?

Food for thought

Under normal conditions, desert rats never drink. Suggest why they drink when given a high protein diet.

22.15

Adaptations to environments (2)

OBJECTIVES

By the end of this spread you should be able to:

- describe the adaptations of xerophytes, hydrophytes, and halophytes to their environments.

Fact of life

The prickly pear cactus, *Opuntia stricta*, a native of Arizona, has been introduced into desert regions throughout the world. By crowding out native species, it has become a serious pest. It was imported into Australian gardens in 1839, and spread into the surrounding countryside. By 1925 it occupied 240 000 km^2 of what was previously prime rangeland on the eastern seaboard. In Australia, the cactus has been controlled successfully by the importation of a moth, *Cactoblastis cactorum*, from Argentina. The caterpillar of the moth eats the cactus.

***Figure 1** Prickly pear cactus (right foreground) and the saguaro cactus (tall pipe-like cactus) growing in the desert of Arizona.*

Adaptations of cereals

Rice, sorghum, and maize form the staple diet of many people in the world. Each cereal is well adapted to survive in its particular environment. Rice is a hydrophyte living in swamps. Its **aerenchyma** (tissue containing large intercellular air spaces) make it buoyant when submerged, and its tolerance to ethanol produced by anaerobic respiration (spread 6.2) enables it to tolerate being submerged in poorly oxygenated water. Sorghum has xerophytic adaptations similar to marram grass (thick cuticle, sunken stomata, extensive root system) enabling it to grow in hot, dry conditions. Maize is a tropical plant and, like sugar cane, has C_4 metabolism enabling it to live in very hot and sunny conditions (see spread 5.6).

Some of the most striking adaptations of plants relate to the availability of water in their environment. In fact, plants are often classified into four groups according to their water supply:

- **Mesophytes** are plants living in areas where water is readily available.
- **Xerophytes** live in areas where water is in short supply.
- **Hydrophytes** are plants in freshwater or very wet environments.
- **Halophytes** occupy salt marshes, estuarine muds, or areas close to the sea, where the high concentration of salts makes obtaining water difficult.

Mesophytes

Most native flowering plants in Britain are **mesophytes**, plants adapted to soils that are neither extremely wet nor extremely dry.

In hot conditions, mesophytes may begin to overheat and suffer **temperature stress**. In well watered soils, they can respond to this by increasing their rate of transpiration which cools the plant. In dry conditions, mesophytes often suffer **water stress**, losing more water to the air than they can regain from the soil. Mesophytes respond to water stress by closing their stomata. Nevertheless, on hot dry days water stress often leads to wilting, caused by loss of cell turgidity. In the short term, wilting lowers the rate of water loss by reducing the surface area of leaves exposed to the air; it may also lower temperature stress by reducing the leaf surface exposed to the Sun. Most mesophytes recover from short periods of wilting, but prolonged water shortages can lead to permanent wilting and death.

Xerophytes

All plants living in hot, dry deserts are xerophytes. They have a number of features, called **xeromorphic features**, which help reduce transpiration and therefore conserve water. Xeromorphic features of cacti such as the saguaro and prickly pear (figure 1) include:

- reduction of leaves to fine spikes, reducing transpiration and protecting the cactus from animal grazers and water robbers, such as small mammals (the stem takes over the function of photosynthesis)
- a stem with a hard, thick epidermis covered with a waxy cuticle
- an ability to fix carbon dioxide at night so that the stomata can be closed during the day (this is due to a special type of metabolism, crassulacean acid metabolism or CAM, described in spread 5.6).

Cacti are **succulent plants**: they are able to store large amounts of water in the leaves or, in this case, the stem. They usually have a shallow but very extensive root system so that they can replace their water stores quickly when it does rain. Saguaro (*Carnegiea gigantea*), the largest cactus, is reported to absorb as much as one tonne of water in a single day. Some plants such as *Acacia* living in the less arid parts of deserts have long taproots that can obtain water from sources deep underground.

As well as conserving water, plants living in hot deserts also have to cope with high temperatures. The waxy surface of xerophyte stems are shiny and reflect sunlight. Saguaro cacti have an additional adaptation: their organ-pipe-like stems point vertically upwards, minimising the surface area exposed to the midday sun.

Deserts are not the only places where xerophytes grow. Marram grass (*Ammophila arenaria*) is a xerophyte that colonises sand dunes. It has an extensive underground network of rhizomes which help to bind sand, stabilising it and preparing it for colonisation by other plants (see spread 22.10). Marram grass leaves have several xeromorphic features (figure 2):

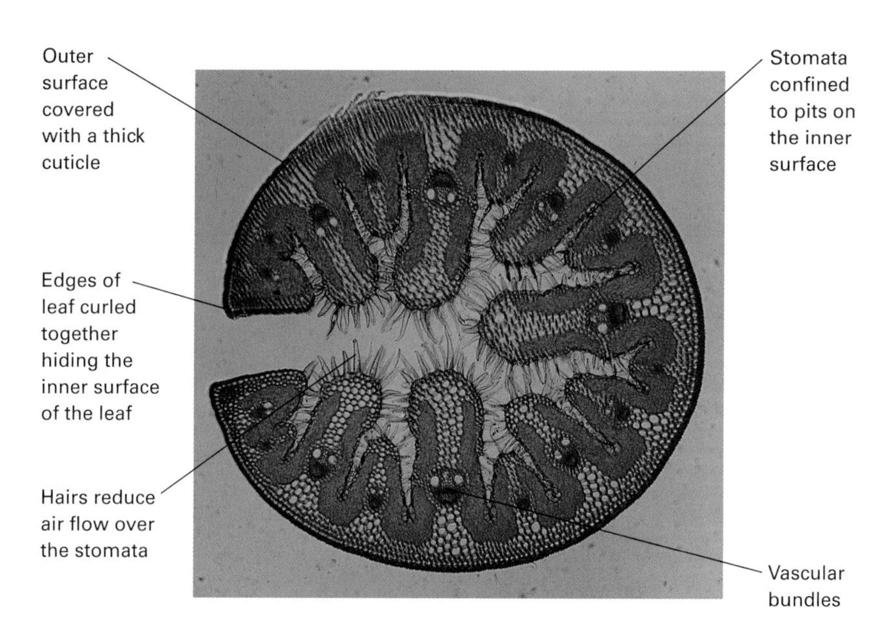

Figure 2 *Transverse section through marram grass (*Ammophila arenaria*).*

- The stomata are sunken in grooves, reduced in number, and confined to the inner surface of the leaf.
- The leaves are able to roll up into a cylindrical shape in conditions of water stress. The rolling traps moist air within the leaf, reducing transpiration.
- A layer of stiff interlocking hairs on the inner epidermis reduces transpiration by trapping air within the leaf.

Plants living in cold or mountain environments also have xeromorphic features. Each winter, soils may become frozen and water is then virtually unobtainable. Plants such as conifers and holly, common in cold climates, often have spiky leaves with a thick epidermis, a waxy cuticle, and stomata in depressions; all features which reduce transpiration losses. The small leaves of many conifers, such as Scots pine (*Pinus sylvestris*), are circular in cross-section which reduces the surface area to volume ratio for transpiration.

Hydrophytes

The bodies of hydrophytes ('water plants') are usually supported by water so they have only small amounts of mechanical and supporting tissues. They have little or no lignified tissue, the xylem is poorly developed, and the root system is either absent or reduced. Hydrophytes have specialised leaves that may be floating, as in water lilies (figure 3), or finely divided, as in Canadian pondweed (*Elodea*). Hydrophyte stems and leaves often store gases in large interconnecting air spaces; this enables them to obtain sufficient oxygen and carbon dioxide from water that is not always fully aerated. As well as storing gases, these air spaces make hydrophytes buoyant when submerged.

Halophytes

Although halophytes ('salt plants') such as the glasswort (figure 4) commonly live in waterlogged estuarine muds and salt marshes, they are faced with a water supply problem: the surrounding water is often very salty. They overcome this problem by actively absorbing salts into their roots. As a result, the roots have a lower water potential than the surrounding water, and water can flow into the plant by osmosis in the usual way. Many halophytes have also evolved xeromorphic features which help them conserve water. Xeromorphic features of the glasswort include a thick epidermis, waxy cuticle, and a reduced number of stomata.

Figure 3 *Water lilies (*Nymphaea*). These hydrophytes have floating leaves, and long leaf stalks attached to rhizomes rooted in the mud of a pond, lake, or slow-moving body of water. Stomata and a waxy cuticle are confined to the upper surfaces of the leaves.*

Figure 4 *Glasswort (*Salicornia*) growing on a salt marsh.*

Quick check

1 a Why do plants living at high altitudes need xeromorphic adaptations?

b Why do hydrophytes have very little lignified tissue?

c Glassworts are edible, salty tasting plants. Why do they accumulate salt in their tissues?

Food for thought

a Suggest why water lilies have a thick epidermis, stomata, and a waxy cuticle on the upper surface of their leaves but a thin epidermis, no stomata, and no cuticle on the lower surface of the leaves.

b Suggest why a typical mesophyte would die if it was transplanted into:
i a desert **ii** a pond
iii a salt marsh.

22.16

Adaptations to environments (3)

OBJECTIVES

By the end of this spread you should be able to:

- describe environmental conditions at high altitude
- describe the effects of high altitude on humans
- explain how humans can become acclimatised to high altitudes.

Fact of life

Llamas are South American camelids which, because they live successfully at high altitudes, are used by physiologists to study adaptations to hypoxic stress (stress due to low oxygen partial pressures). Like Sherpa Tensing (figure 1), llamas have oxygen dissociation curves shifted to the left compared with the curves of low-altitude mammals.

Figure 1 *Sherpa Tensing, who, with Sir Edmund Hillary, was the first to climb Mount Everest.*

Environmental conditions at high altitude

High-altitude environments are potentially dangerous. As altitude increases, air temperature, humidity, and air pressure decrease. With a decrease in the partial pressure of atmospheric gases, the penetration of solar radiation, especially ultraviolet light, increases. This brings with it the risk of skin burns, and snow blindness. Even worse, some wavelengths of ultraviolet light are carcinogenic, so there is a higher risk of developing malignant melanomas (cancerous skin tumours that can be highly invasive).

The temperature drops at a rate of 1°C for every 150 m increase in altitude. This fall, combined with a decrease in relative humidity and an increase in wind speed common at high altitudes, increases the **wind-chill factor**, a measure of the chilling effect of low temperatures on body temperature. A lowering of body temperature can lead to hypothermia, a potentially fatal condition (see spread 8.10). The low temperatures experienced at altitude can also cause frostbite and trench foot. **Frostbite** refers to tissue damage resulting from exposure to very low temperatures. Blood circulation stops or is greatly reduced in the tissues. This can lead to gangrene (local death and decomposition of the tissue). The extent of the frostbite depends on the temperature, the exposure time, and wind-chill factor. For temperatures above freezing, dampness is also an important factor. Prolonged walking through wet terrain in cold environments at high altitude can also cause **trench foot**. This is a nonfreezing cold injury of the foot in which tissue is damaged by cold without being frozen. Nerves, muscles, bone, and cartilage may all suffer long-term damage. The risk of trench foot is greatest among those who are in poor physical condition, dehydrated, ill, or malnourished. Preventative measures include limiting exposure to the cold, taking hot drinks, taking care to keep feet as dry as possible, and an awareness of the early signs of injury (e.g. cold, swollen, and blotchy pink-purple or blanched feet that feel heavy and numb). In addition to contributing to wind chill, the low humidity at high altitude allows water to evaporate more readily from the skin. This can lead to dehydration which can contribute to the development of mountain sickness. In addition, for each 2% loss of body water, there may be a 20% drop in the working capacity of muscles, making activity at high altitudes more difficult.

The most important factor limiting activity at high altitude is the lack of oxygen. With the decrease in barometric pressure with altitude, oxygen

Mountain sickness

High-altitude environments are defined as those higher than 3048 m above sea level. Because of their low oxygen levels, low temperatures, and high wind speeds, they can be extremely hazardous. Anyone ascending too quickly above 2100 m can develop mountain sickness, a condition characterised by shortness of breath, feelings of lassitude, muscular weakness, headaches, rapid pulse, loss of appetite, nausea, vomiting, and sometimes fainting. Those who vomit are in danger of dehydrating and losing mineral salts. This appears to stimulate an increase in the level of ADH which tends to reduce urinary losses of water and salts. The symptoms of mountain sickness become more severe the higher a person ascends and the faster the ascent is made. Individuals differ in their susceptibility, but nearly everyone suffers mountain sickness above 4900 m. The risk appears to be less if you sleep at a lower altitude than the one at which you are active, supporting the mountaineer's adage to 'play high, sleep low'.

Some people who ascend too rapidly, stay too long at high altitudes, or ascend too high, can develop **high-altitude cerebral oedema** (HACE) or **high-altitude pulmonary oedema** (HAPE), two potentially fatal conditions. HACE results from accumulation of fluid in the cranial cavity. Most cases occur above 4300 m. It results in mental confusion that can lead to loss of consciousness and death. HAPE is due to accumulation of fluid in the lungs. It occurs in those who ascend too rapidly above 2700 m. The fluid interferes with ventilation, resulting in shortness of breath and fatigue. Oxygen transport to the brain is impaired, leading to mental confusion and loss of consciousness. Both HACE and HAPE are treated by administering supplemental oxygen and returning the patient to a lower altitude.

partial pressures become lower; this reduces the amount of oxygen that can be taken up by haemoglobin in the blood (see oxygen dissociation curve, figure 3, spread 7.9) and causes **hypoxia** (an inadequate supply of oxygen to respiring tissues). Muscles deprived of oxygen fatigue quickly; an impaired oxygen supply to the brain leads to mental confusion and lowers reaction times.

Altitude acclimatisation

In the short term, exposure to high altitudes causes **hyperventilation** (an increase in the ventilation rate; see spread 7.2) and an increase in submaximal heart rate, which increases cardiac output (spread 7.7), both of which will tend to increase the oxygen supply to respiring tissue. Most people who ascend relatively quickly to altitudes higher than 4000 m suffer mountain sickness to some degree (see box on previous page). However, if a person ascends a mountain slowly over a period of days or weeks, he or she will allow the body to make reversible physiological adjustments so that the body can cope better with the conditions. This process, called **altitude acclimatisation**, includes the following responses:

- an increase in the haemoglobin content and **haematocrit** (volume of red blood cells expressed as a percentage of total blood volume)
- polycythaemia (an increase in the number of red blood cells per mm^3 of blood)
- an initial decrease in plasma volume which gradually returns to normal
- an increase in the number of capillaries supplying muscle
- an increase in myoglobin

Acclimatisation to medium altitudes (1829 m above sea level) takes about two weeks, but acclimatisation to high altitudes (3048 m or more above sea level) may take much longer.

Adaptations of native highlanders

Acclimatisation does not usually make a person as well adapted to high altitudes as someone native to that environment. Compared with native lowlanders, even when acclimatised, native highlanders have a greater depth of breathing, a higher lung capacity, larger tidal volumes, blood with a higher affinity for oxygen, and a higher pulmonary diffusion capacity than lowlanders. (The **pulmonary diffusion capacity** is the volume of gas that can diffuse across the membranes between the alveoli and lung capillaries per minute for each unit of partial pressure difference between the gas inside the alveoli and lung capillaries; the higher the pulmonary diffusion capacity, the greater the rate at which the gas can diffuse from the alveoli to the capillaries.) In addition, native highlanders (figure 1) have oxygen dissociation curves adapted to their environment. All of these factors tend to improve the oxygen transport system in the body.

Sherpa Tensing was a member of the Sherpa people who live in the Himalayas. Among the adaptations which the Sherpa people have to high altitudes is their oxygen dissociation curve which is shifted to the left of the normal curve. This is an advantage because it increases the saturation of haemoglobin with oxygen at the low oxygen partial pressures which occur at these very high altitudes. There is, however, a corresponding disadvantage in that the oxygen is released less readily. Interestingly, llamas and vicunas, mammals which normally live at very high altitudes, also have an oxygen dissociation curve to the left of most other mammals.

In contrast to the Sherpa people, the Quechua Indians living in moderately high altitudes (3000 m to 5500 m) in the Peruvian Andes have an oxygen dissociation curve which has shifted to the right of the normal curve. This means that the oxygen is released more easily, but there is a lower percentage saturation of the haemoglobin with oxygen. Visitors to moderately high altitudes acquire a similar oxygen dissociation curve.

Altitude training

Many elite athletes train at medium altitudes to benefit from the effects of altitude acclimatisation and to enhance their performance in endurance events. Most exercise physiologists believe that the training is of significant benefit only to those who intend competing at high altitude and that it is of little benefit to sea-level competitors. To be effective, altitude training must take place at least 1500 m above sea level for a period of not less than three weeks, with the first week consisting of light exercise only. It is, however, risky for native lowlanders to train at high altitude for very long periods because it may lead to loss of muscle mass and body weight. It takes three to six weeks at sea level to lose all the effects of altitude training and acclimatisation.

Quick check

1 What is the temperature change when someone ascends 300 metres?

2 Define hypoxia.

3 What name is given to reversible physiological adaptations to environmental changes?

Food for thought

Altitude acclimatisation includes an increase in blood volume due to an increased red blood cell count; plasma volume actually decreases. At 4500 m, the blood volume can be as much as 30 per cent more than corresponding values at sea level. With more haemoglobin in the blood, the capacity to carry oxygen is increased. Some athletes increase their haemoglobin content by **blood doping**. They attempt to boost their stamina by injecting themselves with extra red blood cells. Suggest what the potential dangers might be of increasing the density of red blood cells in the blood by blood doping or by altitude acclimatisation.

Summary

The **biosphere** (the living part of Earth) is subdivided into **biomes** such as deserts, tropical rain forests, and temperate grasslands. **Ecology** is the study of the interrelationships between organisms and their environment. The functional unit of ecology is the **ecosystem**, a relatively self-contained **community** of organisms interacting with their **biotic** and **abiotic environment**. Each species within an ecosystem occupies a specific **ecological niche**. Interspecific **competition** may cause the actual or **realised niche** to be much smaller than the potential or **fundamental niche**.
Energy flow in ecosystems can be described in terms of **food chains**, **food webs**, **pyramids of numbers**, **pyramids of biomass**, and **pyramids of energy**. As energy flows from one **trophic** (feeding) **level** to the next, heat is produced therefore there is less energy available for growth and reproduction. Consequently, **gross primary production** and **gross secondary production** are greater than **net primary production** and **net secondary production** respectively. **Population growth** can be described in growth curves, such as the **sigmoid growth curve**. Population growth is limited by **density-dependent factors**, such as food and disease, and **density-independent factors**, such as fluctuations of temperature. When the **carrying capacity** of an ecosystem is exceeded, population density falls. The world population of **humans** seems to be going through an **exponential growth phase**. However, some nations that have become well developed economically are going through a **demographic transition** which will tend to change **population pyramids** (population profiles).
Interrelationships between organisms may be interspecific or intraspecific. Interspecific relationships include **mutualism**, **predation**, **parasitism**, **competition**, **commensalism**, and **allelopathy**. **Predator-prey relationships** are often associated with cyclical changes in population density. Interspecific and intraspecific relationships are involved in the process of **ecological succession**, the sequence of change from the initial colonisation of an area to the establishment of a relatively stable community. **Species diversity** changes during the various **seral stages** of succession, tending to increase with the stability of the ecosystem. Important **nutrient cycles** include those of nitrogen, phosphorus, and carbon. Human activities, such as the production and use of fertilisers, the burning of fossil fuels, and deforestation, affect these cycles and may lead to **eutrophication** and **global warming**. Adaptations of organisms to their environment are illustrated by camels and desert rats which are superbly adapted to conserving water in their desert environments. Plants also show striking adaptations to the availability of water in their environments and can be classified as **mesophytes**, **xerophytes**, **hydrophytes**, and **halophytes** according to their water supply. Human adaptions to high-altitude environments include an increase in haemoglobin content and changes in lung ventilation.

PRACTICE EXAM QUESTIONS

1 The diagram shows part of the food web studied on a rocky shore.

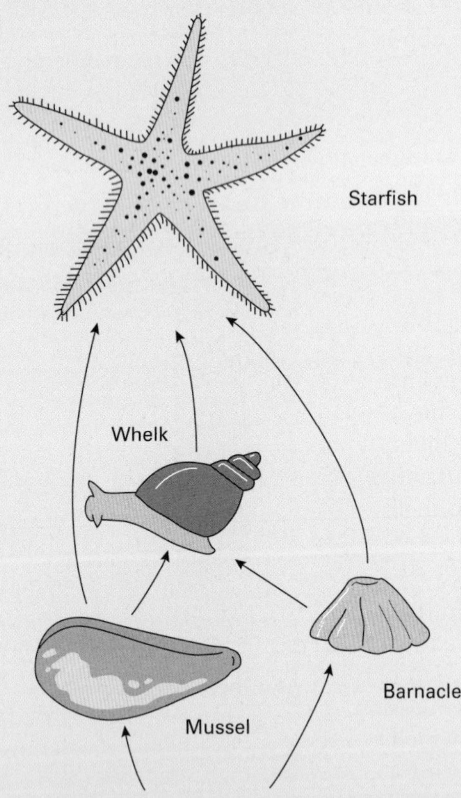

a What do the arrows on the diagram represent? [1]

b All the starfish were removed from a number of experimental areas and then kept out. The number of mussels in these areas increased greatly, until they dominated all other species. Mussels and barnacles live permanently attached to the rock.

Explain **one** way in which the increase in the number of mussels may have led to a reduced number of barnacles. [2]

c The investigators wanted to know the effect of removing all the starfish on the diversity of the rocky shore community and decided to calculate a diversity index. The mean number of species present fell from fifteen to eight when the starfish were removed.

i What other information would the investigator need in order to calculate a change in the diversity index? [1]

ii Both the mean number of species and the standard deviation were calculated. What information does the standard deviation give? [2]

[Total 6 marks]

2 The diagram represents four ways in which changes can take place in the number of organisms in a population.

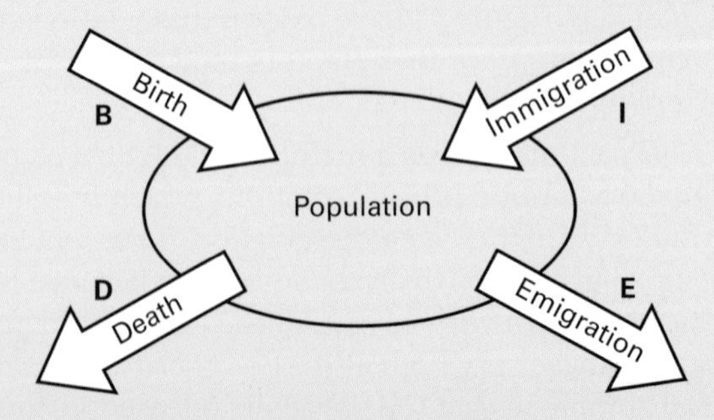

a Explain what is meant by a population. [1]

b Use the letters **B**, **I**, **D** and **E** to write a formula to show:

i a stable population which is in equilibrium;

ii a population which is increasing in size. [2]

c The size of a population may be limited by density-dependent factors or density-independent factors.

Explain how these two types of factor operate to limit population size.

i Density-dependent factors. [1]

ii Density-independent factors. [1]

[Total 5 marks]

3 a The table below shows mean values for primary productivity for four ecosystems: temperate deciduous forest, tropical forest, temperate grassland, and intensively cultivated land in a temperate region.

Ecosystem	Primary productivity / $kJm^{-2}\ yr^{-1}$
Temperate deciduous forest	26 000
Topical forest	40 000
Temperate grassland	15 000
Intensively cultivated land in a temperate region	30 000

i Suggest *two* reasons to account for the higher primary productivity of a tropical forest compared with a temperate forest. [2]

ii Suggest explanations for the difference in primary productivity between temperate grassland and intensively cultivated land. [3]

b Describe how you would estimate the fresh biomass of the producers in a grassland ecosystem. [4]

c Suggest why productivity of an ecosystem is measured in units of energy rather than units of biomass. [2]

[Total 11 marks]

4 The diagram below shows part of the nitrogen cycle.

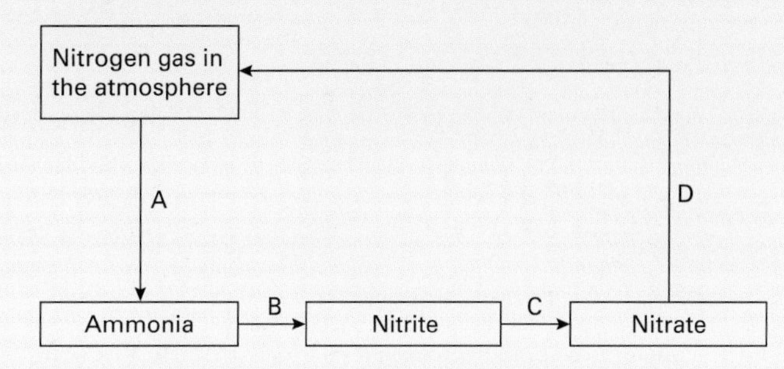

a Name a genus of bacteria which is responsible for each of the reactions A, B, C and D. [4]

b Describe the conditions in which the bacteria responsible for reaction D will thrive. [2]

[Total 6 marks]

5 The diagram shows a number of stages in an ecological succession in a lake.

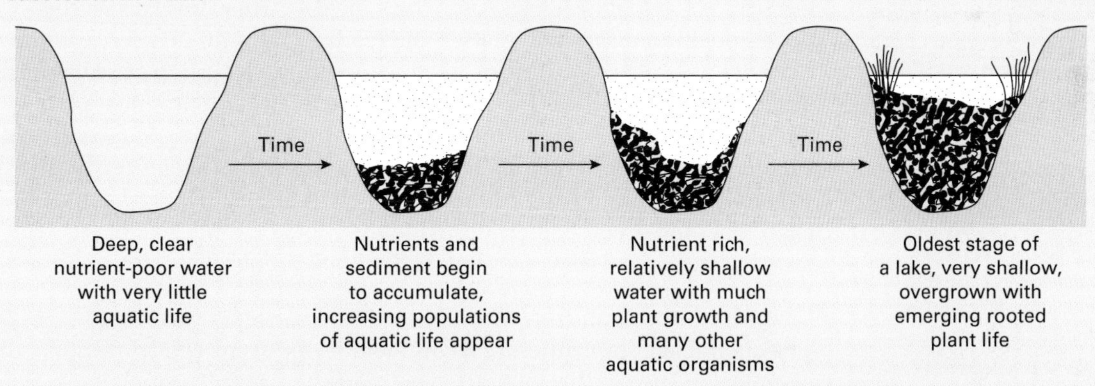

a Use information in this diagram to help explain what is meant by an ecological succession. [2]

b Give **two** general features which this succession has in common with other ecological successions. [2]

c A number of small rivers normally flow into this lake. These rivers flow through forested areas. Explain how deforestation of the area might affect the process of succession in the lake. [2]

[Total 6 marks]

6 a Explain what is meant by an ecosystem. [2]

b What information is required to fully describe the make-up of an ecosystem? [8]

c Comment on the flow of energy through ecosystems and discuss the various ways in which human activity can influence this flow **at all levels** in terrestrial ecosystems. [10]

[Total 20 marks]

23 Applied ecology

Agriculture (1): soils and their management

23.1

OBJECTIVES

By the end of this spread you should be able to:

- describe the main edaphic factors
- distinguish between sandy, silty, clay, and loam soils
- discuss how soil can be managed.

Fact of life

During the 1920s and early 1930s, many people were encouraged to settle in the Midwest of North America. With little thought about how suitable the soil or the climate were for agriculture, the natural prairie grasslands were converted to vast fields of wheat. The soil became exhausted of nutrients, and following dry years in 1934 and 1935, the crop failed. Left exposed to the sun and wind, the soil soon turned to dust leading to its erosion. Huge volumes of topsoil were lost in horrendous dust storms, turning much of the prairies into an uninhabitable area aptly named the Dust Bowl (figure 1). Many thousands of people had to emigrate to areas such as California where they were exploited as cheap labour. This social and economic disaster was caused by mismanagement of soils. Much of the area has now reverted to grassland.

Figure 1 *A farm in the Dust Bowl of the American Midwest during the 1930s.*

Soil, the upper, weathered layer of the Earth's crust, is a vastly underrated human resource. Many people think of it as dirt, but soil, although rarely more than 2m deep, provides the nutrients and water for the plants which, directly or indirectly, provide most of the food we eat. The environmental factors that operate in the soil are called **edaphic factors**. The five most important are physical structure, water content, air content, chemical composition, and soil pH.

Physical structure

The physical structure of a soil depends mainly on the relative proportions of different sized particles in the soil (table 1), and the way they interact with organic matter. The relative proportions of clay, sand, and silt determine **soil texture**. Soil particles are usually glued together with organic matter to make aggregates, collections of particles that act as one material. The distribution, shape, and stability of the aggregate in a soil make up **soil structure**.

Water and air content

The amount of water in a soil depends on its water-holding capacity and the degree to which water that lands on the soil moves down into the soil. Both of these properties depend on the soil's texture and structure. Soil air is present in spaces between the soil particles that are not completely filled with water. The amount of air in a soil therefore depends upon the texture and structure of the soil, and also the amount of water in the soil. Waterlogged soils usually lack oxygen. This slows down the decomposition of organic matter and leads to mineral deficiences. Low oxygen levels also allow anaerobic bacteria to thrive, such as sulphur bacteria which produce hydrogen sulphide (a gas toxic to most plants) and denitrifying bacteria which convert nitrates to nitrogen (see spread 22.12, figure 1). Some plants have evolved mutualistic relationships that overcome the oxygen and mineral deficiences of waterlogged soils. For example, blue-green bacteria provide the roots of some rice plants with oxygen, and mycorrhiza (spread 22.8) supply minerals to their partners.

Chemical composition and pH

The chemical composition of a soil includes nutrient concentrations, the chemical nature of the soil particles (in particular the proportion of clay particles), and the amount of humus in the soil. **Humus** is organic matter that has been broken down by fungi and bacteria. Soils with a high humus content retain water well without preventing drainage through the soil (the downward movement of water by gravity). They therefore tend to be damp but not waterlogged. In contrast, clay also tends to retain water but prevents drainage. Soils rich in humus usually have a higher mineral content. Both humus and clay particles carry negative charges, which prevent positively charged ions from being washed out from the soil by rain (a process called **leaching**), keeping them available for plants.

The hydrogen ion concentration of the soil affects the enzyme systems of soil organisms; pH also affects the degree of dissociation and solubility of ionic substances and thus the availability of mineral ions in the soil.

Soil types

The various edaphic factors interact to create many different soil types, which are usually put into three main categories:

- **Sandy soils** – light soils made up of relatively large particles with large air spaces. Sandy soils gain and lose heat quickly and drain water quickly.
- **Clay soils** and **silty soils** – heavy, often waterlogged soils made up of fine particles with small air spaces. These soils gain and and lose heat slowly and are difficult to work. Clay soils have more than 40 per cent clay particles; silty soils have a smaller proportion of clay particles and tend not to be so waterlogged.
- **Loam soils** – dark soils containing a mixture of sand, silt, and clay particles and usually a high humus content. They have physical properties intermediate between those of the other soil types.

Table 1 *Mineral particles in the soil and their size.*

Particle	Diameter (mm)
Gravel	> 2.0
Coarse sand	0.2–2.0
Fine sand	0.02–0.2
Silt	0.002–0.02
Clay	< 0.002

Soil management

Farmers usually have four main aims when managing the soil:

- to maintain or improve soil structure
- to reduce soil erosion
- to maintain or improve soil fertility
- to control pests and diseases.

Table 2 shows some soil management techniques used to achieve these objectives.

Table 2 *Some soil management techniques. The processes are linked so that, for example, improving the soil structure encourages plant growth and reduces the effects of pests and diseases.*

Technique	Description	Comments
Crop rotation	Farming method in which different crop plants are grown in a field in regular rotation	Prevents the build-up of pests and parasites peculiar to any crop species Minimises the risk of depleting the soil of particular nutrients or water at a particular soil depth because each crop has its own specific requirements. Legumes are usually included in crop rotation because the nitrogen-fixing bacteria in their root nodules add to soil fertility.
Tillage	Mechanical preparation of the soil e.g. turning by ploughing	Suppresses the growth of weeds and promotes a good tilth (i.e. a crumbly and porous soil with a stable, granular structure). Extensive tillage using heavy machinery can compact the soil and be costly in terms of fuel and equipment.
Liming	Addition of agricultural lime to soils. The lime is a mixture of calcium compounds including slaked lime, $Ca(OH)_2$; quicklime, CaO; and limestone, $CaCO_3$.	Increases soil pH, neutralising acidic soils, and promotes clumping of soil particles, thus improving soil structure. Adds carbon dioxide to the atmosphere and may be contributing to global warming
Addition of organic matter	The addition of substances such as manures, straw, and sewage sludge	Promotes humus formation, improving soil structure, and promoting plant growth. Compared with inorganic fertilisers, organic fertilisers are usually more difficult to handle and take a longer time to have an effect. They also have a more variable composition.
Addition of inorganic fertiliser	Addition of granules, pellets, or liquid containing one or more inorganic nutrients (commonly nitrogen, phosphorus, and potassium)	Expensive to make, but easy to handle, quick acting, and reliable. Used appropriately, inorganic fertilisers promote plant growth, but heavy use may damage plants ('burning' or 'scorching' them), and increase soil acidity. Leaching into aquatic environments can lead to eutrophication (spread 23.7).
Drainage and irrigation	Careful management of water supplies	Artificial irrigation can make deserts bloom, but over-use can lower the water table which, in the long term, may lead to increased desertification.

Quick check

1 Distinguish between soil texture and soil structure.

2 Which type of soil (sandy, clay, or loam) would you expect to have the highest nutrient content?

3 What is tillage?

Food for thought

Traditional four-year crop rotations include a year in which red clover is grown in a field for grazing in spring and summer. Suggest why the red clover is ploughed into the soil prior to planting winter wheat the next year. What is the importance of grazing animals in this rotation?

23.2

Agriculture (2): pest control

OBJECTIVES

By the end of this spread you should be able to:

- explain what a pest is
- describe how pests can be controlled.

Fact of life

Herbicide resistance is the most common genetically modified (GM) characteristic in commercial agriculture. Oil rapeseed, a plant cultivated extensively for food and fuel, has recently been genetically modified to optimise the production of useful fatty acids. These plants have also been genetically modified to become resistant to particular herbicides, either by giving the crop a new gene or by switching off an existing gene. In 2007, GM rapeseed made up approximately 87 per cent of Canada's rapeseed crop. It is also grown to a lesser degree in the USA and in certain states in Australia. All GM rapeseed grown commercially throughout the world is herbicide resistant. Supporters argue that herbicide resistance enables weed control to be more efficient and effective; opponents are concerned that it could encourage the over-use of herbicides and lead to the evolution of herbicide-resistant weeds, and that herbicide-resistance genes might flow from commercial plants to weeds.

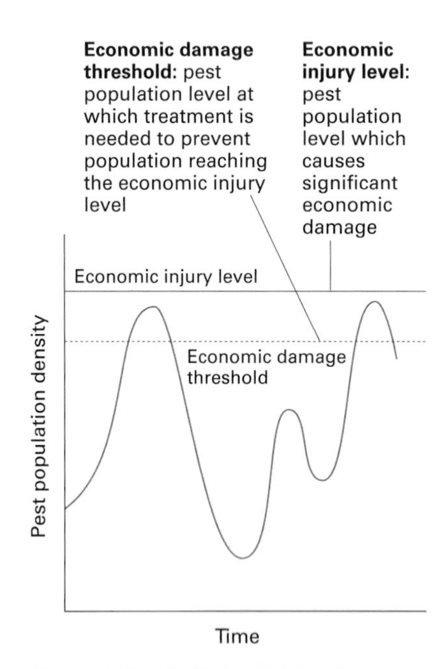

***Figure 1** Graph distinguishing the economic injury level and economic damage threshold of pest populations.*

What is a pest?

A **pest** is any organism that people find undesirable. It may cause harm economically, or affect someone's health. An organism considered a pest in one place or by one person may be beneficial in another; for example, one person's weed is another person's bloom.

Agricultural pests cause economic damage to crops and farm animals. They can have a devastating effect, especially in **monocultures** in which only one crop is grown. These systems are simpler than natural ecosystems and usually lack the predators of pests. Also, growing the same crop on the same land year after year may allow pest populations to build up.

To safeguard crops, most farmers carry out some form of pest control. However, it is rarely possible (or even desirable) to eliminate all pests. Most pests cause significant economic harm only when their population reaches a certain level, called the **economic injury level**. To prevent fast-growing pest populations from reaching this level, control measures have to be started at a lower pest population level, called the **economic damage threshold** (figure 1). Pests may be controlled chemically, biologically, by cultural methods, or by a combination of methods.

Cultural methods of pest control

Cultural methods such as weeding, tillage, and crop rotation are among the most common methods of pest control. Weeding and tillage help to remove weeds; overturning the soil may expose insect pests to predatory birds. Crop rotation (see spread 23.1) often prevents the build-up of pests that occurs in monocultures, but it is only effective when a pest cannot attack successive crops. In Europe, crop rotations that include potato, oil seed rape, and wheat are popular because they discourage the build-up of several potentially damaging pest species.

Crop damage can also be minimised by growing the crop at a particular time in the life cycle of the pest. The crop is sowed or harvested at times when the pest can do least damage, for example, before insect pests lay eggs, or before adults emerge from their dormant state. Infestations of maize crops by the maize weevil (*Sitophilus zeamais*) can be reduced by harvesting the maize before the adult beetle emerges from stores of maize.

Other cultural methods of pest control include:

- removing the remains of crops and badly damaged plants which might harbour pests
- creating physical barriers (for example, apple trees are protected from codling moth caterpillars by putting sticky bands on their trunks)
- covering the soil with organic material (mulching) which prevents light from reaching weeds
- **intercropping** – planting two different crops in the same field; for example, undersowing cereal crops with rye grass provides suitable conditions for ladybirds which control aphids on the cereals.

Chemical control

Toxic chemicals that control pest populations are called **pesticides**. There are hundreds of different types, usually classified according to the pest organism they treat. These include **fungicides**, **herbicides**, and **insecticides**. **Contact pesticides** kill pests without being eaten (for example, by penetrating the cuticle of insects); **systemic pesticides** are taken into a plant and translocated within the plant, and enter the pest when it eats the plant or sap. The ideal pesticide is cheap, effective, specific (it only affects the target species), and is broken down quickly to form harmless substances.

Modern pesticides have played an essential part in feeding an increasing world population, but most have one or more undesirable features. **Broad-spectrum pesticides**, designed to affect a wide range of pests, may also kill harmless organisms or beneficial ones such as the predators of the pest. Other pesticides, such as DDT, are very stable. They persist in the environment and, even if used in very small concentrations, may build up in the food chain (see spread 22.3, figure 1). Most pesticides lose their effectiveness as the pest evolves resistance to them (see spread 20.5). There is a continual search for alternative methods of pest control.

Biological control

Biological control is the control of pests and weeds by other living organisms (including genetically modified organisms), or by biological products. A predator (such as ladybird, figure 2), parasite, or disease-causing organism is usually used to control the pest. The aim is not to eliminate the pest or weed (that would also wipe out the control organism), but to use the control organism to keep the pest or weed population below the economic injury level. Care must be taken to make sure that the control organism does not become a nuisance itself. Cane toads, for example, introduced into Australia to control beetle infestations of sugar cane, have become a serious pest because they kill non-target organisms such as pygmy possums as well as the beetles.

Biological products used as control agents include pheromones and genetically engineered insecticides. Pheromones are chemicals released, for example, by a female insect to attract a mate (see spread 11.12). Synthetic pheromones have been used to lure pests into traps laced with an insecticide.

Integrated pest management

In practice, no one type of pest control is ideal. The Food and Agriculture Organisation of the United Nations recommends considering fully the environmental context and biology of the pest, and then using all suitable methods to maintain the pest population at a level below the economic injury level. This form of pest control is called **integrated pest management**. Pest population densities are monitored, and pesticides used only when the pest reaches the economic damage threshold.

Quick check

1 Define a pest.

2 What is a broad-spectrum pesticide?

Food for thought

Whitefly is a common pest of greenhouse crops that is parasitised by a wasp, *Encarsia formosa*. The temperature requirements of the wasps are more critical than those of the whitefly:

- wasp populations do not survive temperatures below 10 °C
- wasp population growth matches that of the whitefly population above about 18 °C
- wasps reproduce much more rapidly than the whitefly at temperatures above 26 °C

Suggest the implications of these observations on the use of the parasitic wasp to control whitefly populations in greenhouses.

Examples of biological control agents

Bacillus thuringiensis (sometimes abbreviated to Bt) is a bacterium that has been applied as a spray to cabbage plants to infect and kill pest caterpillars. The bacterium has a gene, carried on a plasmid, for the production of a protein that is toxic to the caterpillars. Using **recombinant DNA technology**, this gene has been incorporated into cabbage protoplasts (plant cells with their cell walls removed) which have been cultured to produce fully grown cabbages with a built-in resistance to the caterpillars. When a caterpillar feeds on a cabbage, it takes in a dose of toxin. Genes from different strains of *B. thuringiensis* are being used to genetically engineer other insect-resistant crops including maize, cotton, and potatoes. Such insect-resistant crops are called Bt maize, Bt cotton, and so on, in recognition of their transgenic nature. Advocates for the use of genetic modification hope that insect resistance will reduce the enormous destruction of crops by insects. In Kenya, for example, farmers have reported losing 15–45 per cent of their crop. Anti-GM crop campaigners have expressed concerns about the potential environmental and health risks of using genetically modified insect-resistant crops. These risks include toxic effects on non-target insects such as butterflies, harmful effects on the soil ecosystem, and an evolution of insecticide-resistant pests. In 2011, **GM crops** were grown by around 16.7 million farmers in 29 countries worldwide. No GM crops were grown commercially in the UK, but imported GM products were used in animal feed and in some food products. In 2012, the use of GM crops continues to be a highly controversial and politically sensitive issue with strong arguments made for and against its increase.

Myxomatosis is a viral disease introduced deliberately into Britain in the1950s to control rabbit populations that had become so great that they were causing serious economic and environmental damage. When introduced into Australia in the 1990s, many objected to its use because of the cruel disfiguring effects of the disease on the rabbits, and because the virus could also affect non-target species such as domestic rabbits. Myxomatosis, like other disease control agents, tends to lose its effectiveness as pests evolve resistance. Initially, in Britain it killed over 90% of rabbits infected; it now kills about 50%.

Irradiation is another method of biological control. It involves breeding a pest species such as fruit flies or tsetse flies, separating the males, and sterilising them, for example, by exposure to X-rays. The sterile males are released into the environment to mate with females, which produce infertile eggs.

Figure 2 *Biological control agent: ladybirds have a voracious appetite for aphids.*

23.3

OBJECTIVES

By the end of this spread you should be able to:

- explain how farming has become more intensified
- discuss the advantages and disadvantages of highly intensive farming
- discuss the advantages and disadvantages of removing hedgerows.

Fact of life

Between 1940 and 1970, the introduction of new technology to increase food production was so dramatic that it was called the 'green revolution'. In the UK, food production doubled; in India, wheat production tripled; in parts of Asia rice production increased eight-fold. By taking advantage of improvements in technology, especially biotechnology, food production per hectare continues to increase. The most productive farms tend to be the most intensive and mechanised; they use pesticides, and apply inorganic fertilisers. All require the burning of fossil fuel. To accommodate large machinery, hedgerows, which act as refuges and corridors for wildlife, have been removed. Hedges also act as windbreaks, reducing wind erosion of soils. Because of these 'costs' in fuel and wildlife, some economists question the benefits of the continued 'green revolution'. They argue that increased yields are available only to farmers wealthy enough to finance the changes, and that the 'green revolution' has widened the gap between rich and poor in the world. Others argue that unless farmers use the latest technology and farming techniques, they will be unable to feed the growing world population.

Figure 1 *Highly mechanised farming on land with few hedges brings increased productivity.*

Agriculture (3): crop and animal production

Agricultural production has increased enormously in the last few decades, and seems set to rise even further. These increases have been brought about mainly by the development of new varieties of crops and animals, and more intensive methods of farming.

Selective breeding and genetic engineering

New varieties of farm animals and crops have been produced by artificial selection and genetic engineering. Artificial selection of crops and farm animals has been going on since the dawn of agriculture (spread 20.7). Genetic engineering is a recent innovation which is already having a dramatic impact. Tomatoes have been genetically engineered to improve their flavour and to delay ripening. Spread 18.10 describes how sheep and cattle can be cloned to produce products, and spread 23.2 describes how crops can be genetically engineered to resist pests. Many foods now contain genetically modified organisms, for example, soya beans. Some people believe that these **genetically modified organisms** (GMOs) will help food production keep pace with population growth, and make food cheaper and more readily available. Others believe we already have the capacity to produce sufficient food without GMOs. They think that until we know more about the long-term effects of GMOs on our health and the environment, their use should be stopped or at least restricted.

Intensive farming

Agriculture has become increasingly mechanised, and conditions for growing crops or animals can now be controlled much more than before. On many arable farms machines have replaced labourers almost entirely. Every stage of crop production is carried out by machines, from the preparation of the soil and the sowing of seeds to harvesting and storage. Enormous amounts of pesticide are used to increase crop yields by reducing competition from weeds, and by minimising losses due to pests and diseases. Inorganic and organic fertilisers keep mineral levels high in the soil. Without fertilisers, heavy harvesting of most crops would deplete the soil of minerals. A deficiency of any one mineral limits plant growth. The most common limiting factors are nitrogen, phosphorus, and potassium. Used in the correct amount, fertilisers can increase crop yields greatly, but their application obeys the **law of diminishing returns**: above a certain level, the increase in crop yield for each unit of fertiliser used gets less. Even worse, above a critical level, addition of fertilisers actually reduces crop yield and damages the environment. The correct fertiliser must be chosen for the crop and the soil: crop yield will be boosted only if a fertiliser contains a mineral that has been acting as a limiting factor.

Greenhouse growing and factory farming

Among the most intensive forms of farming are greenhouse cultivation of plants and factory farming of animals. Both these methods grow organisms indoors in a confined space under controlled environmental conditions. Growing plants in greenhouses allows conditions such as light intensity, temperature, carbon dioxide concentrations, and mineral levels to be controlled artificially. Factory farming aims to ensure that food ingested by the animals goes towards production (growth, useful products such as milk, and reproduction). This is acheived by:

- providing a diet that is highly digestible and easily converted to animal products
- minimising unwanted energy expenditure by restricting animal movements and by keeping the animals warm and in the dark
- the addition of artificial growth stimulants to the food
- using young animals (older ones are less efficient at converting food into growth).

Artificial growth stimulants include antibiotics and hormones such as **bovine somatotrophin** (BST), given to cattle. BST is produced by the pituitary gland. As well as stimulating growth, it controls milk production. Since the 1970s, the hormone has been produced artificially in bacteria by recombinant DNA technology (spread 18.9). It has been injected into cows to increase milk production and into beef cattle to help them gain weight. The health of the cattle and of humans drinking the milk or eating the meat appears to be unaffected by the hormone. Antibiotics are given to farm animals mainly to reduce the risk of diseases, but they also boost growth rates. Disease control is especially important in factory farms because of the speed at which a disease can spread in crowded conditions. Factory farming is an emotive topic; some of the points made for and against it are shown in the box below.

Factory farming: for and against

For:

- It is the only way to supply cheap food in the high quantities demanded by the public.
- The conditions are not cruel or stressful; if they were, animal growth and yield would not be maintained.
- Less intensive forms of livestock rearing would require more countryside to be used as farmland.

Against:

- It is cruel and causes stress to animals by keeping them in confined spaces.
- The comfort and well-being of animals should not be sacrificed for financial gain.
- Antibiotics, pesticides, and growth-promoting substances may harm human health and the environment.
- Excessive use of antibiotics in farm animals is contributing to the evolution of antibiotic resistance.
- Intensive cultivation is not cost effective because it relies on the heavy use of fossil fuels, whereas **organic farming** (low-intensity farming without the use of industrially made fertilisers or pesticides) is not dependent on fossil fuels and is healthier for people and the environment.

Genetically modified organisms

A **genetically modified organism** (GMO) is an organism whose genetic make-up has been changed by genetic engineering techniques. For example, recombinant DNA technology is used to combine genes from different species to create a **transgenic organism**. Products from GMOs include human foods and food ingredients, animal feeds, medicines, and fibres.

Ever since the first transgenic animals and crops were created in the 1980s, the use of GMOs has provoked much public debate about costs and benefits, and safety issues. So far, the outcomes of the debates have varied from country to country and this highly emotive topic is likely to remain controversial for many years to come. Those supporting the genetic modification of organisms claim that it is essential for increasing the production of crops and animals to feed an ever-growing global population. As well as increasing yields, genetic modification can improve resistance to disease, pests, and herbicides. It can also enhance taste and increase shelf-life. They argue that prudent and controlled use of genetic modification can improve soil quality and help conserve water and energy. They perceive DNA technology as just another form of artificial selection that farmers have been using for more than 10 000 years to modify the genetics of crops and animals.

Opponents argue that increasing food production by using GMOs is not the best solution to overpopulation; population control measures and improvements in food distribution would be more effective and less risky. They are concerned about potential environmental and human health impacts through unintended gene transfers, such as antibiotic and pesticide resistance, from GMOs to wildlife. They believe that our scientific understanding is insufficient to predict fully all of the potentially harmful effects of genetic modification. Many perceive genetic engineering to produce GMOs as an intolerable tampering with nature. Some do not like the possibility of consuming animal genes in plants and *vice versa*.

Quick check

1 What is BST?

2 Why are animals kept in warm and dark conditions in many factory farms?

3 Hedgerows act as a refuge for many animals, especially in the winter. Give:

a a possible disadvantage

b a possible advantage of this to a farmer.

Food for thought

a Suggest likely advantages and disadvantages of rearing egg-producing chickens on an extended range outdoors rather than intensively in a small space indoors.

b *Agrobacterium tumifaciens* is a bacterium that causes plant tumours. When the bacterium invades a plant (usually at the junction between the stem and root), one of its plasmids enters the plant host nucleus. Plasmid genes are incorporated into host chromosomes, causing the host to secrete opines, special chemicals on which the bacteria feed. Using recombinant DNA technology, genetic engineers have replaced these plasmid genes with a gene that causes a desirable characteristic in the plant. The bacterium can then be used to genetically modify the plant. The method has been used to engineer herbicide resistance into crop plants, such as soya beans. Soya beans are used in 60 per cent of processed foods including margarine, ice cream, breads, breakfast cereals, pasta, and meat substitutes. Suggest advantages and disadvantages of herbicide-resistant crops. What are the advantages and disadvantages of labelling food containing GMOs such as this?

23.4

Fisheries

OBJECTIVES

By the end of this spread you should be able to:

- explain why marine fish population densities vary in different areas of the world
- explain the significance of sustainable yields
- give examples of regulations designed to prevent overfishing.

Fact of life

The annual worldwide catch of fish rose from an estimated 20 million tonnes in 1950 to a peak of 94 million tonnes in 2003. Between 2003 and 2009 it fluctuated around 90 million tonnes. China is the biggest fishing nation. In 2008 it had an annual catch of about 18 million tonnes.

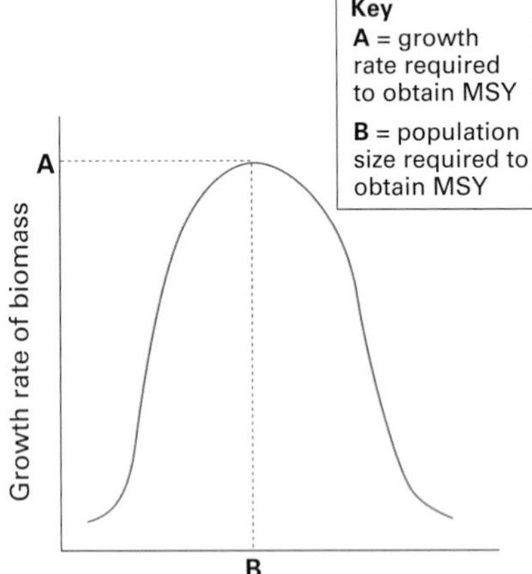

***Figure 1** The maximum sustainable yield (MSY) of a fish population occurs when its growth rate is maximal. To obtain the MSY, the amount of fish harvested and lost in other ways must equal the amount replaced by reproduction and immigration. Harvesting at levels above or below the MSY results in a fall in the growth rate and therefore a fall in the sustainable yield. Note that this graph assumes the fish population has a sigmoid growth curve (spread 22.6, figure 1). Therefore the maximum growth rate occurs during the exponential phase of population growth, when population size is some way short of its maximum limit (the carrying capacity). In theory, it would appear simple to establish MSYs in a fishery. However, to do so we need to know three things: (1) the maximum growth rate of the population; (2) the population size which has the maximum growth rate; (3) the current population size. Unfortunately, this data is incomplete or unreliable for marine fish populations, making it impossible to establish accurate MSYs.*

Fish are an important human resource, supplying around 25 per cent of the protein we eat. The bulk of the world's fish harvest is caught by commercial fleets that use a variety of aids for locating and capturing sea fish. In some cases, as soon as fish are caught they are transferred to a 'mother' factory ship which processes and packs the catches of a dozen or so smaller vessels. Modern fishing is very sophisticated, using the most advanced technology to maximise catches. It has become so efficient that in some areas fish stocks have become severely depleted.

Upwelling zones

Fish are not evenly distributed in the seas. They feed directly or indirectly on phytoplankton, so they are most abundant where phytoplankton productivity is highest. Phytoplankton productivity is directly related to mineral nutrients in the surface waters. Most oceans have a low concentration of nutrients at the surface, so phytoplankton and fish populations are generally low. There are, however, two special situations in which nutrient levels are high. The first is where a combination of wind and currents causes water close to the sea bed to rise. The upwelling carries nutrients into water higher up, where there is sufficient light to support photosynthesis. Approximately half the world's harvest of marine fish are taken in these **upwelling zones**, yet these zones form only about 0.1 per cent of the oceans. One of the richest upwelling zones follows the Humboldt or Peru Current which flows northward along the western coast of South America. The current brings nutrients to the warm, sunny waters off the coast of Peru and Ecuador, supporting an important anchovy fishery. The current is affected by an occasional climatic event called **El Niño** ('the boy child'), because it usually starts at Christmas. During an El Niño, a change in the direction of currents and prevailing winds prevents upwelling, and the fishery fails.

Shallow sea fisheries

Nutrient concentrations are also high in the shallow seas above continental shelves. In these areas, the water is shallow enough for bottom sediments to be disturbed by heavy storms that cause strong wave action on the surface. The seas around the continents are also fed by rivers carrying nutrient-rich sediment that comes from natural erosion and human activities. The most important British fishing ground is the North Sea into which the Rhine and other rivers rich in human effluent flow. The main commercial fish in the North Sea are cod, haddock, plaice, and sole (**demersal** or bottom-feeding fish); and herring and mackerel (**pelagic** fish which feed near the surface).

Fish are a **renewable resource**; that is, a resource which, with proper management, can be used again and again because they are constantly replaced. In a fish farm, fish stocks are replaced by artificial breeding. However, in the oceans fish harvested by humans must be replaced by nature. The size of a fish stock is determined by a number of factors, summarised in the following equation:

$$Sn = Sn - 1 + G + R - M$$

where Sn is the stock size now, expressed as the total biomass of the fish population; $Sn - 1$ is the stock size last year; G is the biomass added by growth of fish already in the stock; R is biomass added by recruitment (new fish); and M is biomass lost by mortality, including harvesting.

If more fish are taken from a population than the number born or migrating into that population (if M exceeds R), the population will decline. This is what appears to have happened to commercial fish stocks in the North Sea. The herring stock, for example, declined from 2 million tonnes in the late 1960s to little more than 100 000 tonnes a decade or so later. The decline was so dramatic that there was a complete ban on herring fishing between 1977 and 1982. Some people believe that pollution or climatic changes helped cause the decline of herring populations, but most scientists agree that overfishing has played an important part.

Overfishing often causes both a fall in the number of fish and a change in its population structure: young, smaller fish become increasingly common as fishing fleets attempt to maintain the size of their catches. Catching large numbers of small fish before they spawn can seriously reduce the ability of the population to reproduce. This may lead to a catastrophic fall in local populations. However, for economic reasons, fishing is likely to stop before it can cause the extinction of a fish population, far less a species.

Sustainable yields

In order not to overfish, fish must not be caught faster than the fish stocks can replenish themselves. This rate of reproduction is known as the **sustainable yield**. It varies with the size of population and its growth rate. The largest amount of a naturally renewable resource, whether it be fish or timber, that can be regularly harvested without causing a decline in stock of that resource is called the **maximum sustainable yield** (**MSY**). The MSY is available only when the growth rate of the resource is at its highest level, and when the harvest is at the correct level to keep the population at its optimal size. Harvests that are too high or too low will result in populations that do not have a maximum growth rate. Therefore, if a population exceeds its optimal size (line **B** in figure 1) by 400 tonnes, the harvest should be 400 tonnes.

Regulations have been made to prevent overfishing or to allow overfished stocks to recover. These include:

- **quotas** – the amount of fish that each country is allowed to catch
- **minimum mesh sizes** – ideally, the mesh size should be large enough to allow small, fast-growing, immature fish to pass through. These fish can then reach maturity, spawn, and help replenish fish stocks.
- **closed seasons** during which fishing is not allowed (usually during the breeding season of a particular fish species)
- **exclusion zones** – areas in which fishing is banned completely.

Fishing is not the only means of obtaining fish for food. **Fish farming**, the deliberate cultivation of fish and shellfish, has been practised in Asia and some European countries for centuries. It has become increasingly important as the demand for fish exceeds the supply from traditional fishing.

Fish farming

Fish farms vary from the simple to the technologically complex. Among the simplest are fish ponds stocked with herbivorous fish such as carp, and enriched with human sewage to stimulate plant growth. At the other end of the spectrum are sophisticated fish farms in which salmon are cultivated. Salmon are produced by selective breeding using eggs and sperm stripped from mature fish. The eggs are incubated in trays in clean running fresh water and kept in the dark until the eggs hatch. The young fish are raised at high population densities in freshwater ponds or tanks. As salmon spend part of their lives in sea water, older fish are transferred to cages in sheltered salt water (figure 2). The characteristics of the fish are artificially selected on commercial grounds, such as fast growth and colour of meat. Environmental conditions such as pH, oxygen content, and temperature of the water are carefully controlled, especially for the larval and young stages. The water is kept clean and fish excrement removed. The fish are fed a highly nutritious diet (usually in the form of pellets) to promote rapid growth, and are given antibiotics to prevent infections. Pesticides are also added to the water to prevent infestations with fish lice (small crustacean parasites). Some fish are kept as brood stock; the remainder are sold. In 2010, over 150 000 tonnes of salmon were farmed in Scotland.

Figure 2 Salmon farming: salmon are kept in cages in a loch and fed with pellets. Uneaten pellets and fish faeces fall through the cage floor onto the bottom sediment. The long-term environmental effects of such farms are unknown.

Quick check

1 Why does the sea off the coast of Peru usually have a large fish population?

2 Assuming a fish population has a sigmoid growth curve, during which phase can the maximum sustainable yield be obtained?

3 What would be the likely consequence of having a mesh size significantly smaller than the minimum set by fisheries regulations?

Food for thought

Salmon farms are not closed systems. Water, along with anything it contains, can leak from freshwater ponds, and salmon can escape from cages into surrounding waters. Suggest why salmon farms might be contributing to the demise of natural stocks of salmon rather than aiding their conservation.

23.5 Deforestation

OBJECTIVES

By the end of this spread you should be able to:

- list the main causes of deforestation
- explain how deforestation may affect the carbon and nitrogen cycles
- discuss the contribution deforestation may make to flooding and desertification
- define sustained development.

Fact of life

Each year, about 120 000 km^2 of forest are thought to be lost, 45 per cent to shifting cultivation (figure 1), 15 per cent each to commercial farming and grazing, 10–15 per cent to roads and dams, and 10 per cent to forestry products such as mahogany furniture and chopsticks.

***Figure 1** Slash and burn in a tropical rainforest in West Africa. **Shifting cultivation** is a form of subsistence farming that causes most of the deforestation in tropical regions. It begins with the clearance of trees by a process called 'slash and burn': some trees and the undergrowth are felled (slashed) but most are burned. The ash from burned trees is used to fertilise the soil. Crops are sown on the cleared ground and cultivation continues until falling yields and recolonisation by wild plants cause the cultivators to move on to another site. After using several different sites, the cultivators may return to the first site to repeat the process. Shifting cultivation requires no artificial fertiliser and can support about 7 people per km^2. There is no serious deterioration of the soil as long as each site is left for 12–15 years to recover. However, speeding up the cycle, for example, because of the pressure of feeding an increasing population, has led to severe soil deterioration and the permanent loss of forest.*

Causes of deforestation and loss of natural forests

At one time, much of the Earth's land surface was covered with forests. **Deforestation**, the permanent removal of trees and undergrowth, happened in the past and continues extensively today, changing the landscape of large areas of the world. Since the advent of agriculture about 10 000 years ago, many forests have been destroyed by burning and felling trees. Until about the late nineteenth century, most of the tropical rainforests were untouched. Since then, more than 40 per cent has gone. The forests have been converted mostly to agricultural land, but also to grassland and scrub. Worldwide, agriculture continues to be the main reason for the loss of natural forests; other reasons include:

- supplying firewood as fuel
- to make room for houses, industrial buildings, roads, and dams
- removal of trees for pulp and paper (mainly in temperate forests)
- cutting trees for timber used in the construction industry
- destruction of trees by atmospheric pollution (acid rain)
- replacement of native trees with fast-growing species such as conifers, eucalyptus, and rubber trees.

Large-scale deforestation is thought to be having a significant effect on global biodiversity, nutrient cycles, and erosion.

Effects of deforestation on biodiversity

Deforestation is having its most dramatic effect on biodiversity in tropical rainforests. Complete replacement of native trees with introduced species (for example, replacing deciduous broad-leaved trees with conifers in parts of Britain), or keeping only a few native species, also leads to a reduction in biodiversity. Tropical rainforests are the most productive biomes on land (spread 22.4), and are believed to be home to about 50 per cent of the world's species. It is impossible to estimate accurately the effect of deforestation on biodiversity because many species living in rainforests are still unknown, but countless organisms are being driven to extinction by the loss of their habitat. Many of the known tropical rainforest species have great human value. For example the rosy periwinkle, *Catharanthus roseus*, has provided the alkaloid anti-cancer drug vincristine, used to treat some forms of leukaemia; other forest plant products have been used as anticoagulants, tranquillisers, and antibiotics. Some of the species that are becoming extinct might have proven to be useful to us as well.

Effects of deforestation on nutrient cycles

Global carbon dioxide levels are increasing (spread 22.13), but the contribution of deforestation to this is complex. Young plants take in more carbon dioxide for photosynthesis than they release in respiration. Removing these plants would probably add to the levels of atmospheric carbon dioxide. However, the amount of carbon dioxide taken up by photosynthesis in established forests of mainly mature trees is virtually the same as that released in respiration by producers, consumers, and decomposers. Therefore, cutting down forests probably has very little effect on global carbon dioxide levels. In contrast, forest burning releases huge amounts of carbon dioxide directly and very quickly into the atmosphere and is probably a major contributor to rising carbon dioxide levels.

The soils beneath deciduous forests are relatively rich, consisting of 10–50% organic material. In contrast, soils under boreal coniferous forests and tropical rainforests are thin and contain no more than 10% organic material. Most of the nutrients such as nitrogen are contained within the living trees. Burning trees in these areas significantly reduces the nitrogen held in the ecosystem. Even if wood ash is added to the soil, the soil soon loses its fertility if crops are grown on it and continually removed. In addition, tree roots bind soil particles together, and the tree canopy prevents rain beating down on the soil. Deforestation therefore causes nutrients to be lost through leaching (dissolving in surface water) and runoff (water running to a stream or river rather than being held in the soil). Experimental forest clearance at Hubbard Brook, New Hampshire, resulted in a 30 per cent increase in runoff and leaching, and an immediate loss of nutrients from the soil causing nitrate levels in streams to increase 40 fold. Eutrophication of the streams, combined with an increase in the amount of light reaching the streams, led to an algal bloom (spread 23.7). After about two years, the soil became heavily eroded and much of it was carried away in streams and rivers: the sediment load of streams in a cleared area was up to 30 times that of a stream in a nearby forested area. Similar effects occur when tropical rainforests are cleared (figure 2). Within a few years, soils in deforested areas may not be fit enough to sustain even subsistence agriculture unless artificial fertilisers and irrigation are used. When forest soils are converted to arable land and inorganic fertilisers added, freshwater pollution is increased through runoff and leaching.

***Figure 2** Soil degradation in this region of the Amazon basin, in Brazil, has been caused by the removal of trees.*

Sustained development

There is a strong case for conserving forests rather than converting them to agricultural land, and for replanting forest in some areas denuded of trees. Trees are among our most important renewable resources. They have many uses including building material, fuel, paper, and packaging. Commercially important forest products include cocoa, Brazil nuts, palm oil, and bananas. Forests make a major contribution to global biodiversity, play an important part in nutrient cycles, and influence climates.

There is general agreement worldwide that forests should have a sustainable development (that is, an economic development that can continue indefinitely), but there is no agreement as to how to bring this about. Any sustainable development would have to include **managed timber production**, with trees removed at a sustainable yield (spread 23.4) so that a forest could supply timber indefinitely. Some land should be designated as '**set-asides**' for wildlife conservation. A sustainable development would also need to balance the conflict between economic development and forest conservation, so that present human populations can satisfy their needs without compromising those of future generations.

Many projects aimed at sustainable development in tropical rainforests include the encouragement of **agroforestry** (the interplanting of trees and crops) and **ecotourism** (attracting visitors to natural forest ecosystems to raise awareness of the need for conservation).

Desertification and deforestation

Desertification, the severe degradation of semi-arid land into desert, has occurred in many parts of Africa and the Middle East. Although many scientists believe that desertification results mainly from climatic changes, deforestation may have contributed to the process, or at least speeded it up, by disrupting the water cycle and soil structure. Reduction in tree cover means reduced transpiration, fewer clouds, and less rainfall in the area. Removing trees also increases the risk of flooding following heavy rains. If the agricultural land becomes heavily populated, it is likely to be overcultivated or overgrazed, and the soil will be less fertile and more easily eroded during periods of drought. Salination, the build-up of soluble salts near the surface of soils, may also occur, usually as a result of upward capillary movement from salty groundwater. The salts may reach a concentration that is toxic to most plants, sterilising the land.

To try and reverse the process of desertification in semi-arid areas, trees are planted and artificial irrigation is used. The process depends on a thorough knowledge of plants suited to semi-arid conditions and the way they interact with soils. One successful project was carried out by the Kenyan Green Belt Movement and involved planting over 7 million trees.

Quick check

1 What is the main cause of deforestation in tropical areas?

2 Why does deforestation of boreal forests and tropical rainforests have far greater long-term effects on soil fertility than deforestation of deciduous forests?

3 How might deforestation increase the risk of flooding?

4 Define sustainable development.

Food for thought

At one time, the Forestry Commission in Britain planted fast-growing conifers in upland areas. The conifer species chosen were hardy enough to tolerate the harsh environmental conditions, and they were often planted in solid blocks to withstand severe weather better. It was also standard practice to plant entire blocks at the same time so that they could all be harvested together. Suggest why these types of plantations were criticised by many people.

The modern forestry practice is to plant trees at different times, producing stands of mixed ages that are harvested by removing small blocks within the forest. Conifer plantations are surrounded with broad-leaved species. Suggest advantages of this new forestry method.

23.6

Air pollution and Global climate change

OBJECTIVES

By the end of this spread you should be able to:

- define pollution
- describe the greenhouse effect
- discuss the significance of changes in atmospheric ozone concentration.

Fact of life

Pollution is not a modern problem. Julius Ceasar (100–44 BC) banned wheeled traffic from the centre of Rome during the daytime in an attempt to reduce noise pollution. In 1273, Edward I introduced the first legislation in England to reduce air pollution by curbing smoke emissions. In 1306, a London manufacturer was tried and executed for disobeying a law forbidding the burning of coal in the city. In 1578, Elizabeth I refused to enter London because of smoke pollution. By 1700, pollution was seriously damaging vegetation and buildings in every major town in Britain.

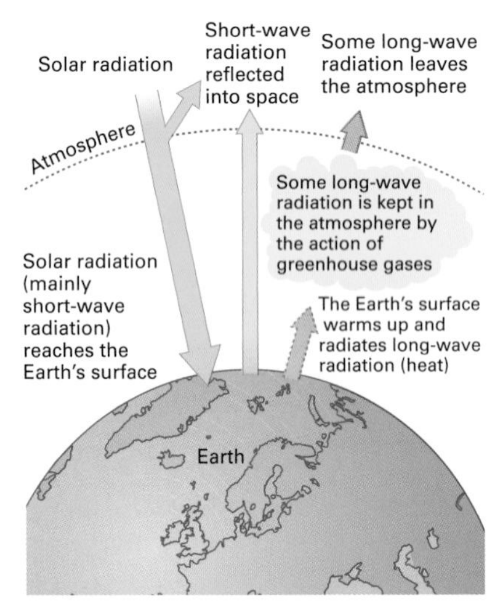

Figure 1 *The greenhouse effect.*

Pollution occurs when substances or energy are released into the environment in amounts large enough to harm anything of human value. A **pollutant** may be physical (for example, noise, heat, and other forms of radiation), chemical (such as heavy metals in industrial wastes), or biological (for example, sewage), and it may harm people, other organisms, or the environment. A pollutant may be a substance of natural origin present in excess (such as volcanic dust or particles of sea salt), but the term is more often used to describe changes brought about by human activities, such as the emission of industrial pollutants, or the discharge of domestic wastes. The pollutant can be in any part of the biosphere: in air, land, or water.

Air pollutants may be gases (such as carbon monoxide from car exhausts; see spread 16.6) or aerosols (solid or liquid particles suspended in the atmosphere). They have many effects on the health of humans and other organisms, and on the natural and built environments. We have already seen how oxides of nitrogen and sulphur emitted as industrial gases can form acid precipitation (spread 2.5). In this spread we shall see how some pollutants can increase the so-called 'greenhouse effect', and how high concentrations of ozone may be both beneficial and harmful.

The greenhouse effect

Solar energy reaches the Earth in the form of short-wave radiation. When that radiation strikes a surface, much of its energy is converted into heat, a form of radiation which has a long wavelength. Carbon dioxide, water vapour, and other gases present in the atmosphere absorb and retain long-wave radiation or reflect it back toward the surface of the Earth. These gases therefore act like panes of glass in a greenhouse, letting light in, but retaining some of the heat before it escapes into space; hence the term 'greenhouse effect' (figure 1).

The retention of heat by the greenhouse effect is a natural process, essential for the evolution of life on Earth. It has been calculated that without it, average surface temperatures would be between –17 and –23 °C; the actual average surface temperature is +15 °C. However, the greenhouse effect appears to be increased by the emission of certain industrial gases, called **greenhouse gases**, the most important of which are carbon dioxide, chlorofluorocarbons, methane, and ozone.

Global climate change

In 1988, the **Intergovernmental Panel on Climate Change** (IPCC) was established to give authoritative and reliable evaluations of the risks of climate change caused by human activity. Thousands of scientists contribute to its work. In 2005, it reported that the average temperature on the Earth's near-surface air and oceans had increased by 0.74 ± 0.18 °C since 1905 and linked this **global warming** to increases in atmospheric greenhouse gases. Physical effects of global warming include rising sea levels, an increase in the melting of ice, and extreme weather events. In 2007, the IPCC declared that the evidence for a warming trend was 'unequivocal' and that humanity has very likely been the driving force in that change.

The precise biological effects of any future global warming are difficult to predict. Up to a certain point, a temperature rise should increase the rate of photosynthesis and primary productivity, but other factors, such as water availability, changes in soil structure, and an increase in insect pests, might become limiting. The distribution of some species such as those on coral reefs and in polar regions are likely to decrease, even to the point of extinction, while the distribution of others such as tropical insect pests are likely to increase.

In 1992, the **United Nations Framework Convention on Climate Change** (UNFCCC) was set up to limit average global temperature increases, to minimise the effects of any resulting climate change, and to cope with whatever impacts occur. Members agreed to reduce greenhouse emissions so that global warming can be limited to below 2.0 °C relative to the pre-industrial level. In 2011, global temperature records indicated that attempts made so far were inadequate to meet this target.

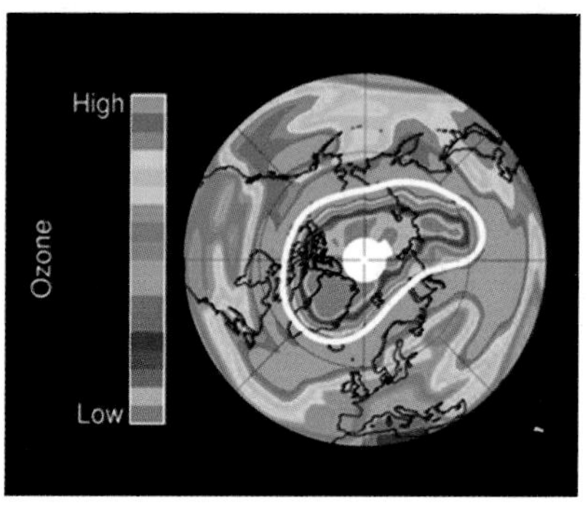

***Figure 2** The first 'hole' in the ozone layer in the Arctic, recorded in April 2011. This 'hole' is similar to that which occurred for the first time over the Antarctic in 1987. According to Dr Markus Rex of the Alfred Wegener Institute for Polar and Marine Research, 'The ozone hole over the Arctic was not only the result of a combination of past pollution due to air pollutants, its development is also connected with long-term changes in the climate system.' A thinner ozone layer leads to an increase in ultraviolet radiation at the Earth's surface that can increase the risk of cataracts and skin cancers in humans, and increase mutation rates in other organisms.*

Greenhouse gases

Water vapour

Atmospheric water vapour comes mainly from evaporation and transpiration. The amount of water vapour in the atmosphere varies, and it has different effects on atmospheric temperatures depending on weather conditions. Water vapour condenses in the atmosphere to form clouds, which are made up of droplets of liquid water. Complete cloud cover can have a cooling effect, preventing some of the Sun's radiation from reaching the Earth's surface. Clear skies allow the Earth's surface to heat up quickly during the day, but also allow it to cool down more rapidly at night. Partial cloud cover can increase surface temperatures by retaining heat radiated from the surface.

Carbon dioxide

Most of the carbon dioxide in the atmosphere is produced naturally by respiration and volcanic activity. Carbon dioxide contributes about 50 per cent of the greenhouse effect. However, carbon dioxide levels have risen steadily since the beginning of the Industrial Revolution, probably as a result of burning fossil fuels and deforestation (spread 22.13).

Chlorofluorocarbons

Chlorofluorocarbons (CFCs) are organic molecules that are not natural but manufactured by humans. They are used as refrigerator coolants in air-conditioning units, as aerosol propellants, and in the manufacture of foam plastics. They are cheap, non-flammable, and non-toxic chemicals, but they are also relatively stable. When released into the atmosphere, CFCs tend to last for a long time (up to 60 years or more). Their concentration in the atmosphere has therefore gradually increased. Although present levels are low (typically less than 0.001 part per million [p.p.m.]), each CFC molecule has an effect several thousands times greater than that of a molecule of carbon dioxide. As well as contributing to the enhanced greenhouse effect, CFCs also affect the ozone layer in the upper part of the atmosphere. CFCs dissociate when exposed to ultraviolet light, releasing highly reactive free chlorine atoms. By a series of chemical reactions, these reduce the ozone concentration. Because of this effect, the use of CFCs is being phased out under the terms of the Montreal Protocol on Substances that Deplete the Ozone Layer, adopted in 1987.

Methane

Methane is produced by anaerobic bacteria living in the mud of boggy habitats, landfill sites, rice paddies, and in the guts of certain insects (such as termites) and ruminants (for example, cattle, sheep, and camels). Human activities such as cattle ranching have increased the release of methane into the atmosphere (a single cow, for example, can produce 200 dm^3 of methane per day).

Ozone

Ozone is the triatomic form of oxygen: its molecule has three oxygen atoms (O_3) rather than the two in the more common form of atmospheric oxygen (O_2). Ozone occurs naturally in the atmosphere mainly by a chemical process involving the action of solar radiation on oxygen gas. At high altitudes, about 20–50 km above the surface, ozone forms a layer (the **ozone layer**, figure 2) which has a concentration of about 1–10 p.p.m. Although high-altitude ozone is a greenhouse gas, its most important effect is to limit the amount of ultraviolet light reaching the Earth's surface. At low altitudes, ozone concentration is usually about 0.01 p.p.m., but it can be boosted by the action of sunlight on pollutants, especially nitrogen oxides in car exhaust fumes. Ozone helps make up the photochemical smog seen in some large cities during certain weather conditions. As a low-altitude pollutant, ozone is harmful. It is chemically very reactive, irritating eyes and respiratory tissues, and it can damage plants. Concentrations greater than 0.1 p.p.m. are considered toxic.

Quick check

1. Under what conditions might a natural substance be regarded as a pollutant?
2. Which of the following gases makes the greatest contribution to the greenhouse effect: carbon dioxide, chlorofluorocarbons, methane, or ozone?
3. At what concentration is ozone toxic?

Food for thought

Suggest a number of different ways in which the conversion of tropical rainforest to a cattle ranch might contribute to air pollution, the enhanced greenhouse effect, and global warming.

23.7

O B J E C T I V E S

By the end of this spread you should be able to:

- describe the main sources of water pollution
- explain how sewage is treated
- discuss the significance of eutrophication

Fact of life

Because soil nitrates are essential for plant growth, artificial nitrogen fertilisers are often added to soils. Unfortunately, about half the nitrate may leach out into the groundwater and, after a delay of up to 30 years, this nitrate can find its way into public water supplies. The nitrate may be converted to nitrite by bacteria in the mouth when the water is drunk. The nitrite is absorbed into the bloodstream and can combine with haemoglobin to form a blue pigment called methaemoglobin. This reduces the ability of the blood to transport oxygen. This can be life-threatening in babies. In the UK, nitrate levels of drinking water are monitored continually to ensure they do not exceed safety levels.

Freshwater pollution

When water vapour is formed by evaporation or transpiration, it is as pure and clean as the distilled water in science laboratories. However, as soon as it enters the atmosphere and continues its journey in the **water cycle** (figure 1), it can become polluted. The source of pollution may be industrial, domestic, or agricultural, and the pollutant may be thermal, nuclear, or chemical (inorganic or organic).

Thermal pollution

Water can become polluted even if no substances are added to it. In many industries, water is used as a coolant. Excess heat is discharged into a nearby waterway, causing thermal pollution. Discharges from power stations, for example, may raise the temperature of rivers and estuaries by several degrees above their normal level. Warm water can carry much less oxygen than cool water. Thermal pollution can therefore kill fish by depriving them of oxygen. It may also cause their death indirectly by encouraging the increased growth of parasites. However, thermal effluent can also have benefits: it increases the growth rate of some commercial shellfish and allows some warm-water organisms, for example tropical prawns, to be cultured in temperate regions such as Britain.

Sewage

Sewage is liquid waste: industrial waste (for example, from abattoirs, factories, and hospitals) and/or domestic waste (including human faeces, urine, and detergents). Sewage is carried through pipes called sewers. Although practically anything can end up in sewage, its typical composition is 99.9% water and dissolved substances and only 0.1% solids. Of the solids, approximately 70% are organic and 30% inorganic (for example, metals and grit). Raw sewage also contains a large number of bacteria, some of which can cause disease (they are pathogenic). Sewage is treated to dispose of all its potentially harmful components. Figure 2 and the box 'Stages in sewage treatment' describe one common type of sewage treatment. However, raw sewage is sometimes released into rivers or discharged into the sea without being fully treated, giving rise to pollution.

Discharge of untreated sewage into a river has an immediate effect on the aquatic environment, causing many changes in both the biotic and abiotic components. Some of these changes are due to specific chemical pollutants (for example, heavy metals such as cadmium from industrial processes, and pesticides from agriculture) and the effect varies according to the chemicals present in the discharge. However, two consequences invariably result from the discharge of large amounts of raw sewage: a decrease in biological oxygen demand, and eutrophication.

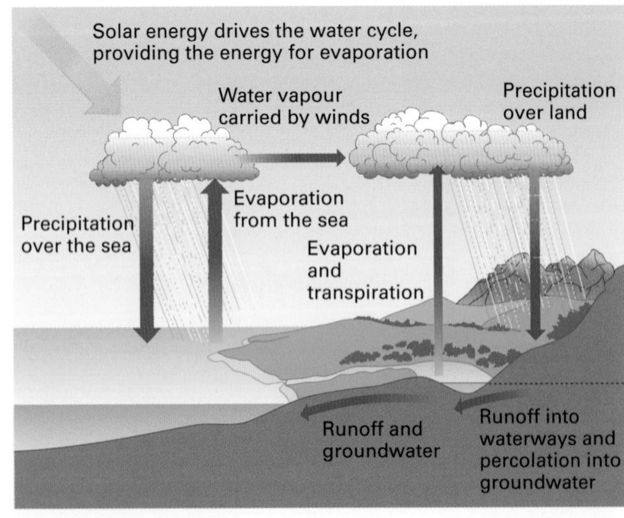

Figure 1 *The water cycle.*

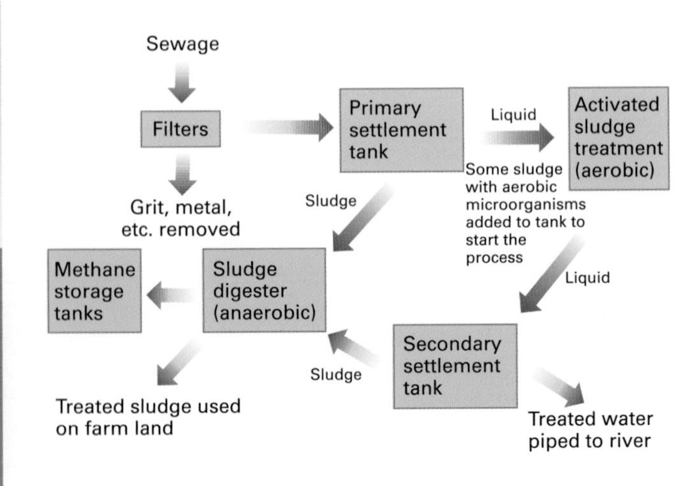

Figure 2 *The activated sludge process: one type of sewage treatment.*

Stages in sewage treatment

Sewage first passes through **filters** which remove solid inorganic matter, and then into **primary settlement tanks** where organic solids are separated as **sludge** from the liquid part. The thick sludge is pumped into **sludge digestion tanks** where, for a period of up to a month, it is digested by anaerobic microorganisms. Sludge digestion produces large amounts of methane gas which can be used as a fuel, and solid sludge, which can be used as organic fertiliser on farmland. The liquid sewage is piped into special tanks called **activated sludge tanks**. These tanks contain aerobic microorganisms which are activated by aeration of the liquid, for example, by pumping compressed air through it continuously. The microorganisms (protoctists and bacteria) feed on the organic material in the liquid and the harmful anaerobic bacteria it contains. In addition to digesting organic material and some harmful bacteria, the aerobic organisms produce a thick mucus which causes most of the remaining organic matter to flocculate (clump together). At the end of the process, the mixture is transferred to a **secondary settlement tank** where the flocculated material settles as a sludge. This can be piped to the sludge digester, leaving behind clean water.

Biological oxygen demand (BOD)

Adding organic material to water stimulates the growth of microorganisms which feed on the material. As the density of microorganisms increases, their demand for oxygen also rises. This demand is called the **biological oxygen demand** (**BOD**) and is measured as the mass (in mg) of oxygen used by 1 dm^3 of water stored in darkness at 20 °C for 5 days. The BOD of unpolluted river water is about 15 mg O_2 dm^{-3}; the BOD of raw sewage is about 100 times higher. Water that is very heavily polluted with raw sewage may become deoxygenated; this can lead to the death of aerobic organisms such as fish.

Eutrophication

Eutrophication occurs when organic material or inorganic nutrients, especially nitrates or phosphates, enter a freshwater habitat, either naturally or as a result of pollution by sewage or agricultural runoff containing fertiliser. The additional nutrients encourage the rapid growth of photosynthesising organisms, especially algae. Dramatic, fast growths of algae are called **algal blooms**. A bloom can smother plants, reduce light intensity in the water, and produce toxins which kill fish. When the algae die, their decomposition by bacteria may lead to the complete deoxygenation of the water, causing the death of aerobic organisms.

Oxygen depletion and eutrophication are not only caused by sewage pollution: they may be caused by any pollutant containing high concentrations of organic or inorganic nutrient, such as fertilisers (inorganic or organic), slurry (animal faeces and urine), or silage effluent which can leach off farmland and pollute water.

Silage and slurry

Silage is a fermented grass product used to feed cattle in winter. To make silage dry, wilted grass is stored at a pH below 5.5 to encourage the growth of beneficial *Lactobacillus* bacteria and discourage contamination by *Listeria* bacteria. If contaminated, silage breaks down into liquid compost which can leak easily from silos and enter aquatic habitats. Silage effluent is up to 200 times more polluting to water courses than raw silage. **Slurry** is a form of manure composed mainly of liquids. On intensive dairy farms, large quantities of slurry are produced, stored in special tanks. Some slurry is spread on fields as fertiliser. If a slurry tank is breached, large volumes of slurry can spill into watercourses where it increases turbidity, smothering large aquatic plants, and can cause severe eutrophication. Slurry is also likely to contain potentially harmful veterinary medicines (e.g. antibiotics) and toxic residues.

Some species, called **indicator species**, have well defined tolerances to water pollution and are used by biologists to monitor water quality (see spread 23.9).

Quick check

1 How might thermal pollution cause the death of fish such as trout?

2 What type of organisms are used during activated sludge treatment?

3 What is an algal bloom?

Food for thought

Figure 3 shows some of the sources of pollutants that affect the quality of river water. Suggest ways of reducing the pollutants or their effects, and consider the following:

- How could reversing the locations at which water is abstracted and discharged affect the output of industrial pollutants?
- In regions with seasonal climates, what effect does time of year have on the concentration of pollutants in water?
- How might vegetation lining the river banks affect pollution levels?

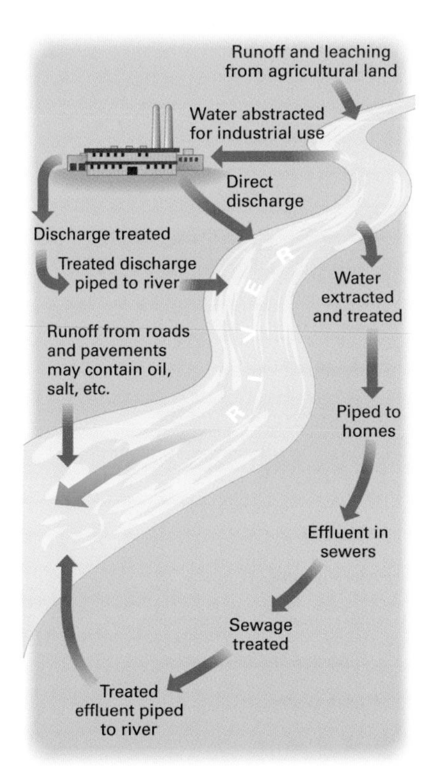

***Figure 3** Sources of water pollution.*

23.8

Marine pollution

OBJECTIVES

By the end of this spread you should be able to:

- list the main types of marine pollution
- discuss the biological effects and treatment of oil pollution.

Fact of life

Over 6 million tonnes of plastics are discarded into the sea each year. Nylon fishing nets, plastic bags, plastic straps, and millions of other objects made from non-biodegradable plastics cause the deaths of countless marine creatures. Some mistake the plastic objects for food. Fish and birds may become poisoned after feeding on small plastic beads resembling the tiny crustaceans that are a normal part of their diet. In sea water, plastic bags can take up the shape of a jellyfish and are often eaten by sea turtles. One scientist extracted enough plastic from the stomach of a leatherback turtle to make a ball over a metre in diameter. In such cases, the animals usually die through starvation because the stomach is clogged up with indigestible plastic which cannot pass through the gut.

There is probably no part of the sea that is completely free from the by-products of human activities. Most of these materials are diluted to such an extent that they are harmless, but others can occur in such large amounts that they are said to be marine pollutants. A **marine pollutant** is defined as a substance or energy introduced by humans into the marine environment which has harmful effects. Table 1 shows some common marine pollutants.

Oil spills

The type of marine pollution that probably captures the greatest public attention is the accidental spillage of huge amounts of oil from large sea tankers. For example, at 20:07 on 15 February 1996, the *Sea Empress* ran aground off Milford Haven, South Wales, releasing 72 000 tonnes of crude oil into the sea. The oil threatened fisheries, tourism, and many important environmental sites on the Pembrokeshire Coast National Park.

The actual environmental impact of such oil spills depends on many factors, including the time of year, the amount of wave action, and the direction in which the oil is carried. Oil may be prevented from reaching environmentally sensitive areas by placing floating booms at strategic places on the sea surface. However, if a slick reaches land, it can reduce the recreational and wildlife value of beaches. Oil driven out to sea probably causes little long-term damage because open water contains fewer species than the coast.

The toxicity of crude oil depends on the age of the spill. In the first few days after a spill, crude oil contains volatile components which are highly toxic. As little as one part per million (p.p.m.) can kill invertebrate larvae, and 1000 p.p.m. can kill fish. Within a few days, most of these volatile compounds evaporate into the atmosphere. Nevertheless, the remaining oil can still damage marine life. Oil on the feathers of sea birds impairs flight and reduces insulation, making birds vulnerable to hypothermia. If the oil is taken into the stomach during preening, it can poison the birds. The *Exxon Valdez* oil spill off the Alaskan coast in 1989 caused the deaths of 500 000 sea birds.

Marine invertebrates at greatest risk are bivalve molluscs such as mussels and cockles which feed by filtering food from the sea water. They filter droplets of oil present in the water as they filter food, and the hydrocarbons are absorbed into their body tissues. The concentrations of hydrocarbons in animals living close to a spill site can increase rapidly by many hundreds of times. Limpets and periwinkles are also susceptible to oil spills because they can take in large amounts of oil when they scrape

Table 1 *Some marine pollutants.*

Pollutant	Comments
Sewage	Increases biological oxygen demand; degradable
Agricultural waste	e.g. from factory farms; degradable
Organic industrial waste	e.g. from food and drink industry; degradable
Nitrate and phosphate fertilisers	Enter nutrient cycle; may contribute to algal blooms; non-persistent
Oil	Degradable; mainly from accidental spillages and cleaning tanker storage tanks
Hot water	e.g. from power stations; heat dissipates quickly soon after entering sea
Heavy metals	e.g. mercury and cadmium, in industrial effluents; persistent
Halogenated hydrocarbons	e.g. PCBs and DDT; persistent
Radioactive materials	e.g. from nuclear power stations; persistent
Mining waste	e.g. china clay waste
Medical waste	e.g. hypodermic needles, blooded bandages, vials of HIV-infected blood

the surface of rocks to obtain food. Limpets are an important component of rocky shore ecosystems; a reduction in their population is often associated with a massive increase in green seaweeds on rocky shores.

Oil does not have to be toxic to reduce the economic value of fisheries. Flesh tainted with oil becomes unacceptable for human consumption and cannot be marketed.

Figure 1 *A guillemot covered with oil. Birds covered in oil have difficulty swimming; many drown. Those that reach land usually ingest or inhale large amounts of oil as they try desperately to remove it from their feathers. The oil damages the bird's digestive system, and it loses body mass and becomes more susceptible to diseases such as pneumonia. Even after being fully cleaned by humans, heavily oiled guillemots that have lost a lot of body mass are extremely unlikely to survive for more than eight days after being released.*

Cleaning up

There are four main ways of dealing with an oil spill at sea.

1. **Leave it alone and let it degrade naturally:** about 25% of crude oil in a spill is volatile and evaporates into the atmosphere within three months; about 60% of crude oil is non-volatile, less dense than water, and floats on the surface where it is broken down by bacteria; the remaining 15% forms globules that sink to the bottom and are broken down slowly.
2. **Mechanical recovery:** this usually involves specialised recovery vessels.
3. **Burning:** this can be 95–98% efficient, but causes black smoke and soot which may pollute coastal towns and cities.
4. **Aerial dispersants:** dispersants are chemicals that reproduce the natural action of waves; they break oil into droplets, stabilise the droplets, and disperse them into the surrounding sea where they may stay. Dispersants prevent slicks from reforming on the surface. They do not actually make oil disappear, but move the oil from the surface down into the water, where it becomes available to naturally occurring marine bacteria which can then degrade it. This degradation can reduce the oil to 5% of its original volume.

Once the oil has reached the shore it may be:

- physically removed by taking off the top layer of sand
- dispersed using detergents or high-pressure hoses
- removed manually by a large workforce using a variety of techniques.

Methods used to clean up an oil spill may speed up, slow down, or have no effect at all on the rate at which the area returns to its unpolluted condition. Inappropriate clean-up procedures can do more harm than good. For example, removing the top layer of sand or applying detergents can destroy shore communities which might have survived the oil. High-pressure hoses can drive oil deeper into the sediment where it can remain for a very long time.

Perhaps the best method of dealing with oil spills is by **bioremediation**. This aims to speed up oil degradation by stimulating naturally occurring bacteria to break down the oil faster. These bacteria use oil as a food source, but they also need oxygen, nitrogen, and phosphorus. Oxygen is usually plentiful, but raising levels of nitrogen and phosphorus by applying fertiliser increases the rate at which the bacteria can feed on oil.

PCBs and the death of seals

In 1988, the seal population of the North Sea suffered a massive mortality. Of 18 000 harbour seals, approximately 12 000 died from a new disease called **phocine distemper virus**. It is thought that the seal population had become infected with the virus after being weakened by the effects of marine pollutants such as nitrogen and phosphorus fertilisers and PCBs (polychlorinated biphenols, chemicals once used as insulators in electrical divices). The pollutants were not at a high enough concentration to cause death, but they may have greatly weakened the immune system of the seals. The highest mortality of seals occurred in areas with the highest levels of PCB pollution. This incident illustrates the complexity of marine pollution. Pollutants do not exist in isolation. They can interact with one another and with other environmental factors. A substance that is harmless on its own can become a deadly pollutant when combined with one or more other factors.

Quick check

1. List two degradable types of marine pollutant.
2. Explain why the addition of a nitrogen fertiliser to sea water can accelerate the clean-up of an oil spill.

Food for thought

Suggest how detergents can have a harmful effect on sea birds.

23.9

Indicator species

OBJECTIVES

By the end of this spread you should be able to:

- explain what an indicator species is
- describe how water quality can be monitored biologically
- discuss the advantages and disadvantages of biological monitoring.

Fact of life

Respiration among aquatic insect larvae varies with oxygen availability. Stonefly nymphs and mayfly larvae have delicate gills with which to acquire oxygen from well aerated waters; chironomid larvae contain haemoglobin to extract oxygen where levels are low – the lower the level, the greater the haemoglobin concentration; rat-tailed maggots (*Eristalis*) have long breathing tubes at the rear through which they can obtain air from above the water surface, allowing them to live in water totally depleted of oxygen (figure 2)

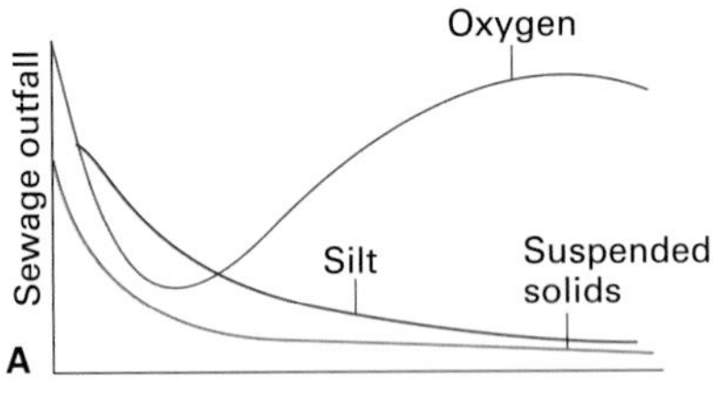

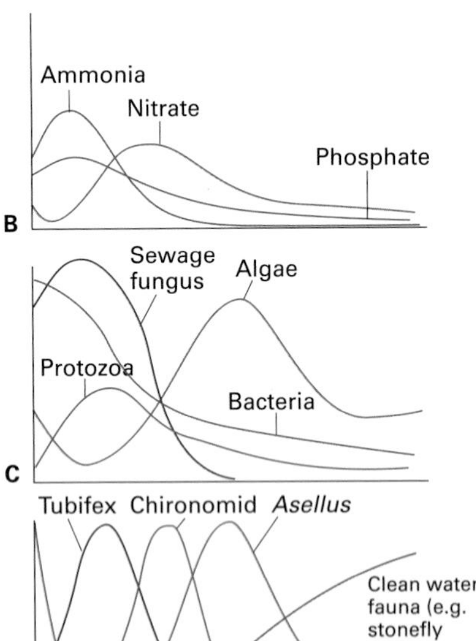

Figure 1 *Changes in the abiotic (**A** and **B**) and biotic (**C** and **D**) conditions of river water downstream of a sewage outfall.*

Before the invention of sophisticated monitoring equipment, coal miners used to take a canary in a cage with them into a mine. Pockets of carbon monoxide are common in mines and can overcome a miner before he is aware of the poisonous gas. Canaries are particularly sensitive to air quality and succumb more quickly to carbon monoxide poisoning than do miners. So a canary falling off its perch was taken as an alarm signal to get out of the mine. The canary was being used as an indicator species to determine whether the conditions in the mine were suitable for the health of the miner.

An **indicator species** is a species that needs a particular environmental condition or set of conditions in order to survive. Whether the species is present, and in what numbers, provides information about the environment. Indicator species are used in a wide range of ecological investigations to find out about both the present and past conditions of, for example, soil and climate. In this spread we shall discuss how indicator species can be used to monitor water and air quality.

Biological monitoring of water quality

The discharge of raw sewage into a river has an immediate impact on the environment at its point of entry and affects the quality of water for some way downstream. Figure 1 shows that changes in the abiotic environment are reflected by changes in the community of aquatic organisms.

The fall in oxygen content is the most dramatic effect of sewage. Fauna requiring clean water and high oxygen levels are absent or in low numbers near the sewage outfall, whereas populations of sewage fungus (a slime consisting of filamentous bacteria, protoctists, and algae, as well as fungi), bacteria, and other organisms that thrive in organic waste are high. Downstream, decomposition of organic matter leads to the release of nitrates which stimulate algal growth, and causes a progressive decrease in oxygen content (known as the 'oxygen sag') of the water. In extreme cases, the oxygen content may fall so low that conditions become anaerobic. There comes a point at which the organic material is completely broken down and the river fauna starts to recover: populations of clean-water species increase and those adapted to the polluted conditions decline. These changes show how, within limits, the activity of aquatic organisms, helped by dilution of the pollutant as it passes downstream, can purify the water. The distance over which the purification takes place varies and is affected by factors such as temperature, volume of water, and the existing population of microorganisms.

Studies of many rivers show that there are characteristic communities of organisms associated with different levels of organic pollution in the river. Clean-water organisms include stonefly nymphs (larvae), mayfly larvae, and caddis fly larvae. These organisms are indicator species of unpolluted, well oxygenated water, and a community containing the

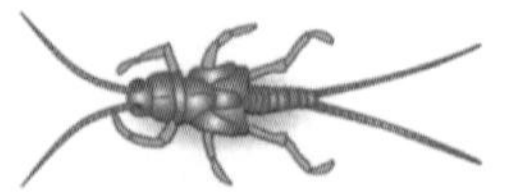

1 Stonefly nymph (up to 30 mm)

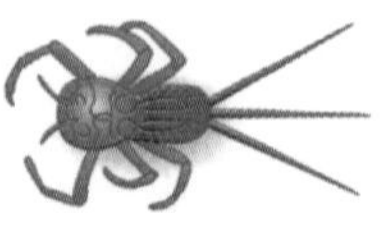

2 Mayfly larva (up to 15 mm)

3 *Asellus* (freshwater louse) (up to about 12 mm)

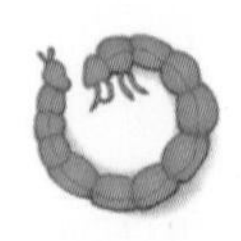

4 Chironomid (bloodworm; a midge larva) (up to 20 mm)

5 Rat-tailed maggot larva (up to 55 mm including tube)

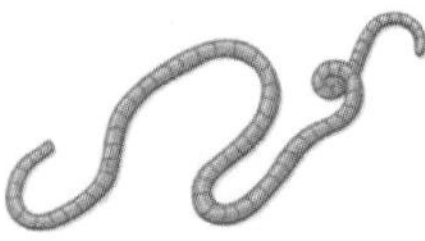

6 Tubifex (sludge worm) (up to 40 mm)

Figure 2 *Biological indicators of organic pollution: these species are arranged in order of the level of pollution they tolerate, where **1** is the least polluted and **6** the most polluted water.*

organisms is an **indicator community** of well oxygenated water. As pollution increases and oxygen content falls, populations of species sensitive to a decrease in oxygen content will fall and those tolerant to organic pollution, such as blood worms, rat-tailed maggots, and tubifex worms, increase (figure 2).

In the UK, samples of freshwater organisms are routinely taken from rivers to monitor levels of organic pollution. A biologist can grade the quality of river water on a scale ranging from clean water with high species diversity to grossly polluted water with no organisms or only a few anaerobic ones. There are a number of **biotic indices** used for this work, based on the diversity and abundance of aquatic organisms present, with particular reference to 'key species' that have known tolerance to organic pollution. Among the simplest is the Trent biotic index which is based on the presence or absence of certain large invertebrates. The Chandler score is a more accurate index, requiring the accurate identification and counting of a number of invertebrate species. Indicator species may also be used to assess levels of inorganic pollutants. For example, chemicals contained within individual indicator species may be analysed to assess levels of inorganic pollutants, such as pesticide organochlorines.

Compared with physicochemical monitoring (measuring physical and chemical conditions), biological monitoring of pollution has advantages and disadvantages.

Advantages of biological monitoring include the following:

- The abundance of organisms depends on the sum of all factors affecting environmental quality; physicochemical factors are usually measured individually.
- The abundance of an organism reflects the effect of continuous exposure to all environmental factors; physicochemical measurements are usually taken at intervals and are often spot checks.
- A brief and damaging pollution incident would continue to affect the abundance of organisms for some time after the event; physicochemical measurements could miss this unless they were continuous.

Disadvantages of biological monitoring include:

- Biological indicators do not reliably identify the precise cause of a pollution incident; physicochemical measurements can identify the precise cause, allowing specific action to be taken.
- The abundance of indicator species may vary naturally, so absence of an organism may not reflect a particular environmental condition; carefully conducted physicochemical measurements do not suffer from this problem.

In most situations, environmental quality is monitored both biologically and physicochemically. Biotic indices are used for routine, continuous monitoring of the environment and when they indicate a problem, physicochemical tests are carried out to identify the precise cause.

Quick check

1 What is sewage fungus, and in what kind of conditions does it occur?

2 Explain why you would not expect aerobic organisms such as mayfly larvae to be found close to a sewage outfall.

3 Why would the absence of stonefly nymph from a stream not necessarily indicate that its water was not of the highest quality?

Food for thought

Suggest why water downstream of sewage outfall has higher population levels of *Escherichia coli* than water upstream of the outfall.

Biological monitoring of air quality

Lichens are commonly used to indicate levels of air pollution. A lichen is a symbiotic association between a fungus and an alga, the fungus providing protection and the alga providing carbohydrates (and sometimes other substances). Lichens live on stable surfaces such as on rocks and trees. Most lichens are particularly sensitive to levels of atmospheric sulphur dioxide, because they have no cuticle or stomata; they absorb water and nutrients directly from the air. Air pollution levels (more accurately, sulphur dioxide levels) are indicated by a biotic index based on lichen species diversity in which abundance includes the number of individuals, extent of coverage, and growth form of the lichens. On the most heavily polluted surfaces, lichens are absent; only algae are present. Lichen diversity and growth increases as the air becomes purer. The highest species diversity occurs on surfaces surrounded by pure air where some species, such as *Usnea* (figure 3), have a particularly luxuriant growth.

Figure 3 *Usnea, an indicator of clean air.*

23.10

OBJECTIVES

By the end of this spread you should be able to:

- give points in favour of conservation
- list the main methods of conservation
- describe an example of conservation in action.

Fact of life

Extinction is a natural and continuous part of the evolution of life on Earth. The rate varies from small to massive. **Mass extinctions** occur when abnormally large numbers of species die out in a very short time. There have been five especially big mass extinctions. The best known is the Cretaceous–Tertiary event when dinosaurs became extinct. The largest event, nicknamed the Great Dying, occurred at the end of the Permian. It wiped out between 50 and 95 per cent of all species. Today, many researchers believe that the Earth is experiencing another mass extinction event as a result of human activities. Although researchers generally agree that biodiversity is declining dramatically, they do not agree about the rate. Different methods are used to estimate the rate, and these give different results.

Figure 1 African elephants, Loxodonta africana, *in their natural habitat. The elephant is regarded as a **keystone species** because many other species in the ecosystem depend on its activities. The African elephant is classified in the IUCN list of threatened species as 'vulnerable to extinction'.*

Biological conservation

Biological conservation aims to maintain the quality of natural environments and their biological resources. Unlike preservation which tries to prevent human interference, conservation involves actively managing biotic and abiotic components to ensure the survival of the maximum number of species and genetic diversity (spread 22.11). In order to survive, populations must have biotic resources (such as food and access to a mate) and abiotic resources (such as a space in which to live, and shelter). A permanent lack of a vital resource will lead to extinction. In many ecosystems, the activities of humans and wildlife are in conflict. To reconcile this conflict, biological conservationists generally adopt a policy of **sustainable development**: development that can continue indefinitely because it is based on exploiting renewable resources while minimising environmental damage, maintaining the overall integrity of the ecosystem, and conserving animal and plant species.

To many people, the reasons for conservation are self-evident: we live in a world enriched by the diversity of its habitats and organisms and these are threatened with extinction. Once extinct, they will be lost forever. To others, conservation is a costly business that can be justified only if it has tangible (for many, this means economic) benefits.

Conserving genetic diversity

Biological conservation includes conserving genetic diversity, which can be measured by assessing differences in nucleotide sequences using **DNA sequencing** techniques. To conserve genetic diversity, ecologists must be able to identify individuals and species. This can be achieved using **DNA profiling** (spread 18.11) and genetic markers. A **genetic marker** is a genetic variation unique to the individual or species that can be identified as a particular nucleotide sequence with a known location on a chromosome. It may be a short nucleotide sequence, such as that surrounding a single base-pair change (**single nucleotide polymorphism**, SNP), or a long one, such as **micro-satellite repeat sequences** (**MRS**s, repeating sequences of a small number [2–5] of base pairs). An SNP occurs when two sequenced fragments of DNA from different individuals contain a difference in a single nucleotide (for example, individuals, one fragment might be GAGCCTA and the other GAGCTTA).

The first DNA profiling to be used routinely involved identifying **restriction fragment length polymorphisms** (**RFLP**s). When a DNA sample is broken into pieces by restriction enzymes, the resulting restriction fragments are separated by gel electrophoresis according to their lengths. In RFLPs, the sequences of nucleotides in samples of DNA are the same except for the locations of restriction sites: they function the same, but when restriction enzymes are used to cut them up, different fragments are created.

Habitat conservation

***In situ* conservation** (conservation in the natural environment) of individual species depends on the conservation of their habitats. Wildlife habitats are continually threatened by agricultural, industrial, and urban development. Such developments can destroy habitats completely or fragment them. **Habitat fragmentation** (the breaking down of a continuous habitat into smaller areas) results in a partial loss of the original habitat, a reduction in habitat patch size, and the potential isolation of habitat patches. Patch isolation can be lessened by the establishment of **wildlife corridors** such as hedges and grassy footpaths between the patches. These allow organisms to move from one patch to another.

The impact of habitat fragmentation depends on the size of the habitat patches that are formed. In the 1980s two ecologists, Robert MacArthur and E. O. Wilson, established a new area of study called **island biogeography** which investigates the factors that affect the species richness of isolated natural communities. In the context of island biogeography, an island is defined as any area of habitat suitable for a given community surrounded by unsuitable habitat. Studies of different islands indicated that the size of habitat is important. Large islands tended to have a greater variety of habitats than small islands, supported more species and had larger populations of each species. And for any particular group of organisms, such as flowering plants or molluscs, species richness tended to increase with the area of an ecosystem up to a maximum, as shown in a **species–area curve** (figure 2). These results had important implications for conservation biology and led to a debate as to which was better for conservation purposes: a single large reserve or several small reserves with a total area equal to that of the single large reserve? This became known as the **SLOSS** (single large or several small) debate. Those in favour of having several small areas argue, among other things, that these could be chosen to have different habitats and ecosystems which would probably increase species diversity. Those in favour of having a single large area argue that having adjoining areas ensures that different communities can interact effectively. They also argue that large reserves are more likely to have populations above the **minimum viable population density** (the lowest population density required to minimise the risk of extinction over a given period of time) than small areas. Much seems to depend on what is meant by 'large'. For conservationists whose primary concern is small annual plants or vegetative perennials, 100 hectares may be regarded as large, whereas for conservationists whose primary concern is large mammals, this would probably be regarded as a tiny area. Most conservation biologists agree that the answer to the debate is case specific and that it is best to conserve as much land as is possible, or at least as much land as is needed to provide everything the target organisms require.

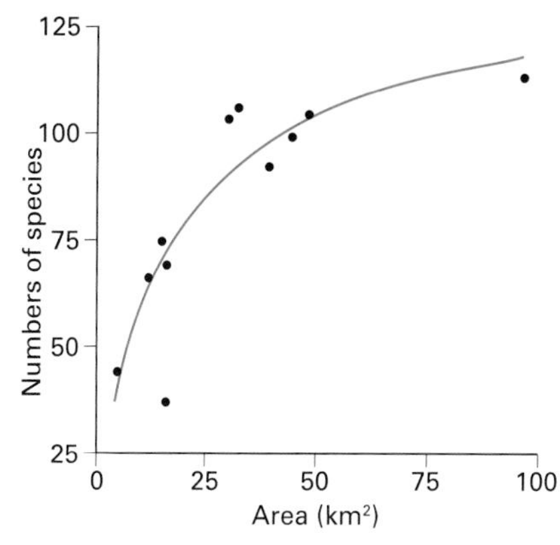

Figure 2 *Species–area curve for benthic (bottom living) freshwater fish in the St Lawrence River, Canada. (Data from Environment Canada)*

International conservation organisations

Wildlife and human activities transcend international boundaries. Effective biological conservation requires international cooperation. For example, the **Convention in International Trade in Endangered Species** (CITES) is an international agreement between governments to ensure that international trade in wildlife and wildlife products, such as ivory, does not threaten the survival of endangered species. Since the 1960s, the **International Union for the Conservation of Nature** (IUCN) has collated wildlife data from a variety of scientific sources and applied precise criteria to assess whether or not a species is endangered (that is, at high risk of extinction in the wild). This work is summarised in the **IUCN Red Lists** of Threatened Species. In 1992, at the Rio Earth Summit, more than 150 government leaders signed the **Rio Convention on Biological Diversity**, committing their governments to promoting sustainable development through the implementation of Agenda 21, a comprehensive plan of action to be taken globally, nationally, and locally by organisations of the United Nations System, by governments, and by major groups in every area in which humans impact on the environment. These commitments were reaffirmed at the World Summit on Sustainable Development (WSSD) held in Johannesburg, South Africa in 2002.

Quick check

1 How have countries such as Kenya taken economic advantage of the aesthetic value of natural environments?

2 What action was taken in 1989 to try to stop the decline of elephant populations in Africa?

Food for thought

Species from every phylum in the animal kingdom are threatened with extinction. Suggest why many conservation schemes focus on saving large mammals.

23.11

Conservation methods

OBJECTIVES

By the end of this spread you should be able to:

- list the main methods of conservation.

Fact of life

The archipelago of the **Galapagos Islands** is one of the most scientifically important and biologically outstanding places on Earth. The islands have a unique flora and fauna which are inextricably linked to Charles Darwin and his theory of evolution (spread 20.1). About 80% of the land birds, 97% of the reptiles and land mammals, 30% of the plants, and 20% of marine species are found nowhere else on Earth. These organisms include the giant Galapagos tortoise, marine iguana, flightless cormorant, and the Galapagos penguin, the only penguin species to be found in the Northern Hemisphere. About 97% of the total land area is National Park. Ever since the islands were discovered in 1535 their wildlife has been threatened by human activities. This included the killing of wildlife, such as the Galapagos tortoise, for food. But potentially most damaging were the organisms introduced to the islands either accidentally or on purpose by humans. Unfortunately, many of these, including goats and blackberries, have been aggressively invasive and detrimental to endemic species. There were no permanent human inhabitants on the islands until the 1830s, but by 2010 the resident population had grown to over 25 000 people and more than 150 000 tourists visited the islands. This influx of people caused so much concern that special conservation measures were implemented to control the spread of invasive species (for example, by limiting human access points and improving quarantine measures) and to limit human residency and tourism (for example, by limiting the number of tourism permits).

Biological conservation may be *in situ* (undertaken in the natural environment) or *ex situ* (undertaken outside the natural environment). As individual species can survive only if they have suitable habitats, the priority of ***in situ* conservation** is to protect and restore natural habitats or to enable new ones to form. This may involve mowing, grazing, or burning to halt succession (spread 22.10), coppicing (the pruning of trees and shrubs to ground level; this allows vigorous regeneration which can provide a sustainable source of timber), or the control of water levels. ***Ex situ* conservation** methods include breeding programmes for endangered species in zoos and botanic gardens. They also include establishing sperm banks and **seed banks** for the long term-conservation of gene pools. This is important not only for wildlife but also for conserving rare breeds of farm animal and different types of agricultural crop.

Where natural habitats are threatened, for example by an urban development, **environmental impact assessments** (EIAs) form an important part of conservation. An EIA is carried out before the development to ensure that the decision-makers consider the effects on the environment. An EIA usually includes an audit of the wildlife and an assessment of the possible positive or negative impacts that the proposed development may have on the wildlife and their environment.

Other conservation methods include:

- reclaiming derelict sites, such as mine waste tips (figure 2)
- protecting endangered species by legislation
- prohibiting the release of non-native species into an area
- controlling pollution, especially in environmentally sensitive areas
- recycling waste
- limiting the exploitation of renewable resources to sustainable yields
- restricting trade in endangered species and their products
- introducing 'closed periods' and exclusion zones for hunting and fishing
- establishing Special Areas of Conservation (SACs), Sites of Special Scientific Interest (SSSIs), National Parks and similar protected areas with restrictions on development.

Ultimately, conservation is the responsibility of each of us. We can all take an active part by supporting conservation organisations, recycling materials, buying 'environmentally friendly' products such as recycled paper, and using the countryside in a way that minimises damage to wildlife.

Figure 1 *The Galapagos islands. Ninety seven per cent of the total land of the archipelago is National Park. A National Park is defined by the International Union for Conservation of Nature as: an extensive area of land that has not been significantly altered by human activities; that contains landscapes, species, or ecosystems of scientific, educational, or aesthetic value; and that is set aside permanently to be managed by the State for conservation. The UK definition of a National Park differs slightly as there are few areas in the UK that have not already been altered by human activity.*

Figure 2 *Soay sheep have been introduced onto reclaimed China clay spoil heaps in Cornwall. China clay (kaolin) mining is an open-cast process. For every tonne of china clay produced, there are between seven and eight tonnes of waste generated. The waste has limited use. Most is dumped to form huge white spoil heaps. Modern spoil heaps are landscaped and seeded with special grasses and nitrogen-fixing legumes, mostly clover, allowing them to be reclaimed. In the early 1970s, Soay sheep were introduced to the pastures formed on reclaimed heaps. Originating from the island of St Kilda, 50 miles off the coast of Scotland, these sheep are light, agile and independent. Their grazing helps to improve the soil and allows the land to be put to commercial use at an early stage.*

Conservation in the countryside

In the UK much of the countryside is used for agriculture and forestry, therefore farmers and foresters play an important role in the conservation of our wildlife. Methods used to support wildlife include:

- retaining hedgerows where they exist and planting new ones where they don't. Although hedgerows cost money and time to create or retain, reduce the land available for planting crops, and make it more difficult to use large machinery, they act as wildlife corridors, form windbreaks which reduce wind erosion of soils, contain refuges for predators that limit the size of pest populations, and make the countryside more aesthetically pleasing by adding variety to the landscape. To be most useful for the conservation of wildlife, hedgerows should be of sufficient length, height, and width to support wildlife, contain some native trees, and be cut at times and at a frequency that suits the wildlife (e.g. nesting birds)
- promoting polyculture instead of monoculture to increase biodiversity and reduce the risk of pest populations becoming established
- using integrated pest management schemes (spread 23.2)
- establishing 'set-aside' areas of land (patches of land on farms and in forests that are set aside for wildlife conservation)
- maintaining indigenous hardwood forests, for example, of oak, ash, and beech which provide a greater diversity of habitats and species compared with softwood forests of introduced softwood species such as larch and spruce
- planting woodland of fast-growing softwoods intermixed with hardwoods to increase biodiversity.

Quick check

1 a Why is mowing, grazing, or burning necessary to conserve grassland and heathland?

b How does coppicing increase species diversity in a woodland?

c Which was the first National Park in England and Wales?

Food for thought

Suggest ways in which you could take positive action to help conserve biological resources.

Summary

Applied ecology, the use of ecological concepts in areas directly relevant to human societies, covers topics such as **agriculture**, **fisheries**, **deforestation**, **pollution**, and **conservation**. In agriculture, the maximisation of crop and animal production requires the intelligent application of biological and ecological concepts to manage **soils**, control pests, and produce new varieties of crops and animals. Soils can be managed by controlling **irrigation**, adding **fertilisers**, and by **rotating crops**, for example. Pests may be controlled by **pesticides**, **biological control agents**, or by **integrated pest management**. New varieties of farm animals and plants can be produced by **selective breeding** and **genetic engineering**. Growth rates can be optimised by intensive methods of fanning, for example, by growing plants in **greenhouses** or animals in **factory farms**. Ecological concepts are applied in fisheries to determine **maximum sustainable yields** and avoid **overfishing**. Applied ecology is playing a key role in understanding how deforestation affects nutrient cycles, both locally and globally.

Pollution includes air pollution, freshwater pollution, and marine pollution. **Air pollution** by **greenhouse gases** may lead to **global warming**. **Freshwater pollution** by nitrate fertilisers and by organic wastes can lead to **eutrophication** and an increase in **biological oxygen demand**. **Marine pollution** may have a number of causes, but one of the most dramatic is **oil spillage**. Effective treatment of oil spills requires sound application of ecological and biological concepts. **Indicator species** can be used to monitor levels of pollution. Lichens, for example, are used to monitor air quality. **Biological conservation** aims to maintain the quality of natural environments and biological resources. The conservation of species, such as the **African elephant**, is fraught with difficulties and requires not only the application of sound ecological concepts, but also the application of sound political and sociological concepts. **Conservation methods** include **coppicing**, control of water levels in an ecosystem, and the designation of conservation areas, such as National Parks and nature reserves.

PRACTICE EXAM QUESTIONS

1 An area of rain forest was cleared for agriculture. Two experimental plots of land were marked out. On one, rice was grown continuously. On the other, crops of rice and beans were alternated. No fertiliser was added to either plot. The graph shows the crop yields for the two plots, over a period of time.

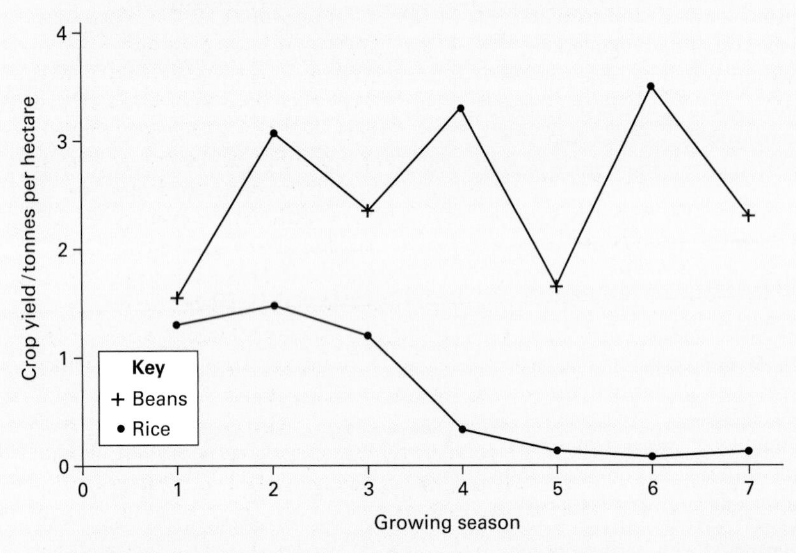

a Explain why:

i the yield fell when rice was grown continuously; [1]

ii the crop yield was maintained when rice and beans were alternated. [2]

b In another investigation, it was found that the biomass of decomposers fell from $54\,g\,m^{-2}$ to $3\,g\,m^{-2}$ when the rain forest was converted to permanent agriculture. Explain how you would expect this fall to influence the recycling of nutrients. [2]

[Total 5 marks]

2 Otters make a comeback

In the 1950s and 1960s there was a dramatic decline in the otter population. Pollution and removal of vegetation from riverbanks wiped them out in almost all of the Midlands and South-East England. In the 1990s, the Government, through conservation organisations such as local wildlife trusts, carried out a number of ambitious projects designed to restore otters to all rivers where they existed before the 1960s. Many of these projects were successful.

a Suggest how pollution might have contributed to otters becoming extinct in many rivers. [8]

b Describe how you would measure the size of a local otter population. [4]

[Total 12 marks]

3 The graph shows the catches of cod from the North-west Atlantic between 1960 and 1990. It also shows the size of quotas which were introduced in the 1970s.

a Why was it necessary to introduce quotas in the 1970s? [1]

b In the 1970s the quota was revised from 400 000 tonnes to 200 000 tonnes. Use evidence from the graph to suggest why this was necessary. [2]

c Give **two** methods, other than quotas, that can be used to regulate fishing. [2]

[Total 5 marks]

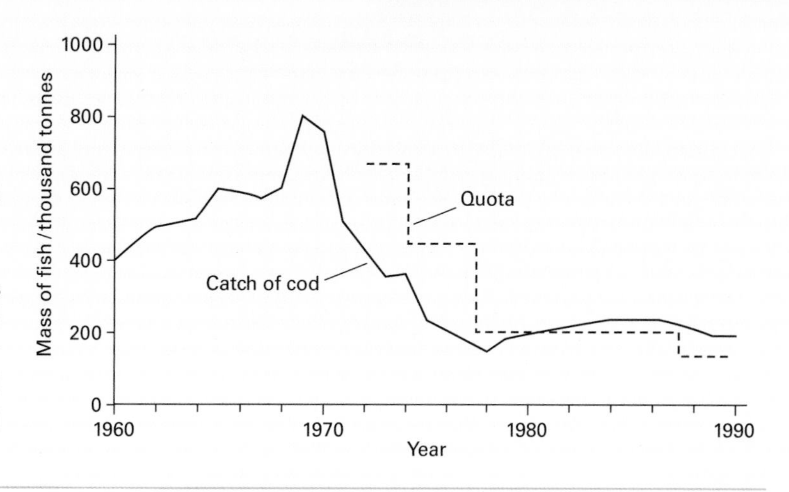

4 Trout and salmon farming have expanded rapidly over the last few years, but the development has not been without environmental costs. Intensive fish farms pollute habitats with suspended solids, nitrates and phosphates, while escaping fish threaten wild fish stocks.

a Suggest **one** source of each of the following from fish farms

i suspended solids

ii nitrates and phosphates [2]

b Suggest **two** ways by which escaped fish may threaten wild fish stocks. [2]

c Explain how pollution with nitrates and phosphates may affect the concentration of oxygen in the surrounding water. [3]

[Total 7 marks]

5 A number of current agricultural practices are of immediate benefit to farmers and growers, but may have long term adverse effects on humans and the environment. For each of the following agricultural practices, state **two** benefits and **two** adverse environmental or human consequences.

a Deforestation. [4]

b Applying nitrogenous fertilisers to crops. [4]

c Burning agricultural waste, such as straw. [4]

d Growing crop plants with genetically engineered resistance to herbicide (e.g. glyphosate). [4]

[Total 16 marks]

6 In this section, 16 marks are allotted for scientific content and 4 marks for orderly presentation and quality of English.

Diagrams may be used to assist your account, but are not essential.

Either a Discuss the role of Government legislation in the conservation of natural habitats in the United Kingdom. [20]

Or b Discuss the principles involved in the conservation of rare species. [20]

[Total 20 marks]

7 Read the following passage.

Decline and fall of the amphibians

Frogs, toads and salamanders are in deep trouble – or so it would seem from the many reports of troubled populations which have emerged in recent years. Some zoologists and conservationists fear that many, if not
5 most, of the world's 4000 or so species of amphibians are in decline, their plight a symptom of global pollution and climate change.
The key factor in the survival of amphibians for more than 300 million years has been their ability to exploit a
10 habitat that is inhospitable to other vertebrates – the temporary pool. By laying eggs in such pools the newly hatched offspring are able to take advantage of a rich but transient flush of food. As the pond dries, the tadpoles metamorphose and emerge onto land, where they occupy
15 a different niche as carnivorous frogs, toads or salamanders. The problem with such a strategy is that it puts the amphibians at the mercy of unpredictable weather conditions. It the rains fail or the pond dries out too early, no young from that season will survive.
20 Some evidence suggests that amphibians are being wiped out by acid rain and the indiscriminate use of pesticides. They are probably most vulnerable to waterbome pollutants and acid rain at the egg and tadpole stages of development. Species vary considerably in the ability to
25 tolerate acidity, but the eggs of most will die if the pH falls much below 4.

(Reproduced from an article by permission of the New Scientist)

a Explain what is meant by 'different niche' (line 15), and how occupying different niches as young and adults helps the survival of amphibians. [3]

b Suggest the sources of the following pollutants described in the passage and explain how they may affect amphibians and their habitat.

i Acid rain.

ii Pesticides. [6]

c Suggest why the egg and tadpole stages may be 'most vulnerable to waterbome pollutants' (lines 22 and 23). [3]

[Total 12 marks]

Appendix

Analysis of biological chemicals

Different **separation techniques** can be used to analyse substances.

Electrophoresis (figure A1) is a technique for separating charged particles. It is used extensively to separate mixtures of proteins or amino acids, or to separate fragments of DNA in DNA profiling (spread 18.11).

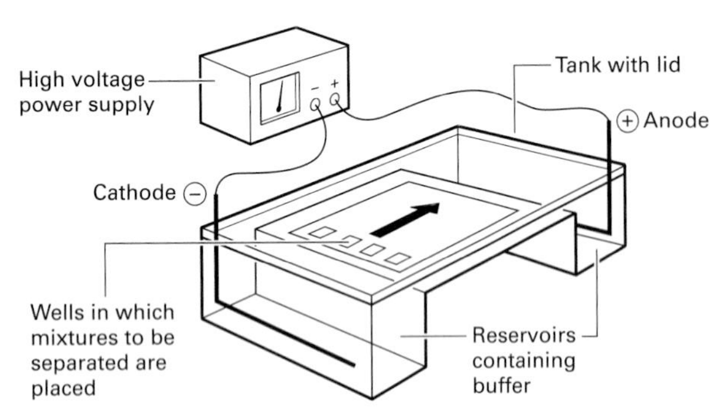

Figure A1 *Electrophoresis. Negatively charged molecules move towards the anode.*

Electrophoresis requires a positively charged electrode (anode), a negatively charged electrode (cathode), and a solution between them capable of conducting electricity. The solution has to be held in a supporting medium. This may be a strip of filter paper or a gel. The supporting medium is soaked in a suitable buffer to keep the pH constant and to prevent any change in the particles' charges (altering the pH of the buffer will alter the charge on an amino acid or protein; spread 2.9). Small amounts of the mixtures to be separated are placed on an appropriate spot on the supporting medium (e.g. a well cut into the gel). A potential difference of about 200 V is applied to the supporting medium for a specific period of time. Particles with a positive charge migrate towards the cathode, and those with a negative charge migrate towards the anode. Once the supporting medium has dried, the final position of the particles can be revealed using a stain. For example, amino acids move on the medium at different rates according to the charge on their R group, and their final position can be detected by spraying the **electrophoretogram** with ninhydrin (a stain which gives an intense blue colour with most amino acids, yellow with proline). The distance each particle has travelled under known conditions can then be measured. If the identity of a particle is unknown, its final position on the gel or filter paper can be compared with positions of standards.

Most laboratories that carry out electrophoresis routinely use a gel such as PAG (polyacrylamide gel) as the supporting medium. The separation of particles in gel electrophoresis depends not only on their charges but also on their size and shape. This is because the gel has tiny pores that enables it to act like a molecular sieve. Thus, charged particles which differ in size separate from each other even if they have the same electrical charge.

A typical analysis of a mixture of proteins begins with treating the mixture with a detergent such as sodium dodecyl sulphate (SDS). The detergent attaches negatively charged ions to the protein molecules. The proteins can then be loaded at the cathode end of the apparatus and will be attracted to the anode. The distance they travel in a given time will depend not on their charge but their size, the smaller molecules moving further than the larger ones. As the distance moved by a protein is normally proportional to its molecular mass, the molecular mass of an 'unknown' protein may be calculated.

In **restriction mapping** (spread 18.9), gel electrophoresis can be used to determine the number and size of the fragments cut at specific restriction sites by different restriction endonucleases. The smaller the fragment, the further it moves along the gel. By using DNA fragments of known size it is possible to estimate the number of base pairs in each test fragment. A fragment may have, for example, 12 kb (12 kilobase pairs, 12 000 base pairs). With the ends of the test DNA marked with a fluorescent dye and the DNA completely digested with two or more restriction endonucleases both singly and together, enough information is gained to make a restriction map (figure A2). When compiling a restriction map, genetic engineers need to take into account the following.

- DNA can be circular or linear.
- Specific restriction endonucleases cut DNA at specific restriction sites.
- Partial digestion produces larger fragments than complete digestion. If DNA is incubated with the restriction endonucleases for long enough to achieve total digestion, larger fragments are cut into smaller ones.
- With linear DNA, the total number of fragments is always one more than the number of restriction sites; with circular DNA, the number of fragments equals the number of restriction sites.
- Similarly sized fragments can mask each other: although they have different base sequences, they can appear as one band on a gel.
- The combined length of the DNA fragments should equal the original length of the undigested starter DNA. If it does not, one or more DNA fragments might be masked.
- When two restriction enzymes are used, it is helpful to look for fragments in one enzyme's pattern that appear to be cut by the other enzyme. It is often good to start with the enzyme that cuts the most fragments, then look for those that are cut again by

the enzyme that cuts the least fragments.

- When linear DNA is digested by two enzymes, the ends of fragments might be free ends (F) of the starter DNA, exposed ends (E1) of the internal parts cut by enzyme 1, or exposed ends (E2) cut by enzyme 2. The ends of a DNA might therefore be: F —— E1; F —— E2; E1 —— E1; E2 —— E2; E1 —— E2 (note: circular DNA has no free ends).
- Determining what kinds of ends are on each fragment provides valuable information about where the restriction sites are for restriction mapping. For example, the free ends could be labelled with a fluorescent dye.

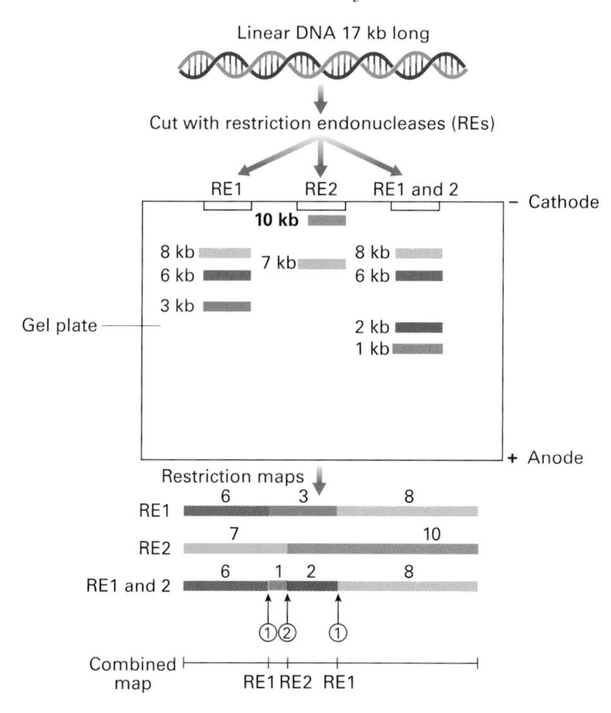

Figure A2 *Restriction mapping. A sample of linear DNA has been digested by two different restriction enzymes, RE1 and RE2, singly and together. The fragments have been separated by gel electrophoresis. All the DNA fragments had the same negative charge and moved towards the anode. Fluorescent markers showed which fragments were at the ends of the original DNA. By using all the information, a restriction map was compiled showing the location of restriction sites 1 and 2, which correspond with the locations on the DNA cut by RE1 and RE2, respectively.*

Chromatography is a technique used to separate mixtures into their components. The technique involves moving the mixture, normally as a gas or liquid (the **mobile phase**), over a **stationary phase** embedded in a relatively inert substrate, such as paper, chalk, or silica gel. Components of the mixture are carried to different parts of the substrate (or pass through the substrate at different rates) and thus separate out. The separation may depend on a number of physical and chemical properties of the component molecules, including their solubility and molecular mass.

There are three basic types of chromatography, named according to the nature of the stationary phase: column chromatography, thin-layer chromatography, and paper chromatography. In **column chromatography**, the mobile phase flows over a supporting matrix such as a gel packed in a glass or metal tube.

Thin-layer chromatography uses a thin layer of silica gel on an inert solid support, usually glass. The mobile phase rises up the thin layer and carries components of the mixture to different parts of the gel. Identification of particular components is achieved by comparing their migration with that of known chemicals.

Paper chromatography is commonly used in schools and colleges to separate mixtures of amino acids, sugars, or plant pigments. Typically, the mixture is dissolved in a suitable solvent, and drops of the resultant solution are repeatedly placed on top of each other to form a small concentrated spot near one end of a paper strip (figure A3). A line is drawn across the paper to mark the position of the spot, and then the end of the paper nearest the spot is dipped into a solvent. The solvent moves up the paper by capillarity, carrying solute molecules with it. As different solute molecules travel at different rates, they are separated from one another. The location of a particular plant pigment may be detected by its colour. The location of particular protein, peptide, and nucleic acid molecules may be detected by their ability to absorb ultraviolet light. The locations of colourless molecules, such as amino acids, require staining (for example, with ninhydrin) to be detected.

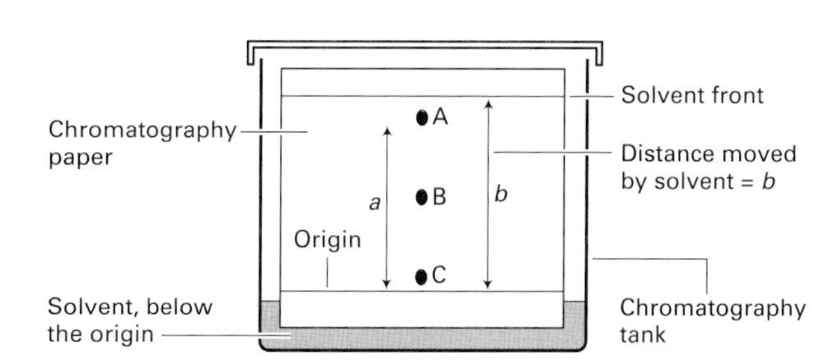

Figure A3 *Chromatography. In this example an unknown mixture is separated into three solutes. Each can be identified by its R_f value. For instance, the R_f value for substance A = a/b.*

Identification of particular solute molecules is usually made on the basis of the distance moved by the solutes and solvent. In any given chromatographic system, the distance moved by a solute relative to the distance moved by the solvent is constant for that solute. This can be expressed as the $\boldsymbol{R_f}$ value (retardation factor) of the solute as shown below:

$$R_f = \frac{\text{distance moved by solute}}{\text{distance moved by solvent front}}$$

Simple paper chromatography may not separate complex mixtures of solutes completely. If this is the case, it may be necessary to perform two-dimensional or **two-way paper chromatography**. This procedure is carried out using a square sheet of chromatography paper. After chromatography has been carried out in the normal way, the paper is removed from the solvent, dried, and rotated 90° so that another chromatographic run can be made, using a second solvent which moves at right angles to the first. As the second solvent has different properties to the first, the partially separated solutes of the first run are usually separated further. The identification of a particular solute can be made by comparison of its position with that of known standard compounds.

Autoradiography is a technique by which large molecules, cell components, tissues, or body organs are radioactively labelled and their image recorded on photographic film, producing an **autoradiograph** or autoradiogram. The technique is based on the fact that beta and gamma rays emitted from radioactive material affect a photographic plate just as light does. If, therefore, a radioactive specimen is placed in contact with or close to a photographic plate, the film is darkened by the ionising radiation from the radioactive parts of the specimen, recording the distribution of radioactivity in the specimen.

Figure A4 is a drawing based on an autoradiograph of a whole plant that has been exposed to carbon dioxide labelled with ^{14}C, a radioactive isotope of carbon. Care has to be taken not to allow the photographic plate to be exposed to light during the procedure. The technique has many applications, including the detection of radioactively labelled compounds on chromatograms.

Figure A4 (courtesy of Horticultural Research International; in Rowland p. 269)

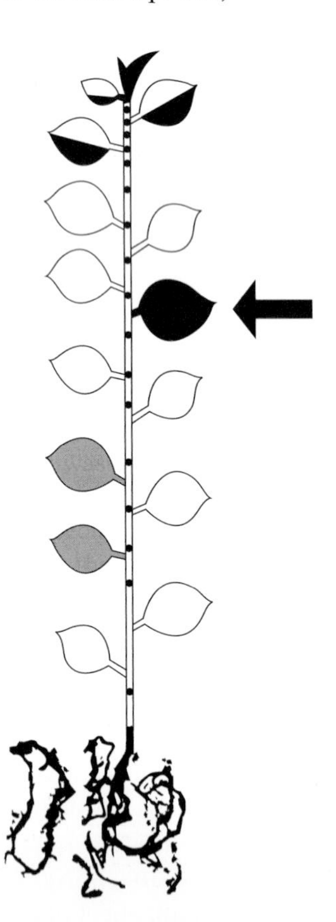

Figure A4 *A diagram based on an autoradiograph of an apple shoot. The leaf marked with an arrow was exposed to radioactively labelled carbon dioxide for 20 minutes. The darkened parts show where the radioactive isotope is concentrated.*

SI units

Generally, in science the International System of Units (SI – for Système International d'Unités) is used. SI units commonly used in biology are:

Physical quantity	Name of unit	Unit symbol
Amount of substance	Mole	mol
Length	Metre	m
Mass	Kilogram	kg
Time	Second	s

Commonly used derived SI units include:

Physical quantity	Name of unit	Unit symbol
Area	Square metre	m^2
Energy	Joule	J
Force	Newton	N
Frequency	Hertz	Hz
Pressure	Pascal	Pa
Volume	Cubic metre	m^3

Prefixes commonly used to denote decimultiples and submultiples of SI units:

Prefix	Multiple	Sign
Pico-	$\times 10^{-12}$	p
Nano-	$\times 10^{-9}$	n
Micro-	$\times 10^{-6}$	μ
Milli-	$\times 10^{-3}$	m
Centi-	$\times 10^{-2}$	c
Deci-	$\times 10^{-1}$	d
Deca	$\times 10$	da
Hecto-	$\times 10^{2}$	h
Kilo-	$\times 10^{3}$	k
Mega-	$\times 10^{6}$	M
Giga-	$\times 10^{9}$	G
Tera-	$\times 10^{12}$	T

Drawings

In examinations, students are encouraged to illustrate their answers. These illustrations can take many forms, including simple sketches, flow diagrams, and coloured drawings. Drawings also play an important part in practical work, providing, for example, an accurate record of an animal or plant tissue being studied. In addition, the act of drawing is an aid to careful observation.

Biological drawings made during practical work should be accurate and clear representations of the specimen. All structures should be drawn in proportion. Rough sketches or elaborately shaded artistic pictures should be avoided. Generally, drawings should be made on plain paper, using a sharp HB pencil. Unless you are instructed otherwise, coloured pencils should not be used.

Students are often instructed to make a drawing of a section of a specimen on a microscope slide. Drawings of such sections usually conform to the conventions described below.

- **Title** The drawing should have an informative title.
- **Size** The drawing should be of an appropriate size – it should cover at least a third of an A4 page; the more detailed the drawing and the more structures it depicts, the larger it should be.
- **Outline** The outline should consist of a single, neat, even and unbroken line.

  It is best to plan the drawing by using small marks to indicate its length and breadth, and a faint outline can be drawn rapidly to show the relative position of the parts.
- **Structures** All structures should be drawn in proportion to each other.

It is important to base the drawing on what is seen, not on what you think you see and certainly not on what is seen in a text book. Illustrations in text books are usually highly stylised diagrams. Simple diagrams can be used to illustrate essays, reports, etc., but they are not usually appropriate as a record of practical work.

- **Shading** This should be kept to a minimum and where used it should be neat (e.g. careful crosshatching).
- **Labels** These should be done neatly and (unless instructed otherwise) in pencil. They should be spaced carefully around the drawing. They should not be written too close to the drawing and never written inside the drawing itself. Each label should be connected to the appropriate part of the drawing with a clear, straight pencil line. Label lines should not cross each other.
- **Magnification/scale** The scale of each drawing and (where appropriate) the magnification at which it was viewed should be given.

- **Low-power (LP) drawings** are drawings of sections viewed using a low power lens of a microscope (×10, ×20, or ×40). The purpose of an LP drawing (figure A5) is to map the different distributions of different layers of tissues and structures present in a section. The drawing, commonly called an LP plan, should not show individual cells.

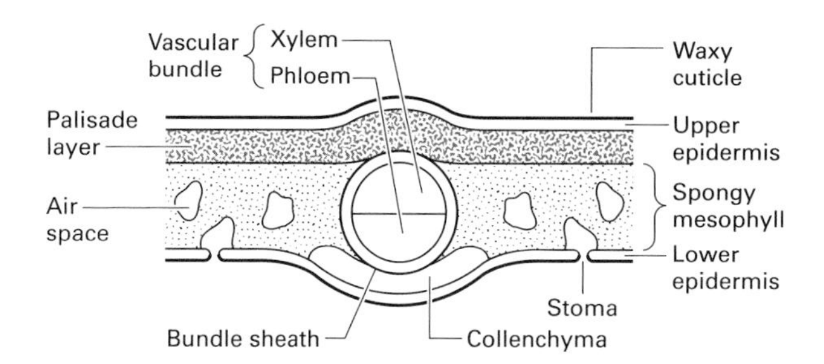

Figure A5 *Low-power plan (x40) of a transverse section through a dictotyledon leaf.*

- **High-power (HP) drawings** are drawings of sections viewed at a magnification of 100× or more. HP drawings (figure A6) are more detailed than LP plans and should show individual cells and, if visible, intracellular structures such as nuclei and chloroplasts. Generally, only a small number of cells need be shown. The cells should be drawn accurately and in proportion to each other. Particular care should be taken when drawing neighbouring cells; they should not appear to interlock and share a common cell membrane or cellulose cell walls. By convention, the cellulose cell walls of plant cells are indicated by a double line, whereas the cell surface membrane of an animal cell is indicated by a single line. It is important to remember that in order to make a clear two-dimensional drawing of a three-dimensional specimen viewed with a microscope, it might be necessary to view the specimen at different depths of focus.

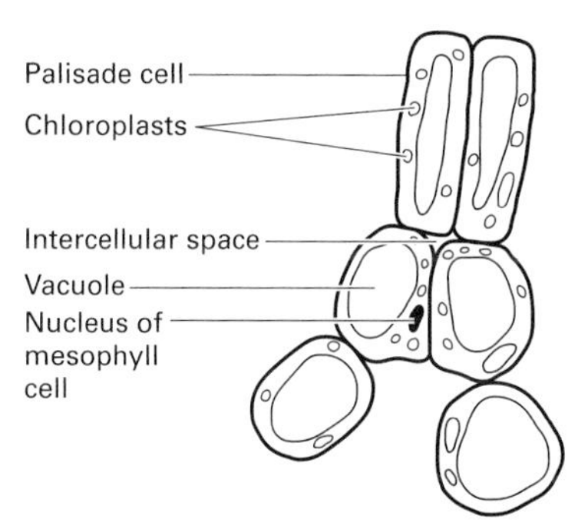

Figure A6 *High-power drawing (x400) of cells in the mesophyll layer of a transverse section of a dicotyledon plant.*

Measuring linear dimensions of objects

The linear dimensions of objects visible to the unaided eye can be measured easily using standard equipment such as rulers and vernier callipers. Smaller objects can be measured using a microscope fitted with an **eyepiece graticule** (a glass or plastic disc with a scale usually subdivided into a hundred divisions). When an object is viewed through the microscope, the scale is superimposed on the object (figure A7). However, the scale on the eyepiece graticule is quite arbitrary and will represent different lengths for different objective lenses. It is therefore necessary to calibrate it; a **stage micrometer** or object micrometer is used for this.

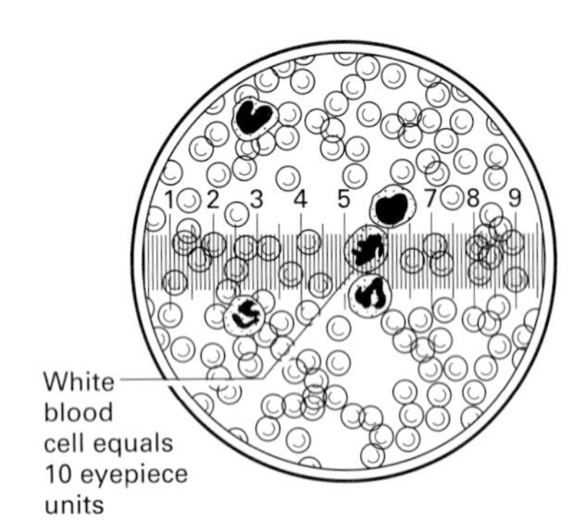

***Figure A7** Measuring the length of blood cells with an eyepiece graticule.*

The stage micrometer has a scale of known length etched on it (usually the scale is 10 mm wide with 100 subdivisions, each 10 μm wide). On looking through the microscope when the stage micrometer is placed on the microscope stage, the arbitrary scale on the eyepiece graticule can be superimposed on the stage micrometer scale. This allows the total length of the scale on the eyepiece graticule to be measured and the length of each subdivision to be determined. If, for example, at ×400 overall magnification, 1 stage micrometer subdivision of 10 μm = 4 eyepiece graticule units, each eyepiece graticule unit = 2.5 μm. The eyepiece scale can now be used to measure objects. However, because the size of the image of the eyepiece graticule changes with each magnification, the calibration of its arbitrary scale has to be repeated for each objective lens of the microscope.

Establishing scales and calculating linear magnifications of drawings, photographs, and electron micrographs On illustrations (drawings, photographs, and micrographs), the scale is usually provided as a magnification factor (e.g. ×400) or in the form of a bar of defined length (e.g. 1 μm). The following method can be used to estimate the dimensions of an object shown in an illustration

1. Use a ruler or vernier callipers to measure the image as it appears in the illustration.
2. If the scale is represented by a bar, measure the bar to find the object's size by proportion. For example, if the image measures 50 mm and a bar representing 1 μm is 20 mm long, then the size of the object is 50/20 × 1 μm = 2.5 μm.
3. If the scale is given as a magnification factor, divide the measurement of the image by this number to obtain the object's actual size. Before doing the calculation, convert all measurements to the same unit (usually μm). Thus, if the image measures 50 mm (= 50 000 μm) and the magnification factor is 20 000, then the size of the object is 50 000/20 000 = 2.5 μm.
4. To calculate the magnification of an illustration of an object of known size, measure the object as it appears on the illustration and divide this measurement by the actual size. For example, if an object of 2.5 μm measures 50 mm (50 000 μm) in an illustration, the magnification will be 50 000/2.5 = × 20 000.

Tests to identify biologically important chemicals

Benedict's test for reducing sugars and non-reducing sugars Typically, a 2 cm^3 solution to be tested is added to an equal volume of Benedict's solution; the solutions are mixed and heated gently. The formation of a coloured precipitate indicates the presence of reducing sugar (for example, glucose); the colour of the precipitate changes from yellow-green to red as the concentration of the sugar increases. Benedict's solution is alkaline, transparent blue, and contains copper(II) sulphate ($CuSO_4$), sodium citrate, and sodium carbonate. Reducing sugars contain either an aldehyde (–CHO) or ketone (C=O) group which can reduce copper(II) ions (Cu^{2+}) to copper(I) ions (Cu^{+}); the copper oxide so formed is a brick-red colour.

When testing for non-reducing sugars (for example, sucrose), any non-reducing sugar present has to be hydrolysed to reducing sugars by boiling the test solution in dilute hydrochloric acid, neutralising the products with sodium hydrogencarbonate (checking the pH with pH paper) and, when the reactants have cooled, carrying out Benedict's test.

Benedict's test can be used qualitatively or semi-quantitatively. Used qualitatively, the test merely indicates the presence or absence of reducing sugars. Used semi-quantitatively, the test (by virtue of the colour of the precipitate produced by the test) can be used to give the relative concentration of reducing sugar in different test solutions, as long as the solutions are treated in exactly the same way.

Iodine/potassium iodide test for starch To test for starch, a few drops of iodine/potassium iodide solution are added to the test substance. This may be either a solid or liquid. A blue-black coloration, resulting from the formation of a polyiodide complex, indicates the presence of starch.

Emulsion test for lipids To test for lipids, absolute ethanol is added to a substance. Distilled water is then added to the mixture after it has been left a sufficient time for lipids to dissolve in the ethanol. When the test mixture is shaken, it appears cloudy if lipids are present because an emulsion consisting of small droplets of lipids in water forms. The test is usually performed on solids or on liquids suspected of being oils. It is not appropriate for watery substances.

Biuret test for proteins To test for proteins, a test sample (if solid) is usually crushed in distilled water and filtered. Sodium hydroxide (or potassium hydroxide) is then added to a known volume of the filtrate a little at a time until the solution becomes clear. A few drops of 1% copper sulphate are then added and mixed. A purple/violet coloration indicates the presence of proteins. The colour change is caused by copper ions forming a complex with adjacent pairs of —CONH— groups in peptide bonds. Similar reactions also occur with $—CSNH_2—$, $—CNNH_2—$ and $—CH_2NH_2—$ bonds which may or may not be part of proteins. The test therefore does not show conclusively that a protein is present, but this is often the most probable explanation for a positive result.

Some useful bacteriological techniques

Culturing bacteria Most bacteriological investigations involve growing pure cultures in a nutrient medium under controlled conditions. **Nutrient media** contain essential raw materials in which microorganisms, cells, or tissues can be grown. These raw materials must include a source of:

- **carbon** (for example, glucose, another simple sugar, or a salt of an organic acid such as sodium ethanoate, commonly known as acetone)
- **nitrogen** (for example, amino acids, peptides, or ammonium salts)
- **mineral salts** (the most commonly required positive ions are those of calcium, potassium, sodium, and iron; the most commonly required negative ions are chloride, phosphate, and sulphate)
- **water** (all active organisms need a source of water)
- **energy** (bacteria may be photoautotrophic, chemotrophic, or heterotrophic; see spread 17.2).

In addition, trace amounts of **growth factors** or **vitamins** (especially some of the B vitamins) are sometimes needed.

Microorganisms may be cultured in a solid medium or a liquid medium. Solid media are commonly made using **agar** (a transparent complex polysaccharide derived from red algae) as a gelling agent into which the nutrients are mixed. Liquid media are used in fermenters (see spread 17.7).

Special nutrient media include selective media and indicator media. A **selective medium** has one or more substances added to inhibit the growth of all but one or a few microorganisms. An example is the addition of an antibiotic to select organisms resistant to it (see replica plating below). An **indicator medium** contains a chemical indicator which enables colonies of one organism to be visibly distinguished from those of another. The indicator usually changes colour due to the activity of particular types of bacteria (see spread 17.4).

Each species of bacteria is adapted to a particular environment and requires a certain set of conditions for maximum growth. Among the environmental factors that need to be considered when culturing bacteria are:

- **pH:** the majority of bacteria grow best at pH 7.4, but some can tolerate very acidic conditions (e.g. *Thiobacillus thiooxidans* can grow in sulphuric acid at pH 0), whereas others thrive in strongly alkaline conditions (e.g. *Vibrio cholerae*)
- **temperature:** bacteria as a group live in a wide range of temperatures, ranging from –5 °C for some polar bacteria to more than 120 °C for bacteria living in or close to hydrothermal vents. For bacteria living in mammals, the optimum is usually about 37 °C with a maximum of 42–43 °C
- **gaseous environment:** bacteria differ in their oxygen requirements. Some species are completely aerobic, others are completely anaerobic. The majority of bacteria species, however, flourish best in an atmosphere that does contain free oxygen, but they can also grow less well in one that does not. The growth of all bacteria benefits from small concentrations of carbon dioxide; a few species require concentrations as high as 10 per cent.

Inoculation This is the introduction of bacteria or other microorganisms into a nutrient medium to start a new culture. It is often carried out using a sterile **inoculating loop**, commonly a nichrome wire loop in a metal handle. The loop can be sterilised repeatedly by heating the wire, loop downwards and almost vertical, in the hottest part of a bunsen flame until the wire becomes red hot. After cooling for about 10 s it is ready for use.

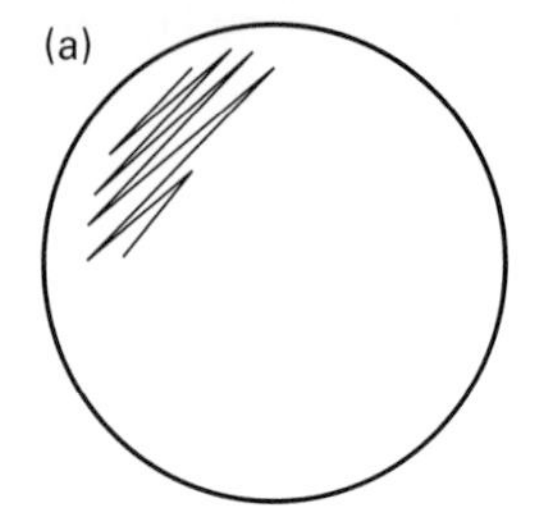

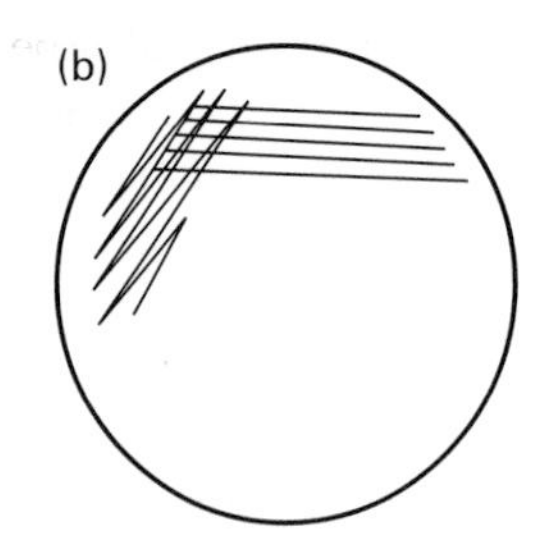

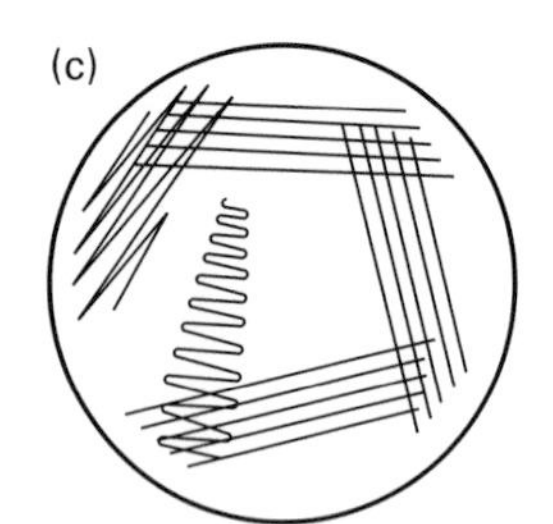

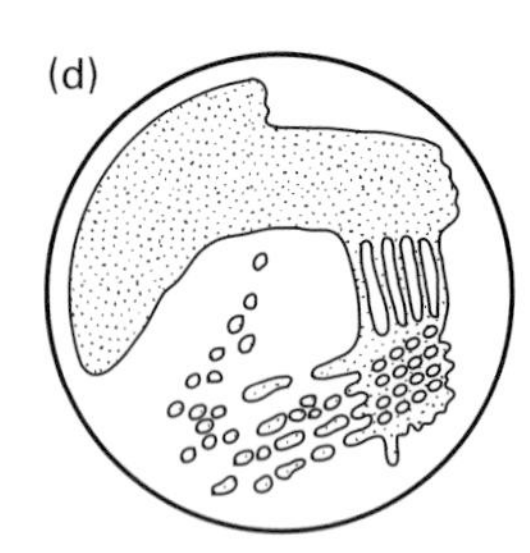

Figure A8 *Streak dilution painting.* ***(a)*** *An initial streak is made from a bacteria sample.* ***(b)*** *The loop is sterilised and used to make small streaks from the first sector.* ***(c)*** *The process is repeated until there are five sets of streaks, with progressively lower concentrations of bacteria.* ***(d)*** *After the plate is incubated at 37 °C, individual bacteria colonies can be seen in the fourth and fifth sets of streaks.*

An inoculating loop can be used to make a streak dilution plate (figure 2, spread 17.4). A sterile loop is dipped into a small sample of bacteria to be streaked. This sample is used as the inoculum and is streaked onto a small sector of the plate. Then, the loop is sterilised before several small streaks are made from the first sector into an adjacent sector. The process is repeated, resterilising the loop between each streaking, until there are about five sets of streaks (figure A8). With each set of streaks, the inoculum should have been diluted so that individual colonies can be distinguished within one or more sectors. As each colony is derived by asexual reproduction of a single individual, a subculture of an individual colony will produce a pure culture (clone).

Stab technique The **stab technique** is used to culture anaerobic organisms or those that grow well in conditions of low oxygen concentration. Typically, a liquid or solid sample is collected on the tip of a straight wire. This is then stabbed vertically into a nutrient agar medium in a test tube. The culture grows out of the stab line. If the sample contains anaerobic organisms, their growth will be confined to the deepest part of the tube (figure A9).

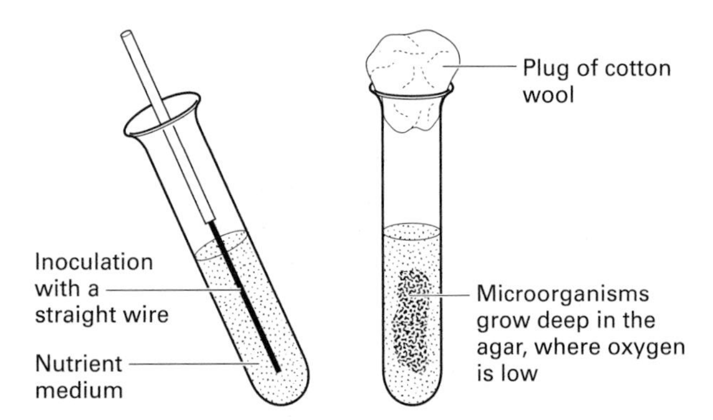

Figure A9 *Stab technique involves inoculation with a straight wire.*

Bioassay to determine the effectiveness of antibiotics or disinfectants **Bioassays** are techniques in which the presence of a chemical is quantified by using living organisms, rather than by carrying out chemical analyses. Bioassays are used to determine the effectiveness of antibiotics or disinfectants on inhibition of bacterial growth. The following technique involves growing a pure culture of bacteria in a Petri dish on which a small disc of filter paper impregnated with a known concentration of antibiotic or disinfectant is placed.

Pure bacteria cultures are usually supplied in test tubes growing on the surface of an agar slope. An inoculation loop can be used to transfer a few loopfuls of the culture onto the surface of nutrient agar in a Petri dish. Alternatively, a pipette can be used to transfer a known volume of a bacteria suspension onto the agar. Typically, an alcohol-sterilised **spreader** (an L-shaped glass instrument) is used to spread out the bacteria as evenly as possible over the agar surface to make a spread plate (figure A10). Once this is done, the filter paper impregnated with antibitotic or disinfectant is placed carefully in the centre of the agar surface. After incubation at 37 °C for 24 hours, any zones of inhibition (regions where bacterial growth have been prevented by the antibiotic or disinfectant) can be seen and measured (figure A11). Results differ for different antibiotics and disinfectants, the same antibiotic or disinfectant on different species of bacteria, and for different strengths of the same antibiotic or disinfectant.

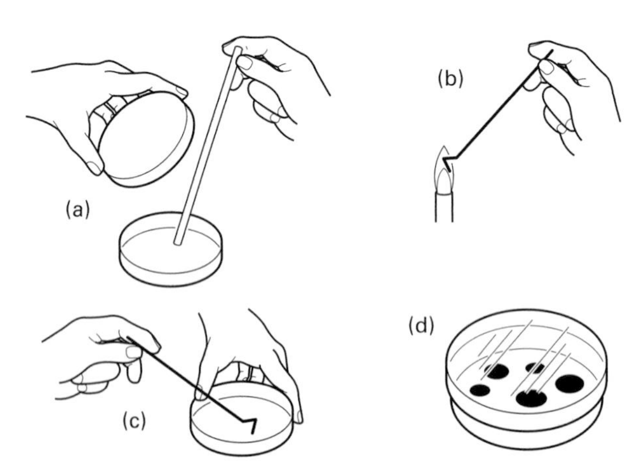

Figure A10 *Making a spread plate.* ***(a)*** *A known volume of bacteria in suspension is transferred onto an agar plate.* ***(b)*** *A glass spreader is dipped in alcohol and then ignited; this sterilises the spreader.* ***(c)*** *The spreader is then used to distribute the bacteria suspension evenly on the agar medium.* ***(d)*** *After incubation, bacteria colonies are distributed across the surface of the plate.*

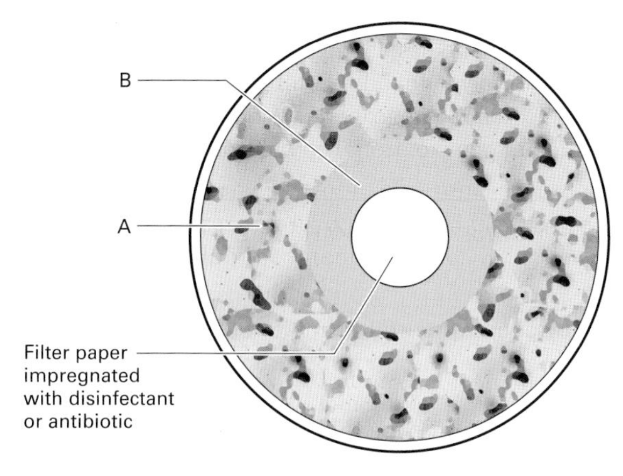

Figure A11 In this bioassay, in region A the bacteria shows continuous growth; in region B, bacterial growth is inhibited.

Replica plating This is a method of transferring bacteria colonies from one agar plate to another while maintaining the same spatial arrangement of the colonies. The transfer is made by gently pressing a sterile pad (usually made of velvet) on to an agar plate on which bacterial colonies can be seen growing. Then the pad is pressed on to sterile agar in a second Petri dish. In this way the bacteria are transferred and grow in exactly the same way in both dishes. Replica plating can be used to identify genetically modified organisms (figure A12).

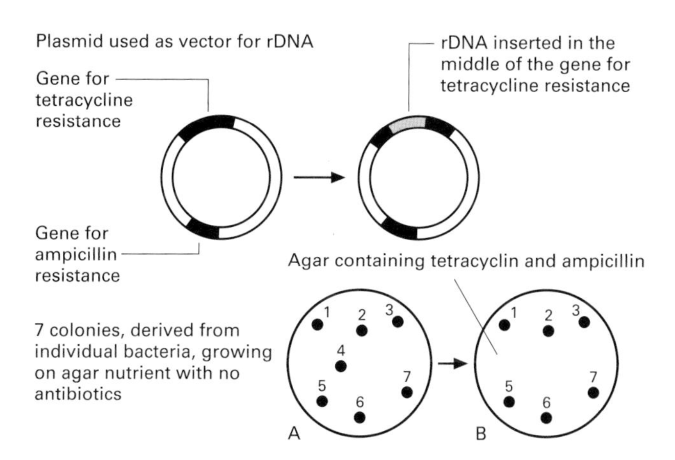

Figure A12 Replica plating. Bacteria in colony 4 must contain the rDNA, as they fail to grow in the agar with tetracycline and ampicillin. Therefore colony 4 in dish A can be used as a source of these bacteria for cloning.

Many of the plasmids used as vectors in genetic engineering (see spread 18.9) have genes for antibiotic resistance, coding for tetracycline resistance and ampicillin resistance. Bacteria with these so-called R-plasmids are able to grow on agar containing both antibiotics. The antibiotic resistance genes act as genetic markers. If recombinant DNA from a donor is inserted into the middle of the tetracycline resistance gene, the gene is made inactive. Consequently, bacteria containing R-plasmids with recombinant DNA fail to grow on agar containing tetracycline and ampicillin; this allows them to be identified using replica plating. In figure A12, bacteria in colony 4 fail to grow in the agar containing tetracycline and ampicillin. Therefore, they must contain the inactive tetracycline-resistance gene with the inserted DNA. Consequently, colony 4 in dish A can be used as a source of these bacteria for cloning.

Measuring population growth of bacteria The growth of a bacteria population can be measured directly by counting the number of cells, or indirectly by measuring a product or effect of their activity that increases with population size (for example, an increase in cloudiness or the production of a gas such as carbon dioxide).

The cloudiness or turbidity of a suspension of bacteria can be measured using a colorimeter or spectrophotometer, a technique known as **turbidimetry**. As the turbidity of the suspension increases, the amount of transmitted light decreases. The percentage of light transmitted can therefore be correlated with turbidity. The **optical density** (a spectrophotometric measurement of light scattered by a suspension at a particular wavelength) of the suspension gives a measure of the decrease in transmitted light. For a given cell type, the relationship between optical density and cell concentration is approximately linear. Under standardised conditions, therefore, the concentration of bacteria cells in a suspension can be determined by optical density. Other indirect measurements include the Resazurin test and methylene blue test for milk freshness (described below).

There are two types of cell count: a **viable count** (the total of living cells only), and a **total count** (the total number of cells, living and dead). Direct total counts can be made using haemocytometers (spread 7.8) or electronic particle counters. (Electrical counters detect small particles due to a change in electrical resistance when they pass through a small aperture in a glass tube.) Viable counts can be made using culture-based counting methods. The most widespread methods use spread plates or pour plates in which a suitable amount of sample is transferred to an agar medium, incubated under appropriate conditions, and the resulting colonies counted. The method is based on the principle that each bacterium will grow into a single colony. Therefore, the number of bacteria in the original sample is equal to the number of colonies after incubation.

Resazurin test and methylene blue test for milk freshness Milk contains carbohydrates, fats, proteins, mineral salts, and water. The main carbohydrate is lactose (milk sugar). If milk is left to stand at room temperature, bacteria break down the lactose to lactic acid. As a result, the milk becomes more acidic and develops a sour taste, making it less palatable to most people. At a constant temperature, the amount of acid produced, and therefore the number of protons (hydrogen ions) in the milk, increases with time. Certain dyes, such as resazurin and methylene blue, change colour as they take up more protons. Their colour change can be used to determine the freshness of milk.

Resazurin is a purple dye which acts as a pH indicator and also as a redox indicator. As it becomes increasingly reduced, it changes from blue in its oxidised state through various shades of pink to colourless. If milk to which resazurin is added is incubated at 37 °C for 30 minutes, its bacterial activity and freshness can be determined: a white-blue coloration indicates a low bacterial count and fresh milk; a pink coloration indicates an intermediate bacterial count and stale milk; and a colourless condition indicates a high bacterial count and sour milk. Similarly, incubation of milk to which methylene blue is added can be used to determine freshness. Methylene blue is a blue dye which becomes colourless when reduced. If the milk is blue after incubation, it is fresh; if it becomes decolorised, it is stale.

Turbidity test for sterilised milk Sterilisation requires milk to be heated to 80 °C or above. At these temperatures, all its albumin becomes denatured and if solutions of inorganic salts or acids are added the albumin separates with the casein. In the turbidity test, heat-treated milk is shaken with ammonium sulphate, filtered, and then the filtrate is heated. Any albumin present in solution reveals itself as a turbidity in the heated filtrate. Turbidity, therefore, indicates inadequate sterilisation. However, if milk has been sterilised adequately, the albumin will have been precipitated and no turbidity will be produced in the heated filtrate.

Iodine/potassium iodide test for assaying carbohydrase activity In this test, a sample suspected of containing microorganisms is cultured on starch agar, a complex solid medium containing starch. When incubation is complete, the starch agar is flooded with the iodine/potassium iodide solution. Carbohydrase activity by microorganisms is assayed by measuring the area of clear zones.

Food

Food additives These may be natural or artificial. Common natural additives include sugar, salt, corn syrup, baking soda, and pepper. Many modern additives, such as vitamins and some flavours, are made in a laboratory, but most of them are exact replicas of naturally occurring substances and the body is unable to distinguish between the natural and artificial forms. The most controversial additives are those which are completely synthetic. Examples are tartrazine, aspartame, and saccharin. **Tartrazine** is an azo dye (a group of synthetic nitrogen-containing chemicals) used to impart a yellow colour to foods and drinks. It has become notorious because of its association with hyperactivity in children. It has also caused allergic reactions and headaches in other sensitive people. **Aspartame** is an artificial sweetener, 200 times as sweet as sucrose. It contains phenylalanine, an amino acid that should be avoided by people with phenylketonuria (see also spread 16.1). **Saccharin** is another artificial sweetener, 300 times as sweet as sucrose. Large doses cause bladder cancer in laboratory animals but, at the time of writing, there is no certain proof that it is carcinogenic in humans. In the USA, foods containing saccharin carry warning labels. Young children and pregnant and lactating women are advised not to use it.

In the European Union, food additives are often given **E numbers**: a set of standard codes which have been approved by the European Union. The main categories of additives are colours (for example E100, curcumin; tartrazine is E102), preservatives (for example E200, sorbic acid); antioxidants (for example E300, L-ascorbic acid); emulsifiers and stabilisers (for example E322, lecithins); and sweeteners (for example E421, mannitol).

Other food additives include:

- acids (for example, citric acid, to give a sour taste)
- anti-caking agents (for example, some phosphates, to help food flow easily)
- anti-foaming agents (for example, oxystearin, to prevent excessive frothing)
- bases (for example, bicarbonate, as a raising agent and acid neutraliser)
- bulking agents (for example, guar gum, adds bulk without adding any calories)
- firming agents (for example, aluminium salts, to retain crispness)
- flavour modifiers (reduce flavour)
- flour improvers (for example, cysteine)
- glazing agents (for example, waxes, to give a polished appearance)
- humectants (for example, glycerol, to prevent foods, such as marshmallow, drying out)
- liquid freezants (for example, liquid nitrogen, to freeze food quickly)
- packaging gases (for example, nitrogen, to control the atmosphere within a package)
- propellants (for example, carbon dioxide, to form an aerosol, forcing food out of containers)
- release agents (for example, silicates, to prevent food sticking to pans)

Table 1 Reference nutrient intakes (RNIs) per day for selected nutrients.

Males	Protein (g)	Vit. B_1 (mg)	Vit. B_2 (mg)	Niacin (mg)	Vit B_6 (mg)	Vit. B_{12} (mg)	Folate (mg)	Vit. C (mg)	Vit. A (μg)	Calcium (mg)	Iron (mg)	Zinc (mg)
1–3 years	14.5	0.5	0.6	8	0.7	0.5	70	30	400	350	6.9	5.0
4–6 years	19.7	0.7	0.8	11	0.9	0.8	100	30	500	450	6.1	6.5
7–10 years	28.3	0.7	1.0	12	1.0	1.0	150	30	500	550	8.7	7.0
11–14 years	42.1	0.9	1.2	15	1.2	1.2	200	35	600	1000	11.3	9.0
15–18 years	55.2	1.1	1.3	18	1.5	1.5	200	40	700	1000	11.3	9.5
19–50 years	55.5	1.0	1.3	17	1.4	1.5	200	40	700	700	8.7	9.5
50+ years	53.5	0.9	1.3	16	1.4	1.5	200	40	700	700	8.7	9.5
Females												
1–3 years	14.5	0.5	0.6	8	0.7	0.5	70	30	400	350	6.9	5.0
4–6 years	19.7	0.7	0.8	11	0.9	0.8	100	30	500	450	6.1	6.5
7–10 years	28.3	0.7	1.0	12	1.0	1.0	150	30	500	550	8.7	7.0
11–14 years	41.2	0.7	1.1	12	1.0	1.2	200	35	600	800	14.8**	9.0
15–18 years	45.0	0.8	1.1	14	1.2	1.5	200	40	600	800	14.8**	7.0
19–50 years	45.0	0.8	1.1	13	1.2	1.5	200	40	600	700	14.8**	7.0
50+ years	45.5	0.8	1.1	12	1.2	1.5	200	40	600	700	8.7	7.0
Pregnancy	+6	+0.1*	+0.3	–	–	–	+100	+10	+100	–	–	–
Lactation												
0–4 months	+11	+0.2	+0.5	+2	–	+0.5	+60	+30	+350	+550	–	+6.0
4+ months	+8	+0.2	+0.5	+2	–	+0.5	+60	+30	+350	+550	–	+2.5

– No increase

* Last trimester only

** Insufficient for women with high menstrual losses

- sequestrants (for example, sodium hydrogen diacetate, to help remove heavy metal from food)
- solvents (for example, glycerol, to dissolve solids in food).

Ecological sampling techniques

Most ecological investigations involve collecting data about the abundance of organisms and their pattern of distribution. Unless the study area is small, it will be necessary to take representative samples. For many purposes, a random sampling procedure should be used to avoid the possibility of bias. In some cases, a systematic method of sampling, such as transect sampling, is more appropriate.

Random sampling Although there are many ways to sample randomly, they are all designed to ensure that each point within the study area has an equal chance of being sampled. For example, if your study site is a conveniently flat and square area measuring 10 m × 10 m, the area could be marked by lengths of string laid out along two edges at right angles to each other. The string could be marked off at 1 m intervals and these intervals numbered 1 to 10. The numbered marks along each 10 m side could then be used to define coordinates of any sampling point within the study area. Pairs of numbers between 1 and 10 drawn at random from a table of random numbers can be used to locate the positions of the sampling point.

If it is not possible to lay out strings to mark a study area, **random walking** can be used to select sampling points. For example, a pair of random numbers can be used to determine the number of paces forward then at right angles (with the spin of a coin determining whether to turn left or right) from starting point to sample point.

Transect sampling This is used to investigate the transition of one community to another, particularly when the communities occur in a linear sequence, for example up a rocky seashore. There are several types of transects including line transects, belt transects, continuous transects, and interrupted transects.

Typically, when a **line transect** is used, a tape or string is laid along the ground in a straight line between two poles to indicate the position of the transect. Sampling is confined strictly to recording all individual organisms actually touching the line. A single line is unlikely to be representative and it is usually necessary to take a series of lines. A **belt transect** is a strip of chosen width through the habitat. It is made by laying two parallel line transects, usually 0.5 m or 1 m apart, between which all individuals are recorded.

A **continuous transect** is a line transect or belt transect in which the whole line or belt is sampled. In an **interrupted transect**, samples are taken at points along the line or belt transect. The points are usually at regularly spaced horizontal or vertical intervals. Vertical intervals are usually the most appropriate when studying a distribution pattern that is likely to be related to height, for example up a rocky shore or mountain.

Once a sampling point has been determined, either at random or systematically, a number of different techniques can be used to take samples. One of the most common is quadrat sampling.

Quadrat sampling There are two main types of quadrats: frame quadrats and point quadrats. A **frame quadrat** is usually a metal, plastic, or wooden frame that forms a known area, such as 0.25 m^2 or 1 m^2. Frame quadrats can be of any regular shape. The most common are square-shaped (figure A13), but circular and even hexagonal frames can be used. However, whatever the shape, the frame quadrat must be of a known area. Frame quadrats can be used to measure population density, percentage cover, or frequency of occurrence of a species within a selected area.

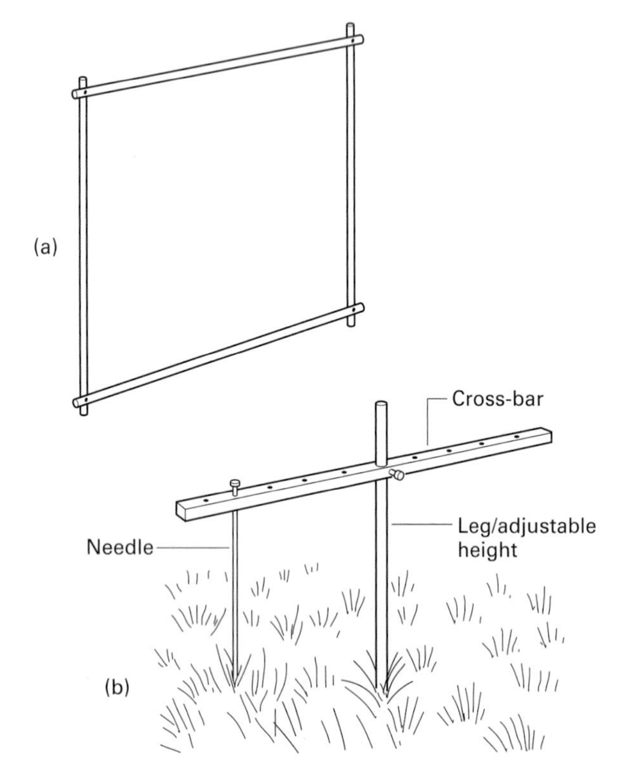

Figure A13 ***(a)*** *A collapsible, square-frame quadrat.* ***(b)*** *A point quadrat.*

To measure **population density**, all individuals present within the chosen quadrats are recorded. It is important to be consistent; for example, record all individuals completely or partially visible. The density of each species can be expressed as an average figure per square metre (m^{-2}) or an extrapolation made to give the total number for the whole study area.

To obtain an accurate estimate of population size, a series of quadrat samples is usually required. The number of samples to be taken can be determined by constructing a **running mean**. A mean is calculated after each replicate. The mean values will fluctuate each time, but will gradually settle down until a point is reached where the addition of another sample has only a very small effect on the mean. When this happens, you can assume the sample size is adequate; sampling more quadrats would be unrewarding and a waste of time.

A frame quadrat divided into a 10 × 10 grid with string can be used to estimate **percentage cover**. For each species being investigated, the number of squares fully occupied, partly occupied, and unoccupied is recorded. Each fully occupied square represents a 1% cover. An estimate needs to be made of how many full squares the partly occupied squares represent so that the total percentage cover can be calculated.

A frame quadrat divided into a 10 × 10 grid can also be used to estimate **frequency of occurrence** of a species. The presence or absence of each species being investigated is recorded for each of the 100 squares. The 'present' score for each species can be used as an index of frequency or in an estimate of abundance. If the 'present' score is expressed as a percentage of the total number of squares, the result will be a **percentage frequency**. A consistent method must be adopted to deal with individuals overlapping the edges of squares. One common method is to include an individual if it touches the top or left-hand sides of a square and to exclude it if it touches the bottom or right-hand sides.

A **point quadrat** (or **pin frame**) is designed for sampling stationary organisms, usually plants. It consists of a frame of adjustable height and a cross bar bearing a number of holes (usually 10) through which a 'pin' such as a knitting needle can be passed (figure A12). The point quadrat is used to record the presence or absence of organisms at a number of points. The needle is passed through each hole in turn and a record made of each species touched by the tip of the needle as it is lowered to the ground.

Point quadrats are usually used to measure percentage cover, the proportion of the ground shaded by aerial parts of organisms. With a 10-needle point quadrat, each 'hit' with a needle represents a 10% cover.

$$\text{Percentage cover} = \frac{\text{hits} \times 100}{\text{hits} + \text{misses}}$$

The record can be confined to organisms which form the canopy and shade other organisms, or it can be made for all organisms touched by each needle. In the latter case, the percentage cover may well exceed 100% because several organisms will overlap and shade the same patch of ground.

Frame quadrats and point quadrats can be used to estimate relative abundance using one of several **abundance scales**. In the **ACFOR** scale each species present at a sample site is assigned to one of the following five categories:

A = Abundant

C = Common

F = Frequent

O = Occasional

R = Rare

When comparing two or more sites, it might be necessary to include an additional category 'N' to represent 'none present'.

ACFOR scales use arbitrary values for the five categories. Consequently, they can be highly subjective. Two different investigators may use different criteria. For example, one investigator may regard a species as 'abundant' if its frequency exceeds 90%, whereas another investigator may regard the species as being abundant if it has a frequency exceeding 80%. In addition, the relative abundance of conspicuous organisms tends to be overestimated and that of inconspicuous ones underestimated. However, ACFOR scales can be made less subjective if the criteria for assigning species to each category are standardised.

Capture-mark-recapture technique The **capture-mark-recapture** is used to estimate the population size of mobile organisms, such as fish. It involves capturing the organisms, marking them in a way that causes no harm, and returning them to their natural environment. After allowing sufficient time for the marked individuals to mix thoroughly with other members of the population, a second sample is taken and the number of marked and unmarked individuals recorded. The population size is estimated using the following equation:

$$\text{Population size (P)} = \frac{\text{number of organisms in first sample} \times \text{number of organisms in second sample}}{\text{number of marked individuals recaptured}}$$

This population estimate is known as the **Lincoln Index**. It is based on the assumption that a second sample containing 10 per cent of the original marked individuals represents 10 per cent of the total population. Reliance on this assumption depends on certain conditions being met:

1. organisms are captured randomly; there is no bias towards a particular group
2. the mark is not lost between the period of release and recapture
3. marking does not hinder the movement of organisms or harm them in any way
4. for the second sample, there is an equal chance of marked and unmarked individuals being caught
5. marked organisms mix randomly with unmarked individuals
6. organisms disperse evenly within the study area
7. during the study period, changes in population size due to immigration, emigration, death, and birth are negligible.

Measuring abiotic factors Many ecological investigations try to relate the distribution and abundance of certain organisms to particular abiotic factors. Among the most commonly measured abiotic factors are temperature, pH, light intensity, humidity, and wind speed.

Temperature can be measured using a standard mercury-in-glass thermometer. Such a thermometer with a thickened bulb and in a protective case can be used in air, soil, and water. To measure microhabitat temperatures it might be necessary to use **thermistor probes** connected to an appropriate meter. Standard domestic **maximum/minimum thermometers** can be used to measure the temperature range over a given period of time which, for many investigations, is more important than a reading at any particular moment.

The **pH** of water samples can be measured directly using pH paper or a pH meter and a probe. Soil pH is best measured using a soil-testing kit which includes tests for pH. Alternatively, you can mix soil with water and measure the pH with a standard probe and meter. To do this, take a sample of soil, place it in a measuring cylinder and add an equal volume of distilled water. Shake vigorously for about 10 minutes and then filter. (If clay particles are present, they can be flocculated by using barium sulphate solution instead of distilled water.) The pH of the filtrate is then measured.

When taking a series of pH readings, remember that pH values are the negative logarithms to the base 10 of hydrogen ion concentration. Consequently, they cannot be averaged in the usual way. The mean of pH 4 and 6 is not 5 but 5.114. To obtain averages, take the mean of the antilogs and then express the result as the log.

Light intensity can be measured at any particular time using **light meters**, such as those used by photographers. However, the readings may not be very useful because light levels fluctuate continuously. In addition, if light intensity is being studied in relation to the distribution of plants, it is more important to measure the total amount of light received over a relatively long period of time, because it is this which most affects plant performance. **Light probes** connected to a computer or data-logger can be used to take continuous measurements of light. Alternatively, it is possible to compare the amount of light received over a long period of time by using **light-sensitive paper**, such as ozalid paper used by printers. This gradually darkens on exposure. For most ecological studies, quality (range of wavelengths) of light is also an important factor. For example, when studying the effect of light intensity on plant distribution, a light probe incorporating a cadmium sulphide cell might be useful, because it is sensitive to the range of wavelengths in the visible spectrum similar to those absorbed by plants.

When considering **wind speed and direction**, it is important to remember that for most ecological studies, degree of exposure is more important than wind speed at any one time. Degree of exposure is a function of wind frequency, speed, and direction. A simple wind-gauge or wind vane measures direction and a simple **anemometer** measures wind speed (figure A14).

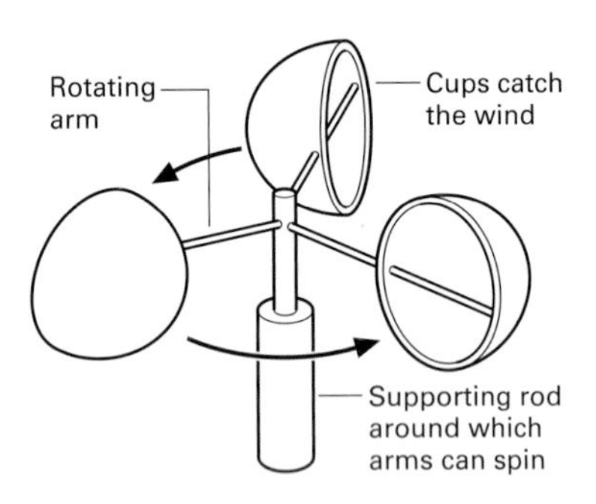

Figure A14 *A cup-shaped anemometer. The arms rotate at a speed proportional to the wind speed.*

Absolute humidity is a measure of the amount of water vapour present in a specified portion of the atmosphere, whereas **relative humidity** is a measure of the moisture content of air at a given temperature relative to air fully saturated with water vapour at the same temperature. For terrestrial organisms, relative humidity is the more important factor. It indicates the readiness with which water vapour can evaporate from the body surface. Relative humidity can be measured using a **hygrometer**. Alternatively, to compare humidities in different microhabitats (for example, rock crevices), cobalt chloride paper can be used. This paper can be easily prepared by immersing strips of filter paper in 5% solution of cobalt chloride and then drying them. There is a close relationship between colour of the paper and air humidity, the paper being blue at low humidities and pale pink at high humidities, with a series of lilacs in between.

When trying to relate the distribution of an organism to an abiotic factor, it is important to measure that factor in the microhabitat of the organism. For example, when studying invertebrates which live in leaf litter, it is necessary to measure the temperature underneath the leaves and not in the air above.

Statistical terms and methods

Normal distribution curve and probability

A **normal distribution curve** has a bell-like shape. It is produced when a population is sampled randomly for a continuously varying character (for example, height in humans) and the frequency of each class of data is plotted graphically (figure A15). The curve of a perfect normal distribution is symmetrical and the mean, median, and mode are equal. Many statistical procedures assume that data have a normal distribution. In a normal distribution 68.27% of the data will lie within ±1 standard deviation of the mean (see below), 95.45% within ±2 standard deviations of the mean, and 99.73% within ±3 standard deviations of the mean.

Probability refers to the likelihood that a given event will occur. It is usually expressed as a value between 0 and 100 per cent, or between 0 (the event has no chance of occurring) and 1.0 (the event is certain to occur).

Measures of central tendency Measures of central tendency include mean, mode, and median. They are used to estimate the average value of a series of values.

The **mean** is obtained by adding all the values together and dividing the total by the number of individual values. Thus, the mean ($\bar{x}$) for values $x_1, x_2, x_3, \ldots, x_n$ is given by

$$\bar{x} = \frac{x_1 + x_2 + x_3 + \ldots + x_n}{n}$$

or

$$\bar{x} = \frac{\Sigma x}{n}$$

where Σ (sigma) means 'the sum of' and $x_1, x_2, x_3, \ldots$ are the individual values.

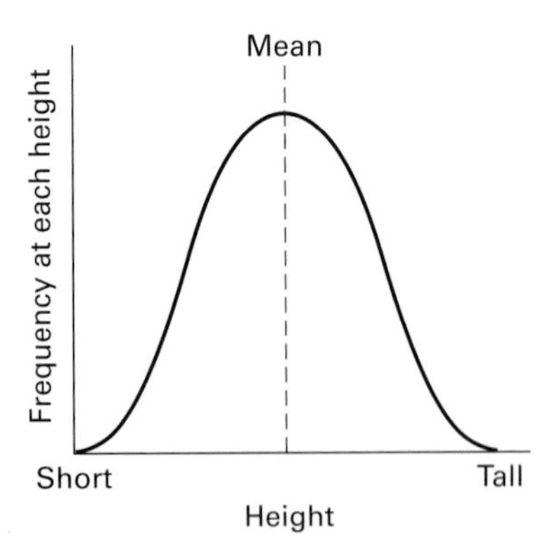

Figure A15 *Normal distribution curve for height, sampled randomly in a population of humans.*

The **mode** is the most frequently occurring value in a set of values. For example, if the number of kittens in 10 litters is 2, 3, 3, 4, 4, 4, 4, 5, 5, 6, the modal value is 4. When a histogram is drawn of a frequency distribution, the mode is the class with the largest area. As columns are usually drawn of equal width, it is normally the class with the tallest column. The mode, therefore, is easy to find providing the distribution is monomodal, i.e. there is only one modal value. In some biological situations, however, distributions are bimodal or even polymodal.

The **median** represents the central or middle value of a set of values. If a series of values is placed in rank order as x_1, x_2, x_3, x_4, and x_5, the median value would be x_3. If there is an even number of values for x, for example x_1 to x_8, the median value is represented as the mean of the two middle values: $(x_4 + x_5)/2$.

Measures of dispersion These are statistical measures of the extent to which a set of values are spread or clustered around a central value (mean, mode, or median). They include range, standard deviation, standard error, and variance.

The **range** is the spread of values of a set of data. The range is the difference between the maximum and minimum values.

The **standard deviation** (s) of a set of values is a measure of the spread of the values from the mean. The mean value is often shown as plus or minus the standard deviation. The standard deviation is calculated using the expression:

$$s = \sqrt{\frac{\Sigma(x - \bar{x})^2}{n}}$$

where s = standard deviation; Σ means 'sum of';

$x - \bar{x}$ = difference between each sample value and the arithmetic mean; n = total number of values.

The standard deviation is the square root of the **variance**.

The **standard error of the mean** is an estimate of the standard deviation of the means of many samples. It is calculated as the standard deviation (s) divided by the square root of the number of individuals in a sample (n), i.e.

$$\text{SE} = s/\sqrt{n}$$

For a sample with values 27, 33, 35, 36, 41, 48, 52, standard deviation = 8.10 and standard error = 3.06.

Variance is another measure of the extent to which values are clustered or dispersed around a mean. The variance of a set of values is calculated using the formula:

$$s^2 = \frac{\Sigma(x - \bar{x})^2}{n}$$

where s^2 = variance; Σ means 'sum of'; $x - \bar{x}$ = difference between each sample value and the arithmetic mean; n = total number of values.

Variance is the square of the standard deviation. A low variance indicates a low dispersion of values about the mean.

Significance tests These include the chi-squared test, and the t-test which uses calculations of standard errors of means. Significance tests are used to determine whether or not differences between two or more sets of data are likely to be real and not attributable to chance or sampling error.

The **chi-squared test** is used to test the significance of deviations between numbers observed in an experiment or investigation and numbers expected from a given hypothesis. The measure of the deviation is called chi squared (χ^2).

The chi-squared test can be used only with data that fall into discrete categories. It is calculated using the following equation:

$$\text{Chi squared} = \text{sum of} \left(\frac{(\text{observed no. in class} - \text{expected no in class})^2}{\text{expected number in class}}\right)$$

The chi-squared value is converted into a probability value using a chi-squared table.

Worked example

The chi-squared test is particularly valuable in genetics to determine whether offspring phenotype ratios fit expected Mendelian ratios (see chapter 19). For example, when Mendel self-crossed heterozygous tall plants, he obtained 787 tall and 277 dwarf plants (figure 3, spread 19.2). On the basis of his law of segregation, the expected ratio is 3:1, and we would expect 798 tall and 266 dwarf plants. When carrying out the test, we begin by assuming there is no difference between the observed numbers and the expected numbers. This assumption of no difference is called the **null hypothesis**. The general formula for calculating chi squared is:

$$\text{Chi squared} = \sum \frac{(O - E)^2}{E}$$

where O = observed result; E = expected result; Σ means 'the sum of'.

Using this formula in this example:

$$\text{Chi squared} = \frac{(787 - 798)^2}{798} + \frac{(277 - 266)^2}{266} = \frac{121}{798} + \frac{121}{266} = 0.606$$

The probability associated with this value can be obtained from chi-squared tables. These tables give chi-squared values against degrees of freedom and probability. **Degrees of freedom** (d.f) is a mathematical term relating to the number of free variables in a system. In this example, d.f. is calculated as $(n - 1)$, where n = the number of categories = the number of phenotypes considered; i.e. d.f. = 2 − 1 = 1.

The probability value for a chi squared of 0.606 at 1 d.f. equals more than 0.5; this means that there is more than a 50% probability (p) that the differences between the observed and expected numbers are due to chance and we accept the null hypothesis (i.e. there is no difference between the observed and expected results and the results of Mendel's experiment are consistent with a 3:1 ratio). The higher the chi-squared value, the smaller the p value. In most biological investigations, if there is more than a 5% chance (p>5%) that the observed result is the same as expected, any deviation is regarded as 'not significant' and might have occurred by chance alone. If, however, there is less than a 5% chance ($p<5\%$) that the observed and expected results are the same, then the deviations are regarded as 'significant'. In this example, we would reject the null hypothesis only if the chi-squared value was more than 3.84.

The **standard error of the sample mean** (often abbreviated to standard error or SE) is used to assess the reliability of the sample mean as an estimate of the true (population) mean. It is estimated using the following equation:

$$SE = s/\sqrt{n}$$

where s = standard deviation and n = sample size.

The equation reflects the fact that SE is directly affected by sample dispersion and inversely related to sample size. This means that the SE will decrease as the number of data values in the sample increases, giving increased precision. In scientific reports, means are often quoted as mean plus or minus standard error for a given sample size. If, for example, in an investigation the mean clutch size of 21 female birds was found to be 5.0 and the standard deviation was 1.28, the standard error would be 0.28. The results could then be summarised as:

$$\text{mean clutch size} = 5.0 \pm 0.280\ (n = 21)$$

The standard error is used to carry out a t-test between two samples; it also allows the calculation of confidence limits for the sample mean.

The ***t*-test** was devised to test the significance of differences between means of two small samples (sample size less than 30) showing normal distribution. The statistic 't' is the difference between the means divided by the standard error of the difference between means:

$$t = \frac{\text{difference between means}}{\text{standard error of difference between the means}}$$

For samples A and B

$$t = \frac{\overline{A} - \overline{B}}{\sqrt{\left((SE_A)^2 + (SE_B)^2\right)}}$$

where $\overline{A}$ = mean of sample A; $\overline{B}$ = mean of sample B; SE_A = standard error of sample A; SE_B = standard error of sample B.

Worked example

An ecologist studying the shell length of mussels at two different tidal levels on a rocky shore obtained the following results:

Sample A: low-shore population	Sample B: upper-shore population
length of shell (cm)	length of shell (cm)
6.6	6.3
5.5	7.2
6.8	6.5
5.8	7.1
6.1	7.5
5.9	7.3

The null hypothesis is that the two samples are from the same population and the difference between their means is not significant. The alternative hypothesis is that the two samples are from different populations and that the difference between their means is significant. To find out if the observed differences between the means is significant or not, the t value can be calculated as follows:

1 *Calculate the two sample means and the differences between them:*

mean for sample A = 6.12 cm
mean for sample B = 6.98 cm
difference between means = –0.87

2 *Calculate the standard deviations of the two samples:*

$s_A = 0.452$ $\quad$ $s_B = 0.434$

3 *Calculate the sample standard errors, square each, add the squares together, then take the square root of this:*

$SE_A = s_A/\sqrt{n_A}$ $\quad$ $SE_B = s_B/\sqrt{n_B}$

$SE_A = 0.452/2.449 = 0.185$ $\quad$ $SE_B = 0.434/2.449 = 0.177$

4 *Calculate the t value from the formula:*

$$t = \frac{\overline{A} - \overline{B}}{\sqrt{\left((SE_A)^2 + (SE_B)^2\right)}}$$

$$t = \frac{\overline{A} - \overline{B}}{\sqrt{\left(0.202^2 + 0.194^2\right)}} = -3.39$$

5 *Calculate the degrees of freedom (d.f):*

d.f. = $(n_A - 1) + (n_B - 1) = 10$

where n_A = number in sample A and n_B = number in sample B.

6 *Use statistical tables to compare the t value with the appropriate critical value:*

In our example, the t value is negative, but this does not matter because only the modulus (absolute value) of t is used. Statistical tables show that a t value of 3.39 exceeds the value for $p = 0.05$ at 10 d.f. The null hypothesis is therefore rejected and we can conclude that the means are different at the 5% level of significance.

(Note that the above is a description of a two-tailed t test for unmatched samples in which the sample size is less than 30. Tests based on the t distribution are also available for one-tailed tests of matched samples.)

Confidence intervals, confidence limits, and confidence levels A **confidence interval** gives an estimated range of values which is likely to include an unknown population parameter. For example, the mean you calculate from a set of values is unlikely to equal the true or population mean. The size of the discrepancy will depend on the size and variability of the sample. The sample mean calculated from a small sample with large dispersion is likely to be very different from the true mean. Conversely, the mean calculated for a large sample with small dispersion will probably be very close to the true mean.

The **confidence level** is the probability associated with a confidence interval. In biology, the most commonly used confidence level is 95%. This means that you can be 95% certain that the true value lies somewhere within the 95% confidence interval.

Confidence limits are the upper and lower values of a confidence interval, that is, the values defining the range of a confidence interval. For a set of values with a normal distribution, the 95% confidence interval for the true mean is defined as plus or minus two standard errors of the sample mean. Therefore, if the mean clutch size of 21 birds equals 5.0 and the standard error equals 0.28, the confidence interval for a 95% confidence level would be 5.0 ± 0.56 and the confidence limits would be 4.44 to 5.56. This means that we can be 95% certain that the true mean clutch size lies within the range 4.44 to 5.56.

The genetic code

The genetic code: the letters U, C, A, and G represent the bases in mRNA, and the abbreviations Phe, Ser, etc. represent amino acids (see table 1, spread 2.9). The codon is read from the 5' end of mRNA to its 3' end; thus phenylalanine (Phe) has the codons UUU and UUC.

First position (5' end)	Second position				Third position (3' end)
	U	C	A	G	
U	Phe	Ser	Tyr	Cys	U
	Phe	Ser	Tyr	Cys	C
	Leu	Ser	Stop	Stop	A
	Leu	Ser	Stop	Trp	G
C	Leu	Pro	His	Arg	U
	Leu	Pro	His	Arg	C
	Leu	Pro	Gln	Arg	A
	Leu	Pro	Gln	Arg	G
A	Ile	Thr	Asn	Ser	U
	Ile	Thr	Asn	Ser	C
	Ile	Thr	Lys	Arg	A
	Met	Thr	Lys	Arg	G
G	Val	Ala	Asp	Gly	U
	Val	Ala	Asp	Gly	C
	Val	Ala	Glu	Gly	A
	Val	Ala	Glu	Gly	G

The Hardy–Weinberg equation

The Hardy–Weinberg equation (spread 20.6) is written as:

$$p^2 + 2pq + q^2 = 1.0$$

where:

- p^2 represents the frequency of the **AA** genotype
- $2pq$ represents the frequency of the **Aa** genotype
- q^2 represents the frequency of the **aa** genotype.

The following example shows how the equation can be used to calculate the genotype frequencies in a human population in which 14 per cent of babies are born with sickle-cell anaemia (spread 16.1). Expressed as probabilities:

$q^2 = 0.14$

$q = \sqrt{0.14} = 0.374$

since $p + q = 1.0$,

$p = 1.0 - 0.374 = 0.626$

$p^2 = 0.626^2 = 0.392$

and $2pq = 2 \times 0.626 \times 0.374 = 0.468$

This means that 46.8 per cent of all the babies born have sickle-cell trait.

Note that the frequency of the homozygous recessive phenotype q^2 needs to be known before the Hardy–Weinberg equation can be used to calculate genotype frequencies. Individuals showing the dominant phenotype may be either heterozygous or homozygous: the frequency of the dominant phenotype is $p^2 + 2pq$. Knowing the frequency of the dominant phenotype (in this case, 86%) does not allow us to calculate directly the frequency of the alleles p and q, or the genotypes. However, it does allow us to calculate the frequency of the recessive phenotype (in this case, 14%), from which we can calculate the frequencies of all the genotypes.

Suggested answers to quick check and exam questions

Please note: the following suggested answers and solutions are intended as a guide for the reader. The respective exam boards have not supplied the answers and solutions to past exam questions nor are they in any way responsible for their accuracy or correctness.

Conventions / abbreviations used:

; denotes new mark point

/ denotes alternative answer for same mark point

eq. equivalent words / answer accepted

AOVP any other valid point

If more answers are given than there are marks, then it is implied that the marks available are up to the given maximum.

Chapter 2 The chemicals of life

Quick Check questions

Spread 2.1

1 atom
2 carbon, hydrogen, oxygen, and nitrogen
3 nucleus
4 neutron

Spread 2.2

1 a ionic bond; b covalent bond
2 carbon is capable of forming four covalent bonds with hydrogen atoms or the equivalent
3 carbon-based molecules can be stable; they can contain a variety of elements; and they can form a great variety of shapes.

Spread 2.3

1 a weak bonds formed by the attraction between a hydrogen atom and an electronegative atom (e.g. oxygen in water)
b sugar and salt have an electrostatic charge; oil does not.
c water inside a plant leaf expands when it freezes, damaging tissue

Spread 2.4

1 a relatively incompressible
b water's transparency enables photosynthesis to take place in aquatic environments.
2 cooling e.g. by sweating

Spread 2.5

1 a solute dissolves in a solvent to form a solution.
2 a solution at pH 6 has 100 times as many protons as a solution at pH 8
3 rain is only one type of precipitation which may be acid (others include mists, fogs, and snow)

Spread 2.6

1 $(CH_2O)_n$
2 glycosidic bond
3 a glucose; **b** sucrose.

Spread 2.7

1 unlike a sugar, polysaccharides are relatively insoluble, not sweet, and cannot be crystallised.
2 a energy storage; **b** structural support in cell walls;
c starch is made of alpha glucose monomers and forms a coiled molecule which can be packed easily for energy storage; cellulose is made of beta glucose monomers and forms straight-chained molecules which can form cross-links to produce a very strong structure.

Spread 2.8

1 glycerol and fatty acids
2 fat has a high proportion of hydrogen.
3 phospholipid has only 2 fatty acids (triglycerides have 3); a phospholipid has a hydrophilic head and a hydrophobic tail (triglycerides are completely hydrophobic).

Spread 2.9

1 a oxygen transport **b** digestion of starch
c control of blood glucose concentrations
2 carbon, hydrogen, oxygen, and nitrogen
3 –CO–NH–
4 an ion which has both a positive and negative charge

Spread 2.10

1 tertiary structure refers to the 3-D shape of each polypeptide chain of haemoglobin whereas quaternary structure refers to the overall shape formed by the four polypeptide chains of each haemoglobin molecule.
2 denaturation.
3 fibrous proteins.

Spread 2.11

1 organic, nitrogen-containing base, five-carbon sugar, and phosphate.
2 by hydrogen bonds.
3 any three cellular reactions requiring free energy (endergonic reactions) such as active transport, contraction of a muscle cell, and synthesis of glucose from carbon dioxide and water.

Exam questions

1 a E; **b** A; **c** B; **d** F; **e** D; **f** C; **g** A, D, E, and F h A.
2 a i A: Glycerol backbone/residue;
B: 3 fatty acid residues.
ii Triglyceride / triacylglycerol
iii Condensation reaction
b i and **ii** Function: long-term energy stores e.g. in seeds
Feature: compact / insoluble in water / releases approximately twice as much energy as one molecule of carbohydrate. Other functions could include insulation / buoyancy / respiratory substrate, etc. Feature must match given function.
3 a As structure shown / eq.(α-1–4 and α-1–6 links must be in the correct positions)

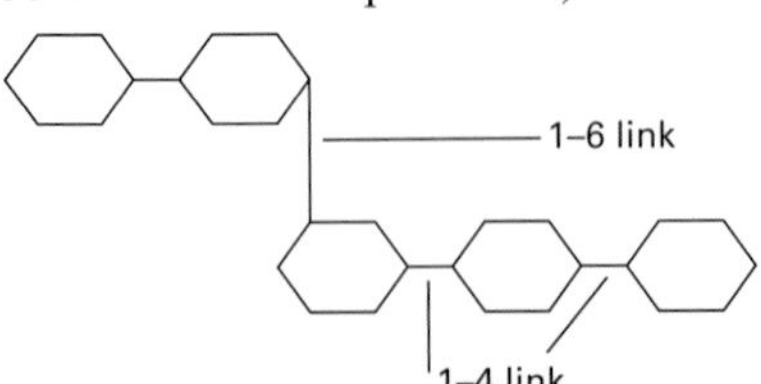

b i Hydrolysis is a chemical reaction in which macromolecules / large molecules are broken down, by the addition of water.
ii Enzymes / hydrolases; acids; heat
iii More 'ends' for enzyme / eq. to attack / more exposed substrate groups

4 Any 4 from the following:

Cellulose	Glycogen
Polymer of β-glucose	Polymer of α-glucose
Structural role / non-storage role	Non-structural role / storage role
The monomers rotate alternately at 180° to each other	No rotation of adjacent monomers
Straight chains, with cross-linking due to hydrogen bonds	Branched chains
β-1–4 linkages	α-1–4 and 1–6 linkages
Found in plants	Found in animals and fungi
Form microfibrils	Does not form microfibrils
Hydroxyl groups project outwards	Hydroxyl groups project inwards
Few organisms can digest cellulose / cellulase is not a common enzyme	Easily digested / hydrolysed by many organisms
Used in paper-making	Few commercial uses

5 a Amylopectin / starch

b Molecule A contains α-glucoses, molecule B contains β-glucoses / monomers in molecule A are linked by α-1–4and α-1–6 links, monomers in molecule B are linked by β-1–4 links.

c i

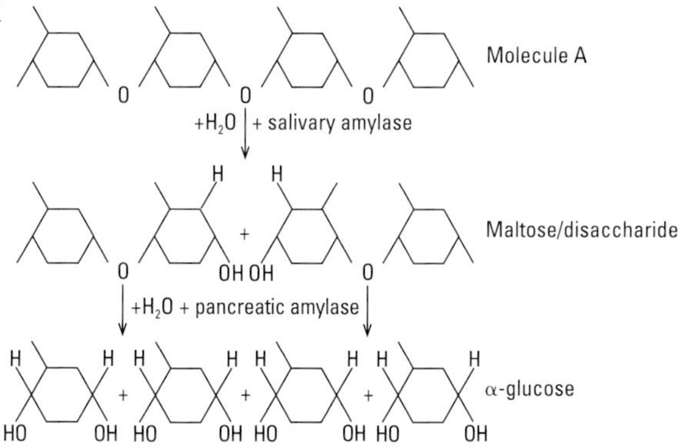

Molecule A

$+H_2O$ + salivary amylase

Maltose/disaccharide

$+H_2O$ + pancreatic amylase

α-glucose

ii The molecular shape / structure of molecule B, caused by β glycosidic bonds, does not fit the active site of the enzyme / amylase / the monomers alternately rotate 180o producing a shape / structure that does not fit the active site of the enzyme / amylase.

6 a i Amino acid

ii Secondary structure / α-helix

iii Hydrogen bond

iv Hydrogen bonds break; the secondary structure / α-helix will unfold / become denatured / coagulate.

v Primary structure / polypeptide chain is folded, usually to form an α-helix or a β-pleated sheet; specific shape is held by many hydrogen bonds

b In the β sheet, polypeptide chains: are arranged in parallel (in the α-helix the chains are helical or spiral as the name suggests) / can not be stretched (α-helices have some ability to stretch) / have a higher tensile strength than α-helices.

c i Fibrous proteins: keratin / collagen / myosin / silk / etc.
Globular proteins: any named enzyme e.g. amylase / any named protein hormone e.g. insulin / antibodies / etc.

ii The folding in on itself of the secondary structure; to form a precise / compact / globular shape; held in place by disulphide / ionic / hydrogen bonds / hydrophobic interactions.

d Polypeptide is linked by peptide bonds (polysaccharide has glycosidic bonds); monomers can be different / choice of 20 amino acids (polysaccharide has identical monomers); sequence of monomers plays a critical role in the final shape / nature of the protein (sequence of monomers in polysaccharide irrelevant as identical); AOVP.

Chapter 3 Metabolic reactions

Quick Check questions

Spread 3.1

1 heat is generated when energy is transformed from one type to another.

2 anabolic

3 energy required to overcome an energy barrier and start a chemical reaction.

Spread 3.2

1 three from: enzymes are **i** specific; they function effectively only in a limited range of **ii** temperature, **iii** pressure, and **iv** pH. Inorganic catalysts are not specific and generally function over a wide range of temperature, pressure, and pH.

2 in the lock-and-key theory the active site is rigid whereas in the induced-fit theory the active site is able to change its shape.

Spread 3.3

1 enzymes can only catalyse reactions in which the substrate shape fits that of the active site.

2 a an increase in temperature **i** increases the kinetic activity of reactants and **ii** increases the rate of denaturation of the enzyme.

b pepsin optimum about pH 3; trypsin optimum about pH 7.8

c pepsin is found in the stomach (acidic conditions); trypsin is found in the duodenum (alkaline conditions).

Spread 3.4

1 a cofactor may be any non-protein substance which enables an enzyme to function effectively; a coenzyme is a non-protein organic substance which enables an enzyme to function effectively (i.e. it is one type of cofactor).

2 reversible, irreversible, competitive, non-competitive.

3 when it reaches a certain concentration, the final product in a series of enzyme-catalysed reactions acts as a non-competitive inhibitor, preventing further production until the concentration of the final product falls below the critical concentration.

Spread 3.5

1 a transferase; **b** hydrolase; **c** oxidoreductase

Spread 3.6

1 advantages include: **i** enzymes catalysed can produce fewer unwanted by-products than inorganic catalysts; **ii** most enzymes work at room temperature and **iii** at atmospheric pressure. Most industrial reactions involving inorganic catalysts require high temperatures and pressures which are expensive to produce.
disdvantages include: **i** work effectively only in a narrow temperature range and **ii** a narrow pH range.

2 alcohol or ammonium sulphate is used to precipitate enzyme from a solution.

3 high temperatures occur during many industrial processes.

Spread 3.7

1 a protein-digesting enzyme obtained from *Bacillus subtilis*.

2 fructose is sweeter than glucose.

3 a transducer converts a biochemical signal into an electrical signal.

Exam questions

1 a Amino / amine group

b i Folding of polypeptide chains in a secondary structure / into a globular shape; folding held by specified bonding / eq.

ii When bonds are broken shape of enzyme is altered / enzyme is denatured; shape of active site is altered; substrate can no longer fit in active site; enzyme activity will be reduced.

2 a Active site

b Shape of substrate exactly fits shape of active site; molecules of other shapes can not fit into active site and therefore can not form enzyme–substrate complexes.

c Competitive inhibitor molecule has similar shape / structure to substrate molecule; it can therefore fit into the enzyme's active site and prevent the substrate entering; enzyme–substrate complexes can not be formed; rate of reaction decreases.

3 a i Substrate used up; therefore less frequent collisions between enzyme and substrate / enzyme–substrate complexes formed less often as concentration of substrate decreased; product may be producing feedback inhibition (end-product inhibition) as it builds up.

ii Turnover number will be higher / doubled / 5000; increased thermal energy will be converted to increased kinetic energy of the enzyme and substrate molecules, increasing the frequency of collisions / formation of enzyme–substrate complexes.

b To show that it was the temperature increase, and no other factor, that increased the turnover number.

4 a Enzymes are specific; the shape of the succinate molecule will fit the shape of the active site of succinate dehydrogenase / to form an enzyme–substrate complex.

b The shape / structure of the malonate molecule is similar to that of succinate / specific example of similar groups given; malonate will also fit into the active site of the enzyme, reducing the number of succinate molecules able to form an enzyme–substrate complex / decreasing the rate of reaction.

5 a Enzyme concentration / temperature / pH

b i Rate of reaction is reduced in the presence of the competitive inhibitor; inhibitor molecules have similar shape to substrate molecules / can fit into active site of enzyme / prevent substrate molecules forming an enzyme–substrate complex by occupying the active site.

ii Number of inhibitor molecules has remained constant, number of substrate molecules has increased; if there are more substrate molecules than inhibitor molecules the substrate molecules have a better chance of filling the active site of the enzyme / effect of inhibitors is reduced / substrate molecules out-compete inhibitor molecules.

c i End-product inhibition

ii Compound **Z** acts as an allosteric / non-competitive inhibitor of enzyme e_1; as **Z** increases, the inhibition increases, less compound **W** will be produced and therefore less compound **Z**; negative feedback mechanism.

6 a Enzymes can be re-used; product and enzyme easily separated / product not contaminated by enzyme / reduces purification processes for product; more thermostable / resistant to changes in temperature / can be used over a wider range of temperatures / temperature of process does not need to be controlled so carefully; same points for pH; allows the process to be run as a continuous process rather than as a batch process; more economical.

b i Adsorption onto an insoluble matrix e.g. resin; entrapment in a gel e.g. alginate, silica; entrapment in microcapsules; covalent bonding onto a matrix e.g. cellulose; cross-linking to another molecule e.g. glutaraldehyde.

ii Use: same concentration / volume of enzyme to immobilise; same concentration / volume of glucose as substrate; control temperature / specified temperature used; control flow rate of substrate over enzyme / contact time (incubation time) of substrate and enzyme / use of stopclock; use of buffers / specified pH; suitable technique for measuring fructose production / glucose reduction; repeats.

c Enzymes more likely to be thermostable / resistant to denaturation at high temperatures; processes using these enzymes can be run at a higher temperature / over a wider range of temperatures.

7 a i Proteins / antibodies / enzymes

ii To detect a specific chemical compound / molecule

b To convert any change brought about by the recognition of a specific molecule by the receptors into an electrical signal which can then be amplified to give a read-out.

c Biosensor detects only glucose / Benedict's detects any reducing sugar; biosensor does not depend on colour changes / a positive result with the Benedict's test is an orange / eq. precipitate which is hard to detect against the red colour of blood; biosensor is faster / does not require heat; biosensor requires less blood; biosensor is more sensitive / will detect smaller changes in blood glucose concentration; biosensor is objective and quantititve / Benedict's is subjective and semiquantitive.

8 a Sequence / position of amino acids determines where bonds will form / exactly how chain will fold; primary structure / chain of amino acids is folded into an α-helix or a β-pleated sheet; to form secondary structure; held in place by many hydrogen bonds; secondary structure is then

folded again to form tertiary structure; other types of bonds e.g. disulphide bonds / ionic bonds are involved in maintaining the precise shape; quaternary structure if more than one polypeptide chain is incorporated into the structure.

b Increase in temperature increases kinetic energy of enzyme / substrate molecules / enzyme / substrate molecules move faster; more collisions between enzyme molecules and substrate molecules occur / more enzyme–substrate complexes form; / converse of first two points for decrease in temperature; at higher temperatures molecules vibrate more; bonds / hydrogen bonds / eq. break; denaturation occurs / enzyme's tertiary structure is altered; shape of active site is altered; substrate no longer fits active site / enzyme–substrate complex can not form.

c Direct inhibitors are usually competitive inhibitors; have a shape similar to the substrate molecule; will fit into active site of enzyme; preventing substrate from entering / forming enzyme–substrate complex; are concentration-dependent / the more inhibitor present, the greater the inhibition / competition for the active site.
Indirect inhibitors are usually non-competitive inhibitors; bind to a site on the enzyme away from the active site; effect is to distort the enzyme molecule and hence the active site; substrate molecules no longer fit into active site / enzyme–substrate complexes not formed

9 i Hydrolases are enzymes which catalyse hydrolysis reactions.

ii Starch is made up of α-glucoses; cellulose is made up of β-glucoses; the active site of amylase is specific to the shape / structure of linked α-glucoses / linked β-glucoses do not fit the active site of amylase.

Chapter 4 Cells

Quick Check questions

Spread 4.1

1 living things are made of cells and these cells have certain things in common.

2 a glycogen granules **b** chloroplast, cell wall, vacuole membrane (tonoplast), vacuole

c cell surface membrane, mitochondria, cytoplasm, and nucleus

Spread 4.2

1 a magnification in a light microscope is varied by changing the power of the glass lenses.

b magnification of an electron microscope is varied by changing the strength of the electromagnets.

2 electron microscopes use beams of electrons which have a shorter wavelength than light, giving electron microscopes a higher resolving power than light microscopes.

3 0.2 μm.

Spread 4.3

1 a mitochondrion; **b** ribosome; **c** lysosome; **d** nucleus.

Spread 4.4

1 cytoplasmic threads passing through small membrane-lined pores in cell walls, connecting one plant cell with another.

2 a nucleoid b A nucleoid has no double-membraned nuclear envelope.

3 eukaryotes have double-membraned organelles.

Spread 4.5

1 during electron microscopy, artefacts may be caused by **i** distortion during dehydration **ii** staining chemicals outside the cell which appear as intracellular structures.

2 freeze-fracturing of cell membranes reveals tiny dots which are thought to be proteins.

3 chilling slows down metabolism and minimises self-digestion of the organelles.

Spread 4.6

1 partial permeability enables cell membranes to control the passage of materials, allowing the passage of some substances while preventing the passage of others.

2 the fluid-mosaic model is so-called because it appears to have the properties of a fluid rather than a solid (the protein component can, for example, move through the phospholipid component), and the protein and lipid components form a pattern like a mosaic made up of small pieces of coloured glass or stone.

3 the carbohydrate component of a glycoprotein, projecting from a cell surface membrane, can act as a receptor site. The shape of the carbohydrate component may fit that of T cells, making the receptor site specific to these cells.

Spread 4.7

1 the net movement of molecules or ions from a region where they are in high concentration to a region where they are in low concentration.

2 surface area; concentration gradient; length of diffusion path.

3 1.5 : 1

Spread 4.8

1 active transport is movement of substances against a concentration gradient using energy derived from the breakdown of ATP, and is associated with an increase in oxygen consumption.

2 active transport uses energy derived from ATP; a decrease in the concentration of oxygen will reduce the amount of ATP generated by aerobic respiration. Diffusion is a passive process not requiring ATP.

3 pinocytosis involves the formation of small vesicles and the entry of liquids or large macromolecules whereas phagocytosis involves the formation of vacuoles and the entry of larger, solid substances.

Spread 4.9

1 0 kPa

2 water moves from A (water potential –100 kPa) to B (water potential –300 kPa).

3 a heavily perspiring athlete should take a hypotonic drink because it will have a higher water potential than body cells so that water will tend to enter body cells from the drink.

Spread 4.10

1 a interphase, prophase, metaphase, anaphase, and telophase.

b interphase

2 mitosis involves one division and the formation of two daughter cells from each parent cell whereas meiosis involves two divisions and the formation of four daughter cells from each parent cell.

Spread 4.11

1 stages of mitosis shown in figure 4.11.2 are, from left to right on the top row, late prophase, anaphase, early prophase, and, from left to right on the bottom row, telophase, interphase, metaphase.

Spread 4.12

1 a meiosis I, metaphase I (first division)
b meiosis I, telophase I (first division)

Spread 4.13

1 multicellularity enables organisms to become larger and also enables their cells to become more specialised.

2 a tissue consists of cells of the same type which act together to carry out a common function, whereas an organ is a group of different tissues that carry out a particular function or functions.

3 a variety of answers are acceptable (see text) e.g.
a i a cheek cell is relatively unspecialised;
ii epithelial tissue **iii** intestine
b i meristematic cell e.g. in the root tip
ii parenchyma tissue **iii** leaf

Exam questions

1

	Plant cell	Animal cell	Prokaryotic cell
nucleolus	✓	✓	✗
plasmid	✗	✗	✓
mitochondrion	✓	✓	✗
cellulose wall	✓	✗	✗

2 a A: cell surface membrane; B: centrioles; C: Golgi apparatus / body; D: rough endoplasmic reticulum.
b Shown / magnified length of mitochondrion is 11 mm (+/– 1mm) / 11 000 µm; drawing has been magnified 12 000 times, so divide by this to get actual length / 11 000/12 000; actual length is 0.9 µm

3 a Ribosomes / polysomes / endoplasmic reticulum
b Same water potential / solute concentration as in mitochondria; prevents osmosis / water entering the mitochondria / mitochondria bursting.
c Pyruvate enters Krebs cycle; hydrogens produced and taken by carrier molecules to electron transport chain; hydrogen picked up by oxygen as the final acceptor at the end of the electron transport chain / hydrogen used to make water and therefore oxygen used up / levels fall.

4 a i **A**: phosphate group; **B**: glycerol residue / backbone; **C**: fatty acids / fatty acid residues
ii Two layers of phospholipid molecules / phospholipid bilayer; hydrophilic / polar phosphate heads of the phospholipids face outwards into the aqueous cytoplasm of the cell and into the aqueous tissue fluid outside the cell; hydrophobic / non-polar hydrocarbon tails face inwards and create a hydrophobic interior to the cell membrane.
b Transport of ions / molecules across the membrane; receptors; cell recognition / antigens / cell markers; enzymes.

5 a $\psi_{cell} = \psi_s + \psi_p$ / = 350 + (–800) kPa; = –450 kPa.
b Water potential is lower / more negative outside the cell; water moves from a high / nearer zero water potential to a lower water potential / down water potential gradient so water will move out of the cell.
c –800 kPa

6 a i –7 Mpa
ii Arrows from A to B, from C to B, and from C to A.
iii Sucrose / solutes reduce water potential; there are fewer 'free' water molecules in sucrose solution / sucrose solution has a lower water potential; than pure water which has a water potential of zero.
b i Glycerol is fat-soluble and dissolves in / passes through lipid layers of both membranes.
ii Sodium ions are transported by channel proteins / sodium pumps; these are not present in the artificial membrane, but are in the biological membrane; facilitated diffusion / active transport of sodium ions may take place in the biological membrane, not in the artificial one.

7 a Sequence of events between one cell division and the next is called the cell cycle; interphase; new materials / organelles synthesised / G1 phase; growth; DNA replicates / S phase; each chromosome becomes two chromatids; G2 / mitochondria / chloroplasts divide; M phase / 4 stages; **prophase**; chromosomes shorten and thicken; centrioles move to poles in animal cells; nucleoli disappear; nuclear membrane disappears; spindle forms; **metaphase**; chromosomes line up around equator; centromeres attached to spindle fibres; **anaphase**; centromeres split; separated chromatids pulled towards poles; **telophase**; chromatids reach poles; chromatin reformed; spindle disintegrates; centrioles replicate; nuclear membrane reforms; nucleoli visible; C phase; cell divides to become two daughter cells; with same chromosome number as parent cell / equal distribution of organelles; cell plate forms in plant cells; G1 phase of interphase / cycle continues; length of cell cycle depends on type of cell / temperature / other factors; named example e.g. onion root tips may divide every 20 hours; AOVP
b Any 4 of the following:

Mitosis	Meiosis
Two daughter cells produced with same number of chromosomes as parent cell	Four daughter cells produced with half the number of chromosomes as parent cell
Daughter cells are genetically identical to / clones of parent cell and each other	Daughter cells show genetic variation to parent cell and to each other
Occurs in the formation of body / somatic / non-sex cells / some spores / gametes in plant / basis of asexual reproduction	Occurs in the formation of gametes / spores / occurs at some stage in the life of sexually reproducing organisms
One main stage	Two stages: meiosis I and meiosis II
Homologous chromosomes do not pair up in prophase	Homologous chromosomes pair up / form bivalents in prophase I
Chiasmata do not form	Chiasmata form
No crossing over	Crossing over
No independent assortment	Independent assortment of chromosomes in metaphase I
AOVP	AOVP

Chapter 5 Photosynthesis

Quick Check questions

Spread 5.1

1 a carbon dioxide **b** oxygen

2 a light-dependent stage occurs in the grana of chloroplasts

b the light-independent stage occurs in the stroma of chloroplasts.

Spread 5.2

1 red and blue

2 the action spectrum shows the relationship, usually plotted as a graph, between the wavelengths of light and the rate of photosynthesis, whereas the absorption spectrum shows the percentage of absorption of light by chlorophyll (or another substance) at different wavelengths of light.

3 the green/yellow parts of the spectrum around 500–550 nm.

Spread 5.3

1 ATP, NADPH, and oxygen

2 oxygen

3 hydroxide ions (OH^-) and protons (H^+)

Spread 5.4

1 ribulose biphosphate, ribulose biphosphate carboxylase (rubisco), and glycerate 3-phosphate.

2 glycerate 3-phosphate

3 NADPH is a reducing agent (it donates hydrogen to GP, reducing it to GALP, a triose phosphate).

4 ATP is used as a source of energy for the reduction of carbon dioxide and the regeneration of RuBP.

Spread 5.5

1 water availability affects many activities in addition to photosynthesis. Water deprivation may kill a plant, but the cause of death may not be connected to photosynthesis.

2 the compensation point for a shade plant is at a lower light intensity than that of a sun plant.

3 light intensity

Spread 5.6

1 a two from wheat, soya beans, and rice.

b sugar cane is called a C_4 plant because it fixes carbon dioxide as a four-carbon compound.

2 at night.

3 a C_4; **b** CAM; **c** C_3

Exam questions

1 a Temperature; carbon dioxide concentration; pH

b Light energy used to split water / photolysis of water; in light-dependent stage; into hydrogen ions / protons, electrons, and oxygen; oxygen released as a waste product via stomata / cut stem;

c $\frac{1}{10^2} = 0.01 \quad \frac{1}{5^2} = 0.04$

number of bubbles at 5 cm $= \frac{20 \times 0.04}{0.01} = 80$

d i lumen; **ii** photosystem II; **iii** photosystem I; **iv** photosystems II and I; **v** photosystem I

e Pump moves protons, by active transport, from the stroma; across the thylakoid membrane into the thylakoid lumen; pump is driven by energy released from excited protons as they pass down an electron transport chain; protons accumulate in lumen forming a concentration gradient; diffuse back across the membrane through chemiosmotic channels / ATP synthase complex; complex converts ADP and phosphate into ATP / photophosphorylation; ATP formed helps drive the light-independent stage.

2 a i Maximise number of algal cells exposed to light

ii Transparent / clear perspex / eq. screen would absorb heat but allow light through / use a 'cold' light source

iii Enzymes inactivated; reactions stopped immediately; so that the radioactive compounds identified and the path taken by radioactive carbon traced

b R_f value calculated and compared with known values for solvent used

c i Glycerate phosphate / GP **ii** X

iii Spot visible after 5 seconds has same position as X

3 a Chloroplasts found in the cytoplasm of plant cells, usually in leaves; numerous in palisade mesophyll cells / nearer upper surface of leaf; some chloroplasts in spongy mesophyll cells / some in guard cells; within cells, chloroplasts may move toward light; allows them to arrange themselves into best position within cell for most efficient absorption of light; chlorophyll and carotenoids / other photosynthetic pigments; located on membranes; on thylakoids / grana; different pigments trap different wavelengths of light, making process more efficient; light, trapped by chlorophyll *a* initiates photophosphorylation / photolysis of water; flattened / fluid-filled thylakoids form stacks / grana; mosaic / tessellated arrangement of chlorophyll in layers increases chance of light striking chlorophyll molecule

b Photolysis of water; by light energy trapped within chlorophyll *a*; in photosystem II; releases protons, electrons, and oxygen; light energy excites electrons from chlorophyll; protons moved from stroma to lumen of thylakoid membrane; by Q cycle; PQ / plastoquinone reduced; diffuses across phospholipid bilayer; as electrons passed onto next carrier in thylakoid lumen, protons released into lumen; energy carried by electrons / from light used to move protons across membrane; protons transferred against a concentration gradient / concentration difference in protons develops across membrane; more energy from light used to reduce NADP; in photosystem I; more protons removed from stroma; transfer of protons to interior of thylakoid creates difference in pH; and strong electrical potential; potential energy stored in proton concentration difference used to form ATP; enzyme complex ATPase synthesises ATP; ATPase in thylakoid membrane, facing the stroma; chemiosmotic mechanism; 3 protons needed for each molecule of ATP formed

4 a i Action spectrum **ii** 450 nm

iii Rate increases from 550 to 625 nm; maximum rate at 625 nm; rate falls gradually from 625 to 675 nm; then steep fall to 700 nm; different photosynthetic pigments absorb different wavelengths; chlorophyll *b* peak absorption at 600–650 nm; chlorophyll *a* peak absorption at 650–675 nm; pigments do not absorb wavelengths above 675 nm.

b i Light-dependent stage

ii Thylakoid membrane / thylakoids / grana

5 a i RuBP: 5C; GP: 3C; Triose phosphate: 3C; Hexose: 6C

ii Stroma

iii Store of RuBP converted to GP; no further ATP / NADPH arriving from light-dependent stage, so GP can not be converted to triose phosphate.

b Photolysis of water provides electrons to replace electrons emitted by chlorophyll molecules when light absorbed; provides protons, which, with electrons, reduce NADP; provides oxygen released as a waste product / used in respiration.

Chapter 6 Respiration

Quick Check questions

Spread 6.1

1 cellular respiration is the process by which energy in food molecules is made available as ATP for an organism to do biological work.

2 without regeneration, the store of ATP would be completely depleted in about 1 second.

3 a an environment with no or very little oxygen.

b an obligate anaerobe can respire only anaerobically whereas a facultative anaerobe can respire either aerobically or anaerobically, depending on the availability of oxygen.

Spread 6.2

1 the cytosol (the fluid part of the cytoplasm)

2 four

3 lactate fermentation.

Spread 6.3

1 a mitochondria. **b** 1

2 oxidative decarboxylation involves removal of carbon dioxide from a molecule coupled with oxidation of the molecule by the removal of hydrogen.

Spread 6.4

1 on the inner membrane (cristae) of mitochondria.

2 the processes along the electron transport system involve oxidation reactions which provide the energy required to make ATP by the phosphorylation (addition of a phosphate group) to ADP.

3 the folding increases the area of the surfaces on which the chemical reactions take place.

Spread 6.5

1 RQ = 36/40 = 0.9

2 36/51 = 0.71

3 protein (assuming only one substrate is being respired).

Exam questions

1 a i Glycolysis **ii** Cytoplasm / cytosol

b i Active transport / ion pumps; phosphorylation / donation of phosphate group to a substrate / named substrate / stage A used as example; anabolic reactions / named example e.g. protein synthesis; muscle contraction / movement; light-independent stage of photosynthesis; DNA replication; cell division; AOVP

ii Stage A

c Stage D

2 a Increases surface area; for enzymes / carrier molecules; associated with ATP synthesis / electron transport chain

b $\text{R.Q.} = \dfrac{\text{volume of } CO_2 \text{ produced}}{\text{volume of } CO_2 \text{ consumed}} = \dfrac{57}{80}$

c i Respiratory substrate is carbohydrate / carbon dioxide produced = oxygen consumed

ii Different respiratory substrates are used / respiratory substrate changes

3 a

Stage of respiration	ATP produced	Carbon dioxide produced	Reduced NAD converted to NAD
glucose → pyruvate	✓		
pyruvate → acetyl coenzyme A		✓	
Krebs cycle	✓	✓	
Electron carrier chain	✓		✓

b i 8 mm^3 oxygen consumed after 10 minutes, therefore rate of consumption $= \dfrac{8}{10}$ mm^3 min^{-1} $= 0.8$ mm^3 min^{-1}

ii $\text{R.Q.} = \dfrac{\text{volume of carbon dioxide produced}}{\text{volume of oxygen consumed}} = \dfrac{8.4}{12.0} = 0.7$

4 a i Mitochondria contain DNA; mitochondria have the biochemical machinery needed for carrying out some protein synthesis.

ii Their DNA only codes for a few of the mitochondrial proteins / they have to rely on nuclear / other DNA to code for the rest of their proteins; they only have the biochemical machinery to carry out *some* protein synthesis / they need to rely on other sources to complete protein synthesis.

iii Liver cells are very active cells / require a lot of energy / many mitochondria needed to provide the energy required

iv Increased surface area; for enzymes / carrier molecules associated with the electron transport chain

b i A: cytoplasm / cytosol; B: matrix of mitochondria; C: inner membranes / cristae of mitochondria

ii Carbon dioxide

iii Stage A: 2 molecules of ATP (net)
Stage B: 2 molecules of ATP
Stage C: 34 molecules of ATP
Note that the above numbers are for molecules of ATP formed directly.

iv Total number of ATP molecules formed from 1 molecule of glucose during respiration = 38; each ATP molecule 'traps' 50 kJ, therefore total of 38 × 50 kJ 'trapped' / = 1900 kJ mol^{-1}; 2881 kJ *could* be made available from glucose / therefore % efficiency $= \dfrac{1900}{2881} \times 100$ / = 65.9%

5 a i A: matrix; B: inner membrane / crista; C: inter-membrane space

ii Arrow pointing to inner membrane

iii Measured length = 79 mm $= \dfrac{79\,000}{70\,000}$ µm;
Actual length = 79 000 = 1.13 µm

b i Aerobic respiration is occurring / oxygen is used up in respiration / oxidative phosphorylation / in the electron transport chain; oxygen combines with hydrogen / is final acceptor for hydrogen; forms water.

ii ATP production stopped / decreased; electron transport can no longer release energy to produce ATP; oxidative phosphorylation stopped; Krebs cycle stopped; hydrogen carriers / NAD can not be recycled; no more oxygen taken up.

Chapter 7 Gaseous exchange and transport in mammals

Quick Check questions

Spread 7.1

1 nasal cavity → pharynx → trachea → bronchi → bronchioles →alveoli

2 the skin would not provide a sufficiently large surface area for gaseous exchange, and gaseous exchange through the skin would also result in excessive water loss.

Spread 7.2

1 a external intercostals **b** internal intercostals

2 tidal volume

Spread 7.3

1 well ventilated and highly perfused with blood to maintain a high concentration gradient; numerous, small, and highly folded to provide a very large surface area; thin (one cell thick) to minimise diffusion distance between the blood in the capillaries and the air in the lungs.

2 0.04 kPa

Spread 7.4

1 mass flow is the bulk movement of substances from one area to another due to differences in pressure.

2 blood can be at a relatively high pressure and distribution of the blood can be controlled.

3 although an artery has a thicker wall than a vein of the same size, the absolute thickness of a vessel will depend on its absolute size. A small artery may have a thinner wall than a large vein.

Spread 7.5

1 superior vena cava → right atrium → tricuspid valve → right ventricle → semilunar valve → pulmonary artery → lung capillaries → pulmonary vein → left atrium → bicuspid valve → left ventricle → semilunar valve → dorsal aorta.

2 left ventricle; it has to pump the blood into the aorta and all the way around the body.

Spread 7.6

1 a atrial systole; pause; ventricular systole; diastole

b 0.3s

2 the main heart pacemaker region is in the sinoatrial node, in the wall of the right atrium near the opening of the superior vena cava.

Spread 7.7

1 a $60 \times 75 = 4500\ cm^3\,min^{-1} = 4.5\ dm^3\,min^{-1}$

b 270 dm^3

2 directly above the base of the thumb on the underside of the wrist where the radial artery passes alongside the radius.

3 the heart rate will increase.

Spread 7.8

1 plasma is the fluid portion of the blood, including proteins, whereas serum is plasma from which proteins (especially fibrinogen) have been removed.

2 haemoglobin; fibrinogen.

3 blood clotting.

Spread 7.9

1 haemoglobin has a non-protein part (haem).

2 four

3 high carbon dioxide concentrations shift the oxygen dissociation curve to the right (the Bohr shift).

Spread 7.10

1 a i three from: increase in rate of breathing; increase in depth of breathing; increase in anaerobic respiration; increase in the activity of intercostals and diaphragm which produce ventilatory movements; increase in amount of anaerobic respiration

ii an increase in cardiac output (produced by an increase in heart rate and stroke volume); shunting of blood from intestines to skeletal muscles; increase in the percentage of oxygen unloaded from haemoglobin (or an improvement in the ability of muscle to extract oxygen).

b high concentration gradient resulting from aerobic respiration of active muscle; a high temperature; high carbon dioxide levels.

Spread 7.11

1 ultrafiltration resulting from a high hydrostatic pressure.

2 red blood cells; platelets; proteins.

Exam questions

1 a i Right lung (left side of diagram) deflated, left lung remains the same size

ii Lungs inflate due to air pressure within thorax / lungs being lower than atmospheric pressure; puncture will cause air pressure inside lungs to be equal to atmospheric / air pressure will no longer be able to drop below atmospheric; right lung can not inflate / right lung collapses.

b i Deoxygenated Hb changes its shape after being oxygenated by one molecule of oxygen, exposing the iron in the haem groups; making subsequent oxygenation easier

ii Curve C

iii Oxygen is released readily; in areas where oxygen concentration is low; haemoglobin molecule is less saturated for the same partial pressure of oxygen.

iv Haemoglobin has a greater affinity for carbon monoxide than for oxygen, therefore haemoglobin prevented from combining with oxygen / tissues deprived of oxygen.

2 a Sinoatrial node / pacemaker initiates heartbeat; wave of excitation spreads across both atria causing virtually simultaneous contraction; wave reaches atrioventricular node; bundle of His / Purkyne fibres; spreads down both ventricles at same time; to base / apex; from where it radiates upwards; ventricles contract simultaneously from base upwards / both will beat / contract at same rate; nerves / named e.g. act on SA node to alter rate; pressure produced by contraction of cardiac muscle; left side of heart has more / thicker muscle; blood ejected from left ventricle at higher pressure / converse for right ventricle.

b Chemoreceptors in aorta / carotid artery; detect changes in blood CO_2 concentration; stretch / pressure receptors in main blood vessels; send impulses to control centre / cardiac inhibitory / excitatory centre; in medulla; impulses from aorta / carotids decrease heart rate / impulses from vena cava increase heart rate; sympathetic nerves from medulla increase heart rate; parasympathetic / vagus nerves decrease heart rate; sympathetic nerves take nerve impulses from excitatory centre / parasympathetic from inhibitory centre to SA node; increase / decrease rate of activity of SA node; adrenaline from adrenal glands; noradrenaline; thyroxine increase heart rate; bind to receptors at SA node.

c Capillary walls permeable to water; high blood pressure at arterial end of capillary; forces water plus small solutes and ions out of capillary; to form tissue fluid; not all tissue fluid returns to capillaries; higher pressure in tissue fluid than lymph; excess drains into lymphatic vessels; now called lymph; fats added; lymph node products / antibodies added.

3 a A: Sinoatrial node

b i Ventricles contract after atria, allowing time for ventricles to fill before contracting.

ii Blood is squeezed out of the ventricles more efficiently / more blood is expelled at each contraction

c Excitatory part of cardiovascular centre stimulated; sympathetic nerves from here stimulate SA node / pacemaker; increase in activity of SA node increases heart rate.

4 a Any six of the following:

Chemical component of blood plasma	Function
Water	Solvent for many dissolved materials; transport medium; provides cells with water; helps regulate blood volume / pressure; AOVP
Carbon dioxide / hydrogen carbonate	Waste product of respiration
Proteins / albumins / globulins	Enzymes; transport of hormones, fat-soluble vitamins, iron, calcium, cholesterol; antibodies; blood-clotting; buffers; AOVP
Hormones / named e.g.	Control of blood sugar level; control of body water content; growth; metabolic rate; AOVP
Sugars / named e.g.	Respiratory substrates; synthesis of other compounds; AOVP
Fatty acids	Respiratory substrates; synthesis of other compounds; AOVP
Glycerol	Synthesis of other compounds; AOVP
Amino acids	Synthesis of proteins; buffers; AOVP
Urea / other excretory products	Removal of excess nitrogen
Vitamins	Required in small quantities for good health
Mineral ions / named e.g.	Regulation of solute potential / pH level in blood; clotting factors; AOVP

b Oxygen diffuses into alveolar capillaries; down a concentration gradient; across capillary endothelium / plasma; combines with haemoglobin in red blood cells; iron atoms within haemoglobin combine with oxygen; in areas of high oxygen concentration / in lung each haemoglobin molecule can combine with 4 oxygen molecules; amount of oxygen that can combine with haemoglobin determined by partial pressure / concentration of oxygen; oxygen dissociation curve shows how much oxygen is carried by haemoglobin / saturation of haemoglobin at different partial pressures of oxygen; mass flow idea.

Carbon dioxide from cells diffuses across tissue fluid / capillary endothelium / plasma; down a concentration gradient; some carbon dioxide transported in solution; as carbonic acid; more carbon dioxide transported combined with protein / as carbaminohaemoglobin compound; amount carried this way depends on amount of oxygen being carried / the less oxygen being carried, the more carbon dioxide can be carried; most carbon dioxide transported as hydrogencarbonate; carbon dioxide combines with water inside red blood cells to form carbonic acid; catalysed by carbonic anhydrase; carbonic acid dissociates into hydrogen and hydrogencarbonate ions; hydrogen ions displace oxygen from haemoglobin; haemoglobinic acid formed; buffering action of haemoglobin; allows large quantities of carbonic acid to be transported to lungs without large shifts in blood pH; negatively charged hydrogencarbonate ions diffuse from red blood cells to plasma; form sodium hydrogen carbonate; positively charged red blood cells attract negative chloride ions from plasma / chloride shift; mass flow idea.

c Capillaries have single layer of endothelial cells; small gaps between these cells make wall permeable to water, small molecules, and ions; capillary networks run close to cells; capillaries in the kidney have pores; slow blood flow through capillaries allows time for material to exchange; material exchange down concentration gradients; in solution; formation of tissue fluid transports material out of capillaries; at arterial end of capillary network; increased solute concentration / more negative solute potential of blood at venous end of capillary bed; draws tissue fluid back into capillary and with it material that has diffused out of cells.

5 a i High partial pressure of oxygen / data given, in lungs causes haemoglobin to pick up oxygen readily / form oxyhaemoglobin / form HbO_8

ii Low partial pressure of oxygen in tissues / data given, due to respiring cells consuming oxygen causes haemoglobin to become less saturated with oxygen / dissociate

b Exercise increases temperature / rate at which carbon dioxide produced / lowers pH; shifts dissociation curve to right / Bohr effect; partial pressure of oxygen in muscle falls; percentage saturation of haemoglobin falls / affinity for oxygen is lowered / increased dissociation

c In absence of oxygen / at low partial pressures of oxygen in tissues / oxygen still used up in respiration; lugworm haemoglobin acts as an oxygen store; when oxygen levels low in tissues it releases oxygen.

6 a i 17–18%

ii Percentage saturation at 43°C and 4 kPa = 18%, difference in saturation = (90 – 18) = 72%;

$$\frac{105 \times 72}{100}; = 75.6 \text{ cm}^3$$

b Increase in temperature associated with increased activity and increased oxygen demand / high temperature would also release more oxygen when activity level high

Chapter 8 Homeostatic control systems

Quick Check questions

Spread 8.1

1 maintenance of a system, such as the internal environment of the body or part of the body, in a steady state.

2 a detectors monitor the actual output of a system

b comparators compare the actual output with the set point.

c effectors restore the output to the set point.

3 negative feedback

Spread 8.2

1 two separate mechanisms controlling deviations in different directions from the set point give a greater degree of control.

2 when blood glucose levels fall below the set point

Spread 8.3

1 cells from the immune system attack beta cells in the islets of Langerhans, destroying a person's ability to secrete insulin

2 insulin injections

3 high-fat diets

Spread 8.4

1 the liver has a double blood supply: the hepatic artery carries oxygenated blood to the liver; the hepatic portal vein delivers blood from the intestines to the liver so that substances absorbed from the gut can be processed before they enter the general circulation.

2 ammonia, carbon dioxide, and water

Spread 8.5

1 a elimination of metabolic waste

b control of the amount of water in the body

2 proximal convoluted tubule, loop of Henlé, distal convoluted tubule

3 lower abdomen, near the groin

Spread 8.6

1 glucose concentration in the glomerular filtrate is about the same as that in the blood in the glomerulus.

2 a a high density of mitochondria; microvilli to increase the surface area for absorption; indented cell surface membranes to form a large area of intercellular spaces bathed with fluid

b to supply ATP for active transport

Spread 8.7

1 a countercurrent multiplier combines a countercurrent exchange mechanism with active secretion of solutes

2 to prevent water from leaving the ascending limb by osmosis (if water followed salts out of the ascending limb, an osmotic gradient could not be created between the fluid in the descending and surrounding tissue fluid).

Spread 8.8

1 a control of the amount of water in the body so that the volume of body fluid can be kept in a steady state

b to conserve water

2 produce more ADH

3 osmoregulation; ionic regulation (regulation of salts); regulation of blood pH and blood pressure

Spread 8.9

1 when a lizard controls its body temperature, it does so by behavioural control mechanisms using external sources of heat gain and loss

2 a mammal is homoiothermic, typically maintaining the body at a temperature higher than that of the environment; this requires the respiration of food to release heat.

3 conduction, convection, and radiation

Spread 8.10

1 a in hyperthermia the body core temperature rises above 41 °C whereas in hypothermia the body core temperature falls below 32 °C

b a dry skin in hot weather may indicate that there is a breakdown in the thermoregulatory mechanism because sweating has stopped

Spread 8.11

1 three from: blood vessels; hot and cold receptors; sweat glands; adipose tissue; hair follicles

2 shunting is the diversion of blood from one region of the body to another

Exam questions

1 a i Blood glucose concentration increases after swallowing due to absorption from gut; detected by pancreas; insulin secreted / released

ii Blood glucose concentration moves away / rises from 'set' level, more insulin is released, decreases blood glucose concentration / blood glucose concentration falls, blood glucose concentration returns to 'set' level.
OR Insulin concentration in blood moves away from 'set' level, brings about a decrease in blood glucose concentration, insulin concentration returns to 'set' level.

b Glucose, a respiratory substrate, is used up more quickly during exercise, so blood glucose concentration falls; glucagon secretion stimulated / glycogen converted to glucose by glucagon / insulin secretion inhibited / insulin converts glucose to glycogen

2 a β cells in the islets of Langerhans

b Increase in blood glucose concentration over the 'set' level is detected by the pancreas; insulin released; decreases blood glucose levels by stimulating uptake by cells / conversion to glycogen; so blood glucose concentration returns to 'set' level; insulin release inhibited

c Insulin is a protein molecule, proteins are digested if swallowed

d Proteins; cell surface membranes

e i Controlling carbohydrates in diet; will reduce blood glucose concentration and hence insulin levels in blood

ii High concentrations of blood glucose are common to both types of diabetes, exercise will help reduce concentration as more glucose will be used up as a respiratory substrate

3 a Water is also reabsorbed

b Descending limb of loop of Henle is permeable to water / water leaves descending limb; less permeable to solutes; so fluid becomes more concentrated; water leaves descending limb by osmosis, due to high solute concentration in medulla; ascending limb is impermeable to water; solutes actively reabsorbed from upper half of ascending limb into tissue fluid; so fluid becomes increasingly dilute.

c ADH makes walls of collecting duct permeable to water, so smaller volume of more concentrated urine produced; absence of ADH makes walls impermeable to water, so larger volume of more dilute urine produced.

4 Glomerulus; water / glucose / urea / uric acid / amino acids / named amino acid / named hormone / named water-soluble vitamin (any 2 of these for the two spaces); ultrafiltration; glucose / amino acid; water

5 a Hypothalamus

b i Via nerves / electrical impulses

ii Arterioles near skin surface; vasodilate; capillary shunts / sphincters increase blood flow through skin capillary networks; more blood flows near surface; more heat lost by radiation

6 a Maintenance of a constant internal environment / eq.

b **Endothermy**: body temperature kept within narrow limits; by physiological mechanisms; e.g. in mammal / bird / named example; **Ectothermy**: body temperature varies with surroundings / environmental temperature; controlled by behaviour / eq.; named ectothermic animal (any other than mammal / bird)

Chapter 9 Heterotrophic nutrition

Quick Check questions

Spread 9.1

1 bracket fungus: saprobiontic; zebra: holozoic herbivore; frog: holozoic carnivore; fanworm: microphagus feeder; mosquito: external parasite (liquid feeder); housefly: liquid feeder

Spread 9.2

1 ingestion, taking food into the gut; digestion, breakdown of food; absorption, movement of nutrients across the gut wall into the blood and lymph; assimilation, incorporation, and using digested food molecules; egestion, elimination of undigested products from the gut.

Spread 9.3

1 tasting and manipulating food

2 salivary amylase; it catalyses the breakdown of starch to maltose

3 the oesophagus lengthens (longitudinal muscle cause it to shorten)

Spread 9.4

1 hydrochloric acid converts pepsinogen to pepsin and provides acidic conditions in which pepsin works most efficiently

2 a stomach wall cells; **b** duodenal lining

3 closure of the pyloric sphincter

Spread 9.5

1 Microvilli are folds in the surface membrane of individual cells whereas villi are folds made of many cells which form finger-like projections in the intestinal wall.

2 a glucose **b** fatty acids and glycerol; **c** amino acids

3 reabsorption of water

Spread 9.6

1 fat

2 carbohydrates, proteins, fats, minerals, vitamins, fibre, and water.

3 A person who has a daily energy expenditure of 12 500 kJ and a PAL of 1.6 will have a BMR of 12 500/1.6 = 7812.5 kJ

Spread 9.7

1 a carbohydrates, fats and proteins, nutrients required in relatively large amounts in the diet

b linoleic acid

c a complete protein contains all the essential amino acids whereas an incomplete protein lacks one or more essential amino acids.

Spread 9.8

1 a iron; **b** calcium, magnesium, phosphorus, fluoride, manganese; **c** iodide

Spread 9.9

1 a cut flesh with a scissors-like shearing action;

b the temporalis closes jaws enabling canine teeth to pierce and grip flesh; **c** hold meat as it is being cut

Spread 9.10

1 a the diastema is a gap which allows freshly cropped food to be separated from the cud

b rumen and reticulum

Exam questions

1 Type of tooth	A Carnassial	B Molar	C Incisor
Named mammal	Cat / dog / any named carnivore	Sheep / cow / any named herbivore	Human / any named omnivore
tooth function	Cracking bones / cutting flesh	Grinding vegetation	Biting

2

Name of enzyme	Site of production	Products of reaction
Lactase	Wall of intestine	Glucose + galactose
Sucrase	Wall of intestine	Glucose + fructose
Amylase	Pancreas	Maltose

3 a Two of salivary glands; pancreas; small intestine

b Salivary glands: unconditioned reflex; food stimulates taste buds / tongue receptors, sensory neurones carry impulses to brain, motor neurones carry impulses from brain to salivary glands, salivary glands produce saliva containing amylase; conditioned reflex may also produce saliva.Pancreas: stomach contents arriving in small intestine stimulates release of CCK / CCK-PZ; this stimulates synthesis of amylase; nervous reflexes stimulate secretion of pancreatic juice containing amylase. Small intestine: amylase not released; site of action is on membrane of epithelium of villi

c Cell surface membrane of epithelial cells of microvilli on villi of mucosa of duodenum

4 a Lacteal absorbs fatty acids and glycerol /

recombined lipid / chylomicrons; lacteal delivers lipids to larger lymph vessels, which then carry lipids to vein / blood system; capillaries absorb glucose / monosaccharides; amino acids; dipeptides; by diffusion / active transport; capillaries deliver absorbed products to hepatic portal vein

b Folds of wall of small intestine, villi, microvilli / brush border increases surface area; for secretion of enzymes; and absorption of digested products; glands in Crypts of Lieberkuhn produce new cells to replace those worn off the villi tips; secrete intestinal juice; lysozyme; goblet cells secrete mucus to lubricate food, helping passage through gut; epithelium is single cell layer, aids diffusion; microvilli have enzymes / named enzymes in cell surface membrane / contact digestion; 'tight junctions' ensure digested food passes through epithelial cells; dense capillary network within villi increases surface area for absorption; good blood flow maintains concentration gradient; smooth muscle in villi allow movement, bringing villi into close contact with digested food; longitudinal and circular muscles / antagonistic muscles in gut wall bring about peristalsis / movement of food along gut / mixing of food

c Pancreatic amylase hydrolyses amylose to maltose; maltase hydrolyses maltose to glucose; most chyme / material from stomach that enters the small intestine contains more amylose than maltose; glucose absorbed by epithelial cells

5 a A: Villi / villus / mucosa; B: submucosa / connective tissue; C: muscle

b Peristalsis; movement / mixing of gut contents / churning food / digested food / to bring villi into close contact with contents

---

Chapter 10 Nervous and hormonal coordination

Quick Check questions

Spread 10.1

1 nerve fibre carrying nerve impulses away from the cell body

2 a nerve impulses **b** hormones

Spread 10.2

1 antidiuretic hormone is made in the hypothalamus (and released from the posterior lobe of the pituitary)

2 a prolactin; **b** thyroxine; **c** androgens (e.g. testosterone)

Spread 10.3

1 a two from oestrogen, progesterone, and testosterone.

b two from adrenaline, glucagon, parathyroid hormone, and insulin.

c cAMP acts as an intracellular chemical messenger activating specific enzyme-catalysed reactions after it has been produced in response to a hormone (first messenger) that attaches onto receptor sites on the cell surface membrane.

Spread 10.4

1 The resting potential is determined by an unequal distribution of charged ions inside and outside a neurone, making the inside negative relative to the outside

2 a resting phase **i** potassium gates closed **ii** sodium gates closed;

b depolarisation phase **i** potassium gates closed **ii** sodium gates open quickly;

c repolarisation phase **i** potassium gates open slowly **ii** sodium gates closed slowly

d undershoot phase **i** potassium gates closed slowly **ii** sodium gates closed.

Spread 10.5

1 a increases **b** remains the same

2 Schwann cell

Spread 10.6

1 a mitochondria generate ATP required fro the synthesis of neurotransmitters **b** calcium **c** acetylcholine

Spread 10.7

1 a brain and spinal cord; **b** the somatic nervous system includes sensory neurones that sense external stimuli and motor neurones which supply skeletal muscle. It is part of the peripheral nervous system (all nervous tissue outside the brain and spinal cord) which also includes the autonomic nervous system.

2 sympathetic nervous system

Spread 10.8

1 a primary receptor consists of a single neurone which is sensitive to a particular stimulus and transmits nerve impulses to another neurone or effector, whereas a secondary receptor consists of a modified epithelial cell sensitive to a particular stimulus and a neurone.

2 interoceptors

3 the amplitude of a generator potential is proportional to the stimulus intensity whereas the amplitude of an action potential is always the same (i.e. it obeys the all or none law).

Spread 10.9

1 a vestibular apparatus consisting of semicircular canals, the utricle and saccule; **b** hammer (malleus); **c** eustachian tube

Spread 10.10

1 a transmits vibrations from stapes to the cochlea

b pitch refers to frequency of sound waves whereas intensity refers to their amplitude;

c detects changes of movement in any direction.

Spread 10.11

1 a the cornea bends (refracts) light as it enters the eye so that the light is focused on the retina **b** lens thickens (becomes more rounded); **c** iris

Spread 10.12

1 cones

2 bleaching occurs: retinal and opsin break apart, altering the permeability of the cell surface membrane of a rod cell; this may lead to an action potential if the light stimulus is high enough.

3 blue, green, and red.

4 Binocular vision is vision with two eyes whereas stereoscopic vision involves using the ability to combine the images formed in the two eyes to produce a single three-dimensional image.

Spread 10.13

1 a protection (cerebrospinal fluid filling spaces

between meninges, cushions the CNS against mechanical disturbance).

b the medulla oblongata contains nerve centres which control vital physiological functions such as breathing, heart rate, and blood pressure.

c cerebrum

Spread 10.14

1 neurosecretions
2 thyrotrophin releasing factor (TRF)
3 metabolic rate and respiration of glucose and fats.

Exam questions

1 a Activates enzyme systems within the cytoplasm in response to adrenaline being detected by cell membrane receptor
b i X: glycogen / other correct example
ii Y: glucose / other correct example
c i Only target cells for adrenaline will have specific protein receptors that recognise / fit adrenaline
ii At each stage in the process / from adrenaline to receptor to Enzyme 1 to cyclic AMP to Substance Y amplification occurs; only a few molecules of Enzyme 1 / adenyl cyclase are needed to activate many molecules of Substance Y / protein kinase / other named example; cascade effect

2 a Motor neurone
b X: axon; Y: dendrite
c Action potential set up / change in potential across axon membrane from negative inside value / –70 mV to positive value / +40 mV / opening of sodium gates / channels in axon membrane / depolarisation / wave of depolarisation along surface of nerve cell
d i Continuous muscle contraction / paralysis
ii Acetylcholine accumulates, continues to depolarise postsynaptic membrane / produce action potential in sarcolemma of muscle cell leading to muscle contraction

3 a Sodium channels / gates in cell surface membrane open, in response to a stimulus which causes a slight depolarisation; sodium ions rapidly enter cytoplasm by diffusion; increased number of positive ions inside membrane; negative resting potential cancelled out; further depolarisation makes potential difference across membrane positive (with respect to outside)
b Potential difference rapidly returns to a negative value
c Myelinated neurone has breaks called nodes of Ranvier; local circuits set up only at these nodes so action potential 'jumps' from node to node / saltatory conduction; faster than the series of smaller local currents in a non-myelinated axon.

4 a A: myelin sheath; B: axon
b Acetylcholine / noradrenaline
c Mitochondrion; ATP produced in aerobic respiration within mitochondria required e.g. for synthesis of neurotransmitter
d Hydrolysis / inactivation of neurotransmitter; to prevent continuous 'firing' of impulses in the postsynaptic nerve / muscle cell

5 a i Calcium; presynaptic; acetylcholine; diffusion; receptor protein; sodium
ii Hydrolysed by enzyme; to choline and ethanoic acid; which are inactive as transmitters
b Adrenergic

6 a i Arrow drawn vertically upwards
ii Absorbs light not detected by receptors; prevents reflection / scattering of light and hence less precise detection by receptors; contains capillary network, supplies receptors with oxygen and glucose.
b i Many rod cells share / synapse with a single relay neurone / synaptic convergence; gives increased sensitivity to low levels of light.
ii Individual cone cells have their own relay neurone; giving greater visual acuity

7

FUNCTION	REGION OF BRAIN
Osmoregulation	Hypothalamus
Control of posture	Cerebellum
Modification of heart rate	Medulla

8 a A: cerebral hemisphere / cerebrum; B: medulla; C: cerebellum
b Regulation of: heart rate; blood pressure; ventilation rate; swallowing; salivation; vomiting; coughing; sneezing

Chapter 11 Locomotion and behaviour

Quick Check questions

Spread 11.1

1 hydroskeleton
2 three of: prevention of water loss; protects soft tissues against mechanical damage; provides mechanical support; acts as an attachment point for muscles and a system of levers for locomotion.
3 compared with a small animal, a dog has low surface area-to-mass ratio; exoskeletons are surface features, therefore the dog would require a very thick exoskeleton to support its relatively high body mass.

Spread 11.2

1 maintain body shape; support soft tissue; act as a system of levers for locomotion; produce blood cells; store minerals
2 cartilage is more flexible and less dense than bone
3 blood vessels, nerves, and lymph vessels
4 33 (7 cervical, 12 thoracic, 5 lumbar, 5 sacral, 4 coccygeal)

Spread 11.3

1 cardiac muscle, smooth muscle, skeletal muscle
2 reduces friction between articulating bones
3 a biceps brachi, **b** triceps

Spread 11.4

1 region between two Z lines of a muscle fibre, acting as the contractile unit of a muscle
2 motor units performing delicate movements are likely to contain slow-twitch fibres whereas motor units performing powerful movements are likely to contain fast-twitch fibres.
3 a more fast-twitch fibres; **b** more slow-twitch fibres.

Spread 11.5

1 a calcium enables myosin heads to bind onto actin;
b tropomyosin blocks the myosin head binding sites;
c acetylcholine acts as a neurotransmitter, depolarising the postsynaptic membrane;
d ATP provides the energy for muscle action
2 temporal summation is the adding together of muscle

tensions produced in a single motor unit at different times, whereas muscle fibre recruitment results from the addition of muscle tensions produced in different motor units at the same time

Spread 11.6

1 gravity and friction
2 myotomes
3 air pressure is greater on the lower surface of the wing, generating lift.

Spread 11.7

1 In walking, one foot is in contact with the ground at all times
2 a cheetah maximises stride length by flexing its back and running on its toes
3 to maintain body posture; to keep muscle primed for action

Spread 11.8

1 receptor, sensory neurone, interneurone, motor neurone, effector (association neurone is absent in monosynaptic reflex arcs).
2 in a reflex action, a particular stimulus always evokes the same response whereas in a fixed action pattern the response to a particular stimulus can be modified by experience or the precise conditions in which the stimulus is presented.
3 positive phototaxis

Spread 11.9

1 Innate behaviour is genetically determined and not modified by experience whereas learning occurs when an animal's behaviour is consistently modified as a result of experience.
2 fatigue or sensory adaptation and generalisation
3 imprinting influences social behaviour, enabling a gosling to follow its mother in order to gain food, protection, and shelter, and, when it matures, to mate and interact positively with other geese.

Spread 11.10

1 in classical conditioning an animal learns to associate a particular stimulus with a reinforcer (unconditioned stimulus) provided by a person, whereas in operant conditioning an animal learns on its own to associate a particular behavioural act with a reinforcer.
2 latent learning helps a rat to become familiar with the area in which it lives

Spread 11.11

1 imitation
2 a dog may solve a detour problem by trial and error whereas a chimpanzee may solve it by reasoning.
3 by rehearsal; by adopting a positive emotional state; by linking new information to information already in the long-term memory

Spread 11.12

1 members of an insect society interact strongly and influence the behaviour of each other whereas members of an aggregation only live in the same place and interact only slightly if at all
2 a substance (a scent) secreted by one organism that stimulates a physiological or behavioural response in another individual of the same species.
3 a distance less than 70 m (no directional information)
b distance (more than 70 m) given by duration of dance and direction given by the angle of the straight run

Spread 11.13

1 head-up display given by gravid female stickleback
2 territories are usually established and maintained by ritualised acts of aggression or signals
3 alpha (or top-ranking) animal

Spread 11.14

1 a zeitgeber is an environmental factor such as light which acts as a cue synchronising a behavioural rhythm to an external cycle (e.g. the daily cycle of dark and light).
2 suprachiasmatic nuclei in the hypothalamus.

Spread 11.15

1 Migrations are regular and in definite directions whereas dispersion takes place in various directions
2 Seasonal migrations ensure maximum food availability at all times
3 Cuckoos do not see their natural parents and migrate at different times to them

Exam questions

1 a i I band
ii I band and H zone
b i Provides energy to break cross-bridges / rotate myosin 'heads' / drive ratchet mechanism; provides energy to pump calcium ions back into sarcoplasmic reticulum; formation of phosphocreatine
ii Calcium ions bind to troponin causing it, and the tropomyosin to which it is attached, to move away from the binding site on actin so myosin 'heads' can attach / switch actin into the 'on' position so that actin and myosin can bind / activate ATPase; release acetylcholine / neurotransmitter from presynaptic membrane so neurotransmitter diffuses across to muscle cell postsynaptic membrane

2 a i 0.02 seconds
ii Impulse caused by stimulus has to reach motor end plate / wave of depolarisation; neurotransmitter has to be released and diffuse across synaptic gap; sarcolemma has to be depolarised and action potential set up in muscle fibre; calcium ions have to be released and bind to troponin causing tropomyosin to twist out of binding site on actin molecule; myosin 'heads' have to attach to exposed binding sites and then change position so actin filaments slide and muscle contracts
b Rapid movement of sodium ions from outside nerve cell surface membrane to inside depolarises axon / generates action potential / makes the potential difference across the membrane positive / changes potential difference from –70 mV to +40 mV; example of positive feedback / initial entry of sodium ions further increases permeability of membrane to sodium ions; decrease in permeability to sodium ions at peak of action potential (linked to increased permeability to potassium ions which move out), starts process of repolarisation / return to original resting potential; action potentials propagated by effect of sodium ions entering axon, creating area of positive charge and a flow of current is set up in a local circuit
c Synapse acts as a 'one-way valve' as neurotransmitter only found at presynaptic end of nerve cell, therefore impulse can only be transmitted from this end to another nerve or muscle cell; receptors for

neurotransmitter only on postsynaptic membrane

3 a Arrival of impulse depolarises pre-synaptic membrane; calcium channels open, resulting in influx of calcium ions; this causes vesicles containing acetylcholine / neurotransmitter to fuse with pre-synaptic membrane, releasing their contents into the synaptic gap; neurotransmitter diffuses across gap; attaches to specific receptor protein on post-synaptic membrane / sarcolemma; change in shape of receptor site opens sodium ion channels in post-synaptic membrane; influx of sodium ions causes depolarisation / end-plate potential of the membrane, and if sufficient, an action potential

b i H zone only contains myosin filaments which are thicker than actin, so appears darker than regions of just actin; dark band contains overlapping actin and myosin filaments, so appears darkest of all; light bands contain only actin filaments, thinner than myosin, so appears lightest of all bands.

ii Light band shortens; as more actin filaments overlap with myosin filaments; H zone shortens as more actin slides past myosin; Z lines, to which actin filaments attached, become closer together / sarcomere shortens; dark band remains the same.

4 a The more times that rats navigated the maze, the faster they got at finding their way out

b Operant conditioning / instrumental conditioning / trial and error learning / associative learning; trial motor activities gave responses which were reinforced by a reward, association of the outcome of a response in terms of reward increased future responses; associative learning efficiency was increased by repetition

c After 3 trials, the greater the amount of food given, the faster the maze was navigated / after only 2 trials, group given least food was slower than groups given more food / after 20 trials, group given 50 pellets was 6 times faster than group given 1 pellet / eq.; larger reward increased / reinforced rats' association of reward with movement through maze / provided increased motivation

5 a A territory is an area held and defended by an organism or a group of organisms of the same species against other organisms; territories are defended from intruders by aggression and/or signals which are often ritualised, especially against members of the same species; ensures that individuals have larger area to find food, when food is scarce, during autumn / winter / when being in pairs is no advantage; in spring / summer, ensures breeding pairs and offspring have sufficient territory for finding food; for nesting / shelter; optimum utilisation of habitat / population spaced out in relation to food supply; reduces risk of infection; heightened defence at breeding time ensures sufficient food for nestlings; ensures female is not approached by other males; gives animals freedom from interference during pair formation; level of aggression shown increases nearer the centre of the territory to protect nest; weaker members of a species, unable to hold / defend a territory may fail to mate / have to live in non-optimum habitat; territorial behaviour ensures only 'fittest' hold a territory, breed, and pass on their genes to the next generation; may act to regulate population size

b Actual fighting would be detrimental to the species; to the individual; song and display are effective as intruder and defender both recognise the ritualised behaviour

c Only specific stimuli act as releasers for aggressive behaviour e.g. red breast of adult male robin; adult males pose more of a threat than juveniles

6 a A specialised form of learning that takes place during a brief, genetically determined period, usually shortly after birth / hatching; mainly in birds and mammals; a particular stimulus becomes permanently associated with a particular response; usually establishes bond between parent and offspring; behaviour is irreversible

b Imprinting effect greatest in ducklings 13–16 hours after hatching; after 16 hours from hatching, imprinting effect decreases sharply / from 1 to 12 hours after hatching imprinting effect gradually increases; ducklings are particularly receptive to stimuli that cause imprinting for only a short period / approximately 3 hours

c They will imprint on first, larger, moving object they see

7 a i Innate behaviour can be described as instinctive, and implies that the animal is genetically programmed with responses to particular stimuli; triggered automatically; in young chaffinches song length and notes used are examples of fixed action patterns / FAPs, as they have not been learned; phrasing is not innate as it must need to be learned; innate behaviour is behaviour that has been modified over many generations / by natural selection; culminating in complex song pattern of chaffinches

ii Imprinting is a specialised form of learning that takes place during a brief, genetically determined critical period in birds (and mammals); chaffinches learn phrasing of song between hatching and leaving the nest; a fairly short period when they must be particularly receptive to this information; a permanent response to a particular stimulus is established; as shown by chaffinches isolated after leaving nest, still being able to develop normal song pattern when exposed to it when eleven months old; imprinting often establishes bond between parent and offspring;

b Allows interaction with others of same species; important in territorial defence; intraspecific competition; intraspecific communication e.g. warning calls; courtship; survival value / being able to breed successfully and pass on genes to next generation

8 a Animal behaviour refers to the observable responses of an animal to the environment around it; responses usually involve movements of all or part of the body, but may include other changes; such as colour changes; 'instinctive' behaviour implies that the animal has inherited responses to particular stimuli; triggered automatically; fixed action patterns / FAPs; behaviour that has been modified over many generations / by natural selection; culminating in complex behaviour patterns; essential in animals with short life span with no time for 'trial-and-error' learning or solitary animals with little opportunity for learning from others of their species; includes taxes / kineses; reflexes / their instincts / social behaviour etc.; 'learned' behaviour involves memory; is an adaptive change in individual behaviour as result of previous experience / environment; includes habituation; classical conditioning e.g. Pavlov's dogs; operant conditioning / trial-and-error learning; exploratory learning; insight learning; imprinting; AOVP

b Knee-jerk reflex; when stretch receptors detect stretch in tendon, impulses are sent via a predetermined pathway along sensory neurone, to spinal cord, to motor neurone, to muscle which contracts; reflex actions are innate / they are unlearned responses determined by inherited nervous pathways; food-begging response of young birds is example of stereotyped behaviour; involves fixed action patterns / FAPs; colour displayed by open mouth of nestling / tapping on red spot on parent's beak are specific stimuli called sign stimuli; parent bird responds by feeding chick; in many insects, sign stimuli activate genetically preprogrammed nervous pathways which make muscles perform appropriate movements; without involving brain; kinesis is a random movement in which rate of movement is related to intensity of stimulus; e.g. woodlice move faster in dry environments than in damp ones; result is that woodlice stay longer in favourable environment; taxis occurs in response to direction of stimulus; woodlice show negative phototaxis; move away from light; result is that they again are more likely to find favourable environment; AOVP

c Advantages: all animals of the same species will have the same set of behaviour patterns e.g. great crested grebes (water birds) have complex set of courtship rituals which both sexes share / complicated courtship dance helps bond pairs; fixed action patterns can be adaptive and can increase the survival chances of organism or offspring; limiting the number of neurones required for a particular response may allow organism to respond appropriately to a wider range of situations, especially in invertebrates with a less developed nervous system; preprogrammed behaviour such as migration in birds e.g. swallows allows complex feeding and breeding patterns that increase survival of the species; natural selection will have caused much instinctive behaviour to be of survival value e.g. parental care; alarm signals; territorial behaviour etc.; AOVP

Disadvantages: once fixed action patterns are initiated, they are usually completed in exactly the same way; e.g. ground-nesting birds retrieve eggs with particular sequence of movements; if egg moves off to one side during sequence, the movements continue regardless; adaptation is very restricted; sign stimuli / releasers produce instinctive behaviour e.g. releaser for robins in breeding season is red colour of another adult male's breast; but robin will attack red items that pose no threat and ignore other birds, without red colouring, that may pose a threat; AOVP

Chapter 12 Animal reproduction

Quick Check questions

Spread 12.1

1 fission, budding, fragmentation

2 advantages of asexual reproduction include: no mate need be found, saving time and energy; large numbers of offspring can be produced quickly; offspring are identical to parents, an advantage in stable environments.

Spread 12.2

1 secretion of a milky white fluid that neutralises acids

2 follicles

3 seminiferous tubules, epididymis, vas deferens, urethra, vagina, uterus, oviduct

4 cilia produce a current which sweeps the oocyte along

Spread 12.3

1 seminiferous tubules

2 primordial germ cells and oogonia

3 Sertoli cells

4 one

Spread 12.4

1 in the oestrus cycle, oestrus (the period of heightened sexual activity coinciding with ovulation) is the most prominent event whereas in the menstrual cycle, menstruation (the discharge of blood and uterine lining) is the most prominent event.

2 **a** first discharge of blood; **b** ovulation

3 progesterone and oestrogen

Spread 12.5

1 mammals are predominantly terrestrial and internal fertilisation ensures sperm do not become dehydrated by coming into contact with air

2 sperm need to be activated by capacitation before fertilisation can occur

3 **a** to enable a sperm to penetrate an oocyte;
b to form a fertilisation membrane, preventing any more sperm from entering the ovum.

Spread 12.6

1 the morning-after pill

2 the condom

Spread 12.7

1 by trophoblast cells digesting their way into the uterine lining

2 fetus

3 ectoderm

Spread 12.8

1 amnion, yolk sac, chorion, allantois

2 chorion

3 to act as an exchange surface, allowing useful substances (e.g. nutrients, certain antibodies, and oxygen) to pass into the fetal circulation and wastes (e.g. carbon dioxide and urea) to pass out; to separate maternal blood from fetal blood and to act as a partial barrier to the passage of some harmful substances; to

act as temporary endocrine gland;

Spread 12.9

1 three from human chorionic gonadotrophic hormone, progesterone, oestrogen, relaxin

2 a relaxes the connective tissue in the bones of the pelvic girdle

b oxytocin stimulates uterine contractions

3 release of amniotic fluid from the womb during the delivery stage

Spread 12.10

1 the effects of prolactin are inhibited by progesterone and oestrogen

2 oxytocin

Spread 12.11

1 growth is a change in dry mass whereas development refers to the progressive changes that take place in an animal from conception to adulthood.

2 pluripotent

3 continuous, allometric, limited

Exam questions

1 a i Type A cells are somatic / diploid / body cells so 100% of cells will have a Y chromosome

ii Type C cells are gametes / sperm / haploid so 50% of cells will have a Y chromosome

b Repeated divisions of germinal epithelial cells / spermatogonia / primary spermatocyte ensure large numbers of cells that can undergo meiotic division in order to produce the vast numbers of sperm required

c Type C cells formed by meiosis in which independent assortment of chromosomes; and crossing over / formation of chiasmata in first division of meiosis will introduce variation

2 a i 38

ii 19

b Cells with more cytoplasm will have more mitochondria; larger food store; cells with less cytoplasm form polar bodies

c Crossing over / formation of chiasmata in first division of meiosis cause rearrangement of some alleles on each chromosome / so that adjacent alleles on a parental chromosome need not be inherited together by the offspring; independent assortment of chromosomes at metaphase I: because orientation of bivalents at the equator of the spindle is random this increases potential variation in gametes

3 a Differences:

Male gametes	**Female gametes**
Smaller, about 2.5 μm diameter / negligible food store	Larger, about 140 μm diameter / has food store
Cell clearly differentiated into head, middle, and tail	More uniform shape, mainly spherical
Motile	Non-motile
Only produced at puberty	Present, though undeveloped, at birth
No polar bodies	Polar bodies
Produced and released in vast numbers	Fewer produced, and many fewer released

AOVP

Similarities: both male and female gametes are: produced by meiosis; in gonads; from germinal epithelium; in three main stages, multiplication, growth, and maturation; final form is as haploid cells; single cells / single nucleus; able to fuse with another gamete

b Finger-like projections / villi of allantochorion (fetal tissue part of placenta) grow into the endometrium (maternal uterine tissue); this vastly increases surface area over which exchange can occur; blood vessels of embryo lie close to maternal blood vessels; endometrium breaks down in region of villi so placental villi are bathed in maternal blood, making exchange more efficient, but still keeping fetal and maternal blood separate

4 a i FSH is released just before menstruation; FSH stimulates growth of follicles; maturing follicles secrete oestrogens, slow increase of this inhibits further production of FSH; oestrogens also stimulate release of LH; LH stimulates growth of follicles; and production of secondary oocyte; also promotes ovulation with FSH in response to rapid increase in oestrogen levels / ovulatory surge; LH promotes development of corpus luteum; stimulates production of progesterone

ii Oestrogens secreted by maturing follicle; stimulate growth and repair of uterine wall; growth of milk-producing tissue in mammary glands; low levels inhibit further production of FSH, so limiting number of follicles maturing at a time; stimulate anterior pituitary gland to secrete LH; increase in oestrogen levels in blood stimulates hypothalamus to secrete GnRH / gonadotrophin-releasing hormone; leading to increase in FSH / LH, bringing about ovulation; corpus luteum secretes small amounts of oestrogen / larger amounts of progesterone which stimulate breast development; prepares uterine wall for possible pregnancy
Progesterone at high blood levels inhibits hypothalamus and pituitary; rapid decrease in levels, due to corpus luteum reducing in size if oocyte not fertilised; triggers menstruation / loss of uterine lining

b Rats: chance encounters will produce offspring / increase population / population can fluctuate in relation to variable food supply
Deer: breeding occurs when most food available for young

5 a i October to April

ii Lambs will be born when grass is growing / food is plentiful; when temperature is milder; higher milk production when ewe grazes on young grass

b Length of daylight hours is an important factor in controlling the timing of breeding season / oestrus; increase in length of daylight hours triggers oestrus; photoperiodic response

6

Hormone	**Secreted by ovaries**	**Reaches highest level in blood before ovulation**
Follicle stimulating hormone (FSH)	✗	✓
Luteinising hormone (LH)	✗	✓
Oestrogen	✓	✓
Progesterone	✓	✗

7 **a i** Labelled diagram of placenta (see Chapter 12); double layer of spongy vascular tissue; chorionic villi of embryo intrude into endometrium of mother; villi contain capillaries of allantois; placental villi bathed in maternal blood / in lacunae but blood of embryo and mother kept separate

ii Oxygen and nutrients / etc. diffuse from maternal to fetal blood; transported in umbilical vein; waste fetal products diffuse from fetal to maternal blood; from umbilical artery; capillaries of allantois in villi form huge surface area for exchange / increased diffusion rate; microvilli of villi increase surface area further; some exchange by active transport; e.g. ions / named ion / amino acids; facilitated diffusion; e.g. glucose; pinocytosis; numerous mitochondria in cells in walls of chorionic villi provide energy for active transport and pinocytosis; cell surface membranes contain protein carrier molecules used in uptake of materials into villi by active transport

b Carbon monoxide; and nicotine; enter mother's blood, cross placenta into baby's circulation; increased likelihood of intrauterine growth retardation; possibly due to reduction of blood flow through placenta / vasoconstrictive effects of nicotine; increased risk of heart abnormalities; carbon monoxide may reduce oxygen-carrying capacity of haemoglobin

---

Chapter 13 Structure and transport in plants

Quick Check questions

Spread 13.1

1 a cuticle **b** the upper surface is more exposed to sunlight, making it hotter than the lower surface.
2 photosynthesis
3 collenchyma consists of living cells with the corners of each cell reinforced by extra cellulose whereas mature sclerenchyma consists of dead cells impregnated with a thick layer of lignin.

Spread 13.2

1 all stems have nodes and internodes
2 support, transport, tissue production, storage of food and water
3 parenchyma cells help support stems by becoming fully turgid and pressing against other cells

Spread 13.3

1 thin-walled, no waxy cuticle, and torpedo shaped
2 a absorption of water and mineral salts;
b mechanical protection of the growth region;
c transport of water

Spread 13.4

1 diffusion
2 mature tracheids have tapered sloping oblique end walls that contain cellulose-lined pits whereas the end walls of mature xylem vessels break down
3 sieve tubes and companion cells

Spread 13.5

1 light intensity
2 as an unavoidable consequence of gaseous exchange
3 tensile strength of water column, adhesion betwen water molecules and the walls of xylem vessels, and cohesion between one water molecule and another.

Spread 13.6

1 higher
2 root pressure
3 the Casparian strip prevents water following the apoplast route

Spread 13.7

1 diffusion
2 by being deciduous (shedding leaves in autumn)
3 a metabolic poison would prevent the production of ATP required for the active transport of potassium, therefore potassium would diffuse out of guard cells, water would leave by osmosis, guard cells would become flaccid and the stoma would close.

Spread 13.8

1 active transport, diffusion, and mass flow
2 xylem and phloem
3 magnesium

Spread 13.9

1 phloem
2 sucrose
3 fluid flows out of the exposed end of a stylet that has penetrated phloem

Spread 13.10

1 loading and unloading of sieve tubes is thought to be an active process carried out by living companion cells.
2 by using different sieve tubes with different sources and sinks

Exam questions

1 a Water potential of soil is higher than that of root cells; due to higher solute concentration in cells; water moves into cells by osmosis; down a water potential gradient; root hairs increase surface area of root / increase rate of water uptake; water passes through cytoplasm in symplastic route; through cell walls in apoplastic route; through vacuolar route; Casparian strip prevents apoplastic movement / forces symplastic route; endodermis secretes ions into xylem; lower water potential in xylem

b Root pressure; caused by secretion of ions from endodermis; adhesion; capillarity; cohesion; between 'sticky' water molecules; maintains an unbroken water column; transpiration; pulls up water

c Stomata able to close; when conditions likely to cause high water loss / when humidity low / temperature high / windy / night; stomata mainly on lower leaf surface; therefore out of direct sunlight; stomata in pits / surrounded by hairs; leaves may curl; saturated air kept near leaf reduces concentration gradient; large number of small pores more efficient at allowing gaseous exchange with less water loss; edge effects / diffusion shells; waxy cuticle on upper leaf surface; relatively impermeable to water; gas exchange surface of mesophyll cells internal, therefore evaporation rate reduced

2 a No photosynthesis when dark, stomata close; conserve water

b i Guard cells have thicker / less elastic wall on side nearest pore / thinner / more elastic wall away from pore; K^+ ions accumulate in guard cells during daylight; decreased water potential draws water into guard cells from subsidiary cells; as guard cells increase in turgidity, only thinner wall can stretch, causing cells to assume a semicircular shape, creating open pore between pairs of guard cells

ii Active transport / carriers / potassium ion pump; energy from ATP; alternative theory: hydrogen ions pumped out of cells, potassium moves in to balance the charges.

3 a i Transpiration rate rose from 0800 to 1400 hours / peaked at 1400 hours / at 51 gh^{-1}; fell from 1400 to 2200 hours; then remained steady / small fluctuations around 6 gh^{-1}

ii Increase in transpiration rate / loss of water vapour due to stomata opening / more stomata opening / opening wider; increase in light intensity; temperature; wind speed; decrease in humidity; converse for decrease in transpiration rate; limiting factors

b Rate of water absorption affected by transpiration rate; water moves from area of high water potential in soil to area of lower water potential in atmosphere; curves are similar shape; transpiration rate peak greater than peak for water absorption; water absorption curve follows / to right of transpiration curve

c Cut shoot from plant under water so no air bubbles; still under water, insert cut end of shoot into potometer, making a water-tight seal; allow to equilibrate; record rate of movement of water in capillary tube of potometer; control environmental conditions / named factor; calculate volume of water absorbed in set time; assume water absorbed is proportional to rate of transpiration; repeat several times to increase reliability; reset water column in capillary tube as necessary; alternative methods

4 a A: sieve tube element; B: companion cell

b i Transport / translocation of products of photosynthesis / sucrose / amino acids / eq.

ii Aphids allowed to feed by inserting stylets into phloem; aphids anaesthetised and bodies removed, leaving stylets behind; fluid collected from stylets; fluid analysed

5 a P: xylem; Q: spongy mesophyll

b i Potassium ion concentration falls in cells Y and Z / decreases by about 6 times in cell Z and by about half in cell Y; rises in cell X / approximately doubles in cell X

ii Active transport of potassium ions into guard cell from cell X; lowers water potential of guard cell; water enters by osmosis; guard cells increase in turgidity; uneven thickening of cell walls cause cells to assume semicircular shape forming open pore between pairs of guard cells

6 a Higher water potential in soil because fewer solutes; lower water potential in cell due to higher concentration of solutes; water enters cell by moving down water potential gradient / from high to low water potential; by osmosis; across cell wall into cytoplasm of root hair / epidermal cells; root hairs increase surface area for water uptake; water travels along cell walls / apoplastic route; through cytoplasm / symplastic route; via vacuolar route; apoplast pathway most important; suberin / Casparian strip in endodermal cells; prevents apoplastic movement / forces symplastic route; endodermis secretes ions into xylem; lower water potential in xylem due to lower / more negative solute potential; due to water moving up the xylem, setting up tension in xylem and thus lowering water potential of its sap; water moves into xylem down the water potential gradient

b Active transport of nitrogen as nitrates; requires ATP; passive uptake / diffusion; taken up in solution; ions move through apoplast only until endodermis; then cross cell surface membrane to enter cytoplasm of endodermal cells; membrane allows selection of which type of ions reach xylem

c Synthesis of amino acids / proteins; poor growth

7 Structure: clear, labelled diagram of stoma; stomata are pores surrounded by banana-shaped guard cells; guard cells contain chloroplasts; thicker / less elastic cell wall on side nearest pore / thinner / more elastic wall away from pore; cellulose microfibrils of cell wall run across width of cell, like hoops on a barrel, forcing cell to elongate rather than expand sideways when turgor increases
Function: gas exchange; for respiration / photosynthesis; transpiration via stomata aids water transport up xylem; evaporation of water helps cool plant; enable plant to regulate balance between the need to obtain CO_2 and to conserve water
Theories: potassium ions accumulate in guard cells in response to light; pumped in by active transport; light may activate ATPase; positive potassium ions balanced electrically by negative ions / malate from starch stored in guard cells; or by negative chlorine ions entering with potassium ions; or hydrogen ions may be pumped out and potassium ions may then enter to balance charge; or sugar accumulates as a photosynthetic product within guard cell (this theory now discounted); water moves in down a water potential gradient, via osmosis, from adjacent cells; as guard cells increase in turgidity, only thinner wall can stretch, causing cells to assume a semicircular shape, creating open pore between pairs of guard cells; theory for closing of stomata: abscissic acid / ABA concentration increases when water loss is high; triggers potassium pump which acts to pump these ions out of guard cells; turgor reduced / stomata close; in darkness, potassium ions leak out of guard cells; water potential increases, water moves out of guard cells, which become flaccid and stoma close

Chapter 14 Reproduction and coordination in flowering plants

Quick Check questions

Spread 14.1

1 a i calyx (sepals) and corolla (petals);
ii petals; **iii** stamens; **iv** carpels

b carpel is a single female sexual organ whereas a pistil is two or more carpels fused into a single structure

2 enclose and protect a flower bud

Spread 14.2

1 four pollen grains grouped together

2 ovule

Spread 14.3

1 increased variation of offspring

2 they both benefit, the bees from obtaining a food source and flowers by having a means of pollination

3 stamens of wind-pollinated plants have to be exposed to the air whereas those of insect-pollinated plants have to be enclosed so that insects have to brush past them

Spread 14.4

1 pollination is the transfer of pollen from one plant to another whereas fertilisation is the fusion of a male gamete with a female gamete

2 angiosperms are the only organisms which have double fertilisation

3 **a** radicle; **b** plumule; **c** cotyledons

Spread 14.5

1 **a** Mediterranean squirting cucumber; **b** sycamore or European maple; **c** coconut; **d** mistletoe.

2 favourable conditions may be short-lived (e.g. in autumn) and dormancy may increase the chances of germination occurring when there is a prolonged period of favourable conditons (e.g. in spring); dormancy increases the time during which seeds may be moved away from parents.

3 increases the chance of seeds germinating only after being buried in soil for a period of time

Spread 14.6

1 water is required for hydrolytic actions to catalyse the breakdown of food stores; water absorption enables seeds to expand quickly, bursting through the seed coat; water causes the embryo to release hormones which stimulate mobilisation of food stores

2 favourable temperatures are required for enzymes essential for germination to work efficiently

3 in hypogeal germination cotyledons remain underground whereas in epigeal germination cotyledons are pushed out of the soil above the surface of the ground

4 coleoptiles protect the plumule during its emergence out of the soil

Spread 14.7

1 an advantage of vegetative propagation is that it can produce large numbers of offspring quickly; one disadvantage of vegetative propagation in habitats that change is that (barring mutations) all offspring are identical to parents

2 **a** in bulbs, food is stored in fleshy leaves whereas in corms food is stored in the swollen rounded base of the stem

b stolons are long aerial stems that grow upward and then bend to the ground whereas runners usually grow along the ground (note that the two terms are sometimes used synonymously)

c a rhizome is a swollen stem which grows horizontally undergound whereas a stem tuber does not grow horizontally

3 perennating organs enable the plants such as bluebells to grow and flower before trees come into leaf

Spread 14.8

1 asexual propagation ensures that desirable qualities of a plant are maintained from one generation to the next; it allows new plants to be produced quickly in large numbers; it involves the minimum of expense and effort.

2 a small piece of a healthy plant which is removed and grown into a new plant

3 totipotency of a cell is its ability to form a whole new plant; it is the basis of micropropagation which uses a small piece of plant tissue to produce a whole new plant

Spread 14.9

1 primary growth involves apical meristems, secondary growth involves lateral meristems; primary growth is concerned with making a plant longer, secondary growth is concerned with making a plant thicker; primary growth occurs in all plants, secondary growth occurs only in some dicots such as trees and shrubs.

2 apical meristems just behind the root tip and shoot tip, and lateral meristems of vascular cambium and cork cambium

3 wood is secondary xylem and its major function is mechanical support

Spread 14.10

1 unlike animal hormones, plant growth substances are not manufactured in special organs and they do not always move away from from their site of production to their target cells

2 auxins promote primary growth and apical dominance; gibberellins stimulate growth of shoots and leaves; cytokinins stimulate cell division; abscisic acid is a growth inhibitor (promotes dormancy); ethene is involved in fruit ripening (other effects in text).

3 antagonistic effects result from two or more substances interacting to reduce each other's action whereas synergistic effects result from two or more substances interacting to give a greater effect than the sum of their individual actions

Spread 14.11

1 nastic responses are independent of the direction of the stimulus whereas a tropic response is related to the direction of the stimulus

2 **a** phytochrome **b** differential growth of cells **c** auxin, a plant growth substance

3 statoliths

Spread 14.12

1 relative length of day and night

2 detection of the photoperiod

3 P_{fr}

Exam questions

1 **a** **i** Embryo / Seed / mature plant

ii Pollen grain; mature embryo sac

iii Pollen grains (microspores); embryo sac; ovum

b Haploid gametophyte generation alternates with diploid sporophyte generation; haploid generation undergoes mitosis to form gametes; gametes fuse to form diploid zygote which grows into diploid sporophyte generation; which undergoes asexual reproduction / meiosis to produce haploid spores and the cycle is repeated

c Between pollen mother cells and pollen grains; between embryo sac mother cell and mature embryo sac

d One of the two male gamete nuclei fuses with the ovum to form a diploid zygote; the other fuses with the two polar / secondary nuclei to form the triploid endosperm

e Any two named bryophytes e.g. *Pellia* / liverwort;

e.g. *Funaria* / moss

2 a Flower has colour / scent / nectar to attract insects; insect lands on wing petals / depresses keel petal; stamens / anthers deposit pollen onto insect's abdomen; insect carries pollen to stigma / carpel of another flower of same species

b Zygote: D, endosperm: G

c Anthers and carpels ripen at different times / dichogamy; e.g. anthers shed pollen before carpels are ripe / protandry; protogyny / stigma withered / ripe before pollen shed; structure of flower may prevent pollen landing on stigma; heterostyly; anthers and carpels may be in different flowers; either on the same plant / monoecious; or different plants / dioecious; self-sterility

3 a Amylase activity on day 2 = 0.4 a.u.
Amylase activity on day 6 = 4.0 a.u.
Increase in activity over the 4 days = 4 – 0.4 a.u. = 3.6
Rate of increase = $\frac{3.6}{4}$ a.u. per day;
= 0.9 arbitrary units day $^{-1}$

b Digestion / hydrolysis of starch; to maltose; maltose is more soluble than starch so can be transported to growth areas / radicle / plumule / embryo; to use as a respiratory substrate

c Amylase activity low until day 2 because little water in barley grain / activity increases as water is absorbed / imbibition; gibberellin is synthesised; diffuses to aleurone layer; stimulates production of amylase / more amylase; results in production of more respiratory substrate

d Starch stores are used up; photosynthesis can now produce carbohydrates

4 a Auxin concentration increases in agar block in side of shoot when shoot is lit from side, compared to in dark; more auxin has moved to side of shoot away from light source; less auxin moves downwards in light; light influences movement / distribution of auxin

b Auxin moves laterally away from light in plant shoots, creating greater concentration of auxin on side away from light; cells stimulated to grow / elongate; resulting in bending towards the light / positive phototropism; more light absorbed increases the rate of photosynthesis

5 a Very low levels / concentrations less than 10^{-2} ppm of auxin increase root growth / low levels of auxin / below 10^{-3} ppm have no effect on shoot; higher levels of auxin / above 10^{-2} ppm decrease root growth / between 10^{-3} and 10 ppm for shoots has opposite effect / increases shoot growth; eq.

b Distilled water applied to oat seedlings

c Hydrogen ion secretion out of cytoplasm into cell walls is stimulated; pH outside cell decreases and promotes loosening of cell wall to allow cell to extend; by swelling due to osmosis / new cell wall material laid down; auxin may bind to receptors on cell surface membrane of epidermal cells which switch on genes to produce enzymes / proteins connected with growth

d Selective action / affect broad-leaved plants / dicotyledons more than cereals / monocotyledons / monocotyledons less sensitive to these substances; cause twisted growth / stop growth

6 a Treatment A; short day plants only flower after exposure to photoperiods of less than a set maximum value / critical day length

b i No flowering; red light is detected / absorbed by phytochrome / P_r / P_{660} converted to P_{fr} / P_{730}, day plants are actually 'long night' plants and require uninterrupted lengths of darkness for flowering,

ii Flowering; in darkness / far-red light absorbed by P_{fr} converted to P_r

Chapter 15 Infectious diseases

Quick Check questions

Spread 15.1

1 health includes mental, spiritual, and physical well-being as well as freedom from disease

2 infectious disease

3 morbidity refers to incidence of disease whereas mortality refers to deaths associated with the disease

Spread 15.2

1 vaccination of everyone within a certain area of the outbreak of a disease

2 HIV is a virus whereas AIDS is the syndrome associated with the virus

3 HIV is transmitted sexually, via infected blood or blood products, via contaminated needles, from mother to fetus across the placenta, from mother to child in breast feeding

4 unlike the smallpox virus, HIV attacks helper T cells; HIV can remain dormant and hidden in the body; HIV is extraordinarily variable, changing the shape of its antigens, making it difficult for the immune system to identify it.

Spread 15.3

1 by contaminated food and water

2 a a poison which afects the intestinal lining

b *Vibrio cholerae*

3 the tuberculosis bacterium is easily transmitted in airborne currents produced by coughing, sneezing, and even speaking

Spread 15.4

1 a *Plasmodium* species **b** *Anopheles* mosquito

2 mosquito nets act as a barrier to mosquitoes, reducing the risk of being bitten by the vector

3 *Trypanosoma*

4 trypanosomes change the surface proteins (the antigens by which they can be identified) on their coat

Spread 15.5

1 cell-mediated response: foreign antigen attaches onto specific helper T cell receptor → helper T cell divides mitotically → resulting cells either remain as memory cells or activate other cells in the immune system
Humoral response: foreign antigen attaches onto specific B cell receptor → B cell divides mitotically → plasma cells formed which secrete one type of antibody which acts specifically against the foreign antigen

2 a body cell which has been antigenically altered may have non-self antigens rather than self antigens, which enables it to be attacked by killer T cells

Spread 15.6

1 a glycoprotein receptor on the surface of B cells; when released, they form specific antibodies

2 neither lymphocytes nor plasma cells are able to divide outside the body
3 humanised monoclonal antibodies are less likely to trigger an immune response

Spread 15.7

1 active immunity
2 antibodies are eventually broken down in the spleen and liver
3 an influenza virus can mutate, changing its surface antigens, so that it may be different to the virus a person was vaccinated against

Spread 15.8

1 **a** autografts and isografts **b** allografts
2 O
3 AB

Spread 15.9

1 a biocidal antibiotic kills microroganisms whereas a biostatic antibiotic inhits the growth and reproduction of microorganisms
2 antibiotics may: weaken bacterial cell walls; inhibit protein synthesis or promote synthesis of abnormal protein; disrupt synthesis of nucleic acids; disrupt cell membranes
3 by natural selection of resistant strains which have arisen by random gene mutations: after exposure to doses of antibiotics that do not kill all bacteria (this commonly occurs when patients do not complete a course of antibiotic treatment) sensitive bacteria die but resistant bacteria survive to pass on their resistance to the next generation

Spread 15.10

1 a proteinaceous infective particle
2 procedures which break down nucleic acids, such as exposure to UV light, do not prevent the development of TSEs

Exam questions

1 a

statement	malaria	cholera	AIDS	TB
causative agent is a bacterium	✗	✓	✗	✓
causative agent is blood-borne	✓	✗	✓	✗
controlled by pasteurisation of milk	✗	✗	✗	✓
sexually transmitted	✗	✗	✓	✗
transmitted by a vector	✓	✗ / ✓ flies can act as vector	✗	✗

b The human immunodeficiency virus (HIV) has a latency period / remains dormant for about eight years on average, no symptoms shown during this period
c HIV widespread in countries where TB is also a serious problem; impaired immune system leaves HIV patients more susceptible to TB; HIV patients do not respond well to some anti-tuberculosis drugs
d Mutant strains of *Mycobacterium* resistant to some antibiotics; number of resistant strains increasing; combination of antibiotics reduce chances of strain multiplying which is resistant to one of the antibiotics

2 a Skin: thick, continuous, keratinised layer reduces entry of bacteria into body; constant shedding of surface cells makes it difficult for bacteria to adhere; acidity of sweat / sebaceous secretions discourage bacterial growth
Tear fluid: contains the enzyme lysozyme, which breaks down bacterial cell walls
Gastric juice: acid / pepsin may digest bacteria
b i Weakened strain induces production of antibodies; by lymphocytes; lymphocyte multiplies rapidly / primary immune response; some cells from this multiplication form memory cells; these survive for many years; infection by TB bacterium will cause memory cells to divide immediately so that their numbers increase faster than numbers of pathogen / secondary immune response
ii Immune system impaired; T lymphocytes / CD4+ lymphocytes / T-helper cells infected by HIV; therefore B lymphocytes / phagocytes not activated
c i Red blood cells from blood of Group A will have A antigens present on cell surface membrane; individuals do not produce antibodies to antigens present on their own blood cells; red blood cells from Group O have neither A nor B antigen present on cell membrane; so Group O individuals produce antibodies a and b; if blood from Group A is transfused into Group O individual antigen A will come into contact with antibody a and cause an immune response resulting in red blood cells clumping together / agglutination; and breaking down / haemolysis
ii Failure will only occur if Group A donates to Group O, 25% risk of picking Group A donor randomly, 75% risk of picking Group O receiver, therefore risk of failure is 75% of 25%, or 19%

3 a A–B is 56 mm = 56 000 μm,

Actual diameter $= \frac{56\,000}{10\,000}$; $= 5.6\,\mu m$

b Lymphocytes have larger nucleus / nucleus with no lobes / agranular cytoplasm
c i Engulf bacteria by phagocytosis; some have antihistamine properties; others produce histamine / heparin
ii Bacterial toxins from ingested bacteria

4 a T cells responsible for cell-mediated response; recognise antigens / non-self cell; divide to form clone of cells; differentiate into different types e.g. T-helper cell / T-cytotoxic / killer cell / T-suppressor cells; correct reference to function of given T cell
b i Rapid rise in first six months; followed by rapid fall over next six months; continued slow fall to two years; rise for next seven years / to year nine; then little change / plateau reached
ii T4 curve starts higher; initial rate of increase slower for T4; slower rate of decline over years 1–2; after year 2 curves have opposite slopes / T4 curve falls as HIV rises; HIV has two peaks / T4 has one
iii HIV particles multiply rapidly; T4 cells multiply in response; kill most of virus particles; T4 cells infected by virus; latency period; more T4 cells killed as virus slowly increases; virus numbers reach a plateau; few T4 cells left

5 a Malaria caused by protoctistan / *Plasmodium*; female mosquito / *Anopheles* bites infected person, sucks up blood containing parasites; mature and fertilise in stomach of mosquito; parasites / zygotes bore into gut wall of mosquito and encyst; large numbers of sporozoites / parasites released into blood; migrate to salivary glands of mosquito; injected into blood, with saliva, when mosquito bites human; mosquito is vector; parasite enters human liver cells and multiplies inside; many parasites / merozoites released from liver; invades red blood cells / more liver cells; pattern repeated / re-invasion every 2–3 days; some develop into gametocytes / male and female stages; remain in blood until taken up by feeding mosquito

b Using insecticides to kill adult mosquitoes; mosquito nets to prevent biting; protective drugs / named drug; drainage of marshes where mosquitoes breed; insecticides / oil on water where mosquitoes breed, to destroy aquatic larvae; biological control, using carnivorous fish to eat larvae

c Insecticides: may be non-specific / affect other insects / beneficial insects; may have long-lasting effects / pass along food chains / affect food webs / accumulate in other species which are food for humans; drugs: may have undesirable side-effects; not all are 100% effective / some strains of *Plasmodium* have developed resistance to some drugs; drainage: has environmental consequences / loss of habitats etc.; oil on water: environmental consequences / affects other air-breathing larvae / oxygen content of water / feathers of visiting birds etc.; AOVP

6 Antigen; B lymphocyte / B / plasma; myeloma cells; clone; pregnancy / disease / named disease e.g. *Chlamydia*, a sexually transmitted disease

7 a Viral antigens activate B lymphocytes; which divide rapidly to produce plasma cells; plasma cells produce antibodies; B-memory cells also produced; virus also engulfed by macrophages / cell-mediated response; this activates T lymphocytes; T-helper / T-killer produced; T-memory cells produced; memory cells retain ability to recognise viral antigen; re-infection by same virus causes memory cells to divide rapidly / secondary immune response

b Antibodies diffuse across placenta from mother; and from breast milk / colostrum

8 Production: mostly produced by fungi; e.g. *Penicillium notatum* produces penicillin / other example; produced by fungus as secondary metabolites / when growth is slowing down; produced commercially by batch culture / fermentation; large-volume fermenter / nutrient medium; optimum temperature / around 24 °C; good oxygen supply; slightly alkaline pH / pH controlled; sterile tank / medium; stirring; monitoring so process stopped when maximum penicillin produced; penicillin typically starts to be produced after 30 hours / reaches maximum after 4 days / production ceases after 6 days; downstream processing / fungal mycelium filtered off; penicillin extracted from liquid filtrate / purified / modified; continuous need to find new antibiotics due to development of antibiotic resistance

Uses: antibiotics kill microorganisms / bactericidal; inhibit growth of microorganisms / bacteriostatic; host's immune system then kills the pathogens; use in farming to increase milk yields / eq.; can cause bacterial cell wall disruption / details; e.g. penicillin / most antibiotics on Gram-positive bacteria; can disrupt bacterial cell membranes by affecting permeability; e.g. tetracyclines / eq. disrupt protein synthesis; by preventing tRNA binding to ribosomes; some antibiotics / penicillin effective on only a few pathogens / narrow-spectrum antibiotics; other antibiotics / chloramphenicol inhibit growth of wide variety of pathogens / broad-spectrum antibiotics

---

Chapter 16 Non-infectious diseases

Quick Check questions

Spread 16.1

1 Huntington's disease
2 e.g. coronary heart disease
3 low oxygen partial pressures

Spread 16.2

1 mast cells
2 dilation of arterioles reduce arterial blood pressure
3 hay fever (allergic rhinitis)

Spread 16.3

1 senescence
2 number of nerve cells in the brain decline and the brain becomes smaller
3 beta amyloid

Spread 16.4

1 load-bearing exercises during adolescence stimulate calcium deposition and strengthens bones
2 HRT provides low doses of oestrogens which reduces demineralisation of bones
3 narrowing of joint space (the space between bones) and irregularities ocur at bone edges

Spread 16.5

1 cancer-causing agent
2 metastasis
3 screening by cervical smears can identify early stages of cancer which are treatable

Spread 16.6

1 carbon monoxide
2 cigarette smoking lowers oestrogen levels and reduces bone mineralisation
3 tars irritate and damage lung tissue both mechanically and chemically

Spread 16.7

1 atherosclerosis is a thickening of the inner layers of arterial walls whereas arteriosclerosis is a loss of elasticity of arterial walls
2 myocardial infarction ('heart attack')
3 age, sex, and family history

Spread 16.8

1 hypertension can develop without symptoms
2 saturated fats
3 e.g. jogging, cycling, swimming

Spread 16.9

1 less than 85% of ideal body mass
2 bulimics are often secretive, they often have a normal body mass, and symptoms may not appear until the order is well-developed
3 obesity occurs when the body mass index exceeds 30

Spread 16.10

1 RNI refers to the amount of nutrient that would satisfy 97.5% of a particular group in the population;

consequently, by definition, it will not be sufficient for 2.5% of the population group.

2 a vitamin A is essential for the formation of retinal, a visual pigment important for vision in low light intensities
  b rickets
  c high blood losses can lead to anaemia, a reduction in the haemoglobin content of blood

Spread 16.11

1 drugs interfere with the normal functioning of the body, whereas foods are required for the normal functioning of the body
2 alcohol can change the mental state and behaviour
3 liver cirrhosis results in loss of healthy liver tissue and can lead to jaundice and death

Exam questions

1 a Wear of cartilage; damage to bone
  b Bones more likely to break due to osteoporosis / brittle bone disease / calcium deficiency; commonest in post-menopausal age-groups; oestrogen levels low / parathyroid no longer inhibited

2 a Thickened / roughened / less smooth / less elastic / fatty deposits under endothelium / irregular raised patches
  b i Blood platelets stick to roughened surface / ruptured; clotting mechanism triggered / clotting factors / thromboxanes released; prothrombin converted to thrombin / fibrinogen converted to fibrin; prostaglandins which inhibit clotting are themselves inhibited
    ii Blood clot blocks flow of blood in artery / region of heart muscle, normally supplied by artery, deprived of adequate blood supply / adequate oxygen / ischaemia; leads to acute chest pains / angina; if condition not treated, affected heart muscle may die / heart attack / myocardial infarction; dead patches of heart muscle increase likelihood of irregular heart beat rhythm / ventricular fibrillation / heart ceases to be an effective pump; other organs, such as the brain, deprived of adequate oxygen / heart attack may be fatal
  c Decrease in saturated fat / cholesterol in diet decreases blood cholesterol concentration / less deposited in arterial walls; decreases number of low-density lipoproteins / LDLs in circulation so less cholesterol deposited; causes rise in ratio of high-density lipoproteins / HDLs to LDLs, so more cholesterol removed from arteries

3 a i Mean resting pulse rate = 70 beats per minute, mean pulse rate after exercise = 105 beats per minute, mean increase in pulse rate = 35 beats per minute;

$$\text{mean \% increase} = \frac{35 \times 100}{70} = 50\%$$

    ii Increased blood flow to muscle cells; bringing increased oxygen / glucose to match increase in rate of respiration / demand for energy release by aerobic respiration
  b i Increased capillary networks in muscles develop with regular exercise and increase oxygen supply; increased numbers of mitochondria in muscle cells; increased amounts of enzymes involved in respiratory pathways; lactic acid removed more quickly; increased amounts of phosphocreatine / muscle glycogen stores / myoglobin; increased cardiac output / increased stroke volume; increase in heart muscle / size of heart chambers
    ii Pulse rate increase after exercise is lower than others in group suggesting he is fitter / heart muscle is stronger / cardiac output is greater / cardiac muscles have better blood supply

4 a Liver cells replaced by fibrous tissue; additional collagen fibres disrupt arrangement of liver cells; liver appears nodular in appearance; texture of liver is harder; blood supply to liver cells is poorer; liver functions less efficient
  b It is measurable; it is not reversible; most heavy drinkers develop this disease so reliable indicator
  c i Alcohol can provide sufficient energy for daily requirements; but does not contain all the nutrients needed for a balanced diet / insufficient vitamins / minerals
    ii Deficiency in vitamin B_1 leads to long-term degenerative changes in brain; loss of memory / inability to learn / confusion / dementia; disturbance of speech / walking / confusion / coma; degeneration of peripheral nerves / myelin sheaths degenerate / loss of nerve function; causes loss of sensation in extremities / feet / hands / difficulty in movement / pain / cramps / numbness / tingling / weakness
  d Positive correlation between mean weekly consumption of alcohol per person per week and percentage of heavy drinkers; correlation coefficient required to calculate strength of relationship; scatter diagram shows that as mean weekly consumption per person per week decreases, so does the percentage of heavy drinkers; where weekly consumption is at low levels / below 25 g per person per week, there are very few heavy drinkers / less than 3% of population / equivalent data quoted; small decreases in mean weekly consumption do not always match claim, e.g. some data shows mean weekly consumption of 125 g per person linked to higher / 22% heavy drinkers compared with mean weekly consumption of 175 g per person linked to 20% heavy drinkers; AOVP
  e Less: crime / vandalism / assault; child neglect / abuse; absenteeism from work; family disputes / marital breakdown; road accidents; homelessness; psychiatric / physical illnesses and hence drain on the National Health Service resources / AOVP

5 a Cancer cells do not respond to signals from other cells / to growth factors in normal way; divide uncontrollably; fail to undergo normal cell death; may have genetic mutations in genes / oncogenes that control cell division
  b Tar in cigarette smoke; asbestos fibres; other named environmental / dietary carcinogens e.g. nitrosamines; viruses / named example e.g. human papilloma virus / linked to switching on oncogenes; X-rays / gamma rays / ionising radiation / ultraviolet light / radon gas; genetic predisposition / family history; age; gender
  c High rates of cell division increase chances of mistake being made during mitosis; producing

mutations in daughter cells

d i Numbers of both types of cells decrease rapidly in first 4 days after treatment / numbers of both cells more than halved; normal cells decrease slightly less than tumour cells / to 400 cells per unit volume compared to 300 for tumour cells; between 4 and 7 days after treatment decrease is halted / no increase / decrease in cell numbers; increase in cell numbers from day 7; more rapid increase for normal cells / reach higher plateau / 800 cells per unit volume compared to 500 for tumour cells; following second dose of drug, both types of cell decrease in number, but tumour cells start from lower number so decrease further to 140 cells per unit volume compared to 320 for normal cells at day 25; day 25 to 28 no increase / decrease in either type of cell; day 28 numbers start to rise again, but more rapidly for normal cells; length of time drug is effective at killing dividing cells is same for both types of cells; but effect is greater on tumour cells

ii Tumour cells have greater rate of division / reduced cell cycle time; drug only kills dividing cells, so greater number of tumour cells will be killed than normal cells / normal cells have longer cell cycle, so fewer cells affected by drug; fewer tumour cells will produce fewer daughter cells / repeated doses decrease overall numbers of parent cells

e i Effect on normal cells would be increased / more normal cells would be killed

ii Recovery of normal cells would be reduced

Chapter 17 Microorganisms and their applications

Quick Check questions

Spread 17.1

1 a core containing DNA or RNA and a protein coat called a capsid
2 reverse transcriptase causes the host cell to make viral DNA using the genetic code contained in viral RNA.
3 aphids feed on plant sap in the phloem; viruses stored in the aphid gut may be regurgitated when the aphid is feeding on another plant, infecting the plant
4 unlike typical living cells, viruses are not made of cells; they have only one type of nucleic acid; they are unable to reproduce.

Spread 17.2

1 archaebacteria tolerate extreme conditions of temperature and pH, similar to those that are thought to have existed when life first evolved; also their ultrastructure and metabolism suggest that they are ancestral to both eubacteria and eukaryotes
2 rod-shaped
3 photoautotrophic bacteria synthesise organic molecules from inorganic molecules using light energy whereas chemoautotrophic bacteria synthesise organic molecules using energy released from oxidation of inorganic chemicals

Spread 17.3

1 asexually by binary fission and sexually by conjugation
2 a bacteriophage gene
3 higher than 72 °C

Spread 17.4

1 spores can survive boiling at 100 °C
2 a single area on an agar plate originates from a single bacterium which has produced a clone of identical individuals.
3 pink staining indicates Gram-negative bacteria therefore streptomycin is more likely to be used

Spread 17.5

1 moulds have thread-like filaments (hyphae) that form a woolly mass (mycelium)
2 a *Mucor* hypha is coenocytic because it consists of a continuous mass of cytoplasm with many nuclei
3 digestive enzymes are secreted onto food in the external environment
4 sporangiospores are asexual whereas zygospores are sexual

Spread 17.6

1 *Penicillium* hyphae have cross-walls called septa whereas *Mucor* hyphae have no septa
2 yeasts are unicellular cells that reproduce by budding
3 budding is asexual reproduction by the formation of one or more outgrowths or buds

Spread 17.7

1 small air bubbles have a larger surface area-to-volume ratio than large air bubbles
2 batch cultures
3 the extraction and purification of a product at the end of a process

Spread 17.8

1 lactate produced during fermentation discourages the growth of microorganisms which would spoil food or cause disease.
2 chymosin coagulates milk as part of curdling
3 Two from *Penicillium roqueforti*, *Penicillium camemberti*, and *Propionibacterium*
4 *Lactobacillus bulgaricus* or *Streptococcus thermophilus*

Spread 17.9

1 foods grown from unicellular microorganisms
2 a methanol **b** carbon dioxide **c** glucose
3 a health food

Spread 17.10

1 carbon dioxide is produced, helping dough to rise and swell
2 malting allows partial germination so that starch can be converted to sugars, and proteins can be converted to amino acids.
3 red wines are produced by allowing must to remain in contact with grape skins for a long time

Spread 17.11

1 acetogenic reactions produce acetate, and hydrogen and carbon dioxide
2 waste disposal and fertiliser production
3 a fuel consisting of a blend of 10% alcohol and 90% petrol

Exam questions

1 a i A
ii Spiral of nucleic acid emerging from viral coat at top of diagram should be labelled
b RNA

c

	protein	ribosomes	mitochondria
Bacteria	✓	✓	✗
Viruses	✓	✗	✗

d Living cells / host required; whole organisms could be infected / grow inside chick embryos / amniotic fluid surrounding embryo inside eggs; cell or tissue culture now more usual; cells surrounded by liquid medium for cell culture; small pieces of plant or animal tissue grown in liquid or on solid medium for tissue culture; cultured material taken from infected organism / or culture inoculated with appropriate virus; monolayer culture / any other valid detail

2 a i A: RNA / ribonucleic acid;
B: capsomere / protein subunits

ii Any two from:

Tobacco Mosaic Virus	T_2 bacteriophage
Capsid has helical structure	Icosahedral head and tail
No tail fibres	Tail fibres / base plate
Simple cylindrical rod shape	More complex structure
RNA	DNA

b Reduction in chlorophyll therefore less light absorbed; less photosynthesis therefore fewer sugars produced; less growth / biomass

3

Feature	Tobacco mosaic virus (TMV)	Human immunodeficiency virus (HIV)
Type of nucleic acid	RNA	RNA
Means of entering host	Transmission by aphids	Sexual contact / infected blood / blood products / transfusion / through placenta from mother to fetus
One sign of infection	Yellow / light green patches on leaves	Swollen glands

4 a A: cocci; B: vibrios; C: rods / bacilli

b i Ribosomes

ii Mesosome

c i Contains murein, not cellulose; nitrogenous polysaccharide, not simple polysaccharide; wall may be coated with additional lipid layers

ii Gram staining

5 a Fungi / Zygomycota; hyphae / mycelium / sporangia

b Hyphae penetrate food source; secrete enzymes; absorb digested products

c Parasite: nutrients already digested / soluble / only has to absorb them; source of nutrients is living / host / *Rhizopus* feeds on dead matter; parasite has detrimental effect on host

6 a i Fungi

ii *Penicillium* has hyphae / mycelium; Saccharomyces has single oval cells

iii Any three of:

Microorganisms (Fungi)	Bacteria
Never contain photosynthetic pigments	Some bacteria contain photosynthetic pigments
Possess organelles surrounded by double membrane / mitochondria / nuclear membrane / eq.	No organelles surrounded by double membrane
Distinct nucleus	No distinct nucleus
Many chromosomes	Circular strand of DNA
Larger ribosomes	Smaller ribosomes
Cell walls contain chitin	Cell walls contain murein

b i Non-essential materials / not needed for growth, synthesised late in growth cycle, such as penicillin

ii Slow growth rate / excretion of waste materials

c Similarity: both secondary metabolites / produced late in growth cycle
Difference: penicillin produced by fungus / *Penicillium* that respires aerobically / alcohol produced by fungus / *Saccharomyces* / yeast respiring anaerobically

d Any four of:

Batch fermentation	Continuous fermentation
Fixed volume of nutrients / nothing added during process	Nutrients / oxygen added during process
Nothing removed until end of process	Products / waste constantly removed during process
Environment changes as fermentation progresses / nutrients / oxygen used up / products build up / heat output increases	Environmental conditions kept constant within fermenter
As conditions become unfavourable, growth rate declines	Growth rate maintained at optimum / maximise profits
May be used for different products	Usually dedicated to one product
Suitable for production of secondary metabolites whose production is not linked to growth	More suitable for production of biomass
Shorter time of running makes it more suitable for unstable strains / reduces risk of mutations	Longer running / less labour intensive
Easier to set up / run / but time in between batches wastes production time	Productivity greater as no time lost between batches / more cost effective

7 a i Provides oxygen for aerobic respiration; stirs mixture

ii Brings bacteria into close contact with nutrients; with oxygen; stops bacteria / nutrients settling ; prevents build-up of waste materials / products in one area

b i See answer to **6d** **ii** See answer to **6d**

8 Ethanol / ethyl alcohol; *Saccharomyces* / yeast; anaerobic; carbohydrate; suitable named plant e.g. potato / named carbohydrate

Chapter 18 Molecular biology of the gene

Quick Check questions

Spread 18.1

1 a nucleoside contains a base and pentose sugar whereas a nucleotide consists of a base, sugar, and phosphate

2 condensation

3 TTAGGC

Spread 18.2

1 a bead-like structure consisting of DNA and histones on chromosomes

2 control the distribution of chromosomes during cell

division

3 telomeres seal the ends of chromosomes; shoelace tips seal the ends of shoelaces

Spread 18.3

1 DNA, nucleotides of 4 different types, DNA polymerase, ATP

2 two bands of DNA would have occured: one containing only ^{14}N, the other consisting only of ^{15}N

3 **a** helicases catalyse the separation of DNA strands
  b DNA binding protein helps keep strands separate during replication
  c DNA polymerase catalyses DNA polymerisation
  d DNA ligase joins pieces of polynucleotide together

Spread 18.4

1 rough-strain *Pneumococcus* are transformed into living smooth-strain bacteria after being mixed with heat-killed smooth-strain *Pneumococcus*

2 RNA and proteins can be destroyed by using the appropriate hydrolytic enzymes, leaving DNA

3 **a** protein is on the coat and DNA is in the core
  b DNA can be labelled with ^{32}P and protein can be labelled by ^{35}S

Spread 18.5

1 phenylketonuria is inherited in a Mendelian manner; lack of this specific enzyme results in the build up of phenylalanine in the blood

2 Beadle and Tatum showed that exposure of *Neurospora* to X-rays caused mutations: some mutant *Neurospora* lacked a single specific enzyme responsible for converting one amino acid to another; this inability was inherited, indicating that the production of each specific enzyme was controlled by a single gene

3 haemoglobin is a conjugated protein that has two types of polypeptide chain, alpha and beta; sickle-cell anaemia is caused by a single mutation resulting in one amino acid being changed in the beta chain, indicating that the alpha and beta chain are controlled by different genes

Spread 18.6

1 the central dogma is the principle that genetic information is transferred only in one direction: from DNA to RNA by transcription and from RNA to proteins by translation (it has now been modified to take into account the transfer of information from RNA to DNA by reverse transcription in some viruses)

2 frame shift

3 his (histamine)

Spread 18.7

1 transcription is the transfer of information from DNA to mRNa whereas translation is the transfer of information from mRNA to a polypeptide chain

2 GCCUUAGCA

3 transport specific amino acids to ribosomes during protein synthesis

Spread 18.8

1 peptide

2 complementary base pairing

3 several ribosomes on the same DNA strand, involved in the synthesis of the same type of polypeptide chains

Spread 18.9

1 a single strand of DNA containing a known sequence of bases and labelled with a radioactive or fluorescent marker

2 **a** restriction endonucleases **b** ligases

3 small rings of DNA contained within some bacteria

Spread 18.10

1 a group of cells, organisms, or genes that are exact copies of each other

2 by recombinant DNA technology

3 it is claimed that Dolly was produced from a non-reproductive cell

Spread 18.11

1 repeating nucleotide sequences found in hypervariable regions

2 DNA profiling is used, like traditional fingerprinting, to help police with their inquiries, and, like a fingerprint, a complete DNA profile is thought to be unique to each person

3 alkaline phosphatase labels genetic probes

Exam questions

1 **a** **A**: phosphate group / phosphoric acid;
  B: deoxyribose / pentose sugar, **C**: organic bases
  b i A G T **ii** A G U
  c i Polypeptide
  ii Each type of tRNA molecule binds with a specific amino acid; have a triplet of bases / anticodon specific to amino acid they carry; anticodon is attracted to codon of mRNA held by ribosome; second tRNA 'docks' into next codon of mRNA, bringing second amino acid; tRNA and mRNA form temporary complex until peptide bond forms between two adjacent amino acids

2 **a** Hydrogen bond
  b i Transcription **ii** Translation
  c i TGC AAT CGA **ii** UGC AAU CGA

3 **a** X: nucleotide; Y: organic base
  b DNA ligase / polymerase; polymerisation / condensation reaction
  c Half the original molecule is kept / new molecule has one original strand and one newly synthesised strand
  d Interphase / S / synthesis phase

4 **a i** 32
  ii Forensic science, identification of individuals from very small samples containing traces of DNA
  b No mRNA involved; new strand of DNA produced, this does not happen in transcription; whole molecule copied, only part of DNA copied in transcription; both strands are copied, only one strand is copied in transcription; only bases A, C, G, and T are involved, in transcription U is also involved
  c Half the original molecule of DNA is kept in each daughter DNA molecule; two daughter molecules of DNA each have one original / parent strand and one strand newly synthesised from free nucleotides

5 **a** Egg cell is haploid / no pairs of chromosomes to divide
  b Embryo cells have same genotype / genome; as formed from one fertilised egg
  c Advantage: known characteristics / high meat yield / fleece quality consistent / breeding season / lambing will be similar / production of human proteins if transgenic animal cloned / AOVP
  Disadvantage: could all succumb to same disease

/ could all be flawed in some characteristic / reduction in gene pool

6 a Transcription is the process by which a complementary mRNA copy is made of the specific region / cistron of the DNA molecule which codes for a polypeptide
Translation is the process by which a specific sequence of amino acids is linked together in accordance with the codons on mRNA
Both processes are involved in protein synthesis. Proteins are required for enzymes to control metabolic pathways, for growth, repair, hormones, muscle contraction, carriage of oxygen, antibodies, blood clotting, etc.

b Transcription: specific region of DNA molecule / cistron unwinds; enzyme breaks hydrogen bonds between complementary base pairs to allow this to happen; bases exposed along strands; only one strand is 'read' / used as a template; complementary 'free' RNA nucleotides attracted to exposed bases; guanine attracts cytosine / eq.; thymine replaced by uracil on RNA; RNA polymerase links RNA nucleotides together to form mRNA molecule; DNA reforms behind RNA molecule; only about 12 base pairs unwind at any one time; length of mRNA formed varies according to the length of polypeptide chain for which it codes
Translation: group of ribosomes attach to mRNA to form a polysome; complementary anticodon of a tRNA–amino acid complex is attracted to the first codon on mRNA; first amino acid is usually methionine; second codon attracts complementary anticodon; ribosome acts as framework to hold mRNA and tRNA–amino acid complexes together; until two amino acids linked by peptide bond; once linked, ribosome moves along mRNA to hold next codon–anticodon complex together; third amino acid linked to dipeptide; continues until polypeptide chain formed; nonsense / stop code reached; polypeptide leaves ribosome; other ribosomes may follow immediately behind first ribosome, along the mRNA, so many identical polypeptides formed at the same time; tRNA released to combine with another amino acid of the same type

---

Chapter 19 Inheritance

Quick Check questions

Spread 19.1

1 heterozygous
2 the genotype refers to the genetic make-up of an organism (i.e. the alleles it has) whereas the phenotype refers to the visible or otherwise measurable characteristics of an organism resulting from an interaction between the genotype and environment
3 continuous variation
4 an agent that causes a mutation (e.g. X-rays)

Spread 19.2

1 The characteristics of a diploid organism are determined by alleles which occur in pairs. Of a pair of such alleles, only one can be carried in a single gamete.
2 Each of a pair of alleles for a particular gene can combine randomly with either of another pair of alleles for a different gene.
3 all the offspring in F_1 generation resulting from a cross between pure breeding dwarf and tall plants were tall but both tall and dwarf plants occurred in the F_2 generation; there were no plants of intermediate height.

Spread 19.3

1 **Tt**
2 **Tt** and **tt** (a ratio of 3:3 = 1:1)
3 Results of a test cross between a homozygous tall plant and dwarf plant

Parent phenotype:	tall × dwarf
genotype	**TT** × **tt**
gametes	all **T** all **t**
offspring genotype	all **Tt**
offspring phenotype	all Tall

Spread 19.4

1 codominance
2 $\mathbf{I^BI^B}$ or $\mathbf{I^BI^O}$
3 dominant lethal alleles which are expressed before an organism has a chance to reproduce will be quickly eliminated

Spread 19.5

1 **RY, Ry**, **rY,** and **ry**
2 a plant of known genotype must be used in the test cross, and the only genotype of which we can be certain is that of the dwarf white-flowered plant because it is recesssive for both characteristics being considered
3 recombinants are offspring that have new combinations of characteristics, different from either parent

Spread 19.6

1 genes which, because they are carried on the same chromosome, tend to be inherited together
2 crossing-over
3 $$\text{COV} = \frac{\text{total number of recombinants}}{\text{total number of offspring}} \times 100$$
$$= \frac{46 \times 100}{427}$$
$$= 10.8\%$$

Spread 19.7

1 human males have different sex chromsomes (X and Y)
2 a person with an XY genotype may develop female sexual organs if Mullerian-inhibiting substance is not produced
3 two Barr bodies may occur in buccal epithelial cells if a person has triple-X syndrome (XXX)

Spread 19.8

1 the X chromsome is much larger than the Y chromosome and carries genes for nonsexual characteristics as well as those concerned with determining sex
2 1 in 4 (25% or 0.25)
3 sex linkage refers to the tendency of certain inherited characteristics occurring more frequently in one sex than the other because the gene for the characteristic is carried on one or other of the sex chromosomes; a sex-linked characteristic may occur in both males and females although it tends to occur more frequently in one sex than the other, whereas sex limitation refers to characteristics that are usually expressed in *only* one or the other sex

Spread 19.9

1 non-separation of one or more homologous

chromosomes during meiosis

2 a amniocentesis can be carried out at about 15–16 weeks of pregnancy whereas chorionic villus sampling can be carried out between weeks 8 and 12;

b amniocentesis carries a lower risk than chorionic villus sampling

3 chorionic gonadotrophin

Spread 19.10

1 CFTP is a carrier protein which transports chloride ions out of cells into the mucus

2 blockage of the pancreatic duct may prevent pancreatic enzymes reaching the duodenum

3 spheres made of lipid

Exam questions

1 a Height of wheat shows bimodal distribution

b i Continuous variation; all possible heights represented with no steps in between / complete gradation shown

ii Environmental variation / named environmental effect / light / water / eq. / genetic variation within wheat strains

2 a Normal sight

b **Nn** / heterozygous; must have **N** as shows night blindness, but must also have **n** as has children with normal vision

c Genotype of individual 10 is **Nn** (**N** as has night blindness, **n** as father is **nn**)
Genotype of individual 11 is **nn** (as has normal vision, which is recessive)
Cross between **Nn** and **nn** results in offspring with possible genotypes: **Nn**, **Nn**, **nn**, **nn** (draw Punnett square to check this)
Phenotype of offspring: 50% with night blindness, 50% normal vision; of the offspring with night blindness, half could be male, half female, so probability of girl with night blindness is 25% / 0.25 / ¼

3 a i One, gamete is haploid and therefore only has one copy of each chromosome, carrying the gene for fruit colour

ii Two, leaf cell is diploid, with two copies of each chromosome, and therefore can carry two alleles for fruit colour, both the same, in this case

b i **Rr Tt** / **Tt Rr**

ii Tall, with yellow fruit

iii Parent 1 genotype: **Rr**; parent 2 genotype: **rr** (**T/t** alleles can be ignored)

gametes	R	r
r	Rr	rr
r	Rr	rr

Offspring phenotypes: 2 red, 2 yellow; 50% of offspring will have yellow fruit

4 a Gene is a specific sequence of bases on DNA / occurs at a specific locus on DNA; coding for a particular polypeptide; allele is a different form of a gene; only one allele present at a locus / alleles separated at meiosis

b i 1: $I^O I^O$; 2: $I^A I^B$; 4: $I^A I^O$; 5: $I^B I^O$; 6: $I^B I^O$;

ii Blood group could be either group A or group B; as genotype could be $I^A I^O$, giving group A, or $I^B I^O$, giving group B; either IA or IB inherited from parent 2; only I^O could be inherited from parent 1

5 a i $\mathbf{X^B X^b}$

ii Two X chromosomes needed to express both black and ginger alleles, males have only one X, no alleles for coat colour on Y

b

	black female	ginger male
Parental genotypes:	$\mathbf{X^b\ X^b}$	$\mathbf{X^B\ Y}$
Gamete genotypes:	$\mathbf{X^b}$	$\mathbf{X^B}$ and **Y**
Offspring genotypes:	$\mathbf{X^B\ X^b}$	$\mathbf{X^b\ Y}$

Percentage of tortoiseshell kittens = 50%

6 a i

Parental phenotypes:	normal / carriers
Parental genotypes:	both $\mathbf{Hb^A\ Hb^S}$
Genotypes of gametes:	$\mathbf{Hb^A}$ and $\mathbf{Hb^S}$
Genotypes of offspring:	$\mathbf{Hb^A\ Hb^A}$, $\mathbf{Hb^S\ Hb^S}$, $\mathbf{Hb^A\ Hb^S}$ $\mathbf{Hb^A\ Hb^S}$

Phenotypes of offspring: 25% normal, 25% sickle-cell disease; 50% normal (carriers)

ii 1 in 4 / 25%

b Individuals homozygous for sickle-cell anaemia will have the disease, sickle-cell anaemia is fatal, therefore low numbers in the population; heterozygotes do not have the disease, but are carriers; little disadvantage in being heterozygous under normal conditions; being heterozygous confers an advantage as it confers some resistance to malaria

7 a A different / alternative form of a gene

b i In metaphase I of meiosis, independent assortment of chromosomes occurs; homologous pairs arrange themselves randomly relative to the orientation of other bivalents, so when they are pulled apart, entirely new genetic combinations of alleles will arise in gametes

ii 25% probability / $\frac{1}{4}$ / 0.25

Chapter 20 Evolution

Quick Check questions

Spread 20.1

1 the change over successive generations of the genetic composition (allele frequency of a population) that may result in the formation of new species from pre-existing species

2 neo-Darwinism incorporates new scientific evidence, particularly from genetics and molecular biology

Spread 20.2

1 fossils such as those of *Archaeopteryx* lie between two present-day types of animals and provide strong support that the two types evolved from a common ancestor

2 two from: the genetic code is almost universal, respiration follows similar metabolic pathways in all organisms, and ATP is the universal energy currency

Spread 20.3

1 analogous

2 a convergent evolution **b** parallel evolution

3 e.g. pentadactyl limb in mammals, Australian marsupials, or Darwin's finches

Spread 20.4

1 ability to pass on alleles to subsequent generations; the fittest individual in a population is the one that produces the largest number of offspring

2 a disruptive selection; **b** intermediates would be at a selective disadvantage because they would be easily seen against either a green or brown background.

Spread 20.5

1 heavy-metal tolerant plants are less competitive in unpolluted areas and rarely survive

2 a increased frequency of alleles
b no change in the rate of mutation

Spread 20.6

1 in malaria-infested areas, heterozygotes $\mathbf{Hb^AHb^S}$ are fitter than either homozygote: they are much less likely to suffer malaria than those homozygous for the normal allele ($\mathbf{Hb^AHb^A}$ genotypes) and they do not suffer sickle-cell anaemia like the $\mathbf{Hb^SHb^S}$ genotypes

2 when the population is large, mating is random, no mutations occur, there is no net immigration into or out of the area

3 4.17%

Spread 20.7

1 directional selection

2 a inbreeding reduces genetic diversity;
b outbreeding increases genetic diversity

3 harmful recessive alleles may be less likely to be present in the homozygous condition and some allele combinations may interact positively

Spread 20.8

1 the two species may not normally interbreed because, for example, they have different courtship dances

2 mechanical isolation

3 allopatric speciation occurs in two or more demes which are geographically isolated whereas sympatric speciation occurs in two or more demes living in the same geographical location.

Spread 20.9

1 a colchicine **b** colchicine allows chromosome replication to take place but prevents cell division

2 the two sets of chromosomes in the hybrid are not homologous (one set coming from *S. maritima* and one set from *S. alterniflora*), therefore meiosis can not take place

Spread 20.10

1 a associated with a shorter and flatter nose which has allowed the evolution of stereoscopic vision
b gives a powerful grip
c results in reduced number of offspring associated with increased parental care
d allows increased mobility of forearm

Spread 20.11

1 monkeys have insufficient mobility of the upper limb to perform full brachiation

2 ape big toes are opposable and not aligned parallel with the other toes, therefore they are not well adapted to supporting the whole of the body weight

Spread 20.12

1 Neanderthals had a larger brain than modern humans

Spread 20.13

1 a increased powers of verbal communication (speech)
b more upright posture
c produce a power grip or a precision grip
d greater development of facial muscles for nonverbal communication and improved stereoscopic vision

Exam questions

1 a If 96% of the population are black moths, then 4% mustbe light-coloured moths.
'Light' moths are homozygous recessive, **bb**, as black is the dominant allele for colour.
Frequency of homozygous recessive is given by q^2 in the Hardy–Weinberg principle, therefore, $q^2 = 4\%$ or 4 in 100 = 0.04;
If $q^2 = 0.04$, then $q = 0.2$;
Frequency of both alleles, $p + q = 1$, where p represents the dominant allele, **B**.
Therefore, p, frequency of the dominant allele, **B**, = 0.8

b Some black moths were heterozygous, so two heterozygous moths breeding together would have a 1 in 4 probability of producing the recessive light-coloured form

c In rural populations, the light-coloured form that predominates is homozygous recessive; the dominant allele that produces black moths can not occur in the population unless a new mutation arises; or stray moths from the urban population migrate into rural population

2 a Resistance: any inherited characteristic of an organism which reduces the effect of an adverse environmental factor / named example such as pest / pesticide / salinity / etc.; adverse factor does not cause the mutations which produce the resistant alleles; alleles for resistance already present in gene pool of population, although at low frequency; or new alleles arise randomly by mutation; presence of adverse factor creates strong selection pressures for resistant allele; allele for resistance may initiate production of enzyme to break down drug / pesticide / virus / eq.; or induce enzymes that allow other metabolic pathways to be utilised, by-passing effect of adverse factor; non-resistant strains destroyed; leaving less competition for resistant strains; resistance can be introduced into organisms by transferring genes from resistant species / from susceptible species; resistance usually under the control of one or two gene loci; credit for other details;

b Drug resistance: some strains of bacteria have developed resistance to drugs / antibiotics; *Staphylococcus aureus* / eq.; allow bacteria to survive in the presence of methycillin / penicillin / other named antibiotic; gene conferring resistance can be transferred to other bacteria making them resistant;
Pesticide resistance: some forms of insects such as mosquitoes / *Anopheles* / *Aedes* / other named example; have developed resistance to insecticides such as DDT / other named example; rats / other named example; have developed resistance to warfarin / other named example;
Herbicide resistance: groundsel / *Senecio vulgaris* / other named example; is a major pest as a weed, competing with cash crops for resources; triazine herbicides bind to thylakoid membranes of chloroplasts / prevent photosynthesis; resistant plants may possess geneticallycontrolled mechanisms to prevent herbicide binding;
Disease resistance: some rabbits have developed resistance to myxomatosis; virus rendered ineffective by mutated gene in some cases; in others,

behaviour of rabbit is altered by mutant gene, causing it to spend more time out of its burrow and therefore less chance of being infected by fleas from other rabbits; resistance to rust in flax;
Resistance to heavy metals: heavy metals / tin / copper / lead / nickel toxic to most plants; some varieties of grasses / *Festuca ovina* / *Agrostis tenuis* / other named example; have developed resistance to toxicity / can survive high levels of these metals; so can colonise where other plants can not; AOVP

3 Original population containing common ancestor shared a gene pool; groups from original population became geographically isolated; by physical barriers such as land masses moving away from each other / oceans; separation of groups stops flow of genes between them; each group / deme may then evolve in different ways / adaptive radiation; so that even if re-united, they would not be able to breed successfully; allopatric speciation

4 a i 63

ii Odd number of chromosomes; sets of chromosomes from different species are not homologous; bivalents can not form during meiosis; meiosis can not occur / gametes / pollen not formed

b Chromosomes replicate without cell division; during mitosis, chromosomes do not separate / all go to one pole; non-disjunction; diploid gametes fuse; to give tetraploid / 4n hybrid

5 a i Group of very closely related forms, subgroups of the same family; may be further separated into species; examples: *Homo* (humans) / *Canis* (dogs, coyotes, and wolves) / other generic name

ii Group of organisms that can breed together to produce fertile young; very similar in structure / chemical makeup / DNA sequences / behaviour; examples: *Homo sapiens* (modern humans) / *Canis familiaris* (domesticated dog)

b i Geographical isolation / by physical barrier / valleys stop flow of genes between them; each species may then evolve in different ways / adaptive radiation due to differences in environments

ii Natural selection; directional selection / selection favours one extreme of an inherited characteristic / shifts population mean for characteristic; stabilising selection / selective pressures favour the mean / selects against extremes; disruptive selection / favours extremes of variance / divergence of population into two subpopulations

c Interspecific competition reduced; sufficient food for both species

d **A** and **B** are more closely related than **A** and **D**; therefore may share more characteristics / have similar dietary requirements / have similar shelter / territory / breeding / eq. needs

6 a Relative dating by comparing other fossils from the sedimentary rock that the bone was found in with fossils from elsewhere of known age; absolute dating of volcanic rock by radioactive dating using uranium–lead / thorium / fission track method; using potassium–argon method; radioactive carbon dating of fossil bones.

b Anatomical similarities between modern human and human ancestor bones allow skeleton to be put together; known ratios between / proportions of two or more bones allow estimation of size of missing bones; orientation / size of heads of bones / femurs / eq. allow estimation of joint movement / angle of that part of the skeleton / overall height; position of bones in relation to each other / e.g. backbone to base of skull / shape of hip bones allow estimation of posture; ridges on bones allow estimation of degree of muscle attachment; and hence facial features; computer programmes can now add muscle, fat, and skin layers to create approximation of appearance from data generated by detailed study of bones; fossil footprints give information about stance; Difficulties: fossil record incomplete; unlikely to have every bone for each specimen / have to estimate appearance of missing bones; human fossils are rare; those found represent fraction of total population / may be atypical specimens; may be geographically biased / some areas may conserve fossils more easily; fossil bones may be distorted / crushed / difficult to interpret; no fossils of soft parts of early humans / AOVP

Chapter 21 The variety of living things

Quick Check questions

Spread 21.1

1 the five-kingdom classification aims to be natural but is in part artificial

2 *Panthera tigris*

3 a Protoctista, Fungi, Animalia, and Plantae

b Prokaryotae, Protoctista, Fungi, and Plantae

Spread 21.2

1 DNA which is not incorporated into chromosomes

2 a prokaryote flagella have no internal microtubuluar structure; eukaryote flagella have a typical 9+2 microtubular structure

b prokaryote ribosomes are smaller (70S) than those of eukaryotes (80S)

Spread 21.3

1 a euglenoids have a different biochemistry

b oomycetes have undulipodia and cellulose cell walls

c Chlorophyta are green whereas Rhodophyta are red

Spread 21.4

1 a osmoregulation **b** locomotion and feeding

2 a ATPase catalyses the breakdown of ATP so that energy can be released for movement

b the flexible pellicle of *Euglena* is made of protein, not cellulose.

Spread 21.5

1 saprotrophically, digesting their food extracellularly and absorbing the digested products

2 fungi cell walls contain chitin, not cellulase

Spread 21.6

1 In all land plants the gametophyte generation is haploid and reproduces asexually, whereas the sporophyte generation is diploid and reproduces sexually.

2 bryophyte leaves are haploid whereas true leaves of fern are diploid.

3 gametophyte

Spread 21.7

1 reduce water loss

2 many angiosperms are insect-pollinated whereas most

conifers are wind-pollinated

3 a monocot leaf has parallel veins whereas the veins of a dicot leaf usually form a net-like pattern

Spread 21.8

1 nervous coordination

2 an organism with bilateral symmetry has only one plane which passes through the centre of the body, whereas a radially symetrical organism has more than one plane which passes through the centre of the body.

Spread 21.9

1 **a** a stinging cell
  b a polyp is a sessile or sedentary asexually reproducing stage whereas a medusa is a free-swimming sexually reproducing stage
  c ectoderm, mesoglea, and mesoderm
  d mesoglea

Spread 21.10

1 platyhelminths are triploblastic and have an organ level of organisation whereas cnidarians are diploblastic and have a tissue level of organisation

2 mesoderm

Spread 21.11

1 tapeworms have no gut, liverflukes have a gut; the tapeworm body is divided into proglottids, liver flukes have no proglottids

2 loss of organ systems and functions not needed for a parasitic mode of life

Spread 21.12

1 there are a fixed number of earthworm segments and they are all of the same age; the number of proglottids vary as do their age

2 the body wall musculature is separated from the gut musculature by the coelom

3 earthworms have a relatively low level of activity and their cylindrical body provides a large enough surface area-to-volume ratio for diffusion to satisfy their requirements for gaseous exchange.

Spread 21.13

1 **a** annelids and arthropods are triploblastic, coelomate, metamerically segmented, and have both a mouth and anus.
  b arthropods have jointed limbs and an exoskeleton; annelids do not.

2 **a** millipedes have two pairs of limbs on each segment whereas centipedes have one pair on each segment
  b insects have three pairs of walking legs whereas arachnids have four pairs.

Spread 21.14

1 **a** uric acid **b** urine produced by mammals is watery and its production results in a loss of body water whereas uric acid produced by insects is relatively solid and results in little water loss.

2 at least three

3 the fore-wings and hind-wings are hitched together

Spread 21.15

1 tracheoles

2 a holometabolous life cycle involves a complete metamorphosis whereas a hemimetabolous life cycle involves incomplete metamorphosis

3 Metamorphosis reduces competition between different developmental stages, and allows each stage to become specialised for particular functions

Spread 21.16

1 **a** nematodes have a gut with a mouth and anus, platyhelminths (if they have a gut) only have a mouth
  b shells of gastropods have a single valve whereas those of pelecypods have two valves (i.e. they are bivalves)
  c cnidarians and echinoderms are both multicellular and generally exhibit radial symmetry

Spread 21.17

1 a flexible rod running along the back

2 fins of cartilaginous fish are fleshy whereas fins of bony fish are supported by fin rays

3 water flows along the external surface of the gills in the opposite direction to the blood blowing in the gill lamellae

Spread 21.18

1 **a** uric acid **b** uric acid is relatively solid, allowing reptiles and birds to conserve water

2 **a** reptiles and birds **b** birds and mammals

Exam questions

1 **a i** Complex family / taxonomic tree, with different levels; largest group / kingdom at top, smallest group / species / subspecies at bottom; organised according to diagnostic features / denoting closeness of relationships / evolutionary history
  ii Based on evolutionary relationships / history / suggesting how species arose by adaptation of existing forms; organisms belonging to one group believed to have a common ancestor / indicating closeness of relationship / comparative anatomy / biochemistry / embryology / DNA sequencing / behaviour etc. used to group organisms
  b Missing words, from top to bottom: phylum; order; genus; species

2 **a i** Plants / Plantae
  ii Organisms in kingdom **B** possess cell wall; large vacuole containing cell sap; cellulose as a structural material
  b D **c E**

3 **a i** **A**: mitochondrion / outer membrane of mitochondrion; **B**: rough endoplasmic reticulum / ribosome on RER
  ii Distinct nucleus; nuclear membrane / nuclear envelope; smooth endoplasmic reticulum; Golgi body
  b Diameter at **X** = 16.5 mm (accept 16 / 17) 23.5 mm scale bar (accept 23 / 24) represents 5 μm

Therefore, diameter at **X** = $\frac{5 \times 16.5}{23.5} = 3.5\,\mu m$

4 **a i** Cnidaria
  ii Platyhelminthes
  b i Triploblastic with coelom / body cavity in mesoderm
  ii Coelom allows independent movement / function of organs / division of labour; more tissues / organs / systems can develop

5 **a** Suckers allow attachment to gut wall of host; prevent tapeworm being swept out of gut by movement of contents within lumen / peristalsis
  b i Parasitic nutrition; host digests food; digested products absorbed through body wall of tapeworm; ref. to contact digestion by tapeworm

ii *Rhizopus* is a saprobiont / saprophyte; secretes extracellular enzymes to digest its own food

6

Organism	Phylum	One visible external feature
A	Angiosperms / Angiospermophyta	Flower / petal / sepal / stigma / anther
B	Mollusca	Shell / muscular food
C	Arthropoda	Jointed legs / appendages / exoskeleton

7 a Label line to exoskeleton, immediately above muscle B

b i Antagonistic

ii Muscle **A** contracts, muscle **B** relaxes, wing lowered; muscle **A** relaxes, muscle **B** contracts, pulls exoskeleton attached to base of wing inwards, wing raised

c Allows muscle **B** / elevator muscle to pull wing upwards; some energy from muscle contraction is stored in specialised pads / resilin, located in the wing pivots of the exoskeleton, as wing passes the 'click' point, stored energy is released, producing powerful downstroke

8 a i Diffusion / through air until cell surface reached / down a concentration gradient / passes through small openings / spiracles and stomata

ii Carbon dioxide diffuses from cells into the air in the tracheal system in insects, or via haemocoel / exoskeleton; carbon dioxide from fish will diffuse from blood and dissolve in the water passing over the gills

b One cycle takes 0.6 s, therefore ventilation rate

$= \frac{1.0}{0.6} = 1.67$ cycles per second

= 100 cycles per minute

c Pressure in opercular cavity is less than that in the buccal cavity for first part of cycle / in first 0.4 s, therefore water moves from buccal cavity over the gills to opercular cavity down pressure gradient; after 0.3 s buccal cavity expands, lowering pressure, causing water to be drawn in through mouth, water can not enter through opercular cavity as operculum shuts

d Mouth / entrance to oesophagus closes; floor of buccal cavity raised, increasing pressure inside cavity

Chapter 22 Environmental biology

Quick Check questions

Spread 22.1

1 dormant bacterial and fungal spores

2 temperature and precipitation

Spread 22.2

1 A community consists of populations of different species whereas a population consists of individuals of the same species

2 Abiotic factors result from the non-living part of an ecosystem, whereas biotic factors result from the activities of living organisms.

Spread 22.3

1 a arrows represent the direction of energy flow

b filamentous algae, planktonic algae, and water lilies

c the rate of energy flow decreases at each higher tropic level

Spread 22.4

1 net primary production = gross primary production – respiration

2 Net primary productivity is used to compare efficiencies because only the energy stored in plant tissues is available to consumers

3 percentage of energy passing from one trophic level to another

Spread 22.5

1 profession

2 fundamental niche

3 if two species had identical niches, competition would eventualy result in the exclusion of one.

Spread 22.6

1 the population grows exponentially

2 the maximum population size that can be sustained over a relatively long period by a particular environment

3 because a change in temperature will affect the same proportion of the population regardless of its density

Spread 22.7

1 development of agriculture

2 change in a human population from having high birth rate and high death rate to having low birth rates and low death rates

3 a high percentage of the population of less well-developed countries is younger than fifteen years and therefore includes many young women who are approaching child-bearing age

Spread 22.8

1 competition has harmful effects on both populations involved

2 predators kill their prey to obtain nourishment, parasites do not

3 a the plant obtains soluble mineral nutrients from the activities of mycorrhiza;

b the mycorrhiza obtain food (e.g. sucrose) from the plants

Spread 22.9

1 a as predators have evolved better hunting strategies, prey have evolved better defensive strategies

b populations were estimated from furs traded with the Hudson Bay Company

c on islands with no lynx predators, the snowshoe hare population still oscillates

Spread 22.10

1 secondary succession usually occurs where soil, often containing seeds, is present

2 pioneer species, through their activities, may create a soil suitable for the next stage of succession

Spread 22.11

1 D = 756/158 = 4.78

2 a an estuarine mudflat; b a woodland with high species diversity

Spread 22.12

1 a ammonification; b nitrogen fixation; c nitrification; d denitrification

Spread 22.13

1 carbon dioxide fixation

2 three from: limestone quarrying and cement making; burning fossil fuels; breathing; deforestation.

Spread 22.14

1 the camel's hump stores fat which acts as an insulator, protecting underlying tissue from heat; it also acts as a food store

2 a oxidation or metabolic water obtained from respiration of seed **b** the long loop of Henlé gives desert rats a more efficient countercurrent exchange mechanism, enabling them to produce a more concentrated urine than that of the beaver **c** water exhaled is absorbed by dry seeds which are then eaten.

Spread 22.15

1 a plants at high altitudes have xeromorphic adaptations to reduce transpiration losses when soil is frozen and free water is unavailable

b hydrophytes are supported by water

c salt is actively absorbed in tissues of glassworts to create a concentration gradient for the uptake of water by osmosis

Spread 22.16

1 2 °C

2 Hypoxia is an inadequate supply of oxygen to respiring tissues

3 Acclimatisation

Exam questions

1 a Flow of energy through trophic levels

b Larger population of mussels will consume more microscopic plants and animals, reducing food supply for barnacles / larger population of mussels will provide more food for whelks, therefore whelk population increased and more barnacles eaten

c i Number of organisms of each species

ii How variable the data is / spread around the mean; the reliability of a sample mean is greatest if it is large and the standard deviation is small / the smaller the standard deviation, the more reliable the mean

2 a All the individuals of a given species in a particular habitat

b i $B + I = D + E$

ii $B + I > D + E$ / $(B + I) - (D + E) = >0$ / eq.

c i Biotic factors such as competition for resources, predation, and parasitism increase with increasing population density and therefore act to limit population size

ii Abiotic / environmental factors such as water availability, light intensity, and temperature changes will act on the population, possibly unfavourably, whatever the density, and therefore limit population size

3 a i Tropical rainforest has higher / more consistent temperatures / all year; higher plant density; more light energy / greater light intensity; more water available / higher rainfall; more evergreen plants / fewer deciduous plants

ii Intensively cultivated land: crop varieties selected for high yield; monoculture / crops all same type of plant; pests / diseases all controlled; fertilisers used to maximise yield; irrigation; new crops planted immediately after harvest

b Use random numbers / eq. to place quadrat of given size; remove all plants from quadrat; repeat many times, each time placing quadrat randomly; remove soil / animals from plants; weigh to find fresh mass per known quadrat area; find mean value and multiply appropriately to find mass per area of grassland; or count number of plants in quadrat area; remove one plant; weigh and multiply by number of plants

c Water content of biomass varies; biomass includes inorganic components; productivity in energy units is truer reflection of energy capture by producers

4 a A: *Rhizobium* / *Azotobacter* / *Clostridium* / *Bacillus* / *Klebsiella* / etc.;
B: *Nitrosomonas* / *Nitrococcus* / etc.;
C: *Nitrobacter* / etc.;
D: *Pseudomonas* / *Bacillus* / *Thiobacillus* / etc.

b Anaerobic; water-logged soil / bogs / marshes / eq.

5 a Change / often an increase in species / diversity of organisms present; resulting in a change to their environment, which benefits other species, for example addition of nutrients to the water in the lake

b Increase in number of species / diversity; increase in numbers of organisms / biomass; increase in complexity of organisms; increase in nutrients / minerals / soil available; decrease in space available for new species; more interspecific competition as conditions become less harsh; AOVP

c Removing forest cover may increase soil erosion; more sediment deposited in lake may speed up succession; as more sediment for plants to grow in / more minerals; AOVP

6 a All the biotic and abiotic components in a natural / self-contained unit, through which energy flows and nutrients cycle

b Species present; numbers of each species / index of diversity; trophic level of each species; which are producers; herbivores; secondary consumers; tertiary consumers; top carnivores; decomposers; food chains / webs; biomass; energy input into system / amount of light available; amount of light absorbed / gross primary productivity; net primary productivity; other climate details: temperature; humidity; availability of water / rainfall; pH of water; salinity of water; type of soil / geological material; mineral availability; AOVP

c Some solar energy is captured by photoautotrophs / producers; in photosynthesis; some is reflected / not all is absorbed / not all is of useful wavelengths; about 1% of incoming useful light is converted into gross primary productivity / GPP; cultivated crops may achieve higher levels of GPP and net primary productivity during growing season; amount of light falling on producers can be decreased by selective planting e.g. coniferous woodlands will not only block light from forest floor, but tall trees will cast shadows on adjacent land; deforestation drastically reduces numbers of producers / amount of light absorbed; contributes to soil erosion and hence even fewer producers; air pollution / acid rain can decrease numbers of producers; affect other trophic levels; ground-level ozone pollution can affect crops detrimentally; desertification can result in formerly productive land becoming useless; use of greenhouses and lighting of certain wavelengths / red, blue can increase absorption of light / increase photosynthesis; covering ground with black plastic / eq. to reduce weed growth will decrease light energy absorbed by plants, but will increase thermal energy of soil; removal of crops reduces food energy available for herbivores / food chain; herbivores / primary consumers only able to make use of some of energy trapped in producers / indigestible cellulose cell walls / not all of plant eaten etc.; herbivores eaten by humans / rabbits / cows etc.; / by-passing rest of food chain, so their biomass is not available to other members of food web; intensive farming / high density of cows / pigs / eq. will produce considerable biomass from excretory products; encouraging decomposer food chains; secondary consumers / carnivores obtain only small percentage of energy taken in by primary consumers; losses of heat energy from respiration; indigestible bones / eq.; predators e.g. foxes, may be hunted for sport; other large secondary consumers may be killed as they interfere with human activities e.g. deer killed in road accidents or from ignorance e.g. grass snakes; or for other reasons e.g. badgers culled because of TB link; reduces energy available for tertiary consumers; burning dead organisms, including humans; forest fires started by humans make less energy available for decomposers and detrivores; spraying fields with cocktails of chemicals may have effects on soil decomposers, making the recycling of minerals slower; AOVP

Chapter 23 Applied ecology

Quick Check questions

Spread 23.1

1 Soil texture is determined by the relative proportions of silt, sand, and clay particles, whereas soil structure refers to the distribution, shape, and stability of aggregates.
2 loam
3 mechanical preparation of soil

Spread 23.2

1 A pest is any undesirable organism.
2 Broad-spectrum pesticides affect a wide range of pests.

Spread 23.3

1 Bovine somatotrophin, a hormone which stimulates growth
2 warm and dark conditions minimise energy expenditure by minimising activity
3 **a** hedgerows can harbour weeds, diseases, and pests
b hedgerows can act as refuges for predators of pests

Spread 23.4

1 The sea off Peru is usually the site of upwelling which supplies nutrients for plankton on which fish depend for food.
2 log (exponential) phase
3 small fish will be caught in a net with small mesh size before they have a chance to reproduce

Spread 23.5

1 forests have been converted to agricultural land in tropical areas
2 most nutrients in a boreal or tropical forest are contained within living trees whereas in deciduous forests most nutrients are contained within the soil.
3 Deforestation may increase runoff, thereby increasing the risk of flooding
4 sustainable development is economic development which can continue indefinitely

Spread 23.6

1 when substances or energy are released into the environment in amounts high enough to harm anything of human value
2 carbon dioxide
3 ozone concentrations greater than 0.1p.p.m. are considered toxic

Spread 23.7

1 thermal pollution warms water, reducing its oxygen-carrying capacity, killing fish by depriving them of oxygen
2 aerobic microorganisms
3 a dramatic, fast growth of algae

Spread 23.8

1 Two from: sewage, agricultural wastes, organic industrial wastes
2 nitrate fertiliser may accelerate oil-spill clean-ups by stimulating the growth of naturally occurring bacteria which break down the oil.

Spread 23.9

1 sewage fungus is a slime consisting of filamentous bacteria, protoctists, and algae, as well as fungi; it thrives in areas depleted of oxygen and rich in organic material
2 areas near a sewage outfall are low in oxygen, making them unsuitable for mayfly larvae
3 abundances of organisms vary naturally, and they may be absent due to natural causes unrelated to pollution

Spread 23.10

1 ecotourism has been used to take economic advantage of the aesthetic value of natural environments
2 a worldwide ban on the trade in ivory.

Spread 23.11

1 **a** in most environments, grassland and heathland will develop into scrub and woodland without mowing, grazing, or burning
b by providing a wide range of microhabitats
c The Peak District

Exam questions

1 **a i** Rice removes minerals from soil / minerals not replaced
ii Beans are legumes / have root nodules containing nitrogen-fixing bacteria; after decomposition of bean plants, nitrates available to rice
b Recycling of 'nutrients' / minerals would be decreased as decomposers needed to break down organic material and their numbers have fallen drastically; if organic material not broken down, then less nitrates will be available for plants
2 **a** Pollution is most likely to have affected the otters' food and habitat; dumping of rubbish in rivers; sharp objects could have cut them and wounds become infected; plastic can rings / bags could have harmed / suffocated young otters; poisons / toxins released into river; e.g. copper / zinc / lead / mercury / cyanide; will kill fish / otters' food

supply; oil spills less likely in fresh water, but could happen on a small scale, could affect the otters directly by affecting the waterproofing of their fur; eutrophication; leaching of excess minerals / fertilisers from farm land entering river; could have caused algal blooms; aerobic decomposer bacteria multiply rapidly when algae die; depriving water of oxygen; killing fish (and other organisms) and reducing otters' food supply; thermal pollution from factories, returning warm water to rivers, could have similar effect; also alter the species of fish present to some less desirable to otters; acid rain could lower pH of river water, killing fish; possibly also having direct effect on otters; sewage entering rivers could carry diseases affecting otters; may create a biological oxygen demand / BOD, killing fish; AOVP

b Capture several otters in humane way and mark / tag / ring them in some way that will not cause harm; release them at same place where caught; capture another sample some time later and count how many of these are tagged; calculate size of population on the assumption that the proportion of tagged to untagged individuals in second sample is the same as the proportion of tagged to untagged individuals in the population as a whole; use Lincoln index to calculate this

3 a Mass of fish being caught was decreasing steeply, suggesting that numbers of cod in the Atlantic were also falling

b Despite introduction of quotas, mass of fish being caught continued to fall, suggesting that stocks in the Atlantic had not recovered; when quotas were introduced just under 700 thousand tonnes of fish were being caught, and after two years of the initial quota, this had fallen to under 400 thousand tonnes, so further action needed

c Have some areas of the ocean that are 'out of bounds' / from which fish are not allowed to be taken; have regulations / laws that insist on young fish / pregnant females being returned to the water if caught; regulations about size of holes in nets so younger / smaller fish can escape

4 a i Faecal matter from fish

ii Excretory material from fish contains ammonia / dead fish / excess fish meal

b Compete for food supplies; wild fish, or their young, may be eaten

c Increase in minerals / nitrates / phosphates causes algal bloom / algae / small water plants to multiply rapidly / eutrophication; so much growth occurs that light prevented from reaching plants / algae below surface which then die; massive increase in aerobic, decomposer bacteria decreases concentration of oxygen in water

5 a Benefits: wood used as fuel; land cleared for agriculture to grow more food; profit made from selling timber for export;
Adverse consequences: burning / decay of timber releases additional carbon dioxide / contributes to the greenhouse effect; fewer producers to absorb carbon dioxide / contributes to the greenhouse effect; soil may be eroded; eventual desertification; silting up of rivers etc.; heavy rain no longer absorbed by forests leads to increased risk of flooding; loss of biodiversity / habitats / species; AOVP

b Benefits: increased productivity / total yield of farmland; increased growth rate of crops; poor land can be used for agriculture;
Adverse consequences: eutrophication / soluble excess fertiliser increases algal growth in rivers / lakes, decomposer organisms thrive when this decays, reducing oxygen content and killing aerobic population in water; excess fertiliser wastes money / reduces profits; too little fertiliser may not produce enough increase in yield to pay for cost of fertiliser

c Benefits: quick / easy clearance of waste; destroys fungal spores / parasitic eggs / eq.;
Adverse consequences: increased amounts of carbon dioxide, adding to greenhouse effect; less humus / organic material to return to, and improve the quality of, soil

d Benefits: weeds controlled more easily / less labour / cost involved in overcoming weeds; increasing crop yield as less competition for resources / eq.; increasing profits; Adverse consequences: herbicide resistance may theoretically spread to weeds; consumers may dislike the idea of genetically engineered foods; may be more expensive than normal crop plants

6 see spreads 23.10 and 23.11

7 a An alternative way of life, or role in the community; tadpoles exploit a different food supply to the adults; young and adults not in competition for food / shelter / space

b i Nitrogen oxides / sulphur dioxide; from combustion of oil / coal / fossil fuels; form dilute nitric acid / sulphuric acid when dissolved in rain; may travel some distance, carried by wind, so source of acid rain in given passage may be from industry in another country; low pH in lakes may directly harm / affect eggs; harm tadpoles / irritate their gills; heavy metals may increase in concentration, poisoning the tadpoles; if young affected, overall population of amphibians may decrease

ii Over-use of pesticides in farming; may result in excess being washed into ponds; some non-specific pesticides could act directly on eggs / tadpoles / adults; long-lasting pesticides may enter food chain and accumulate in tadpoles / adults; effects on food chain could alter populations and affect food supply of tadpoles; some herbicides kill all vegetation, so could destroy adult's habitat / food source for tadpole in herbivorous stage

c Adult amphibians live on land, apart from breeding season, so are less likely to be vulnerable to pollutants in water; eggs / tadpoles are aquatic for all of their existence, so have greater exposure to pollutants than adults; soluble material may enter soft / unshelled eggs by diffusion; water passing over tadpole gills comes into close contact with circulatory system, so harmful material, dissolved in water could diffuse into tadpole's blood; amphibians have thin skins, allowing pollutants to diffuse easily across tadpole skin

Index

F

Acknowledgements

The publisher would like to thank the following for permission to reproduce examination questions:

AQA – Assessment and Qualifications Alliance

- AEB: p39q5, p54q1,3, p55q8,9, p82-3q3,6, p99q5, p110q2,3, p134-5q2,5, p158-9q1,5, p180q3, p210q1, p210-11q3, p242q1,2, p266-7q1,3,5, p288q1, p315q4, p338q3, p362q1, p412q2, p434-5q2,5,7, p463q6, p500-1q1,2,4, p560q1
- NEAB: p54-5q4,5,6,7, p83q4, p98-9q2, p135q3, p158-9q3, p181q4, p211q5,6, p242-3q3,4,5,7, p266-7q2,4, p288q2, p315q5, p362q2,3, p386-7q1,2,4,7, p412-13q1,4,5, p434-5q1,3,6, p462-3q3,7, p500-1q3,7,8, p536-7q1,2,5, p561q3,4,7

Council for the Curriculum, Examinations, and Assessment (Northern Ireland)

- CCEA: p39q4, p82q1, p83q7, p211q4,7, p315q6, p338q2, p462q1

Edexcel Foundation

- London: p38q2, p54q2, p82q2, p83q5, p99q4, p110-11q1,5, p135q5, p159q4,6, p180-1q2,5, p211q8, p267q6, p289q3,4,5, p314-15q2,3, p339q4,6,7,8, p387q3,5,8, p413q3, p435q4, p463q4, p501q5,6, p537q3,4

OCR – Oxford, Cambridge and RSA Examinations

- OCR: p98q1, p561q5
- Cambridge (UCLES): p111q4, p267q7, p363q4,5, p387q6, p462-3q2,5
- Oxford: p38-9q1,3
- Oxford & Cambridge Schools: p39q6, p134q1, p158q2, p243q6, p314q1, p338q1, p561q6

Welsh Joint Education Committee

- WJEC: p99q3, p135q4, p180q1, p210q2, p243q8, p289q6,7, p339q5, p413q6, p537q6

Text extracts on the section openers are taken from the following publications:

p. 11	OUP. Reprinted from *Introduction to the Study of Man* by J. Z. Young by permission of OUP.
p. 85	*The Origin of Species* by Charles Darwin (available in several editions)
p. 315	*The Compleat Angler* by Izaak Walton (available in several editions)
p. 389	Reprinted from *A Guided Tour of the Living Cell* by Christian de Duve, published by Scientific American Books (W. H. Freeman) in 1984
p. 503	Reprinted from *Gerald Durrell: The Authorized Biography* by Douglas Botting, published by Harper Collins, 1999

Cover image by: Eye of Science/Science Photo Library

Chapter 1
1.1.1: A.N.T. Photo Library/NHPA/Photoshot; 1.1.2: Ronald Grant Archive.

Chapter 2
2.1.1: Steve Turner/Getty Images; 2.3.5: David Wrobel/Visuals Unlimited/Science Photo Library; 2.3.1: David Hardy/Science Photo Library; 2.4.2: David Scharf/Science Photo Library; 2.4.2: Science Photo Library; 2.4.2: Stephen Dalton/NHPA/Photoshot; 2.7.3: Jerzy Gubernator/Science Photo Library; 2.7.6: DIOMEDIA; 2.7.4: CNRI/Science Photo Library; 2.10.3: Science Photo Library.

Chapter 3
3.1.1: Telegraph Colour Library; 3.2.3: Science Photo Library; 3.5.1: Telegraph Colour Library.

Chapter 4
4.1.1: Biophoto Associates/Getty Images; 4.1.2b: Biophoto/Professor Leedale; 4.1.3b: Trip Photo Library; 4.2.1: Science Photo Library; 4.3.2a: Science Photo Library; 4.3.2b: Science Photo Library; 4.3.4: Science Photo Library; 4.3.5: Biophoto/Professor Leedale; 4.3.6: Science Photo Library; 4.5.1b: Science Photo Library; 4.6.2: Biophoto/Professor Leedale; 4.8.5: Oxford Scientific Films; 4.10.1: Telegraph Colour Library; 4.12.3: Biofotos/Heather Angel; 4.13.4: Biophoto/Professor Leedale.

Chapter 5
5.1.2: Biophoto/Professor Leedale; 5.1.3: Science Photo Library; 5.6.2a&b: Oxford Scientific Films.

Chapter 6
6.4.2: Biophoto/Professor Leedale; 6.5.1: Popperfoto/Getty Images; 6.5.3: Dr Jeremy Burgess/Science Photo Library.

Chapter 7
7.1.1: Ardea; 7.5.2-7.9.1: Science Photo Library; 7.10.1: Popperfoto/Getty Images; 7.11.1: Science Photo Library.

Chapter 8
8.1.1a: Telegraph Colour Library; 8.1.1b: Science Photo Library; 8.3.1: Popperfoto/Getty Images; 8.3.2: Science Photo Library; 8.5.1: Science Photo Library; 8.7.1: Telegraph Colour Library; 8.11.1: Telegraph Colour Library.

Chapter 9
9.1.1b: Peter Menzel/Science Photo Library; 9.1.2a: Biofotos/Heather Angel; 9.1.2b: Oxford Scientific Films; 9.1.2c: Telegraph Colour Library; 9.1.2d: Biofotos/Heather Angel; 9.1.2e: Oxford Scientific Films; 9.1.2f: Oxford Scientific Films; 9.4.2: Science Photo Library; 9.6.2: Science Photo Library; 9.7.2r: Biophoto/Professor Leedale; 9.7.2l: Telegraph Colour Library; 9.9.2: Corbis.

Chapter 10
10.1.4: Corbis; 10.6.1: Corbis; 10.7.2: Science Photo Library; 10.9.1: Oxford Scientific Films; 10.13.1: Tony Stone/Getty Images.

Chapter 11
11.1.1: Science Photo Library; 11.2.1: Trip Photo Library; 11.3.1a: Biophoto/Professor Leedale; 11.3.1b: Science Photo Library; 11.3.1c: Oxford Scientific Films; 11.4.1: Corbis; 11.6.3: Biofotos/Heather Angel; 11.8.1: Science Photo Library; 11.8.3a: Ardea; 11.9.1: Central Press/Getty Images; 11.9.2: Corbis; 11.10.1: Hulton Getty Images; 11.11.1: Suzi Eszterhas/Getty Images; 11.11.1: Hulton Getty Images; 11.12.1: Science Photo Library; 11.13.3: © stagewestphoto/Fotolia; 11.13.2: Simon Fraser/ Science Photo Library; 11.13.4: Oxford Scientific Films; 11.14.1: Bruce Coleman; 11.14.2: Science Photo Library; 11.15.1: Jeffrey Lepore/Science Photo Library; 11.15.2: Andrew Howe/Getty Images.

Chapter 12
12.1.1: Biophoto/Professor Leedale; 12.1.2-5.1: Oxford Scientific Films; 12.5.3: Science Photo Library; 12.10.3: Biofotos/Heather Angel.

Chapter 13
13.1.1(both): Geoscience; 13.2.1-2: Trip Photo Library13.2.4: Philip Harris; 13.3.2tl: Oxford Scientific Films; 13.3.2: Biophoto/Professor Leedale; 13.3.4: Science Photo Library; 13.4.1: Science Photo Library; 13.6.1: Heather Angel; 13.6.1: Feng Wei Photography/Getty Images; 13.6.2: Dennis Drenner/Visuals Unlimited ; 13.7.1: Science Photo Library; 13.8.1a: Nigel Cattlin; 13.9.2a: Holt Studios; 13.9.2b: Science Photo Library; 13.9.2c: Science Photo Library.

Chapter 14
14.1.1: Biofotos/Heather Angel; 14.1.4: Biophoto/Professor Leedale; 14.3.1a: Science Photo Library; 14.3.1b: Science Photo Library; 14.3.2a: Science Photo Library; 14.3.2b: Biofotos/Heather Angel; 14.5.1: Biofotos/Heather Angel; 14.5.2: Oxford Scientific Films; 14.5.3: Oxford Scientific Films; 14.8.1: Biofotos/Heather Angel; 14.9.1: Biofotos/Heather Angel; 14.9.2: Dr. Keith Wheeler/Science Photo Library; 14.9.3: Science Photo Library; 14.10.1: Holt Studios; 14.10.2: Holt Studios; 14.11.1: Oxford Scientific Films/C G Gardener; 14.11.1: Power and Syred/ Science Photo Library; 14.12.3: Biofotos/Heather Angel.

Chapter 15
15.1.1: Oxford Scientific Films; 15.2.1-5.2: Science Photo Library; 15.6.1: CNRI/Science Photo Library15.6.1: Science Photo Library; 15.6.4: Science Photo Library; 15.7.1: Science Photo Library; 15.9.1: Telegraph Colour Library; 15.10.1: Science Photo Library.

Chapter 16
16.1.1: Redferns;
16.1.2a: Oxford Scientific Films; 16.1.2b: Oxford Scientific Films; 16.2.1: Science Photo Library; 16.2.3: Science Photo Library; 16.3.2: Science Photo Library; 16.4.3: Science Photo Library; 16.5.2: Trip Photo Library; 16.5.2: Trip Photo Library; 16.6.1: National Medical Slidebank; 16.6.2: Biophoto/Professor Leedale; 16.7.1: Wellcome Trust/National Medical Slide Bank; 16.7.1a: Frederick C. Skvara, MD/ Visuals Unlimited, Inc./ Science Photo Library; 16.7.1b: Carolina Biological Supply, Co/Visuals Unlimited, Inc; 16.9.1: Ed Young/ Science Photo Library; 16.9.1: Science Photo Library; 16.10.2: Biophoto/Professor Leedale; 16.11.1: Tony Stone.

Chapter 17
17.1.3a: Holt Studios; 17.1.3b: Science Photo Library; 17.2.1: Telegraph Colour Library; 17.2.3: Biophoto/Professor Leedale; 17.2.3a: Michael Abbey/ Science Photo Library; 17.3.1: Science Photo Library; 17.3.2: Science Photo Library; 17.4.1: Science Photo Library; 17.4.3: Science Photo Library; 17.4.5a: John Durham/ Science Photo Library; 17.4.5a: Dr.Rosalind King/ Science Photo Library; 17.4.5a: Science Photo Library17.4.5b: Science Photo Library17.5.1a: Eye of Science/ Science Photo Library; 17.5.1a: Premaphotos Wildlife17.5.1b: Science Photo Library17.5.1b: Power and Syred/ Science Photo Library; 17.5.1c: Bruce Coleman; 17.6.1a: Biophoto/Professor Leedale; 17.6.1b: Science Photo Library; 17.6.2: Science Photo Library; 17.7.3: Science Photo Library; 17.8.1: Anthony Blake Photo Library; 17.8.2: Anthony Blake Photo Library; 17.10.1: Ronald Sheridan Photo Library; 17.11.2: Panos Photo Library.

Chapter 18
All images by Science Photo Library.

Chapter 19
19.2.4: Science Photo Library; 19.4.4: Oxford Scientific Films; 19.5.1: Corbis; 19.7.1: Oxford Scientific Films; 19.7.3: Science Photo Library; 19.9.1: Science Photo Library; 19.9.2: Science Photo Library; 19.9.4: Science Photo Library;

Chapter 20
20.1.1: Science Photo Library; 20.1.2: Telegraph Colour Library; 20.2.1: Science Photo Library; 20.3.1a: Biofotos/Heather Angel; 20.3.1b: Biofotos/Heather Angel; 20.5.1: Chris Mattison/FLPA; 20.5.1: Biofotos/Heather Angel; 20.5.2a: Science Photo Library; 20.5.2b: Bruce Coleman; 20.7.1: Corbis; 20.7.2: Gerard Lacz/Rex Features; 20.7.2: Oxford Scientific Films; 20.8.1a: Biofotos/Heather Angel; 20.8.1b: Biofotos/Heather Angel; 20.9.2a: Holt Studios; 20.9.2b: Science Photo Library; 20.9.2c: Biofotos/Heather Angel; 20.10.1: Biofotos/Heather Angel; 20.11.1a: www.visualphotos.com; 20.11.1b: www.visualphotos.com; 20.11.1c: Tom Mchugh/SCIENCE PHOTO LIBRARY; 20.11.1: Gerry Willis ENP Images; 20.12.1: Science Photo Library; 20.12.2: Natural History Museum.

Chapter 21
21.1.1: Science Photo Library; 21.2.2: Dr Fishelson; 21.3.1: Oxford Scientific Films; 21.4.1: Science Photo Library; 21.4.1: Shutterstock; 21.5.2: Biofotos/Heather Angel; 21.6.4: Science Photo Library; 21.7.2: Topham Picture Library; 21.7.2: Shutterstock; 21.8.1: Oxford Scientific Films; 21.9.4: Biofotos/Heather Angel; 21.11.3: Science Photo Library; 21.12.1a: Biofotos/Heather Angel; 21.12.1b: Biofotos/Heather Angel; 21.12.1c: Science Photo Library; 21.14.1: Telegraph Photo Library; 21.15.3: Biofotos/Heather Angel; 21.16.1: Science Photo Library; 21.16.2: Geoscience; 21.16.3: Oxford Scientific Films; 21.16.4: Telegraph Photo Library; 21.16.5: Biofotos/Heather Angel; 21.17.1: Ardea; 21.17.2: NHPA; 21.18.1: Bruce Coleman; 21.18.2: Oxford Scientific Films; 21.18.3: Biofotos/Heather Angel; 21.18.4: Ardea; 21.18.5: Telegraph Photo Library.

Chapter 22
22.1.1: Biofotos/Heather Angel; 22.2.1: Science Photo Library; 22.8.2: Biofotos/Heather Angel; 22.9.3: Bruce Coleman; 22.10.1: Bruce Coleman; 22.10.4: NHPA/Photoshot; 22.10.4: Biofotos/Heather Angel; 22.11.2: Science Photo Library; 22.12.2: Biofotos/Heather Angel; 22.14.1: Biofotos/Heather Angel; 22.14.2: Ardea; 22.14.3: Bruce Coleman; 22.15.1: Science Photo Library; 22.15.2: Biofotos/Heather Angel; 22.15.3: Biofotos/Heather Angel; 22.15.4: Biofotos/Heather Angel; 22.16.1: Popperfoto/Getty images; 22.16.1: Baron/Getty Images.

Chapter 23
23.1.1: Corbis; 23.2.2: Science Photo Library; 23.3.1: Science Photo Library; 23.4.2: Science Photo Library; 23.5.1: Oxford Scientific Films; 23.5.2: NHPA; 23.6.2: Science Photo Library; 23.8.1: Oxford Scientific Films; 23.9.3: Biofotos/Heather Angel; 23.10.1: Ardea; 23.11.1: NHPA; 23.11.2: Gerard Lacz/Rex Features; 23.11.2: Biofotos/Heather Angel.

Additional Artwork by Mark Walker